MW00837655

TOPOLOGICAL FLUID MECHANICS

INTERNATIONAL UNION OF THEORETICAL AND APPLIED MECHANICS

TOPOLOGICAL FLUID MECHANICS

Proceedings of the IUTAM Symposium

Cambridge, UK 13–18 August 1989

Edited by H. K. MOFFATT and A. TSINOBER

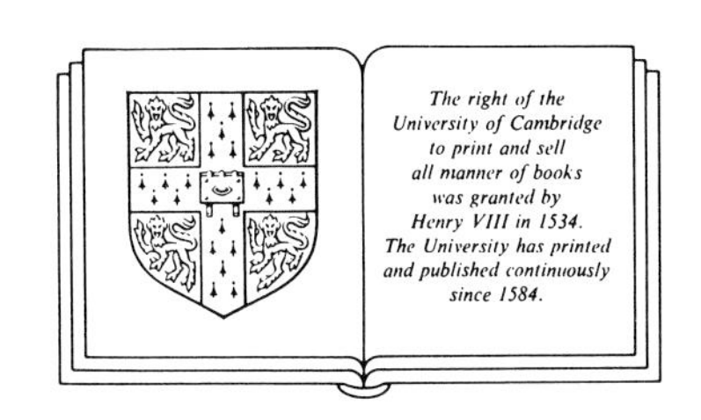

CAMBRIDGE UNIVERSITY PRESS
Cambridge
New York Port Chester
Melbourne Sydney

Published by the Press Syndicate of the University of Cambridge
The Pitt Building, Trumpington Street, Cambridge CB2 1RP
40 West 20th Street, New York, NY 10011, USA
10 Stamford Road, Oakleigh, Melbourne 3166, Australia

First published 1990

Printed in Great Britain at the University Press, Cambridge

British Library cataloguing in publication data

Topological fluid mechanics: proceedings of the IUTAM Symposium, Cambridge UK, 13–18 August 1989.
1. Western bloc countries. Imports from Soviet Union: Natural gas & petroleum. Politico–economic aspects
I. International Union of Theoretical and Applied Mechanics II. Moffatt, H. K. III. Tsinober, A.
532.'001'51

ISBN 0–521–38145–2

Library of Congress cataloguing in publication data available

ISBN 0 521 38145 2

International Committee

V.I. ARNOL'D	USSR
U. DALLMANN	FRG
U. FRISCH	France
P. GERMAIN (President IUTAM)	France
H. HASIMOTO	Japan
F. HUSSAIN	USA
H.K. MOFFATT (Chairman)	UK
E.N. PARKER	USA
A.E. PERRY	Australia
A. TSINOBER	Israel

Local Organising Committee

H.K. MOFFATT (Chairman)
K. BAJER
R.E. BRITTER
D.G. DRITSCHEL
M. GASTER
P.H. HAYNES
J.C.R. HUNT
SARAH KIRKUP (Secretary)
A.J. MESTEL
M.R.E PROCTOR
W.E. WARD

This Symposium was sponsored by the International Union of Theoretical and Applied Mechanics, with additional financial support which is gratefully acknowledged from the following organisations:

- The Royal Society
- The London Mathematical Society
- Trinity College Cambridge
- US Air Force European Office of Aerospace Research and Development (EOARD)
- Office Naval Research Europe (ONREUR)
- ALCAN International Ltd.
- British Aerospace plc
- Cambridge Environmental Research Consultants Ltd.
- ICI Engineering
- Rolls Royce plc
- Schlumberger Cambridge Research Ltd.

Contents

VI: Vortex Interaction and Reconnection 467

VII: Homogeneous Turbulence 547

Preface

The topic of this meeting was chosen in response to the developing interest in aspects of fluid mechanics and of magnetohydrodynamics that can properly be described as topological, rather than exclusively analytical, in character. On the one hand, there are purely kinematic problems such as the classification of possible streamline structures in three dimensions; the deformation of convected lines and surfaces in a prescribed flow field; and the relation between Lagrangian and Eulerian properties for both laminar and turbulent flows. On the other hand, there are dynamical problems such as the treatment of changes in flow topology associated with symmetry-breaking instability; topological invariants (e.g. the helicity invariant) associated with the Euler equations; the manner in which topological constraints may be broken (rapid reconnexion, or cut-and-connect mechanisms) in high Reynolds number flow; the influence of helicity fluctuations, or more generally of any departure from reflexional symmetry, in the fundamental dynamics of turbulence; and the manner in which singularities of the Euler equations may develop within a finite time. Underlying all these problems is a desire to identify structures (if any) that are characteristic of fully developed turbulent flows, with a view to constructing an improved statistical theory of turbulence.

Between these kinematic and dynamical aspects lie a fascinating range of problems that arise in the magnetohydrodynamics (MHD) of highly conducting fluids: relaxation to magnetostatic equilibrium under the constraint of conserved field topology; formation of discontinuities during such relaxation; spontaneous magnetic field growth (i.e. dynamo action) due to fluid motion either with or without helicity; and, particularly, the phenomenon of 'fast dynamo action', tantalisingly difficult to describe mathematically, and yet almost certainly the root cause of solar and stellar magnetism.

In focussing on these topics, the Symposium stimulated a fruitful interaction between, on the one hand, fluid dynamicists working on fundamental mechanisms of vorticity dynamics and turbulence, and on the other, plasma physicists with expertise in problems of magnetic field topology and reconnexion processes. The power of topological arguments remained at the forefront of the discussion throughout the week, emphasis being placed on global, rather than local, characteristics of field and flow phenomena.

There were 119 registered participants at the Symposium, representing 15 countries. Accommodation was provided at Pembroke College, Cambridge. The scientific programme consisted of Lecture sessions, supplemented by three Poster-discussion sessions. All of the papers presented have been refereed, and where necessary revised, and all are included in this Volume. We apologise for its consequent size, but we believe that it encompasses an impressive body of work which reflects a widespread desire to understand qualitative properties of fluid-dynamical systems as an essential preliminary to quantitative and computational analysis.

H.K.M.
A.T.
30 September 1989

Opening Address

P. GERMAIN, PRESIDENT OF IUTAM

Académie des Sciences, Paris VI, France

Mr. Chairman, dear colleagues and friends, it is a great pleasure for me to have the opportunity to welcome you this morning on behalf of IUTAM, the International Union for Theoretical and Applied Mechanics, and more specifically on behalf of its Bureau. As you know, IUTAM symposia are the most regular manifestations of the scientific activities of the Union. Every year, about nine Symposia on topics of special interest in the field of mechanics are carefully selected by the General Assembly from among approximately twice as many proposals, following on the advice and recommendations of the Symposia Panels and of the Bureau. One may be sure that each of these Symposia which gathers together the best specialists of the field makes a complete review of the most recent achievements of the subject. Looking back on the collection of volumes published over more than forty years, which contain all the papers presented during IUTAM Symposia, provides a good view of the main trends of our discipline and of their evolution during this period. Each IUTAM Symposium is generally a meeting of great scientific significance. This is why IUTAM is always grateful to colleagues who agree to organize a Symposium, to the scientific committees which advise on selection of participants and last, but not least, to the authors of papers and to all participants in the discussions.

Today, I have special reason, as President of IUTAM, to express my deep satisfaction and to deliver my warmest congratulations. The reasons lie in the originality of the topic, in its exceptional interest, and in the outstanding scientific and practical preparations. The proposal made by my colleague and friend, Keith Moffatt, to call this Symposium in Cambridge, England, and to take the responsibility for its organization, was unanimously welcomed: "Topological Fluid Mechanics" is certainly a quite new title among all the titles of former IUTAM Symposia; and today a most appropriate one.

Obviously, scientists working in mechanics are fascinated by the facilities offered by modern computers, these powerful tools which every day increase their capacities and open new opportunities. The engineer, provided he has the competence, the time and the money, may compute every flow and hence the pressures and forces on immersed bodies. But these numerical results, which require extremely sophisticated numerical analysis and very extensive analytical and functional study of the problem to be solved, even if they are correct and precise, do not automatically provide a good understanding of the flow which is computed. Quantitative results gain in their significance when coupled with a clear understanding of qualitative properties. A meeting devoted to topological fluid mechanics, mainly concerned with these qualitative properties of the flows, is therefore particularly welcome. First, on account of the intrinsic aesthetic character, always attractive to scientists, of the concepts, reasonings, and results which will be presented; secondly, to review and improve the tools and ideas which allow us to extract full benefit from complex computations thanks to good interpretation, and also to enable more effective computation in the future; and thirdly, to improve our capacity to visualise and comprehend the structure of complex three-dimensional flows.

This goal is related with the ideals of our forefathers, like Lagrange who, after the formulation of the laws of mechanics by Newton, worked to clarify and to stress their geometrical character. Analytical mechanics has developed successfully along these lines. Representations of the motion of a system have realized this ideal through intrinsic mappings on manifolds with a riemannian or a symplectic geometry and on the bundles they generate. May I emphasize that the language of fluid mechanics has often been used to visualize a continuous group of motions or mappings. In some of these representations, a motion may be a streamline, in others a vortex line. Invariants, phase portraits, Poincaré sections, basins of attraction, singular points on manifolds each with its special neighbourhood, and strange attractors with fractal structure are all qualitative concepts which are familiar in analytical mechanics and which we shall meet often during this Symposium; but here, in the context of real fluid mechanics.

The IUTAM Symposium on topological fluid mechanics held in Cambridge is therefore an important scientific event for our International Union. It follows the exceptional IUTAM Symposium on "Fluid Mechanics in the spirit of G.I. Taylor" which took place in 1986 in Cambridge under George Batchelor's Chairmanship, and is in similar spirit, but perhaps with more theoretical and geometrical flavour. Here in Cambridge where, three hundred years after Newton, mechanics still finds its best inspiration, I have particular pleasure in opening this Symposium and in wishing you a fruitful week of scientific interaction.

The Topological (as opposed to the analytical) Approach to Fluid and Plasma Flow Problems

H.K. MOFFATT

Department of Applied Mathematics and Theoretical Physics, University of Cambridge, Silver Street, Cambridge CB3 9EW, UK

1. INTRODUCTION

This talk is by way of introduction to the whole meeting, and I shall therefore focus on one or two facts and conjectures, which will be explored in more detail in some of the subsequent lectures. My first assertion is that topological, rather than analytical, techniques and language provide the natural framework for many aspects of fluid mechanical research that are now attracting intensive study. The reason for this is very clear: in a fluid flow, in which the velocity field is a continuous function of position and time, the particle paths may be defined by a function $\mathbf{x}(\mathbf{a}, t)$, representing the position at time t of the particle whose initial position is $\mathbf{a}$. For finite t, the function $\mathbf{x}(\mathbf{a}, t)$ is continuous and invertible, and the inverse function $\mathbf{a}(\mathbf{x}, t)$ is also continuous. In these circumstances, any structure (whether described by a scalar, vector or tensor field) which is convected with the flow will have topological properties that are invariant in time (although its geometrical properties are by no means invariant, and indeed in general become exceedingly complex, in time). The simplest such structure is a patch of dye, passively convected with the flow. If the patch is topologically spherical at $t = 0$, then it remains so for all finite t; if it is topologically toroidal, then it remains toroidal, and so on.

The qualification 'at finite t' is important here, because, as t tends to infinity, the mapping $\mathbf{a} \to \mathbf{x}(\mathbf{a}, t)$ induced by a continuous velocity field may become discontinuous. For example, consider a system consisting of two viscous fluids, of densities ρ_1 and ρ_2, the lighter fluid lying above the heavier fluid (e.g. oil on water). Suppose that a drop of oil is introduced in the water, so that it rises under buoyancy forces towards the interface. The velocity field in the entire system is continuous for all t, but the gap between the drop and the overlying layer of oil tends to zero as t tends to infinity so that the induced mapping becomes discontinuous in the limit.

The topological invariance referred to above is represented by invariance of a certain family of integrals involving the convected field. For a scalar field $\theta(\mathbf{x},t)$ satisfying the Lagrangian conservation equation $D\theta/Dt = 0$, these integrals are rather trivial:

$$\int F(\theta)dV = \text{cst}. \tag{1}$$

where $F(\theta)$is any function of θ. The topology of the θ field is characterised by the family of surfaces $\theta = \text{cst}$,and the choice $F(\theta) = \delta(\theta - \theta_c)$ focusses attention on one such surface. The total family of invariants (1) guarantees that the topology of the field θ is conserved.

2. HELICITY INVARIANTS

The situation is less trivial and much more interesting, when we consider convected vector fields. These can be either the gradient of a scalar field ($\mathbf{G} = \nabla\theta$) or the curl of a vector field ($\mathbf{B} = \text{curl}\mathbf{A}$) or a combination of these. The prototype field of the latter type is the magnetic field in a perfectly conducting fluid which satisfies the frozen-field equation

$$\frac{\partial \mathbf{B}}{\partial t} = \text{curl}(\mathbf{u} \wedge \mathbf{B}) \tag{2}$$

or the equivalent Lagrangian equation

$$\frac{D}{Dt}\left(B_i(\mathbf{a},t)\frac{\partial a_j}{\partial x_i}\right) = 0 \tag{3}$$

where $B_i(\mathbf{a},t)$ represents the magnetic field at the current position of the fluid particle which was initially at position $\mathbf{a}$. The invariant integral that characterises the topology of $\mathbf{B}$ was discovered by Woltjer 1958, and is now known as the magnetic helicity $\mathcal{H}_M$ (sometimes denoted K in the plasma physics literature). We define this as follows: for simplicity, suppose that the fluid is contained in a simply connected domain, and let S be any closed surface within this domain, moving with the fluid, on which $\mathbf{B} \cdot \mathbf{n} = 0$ (a condition that clearly persists if it holds at some initial instant). Let V be the volume inside S. Then the magnetic helicity of the field within V is defined by

$$\mathcal{H}_M(V) = \int_V \mathbf{A} \cdot \mathbf{B} dV, \tag{4}$$

where $\mathbf{A}$ is an arbitrary magnetic potential for $\mathbf{B}$. Note that, under the assumed conditions, this integral is gauge invariant. $\mathcal{H}_M(V)$ is also invariant in time, and this is entirely a consequence of equation (2).

It is important to note first that the *helicity density* $\mathbf{A} \cdot \mathbf{B}$ is *not* invariant in time, although its value following a fluid particle can be forced to be time invariant by choosing a particular gauge for $\mathbf{B}$ (as recognised by Elsasser 1946 in work foreshadowing that of Woltjer); and secondly, that to every magnetic surface, there corresponds an integral invariant of the form (4), and that if there is an infinite family of such surfaces, then we have a corresponding infinite family of invariants. If however the magnetic field is 'space-filling' in any subdomain $\hat{D}$, then there is only one helicity integral for that subdomain.

Note also that the family of invariants (4) exists even if the field $\mathbf{B}$ is not dynamically passive. The velocity field $\mathbf{u}$ in equation (2) may be allowed to depend in any complex nonlinear manner on $\mathbf{B}$, but this does not affect the proof of the invariance of the integrals (4). There are a number of important physical circumstances in which this type of nonlinearity arises. For example, flows can be driven by a 'magnetic buoyancy instability' in which the velocity field $\mathbf{u}$ is quadratically related to $\mathbf{B}$, so that the right-hand side of equation (2) becomes cubic in $\mathbf{B}$. Despite the nonlinearity of the problem, the family of invariants (4) still exists.

The topological significance of these invariants was recognised by Moffatt (1969), through consideration of the trivial case in which the magnetic field is confined to two closed flux tubes of vanishingly small cross-section. Taking V to be the whole domain, $\mathcal{H}_M$ can be explicitly evaluated, with the result

$$\mathcal{H}_M = \pm n \Phi_1 \Phi_2, \tag{5}$$

where Φ_1 and Φ_2 are the two fluxes, and n is the (Gauss) winding number of the two tubes relative to one another. This result of course establishes a very clear relationship between the frozen field equation (2) and the fundamental topological concept of linkage. The relationship has been cemented by Arnol'd (1974) who showed that, in an asymptotic sense, helicity still represents linkage even when the field lines are not closed, but wander chaotically in the fluid domain; Arnol'd's identification of helicity with an asymptotic form of the Hopf invariant provides a powerful bridge between fluid mechanics and topology.

3. HELICITY ASSOCIATED WITH THE VORTICITY FIELD

It was this topological interpretation of helicity that led to the immediate realisation that there must exist an analogous helicity invariant, namely

$$\mathcal{H} = \int \mathbf{u} \cdot \boldsymbol{\omega} dV \tag{6}$$

corresponding to the frozen field Euler evolution of the vorticity field $\boldsymbol{\omega}$. This is the counterpart of Kelvin's (1869) circulation theorem, and it is noteworthy that Kelvin himself recognised that knotted vortex lines would have invariant topology, under evolution governed by the Euler equations. Kelvin developed his 'vortex theory of atoms' on this basis, in collaboration with P.G. Tait, who was thereby motivated (Tait 1898) to classify and catalogue all knots of increasing order of complexity. The vortex theory of atoms turned out to be misconceived, but the catalogue of knots remains as the cornerstone of an established branch of topology, in which there has been a recent renewed upsurge of interest.

In the fluid mechanical context, helicity, like energy, is conserved only for ideal fluid flow, and helicity generally changes under the influence of viscous effects. Change of helicity is associated with diffusion and reconnection of vortex lines, thus changing the topology of the vorticity field. Note that, whereas viscosity always leads to dissipation of energy, it can be responsible for the production, as well as the destruction, of helicity. The interaction of two vortex rings provides the prototype problem, in which the evolution of the total helicity of the flow provides an indicator of viscous interaction and reconnection.

The fact that energy is an 'inviscid invariant' is of course fundamental to the Kolmogorov theory of turbulence, involving a cascade of energy from large scales to small scales. The existence of a second robust inviscid invariant, namely the mean helicity, has naturally raised the question of the influence that non-zero mean helicity may have on this type of cascade process. The influence of helicity in turbulence has excited some controversy in the recent literature. Whatever the outcome of this controversy, we are faced with a problem: if helicity *does* affect the cascade process, then we have to understand exactly how it does so; if helicity *does not* affect the cascade process, then equally we have to understand how the fluid behaves in such a way as to respect one inviscid invariant (namely energy) and ignore another (namely helicity).

4. HELICITY AND THE SPONTANEOUS GROWTH OF MAGNETIC FIELDS

Helicity has long been known to be of fundamental significance in the context of dynamo theory, that is the theory of the spontaneous growth of magnetic fields in electrically conducting fluids in motion. When the flow is turbulent, magnetic field fluctuations are generated, and these interact with the turbulence to provide a mean electromotive force $\mathcal{E}$, which can in principle be expanded as a series in the mean field $\mathbf{B}$ and its derivatives:

$$\mathcal{E} = \alpha \mathbf{B} - \beta \nabla \wedge \mathbf{B} + \dots . \tag{7}$$

The coefficients α, β, etc. in this equation are determined exclusively by the statistical properties of the turbulence, and by the physical properties of the fluid, particularly its magnetic diffusivity η. The leading coefficient α is of particular importance, since it is this term which gives rise to dynamo instability. This 'α-effect' was anticipated in the early work of Parker (1955), and again in the nearly axisymmetric dynamo model of Braginskii (1964), but it reached maturity with the work of Steenbeck, Krause & Rädler (1966), who recognised explicitly the relationship between α and the underlying helicity of the flow. Evidently, α is a pseudo-scalar quantity (i.e. one which changes sign under change from a right-handed to a left-handed frame of reference) and, insofar as helicity is the simplest measure of lack of reflexional symmetry in a turbulent flow, such a relationship is to be expected. However, helicity is not the *only* measure of lack of reflexional symmetry, and, only in the large η (i.e. low magnetic Reynolds number) limit is the relationship between α and helicity straightforward; in this limit, α may be expressed as a weighted integral of the helicity spectrum function:

$$\alpha = -\tfrac{1}{3\eta} \int k^{-2} H(k) dk. \tag{8}$$

The mean field equation then takes the form

$$\frac{\partial \mathbf{B}}{\partial t} = \alpha \nabla \wedge \mathbf{B} + \eta \nabla^2 \mathbf{B}, \tag{9}$$

and unstable modes of Beltrami form ($\mathbf{B} = K \nabla \wedge \mathbf{B}$) exist provided

$$|\alpha K| > \eta K^2. \tag{10}$$

There are great difficulties in the mean field theory in the alternative *low* diffusivity limit, which have not yet been fully resolved. This limit is very important in astrophysical contexts, and it is customary to suppose that both α and β are determined in order of magnitude by the velocity and length scales (u_0, l_0) of the turbulence (and independent of the magnetic diffusivity) in this limit, i.e.

$$\alpha \sim u_0 \quad , \quad \beta \sim u_0 l_0. \tag{11}$$

The maximum growth rate for the mean field occurs on a length scale of order β/α (according to equation (9)), and we are now faced with the difficulty that, if the estimates (11) are correct, then the mean field instability progresses most efficiently on the scale l_0 of the turbulence itself. This cuts at the heart of mean field theory, which relies on a separation of scale between fluctuating and mean quantities. (The only escape would be if a small numerical coefficient ϵ were to appear in the expression for α, i.e. $\alpha = \epsilon u_0$.)

Current efforts are being increasingly directed at the 'fast' dynamo problem, in which it is assumed from the outset that the dominant scale of the magnetic field is not much greater than that of the velocity field. In fact, it may be very much *smaller*: a generic feature of 'fast' dynamos (for which, by definition, the growth rate is independent of magnetic diffusivity η as this tends to zero) is that the scale of the magnetic field must nearly everywhere be of order $\eta^{\frac{1}{2}}$, and this means of course that the field is nondifferentiable nearly everywhere as η tends to zero. The beginnings of such structures have emerged from the numerical simulations of the ABC dynamo by Galloway & Frisch (1986), which emphasise the central role played by the saddle points of the flow, where maximal stretching of the magnetic field takes place.

These magnetic structures presumably have their counterpart in the vorticity field structures of turbulent flow. Vortex stretching is like magnetic field stretching, but of course the velocity field is itself determined (in a nonlocal way) by the vorticity field. The question of whether the vorticity field can become singular, under Euler evolution, has attracted intense interest, and quite rightly so, because it is crucial to the understanding of the turbulent process. If a singularity *does* form, then its structure is of seminal importance for the process of energy dissipation; if it does *not* form, as now appears to be the case (see Pumir & Siggia, this volume) then there is some mechanism at work which tends to suppress nonlinear transfer of energy to very high wave numbers.

5. MAGNETIC RELAXATION: AN UNCONVENTIONAL ROUTE TO THE DETERMINATION OF EULER FLOWS

The concept of a fluid that is perfectly conducting, but nevertheless viscous, is a mathematical abstraction, rather than a physical reality, but has nevertheless proved remarkably fruitful at a fundamental level. In such a fluid, magnetic lines of force are frozen in the fluid, so that their topology is conserved (apart from the possibility that discontinuities may form by the 'squeeze film' process described in the introduction) but at the same time, energy is dissipated by viscosity. In these circumstances, a magnetic field will relax to a magnetostatic equilibrium state compatible with its initial topology. The formation of discontinuities (i.e. current sheets) is an inescapable part of this process, and this will occur even when the initial field and all its derivatives are continuous.

The magnetostatic equations are exactly analogous to the steady Euler equations (with magnetic field now analogous to velocity, not vorticity). This means that to every magnetostatic equilibrium, there corresponds a steady Euler flow. Mag-

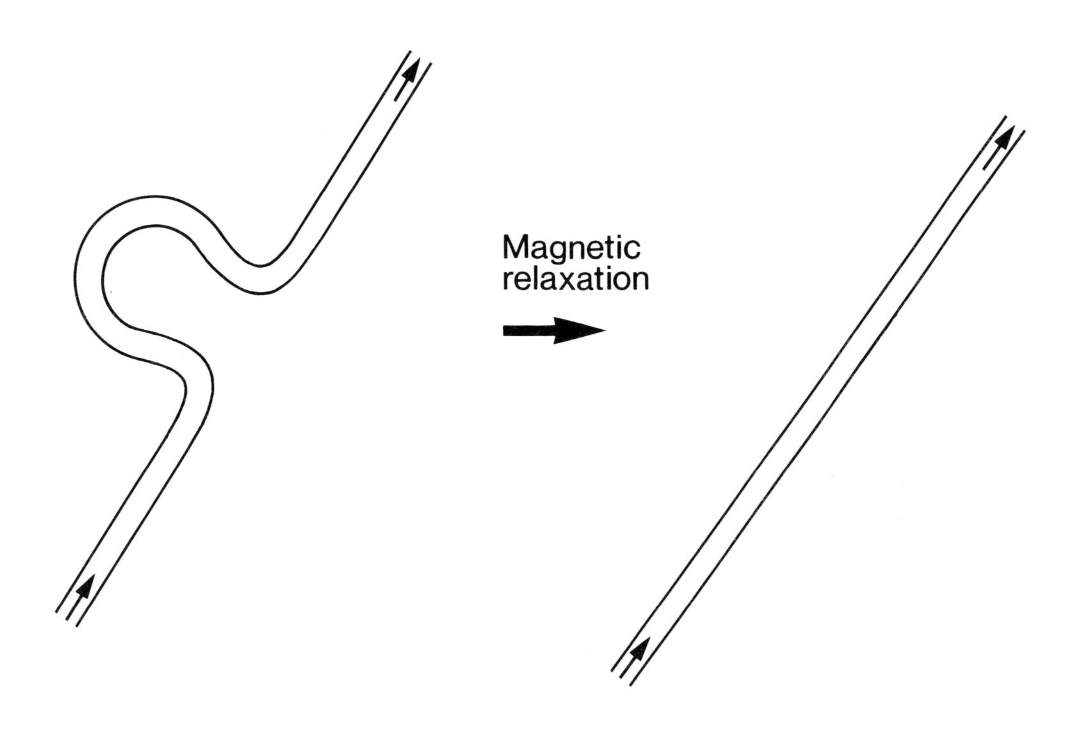

FIGURE 1. Relaxation of a magnetic flux tube in a perfectly conducting, viscous fluid.

netic relaxation therefore provides a route to the determination of Euler flows whose streamline topology may be prescribed in advance. This approach is quintessentially topological, rather than analytical in character, and it has proved powerful in establishing the existence of flows with particular topological properties (Moffatt 1985, 1986). Helicity plays a central role also in the magnetic relaxation problem, since through a combination of Schwarz and Poincaré inequalities (Arnol'd 1974) the magnetic energy of a field configuration is bounded below by a positive constant when the helicity of the configuration is non-zero. In the more esoteric situation when the helicity is zero, but there is nevertheless a higher order linkage present, the magnetic energy is still bounded below (Freedman 1988) but an estimate of this lower bound has yet to be found.

Alternative 'artificial' relaxation processes may be devised which conserve vorticity, rather than streamline, topology (see Vallis et al, and Carnevale & Vallis, this volume). These techniques appear to have tremendous potential, constrained at present only by the limits of computational power.

6. CONNECTIONS AND RECONNECTIONS

Consider a magnetic flux tube with an Ω-shaped kink in it. If the fluid is perfectly conducting, but viscous, the Lorentz force associated with the kink will cause this to relax to rectilinear form (figure 1). If the viscous effect is weak, there will of course be Alfven oscillations involved (as for a weakly damped pendulum); if viscosity is strong then these oscillations do not occur.

Suppose now that we start with two magnetic flux tubes of equal strength in the configuration of figure 2: each tube has an Ω-shaped kink, and these are 'connected'; the axes of the two tubes far from these kinks are non-parallel and do not intersect. Now the tubes obviously interfere with each other during the relaxation process, which for simplicity we assume to be dominated by viscosity. A field discontinuity tends to form, and if η is small, but non-zero, field diffusion and reconnection of field lines must occur.

This reconnection can occur in two ways (figures 2(c), (d)) or more probably as a mixture of these, each tube being in effect forced to bifurcate. Note that, whereas in case (c) the relaxation asymptotes to a state with two non-intersecting rectilinear flux tubes, in case (d), relaxation will continue indefinitely (there being initially an infinite reservoir of magnetic energy).

Consider now the situation when we have *vortex* tubes, rather than magnetic flux tubes, in the configurations of either figure 1 or 2. Suppose for example that the kinks have the form of Hasimoto solitons (Hasimoto 1972). The soliton of figure 1 can propagate along the vortex tube. The solitons of figure 2 may try to propagate but will very soon interfere with each other in a manner that can only be resolved by viscous diffusion and reconnection. Again reconnection can proceed to the configuration (c) or (d), or a mixture of these. The configuration (c) is relatively stable, while the configuration (d) will continue to evolve rapidly due to the twisted configuration of the tubes.

Examples of this kind may serve as prototypes of the reconnection processes associated with Joule dissipation in the MHD context and with viscous dissipation in high Reynolds number turbulence [see Melander & Hussain; Kerr et al; and Kida et al; this volume; also Meiron et al 1989].

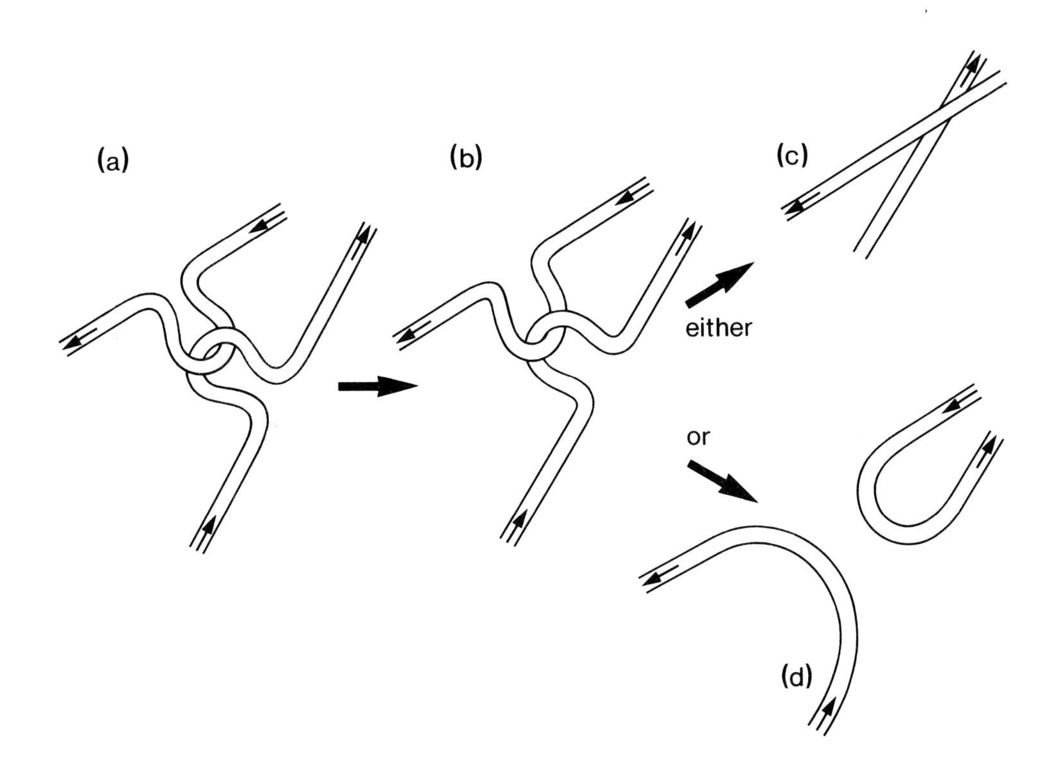

FIGURE 2. Relaxation of two connected flux tubes. A discontinuity forms (*b*), and reconnection occurs to the configuration (*c*) or the configuration (*d*), or to a mixture of these. The figure may also be interpreted in terms of vortex tubes supporting Hasimoto solitons.

7. CONCLUDING REMARKS

In this introductory survey, I have endeavoured to cover a range of topics in which topological, rather than analytical, considerations play a critical part. Helicity, being the natural, and simplest, measure of topological complexity of a convected vector field, plays a prominent role. In turbulent flow, it is the mean helicity that is an inviscid invariant. When this mean helicity is zero, then, as shown by Levich & Tsinober (1983), there still exists an integral invariant characterising helicity fluctuations. Current experimental and numerical investigations of the role of helicity fluctuations in turbulent flow are greatly to be welcomed, and it is to be hoped that these will shed new light on the fundamental mechanisms of turbulence.

The method of magnetic relaxation, and analogous relaxation techniques which conserve either streamline or vorticity topology, seems to hold great promise as a means of determining Euler flows of highly complex form. Tangential discontinuities (i.e. vortex sheets, or vortex gradient sheets, depending on the type of problem considered) may appear naturally in the relaxed fields, even if the initial fields are smooth.

For relaxation problems of this kind an important unsolved problem presents itself: what is the appropriate 'minimal' function space that contains all asymptotically relaxed fields starting from smooth initial conditions? An answer to this question would characterise the 'typical' degree of irregularity of *steady* solutions of the Euler equations, and would represent one step towards analysis of structures that develop under *unsteady* conditions and that may be expected to be generic for the problem of turbulence.

REFERENCES

ARNOL'D V.I. 1974 Proc. Summer School in Differential Equations, Erevan, Armenian SSR Acad. Sci [English translation: Sel. Math. Sov. **5**, 327-345 (1986)]

BRAGINSKII S.I. 1964 Sov. Phys. JETP **20**, 726-735

ELSASSER W.M. 1946 Phys. Rev. **69**, 106

FREEDMAN M.H. 1988 J. Fluid Mech. **194**, 549-551

GALLOWAY D. & FRISCH U. 1986 Geophys. Astrophys. Fluid Dyn. **86**, 53-83

HASIMOTO H. 1972 J. Fluid Mech. **51**, 477-485

KELVIN, LORD (then W. THOMSON) 1869 Trans. Roy. Soc. Edin. **25**, 217-260

LEVICH E. & TSINOBER A. 1983 Phys. Lett. **93a**, 293-297

MEIRON D.I., SHELLEY M.S., ASHURST W.T. & ORSZAG S.A. (1989) "Numerical studies of vortex reconnection", in: Mathematical Aspects of Vortex Dynamics, ed A.E. Caflisch, SIAM, pp. 183-199.

MOFFATT H.K. 1969 J. Fluid Mech. **35**, 117-129; 1985 **159**, 359-378; 1986 **173**, 289-302.

PARKER E.N. 1955 Astrophys. J. **122**, 293-314

STEENBECK M., KRAUSE F. & RADLER K.-H. (1966) *Z. Naturforsch.* **21a**, 369-376. [English translation: Roberts P.H. & Stix M. (1971), Tech. Note 60, NCAR, Boulder, Colorado.]

TAIT P.G. 1898 Scientific Papers I. Cambridge University Press

WOLTJER L. 1958 Proc. Nat. Acad. Sci. USA., **44**, 489-491

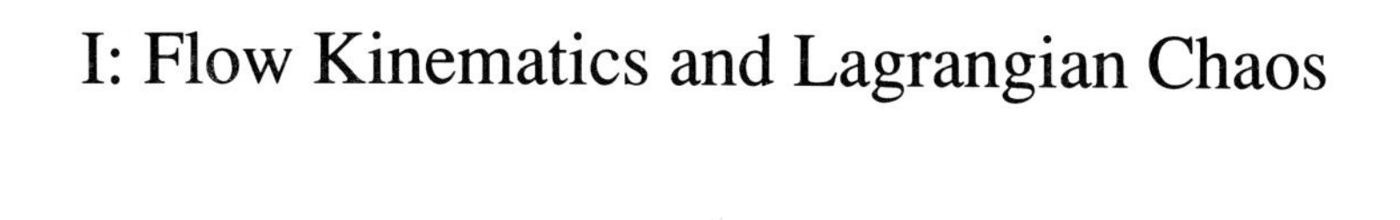

I: Flow Kinematics and Lagrangian Chaos

Kinematics of Chaotic Mixing: Experimental and Computational Results

J.M. OTTINO

Department of Chemical Engineering, University of Massachusetts, Amherst, USA

1 INTRODUCTION

The study of mixing of passive tracers by chaotic motions provides a beautiful example of interplay between analysis, computations, and laboratory experiments. The conceptual framework rests on a kinematical foundation and the analysis is influenced by recent developments in dynamical systems and deterministic chaos (Ottino 1989a). Studies based on rather simple flows display a wealth of intricate behavior and provide a glimpse into the inner workings of more complex processes such as those encountered in nature and industrial applications.

It has been known for quite sometime that mixing of passive tracers is basically stretching and folding and that stretching and folding is the fingerprint of chaos. However, the connection between these two issues is rather recent and its implications are just being exploited (Ottino 1989b). Not all stretching and folding leads to chaos. In order for the flow to be chaotic it has to produce a *Smale horseshoe map* [Smale 1967, Moser 1973; this construction can be taken to be as the *definition* of chaos]. To produce a horseshoe the flow must be capable of deforming a region of fluid and returning it--stretched and folded--to its original location; furthermore the striations created by the reverse flow have to intersect the striations of the forward flow at an angle greater than approximately 19° (Chien, Rising & Ottino 1986). Repeated applications of a Smale horseshoe lead to a layered structure--very much like puff pastry--consisting of folds within folds. The chaos however, need not be widespread and might be confined to regions of small extent; where folds do not invade, islands form (see Figure 1). It is apparent that this picture cannot occur in *steady* two-dimensional flows; however, such a picture is likely in unsteady two-dimensional flows and steady three-dimensional flows.

These kinds of concepts have produced a sort of revival of kinematical studies and an awareness of the role played by topology in the understanding and analysis of flows. The key idea is that Eulerian velocity field, $\mathbf{v}(\mathbf{x}, t)$, regarded as a dynamical system,

$$(d\mathbf{x}/dt)_{\mathbf{X}} = \mathbf{v}(\mathbf{x}, t) \qquad \text{with} \qquad \nabla.\mathbf{v} = 0 \tag{1}$$

admits chaotic particle trajectories for relatively simple right-hand sides, $\mathbf{v}(\mathbf{x}, t)$ ($\mathbf{X}$ denotes the position of $\mathbf{x}$ at t=0). The exploration of this possibility in steady flows can be traced back to Arnold and Hénon and studies of Euler flows (Arnold 1965, Hénon 1966); the recognition that time-periodic two-dimensional flows can produce chaos is due to Aref (1986). However, for the purposes of this presentation we will restrict ourselves to Stokes flows since this is the only case for which comprehensive experiments have been carried out (e.g., Ottino 1989a,b, Chaiken et al. 1986).

The solution of equation (1) with $\mathbf{x}=\mathbf{X}$ at time t=0, gives the *flow* or *motion*,

$$\mathbf{x} = \Phi_t(\mathbf{X}) \qquad \text{with} \qquad \mathbf{X} = \Phi_{t=0}(\mathbf{X}) \tag{2}$$

i.e., the particle $\mathbf{X}$ is mapped to the position $\mathbf{x}$ after a time t. The best understood case corresponds to time-periodic velocity fields, $\mathbf{v}(\mathbf{x}, t) = \mathbf{v}(\mathbf{x}, t + T)$; in such a case, it is standard practice to deal with a mapping rather than with the entire flow.

The basic question is the following: What are the conditions under which a deterministic flow $\mathbf{v}(\mathbf{x}, t)$ is able to produce widespread and efficient stretching and folding throughout the space occupied by the fluid ? Equivalently, what is the 'best' way--assuming that a mixing measure has been adopted--to achieve a desired degree of mixing ? If one had an explicit expression for the flow (equation 2) many possibilities for analysis are open (see Ottino 1989a). However, the motion is almost never obtainable in closed form. A partial answer to these questions is possible in the case of two-dimensional time-periodic and three-dimensional spatially periodic flows.

2 TIME PERIODIC AND SPATIALY PERIODIC FLOWS

The simplest flows are steady two-dimensional velocity fields and variations thereof, such as the so-called *duct flows*:

$$dx/dt = \partial\psi/\partial y \ , \ dy/dt = -\partial\psi/\partial x \ , \ \psi = \psi(x, y) \ , \ dz/dt = v_z(x, y) > 0 \tag{3}$$

The flow in the x-y plane is characterized by the streamfunction $\psi(x, y)$. If the flow is bounded, the x-y flow can be divided into regions of closed streamlines. Let $T(\psi)$ denote the period corresponding the streamline ψ; *i.e.*, the time it takes for a fluid particle to return to its initial position in the x-y plane. The contribution to the stretching produced by the two-dimensional flow is given by

$$\mathbf{dx}_{nT} = \mathbf{dX}.\,[\,\mathbf{1}\text{-}\,(\mathrm{d}T/\mathrm{d}\psi)(\nabla\psi)\mathbf{v}\,]^{n}\ , \tag{4}$$

where $\mathbf{dX}$ represents the initial length of the material filament and $\mathbf{dx}_{nT}$ its state after a time nT. The length of $\mathbf{dx}$ increases as t for long times. In a duct flow the filament can stretch along the z-axis and the total stretch goes as t^2 for long times (Franjione 1989). Some concepts based on dynamical systems suggest how to improve upon this situation. The first possibility is to make $\psi(x, y, t) = \psi(x, y, t{+}T)$; the second is to make ψ spatially periodic, *i.e.*, $\psi(x, y, z) = \psi(x, y, z{+}L)$. Both possibilities lead to exponential stretching in at least some region(s) of the flow.

Dynamical systems concepts provide valuable guidance; however, only part of 'chaos theory' is useful and our understanding of mixing is far from complete. There are two main aspects which distinguish mixing from more conventional studies in dynamical systems and chaos. The first aspect is that in mixing one is interested primarily in *rate processes*, rather than asymptotic structures and long time behavior; the second is that the *perturbations from integrability are large*, since this is (roughly) when the best mixing occurs. The first aspect limits the application of asymptotic tools such as Poincaré sections [when they can actually be used; in many cases, such as the system of Figure 4, they are of very limited utility]; the second limits the applicability of perturbative techniques such as the Melnikov method (Wiggins 1988). There is also a computational counterpart to these issues. Conventional studies are based primarily on an Eulerian picture of the velocity field and often a visualization of the field in terms of discrete nodes is sufficient, especially in steady flows. However, in mixing one needs a Lagrangian description and this fact alone requires a substantially better degree of resolution of the velocity field itself (for example this is crucial in the case of fast reaction between two fluids which are controlled by the amount of area between the fluids; note that this does not mean that we cannot capture details such as folds and islands; see Figure 1). It is therefore clearly desirable to devise techniques valid for large perturbations which do not rely on an exact solution for the velocity field. One of our objectives of this paper is to highlight the role played by symmetries in the understanding of chaotic mixing (for related applications of symmetries see Beloshapkin *et al.* 1989).

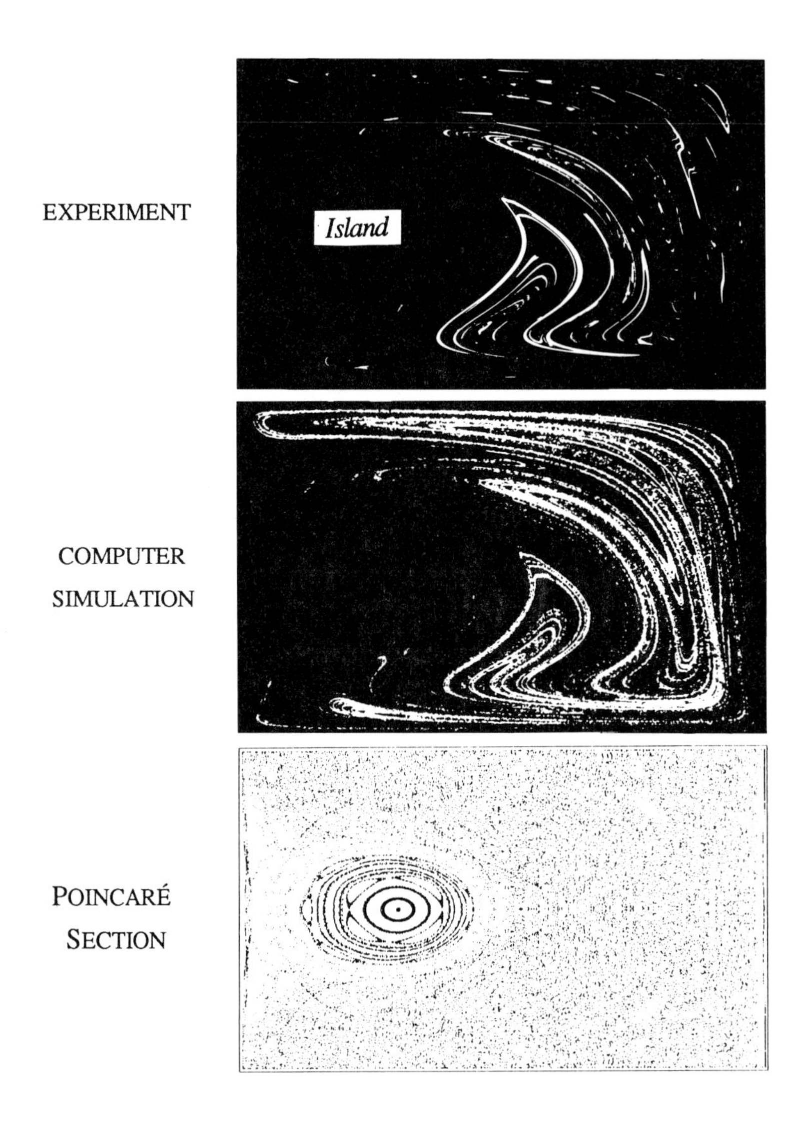

Figure 1. Comparison between a typical experiment, a computer simulation, and its corresponding Poincaré section, for a cavity flow (Leong 1989). The top two patterns are generated in just 8 periods. The Poincaré section, $O(10^3)$ periods, reveals a complex structure within the island, involving smaller islands, as well as a chaotic web. The rate of rotation in the smaller islands is very slow, and the green marker is largely unaffected by this structure (see also color plates).

In order to illustrate the main concepts consider a cavity flow operating under Stokes flow (Leong and Ottino 1989). The flow consists of a rectangular region capable of producing a two-dimensional velocity field in the x-y plane (Figure 1). The top and bottom walls can be moved in a steady or time-dependent manner (the top wall moves from left to right, the bottom wall moves from right to left). If the flow is time-periodic then it is convenient to record the location of initial conditions at times nT (n=1, 2, 3,...). In the best possible scenario it would then be possible to compute the evolution of the system by means of a nonlinear mapping $f^n(\mathbf{x}_0)=\mathbf{x}_n$. The analysis would probably start by computing the location and character of the periodic points, $f^n(\mathbf{p})=\mathbf{p}$, and their associated stable and unstable manifolds (see Table 7.1 in Ottino 1989a). This problem is very complex in its own right, however, to make matters worse, the mapping f is usually not known explicitly. The remarkable thing is that one does not have to know the map f explicitly in order to understand mixing flows from an experimental viewpoint. Much can be done by exploiting the symmetry of the velocity field and the problem is particularly simple for the case of creeping flows such as those in a time-periodic cavity.

3. SYMMETRIES

A map f with inverse f^{-1} is said to be *symmetric* if there exists a map $\mathbf{S}$, with $\mathbf{S}^2 = \mathbf{1}$, such that f and its inverse are related by $f = \mathbf{S} f^{-1} \mathbf{S}$. The set of points $\{\mathbf{x}\}$ such that $\{\mathbf{x}\} = \mathbf{S}\{\mathbf{x}\}$ is called the *fixed line* of $\mathbf{S}$. As an illustration consider the case of cavity flows. Assume that the flows consist of an arbitrary sequence of discontinuous flows induced by top and bottom wall motions; further assume that both walls are moved at the same speed for the same time t, that is they suffer the same displacement, and that the sense of rotation induced by the displacement is co-rotational (the possibilities are endless; virtually almost all these conditions can be relaxed; the motions can be counter-rotational, the sequence can be continuous rather than discontinuous, etc.). The motion corresponding to a wall displacement D is called a *period.* Let the motion induced by moving the top wall during a time t be denoted $\mathbf{T}_t$ (*i.e.*, a set of particles $\{\mathbf{x}\}$ is mapped to $\mathbf{T}_t\{\mathbf{x}\}$ at time t). Similarly, denote by $\mathbf{B}_t$ the motion induced in the cavity by moving the bottom wall. Since the system operates in the creeping flow regime the system is characterized by a displacement per period, D, and the specification of the sequence of top and bottom displacements. It is easy to show that both the top and bottom wall motions are symmetric with respect to the y-axis,

$$\mathbf{T}_t = \mathbf{S}_y \mathbf{T}_t^{-1} \mathbf{S}_y \quad , \quad \mathbf{B}_{t'} = \mathbf{S}_y \mathbf{B}_{t'}^{-1} \mathbf{S}_y \tag{5}$$

where $\mathbf{S}_y$ denotes the map $(x, y)\rightarrow(-x, y)$. In this case the fixed line of $\mathbf{S}_y$ is the y-axis itself (see Figure 2). Furthermore if $t=t'$ then the top and bottom motions are related by

$$\mathbf{T}_t = \mathbf{S}_y \mathbf{B}_t^{-1} \mathbf{S}_y \quad , \quad \mathbf{T}_t = \mathbf{R}\ \mathbf{B}_t \mathbf{R} \qquad (6)$$

where $\mathbf{R} \equiv \mathbf{S}_y \mathbf{S}_x = \mathbf{S}_x \mathbf{S}_y$ denotes a 180° rotation $(x, y)\rightarrow(-x, -y)$. The symmetries of composition of flows can be deduced by using rules from map algebra. (Franjione, Leong & Ottino 1989). For example, the flow $\mathbf{F}_{2t} \equiv \mathbf{B}_t \mathbf{T}_t$; *i.e.*, a periodic sequence of top and bottom wall motions with equal displacement, has symmetry with respect to the x-axis,

$$\mathbf{F}_{2t} = \mathbf{S}_x \mathbf{F}_{2t}^{-1} \mathbf{S}_x$$

whereas the flow $\mathbf{G}_{2t} \equiv \mathbf{B}_{t/2} \mathbf{T}_t \mathbf{B}_{t/2}$; *i.e.*, a series of top and bottom wall motions, but where observations are made half-way through the bottom wall displacement, has y-symmetry instead. A flow can have more than one symmetry. For example, the $\mathbf{F}_{2t}$ flow can be written also as

$$\mathbf{F}_{2t} = \mathbf{S}^* \mathbf{F}_{2t}^{-1} \mathbf{S}^*$$

where the symmetry $\mathbf{S}^*$ is given by $\mathbf{S}^* \equiv \mathbf{S}_y \mathbf{T}_t = \mathbf{T}_t^{-1} \mathbf{S}_y$. The fixed line of $\mathbf{S}^*$ is not obvious; it corresponds to the mapping of the y-axis with $\mathbf{T}_{t/2}^{-1}$; that is, the inverse of the top wall flow with half the displacement. In this case the symmetry line is a curve.

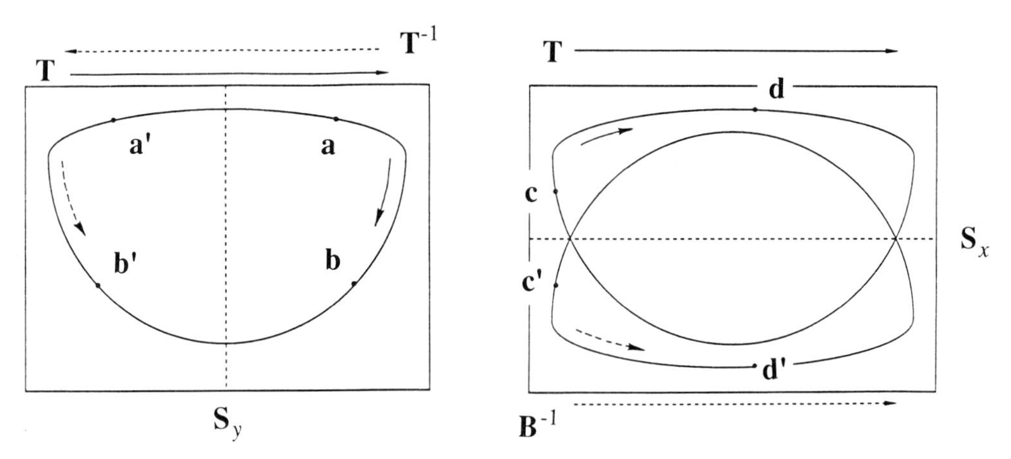

Figure 2. Schematic representation of symmetries in cavity flow: According to equation (5) the point **a** is mapped into **b** by **T**; equivalently **a** is mapped into **a'** by $\mathbf{S}_y$, **a'** into **b'** by $\mathbf{T}^{-1}$, and finally **b'** to **b** by $\mathbf{S}_y$ (see left figure). According to equation (6) the point **c** is mapped to **d** by **T**; equivalently **c** can be mapped to **c'** by $\mathbf{S}_x$, **c'** to **d'** by $\mathbf{B}^{-1}$, and finally **d'** to **d** by $\mathbf{S}_x$ (see right figure).

Extensive experimental studies in two-dimensional time-periodic flows show that there are large dynamic structures, called islands, which remain segregated even after long times (see Figure 1). Islands are routinely observed in computer experiments, and theory indicates that elliptic points are surrounded by invariant KAM (Kolmogorov-Arnold-Moser) surfaces that isolate the contents of the islands from the rest of the fluid; however, the theory is valid for small deviations from nonchaotic (regular) behavior. If the perturbation increases the KAM curves can develop 'holes', becoming what is called a *cantori*, which allow slow leaking of material placed in the interior of the island (experimentally it would be very hard to distinguish between a KAM curve and a cantori, Leong & Ottino 1989).

Islands preclude mixing. However, unless one has an explicit expression for the mapping, theory to date is insufficient to predict the evolution and location of islands even in relatively simple flows. Can the restriction imposed by islands be somehow eliminated even if an analytical description for the flow *is not* known ? Let us consider the following problem: Suppose that we have an initial condition to be mixed. The objective is to accomplish the best possible mixing using some allotted amount of displacement which we can use any way we want. We can accomplish good mixing by using a time periodic sequence of top and bottom motions. However, if we select a 'bad' D large islands will form (we should point out also that we do not know how 'large' the islands are going to be; if we are lucky they might be small). Two questions immediately follow: (i) how quickly can we *recognize* that the flow contains a large island ? (or islands of various sizes forming at different rates), and (ii) how can we destroy the island ? However, even if quickly identified and 'broken', other islands will form and these have to be destroyed too. Can islands be systematically prevented from forming ? The answer seems to be yes (Franjione, Leong & Ottino 1989). One possibility is to exploit the symmetries of the system. However, as soon as a time periodic sequence is chosen the flow contains symmetries and islands are restricted to lie on axes of symmetries or in pairs about the axis. Thus for example, the flow ...**BT BT BT** has x-symmetry and islands lie on the x-axis or in pairs about the x-axis. A way to prevent the formation of the island is to follow the flow $\mathbf{BT} = \mathbf{Q}_0$ by an identical flow which is rotated by 180°. If the first flow is 'rotated' immediately after the application of top and bottom motions, the new flow now reads $[\mathbf{R}(\mathbf{BT})\mathbf{R}]\ \mathbf{BT} = \mathbf{Q}_1 = \mathbf{TB\ BT}$ and the placement of the possible islands is switched (there is no fundamental reason to rotate the flow after just one cycle; a rotation after two cycles would be $[\mathbf{R}(\mathbf{BT\ BT})\mathbf{R}]\ \mathbf{BT\ BT}$). If the process is repeated we obtain $[\mathbf{R}(\mathbf{TB\ BT})\mathbf{R}]$ $\mathbf{TB\ BT} = \mathbf{Q}_2 = \mathbf{BT\ TB\ TB\ BT}$. It is interesting to note that the sequence generated is self-similar; *i.e.*, if $\mathbf{BT} \rightarrow \mathbf{B}'$ and $\mathbf{TB} \rightarrow \mathbf{T}'$, the sequence of $\mathbf{T}'$ and $\mathbf{B}'$ is identical to the original sequence. The usefulness of these ideas has been verified experimentally (Franjione, Leong & Ottino 1989).

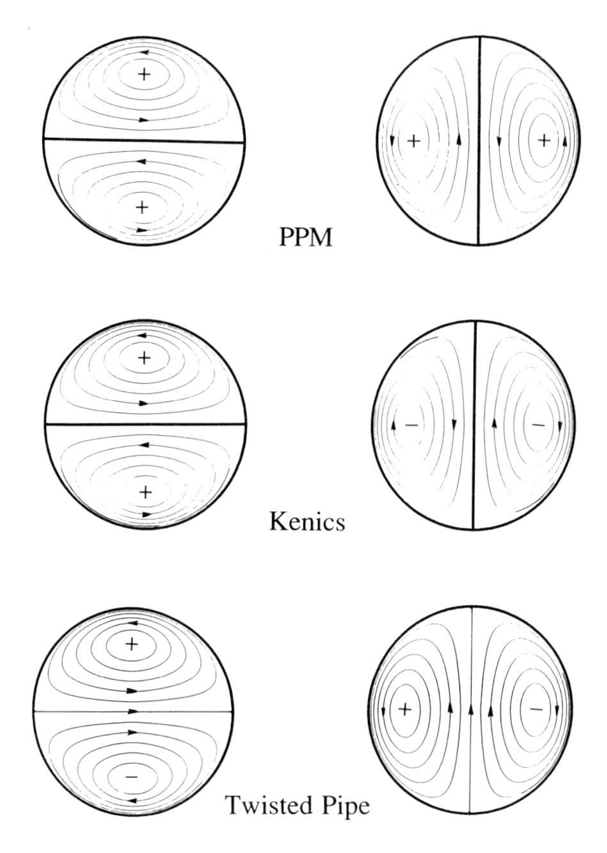

Figure 3. Comparison of cross sectional flows in PPM, twisted pipe, and Kenics^R mixer.

Ideas based on symmetries apply to spatially periodic duct flows as well. Two such systems are the partitioned-pipe mixer (PPM) system and the Kenics^R mixer. The PPM consists of a pipe partitioned into a sequence of semi-circular ducts by means of rectangular plates placed orthogonally to each other (the system can be regarded as an idealized version of a static mixer or a model porous medium). The fluid is forced under Stokes flow through the pipe by means of an axial pressure gradient while the pipe is rotated about its axis relative to the assembly of plates, thus resulting in a cross-sectional flow in the cross-sectional plane in each semi-circular element. Neglecting developing flows, a fluid particle jumps from streamsurface to streamsurface in between adjacent elements. Thus the flow consist of volume preserving flow (in the ducts) followed by an area preserving subdivision. The Kenics mixer is a special type of static mixer (Middleman 1977). This system consists of a tube with internal helical subdivisions, elements, of alternating right hand and left hand pitches. Ideally, after each element the streams are subdivided into two and after n-elements the striation thickness is reduced by 2^{-n}. Another spatially periodic system is the *twisted pipe* of Jones, Thomas & Aref (1989). The system consists of a periodic array of pipe bends; within each bend there is secondary flow consisting of two identical counter-rotating vortices. In a rough sense the streamline pattern in the cross section of all these systems is the same. What is different however, is the sense of rotation of the two vortices and how the sense of rotation changes from element to element (Figure 3). Evidently the symmetry properties of all these systems differ in a fundamental sense and drastic differences in mixing behavior are expected.

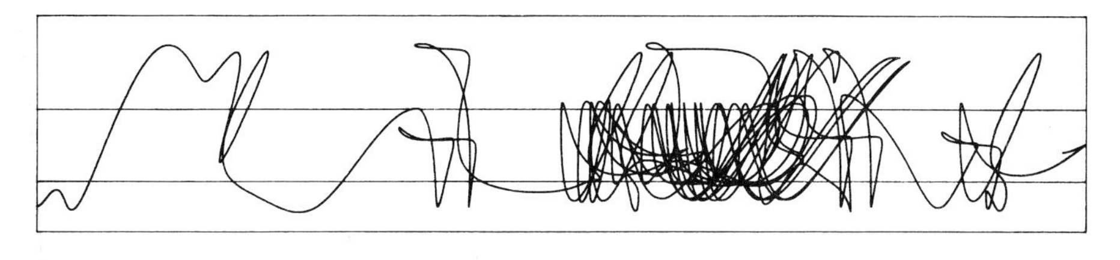

Figure 4. The eccentric helical annular mixer is a case of a continuous time-periodic flow. The cross section of this system corresponds to a combination of a the journal bearing flow and the axial flow pressure driven Poiseuille flow. The figure shows a lateral view (top) of a streakline injected into the flow (Kusch 1989).

The understanding of continuous time-periodic flows is substantially less developed. The ideas of symmetry considered above apply to projections of the cross sectional flow. However, substantial complexity is developed in the axial sense (Figure 4) and in fact this complexity is probably the most important aspect of continuous mixers from an experimental and practical viewpoint since it is connected with the age distribution of fluid elements in the mixer. Some understanding of these issues can be obtained by means of pathlines and streaklines, axial dispersion of tracer particles, and stretching of material lines. However, experimental studies in this class of systems are just starting (Kusch 1989) and our understanding is a bit less developed than that of two-dimensional or spatially periodic flows.

Acknowledgement: The work presented here was supported by DOE-Office of Basic Energy Sciences, the Donors of the Petroleum Research Fund, administered by the American Chemical Society, and the Materials Research Laboratory of the University of Massachusetts. I would like to express my gratitude to J.G. Franjione, H.A. Kusch, and C.W. Leong for their assistance during the writing of this paper.

REFERENCES

AREF, H 1984. Stirring by chaotic advection, *J. Fluid Mech.*, **143**, 1-21

ARNOLD, V.I. 1965. Sur la topologie des écoulements stationnaires des fluids parfaits. *Comptes Rendus Acad. Sci. Paris*, **A261**, 17-20

BELOSHAPKIN, V.V., CHERNIKOV, A.A., NATENZON, M. YA., PETROVICHEV, B.A., SAGDEEV, R.Z. & ZASLAVSKY, G.M. 1989. Chaotic streamlines in pre-turbulent states. *Nature*, **337**, 133-7

CHAIKEN, J., CHEVRAY, R., TABOR, M. & TAN, Q.M. 1986. Experimental study of Lagrangian turbulence in Stokes flow. *Proc. Roy. Soc. London*, **A408**, 165-74

CHIEN, W.-L., RISING, H., OTTINO, J.M. 1986. Laminar mixing and chaotic mixing in several cavity flows. *J. Fluid Mech.*, **170**, 355-77

FRANJIONE, J.G., LEONG, C.W. & OTTINO, J.M. 1989. Symmetries within chaos: a route to effective mixing, *Phys. Fluids A.*, to appear

FRANJIONE, J.G. 1989. Ph.D. Thesis, Dept. of Chemical Engineering, University of Massachusetts, Amherst, in progress

HÉNON, M. 1966. Sur la topologie des lignes de courant dans un cas particulier. *Comptes Rendus Acad. Sci. Paris*, **A262**, 312-4

JONES, S.W., THOMAS, O.M. & AREF, H. 1989. Chaotic advection by laminar flow in a twisted pipe. *J. Fluid Mech.*, in press.

KUSCH, H.A. 1989. Ph.D. Thesis, Dept. of Chemical Engineering, University of Massachusetts, Amherst, in progress.

LEONG, C.-W. 1989. Ph.D. Thesis, Dept. Chemical Engineering, University of Massachusetts, Amherst, in progress

LEONG, C.-W. & OTTINO, J.M. 1989. Kinematics of mixing in slow flows: an experimental study of chaotic mixing in a time-periodic cavity flow. *J. Fluid Mech.*, to appear

MIDDLEMAN, S. 1977. *Fundamentals of polymer processing*. New York: McGraw-Hill

MOSER, J. 1973. *Stable and random motion in dynamical systems*. Princeton: Princeton University Press

OTTINO, J.M., LEONG, C.W., RISING, H. & SWANSON, P.D. 1988. Morphological structures produced by mixing in chaotic flows. *Nature*, **333**, 419-25

OTTINO, J.M. 1989a. *The kinematics of mixing: stretching, chaos, and transport*, Cambridge: Cambridge University Press

OTTINO, J.M. 1989b. The mixing of fluids. *Scientific American*, **260**, 56-67

SMALE, S. 1967. Differentiable dynamical systems. *Bull. Amer. Math. Soc.*, **73**, 747-817

WIGGINS, S. 1988. *Global bifurcations and chaos: analytical tools*, New York: Springer

Lagrangian Chaos and Transport of Scalars

Bruno Eckhardt

Fachbereich Physik der Philipps-Universität, Renthof 6
3550 Marburg, Federal Republic of Germany

1 INTRODUCTION

The evolution of passive scalars advected by simple flows can be very complicated (Aref 1984). Nearby points can separate exponentially and material lines are stretched and convoluted to produce fractal like images (see the beautiful experiments of Chaiken *et al* 1986 and Chien *et al* 1987). Eventually, once the concentration gradients become too large, diffusion takes over and smoothens the structures. These different regimes call for different descriptions.

For short times, the effects of advection and diffusion can be discussed in some detail for Gaussian initial distributions. A direct link between the evolution of variances and the linearization of the flow field around the Lagrangean paths can be found (section 2).

For longer times, this approximation breaks down and another representation is needed: Markovian transport models. These are illustrated using a simple model of Solomon and Gollub (1988a,b) for their experiments on diffusion in oscillatory Rayleigh-Bénard flow (section 3).

I conclude with a summary and a comment on the kinematic dynamo in section 4.

2 SHORT TIME BEHAVIOUR

The time evolution of a dynamically passive scalar field $\phi(\mathbf{x}, t)$ advected by a given flow $\mathbf{v}(\mathbf{x}, t)$ and subject to diffusion characterized by a diffusion constant D is governed by

$$\frac{d\phi}{dt} = \frac{\partial\phi}{\partial t} + (\mathbf{v}\cdot\nabla)\phi = D\Delta\phi. \tag{1}$$

The relative importance of advection ($(\mathbf{v}\cdot\nabla)\phi$) over diffusion ($D\Delta\mathbf{v}$) is measured by the dimensionless *Peclet* number $Pe = ul/D$, where u and l are typical (external) velocity and length scale of the flow. Large Pe (as e.g. in the experiments of Gollub and Solomon discussed below, where $Pe \approx 10^6$) corresponds to advection dominated evolution.

Neglecting the rhs of eq (1) for this case, a solution is easily found:

$$\phi(\mathbf{x}(\mathbf{x}_0,t),t) = \phi(\mathbf{x}_0,0) \tag{2}$$

where $\mathbf{x}(\mathbf{x}_0,t)$ denotes a Lagrangian path which, starting from $\mathbf{x}_0$ at time zero, has reached $\mathbf{x}$ at time t.

The effect of diffusion can be taken into account in a Gaussian approximation,

$$\phi(\mathbf{x},t) = \frac{const}{(\det\,\Gamma)^{1/2}} \exp\left(-\frac{1}{2}(\mathbf{x}-\mathbf{q})^T\Gamma^{-1}(\mathbf{x}-\mathbf{q})\right). \tag{3}$$

The center of the Gaussian, $\mathbf{q}(t)$, moves along a Lagrangian path

$$\dot{\mathbf{q}}(t) = \mathbf{v}(\mathbf{q}(t),t). \tag{4}$$

Expanding the velocity field linearly around the path,

$$\mathbf{v}(\mathbf{x},t) = \mathbf{v}(\mathbf{q}(t),t) + \mathcal{L}(\mathbf{x}-\mathbf{q}) + O((\mathbf{x}-\mathbf{q})^2) \tag{5}$$

with

$$\mathcal{L}_{i,j} = \frac{\partial v_i}{\partial x_j}(\mathbf{q}(t)) \tag{6}$$

one finds an equation for the matrix of variances Γ,

$$\dot{\Gamma}(t) = 2D + \Gamma\mathcal{L}^T + \mathcal{L}\Gamma. \tag{7}$$

This equation can be solved in closed form:

$$\Gamma(t) = A(\Gamma(0) + 2D\int_0^t (A^TA)^{-1}d\tau)A^T \tag{8}$$

where

$$\dot{A}(t) = \mathcal{L}A \tag{9}$$

solves the linearized equations of the flow with initial condition $A(0) = 1$. For a divergence free velocity field, $\mathrm{Tr}\mathcal{L} = 0$, whence $\det A = 1$.

In the simplest cases, $\mathcal{L}$ has no component in the direction of the motion; spreading along the paths is then entirely due to diffusion. In a plane perpendicular to the

Lagrangian path (taken to be the (x, y)-plane), one has an interplay between diffusion and advection. One can distinguish three cases according to the eigenvalues of the matrix $\mathcal{L}$:
(i) Elliptical case: Here, $\mathcal{L}$ has imaginary eigenvalues and the velocity field is similar to a rotation, e.g. $\mathbf{v} = (-y, x)$. The Gaussian spreads linearly with time and rotates with the flow.
(ii) Parabolic case (shear flow): Both eigenvalues are zero, e.g. $\mathbf{v} = (\alpha y, 0)$. Now there is a cubic component in the variances:

$$\Gamma(t) = \begin{pmatrix} \frac{2}{3}D\alpha^2 t^3 + \alpha^2\Gamma_{22}t^2 + 2(D + \alpha\Gamma_{12})t + \Gamma_{11} & D\alpha t^2 + \alpha\Gamma_{22}t + \Gamma_{12} \\ D\alpha t^2 + \alpha\Gamma_{22}t + \Gamma_{12} & 2Dt + \Gamma_{22} \end{pmatrix}. \quad (10)$$

(Here Γ_{ij} denote the initial values at time zero.) This anomalous diffusion law $\sigma \approx t^{3/2}$ has been discussed by Rhines and Young (1983). It is observed on times long enough for the cube to dominate, but short enough so as not to invalidate the approximation.
(iii) Hyperbolic case (straining field): The eigenvalues of L are real ($\mathbf{v} = \lambda(y, x)$) and the variances grow exponentially:

$$\Gamma(t) = \begin{pmatrix} \Gamma_{11} + (\Gamma_{12} + D/\lambda)\sinh(2\lambda t) & (\Gamma_{12} + D/\lambda)\cosh(2\lambda t) - D/\lambda \\ (\Gamma_{12} + D/\lambda)\cosh(2\lambda t) - D/\lambda & \Gamma_{22} + (\Gamma_{12} + D/\lambda)\sinh(2\lambda t) \end{pmatrix}. \quad (11)$$

Now variances double within a time of the order of $t_* = \log 2/2\lambda$.

The conclusion from this section is that the (in-)stability of the particle trajectories affects the spreading of passive scalars. In particular the hyperbolic stretching near stagnation points (or X-points) gives rise to exponentially fast spreading. However, once the Gaussian becomes sufficiently broad, it no longer is a useful approximation. To describe the evolution on longer times, Markovian transport models have been used succesfully.

3 MARKOVIAN TRANSPORT MODELS

3.1 General Remarks

The basic idea of these kinds of models is to divide up the fluid volume into cells and to approximate the motion between cells by a random walk. Of course, a good choice of the cells, a justification for the neglect of correlations between steps and an estimate of transition rates are needed.

The first problem, the choice of the cells, has a long history, going all the way back to the thirties when Wigner, Eyring and others discussed rates of chemical reactions. There, the 'cells' are the initial and final states and interest focussed on the boundary between them, the so-called 'transition state'. From the vantage point of dynamical systems, these theories have been completed (in two degrees of freedom) in the work of Pechukas, Pollak and Child (see Pechukas 1981 and Pollak 1985 for details and references). Parallel developments include Channon and Lebowitz (1980), Bensimon and Kadanoff (1984), MacKay, Meiss and Percival (1984, 1987) and MacKay and Meiss (1986). This theory will be illustrated for a simple but very instructive model below.

Decay of correlations is to a large extend directly related to sensitive dependence on initial conditions. Because of the exponential instability, a cloud of particles will rapidly spread out more or less uniformly over phase space, thus quickly forgetting its initial condition (Bunimovich 1985). Memory effects can partly be accounted for by subdividing the cells further.

3.2 Oscillatory Rayleigh-Bénard Convection

Solomon and Gollub (1988a,b) have measured diffusion rates in stationary and oscillatory Rayleigh-Bénard convection. They suggest to approximate the flow by a planar model, $\mathbf{v} = (-\partial\Phi/\partial z, 0, +\partial\Phi/\partial x)$ with a streamfunction

$$\Phi(x, z, t) = \cos(x - x_c(t))\cos(z). \tag{12}$$

This model does not satisfy the proper boundary conditions, it ignores 3-d effects, and models rather poorly the complicated oscillatory instability (because of the Prandtl number of 4.7, one has non-uniform oscillations of blobs, see Bolton, Busse and Clever 1986). However, it nicely illustrates the qualitative features. Similar models have been studied by Broomhead and Ryrie (1988) and Knobloch and Weiss (1987).

In the stationary case, $x_c = 0$. Time is scaled such that at the center of the rolls ($x = 0, \pi$; $z = 0$) the period of rotation is 2π. Lengths are scaled so that the thickness of the layer is π and the wavelength of the rolls is 2π. The second instability is mimicked by an oscillation of the centers of the rolls, $x_c = \epsilon\cos(\omega t)$.

In the stationary case, there are hyperbolic fixed points (stagnation points) of the flow at $x = \pi/2, 3\pi/2$; $z = \pm\pi/2$. Their stable and unstable manifolds connect smoothly to form the boundary between cells: a layer of fluid that asymptotically moves from one stagnation point to another (also known as saddle connection). The period of rotation changes across streamlines. This shear is responsible for the larger diffusivity as calculated by Shraiman (1987) and Rosenbluth et al (1987) and measured by Solomon and Gollub (1988a). However, before this asymptotic regime is reached, there is a transient, anomalous regime because of the parabolic

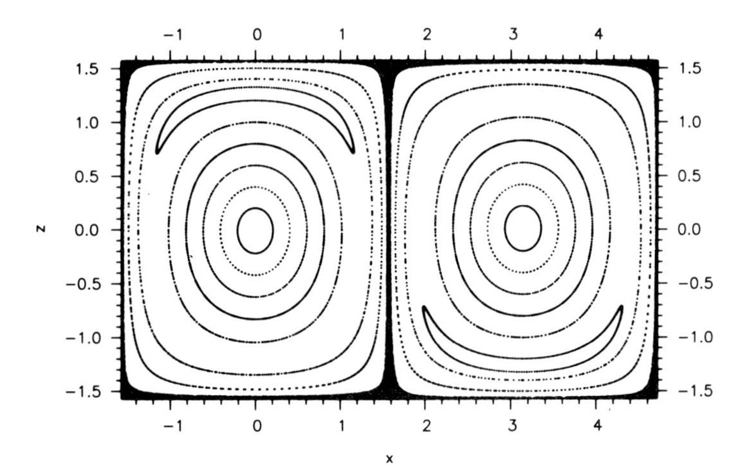

Fig. 1. Stroboscopic map of trajectories in the model for oscillatory Rayleigh Bénard convection. The positions (x, z) are recorded at every period of the oscillations. The parameters are $\epsilon = 0.01$, $\omega = 0.6$.

spreading indicated above (Eq (10)). This, too, has been verified experimentally (Cardoso and Tabeling 1988).

As soon as time dependence sets in, chaos is observed (see the stroboscopic map in Figure 1). The mechanism responsible for chaos is a clustering of resonances (Chirikov 1979). As one approaches the boundary, the period of rotation diverges logarithmically. Therefore, one always has initial conditions for which the rotation within a roll and oscillation are in resonance. A theorem of Birkhoff (Lichtenberg and Liebermann 1981) then says that this streamline breaks up into a chain of elliptic and hyperbolic fixed points. The original stagnation points on the upper and lower surface now move periodically with the oscillations (first order perturbation theory describes their location and phase very well). Their stable and unstable manifolds no longer connect smoothly, but cross in an infinite number of *heteroclinic* points (see Figure 2).

To set up a transport model in the form of a random walk between cells, we follow Channon and Lebowitz (1980) and take cells formed from the following lines: the top and bottom surface between the fixed points and the segments of the stable and unstable manifolds to the primary heteroclinic points, eg. point A_0 in Figure 2. (In Figure 2 only two fixed points and their manifolds are shown. There is another pair near $x \approx -\pi/2$, $z = \pm\pi/2$). After one iteration the point A_0 will be mapped onto A_1. The segment of the unstable manifold between A_0 and A_1 crosses the stable

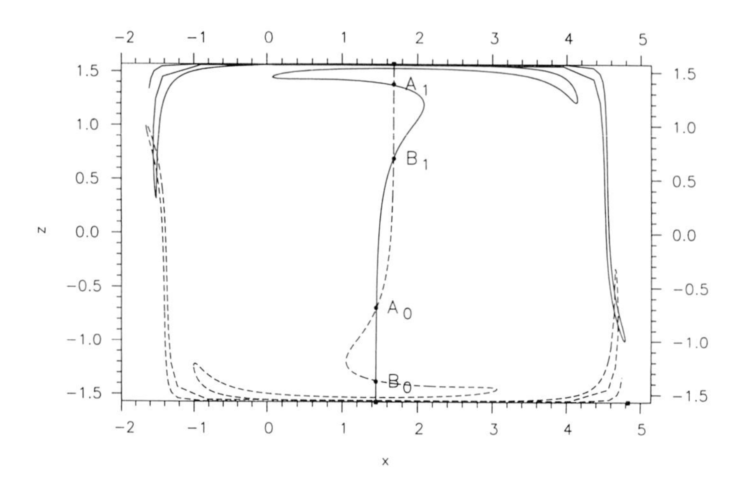

Fig. 2. Stable and unstable manifolds of two stagnation points. The continuous line is one of the unstable manifolds of the lower fixed point, and the dashed line is the stable manifold of the upper fixed point. The area bounded by the segments of the stable and unstable manifolds connecting the points $A_0 B_1$ equals the area bounded by the arcs between $B_1 A_1$. This is the area exchanged with the roll to the right.

manifold once in B_1. There are two segments of manifolds connecting A_0,B_1 and B_1,A_1, enclosing an area in phase space. Comparing the new cell with the old one, one observes that the area between A_0,B_1 has been mapped outside the cell, but the area between B_1,A_1 has been mapped inside. By volume conservation, the two areas are the same. The Markovian assumption is that particles that have entered a cell are rapidly randomized, so that there are no correlations between consecutive maps of the cells.

This model can be refined by dividing up cells further into regions that map out after $2, 3, \ldots, n$ iterations (see Dana, Murray and Percival 1989).

With these preparations, we can now compute transport rates. There is an accurate way of estimating the area between the segments as differences of heteroclinic orbits (see Pollak and Child 1980 and MacKay and Meiss 1986 for details). A calculation of Melnikov type (Chirikov 1979, ch. 4.4) gives an analytical formula for the effective diffusion constant

$$D_* = \frac{\epsilon\omega}{\pi \cosh(\pi\omega/2)} \tag{13a}$$

However, for this formula to be applicable, one needs that the oscillations are fast compared to the rotations. Numerical experiments confirm the linear dependence of D_* on the amplitude of the oscillations, but are at variance with the exponential law in frequency. The linear analysis of the fixed points shows that they are shifted by $\epsilon\omega/(1+\omega^2)$ relative to their unperturbed z-positions. Figure 2 suggests that this shift may be the dominant effect in determining the area enclosed by A_0,B_1, prediciting a diffusion rate

$$D_* = \frac{\epsilon\omega}{\pi(1+\omega^2)}. \tag{13b}$$

This is in reasonable agreement with the numerical data for large frequencies.

If the oscillations are much slower than the rotations, then it is useful to change coordinates to a frame of reference moving with the oscillations, $x' = x - x_c(t)$. In these coordinates, the streamfunction reads

$$\Phi(x', z, t) = \cos x \cos z + z\epsilon\omega \sin(\omega t). \tag{14}$$

The essentially new feature is that one has not only transitions to neighbouring cells but over many cells. For fixed ωt, this stream function is the one used by Moses and Steinberg (1988) to model mass transport in travelling waves in binary mixtures. Their experiments show the long range transport rather clearly. An analysis of transport in this regime will be presented elsewhere.

3.3 Anomalous diffusion due to trapping near islands

There is plenty of numerical evidence for slow algebraic decay of correlations in chaotic Hamiltonian systems (representative references are Karney 1983 and Shepelyansky and Chirirkov 1988). The favorite explanation is based on 'cantori', aperiodic chains of hyperbolic fixed points left behind in phase space by tori after break up. Similar to the ordinary, periodic hyperbolic fixed points discussed above, one can associate stable and unstable manifolds to them. MacKay, Meiss and Percival (1984) argue that near criticality (i.e. in a parameter range where a torus has just been destroyed), these cantori are the strongest barriers to transport. These cantori surround the islands in the chaotic sea, which in turn have islands with cantori around them and so forth. Thus there is an infinite hierachy of cells (bounded by cantori) with transport between them. Assuming selfsimilarity, one can indeed derive algebraic decay, however, with still some uncertainty in the exponent (see Meiss and Ott 1986 for further discussions). Pasmanter (1988) discusses an application of this anomalous diffusion to estuaries. In numerical simulations their influence is sometimes easily detected, see Figure 3.

In this model, it is assumed that arbitrarily small cells can be realized (transport on long times is dominated by smaller and smaller cells). However, diffusion cuts

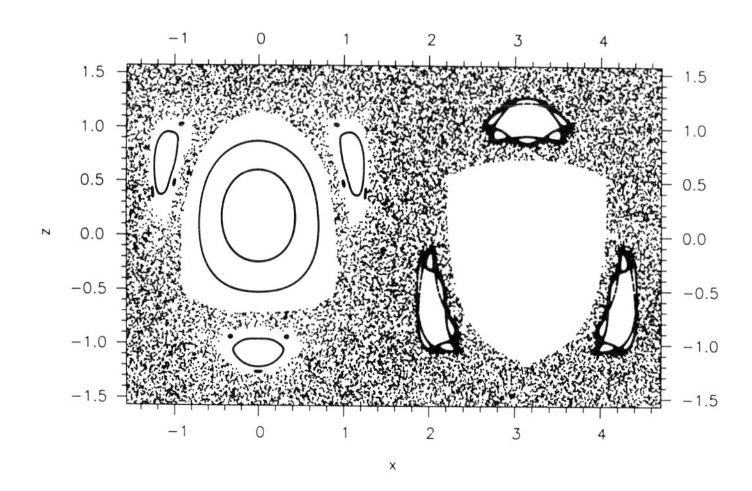

Fig. 3. An example of oscillatory flow with noticable trapping near islands. All points in the chaotic region are generated from one initial condition. The parameters are $\omega = 2$, $\epsilon = 0.3$.

off the hierachy of scales at some level, hence limits the time interval over which this algebraic decay may be observed. For the experiment of Solomon and Gollub I have estimated that diffusion takes over after about 100 periods of rotation of the rolls. This is not enough to see the influence of cantori.

3.4 Realistic Boundary Conditions

The model discussed thus far has free-free boundary conditions on the top and bottom surface. More realistic are rigid-rigid boundary conditions. Then the structure of fixed points is slightly different. The two fixed points on the boundary survive, but two new periodic solutions are created from the separatrix near the top and bottom fluid layer. It is their stable and unstable manifolds that can now be used to divide phase space.

If the side walls are taken into account, then the division between cells breaks down already for stationary flows (see Kessler, Dallmann and Oerterl 1984 and Arter 1985 for particle paths). It is not clear how to generalize the transport models to three degrees of freedom.

4 CONCLUSIONS

In the previous two section I have summarized a nonlinear dynamicists view of transport of scalars in Hamiltonian systems. The essential idea is the introduction of Markov models based on segments of stable and unstable manifolds. A Melnikov type calculation allows to derive the effective diffusion constant eq (13a), to be corrected by the motion of the fixed points, eq (13b). The linear dependence on the amplitude has been verified in numerical simulations. That it has been observed in the experiment of Solomon and Gollub (1988b) as well shows the robustness of this feature to details of the flow. Two problems deserve further study: the transport in the adiabatic regime (oscillations much slower than rotations) and transport in fully 3-d flows.

Another interesting aspect not yet fully explored is the existence of correlations, as e.g. illustrated in Figure 3. In fluids they may not be observable, but they may be relevant for the kinematic dynamo problem (which first requires an extension of the above models to transport of vectors). The evolution of a magnetic field **B** in a conducting fluid with (prescribed) velocity field **u** is governed by

$$\frac{\partial \mathbf{B}}{\partial t} + (\mathbf{u}\nabla)\mathbf{B} - (\mathbf{B}\nabla)\mathbf{u} = \frac{1}{R_m}\Delta\mathbf{B} \tag{15}$$

(with R_m the magnetic Reynolds number). The terms are grouped such that the lhs of the equation has a simple solution, containing the Jacobian of the mapping,

$$\mathbf{B}(\mathbf{x}, t) = A\mathbf{B}(\mathbf{x}_0(\mathbf{x}, t), 0) \tag{16}$$

where A is the matrix of eq (9). As discussed before, for hyperbolic flow, this implies exponential growth of the field components. This is counteracted by diffusion (which tends to smear out things exponentially almost as fast as they built up) and by the mixing of components due to the remaining terms.

Instability can win over diffusion, if there is an increased probability of being in an unstable region (Bayly 1986), that is to say, if motion is correlated. One cause of correlation (trapping in selfsimilar island structures) has been mentioned before. However, as discussed by Ruelle (1986), even motion in strongly chaotic systems can be correlated showing resonances. In numerical simulations, these resonances show up very much like the trapping near islands in Figure 3: trajectories visit certain regions in phase space preferentially. Work on a quantitative theory of these resonances is in progress (Eckhardt, Gomez-Llorente and Pollak 1989), but it is already becoming clear that what looks like homogeneous chaos on long time scales may be well organized and coherent on intermediate time scales. This is perhaps the mechanism by which Lagrangian chaos leads to fast dynamos.

5 ACKNOWLEDGEMENTS

It is a pleasure to thank H Aref, J Gollub, T Solomon, V Steinberg for discussions and S Großman for a critical reading of the manuscript.

6 REFERENCES

Aref, H., 1984 Stirring by chaotic advection, *J. Fluid Mech.*, **146**, 1–21

Arter, W., 1985, Nonlinear Rayleigh-Bénard convection with square planform, *J. Fluid Mech.*, **152**, 391–418

Bayly, B.J., 1986 Fast magnetic dynamos in chaotic flows, *Phys. Rev. Lett.*, **57**, 2800–2803

Bensimon, D., and L.P. Kadanoff, 1984 Extended chaos and disappearance of KAM trajectories, *Physica*, **13 D**, 82–89

Broomhead, D.S. and S.C. Ryrie, 1988 Particle paths in wavy vortices, *Nonlinearity*, **1**, 409–434

Bunimovich, L.A., 1985 Decay of correlations in dynamical systems with chaotic behavior, *Sov. Phys. JETP*, **62**, 842–852

Bolton, E.W., F.H. Busse and R.M. Clever, 1986 Oscillatory instabilities of convection rolls at intermediate Prandtl numbers, *J. Fluid Mech.*, **164**, 469

Cardoso, O. and P. Tabeling, 1988 Anomalous diffusion in a linear array of vortices, *Europhys. Lett.*, **7**, 225–230

Chaiken, J., R. Chevray, M. Tabor and Q.M. Tan, 1986 Experimental study of Lagrangian turbulence in a Stokes flow, *Proc. R. Soc. (London)*, **A 408**, 165–174

Channon, S.R. and J.L. Lebowitz, 1980 Numerical experiments in stochasticity and homoclinic oscillation, *Ann. NY Acad. Sci*, **357**, 108–118

Chien, W.-L., H. Rising and J.M. Ottino, 1987 Laminar mixing and chaotic mixing in several cavity flows, *J. Fluid Mech.*, **170**, 355–377

Chirikov, B.V., 1979 A universal instability of many-dimensional oscillator systems, *Phys. Rep.*, **52**, 263-379

Dana, I., N.W. Murray and I.C. Percival, 1989 Resonances and diffusion in periodic Hamiltonian maps, *Phys. Rev. Lett.*, **62**, 233–236

Eckhardt, B., J. Gomez-Llorente and E. Pollak, 1989 submitted for publication

Karney, C.F.F., 1983 Long-time correlations in the stochastic regime, *Physica*, **8D**, 360-380

Kessler, R., U. Dallmann and H. Oertel, 1984 Nonlinear transitions in Rayleigh-Bénard convection, in *Turbulence and chaotic phenomena in fluids*, T. Tatsumi (ed), (Amsterdam: Elsevier Science), pg. 173–178

Knobloch, E. and J.B. Weiss, 1987 Chaotic advection by modulated travelling waves, *Phys. Rev. A*, **36**, 1522–1524

Lichtenberg, A.J. and M.A. Lieberman, 1981 *Regular and Stochastic Motion*, (Berlin: Springer)

MacKay, R.S. and J.D. Meiss, 1986 Flux and differences in action for continuous time Hamiltonian systems, *J. Phys. A: Math. Gen.*, **19**, L225–L229

MacKay, R.S., J.D. Meiss and I.C. Percival, 1984 Transport in Hamiltonian systems, *Physica*, **13 D**, 55–81

MacKay, R.S., J.D. Meiss and I.C. Percival, 1987 Resonances in area preserving maps, *Physica*, **27 D**, 1–20

Meiss, J.D., and E. Ott, 1986 Markov tree model of transport in area preserving maps, *Physica*, **20 D**, 387–402

Moses, E. and V. Steinberg, 1988 Mass transport in propagating patterns of convection, *Phys. Rev. Lett.*, **60**, 2030–2033

Pasmanter, R.A., 1988 Anomalous diffusion and anomalous stretching in vortical flows, *Fluid Dyn. Res.* **3**, 320–326

Pechukas, P. 1981 Transition state theory, *Ann. Rev. Phys. Chem.*, **32**, 159–177

Pollak, E., 1985 Periodic orbits and the theory of reactive scattering, in *Theory and Models of Chemical Reactions*, vol III, M Baer (ed), (Boca Raton: CRC)

Pollak, E. and M. Child, 1980 Classical mechanics of a collinear exchange reaction, *J. Chem. Phys.*, **73**, 4373–4380

Rhines, P.B. and W.R. Young, 1983 How rapidly is a scalar mixed within closed stream lines? *J. Fluid Mech.*, **133**, 133–145

Rosenbluth, M.N., H.L. Berk, I. Doxas and W. Horton, 1987 Effective diffusion in laminar convective flows, *Phys. Fluids*, **30**, 2636–2647

Ruelle, D., 1986 Resonances of chaotic dynamical systems, *Phys. Rev. Lett.*, **56**, 405–407

Shepelyansky, D.L. and B.V. Chirikov, 1988 Renormalization chaos and motion statistical properties, *Phys. Rev. Lett.*, **61**, 1039

Shraiman, B.I., 1987 Diffusive transport in a Rayleigh-Bénard convection cell, *Phys. Rev. A*, **36**, 261–267

Solomon, T.H. and J.P. Gollub, 1988a Passive transport in steady Rayleigh-Bénard convection, *Phys. Fluids*, **31**, 1372–1379

Solomon, T.H. and J.P. Gollub, 1988b Chaotic particle transport in time-dependent Rayleigh-Bénard convection, *Phys. Rev. A*, **38**, 6280–6286

On Chaotic Flow around the Kida Vortex

L.M. Polvani and J. Wisdom
Massachusetts Institute of Technology, Cambridge MA 02139

The ordinary accounts of this vortex
in no way prepared me for what I saw.

Edgar Allan Poe

The "Kida vortex," a two-dimensional elliptical patch of constant vorticity embedded in a uniform quadratic background shear flow, has recently been the subject of a number of studies (Kida 1981, Dritschel 1989a, Meacham *et al.* 1989). These have concentrated mostly on understanding the linear stability of this solution and its nonlinear evolution under perturbations. In this paper we report on a new and surprisingly interesting aspect of the Kida vortex: chaotic advection is induced by the vortex in the surrounding fluid over regions much larger than the vortex itself, even for relatively small values of the background shear.

We start by presenting the Kida solution and describing the possible behaviors. We then proceed to derive the equations of motion for a passive tracer advected by the velocity field of the Kida vortex, and present some Poincare sections. The growth of the chaotic regions is studied by applying simple resonance overlap ideas. We conclude with a discussion of the implications of this finding for our understanding of two-dimensional turbulent fields.

1. THE KIDA VORTEX

The flow induced by an elliptical region of constant vorticity embedded in a quadratic shear flow is an exact time-dependent nonlinear solution of the Euler equations in two dimensions (Kida 1981). The quadratic form of the background is interesting because it can be thought of as the beginning of a local Taylor series expansion of the complicated velocity field in which the vortex is located. Since a shear flow can be decomposed as the sum of a rotation ω and a strain s, the streamfunction associated with the background field can be written:

$$\Psi_B = (1/4)\ (\omega + s)\ x^2\ +\ (1/4)\ (\omega - s)\ y^2, \tag{1}$$

where (x, y) are inertial Cartesian coordinates; ω and s may be functions of time. The result of Kida (1981) is that an elliptical vortex patch embedded in (1) remains elliptical for all times. The flow field is then entirely determined by the aspect ratio λ of the ellipse (we have chosen the convention $0 < \lambda < 1$) and its orientation, specified by the angle φ between the x-axis and the major axis of the ellipse. The evolution of

λ and φ is governed by the following system:

$$\frac{d\varphi}{dt} = \Omega_K + \frac{1}{2}[\omega + s\,\Lambda\,\cos 2\varphi] \qquad \text{and} \qquad \frac{d\lambda}{dt} = -s\,\lambda\,\sin 2\varphi\,, \tag{2}$$

where $\Omega_K = \lambda/(1+\lambda)^2$ is the angular velocity of an elliptical vortex patch in the absence of shear (a Kirchhoff vortex), and $\Lambda = (1+\lambda^2)/(1-\lambda^2)$. All variables in (1) and (2) are non-dimensional, with time and length scales chosen so that the patch has vorticity 1 and area π. Although φ and λ are not canonically conjugate variables, (2) can be shown to be a Hamiltonian system, and trajectories in (φ, λ) space are integrable. Despite their apparent simplicity, (2) give rise to a rich variety of behaviors, even for the case of constant ω and s to which we hereinafter restrict our discussion. An example is illustrated in Figure 1, where the (φ, λ) orbits for $\omega = -.2$ and $s = 0.1$ are plotted. Depending on their aspect ratio vortices either rotate or "librate" around stationary solutions.

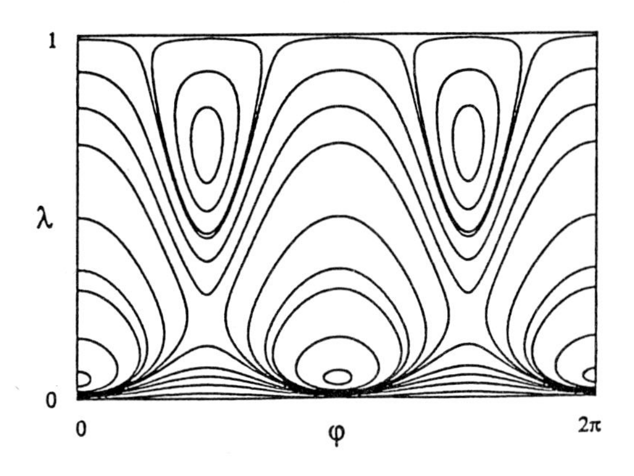

Figure 1: The (φ, λ) orbits for Kida vortices at $\omega = -0.2$ and $s = 0.1$, in region D of Figure 2.

Following Meacham *et al.*, it is simplest to classify the Kida solutions by dividing the (s, ω) parameter space into different regions, depending on the number of stationary solutions of (2); the latter were originally determined by Moore & Saffman (1971). Note first that (2) is invariant under the transformation $s \to -s$ and $\varphi \to \varphi + \frac{\pi}{2}$; thus only the $s > 0$ half-plane need be considered.

The whole phenomenology of the Kida solutions, synthesized in Figure 2, is physically understood as follows. For a given rotation ω, if the strain s is sufficiently large vortices of all aspect ratios will be sheared away: this is region C, where the only stationary solutions of (2) are for $\lambda = 0$. Conversely, for a given s all vortices will survive the straining provided the rotation is large enough, in either sense (these are regions A and E). In region B the rapidly rotating vortices (i.e. the more circular ones, c.f. the expression for Ω_K above) survive the straining although the background streamlines are hyperbolic, while the slow elongated vortices are sheared away. The richest behavior is found in region D, of which an example is given in Figure 1.

Figure 2: Phenomenology of the Kida solutions in the (ω, s) plane. The letters designate the regions described in the text.

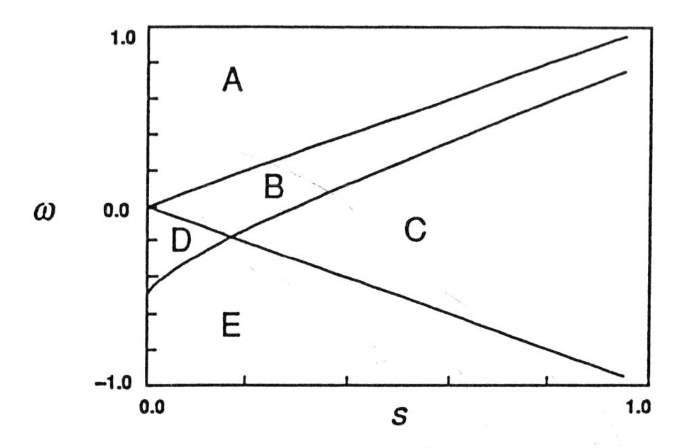

The linear stability of the Kida solutions to two-dimensional normal mode perturbations has recently received much attention (Dritschel 1989a, Meacham *et al.* 1989). The essential result is the following: in addition to a Love-type instability similar to the one of the Kirchhoff vortex, resonance between the frequency of the (φ, λ) motion and the frequency of the disturbances on the elliptical boundary leads to a Mathieu-like parametric instability. However, the unstable bands become vanishingly thin and are associated with extremely small growth rates in the limit $s \to 0$. Moreover, the relevance of this linear analysis to the fully nonlinear evolution is unclear, since the instability often manifests itself only through the emergence of very thin filaments while the vortex retains its elliptical shape for long times. All results presented in this paper are for cases that are linearly stable.

2. CHAOTIC ADVECTION AROUND KIDA VORTICES

Having reviewed the Kida solutions from an *Eulerian* viewpoint, we now present new results regarding the *Lagrangian* motion of passive tracer advected by a Kida vortex. The total streamfunction $\Psi_T(x, y, t) = \Psi_B(x, y) + \Psi_V(x, y, t)$, is the sum of the quadratic background flow (1) and the streamfunction Ψ_V of a Kirchhoff vortex of aspect ratio $\lambda(t)$ and inclination $\varphi(t)$. In Cartesian coordinates, the position (x, y) of a particle evolves according to:

$$\frac{dx}{dt} = -\frac{\partial \Psi_T}{\partial y} \quad \text{and} \quad \frac{dy}{dt} = \frac{\partial \Psi_T}{\partial x}. \tag{3}$$

These are Hamilton's equations. The Hamiltonian Ψ_T contains explicit periodic time dependence, which suggests the presence of chaotic advection. Inside the vortex, however, Ψ_V is merely quadratic in x and y (Lamb 1945), so that (3) is linear with periodic coefficients; thus motion in the interior of the vortex is integrable.

To integrate (3) for the exterior field, we have found it easier to work in elliptical coordinates (ρ, ϑ) rotating and stretching with the vortex, since the analytical expression for Ψ_V is known in terms of ρ and ϑ, and there is no simple way of expressing (ρ, ϑ) in terms of (x, y). The conversion between the two coordinate systems, the combination of an elliptical transformation and a rotation, is given by:

$$\begin{aligned} x &= c \cosh\rho \,\cos\vartheta\, \cos\varphi \;-\; c\,\sinh\rho\, \sin\vartheta\, \sin\varphi \\ y &= c \cosh\rho\, \cos\vartheta\, \sin\varphi \;+\; c\,\sinh\rho\, \sin\vartheta\, \cos\varphi, \end{aligned} \tag{4}$$

where $c^2 = (1-\lambda^2)/\lambda$ depends on time through λ. Substitution of (4) into (3) yields, after some algebra, the evolution equations for ρ and θ:

$$\begin{aligned} \frac{d\rho}{dt} &= \left(\frac{c^2h^2}{2}\right)\left[-\left(\Omega_K \sin 2\vartheta \;+\; 2c^{-2}\frac{\partial\Psi_V}{\partial\vartheta}\right) \;+\; \frac{1}{2}\, s\, \mathcal{F}(\rho,\vartheta,\varphi)\right] \\ \frac{d\vartheta}{dt} &= \left(\frac{c^2h^2}{2}\right)\left[-\left(\Omega_K \sinh 2\rho \;-\; 2c^{-2}\frac{\partial\Psi_V}{\partial\rho}\right) \;+\; \frac{1}{2}\, s\, \mathcal{G}(\rho,\vartheta,\varphi)\right], \end{aligned} \tag{5}$$

with

$$\begin{aligned} \mathcal{F} &= (\cosh 2\rho\, \sin 2\vartheta\, \cos 2\varphi + \sinh 2\rho\, \cos 2\vartheta\, \sin 2\varphi) \\ &\quad - \Lambda\,(\cos 2\varphi\, \sin 2\vartheta + \sinh 2\rho\, \sin 2\varphi) \\ \mathcal{G} &= (\sinh 2\rho\, \cos 2\vartheta\, \cos 2\varphi - \cosh 2\rho\, \sin 2\vartheta\, \sin 2\varphi) \\ &\quad + \Lambda\,(\sin 2\varphi\, \sin 2\vartheta - \sinh 2\rho\, \cos 2\varphi)\,, \end{aligned}$$

and

$$c^2h^2 = \left(\cosh^2\rho \;-\; \cos^2\vartheta\right)^{-1}.$$

We have used a Bulirsch-Stoer algorithm to numerically integrate the autonomous 2-degree of freedom Hamiltonian problem constituted by (2) and (5), instead of solving (3) directly. Notice from (2) that for $s = 0$ the ellipse rotates at constant angular velocity with constant aspect ratio; in this case the Lagrangian fluid motion is integrable and, in the corotating frame, the streamlines are time-independent and appear as illustrated in Figure 3. It is important to note the presence of two saddle points S_1 and S_2 as well as two centers (usually referred to as 'ghost vortices'). For the weakly perturbed system, the chaotic zones are found to appear surrounding the separatrices (represented by the dotted lines in Figure 3), as is expected for weakly perturbed Hamiltonian systems.

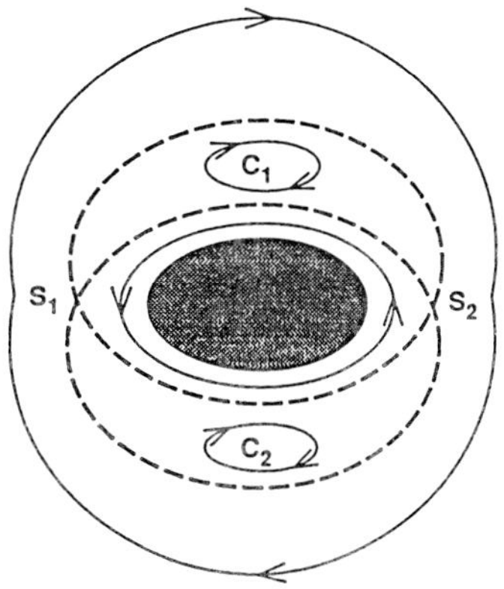

Figure 3: The Kirchhoff vortex streamfunction in the corotating frame. The solid lines are particles trajectories, the dashed lines are the separatrices, S_1 and S_2 the saddle points and C_1 and C_2 the 'ghosts'.

The important result of this study is that for small values of s and ω remarkably large chaotic zones are found surrounding the entire vortex. We show a typical example in Figure 4, were the surface of section for a Kida vortex at $\omega = -0.2$ and $s = 0.1$ is presented. The aspect ratio for this ellipse oscillates in the range $0.377431\ldots < \lambda < 0.9$. The figure is obtained by plotting a dot corresponding to the position of a single particle every period of the (φ, λ) motion, when $\varphi = 0$. In this and subsequent figures, the ellipse inside the chaotic zone coincides with the boundary of the vortex. We have found that chaotic zones are often very close to the vortex boundary. In view of this, it is not surprising that the behavior of vorticity filaments surrounding Kida vortices has been found to be extremely complicated (Dritschel 1989a). The presence of chaotic zones surrouding elliptical vortices may also provide an efficient mechanism for vortex stripping.

Another instance of great interest is the case $|\omega| = |s|$, for which the backround flow is a pure shear (an unbounded Couette flow) . The case $\omega = -s = 0.5$ is presented in Figure 5, for a vortex of aspect ratio $\lambda = 0.8$ at $\varphi = 0$. The chaotic zone appears to extend to arbitrarily large values of x, and is confined in y around the vortex. These results suggest that the elliptical vortices formed in unstable mixing layers may actually play a very important role in the mixing process.

To understand how the large chaotic regions emerge as the strength of the perturbing strain is increased, we have applied some simple ideas from the theory of overlapping resonances problem and found two main types of overlap that lead to large chaotic zones. The purpose of this analysis is by no means a detailed investigation of all the resonance overlaps in the Kida problem, but a simple approach that would account for the existence of large chaotic zones at small values of s and ω.

For case $\omega < 0$, one of the main resonances is associated with the 'ghost' of the exterior unperturbed Kirchhoff flow Ψ_V (see Figure 3), for which the period of the Lagrangian trajectory is identical to the period of the (φ, λ) motion; we designate this the 'Kirchhoff resonance'. The other main resonance is associated with the streamline pattern due to the superposition of the background quadratic shear with the equivalent monopolar velocity field of the vortex. We then consider the streamfunction Ψ_{pv} obtained by the superposition of a point vortex of strength π and the background flow (1), i.e. $\Psi_{pv} = (1/2)\ln r + \Psi_B$, where r is the radial distance from the origin. Such a streamfunction has an associated streamline pattern geometrically identical to the one shown in Figure 3, with the exception that it is rotated by $\pi/2$. The resonance of interest is the one associated with the 'ghost' vortex of Ψ_{pv}. It is a simple matter to find the radial distances r_{pv} and r_K of the resonances associated with Ψ_{pv} and Ψ_V (Polvani *et al.* 1989), and the respective half-widths Δr_{pv} and Δr_K, although the computation has to be performed numerically since transcendental equations need to be solved. According to the simple resonance overlap criterion we expect to observe large chaotic regions if

$$\Delta r_{pv} + \Delta r_K > r_{pv} - r_K \,. \tag{6}$$

As an illustration, the case $\omega = -0.1$ and $\lambda = 0.9$ is presented in Figure 6, where four surfaces of section are shown for increasing values of the perturbation

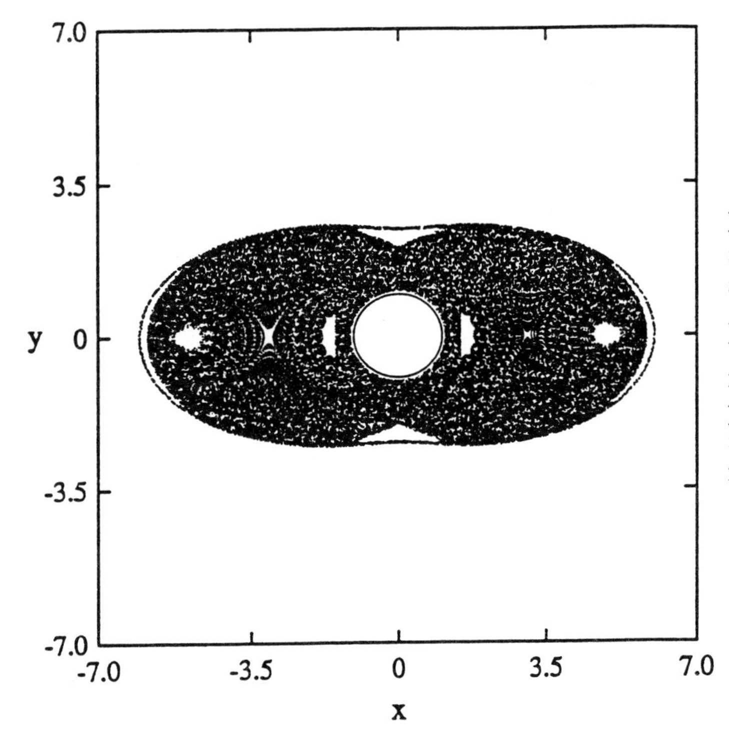

Figure 4: Surface of section for the Kida vortex at $\omega = -0.2$, $s = 0.1$ and $\lambda = 0.9$ at $\varphi = 0$. The figure was obtained with a single initial condition; 70,000 points are plotted. The aspect ratio varies in the interval $0.377431 < \lambda < 0.9$.

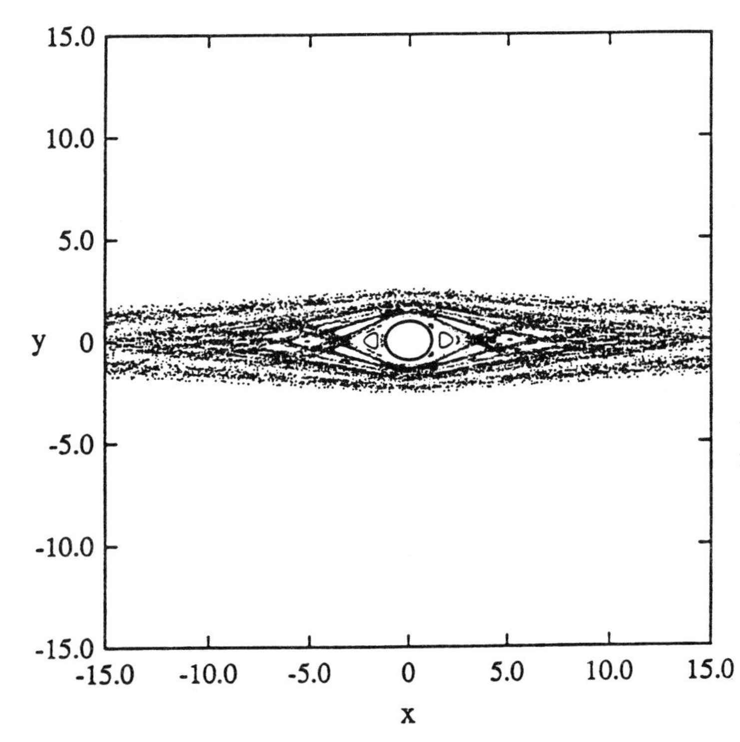

Figure 5: Surface of section for the Kida vortex at $\omega = 0.5$, $s = -0.5$ and $\lambda = 0.8$ at $\varphi = 0$. The background field is a simple shear flow of the form $u = (1/2)\,y$. The chaotic region appears to extend to arbitrarily large values of x. The aspect ratio for this vortex varies within $0.384535 < \lambda < 0.9$.

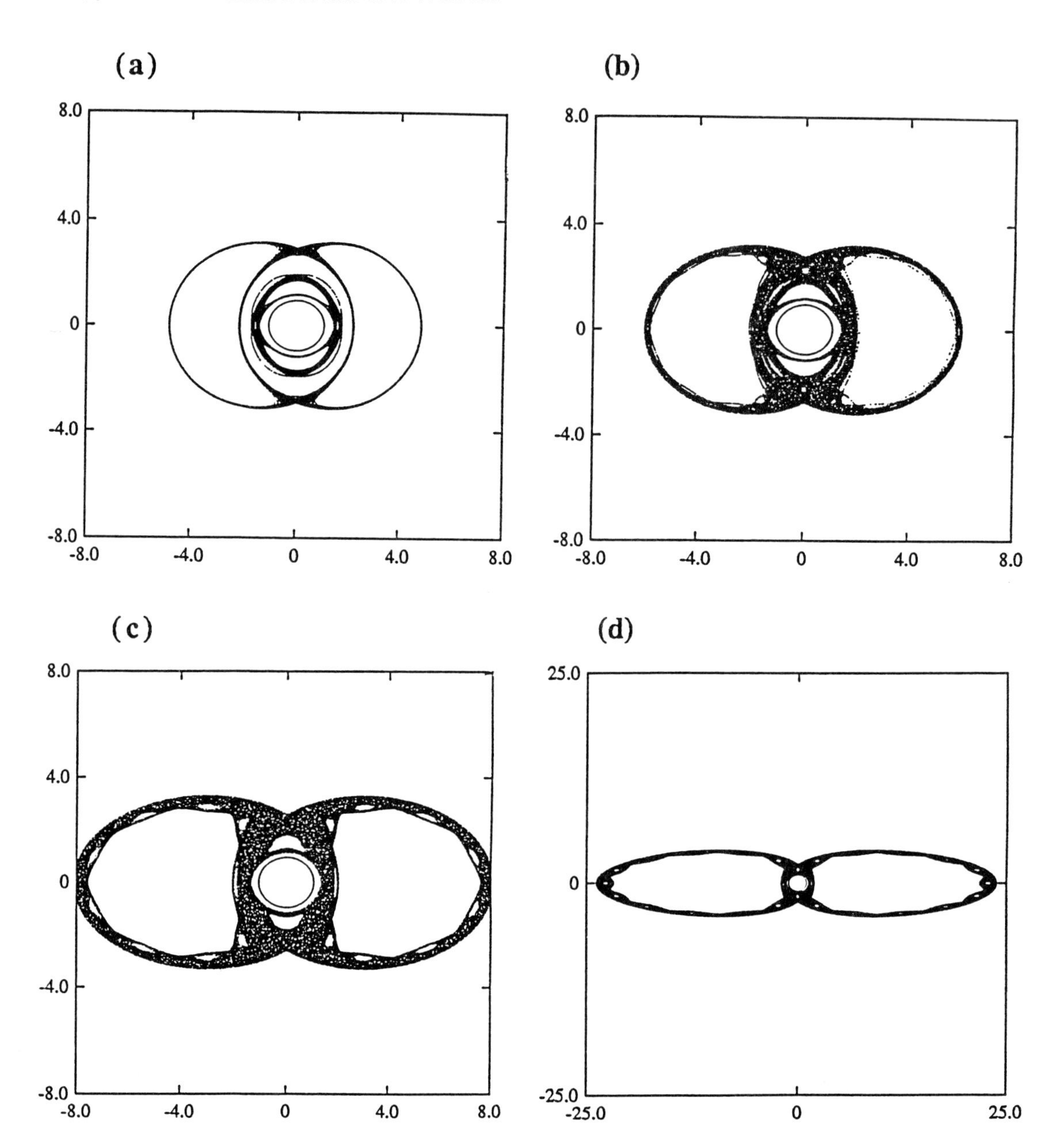

Figure 6: Surfaces of section at increasing values of the strain parameter s for Kida vortices at $\omega = -0.10$ and $\lambda = 0.9$ at $\varphi = 0$. (a) $s = 0.015$, (b) $s = 0.03$, (c) $s = 0.05$ and (d) $s = 0.09$. Each figure contains between 50,000 and 100,000 points.

parameter s. Resonance overlap occurs at $s = 0.0276$. At $s = 0.015$ (Fig. 6a) the two chaotic regions associated with the above mentioned resonances appear to be well separated. They have merged into a single chaotic zone at $s = 0.03$ (Fig. 6b). The transition compares favourably with the resonance overlap prediction (6). As increased the chaotic regions become larger. The section at $s = 0.05$ is shown in Figure 6c. It should be noted that the aspect ratio for this vortex varies in the range $0.696 < \lambda < 0.90$. This shows that even rather circular vortices may possess very large chaotic regions in the presence of extremely small straining fields. As the background approaches a pure shear (i.e. close to the line $|\omega| = |s|$), the chaotic zone becomes truly enormous (see Fig. 6d for the case $s = 0.09$), and are elongated in the direction of the shear.

When $\omega > 0$ the resonance associated with Ψ_{pv} cannot exist, and the large chaotic zones are produced by the overlap of resonances of the full Hamiltonian. Since the actual expansion of Ψ_T as a sum of resonances is rather complicated, we have preferred a simple numerical approach to determine both the location and width of the primary resonances. To the nth resonance, associated with the orbit that has a period $T_n = nT_{\varphi\lambda}$ (where $T_{\varphi\lambda}$ is the period of the (φ, λ) motion obtained from (2)), corresponds the frequency $\omega_n = \omega_{\varphi\lambda}/n$. To locate the resonances in the (x, y) plane we have used a Nelder-Mead scheme that minimizes the return distance on the section after $n\ T_{\varphi\lambda}$ periods. The frequency half-width $\Delta\omega_n$ associated with the nth resonance is determined by relating it to the frequency α_n of small oscillations around that resonance, which we compute by means of Floquet multipliers. Since the Kida vortex has a 4-fold symmetry, the nth resonance can be associated with a pendulum-like Hamiltonian of the form $H_n = \frac{1}{2}Ap^2 - B\cos(2n\vartheta - 2\omega_{\varphi\lambda}t)$, where p is canonically conjugate to the elliptical angle ϑ and zero at the resonance center. The actual values of A and B are immaterial. It is simple to show that for H_n the half-width Δw_n is related to α_n by the identity $\Delta\omega_n = \alpha_n/n$. Again we expect overlap when the simple criterion $(\Delta\omega_{n_1} + \Delta\omega_{n_2}) > (\omega_{n_1} - \omega_{n_2})$ is satisfied.

An example of this type of resonance overlap is given in Figure 7, where the surfaces of section for $\omega = 0$ and $\lambda = 0.5$ (at $\varphi = 0$) are shown at four values of s. At $s = 0.002$ (Fig. 7a) the chaotic regions associated with the Kirchhoff ($n = 1$) and $n = 2$ resonances are shown (50,000 points are plotted for each), and are not joined. The resonance overlap criterion predicts a value of $0.01 < s < 0.015$ for overlap. We find the $n = 1$ and $n = 2$ chaotic zones merge for s between 0.05 and 0.10; the latter case is shown in Figure 7b. The agreement with the resonance overlap prediction is satisfactory. In Figure 7c we have computed the chaotic regions for the first 7 primary resonances at $s = 0.02$; most of them are rather thin, and only the first two have merged (the figure contains approximately 300,000 points). Finally, the surface of section section for $s = 0.03$ (Fig. 7d) shows that most of them have by then fused into a single chaotic region; it is quite remarkable that the variations in aspect ratio and angular velocity for this vortex are very small, with $0.43525148\ldots < \lambda < 0.5$ and the a period differing from the Kirchhoff value by less than a half of one percent. Also of great interest is the fact that the chaotic zone is actually open (see the top of

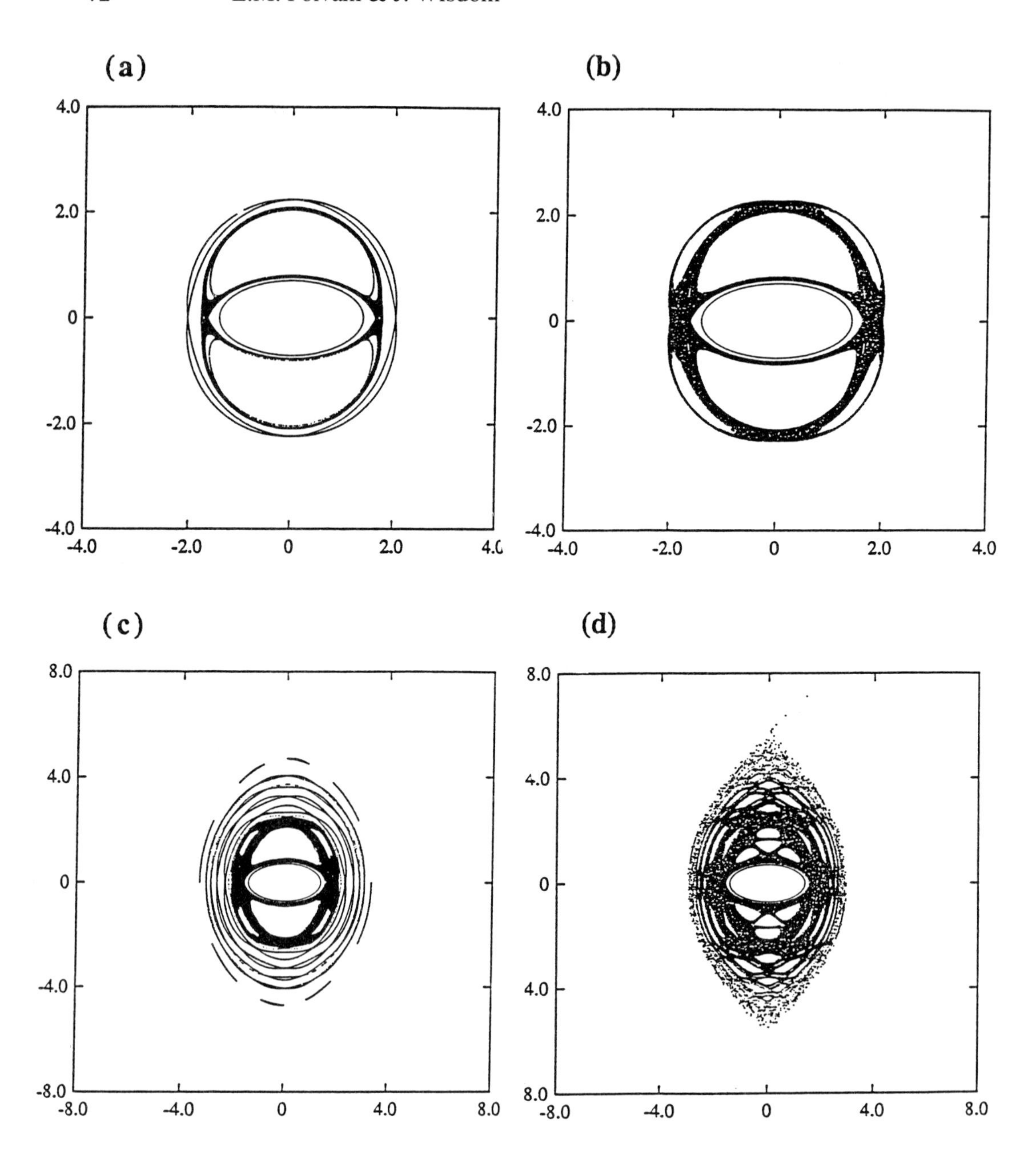

Figure 7: Surfaces of section at increasing values of the strain parameter s for Kida vortices at $\omega = 0$ and $\lambda = 0.5$ at $\varphi = 0$. (a) $s = 0.002$, (b) $s = 0.01$, (c) $s = 0.02$ and (d) $s = 0.03$. Notice that in (d) the particle actually escapes (see top of the figure) from the vicinity of the vortex.

Figure 7d), and the particle eventually escapes in the direction of the straining flow. In general we have found that for open chaotic zones, which exist only in region B of Figure 2, escapes occur very quickly (in a few $T_{\varphi\lambda}$ periods) unless $|\omega|$ is very close to $|s|$.

3. IMPLICATIONS FOR TWO-DIMENSIONAL TURBULENCE

Beyond their intrinsic value as the only presently known analytic nonlinear time-dependent solution of the Euler equations in two dimensions, Kida vortices are of great interest as the simplest model for the evolution of a vortex embedded in a complicated velocity field. This situation is encountered in high-Reynolds-number numerical simulations of two-dimensional turbulence, where strong isolated coherent vortices are known to emerge spontaneously from initially turbulent fields (McWilliams 1984). Recent results at very high resolution (Dritschel, 1989a) indicate that the vortices of 2D turbulence not only retain their ellipticity but are progressively stripped of the outer (lower) layers of vorticity, yielding elliptical structures with large vorticity gradients, for which a vortex patch is a reasonable first approximation.

In as much as the influence of the many distant vortices present in the field can be approximated by a uniform and constant quadratic shear, the discovery of very large chaotic zones surrounding Kida vortices suggests that the coherent structures of 2D turbulence may in fact be the *mixers* of the low levels of vorticity – which have been shown to behave not unlike a passive tracer (Babiano *et al.*, 1986). In particular, the results presented here for the Kida problem seem to provide a rationale for the recent observations of Benzi *et al.* (1988), who have discovered, in high-resolution spectral simulations of 2D turbulence, that the largest local exponential divergence of Lagrangian trajectories is found predominantly in the regions surrounding the coherent structures, while the motion in their interior is very stable.

Since the mixing we have described here is directly attributable to the dynamics of the homoclinic tangle (see, for instance, Rom-kedar *et al.* 1989) arising from the presence of hyperbolic points in the flow field surrounding the vortex, there is every reason to believe that similar behavior is generic to any two-dimensional vortex with a non-axisymmetric and time-dependent streamfunction. Among the many applications of interest, we would like to draw attention the stratospheric polar vortex, for which the advection of chemical tracers is a question of great current interest.

REFERENCES

BABIANO, A., BASDEVANT, C., LEGRAS, B. & SADOURNY, R. 1986 Vorticity and passive scalar dynamics in two-dimensional turbulence. *J. Fluid Mech.* **183**, 379-397.

BENZI, R. , PATERNELLO, S. & SANTANGELO, P. 1988 Self-similar coherent structures in two-dimensional decaying turbulence. *J. Phys. A* **21**, 1221-1237.

DRITSCHEL, D. 1989a The stability of elliptical vortices in an external straining flow. *J. Fluid Mech., sub judice.*

DRITSCHEL, D. 1989b Contour Dyamics and Contour Surgery. *Comp. Phys. Rep* **10**, 77-146.

KIDA, S. 1981 Motion of an elliptic vortex in a uniform shear flow. *J. Phys. Soc. Japan*, **50**, 3517-3520.

LAMB, Sir Horace, *Hydrodynamics*, Dover, N.Y.(1945).

MCWILLIAMS, J.C. 1984 The emergence of isolated coherent vortices in turbulent flow. *J. Fluid Mech.* **146**, 21-43.

MEACHAM, S.P., FLIERL, G.R. & SEND, U. 1989 Vortices in shear. To appear in *Dyn. Atoms. and Ocean.*

MOORE, D.W. & SAFFMAN, P.G. 1971 The structure of a line vortex in an imposed strain. In *Aircraft Wake Turbulence and its Detection*, p. 339, Plenum, N.Y. (1971).

POLVANI, L.M., FLIERL, G.R. & ZABUSKY, N.J. 1989a Filamentation of coherent vortex structures via separatrix crossing: a quantitative estimate of onset time. *Phys. Fluids A* **2**,181-184.

ROM-KEDAR, V., LEONARD, A. & WIGGINS, S. 1989 An analytical study of transport, mixing and chaos in an unsteady vortical flow. *J. Fluid Mech.*, submitted.

Fluid Webs with Lagrangian Chaos and Intermittency

A.A. CHERNIKOV, R.Z. SAGDEEV &
G.M. ZASLAVSKY

Space Research Institute, Profsoyuznaya 84/32,
Moscow 117810, USSR.

1. INTRODUCTION

The aim of this report is to describe the existence of a fairly generic phenomenon which appears in fluid stochastic webs, i.e. in finite measure domains of fluid separating cells of a pattern and full of chaotic streamlines. The reason is that three-dimensional periodic or quasiperiodic patterns produce equations for streamlines which correspond to dynamical systems with chaos. From a very general point of view, the boundary between neighbouring cells of a pattern must be unstable. The instability turns out to create stochastic layers between cells with chaotic streamlines. Connected stochastic layers make a network in 3D-space of finite thickness. This is a fluid stochastic web. All streamlines inside the web are of chaotic form. So chaotic particle motion is possible inside a web of regular form. This random walking of a streamline or of a fluid particle is a form of Lagrangian turbulence.

There are several examples for which Lagrangian turbulence has been recognized (Arnold 1979, Dombre et al 1986, Aref 1984), and several recent articles (Zaslavsky et al 1988, Beloshapkin et al 1989, Ottino et al 1988). So-called ABC-flow (Arnold-Beltrami-Childress) will be of special interest for us. The concept of stochastic webs (Zaslavsky et al 1986) will be applied to a stationary incompressible inviscid flow with Beltrami property in the form

$$\mathrm{rot}\ \mathbf{V} = \pm\mathbf{V} \tag{1}$$

and with symmetry or quasisymmetry (Zaslavsky et al 1988). The significance of (1) for turbulence has been shown in different ways (Tsinober & Levich 1983, Moffatt 1985, Levich & Tsinober 1983).

It appears that Lagrangian turbulence is a carrier of the pattern and its symmetry. So the advection of particles is not a usual diffusion process, but turns out to be strongly intermittent with Levy flights and with anomalous transport.

2. Q-FLOWS

The ABC-flow may be generalized by introducing 'Q-flows' (flows with quasisymmetry, which includes flows with periodical symmetry as a particular case) (Zaslavsky et al 1988, Beloshapkin et al 1989, Ottino et al 1988). Let the function $\psi = \psi(x, y)$ be any solution of the equation

$$\Delta\psi + \psi = 0 \tag{2}$$

for some given boundary condition, where $\Delta \equiv \partial^2/\partial x^2 + \partial^2/\partial y^2$. Then the flow

$$\begin{aligned} V_x &= -\frac{\partial\psi(x,y)}{\partial y} + \epsilon \sin z \\ V_y &= \frac{\partial\psi(x,y)}{\partial x} - \epsilon \cos z \\ V_z &= \psi(x,y) \end{aligned} \tag{3}$$

satisfies the Beltrami property (1) and the incompressibility condition

$$\text{div } \mathbf{V} = 0. \tag{4}$$

If

$$\psi \equiv \psi_q = \sum_{j=1}^{q} \cos \left[x \cos \frac{2\pi}{q} j + y \sin \frac{2\pi}{q} j \right] \tag{5}$$

where q is an integer, then (2) is satisfied, and so (3) determines a stationary Q-flow.

The function ψ_q coincides with the stream function for Kolmogorov flow if $q = 1, 2$, and $\psi_1 = -\psi_2 = \cos x$, or with the stream function for Rayleigh-Benard convection if $q = 3, 6$ and

$$\tfrac{1}{2}\psi_6 = \psi_3 = \cos x + \cos \tfrac{1}{2}(x + \sqrt{3}y) + \cos \tfrac{1}{2}(x - \sqrt{3}y)\,. \tag{6}$$

For $q = 4$ and

$$\psi_4 = \cos x + \cos y \tag{6$'$}$$

the velocity field (3) corresponds to the ABC-flow. If $q \neq 1, 2, 3, 4, 6$, then Q-flow has quasicrystal symmetry (non-periodic rotational q-fold symmetry) in the (x, y)-plane. The possible existence of 5-fold symmetry in hydrodynamics was discussed in Beloshapkin et al (1988) and Yakhot et al (1986).

3. STREAMLINES FOR Q-FLOWS

The equations

$$\frac{dx}{V_x} = \frac{dy}{V_y} = \frac{dz}{V_z} \tag{7}$$

determine the streamlines of the velocity field $\mathbf{V}(x, y, z)$. A more convenient description of the system (7) in the form

$$dx/dz = V_x/V_z; \quad dy/dz = V_y/V_z \tag{8}$$

shows that we are dealing with a 'nonstationary' problem for a dynamical system with two-dimensional phase space. The variable z plays the role of time. It is sometimes possible to represent the system (8) in Hamiltonian form and then the well-developed apparatus of the theory of dynamical systems begins to work to its full extent. The streamlines can wind on invariant surfaces $z = z(x, y)$ or behave irregularly or stochastically in space. In particular, in two-dimensional fields with no dependence of (V_x, V_y) on z the streamlines are always regular and in the three-dimensional case, generally speaking, some of the lines always behave stochastically if the z-dependence is periodic.

For Q-flows of the form (3), the system (8) takes the form

$$\begin{aligned} \frac{dx}{dz} &= \frac{1}{\psi}\left(-\frac{\partial\psi}{\partial y} + \epsilon \sin z\right); \\ \frac{dy}{dz} &= \frac{1}{\psi}\left(\frac{\partial\psi}{\partial x} - \epsilon \cos z\right). \end{aligned} \tag{9}$$

This is a generalized Hamiltonian form. To get the usual Hamiltonian representation, we introduce a nonlinear transformation

$$p \equiv p(x, y) = \int_0^y dy' \psi(x, y'). \tag{10}$$

Then equations (9) can be written

$$dx/dz = -\partial H/\partial p; \quad dp/dz = \partial H/\partial x \tag{11}$$

with Hamiltonian

$$H(p, x, z) = \psi(x, y(p, x)) - \epsilon y(p, x) \sin z - \epsilon x \cos z, \tag{12}$$

which depends periodically on the 'time' z.

Two examples: there is an explicit formula for $y(p,x)$ for $q = 4$ and $q = 3$ (Chernikov et al 1988). For $q = 4$ and ψ_q from (6), eqs. (10) - (12) give

$$H = \cos x + \cos y(x,p) - \epsilon y(x,p) \cos z - \epsilon x \sin z, \tag{13}$$

where

$$y(x,p) = \frac{p}{\cos x} + 2\sum_{n=1}^{\infty} \frac{(-1)^n}{n} J_n \left[\frac{n}{\cos x}\right] \sin \left[\frac{np}{\cos x}\right].$$

If $q = 3$ and ψ_3 is given by (6), then

$$y(x,p) = \frac{p}{\cos x} + \frac{4}{\sqrt{3}}\sum_{n=1}^{\infty} \frac{(-1)^n}{n} J_n \left[\frac{n \cos(x/2)}{2 \cos x}\right] \sin \left[\frac{\sqrt{3}np}{2 \cos x}\right]$$

4. STOCHASTIC WEB

This part contains our central result for small ϵ. The Q-flow (3) divides the whole 3D-space in convective cells and creates a 3D-pattern. The pattern has rotational symmetry in the (x,y)-plane and helical symmetry along the z-axis. If $q = 4$, then the symmetry is cubic. The cells are separated by some connected thin layers of finite measure which we describe as the 'liquid web'. In rough approximation, the flow in the web is stochastic, and in the cells is regular. Because of the connectedness of the web, unlimited advection of passive particles is possible. It may be said that the liquid stochastic web makes an invariant skeleton which appears in a pattern of the flow.

Chaotic streamlines for the ABC flow were mentioned in Arnold (1979) and analysed in Dombre et al (1986). The web concept for Q-flows was introduced by Zaslavsky et al (1988). In this reference, a method for solving the problem of the width of a web was proposed. The method is based on the theory of stochastic layers for Hamiltonian systems (Zaslavsky 1985). With $\epsilon = 0$, the $2D$-case) the streamline Hamiltonian (13) describes a family of cylindrical surfaces corresponding to different values of the 'energy' integral $H = E$, where the integral is simultaneously the stream function. There are special separatrix surfaces which are strongly affected by even a small perturbation. Their section by the plane $z = \text{cst}$ gives a connected network (for $q = 3,4,6$) or disconnected networks (for $q = 5,7,\ldots$). These networks are destroyed by perturbation if $\epsilon \neq 0$, and a connected stochastic web is formed in their place.

It has been shown (Zaslavsky et al 1988, Chernikov et al 1988) that the thickness of stochastic webs is of order ϵ. Some typical examples of webs sections are given in figures 1-4, which show both closed curves and a 'stochastic sea'. Closed curves appear as the section of invariant surfaces by the plane $z =$ cst (figures 1-3) or $x =$ cst (figure 4). The stochastic sea is formed by the points of section of one chaotic streamline on the same plane.

But the real situation is much more complicated than this. Inside the web there are islands of stability (see figures 1-4) which are filled with invariant surfaces like tubes or quasiperiodically meandering tubes. Between the tubes small regions of chaos exist separated from the stochastic web. This shows up in a very specific advection process along the stochastic web.

5. LEVY FLIGHT OF STREAMLINES

Let us consider the generic case when all directions are equivalent in some sense, i.e. $\epsilon \sim 1$. Then the web width is of order one and there are gaps of order one in the web. The process of random walking of streamlines along the web is not like the usual diffusion process (Zaslavsky et al 1986) but rather like the Levy flight (Montroll & Schlesinger 1984).

Let $R(t)$ be the distance between the position of the point of crossing the plane (x, y) of a streamline and its initial position. Here 'time' t denotes number of crossings since the start. If $< R(t) >$ is averaging over many streamlines with different initial conditions inside the web, then asymptotically for large t

$$< R(t) > \sim t^{\alpha} \quad \text{for some } \alpha. \tag{14}$$

For a usual diffusion process, $\alpha = \frac{1}{2}$. For Levy flight $0 < \alpha < 1$. The value $\alpha = 1$ is for free motion ($R =$ cst. t) and the value $\alpha = 0$ denotes the capture of streamlines in some finite domain. The situations of α close to one and α close to zero are common for chaotic properties of the dynamical system (9).

The situation as described is clearly seen in figures 5, 6 for $q = 3$. The diffusional spot in figure 5 does not indicate visible jumps or flight. Numerical simulation (Chernikov et al., to be published) shows that for this case, $\alpha = 0.5957$, which is close to the usual diffusion process. In figure 6 the situation is very different. Long tracks indicate almost free motion along boundaries of certain invariant tubes. This explains the larger value $\alpha = 0.782$ for this case. Another important thing is seen in figures 5, 6. Changing the perturbation parameter ϵ shows itself in reconstruction of the phase plane. Due to a bifurcation, new islands appear and create (or change) possibilities for Levy flight. Some simple estimates can be made for the Levy flight transport process (Zaslavsky et al 1989). These are based on the multifractal properties of the chaos of streamlines. Let $P(R; t)$ be the probability density of finding a value R in the interval $(R, R + dR)$ at time t. Then

$$< R >_t = \int dR\ P(R;t) = \int dR \int dk\ \exp(-ikR)\ P(k;t) \tag{15}$$

where $P(k;t)$ is the Fourier transform of $P(R;t)$. For the usual Levy flight

$$P(k;t) = \exp\ (-C\ k^\beta t), \quad (k > 0) \tag{16}$$

where C is a constant and β is an exponent connected with the fractal dimension of the random walk (Montroll & Schlesinger 1984). For a multifractal motion, the expression (16) may be generalized:

$$P(k;t) = \exp\left(-t \int_{\beta_1}^{\beta_2} d\beta\ \rho(\beta)\ C(\beta)\ k^\beta\right),\ (k > 0) \tag{17}$$

where β_1, β_2 are minimal and maximal exponents $(1 \leq \beta_1 < \beta_2 \leq 2)$ and $\rho(\beta)$ is the statistical weight of the exponent β. From (15), (17), we obtain

$$< R >_t \sim t^{1/\bar{\beta}}, \tag{18}$$

where $\bar{\beta}$ is some intermediate value of $\beta(\beta_1 < \bar{\beta} < \beta_2)$. If there is a discrete spectrum for β, i.e.

$$\rho(\beta) = \sum_s \rho_s \delta(\beta - \beta_s),$$

then the expression (17) determines different intermediate asymptotics for different time-scales:

$$< R >_t \sim t^{1/\beta_s}. \tag{19}$$

[As $t \to \infty$ then $\beta_s \to \beta_{\text{min}}$.] The formula (19) shows why different possibilities for transport laws may exist. Changing ϵ changes the phase portrait of streamlines; this results in the modification of the multifractal spectra and in evolution of the exponent $\bar{\beta}$ in (18) or of the exponents β_s in (19).

6. CONCLUSION

It seems that quasisymmetric periodic flows represent a rather general pattern phenomenon. Q-flow provides a good example of the situaton which may precede the onset of turbulence. Streamline structure may be investigated by the general methods of dynamical systems. These show the existence of liquid stochastic webs – a form of organization of a pattern. Chaotic wandering of streamlines inside the web implies that fluid particles have a disordered spatial motion, usually called

Lagrangian chaos. This Lagrangian chaos reveals itself in very strong intermittency properties, associated with Levy flight of fluid particles.

REFERENCES

AREF H. (1984) *J. Fluid Mech.* **143**, 1-21.

ARNOLD V.I. (1979) Mathematical Methods in Classical Mechanics. Nauka, Moscow.

BELOSHAPKIN V.V., CHERNIKOV A.A., SAGDEEV R.Z. & ZASLAVSKY G.M. (1988) *Phys. Lett.* **A133**, 395-398.

BELOSHAPKIN V.V., CHERNIKOV A.A., NATENZON M.Ya., PETROVICHEV B.A., SAGDEEV R.Z. & ZASLAVSKY G.M. (1989) *Nature* **337**, 133-137.

CHERNIKOV A.A., SAGDEEV R.Z., USIKOV D.A. & ZASLAVSKY G.M. (1988) In: Sov. Sci. Rev. C Math. Phys. Vol. 8, Ed. Ya.G. SINAI. Harwood Acad. Publ.

CHERNIKOV A.A., PETROVICHEV B.A., ROGALSKY A.A., SAGDEEV R.Z. & ZASLAVSKY G.M. (to be published).

DOMBRE T., FRISCH U., GREEN J.M., HENON M., MEHR A. & SOWARD A.M. (1986) *J. Fluid Mech.* **167**, 353-391.

LEVICH E. & TSINOBER A. (1983) *Phys. Lett.* **A93**, 293-297.

MOFFATT H.K. (1985) *J. Fluid Mech.* **159**, 359-378.

MONTROLL E.W. & SCHLESINGER M.F. (1984) In: Studies in Statistical Mechanics Vol. 11, p. 5. Eds. Lebowitz J. & Montroll E.W. North Holland.

OTTINO J.M., LEONG C.W., RISING H. & SWANSON P.D. (1988) *Nature* **333**, 419-425.

TSINOBER A. & LEVICH E. (1983) *Phys. Lett.* **A99**, 321-324.

YAKHOT V., BAYLY B.J. & ORSZAG S.A. (1986) *Phys. Fluids* **29**, 2025-2027.

ZASLAVSKY G.M. (1985) Chaos in dynamic systems, Harwood.

ZASLAVSKY G.M., ZAKHAROV M.Yu, SAGDEEV R.Z., USIKOV D.A. & CHERNIKOV A.A. (1986) *Zh. Exsp. Teor. Fiz.* **91**, 500-516.

ZASLAVSKY G.M., SAGDEEV R.Z. & CHERNIKOV A.A. (1988) *Zh. Exsp. Teor. Fiz.* **94**, 102-115.

ZASLAVSKY G.M., SAGDEEV R.Z., CHAIKOVSKY D.K. & CHERNIKOV A.A. (1989) *Zh. Exsp. Teor. Fiz.* **95**, 1723-1733.

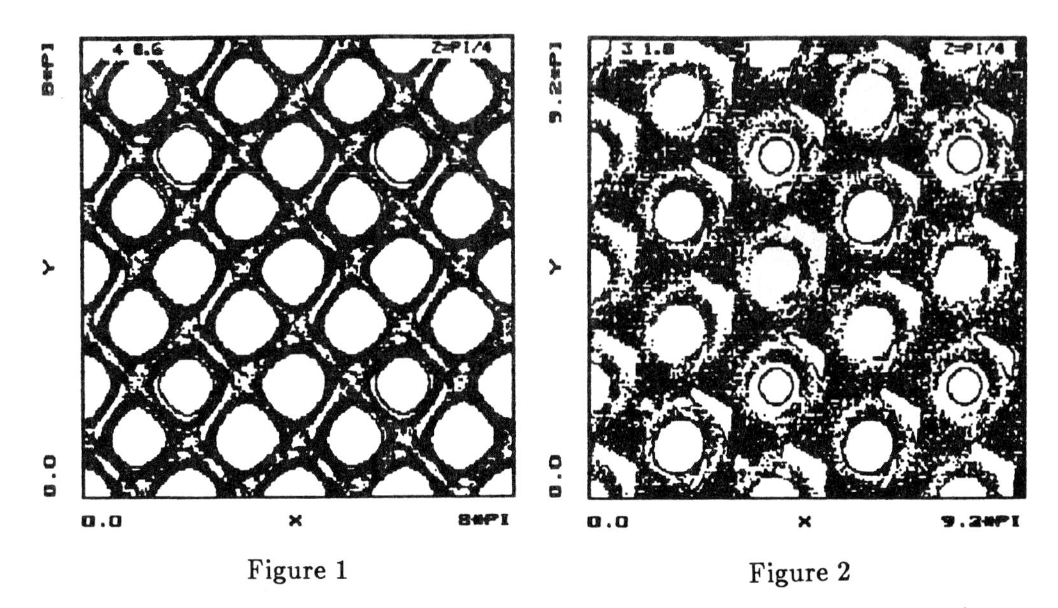

Figure 1

Figure 2

Figure 1. Square stochastic web of streamlines: $q = 4; \epsilon = 0.6; z =$ cst $= \pi/4$; the size of the square is $8\pi \times 8\pi$.

Figure 2. Hexangonal stochastic web on (x, y)-plane: $q = 3; \epsilon = 1.0$; the size of the square is $9.2\pi \times 9.2\pi$.

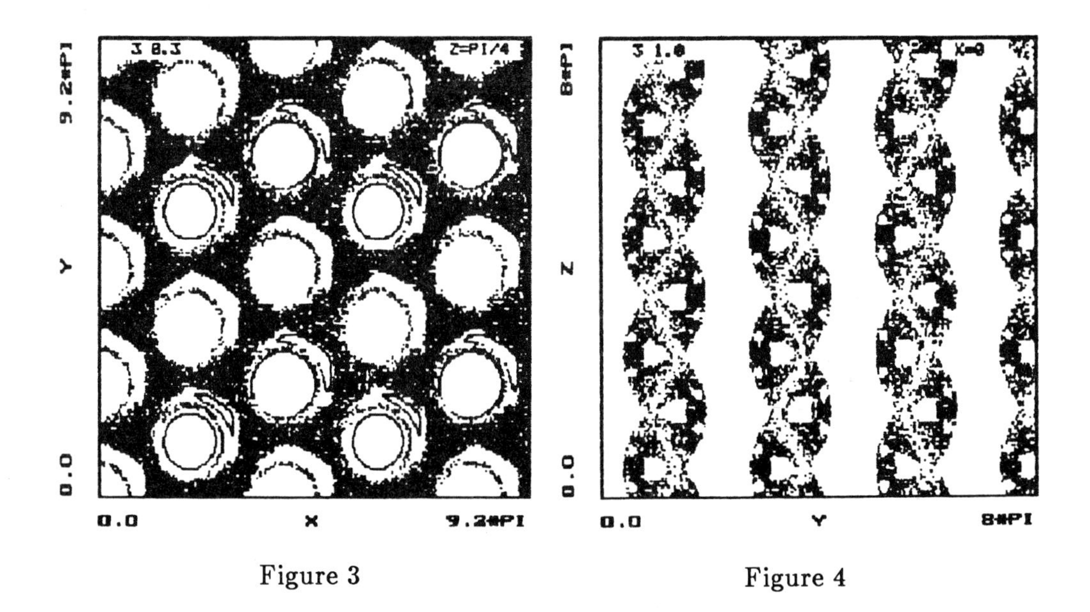

Figure 3

Figure 4

Figure 3. The same as figure 2 but for $\epsilon = 0.3$.

Figure 4. The same as figure 2 but for (y, z)-plane and size of square $8\pi \times 8\pi$.

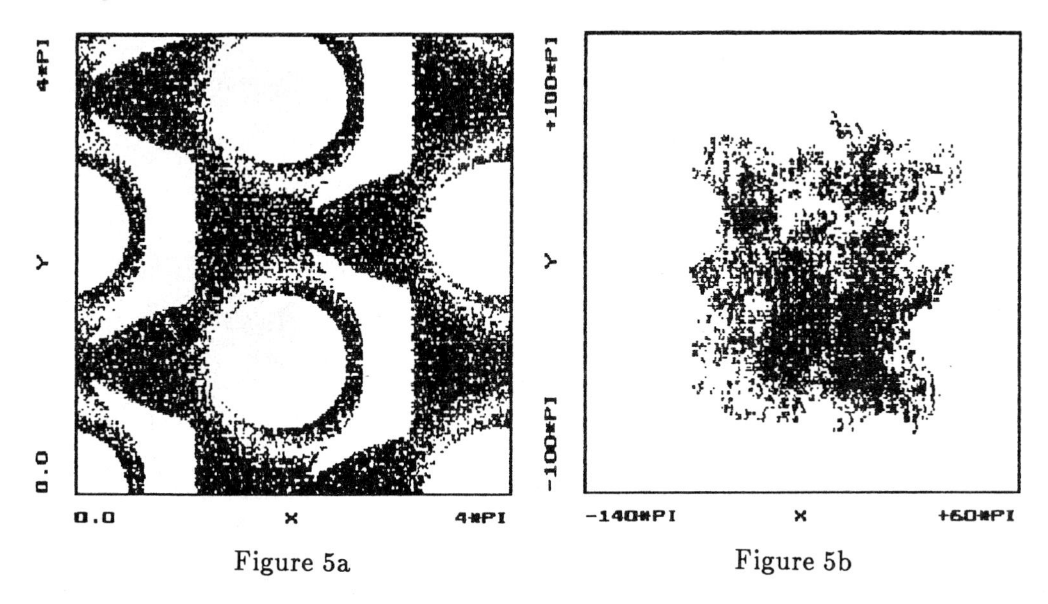

Figure 5a

Figure 5b

Figure 5. Streamlines sections by the plane $z =$ cst for $q = 3, \epsilon = 0.7$ for one cell of the web (a) and the overall picture of random walking on the (x, y)-plane (b).

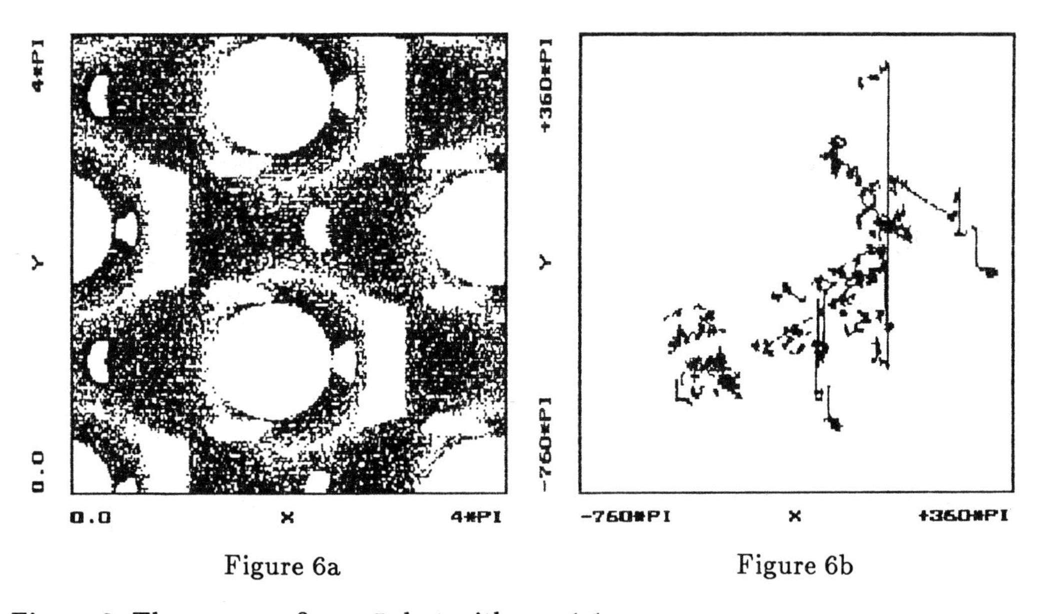

Figure 6a

Figure 6b

Figure 6. The same as figure 5, but with $\epsilon = 1.1$.

Numerical Simulation of the Lagrangian Flow Structure in a Driven-Cavity

KATSUYA ISHII & REIMA IWATSU

Institute of Computational Fluid Dynamics, 1-22-3 Haramachi, Meguro-ku, Tokyo, Japan

1. INTRODUCTION

It is well known that three-dimensional steady flows with a simple Eulerian representation can have a chaotic Lagrangian structure. The most famous example is an ABC flow which is a three-dimensional steady solution of the Euler equations characterized by the Beltrami property(Arnold 1965, 1980, Dombre et al. 1986). Recently, Zaslavskiĭ et al.(1988) studied other classes of the Euler solutions with the Beltrami property. Another type of the flow with a chaotic Lagrangian structure was reported by Arter(1983) which is formed in a truncated model of the Rayleigh-Bénard flow with a square planform to the second order of $(R - R_c)^{1/2}$, where R and R_c are the Rayleigh number and the critical Rayleigh number, respectively. The previous studies on the Lagrangian flow structure mentioned above are based on analytic representations of the velocity fields. However, in many viscous three-dimensional flows of practical importance, the velocity vector fields do not have any analytical representations: consequently, little knowledge has been obtained on the characteristics of the Lagrangian structure of these flows.

In this paper, we study the Lagrangian structure of viscous steady flows in a lid-driven cubic cavity at various Reynolds numbers. Although the boundary conditions of this problem are simple, the velocity field does not have any analytic representation. Hence, we obtain an approximate velocity field by using a finite difference method and interpolation of its solution. The obtained velocity field gives a dynamical system of fluid particles whose orbits represent approximate streamlines. This dynamical system of streamlines will be studied by using the Poincaré section method.

2. VELOCITY VECTOR FIELD

Consider three-dimensional flows in a square cubic box driven by its sliding upper wall. We choose the Cartesian coordinates (x, y, z) as shown in Fig. 1 so that the cubic box is located at $0 \leq x \leq 1$, $0 \leq y \leq 1$, and $0 \leq z \leq 1$. It is assumed that the plane of $z = 1$ slides in the positive x direction with the velocity of unity.

The governing Navier-Stokes equations expressed in the non-conservation form are given as

$$\mathrm{div}\mathbf{V} = 0 \tag{1}$$

$$\frac{\partial \mathbf{V}}{\partial t} + (\mathbf{V}\cdot \mathrm{grad})\mathbf{V} = -\mathrm{grad}p + \frac{1}{Re}\Delta \mathbf{V} \tag{2}$$

where $\mathbf{V} = (u, v, w)$ is the velocity vector, p is the pressure, and Re is the Reynolds number.

The numerical finite difference technique adopted here is a modified MAC method, which was originally developed by Harlow and Welch(1965) and improved by Kawamura et al.(1985, 1986). The simple geometry of a square cubic cavity allows us to exploit an efficient orthogonal grid systems (ξ, η, ζ). In order to resolve rapid changes of velocities in the boundary layer, a one-dimensional grid clustering transformation was applied. For instance, in the x direction, we have

$$x = \frac{1}{2b}(\tanh(ah) + b) \tag{3}$$

$$h = -1 + 2\frac{\xi - 1}{\xi_{max} - 1} \qquad (\xi = 1, 2, \ldots, \xi_{max})$$

where $b = \tanh(a)$. Similar transformations were effectuated in the y and z directions. The values of ξ_{max} and b were respectively set at 81 and 0.7 in the steady flows at relatively low Reynolds numbers considered in this paper. (For the flow at $Re \geq 2000$ shown in Fig.2, we choose $b = 0.965$.) In the gird systems with $b = 0.7$, the minimum and maximum spacings between two adjacent grid points were 8.019×10^{-3} and 1.548×10^{-2}, respectively.

The continuous velocity vector field in the cubic cavity was obtained by using the first-order interpolation of the velocity values in the calculated coordinate systems (ξ, η, ζ)(Shirayama 1988). The first derivatives of this velocity field are piecewise continuous functions of $\mathbf{x}$. The values of their discontinuities depend on the smoothness of the finite difference solution.

The streamline in a steady flow is identified with the orbit of a fluid particle which is determined by

$$\frac{d\mathbf{x}}{dt} = \mathbf{V}(\mathbf{x}). \tag{4}$$

We used the second order time integration to solve eq. (4). The typical CPU time for each case was about 10 hours on supercomputers NEC SX2 and FUJITSU VP400, which are located at iCFD.

3. RESULTS

We calculated the flows in a driven-cavity by the finite difference method over a wide range of Reynolds number $100 < Re < 10000$(Iwatsu et al. 1989). Steady state solutions were attained when $Re \leq 2000$. The values of the total kinetic energy of the flow fields, $\frac{1}{2}\int \mathbf{V}^2 \mathrm{dV}$, at different Reynolds numbers are shown in Fig. 2. Since we choose the lid-driven velocity as the characteristic velocity, the value of the total energy decreases as the Reynolds number increases. As shown in Fig. 2 the decay rate in the kinetic energy increases beyond the critical Reynolds number, Re_c, between 2000 and 3000. This trend is accompanied by the steady-unsteady transition in the global structure of the driven-cavity flow, which will play an important role in transition to turbulence of the flow field.

In this paper, we focus our attention on the steady flow fields at $100 < Re < 400$. It is noted that no changes occur in the kinetic energy-Reynolds number relation in this Reynolds number range as shown in Fig. 2. The streamlines at $Re = 100$ in the plane of symmetry at $y = 0.5$ and the planes near the side walls at $y = 0.008$ and $y = 0.992$ are displayed in Fig. 3(a). We find in Fig. 3(a) that a large eddy occupies the large volume of the cubic box except for the corner regions, which is reported by experiments and the other calculations. (Ku, Hirsh and Taylor 1987, Hwang and Hyunh 1987, Koseff and Street 1984). The streamlines on the planes near the side walls at $y = 0$ and $y = 1$ approach to one point with a spiral motion. The spiral streamlines on the plane of symmetry at $y = 0.5$ moves from a fixed point to the walls. This fixed point is located at $(x, y, z) = (0.608, 0.5, 0.765)$. The projected velocity (v, w) on the plane of $x = 0.608$ is shown in Fig. 3(b). Four secondary eddies are observed, which are divided by stable and unstable manifolds of the saddle point of $(y, z) = (0.5, 0.765)$. Typical streamlines based on this velocity field are shown in Figs. 4(a) and (b), and two streamlines originated near the focus on the side walls are shown in Fig. 4(c). It is found in Fig. 4(c) that there exists a fixed point on the plane of $y = 0$ at which the derivative matrix $\partial\mathbf{V}/\partial\mathbf{x}$ has one negative eigenvalue, a complex eigenvalue with a positive real part and its conjugate. Two streamlines in Fig. 4(a) appear to be closed curves and the streamline in Fig. 4(b) appears to form a torus.

In order to perform an extensive study of these behavior, we shall resort to the method of Poincaré sections in which the three-dimensional dynamical system is represented by the successive intersections of streamlines with one or several planes. Figure 5(a) shows a Poincaré section of the streamlines at $x = 0.5$ at $Re = 100$. Since the flow is symmetrical with respect to the plane of $y = 0.5$, this map is of symmetry with respect to the plane of $y = 0.5$. We can also find the upper and lower sets of points in Fig. 5(a). The points of upper set represents the location where streamlines traverse the plane of $x = 0.5$ in the positive direction. In other words, the fluid particles have the positive values of u at these points. The points of lower set are the intersections of the streamlines in the negative direction. The Poincaré sections of a streamline form a closed curve in many cases, which signifies that this streamline is located on a torus. The fluid particles on these streamlines have two frequencies, which respectively correspond to the basic motion of the $x-z$ plane in Fig. 3(a) and the secondary motion in Fig. 3(b). Among the observed tori, the basic frequency of the largest torus ω_1 is 0.82 and the second frequency ω_2 is 0.15, while they are found to be $\omega_1 = 0.91$ and $\omega_2 = 0.18$ for the smallest torus. According to the theory of dynamical system, the ratio of two frequencies of the fluid particle moving on a torus must be irrational. We also find a fixed point, five periodic points and six islands in Fig. 5(a). The fixed point represents a closed streamline in Fig. 4(a). The remaining two maps represent the streamlines with two frequencies whose rational ratio are 5:1 and about 6:1, respectively. The streamlines in the outer region is chaotic and we cannot get a definite mapping.

Figures 5(b-d) show the Poincaré sections at $Re = 200$, 300, and 400, respectively. These figures show that the area of tori becomes small as the Reynolds number increases. In particular, in the Poincaré section at $Re = 400$ we cannot find any closed curves formed by the intersections. The two frequencies and its ratio of the streamlines observed on a torus at $Re = 100$, 200, and 300 are given in Table 1. While the basic frequency gradually decreases, the second frequency rapidly increases as the Reynolds number increases. Several island structures obtained in Figs. 5(b,c) are attributed to the rational ratio of two frequencies.

The resonance of $3:1$ around $Re = 220$ is quite different from the other resonances. Figures 5(e-g) show the left halves of the Poicaré sections at $Re = 219$, 225, and 235. In particular, we cannot observe any tori around the central closed streamline at $Re = 225$. These phase portraits under the resonance of $3:1$ are surprisingly similar to those in Hamiltonian systems with two degrees of freedom described by Arnold et al.(1988). The stable resonances of $5:2$, $7:3$, and so on, can be observed as the Reynolds number gradually increases beyond the resonance of $3:1$. However, the resonance of $2:1$ is unstable and breaks the structure of tori around the central closed orbit shown in Fig. 5(h) at $Re = 335$. After the resonance of $2:1$ occurs, any definite structure around the closed orbit cannot be obtained.

The three-dimensional behavior of a streamline in a relatively short time period at $Re = 400$ are shown in Figs. 6(a,b). The local streamlines at $Re = 1000$ are depicted in Figs. 7(a,b), which show similar behavior. These figures show that the local (Euler) structure of the velocity field may not change significantly. The position of the fixed points near the center of the box are shown in Fig. 8. The fixed point moves toward the geometrical center of the box as the Reynolds number increases.

In Fig. 7(b), we can find a secondary eddy region near the upstream bottom corner $(1, y, 0)$. This and another secondary eddy near the downstream bottom corner $(0, y, 0)$ are observed in all the calculated steady flows, which have been extensively studied for the two-dimensional flow. The centers of these secondary eddies on the plane of symmetry at $y = 0$ are fixed points at which $\partial \mathbf{V}/\partial \mathbf{x}$ has one positive eigenvalue, a complex eigenvalue with negative real part and its conjugate.

4. DISCUSSION AND CONCLUDING REMARKS

It is shown in §3 that the Lagrangian structures of the approximate velocity fields based on the finite difference solution are similar to those of the Hamiltonian dynamical system with two degrees of freedom. Overall characteristics of the structures change smoothly with the Reynolds number. In addition, since the structure of invariant tori in the Hamiltonian system with two degrees of freedom is stable under small disturbances (Arnold 1980,1988), these Lagrangian characteristics of the approximate velocity fields will be preserved in the presence of small disturbances which conserves mass. Therefore, we may conclude that the exact solutions of the Navier-Stokes equations will have the same Lagrengian structure as the present approximate solution at moderate Reynolds numbers.

When $Re = 0$, the Navier-Stokes equations are reduced to linear equations which are known as the Stokes equations. Therefore, the steady velocity field in a driven-cavity at $Re = 0$ must satisfy the following relations:

$$\begin{aligned} u(0.5 - x, y, z) &= u(0.5 + x, y, z) \\ v(0.5 - x, y, z) &= -v(0.5 + x, y, z) \\ w(0.5 - x, y, z) &= -w(0.5 + x, y, z) \end{aligned}$$

$$(0 < x < 0.5, 0 < y, z < 1)$$

All the streamlines in this flow field are symmetrical with respect to the plane of $x = 0.5$, so that the transverse streamlines through the plane of $x = 0.5$ are closed curves whose Poincaré sections at $x = 0.5$ are points. The fluid particle on this streamline has only one frequency ω_1. In addition, the fixed points in the Stokes flow field must form a line on the plane of symmetry at $x = 0.5$, though we may have the other fixed points in the regions of small eddies near the bottom corners of the cavity. This structure is clearly unstable under disturbance.

Because the Lagrangian structures mentioned in §3 are similar to those of the Hamiltonian dynamical system, it is assumed that we can construct the "effective" Hamiltonian of this three-dimensional viscous flow in the primary eddy region as the streamfunction of two-dimensional flows (Whittaker, 1927). Under this assumption, as the Reynolds number increases, the incompressible flow field with an infinite number of closed streamlines at $Re = 0$ bifurcates the flow, which consists of chaotic regions and invariant tori with a small second frequency ω_2 which vanishes at $Re = 0$. This conjecture is supported by the behavior of the second frequency ω_2 in Table 1.

We have successfully obtained the Lagrangian structures of the driven-cavity flows at various Reynolds numbers. While the transition to unsteady flow takes place at $2000 < Re_c < 3000$, the transition to Lagrangian chaos occurs at $300 < Re_{cL} < 400$. For $Re < 400$, we have observed a similar structure to a Hamiltonian system with two degrees of freedom that has invariant tori, islands and chaotic regions. This structure develops from a definite structure at $Re = 0$ with increasing Reynolds number. However, analysis on the relationship between the Euler and Lagrangian turbulence is left as a future investigation.

The authors would like to thank Professor T. Kambe of University of Tokyo and Professor K.Kuwahara of ISAS for their valuable suggestions and discussions.

REFERENCES

Arnold,V.I.1965 *C.R.Acad.Sci.Paris* **261**, 17

Arnold,V.I.1980 *Mathematical Method of Classical Mechanics.* Springer (original in Russian: Nauka, 1974).

Arnold,V.I.,Kozlov,V.V. and Neishtadt,A.I. 1988 *Dynamical Systems III.* Springer (original in Russian: VINITI, 1985) Ch.7 259-260.

Arter,W. 1983 *Phys. Letter* **97a**, 171.

Dombre,T., Frish,U., Greene,J.M., Hénon,M., Mehr,A., and Sward,A.M. 1986 *J. Fluid Mech.***167**, 353.

Harlow,F.H. and Welch, J.E. 1965 *Phys. Fluids* **8** 2182.

Hwang,D.P., and Huynh,H.T. 1987 *in Proceedings of the International Conference of Laminar and Turbulence Flow.*

Iwatsu, R., Ishii, K., Kawamura T., Kuwahara K. and Hyun, J.M. 1989 *AIAA paper* **89-0040**.

Kawamura T. and Kuwahara K. 1985 *AIAA paper* **85-0376**.

Kawamura T. and Kuwahara K. 1986 *Fluid Dyn. Res.* **1** 145.

Koseff, J.R., and Street, R.L. 1984 *J.Fluids Eng.* **106** 385.

Ku, H.C., Hirsh, R.S. and Taylor, T.D. 1987 *J. Comput. Phys.* **70** 439.

Shirayama, S. 1989 *AIAA paper* **89-1803**.

Whittaker, E.T. 1927 *A Treatise on the Analytical Dynamics of Particles and Rigid Bodies.* Cambridge §122 284.

Zaslavskiĭ,G.M., Sagdeev,R.Z., and Chernikov,A.A. 1988 *Sov. Phys. JETP* **67** 270-277.

Re	ω_1	ω_2	ω_1/ω_2
100	0.91 − 0.82	0.18 − 0.15	4.92 − 5.43
200	0.80 − 0.77	0.25 − 0.22	3.24 − 3.48
300	0.74 − 0.74	0.31 − 0.30	2.35 − 2.49

Table.1 The first and second frequencies of the observed tori (the left value corresponds to the smallest torus and the right is the largest torus)

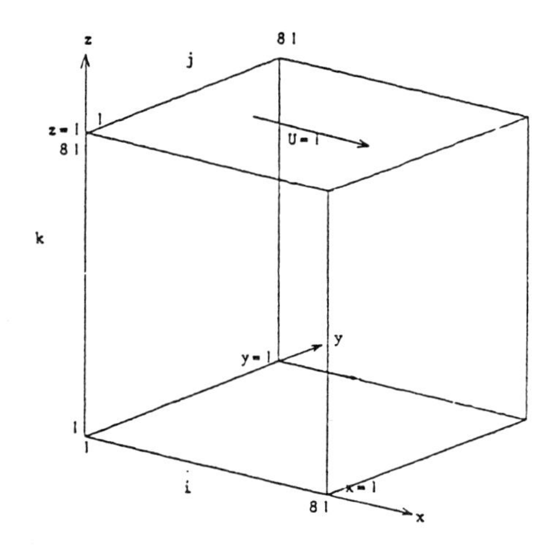

Fig.1 Flow configuration and coordinate system.
Boundary condirions
$(u, v, w) = (1, 0, 0)$ at z=1,
$(u, v, w) = (0, 0, 0)$ on the other surfaces.

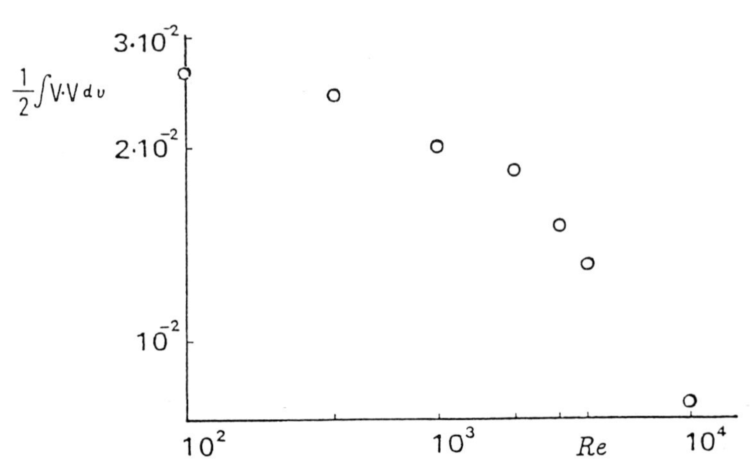

Fig.2 Total kinetic energy versus Reynolds number.

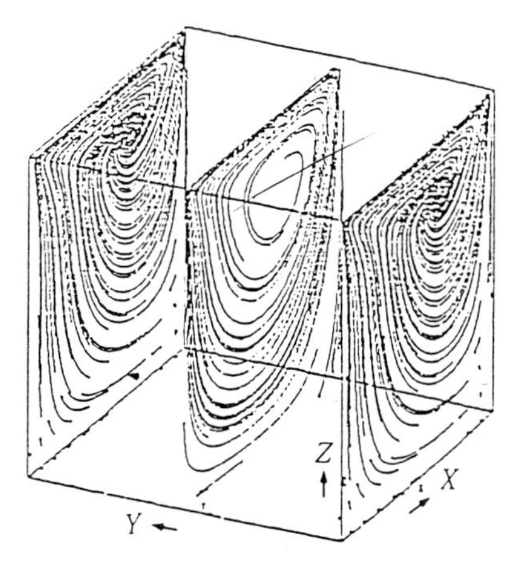

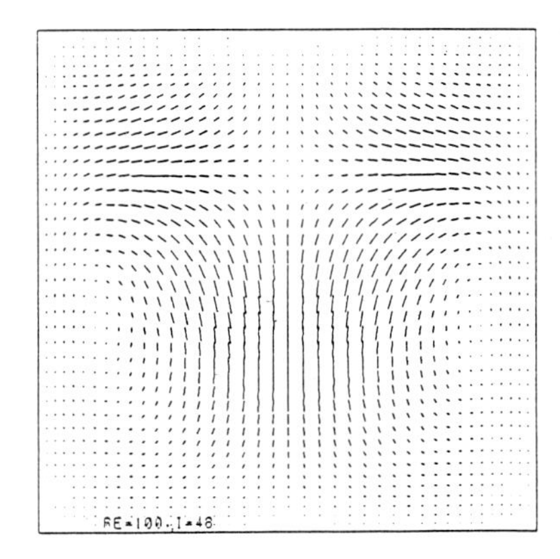

(a) Streamlines on the planes of $y = 0.01$, 0.5 and 0.99. (b) projected velocity (v, w) on the plane of $x = 0.608$.

Fig.3 Flow field at $Re = 100$.

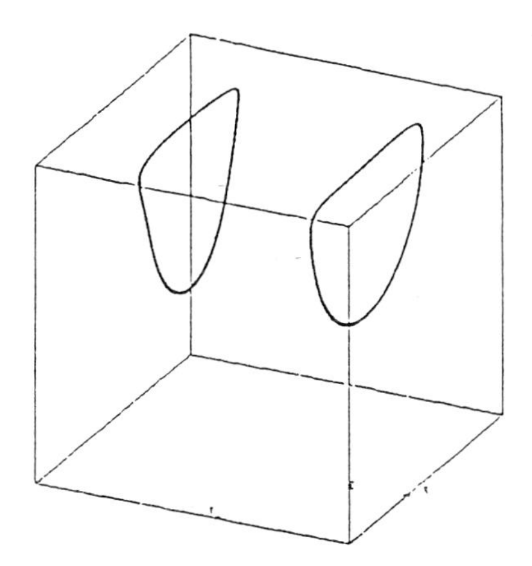

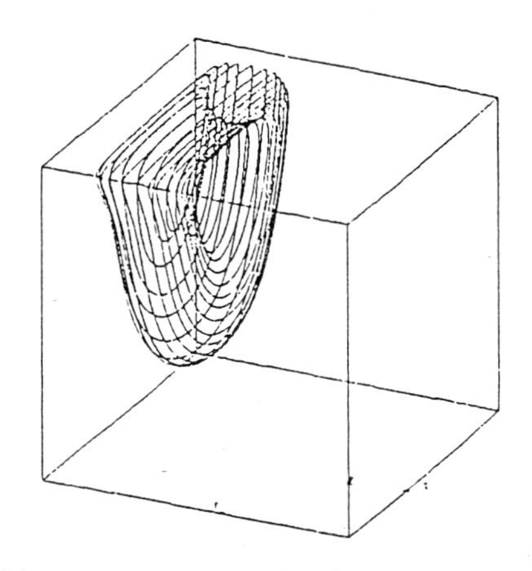

(a) a streamline started at $(0.5, 0.5 \mp 0.27, 0.45)$. (b) a streamline started at $(0.5, 0.64, 0.33)$.

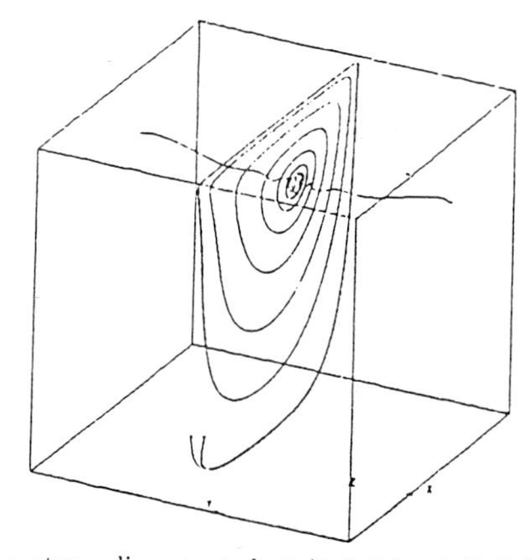

(c) a streamline started at $(0.5, 0.5 \mp 0.49, 0.80)$.

Fig.4 A typical streamline at $Re = 100$.

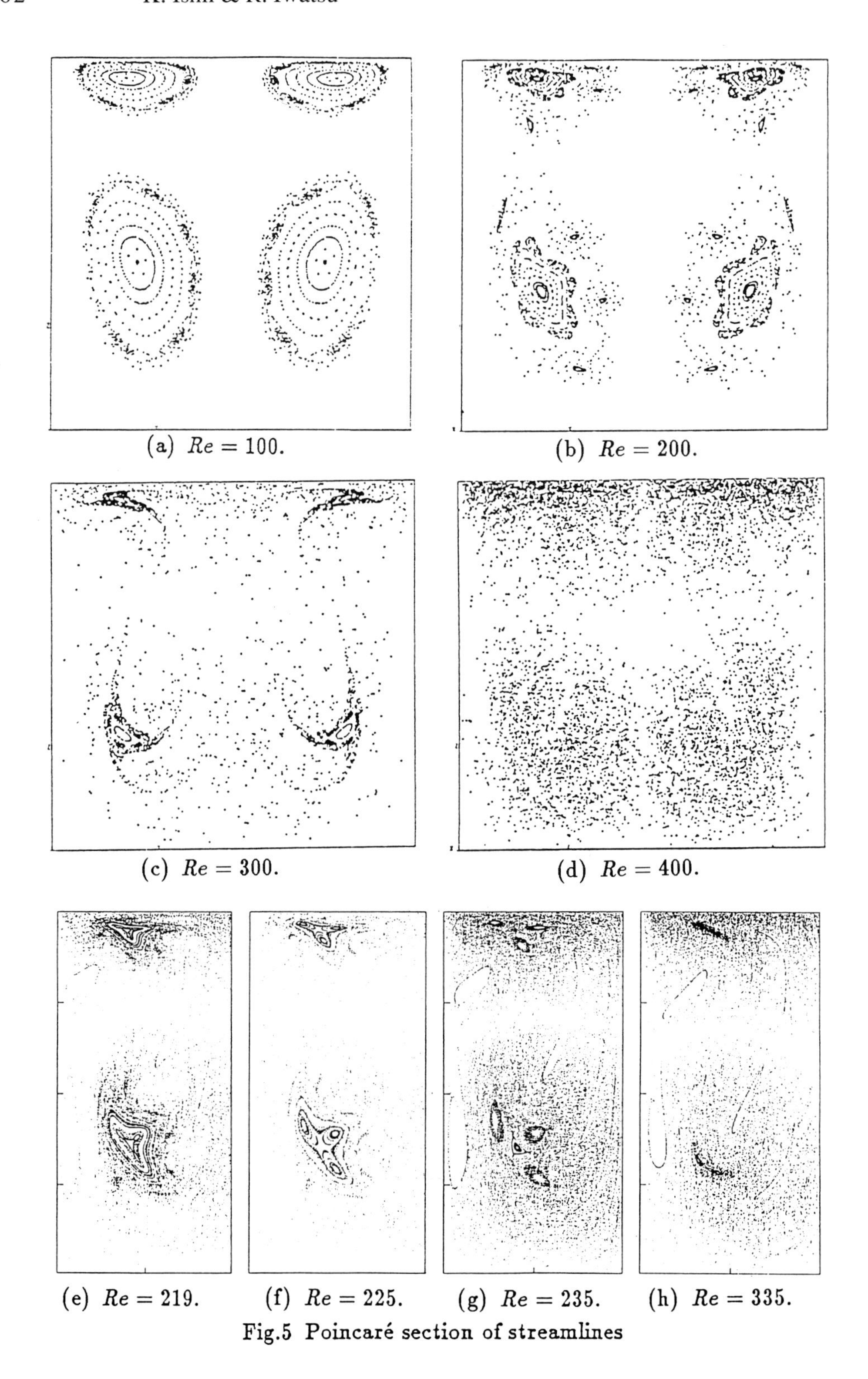

(a) $Re = 100.$ (b) $Re = 200.$

(c) $Re = 300.$ (d) $Re = 400.$

(e) $Re = 219.$ (f) $Re = 225.$ (g) $Re = 235.$ (h) $Re = 335.$

Fig.5 Poincaré section of streamlines

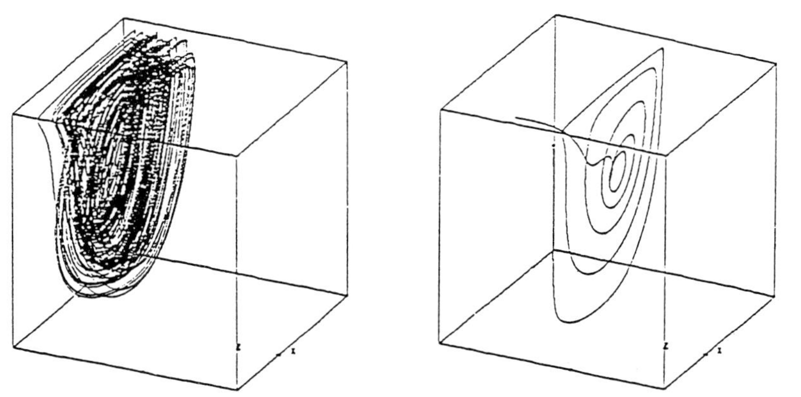

(a) a streamline started at (0.5, 0.77, 0.44). (b) a streamline started at (0.64, 0.99, 0.71).

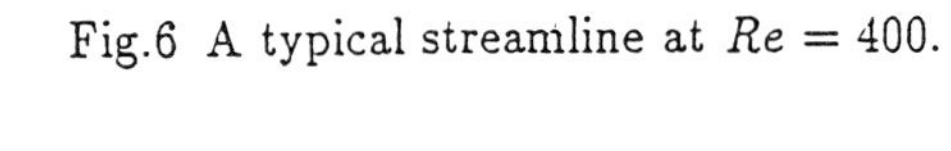

Fig.6 A typical streamline at $Re = 400$.

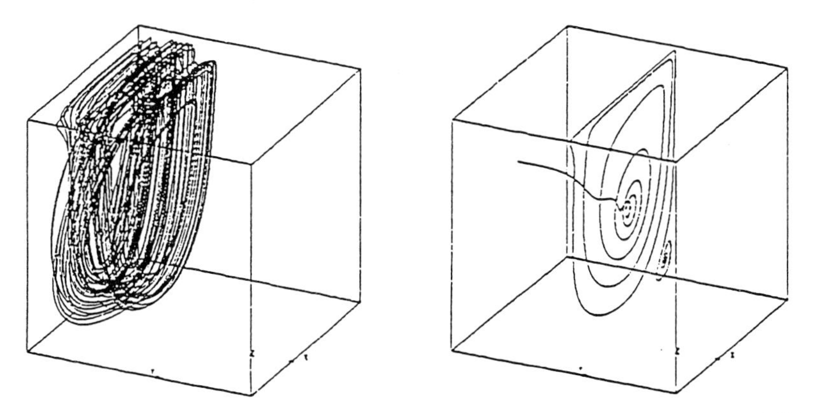

(a) a streamline started at (0.5, 0.64, 0.14). (b) a streamline started at (0.5, 0.99, 0.59).

Fig.7 A typical streamline at $Re = 1000$.

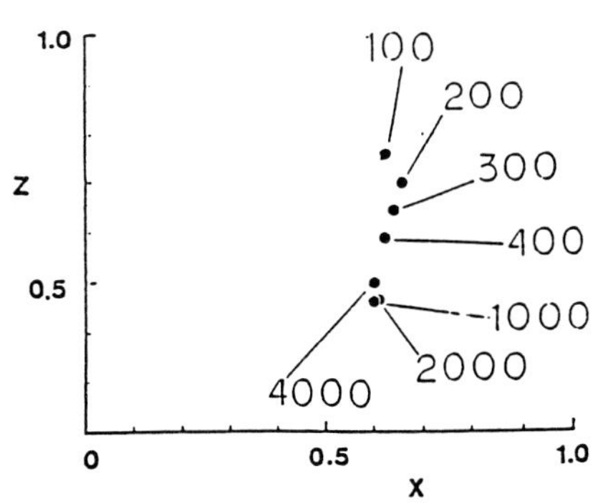

Fig.8 The position of the fixed point near the center of the symmetric plane $y = 0.5$ at different Reynolds numbers.

Streamline Coordinates, Moving Frames, Chaos and Integrability in Fluid Flow

J.J. FINNIGAN

CSIRO Centre for Environmental Mechanics, Canberra, Australia

1 INTRODUCTION

There are significant advantages to writing flow equations in streamline coordinates or, more precisely, in an orthonormal moving frame aligned locally with the streamlines. In such equations, dependent variables are precisely those quantities that are measured in rectangular Cartesian coordinates aligned with the streamlines, while the independent variables are simply distance along the streamline and along two sets of orthogonal trajectories. In experimental studies of distorted flow fields, all that is required to use such equations is that measured quantities be rotated, post facto, into the local streamline frame, a trivial operation but one to which, at least in the case of atmospheric field studies of flow over hills, there exists no simple alternative.

A more fundamental motivation is the hope that by expressing the equations of motion in intinsic form, we may discern simplifications and parametrizations that are universal and not peculiar to a particular type of flow. An example of the insights that can arise is provided by the analysis by Finnigan et al. (1989) of turbulent boundary-layer flow over a two-dimensional ridge. There, the geometric parameters related to the flow structure that appear naturally in the streamline flow equations proved to be the appropriate parameters to interpret data from a range of situations.

In the following sections we shall outline the derivation of intrinsic streamline flow equations in the general, three-dimensional case and show how the appearance of streamline chaos limits the range of applicability of this concept. We shall also see how the same techniques provide a direct approach to the equally important topic of the onset of chaotic advection, that is, of Lagrangian turbulence.

2 FLOW EQUATIONS IN THE ORTHONORMAL MOVING FRAME

Let $\underline{e}_i$ be a right-handed, orthonormal basis in E^3 and let $\underline{u}$ be the velocity vector and $\underline{\underline{T}}$ the stress tensor of an incompressible, adiabatic fluid flow. The equation of motion of this fluid is

$$\frac{\partial \underline{u}}{\partial t} = - \nabla . \underline{\underline{T}} \qquad (1)$$

$\underline{u}$ and the second order tensor $\underline{T}$ may be expanded in the basis $\underline{e}_i$ as follows:

$$\underline{u} = u^i \underline{e}_i = u^1 \underline{e}_1 + u^2 \underline{e}_2 + u^3 \underline{e}_3 \qquad (2a)$$

$$\underline{T} = \tau^{ij}\, \underline{e}_i \otimes \underline{e}_j = \tau^{11}\, \underline{e}_1 \underline{e}_1 + \tau^{12} \underline{e}_1 \underline{e}_2 + \ldots + \tau^{33}\, \underline{e}_3 \underline{e}_3\,, \qquad (2b)$$

where the usual convention of summation over repeated indices is assumed.

We now define ∂_i the directional derivative in the direction $\underline{e}_i$ as $<\nabla\underline{A}, \underline{e}_i>$, where ∇ is the grad operator, $\underline{A}$ a tensor of any order and $<\underline{f}, \underline{g}>$ denotes the inner product of two tensors, $\underline{f}$ and $\underline{g}$. We define $\nabla.\underline{T}$ by

$$\nabla.\underline{T} = \partial_j [\underline{e}_i \otimes \underline{e}_j\ \tau^{ij}] \qquad (3)$$

Substituting equations (2) and (3) into (1), we obtain

$$\frac{\partial}{\partial t}\, u^i\, \underline{e}_i = -\, \underline{e}_i \left\{ \partial_j \tau^{ij} + \tau^{\alpha j}\, \Gamma^i_{\alpha j} + \tau^{i\alpha}\, \Gamma^j_{\alpha j} \right\}, \qquad (4)$$

where $\Gamma^i_{jk} = <\underline{e}_i, \partial_k \underline{e}_j>$, the i'th component of the rate of change of the j'th base vector in the k direction. Γ^i_{jk} is called the connection coefficient. Equation (4) has three components when referred to the basis $\underline{e}_i$. Remembering that $<\underline{e}_i, \underline{e}_j> = \delta_{ij}$, its k'th component is:

$$\frac{\partial u^k}{\partial t} + u^i \Gamma^k_{it} = -\, \partial_j \tau^{kj} - \tau^{\alpha j}\, \Gamma^k_{\alpha k} - \tau^{k\alpha}\, \Gamma^j_{\alpha j} \qquad (5)$$

and Γ^k_{it} are the components of the time rate of change of $\underline{e}_i$. Evidently all that is required to write flow equations in terms of an arbitrary orthonormal basis $\underline{e}_i$ is knowledge of the connection coefficients that is, of the derivatives of $\underline{e}_i$ along the vector lines tangent to $\underline{e}_i$. Terms involving the connection coefficients appear in the equations whenever the basis varies in time or space. The best-known consequence of a time-varying basis is the Coriolis "force" that appears in flow equations referred to rotating axes. For the rest of this paper, however, we shall confine ourselves to discussion of streamline patterns that do not vary in time; for example, in turbulent flows we shall use the time mean, not the instantaneous velocity field.

To fulfil the requirements outlined in the introduction, we must specify an orthonormal frame with $\underline{e}_1$ everywhere tangent to the streamline. Therefore, $\underline{e}_1$ is given by

$$\underline{e}_1 = \underline{u}/|\underline{u}| = \underline{u}/Q \qquad (6)$$

and the $\underline{x}^1$ lines are the vector lines of the $\underline{e}_1$ field.

The next step is to specify $\underline{e}_2$ and $\underline{e}_3$. In axially symmetric and two-dimensional flow fields it is possible to define a true coordinate system, where the coordinate surfaces are constant surfaces of the appropriate stream function and a pseudo-potential function (Finnigan, 1983). This coordinate system can be used to furnish the orthonormal moving frame we require by applying a normalizing procedure to the coordinate basis. In that case $\underline{e}_2$ and $\underline{e}_3$ turn out to be, respectively, the principal normal and binormal vectors to the streamline. In axially symmetric and two-dimensional flows the streamlines are plane curves.

This is not the case for more general flows; nevertheless, for consistency, we will continue to identify $\underline{e}_i$ with the tangent, principal normal and binormal in the general case, where the streamline $\underline{x}^1$ may be a twisted curve.

We define the principal normal, $\underline{e}_2$, by the first of the Serret-Frenet equations:

$$\partial_1 \underline{e}_1 = \underline{e}_2/R \tag{7}$$

The binormal, $\underline{e}_3$, is orthogonal to $\underline{e}_1$ and $\underline{e}_2$ so that $\underline{e}_i$ form a right-handed triple.

$$\underline{e}_3 = \underline{e}_1 \wedge \underline{e}_2 \tag{8}$$

The $\underline{x}^2$ and $\underline{x}^3$ lines are the vector lines of the principal normal and binormal fields. The basis and the coordinate lines $\underline{x}^i$ are therefore determined completely by the velocity field. A basis consisting of the unit tangent, principal normal and binormal to a curve is sometimes called the Frenet frame.

In the following sections we shall attempt to determine the connection coefficients in terms of geometric parameters of the fields of vector lines $\underline{x}^i$ and then to relate these parameters to properties of the primary velocity field, specifically to its divergence and curl. In this way we hope to produce a set of flow equations which are self-consistent and closed in the sense that the connection coefficients appearing in (5) are expressed as functions of the dependent and independent variables only.

3 DETERMINING THE CONNECTION COEFFICIENTS

Since the basis is orthonormal

$$\langle \underline{e}_i, \underline{e}_j \rangle = \delta_{ij} \tag{9}$$

hence

$$\Gamma^i_{kj} = -\Gamma^j_{ik} \;; \qquad \Gamma^i_{ik} = 0 \tag{10}$$

and there are only nine independent connection coefficients.

Variation of $\underline{e}_i$ in the $\underline{x}^1$ direction is described by the Serret-Frenet equations of which (7) was the first member. The complete equations are:

$$\partial_1 \underline{e}_1 = \underline{e}_2/R \;; \quad \partial_1 \underline{e}_2 = -\underline{e}_1/R + \underline{e}_3/\sigma \;; \quad \partial_1 \underline{e}_3 = -\underline{e}_2/\sigma \tag{11a,b,c,}$$

$1/\sigma$ is called the torsion of the curve. It can be interpreted as the rate at which $\underline{e}_2$ rotates towards $\underline{e}_3$ in the plane spanned by $\underline{e}_2$ and $\underline{e}_3$ (the normal plane) with motion in the $\underline{x}^1$ direction. In the same way, $1/R$ can be regarded as the rate at which the tangent, $\underline{e}_1$, rotates towards $\underline{e}_2$ in the plane spanned by $\underline{e}_1$ and $\underline{e}_2$ (the osculating plane) with motion in the $\underline{x}^1$ direction. A curve is determined to within its position in space by $1/R(x^1)$ and $1/\sigma(x^1)$, the curvature and torsion parametrized by arc length x^i, where $x^i = |\underline{x}^i|$. Cartesian position coordinates can be recovered by integrating $1/R$ and $1/\sigma$ from some point on a streamline whose Cartesian coordinates are known. $1/R$ and $1/\sigma$ are consequently known as the intrinsic parameters of a space curve (Eisenhart, 1940).

The connection coefficients Γ^i_{j1} can be read off directly from equation (11a,b,c)

$$\Gamma^2_{11} = -\Gamma^1_{21} = 1/R \ ; \quad \Gamma^3_{21} = -\Gamma^2_{31} = 1/\sigma \ ; \quad \Gamma^3_{11} = -\Gamma^1_{31} = 0 \qquad (12a,b,c)$$

Six independent equations are required to determine the remaining connection coefficients, Γ^i_{j2} and Γ^i_{j3}. The required equations are the structure equations of the basis in E^3. Let $\underline{d}$ denote the exterior derivative operator, then

$$\underline{d}\ \underline{e}_j = \underline{\omega}^i_j \underline{e}_i \qquad (13)$$

where the exterior derivative of the base vectors is expanded in the basis with one-form coefficients ω^i_j. Evidently

$$\underline{\omega}^i_j(\underline{e}_k) = \Gamma^i_{jk} \ . \qquad (14)$$

By Poincaré's lemma,

$$\underline{d}(\underline{d}\ \underline{e}_j) = 0 = \underline{d}\ (\underline{\omega}^i_j, \underline{e}_i)$$

$$d\underline{\omega}^i_j \underline{e}_i = \omega^k_j \wedge (\omega^i_k \underline{e}_i)$$

where $\wedge$ denotes exterior multiplication and, since $\underline{e}_i$ are linearly independent,

$$d\underline{\omega}^i_j = \underline{\omega}^k_j \wedge \underline{\omega}^i_k \qquad (15)$$

Equations (15) are the structure equations.

Combining (14) and (15), we obtain nine independent equations for the connection coefficients. We shall discard those that contain derivatives of the unknown connection coefficients in the $\underline{x}^2$ and $\underline{x}^3$ directions. This leads to the following set of six coupled, first order, inhomogeneous ordinary differential equations:

$$\partial_1 \begin{bmatrix} \Gamma_1 \\ \Gamma_2 \\ \Gamma_3 \\ \Gamma_4 \\ \Gamma_5 \\ \Gamma_6 \end{bmatrix} + \begin{bmatrix} -\Gamma_1 & 0 & -(\Gamma_4 + 1/\sigma) & -1/\sigma & 0 & 0 \\ 0 & -\Gamma_2 & (1/\sigma - \Gamma_4) & 1/\sigma & 0 & -1/R \\ 1/\sigma & -1/\sigma & -1/La & 0 & 0 & 0 \\ 1/\sigma & -1/\sigma & 0 & -1/La & 1/R & 0 \\ 0 & 0 & 0 & (\Gamma_6 - 1/R) & -\Gamma_1 & 1/\sigma \\ 0 & 1/R & \Gamma_5 & 0 & -1/\sigma & -\Gamma_2 \end{bmatrix} \begin{bmatrix} \Gamma_1 \\ \Gamma_2 \\ \Gamma_3 \\ \Gamma_4 \\ \Gamma_5 \\ \Gamma_6 \end{bmatrix}$$

$$+ \begin{bmatrix} \partial_2(1/R) - (1/R)^2 \\ 0 \\ \partial_3(1/R) \\ 0 \\ 1/\sigma R - \partial_2(1/\sigma) \\ \partial_3(1/\sigma) \end{bmatrix} = 0 \qquad (16)$$

where, for clarity, we have written

$$\Gamma_1 = \Gamma^1_{22}; \quad \Gamma_2 = \Gamma^1_{33}; \quad \Gamma_3 = \Gamma^1_{23}; \quad \Gamma_4 = \Gamma^1_{32}; \quad \Gamma_5 = \Gamma^3_{22}; \quad \Gamma_6 = \Gamma^2_{33} \ .$$

and $1/La = 1/Q\ \partial_1 Q$. The appearance of $1/La$, the inverse of the 'e-folding' distance of streamwise acceleration, will be explained shortly.

Solving equation (16) involves finding six 'integrals of the motion'. These are independent functions of the Γ's that are constant along $\underline{x}^1$. In comparable problems in classical mechanics, it is often possible to obtain some of the integrals of the motion by the application of general principles like the conservation of momentum and energy. We shall pursue an analogous course by examining the properties of the velocity field that are invariant under coordinate translation or rotation.

The variation of a given velocity field in space is described by the deformation tensor D_{ij} where,

$$D_{ij} = \partial u_i/\partial y_j \tag{17}$$

and u_i are the components of the velocity vector $\underline{u}$ in the rectangular Cartesian coordinate frame y_i. D_{ij} may be split into symmetric and skew-symmetric parts,

$$D_{ij} = \tfrac{1}{2}e_{ij} + \tfrac{1}{2}s_{ij} = \tfrac{1}{2}(\partial u_i/\partial y_j + \partial u_j/\partial y_i) + \tfrac{1}{2}(\partial u_i/\partial y_j - \partial u_j/\partial y_i) \tag{18}$$

e_{ij} is the rate of strain tensor while the elements of s_{ij}, the rotation tensor, are Ω_i, where Ω_i are the components in y_i of the vorticity vector, $\underline{\Omega} = \text{curl}\ \underline{u}$.

e_{ij}, a real, symmetric third-rank tensor, has three scalar invariants, the Cayley-Hamilton invariants. These are the coefficients of the characteristic equation of e_{ij}. Two of these invariants have no simple physical meaning (Dishington, 1960) but the third is the trace of e_{ij}, where

$$\text{tr}\ e_{ij} = e_{ii} = \nabla.\underline{u} \tag{19}$$

In an incompressible flow, $\nabla.\underline{u} = 0$ and, if we express $\nabla.\underline{u}$ in the $\underline{e}_i$ basis, we obtain

$$\nabla.\underline{u} = 0 = \partial_1 Q + (\Gamma^2_{12} + \Gamma^3_{13})Q$$

hence,

$$1/La - \Gamma^2_{12} - \Gamma^3_{13} = 1/Q\ \partial_1 Q - \Gamma^2_{12} - \Gamma^3_{13} = 0 \tag{20}$$

Equation (20) can be regarded as an integral of the motion.

Moving to the skew symmetric part of the deformation tensor, the vorticity vector, curl $\underline{u}$, may be expressed in the $\underline{e}_i$ basis as

$$\text{curl}\ \underline{u} = QA\underline{e}_1 + \partial_3 Q\underline{e}_2 + (Q/R - \partial_2 Q)\underline{e}_3 \qquad \text{(Truesdell, 1954)} \tag{21}$$

The quantity A is called the abnormality of the field of vector lines. It is the ratio of two of the quadratic scalar invariants of the velocity and vorticity fields, the mean helicity and the kinetic energy per unit mass.

$$A = \frac{\langle \underline{u}, \text{curl } \underline{u} \rangle}{\langle \underline{u}, \underline{u} \rangle} = \langle \underline{e}_1, \text{curl } \underline{e}_1 \rangle \tag{22}$$

while

$$\text{curl } \underline{e}_1 = (\Gamma^1_{23} - \Gamma^1_{32})\underline{e}_1 + \Gamma^2_{11}\underline{e}_3 = (\Gamma^1_{23} - \Gamma^1_{32})\underline{e}_1 + \underline{e}_3/R \tag{23}$$

where

$$A - (\Gamma^1_{23} - \Gamma^1_{32}) = 0 \tag{24}$$

We may regard (24) as a second integral of the motion. Equations (21) and (23) provide an immediate connection between the geometric parameter $1/R$ and the component of the vorticity in the $\underline{x}^3$ direction. The relationship between the torsion $1/\sigma$ and A is less direct, as we see now. Two of the parameters appearing in the coefficient matrix of (16) can be identified with the abnormalities of the fields of $\underline{x}^2$ and $\underline{x}^3$ lines, An and Ab, respectively.

$$An = \langle \underline{e}_2, \text{curl } \underline{e}_2 \rangle = \Gamma^1_{23} - \frac{1}{\sigma} = \Gamma_3 - \frac{1}{\sigma} \tag{25}$$

$$Ab = \langle \underline{e}_3, \text{curl } \underline{e}_3 \rangle = -\left[\Gamma^1_{32} + \frac{1}{\sigma}\right] = -\left[\Gamma_4 + \frac{1}{\sigma}\right] \tag{26}$$

Combining (24), (25) and (26) to obtain

$$A - An - Ab = 2/\sigma \tag{27}$$

we see that the connection between the abnormality and the torsion of the field of streamlines is mediated by the abnormalities of the fields of orthogonal trajectories, $\underline{x}^2$ and $\underline{x}^3$.

The question of which parameter combinations allow a coupled dynamic system like (16) to be integrated, that is, the remaining four integrals of the motion to be found, has not yet received a complete answer. In a recent survey article, Tabor (1984) suggests that a general approach could be based on Painlevé tests, that is, the search for complex x^1 singularities of the solution trajectories in the six dimension state space of (16). This approach has been used by Dombre et al. (1986) in their study of Lagrangian particle trajectories of ABC flows. In this paper we will investigate instead, a heuristic approach based upon the connection between (16) and the conjugate dynamical system describing particle motion along steady streamlines. This method is motivated by the nature of those simplifications of the coefficient matrix of (16) that allow integration by inspection. To introduce this approach we shall first survey the flow fields for which solutions to (16) may be found in this way.

4 THE INTEGRABILITY OF THE STRUCTURE EQUATIONS

The most complicated flow for which (16) can be integrated by inspection has An = Ab = 0.

Here, in virtue of (25) and (26),

$$\Gamma_3 = \Gamma^1_{23} = 1/\sigma \quad ; \quad \Gamma_4 = \Gamma^1_{32} = -1/\sigma \tag{28}$$

whence $A = 2/\sigma$

Manipulation of (16) in combination with (28) and the integrals of the motion (20) and (24) yields expressions for all of the connection coefficients. No quadratures need to be performed explicitly.

The successively simpler flows with An = Ab = A = 0; with axial symmetry and with symmetry under reflection in the $\underline{x}^1 - \underline{x}^2$ plane (2-D flow) all allow (16) to be integrated directly and produce simple expressions for the connection coefficients. We shall return to consider their flow equations below. For the moment we note that the property that distinguishes obviously integrable from nonintegrable solutions of the structure equations is whether An and Ab differ from zero.

In systems that are nonintegrable, solution trajectories are exponentially sensitive to initial conditions. Such systems are termed chaotic. If the phase or state space of such a system contracts at the same time as solution trajectories are diverging exponentially, the manifold of trajectories must be a fractal. The solutions of the Lorenz equations, for example (Lorenz, 1963), lie on "strange" attractors of fractional dimension (Yorke and Yorke, 1981).

It is useful at this point to formalize the concept of the dynamical system - the fluid flow - that is conjugate to the structure equations in that the coefficients of (16) are determined by that flow field. If y_i are Cartesian coordinates of a point in the steady velocity field $\underline{u}(y_i)$, we may write:

$$u_1 = \frac{dy_1}{dt} = f_1(y_i) \;;\quad u_2 = \frac{dy_2}{dt} = f_2(y_i) \;;\quad u_3 = \frac{dy_3}{dt} = f_3(y_i) \qquad (29a,b,c)$$

The particle trajectory system may also be written in terms of the intrinsic parameters of the streamline, Q, 1/R and $1/\sigma$, in order to show the connection between the fluid flow and structure equations more directly, but (29) is the more familiar form.

Equation (29) describes the particle trajectories, which, in a steady flow, are the streamlines, while the connection coefficients Γ^i_{ij} may be reinterpreted geometrically as the rate at which tangent vectors to the streamlines rotate relative to one another with movement along the $\underline{x}^i$ lines. The structure equations (16), therefore, provide a description of the way the streamlines are separating as the flow proceeds.

The identification of nonintegrability with chaotic trajectories finds an immediate physical interpretation in the flow system (29). Since this flow is incompressible, its state space (real space) does not shrink and the solution trajectories (streamlines) lie on integral manifolds (stream surfaces). However, if the system is not integrable, streamlines are chaotic and the stream surfaces correspondingly convoluted.

Stream surfaces in a fluid flow are the integral surfaces of the field of tangent planes spanned by $\underline{e}_1$ and the unit tangent vectors, $\underline{n}$, to some orthogonal trajectories of the streamlines. A particular orthogonal trajectory will lie on the stream surface. The condition for the existence of such an integral surface is that

$$\underline{\omega} \wedge \underline{d}\,\underline{\omega} = 0 \tag{30}$$

where the one form $\underline{\omega}$ corresponds to the plane spanned by $\underline{e}_1$ and $\underline{n}$. Equation (30) is the Frobenius integration condition (Arnold, 1978; Spivak, 1979). $\underline{n}$ may be expressed in terms of $\underline{e}_2$ and $\underline{e}_3$ as

$$\underline{n} = \underline{e}_2 \sin\theta + \underline{e}_3 \cos\theta \tag{31}$$

then

$$\underline{\omega}(\underline{e}_2) = \sin\theta; \quad \underline{\omega}(\underline{e}_3) = \cos\theta \; . \tag{32}$$

When $\theta = 0$, $\underline{n} = \underline{e}_3$ and (33) can be expressed as $\underline{\omega} \wedge \underline{d}\,\underline{\omega} = \mathrm{Ab} = 0$. Similarly, when $\theta = \pi/2$, $\underline{n} = \underline{e}_2$ and we have $\underline{\omega} \wedge \underline{d}\,\underline{\omega} = \mathrm{An} = 0$. Hence the vanishing of An implies that the $\underline{x}^2$ curves lie on stream surfaces; when Ab vanishes, $\underline{x}^3$ curves lie on stream surfaces. Clearly, the possibility of integrating the structure equations is related to the character of the stream surfaces in the conjugate fluid flow.

In the general case (31), the Frobenius condition (30) can be expressed as an equation for θ:

$$\frac{d\theta}{dx} = (\Gamma^1_{22} - \Gamma^1_{33}) \sin\theta \cos\theta - \mathrm{An} \sin^2\theta - \mathrm{Ab} \cos^2\theta \tag{33}$$

Equation (33) is the basis of a heuristic approach to investigating the integrability of the flow field and its associated structure equations. If a given velocity field has plane strain and rotation descriptors, $(\Gamma^1_{22} - \Gamma^1_{33})$, An and Ab such that solutions to (33) cannot be found in some region of the flow, then in that region, the streamlines will not form smooth stream surfaces but will be chaotic. We conjecture that such values of $(\Gamma^1_{22} - \Gamma^1_{33})$, An and Ab mean that in this region the structure equations will not be integrable and streamline moving frame equations cannot be used.

Lack of space forbids a detailed discussion of (33) here but the following points must be made. If $\mathrm{An}(x^1)$, $\mathrm{Ab}(x^1)$ and $(\Gamma^1_{22} - \Gamma^1_{33})\,(x^1)$ are chosen as certain smoothly varying, differentiable functions that correspond to a flow that has $A \neq 0$ and is a solution of the Euler equations, (33) satisfies a Lipschitz condition signifying that a solution to the Cauchy initial value problem exists. However, numerical integration of (33) with these choices of coefficient reveals random, aperoidic dependence of θ on x^1. In this sense (33) behaves in the same way as the archetypal nonlinear equation studied by Feigenbaum (1983). Work is continuing on this question of what exactly is meant by "solutions to equation (36) cannot be found", in particular we wish to follow a streamline in the chaotic region of the ABC flow in order to investigate the associated behaviour of the θ equation.

We have considered the question of the integrability and associated chaotic advection of a field of streamlines. Of equal importance in practical situations is the degree of mixing that occurs. This question can be answered by considering the Lyapunov stability of the flow field system (29). Such an approach will not, however, produce general criteria but will be peculiar to the particular flow investigated. 54Since the form of the structure equations does not depend upon the conjugate flow field, we should look instead at the Lyapunov stability of (16). Once again lack of space forbids a detailed treatment in the present paper. We can, however, make the following related observation following Lorenz (1963).

If (16) is linearized about a basic state Γ_{i0} at $x^1 = x^1_0$, then $\partial_1 V_0$, the change with motion along $\underline{x}^1$ of a small volume of the six-dimensional state space of Γ_i, is given by the trace of the coefficient matrix of the linearized structure equations:

$$\partial_1 V_0 = -\left[3\Gamma_{10} + 3\Gamma_{20} + \Gamma_{50} + \Gamma_{60} + 2/\mathrm{La}\right] \tag{34}$$

If we are considering departures from an initial state of parallel rectilinear flow such that the $\underline{x}^i$ lines are orthogonal straight lines, then $\Gamma_{i0} = 0$ and

$$\partial_1 V_0 = -2/\mathrm{La} \tag{35}$$

This criterion is essentially the same as the one obtained by Zak (1986) characterizing the orbital instability of a general nonlinear system. It is interesting to observe that the criteria for relative separation of flow lines involves $1/L_a = \left[\Gamma^1_{22} + \Gamma^1_{33}\right]$, the divergence of the flow, and the related quantities that appear in (34) but not Γ_3 and Γ_4. In contrast, the criterion for chaotic behaviour, (33), interpreted as failure of streamlines to form smooth stream surfaces, involves the relative rotations or helicity of the flow field, expressed as An and Ab, and the tendency of the flow lines to converge in one direction normal to $\underline{e}_1$ while simultaneously diverging in an orthogonal direction, the property characterized by $\Gamma^1_{22} - \Gamma^1_{33}$.

5 PROPERTIES OF THE INTEGRABLE SYSTEMS

Thus far, only the two-dimensional streamline equations have been used in analysis of data (Finnigan and Bradley, 1983; Finnigan, 1988; Finnigan et al., 1989). Both 2-D and axially symmetric flows are special cases of complex lamellar flow: A = 0. They have the added restriction that An = Ab = 0 and special symmetry as mentioned earlier. The streamwise ($\underline{x}^1$ direction) and normal ($\underline{x}^2$ direction) mean momentum equations of axially symmetric flow are

$$\partial_1(P/\rho + \tfrac{1}{2}Q^2) = -\partial_1\tau^{11} - \partial_2\tau^{12} + (\tau^{11} - \tau^{22})/\mathrm{La} + \tau^{12}(2/R + 1/r\ \partial_2 r) + (\tau^{22} - \tau^{33})\ 1/r\ \partial_1 r \tag{36a}$$

$$\frac{Q^2}{R} + \partial_2 P/\rho = -\partial_1\tau^{12} - \partial_2\tau^{22} + (\tau^{11} - \tau^{22})/R + \tau^{12}(2/\mathrm{La} + 1/r\ \partial_1 r) - \tau^{22}\ .\ 1/r\ \partial_2 r \tag{36b}$$

ρ is the density and r is the radius of curvature of the $\underline{x}^3$ lines, which are circles about the axis of symmetry and can be replaced by an expression involving the integral of a function of $1/R(x^1)$ along $\underline{x}^1$ in order to close (36)a and b.

The two-dimensional equations follow by setting $1/r = 0$ in (36). Analysis of turbulent flow over two-dimensional ridges based on this limit of (36) reveals that $1/R$ and $1/La$ are natural scaling parameters of turbulent stresses, which may be regarded as generated by rates of mean strain. These are succinctly described in this intrinsic form. In this sense moving to the streamline moving-frame description led to the simplifications that were anticipated in the introduction.

In more complicated situations, where two-dimensional and axial symmetry are absent, mean rates of strain in intrinsic form include the torsion $1/\sigma$. A full analysis of such a system remains to be performed but work by Kao (1987) and Aref et al. (1988) emphasizes the extreme sensitivity of turbulent stress and scalar fluxes to torsion in such flows.

6 CONCLUSIONS

In order to write flow equations in the Frenet frame aligned with steady streamlines, we must determine nine independent connection coefficients. Three of these follow immediately from the Serret-Frenet equations of a space curve - the streamline - but the remaining six must be determined from the structure equations of Euclidean three-space, E^3. These furnish a set of six coupled, first-order ode's, the independent variable being x^1, arc length along the streamline. By considering the conjugate flow field, we are led to conjecture that integrability of the structure equations, which we define as solution trajectories lying on an integral manifold, depends upon the integrability of the dynamical system describing particle movement in the steady flow field. In short, streamlines must form smooth stream surfaces if the structure equations are to be integrable.

This direct connexion is made plausible by the geometric interpretation of connection coefficients as the rate at which the unit tangents to streamlines rotate apart as the flow proceeds. The intrinsic streamline representation in turn allows the integrability condition to be formulated in a general way, independent of the properties of a particular flow. The diagnostic for nonintegrability reduces to a first-order ode in x^1 that has apparently stochastic behaviour (in the Feigenbaum (1983) sense) wherever smooth stream surfaces do not exist. We may speculate, therefore, that the particular combination of flow parameters - $(\Gamma^1_{22} - \Gamma^1_{33})$, An, Ab - that comprise the coefficients of (33) are those that determine the onset of chaotic advection.

In cases where the structure equations can be integrated, streamline flow equations can be found, in which length scales describing flow divergence or acceleration, vorticity and helicity arise naturally. These are known to be parameters upon which real flows exhibit sensitive dependence.

7 REFERENCES

Aref, H., Jones, S.W. and Thomas, O.M. (1988). Computing particle motions in fluid flows. *Computers in Physics*, ***2***, 22-27.

Arnold, V.I. (1978). *Mathematical Methods of Classical Mechanics*. Springer-Verlag, New York, 462 pp.

Dishington, R.H. (1960). Rate of strain invariants in the kinematics of continua. *Phys. Fluids*, ***3***, 482.

Dombre, T., Frisch, U., Green, J.M., Hénon, M., Mehr, A. and Soward, A.M. (1986). Chaotic streamlines in the ABC flows. *J. Fluid Mech.* ***167***, 353-391.

Eisenhart, L.P. (1940). *An Introduction to Differential Geometry*. Princeton University Press, Princeton, NJ.

Feigenbaum, M.J. (1983). Universal Behavior in Non-linear Systems. *In* Order in Chaos: Proceedings of Int. Conf. on Order in Chaos, Los Alamos, 24-28 May, 1982. pp. 16-39.

Finnigan, J.J. (1983). A streamline co-ordinate system for distorted, two-dimensional shear flows. *J. Fluid Mech.* ***130***, 241-258.

Finnigan, J.J. (1988). Air flow over complex terrain. In *Flow and Transport in the Natural Environment: Advances and Applications*. Springer-Verlag, Heidelberg, pp. 183-229.

Finnigan, J.J. and Bradley, E.F. (1983). The turbulent kinetic energy budget behind a porous barrier: An analysis in streamline co-ordinates. *J. Wind Eng. and Ind. Aerodyn.* ***15***, 157-168.

Finnigan, J.J., Raupach, M.R., Bradley, E.F. and Aldis, G.K. (1989). A wind tunnel study of turbulent flow over a two-dimensional ridge. *Boundary Layer Meteorol.* (in press).

Kao, H.C. (1987). Torsion effect on fully developed flow in a helical pipe. *J. Fluid Mech.* ***184***, 335-356.

Lorenz, E.N. (1963). Deterministic nonperiodic flow. *J. Atmos. Sci.* ***20***, 130-141.

Spivak, M. (1979). *A Comprehensive Introduction to Differential Geometry*. Publish or Perish Inc., Berkeley, CA.

Tabor, M. (1984). Modern dynamics and classical analysis. *Nature*, ***310***, 277-282.

Truesdell, C.A. (1954). *The Kinematics of Vorticity*. Indiana University Press, Bloomington.

Yorke, J.A. and Yorke, E.D. (1981). Chaotic behaviour and fluid dynamics. In *Hydrodynamic Instabilities and the Transition to Turbulence* (Topics in Applied Physics, Vol. 45) Springer-Verlag, Berlin, pp. 77-95.

Zak, M. (1986). Criteria of chaos in non-linear mechanics. *Int. J. Non-linear Mech.* ***21***, 175-182.

The Geometry of Lagrangian Orbits

E. DRESSELHAUS and M.TABOR

Department of Applied Physics
Columbia University

1. INTRODUCTION

The stirring and mixing of passive scalars is determined by the Lagrangian history of the fluid particles. The efficiency of these processes depends on the geometric properties of the orbits. Thus the "stretching and folding" of a line element that results from chaotic particle advection in the neighbourhood of a hyberbolic fixed point leads to efficient stirring. The recent recognition that very simple flow fields can generate highly chaotic fluid particle orbits — a phenomenon variously termed "chaotic advection" or "Lagrangian turbulence" — has stimulated a new and significant research activity (Aref (1984), Aref & Balachandar (1985), Chaiken et.al.(1986,1987), Khakhar et.al.(1986), Ottino et.al.(1988), Rom-Kedar et.al.(1989)). For example, Aref and coworkers and Ottino and coworkers have explored a variety of two-dimensional flow configurations for which chaotic advection can occur with the latter emphasizing the significance of these results to mixing technologies. Wiggins and coworkers (see Rom-Kedar et. al.) have concentrated on developing detailed mathematical models of the associated transport processes.

In all the above mentioned works the emphasis has been almost exclusively on <u>two-dimensional</u> flows at very low Reynolds number. The underlying idea is, of course, that for 2–D incompressible flows a stream function $\psi = \psi(x,y,t)$ can be introduced and hence used to describe the fluid particle motions as the Hamiltonian dynamical system

$$\dot{x} = \frac{\partial \psi}{\partial y}, \quad \dot{y} = -\frac{\partial \psi}{\partial x} \tag{1}$$

For unsteady flows, i.e. explicitly time dependent ψ, the particle dynamics is generically chaotic. By drawing on the extensive understanding of Hamiltonian dynamics the observed chaotic advection phenomena can be analyzed (and predicted) in detail.

However, almost completely lacking to date has been an investigation of chaotic advection in <u>three dimensions</u> and the "fate" of chaotic advection at <u>higher Reynolds</u>

number. (This latter problem might glibly be termed the interaction between "Eulerian" and "Lagrangian" turbulence.) These two issues must, to some extent, go hand in hand since turbulent flows are essentially three dimensional.Furthermore any discussion of mixing and stirring at higher Reynolds number must inevitably probe the small-scale structures of these flows since it is these structures that induce the strongest gradients which in turn will result in the greatest stretching and folding of material lines. In this paper we suggest some ideas and models for exploring these ideas at low and high Reynolds number with special emphasis on characterizing the geometric properties of Lagrangian orbits.

2. CHAOTIC ADVECTION IN 3 DIMENSIONS: THE GEOMETRY OF FLUID PARTICLE ORBITS.

In three dimensions the fluid particle trajectories are governed by the non-Hamiltonian dynamics

$$\dot{x} = u(x,y,z,t)$$

$$\dot{y} = v(x,y,z,t) \tag{2}$$

$$\dot{z} = w(x,y,z,t)$$

where u, v, w are the three components of the velocity field obtained by solving by the relevant Navier-Stokes equation. In three dimensions chaotic motion is still possible for stationary flows, i.e. the right hand side of (2) is time independent. A valuable model of a stationary 3–D flow field is the ABC system studied by Dombre et.al.(1986) which takes the form

$$u = A \sin z + C \cos y$$

$$v = B \sin x + A \cos z \tag{3}$$

$$w = C \sin y + B \cos x$$

This is a solution to the (time-independent) Euler equations and is an example of a Beltrami flow, i.e. a flow for which $\omega \times \mathbf{u} = 0$, where ω is the vorticity vector. Beltrami flows, or to be more precise, regions with the Beltrami property are believed to play a significant role in describing turbulent flows (Pelz et.al.(1985)). The detailed study of (3) by Dombre et al. reveals the expected generic mixture of regular and chaotic orbits and a complicated stable/unstable fixed point structure. A discrete time mapping version of the ABC flow has been studied by Feingold et.al.(1988).

In the case of 2–D advection the possible morphologies of evolving particle distributions were predicted some time ago by Berry et.al.(1979) to be either whorls, which are tight curling structures due to the presence of elliptic fixed points, or tendrils , which are flailing, exponentially growing structures due to hyperbolic fixed points. The latter, with their essential combination of stretching and folding characteristics, are responsible for efficient mixing in 2 dimensions. However,there is, to date, no corresponding classification of the possible morphologies for 3–D flows. This is far more difficults than for the 2–D case since the structure and intersections of the stable and unstable manifolds in 3–D phase space is extremely complicated (see, for example, the discussion of null-null lines in Dombre et.al.). One approach to studying this structure will (inevitably) involve detailed numerical studies using three-dimensional graphics facilities.

We believe that a significant insight into the 3–D problem can be provided by a study of the geometry of particle orbits, i.e. their curvature and torsion and various time averages and correlations thereof. A first step in this direction is a recent study by Dresselhaus and Tabor (1989) of the persistence of strain. This quantity is simply

$$\sigma^2 = \frac{1}{2}\text{Trace } \mathbf{A}^2 \tag{4}$$

where $\mathbf{A}$ denotes the velocity gradient tensor; $(\mathbf{A})_{ij} = \partial u_i/\partial x_j$. The persistence of strain gives a simple and compact measure of the straining and rotational components of the motion since it is easily shown that

$$\sigma^2 = \frac{1}{2}\sum_i (s_i^2 - \omega_i^2) \tag{5}$$

where s_i and ω_i are the eigenvalues of the rate-of-strain and vorticity tensors respectively. In 2 dimensions σ^2 is the Gaussian curvature of the stream function. Clearly if $\sigma^2 < 0$ the motion is rotation dominated and the associated morphology will be whorl-like whereas for $\sigma^2 > 0$ the motion is strain dominated with a tendency for tendril formation. The study of σ^2 in the Lagrangian frame of fluid dynamics appears to be a new concept. By taking the time average of the real and imaginary parts of σ itself (i.e. $\sigma = (\text{Trace } \mathbf{A})^{1/2}$) along a Lagrangian orbit the ratio

$$\chi = \frac{\sigma_{Re}}{\sigma_{Im}} \tag{6}$$

gives a simple measure of stretch-fold ratio of particle orbits. We believe that such a quantity is relevant to quantifying mixing efficiency . Indeed, σ^2 and χ may be more valuable than the traditional Lyapunov exponents since the latter are,in effect, the long time average of the eigenvalues of the real part of the tangent map and hence contain no "folding" information. In addition σ^2 and χ are also significant as statistical quantities at higher Reynolds number (see section 3). Apart from its fluid dynamical significance we have shown (Dresselhaus and Tabor (1989)) that the quantities σ^2 and χ can also be used to provide a geometrical description of the invariant sets of dynamical systems in general since here **A** is simply the tangent map.

In addition to the persistence of strain a study of the orbital curvature and torsion (the former is very close, in 2-D, to the persistence-of-strain) should provide a rather refined description of fluid particle advection (and dynamical systems in general) that will be especially valuable in 3-D flows. In 2-D the stretching and foldings of tendrils gives an immediate picture of the striation of an advected quantity whereas in 3-D the extra dimension allows for the additional geometric complexity of "twisting" that can be quantified by studying the torsion. In fluid dynamics the study of curvature and torsion, in the context of vortex filaments, was introduced by Betchov (1965). Here, the novel feature is to study the curvature and torsion for Lagrangian orbits. In dynamical systems these quantities have not been studied extensively since they are metric dependent; for example helical flow on a torus is isomorphic to straight line flow on the unit square (with identified edges). However, this lack of invariance should not be a deterrent since for a given system, especially fluid flows in the cartesian metric, they can provide a rich geometric classification of different orbits.

For an orbit $\mathbf{x}(t) = (x(t),y(t),z(t))$ the calculation of its curvature and torsion is standard. On parameterizing the orbit in terms of arc length, s, one obtains the standard Frenet equations (here prime denotes derivative with respect to s)

$$\mathbf{u}' = \kappa\mathbf{p}$$

$$\mathbf{p}' = -\kappa\mathbf{u} + \tau\mathbf{b} \qquad (7)$$

$$\mathbf{b}' = -\tau\mathbf{p}$$

where $\mathbf{u}$ is the unit tangent vector and $\kappa(s) = |\mathbf{u}'(s)|$ the curvature; $\mathbf{p}$ and $\mathbf{b}$ are the unit normal and unit binormal vectors respectively and $\tau(s) = -\mathbf{p}(s).\mathbf{b}'(s)$ is the torsion. Explict expressions for κ and τ are easily obtained in terms of the total "time" derivatives of $\mathbf{x}$. Notions of curvature and torsion are easily generalized to curves in higher dimensional spaces. In addition to the information contained in κ and τ individually we envision that various ratios thereof, for example a screw-twist ratio, $\xi = \tau/\kappa$, analogous to the stretch-fold ratio χ, may provide further quantification of mixing efficiencies.

Preliminary investigation of these quantities have been carried out for a variety of flow fields such as the Lorenz system and the ABC flow. Especially instructive are studies of the "stretch-fold-twist" (STF) flow proposed by Moffatt (Bajer & Moffatt (1989). The general form of this bounded spherical flow takes the form $\mathbf{u} = \mathbf{U} + \mathbf{V} + \mathbf{W}$ where

$$\mathbf{U} = \mathbf{a}(1 - 2r^2) + (\mathbf{a}.\mathbf{x})\mathbf{x}$$

$$\mathbf{V} = \Omega \times \mathbf{x} \tag{8}$$

$$\mathbf{W} = - \mathbf{x} \times \mathbf{T}$$

Here $\mathbf{a} = (0, a, 0)$, $r^2 = \mathbf{x}.\mathbf{x}$, $\Omega = (0, \omega, 0)$ and $\mathbf{T} = T_{ij}\, x_i x_j$; with a typical choice for T_{ij} being $T_{ij} = \lambda\delta_{11}$. The parameters ω and λ characterize the rotation and twist of the velocity field as a whole. (It is most important to differentiate between the geometric properties of the flow field itself, i.e. the Eulerian frame and those of the individual orbits, i.e. the Lagrangian frame - the two are not necessarily related!). Numerical studies of the particular form studied by Bajer and Moffatt (their eqn (1.7)) show a striking increase in the spectral complexity of κ and τ as the rotation parameter ω is increased. Along a given orbit the biggest fluctuations in κ and τ are observed as the orbit swings past the various fixed points of the flow field. A careful quantification of this behaviour as a function of fixed point structure is now required. Here we present some typical (and preliminary) results illustrating the behaviour of χ, the stretch-fold ratio, and ξ, the twist-screw ratio, for the STF flow. In figure(1) we show, for fixed λ (λ = 1.0), the behaviour of χ, for a typical chaotic orbit, as a function of increasing ω. As ω increases the flow field becomes rotation dominated and this is reflected in the behaviour of the individual (Lagrangian) orbit by (overall) a corresponding decay in χ.

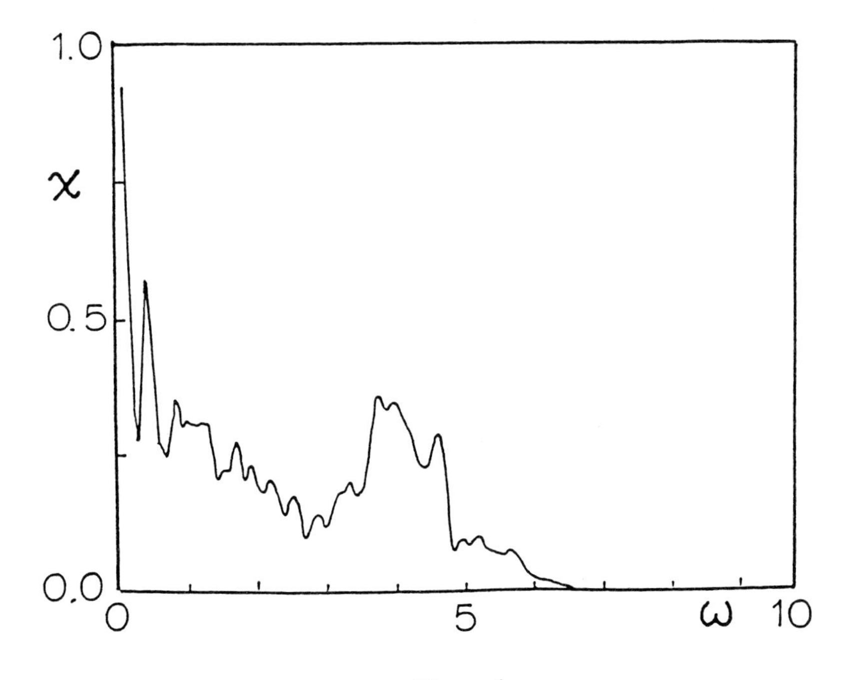

Figure 1

In figure(2) we show a corresponding plot of ξ. Here we observe that the biggest fluctuations in this quantity (even allowing for rather large standard deviations) occur in the ω range that Bajer & Moffatt (1989) found to correspond to the transition from predominantly chaotic to predominantly integrable motion. It is worth noting that the geometrical quantities, σ^2, χ, ξ etc. are trivial to compute compared to Lyapunov exponents.

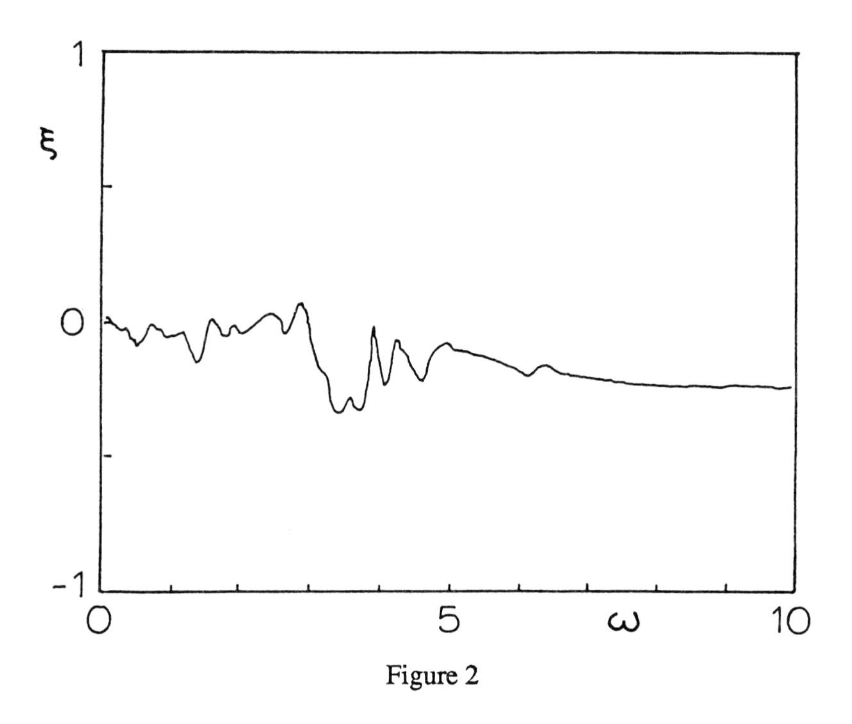

Figure 2

3. CHAOTIC ADVECTION AND LAGRANGIAN QUANTITIES AT HIGHER REYNOLDS NUMBER

The Lagrangian description of turbulence is a venerable problem. A substantial older literature due to Batchelor, Kraichnan, Corrsin and many others is available (see, for example, Monin and Yaglom (1975)). All these theories are essentially statistical. In our problem of chaotic advection at higher Reynolds number (Re) we still need to retain some detailed dynamical information. Since this is missing from the older works there is much scope for new research. A key question, for example, is: how is mixing efficiency affected by increasing Re? One way of rephrasing this question is to ask : how do Lyapunov exponents scale with Re?

In order to illustrate some of the concepts involved in answering this difficult question the following, heuristic, argument should be noted. Two, infintesimally close particles will only separate significantly over regions of space for which the associated velocity gradient (i.e. straining field) is well correlated. The largest scale over which this can occur in (fully developed) turbulent flow is, in fact, the Kolmogorov micro-scale. This length scale scales as $Re^{-3/4}$. Of course as this length scale becomes smaller, the associated gradient increases. This gradient (the micro-scale frequency) scales as $Re^{1/2}$. Considering that a Lyapunov exponent is a long-time average of particle separation the competition between these two effects as a function of Re is clearly very complicated. Following on from our discussion of chaotic advection in 3–dimensions we believe that, in addition to Lyapunov exponents, the scaling properties of the geometrical quantities such as persistence of strain, curvature and torsion will be extremely significant.

At the moment there is little in the way of convincing scaling laws for Lagrangian auto–correlations — expecially for gradient quantities. Properties of curvature and torsion auto–correlations and cross–correlations are completely unknown

The key quantities to study are clearly time correlations. The simplest of these is the velocity auto-correlation. This can either be in the Eulerian frame, namely

$$C^{(E)}_{ij}(\tau) = \langle u_i(\mathbf{x},t)u_j(\mathbf{x},t+\tau)\rangle \tag{9}$$

or the Lagrangian frame, namely

$$C^{(L)}_{ij}(\tau) = \langle u_i(\mathbf{a},t)u_j(\mathbf{a},t+\tau)\rangle \tag{10}$$

where **a** denotes a fluid particle label. In the theory of isotropic, homogeneous turbulence the study of time correlations, as opposed to space correlations, has met with only modest success. Assuming a Kolmogorov cascade it follows from simple scaling arguments (if they exsist for time correlations) that the frequency spectrum (i.e. the Fourier transform of $C(\tau)$) of either correlation behaves as (Tennekes & Lumley (1972))

$$\phi(\omega) = \varepsilon\, \omega^{-2} \tag{11}$$

where ε is the (universal rate) of energy dissipation. However, if one allows, in the Eulerian frame, for the possibility that the large scale, energy containing, eddies "sweep" the smaller eddies (this is sometimes referred to as the "random Taylor hypothesis") the spectrum changes to (Tennekes (1975))

$$\phi(\omega) = \varepsilon^{2/3} u_0^{2/3}\, \omega^{-5/3} \tag{12}$$

which is non-universal since it now depends on some typical large scale velocity u_0. It seems reasonable to suggest that the sweeping result, $\omega^{-5/3}$, corresponds to the Eulerian spectrum and the non-sweeping result, ω^{-2}, corresponds to the Lagrangian spectrum (by definition there cannot be sweeping in the Lagrangin frame). One consequence of this is that the ratio of some Lagrangian (micro) time-scale to the corresponding Eulerian time-scale scales as $Re^{1/4}$. Thus along a Lagrangian orbit the velocity of a particle becomes ever better correlated relative to its Eulerian counterpart as a function of increasing Reynolds number. However, the notion of sweeping is not without controversy. A recent paper by Yakhot et.al.(1988) using newly developed renormalization group (RNG) methods suggests that there is, in fact, no sweeping. In other words: the large and small scale motions cannot be decorrelated with the result that the Eulerian spectrum scales more like ω^{-2}. However, a recent investigation by Nelkin and Tabor (1989), using simpler statistical ideas, reinforces the validity of the sweeping hypothesis which, apart from the scaling issue, also indicates possible defficiencies with the above mentioned RNG methods.

A relatively uncharted area, of great importance to the ideas of orbital geometry discussed here, is that of time correlations for velocity gradient quantities. Since these are greatest for the smallest scales in the flow the behaviour of the associated correlations will probe the properties at this end of the spectrum (intermittency effects will inevitiably become very important here). An especially significant quantity is the persistence of strain auto-correlation in the Lagrangian frame, namely

$$P(\tau) = \langle \sigma^2(\mathbf{a},t)\, \sigma^2(\mathbf{a},t+\tau) \rangle \tag{13}$$

Such an auto–correlation (along a Lagrangian orbit) will determine the deformation of a (small) elastic body in a turbulent flow: if the auto–correlation time is long the body will be significantly stretched. (This is fundamental for an understanding of the role of polymer molecules in drag reduction (Tabor & de Gennes (1986)). It is easy to show that σ^2 has a simple fluid dynamical interpretation, namely

$$\sigma^2 = \nabla.(u.\nabla u) = - \nabla^2 p \tag{14}$$

where p is the pressure field(here we are setting the density,ρ, to unity). In many numerical simulations this Poisson equation is used to compute a velocity field from a given pressure field. Thus it may be fairly straight forward to compute the two-point, one-time and (maybe) the one-point two-time Eulerian correlations. Numerical studies of the Lagrangian quantity (14) will, again, be much more difficult.(Recent simulations (Yeung & Pope (1988)) corresponding to a Reynolds number of several thousand indicates that σ^2 - there termed the "psuedo dissipation" - has a lognormal distribution). In addition, scaling laws and properties - both at a theoretical and numerical level - of the persistence of strain autocorrelation in the Lagrangian frame will require new insights into the nature of pressure fluctuations.

REFERENCES

Aref, H. 1984 J. Fluid Mech. 143, 1.

Aref, H. & Balachandar, S. 1985 Phys. Fluids 29, 3515.

Bajer, K. & Moffatt, H.K. 1989 "On a class of steady confined Stokes flows with chaotic streamlines" submitted to J.Fluid.Mech.

Berry, M. V. Balazc, N. L., Tabor, M. & Voros, A. 1979, Ann. Phys. N.Y. 122, 26.

Betchov, R. 1965 J. Fluid Mech. 22, 471.

Chaiken, J., Chevray, R., Tabor, M. & Tan, Q. M. 1986 Proc. Roy. Soc. A 408, 165.

Chaiken, J., Chu, C. K., Tabor, M. & Tan, Q. M. 1987 Phys. Fluids 30, 687.

Dombre, T., Frisch, U., Greene, J. M., Henon, M., Mehr, A. & Soward, A. M. 1986 J. Fluid Mech. 167, 353.

Dresselhaus, E. & Tabor,M. 1989 J.Phys.A 22, 971.

Feingold, M., Kadanoff, L. P. & Piro, O. 1988 J. Stat. Phys. 50, 529.

Khakhar, D. V., Rising, H. & Ottino, J. M. 1986 J. Fluid Mech. 172, 419.

Monin, A. S. & Yaglom, A. M. 1975, "Statistical Fluid Mechanics," Vols I & II, MIT press.

Nelkin, M. & Tabor, M. 1989 "Time correlations and random sweeping in isotropic turbulence," submitted to Phys.Fluids.A.

Ottino, J. M., Leong, C. W., Rising, H. & Swanson, P. D. 1988 Nature, 333, 419.

Pelz, R., Yakhot, V., Orstag, S. A., Shtilman, L. & Levich, E. 1985 Phys. Rev. Lett. 54, 2505.

Rom–Kedar, V., Leonard, A. & Wiggins, S. 1989 "An Analytical Study of Transport, Mixing and Chaos in an Unsteady Vortical Flow" (preprint).

Rom–Kedar, V. & Wiggins, S. 1989 "Transport in Two-Dimensional Maps," Arch. Rat. Mech. & Anal. (in press).

Tennekes, H. 1975 J. Fluid Mech. 67, 561.

Tennekes, H. & Lumley, J. L. 1972 "A First Course in Turbulence," MIT press.

Yakhot, V., Orszag, S. A. & She, Z. S. 1989 Phys. Fluids A1, 184.

Yeung, P. K. & Pope, S. B. 1988 "Lagrangian Statistics from Direct Numerical Simulations of Isotropic Turbulence," Cornell University preprint FDA-88-16.

Scalar Field Topology in Turbulent Mixing

CARL H. GIBSON

Departments of Applied Mechanics and Engineering Science and Scripps Institution of Oceanography, R-010, University of California at San Diego, La Jolla, CA 92093-0411, USA

ABSTRACT

The microstructure of scalar fields like temperature mixed by turbulence depends on the interaction of the rate-of-strain tensor field with scalar extremum points and extremal lines, produced by turbulent convection, where the scalar gradient is zero, minimal or maximal. A topological description of the interaction process is crucial to understanding the turbulent mixing process. Mixing of magnetic fields and self mixing of the vorticity field of the turbulence itself may also be formulated in terms of the interaction of the rate-of-strain tensor field with similar topological features of these vector fields.

1 INTRODUCTION

Turbulence and turbulent mixing processes are notoriously complex and demand clear physical models for progress in understanding. Because the processes are intrinsically four dimensional, sometimes involving several coupled scalar and vector fields governed by nonlinear equations, topological descriptions of the coupled fields as they evolve in time are virtually mandatory either to fully comprehend and visualize the significance of mathematical or numerical solutions of model problems, to interpret laboratory data, or to formulate improved physical models. Attempts to analyze turbulence processes by purely mathematical descriptions, or by dimensional analyses devoid of clear physical models, generally produce marginally reliable, or incorrect, results.

2 PASSIVE SCALAR MIXING THEORIES

An example is the partially correct Oboukhov (1949) - Corrsin (1951) extension of Kolmogorov's universal similarity hypotheses to scalar mixing. By analogy to the inertial-viscous damping scale of turbulence at the length scale $L_K \equiv (\nu^3/\varepsilon)^{1/4}$ predicted by Kolmogorov (1941), Oboukhov and Corrsin independently predicted a scalar inertial subrange with the scalar spectrum ~ $k^{-5/3}$, which is correct, and an inertial-diffusive damping of turbulent scalar field fluctuations at the analogous scale $L_C \equiv (D^3/\varepsilon)^{1/4}$, which is incorrect, where ν is the kinematic viscosity, ε is the viscous dissipation rate, and D is the diffusivity of the scalar. Batchelor (1959) realized that the latter result was incorrect for the case of weakly diffusive scalars with Prandtl number $Pr \equiv \nu/D \gg 1$ because the mixing at scales $L \approx L_C$ smaller than the Kolmogorov scale L_K should occur by a uniform straining mechanism depending on the rate-of-strain parameter $\gamma \equiv (\varepsilon/\nu)^{1/2}$ with a strain-diffusive damping length scale of $L_B \equiv (D/\gamma)^{1/2}$ rather than at L_C. Batchelor (1959) used the Fourier transformed scalar conservation equation to predict the absolute level and form of the spectrum for large Pr scalar fields like temperature and salinity in water. The Batchelor (1959) universal spectral form has now been confirmed by a variety of studies in the laboratory and ocean, and stands as one of very few successful analytic predictions of turbulence theory by mathematical analysis.

1.1 The Batchelor (1959) Wavecrest Compression Model

The Batchelor (1959) theory for $Pr \gg 1$ is based on the model that because γ is uniform on scales smaller than L_K but larger than L_C, the wavecrests of Fourier elements will be convected together by the uniform rate-of-strain tensor $\overleftrightarrow{e}$ along the compression principal axes, where $e_{ij} = (u_{i,j} + u_{j,i})/2$. This model fails for the case of $Pr \ll 1$ where both possible diffusive damping scales L_C and L_B are larger than the scale of uniform strain L_K. The smallest Fourier wavecrest separation for such strongly diffusive scalars should be much longer than the scale of uniform straining L_K. Therefore, Batchelor et al. (1959) conclude that γ should be irrelevant to the turbulent mixing of small Pr scalars and predict that the smallest scale fluctuations for $Pr \ll 1$ will be at $L \approx L_C$ which is independent of ν and γ, rather than at L_B. The model is shown schematically in Figure 1.

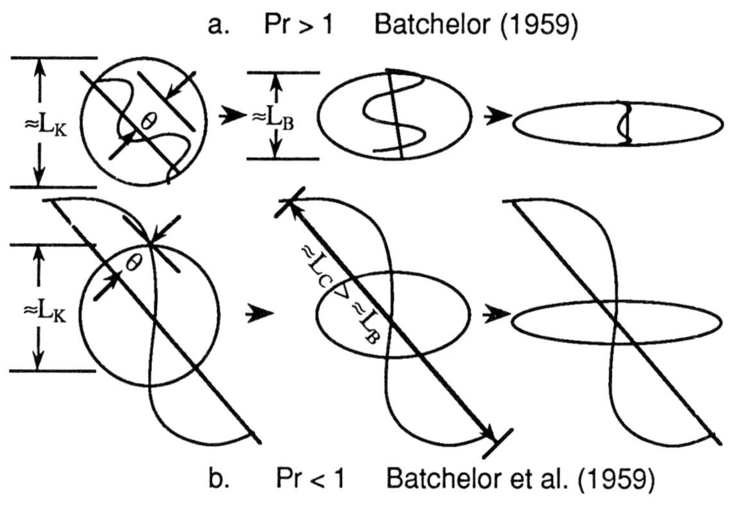

Figure 1. Wave crest compression model of Batchelor (1959), Fig. 1a, for $Pr \gg 1$; and Fig. 1b, for $Pr \ll 1$, from Batchelor et al. (1959). In Fig.1a a scalar Fourier element with wavelength smaller than the Kolmogorov scale L_K is convected to higher wavenumbers by the straining of a spherical fluid element, and is damped by diffusion when it reaches scales less than the Batchelor scale L_B. In Fig. 1b the local straining has no effect on the smallest scalar fluctuations because the wavecrests are separated by scales $L \geq L_C$ larger than L_K.

However, measurements of turbulent temperature fluctuations in mercury with $Pr \approx 0.02$ by Clay (1973), numerical simulations of Ashurst et al. (1988), Kerr (1985) with $Pr = 0.1$, 0.2 and 1.0 and numerical simulations of Gibson et al. (1988) with $Pr = 0.001$ to 0.2

indicate that the rate-of-strain γ is indeed relevant to the smallest scale mixing, even for Pr « 1. The evidence is that all spectra show diffusive cutoffs at $k \approx L_B^{-1}$, and the strain-rate-scalar dissipation correlation coefficient $\Sigma \equiv \left(\overline{u_{,x}T_{,x}{}^2}\right)\left(\overline{u_{,x}{}^2}^{1/2}\ \overline{T_{,x}{}^2}\right)$ is consistently non-zero for Pr < 1, with values about -0.5 as observed for Pr ≥ 1 showing the enhancement of local temperature gradients $T_{,x}$ by negative streamwise velocity gradients $u_{,x}$; that is, compressive straining, in the streamwise direction x (see Figure 7, §2).

The Batchelor et al. (1959) mixing model has been extended to strongly diffusive magnetic fields by Moffatt (1961, 1962), Golytsyn (1960) and Kraichnan & Nagarajan (1967) who reproduce for the magnetic field spectrum ϕ_B the strong diffusive cutoff $\phi_\theta \sim k^{-17/3}$ of the scalar spectrum for Pr « 1 at $k \approx L_C^{-1}$ predicted by Batchelor et al. (1959), as shown in Figure 2. It seems likely that if scalar fields with Pr « 1 are affected by the rate-of-strain of turbulence, then vector fields will also be affected, as discussed by Gibson (1988).

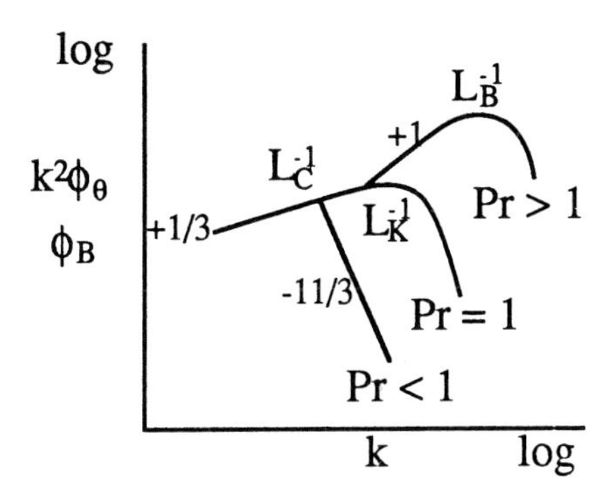

Figure 2. Spectral forms for scalar dissipation spectrum $k^2\phi_\theta$ and magnetic field spectrum ϕ_B for various Pr ranges according to wavecrest compression model of Fig. 1.

The critical limitation of the Fourier element wavecrest compression model of Fig. 1 is that it lacks phase information present in an actual scalar field mixed by turbulence. A topological-time view of the turbulence-scalar fields interaction reveals possible new mechanisms for γ-induced stirring and mixing not indicated by the evolution of individual Fourier elements.

1.2 The Gibson (1968a,b) Model of Straining Scalar Topology

Gibson (1968a,b) focuses on the evolution of isoscalar surfaces and topological features of scalar fields associated with points, lines and surfaces with minimum, maximum and zero gradient, based on the form of expressions derived for the velocity of the isoscalar surfaces and the points of minimum and maximum scalar values. For a dynamically passive scalar field θ with mean zero that obeys the equation

$$\frac{\partial\theta}{\partial t} + \vec{v}\cdot\nabla\theta = D\nabla^2\theta \tag{1}$$

where t is time, $\vec{v}$ is velocity and D is the diffusivity of θ, the velocity $\vec{v_\theta}$ of an isothermal surface at point $\vec{x}$ with molecular diffusivity D is given by

$$\vec{v}_\theta = \vec{v} - \frac{D\nabla^2\theta}{|\nabla\theta|}\vec{g}\,;\,\vec{g} \equiv \frac{\nabla\theta}{|\nabla\theta|}\ . \tag{2}$$

Equation (2) shows that isothermal surfaces move with the fluid velocity but also diffuse with respect to the fluid in the direction of the scalar gradient unit vector $\vec{g}$. The magnitude of the diffusion velocity on scale L from (2) is D/L, which equals the turbulence velocity $(\varepsilon L)^{1/3}$ from Kolmogorov's second universal similarity hypothesis at the scale $L_C = (D^3/\varepsilon)^{1/2}$. Thus turbulence acting on a uniform scalar gradient can only wrinkle the gradient but not overturn it on scales smaller than L_C, as shown by Gibson (1968a) and Figure 3. Since eddy scales $L > L_K > L_C$ for $Pr > 1$, all eddies can cause gross distortions, but for $Pr < 1$ a range of scales $L_C > L > L_K$ exists for which some eddies cannot. Because the overturning time scale for turbulent eddies is smallest for the smallest eddies, the first production of zero gradient features by turbulence will be at scales L_K for $Pr > 1$ and at scales L_C for $Pr < 1$. Subsequently larger eddies can produce such features, but at a slower rate. Zero gradient topologies produced by the larger eddies will have larger scalar contrasts between extremum values and their associated saddle points, so they will persist longer than smaller amplitude, smaller scale features.

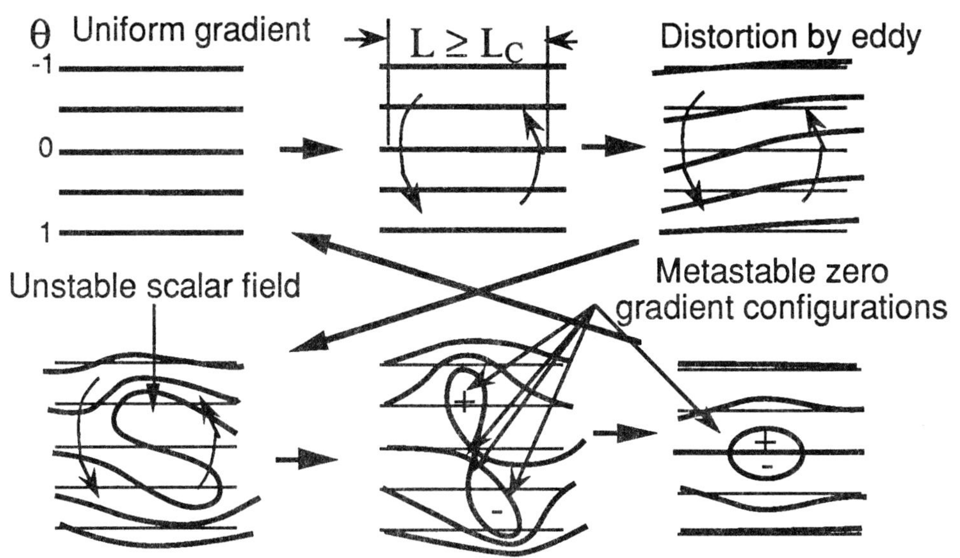

Figure 3. Schematic of scalar topology mixing model. Starting from a uniform downward scalar gradient at top left, a temporary isolated eddy of scale $L \geq L_C$ produces zero gradient extrema, saddle points and finally a doublet with a saddle line before diffusive relaxation to the original state. Only these five zero gradient configurations are "metastable" in the sense that they have finite possible lifetimes.

From (2) the initial velocity of the isoscalar surfaces in Fig. 3 will be the fluid velocity of the eddy since $\nabla^2\theta = 0$ if $\nabla\theta$ is constant. However, as the surfaces are distorted by convection, diffusion velocities develop in those directions required to restore the uniform gradient. Since the convection velocity of the disturbing eddy is larger than the induced diffusive velocities, by hypothesis, the isoscalar surfaces will become grossly distorted, and therefore unstable, and the initially singly connected isoscalar surfaces will split up to become multiply connected with a virtually unique topological signature of turbulent, and fossil turbulent, scalar fields. Once a scalar field has been scrambled by turbulence, it must unscramble itself by molecular diffusion, and this may take much longer than it takes for the turbulence which did the scrambling to be damped. This is particularly true for high Reynolds number, high Froude number turbulent events in stably stratified fluids such as the ocean and atmosphere where microstructure "fossils" of turbulence persist in several scalar fields long after buoyancy forces have converted the turbulent kinetic energy to internal waves. Gibson (1986) suggests that most temperature microstructure sampled in the ocean is fossil turbulence at the largest scales, and sometimes all scales, of the observed temperature fluctuations.

Temporarily stable (metastable) zero gradient features shown such as extremum points and saddle points will generally not occur in a fluid without turbulence since their formation requires either the gross isoscalar surface distortions produced by turbulence eddies larger than L_C, with consequent scalar instabilities, or artificial production by local scalar sources or sinks. According to the turbulent mixing theory of Gibson (1968a,b), Gibson et al. (1988) and Gibson (1988), these topologies are crucially important to the smallest scale mixing processes for turbulent scalar or vector fields, especially for strongly diffusive properties with $Pr \ll 1$. The mixing mechanism proposed depends on the special interaction of the rate-of-strain tensor field with features of the scalar field which have small or zero scalar gradients. Such features tend to follow the fluid motion, and therefore respond to the local rate-of-straining, as shown by the following expression for the velocity $\vec{v_0}$ of a zero (or minimum) gradient scalar feature, derived by Gibson (1968a)

$$\vec{v_0} = \vec{v} - D\left[\left(\frac{\theta_{,jj1}}{\theta_{,11}}\right), \left(\frac{\theta_{,jj2}}{\theta_{,22}}\right), \left(\frac{\theta_{,jj3}}{\theta_{,33}}\right)\right] \tag{3}$$

where subscripted commas indicate partial differentiation and repeated indices are summed over the three possible values. In (3) the coordinate axes are aligned with the local principal axes of the $\theta_{,ij}$ tensor. At extremum points, and on minimal gradient lines, the components of the diffusion velocity vector on the right of (3) tend to be zero or small. At such points, the tensor principal values in the denominators of the diffusion velocity components are maximum and the numerators are zero if the (scalar distribution near) the extremum is symmetric. If the extremum (or the extremal line) is not symmetric, the direction of the extremum diffusion velocity $\mathbf{v_0}$ is always toward a point of greater symmetry as shown in Figure 4.

Figure 4. An unsymmetric scalar fluctuation is shown at time t = 0. The direction of the zero gradient point velocities, from (3), are in directions to produce greater symmetry about the peaks. The scalar distributions about the extrema are more symmetric and damped by diffusion at a later time t > 0, as shown, and the point velocities $\mathbf{v_0}$ have decreased. When the distributions are symmetric the extrema are convected with the fluid velocity.

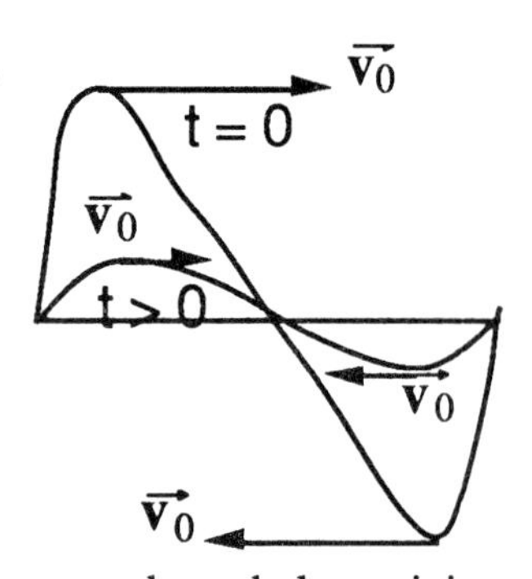

These zero gradient points and minimal gradient lines are important to the turbulent mixing process because they tend to move with the fluid so they can respond to the local fluid motion $\vec{u} = \overleftrightarrow{\Omega} \cdot \mathbf{d}\,\vec{\mathbf{x}} + \overleftrightarrow{\mathbf{e}} \cdot \mathbf{d}\,\vec{\mathbf{x}}$ at a position $\mathbf{d}\vec{\mathbf{x}}$ away from the extremum which consists of spinning by the local rotation tensor $\overleftrightarrow{\Omega}$ and stretching by the local rate-of-strain tensor $\overleftrightarrow{\mathbf{e}}$. Consider an initially symmetric maximum point. It moves with the local fluid velocity, but rapidly will become distorted in response to the principal axes of $\overleftrightarrow{\mathbf{e}}$. The effects of the local spinning presumably will be like a random convective noise or an enhanced transverse diffusivity. Local spin effects have not been precisely modelled, but are not expected to have any qualitative effect on the stain induced development of minimal gradient lines except to randomly deflect their trajectories.

From every extremum where $\overleftrightarrow{\mathbf{e}} \neq 0$ a minimal scalar gradient line will develop along the stretching axis, and a maximal scalar gradient line will develop along the compression principal axis of $\overleftrightarrow{\mathbf{e}}$. These lines will propagate rapidly into the fluid and will always tend to be aligned with the principal axes of $\overleftrightarrow{\mathbf{e}}$. Consequently the compressive axes of $\overleftrightarrow{\mathbf{e}}$ will be strongly correlated and aligned (with maximum mixing rates) over long distances $L \gg L_K$ along these maximal gradient lines and the stretching axes of $\overleftrightarrow{\mathbf{e}}$ will be strongly correlated and aligned (with minimum mixing rates) over long distances $L \gg L_K$ along the minimal gradient lines, contrary to the wavecrest compression model of §1.2 which assumes decorrelation of mixing and straining for scales larger than L_K. The existence of and location of the maximal gradient lines in the fluid reflect the history of their development, and this depends on the local straining. These lines also reflect the expression for the diffusive velocity $\vec{v}_D = -D\left[\left(\nabla^2\theta\right)/\left|\nabla\theta\right|\right]\vec{g}$ of isoscalar surfaces in (2), since $\vec{v}_D$ decreases for increasing gradient magnitude, so that nearby isoscalar surfaces will be more strongly influenced by convection where the scalar gradient magnitude is largest.

Figure 5 summarizes the scalar topology mixing model. As shown by Gibson (1968a), for $Pr < 1$ the lifetime of the zero gradient features produced by a turbulent eddy from a region of uniform scalar gradient is at least of order L_C^2/D, which is larger than the time $(\nu/\varepsilon)^{1/2}$ it takes for new features to form by strain splitting by a factor of $Pr^{-1/2} > 1$. Therefore, many extrema will form by stretching induced splitting from each one formed by the mechanism shown in Fig. 3. Each splitting results in the formation of a new saddle point of the same sign as the new extremum. Note that in three dimensions, saddle points have signs because there are three values of the eigenvalues of the tensor $\theta_{,ij}$. For a maximum point, all three values are negative, and for the associated saddle point two values are negative and one is positive.

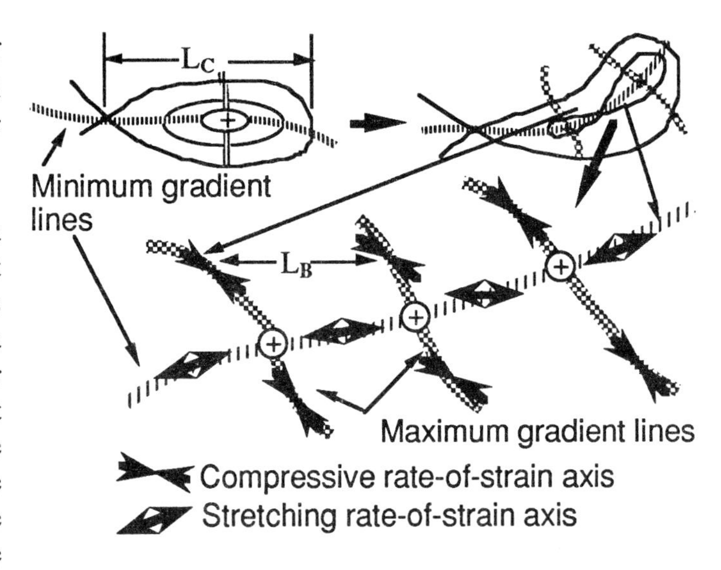

Figure 5. Production of extrema, minimal and maximal gradient lines, for Pr « 1.

At top left of Fig. 5 a maximum zero gradient point, indicated by a plus sign, is isolated from a region of uniform scalar gradient by a turbulent eddy of scale equal to the minimum size L_C. The extremum is split by the straining as shown on the right at a later time. A minimal gradient line in light hatching connects the extremum with its associated saddle point, and tends to be aligned with the stretching axes of the local rate-of-strain tensor along the line, a segment of which is shown later in the closeup below by heavy double arrows in opposite directions. The extremum will split several times as the straining induces eccentricity, and therefore instability, of the subsequent scalar extrema. Maximal gradient lines emanate from the extrema, shown by heavier hatched lines, and tend to be aligned with the compression axes of the local rate-of-strain tensor, shown by double arrows pointing together. The separation of maximal and minimal gradient lines is of order the Batchelor length scale L_B and their lengths are of order at least L_C according to the theory, as shown.

Because of the causal relationship existing between the location of the minimal and maximal gradient lines in scalar mixing and the directions of the stretching and compression principal axes of the rate-of-strain tensor, it is reasonable to expect that a strong correlation will exist between them, and that the rate-of-strain will be an important factor in determining the highest wavenumber portion of the scalar spectrum for turbulent mixing of arbitrary Prandtl number scalar fields. This assumption is the basis of the universal similarity hypothesis proposed by Gibson (1968a) that the smallest scale structures should depend only on the scalar variance dissipation rate $\chi \equiv 2D(\nabla\theta)^2$, D and γ. The spectral consequences of this hypothesis are given by Gibson (1968b) and are shown in Figure 6. The Batchelor et al. (1959) inertial diffusive subrange is shifted from $k \approx L_C^{-1}$ to $k \approx L_B^{-1}$, and a new strain-diffusive subrange, with $\phi_\theta \sim k^{-3}$ is predicted, with a universal exponential cutoff at $k \approx L_K^{-1}$.

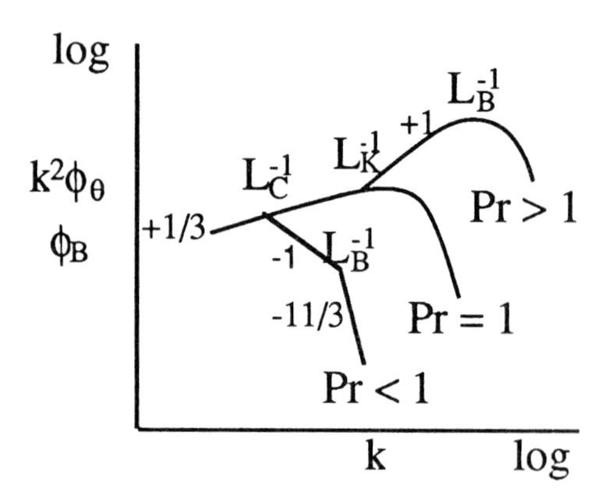

Figure 6. Spectral forms for scalar dissipation spectrum $k^2\phi_\theta$ and magnetic field spectrum ϕ_B for various Pr ranges according to straining scalar topology model of Fig. 5. The -11/3 subrange of Fig. 2 has been shifted from L_C^{-1} to L_B^{-1} with a new -1 subrange in between, from Gibson (1968b).

As shown by Wyngaard (1971) the parameter Σ is very sensitive to the highest wavenumber portion of the scalar spectrum, by the equation

$$\Sigma = -(4/5)\sqrt{15}\left(\int_0^\infty k^4\phi_\theta dk\right)_B \tag{4}$$

where the B subscript on the integral indicates normalization by the Batchelor length, time and scalar scales L_B, $T_B \equiv \gamma^{-1}$, $\Sigma_B \equiv (\chi T_B)^{1/2}$, respectively. The stronger cutoff for the wavecrest compression model of Fig. 2 causes Σ to approach zero for Pr « 1. This is inconsistent with Σ measured for turbulent mercury temperature fluctuations by Clay (1973) and numerical simulations of Kerr (1985) which give $\Sigma \approx -0.5$, as shown by Gibson et al. (1988). A higher wavenumber cutoff at $k \approx L_B^{-1}$ such as the Pr « 1 spectrum of Fig. 6 is required to give such Σ values, as shown in Figure 7. Substituting into (4) with $\Sigma = -0.5$ gives a transition at $k \approx 0.4\ L_B^{-1}$ between the -11/3 and -1 subranges of $k^2\phi_\theta$, which is just what is observed, as shown by Gibson et al. (1988, Fig. 1).

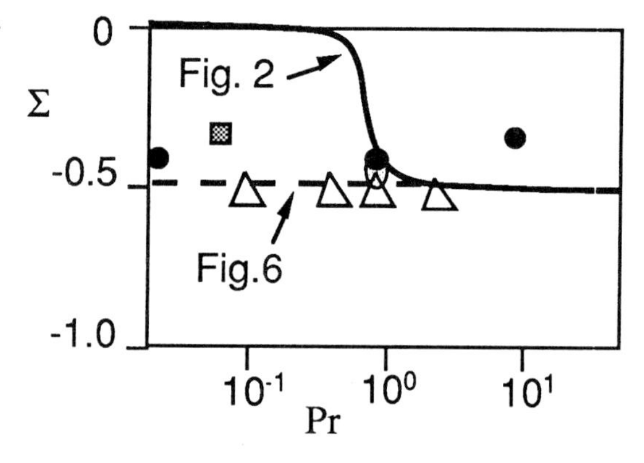

Figure 7. Σ versus Pr, from Fig. 3 of Gibson et al. (1988). Measured laboratory values: dark circles; turbulent temperature in mercury, air and water, from Clay (1973), square; plasma electron density, Granatstein et al. (1971), oval; air temperature from Boston (1970). Triangles are from numerical simulations of Kerr (1985). Solid line is from spectra of Fig. 2, dashed line from Fig. 6, computed using Equation (4).

3 SUMMARY AND CONCLUSIONS

Topological modelling is used to understand, visualize and explain the evidence from laboratory and numerical simulations that the local rate-of-strain determines the smallest scales of both strongly and weakly diffusive passive scalar fields mixed by turbulence. Turbulent convection produces gross distortions in an initially uniform scalar gradient field with singly connected isoscalar surfaces: the equilibrium condition of scalar fields in the absence of turbulent convection. The distorted isoscalar surfaces are diffusively unstable and split up to form extrema and associated saddle points. Further generation and evolution of scalar extremum points, saddle points, and maximal and minimal gradient lines by local straining produces the complex topology characteristic of turbulent scalar fields. For all Prandtl number scalars, local straining produces minimal gradient lines emanating from the extrema, and splitting of the extrema. This increases the numbers of maximal gradient lines where the actual mixing takes place. Thus the minimal gradient lines reflect stirring and the maximal gradient lines reflect mixing, both induced by the rate-of-strain tensor. This basic rate-of-strain induced turbulent mixing mechanism applies independent of Pr, and is consistent with the laboratory and numerical simulation evidence.

4 ACKNOWLEDGEMENTS

This work was carried out under the auspices of the Society for Statistical Geometry with financial support from NSF and ONR.

5 REFERENCES

ASHURST, W. T., A. R. KERSTEIN, R. M. KERR and C. H. GIBSON 1987 Alignment of Vorticity and Scalar Gradient with Strain Rate in Simulated Navier-Stokes Turbulence. *Phys. of Fluids* **30**, 2343-2353.

BATCHELOR, G. K. 1959 Small-scale variation of convected quantities like temperature in turbulent fluid. Part 1. General discussion and the case of small conductivity. *J. Fluid Mech.* **5**, 13-133.

BATCHELOR, G. K., HOWELLS, I. D. & TOWNSEND, A. A. 1959 Small-scale variation of convected quantities like temperature in turbulent fluid. Part 2. The case of large conductivity. *J. Fluid Mech.* **5**, 134-139.

BOSTON, N. E. 1970 An investigation of high wave number temperature and velocity spectra in air. Ph. D. thesis. University of British Columbia.

CLAY, J. P. 1973 Turbulent mixing of temperature in water, air and mercury. Ph. D. thesis, University of California at San Diego.

CORRSIN, S. 1951 On the spectrum of isotropic temperature fluctuations in isotropic turbulence. *J. Appl. Phys.* **22**, 469-473.

GIBSON, C. H. 1968a Fine structure of scalar fields mixed by turbulence, I. Zero gradient points and minimal gradient surfaces. *Physics of Fluids* **11**, 2305-2315.

GIBSON, C. H. 1968b Fine structure of scalar fields mixed by turbulence, II. Spectral theory. Physics of Fluids **11**, 2316-2327.

GIBSON, C. H. 1986 Internal waves, fossil turbulence, and composite ocean microstructure spectra, *J. Fluid Mech.* **168**, 89-117.

GIBSON, C. H. 1988 Isoenstrophy points and surfaces in turbulent flow and mixing. *Fluid Dynamics Research.* **3**, 331-336.

GIBSON, C. H., W. T. ASHURST and A. R. KERSTEIN 1988 Mixing of strongly diffusive passive scalars like temperature by turbulence. *J. Fluid Mech.* **194**, 261-293.

GOLYTSYN, G. S. 1960 Fluctuations of the magnetic field and current density in a turbulent flow of a weakly conducting fluid. *Dokl. Akad. Nauk SSSR* **32**, 315.

GRANATSTEIN, V. L., BUCHSBAUM, S. J. & BUGNOLO, D. S. 1966 Fluctuation spectrum of a plasma additive in a turbulent gas. *Phys. Rev. Lett.* **6**, 504.

KERR, R. M. 1985 Higher-order derivative correlations and the alignment of small-scale structures in isotropic numerical turbulence. *J. Fluid Mech.* **153**, 31-58.

KRAICHNAN, R. H. & NAGARAJAN, S. 1967 Growth of turbulent magnetic fields. *Physics of Fluids* **10**, 859.

KOLMOGOROFF, A. N. 1941 The local structure of turbulence in incompressible viscous fluid for very large Reynolds number. *Dokl. Akad. Nauk SSSR* **30**, 301.

MOFFATT, K. 1961 The amplification of a weakly applied magnetic field by turbulence in fluids of moderate conductivity. *J. Fluid Mech.* **11**, 625.

MOFFATT, K. 1962 Intensification of the earth's magnetic field by turbulence in the ionosphere. *J. Geophys. Res.* **67**, 307.

OBOUKHOV, A. M. 1949 Struktura temperaturnovo polia v turbulentnom potoke. Izv. Akad. Nauk SSSR Ser. Geofiz. **3**, 59.

WYNGAARD, J. C. 1971 The effect of velocity sensitivity on temperature derivative statistics. *J. Fluid Mech.* **48**, 783-769.

Algorithms for Classification of Turbulent Structures

A.A. WRAY

NASA Ames Research Center

J.C.R. HUNT

University of Cambridge

1 INTRODUCTION

It has become clear in recent years that turbulent flows of all types contain 'structures' with various degrees of coherence, and that these entities are created, destroyed, and otherwise evolve at identifiable events within the flow, probably thereby accounting for a large measure of the important statistical properties of the turbulence. Examples of such events might be high values of filtered vorticity (Hussain 1986) or high velocity, perhaps in combination with straining (Adrian & Moin 1988), or high Reynolds stress (Blackwelder & Kaplan 1976). We are developing algorithms for detecting these structures in computational turbulence. The simulations are of incompressible isotropic turbulence, with a steady force applied to achieve statistical stationarity, as was done in Hunt, Buell, & Wray (1987).

We identify four classes of flow structure: "eddies", which should correspond to the intuitive concept of a circulating rotational region in which fluid particles typically have a long residence time; "convergence zones" (or, more properly, convergence-divergence zones) which contain fluid undergoing significant irrotational strain; "shear zones", which are regions of significant vorticity but without a circulating flow pattern; and "streams", in which the fluid is moving fairly rapidly but without significant distortion or rotation.

We use pointwise tests of $\mathbf{u}, p$, and their derivatives to define the classes. An advantage of using the pressure is that it brings in a broad region of information about the flow field, owing to its elliptic dependence on $\mathbf{u}$. (Other elliptic quantities, such as partitions of the pressure defined by subdividing the rhs of the pressure Poisson equation, might also prove useful.) Combinations of simple inequalities of these quantities provide a very efficient algorithm for classification; we hope to show here that the classification is a useful one.

These flow regions are studied because they are important sites for certain processes occurring in the flow, for example kinetic energy production and dissipation, enstrophy production and dissipation, and mixing and chemical reactions. In particular, there is evidence that in reacting flows fast reactions are concentrated in locations where streamlines converge (Leonard & Hill 1988), i.e., in convergence zones, and slow reactions in recirculating eddying regions (Broadwell & Breidenthal 1982), i.e., in eddies. Eddy regions are also of importance in certain flame and combustion problems (Peters 1988). Furthermore, recent experimental and computational research (e.g. Hunt et al. 1988; Maxey 1987; Chung & Troutt 1988; Fung & Perkins 1989) has shown that low-density particles tend to concentrate in low-pressure regions, possibly "eddies", while denser particles, especially if buoyancy forces are important, tend to concentrate in streams between eddies. Also, those particles that are entrained into eddying regions can remain for long periods. Several other investigators have been studying strong eddying or vortical regions using various criteria for these regions, such as low pressure, strong rotational motion as defined by the local deformation tensor (Herring 1988; Perry & Chong 1987), or high values of the vorticity of the filtered velocity field (Hussain 1986).

2 NUMERICAL METHOD

The events, structures, and processes characterizing the four flow zone types are of necessity statistical in nature, and to obtain good statistics we prefer to use a statistically stationary velocity field rather than a decaying one, which for isotropic turbulence requires that an energy-input mechanism be added to the Navier-Stokes system.

The method used in these computations is to introduce a *steady* but spatially nonuniform force field $\mathbf{F}(\mathbf{x})$ at the largest scales of the flow, which induces a mean velocity field in time at these scales. This mean flow is unstable, allowing instabilities to grow and energy to be transferred to smaller-scale motions. (One can also let the force be time-varying (Eswaran & Pope 1988)). If the initial conditions are random and the Reynolds number high enough, the force field will maintain the turbulence with statistical properties determined by $\mathbf{F}$ and the viscosity ν.

The computations were performed on a cubical $N \times N \times N$ mesh with periodic boundary conditions, where $N = 64$. The power spectrum of the force is chosen to have contributions only from values of $k = \sqrt{2}$. For each $\mathbf{k}$ in the $k = \sqrt{2}$ sphere, the amplitude of the forcing was set so that $\hat{F}_i(\mathbf{k})\hat{F}_i^*(\mathbf{k}) = \alpha^2$, and α and ν were chosen to obtain a microscale Reynolds number $Re_\lambda \approx 25$. In practice $\mathbf{F}$ can only be made approximately isotropic, since there are only a finite number of Fourier components on a discrete mesh. We constrain the moments of the Fourier coefficients up to second order, averaged over spherical shells in $\mathbf{k}$-space, to have their isotropic values, and require the single third-order moment $\text{avg}(\hat{F}_1\hat{F}_2\hat{F}_3) = 0$. $\mathbf{F}$ is further required to be solenoidal so as to not generate large pressure fluctuations.

The 3-D Navier-Stokes equations were solved under the above conditions using the spectral code of Rogallo (1981) modified to include the body force $\mathbf{F}$. It was found that $\langle u_j u_j \rangle$ initially increased or decreased, but eventually oscillated around a stationary value. Then a time series $\hat{\mathbf{u}}(\mathbf{k}, t)$ was collected on a time interval of 10 time steps, or about 1 minimum Eulerian time scale $\tau_{\min}^{(E)} \sim \ell_{\text{Kol}}/u_0$. In our computational units, $u_0 \approx 1.2$ and $L_1 \approx 0.63$; the Taylor microscale $\lambda \approx 0.416$.

3 ZONE ALGORITHMS

Denoting by II the second invariant of $\partial u_i/\partial x_j$, the characterization of the four structure classes which we have used is

$$\begin{aligned}
&\text{Eddy if}: && II < -II_{\text{E}} && \text{and} && p < -p_{\text{E}} && (1a)\\
&\text{Convergence zone if}: && II > +II_{\text{C}} && \text{and} && p > +p_{\text{C}} && (1b)\\
&\text{Shear zone if}: && II < -II_{\text{E}} && \text{and} && -p_{\text{E}} < p < +p_{\text{C}} && (1c)\\
&\text{Stream if}: && |II| < II_{\text{S}} && \text{and} && |\mathbf{u}| > u_{\text{S}} && (1d)
\end{aligned}$$

These criteria are made mutually exclusive by choosing $II_{\text{S}} \leq \min(II_{\text{E}}, II_{\text{C}})$, but they are not exhaustive, and much of a given turbulent flow field remains unclassified.

These mathematical characterizations cannot, of course, be rigorously derived, particularly given the fact that the zone classes themselves are fuzzy concepts. The criteria (1) were arrived at by a combination of intuition and numerical optimization. The optimization consists of observing a tentative classification against a background of various flow quantities, such as velocity vectors, pressure or vorticity fields, eigenvalues and invariants of the strain and deformation tensors, etc. The process of optimization over our low-Reynolds-number isotropic simulations has led us to the following values for the parameters appearing in (1):

$$\begin{aligned}
&II_{\text{E}} = \frac{1}{2} II_{\text{rms}} \qquad p_{\text{E}} = \frac{1}{2} p_{\text{rms}} \qquad II_{\text{C}} = II_{\text{rms}} \qquad p_{\text{C}} = p_{\text{rms}}\\
&II_{\text{S}} = \min(II_{\text{E}}, II_{\text{C}}) \qquad u_{\text{S}} = |\mathbf{u}|_{\text{rms}}
\end{aligned} \tag{2}$$

where the pressure and velocity rms values are taken relative to the global means.

Eddies clearly are regions where vorticity dominates irrotational strain, so that it is natural to have a condition such as the first one in (1a) involving the second invariant $II = s_{ij}s_{ij} - \frac{1}{2}|\boldsymbol{\omega}|^2$, where $s_{ij} = \frac{1}{2}(u_{i,j} + u_{j,i})$. However, we wish to define eddies as regions of *circulating* flow and thus choose them to be regions of low pressure, so that streamlines tend to curve around them (Hunt, Wray, and Moin 1988). It is the elliptic dependence of the pressure on $\mathbf{u}$ that allows us to make such a condition local in $\mathbf{x}$. Similarly, convergence zones (1b) are regions in which the dominant deformation is irrotational strain, but we also wish them to have converging-diverging streamlines and so require them to contain a high pressure. Shear zones (1c) are defined to be vortical but *without* circulating flow and so are required to have moderate pressures. Streams (1d) are regions of *weak* deformation but substantial speed.

The characterizations (1) are computationally very convenient and efficient, but the result of this convenience and efficiency is some physical shortcomings. Clearly the pressure conditions would better refer to a pressure difference across an eddy or convergence zone rather than to a difference between the local pressure and the global rms value. Similarly, the conditions on II should refer to local values inside and outside the region in question. Finally, the speed condition in the stream characterization should be a local relative one. In defense of the rough approximations (1) we can say that, at the low Reynolds number ($Re_\lambda \approx 25$) in these simulations, there are normally only a few dozen or so of each type of region in the computational domain at any time, and no real "hierarchy" of, say, eddies within streams within eddies, and so on, exists. Furthermore, only about one-half the total volume is classified as any of the four zone types (1), and most of that is in the stream regions, so that the global rms values are not bad approximations to the magnitudes of the pressure and II outside any particular classified region. Finally, algorithms which use the more accurate local concepts are computationally *much* more expensive and so would reduce the utility of the method for analysis of large complex databases.

For eddies in particular, the criterion (1a) seems to identify intuitively correct-looking regions more reliably than criteria using the character of the eigenvalues of the local deformation tensor (Perry & Chong 1987). It is encouraging that the characterizations do not seem to be strongly dependent on the values of the parameters (namely the coefficients of the rms quantities above). Furthermore, additional intuitive ideas about the nature of the identified structures seem to hold; for instance, convergence zones correlate very strongly with negative III, the third invariant of s_{ij}, ("pancake" straining). Such qualitative statistical conclusions are also not strongly dependent on the zonal parameters.

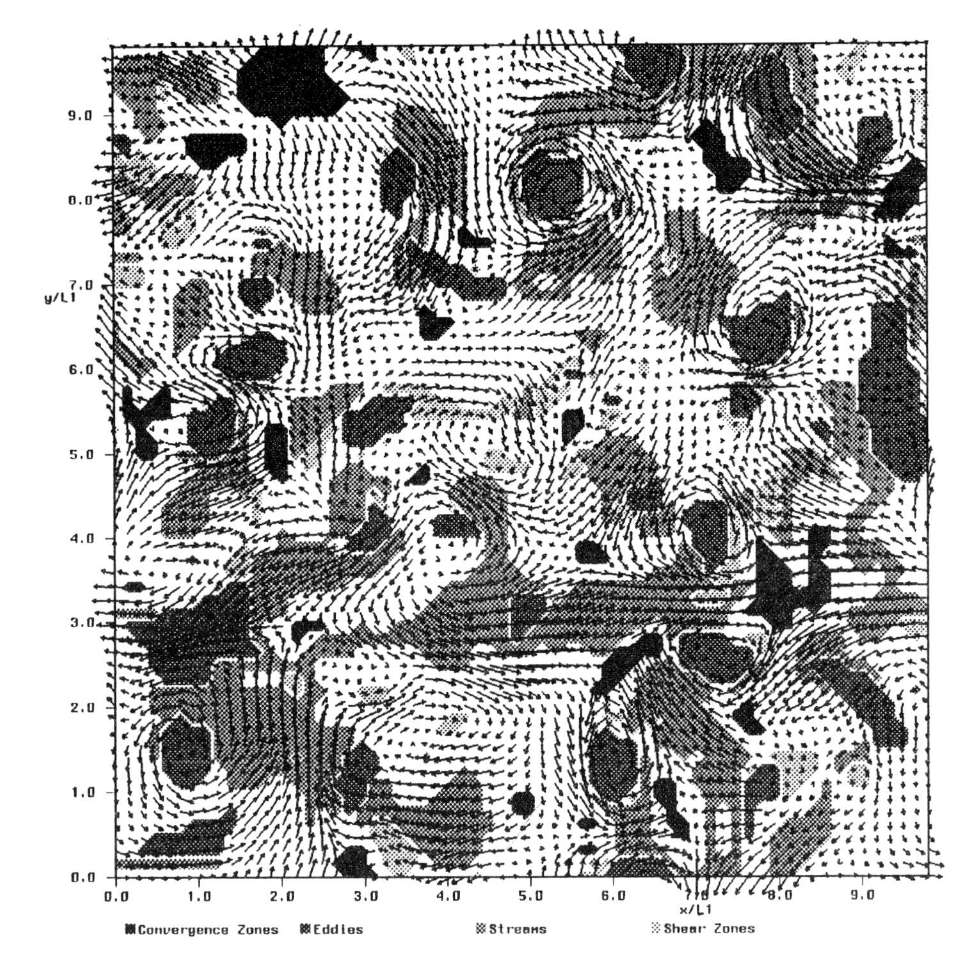

FIGURE 1. Zone classification in a typical plane; see also the corresponding colour plate. (See also Colour Plates)

4 GEOMETRIC PROPERTIES OF THE ZONES

Figure 1 shows a typical plane of the 64^3 flow field at a single time. The arrows are the projection of the fluid velocity on the plane. The lightest gray shading is used for those areas classified as shear zones, next-to-lightest gray marks streams, next-to-darkest gray denotes eddies, and the darkest gray is used for convergence zones. The colour plate shows these zones much more clearly: there blue denotes shear zones; yellow, streams; green, eddies; and red, convergence zones.

There are several prominent eddies shown in by the swirling structure of the velocity vectors, for example at $x \approx 1.0$, $y \approx 1.5$ and at $x \approx 5.5$, $y \approx 8$. The zone algorithms have correctly identified eddies at these locations in figure 1. Other areas, for instance the elongated region at $x \approx 9.5$, $y \approx 5 - 6.5$, do not show swirling flow *in this plane*. However, by looking at the $y - z$ plane at $x = 9.5$, one sees a swirling motion in that plane. Indeed, we have found that, in nearly every case, there is a circulating flow associated with regions for which (1a) is true.

Some of the shear zones shown in figure 1 tend to lie nearby the eddies and seem to be regions of shear flow which are a part of the larger-scale eddy structure but without particularly low pressures or strongly curving flow patterns. An example of such a structure is at $x \approx 7.5$, $y \approx 3.3 - 4$. Other shear zones are isolated from eddies, and are either weak swirls, as at $x \approx 5$, $y \approx 4.9$, or are simply areas of local shear flow, such as the small area at $x \approx 1.8$, $y \approx 4.3$.

Stream zones (next-to-lightest gray in figure 1) are often associated with an even larger-scale motion surrounding and connecting the eddy regions. The motion is, by condition (1d), faster than average in speed, weakly strained, and nearly irrotational. Such strong motion is presumably induced by the eddies. Streams are the largest structures described by the criteria (1), occupying approximately 25% of the total volume, compared with about 13% for the eddies, 6% for the shear zones, and 4% for the convergence zones, so that about 48% of the total volume is classified.

The convergence zones (darkest gray in figure 1) correspond graphically to areas where the velocity vectors are either converging or diverging, such as at $x \approx 8$, $y \approx 3.5$ and $x \approx 2 - 3$, $y \approx 9 - 10$. These seem to be generally the collision points of streams, usually an oblique collision, as at $x \approx 6 - 6.5$, $y \approx 2.1 - 2.6$, but sometimes nearly head-on. The stream criterion of low deformation disallows stream regions continuing right up to convergence zones, but they often come quite close.

The overall geometric picture, certainly highly simplified but one we believe instructive, is that of eddies, sometimes flanked by less-swirling shear zones, pumping fluid along the streams, which may collide to form convergence zones, deflecting the streams away and back toward the eddies. Other shear zones lie along the sideways juxtaposition of streams and may be progenitors of eddies.

While the criteria (1) seem to distinguish geometric structures in a reasonable way, we have found as well that the kinematic and dynamic natures of the flow regions differ in interesting and important ways, as the following discussion will describe.

5 KINEMATIC AND DYNAMIC PROPERTIES OF THE ZONES

We want to assess how well the zones isolate and distinguish various features of the turbulent dynamical processes. We do this by looking at the statistics of various quantities taken over the four zone types and compare them with each other and with the overall statistics. Though the sample size for a given zone type at one time step is rather small and the statistics therefore rather noisy, we time-average over approximately 10-20 steps, each 100 time-steps apart. The resulting averages are smooth enough to have some confidence in their validity.

As noted above, the fractions of the total volume occupied by the four zone types are: convergence zones 4%, eddies 13%, streams 25%, and shear zones 6%. However, important dynamical quantities tend to be concentrated in one or more of the zones.

The streams dominate energy contributions with 42% of the total compared to their 25% of the volume. The eddies have just about an average concentration of energy, and the others lower than average. Eddies and shear zones dominate contributions to the total enstrophy with 37% and 12%, respectively, that is, between two and three times the average concentration. Streams contribute 15% and convergence zones 2%, or significantly below-average enstrophy.

The contribution to the total strain rate squared, $\int s_{ij}s_{ij}dV$, is 11% from the convergence zones; in other words, convergence zones have a much higher than average concentration of s^2. The other zones have lower than average concentrations: eddies contribute 11% of the total, streams 17%, and shear zones 5%. The conditions (1) clearly define zones likely to have such a distribution of s^2.

The contribution to the total enstrophy growth rate, $d(\int |\omega|^2 dV)/dt$, is 30% from the eddies and 15% from the shear zones, while the convergence zones have about an average concentration (5% contribution vs. 4% volume) and the streams much lower than average (11% contribution vs. 25% volume). Because the enstrophy production term, $s_{ij}\omega_i\omega_j$, is quadratic in the vorticity, it is not surprising that it is concentrated in the eddies and shear zones, especially since their irrotational strains tend to be only a little below average. Likewise, the strain factor in this term enhances a convergence zone's enstrophy growth rate while the vorticity factors must reduce it there. It is interesting that the shear zones have a higher ratio (1.25) of enstrophy production to enstrophy than do the eddies (0.81); perhaps the shear zones are areas of vorticity growth and the eddies mainly zones of more nearly steady-state and decaying vorticity.

The contributions to the mean value of the third invariant of s_{ij}, III, are 22% from the convergence zones and smaller than average concentrations for the other zones: eddies contribute 7% of the total, streams 9%, and shear zones 4%. All the zones contribute a net negative III, that is, all the zones have pancake straining on the average, though except in the convergence zones the magnitude of III is below average; in the unclassified zones one has occasional sausage straining.

The contributions to total energy dissipation rate are 7% in convergence zones, 25% in eddies, 16% in streams, and 9% in shear zones. Thus, streams have a significantly below-average energy dissipation rate, while the eddies and convergence zones have close to twice the average rate and shear zones about 50% more than the average. This result certainly agrees with the usual intuitive picture.

To summarize these contributions to mean dynamical measures, we find that convergence zones, which tend to be quite small in size, have large values of strain rate, and this strain is very predominantly of the pancake type, more so even than the strain in all regions, which is itself predominantly pancaking. No other zones are strongly straining. The eddies and shear zones contribute large amounts of enstrophy growth rate due to vortex stretching, indicating good alignment of strain and vorticity there, since the strain rates in those zones are below average. Energy is concentrated in the streams and energy dissipation in convergence zones and eddies and to a lesser extent in shear zones.

Probability density functions (PDF's) of various quantities over the zones are also instructive as to the dynamics of the regions. The velocity PDF's are all very nearly gaussian, except in the streams where the condition on the speed in (1d) causes a large flat region in the PDF from $-u_S$ to $+u_S$. As one would expect from the energy concentrations, the velocity PDF in eddies is essentially identical to the overall one, while the convergence zones and shear zones have narrower, i.e., less energetic, velocity PDF's.

The PDF's of the longitudinal derivatives ($\partial u/\partial x$, $\partial v/\partial y$, $\partial w/\partial z$) have the usual skewed appearance in the eddies, streams, and shear zones, where they look basically like the overall PDF but are slightly narrower, consistent with these zones' below-average contribution to III. The mean value of a derivative in each zone is of course statistically zero by homogeneity of the field and of the zone criteria, but the skewnesses are all negative. The convergence zones on the other hand, which contribute heavily to III and are more pancaking than average, have a highly unsymmetrical, double-humped PDF shown in figure 2. This shape gives insight into the enstrophy generation process as measured by the velocity derivative skewness. That part of the curve on the positive x-axis, which contributes to positive skewness, while weaker in mean-cube than the negative side, has a much higher peak than the broad negative-x part of the distribution. Thus elongational straining has higher *maximum* values than the flattening, but the flattening type has a broader *range* of values such that its total contribution to the mean-cube is greater. There seem

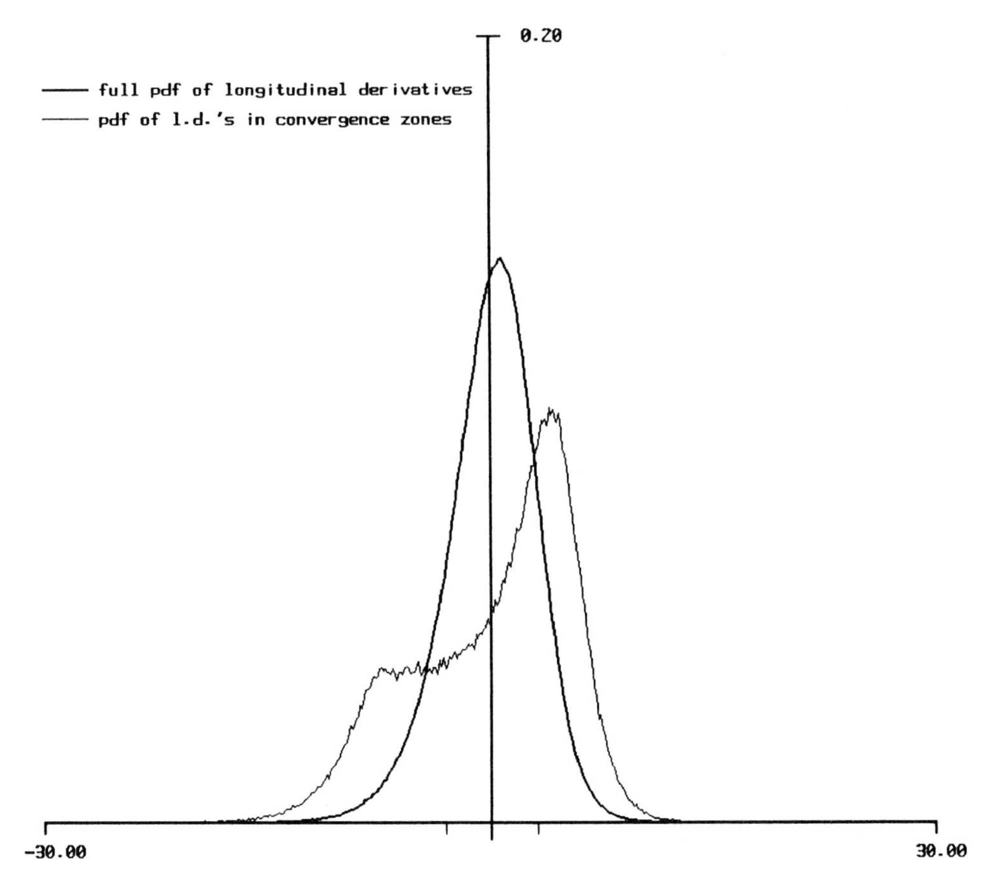

FIGURE 2. PDF's of longitudinal derivatives overall and in convergence zones.

to be distinct mechanisms at work here creating the two types of strain and two types of PDF of longitudinal derivative. Convergence zones could be divided into elongating and flattening types to try to separate the mechanisms.

All the zonal PDF's of both longitudinal and transverse derivatives display exponential tails for very large magnitudes of the derivatives. These tails appear to have the same asymptotic slope in all zones except in streams, where the slope is smaller; the sample is very poor for such extreme values, however, and these conclusions are tentative.

The PDF's of the stretching, $s_{ij}\omega_i\omega_j$, are all very sharply peaked at 0 with positive means and long tails which decay more slowly than exponential. The tails are shorter and steeper for negative stretching than for positive. These PDF's in eddies, and to a lesser extent in shear zones, have much longer tails than in the other zones. Energy disspation PDF's also have very much longer tails in the eddy regions than elsewhere. Enstrophy dissipation PDF's have slightly longer tails in the eddies, but the other zones' distributions for this quantity have quite long tails as well.

The dynamical picture that emerges from these observations is that vortex stretching and energy and enstrophy dissipation are both very intermittent processes, even at this low Reynolds number, and that eddy regions in particular are subject to rare strong bursts of such events. The lack of bursts of energy dissipation in streams must mean that their velocity fields remain smooth (imposed by condition (1d) on *II*), though sharp gradients can appear in the vorticity, as the enstrophy dissipation PDF's for streams show. These bursts in streams could be occurring on their edges where vorticity gradients might be occasionally high.

REFERENCES

Adrian, R.J. & Moin, P. 1988, *J. Fluid Mech.*, **190**, 531-559.
Blackwelder, R.F. & Kaplan, R.E. 1976, *J. Fluid Mech.*, **76**, 89-112.
Broadwell, J.R. & Breidenthal, R.E. 1982, *J. Fluid Mech.*, **125**, 397-410.
Chung, J.N. & Troutt, T.R. 1988, *J. Fluid Mech.*, **186**, 199-222.
Eswaran, V. & Pope, S. B. 1988, *Computers & Fluids* **16**, No. 3, 257-278.
Fung, J. & Perkins, R.J. 1989, *Proc. 2nd European Turbl. Conference*, Springer-Verlag, Berlin .
Herring, J. 1988, *Proc. 17th Int. Cong. Theor. Appl. Mech.*, Grenoble.
Hunt, J.C.R., Buell, J. & Wray, A.A. 1987, *Proc. CTR Summer Program*, 77-94.
Hunt, J.C.R., Wray, A.A., & Moin, P. 1988, *Proc. CTR Summer Program*, 193-208.
Hunt, J.C.R. & Auton, T.R., Sene, K., Thomas, N.H. & Kowe, R. 1988, *Proc. Conf. Transient Phenomena in Multiphase Flow*, Hemisphere (in press).
Hussain, A.K.M.F. 1986, *J. Fluid Mech.*, **173**, 303-356.
Leonard, A.D. & Hill, J.C. 1988, *Proc. CTR Summer Program.*
Maxey, M.R. 1987, *J. Fluid Mech.*, **174**, 441-465.
Perry, A.E. & Chong, M.S. 1987, *Ann. Rev. Fluid Mech.*, 125-155.
Peters, N. & Williams, F.A. 1988, *22nd Symposium on Combustion*, Seattle.
Rogallo, R.S. 1981, *Numerical experiments in homogeneous turbulence*, NASA TM 81315.

Chaotic Advection by a Point Vortex in a Semidisk

HISASHI OKAMOTO[1] & YOSHIFUMI KIMURA[2]

[1]University of Tokyo, Japan
[2]Center for Nonlinear Studies, Los Alamos National Laboratory, USA.

§1. Introduction. We consider the motion of a particle which is advected by a point vortex in a semi-disk. The purpose of this paper is to show how the motion of the advected particle changes from a periodic one to a chaotic one. We actually present an alternative perspective to what is observed in [**1**], where it is shown, by numerical computations, that two point vortices in a semi-disk behave chaotically if the energy of the orbits are sufficiently high, while they move quasi-periodically if the energy is low. One of the points in [**1**] is : even two vortices give rise to chaos if they are confined in a semi-disk, while three vortices are necessary to cause a chaos in the case of a full-disk and four vortices necessary in the case of the whole plane.

In this paper, we present a mathematical framework which we believe to give a clearer understanding of the dynamical system governing two vortices. In this framework, we obtain differential equations which depend on a certain parameter $\alpha \in [-1,1]$. The differential equation studied in [**1**] is the one given here with $\alpha = -1$. It is therefore important to understand the structural change of the phase portrait as α runs in $[-1,1]$. As a first step toward this, we consider in this paper the case where $\alpha = 0$. Our method is classical: the Poincaré map. We study the transition from periodic motions to chaotic ones.

§2. The equation and its nondimensionalization. In this section we write the governing equation and suitably nondimensionalize it. We put

$$D_R = \{z \in \mathrm{C}; |z| < R, \mathrm{Im}(z) > 0\},$$

which is an open semidisk of radius R in the complex plane. Suppose that there are two point vortices $z(t)$ and $w(t)$ $(-\infty < t < \infty,\quad z, w \in D_R)$. Let κ_1 and κ_2 denote the intensity of the vortices z and w, respectively. Then the motion of these two vortices in D_R are governed by the following (2.1,2) (see [**1**]) :

$$(2.1)\quad \dot{z} = \frac{-i}{2\pi}\left[\frac{\kappa_1}{\overline{z}-z} + \frac{\kappa_1}{\overline{z}-\frac{R^2}{z}} - \frac{\kappa_1}{\overline{z}-\frac{R^2}{\overline{z}}} - \frac{\kappa_2}{\overline{z}-\overline{w}} + \frac{\kappa_2}{\overline{z}-\frac{R^2}{w}} + \frac{\kappa_2}{\overline{z}-w} - \frac{\kappa_2}{\overline{z}-\frac{R^2}{\overline{w}}}\right],$$

$$(2.2)\quad \dot{w} = \frac{-i}{2\pi}\left[\frac{\kappa_2}{\overline{w}-w} + \frac{\kappa_2}{\overline{w}-\frac{R^2}{w}} - \frac{\kappa_2}{\overline{w}-\frac{R^2}{\overline{w}}} - \frac{\kappa_1}{\overline{w}-\overline{z}} + \frac{\kappa_1}{\overline{w}-\frac{R^2}{z}} + \frac{\kappa_1}{\overline{w}-z} - \frac{\kappa_1}{\overline{w}-\frac{R^2}{\overline{z}}}\right],$$

where the dot means differentiation with respect to time t. We change the variables to nondimensional ones by $z \to Rz$, $w \to Rw$, $t \to 2\pi R^2 t/\kappa_1$. Then we have

$$(2.3)\quad \dot{z} = \frac{-i}{\overline{z}-z} + \frac{-i}{\overline{z}-1/z} + \frac{i}{\overline{z}-1/\overline{z}} + \frac{\alpha i}{\overline{z}-\overline{w}} + \frac{-\alpha i}{\overline{z}-1/w} + \frac{-\alpha i}{\overline{z}-w} + \frac{\alpha i}{\overline{z}-1/\overline{w}},$$

$$(2.4)\quad \dot{w} = \frac{-\alpha i}{\overline{w}-w} + \frac{-\alpha i}{\overline{w}-1/w} + \frac{\alpha i}{\overline{w}-1/\overline{w}} + \frac{i}{\overline{w}-\overline{z}} + \frac{-i}{\overline{w}-1/z} + \frac{-i}{\overline{w}-z} + \frac{i}{\overline{w}-1/\overline{z}},$$

where $\alpha = \kappa_2/\kappa_1$. These are the equations which we wish to analyse. Note that the phase space of this dynamical system is $(D_1 \times D_1) \setminus \{(z,w); z = w\}$ and that the only α appears as a nondimensional parameter running from $-\infty$ to $+\infty$.

Remark 1. It is enough to consider only $-1 \leq \alpha \leq 1$. For, if $G(\alpha, z, w)$ denotes the right hand side of (2.3), then the right hand side of (2.4) is $\alpha G(1/\alpha, w, z)$. This implies that the dynamics of (α, z, w) is the same as $(1/\alpha, w, z)$, if we change the time scale.

In [**1**] orbits of (2.3,4) are numerically computed in the case of $\alpha = -1$. Some of them with a high energy are chaotic, i.e., they have continuous power spectra. On the other hand, as far as the authors know, no chaotic motion has been found if α is positive. Accordingly it is important to consider the structural change of the phase portrait as α runs from -1 to $+1$. For instance, we should determine where in the parameter space chaotic motions appear and where they do not. In this paper we consider the case of $\alpha = 0$, which enables us to use a mathematical theory. When $\alpha = 0$, we have

$$(2.5)\qquad \dot{z} = \frac{-i}{\overline{z}-z} + \frac{-i}{\overline{z}-1/z} + \frac{i}{\overline{z}-1/\overline{z}}$$

and

$$(2.6)\qquad \dot{w} = \frac{i}{\overline{w}-\overline{z}} + \frac{-i}{\overline{w}-1/z} + \frac{-i}{\overline{w}-z} + \frac{i}{\overline{w}-1/\overline{z}}.$$

The meaning of this system is that the intensity of w is infinitely small compared with that of z. Therefore z moves irrelevantly to w, while the motion of w is influenced by z. Note that (2.5) is independent of w. We may alternatively say that w moves as a passive particle in a vector field created by z. We now prove some elementary properties of (2.5,6). We introduce Hamiltonians

$$H(z) = \frac{1}{2}\log\frac{|1-z^2|}{|1-z\overline{z}||z-\overline{z}|} \quad \text{and} \quad \widetilde{H}(w,t) = \frac{1}{2}\log\frac{|w-z(t)||w-1/z(t)|}{|w-\overline{z(t)}||w-1/\overline{z(t)}|}.$$

Then (2.5,6) are written as the following Hamiltonian systems, respectively:

$$\dot{z} = 2i\frac{\partial H}{\partial \overline{z}}, \tag{2.7}$$

$$\dot{w} = 2i\frac{\partial \widetilde{H}}{\partial \overline{w}}. \tag{2.8}$$

PROPOSITION 1. *The system (2.7) is completely integrable and has a unique equilibrium:*

$$z = i\sqrt{\sqrt{5}-2} \tag{2.9}$$

Other orbits of (2.7) are periodic ones which surround this equilibrium, see Figure 1.

PROOF: The essential part of the proof is given in [**2**]. We, however, give a complete proof in our framework. Let us use the polar coordinates (I, σ) defined by $\sqrt{2I}e^{i\sigma} = z$. Then, by the definition of the Hamiltonian, we have

$$e^{-4H} = \frac{(1-2I)^2 8I\sin^2\sigma}{4I^2+1-4I\cos 2\sigma}. \tag{2.10}$$

We now introduce some symbols. We put

$$A = e^{-4H}, \qquad \xi = 1-2I, \qquad f(\xi) = -4\xi^3 + (4-A)\xi^2 + 4A\xi - 4A.$$

Then (2.10) is rewritten as :

$$\cot\sigma = \frac{\sqrt{f(\xi)}}{\sqrt{A}\xi}, \tag{2.11}$$

This equation defines a family of closed curves in D_1. If we regard the right hand side of (2.10) as a function of (I, σ), then we see that it has one and only one maximum at $\sigma = \pi/2$, $I = (\sqrt{5} - 2)/2$. At this point A takes it maximum value $10\sqrt{5} - 22$. If $0 < A < 10\sqrt{5} - 22$, then (2.11) defines a closed curve enclosing the point (2.9) inside it. On these curves, the motion of z is described as follows. Taking the real part of (2.5) multiplied by $\overline{z}$, we have

$$\dot{I} = \frac{1}{2}\cot\sigma - \frac{4I\sin\sigma\cos\sigma}{4I^2 + 1 - 4I\cos 2\sigma} \tag{2.12}$$

By (2.11,12) we have $\dot{\xi} = (A - \xi^2)\sqrt{f(\xi)}/(\sqrt{A}\xi^3)$. This equation defines the time evolution of the vortex $z(t)$ on the closed curves given by (2.11). We can solve this equstion by means of elliptic functions and see that the solutions are periodic.

By the periodicity of z, the equation (2.8) is a system whose Hamiltonian depends periodically on t.

PROPOSITION 2. *The differential equation (2.6) is definable on the boundary of D_1. The boundary of D_1 is invariant with respect to the flow given by (2.6). $w = 1, -1$ are unstable equilibria.*

PROOF: The right hand side of (2.6) is equal to the following

$$\frac{i(\overline{z} - z)(1 - |z|^2)(1 - \overline{w}^2)}{\{\overline{w}^2 - (z + \overline{z})\overline{w} + |z|^2\}\{\overline{w}^2|z|^2 - (z + \overline{z})\overline{w} + 1\}}. \tag{2.13}$$

It is clear from (2.13) that $w = -1, +1$ are equilibria. On the boundary circumference, we have $w = e^{i\gamma}$ $(0 \leq \gamma \leq \pi)$. In this case, (2.13) is equal to $c(e^{2i\gamma} - 1) = 2c\sin\gamma i e^{i\gamma}$, where $c \in \mathbf{R}$. This means the vector field is tangent to the boundary. Similarly it is tangent in the case of $w \in [-1, 1]$, since (2.13) $\in \mathbf{R}$ when $w \in \mathbf{R}$. Therefore the boundary of D_1 is an invariant set.

Thus the equation (2.5,6) has nice properties which (2.3,4) with $\alpha \neq 0$ does not share. Notice that (2.4) can not be defined on the boundary for $\alpha \neq 0$. Although (2.5,6) are simple, it is connected through α to the equation considered in [**1**] .

Since a similar problem is considered in Aref and Pomphrey [**5**,6], we would like to mention our motivation here. In [**5**,6], they consider the motion of a passive vortex stirred by three identical vortices. Since this problem is a special case of three vortices with different intensities, it seems to us that our problem is simpler than theirs. Note that the vortex z by which the motion of w is take place, can move periodically or stationary and there is no motion of other kind. On the other hand, three vortices can move with more varieties, e.g., they can collide ([7]).

Suppose that z is the equilibrium (2.9). Then (2.8) is independent of time, which implies that the Hamiltonian $\widetilde{H}$ is constant along individual orbits. Consequently (2.8) is completely integrable and the orbits of (2.8) consist only of closed Jordan curves defined by $\widetilde{H}(w, i\sqrt{\sqrt{5}-2}) = constant$. Furthermore, they occupy the whole phase space of (2.8) except for the boundary (see Figure 2). If the initial position of z is placed slightly apart from (2.9), then z moves on a small closed curve surrounding the equilibrium. In this case $\widetilde{H}$ is no longer independent of time and complicated orbits may appear. Let T be the period of z. Then we can obtain a Poincaré map in a usual way:

$$f : w(0) \to w(T). \tag{2.14}$$

We give in APPENDIX a theorem by which the map (2.14) becomes well-defined in $\Omega \equiv \overline{D_1} \setminus \{z(0)\}$. This is equivalent to saying that

$$\text{if} \quad w(0) \neq z(0), \text{ then } \quad w(t) \neq z(t) \text{ for all } \quad t.$$

If this is proved, it is clear that the map (2.14) is one-to-one, onto and continuous. Furthermore it preserves the area. Although our "proof " is not complete, we think the account in APPENDIX is a strong evidence of the correctness of the theorem.

We now examine the properties of the Poincaré map. It is enough to consider the case where $z(0) = iq + i\sqrt{\sqrt{5}-2} \quad (0 < q < 1 - \sqrt{\sqrt{5}-2})$. Let the mapping be denoted by f_q when $z(0) = iq + i\sqrt{\sqrt{5}-2}$. Several orbits are drawn on each figures 3-8. Figure 3, $\cdots$, 8 correspond to $q = 0.01, 0.05, 0.1, 0.25, 0.3, 0.4$, respectively.

It should be noticed that there is a fixed point in a lower part of the imaginary axis and that it is enclosed by a layer of closed curves. This shows that there is a periodic orbit which has exactly the same period as that of $z(t)$ and that it is stable. Some topological argument shows that there must be an unstable fixed point. Figure 1 shows that the unstable fixed point is on the upper side of the imaginary axis and that the stable fixed point are connected to the unstable one by a homoclinic orbit. We can observe that the region occupied by the invariant circles reduces and the islands grows up in accordance with the increase of q. We also notice that, even in the case of a large q, there are KAM tori around the point $z(0)$. The reason is that, when $w(0)$ is close to $z(0)$, the interaction of w with the boundary is negligibly small compared with the interaction between w and z (see the definition of F_0 in the APPENDIX).

Conclusion. Our equation (2.6), despite its simple appearance, exhibits chaotic orbits. It seems to the authors that ours is one of the simplest equation among the chaos-displaying vortex systems. As is shown in [4], streamlines of a stationary 3-D Euler flow can be chaotic. Our example shows that 2-D time-periodic Euler flow may have chaotic trajectories of particles.

APPENDIX 1. Here we prove :

THEOREM A. *For any tubular neighborhood N of $O = \{(z(t),t); 0 \le t < T\}$, there is an invariant torus such that it lies in N and that O lies inside it*

The precise meaning of this theorem is as follows: The phase space of (2.8) is $\bigcup_{0\le t\le T}(\overline{D_1}\setminus\{z(t)\})$, where the sections $t=0$ and $t=T$ are identified. Therefore it is homeomorphic to $(\overline{D_1}\setminus\{z(0)\})\times S^1$, where S^1 is a circle. Note that $\{z(0)\}\times S^1$ corresponds to the orbit of z. The above theorem asserts that all the neighborhood of $\{z(0)\}\times S^1$ has an invariant torus which contains $\{z(0)\}\times S^1$ inside.

FORMAL PROOF OF THEOREM A: Let us introduce $U+iV = u+iv-z(t)$ where $u+iv=w$. Then (2.8) is rewritten as

$$\dot{U} = -\frac{\partial K}{\partial V}, \qquad \dot{V} = \frac{\partial K}{\partial U}, \tag{A.3}$$

where we have put

$$K(U,V,t) = \frac{1}{2}\log\frac{|U+iV||U+iV+z-1/z|}{|U+iV+z-\overline{z}||U+iV+z-1/\overline{z}|} + V\mathrm{Re}(\dot{z}) - U\mathrm{Im}(\dot{z}).$$

Note that the right hand side depends on t through $z=z(t)$. If we define $K_0(U,V)$ and $K_1(U,V,t)$ by $K_0(U,V) = \frac{1}{4}\log(U^2+V^2)$, $K_1(U,V,t) = K(U,V,t) - K_0(U,V)$ then, K_1 is continuous on $\overline{D_1}$, the closure of D_1. Note that the orbits of $\dot{U} = -\frac{\partial K_0}{\partial V}$, $\dot{V} = \frac{\partial K_0}{\partial U}$, are simply the circles about the origin. We attempt to apply the KAM theory to the Hamiltonian system (A.3). Let $\epsilon > 0$ be a small parameter. We introduce canonical variables (p,q) by $p = \frac{U^2+V^2}{2\epsilon^2}$, $q = \arg(U+iV)$. We further change t to $\epsilon^2 t$. Then (A.3) becomes :

$$\dot{p} = -\frac{\partial F}{\partial q}, \qquad \dot{q} = \frac{\partial F}{\partial p}, \tag{A.4}$$

where we have put

$$(A.5) \qquad F = F_0(p) + F_1(p,q,t,\epsilon), \quad \text{with} \quad F_0(p,q,t,\epsilon) = \frac{1}{4}\log p,$$

$$F_1(p,q,t,\epsilon) = \frac{1}{2}\log\frac{|\epsilon\sqrt{2p}e^{iq} + z(\tau) - 1/z(\tau)|}{|\epsilon\sqrt{2p}e^{iq} + z(\tau) - \overline{z(\tau)}||\epsilon\sqrt{2p}e^{iq} + z(\tau) - 1/\overline{z(\tau)}|}$$

$$+ \epsilon\sqrt{2p}\sin q\mathrm{Re}(\dot{z}(\tau)) - \epsilon\sqrt{2p}\cos q\mathrm{Im}(\dot{z}(\tau)),$$

where $\tau = \epsilon^2 t$. These are defined on $q \in \mathbf{R}/2\pi\mathbf{Z}$ and $p \sim 1$. In this setting we wish to use Theorem 2 in Arnold [**3**]. This theorem garantees the existence of invariant tori for $\epsilon > 0$ which is close to unperturbed torus $p = p_0(\in [1/2, 2])$ where p_0 is sufficiently incommensurable. There is, however, one difficulty that the slowly changing parameter is ϵt in [**3**], while it is $\epsilon^2 t$ in (A.5). We hope that this diffuiculty is overcome if we follow the method of [**3**] in detail. Accordingly we are satisfied by the form (A.4) and stop here rather than pursuing rigorous proof, which seems to require a formidable calculation.

References

1. Y. Kimura and H. Hasimoto, J. Phys. Soc. Japan **55** (1986), 5–8.
2. Y. Kimura, Y. Kusumoto and H. Hasimoto, J. Phys. Soc. Japan **53** (1984), 2988–2995.
3. V.I. Arnold, Soviet Math. Dokl. **3** (1962), 136-140.
4. V.I. Arnold, "Mathematical Methods of Classical Mechanics," Springer Verlag, New York, Heidelberg, Berlin, 1978.
5. H. Aref and N. Pomphrey, Proc. R. Soc. London A **380** (1982), 359–387.
6. H. Aref and N. Pomphrey, Phys. Lett. A **78** (1980), 297–300.
7. H. Aref, Ann. Rev. Fluid Mech. **15** (1983), 345–389.

Keywords. point vortex, homoclinic orbit, Chaos

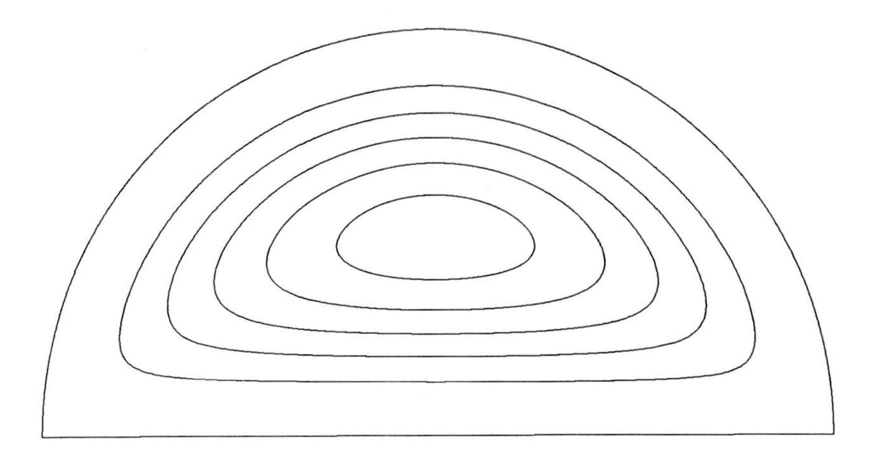

Figure 1. Orbits of $z(t)$

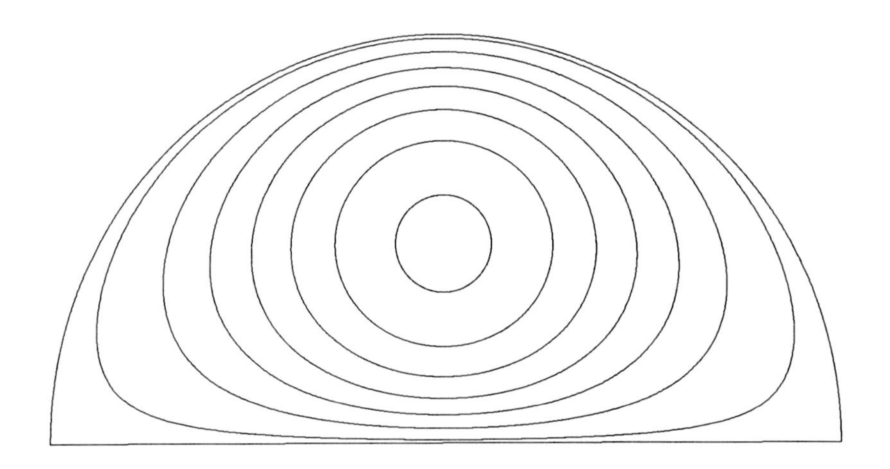

Figure 2. Orbits of $w(t)$ when $z \equiv \sqrt{\sqrt{5} - 2}\, i$

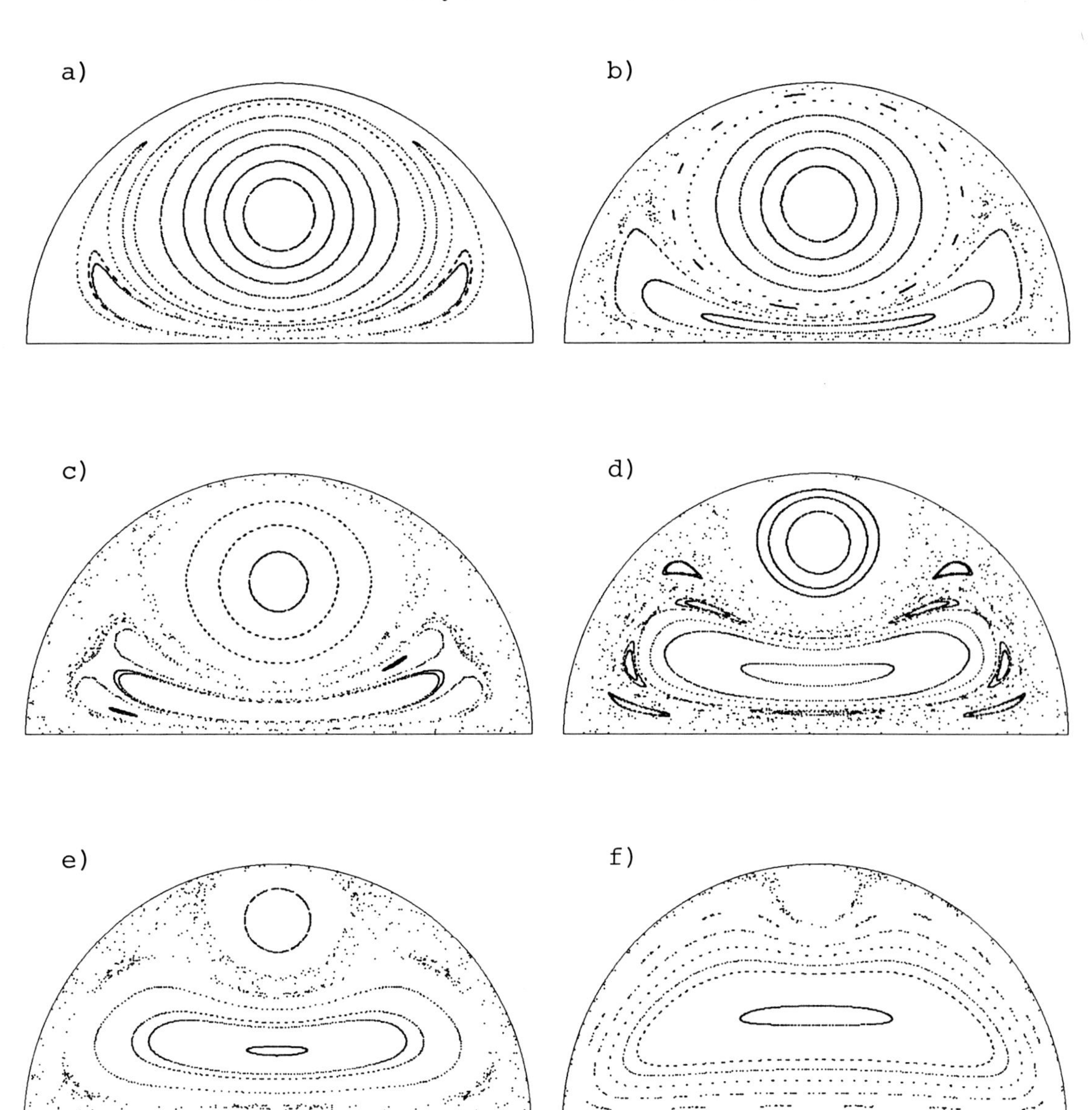

Figure 3. Poincaré map:

a) $q = 0.01$, b) $q = 0.05$, c) $q = 0.1$, d) $q = 0.25$, e) $q = 0.3$, f) $q = 0.4$

II: Dynamo Theory

Self-Excitation and Reconnections of Magnetic Field in Random Flow

S.A. MOLCHANOV, A.A. RUZMAIKIN & D.D. SOKOLOFF

Institute of Terrestrial Magnetism, Moscow, USSR

1. INTRODUCTION

The problem of self–excitation of magnetic field in moving conducting medium is important in two respects. First, this process produces and maintains magnetic fields of the Earth, the Sun, galaxies and other astronomical objects. Second, description of this phenomenon hardly fits the frameworks of standard methods of mathematical physics and necessitates further development of novel mathematical and physical ideas which can be applied in other fields of science as well.

The peculiarity of description of magnetic field self–excitation in a given flow, known as the kinematic dynamo problem, is as follows. The self–excitation is possible only when inductive action of motions strongly dominates over dissipation, i.e. when the so–called magnetic Reynolds number is large. Usually, this is the case in cosmic environments. On the other hand, if magnetic Reynolds number is infinitely large, the field is frozen into the medium and magnetic lines follow motions of the fluid particles. In other words, magnetic flux through any comoving contour is conserved. It would seem that the frozenness is incompatible with generation of magnetic field: the one and the same magnetic flux both should be conserved due to frozenness and grow exponentially due to generation.

One solution of this dilemma consists in introduction of a weak but finite dissipation which amounts to allowance for slow changes of the flux through a given comoving contour. If this is the case, the self–excitation process is called a *slow dynamo*: the growth rate of magnetic field is determined by magnetic Reynolds number and tends to zero when the latter tends to infinity (Zeldovich and Ruzmaikin, 1980). Although this mechanism of field growth can be essential in some objects like powerful breeder reactors with the liquid–metal cold stream or jets emanating from active galaxies, its characteristic times generally exceed characteristic times of magnetic field variations in most astronomical objects.

However, there exists one more way to self–excitation of magnetic field. This process has been demonstrated ***impromptu*** by Ya.B. Zeldovich in the course of discussion at symposium in Krakow (Poland) in 1972 (this shows again how useful are symposia). The point is that what is conserved for $R_{\mathrm{m}} = \infty$, is the magnetic flux through a comoving (Lagrangian) contour while one is interested in exponential growth of magnetic flux through a fixed (Eulerian) initial contour. The latter can occur when the separation between Eulerian and Lagrangian contours grows in time at exponential rate. Such exponential divergence is typical of random, turbulent fluid flows and is also reproduced by stationary flows with chaotic streamlines. In this case the magnetic dissipation serves as a relatively weak effect which smooths out the field irregularities but does not affect the field growth rate in the limit of large magnetic Reynolds numbers. Such self–excitation process has been called the *fast dynamo* (Zeldovich *et al.*, 1983).

Some types of random flows have been found now, which act as fast dynamos in the sense of growth of the mean magnetic field (Moffatt, 1978; Krause and Rädler, 1980) or its correlation function (Kraichnan and Nagarajan, 1967; Léorat *et al.*, 1981). Among such flows is the renovating flow in which the memory of the velocity field is lost at every moment proportional to a given renovation time. Strictly speaking, these flows are not statistically stationary but they approximately reproduce properties of stationary flows when considered at periods whose duration is much longer than the renovation time; the renovation time itself plays the role of the memory time (Molchanov *et al.*, 1985).

However, the situation with the fast–dynamo theory turns out to be much more complicated than it could be perceived beforehand. Generation of magnetic field by as random flow brings about the phenomenon of intermittency which manifests itself by highly inhomogeneous, in space and time, growth of magnetic field. Widely spaced anomalously strong magnetic concentrations appear against the background of typical magnetic fields; these concentrations can make the main contribution to the averaged magnetic energy. Therefore, the analysis of the fast growth of averaged parameters of magnetic field should be supplemented by investigation of behavior of its typical realization. Such results are also accessible and a relevant technique has been developed by Molchanov *et al.* (1984) who have proved that the so–called *dynamo theorem* which establishes the possibility of exponential growth of a typical realization of magnetic field embedded in a sufficiently smooth renovating random flow.

Another problem consists in inclusion into the theory of a weak magnetic dissipation. Even in those cases which are described by the dynamo theorem, the growth rate of magnetic field rapidly decreases when magnetic diffusivity increases. It would seem that for large magnetic Reynolds numbers the difference between the growth rate of magnetic field and its limiting value is proportional to inverse magnetic Reynolds number. However, there exist explicit examples for which this difference is proportional to a certain power of the logarithm of R_m. However, in random flows of other types a stronger influence of weak magnetic diffusion is possible. In such flows, the growth rate of a typical realization of magnetic field can differ from the growth rate obtained in neglect of the dissipation. This implies that the characteristic time of transformation of kinetic energy into heat coincides with the characteristic time of hydrodynamic motions and is independent of magnetic diffusivity (in the limit of large magnetic Reynolds numbers). This phenomenon, which can be called the *fast dissipation*, is no less interesting than the fast dynamo because it can serve as an effective mechanism of energy transformation in cosmic environments. It is important now to find out which kinds of random flows act as fast dynamos and which give rise to the fast dissipation (possibly preserving the fast growth of magnetic field). Below we discuss this problem further.

We note that one more type of generation of magnetic field has been proposed, in which the fast growth of the field is due to tangled streamlines of a stationary velocity field or singular gradients of the velocity field (Arnold *et al.*, 1981; Galloway and Frisch, 1981; Soward, 1988).

2. The Dynamo Maps

The physical mechanism of self–excitation and basic properties of growing magnetic fields can be clarified without exact solution of MHD equations. Consider a closed magnetic tube of the strength H, length L and cross–section S embedded in a flow of conducting incompressible fluid. The basic scenario of amplification of magnetic field has been outlined by Ya.B. Zeldovich as follows. First, the tube is stretched to twice its initial diameter so that, due to incompressibility and conservation of magnetic flux $\Phi = HS$, the length L is transformed to $2L$, S transforms to $S/2$ and H transforms to $2H$. After that the tube is twisted into a figure eight shape and the two halves are superposed. At this stage we have $2L \to L$, $S/2 \to S$ and $2H \to 2H$. Of course, magnetic flux through a Lagrangian contour is conserved and what doubles is that through the Eulerian contour which initially encircles the tube once and through which the tube passes twice after the amplification cycle: $\Phi \to 2\Phi$. As a result, both the magnetic field and magnetic flux are doubled while the initial shape of the tube is restored (strictly speaking, this requires some magnetic diffusion which can wash out the irregularities resulting from twisting and folding of magnetic lines). When this procedure is repeated, magnetic field grows with the characteristic time $T/\ln 2$ where T is the duration of one cycle.

In this dynamo, magnetic field is considered as a field embedded in three–dimensional space. This process can be modeled by a simple one–dimensional discrete baker's map which, however, admits discontinuous transformation of magnetic field (Fig. 1). Consider magnetic lines on a plane which are directed along the y–axis and whose density varies along the x–axis, *i.e.* consider a one–dimensional model (dependence on x) of three–dimensional process of magnetic field generation. The remaining coordinate, y is essential for account for the vectorial nature of magnetic field.

The Zeldovich dynamo is sufficiently simple and can serve as an illustration of the mechanism of self–excitation of magnetic field but it is too simplified when one wishes to obtain a realistic picture of this mechanism. The model of this dynamo, based on theory of maps, which has proved to be very useful in modern hydrodynamics, allows to take into account more delicate details and to make the picture more realistic while preserving the simplicity of analysis. The first step in this direction has been made by Finn and Ott (1988) (see also Childress, 1988) who note that the stretching can be inhomogeneous so that the fraction μ of the tube length is stretched to the length L, and so does the remaining part of the tube. As a result, after many generation cycles for $\mu < 1/2$ the field distribution becomes very rugged, fractal. This fact reflects a real process of magnetic field tangling and requires inclusion of magnetic diffusion which smooths small–scale inhomogeneities of magnetic field [see Molchanov *et al.* (1989) for details].

3. An Ensemble of Dynamo Maps

In order to introduce further realism into this picture, one should consider not only baker's map or some similar map of a single square but rather the maps which transform an ensemble of such squares. Suppose that the magnetic tube folding occurs in each square but only the fraction p of magnetic tubes are twisted and magnetic fields are amplified while in the fraction $q = 1 - p$ the fields in two halves of the final ring become antiparallel (due to absence of twisting). Such process is

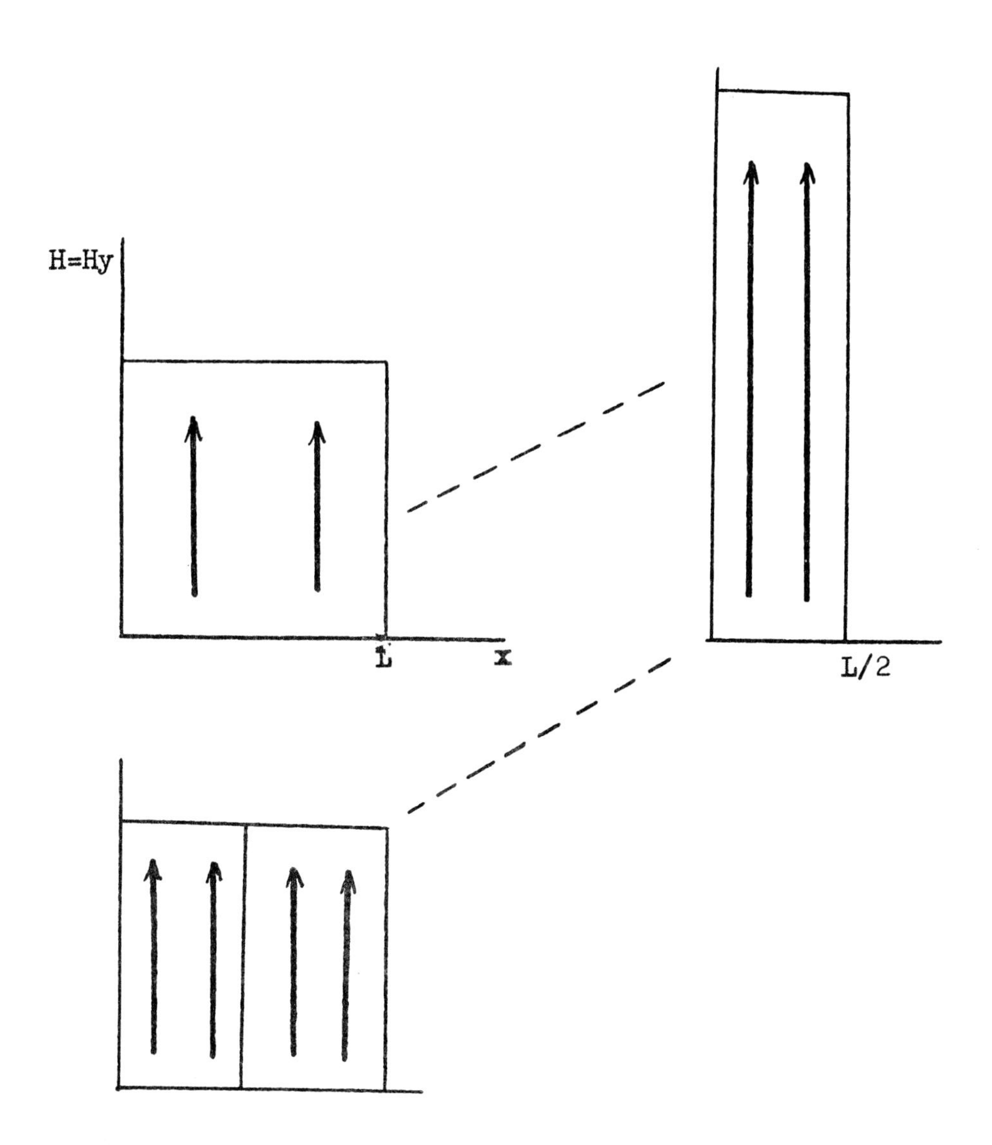

Fig. 1. Baker's map doubles the number of arrows within a square thus imitating the Zeldovich dynamo.

illustrated by Figs. 2 a,b,c where the initial and final distributions of magnetic field are shown. After n cycles, magnetic field in every half of a square is amplified from H_0 to $H_0 2^n$. However, in many squares the fields in neighbouring half–squares are antiparallel. Thus, non–intermittent but alternating magnetic field is generated in the considered random medium (the field growth rate is uniform throughout the considered volume).

Now let us take into account the dissipation presuming that magnetic field in every square is averaged after every cycle. This drastically changes the spatial distribution of magnetic field. Now the magnetic field in individual square is given by

$$H_n = H_0 2^n \theta_1 \theta_2 \ldots \theta_n, \qquad (1)$$

where θ_i are independent random numbers whose values are unity with the probability p and zero with the probability q. It is clear that for $n \to \infty$ the magnetic field vanishes within most squares except the rare ones in which all numbers θ_i are unity. [Random sets θ_i in different cells can be considered statistically independent. Therefore, at every stage of evolution of an infinite space of squares always exist the squares in which magnetic field differs from zero and, moreover, is very large, $H = H_0 2^n$. Due to chance, all n number θ_i are unity in all these squares.] The average magnetic field in this medium is given by

$$<|H_n|> = 2^n p^n.$$

The higher moments grow as

$$<|H_n|^k> = 2^{n^k} p^n,$$

Thus, the growth rate of the k'th moment of magnetic fields as follows:

$$\gamma_k = \lim_{n \to \infty} \frac{\ln <|H_n|^k>}{k^n} = \ln 2 - \frac{|\ln p|}{k}, \qquad (2)$$

where k is any non–zero number. At each fixed point of the considered medium the typical realization of magnetic field decays [see (1)] while higher moments grow at the rate which increases with the moment order. The dissipation, which is active within individual squares, destroys at each cycle the field within the fraction q of the volume and after n cycles the field disappears in the fraction $1-p^n$ of the volume. The fast dissipation of magnetic energy arises owing to development of magnetic structures shown in Fig. 2d which involve numerous reconnections of magnetic lines. However, a strong field survives within small widely spaced islands (intermittency).

Now consider another way of introduction of randomness. Let the folding be such that neighboring magnetic lines always become parallel. However, instead of θ_i we allow for larger generally nonvanishing stretching factors, so that multiple figures eight may appear and stretching by the factor three, four or more has finite

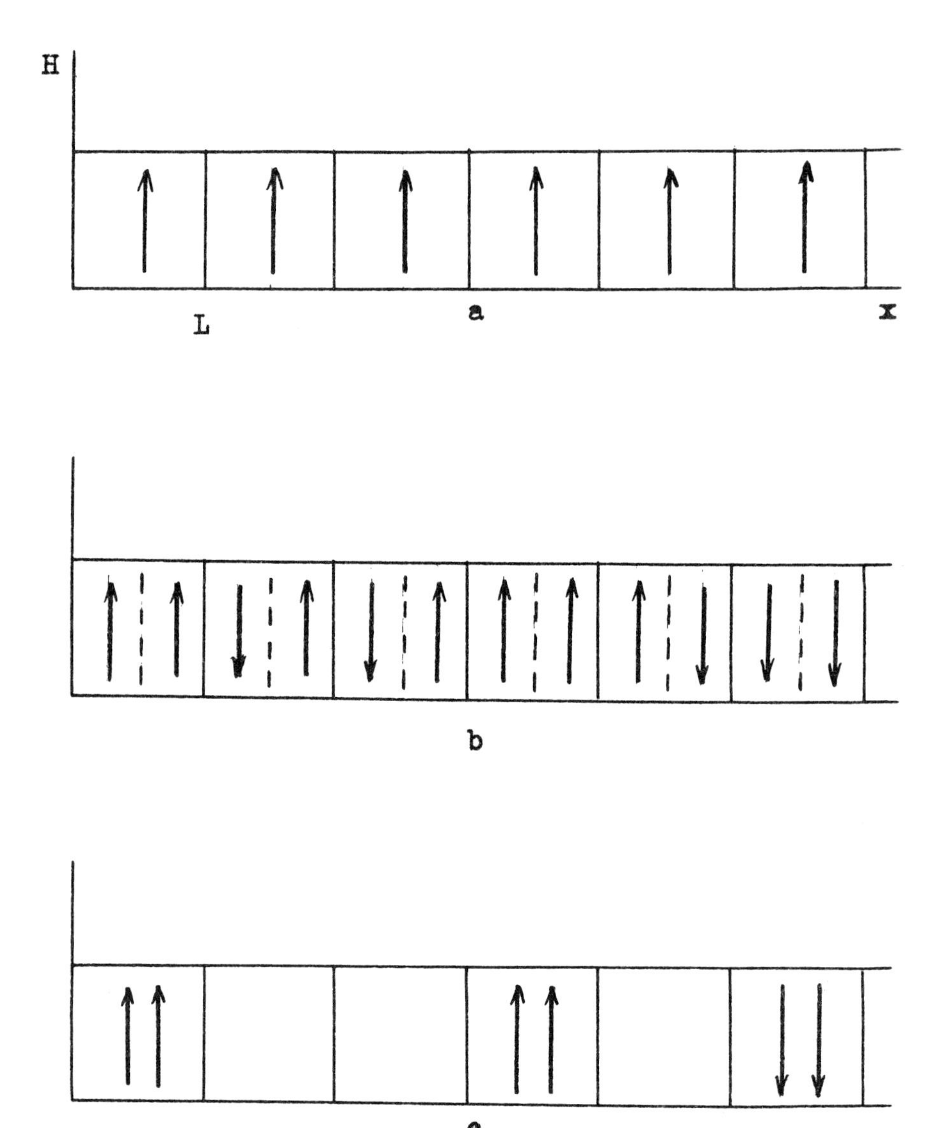

Fig. 2. Shown is the random medium consisting of squares whithin each of which the bakers's map acts. In (b) the field is twice as strong as in (a) but the field direction is arbitrary. As a result, introduction of diffusion (c) leads to fast dissipation of the field.

probability, followed by folding of the corresponding order. Then in a given square the magnetic field is amplified by a random factor after one cycle (Fig. 3 a,b). In neglect of dissipation, after n cycles the magnetic field in a given square becomes equal to

$$H_n = H_0 \xi_1 \xi_2 \ldots \xi_n,$$

where ξ_i are random numbers which acquire the values 1,2,3,.. with finite probability. It can be easily verified (see Zeldovich *et al.*, 1987) that a typical realization of this magnetic field grows while statistical moments of this field grow at even larger, progressively increasing growth rates (the detailed estimates depend on statistical properties of the numbers ξ_i). This means that intermittency arises even for zero dissipation. The intermittency manifests itself in anomalously rapid growth of magnetic field at positions of anomalously rapid amplification of magnetic field, where the loops are stretched very strongly. Of course, magnetic diffusion will smear these inhomogeneities but it cannot result in catastrophic annihilation of oppositely directed magnetic tubes, typical of the last example, because the field peaks are separated very widely. The structure of inhomogeneities is shown in Fig. 3c. In this example we have a fast generation of intermittent magnetic field; the growth rate continuously changes with magnetic diffusivity so that there exists the limiting value $\gamma(\nu_m) \to \gamma_0$ for $\nu_m \to 0$.

Of course, both these examples are schematic and each one reflects only one aspect of the field self–excitation in a real turbulent flow. The generation mechanism and the role of weak magnetic diffusivity in renovating random flows (see Molchanov *et al.*, 1984, 1985) are close to the second example. It is interesting to find realistic flows which reproduce the properties of the first example and lead to fast dissipation.

3. The Role of Three–Dimensionality of the Flow

The effects of fast generation can prevail over fast dissipation in a real flow only when reconnections are rare. The simplest cause of this prevalence can be the dimensionality of the flow. It is clear that in two–dimensional flow the figure–eight magnetic rope cannot be produced by continuous deformation. The quasi–two=dimensional maps considered above are useful only because they are discrete and magnetic lines are cut. For two–dimensional motion, magnetic tubes can be stretched ensuring a temporal growth of the field but earlier or later oppositely directed segments of the tube meat each other and what appears is the fastly annihilating structure shown in Fig. 2d. Note that the smaller is magnetic diffusivity, the faster is dissipation of such structures when they overlap (see Zeldovich *et al.*, 1984).

In three dimensions, the approaching tubes can be neither parallel nor antiparallel. They can meet each other obliquely, with an arbitrary angle λ between them, so that the annihilation is never total and the resulting structures resemble those shown in Fig. 3c.

In a real turbulent flow, the process of folding of magnetic tubes is accompanied by decrease in the scale of magnetic field due to divergence of closely spaced particles, which complicates introduction of magnetic diffusion. Rigorous estimates in this field, which are based on model examples analogous to random

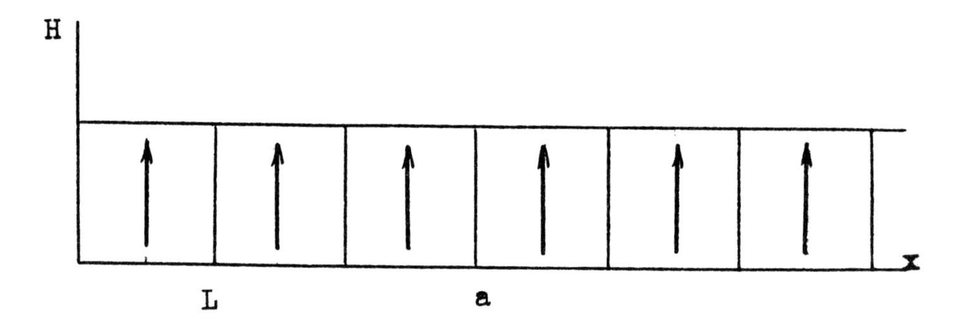

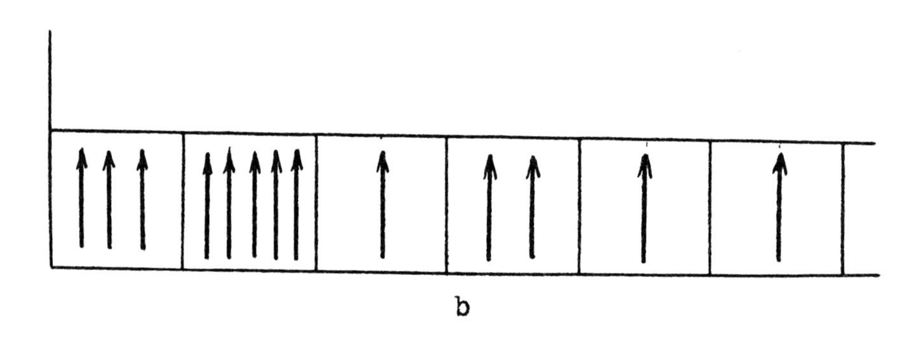

Fig. 3. In this medium magnetic field is arbitrarily amplified between (a) and (b). Magnetic diffusion only weakly affects the resulting structures (c).

maps considered here (see Zeldovich *et al.*, 1988), essentially rely on multiple differentiability of the velocity field. In physical terms , in prove of the magnetic field growth it is implicitly presumed that magnetic diffusivity is smaller than the hydrodynamic one. In this case the generation of magnetic field by a turbulent flow occurs due to the motions which determine the correlation properties of the flow, *i.e.* those at the Taylor microscale, rather than the inertial–range motions. Meanwhile, the role of the latter range is essential when the hydrodynamic diffusivity is smaller than the magnetic one. The energy–range flow then is not smooth but has only fractional number of Hölder derivatives; for Kolmogorov turbulence, this number is 1/3 (see, e.g., Zeldovich and Sokoloff, 1985). In this situation the generated magnetic field is very rugged and we can hypothesize that a typical realization of magnetic field decays and only the averaged characteristics can grow.

References

Arnold, V.I., Zeldovich, Ya.B., Ruzmaikin, A.A. and Sokoloff, D.D., 1981, Magnetic field in a stationary flow with stretching in a Riemannian space, *Sov. Phys. JETP*, **56**, 1083–1086.

Childress, S., 1988, Order and disorder in planetary dynamos, *Summer Study Program in Geophys. Fluid Dynamics*, WHO, Mass.

Finn, J. and Ott, E., 1988, Chaotic flows and fast magnetic dynamo, *Phys. Rev. Lett*, **60**, 760–763.

Galloway, D. and Frisch, U., 1981, A numerical investigation of magnetic field generation in a flow with chaotic streanlines, *Geophys. Astrophys. Fluid Dyn.*, **29**, 13–18.

Kraichnan, R.H. and Nagarajan, S., 1967, Growth of turbulent magnetic fields, *Phys. Fluids*, **10**, 859–870.

Krause, F. and Rädler, K.–H., 1980, *Mean–Field Magnetohydrodynamics and Dynamo Theory*, Pergamon Press, Oxford.

Léorat, J., Pouquet, A. and Frisch, U., 1981, Fully developed MHD turbulence near critical Reynolds number, *J. Fluid Mech.*, **119**, 419–444

Moffatt, H.K., 1978, *Magnetic Field Generation in Ealactrically Conducting Fluids*, Cambridge Univ. Press, Cambridge.

Molchanov, S.A., Ruzmaikin, A.A. and Sokoloff, D.D., 1984, A dynamo theorem, *Geophys. Astrophys. Flud Dyn.*, **30**, 241–259.

Molchanov, S.A., Ruzmaikin, A.A. and Sokoloff, D.D., 1989, Magnetic intermitency, in *Nonlinear Waves*, Springer, Berlin.

Soward, A.M., 1987, Fast dynamo action in a steady flow, *J. Fluid Mech.*, **180**, 267–295.

Zeldovich, Ya.B. and Ruzmaikin, A.A. 1980, Magnetic field in conducting fluid movving in two dimensions, *JETP*, **78**, 980–986.

Zeldovich, Ya.B. and Sokoloff, D.D., 1985, Fractals, self–similarity, intermediate asymptotics, *Usp. Fiz. Nauk*, **146**, 493–506.

Zeldovich, Ya.B., Ruzmaikin, A.A. and Sokoloff, D.D., 1983, *Magnetic Fields in Astrophysics*, Gardon and Breach, New York, London, Paris.

Zeldovich, Ya.B., Molchanov, S.A., Ruzmaikin, A.A. and Sokoloff, D.D., 1984, Kinematic dynamo in the linear velocity field, *J. Fluid Mech.*, **144**, 1–11.

Zeldovich, Ya.B., Molchanov, S.A., Ruzmaikin, A.A. and Sokoloff,

D.D., 1985, Intermittency of passive fields in random media, *JETP*, **89**, 2061–2072.
Zeldovich, Ya.B., Molchanov, S.A., Ruzmaikin, A.A. and Sokoloff, D.D., 1987, Intermittency in random media, *Usp. Fiz. Nauk*, **152**, 3–32.

Dynamo Action at Large Magnetic Reynolds Number in Spatially Periodic Flow with Mean Motion

A.M. SOWARD[1] & S. CHILDRESS[2]

[1]Department of Mathematics and Statistics, The University, Newcastle upon Tyne NE1 7RU, UK
[2]Courant Institute of Mathematical Sciences, 251 Mercer Street, New York University, NY 10012, USA

Kinematic dynamo action at large magnetic Reynolds number is considered for a 2-dimensional steady spatially periodic flow. Magnetic boundary layers are triggered at the X-type stagnation points, which provide the sites for dynamo activity. With the addition of mean motion, the streamlines downstream of the stagnation points, on which the layers lie, may be dense in finite regions of the flow. This introduces a new generic feature for advection-diffusion by steady flow. It is a first step towards analysis of flows with chaotic streamlines, which are believed to be likely candidates for fast dynamos.

1. BACKGROUND AND PROBLEM

We consider the self-excitation of magnetic field $\underset{\sim}{B}(\underset{\sim}{x},t)$ caused by the steady flow $u(\underset{\sim}{x})$ of a highly electrically conducting fluid. Magnetic induction is governed by

$$\partial\underset{\sim}{B}/\partial t = \underset{\sim}{\nabla}\times(\underset{\sim}{u}\times\underset{\sim}{B}) + R^{-1}\nabla^2\underset{\sim}{B} , \qquad \underset{\sim}{\nabla}\cdot\underset{\sim}{B} = 0 , \tag{1}$$

in which the magnetic Reynolds number is assumed to be large, $R \gg 1$. In this limit the nature of the inductive process is sensitive to the strealine topology. Much attention has focussed on the class of spatially periodic flows defined by

$$\underset{\sim}{u} = \bar{\underset{\sim}{u}}_H + \underset{\sim}{u}' , \qquad \bar{\underset{\sim}{u}}_H = \epsilon(\sin(\ell z),\cos(\ell z),0), \tag{2a}$$

$$\underset{\sim}{u}' = (\partial\psi'/\partial y,-\partial\psi'/\partial x,K\psi'), \quad \psi' = \sin x\sin y+\delta\cos x\cos y , \tag{2b}$$

where K, ℓ, ϵ and δ are constants. With $\epsilon = 0$, the motion defined by $\underset{\sim}{u}'$ lies entirely on the streamsurfaces ψ' = constant. When ϵ and ℓ are non-zero, the complete motion is no longer integrable and regions with chaotic particle paths ensue. This is exemplified by the special case

$K = \ell = \sqrt{2}$, which after a coordinate transformation defines the well-known ABC-flows discussed by Dombre et al (1986).

Solutions of (1) may be sought in the form

$$\underset{\sim}{B} = \mathrm{Re}[e^{pt+inz}\hat{\underset{\sim}{B}}(\underset{\sim}{x};n,R)] \tag{3}$$

for various values of Floquet exponent n, where the complex vector $\hat{\underset{\sim}{B}}$ has the same spatial periodicity as the motion. In the case of ABC-flows, Arnold and Korkina (1983) and later Galloway and Frisch (1984,1986) have obtained numerical solutions and determined the value of the growth rate p by employing modal expansions. Despite the large values of R considered, of order 10^3, no clear asymptotic regime was reached. Analytic results, on the other hand, have been largely limited to the integrable cases of the motion (2), and obtained by boundary layer methods. The following three cases have received particular attention:

(i) **The case $\epsilon = \delta = 0$**

Like the motion, the vector

$$\hat{\underset{\sim}{B}}(\underset{\sim}{x}_H) = \bar{\underset{\sim}{B}}_H + \underset{\sim}{b}' \qquad [\underset{\sim}{x}_H = (x,y)] \tag{4}$$

is z-independent. The contribution $\bar{\underset{\sim}{B}}_H$ is a constant complex vector lying in the horizontal xy-plane, while $\underset{\sim}{b}'$ is spatially periodic. Roberts (1972) investigated this complete problem numerically. On the other hand, for sufficiently small n, $\hat{\underset{\sim}{B}}$ may be determined correct to leading order by setting $p = n = 0$. Then the mathematical problem reduces to determining the steady 2-D magnetic field structure which arises when there is an applied mean field

$$\bar{\underset{\sim}{B}}_H = (\partial\bar{A}/\partial y, -\partial\bar{A}/\partial y, 0) \ , \quad \bar{A} = \bar{B}_x y - \bar{B}_y x \ . \tag{5}$$

Magnetic flux is expelled from the closed eddies and is concentrated in sheets of width of order $R^{-1/2}$ on the cell boundaries $\psi' = 0$. Childress (1979) formulated the boundary layer problem and considered the magnitude of the 2-D isotropic alpha-tensor, which ensues (cf (15) below). Soward (1987) showed that the result is valid provided that $n \ll R^{1/2}(\ell nR)^{-1}$ and obtained the appropriate modifications valid when

$n = O[R^{1/2}(\ell nR)^{-1}]$. These short vertical length scale modes have the fastest growth rates and are 'almost' fast dynamos (see also Soward, 1989). Nevertheless, the important features of the dynamo process are isolated by the small n-limit and so the Childress (1979) approach is adopted in the cases (ii) and (iii) which follow.

(ii) **The case $\epsilon = 0$, $\delta = O(R^{-1/2})$**

One way to break the symmetry of the case (i) flow is to relax the condition $\delta = 0$. So instead of swirling vortices filling the squares, $\Pi_{m,n} = [m\pi,(m+1)\pi]\times[n\pi,(n+1)\pi]$ (integer m,n), the closed streamline region (of the horizontal motion $\underset{\sim}{u}_H$) contracts to form rows of cat's eyes. Their boundaries are the separatrices, which connect the 'junction' streamlines at the opposite corners $(m\pi,n\pi)$, $((m+1)\pi,(n+1)\pi)$ of the squares $\Pi_{m,n}$. Outside the cat's eyes, in the 'channel' regions $|\psi| < \delta$, the $\underset{\sim}{u}_H$-streamlines are open. When $1 \gg \delta \gg R^{-1/2}$, magnetic boundary layers triggered at the 'junction' streamlines are confined close to the separatrices, $|\psi| - \delta = O(R^{-1/2})$. Magnetic flux is expelled from the cat's eyes, while the nature of the mainstream 'channel' solutions depends on the orientation of the mean field (5), and the resulting 2-D alpha-tensor is strongly anisotropic. When the mean field is **parallel** to the 'channel' direction (1,1), the magnetic field aligns itself with the flow and there is very little inductive effect. On the other hand, when the mean field lies in the direction (1,-1) **perpendicular** to the 'channels', field lines are stretched out along the 'channels' and intensified by a factor of order R leading to a strong inductive effect. The growth rate, which depends on the geometric mean of these effects, is $p = O[KnR^{-1/2}\sqrt{R^{1/2}\delta}]$. There are upper limits on the size of n and, as in case (i), 'almost' fast modes only occur when the 'channel' and boundary layer widths are comparable, $\delta = O(R^{-1/2})$.

(iii) The case $\epsilon = O(R^{-1/2})$, $\ell \to 0$, $\delta = 0$

In cases (i) and (ii) motion is confined to stream surfaces. When $0 < \epsilon << 1$, $\ell \neq 0$, a 'web' of chaotic particle paths forms close to the boundaries of the square $\Pi_{m,n}$ (see Zaslavskii et al, 1988). Since exponential stretching of line elements is no longer confined to the separatrices (see also case (iv)), it is often argued that such dynamos may be fast. As a tentative step in that direction, Soward and Childress (1989) considered the double limit $\ell \to 0$, $\ell z \to \theta_0$ (a constant) paying particular attention to mean flows

$$\bar{\underset{\sim}{u}}_H = \epsilon(M,N,0)/(M^2+N^2)^{1/2} \qquad (\epsilon << 1) \;, \tag{6}$$

with rational tangents, $\tan\theta_0 = M/N$, where M,N are relatively prime integers. The stream function for the horizontal motion is

$$\psi = \bar{\psi} + \psi' \;, \qquad \bar{\psi} = \bar{u}_x y - \bar{u}_y x \;. \tag{7}$$

The closed streamline flow is confined to eddies $\mathcal{D}^{(\pm)}_{m,n}$ [+,ℓ even; −,ℓ odd; ℓ = m+n] inside the squares $\Pi_{m,n}$. Their boundaries are terminated at single X-type stagnation points located at

$$[(m + \tfrac{1}{2}\{1+(-1)^{\ell}\})\pi,\ (n + \tfrac{1}{2}\{1-(-1)^{\ell}\})\pi] \;, \tag{8a}$$

where ψ takes the values

$$\psi^{(\pm)}_{m,n} = [M(n+\tfrac{1}{2}) - N(m+\tfrac{1}{2})](\Delta\psi) \pm \tfrac{1}{2}\psi^{(c)} + O(\epsilon^2) \tag{8b}$$

with

$$\psi^{(c)} = L(\Delta\psi) \;, \qquad \Delta\psi = \epsilon\pi/(M^2+N^2)^{1/2} \;, \qquad L = M + N \;. \tag{8c}$$

The eddy structure about the square $\Pi_{0,0}$ is illustrated in the figure. In the limit $\epsilon << 1$, the streamlines through the X-type stagnation points form the boundaries, $\psi[= \psi^{(\pm)}_{m,n}] = k(\Delta\psi)$ (integer k), of spatially peridoic 'channels' carrying fluid flux $\Delta\psi$. A periodicity section of channel starting at (x_0,y_0) ends at $(x_0,y_0) + (2\pi/\tau)(M,N)$, where

$$\tau = 2,\ L \text{ even}, \qquad \tau = 1,\ L \text{ odd} \;. \tag{9}$$

The circulation, $\int \underset{\sim}{u}\cdot d\underset{\sim}{x}$, along each side of the squares $\Pi_{m,n}$ is 2, and the total circulation over a periodicity section of 'channel' is

$$\Gamma_k = (8/\tau)\Delta_k L \quad \text{on} \quad D_k: (k+1)(\Delta\psi) > \psi > k(\Delta\psi) \;, \tag{10a}$$

where

$$\Delta_k = 1 + (-1)^k(\tau-1)/L \;. \tag{10b}$$

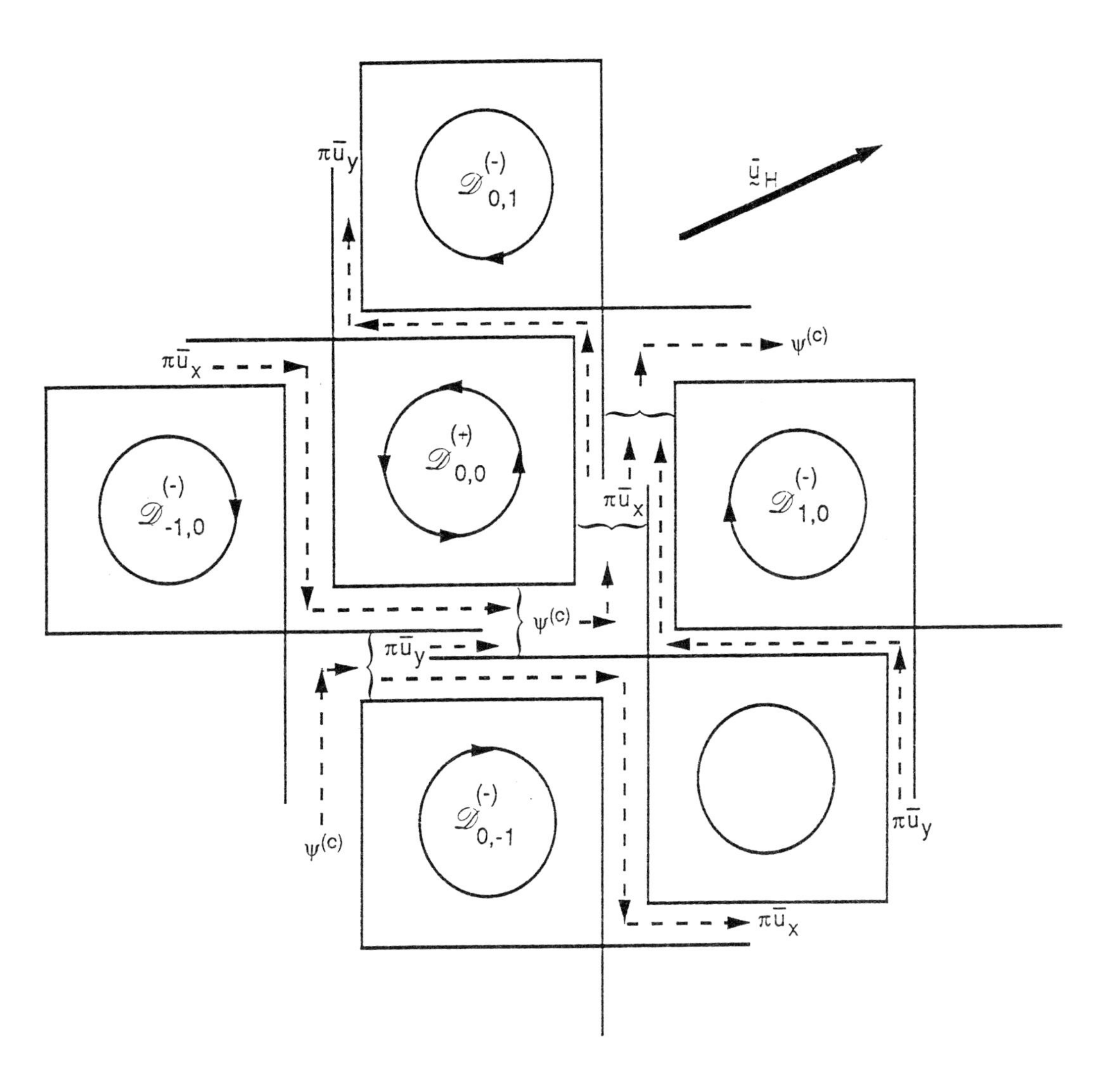

Like the cat's eye configuration of case (ii), boundary layers are triggered at the X-type stagnation points. They penetrate a distance $R^{-1/2}$ into the eddies. In the 'channels', on the other hand, they thicken over the length of the periodicity section (see (10a)) to a width $O(R^{-1/2}L^{1/2})$. Consequently mainstream 'channel' solutions only exist when the 'channel' width $\Delta\psi$ is large compared to the 'channel' boundary layer width. In terms of β, the ratio of the gap between eddies ϵ to eddy boundary layer width $R^{-1/2}$, the condition is

$$\beta \equiv \epsilon R^{1/2} >> [L(M^2+N^2)]^{1/2} , \tag{11}$$

which can be met at fixed M,N for sufficiently large R. On the other hand, at fixed small ϵ and large R, (11) is only met by a discrete set of rational tangents M/N. In this respect, the irrational limit of $\tan\theta_0$ exhibits an important feature of flows with chaotic particle paths, namely that streamlines on which exponential stretching occurs are dense ($\Delta\psi=0$) in a finite region of the flow. Solutions are described in Section 2 below.

(iv) The case $\epsilon = 1/\sqrt{2}$, $\ell = K = \sqrt{2}$, $\delta = 0$

We mention briefly the symmetric ABC-flow. Childress (1979) suggested how the corresponding 3-D isotropic alpha-tensor might be calculated. Childress and Soward (1985) attempted the calculation but the difficulties they encountered with the chaotic particle paths relate to the irrational limit mentioned above. Specifically boundary layers are triggered at stagnation points and the two-dimensional manifolds on which they lie are probably dense in the chaotic regions of the flow. The solution to this dynamo problem in the limit $R \to \infty$ remains unsolved.

2 RESULTS

We consider case (iii) and seek steady 2-D solutions

$$\underset{\sim}{B} = (\partial A/\partial y, -\partial A/\partial x, KB) \tag{12}$$

of (1) subject to the periodicity conditions

$$A(m\pi+x, n\pi+x) = \bar{A}_{m,n}+A(x,y), \qquad B(m\pi+x, n\pi+y) = B(x,y) \tag{13a}$$

for integer m,n of even sum $\ell(=m+n)$, where

$$\bar{A}_{m,n} = (n\bar{B}_x - m\bar{B}_y)\pi , \quad \bar{B} = 0 \tag{13b}$$

and the bar denotes the horizontal average. The z-component

$$-\bar{E} = (\bar{\underset{\sim}{u}}_H \times \bar{\underset{\sim}{B}}_H)_z \tag{14}$$

of the electromotive force is constant, while the mean of its horizontal components

$$K\bar{\underset{\sim}{E}}_H = \overline{[\underset{\sim}{u}\times\underset{\sim}{B}]}_H = K\overline{[(\psi-\bar{\psi})\underset{\sim}{\nabla}A - B\underset{\sim}{\nabla}\psi]} = \underset{\approx}{\alpha}\cdot\bar{\underset{\sim}{B}}_H \tag{15}$$

is linearly related to $\bar{B}_H$ by the alpha-tensor, α. The scalars A and B satisfy

$$u_H\cdot\nabla A - R^{-1}\nabla^2 A = \bar{E} , \tag{16a}$$

$$u_H\cdot\nabla B - R^{-1}\nabla^2 B = B_H\cdot\nabla\psi' \equiv -u_H\cdot\nabla A + \bar{u}_H\cdot\nabla A . \tag{16b}$$

Their solutions are used to calculate $\bar{E}_H$ from (15) and hence α.

Solutions of (16) at large R are obtained by using the stream function ψ and the circulation, $\sigma = \int u\cdot dx$, as coordinates. The key idea in the wide 'channel' limit (11) is to average the mainstream 'channel' solutions over the periodicity interval $(2\pi/\tau)(M,N)$. It leads to

$$\left.\begin{array}{l} R^{-1}\partial^2 A/\partial\psi^2 \\ \\ R^{-1}\partial^2 \mathcal{B}/\partial\psi^2 \end{array}\right\} = \left[\frac{\partial A}{\partial\sigma}\right]_k = \frac{2}{\Gamma_k\tau}\bar{A}_{M,N} \qquad \left\{\begin{array}{l} \partial A/\partial\psi = (\partial A/\partial\psi)(\psi), \\ \\ B+\bar{\psi}(\partial A/\partial\psi) = \mathcal{B}(\psi) \end{array}\right. \tag{17}$$

on each channel D_k, and

$$\left.\begin{array}{l} \pm R^{-1}\partial A/\partial\psi = (\pi^2/8)\bar{E} = (1/8)(\Delta\psi)\bar{A}_{M,N} \\ \partial\mathcal{B}/\partial\psi = 0 \end{array}\right\} \tag{18}$$

at the outer edge of each eddy $\mathcal{D}^{(\pm)}_{m,n}$. The solutions of (17) and (18) are fixed by matching conditions across the 'channel' and eddy boundaries. Here we summarise briefly the results for mean magnetic fields <u>**parallel**</u> $(-\bar{E} = [\bar{u}_H\times\bar{B}_H]_z = 0)$ and <u>**perpendicular**</u> $(\bar{u}_H\cdot\bar{B}_H = 0)$ to the mean motion.

In the <u>**parallel**</u> case, we obtain

$$\left.\begin{array}{l} \partial A/\partial\psi = \Delta_{k+1}(\bar{u}_H\cdot\bar{B}_H)/|\bar{u}_H|^2 \\ B = (\psi-\bar{\psi})(\partial A/\partial\psi) \end{array}\right\} \quad \text{on } D_k , \tag{19a}$$

and

$$\partial A/\partial\psi = B = 0 \qquad \text{on } \mathcal{D}^{(\pm)}_{m,n} . \tag{19b}$$

To the same order of accuracy the mean electromotive force vanishes, $K\bar{E}_H = 0$. In the <u>**perpendicular**</u> case, we obtain

$$\left.\begin{aligned} \partial A/\partial\psi &= \tfrac{1}{4}\pi^2 R\bar{E}(\psi-\psi_{k+\frac{1}{2}})/(\Delta_k\psi^{(c)}) \\ B-B^{(s)} &= (\pi^2/8)R\bar{E}[(\psi-\bar{\psi})^2 - (\psi_{k+1}-\bar{\psi})(\psi_k-\bar{\psi})]/(\Delta_k\psi^{(c)}) \end{aligned}\right\} \text{ on } D_k \tag{20a}$$

and

$$\left.\begin{aligned} \partial A/\partial\psi &= \pm(\pi^2/8)R\bar{E} \\ B-B^{(s)} &= (\psi^{(\pm)}_{m,n}-\bar{\psi})(\partial A/\partial\psi) \end{aligned}\right\} \quad \text{on } \mathscr{D}^{(\pm)}_{m,n} , \tag{20b}$$

where

$$B^{(s)} = -(\pi^2/16)R\bar{E}\psi^{(c)} , \qquad \psi_k = k(\Delta\psi) . \tag{21}$$

The resulting electromotive force is

$$K\bar{\underset{\sim}{E}}_H = (K\pi^2/16)R[1-(3\Lambda L^2)^{-1}]\psi^{(c)}\bar{\underset{\sim}{u}}_H \times (\bar{\underset{\sim}{u}}_H\times\bar{\underset{\sim}{B}}_H) , \tag{22a}$$

where

$$\Lambda = 1 - (\tau-1)L^{-2} . \tag{22b}$$

The combined results for <u>parallel</u> and <u>perpendicular</u> mean fields imply that the 2-D alpha-tensor defined by (15) is

$$\underset{\sim}{\alpha} = -\alpha_0[\underset{\sim}{I} - \epsilon^{-2}(\bar{\underset{\sim}{u}}_H\bar{\underset{\sim}{u}}_H)] , \tag{23a}$$

where

$$\alpha_0 = [1 - (3\Lambda L^2)^{-1}]\alpha_0^{(irr)} \tag{23b}$$

and

$$K^{-1}\epsilon^{-3}R^{-1}\alpha_0^{(irr)} = (\pi^3/16)(|\sin\theta_0| + |\cos\theta_0|) . \tag{23c}$$

Here $\alpha_0^{(irr)}$ is the limiting value of α_0 for irrational tangents ($L\to\infty$). Their values for low values of M,N are listed in the table. To test the results, the boundary layer problem was also solved numerically for $\beta = O(1)$ and the quantity

$$K^{-1}\tilde{\alpha}^{(num)} = -\beta^{-3}R^{\frac{1}{2}}(\bar{\underset{\sim}{E}}_H\cdot\bar{\underset{\sim}{B}}_H)/|\bar{\underset{\sim}{B}}_H|^2 \tag{24}$$

was calculated for <u>perpendicular</u> fields. The results for $\pi\beta = 30$ and 60 are listed in the table, together with the interpolated values at $(\pi\beta)^{-1} = 0$, on the basis of a linear dependence on β^{-1}. Good agreement is obtained with (23b) for the smallest values of M and N, which satisfy the condition (11).

L-ODD	$\pi\beta = 30$	$\pi\beta = 60$	$\beta \to \infty$	Asymptotic value	Irrational asymptotic value
(M,N)	$K^{-1}\tilde{\alpha}^{(num)}$	$K^{-1}\tilde{\alpha}^{(num)}$	$K^{-1}\tilde{\alpha}_0^{(0)}$	$K^{-1}\epsilon^{-3}R\alpha_0$	$K^{-1}\epsilon^{-3}R\alpha_0^{(irr)}$
(0,1)	1.255	1.273	1.291	1.292	1.938
(1,2)	2.436	2.469	2.502	2.503	2.600
(2,3)	2.549	2.608	2.667	2.652	2.687
(1,4)	2.243	2.272	2.302	2.319	2.350
(3,4)	2.593	2.640	2.687	2.694	2.712
(2,5)	2.387	2.455	2.522	2.502	2.520
(1,6)	2.102	2.168	2.234	2.215	2.230
L-EVEN					
(1,1)	2.339	2.387	2.435	2.436	2.740
(1,3)	2.300	2.346	2.391	2.397	2.452
(1,5)	2.150	2.202	2.254	2.259	2.281

3 CONCLUSIONS

The comparison between the numerical and analytic results highlight the significance of condition (11). They say that given an irrational tangent, the eigenfunction responds indefinitely revealing new structures on ever-shortening length scales as $R \to \infty$. The encouraging feature of our problem is that α_0 converges on to the smooth function $\alpha_0^{(irr)}$ as $L \to \infty$; the irrational limit. For generic 3-D flows exhibiting Lagrangian chaos, we must also expect this indefinite response as $R \to \infty$. The hope is that p tends to a well-defined limit as $R \to \infty$.

Finally, we speculate on the dynamo solution, which occurs in case (iii) when $0 < \ell << 1$ and the slow mean flow $\bar{\underset{\sim}{u}}_H$ alters its direction with z on the long length scale ℓ^{-1}. If we consider modes (3) with $n = O(\ell)$, the main limitation on the validity of our theory stems from the vertical z-motion $K\psi'$ in the 'channels'. It leads to an additional term in the equation (17) for A of order $iK\epsilon\ell A$ which causes A to decay horizontally on the length $(K\epsilon Rn)^{-½}$. This is comparable with the channel width, when

$$n = O[(K\epsilon\beta^2)^{-1}] = O[(K\epsilon^3 R)^{-1}] \, , \tag{25}$$

which provides an upper bound on n. For $\bar{\underset{\sim}{u}}_H$ varying its orientation, the alpha-tensor (23a) is similar to one considered earlier by Soward and Childress (1986). They found that the corresponding growth rate is

$$p = O(\alpha_0 n) \tag{26}$$

which by (23b,c) and (25) is O(1). Of course, this is only an estimate and possibly like cases (i) and (ii) the dynamo is only 'almost' fast. Note, however, that, when ℓ is fixed and $R \to \infty$, the above estimates are inapplicable. Nevertheless, the small regions of Lagrangian chaos, which the above estimates ignore, are likely to be sufficient to render the dynamo fast. Certainly Vishik (1988) has shown that, without exponential stretching of material line elements, fast dynamo action is impossible. He has also indicated (Private Communication) that when exponential stretching is limited to regions of zero measure like in our cases (i), (ii) and (iii) (for rational tangents) fast dynamo action probably cannot occur.

ACKNOWLEDGEMENTS

One of us, A M Soward, has benefited from interaction with Dr M M Vishik, while a guest of the Institute of the Physics of the Earth of the Academy of Sciences of the USSR, Moscow, during March and April 1989. Both of us are very grateful to Dr M M Vishik for discussions.

REFERENCES

Arnold, VI & Korkina, EI (1983). Vest.Mosk.Un.Ta.Ser. 1, *Math.Mec.* 3, 43-46 (in Russian).

Childress, S (1979). *Phys.Earth Planet.Int.* 20, 172-180.

Childress, S & Soward, AM (1985). *Chaos in Astrophysics.* JR Buchler, JM Perdang and EA Siegel, eds. NATO ASI Series C: Mathematical and Physical Sciences, Volume 161, pp223-244.

Childress, S & Soward, AM (1989). Scalar transport and alpha-effect for a family of cat's eye flows. *J.Fluid Mech.* (to appear).

Dombre, T, Frisch, U, Greene, JM, Hénon, M, Mehr, A & Soward, AM (1986). *J.Fluid Mech.* 167, 353-391.

Galloway, DJ & Frisch, U (1984). *Geophys.Astrophys.Fluid Dynamics* 29, 13-18.

Galloway, DJ & Frisch, U (1986). *Geophys.Astrophys.Fluid Dynamics* 36, 53-83.

Roberts, GO (1972). *Phil.Trans.Roy.Soc.Lond.* A271, 411-454.

Soward, AM (1987). *J.Fluid Mech.* 180, 267-295.

Soward, AM (1989). *Geophys.Astrophys.Fluid Dynamics* (to appear).

Soward, AM & Childress, S (1986). *Adv.Space Res.* 6, 7-18.

Soward, AM & Childress, S (1989). *Phil.Trans.Roy.Soc.Lond.A* (to appear).

Vishik, MM (1988). *Izvestiya AN USSR, Fisika Zemli*, No 3, 3-12.

Zaslavskii, GM, Sagdeev, RZ & Chernikov, AA (1988). *Zh.Eksp.&Teor.Fiz.* 94, 102-115. [English Translation: *Sov.Phys.JETP*, 67, 270-277 (1988)].

Generation of Steady and Oscillatory Magnetic Fields Through Inverse Cascades in Alpha-Effect Dynamos

B. Galanti[1], A.D. Gilbert[1,2], P.L. Sulem[1,3]

[1] - CNRS, Observatoire de Nice, BP 139, 06003 Nice-Cedex, France

[2] - D.A.M.T.P., Silver St., Cambridge CB3 9EW, U.K.

[3] - School of Mathematical Sciences, Tel Aviv University, Israel

1. INTRODUCTION

A small-scale helical flow acting on a large-scale magnetic field generates a mean electromotive force (emf) which can lead to a dynamo action (Parker 1955, Steenbeck, Krause and Rädler 1966). Saturation of the growing magnetic field may result from interaction with the large-scale velocity field induced by the Lorentz force. Hewever, it is possible that these two fields are approximately in MHD equilibrium, having negligible non-linear interactions. In this case, saturation occurs by a second mechanism: a strong magnetic field may modify the motion at small scales so as to reduce the mean emf and the alpha effect (Kraichnan 1979, Meneguzzi, Frisch and Pouquet 1981). The condition of MHD equilibrium holds when the large-scale fields are one-dimensional, that is depend on only one cartesian coordinate. Various such models were considered in situations where there is no motion at large scales (Krause and Rädler 1980, Zeldovich, Ruzmaikin and Sokoloff 1983, Krause and Meinel 1988, Gilbert and Sulem 1989). The main conclusion of the last reference is the development of an inverse cascade of magnetic field, which ultimately saturates in a stationary state dominated by the largest available scale (Sections 2 and 3).

In Section 4, we consider a dynamo in which the magnetic field interacts with a large-scale velocity field. Since in one dimension, the only interactions between the large scales are those mediated by the small scales, we require that the latter be sensitive to a large-scale velocity field. This effect, referred to as an AKA-type effect (Sulem, She, Scholl and Frisch 1989), necessitates lack of Galilean invariance and can lead to the development of a large-scale velocity field (Frisch, She and Sulem 1987). When a dynamo is coupled to an AKA instability, it can lead to interesting dynamical behaviour for the large-scale fields (Galanti, Sulem and Gilbert 1989).

2. EQUATIONS GOVERNING THE LARGE-SCALE DYNAMICS

The basic MHD equations at unit Prandtl number are written in the form:

$$\begin{aligned} \partial_t \mathbf{u} &= \nabla \cdot (\mathbf{bb} - \mathbf{uu}) - \nabla p + \nu \nabla^2 \mathbf{u} + \mathbf{f} \\ \partial_t \mathbf{b} &= \nabla \times (\mathbf{u} \times \mathbf{b}) + \nu \nabla^2 \mathbf{b} \\ \nabla \cdot \mathbf{u} &= \nabla \cdot \mathbf{b} = 0, \end{aligned} \tag{1}$$

where $\mathbf{f}(\mathbf{x}, t)$ is a prescribed solenoidal body force. We define a Reynolds number $R = l^3 f/\nu^2$, based on the characteristic scale l, and strength f, of the body force. We shall be studying the case of small Reynolds number, $R \ll 1$. For certain body forces, the resulting "basic flow" is known to be susceptible to dynamo instabilities (see Moffatt 1978) and AKA instabilities (Frisch et al. 1987) which generally occur on large spatial scales, $O(R^{-2} l)$, and evolve over long time scales, $O(R^{-4} l^2/\nu)$, compared with those of the basic flow.

The multiple-scale expansion describing the non-linear saturation of the instabilities is explained in Gilbert and Sulem (1989). Here we just recall the equations governing the large scale velocity $\mathbf{U}$ and magnetic field $\mathbf{B}$, assumed to be strong compared to the amplitude of the basic flow. For this assumption to be valid requires that these fields be in MHD equilibrium, interacting only by their effects on the mean stress and emf. They then must satisfy

$$\begin{aligned} \partial_t \mathbf{U} &= \nabla \cdot \mathbf{S} - \nabla P + \nu \nabla^2 \mathbf{U} \\ \partial_t \mathbf{B} &= \nabla \times \boldsymbol{\mathcal{E}} + \nu \nabla^2 \mathbf{B} \\ \nabla \cdot \mathbf{U} &= \nabla \cdot \mathbf{B} = 0, \end{aligned} \tag{2}$$

together with the MHD equilibrium condition,

$$\nabla \cdot (\mathbf{BB} - \mathbf{UU}) - \nabla P' = 0 \quad , \quad \nabla \times (\mathbf{U} \times \mathbf{B}) = 0. \tag{3}$$

In eqs. (2) and (3), P and P' refer to different order contributions to the large-scale pressure. Furthermore the mean stress tensor $\mathbf{S}$ and the mean emf $\boldsymbol{\mathcal{E}}$ are non-linear functions of $\mathbf{U}$ and $\mathbf{B}$, which may be calculated by solving the equations for the small scale-fields, in the form:

$$S_{ij} = \int d^3\mathbf{k}\, d\omega\, \frac{(\mathbf{k} \cdot \mathbf{B})^2 - (\mathbf{k} \cdot \mathbf{U} - \omega)^2 - \nu^2 k^4}{|Q|^2} \hat{f}_i^*(\mathbf{k}, \omega) \hat{f}_j(\mathbf{k}, \omega), \tag{4}$$

$$\mathcal{E}_i = \int d^3\mathbf{k}\, d\omega\, \frac{-\nu k_i (\mathbf{k} \cdot \mathbf{B})}{|Q|^2} h^f(\mathbf{k}, \omega). \tag{5}$$

Here $\hat{\mathbf{f}}(\mathbf{k}, \omega)$ is the Fourier transform of the body force and we define $h^f(\mathbf{k}, \omega) = \hat{\mathbf{f}}^*(\mathbf{k}, \omega) \cdot i\mathbf{k} \times \hat{\mathbf{f}}(\mathbf{k}, \omega)$ to be the helicity in each Fourier mode of the body force. The quantity, Q, is defined by $Q = (\nu k^2 + i(\mathbf{k} \cdot \mathbf{U} - \omega))^2 + (\mathbf{k} \cdot \mathbf{B})^2$.

3. INVERSE CASCADE IN A HELICAL FLOW

A simple example of body force leading to an alpha-effect dynamo is (Roberts 1972)

$$\mathbf{f}(\mathbf{x},t) = (f\sin(k_0 y), f\cos(k_0 x), f[\sin(k_0 x) + \cos(k_0 y)]), \tag{6}$$

which may be viewed as the sum of two Beltrami waves. The linearly most unstable modes depend solely on the z coordinate. We shall examine only such one-dimensional fields, $\mathbf{U} = \mathbf{U}(z,t)$ and $\mathbf{B} = \mathbf{B}(z,t)$, with $U_3 = B_3 = 0$, for which conditions (3) of MHD equilibrium at large scales are trivially satisfied. Furthermore the equation for $\mathbf{U}$ reduces to a diffusion equation and we may take $\mathbf{U} \equiv 0$. We are thus left with the following equations for the magnetic field components:

$$\begin{aligned} \partial_T \bar{B}_1 &= -\alpha\partial_Z \left[\bar{B}_2/(\bar{B}_2^2+1)^2\right] + \partial_Z^2 \bar{B}_1 \\ \partial_T \bar{B}_2 &= \alpha\partial_Z \left[\bar{B}_1/(\bar{B}_1^2+1)^2\right] + \partial_Z^2 \bar{B}_2. \end{aligned} \tag{7}$$

Here we have made a convenient change of variables; the length L is a characteristic scale of the domain and we have defined, $Z = z/L, T = \nu t/L^2, \bar{\mathbf{B}} = \mathbf{B}/k_0\nu$. The parameter, $\alpha = R^2 L k_0$, controls the number of linearly unstable magnetic field modes, being the ratio of the size of the domain to the scale of neutrally stable magnetic field modes.

We present here an integration of eq.(7) in a 2π-periodic domain, for $\alpha = 20$, starting with a weak magnetic field (Fig.1). Magnetic modes with wave-numbers in the range $0 < |k| < 20$ are linearly unstable. At early times, the dynamics is linear and the magnetic field is dominated by the growth of the linearly most unstable modes with $|k| = 10$. At larger amplitudes, energy is transferred to successively larger scales in an *inverse cascade*, produced by merger and destruction of peaks in the magnetic field components. Fig.2 shows the evolution of the energy in modes with $|k| \leq 5$; we see that the energy spectrum becomes dominated by successively larger scales. Note how the field remains in a slowly evolving state with two peaks for a long period, before a sudden transition to a state with one peak, and negligible energy in modes with $|k| = 2$.

Eventually the field saturates in a stationary state; the modes with $|k| = 1$ dominate the spectrum. All the even modes have decayed to zero, as may be seen from the symmetry of the final state. This behaviour is consistent with the study of Krause and Meinel (1988), who examine the stability of steady solutions of non-linear alpha effect dynamo equations in a domain which is a plane layer. They find that only solutions of largest scale in the layer are stable and discuss their symmetry properties.

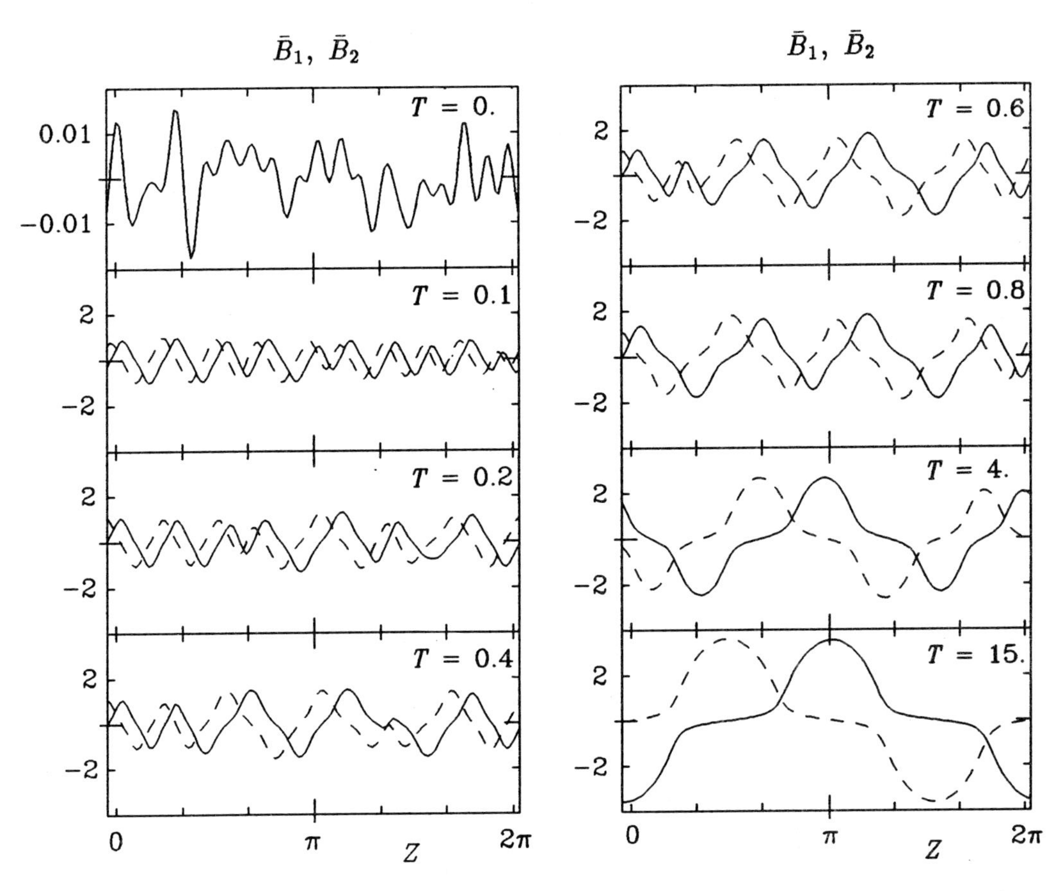

Fig.1 : Evolution of a magnetic field for the body force (6) and $\alpha = 20$; the solid line marks the component $\bar{B}_1$, the dashed line, $\bar{B}_2$. The initial conditions for this run have $\bar{B}_1 = \bar{B}_2$.

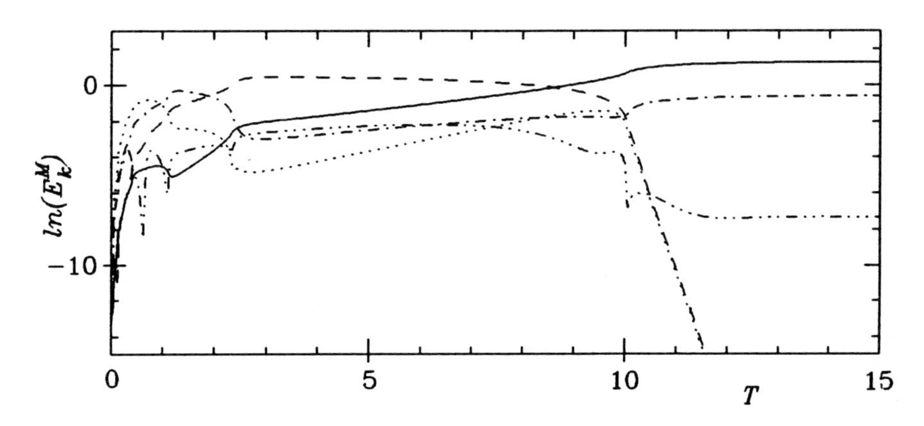

Fig.2 : Evolution of the magnetic energy modes E_k^M for $|k| = 1$ (—), $|k| = 2$ (- - -), $|k| = 3$ (-.-.-), $|k| = 4$ (.....) and $|k| = 5$ (-...-) for the body force (6), with $\alpha = 20$.

4. COUPLED DYNAMO AND AKA INSTABILITIES

A simple mechanism to generate a large-scale velocity field in a one-dimensional model consists of stirring the basic (small-scale) flow with a body force which in addition to a dynamo instability, gives a linear AKA instability (Frisch et al. 1987). A simple example is

$$\begin{aligned} f_1 &= f\cos(k_0 y + \omega_0 t) \quad ; \quad f_2 = f\cos(k_0 x - \omega_0 t); \\ f_3 &= c(f_1 + f_2) + f\{\sin(k_0 x - \omega_0 t) - \sin(k_0 y + \omega_0 t)\}. \end{aligned} \tag{8}$$

Here c is a free parameter governing the strength of the AKA instability. Using the same units as in Section 3 with, in addition, $\bar{\mathbf{U}} = \mathbf{U}/k_0\nu$, Eqs.(2) become

$$\begin{aligned} \partial_T \bar{B}_1 &= -4\alpha\partial_Z \frac{\bar{B}_2}{(\bar{B}_2^2 - (\bar{U}_2+1)^2+1)^2 + 4(\bar{U}_2+1)^2} + \partial_Z^2 \bar{B}_1, \\ \partial_T \bar{B}_2 &= 4\alpha\partial_Z \frac{\bar{B}_1}{(\bar{B}_1^2 - (\bar{U}_1-1)^2+1)^2 + 4(\bar{U}_1-1)^2} + \partial_Z^2 \bar{B}_2, \\ \partial_T \bar{U}_1 &= 4c\alpha\partial_Z \frac{\bar{B}_2^2 - (\bar{U}_2+1)^2 - 1}{(\bar{B}_2^2 - (\bar{U}_2+1)^2+1)^2 + 4(\bar{U}_2+1)^2} + \partial_Z^2 \bar{U}_1, \\ \partial_T \bar{U}_2 &= 4c\alpha\partial_Z \frac{\bar{B}_1^2 - (\bar{U}_1-1)^2 - 1}{(\bar{B}_1^2 - (\bar{U}_1-1)^2+1)^2 + 4(\bar{U}_1-1)^2} + \partial_Z^2 \bar{U}_2. \end{aligned} \tag{9}$$

The parameters α and $c\alpha$ control the number of linearly unstable magnetic and kinetic modes respectively. Various regimes and transitions are observed according to their values. We discuss here only a few typical ones, for $c = 1$, a detailed investigation being presented elsewhere (Galanti et al. 1989).

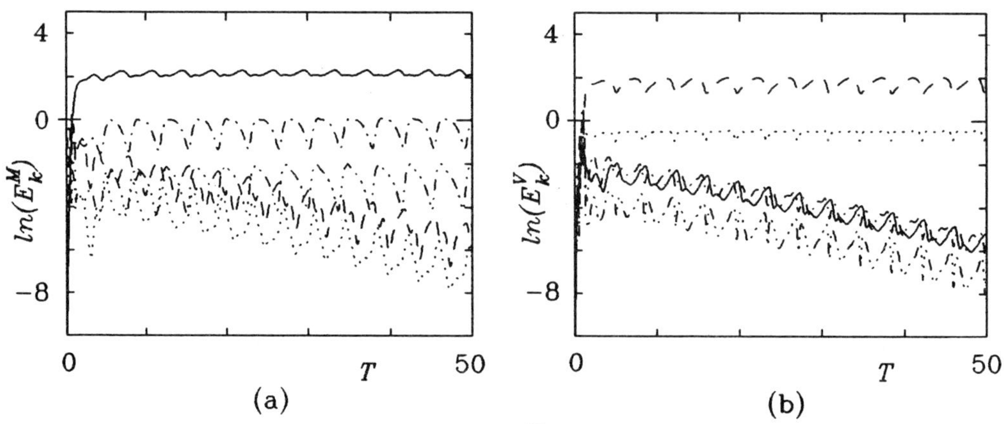

Fig.3 : Evolution of the magnetic energy modes E_k^M (a) and velocity energy modes E_k^V (b) for the body force (8) with $\alpha = 14$, $c = 1$. Same line styles as in Fig.2.

When α crosses a critical value located between 8 and 10, the long time behaviour of the solution bifurcates from a steady to a time-periodic state. The development of

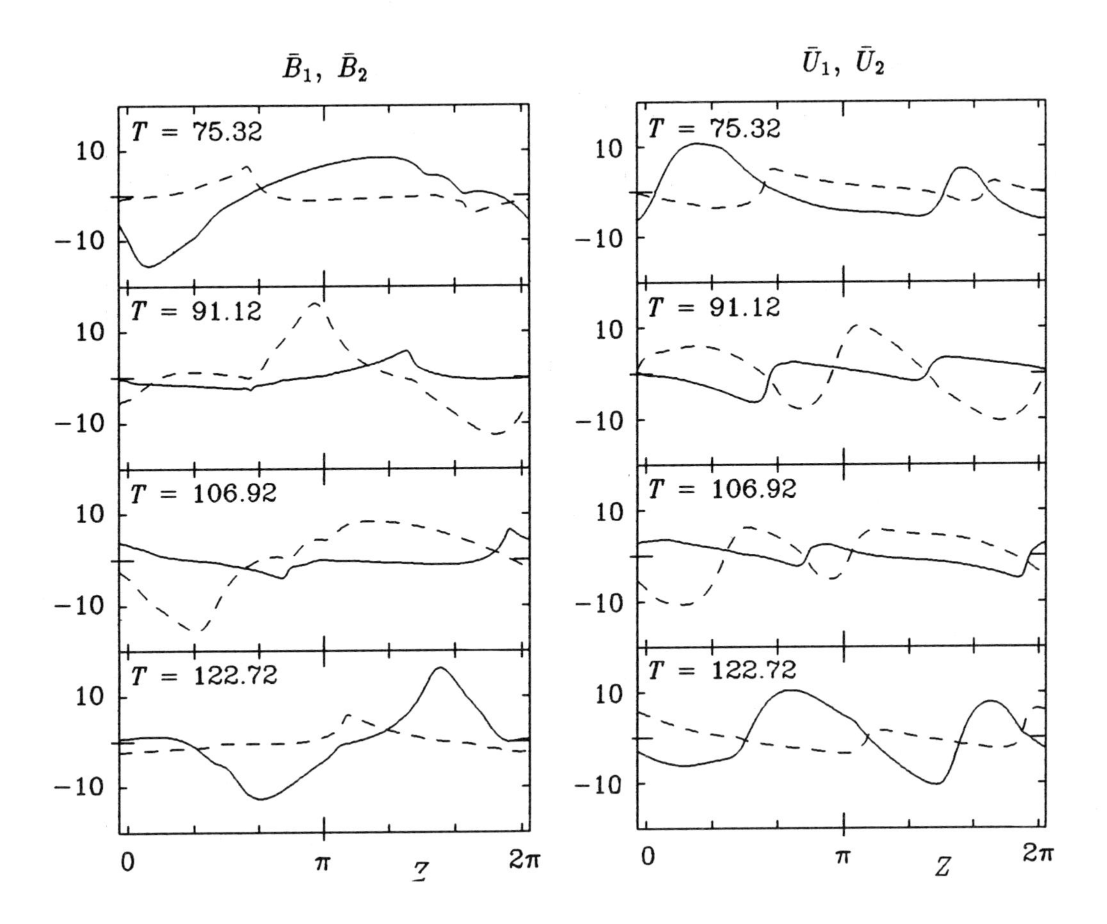

Fig.4 : Profiles of the magnetic and velocity fields in a period for the body force (8) with $\alpha = 30.5$, $c = 1$.

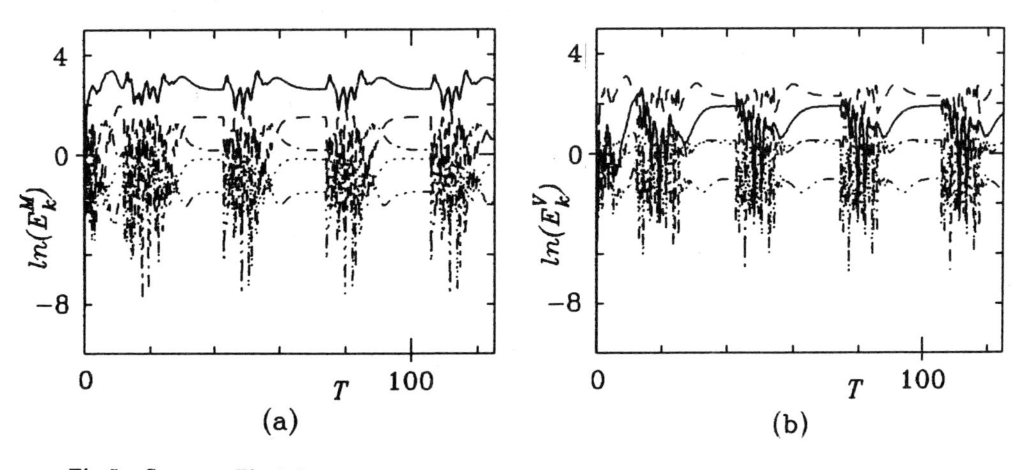

Fig.5 : Same as Fig.3 for $\alpha = 30.5$.

this periodic regime is illustrated in Fig.3 which displays the energy in the leading kinetic and magnetic modes of the solution for $\alpha = 14$. After a very short transient during which the linearly most unstable modes ($k = 7$) are dominant, an inverse cascade develops. Ultimately, the velocity and the magnetic field stabilize in a periodic state. Note that the magnetic field displays the same spatial symmetry as in the example studied in Section 3, while the velocity is π-periodic in space. The next bifurcation is anticipated by the following phenomenon: for increasing α's, the convergence to the asymptotic symmetric state becomes slower. As seen in Fig.3, the energy of the asymptotically irrelevant modes decays exponentially in time. The decay rates of these modes decrease when α increases. When they vanish at $\alpha \approx 14.6$, the long-time solution bifurcates to a non-symmetric state. Near the threshold, the solution is steady but it becomes time-periodic at a critical value of α located between 29.25 and 29.50. This second periodic regime is illustrated in Fig.4 which corresponds to $\alpha = 30.5$. We observe that the Fourier modes display sudden transitions between time intervals of very quiet evolution and bursts of violent oscillations. The solution in the physical space is shown in Fig.5 at various times during a period. For higher values of α, the solution is no longer periodic. Figs.6 show the leading Fourier modes for $\alpha = 32$. The solution looks chaotic: however, intervals of quiet evolution still survive. A precise analysis would require significantly longer integrations.

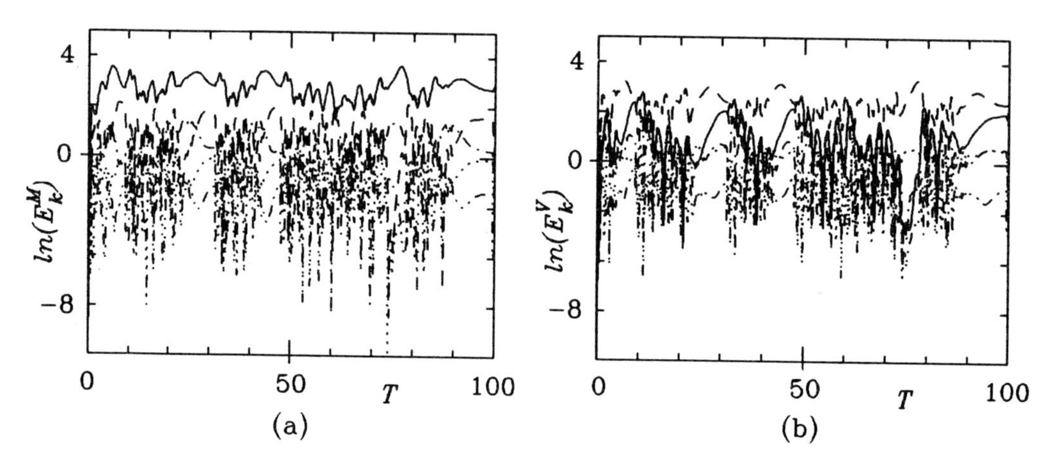

Fig.6 : Same as Fig.3 for $\alpha = 32$.

The above solutions obtained by starting with seed velocity and magnetic fields appear to be globally stable. If the magnetic field is initially zero, a large scale velocity field is subject only to the AKA instability and converges to a steady state with a 2π-period in space. If a seed magnetic field is then introduced in the system, this velocity profile is destabilized and the system evolves to the same final state as that reached from infinitesimal initial velocity and magnetic fields. In particular, the velocity may become π-periodic in space.

5. CONCLUDING REMARKS

We have reported on the non-linear saturation of large-scale instabilities which develop in alpha-effect dynamos, both with and without the presence of an AKA effect. We restricted ourselves to the case of one-dimensional large-scale fields which interact only through their back-reaction on the small-scale fields and the resulting non-linear modification of the alpha and AKA transport effects. As a consequence the large-scale fields saturate at a level significantly larger than the amplitude of the basic flow. If there is an alpha effect, but no AKA effect and no large-scale velocity field, we observe an inverse cascade of magnetic field to the largest scales of the system and saturation in a steady state. When both AKA and alpha effects are present, a large-scale velocity field is excited and coupled to the magnetic field. We again see an inverse cacade, but now the ultimate state of the system may be steady, time-periodic or chaotic.

Now suppose that the large-scale fields are allowed to be three-dimensional, that is, depend on all spatial coordinates. Saturation may occur either by the reduction of alpha or AKA effects seen above, or by direct coupling, when the large-scale fields fail to be in MHD equilibrium. In the latter case the fields would saturate at a significantly lower level. The question of the saturation mechanism for dynamo instabilities in three dimensions is currently under investigation. Results were obtained recently in the case of a pure AKA instability in three dimensions (Galanti and Sulem 1989). The inverse cascade is still observed, however the saturation level of the instability suggests that the dynamics of the large-scale flow are dominated by direct couplings.

REFERENCES

Frisch, U., She, Z.-S. & Sulem, P.-L. 1987 *Physica*, **28D**, 382-392.

Galanti, B., Sulem P.L. & Gilbert, A.D. 1989, in preparation.

Galanti, B. & Sulem P.L. 1989, in preparation.

Gilbert, A.D. & Sulem P.L. 1989 "On inverse cascades in alpha effect dynamos," *Geophys. Astro. Fluid Dyn.*, in press.

Kraichnan, R.H. 1979*Phys. Rev. Lett.*, **42**, 1677-1680.

Krause, F. & Meinel, R. 1988 *Geophys. Astrophys. Fluid Dyn.*, **43**, 95-117.

Krause, F. & Rädler, K.-H. 1980 "Mean-field magnetohydrodynamics and dynamo theory," Pergamon Press.

Meneguzzi, M., Frisch, U. & Pouquet, A. 1981 *Phys. Rev. Lett.*, **47**, 1060-1064.

Moffatt, H.K. 1978 "Magnetic field generation in electrically conducting fluids," Cambridge University Press.

Parker, E.N. 1955 *Atrophys. J.*, **122**, 293-314.

Pouquet, A., Meneguzzi, M. & Frisch, U. 1986 *Phys. Rev.*, **33A**, 4266-4276.

Roberts, G.O. 1972 *Phil. Trans. Roy. Soc.*, **A271**, 411-454.

Steenbeck, M., Krause, F. & Rädler, K.-H. 1966 *Z. Naturforsch*, **21a**, 369-376.

Sulem, P.-L., She, Z.-S., Scholl, H. & Frisch, U. 1989 *J. Fluid Mech.*, **205**, 341-358.

Zeldovich, Ya.B., Ruzmaikin, A.A. & Sokoloff, D.D. 1983 "Magnetic fields in astrophysics," Gordon and Breach.

The Helical MHD Dynamo

A. GAILITIS

Institute of Physics, Latvian SSR Academy of Sciences,
229021 Riga, Salaspils, USSR

ABSTRACT

The helical dynamo is the simplest of the numerous types of MHD dynamos. Due to its symmetry, the model can be semi-analytically explored. As there is no need for the common assumption $R_m \gg 1$, simple generating configurations can be analysed. The critical R_m appears to be the lowest one known. The main theoretical problem derives from the $z, -z$ asymmetry. As a result, the neutral curve for an infinite model does not correspond to the limit of a real self-excitation margin for a long but finite model. A one-way helical stream requires some entrance AC excitation all the time (convective type of generation). We focus our attention on a helical stream surrounded by a counter-flow so that instead of convective generation we have genuine self-excitation. A special technique is developed to search for self-excitation. The generated field pattern is presented.

The counter-flow scheme was used in a laboratory liquid sodium experiment. At subcritical conditions the model response to external excitation agrees with our theory. The critical flow rate was not reached because of mechanical problems. We hope to achieve self-excitation after mechanical vibration is eliminated.

1. FORMULATION

Many motions of electrically uniform fluid (σ = const) can generate a magnetic field. The simplest dynamo model is an endless helical stream maintaining full electrical contact with its immobile surroundings. In the original Ponomarenko (1973) model the stream moves uniformly as a solid cylinder ($v_r = 0$, $v_\phi = \omega r$, $v_z =$ const, Figure 1).

The generated field depends on t, ϕ and z exponentially:

$$B_r \pm iB_\phi = B_\pm(r) \cdot \exp\,(pt + ikz + im\phi) \tag{1}$$

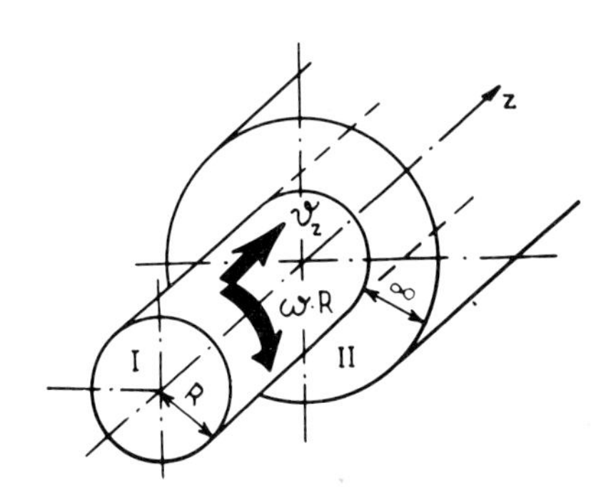

FIGURE 1. The Ponomarenko dynamo

Below we consider one, two, and more coaxial cylinders moving one within another with different constants ω and v_z. In each cylinder, the $B_\pm(r)$, as a consequence of Maxwell equations, are Bessel functions I_m and K_m of complex argument.

Continuity of magnetic and electrical fields and the condition $B_\pm(\infty) = 0$ lead to a secular equation of the form

$$F(R_m, k_r + ik_i, p_r + p_i, \text{ geom. param.}) = 0 \tag{2}$$

The magnetic Reynolds number $R_m = \sigma\mu_0 v_{max} R$ is determined by the maximum velocity v_{max} within the inner cylinder and its radius R (below $R = 1$ and $\sigma\mu_0 = 1$).

2. THE PONOMARENKO MODEL

With one moving cylinder (Figure 1) the equation (2) is rather simple:

$$\omega(1/R_+ - 1/R_-) = 2i \tag{3}$$

where

$$R_\pm = qI_m(q)/I_{m\pm1}(q) + sK_m(s)/K_{m\pm1}(s)$$
$$s = (k^2 + p)^{\frac{1}{2}}, \qquad q = (k^2 + p + i(m\omega + kv_z))^{\frac{1}{2}}$$

The complex equations (2) or (3) are exact but transcendental (Bessel functions) and without any exact solution. For $R_m \gg 1$, the Bessel functions can be replaced by their asymptotic expressions. In this limit, (3) was solved by Ponomarenko (1973). An equivalent WKB approach for the case of radially dependant ω and v_z was developed by Ruzmaikin & Sokolov (1988), Gilbert (1988), and others.

With experiments in mind, we consider the generation threshold where the condition $R_m \gg 1$ is not valid. To obtain a relation between two of the five arguments named in (2), keeping the others fixed, we use numerical iteration (Gailitis & Freiberg 1976). In Figure 2, the disconnected neutral curve ($k_i = p_r = 0$) looks like the neutral mechanical stability curve for a laminar boundary layer with an inflected profile. Inside, $\gamma = p_r > 0$, the field is growing while outside it is decaying. An AC (rotating) field is generated, whose frequency $\Omega = p_i$ (dashed lines) varies along the curve from high on the upper branch to low on the lower one. At $R_m \gg 1$, on the upper branch the field is close to being frozen within the cylinder, while on the lower branch it is in immobile surroundings. The leftmost point on the neutral curve is the critical R_m^*. Figure 2 is for the main mode $m = 1$. Confirming Cowling's theorem, no axisymmetric solution ($m = 0$) exists. The higher modes ($m = 2, 3, \ldots$) occur for much higher R_m.

2.1 Convective Nature of the Ponomarenko Model

Inside the neutral curve (Figure 2) field perturbations grow. But there is no self-excitation. A growing field moves axially with a non-zero group velocity $v_{gr} = -dp_i/dk_r$. Hence it exists within the finite model for a limited time only. The temporal growth ends with the last primordial perturbation leaving the model. Any further field must be supported by some persistent AC perturbation at the entrance. For a suitable frequency the model acts as an amplification line with the exit field much higher than the entrance one. This convective type of fluid instability is common in boundary layers, channels and many other cases when the z-direction is not equivalent to $-z$. When amplification is high, the convective instability gives the same result as real self-excitation. The laboratory dynamo cannot be very long and so the Ponomarenko model must be modified.

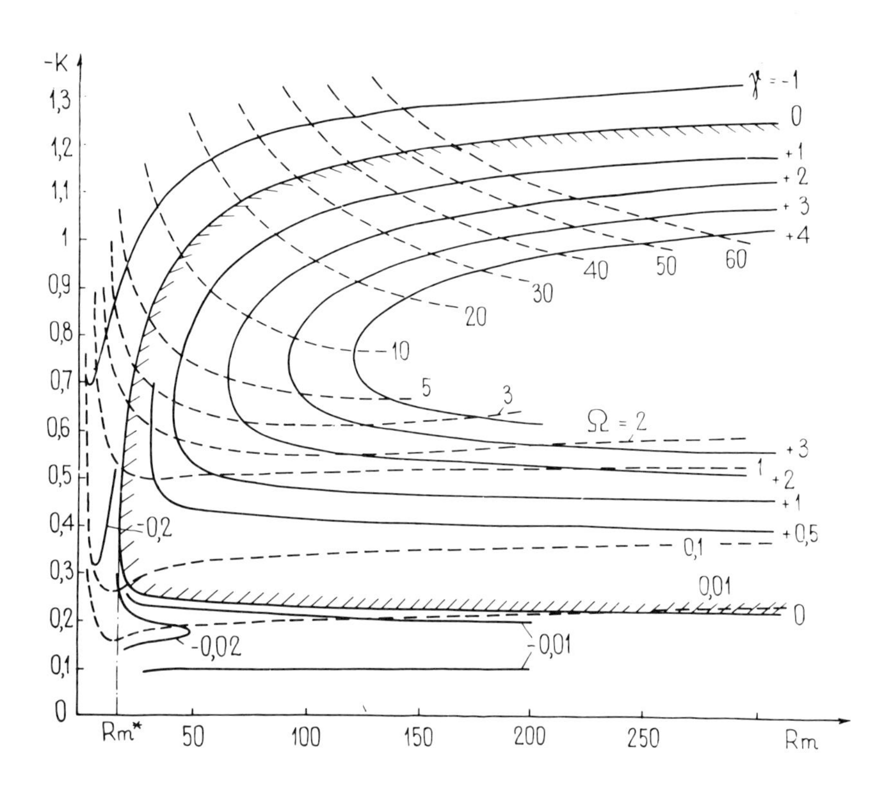

FIGURE 2. Ponomarenko dynamo. Neutral curve (hatched), growth rate (solid) and frequency (dashed) lines. $v_z/\omega = 1.3$.

2.2 Ponomarenko Model Response to External Excitation

The above critical $R_m = 17.7$ is low enough to be reached in the laboratory. Nevertheless, high hydraulic power is needed to move the fluid conductor. Hence the first stage must be a subcritical experiment to predict an experimental value for R_m^* using the magnetic field response to an external AC excitation (Figure 3). At the correct AC frequency, the $1/B$ curves should reach the R_m axis at R_m^*. At some detuning these curves are displaced. Measurement in the centre of the model can indicate R_m^* much more clearly than can measurements outside.

3. THE COUNTERFLOW MODEL

A helical stream surrounded by an axial flow reversing the whole central flow rate can self-excite. There are two critical values of R_m (Figure 4). The first critical $R_m^*(= 14.28)$ remains as before the leftmost point of the neutral curve

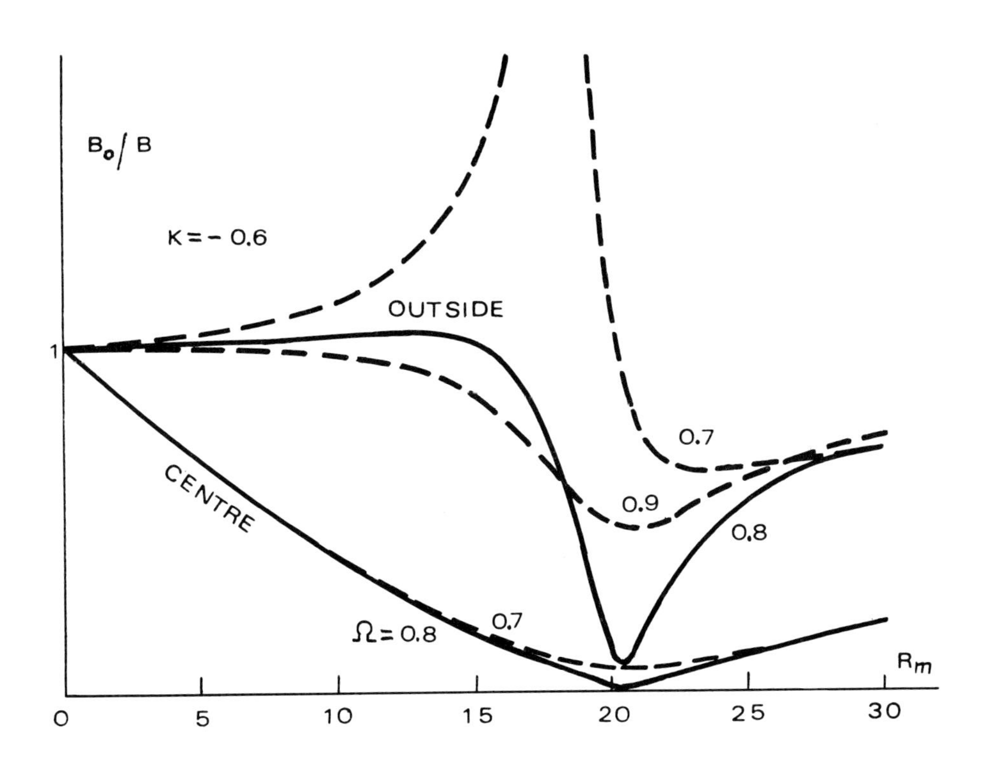

FIGURE 3. Ponomarenko model under external excitation. Computer simulated field measurements: $r = 0$ and $r = 3.01$. $0 \leq r < 1$: helical stream with $\omega = v_z$; $1 < r < 3$: immobile conductor; $r = 3$: helical coil with $j =$ exp $(i\Omega t + ikz + i\phi)$; $r > 3$: insulator.

$k_i = p_r = 0$. The new critical $R_m^{**}(= 15.74)$ is the leftmost point of an envelope of the family of curves $k_i = const, p_r = 0$. Between the two critical values of R_m, generation is convective. At the leftmost point of the envelope $dR_m/dk = -(\partial F/\partial k)/(\partial F/\partial R_m) = 0$. Because of (2) there is a zero of the complex group velocity too: $dp/dk = -(\partial F/\partial k)/(\partial F/\partial p) = 0$. A long model starts to generate without any entrance support. The Ponomarenko model has no envelope and hence no absolute generation.

Finite models start to generate at some higher R_m. For any such R_m, a curve $k_i = const$ has equal frequencies on the lower and the upper branches as marked by crosses in Figure 4. Together, the two solutions form a deformed standing wave (Figure 5). The knot distance $L = 2\pi/(k_1 - k_2)$ is an estimate for a model generating length (Figure 6).

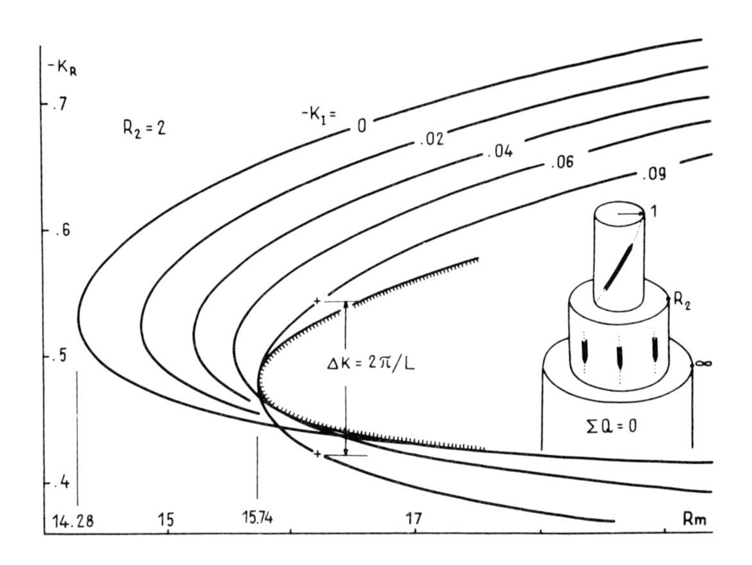

FIGURE 4. Envelope indicates an absolute generation. $\omega = v_z$.

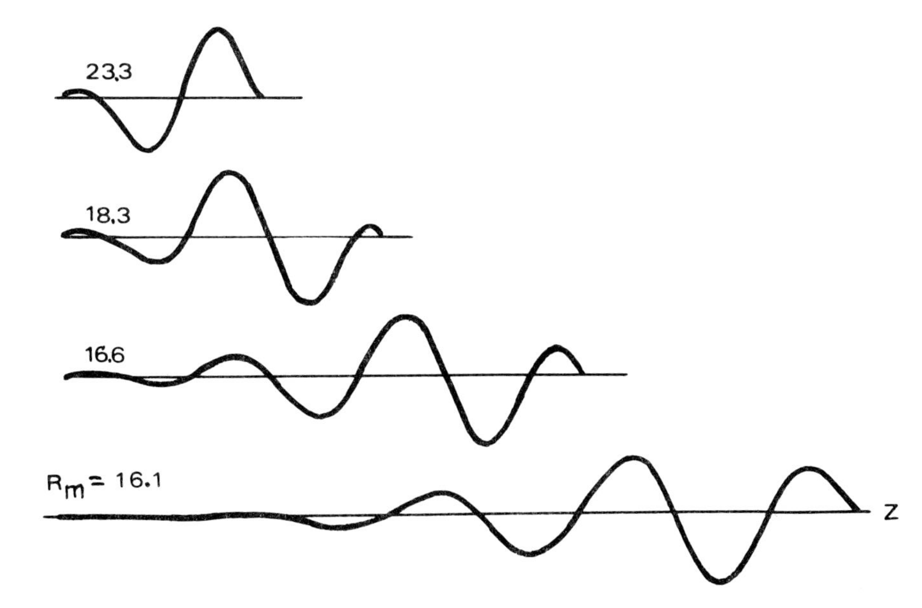

FIGURE 5. Axial profile of the generated field

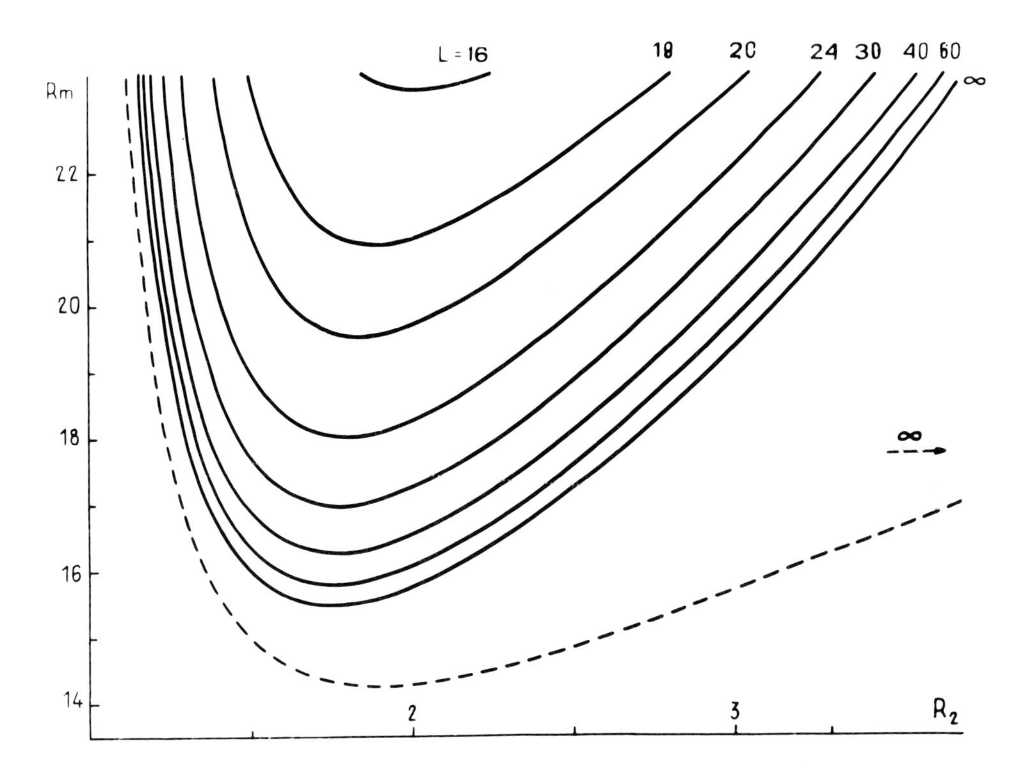

FIGURE 6. Generation R_m depends on model length L and outer radius of reversed flow, R_2. The dashed line is the convective generation margin

4. HELICAL MHD DYNAMO MODEL EXPERIMENT

We are using the counter-flow scheme for a laboratory MHD experiment (Gailitis et al 1987). The model is all welded from stainless steel and is filled, except for an open measuring channel 1, with liquid sodium (Figure 7). Sodium passes through the model from entrance 2 to exit 10. In the helical labyrinth 3, the sodium stream is accelerated and made helical: axial motion and rotation round the axis. When passing through the main channel 4, the sodium rotation is maintained by inertia only. Here, both the calculation and air test indicate a 20 - 25% loss of angular momentum (Figure 8). Rotation ends in the twelve-wall labyrinth 5 of the reverse system; hence the flow is axial in the counterflow channel 6.

The counter flow channel is separated from the main channel and from an immobile sodium volume 7 by two thin (1mm) coaxial cylinders, providing good electrical contact in sodium. The electrodynamics of the field generation is calculated for the real electrical resistance of the inner cylinder walls as well as for

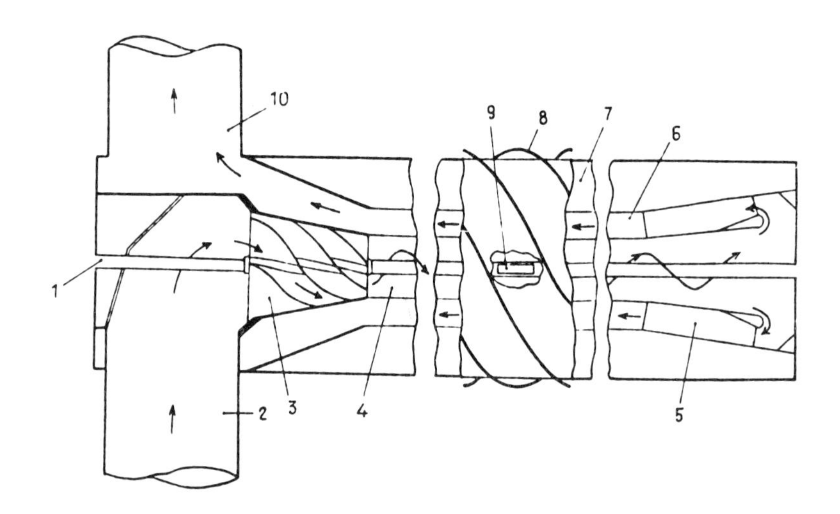

FIGURE 7. Experimental model. Length 3m, diameter 0.5m, empty 400 kg.

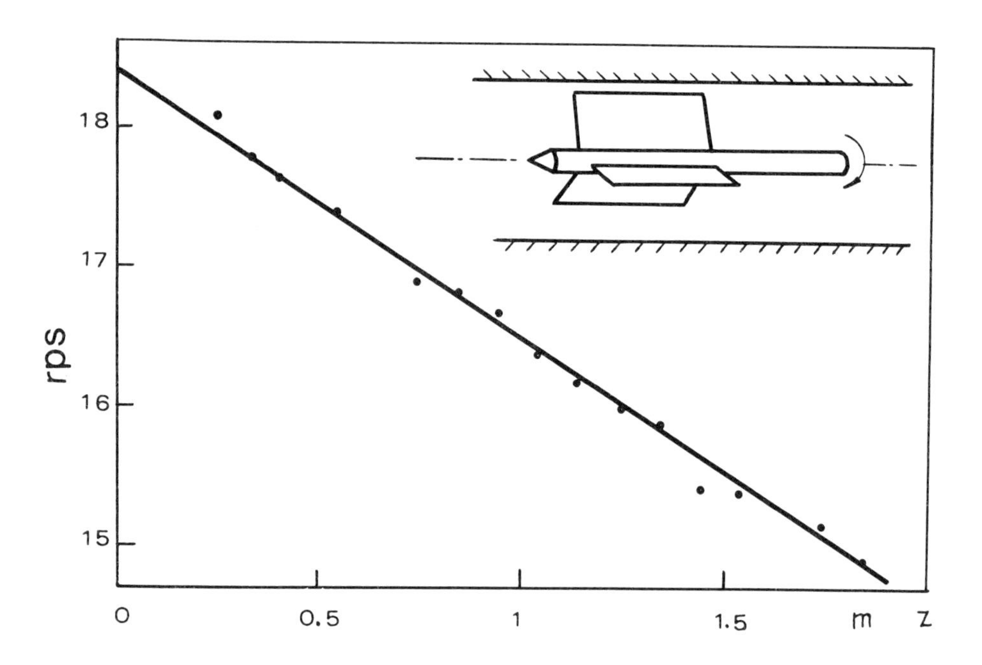

FIGURE 8. Air test. Rotation of a tiny mill inserted in the main channel 4. Measuring channel 1 withdrawn.

the insulator outside the model and inside the measuring channel. All the dimensions are set at optimum to have the calculated critical *Rm* as low as possible.

To start the experiment at subcritical flow rates, as proposed in chapter 2, the whole model was located in a helical three-phase excitation winding 8 linked with a low-frequency (0 - 10 Hz) generator. In the channel 1, we placed a coil 9 sensitive to a transverse magnetic field and connected it to a voltmeter. To find the maximum signal U, we tuned the three-phase generator frequency. The inverse signal $1/U$ versus flow-rate for three frequencies is plotted in Figure 9. The critical flow-rate, as in Figure 3, is actually the furthest left point where th $1/U$ curves cross the coordinate line. In Figure 9, the crossing point is not reachec as it seems to be even lower than the theoretical value (*).

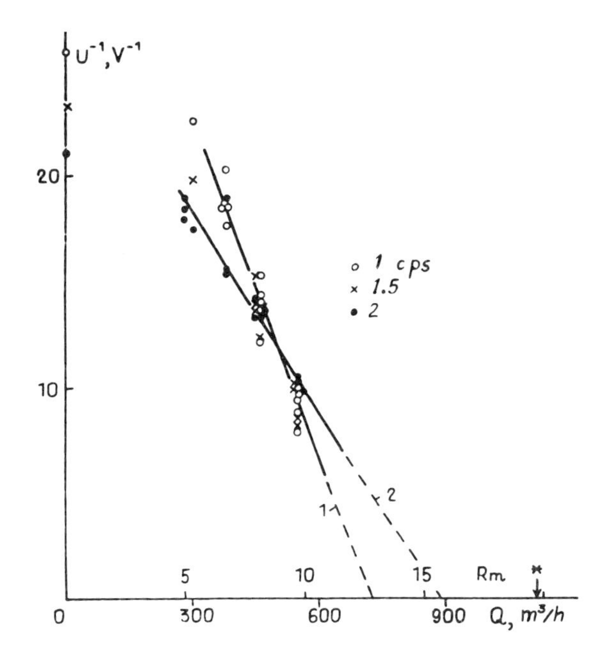

FIGURE 9. Model's response to external excitation

The experiment was carried out at 200°C; the flow-rate of the pump used can be up to 1200 m^3/h. The experiment was interrupted at 660 m^3/h before reaching self-excitation when our model was damaged by unexpected mechanical vibration.

At subcritical conditions model response to external excitation agrees with theory. We hope to reach self-excitation after we can eliminate the mechanical vibration.

REFERENCES

GAILITIS A. & FREIBERG J., (1976), On the theory of helical MHD dynamo. Magn. Gidrodin., No. **2**, 3 [Magnetohydrodynam. Vol. 12, 127].

GAILITIS A.K., KARASEV B.G., KIRILLOV I.A., LIELAUSIS O.A., LUZHANSKII S.M., OGORODNIKOV A.P., PRESLITSKII G.V. (1987) Liquid metal MHD dynamo model experiment, *Magn. Gidrodin* No. **4**, 3.

GILBERT A. (1988) Fast dynamo action in the Ponomarenko dynamo. *Geophys. Astrophys. Fluid Dyn.*, **44**, 241-258.

PONOMARENKO Y.B. (1973) On the theory of hydromagnetic dynamo. *Zh. Prikl. Mech. Tech. Fiz (USSR)*, **6**, 47 -51.

RUZMAIKIN A., SOKOLOFF D., SHUKUROV A., (1988). Hydromagnetic screw dynamo. *J. Fluid Mech.* v. **197**, 39 - 56.

Relations between Helicities in Mean-field Dynamo Models

K.-H. RÄDLER & N. SEEHAFER

Zentralinstitut für Astrophysik Potsdam der AdW der DDR, Potsdam, 1560, GDR.

ABSTRACT

Within the framework of mean-field dynamo theory some results concerning the current helicities of fluctuating and mean magnetic fields are derived. Using the second order correlation approximation a relation between the current helicity of the fluctuating magnetic field and parameters determining the α-effect is established. On this basis it is shown that the energy stored in the mean magnetic field of an α^2-dynamo is prevented from decaying only if, at least in some region, the current helicities of the fluctuating and the mean magnetic fields have opposite signs and the modulus of the former exceeds that of the latter. Results of an analysis of magnetic field configurations in solar active regions are also presented and discussed with reference to current helicity and the α-effect.

1. INTRODUCTION

In investigations of dynamo processes which are believed to be responsible for the magnetic fields of cosmical bodies, the mean-field approach has proved useful (see, e.g., Krause & Rädler, 1980). In this approach the magnetic flux density, $\mathbf{B}$, in an electrically conducting fluid as well as the velocity, $\mathbf{u}$, of the fluid motions are understood as superpositions of mean parts, $<\mathbf{B}>$ and $<\mathbf{u}>$, which are defined by a proper averaging procedure, and fluctuating parts, $\mathbf{B}'$ and $\mathbf{u}'$. The mean magnetic flux density $<\mathbf{B}>$ inside the fluid is governed by the equation

$$\eta \triangle <\mathbf{B}> + \text{curl } (<\mathbf{u}> \times <\mathbf{B}> + \boldsymbol{\epsilon}) - \partial <\mathbf{B}> / \partial t = 0. \tag{1}$$

Here the magnetic diffusivity η of the fluid is assumed to be constant. $\boldsymbol{\epsilon}$ is the mean electromotive force caused by fluctuations,

$$\boldsymbol{\epsilon} = <\mathbf{u}' \times \mathbf{B}'>; \tag{2}$$

angular brackets always indicate averages. For a wide range of reasonable assumptions the quantity ϵ, when represented in Cartesian coordinates, has the form

$$\epsilon_i = a_{ij} < \mathbf{B} >_j + b_{ijk}\, \partial < \mathbf{B} >_j / \partial x_k. \tag{3}$$

The tensorial coefficients a_{ij} and b_{ijk} depend on $< \mathbf{u} >$ and $\mathbf{u}'$. Within the framework of kinematic theory, to which we restrict ourselves in this paper, they are independent of $< \mathbf{B} >$. The first term on the right-hand side describes the α-effect, the second term various effects like that of the turbulent magnetic diffusivity. In the simple case in which $< \mathbf{u} >$ is equal to zero and $\mathbf{u}'$ corresponds to isotropic turbulence, relation (3) reduces to

$$\epsilon = \alpha < \mathbf{B} > - \beta \operatorname{curl} < \mathbf{B} >, \tag{4}$$

with scalar coefficients α and β which are determined by $\mathbf{u}'$. The first term on the right-hand side then describes the ideal, that is isotropic, α-effect and the second term an effect which is completely covered by introducing a turbulent magnetic diffusivity.

In dynamo processes the helicities of both the motion and the magnetic field are of interest. In the following we pay particular attention to the *current helicity*, $\mathbf{B} \cdot \operatorname{curl} \mathbf{B}$, the mean value of which, $< \mathbf{B} \cdot \operatorname{curl} \mathbf{B} >$, can be represented as the sum

$$< \mathbf{B} \cdot \operatorname{curl} \mathbf{B} > = < \mathbf{B} > \cdot \operatorname{curl} < \mathbf{B} > + < \mathbf{B}' \cdot \operatorname{curl} \mathbf{B}' > \tag{5}$$

of two contributions resulting from the mean and fluctuating magnetic field. In this paper we show that there is, at least in some approximation, a simple connection between the current helicity $< \mathbf{B}' \cdot \operatorname{curl} \mathbf{B}' >$ of the fluctuating magnetic field and the α-effect coefficients a_{ij}, or α, which are in turn related to the kinematic helicity of the fluctuating motions. Using this result we study the current helicities $< \mathbf{B}' \cdot \operatorname{curl} \mathbf{B}' >$ and $< \mathbf{B} > \cdot \operatorname{curl} < \mathbf{B} >$ of the fluctuating and mean magnetic fields in α^2-dynamos. It is shown that the energy of the mean magnetic field can only be maintained or grow if there is a region in which the signs of $< \mathbf{B}' \cdot \operatorname{curl} \mathbf{B}' >$ and $< \mathbf{B} > \cdot \operatorname{curl} < \mathbf{B} >$ are different and, in addition, $| < \mathbf{B}' \cdot \operatorname{curl} \mathbf{B}' > |$ exceeds $| < \mathbf{B} > \cdot \operatorname{curl} < \mathbf{B} > |$. Furthermore, we present some results on the current helicity in active regions on the Sun which have been derived from observational data, and discuss these in the light of our findings concerning current helicity and the α-effect.

2. CURRENT HELICITY OF THE FLUCTUATING MAGNETIC FIELD AND α-EFFECT

In order to derive a relation between the current helicity $< \mathbf{B}' \cdot \text{curl}\, \mathbf{B}' >$ of the fluctuating magnetic field and the α-effect coefficients a_{ij}, or α, we restrict ourselves to the case in which $< \mathbf{u} >$ is zero and $\mathbf{u}'$ so small that the second order correlation approximation applies. Then we have

$$\eta \triangle \mathbf{B}' - \partial \mathbf{B}'/\partial t = - \text{ curl } (\mathbf{u}' \times < \mathbf{B} >). \tag{6}$$

For the determination of $\boldsymbol{\epsilon}$ and also of $< \mathbf{B}' \cdot \text{curl}\, \mathbf{B}' >$ it is useful to subject $\mathbf{B}'$ and $\mathbf{u}'$ to a Fourier transformation of the form

$$F(\mathbf{x}, t) = \int \int \hat{F}(\mathbf{k}, \omega) e^{i(\mathbf{k}\cdot\mathbf{x} - \omega t)} d^3\mathbf{k}\, d\omega. \tag{7}$$

Then (6), with $< \mathbf{B} >$ taken as constant, reduces to

$$\hat{\mathbf{B}}_i' = i\, \epsilon_{ijk}\, \epsilon_{klm} \frac{k_j \hat{u}_l' < \mathbf{B} >_m}{\eta k^2 - i\omega}. \tag{8}$$

We further assume that $\mathbf{u}'$ describes a homogeneous and steady field of turbulence. This implies that

$$< \hat{u}_i'(\mathbf{k}, \omega) \hat{u}_j'(\mathbf{k}', \omega') > = \hat{Q}_{ij}(\mathbf{k}', \omega')\ \delta(\mathbf{k} + \mathbf{k}')\ \delta(\omega + \omega'), \tag{9}$$

where $\hat{Q}_{ij}(\mathbf{k}, \omega)$ is the Fourier transform of the correlation tensor $Q_{ij}(\boldsymbol{\xi}, \tau)$ defined by

$$Q_{ij}(\boldsymbol{\xi}, \tau) = < u_i'(\mathbf{x}, t)\ u_j'(\mathbf{x} + \boldsymbol{\xi},\ t + \tau) > \tag{10}$$

A straightforward calculation of $\boldsymbol{\epsilon}$ using (8) and (9) provides us with

$$a_{ij} < \mathbf{B} >_i < \mathbf{B} >_j = -\eta \int \int \frac{(\mathbf{G}(\mathbf{k}, \omega) \cdot < \mathbf{B} >)\, (\mathbf{k} \cdot < \mathbf{B} >)\, k^2}{(\eta k^2)^2 + \omega^2} d^3k\, d\omega. \tag{11}$$

In an analogous way we find

$$< \mathbf{B}' \cdot \text{ curl } \mathbf{B}' > = \int \int \frac{(\mathbf{G}(\mathbf{k}, \omega) \cdot < \mathbf{B} >)\, (\mathbf{k} \cdot < \mathbf{B} >)\, k^2}{(\eta k^2)^2 + \omega^2} d^3k\, d\omega. \tag{12}$$

In both cases G is given by

$$G_i(\mathbf{k},\omega) = -i\ \epsilon_{ijk}\hat{Q}_{jk}(k,w). \tag{13}$$

$\mathbf{G}(\mathbf{k},w)\cdot\mathbf{k}$ is just the Fourier transform of the kinematic helicity spectrum function $H(\boldsymbol{\xi},\tau)$ defined by

$$H(\boldsymbol{\xi},\tau) = < \mathbf{u}'(\mathbf{x},t)\cdot\ \text{curl}\ \mathbf{u}'(\mathbf{x}+\boldsymbol{\xi},\ t+\tau) > . \tag{14}$$

We note that (11) can also readily be derived from a relation mentioned by Krause and Rädler (1980, eq. 7.1), and (12) from a result by Bräuer and Krause (1972, eq. 14) or, for incompressible fluids, from a result by Rüdiger (1974, eq. 26).

Comparing (11) and (12) we see that

$$< \mathbf{B}'\cdot\ \text{curl}\ \mathbf{B}' > = -(1/\eta)a_{ij} < \mathbf{B} >_i < \mathbf{B} >_j . \tag{15}$$

For the special case of isotropic turbulence, for which $a_{ij} = \alpha\delta_{ij}$, we have simply

$$< \mathbf{B}'\cdot\ \text{curl}\ \mathbf{B}' > = -(\alpha/\eta) < \mathbf{B} >^2 . \tag{16}$$

Relations of this type have already been given by Keinigs (1983) and by Mattheaus et al (1986).

As is well-known, the sign of α is, as a rule, opposite to that of the kinematic helicity of the fluctuating motions. Hence, as expected, the sign of the current helicity of the fluctuating magnetic field coincides with that of the kinematic helicity of the motions responsible for them.

3. CURRENT HELICITIES IN AN α^2-DYNAMO

Let us now consider a fluid body surrounded by free space and assume that the mean magnetic field inside this body is governed by (1) and continues as an irrotational field in the external space. Then we have

$$\frac{d}{dt}\int_\infty \tfrac{1}{2} < \mathbf{B} >^2 dV = -\int \Big(\eta\ \text{curl}^2 < \mathbf{B} > + \\ + < \mathbf{u} > \cdot(\ \text{curl}\ < \mathbf{B} > \times < \mathbf{B} >) +\ \text{curl}\ < \mathbf{B} > \cdot \boldsymbol{\epsilon}\Big)\, dV, \tag{17}$$

where the integral on the left is over all space and that on the right over the fluid body. The energy stored in the mean magnetic field is maintained, or it grows, if the integral on the right is zero, or negative.

We restrict attention to α^2-dynamos and assume that ϵ is given by (4) with non-negative β. Thinking of a proper frame of reference we further put $<\mathbf{u}>=0$. We require that the energy of the mean magnetic field does not decay, that is,

$$\int \left(\alpha <\mathbf{B}> \cdot \text{ curl } <\mathbf{B}> - (\eta+\beta) \text{ curl}^2 <\mathbf{B}>\right) dV \geq 0. \tag{18}$$

This implies that there must be sufficiently extended regions of the fluid in which the signs of α and $<\mathbf{B}> \cdot \text{ curl } <\mathbf{B}>$ coincide. As long as (16) applies, in these regions the signs of the current helicities $<\mathbf{B}> \cdot \text{ curl } <\mathbf{B}>$ and $<\mathbf{B}' \cdot \text{ curl } \mathbf{B}'>$ of the mean and the fluctuating magnetic fields are different.

In order to deduce a further result concerning the current helicities we start again from (18) and introduce

$$<\mathbf{B}' \cdot \text{ curl } \mathbf{B}'> = -f <\mathbf{B}> \cdot \text{ curl } <\mathbf{B}> \tag{19}$$

with a factor f depending on the space coordinates. In this way we arrive at

$$\int f\left((<\mathbf{B}> \cdot \text{ curl } <\mathbf{B}>)^2 / <\mathbf{B}>^2\right) dV \geq \int (1+\beta/\eta)(\text{ curl } <\mathbf{B}>)^2 dV \tag{20}$$

According to a mean value theorem of integral calculus, f takes somewhere in the fluid volume a value f^* such that

$$\int f \frac{(<\mathbf{B}> \cdot \text{ curl } <\mathbf{B}>)^2}{<\mathbf{B}>^2} dV = f^* \int \frac{(<\mathbf{B}> \cdot \text{ curl } <\mathbf{B}>)^2}{<\mathbf{B}>^2} dV. \tag{21}$$

From (20) and (21) we have

$$f^* \geq \int (1+\beta/\eta)(\text{ curl } <\mathbf{B}>)^2 dV \Big/ \int \left((<\mathbf{B}> \cdot \text{ curl } <\mathbf{B}>)^2 / <\mathbf{B}>^2\right) dV \tag{22}$$

and therefore

$$f^* \geq 1. \tag{23}$$

Assuming continuity of all relevant quantities, we conclude that the energy of the mean magnetic field can only be prevented from decaying if there is some region in the fluid where the signs of $< \mathbf{B}' \cdot \text{curl}\, \mathbf{B}' >$ and $< \mathbf{B} > \cdot\, \text{curl} < \mathbf{B} >$ differ and, in addition, $| < \mathbf{B}' \cdot \text{curl}\, \mathbf{B}' > | > | < \mathbf{B} > \cdot\, \text{curl} < \mathbf{B} > |$.

4. CURRENT HELICITY IN SOLAR ACTIVE REGIONS

Let us now proceed to an application of our relations (15) and (16) connecting the current helicity of fluctuating magnetic fields and the α-effect to solar phenomena. The solar magnetic fields are attributed to an $\alpha\omega$-dynamo operating in the convection zone of the Sun, that is, below the visible surface. In the observable atmosphere, besides weak background fields, strong magnetic fields are found in active regions. These regions are believed to result from the emergence of magnetic flux ropes which have broken away from the predominantly toroidal field below the surface and carried up by magnetic buoyancy. Above the photosphere (the thin layer at the base of the atmosphere which represents the surface of the Sun in white light), the magnetic energy density in active regions dominates, except for explosive events (such as flares) over the thermal, kinetic and gravitational energy densities. Therefore a quasi-static equilibrium of the plasma may be assumed with a force-free magnetic field. Of course, there is some evolution of the configuration, which is induced by plasma motions in or below the photosphere. However, these motions are slow compared to the Alfven velocity in the superphotospheric plasma, that is, small compared to the velocity of the upward propagation of disturbances caused by them, so that each state may be considered as an equilibrium state (cf. Low, 1982). Since then the magnetic field above the photosphere should be force-free, we have there

$$\text{curl}\, \mathbf{B} = \alpha_{ff} \mathbf{B}, \tag{24}$$

for some pseudo-scalar α_{ff}. The sign of the current helicity $\mathbf{B} \cdot \text{curl}\, \mathbf{B}$ coincides with that of α_{ff}.

Seehafer (1989) has compiled 16 active regions for which the factor α_{ff} was estimated. Using observed photospheric magnetic fields as boundary data force-free fields with constant α_{ff} in the volume above the photosphere were calculated and α_{ff} was varied until an optimum coincidence of the calculated field line configrations with observed superphotospheric structures believed to be aligned with the field was obtained. Of the 16 regions, which belonged to the activity cycles 20 (beginning 1965) and 21 (beginning 1976), 12 lay in the northern and 4 in the southern hemisphere. For 11 of the 12 regions in the northern hemisphere a_{ff} was negative, for one positive. In the southern hemisphere, in 3 cases α_{ff} was positive, in one case a change of the sign of α_{ff} within the region was suggested. Thus we are led to conclude that in solar active regions the current helicity is predominantly negative in the northern and positive in the southern hemisphere (though further investigations are needed to firmly establish this result).

In the traditional mean-field concept of the solar dynamo the mean magnetic field does not reflect the magnetic fields of the individual active regions. Although these fields may contribute to the mean field, they are presumably mainly fluctuating fields. Their helicities then have to be interpreted in the sense of $< \mathbf{B}' \cdot \mathrm{curl}\, \mathbf{B}' >$. Using the relation (15) we thus conclude that $a_{ij} < \mathbf{B} >_i < \mathbf{B} >_j$ is predominantly positive in the northern and negative in the southern hemisphere. If the α-effect is assumed to be isotropic we may replace (15) by (16) and conclude that α is mainly positive in the northern and negative in the southern hemisphere. This corresponds to the usual picture used in many solar dynamo models (see, e.g., Krause and Rädler , 1980). Of course, the α-effect in the Sun deviates from isotropy. Taking this into account and assuming that the toroidal component of the mean magnetic field is large compared to the other components, we may conclude that $\alpha_{ij} < \mathbf{B} >_i < \mathbf{B} >_j$ is approximately equal to $\alpha_{\varphi\varphi} < \mathbf{B} >^2$ where $\alpha_{\varphi\varphi}$ is just that component of the α-tensor which is responsible for the regeneration of the poloidal from the toroidal field. Clearly, the sign rule formulated above for the scalar α then applies to $\alpha_{\varphi\varphi}$, too.

REFERENCES

BRÄUER, H. & KRAUSE, F. (1972), *Astron. Nachr.* **294**, 179.

KEINIGS R.K. (1983), *Phys. Fluids* **26**, 2558.

KRAUSE F. & RÄDLER K.-H. (1980), Mean-Field Magnetohydrodynamics and Dynamo Theory, Akademie Verlag, Berlin, and Pergamon Press, Oxford.

LOW B.C. (1982), *Rev. Geophys. Space Phys.*, **20**, 145.

MATTHAEUS W.H., GOLDSTEIN M.L. & LANTZ S.R. (1986), *Phys. Fluids* **29**, 1504.

RÜDIGER, G., (1974), *Astron. Nachr.* **295**, 275.

SEEHAFER N. (1989), *Solar Phys.*, submitted.

Boundary Effects on the MHD Dynamo in Laboratory Plasmas

Y.L. HO* & S.C. PRAGER

University of Wisconsin, Madison, Wisconsin USA

1 INTRODUCTION

One noteworthy recent result of laboratory plasma physics is the demonstration of an effect, related to dynamo processes, in which confining magnetic field is generated partially by fluctuations within the plasma.[1] The laboratory magnetic field generation problem is effectively treated in a cylindrical plasma using resistive magnetohdrodynamics. Consider, for the moment, a one-dimensional steady-state configuration in which the magnetic field depends upon radius only and consists of axial and azimuthal components B_z and B_θ. In such a 1D system, the axial field must be unidirectional. If B_z were to change direction with radius, then the radius at which B_z passes through zero would contain an azimuthal current density, $j_\theta = -dB_z/dr$, parallel to the magnetic field. From the parallel component of Ohm's law, $\mathbf{E} + \mathbf{v} \times \mathbf{B} = \eta \mathbf{j}$, we see that current would require a nonzero azimuthal electric field. Faraday's law then requires a time-varying axial magnetic flux to generate the electric field; hence, a steady state with reversed axial magnetic field is impossible. It is easy to show that the impossibility of a steady-state reversed field extends to a 2D magnetofluid, such as an axisymmetric torus.

The generation of the reversed axial magnetic field, i.e., the axially symmetric field beyond the radius at which B_z is zero, is often referred to as a dynamo-like problem. The axisymmetric field is considered to be generated by higher dimensional fields, or spatial fluctuations. In the sense that a spatial mean field (uniform along the z direction) is generated by three-dimensional magnetic and velocity fields, this problem is similar to the astrophysical dynamo. It differs in that only a portion of the magnetic field is self-generated and, of course, it is amenable to detailed comparison of theory with controlled experiments. The effect also is important to the fusion reactor concept known as the reversed field pinch.

Past workers have shown by MHD computation that local plasma current is generated by the mean component of the fluctuation-induced **v** x **B** electric field.(2-5) For a variety of initial conditions, the plasma arranges its topology, by magnetic reconnection, to establish a unique reversed field steady state. Analytical insight is provided by quasilinear theory(6) and variational approaches.(7,8) The majority of the past work, both computational and analytical, has included a perfectly conducting cylindrical wall at the plasma surface. Our purpose here is to examine the role of the boundary condition in this laboratory dynamo-like process. The boundary is expected to be a key ingredient in the process. Its variation will help elucidate the dynamics of the field generation. Recent experiments have switched from the highly conducting walls of the past several decades, to walls with short electrical penetration times.

We examine the boundary effect through 3D compressible resistive MHD computation.(9) The code is described in section 2. In section 3 we briefly review results with the perfectly conducting, close-fitting boundary. In section 4, we present computational results with various boundary conditions, including both resistive and distant walls. It is seen that the field generation and final state of the plasma is strongly dependent on the boundary. In section 5 we interpret the results through magnetic helicity conservation. In a driven system the magnetic helicity of the plasma is a balance between injection and dissipation. Variation of the boundary alters the dissipation and, thereby, the injection rate must be correspondingly altered to maintain a steady-state.

2 CODE DESCRIPTION

The initial value code solves the full, compressible MHD equations for a pressureless cylinder periodic in z.(10) The equations, in dimensionless form are

$$\partial \mathbf{A}/\partial t = S(\mathbf{v} \times \mathbf{B}) - \eta \mathbf{j},$$

$$\partial \mathbf{v}/\partial t = -S(\mathbf{v} \cdot \nabla \mathbf{v}) + S(\mathbf{j} \times \mathbf{B}) + \nu \nabla^2 \mathbf{v} \tag{1}$$

where S is the Lundquist number, and time, velocity, and length are normalized to the resistive diffusion time ($\tau_R=\mu_o a^2/\eta$), Alfven speed, and plasma radius, a. Viscosity (normalized to a^2/τ_R) and mass density are held constant in space and time. Resistivity is normalized to its central value and, to resemble experiments, is taken to vary as $(1 + 9r^{30})^2$. The Lundquist number is taken as 6×10^3. The algorithm is finite-differenced radially and pseudospectral in the other two dimensions. The semi-implicit algorithm is used to remove the severe time step restrictions imposed by Alfven waves, thus permitting longer time scale nonlinear phenomena to be tracked economically.

We employ 43 axial modes (axial mode numbers n = -21 to 21), 3 azimuthal modes (azimuthal mode number m=0 to 2), 127 radial points and set the aspect ratio $L/2\pi a = 2.5$. The intent of the calculation is not to investigate small-scale turbulence, but to determine the effect of nonlinear, relatively long wavelength fluctuations on the evolution of the mean, symmetric quantities.

The boundary conditions at r = a are as follows. At the plasma surface is a resistive shell of resistivity η_s and thickness Δ. The shell is modeled by a jump condition on the radial magnetic field,(11)

$$\partial B_r/\partial t = a/\tau_s\, [\partial B_r/\partial r] \qquad (2)$$

where [ ] denotes a jump across the shell (assumed to be thin) and $\tau_s = \Delta a \mu_o/\eta_s$ is the shell penetration time. The mean velocity (with m=n=0) is given by $v_\theta = v_z = 0$ and $\mathbf{v}_r = \mathbf{E} \times \mathbf{B}/B^2$, where **E** and **B** are the mean electric and magnetic fields. The nonsymmetric velocity (m or n nonzero) and the radial current density vanishes at r=a. The vacuum fields, between the resistive shell and perfectly conducting wall, are calculated analytically and matched to the numerical plasma solution by eq. (2).

3 PERFECTLY CONDUCTING BOUNDARY

Before varying the boundary condition, we briefly review the case of the close-fitting perfectly conducting boundary. The code is initialized with magnetic energy uniformly distributed in m and n (except for the m=n=0 component which is large). The total axial magnetic flux and axial current are held constant during the run. Several key features of the laboratory dynamo are manifest in the steady state which evolves in about one-fifth of the resistive diffusion time. The mean magnetic fields (the m=n=0 component) evolve to the state shown in Fig. 1. The axial field is reversed at the edge. This reversed field is generated by the three-dimensional fields, as is captured in the component of the mean-field Ohm's law parallel to the mean field,

$$E + E_f = \eta j \tag{3}$$

where $E = \mathbf{E} \cdot \langle \mathbf{B} \rangle / \langle B \rangle$ is the parallel electric field, j is the mean parallel current density, $E_f = S\langle \mathbf{v} \times \mathbf{B} \rangle \cdot \langle \mathbf{B} \rangle / \langle B \rangle$ is the parallel fluctuation-induced electric field, and $\langle \ \rangle$ denotes the mean quantity obtained by averaging over θ and z. The three quantities of Ohm's law are plotted in Fig. 2. In steady state, E is the externally produced field. From Fig. 2 we see that it is of a shape to produce a peaked current density profile and by itself could not sustain the current density in outer region of reversed axial magnetic field. The fluctuation field E_f, on the other hand, is in a direction to drive current on the edge to sustain the reversed magnetic field and thereby produce the dynamo. Over the inner half of the cylinder E_f suppresses the current. The net effect of E_f is to flatten the current density.

Fourier decomposition of the fields in θ and z indicates that the dominant modes are m=1 and a band of n values between roughly 5 and 10, as seen in Fig. 3.. These modes satisfy the relation $\mathbf{k} \cdot \langle \mathbf{B} \rangle = 0$. Such modes are resonant with the mean magnetic field, that is, the mode amplitude is constant along the mean field lines. In this case, the spatial perturbations cause the field lines to wander radially and, if the modes are sufficiently plentiful, to produce a chaotic field pattern. In the absence of the fluctuations the field lines would spiral on circular surfaces. The fluctuations cause the lines to break and reconnect. It is the resistive reconnection of the field that permits the plasma to alter its topology as it

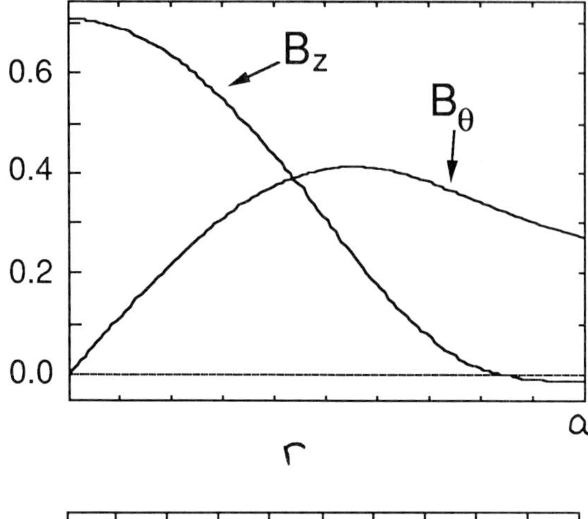

Fig. 1: Mean magnetic field for close-fitting perfectly conducting boundary.

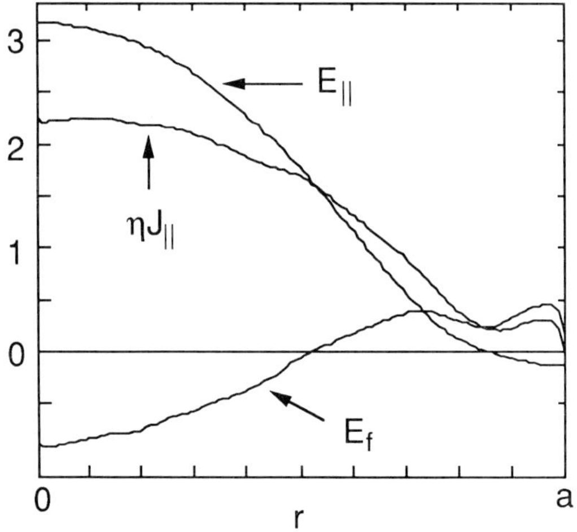

Fig. 2: Terms in the Ohm's law, Eq. (3), for close-fitting perfectly conducting boundary

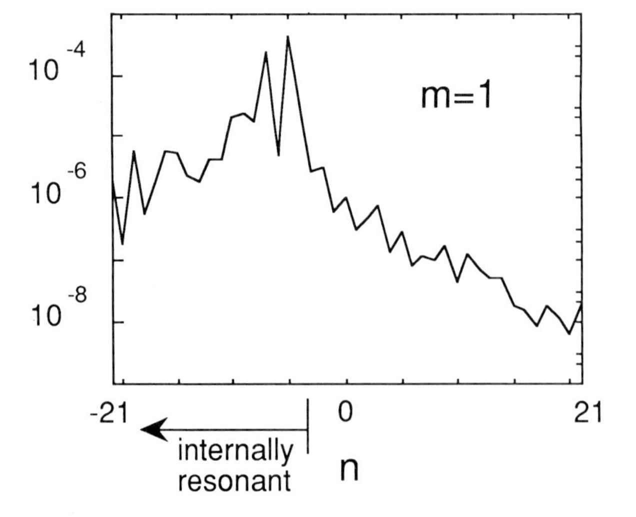

Fig. 3: Final n spectrum of magnetic energy for m = 1 components.

evolves in time.

Often analytic linear stability analysis, treating the mean field as an equilibrium state, is used to compare with the nonlinear evolution. The modes that dominate the nonlinear computation are generally linearly unstable "tearing" modes, driven unstable by the current density gradient. The fluctuation-induced electric field, E_f, diminishes the gradient so as to stabilize the modes.

The magnetic field fluctuations are accompanied by flow with kinetic energy about ten times smaller than the fluctuating magnetic energy. This corresponds to a typical flow speed in the m=1 mode of about 0.2% of the Alfven speed. The laboratory dynamo effect is similar to the previously studied astrophysical dynamo in that the correlated flow and field fluctuations may generate a lower dimensional magnetic field by the **v** x **B** effect. However, there are two differences. First, the laboratory field is only partially self-generated by the flow and requires an externally applied magnetic and electric field for the remainder. Second, the energy source of the fluctuations is the applied electric field in the laboratory case.

The k spectrum of magnetic field at the plasma surface and the radial structure of the mean field are consistent with experimental observations.[12] Details of the internal dynamics await further experimentation.

4 RESISTIVE AND DISTANT BOUNDARIES

To illustrate the effect of the boundary on the dynamo, we consider the case in which the perfectly conducting wall is separated from the plasma surface. To track the dependence on the wall position, we expand the wall slowly during one run, while fixing the total axial current and resistive shell radius. The fields are maintained in a quasi-steady state with $\nabla \times \mathbf{E}$ approximately zero. Now that the radial magnetic field can be nonzero at the plasma surface, the fluctuation amplitude increases significantly, as indicated in the electric field plot in Fig. 4, for the case that the wall radius exceeds the plasma radius by 15%. Although the amplitude increases, the azimuthal and axial k spectra of the fluctuations remains mostly unchanged. The radial profile of the fluctuation-induced electric field does change somewhat in that the inner region of current-suppressing electric field increases. The consequent change in the mean quantities are two-fold. First, in

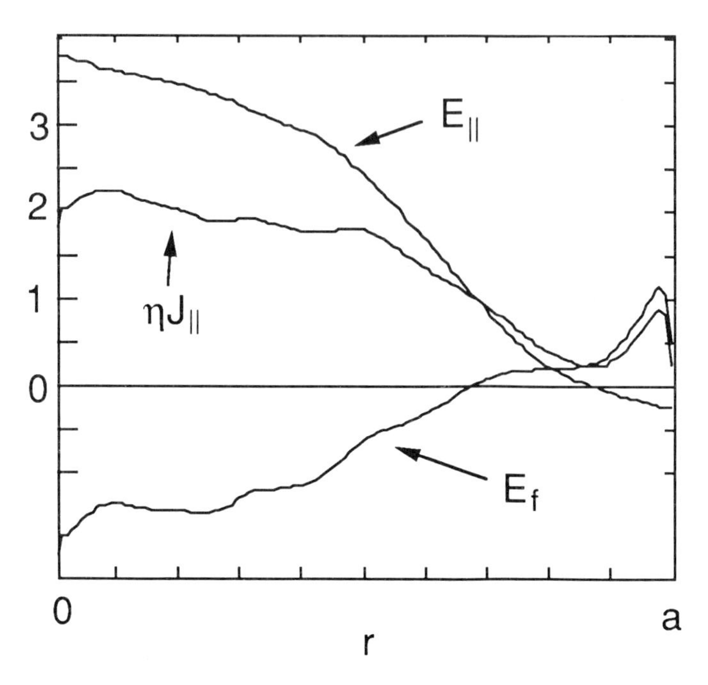

Fig. 4: Terms in Ohm's law for the case with conducting wall radius 15% greater than the plasma radius.

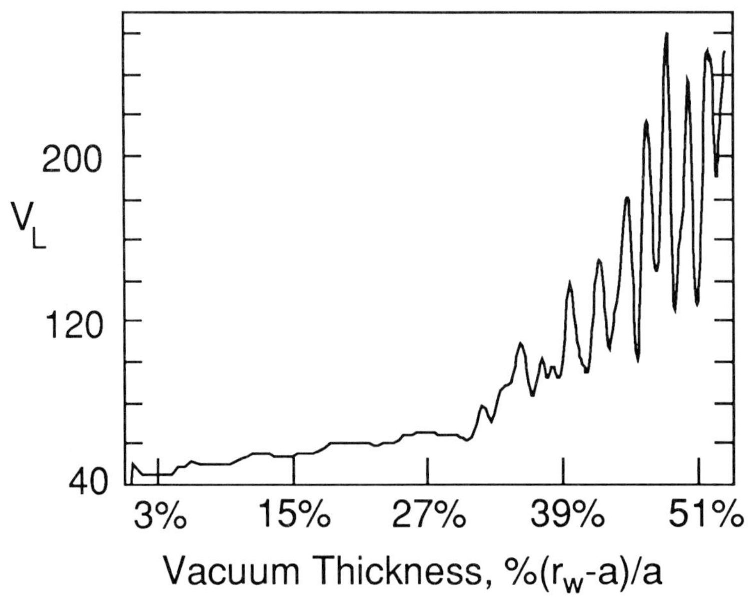

Fig. 5: Axial voltage vs conducting wall position
r_W = w = radius. a = plasma radius.

order to hold the current constant as the current-suppressing electric field E_f increases, the axial electric field, or the axial voltage $V_z = \int\langle E_z\rangle dz$, must increase. The increase in axial voltage is shown in Fig. 5 as a function of wall position. Second, the increase in the magnitude of E_f at the edge increases the azimuthal current density and therefore deepens the axial field reversal (increases B_z at the edge). Both effects, the voltage increase and the increased field reversal, are observed in experiments in which a highly conducting shell is replaced by a distant shell.(13)

To test whether the dynamo sustainment of field reversal is possible without a conducting boundary, we remove the conducting wall to a radius of 10 plasma radii. The plasma remains surrounded by a resistive shell. We find that in one shell penetration time the fluctuating magnetic and kinetic energy increases by one and two orders of magnitude, respectively (Fig 6). The fluctuation electric field, E_f, is now quite large (Fig. 7) and a large axial voltage is required to maintain the current constant, as shown in Fig. 6. The m spectrum is still dominated by internally resonant m=1 modes which cause reconnection. The m=0 contribution is also significant. Removal of the m=0 modes from the evolution leads to enhanced m=1 modes and enhanced axial voltage. Thus, the m=0 modes probably facilitate energy transfer from the linearly unstable m=1 modes(11,14) to the linearly stable m=2 (and m=1) modes which resistively and viscously dissipate the energy. The m=0 modes broaden the n spectrum of the m=1 modes (by m=0/m=1 coupling) which then yields more efficient coupling to the m=2 modes.

The code was not run sufficiently long to establish whether the axial voltage saturates; however, it appears that the sustainment of the magnetic configuration with the reversed axial magnetic field is difficult (i.e., requires large voltage), and perhaps impossible. These results are also not inconsistent with experiments with resistive shells in which the plasma terminates in several shell penetration times, despite the application of axial voltage.(15)

5 MAGNETIC HELICITY DISSIPATION

The effect of alteration of the boundary can be interpreted in terms of the conservation of magnetic helicity,(16,17) $H_m = \int \mathbf{A}\cdot\mathbf{B}dV$, contained within the conducting wall. For the case in which the wall is displaced from the plasma surface, the volume includes the

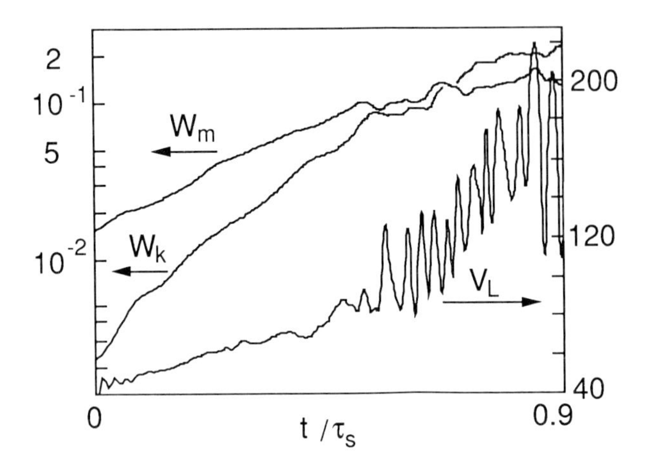

Fig. 6: Time dependence of fluctuating magnetic energy, fluctuating kinetic energy, and axial voltage for conducting wall radius equal to ten plasma radii.

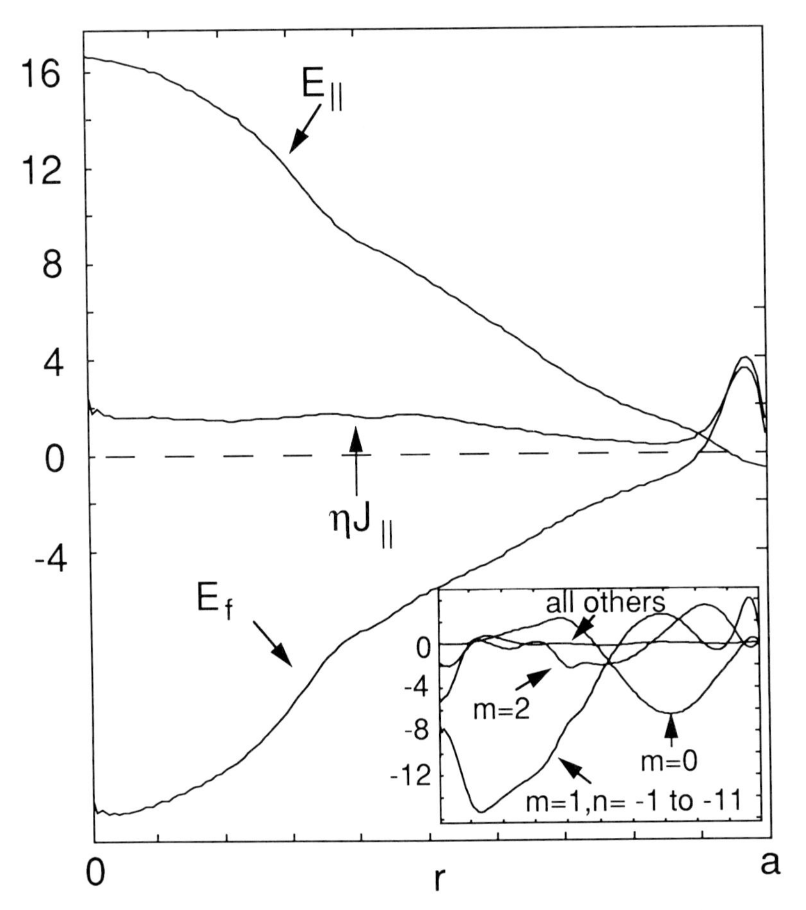

Fig. 7: Ohm's law terms for $r_w = 10a$ at t = one shell time. Inset is E_f vs r for various spectral bands.

vacuum region as well as the plasma volume. From Faraday's and Ohm's laws the helicity change in time is given by

$$dH_m/dt = 2V_z\Phi_z - 2\int \eta\, \mathbf{J}\cdot\mathbf{B}\, dV - 2\int \chi\, \mathbf{B}\cdot d\mathbf{S}, \tag{4}$$

where Φ_z is the total axial magnetic flux within the conducting wall and χ is the electrostatic potential. The expression indicates that helicity changes by the application of axial voltage, by resistive dissipation, and by the intersection of the magnetic field lines with a charged surface. If we write the mean quantities as $\mathbf{J}_o,\mathbf{B}_o$ and the fluctuating quantities as $\mathbf{j},\mathbf{b}$, then the helicity balance equation in steady state is

$$V_z\Phi_z = \int \eta\, \mathbf{J}_o\cdot\mathbf{B}_o dV + \int \eta\, \mathbf{j}\cdot\mathbf{b} dV - \int \chi\, \mathbf{b}\cdot d\mathbf{S}. \tag{5}$$

The left hand side represents helicity injection into the volume within the conducting wall. It is balanced by dissipation due to resistive damping of the mean and fluctuating fields, and by surface loss of helicity. We have numerically evaluated the four terms of eq. (5) for various positions of the conducting wall, as shown in Table I. For the case that the conducting wall is at the plasma surface the helicity is lost mainly through resistive dissipation of the mean fields. Since the magnetic field is tangent to the conducting surface, the surface helicity loss is zero. As the wall is removed, the increase in voltage (Fig. 5) represents an increased helicity injection rate. Since the current is held constant the mean quantities, and their resistive dissipation, are roughly unchanged. The fluctuations increase as the wall is retracted, but for the case considered in the Table, remain small compared to the mean quantities. The increase in helicity dissipation is mainly due to an increase in the surface helicity loss. The surface potential is caused by obstruction of current flow by the resistive shell which is penetrated by the fluctuating magnetic flux. Surface charges accumulate and are calculated by requiring continuity of the electric field parallel to the resistive surface.

TABLE I. Evaluation of terms appearing in helicity balance equation (eq.5)

%Vacuum	$V_L\Phi_z$	$\int \eta \mathbf{J} \cdot \mathbf{B}\, d^3\mathbf{r}$	$\int \eta \mathbf{j} \cdot \mathbf{b}\, d^3\mathbf{r}$	$-\int_S \chi\, \mathbf{b} \cdot d\mathbf{s}$
0%	24.5	24.9	0.5	0
15%	32.9	25.1	1.4	6.2
45%	43.6	25.5	2.6	13.1

6 SUMMARY

In recent laboratory experiments, a dynamo-like mechanism has been demonstrated in which a portion of the axisymmetric component of the magnetic field is believed to be sustained by 3D spatial fluctuations in the field and flow. With a conducting shell at the plasma surface, past MHD computation shows that sustainment arises from fluctuations which cause magnetic reconnection. If the conducting wall is retracted from the plasma surface, the fluctuations are amplified and the dynamo sustainment is still active for the times studied, but an increased energy input to the plasma is required through the applied electric field. The retraction of the conducting wall enhances the helicity dissipation rate by the intersection of the fields with the resistive surface which bounds the plasma. This enhanced helicity dissipation is balanced by the helicity injection that accompanies the increased applied electric field.

This work has been supported by the U.S. Department of Energy.

* Present Address: Science Applications International Corporation, San Diego, California, USA.

REFERENCES

1. See, for example, E. J. Caramana and D.A. Baker, Nucl. Fusion **24**, 423 (1984).

2. A.Y. Aydemir and D.C. Barnes, Phys. Rev. Lett. **52**, 930 (1984).

3. H.R. Strauss, Phys. Fluids **27**, 2580 (1984);.

4. E.J. Caramana and D.D. Schnack, Phys. Fluids **29**, 3023 (1986).

5. J. Dahlburg, D. Montgomery, G.D. Doolen and L. Turner, J. Plasma Physics **40**, 39 (1988).

6. E. Hameiri and A. Bhattacharjee, Phys. Fluids, **30**, 1743 (1987).

7. J.B. Taylor, Phys. Rev. Lett., **33**, 1139 (1974).

8. D. Montgomery. L. Phillips and M.L. Theobold, submitted to Phys. Rev. A.

9. Y.L. Ho, D.D. Schnack and S.C. Prager, Phys. Rev. Lett., **62**, 1507 (1989).

10. D.D. Schnack, d.C. Barnes, Z. Mikic, D.S. Harned and E.J. Caramana, J. Comput. Phys. **70**, 330 (1987).

11. C.G. Gimblett, Nucl. Fusion, **26**, 617 (1986).

12. See, for example, I.H. Hutchinson M. Malacarne, P.Noonan and D. Brotherton-Ratcliffe, Nucl. Fusion **24**, 59j (1984).

13. B. Alper and R.J. La Haye, 16th European Conference on Contr. Fus. and Plasma Phys., Venice, Vol. 13B, Part II, 753 (1989).

14. Y.L.Ho and S.C Prager, Phys. Fluids **31**, 1673 (1988).

15. B. Alper et. al., Plasma Physics and Controlled Fusion, **31**, 205 (1989).

16. T.R. Jarboe and B. Alper, Phys. Fluids **30**, 1177 (1987).

17. H.Y.W. Tsui, Culham Laboratory Report No. CLM-P819.

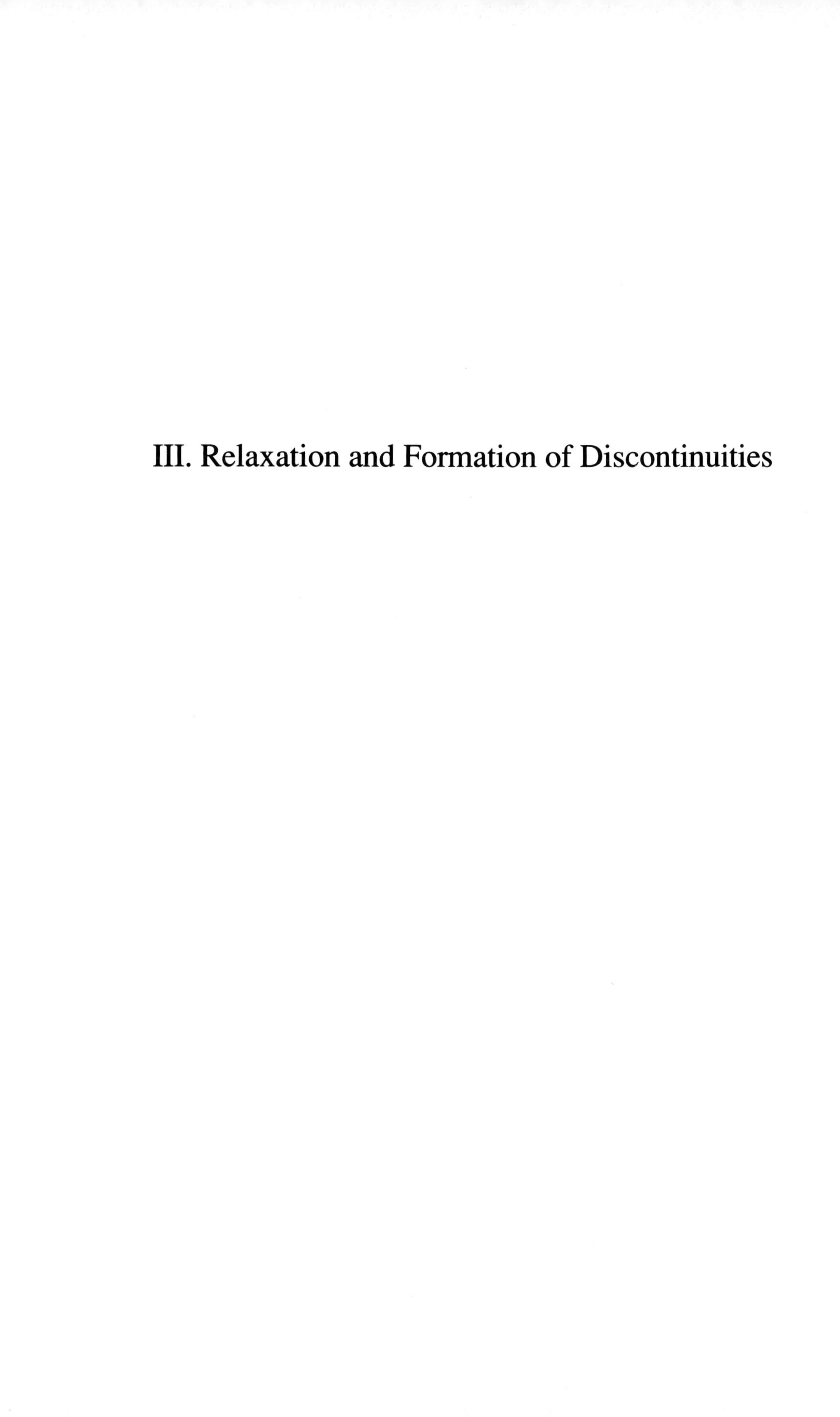

III. Relaxation and Formation of Discontinuities

High Velocities in Quasi-Static Magnetic Singularities

E.N. PARKER

Enrico Fermi Institute and Depts. of Physics and Astronomy
University of Chicago, Chicago, Illinois 60637 USA

1 INTRODUCTION

Tangential discontinuities (current sheets) are an intrinsic part of the static equilibrium of a magnetic field with any but the simplest topology. The curious feature of the tangential discontinuity (TD) is that an arbitrarily slow quasi-static deformation of the equilibrium field causes a thin layer of field and fluid to slide across the TD at the characteristic Alfven speed. Only the thickness of the layer sliding across the TD declines with declining overall deformation rate. Simple estimates suggest the possibility that the thin high speed layer of field and fluid may be a significant part of the high speed small-scale (generally unresolved) plasma motions that contribute (along with Stark broadening) to the extreme line widths observed in solar flares.

To understand the origin of the high speed sheets of field and fluid, consider the manner in which the TD arises in deformed magnetic field. Employing the standard scenario (Parker, 1972, 1979), consider a magnetic field B_0 extending uniformly in the z-direction through an infinitely conducting fluid from the boundary plane $z = 0$ to the boundary plane $z = L$. The footpoints of the field are held fixed at $z = 0$ and the fluid is moved about with the continuous time dependent velocity field

$$v_x = +kz\partial\psi/\partial y, v_y = -kz\partial\psi/\partial x, v_z = 0, \tag{1}$$

throughout $O \leq z \leq L$, where $\psi = \psi(x, y, kzt)$. The magnetic field is carried with the fluid so that after a time t,

$$B_x = +B_0\ kt\partial\psi/\partial y, B_y = -B_0\ kt\partial\psi/\partial x, B_z = B_0. \tag{2}$$

(Parker, 1986). The successive mixing patterns $\psi(x, y, kLt)$ introduced at $z = L$ appear as successive winding patterns along the field from $z = 0$ to $z = L$. Note that (B_x, B_y, B_z) is a continuous function of position provided only that $\nabla\psi$ is a continuous function of position. After a time $t = \tau$ the fluid motion is stopped and the fluid is released so that the field is free to seek its own equilibrium.

It should be noted that, if the fluid pressure applied at the boundaries is not uniform, then there is a flux bundle connecting from a footpoint (x_0, y_0) at $z = 0$

to a footpoint (x_L, y_L) at $z = L$ with different applied fluid pressures. The result is fluid motion along the lines of force from $z = 0$ to $z = L$ for as long as $p(x_0, y_0)$ at $z = 0$ is not equal to $p(x_L, y_L)$ at $z = L$. This is an interesting situation (cf. Montesinos and Thomas, 1989) but not central to the purpose of this paper. So suppose that the fluid pressure applied at $z = 0, L$ is uniform, with a value p_0. Since $\mathbf{B} \cdot \nabla p = 0$ in static equilibrium, it follows that $p = p_0$ throughout $O \leq Z \leq L$ and the magnetic field is force-free, with

$$\nabla \times \mathbf{B} = \alpha \mathbf{B}, \qquad \mathbf{B} \cdot \nabla \alpha = 0 \tag{3}$$

where $\alpha = \mathbf{B} \cdot \nabla \times \mathbf{B}/B^2$ represents the torsion in each elemental flux bundle. The set of all continuous solutions to the force-free equilibrium equations is sufficiently limited in topology that most field topologies require solutions with internal TD's. That is to say, a continuous stream function providing a change in the topology of the mixing pattern along the field produces a field with a change in the topology of the winding pattern along the field. However, $\mathbf{B} \cdot \nabla \alpha = 0$ decrees that the torsion in the field cannot change along any individual line of force extending through two or more winding patterns. This incompatibility is evaded by the development of internal TD's (Parker, 1972, 1986, 1988e) across which the torsion has the form of a delta function with either sign. The TD does not violate $\mathbf{B} \cdot \nabla \alpha = 0$ because the TD contains no magnetic flux.[1]

The initial field, described by equation (2), is continuous, and the TD's develop as the field relaxes to the lower energy state described by equation (3), while preserving the topology of the initial intertwining of the lines of force given by equation (2). The individual TD develops during the relaxation process when one flux bundle pushes its way into a neighboring flux bundle, shoving apart the lines of force and generally intruding into the neighboring region of field. In this way the intruding flux bundle pushes away the field with which it is continuous and comes into contact with field that was not initially (at time $t = \tau$) contiguous with it. There is generally a discontinuity in field direction, i.e. a TD, at the contact surface. In fact the discontinuity in direction occurs in a region of enhanced field pressure, so that the directional discontinuity is enhanced by the refraction of the lines of force - the *surface rotation effect* (Parker, 1987).

[1]It is readily shown that in the limit of a large number n of uncorrelated winding patterns along the field introduced by $\psi(x, y, kzt)$ over the time τ, α declines asymptotically to zero as $n^{-1/2}$, and the field approaches a potential form, described by Laplace's equation. The potential field admits no internal torsion, whereas the field is mixed by ψ into topologies with significant torsion along each line of force. Again the field evades this contradiction by developing internal surfaces of TD with the field described by Laplace's equation throughout the regions of continuity between the surfaces of TD (Parker, 1989a).

2 THE OPTICAL ANALOGY

The formal mathematical optical analogy is a convenient formulation of the equilibrium equation (3) for illustrating the occurrence of gaps in the field through which the intrusion of flux bundles causes TD's (Parker, 1989b). The essential point for the optical analogy is that the two dimensional magnetic field in any local flux surface is expressible as the gradient of a scalar function ϕ of position on the flux surface. Noting that the flux surface generally forms a non Euclidean two dimensional space, write the diagonal terms of the metric tension as h_1^2, h_2^2 so that B_i can be expressed as

$$B_1 = -(1/h_1)\partial\phi/\partial x_1, \; B_2 = -(1/h_2)\partial\phi/\partial x.$$

in the flux surface, where (x_1, x_2) represents an orthogonal coordinate system. Note that the divergence of the field in the flux surface is generally nonvanishing, so ϕ generally does not satisfy Laplace's equation. The lines of force are given by

$$Bh_1 dx_1/ds = -(1/h_1)\partial\phi/\partial x_1, \tag{4}$$

$$Bh_2 dx_2/ds = -(1/h_2)\partial\phi/\partial x_2 \tag{5}$$

where ds is an element of arc length along the line of force.

We recognize equations (4) and (5) as having the same form as the equations for the optical ray path of a wave exp $i\phi$ in a medium with index of refraction B. It follows from Fermat's principle that the magnetic line of force connecting two fixed points P and Q minimizes the integral of B

$$\delta \int_P^Q ds\ B = 0$$

so that the path is describable by the appropriate Euler equation. Using cartesian coordinates (x, y) in a flat space, the Euler equation can be written in the form

$$y''/(1 + y'^2) = \partial \ell n B/\partial y - y' \partial \ell n B/\partial x,$$

where the ray path is $y = y(x)$ and the prime indicates differentiation with respect to x. The essential point is that the lines of force are refracted around (i.e. the lines are concave toward) regions of enhanced B. Thus a region of enhanced B may exclude the lines of force, producing a gap in the flux surface, unless the field lines are fixed at sufficiently nearby boundaries. In optical language, a region of enhanced B midway between the boundaries creates a gap, or zone of exclusion, in the otherwise continuous pattern of rays when the separation λ of the boundaries is in excess of four times the effective local length f of the region of enhanced index of refraction, B. When $\lambda > 4f$ the shortest optical path lies around, rather than

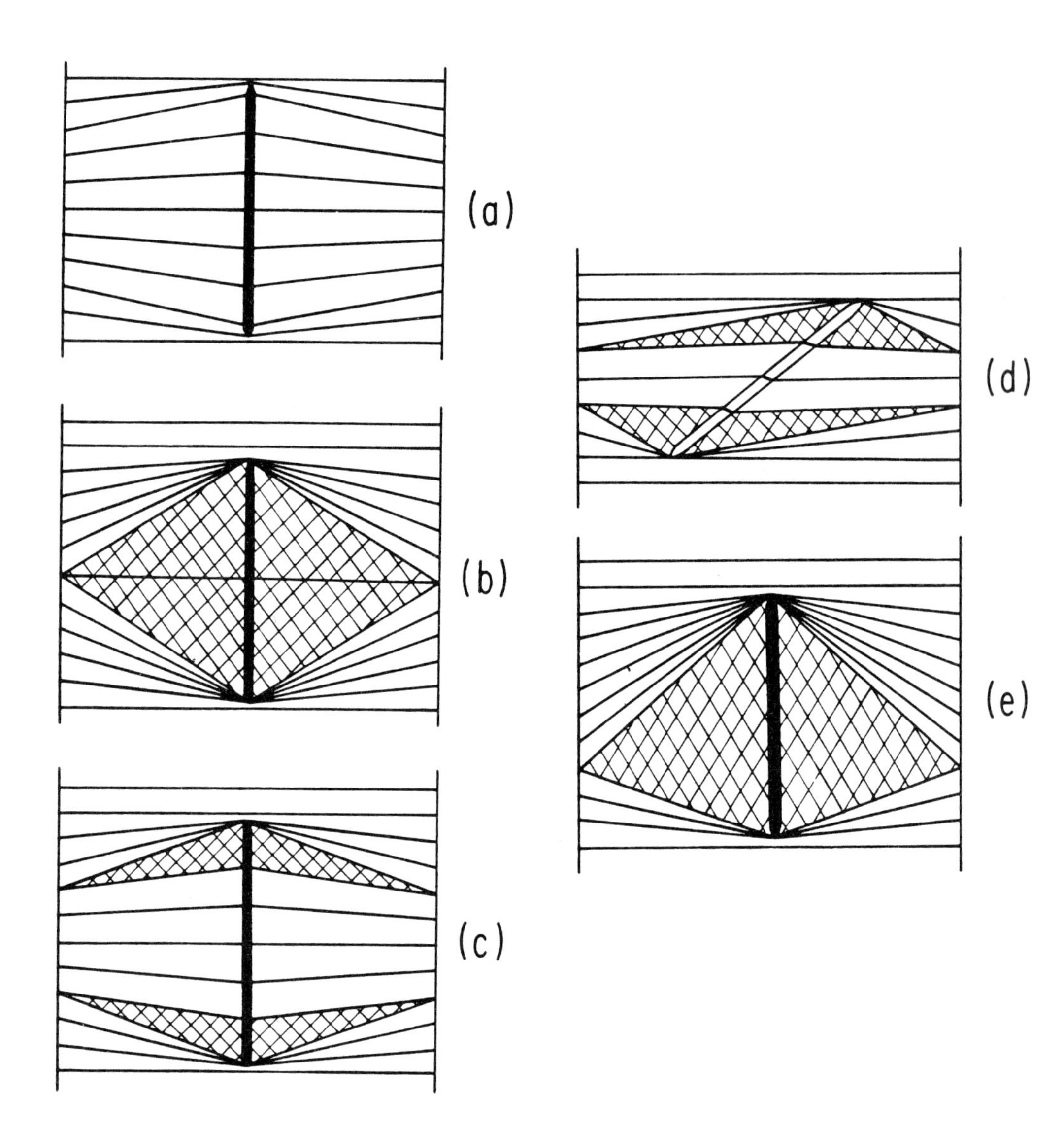

FIGURE 1: A schematic drawing of the optical ray paths or lines of force across a region of enhanced index of refraction B indicated by the vertical bar up the middle of each diagram. (a) shows the field pattern for a relatively weak enhancement, (b) a strong enhancement, (c) a flat topped enhancement, (d) an oblique flat topped enhancement, and (e) a strong enhancement as in (b), displaced upward relative to the field across it.

through, the region of enhanced B. Figure 1(a) sketches the lines of force extending across the region of enhanced B indicated by the heavy bar up the middle when $\lambda < 4f$ and Figure 1(b) shows the lines when $\lambda > 4f$. The cross hatched region represents the zone of exclusion, across which there are no paths. Figure 1(c) shows the interesting situation in which the maximum in B is sufficiently broad that the effective focal length for the rays near the optical axis exceeds $\lambda/4$, while at greater distance from the optical axis, the focal length is less than $\lambda/4$. The result is a band of transmission across the middle of the region with a gap on either side. Fig. 1(d) shows the same situation as Figure 1(c) except that the general field direction is oblique to the optical axis, illustrating the simple refraction of the lines in the region of enhanced B. It is this refraction that gives rise to the surface rotation effect, mentioned earlier.

The effect with which we are concerned here is best understood by close attention to Fig. 1(b). The lines of force are excluded from the cross hatched region in the symmetric manner shown in Fig. 1(b) if the enhancement of B is increased from zero without moving relative to the field. But suppose that the region of enhanced B is subsequently displaced upward in the figure. The field is distorted to the asymmetric form shown in Fig. 1(e). The field is pushed upward ahead of the enhanced B until finally the optical pathlength around the upper end becomes longer than the optical path through the region of enhanced B. At that point the shortest available optical path is around the lower end of the enhanced B. The line jumps from its position at the upper end to the path around the lower end. The jump takes place at the Alfven speed in the deflected component of the field, no matter how slowly the enhanced B is moving upward across the field.

The effect is readily understood in terms of the Maxwell stresses. The tension along the lines of force tends to pull the lines into the region of enhanced B, but is opposed by the excess magnetic pressure. The optical analogy is simply the formal calculation of the result of the contending tension and pressure. The upward motion of enhanced B and the associated gap in Figure 1(e) increases the component of the tension pulling the field above the gap into the enhanced B. When the field is finally stretched enough to pull the lines into the region of enhanced B, it goes with a rush because the pressure gradient opposing the tension diminishes. Upon crossing the optical axis the pressure gradient pushes downward (in Figure 1(e)) and there is no equilibrium where the field might stop until the line of force comes to the lower end of the gap, below the region of enhanced B.

The dash across the region of enhanced B occurs at the characteristic Alfven speed in the transverse component of the field because it is the tension in the field, no longer balanced by the gradient in the magnetic pressure, that pulls the field across.

Formal mathematical examples of refraction patterns of the field subject to a variety of forms of pressure variation are available in the literature (Parker, 1989c).

Formal mathematical examples of the dynamical motions of the field and fluid sliding, or snapping, across the gap are also available (Parker, 1989d), for the interested reader. The calculations show in detail the sliding of the field across the gap, along the lines described from the basic physical considerations noted above.

3 PROGRESSIVE WINDING OF THE FIELD

Imagine the consequences of the slow progressive winding and wrapping of the lines of force throughout $O < z < L$ caused by the continual swirling and shuffling of the footpoints at $z = L$, with the field throughout $O < z < L$ near force-free equilibrium all the time. The progressive wrapping and mixing of the flux bundles, each exerting an inhomogeneous pressure on the closed packed neighboring bundles encountered along its length, produces gaps and TD's at various locations in the field. The continuing winding of the field causes the pressure maxima applied to each TD surface to drift slowly across that surface. The result is illustrated in Figure 1(e), with a steady flow of fluid and field from the bow to the stern of each moving region of enhanced B. It follows that each TD in the slowly evolving field configuration has a thin layer of fluid and field sliding over it. The TD is no longer the simple discontinuity of the static equilibrium. It becomes a dynamical entity, combining a current sheet with a high speed, usually very thin, sheet of fluid. Resistive and thermal instabilities, and magnetic reconnection may be profoundly altered by the presence of the rapidly moving thin sheet of fluid and field.

4 THE SOLAR CORONA AND SOLAR FLARES

It is an interesting question whether the fluid sliding across the TD's in the bipolar magnetic fields of the solar X-ray corona may be observable as high speed jets of plasma. The mean magnetic field strength $B \cong 10^2$ gauss provides an Alfven speed of about 2×10^3 km/sec in the typical coronal density of 10^{10} atoms/cm^3. We have estimated (Parker, 1983) that the deformation of the field by the random motion (with a velocity $v \cong 0.5$ km/sec) of the footpoints of the field proceeds to the point where the rms transverse field component $B_\perp$ is of the order of $\frac{1}{4}B$, or 25 gauss. The Alfven speed V in $B_\perp$ is then about 500 km/sec. So we expect plasma motions in excess of one hundred km/sec. The problem is that the rate of deformation of the field is so slow ($v \cong 0.5$ km/sec) that the high speed sheet of fluid is too thin to be observable (Parker, 1989d). The thickness h may be estimated from conservation of magnetic flux.

If a layer of field of thickness h_0 moving at velocity v flows with velocity V with a thickness h, then conservation of magnetic flux requires $vh_0 = Vh$, assuming that the field strength is about the same in both places. If V=500 km/sec and v=0.5 km/sec, the result is $h = 10^{-3}h_0$. So even if h_0 is as large as the characteristic granule size, of the order of 10^3 km, h is only 1 km, providing only 10^{15} atoms/cm^2 in the line of sight.

It has been suggested that the sudden reconnection at one TD in a magnetic field containing many TD's (pushed close to the threshold for rapid reconnection by some large-scale deformation of the field) may disturb the field sufficiently to trigger a burst of reconnection in several of the TD's. This coordinated dissipation at a large number of otherwise quiescent TD's may be the major energy release in a solar flare (Parker 1988), although probably not the most intense concentrated energy release. If this is indeed the case, one can imagine that in a flare the interwoven field containing the TD's may evolve rapidly, rather than slowly, during the explosive phase with v of the order of 10^2 km/sec or more. In that case h may be 10^2 times larger, i.e. of the order of 10^2 km, providing 10^{17} atoms in the line of sight. It appears, then, that the curious phenomenon of the high speed sheets of fluid and field sliding across the TD's in a flare region may be a source of plasma jetting within the flare. The jetting of plasma at sites of rapid reconnection is probably the major effect, but it is conceivable that the high speed sheets may participate too. The relative proportions depend upon how much massaging of the quasi-static TD's occurs before a burst of rapid reconnection reduces the TD below the threshold for active dissipation.

Whatever the role of the high speed sheets of fluid in contributing to the observed jetting of the fluid in the active corona and in the flare, it is clear that the sheets greatly complicate the dynamics of the magnetic reconnection at the TD's.

This work was supported in part by the National Aeronautics and Space Administration under NASA Grant NGL 14-001-001.

REFERENCES

Montesinos, B. and Thomas, J.H., "Siphon flow in isolated magnetic flux tubes. II. Adiabatic flows," *Astrophys. J.* **337**, 977-988 (1989).

Parker, E.N. "Topological dissipation and the small-scale fields in turbulent gases" *Astrophys.* **174**, 499-510 (1972).

Parker, E.N. Cosmical Magnetic Fields, Clarendon Press, Oxford (1979) pp. 359-378.

Parker, E.N. "Magnetic neutral sheets in evolving fields. II. Formation of the solar corona" *Astrophys. J.* **264**, 642-647 (1983).

Parker, E.N. "Equilibrium of magnetic fields with arbitrary interweaving of the lines of force. I. Discontinuities in the torsion" *Geophys. Astrophys. Fluid Dyn.* **34**, 243-264 (1986).

Parker, E.N. "Magnetic reorientation and the spontaneous formation of tangential discontinuities in deformed magnetic fields" *Astrophys. J.* **318**, 876-887 (1987).

Parker, E.N. "Stimulated dissipation of magnetic discontinuities and the origin of solar flares" *Solar Phys.* **111**, 297-308 (1988).

Parker, E.N. "Spontaneous tangential discontinuities and the optical analogy for static magnetic fields. I. Force-free fields, potential fields, and discontinuities" *Geophys. Astrophys. Fluid Dyn.* **45**, 159-164 (1989a).

Parker, E.N. "Spontaneous tangential discontinuities and the optical analogy for static magnetic fields. II. The optical analogy" *Geophys. Astrophys. Fluid Dyn.* **45** 169-182 (1989b).

Parker, E.N. "Sponetaneous tangential discontinuities and the optical analogy for static magnetic fields. III. Zones of exclusion " *Geophys. Astrophys. Fluid Dyn.* **46** (in press) (1989c).

Parker, E.N. "Spontaneous tangential discontinuities and the optical analogy for static magnetic fields. IV. High speed fluid sheets" *Geophys. Astrophys. Fluid Dyn.* (in press) (1989d).

Parker, E.N. "Spontaneous tangential discontinuities and the optical analogy for static magnetic fields. V. Formal integration of the force-free field equations" *Geophys. Astrophys. Fluid Dyn.* (in press) (1989e).

Current Sheet Formation in Magnetostatic Equilibria

ELLEN G. ZWEIBEL
Department of Astrophysical, Planetary and Atmospheric Sciences, University of Colorado, Boulder, CO 80309, U.S.A.

MICHAEL R.E. PROCTOR
Department of Applied Mathematics and Theoretical Physics, University of Cambridge, Silver Street, Cambridge CB3 9EW, U.K.

The quasi-static evolution of two-dimensional magnetostatic equilibria is examined in the case where there is a separatrix field line separating regions of different fieldline connectivity. It is shown that in general there will be a current sheet on this separatrix for arbitrarily small displacements of the footpoints. A nonlinear analysis confirms the main results of the linearized theory.

1 INTRODUCTION

The magnetic Reynolds number in astrophysical situations is generally so large (e.g. $10^{10} - 10^{12}$ for solar magnetic loops and $10^{19} - 10^{21}$ in the interstellar gas) that resistive processes such as magnetic fieldline reconnexion and Ohmic heating can be ignored. On the other hand, resistive processes do become important if the magnetic field becomes concentrated or sheared on scales which are much smaller than the gross size of the system, so that the effective magnetic Reynolds number is correspondingly reduced.

Parker (1972) proposed that when the evolution of a magnetofluid is driven by slow motions of its boundary, the magnetic field can become discontinuous even if the boundary motions are smooth. Although actual discontinuities cannot form in a system with nonzero resistivity, it is widely accepted that if the nonresistive theory predicts a discontinuity then restoring resistivity would lead to a highly concentrated current layer at the same location as the discontinuity. Parker's proposal has stimulated a great deal of interest and controversy, as can be seen from some of the contributions to this Symposium.

The purpose of this paper is to consider a special case of the general problem, namely systems which have an invariant direction, and to focus on the rôle of magnetic fieldline separatrices and on comparison between what can be learnt from considering infinitesimal as against finite-amplitude perturbations. The calculations presented generalise to twisted fields, nonzero gas pressure and finite amplitude the work of Moffatt (1987), who considered small perturbations of potential fields. In Section 2 we consider perturbations to a two-dimensional equilibrium system, including the effects of magnetic shear, gas pressure and gravity. In Section 3 we consider distortions of arbitrary amplitude under the assumption that they evolve quasistatically. Section 4 is a summary and discussion.

2 PERTURBATIONS TO A TWO-DIMENSIONAL SYSTEM

We consider solutions of the magnetostatic equation

$$(\nabla \times \mathbf{B}) \times \mathbf{B} - \nabla p = 0 \tag{1}$$

in the halfspace $z > 0$. The boundary $z = 0$ is taken to be rigid and perfectly conducting. For the present we ignore gravity; the effect of nonzero gravitational acceleration will be discussed below.

If $\mathbf{B}$ and p are assumed to be independent of x then we can write $\mathbf{B}$ as the curl of a vector potential $\mathbf{A}(y,z) = \hat{\mathbf{x}}A(y,z)$. It follows that $\mathbf{B}\cdot\nabla A = 0$. The magnetostatic equation (1) can be rewritten in terms of A as

$$\nabla A(\nabla^2 A) + \nabla(p + \frac{B_x^2}{2}) + \hat{\mathbf{x}}\,\hat{\mathbf{x}}\cdot(\nabla B_x \times \nabla A) = 0. \tag{2}$$

Equation (2) shows that both p and B_x must be functions of A, while

$$\nabla^2 A + \frac{d}{dA}(p + \frac{B_x^2}{2}) = 0. \tag{3}$$

We will be interested in solutions of equation (3) in which almost all fieldlines have two endpoints anchored in the surface $z = 0$.

Now suppose that the field undergoes a small ideal displacement $\boldsymbol{\xi}$ which takes it to a new equilibrium. The displacement is specified on the boundary $z = 0$ and elsewhere in the fluid may be regarded as the outcome of a quasi-static relaxation process. We require $\boldsymbol{\xi}$ to be independent of $\hat{\mathbf{x}}$. The Eulerian displacements of A, p and $\mathbf{B}$ are

$$\delta\mathbf{B} = \nabla \times (\hat{\mathbf{x}}\delta A) + \hat{\mathbf{x}}\delta B_x = \delta\mathbf{B}_\perp + \hat{\mathbf{x}}\delta B_x, \tag{4a}$$

$$\delta A = -(\boldsymbol{\xi} \cdot \nabla) A, \tag{4b}$$

$$\delta p = -(\boldsymbol{\xi} \cdot \nabla) p - \gamma p (\nabla \cdot \boldsymbol{\xi}) = \delta A \frac{dp}{dA} - \gamma p (\nabla \cdot \boldsymbol{\xi}), \tag{4c}$$

$$\delta B_x = -(\boldsymbol{\xi} \cdot \nabla) B_x + (\mathbf{B} \cdot \nabla) \xi_x - B_x (\nabla \cdot \boldsymbol{\xi}) = \delta A \frac{dB_x}{dA} + (\mathbf{B} \cdot \nabla) \xi_x - B_x (\nabla \cdot \boldsymbol{\xi}). \tag{4d}$$

If we linearize equation (1) and substitute for δp and $\delta \mathbf{B}$ from equations (4), the result is a set of three coupled second order partial differential equations for the three components of $\boldsymbol{\xi}$. In at least some cases these equations do not have a unique solution. For if the unperturbed solution to equation (1) is marginally stable then there is at least one displacement $\boldsymbol{\xi}_h$, vanishing on the boundary $z = 0$, which takes the system to a new equilibrium. A constant multiple of $\boldsymbol{\xi}_h$ can always be added to a solution of the problem with inhomogeneous boundary conditions, so the solution in this case is clearly not unique. Unfortunately the theory of linear partial differential equations does not offer much guidance about the existence, uniqueness and regularity on non-standard boundary value problems of this type. In view of these uncertainties we will assume in what follows that at least one solution exists, and learn what we can about its properties. If we substitute equations ($4b, d$) into the x component of equation (2), and linearize, the result is

$$\nabla((\mathbf{B} \cdot \nabla)\xi_x - B_x(\nabla \cdot \boldsymbol{\xi})) \times \nabla A = 0,$$

so that

$$(\mathbf{B} \cdot \nabla)\xi_x - B_x(\nabla \cdot \boldsymbol{\xi}) = f(A). \tag{5}$$

The other components of equation (2) become

$$\nabla \delta A (\nabla^2 A) + \nabla A (\nabla^2 \delta A) + \nabla(\delta p + B_x \delta B_x) = 0, \tag{6}$$

which may be rewritten using equations (3)-(5) as

$$\nabla A \Big(\nabla^2 \delta A + \frac{d}{dA}(f B_x) + \delta A \frac{d^2}{dA^2}(p + \frac{B_x^2}{2}) \Big) = \nabla(\gamma p \nabla \cdot \boldsymbol{\xi}). \tag{7}$$

Equation (7) may be satisfied in two possible ways. Either $\gamma p(\nabla \cdot \boldsymbol{\xi})$ must be a function of A or the right and left hand sides of equation (7) must vanish independently. We consider the former possibility first, in which case

$$\nabla \cdot \boldsymbol{\xi} = h(A). \tag{8}$$

Then combining equations (5) and (8) and integrating along a contour of constant A, say $A = \alpha$, with endpoints $P(\alpha)$ and $Q(\alpha)$ on $z = 0$, we find

$$\big(f(\alpha) + B_x(\alpha) h(\alpha)\big) \int_{A=\alpha} \frac{ds}{|\nabla A|} = \xi_x\big(Q(\alpha)\big) - \xi_x\big(P(\alpha)\big). \tag{9}$$

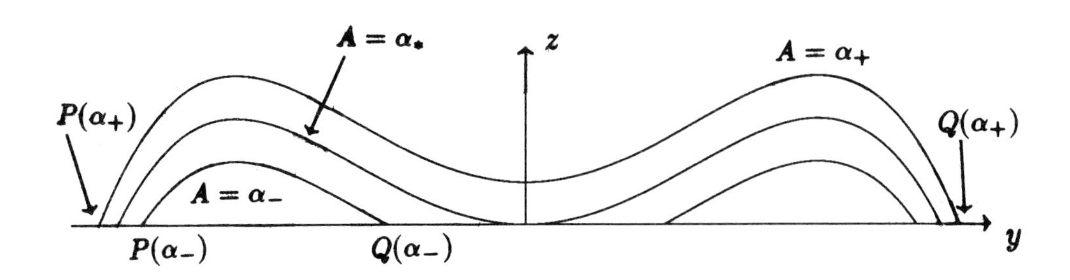

FIGURE 1. Sketch of the field line topology, showing the notation used.

The right hand side of equation (9) is given by the boundary conditions on $\boldsymbol{\xi}$, while the contour integral on the left hand side is to be evaluated from the known unperturbed fieldline geometry. Therefore the combination $f(A) + B_x(A)h(A) = (\mathbf{B} \cdot \nabla)\xi_x$ is determined by equation (9).

Now consider a magnetostatic equilibrium with surfaces of constant A as shown in Figure 1. The surface $A = \alpha_*$ divides the space into two topologically distinct regions; for $A < \alpha_*$ (say) fieldlines do not cross the z-axis and are single humped, while for $A > \alpha_*$ they do cross the z-axis and have a double-humped shape. We assume here that the z-axis is a plane of symmetry, but our arguments can easily be adapted when this is not the case.

Let us evaluate $f(A) + B_x(A)h(A)$ from equation (9) for two fieldlines labelled α_+ and α_- which are just inside and just outside the separatrix ($\alpha = \alpha_*$), respectively. If $f(A) + B_x(A)h(A)$ is continuous across the surface $A = \alpha_*$ then it can be calculated either from α_+ as α_+ approaches α_* or from α_- as α_- approaches α_*. That is, we must have

$$\lim_{\alpha_+ \to \alpha_*} \left[\frac{\xi_x(Q(\alpha_+)) - \xi_x(P(\alpha_+))}{\int\limits_{A=\alpha_+} \frac{ds}{|\nabla A|}} \right] = \lim_{\alpha_- \to \alpha_*} \left[\frac{\xi_x(Q(\alpha_-)) - \xi_x(P(\alpha_-))}{\int\limits_{A=\alpha_-} \frac{ds}{|\nabla A|}} \right]. \tag{10}$$

It is clear from the geometry of the problem that as α_+ and α_- tend to α_*,

$$\int\limits_{A=\alpha_+} \frac{ds}{|\nabla A|} \to 2 \int\limits_{A=\alpha_-} \frac{ds}{|\nabla A|}. \tag{11}$$

If we assume that the displacement on the boundary $z = 0$ is continuous, then $\xi_x(P(\alpha_+))$ and $\xi_x(P(\alpha_-))$ converge to the same value as the limits are taken. On the other hand, $Q(\alpha_-)$ and $Q(\alpha_+)$ are far apart and there is no reason why $\xi_x(Q(\alpha_+))$ and $\xi_x(Q(\alpha_-))$ should have any particular relationship. Thus equation (10) will not be satisfied, and $f(A) + B_x(A)h(A)$ will be discontinuous across the separatrix, which implies the existence of a current sheet there. To see this, we note that the Lagrangian perturbation of the total (gas plus magnetic) pressure, $\Delta\Pi$, must be continuous across the separatrix (Bernstein *et al.* 1958). Bearing in mind that the Lagrangian perturbation ΔM of any quantity M is related to its Eulerian counterpart δM by $\Delta M = \delta M + (\boldsymbol{\xi}\cdot\nabla)M$, the Lagrangian pressure balance condition is

$$\nabla A \cdot \nabla \delta A + (\boldsymbol{\xi} \cdot \nabla)\left(\frac{|\mathbf{B}_\perp|^2}{2}\right) + B_x f - \gamma p h = \Delta\Pi \quad \text{is continuous.}$$

Now if $f(A) = \Delta B_x$ is discontinuous then there is not only a discontinuity in the magnetic shear, but there must also be a discontinuity in the perturbed pressure Δp and/or $\nabla\delta A$. If $h = -\Delta p/\gamma p$ is discontinuous then there is likewise a discontinuity in ΔB_x or $\nabla\delta A$. Therefore the discontinuity in $p + B_x h = (\mathbf{B}\cdot\nabla)\xi_x$ requires a current sheet if the system is in equilibrium. There is one important exception to this requirement. If ξ_x is chosen to be an antisymmetric function of y, then $\xi_x(Q(\alpha_-)) \to 0$ as $\alpha_- \to \alpha_*$ while $Q(\alpha_+) = -P(\alpha_+)$. It can then easily be shown (bearing (11) in mind) that the necessary condition (10) holds in this case, so that $f + B_x h$ is indeed continuous across the separatrix. Of course, the displacements will not satisfy such a symmetry condition in general.

We now return to the alternative way to satisfy equation (7), namely that the left and right hand sides vanish independently. Unless the pressure vanishes, this requires $p(\nabla\cdot\boldsymbol{\xi}) = K$ where K is a constant. Having a definite expression for $\nabla\cdot\boldsymbol{\xi}$, we can integrate equation (5) along a fieldline to find $f(A)$. Next we solve for δA (which is given on the boundary $z = 0$) using the condition that the left hand side of equation (7) vanishes. Once we know δA, we have found the projection of (ξ_y, ξ_z) in the direction perpendicular to $\mathbf{B}_\perp$ (cf. equation (4*b*)). The parallel component is found by integrating $\nabla\cdot\boldsymbol{\xi} = K/p$. But the parallel component is given at both endpoints of a fieldline, and we cannot satisfy both these conditions by integrating a first order equation. Thus requiring that the left and right hand sides of equation (7) vanish separately will not in general lead to self-consistent solutions.

The picture changes if $p = 0$. In this case there seems to be no way to determine $\nabla\cdot\boldsymbol{\xi}$, because the component of (ξ_y, ξ_z) parallel to $\mathbf{B}_\perp$ is undetermined (Zweibel & Li 1987). The case $p = 0$ is thus a singular limit.

Finally, we discuss the effect of gravity. For simplicity, suppose that the gravitational acceleration $\mathbf{g} = -g\hat{\mathbf{z}}$ where g is constant and that in equilibrium the gas temperature T is uniform. Then equation (1) becomes

$$(\nabla \times \mathbf{B}) \times \mathbf{B} - \nabla p - \rho g \hat{\mathbf{z}} = 0 \tag{12}$$

where $\rho = \mu p/kT$. We can show that B_x is again a function of A, while

$$p = R(A)e^{-z/H}, \tag{13}$$

where $H = kT/\mu g$ is the constant scale height of the atmosphere. Equation (3) is replaced by

$$\nabla^2 A + e^{-z/H}\frac{dR}{dA} + \frac{d}{dA}\frac{B_x^2}{2} = 0. \tag{14}$$

If we now displace the fluid by a small amount $\boldsymbol{\xi}$, the Eulerian density perturbation is given by

$$\delta\rho = -(\boldsymbol{\xi}\cdot\nabla)\rho - \rho(\nabla\cdot\boldsymbol{\xi}), \tag{15}$$

In view of equation (13), only the first equality in equation (4c) remains valid. Equation (5) still holds, but equation (7) becomes

$$\nabla A\left(\nabla^2 \delta A + \frac{d}{dA}(B_x f) + \delta A \frac{d^2}{dA^2}\left(\frac{B_x^2}{2}\right) + \delta A e^{-z/H}\frac{d^2 R}{dA^2}\right) +$$

$$\hat{\mathbf{z}} e^{-z/H}\frac{(\gamma-1)}{H}R(\nabla\cdot\boldsymbol{\xi}) = e^{-z/H}\nabla\left(\gamma R(\nabla\cdot\boldsymbol{\xi}) - \frac{\xi_z}{H}R\right). \tag{16}$$

Evidently we cannot conclude that $\nabla\cdot\boldsymbol{\xi}$ is a function of A when gravity is significant.

3 NONLINEAR DISPLACEMENTS

We now reconsider the two-dimensional problem posed in Section 2, but now permitting displacements of arbitrary amplitude. We assume that the fluid velocity is always highly subsonic and subalfvenic so that the system evolves through quasi-static equilibria. Then the relevant equations are equation (1) together with

$$\frac{DA}{Dt} = 0; \qquad \frac{DB_x}{Dt} = (\mathbf{B}\cdot\nabla)v_x - B_x\nabla\cdot\mathbf{v}, \tag{17a,b}$$

$$\frac{D\rho}{Dt} = -\rho\nabla\cdot\mathbf{v}; \qquad p = K(A)\rho^\gamma, \tag{17c,d}$$

where $\mathbf{v}(y,z,t)$ is the fluid velocity and equation (17d) expresses the conservation of entropy on magnetic flux surfaces.

As in Section 2, equation (1) implies that B_x and p must remain functions of A as the system evolves, so that $(B_x, p) = \big(B_x(A,t), p(A,t)\big)$ where B_x and p will also depend implicitly on t through A. For any quantity $S(A,t)$

$$\frac{DS}{Dt} = \frac{\partial S}{\partial t} + \frac{\partial S}{\partial A}\frac{DA}{Dt} = \frac{\partial S}{\partial t} \tag{18}$$

by virtue of equation (17*a*). Using equation (18) in equation (17*c*) we can see that

$$\nabla \cdot \mathbf{v} = -\frac{D}{Dt}(\ln \rho) = h(A,t); \tag{19}$$

thus equation (17*b*) yields

$$(\mathbf{B} \cdot \nabla) v_x = g(A,t) \tag{20}$$

for some g, h.

If we integrate equation (20) along a fieldline $A = \alpha$ with endpoints P and Q as in Section 2, we find

$$g(\alpha,t) = \big(v_x(Q,t) - v_x(P,t)\big) \Big/ \int_P^Q \frac{ds}{|\nabla A|}. \tag{21}$$

In contrast to equation (9), which was expressed in terms of the *unperturbed* potential A, equation (21) needs to be interpreted cautiously since the denominator is undetermined at this stage. Even if a sensible solution can be found for $g(\alpha,t)$, this does not guarantee the existence of a solution to the problem posed at the start of the section, since it is still necessary to solve equation (1) in the (y,z) plane. Let us, however, suppose that there *is* a solution without current sheets, and see what restrictions this imposes on the possible footpoint velocities v_x.

If there is a solution with the topology of Figure 1 for which A and its derivatives are continuous, and if B_x is also continuous at all points, then for any given time t the denominator of equation (21) will change continuously as α approaches α_* from above or below. For $\alpha < \alpha_*$ the contour $\alpha =$ const. has two disjoint portions, and the limiting value $g(\alpha_*, t)$ may be calculated from *either* portion (so that both must give the same answer) as $\alpha \uparrow \alpha_*$. As $\alpha \downarrow \alpha_*$, on the other hand, $g(\alpha_*, t)$ is given as the limit of (21) evaluated along a single contour. If, as is supposed, g is continuous at $\alpha = \alpha_*$ then these three limits must all be identical. There is no reason why this should be the case in general; while we have not been able to find necessary conditions for nonexistence, it is easy to exhibit large classes of footpoint displacement for which g cannot be continuous. In particular, there can be no continuous solution for any footpoint motion v_x such that $\big(v_x(P_1) - v_x(0)\big)\big(v_x(P_2) - v_x(0)\big) > 0$, where P_1 and P_2 are the values of y at the endpoints of the separatrix field line. This is because the limiting values of g derived from each of the two disjoint portions will have *opposite sign* (and be nonzero) and so cannot be the same. We conjecture that general footpoint motions will lead to a similar singularity. Now we can write

$$g(A,t) = \frac{\partial}{\partial t}B_x(A,t) - \frac{1}{\gamma}B_x(A,T)\frac{\partial}{\partial t}(\ln p), \tag{22}$$

while the x component of the current, $j_x = -\nabla^2 A$, satisfies

$$j_x = \frac{\partial}{\partial A}\left(p + \frac{B_x^2}{2}\right). \tag{23}$$

Thus if g is discontinuous, and B_x continuous (by supposition), then it would seem that $\partial p/\partial t$ is discontinuous across $A = \alpha_*$ and therefore that j_x is singular there. Thus, at least for footpoint motions as described above, we conclude that linearized theory gives the correct indication of a singularity. For infinitesimal motions, of course, the singularity always occurs in j_x; however it does seem possible for finite displacements that there could be instead a discontinuity in B_x (this does not have to occur at every point of the separatrix; it might be possible to have a segment of current sheet just above the separatrix point (along $y = 0$ in Figure 1).

4 DISCUSSION AND CONCLUSIONS

We have considered the quasistatic evolution of two-dimensional magnetic equilibria under displacements of the boundary which preserve the invariant horizontal direction. We have shown that if the equilibrium has a separatrix – *i.e.* a surface across which the connectivity of the fieldlines to the booundary changes discontinuously – then smooth motions of the boundary will produce a tangential discontinuity of the magnetic field across the separatrix. Moffatt (1987), Aly (1987) and Low & Wolfson (1988) have made similar arguments for equilibria without pressure or magnetic shear.

We have not shown that the problem as posed has a unique solution. Nor have we shown that current sheets do not form in equilibria without separatrices (or neutral points). Our demonstration of a discontinuity does not apply when gravitational acceleration is included (gravity is normally neglected in structures which are shallower than a thermal scale height). It is not known whether solutions are smooth in this case.

The outstanding problem left by this work (apart fom questions of uniqueness and existence) is the extension to three dimensions. Our method fails when there is no invariant direction and the field cannot be described in terms of the vector potential A.

It is interesting that the results can be obtained from a linearized analysis, and that the results are consistent with nonlinear theory. This is not the case for the problem of a uniform vertical field stretched between two conducting plates. Small perturbations of this equilibrium lead to smooth interior fields (Zweibel & Li 1987). Parker (1986) has argued that discontinuities will develop from finite displacements.

Finally, we note that thermal pressure is an important qualitative ingredient in the problem, even though it may be quantitatively small. Only in the presence of pressure can we conclude that $\nabla \cdot \boldsymbol{\xi}$ or $\nabla \cdot \mathbf{v}$ must be functions of A, as is necessary in establishing equations (9) and (21).

EGZ thanks the Universities of Cambridge and Chicago, where part of this work was carried out, for their hospitality. We are also happy to acknowledge useful discussions with N.R.Lebovitz, H.K.Moffatt and E.N.Parker. Material support was provided by NASA, the National Science Foundation, the Science and Engineering Research Council of Great Britain and Trinity College, Cambridge.

REFERENCES

Aly, J.J. 1987, in *Proc. Workshop on Interstellar Magnetic Fields, Schloss Ringberg (FRG)*, September 8-12, 1986 (Berlin, Springer-Verlag), p.240.

Bernstein, I.B., Frieman, E.A., Kruskal, M.D. & Kulsrud, R.M. 1958, *Proc. Roy. Soc.* **A244**, 17.

Low, B.C. and Wolfson, R. 1988 *Astrophys. J.* **324**, 574.

Moffatt, H.K. 1987, in *Advances in Turbulence*, ed. J. Comte-Bellot & J. Mathieu (Berlin, Springer-Verlag), p.228.

Parker, E.N. 1972 *Astrophys. J.* **174**, 499.

Parker, E.N. 1986 *Geophys. Astrophys. Fluid Dyn.* **34**, 243.

Zweibel, E.G. and Li, H.-S. 1987 *Astrophys. J.* **312**, 423.

Minimum Dissipation States and Vortical Flow in MHD

D. MONTGOMERY[1], J.P. DAHLBERG[2], L. PHILLIPS[1] & M.L. THEOBALD[1]

[1]Dartmouth College, Hanover New Hampshire 03755, USA
[2]U.S. Naval Research Laboratory, Washington, DC 20375, USA

ABSTRACT

For more than thirty years, discussion of the confined states of a magnetized, current-carrying plasma column has been dominated by a paradigm of an ideal axisymmetric MHD equilibrium with no steady-state flow velocity. Anything else that has happened has been interpreted as a consequence of "instabilities," regarded as perturbations of such equilibria. We question this way of viewing the problem, and suggest that it has only been possible to sustain it this long because of the absence of good internal diagnostics for the relevant magnetohydrodynamic variables. Electrically-driven, magnetically-supported, cylindrical magnetofluids seem naturally to relax into states which contain a pair of low-mode-number, counterrotating, helical vortices. The current channel and magnetic topology are the result of an axisymmetric current distribution plus a smaller helical component. The evidence for the configuration is numerical and analytical. The poloidal flow is such as to connect the hot geometrical center of the plasma with the cooler perimeter. The flow thus becomes a candidate for a dominant heat loss mechanism. Ideal and slightly non-ideal configurations may be as different in MHD as they are in hydrodynamics, and the former may play an equally minor role in practice.

* * * * * * *

One of the more familiar phenomena of plasma physics is the "kinking" of a magnetohydrodynamic (MHD) fluid, when a current is forced to flow along a dc magnetic field. The term refers to a helical spatial dependence of the magnetic field, as distinct from the helical shape the field lines have in the presence of axisymmetry. Kinking is usually presented [e.g., Bateman, 1978] as the result of an instability which develops on an axisymmetric, static current channel. This current channel, in turn, is regarded as the natural "equilibrium" configuration of a plasma when an axial electric field, or voltage drop, forces a current to flow along an axial magnetic field. This way of viewing the problem has dominated the MHD of fusion research, and also plays a paradigmatic role in astrophysics and space plasma physics. It has been difficult to think outside this framework.

However, a case can be made that a kinked, rather than an axisymmetric, MHD current channel is in fact the natural configuration of a magnetofluid driven by a voltage applied along a dc magnetic field, any time the resistivity and viscosity are small enough [Montgomery, Phillips and Theobald, 1989]. The ratio of helical to axisymmetric components can vary, depending upon circumstances, and the helical deformations are accompanied by a finite vortical flow velocity. It seems to be that it is the helically-deformed configuration, not the axisymmetric one, to which the plasma will naturally relax after a turbulent formation phase. It is likely that it is this helical state with vortical flow that should be the object of stability studies, if any state is. It is our best candidate for a class of *resistive*, as contrasted with *ideal*, equilibria.

This possibility was first suggested by a series of three-dimensional MHD numerical computations [Dahlburg, et al., 1985; 1986a,b; 1987; 1988]. These computations concerned currents driven in cylinders with periodically identified ends and conducting walls that bounded incompressible magnetofluids supported by axial dc magnetic fields. These included calculations which were driven and undriven; Strauss approximation and full MHD; with and without local temperature variation and temperature-dependent transport coefficients. In every case, partially-helical magnetic and current channels developed, accompanied by a few percent of kinetic energy of flow, concentrated mostly in a pair of counterrotating helical vortices of low axial and azimuthal mode number, the same mode numbers as for the magnetic variables. For some time, explanations of these partially helical states were sought in instabilities, in numerical effects, and in attempted extensions and modifications of Taylor's "minimum energy" principle [Taylor, 1974, 1986].

For several years, our work had involved turbulent initial value problems in which the various global ideal invariants would decay at unequal rates [Montgomery, et al., 1978, Matthaeus and Montgomery, 1980, Ting, et al., 1986, for example]. Extremal states in which rapidly-decaying invariants decay almost to the minimum possible values, compatible with the numerical values of the slowly-decaying invariants, often resulted. These processes were variously dubbed "selective decay" and "dynamic alignment," depending upon which quantity decayed rapidly relative to which other. None of these explanations seemed quite satisfactory, particularly for the driven, steady-state problem. For a time, it seemed that perhaps relaxation back to axisymmetry was being thwarted by a lack of temperature dependence of the transport coefficients and a self-consistent evolution of the heat flow. Eventually, the calculations reached the point [Theobald, et al., 1989] of allowing local temperature dependences, expecting to see "sawtooth oscillations" accompanied by periodic relaxation to axisymmetry, according to a scenario proposed by Kadomtsev [1975].

We did see the sawteeth, but not a relaxation to axisymmetry. The helical current channel again formed, and then executed oscillations in temperature and kinetic activity that in many ways looked convincingly like laboratory sawteeth. The principal novelty was that the pulsating was being done by a helical, rather than an axisymmetric, current channel. At that point, we gave up expecting to see axisymmetric states and sought explanations outside the framework of zero-flow equilibria.

At present, the most convincing explanation we have been able to put forward was one in terms of the *minimum rate of energy dissipation.* This is an only partially proved but useful 19th century principle which is apparently a simple ancestor of the more contemporary, more elaborate, and more controversial principle of minimum entropy production rate [e.g., Prigogine, 1947, Keizer, 1987] about which we shall say no more here. The minimum dissipation rate principle was apparently first articulated by Kirchoff in 1848 [see, in particular, Jaynes, 1980], for rigid but spatially non-uniform electrical conductors. It can be proved that a given distribution of electrostatic potential over the boundary of a material with conductivity $\sigma(\mathbf{x}) \equiv 1/\eta(\mathbf{x})$ leads to a current distribution which minimizes the total Ohmic dissipation R_j, where $R_j \equiv \int \eta \mathbf{j}^2 d^3x = \int \sigma(\nabla\phi)^2 d^3x$, with current density $\mathbf{j}$ and electrostatic potential ϕ. Later in the century, Helmholtz, Korteweg, and Lord Rayleigh (see Lamb's *Hydrodynamics*, 1945) showed that the various elementary incompressible hydrodynamic flows could be calculated by minimizing the viscous dissipation rate R_ω, where $R_\omega \equiv \int \rho_0 \nu \boldsymbol{\omega}^2 d^3x$ (ρ_0 = mass density, considered uniform, ν = kinematic viscosity, and $\boldsymbol{\omega} = \nabla\times\mathbf{v}$ = vorticity, $\mathbf{v}$ = fluid velocity). For other circumstances (e.g., turbulence), the flows are not at present calculable as minimum dissipation states in any recognizable sense; but for either electrically or mechanically driven states with not too high Lundquist and Reynolds numbers, there seems to be reason to believe that the elementary states are predicted by this minimum principle.

All systems, turbulent or not, have a universal tendency to relax toward thermal equilibrium and spatial uniformity. Systems driven mechanically or electrically at the boundary are prohibited from reaching spatial uniformity but presumably would like to get as close as they can. Since $\boldsymbol{\omega} = \nabla\times\mathbf{v}$ and $\mathbf{j} = \nabla\times\mathbf{B}$ ($\mathbf{B}$ is the magnetic field), R_ω and R_j are good mean square measures of the magnitudes of the gradients of $\mathbf{v}$ and $\mathbf{B}$, which for incompressible flow are both divergenceless fields. There seem to be no known counter examples that would indicate anything other than minimal dissipation rates as long as the systems are not driven too hard. (In particular, the minimization of R_ω is *not* restricted to Stokes flow, as is sometimes thought; it works for rotating Couette flow.)

It seems natural to attempt to find the elementary states of driven, dissipative MHD by minimizing the combined MHD dissipation integral, $R_\omega + R_j$. It should be kept in mind that we probably do not yet know the elementary states of driven, resistive, viscous magnetofluids, in most cases. Except for Hartmann flow and perhaps one or two other examples, most of the explicit, bounded, steady-state solutions to the MHD equations that we have are zero-flow solutions of the *ideal* MHD equations ($\nu = 0$, $\eta = 0$). These are of course very numerous, but for the following reasons, very likely irrelevant to the non-ideal case.

Consider the induction equation of MHD,

$$\frac{\partial \mathbf{B}}{\partial t} = -\nabla\times\mathbf{E} = \nabla\times(\mathbf{v}\times\mathbf{B} - \eta\mathbf{j}) = \nabla\times(\mathbf{v}\times\mathbf{B}) + \eta\nabla^2\mathbf{B}, \tag{1}$$

where we are now in dimensionless units, and η is now the magnetic diffusivity; $\mathbf{E} = -\mathbf{v}\times\mathbf{B} + \eta\mathbf{j}$ is the electric field; and $\nabla\times\mathbf{j} = -\nabla^2\mathbf{B}$, with η assumed constant and uniform. If $\partial/\partial t = 0$ and $\mathbf{v} = 0$, the induction equation reduces to $\nabla\times\mathbf{j} = 0$. This implies $\mathbf{j} = \nabla\psi$, where ψ is a scalar solution of Laplace's equation, $\nabla^2\psi = 0$. Solutions to Laplace's equation are determined mainly by boundary conditions. If these are axisymmetric, the only remaining static zero-flow current profile is $\mathbf{j} = j_0\hat{e}_z$, where j_0 is constant and $\hat{e}_z$ is in the direction of the cylinder axis. Certainly *not* permitted are the axisymmetric ideal current profiles $\mathbf{j} = \mathbf{j}^{(0)} = (0, j_\varphi(r), j_z(r))$, which are popular in fusion research, with the dependence on r (cylindrical coordinates now) being essentially arbitrary! Resistive equilibria are very much harder to come by than ideal ones.

A heretofore acceptable answer to the observation that such ideal current profiles as $\mathbf{j}^{(0)}$ fail to satisfy $\nabla\times\mathbf{j}^{(0)} = 0$ has been that "η is very small." However, the next step of the game is then to *linearize* the MHD equations, including eq. (1), about this ideal axisymmetric equilibrium, writing $\mathbf{B} = \mathbf{B}^{(0)} + \epsilon\mathbf{B}^{(1)} + \ldots$, $\mathbf{j} = \mathbf{j}^{(0)} + \epsilon\mathbf{j}^{(1)} + \ldots$, $\mathbf{v} = \epsilon\mathbf{v}^{(1)} + \ldots$, where $\nabla\times\mathbf{B}^{(0)} = \mathbf{j}^{(0)}$, and the first order perturbative fields are written with a formal expansion parameter ϵ. The $\eta\nabla\times\mathbf{j}^{(1)}$ is then retained in the linearized version of eq. (1), even though $\eta\nabla\times\mathbf{j}^{(0)}$ has been discarded from the zeroth order and never reappears at any order in the perturbation theory. Following this procedure, many "resistive instabilities" have been calculated and their "nonlinear saturation" explored. Our thinking about current-carrying magnetically-supported plasmas has been to an extraordinary degree conditioned by this way of viewing the problem, and sophistication has been brought to bear in developing this assumption that is lacking in the assumption itself.

We may ask what happens if we treat η as small but non-zero. In what sense, then, do the static, zero-flow solutions of eq. (1) approximate the ideal equilibria as $\eta \to 0$? The answer is, apparently not at all, except for the one constant current profile $\mathbf{j} = j_0\hat{e}_z$, where j_0 is a constant, proportional to the applied axial electric field.

So what states are achieved if an axial electric field is applied to a periodic cylinder of conducting magnetofluid in the presence of boundary conditions? It depends of course on the boundary conditions and these are still matters of some discussion. Two popular electrical boundary conditions are the following. (1) Put a rigid, perfectly-conducting boundary at $r = a$, say. This requires $\mathbf{B}\cdot\hat{n} = 0$ and $\mathbf{j}\times\hat{n} = 0$, where $\hat{n} = \hat{e}_r$ is the unit normal to the conductor. (2) Alternatively, put a perfect conductor coated with a thin layer of insulating dielectric at $r = a$. This leaves $\mathbf{B}\cdot\hat{n}$ alone but changes $\mathbf{j}\times\hat{n} = 0$ to $\mathbf{j}\cdot\hat{n} = 0$. The following remarks are confined to case (2), even though more numerical computations have been done for case (1), because case (2) is analytically the more tractable. (Neither assumption represents the laboratory situation very well).

There is a solution $\mathbf{j} = j_0\hat{e}_z$, where $j_0 = E_0/\eta$ and E_0 is the applied electric field. This is the minimum dissipation state *if* we assume at the outset that there is no flow. This seems like a reasonable assumption, since R_ω is positive semi-definite. But an unexpected result is [Montgomery, et al., 1989] that it is possible to *lower* the total dissipation rate below $\mathbf{v} = 0$ minimum by permitting a finite $\mathbf{v}$. The discussion of the $\mathbf{v} \neq 0$ state, which has at this point been given only in perturbation theory, will be given presently. It has not yet been possible to solve the general variational problem of minimizing $R_j + R_\omega$ with $\mathbf{v}$ and $\mathbf{B}$ both free to vary, though this is under investigation.

The difficulty lies in the nature of the constraints that need to be taken into account [Montgomery and Phillips, 1988]. One constraint is the constant rate of supply and dissipation of magnetic helicity, $<\mathbf{j}\cdot\mathbf{B}> =$ const., which the electric field imposes. This is a relatively straightforward constraint to take into account, but a much more difficult one is the full generalized Ohm's law, which amounts to a *pointwise* constraint relating $\mathbf{v}$ to $\mathbf{B}$ [Montgomery, et al., 1989]:

$$0 = -\eta\mathbf{j} + \mathbf{v}\times\mathbf{B} + E_0\hat{e}_z + \nabla\tilde{\Phi}. \tag{2}$$

Here, E_0 is the spatially averaged value of the electric field and $\nabla\tilde{\Phi}$ is what is left over in the electric field and has zero spatial average. $\tilde{\Phi}$ is obtained from the Poisson equation

$$\nabla^2\tilde{\Phi} = -\nabla\cdot(\mathbf{v}\times\mathbf{B}). \tag{3}$$

$\tilde{\Phi}$ must be expressed in terms of $\mathbf{v}$ and $\mathbf{B}$ *before* the variation is carried out. The fact that (2) must be satisfied at each spatial point, with $\tilde{\Phi} = \tilde{\Phi}(\mathbf{v},\mathbf{B})$, greatly complicates the variational problem beyond the level of the inclusion of simple integral constraints of the usual global variety; it moves the problem into a rather unfamiliar area of the calculus of variations.

What can be done is to construct a perturbation-theoretic solution which lowers the total dissipation rate below the $\mathbf{v} = 0$ minimum value, $\eta j_0^2 = E_0^2/\eta$ per unit volume. One goes up to terms of third order in the series expansion

$$\mathbf{B} = \mathbf{B}^{(0)} + \epsilon\mathbf{B}^{(1)} + \epsilon^2\mathbf{B}^{(2)} + \ \ldots \tag{4a}$$

$$\mathbf{v} = \epsilon\mathbf{v}^{(1)} + \epsilon^2\mathbf{v}^{(2)} + \ \ldots \tag{4b}$$

where the small parameter ϵ now indicates the smallness of the helical contributions $\mathbf{B}^{(1)}$, $\mathbf{v}^{(1)}$ relative to the axisymmetric $\mathbf{B}^{(0)}$.

Here, $\mathbf{B}^{(0)} = B_0\hat{e}_z + (rj_0/2)\hat{e}_\varphi$, and $\nabla\times\mathbf{B}^{(0)} = j_0\hat{e}_z$, with j_0 and B_0 both constant and uniform. This zeroth order is the state of minimum dissipation if $\mathbf{v} = 0$.

The first order solution $\mathbf{v}^{(1)}$, $\mathbf{B}^{(1)}$ is the linear solution indicated by Storer [1983]. It is

$$\mathbf{B}^{(1)} = \lambda\nabla\times\hat{e}_z\rho + \nabla\times(\nabla\times\hat{e}_z\rho), \tag{5}$$

where $\rho = J_m(\gamma r)\exp(im\varphi - ik_nz)\exp(-i\Omega t)$. J_m is the Bessel function of order m; m is the poloidal mode number ($m = 0, \pm1, \pm2, \ldots$); $k_n = 2\pi n/L_z$, where n is the axial mode number ($n = 0, \pm1, \pm2, \ldots$) and L_z is the periodicity length in

z; $\lambda^2 = \gamma^2 + k_n^2$, and is determined by the requirement that $B_r^{(1)} = 0$ at $r = a$. $\mathbf{v}^{(1)}$ is given in terms of $\mathbf{B}^{(1)}$ by

$$\mathbf{v}^{(1)} = \frac{(\Omega + i\eta\lambda^2)}{k_n B_0 - (mj_0/2)} \mathbf{B}^{(1)} \tag{6}$$

provided Ω is determined by

$$\Omega = -\frac{i(\eta+\nu)\lambda^2}{2}\left\{1 \pm \left[1 - 4\frac{(f + \nu\eta\lambda^2)}{(\eta+\nu)^2\lambda^4}\right]^{1/2}\right\}, \tag{7}$$

where $f \equiv ((mj_0/2) - k_n B_0)\,[(mj_0/2) - k_n B_0 + k_n j_0/\lambda]$.

At threshold in which the first unstable normal mode appears as j_0 is raised, $\Omega = 0$ and the first-order solution is static. It then generates higher order solutions in the expansion, through the nonlinear terms.

The second-order solution has $\mathbf{v}^{(2)} = 0$,

$$\mathbf{j}^{(2)} = \frac{1}{\eta}\mathbf{v}^{(1)} \times \mathbf{B}^{(1)} \tag{8}$$

and $\nabla \times \mathbf{B}^{(2)} = \mathbf{j}^{(2)}$. Real parts of (5) and (6) must be used to go above first order. Both $\mathbf{j}^{(2)}$ and $\mathbf{B}^{(2)}$ are axisymmetric functions of r only, and have only $\hat{e}_\varphi$ and $\hat{e}_z$ components. The direction of $\mathbf{j}^{(2)}$ is such as to oppose $j_0\hat{e}_z$, and the total toroidal current is depressed below its zeroth order value. This is adequate to reduce the overall dissipation, Ohmic plus viscous, below its zeroth order value; for, considering the total dissipation (dimensionless units) through second order,

$$R_\omega + R_j = \eta < (\mathbf{j}^{(0)})^2 > + 2\eta < \mathbf{j}^{(0)} \cdot \mathbf{j}^{(2)} >$$

$$+ \eta < (\mathbf{j}^{(1)})^2 > + \nu < (\boldsymbol{\omega}^{(1)})^2 > + \ldots, \tag{9}$$

$2\eta < \mathbf{j}^{(0)} \cdot \mathbf{j}^{(2)} >$ is < 0 and is exactly twice the magnitude it needs to have in order to be able to compensate for the extra dissipation provided by the square of the first-order terms. A nonlinear solution with (so far arbitrary) amplitude for $\mathbf{B}^{(1)}$ results.

Thus, allowing a small amount of the non-axisymmetric solution (6) and (8), with the first linearly unstable pair of (m, n) numbers, will lower the dissipation below the level of the axisymmetric state. An amplitude limitation is not provided at this order in perturbation theory; for any multiple of $\mathbf{v}^{(1)}$, $\mathbf{B}^{(1)}$ is also a solution, if the square of that multiple is used to multiply the second-order solution. All we can say at this order in perturbation theory is that the dissipation has been lowered. It probably requires quartic terms in $R_\omega + R_j$, at least, to provide a positive-definite contribution to $R_\omega + R_j$ that will eventually dominate, as the overall amplitude of the nonaxisymmetric part increases, and thus determines a limiting amplitude. This problem is presently under investigation, but the fourth-order perturbation theory is difficult.

What can be done is to make graphical comparisons between the first three orders of the analytical solution (eqs. (5)-(8)) and similar displays constructed from the numerical solutions of the MHD and/or Strauss equations. The geometrical parameters (a, L_z, etc.) and the mean magnetic field B_0 may be chosen as similarly as the difference between the square and circular boundary will allow ("minor radius" $= a =$ half the width of the square), and then try to choose the dominant m, n number pair and overall amplitude to match the results of the computation. This has been done in Montgomery, Phillips, and Theobald [1989], where a comparison of the $m = n = 1$ analytic solution has been made with the results of a driven, 3D, Strauss equations computation [Theobald, et al., 1989]. The results are somewhat striking, even though the conditions of the two solutions are far from identical. In particular, the double vortex structure and the partially helical current channel are notably similar. The similarity seems too close to be accidental.

We are at present experimenting with a calculational scheme which shows promise of determining the amplitudes of the helical components non-perturbation-theoretically. These results will be reported elsewhere (Montgomery, 1989).

A major obstacle to a completed theory at this point lies in the area of boundary conditions. The analytical theory would feel most satisfactory if the following sets of boundary conditions, magnetic and mechanical, had been imposed. (1) *Magnetic boundary conditions* should be $\mathbf{B} \cdot \hat{n} = 0$, for a perfect conductor at $r = a$, and either $\mathbf{j} \times \hat{n} = 0$ (for a bare perfect conductor) or $\mathbf{j} \cdot \hat{n} = 0$ (for a perfect conductor coated with a thin layer of insulating dielectric). These boundary conditions may be considered to have been adequately satisfied, though it is the case that $\mathbf{j} \times \hat{n} = 0$ in the computations while the analytical solutions (so far) have had $\mathbf{j} \cdot \hat{n} = 0$. (2) *Mechanical boundary conditions* should be $\mathbf{v} \cdot \hat{n} = 0$ at $r = a$ and either $\hat{n} \times \mathbf{v} = 0$ (no slip) or $\hat{n} \cdot \boldsymbol{\sigma} = 0$ (stress free, where $\boldsymbol{\sigma}$ is the viscous stress tensor) at $r = a$. In fact, the $\mathbf{v} \cdot \hat{n} = 0$ is satisfied both in the computations and in the analytical solution, but the tangential boundary conditions are stress free in the computations and $\boldsymbol{\omega}\cdot\hat{n}=0$ in the analytical solution (both have finite $\mathbf{v}\times\hat{n}$). It will be a far from trivial matter to bring the boundary conditions into line.

Neither set of boundary conditions actually represents the wall conditions very well in the current generation of plasma experiments. The walls are to a significant extent chemically interactive with a hot plasma, and neither perfectly conducting nor perfectly insulating, compared with standard expressions for plasma conductivities. Since most present experiments are targeted towards hot plasmas, there tend to be enormous temperature drops between the centers of the discharges and the walls. Plasma dissipation coefficients typically fall off with increasing powers of the temperature (e.g., conductivity $\sim T^{3/2}$ and shear viscosity $\sim T^{-1/2}$, in many cases), which means that in a classic thermodynamic sense, the wall region is likely to be far more dissipative than the interior and will constitute a boundary layer. It remains to be seen whether boundary layer theory can be adapted so as to match analytical solutions like those we have displayed to no-slip or stress-free mechanical boundaries.

We are currently exploring the possibilities for minimizing the dissipation using an expansion of the fields in eigenfunctions of the curl (Chandrasekhar-Kendall functions). The method shows promise and will be reported in detail elsewhere.

ACKNOWLEDGEMENT

This work was supported in part by NASA Grant NAG-W-710 and U.S. Department of Energy Grant DE-FG02-85ER51394, and in part under the auspices of the U.S. Department of Energy at Los Alamos National Laboratory.

REFERENCES

Bateman, G. 1978 *MHD Instabilities.* MIT Press, Cambridge (MA).

Dahlburg, J.P., Montgomery, D. and Matthaeus, W.H. 1985 *J. Plasma Phys.* **34**, 1.

Dahlburg, J.P., Montgomery, D., Doolen, G.D. and Matthaeus, W.H. 1986a *J. Plasma Phys.* **35**, 1.

Dahlburg, J.P., Montgomery, D., Doolen, G.D. and Turner, L. 1986b *Phys. Rev. Lett.* **57**, 428.

Dahlburg, J.P., Montgomery, D., Doolen, G.D. and Turner, L. 1987 *J. Plasma Phys.* **37**, 299.

Dahlburg, J.P., Montgomery, D., Doolen, G.D. and Turner, L. 1988 *J. Plasma Phys.* **40**, 39.

Jaynes, E.T. 1980, *Ann. Rev. Phys. Chem.* **31**, 579.

Kadomtsev, B.B. 1975 *Fiz. Plasmy* **1**, 710 (*Sov. J. Plasma Phys.,* **1**, 389).

Keizer, J. 1987 *Statistical Thermodynamics of Nonequilibrium Processes* (Springer-Verlag, New York).

Lamb, H. 1945 *Hydrodynamics*, 6th ed., Dover, New York, pp. 617-619.

Matthaeus, W.H. and Montgomery, D. 1980 *Proc. Int. Conf. on Nonlinear Dynamics, New York, 1979* (Ann. N.Y. Acad. Sci. **357**, 203).

Montgomery, D., Phillips, L. and Theobald, M.L. 1989 *Phys. Rev.* **A40**, 1515.

Montgomery, D. and Phillips, L. 1989 *Physica D* (in press).

Montgomery, D. and Phillips, L. 1988 *Phys. Rev. A* **38**, 2953.

Montgomery, D., Turner, L. and Vahala, G. 1978 *Phys. Fluids* **21**, 757.

Montgomery, D. 1989 in *Trends in Theoretical Physics*, **I**, ed. by P.J. Ellis and Y.C. Tang (New York, Addison-Wesley; to appear).

Prigogine, I. 1947 *Etude Thermodynamique des Phénoménes Irréversibles* (Dunod, Paris).

Storer, R.G. 1983 *Plasma Phys.* **25**, 1279.

Taylor, J.B. 1974 *Phys. Rev. Lett.* **33**, 1139.

Taylor, J.B. 1986 *Revs. Mod. Phys.* **58**, 741.

Theobald, M.L., Montgomery, D., Doolen, G.D. and Dahlburg, J.P. 1989 *Phys. Fluids* **B1**, 766.

Ting, A.C., Matthaeus, W.H. and Montgomery, D. 1986 *Phys. Fluids* **29**, 326.

Island Formation in Magnetostatics and Euler Flows

CHRIS C. HEGNA & A. BHATTACHARJEE

Department of Applied Physics, Columbia University

"A not uncommon phenomenon in science is the rapid transition, without apparent cause, from one state in which a concept is almost universally disbelieved and rejected to another state in which it is so transparent as to require no comment and exhibit no visible history. Sometimes there may be a transitional period during which both states coexist simultaneously (and schizoidally) in the scientific community. It is my belief that the subject of this paper is either in the course of such a transition or close to it (but in which state it lies preponderantly, I am not willing to guess)."

H. Grad in *Proceedings of the Workshop on Mathematical Aspects of Fluid and Plasma Dynamics*, Trieste, May 30–June 2, 1984 (Università Degli Studi Di Trieste, Facultà Di Scienzi, Istituto Di Mechanica), pp 253–282.

1. INTRODUCTION

The problem of existence and structure of magnetostatic equilibria has been a preoccupation of plasma physicists for almost three decades (Kruskal and Kulsrud, 1958; Grad, 1967). Apart from the fundamental interest in this problem, it is of considerable practical importance for the confinement of thermonuclear plasmas in toroidal devices. The equations governing magnetostatic equilibria are,

$$\mathbf{J} \times \mathbf{B} = \nabla p \quad , \tag{1}$$

$$\nabla \times \mathbf{B} = \mathbf{J} \quad , \tag{2}$$

$$\nabla \cdot \mathbf{B} = 0 \quad , \tag{3}$$

where $\mathbf{J}$ is the current density, $\mathbf{B}$ is the magnetic field and p is the plasma pressure. When the toroidal device is axisymmetric, the magnetic field-lines, which constitute a Hamiltonian system, lie on flux surfaces which are topologically toroidal. However, in the absence of a direction of symmetry, the equations of field-line flow are not generally integrable. Even the vacuum field, which obeys the condition $\nabla \times \mathbf{B} = 0$ and is analytically soluble in toroidal coordinates (Morse and Feshbach, 1953), does not generally lie on nested surfaces. We may be tempted to sidestep the problem of non-existence of surfaces by asserting that because $\mathbf{B} \cdot \nabla p = 0$, field-lines must lie on a smooth family of equilibria with nested p-surfaces. However, the fallacy in this assertion can be seen by taking the $p \to 0$, $\mathbf{J} \to 0$ limit of equation (1). This limit is uniform and unique, and its existence would require the vacuum field to lie on nested flux surfaces, which is in clear contradiction with the known property of non-integrability of the vacuum field. This counter-example also indicates the singular nature of the low-pressure expansion around a three-dimensional vacuum field (Grad, 1984).

Undaunted by the formidable mathematical difficulties of the problem of non-existence of magnetic surfaces, plasma physicists have searched for solutions which "improve" the quality of flux surfaces in vacuum. A typical Poincaré section of field-line flow shows the appearance of islands and stochastic regions which preclude the possibility of a dense set of flux surfaces, and the purpose of the search is to eliminate the islands and stochastic regions as much as possible. This makes practical sense, and a way to achieve it is to "compensate" for symmetry-breaking in a vacuum field by imposing external magnetic perturbations (Cary, 1984). The mathematical procedure uses a variant of Hamiltonian perturbation theory (Cary and Littlejohn, 1983) which can be carried out in principle, order by order (but usually stops at the first order). The procedure does not converge to an integrable field, but merely postpones the divergence to some higher order which is argued to be unimportant for practical purposes. The result of the exercise is a "visible" reduction in the size of islands and chaotic regions over a wide region of a toroidal plasma. This procedure raises an interesting question: if we begin with a nearly-integrable, vacuum magnetic field which lies on nested flux surfaces, what is the effect of a small scalar plasma pressure on the equilibrium? This question can be posed in a somewhat more general way: if we postulate, a priori, that a three-dimensional magnetic field lies on nested surfaces, what is the structure of the solutions of the magnetostatic equations? We show in Section 2 that the magnetostatic equations can be inverted exactly to obtain a formula for the projection of $\mathbf{J}$ parallel to $\mathbf{B}$ (represented by $J_{\parallel}$). This formula [equation (21)] shows that $J_{\parallel}$ is singular at rational surfaces. To find the constant multiplying the δ-function, we have to resolve the singularity at a rational surface. This can be done by a boundary-layer analysis (Cary and Kotschenreuther, 1985; Hegna and Bhattacharjee, 1989) in which the parallel current in the "outer layer" is given by equation (21), the "inner layer" is localised at a particular rational surface, and the effect of coupling to other rational surfaces is neglected. The latter is a valid approximation when the island-width is not too large. The asymptotic matching between the inner and outer layers, briefly described in Section 3, determines the constant multiplying the δ-function in equation (21). In order to complete the calculation, it is necessary to determine the pressure in the neighborhood of the island. Clearly, equations (1) – (3) do not determine the pressure, except for the requirement that the pressure be constant on a magnetic surface. In an almost static, steady-state plasma which has a small but finite resistivity η , an important additional constraint is provided by Ohm's law,

$$-\nabla\phi + \mathbf{v}\times\mathbf{B} = \eta\mathbf{J} \quad , \tag{4}$$

where ϕ is the electrostatic potential (assumed to be single-valued), and $\mathbf{v}$ is the small diffusion velocity. It is not difficult to show that equation (4) leads to a diffusion equation for p [Kruskal and Kulsrud, 1958; Cary and Kotschenreuther, 1984; Hegna and Bhattacharjee, 1989], and thus provides an important ingredient for completing the calculation. Writing $\mathbf{J}_{\parallel} = Q\mathbf{B}$, we note that equation (4) implies,

$$Q = -\frac{1}{\eta B^2}\,\mathbf{B}\cdot\nabla\phi \quad , \tag{5}$$

which suggests that the presence of a small but finite η is crucial in resolving the singularity in the parallel current. Integrating equation (5) over a surface bounded by magnetic field-lines, we obtained the solubility condition,

$$\eta\int d\tau\, QB^2 = 0 \quad . \tag{6}$$

By integrating Ampere's Law $\nabla\times\mathbf{B} = \mathbf{J}$ under a few simplifying approximations discussed in Section 3, it is then possible to obtain an expression for the nonlinear island-width.

The expression for the island-width is found to depend on the sign of the quantity D_R , which determines the stability of a toroidal plasma with respect to resistive interchange modes, and was first introduced in the stability literature by Glasser, et al. (1975). Why should the island-width in a three-dimensional equilibrium calculation depend on a stability criterion for axisymmetric plasmas? The answer is to be found in the resemblance between two-dimensional axisymmetric equilibria with saturated helical instabilities and intrinsically three-dimensional configurations. If a two-dimensional equilibrium is unstable, a perturbation grows from the equilibrium until a nonlinear process saturates the mode and alters the equilibrium. A way to think about a three-dimensional equilibrium is a two-dimensional equilibrium with an intrinsic symmetry-breaking perturbation; if the equilibrium is unstable, it will have the same island structure as the axisymmetric device with saturated instabilities.

The viewpoint we have presented above sets the stage for our consideration of three-dimensional Euler flows. Moffatt (1985, 1986) has recently used a method of magnetic relaxation to demonstrate the existence of magnetostatic equilibria, and by analogy, of steady Euler flows. The analogy is immediately apparent if we write down the equations,

$$\mathbf{u} \times \omega = \nabla h \quad , \tag{7}$$

$$\nabla \times \mathbf{u} = \omega \quad , \tag{8}$$

$$\nabla \cdot \mathbf{u} = 0 \quad , \tag{9}$$

which describe the steady, inviscid flow of an incompressible fluid of constant density. (Here $h = p/\rho + (1/2)u^2$, ρ is the density, $\mathbf{u}$ is the velocity and ω is the vorticity.) The analogy between the variables $\mathbf{B} \leftrightarrow \mathbf{u}$, $\mathbf{J} \leftrightarrow \omega$, and $p \leftrightarrow h_0 - h$ (for some constant h_0) is obvious. This means that the component of ω parallel to $\mathbf{u}$ ($\omega_{\|}$) in a toroidal sheared flow has the same singularity as $\mathbf{J}_{\|}$ in a torus. Equation (25) in Section 2 shows this singularity in the vicinity of a rational surface explicitly.

The boundary-layer analysis for Euler flows can be set up in analogy with corresponding analysis for magnetostatic equilibria, but only up to a point. A fundamental distinction between magnetostatics and Euler flows lies in the constraints which determine p and h, respectively. Whereas in magnetostatics, a small but finite dissipation is introduced via Ohm's law, the natural source of dissipation in steady-state Euler flows is viscosity via the Navier-Stokes equation, which can be written in the form,

$$\mathbf{u} \times \omega = \nabla h - \nu \nabla^2 \mathbf{u} \quad , \tag{10}$$

where the constant coefficient ν is taken to be small. The curl of equation (10) gives

$$\nabla \times (\mathbf{u} \times \omega) = -\nu \nabla^2 \omega \quad , \tag{11}$$

which should be contrasted with the curl of equation (4), given by,

$$\nabla \times (\mathbf{v} \times \mathbf{B}) = -\eta \nabla^2 \mathbf{B} \quad . \tag{12}$$

Equations (11) and (12) suggest that the analogy $\mathbf{B} \leftrightarrow \mathbf{u}$ which holds for equations (1) – (3) and (6) – (8) does not carry over to equations (11) and (12) for which it would be more appropriate to consider the analogy $\mathbf{B} \leftrightarrow \omega$. Moffatt (1986) has shown that though magnetostatic equilibria and steady Euler flows are analogous, their linear stability properties have vital differences, a point we will return to later.

The projection of equation (10) along $\mathbf{u}$ gives,

$$\omega^2 - \nabla^2 h = -\frac{1}{\nu} \mathbf{u} \cdot \nabla h \quad . \tag{13}$$

Integrating equation (13) over a volume bounded by velocity field-lines, we get the solubility condition,

$$\nu \int d\tau \, [\nabla^2 h - \omega^2] = 0 \quad . \tag{14}$$

Equation (14) is nothing but a statement of conservation of energy (Landau and Lifshitz, 1986) in a steady-state. In a steady-state, the time-rate of change of the kinetic energy of a fixed volume of fluid vanishes. Hence, the flux of energy into a given volume must be balanced by the energy dissipation within the volume.

As shown in Section 4, the differences between equations (5) and (13) (or equivalently, between equations (6) and (14)) lead to important differences between the expressions for the island-width in toroidal magnetostatic equilibria and Euler flows. The island-width for magnetostatic equilibria, as stated earlier, can be related to the stability criterion with respect to resistive interchange modes, and can be made innocuously small by arranging the plasma configuration to be stable with respect to interchange modes. However, the expression for the island-width for Euler flows suggests that it would be very difficult to avoid large islands in sheared, toroidal flows. The fluid equilibria are thus more fragile than their plasma counterparts, and will generally tend to exhibit spatial chaos due to island overlap. In this respect, our result lends indirect support to Moffatt's conjecture that "topologically non-trivial Euler flows" are "unstable" (Moffatt, 1986). Other possible implications of our result will be considered in the Conclusion of this paper.

2. TANGENTIAL DISCONTINUITIES IN MAGNETOSTATIC EQUILIBRIA AND EULER FLOWS

In a toroidal plasma, the magnetic field **B** is taken to lie on flux surfaces far from the rational surface of interest. Each flux surface is labeled by Φ, where $2\pi\Phi$ is the toroidal flux through the surface. The poloidal and toroidal angles are described by θ and ϕ, respectively. **B** can be represented in the contravariant form,

$$\mathbf{B} = \nabla\Phi \times \nabla(\theta - \iota\!\!\!-\, \phi) \quad , \tag{15}$$

where $\iota\!\!\!- = \iota\!\!\!-(\Phi)$ is the rotational transform, which measures the twist of the magnetic field-line. Equation (15) is reminiscent of the well-known Clebsch representation for divergence-free vector fields. In the covariant basis, **B** is represented by (Boozer, 1981),

$$\mathbf{B} = g(\Phi)\nabla\phi + I(\Phi)\nabla\theta + \beta(\Phi, \theta, \phi)\nabla\Phi \quad . \tag{16}$$

The Jacobian $\mathcal{J} \equiv (\nabla\Phi\cdot\nabla\theta \times \nabla\phi)^{-1}$ can be obtained from the scalar product of equations (15) and (16);

$$\mathcal{J} = \frac{g + \iota\!\!\!-\, I}{B^2} \equiv \frac{\gamma(\Phi)}{B^2} \quad . \tag{17}$$

$\mathcal{J}$, as well as all (single-valued) equilibrium quantities, can be expressed as a Fourier series. In particular,

$$\mathcal{J} = \sum_{m,n} \mathcal{J}_{mn}(\Phi) \exp(im\theta - in\phi) \quad . \tag{18}$$

From equation (1), we obtain,

$$\mathbf{J}_\perp = \frac{\mathcal{J}}{\gamma} p' \nabla\Phi \times [g\nabla\phi + I\nabla\theta] \ , \tag{19}$$

and

$$\mathbf{J}_\parallel = Q\mathbf{B} = \sum_{m,n} Q_{mn}(\Phi) \exp(im\theta - in\phi) \ \mathbf{B} \quad . \tag{20}$$

From equation (2), which implies that $\nabla\cdot\mathbf{J} = 0$, we obtain the general solution,

$$Q_{mn} = -p' \mathcal{J}_{mn}\left(\frac{1}{\bar{\iota} - n/m} - \frac{I}{\gamma} \right) + \hat{Q}_{mn}\, \delta(\Phi - \Phi_0) \quad . \tag{21}$$

Here $\Phi = \Phi_0$ at the rational surface $\bar{\iota} = n/m$. Clearly, Q_{mn} is singular at the rational surface. However, the amplitude of the current sheet $\hat{Q}_{mn}$, which constitutes a tangential discontinuity, is not specified by the "exterior" solution.

Equation (21) is a slight generalization of our earlier result (Hegna and Bhattacharjee, 1989), and shows that the parallel current $\mathbf{J}_\parallel$ is intrinsically singular. This result was first obtained by Cary and Kotschenreuther (1985) who used a linear expansion around a vacuum field. Our demonstration of the singularity shows that it is not an artifact of linearization. ["The linear expansion," cautioned Grad, "is inherently unreliable; it does not easily distinguish pathology (nonexistence) from the mere inadequacy of linear representation for an essential nonlinearity such as a helical island. The appearance of a singularity in a linear solution, in principle, casts doubt on the linearization only." (Grad, 1984) Grad's remark, which is worth bearing in mind, should not be interpreted in the context of our present discussion to mean that the analysis of Cary and Kotschenreuther is flawed. Though they used the linear approximation in the exterior region, which restricts the realm of validity of their result, they were careful to avoid linearization in the vicinity of a rational surface where magnetic islands are formed.]

By analogy with magnetostatics, we can write,

$$\mathbf{u} = g^E(\Phi^E)\nabla\phi + I^E(\Phi^E)\nabla\theta + \beta^E(\Phi^E,\theta,\phi)\nabla\Phi^E \quad , \tag{22}$$

where the superscript E designates the analogous functions for Euler flows. From equations (7) and (8), it follows that,

$$\omega_{\perp} = \frac{\ell^E}{\gamma^E} h' \nabla\Phi^E \times [g^E \nabla\phi + I^E \nabla\theta] \quad , \tag{23}$$

and

$$\omega_{\parallel} = Q^E \mathbf{u} = \sum_{m,n} Q^E_{mn}(\Phi) \exp(im\theta - in\phi) \ \mathbf{v} \quad , \tag{24}$$

where

$$Q^E_{mn} = h' \ell^E_{mn} \left(\frac{1}{\iota^E - n/m} - \frac{I^E}{\gamma^E} \right) + \hat{Q}^E_{mn} \, \delta(\Phi^E - \Phi^E_{mn}) \quad . \tag{25}$$

Thus, $\omega_{\parallel}$ has a singular structure exactly analogous to $\mathbf{J}_{\parallel}$.

The singularity in $\mathbf{J}_{\parallel}$ (or $\omega_{\parallel}$) gives rise to a discontinuity in $\mathbf{B}$ (or $\mathbf{u}$) at the rational surface. This discontinuity is measured by the parameter Δ' (or $\Delta^{E\prime}$) which is used to pick out the surface current (or vorticity), and subsequently matched to the current (or vorticity) obtained from the inner region. We consider the formation of a magnetic island at the rational surface $\iota = n_r/m_r$. It is convenient to transform to the new angle coordinates, $\alpha = \theta - (n_r/m_r)\phi$, $\zeta = \phi$. Introducing the vector potential $\mathbf{A}$, defined by $\mathbf{B} = \nabla \times \mathbf{A}$, we write in the new basis,

$$\mathbf{A} = \sum_{m,n} [A_{\alpha mn}(\Phi)\nabla\alpha + A_{\zeta mn}(\Phi)\nabla\zeta] \exp[im\alpha + i\{m(n_r/m_r - n\}\zeta], \tag{26}$$

where we have adopted the gauge $A_{\Phi} = 0$. At the rational surface, the resonant symmetry-breaking field is $\partial_{\alpha} \bar{A}_{\zeta}$, where the overbar indicates an average over ζ . The parameter Δ' is defined for such resonant helicity by the relation,

$$\Delta' = [\partial_{\Phi} \bar{A}_{\zeta}(\Phi_r^+) - \partial_{\Phi} \bar{A}_{\zeta}(\Phi_r^-)] \,/\, \bar{A}_{\zeta}(\Phi_r) \quad , \tag{27}$$

where $\Phi = \Phi_r$ at the rational surface. Subsequently, we use the single-harmonic approximation in which the resonant amplitude $m = m_r$ makes the dominant contribution to Δ' . By integrating $\nabla \times \nabla \times \mathbf{A} = \mathbf{J}$ across the rational surface, we obtain (Hegna and Bhattacharjee, 1989),

$$\Delta' A_{\zeta} = - g \, \overline{|\nabla\Phi|^{-2}} \; \overline{\hat{Q}} \quad , \tag{28}$$

which relates the amplitude of the singular current at the rational surface to Δ', required for the asymptotic matching between the exterior and inner regions. The analogous relation for Euler flows is obtained by attaching superscripts E to the appropriate equilibrium quantities.

3. THE INNER REGION AND THE ISLAND EQUATION

In the neighborhood of the rational surface, **B** (or **u**) -lines do not lie on surfaces of constant Φ (or Φ^E), but on level surfaces of a new invariant function, approximately given by,

$$\psi = (\iota'/2)(\Phi - \Phi_r)^2 - A_\zeta \quad , \tag{29}$$

where ι' is evaluated at $\Phi = \Phi_r$. For a magnetic island dominated by one harmonic, the island size is,

$$\delta\Phi = 2\,|\,\bar{A}_\zeta / \iota'\,|^{1/2} \tag{30}$$

the extent of which in the rotational transform is,

$$\delta\iota = 2\,|\,\bar{A}_\zeta\, \iota'\,|^{1/2} \tag{31}$$

Analogous expressions hold for Euler flows. The main purpose of the boundary-layer analysis, which has been discussed elsewhere (Hegna and Bhattacharjee, 1989), is to obtain an expression for $\bar{A}_\zeta$. $\bar{A}_\zeta$ depends on amplitude of the singular current, which can be obtained from the matching condition (28) and the solubility condition (6). We get

$$\Delta' \approx -(\delta\iota\, D_R) / A(\Phi_r) \quad , \tag{32}$$

where D_R is defined by Glasser, et al. (1975). The island width is given by,

$$\delta\iota = r/2 + \sqrt{(r/2)^2 + C} \quad , \tag{33}$$

where $r \approx D_R/m_r$ and $C \approx (\beta_t / m_r^2 \varepsilon^2)(\mathcal{J}_{m_r n_r} / \mathcal{J}_{00})$; here β_t is the plasma beta (typically a few percent for fusion plasmas) and ε is the ratio of the minor and major radii of the torus. It can be shown readily that if the plasma is unstable to resistive interchange modes, i.e., $D_R > 0$, $\delta\iota$ grows linearly with D_R. More importantly, the island width scales as m^{-1}, and the mean density of islands with $m < M$ scales as M^2 ; hence, island overlap and field-line stochasticity occurs for arbitrarily low β_t (Cary and Kotschenreuther, 1985).

If on the other hand, $D_R < 0$, the island width saturates at a very low value determined by the constant C which can be arranged to be small.

The corresponding result for Euler flows is very different, primarily because the solubility condition (14) is very different from the condition (6) for magnetostatics. The island width for Euler flows is seen to be,

$$\delta\text{ł} \approx (r^E)^{2/3} \tag{34}$$

where $r^E \approx |D_E|^{1/2} / m_r$, and D_E depends on h' and $\text{ł}^{E'}$. Since the island width scales as $m^{-2/3}$, island overlap and Beltramization is expected to occur for arbitrarily low h . Whereas for magnetostatic equilibria, it is possible to make the island width small by a choice of equilibrium profiles which make $D_R < 0$, it appears almost impossible to do the same for Euler flows.

4. CONCLUSIONS

The main aim of this paper has been to investigate the relationship between three-dimensional, sheared, toroidal magnetic fields and Euler flows. We have found that there is a strong analogy between magnetic fields and Euler flows in the context of island formation, but that the analogy extends only up to a point. Whereas the island width in a plasma can be related to the resistive interchange stability criterion, and can be made small by arranging the plasma configuration to be stable to interchange modes, the islands in Euler flow will tend to overlap and cause Beltramization.

Moffatt has conjectured that unstable Euler flows with tangential discontinuities may be "regarded as fixed points in the function space in which solutions of the unsteady Euler equations evolve" (Moffatt, 1985). While our work says nothing about this conjecture, it does suggest that if the conjecture is true, Beltramization of the flow will tend to occur. Recent simulations of decaying isotropic turbulence seem to confirm a certain tendency to Beltramization, but this does not seem to necessarily drive the depression of the mean-square value of the nonlinear term $\{ |\mathbf{u} \times \omega - \nabla h| \}^2$ in the Navier-Stokes equation (Kraichnan and Panda, 1988). These simulations point to the importance of time-variations excluded from our equilibrium analysis which, however, is carried out in a geometry more complex than that of the simulation. It is plausible that the mechanism of island overlap presented in this paper may Beltramize turbulent flows, even though these flows may never reach a true equilibrium state.

ACKNOWLEDGMENTS

This work was supported by the U. S. Department of Energy Grant No. DE-FG0286ER-53222.

REFERENCES

Boozer, A.H. 1981 Phys. Fluids 24, 1999.

Cary, J.R. 1984 Phys. Fluids 27, 119.

Cary, J.R. and Kotschenreuther, M. 1985 Phys. Fluids 28, 1392.

Cary, J.R. and Littlejohn, R.G. 1983 Ann. Phys. 151, 1.

Glasser, A.H., Greene, J.M. and Johnson, J.L. 1975 Phys. Fluids 18, 875.

Grad, H. 1967 Phys. Fluids 10, 137.

Grad, H. in *Proceedings of the Workshop on Mathematical Aspects of Fluid and Plasma Dynamics*, Trieste, May 30–June 2, 1984 (Università Degli Studi Di Trieste, Facultà Di Scienzi, Instituto Di Mechanica), pp 253–282.

Hegna, C.C. and Bhattacharjee, A. 1989 Phys. Fluids B1, 392.

Kraichnan, R.H. and Panda, R. 1988 Phys. Fluids 31, 2395.

Kruskal, M.D. and Kulsrud, R.M. 1958 Phys. Fluids 1, 265.

Landau, L.D. and Lifshitz, E.M. 1986 *Fluid Mechanics* (Pergamon Press).

Moffatt, H.K. 1985 J. Fluid Mech. 159, 359.

Moffatt, H.K. 1986 J. Fluid Mech. 166, 359.

Morse, P.M. and Feshbach, H. 1953 *Methods of Theoretical Physics* (McGraw-Hill, New York).

Steady Motion with Helical Symmetry at Large Reynolds Number

S. CHILDRESS, M. LANDMAN & H. STRAUSS

Courant Institute of Mathematical Sciences, New York University,
251 Mercer Street, New York, NY 10012, USA

1. INTRODUCTION

One of the early applications of topological ideas to fluid flows concerned the role of streamline pattern on the distribution of vorticity in steady flows at large Reynolds number. Prandtl (1904) noted that in two-dimensional incompressible flow a regular pattern of closed nested streamlines would support a constant vorticity in the limit of infinite Reynolds number, provided that the viscous stresses remained negligible. Subsequently Batchelor (1956a) rederived and strengthened this result, and considered analogous results for axisymmetric flow. He also applied the 2D result to bluff-body wakes (Batchelor 1956b), and there have been many subsequent studies of 2D inviscid flows with piecewise-constant vorticity, see e.g. Lagerstrom (1975) and Moore and Saffman (1988) for references to this literature.

The Prandtl-Batchelor theory thus aims to select steady Navier-Stokes limits for large Reynolds number from a much larger collection of steady inviscid flows, without regard to their stability to time-dependent perturbations. Little is known concerning the resulting family of limits, however, even in the 2D case. Related methods can be applied to problems of advection and diffusion of scalar and vector fields, see e.g. Rhines and Young (1983) and Childress and Soward (1989).

The object of the present paper is to apply Batchelor's method to flows with helical symmetry. This problem is of interest as an extension of Batchelor's results, since both 2D and axisymmetric flows can be view as limits of helically-symmetric flows. It is also of intrinsic interest when applied to vortical fields. Helical vortex tubes can stretch while maintaining the symmetry, even though the representation of the fluid is essentially two-dimensional. Axisymmetric structures such as vortex rings also undergo vortex stretching, but, as we have noted, the rings are here regarded as a special case of helical structures.

The fact that streamlines within a bounded region need not close in three dimensions introduces complications into any extension of the Prandtl-Batchelor theory. This point was made by Blennerhassett (1979), who studied flows of the form $(\psi_y(x,y), -\psi_x(x,y), W(x,y))$, and gave an extension of Batchelor (1956a) to this case. With helical symmetry there is a natural reduction of the various line integrals to two dimensions, and Blennerhassett's (1979) result will be recovered below as the two-dimensional limit for small pitch angle.

2. HELICAL SYMMETRY

Helical geometry has been used extensively in plasma physics, where the notation we adopt has been introduced, see e.g. Park *et al.* (1984). Let (r,θ,z) be cylindrical polar coordinates in three dimensions. We say that a scalar function of r,θ,z,t has *helical symmetry* provided that it is a function of r, ϕ, and t, where $\phi \equiv \theta + \alpha z$, α being a positive real parameter. We define the (non-unit) vector **h** by

$$\mathbf{h} \equiv \frac{r\nabla r \times \nabla \phi}{1+\alpha^2 r^2}. \tag{1}$$

We may write this as $\mathbf{h} = h^2(\mathbf{i}_z - \alpha r \mathbf{i}_\theta)$ where

$$h^2 = \frac{1}{1+\alpha^2 r^2}. \tag{2}$$

We note the following properties of **h**,

$$\nabla\cdot\mathbf{h} = 0, \quad \nabla\times\mathbf{h} = -2\alpha h^2 \mathbf{h}, \tag{3}$$

which shows that **h** is a Beltrami field, hence is itself a steady Euler flow.

The velocity field **v** and the vorticity field $\omega \equiv \nabla\times\mathbf{v}$ are both divergence-free vector fields in three dimensions. We say that these fields have helical symmetry if they can be written in the form

$$\mathbf{v} = v\mathbf{h} + \nabla u \times \mathbf{h}, \tag{4a}$$

$$\omega = \zeta\mathbf{h} + \nabla\psi\times\mathbf{h}, \tag{4b}$$

where the four scalar functions all have helical symmetry. We shall refer to the second terms on the right of (4) as the *transverse* part of the vector. It is not difficult to verify, from (1)-(3) and the helical symmetry of scalars, that (4) indeed defines solenoidal fields. The connection between the two fields is obtained from

$$\nabla\times\mathbf{v} = \omega = -h^{-2}\mathbf{h}(\nabla\cdot h^2\nabla u + 2\alpha h^4 v) - \mathbf{h}\times\nabla v \tag{5}$$

from which there follows

$$\zeta = -h^{-2}(\Delta^* u + 2\alpha h^4 v), \tag{6a}$$

$$\Delta^* u \equiv \nabla\cdot(h^2\nabla u) = r^{-2}h^2\frac{\partial^2 u}{\partial\phi^2} + r^{-1}\frac{\partial}{\partial r}\left[rh^2\frac{\partial u}{\partial r}\right], \tag{6b}$$

$$\psi = v + \text{constant}. \tag{6c}$$

In deriving (5), an intermediate step utilizes a calculation in cylindrical coordinates to yield

$$\nabla\times(\nabla u\times\mathbf{h}) = -\mathbf{h}h^{-2}\Delta^* u + (\mathbf{h}\cdot\nabla\frac{\partial u}{\partial r})\mathbf{i}_r + (\mathbf{h}\cdot\nabla\frac{\partial u}{\partial z})\mathbf{i}_z + (r^{-1}\mathbf{h}\cdot\nabla\frac{\partial u}{\partial\theta})\mathbf{i}_\theta.$$

The result then follows from the helical symmetry of scalars.

3. FORM OF THE EQUATIONS FOR THE SCALAR FUNCTIONS

To obtain the equations of motion of an incompressible perfect fluid we need to exhibit equations for v and u. These are obtained by taking the dot product of **h** with the momentum and vorticity equations. Now $\mathbf{h}\cdot\nabla$ of any scalar with helical symmetry is zero. In order to allow application to Poiseuille flow (discussed using helical geometry by Landman 1989), as well as comparison with Blennerhassett (1979), we allow a uniform pressure gradient along the z-direction and thus set $\mathbf{h}\cdot\nabla p = -f_0 h^2$ where f_0 is $1/\rho$ times the constant pressure drop per unit length in z. The dot products with $\mathbf{h}$ yield

$$h^2 \frac{\partial v}{\partial t} = \mathbf{h}\cdot(\mathbf{v}\times\omega) + h^2 f_0, \tag{7a}$$

$$h^2 \frac{\partial \zeta}{\partial t} = \mathbf{h}\cdot\nabla\times(\mathbf{v}\times\omega). \tag{7b}$$

Computing the first of these, using (4) and the helical symmetry, we have

$$h^2 \frac{\partial v}{\partial t} = (\mathbf{h}\times\mathbf{v})\cdot\omega = -h^2 \nabla u\cdot(\mathbf{h}\times\mathbf{v}) = h^2 \mathbf{h}\cdot(\nabla u \times \nabla v),$$

and therefore

$$\frac{\partial v}{\partial t} = \mathbf{h}\cdot(\nabla u\times\nabla v) + f_0. \tag{8}$$

For (7b) we obtain

$$\mathbf{v}\times\omega = (\nabla u\times\nabla\psi)\cdot\mathbf{h}\,\mathbf{h} + h^2(v\nabla\psi - \zeta\nabla u),$$

and therefore

$$\mathbf{h}\cdot\nabla\times(\mathbf{v}\times\omega) = -2\alpha h^4(\nabla u\times\nabla\psi)\cdot\mathbf{h} + [\nabla(h^2 v)\times\nabla\psi]\cdot\mathbf{h} - [\nabla(h^2\zeta)\times\nabla u)\cdot\mathbf{h},$$

where we have used the Beltrami property of $\mathbf{h}$. Thus

$$\frac{\partial}{\partial t}(\Delta^* u + 2\alpha h^4 v) = 2\alpha h^4(\nabla u\times\nabla\psi)\cdot\mathbf{h} - [\nabla(h^2 v)\times\nabla\psi)\cdot\mathbf{h} + [\nabla(h^2\zeta)\times\nabla u)]\cdot\mathbf{h}. \tag{9}$$

Subtracting $2\alpha h^4$ times (8) from (9) and using (6c) we then obtain

$$\frac{\partial \Delta^* u}{\partial t} = \alpha^2 h^4[r\nabla r\times\nabla(v^2)]\cdot\mathbf{h} + [\nabla(h^2\zeta)\times\nabla u)]\cdot\mathbf{h} - 2\alpha h^4 f_0. \tag{10}$$

Euler flows with helical symmetry are thus determined by (4),(6),(8), and (10).

Next, consider the Navier-Stokes equations for incompressible flow. For (8) we must add to the right-hand side

$$\eta h^{-2}\mathbf{h}\cdot\nabla^2\mathbf{v} = -\eta h^{-2}\mathbf{h}\cdot\nabla\times\omega.$$

Here η is the kinematic viscosity of the fluid. If we use an expression analogous to (5) for $\nabla\times\omega$, we see that

$$-\mathbf{h}\cdot\nabla\times\omega = \Delta^*\psi + 2\alpha h^4\zeta,$$

so that (8) is replaced by

$$\frac{\partial v}{\partial t} = \mathbf{h}\cdot(\nabla u\times\nabla v) + f_0 + \eta h^{-2}(\Delta^* v + 2\alpha h^4\zeta). \tag{11}$$

Similarly the viscous vorticity equation can be computed including again the multiple of the momentum component. We must compute

$$V \equiv \mathbf{h}\cdot\nabla\times\nabla\times\omega + 2\alpha h^4\mathbf{h}\cdot\nabla\times\omega.$$

Since

$$\nabla\times\omega = (\nabla\times\omega)\cdot\mathbf{h}\,\mathbf{h} + \nabla\zeta\times\mathbf{h}$$

we have

$$V = -2\alpha h^4(\nabla\times\omega)\cdot\mathbf{h} - \Delta^*\zeta + 2\alpha h^4\mathbf{h}\cdot(\nabla\times\omega) = -\Delta^*\zeta.$$

Thus the Navier-Stokes form of (10) is

$$\frac{\partial\Delta^*u}{\partial t} = \alpha^2h^4[r\nabla r\times\nabla(v^2)]\cdot\mathbf{h} + [\nabla(h^2\zeta)\times\nabla u)\cdot\mathbf{h} - 2\alpha h^4 f_0 - \eta\Delta^*\zeta. \tag{12}$$

It is a straightforward exercise to show that (11), (12) tend to the equations for flow independent of z as $\alpha \to 0$, and that (11), (12) tend to the equations for axisymmetric flow in limit $\alpha \to \infty$, f_0 fixed, with $v/\alpha r$ and u/α tending to a swirl component of velocity and the Stokes stream function, respectively.

4. STEADY EULER FLOWS

A treatment of symmetric steady flow in a general orthogonal curvilinear coordinate system is given by Greenspan (1968, p. 216). We look for general steady solutions of (8), (10). The constant pressure gradient f_0 is no taken to be zero, since $f_0 = O(\eta)$. Because of the helical symmetry, a flow which appears to rotate with constant angular velocity in one coordinate frame, will be steady in another frame moving with constant speed relative to the first in the direction of the z-axis. Thus we need not introduce rotating coordinates for the general steady case.

From (8) we see that, in steady flow with no pressure gradient,

$$v = F(u) \tag{13}$$

where F is an arbitrary function. From (10) it then follows that

$$h^2\zeta = h^2\, FF'(u) + G(u) \tag{14}$$

where G is another arbitrary function. Note that, in the axisymmetric limit $\alpha \to \infty$ (see above), the term in (6a) involving 2α is negligible. With

$$u = \alpha\tilde{u},\; F(u) = \alpha\tilde{F}(\tilde{u}),\; G(u) = \frac{1}{\alpha}\tilde{G}(\tilde{u}), \tag{15a}$$

we obtain from (14), in the axisymmetric limit

$$[r\frac{\partial}{\partial r}\frac{1}{r}\frac{\partial}{\partial r} + \frac{\partial^2}{\partial z^2}]\,\tilde{u} + \tilde{F}\tilde{F}'(\tilde{u}) + r^2\,\tilde{G}(\tilde{u}) = 0. \tag{15b}$$

This is the classical equation for steady axisymmetric steady swirling flow, see e.g. Batchelor (1967, eqn. (7.5.11)).

5. STEADY FLOW AT LARGE REYNOLDS NUMBER

To study the Prandtl-Batchelor theory with helical symmetry, we assume that the surfaces $u =$ constant form a nested family of tubes not containing the axis $r = 0$. In the r–ϕ plane we thus restrict attention to a domain D_0 not containing the origin, and to a family of nested closed curves $C(u)$, $u_1 \le u \le u_2$, the level curves of u, bounding simple domains $D(u)$. Within this domain the viscous forces are assumed to be small, in the specific sense that the inviscid limits (13),(14) differ from the formal viscous solution asymptotic for small η by a term which is $o(1)$ as $\eta \to 0$ uniformly over the domain. Although it is not obvious that both streamlines and vortex lines should be so situated, this is indeed a consequence of our assumptions, since from (6c) and (13) we see that the transverse parts of the vector fields have the same level curves.

From Batchelor (1956a) we can extract the following basic calculation under the above assumptions: The line integral of $\mathbf{v}\times\omega$ along a streamline S or vortex line V must vanish identically since this vector is orthogonal to the local tangent vector in either case. From the steady Navier-Stokes equations it then follows that

$$\eta \int_S \nabla\times\omega\cdot\mathbf{ds} + [H] = 0, \tag{16}$$

where $[H]$ is the jump in the value of the Bernoulli function $H \equiv |\mathbf{v}|^2/2 + p/\rho$. Since the flow has helical symmetry, we can restrict the integral to one circuit of the level curve, so that

$$\frac{\eta}{T}\int_0^T (\nabla\times\omega)\cdot\mathbf{v}dt = \frac{1}{T}\int_0^T f_0\, \dot{z}\, dt, \tag{17}$$

where T is the time required to orbit the contour u = constant. Given the helical symmetry we can then express (17) as a line integral in the r–ϕ plane around a contour u = constant:

$$\eta \int_{C(u)} (\nabla\times\omega)\cdot\mathbf{v}\,(|\nabla_2 u|)^{-1} ds_2 = f_0[\langle h^2\rangle F - \frac{1}{\alpha}\int h^2 d\phi]. \tag{18}$$

Here $ds_2 \equiv dr + r d\phi$ in the r–ϕ plane and $\nabla_2 \equiv (\frac{\partial}{\partial r}, \frac{\partial}{r\partial\phi})$. Note that $\nabla_2 u$ is a gradient of a function of r and ϕ.

To establish (18) we may express the integrand in Lagrangian variables of the particle path. Thus, writing the integrand of the left-hand side of (17) as $Q(r,\phi)$, the Lagrangian form is

$$\frac{\eta}{T}\int_0^T Q(r(t), \theta(t) + \alpha z(t))dt,$$

where

$$\frac{dz}{dt} = h^2 v - \alpha r h^2 \partial u/\partial r,\ r\frac{d\theta}{dt} = -h^2\alpha r v - h^2\partial u/\partial r,\ r\frac{dr}{dt} = \partial u/\partial\phi.$$

We see that this takes the form

$$\frac{\eta}{T}\int_0^T Q(r(t),\phi(t))dt$$

where

$$r\frac{d\phi}{dt} \equiv v_\phi = -\partial u/\partial r.$$

Thus u is a streamfunction which generates the flow along the contour C(u) as

$$(v_r, v_\phi) = (r^{-1}\partial u/\partial\phi, -\partial u/\partial r).$$

This yields (18) as the appropriate line integral, where we define, for any function $Q(r,\phi)$,

$$<Q> \equiv \int \frac{Q}{|\nabla_2 u|} ds_2. \tag{19}$$

Note the essential role played here by the z-components in (4a). It is tempting to think of the term $\nabla u \times \mathbf{h}$ in (4a) as "flow around the contour", which would then yield a speed which is augmented by a factor h. This is however not the speed seen on the projection of the curve C(u) when the three- dimensional helical motion is projected down.

In a similar way we obtain, from the integral over a vortex line,

$$\eta \int_{C(u)} (\nabla\times\omega)\cdot\omega\,(F'(u)|\nabla_2 u|)^{-1} ds_2 = \frac{f_0}{F'(u)}[<h^2\zeta> - \frac{F'(u)}{\alpha}\int h^2 d\phi]. \tag{20}$$

Then we may write (18) and (20) as

$$<h^2\nabla u\cdot\nabla\zeta - v(\nabla\cdot h^2\nabla v + 2\alpha h^4\zeta)> = \frac{f_0}{F'(u)}[<h^2\zeta> - \frac{F'(u)}{\alpha}\int h^2 d\phi], \tag{21a}$$

$$<h^2\nabla u\cdot\nabla\zeta - (F')^{-1}\zeta(\nabla\cdot h^2\nabla v + 2\alpha h^4\zeta)> = \frac{f_0}{\eta F'}[<h^2\zeta>F'(u) - \frac{1}{\alpha}\int h^2 d\phi]. \tag{21b}$$

Comparing (21a,b) using (14) we see that

$$<h^{-2}\nabla\cdot h^2 F'\nabla u + 2\alpha(FF'h^2 + G)> = -\frac{f_0}{\eta}<1>. \tag{22}$$

To evaluate (22), we first note that

$$<h^{-2}\nabla\cdot(h^2 F'\nabla u)> = <\nabla\cdot(F'\nabla u)> - <2r\alpha^2 h^2 \partial u/\partial r F'>, \tag{23}$$

For the first term on the right of (23), we rewrite the term as $\nabla_2\cdot F'\mathbf{g}$ and use the identity

$$<\cdot> = \frac{d}{du}\iint_{D(u)} (\cdot)\, r dr d\phi. \tag{24}$$

This identity follows from writing an area element as dsdn. Note that (24) implies that $<1> = A'(u)$ in (22), where A(u) is the area within the contour of constant u. Applying (24) to $\nabla_2\cdot F'\mathbf{g}$ we obtain

$$<\nabla\cdot(F'\nabla u> = \frac{d}{du}[<(\nabla u)^2>F'], \tag{25}$$

since it is seen that

$$\nabla_2 u\cdot\mathbf{g} = (\nabla u)^2 = (rh)^{-2}u_\phi^2 + u_r^2.$$

For the second term on the right we use $-\partial u/\partial r = r d\phi/dt$. Using (2) and the fact that C(u) is closed and does not contain the origin, we obtain

$$-<2r\alpha^2 h^2 \partial u/\partial r> = -2<h^2 d\phi/dt> = -2\int_{C(u)} h^2 d\phi. \tag{26}$$

Thus, (22) takes the form

$$[\gamma F']' - 2\int_{C(u)} h^2 d\phi\, F' + 2\alpha FF'<h^2> + 2\alpha GA' = -\frac{f_0 A'}{\eta}, \tag{27a}$$

where γ is a circulation integral,

$$\gamma \equiv <(\nabla u)^2>. \tag{27b}$$

We may study (21a) in a similar way. The equation which results is

$$\Gamma(F')^2 - FF'\Gamma' + \gamma G' + 2 \int_{C(u)} h^2 d\phi - 2\alpha F^2 F' <h^4>$$

$$- 2\alpha FG <h^2> = \frac{f_0}{\eta}[<h^2>F - \frac{1}{\alpha}\int h^2 d\phi]. \tag{28a}$$

where

$$\Gamma \equiv <h^2(\nabla u)^2>. \tag{28b}$$

Equations (27) and (28) are the main results of this paper. Given the curves C(u), we may in principle evaluate the various terms in these equations and compute F(u), G(u) by solving ordinary differential equations. Then (6) and (14) can be used to compute $u(r,\phi)$. A solution is obtained when the computed contours equal the starting contours. Since the very form of the equations depends upon the contour geometry, we refer to (6), (14), (27), (28) as a *generalized differential equation* (or GDE). The term "queer differential equation" is also sometimes used. This kind of system arises naturally in plasma physics, as was emphasized by Grad, see Grad *et al.* (1975). The flows with helical symmetry reveal that the Prandtl-Batchelor formalism produces a quite complicated GDE in the general case.

6. SPECIAL CASES

For two-dimensional motion with $v_z = 0$, the steady flows considered by Batchelor (1956a) are determined by the partial differential equation $\nabla^2 u = f(u)$ for the streamfunction u. The Prandtl-Batchelor theorem for this case implies that f should be constant over a region of closed streamlines. Specification of the function f(u) thus determines the equation apart from the contours C(u), and so this problem does not involve a GDE. An related example of a GDE (analogous to the Grad-Shafranov equation, see Laurence and Stredulinsky, 1985) is $\nabla^2 u = f'(u)/A'(u)$, since the dependence of area upon u is not known *a priori*.

To take the two-dimensional limit of (27),(28), now allowing v_z to be nonzero, we let $\alpha \to 0$. It is easily seen (since $h^2 \to 0$ and $\int_{C(u)} d\phi = 0$) that (27) and (28) reduce to

$$(\gamma F')' = -\frac{f_0 A'}{\eta}, \quad \Gamma(F')^2 - FF'\Gamma' + \gamma G' = \frac{f_0 A' F}{\eta}. \tag{29}$$

The z-component of vorticity is now $FF' + G \equiv H'(u)$, and we deduce from (29) that

$$\gamma H'' = 0. \tag{30}$$

Thus vorticity is constant over the region of closed streamlines, $H' \equiv K = \text{constant}$. From the first of (29), together with Stokes' theorem in the form $\gamma(u) = K\,A(u)$, we obtain Blennerhassett's (1979) result

$$F(u) = -\frac{f_0 u}{\eta K} + C\int \frac{1}{\gamma} du + D \tag{31}$$

where C and D are arbitrary constants. Note that (31) determines F(u) once the equation $\nabla^2 u = -K$

has been solved subject to boundary conditions, so we have a PDE, not a GDE. Batchelor's (1956a) 2D result corresponds to $f_0 = F = 0$.

Consider next axisymmetric swirling flow. It is also not difficult to show, using the substitutions (15a), that (29) is again obtained in the limit of infinite α with $f_0 = 0$, Γ being assigned its asymptotic form. In fact since $\gamma = 0$ at the stagnation point within D_0, we see that any regular F must be a constant ($\tilde{F}$ being r times the swirl velocity), and therefore G is also a constant, which is the Batchelor limit for swirling flow. Again the contours are determined by a PDE.

7. SLENDER TUBES

We briefly consider a simple application of (27),(28) to slender vortex tubes. The existence of steady solutions of this kind has been established by Adebiyi (1981). Suppose that $f_0 \equiv 0$ and that D_0 is a small patch in the r–ϕ plane, in the sense that h is approximately constant, $h \approx h_0$, on the domain. Then (27), (28) may be written, approximately,

$$[\gamma F']' + 2\alpha A'[h_0^2 FF' + G] \tag{32a}$$

$$h_0^2 \gamma (F')^2 - FF' h_0^2 \gamma' + \gamma G' - 2\alpha A' h_0^2 F[h_0^2 FF' + G]. \tag{32b}$$

Using (32a) to eliminate $h_0^2 FF' + G$ from the last term of (32b)and factoring out γ, we obtain an equation which may be integrated once, yielding

$$h_0^2 FF' + G = \text{constant} = -K. \tag{33}$$

Using (33) in (32a), integrating, and then using the fact that $\gamma = 0$ at the stagnation point within D_0, we obtain

$$\gamma F' = 2\alpha K A(u). \tag{34}$$

From (6) and (14) we then have the the equation for u in the approximate form

$$\nabla_2^2 u + 2\alpha h_0^2 F = K/h_0^2. \tag{35}$$

In (35) we have neglected a term coming from differentiation of h^2 as negligible for slender vortex tubes. If we approximate the tube as circular. with the radius of a contour denoted by R, then (34), (35) may be integrated to yield the family of profiles

$$u(R) = -\frac{1}{8}\alpha^2 h_0^2 R^4 + \frac{1}{4}(K/h_0^2 - 2\alpha h_0^2 C_1) R^2 + C_2, \tag{36}$$

where $C_{1,2}$ are arbitrary constants.

We thank Trevor Stuart for reference to the paper by Blennerhassett.

REFERENCES

Adebiyi, A. (1981). On the existence of steady helical vortex tubes of small cross-section. *Q. J. Mech. Appl. Math.* **34** , 153-177.

Batchelor, G.K. (1956a). On steady laminar flow with closed streamlines. *J. Fluid Mech.* **1** , 177-190.

Batchelor, G.K. (1956b). A proposal concerning laminar wakes behind bluff bodies at large Reynolds number. *J. Fluid Mech.* **1** , 388-398.

Batchelor, G.K. (1967). *An Introduction to Fluid Mechanics.* Cambridge University Press.

Blennerhassett, P.J. (1979). A three-dimensional analogue of the Prandtl-Batchelor closed streamline theory. *J. Fluid Mech.* **93** , 319-324.

Childress, S. and Soward, A.M. (1989). Scalar transport and alpha-effect for a family of cat's-eye flows. In press, *J. Fluid Mech.*

Grad, H., Hu, P.N., and Stevens, D.C. (1975). Adiabatic evolution of plasma equilibria. *Proc. Nat. Acad. Sci.* **72** , 3789-3793.

Greenspan, H.P. (1968). *The Theory of Rotating Fluids.* Cambridge University press.

Lagerstrom, P.A. (1975). Solutions of the Navier-Stokes equations at large Reynolds number. *SIAM J. Appl. Math.* **28** , 202-214.

Landman, M.J. (1989). Time dependent helical waves in rotating pipe flow. Submitted to *J. Fluid Mech.*

Laurence, P., and Stredulinsky, E.W. (1985). A new approach to queer differential equations. *Comm. Pure Appl. Math. XXXVIII, 333-355.*

Moore, D.W., Saffman, P.G., and Tanveer, S. (1988). The calculation of some Batchelor flows: The Sadovskii vortex and rotational corner flow. *Phys. Fluids* **31** , 978-990.

Park, W., Monticello, D.A., and White, R.B. (1984). Reconnection rates of magnetic fields including the effects of viscosity. *Phys. Fluids* **27** , 137-149.

Prandtl, L. (1904).Über flüssigkeitsbewegung bei sehr kleiner reibung International Mathematical Congress, Heidelberg, 484-491; see Gesammelte Abhandlungen II (1961), 575-584.

Rhines, P.B. and Young, W.R. (1983). How rapidly is a passive scalar mixed within closed streamlines? *J. Fluid Mech.* **133** , 133-145.

General Magnetic Reconnection in 3D Systems

A. OTTO[1], M. HESSE[2] & K. SCHINDLER[1]

[1]Ruhr-Universität Bochum, FRG
[2]Los Alamos National Laboratory, USA

The topological aspects of magnetic reconnection are well understood in two-dimensional plasma systems. However new developments in theory and observation revealed that certain notions used in the 2D case are structurally unstable, such that a revision of the concept of magnetic reconnection for 3D systems became necessary. A notion of general magnetic reconnection (GMR) that avoids the structural instability is based on the violation of the line conservation property due to a strongly localized resistivity. In this framework the cases where the magnetic field contains magnetic nulls form a special subclass. Here we concentrate on magnetic fields without singularities and show that the key quantity is the component of the electric field parallel to the magnetic field. An abstract formulation in terms of topological notions involves both the magnetic and the plasma flow field.

A potential application of GMR that is of interest for space plasma physics is the flux transfer event (FTE) occurring at the boundary of the earth' s magnetosphere. FTE's seem to play an important role for the plasma and energy transport from the solar wind into the magnetosphere.

We have carried out three-dimensional resistive MHD computations, that apply to the formation of FTE's as a general magnetic reconnection process. Special emphasis is given to the role of the parallel electric field. Further the time development of the magnetic interconnection of fluid elements and the local magnetic opening of a previously closed magnetospheric boundary will be addressed.

1. INTRODUCTION

Magnetic reconnection is widely accepted as a major process with respect to the plasma transport in many astrophysical and laboratory systems. The range of applications of magnetic reconnection includes the field of geomagnetic activity (*Hones* 1979), eruptive processes on the sun (*Priest* 1989) and in general the interaction regions of plasmas of different origin (e.g. planetary and interplanetary plasmas, interplanetary and interstellar plasmas, pulsar magnetopauses etc). With respect to geomagnetic activity reconnection is expected to occur at the sunward boundary (steady state reconnection: *Dungey* 1959; formation of flux transfer events: *Russel and Elphic* 1979) and in the geomagnetic tail (formation of plasmoids: *Schindler* 1974; *Otto* 1989).

In the past the process of magnetic reconnection in most of these applications has been discussed by topological arguments based on a two-dimensional reconnection geometry (*Vasyliunas* 1975). Central properties of two-dimensional reconnection theory are the existence of neutral lines and separatrices. However (*Schindler et al.* 1988) have shown that a concept based on neutral lines is structurally unstable in three dimensions. Thus a revision of the concept of magnetic reconnection became necessary (*Schindler et al.* 1988; *Hesse and Schindler* 1988; *Birn et al.* 1989). A basic property of general magnetic reconnection (GMR) is the violation of line conservation due to a strongly localized resistivity. Based upon line conservation it has been shown (*Hesse and Schindler* 1988) that the electric field parallel to the magnetic field in a localized domain D_r is an important quantity of magnetic reconnection with respect to the global topology. Another approach to define magnetic reconnection involves singularities of the magnetic field (*Greene* 1988). In section 2 we will outline the concept of GMR and compare it with the approach that involves magnetic nulls, where special emphasis is given to the transformation properties of both concepts.

The generic case of magnetic reconnection in most plasma systems should be three-dimensional. However with respect to astrophysical plasmas even with in situ observations by satellites it is difficult to determine the three-dimensional structure of a reconnection process. A potential application of GMR has been provided by the observation of flux transfer events (FTE's) (*Russel and Elphic* 1978; *Paschmann et al.* 1982; *Saunders et al.* 1984) at the boundary between the interplanetary and the terrestrial magnetic field (magnetopause). FTE's typically exhibit a bipolar signature of the magnetic field normal to the magnetopause together with a particle population which contains solar wind as well as magnetospheric particles. Thus *Russel and Elphic* (1978) concluded that magnetic flux of the solar wind region must have been connected to magnetospheric flux. The intermittent nature of FTE's is a strong indication that they are produced by a reconnection process that is localized in space and time, which necessarily leads to three-dimensional structures. Estimates of the total reconnected flux together with occurrence rate of FTE's indicate that these events provide a major transport mechanism at the interface between solar wind and magnetospheric plasma.

A number of models have been suggested to describe the formation (*Lee and Fu* 1985; *Scholer* 1988) and/or the interior structure (*Sonnerup* 1987) of flux transfer events. Naturally different reconnection mechanisms lead to different topological structures of FTE's (single X-line reconnection: *Scholer* 1988; multiple X-line reconnection: *Lee and Fu* 1985). We remark that already the terminology indicates that most of the existing models have only been worked out in two dimensions (Note that neutral lines are structurally unstable in three dimensions).

There is a number of three-dimensional simulations (*Ogino et al.* 1989) which are aimed on the global interaction of the solar wind plasma with the magnetosphere. Since basic instabilities (resistive tearing) which may lead to magnetic reconnection involve rather small length scales it is doubted whether these global simulation provide sufficient resolution to analyse the reconnection process.

In section 3 we will present a three-dimensional MHD simulation of the formation of FTE's which is based upon a strongly localized resistivity. The reconnection process and the topological aspects of the developing structures will be discussed in the framework of GMR where the violation of magnetic line conservation due to a parallel electric field is a key quantity.

2. GENERAL MAGNETIC RECONNECTION

The concept of magnetic reconnection was developed largely for systems where one of the space coordinates is ignorable, such that the electromagnetic fields and electric charge and current densities depend on two space coordinates only. In those cases one can tie magnetic reconnection to the topological structure of the magnetic field in the poloidal plane, i.e. perpendicular to the displacement defining the invariance.

We illustrate this procedure in terms of translational invariance with respect to the cartesian z-coordinate assuming that the toroidal field component B_z vanishes. Suppose the magnetic field $\mathbf{B}$ has a hyperbolic point singularity $\mathbf{B} = 0$ at the origin (Fig. 1). The line ($x = 0$, $y = 0$, z) is usually referred to as an X-type neutral line. Obviously, an entirely ideal flow obeying ideal Ohm's law

$$\mathbf{E} + \mathbf{v} \times \mathbf{B} = 0 \tag{1}$$

would imply that the electric field $\mathbf{E}$ vanishes at the neutral line.

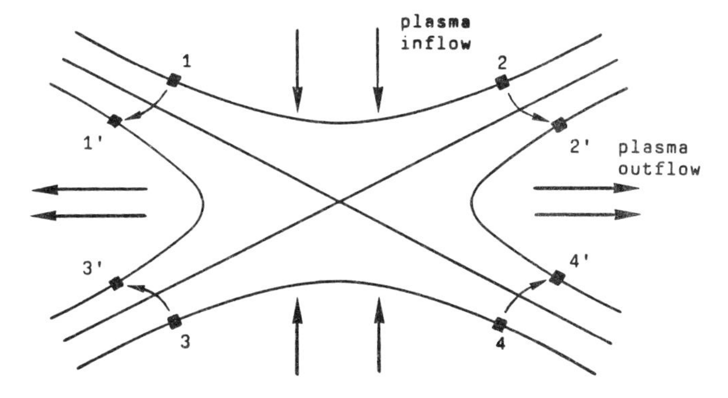

Figure 1: Sketch of a two-dimensional magnetic reconnection geometry with a hyperbolic X-type neutral line

Also, (1) implies the property of magnetic line conservation, expressing the fact that two plasma elements that are connected by a magnetic field line at any time. In fact, the precise condition for magnetic line conservation is $\mathbf{B} \times \nabla \times (\mathbf{E} + \mathbf{v} \times \mathbf{B})) = 0$. Reconnection is based on physical processes that violate (1) in a small region around the X-line such that $E_z \neq 0$. The presence of that non-ideal region implies a global effect, in that plasma elements can cross the separatrix in the ideal region (see Fig. 1), which would be excluded in the case of strictly ideal flow in the entire domain. Since in the ideal flow region the concept of magnetic flux frozen into moving plasma is appropriate one may visualize that the plasma moving across a separatrix carries magnetic field lines that change their topological structure, when instantaneously they become the separatrix .

The field lines carrying the plasma elements 1,2 and 3,4 (Fig. 1) can be regarded to reconnect to from the field lines carrying the plasma elements 1',3', and 2',4' at some later time. It is consistent with these properties to tie the concept of magnetic reconnection either to the plasma flow across a separatrix (*Vasyliunas* 1984) or to the presence of a toroidal electric field component (E_z in the present example) at the X-line (*Sonnerup* 1984). In the latter picture the notion of a 'separatrix' generalizes the X-type neutral lines to cases with $B_z \neq 0$.

Although these concepts of magnetic reconnection have been remarkably successful to understand a number of processes in two-dimensional geometries, difficulties arise when one tries generalize them to three-dimensional field configurations. The reason is simply that an X-type neutral line, which is a structurally stable object in two dimensions, looses this property in three dimensions. There exists arbitrarily small field perturbations that eliminate the X-line. (In two dimensions with $B_z = 0$, where the signature of the X-line is an X-type neutral point of the poloidal field, an arbitrary smooth perturbation can only shift but not eliminate the X-point.) This structural instability also applies to separators. Therefore, it has been suggested that magnetic reconnection should be defined in terms of the violation of magnetic line conservation alone:

In this picture magnetic reconnection is the breakdown of the property of magnetic line conservation as the result of a localized non-idealness (*Axford* 1984; *Schindler et al.* 1988). In the presence of a neutral line this definition reduces to the earlier definition based on magnetic field topology. Because the structural instability of neutral lines we include however, that the generic case of magnetic reconnection corresponds to cases with $\mathbf{B} \neq 0$. (A remark on isolated neutral points is made further below.)

For cases with $\mathbf{B} \neq 0$ the relevant property is the electric field component $E_{\|}$ parallel to the magnetic field inside the small non-ideal domain. If the integral $\int E_{\|}\, ds$ extended over the field line sections situated inside the non-ideal domain is different from zero on a finite measure set of field lines, the reconnection process is global in the same sense discussed above for the case of an X-type neutral line. There are also implications regarding magnetic helicity. The most important result seems to be redistribution of helicity among flux tubes (*Wright and Berger* 1989).

According to (*Greene* 1988) reconnection might occur at isolated magnetic neutral points or 'nulls'. There is an additional test, that a concept of magnetic reconnection must pass before it can be finally accepted: It should be possible to attribute some Lorentz-invariant meaning to reconnection. In other words it should not depend on the inertial frame of a given observer whether or not reconnection takes place. It is obvious that the notion of reconnection for $\mathbf{B} \neq 0$ as discussed above satisfies this criteria (*Hesse and Schindler* 1988) because $\mathbf{E} \cdot \mathbf{B}$ is a Lorentz-invariant quantity.

Although in many cases the existence of a magnetic null seem undebatable between different observes, the location of the null is relativistically not invariant. Two observers in different inertial systems will see the null at different spatial locations. This is an immediate consequence of the general transformation of electromagnetic fields **E** and **B**. However the presence or absence of non-ideal plasma behaviour is a matter of particle dynamics which is based on Lorentz covariant equations. Thus it seems that from a relativistic point of view magnetic nulls may not be a suitable signature for identifying magnetic reconnection.

3. THREE-DIMENSIONAL MHD SIMULATION OF MAGNETIC RECONNECTION

For a better understanding of general magnetic reconnection we have carried out three-dimensional resistive MHD computations. As an example the results of one such simulation will be presented in this section. These results provide some insight into the dynamical evolution of a 3D reconnection process as well as an illustration of the topological aspects involved in general magnetic reconnection. We have already mentioned in the introduction that a potential application of 3D reconnection might be flux transfer events at the earth's magnetopause. Thus it is interesting whether the results of such computations would explain the observed FTE signatures. The basic equations considered for the simulation are

$$\frac{\partial \rho}{\partial t} = -\nabla \cdot (\rho \mathbf{v}) \tag{2}$$

$$\frac{\partial \rho \mathbf{v}}{\partial t} = -\nabla \cdot (\rho \mathbf{v}\mathbf{v} + \frac{1}{2}(\mathbf{p} + \mathbf{B}^2)\mathbf{1} - \mathbf{B}\mathbf{B}) \tag{3}$$

$$\frac{\partial \mathbf{B}}{\partial t} = \nabla \times (\mathbf{v} \times \mathbf{B} - \eta \nabla \times \mathbf{B}) \tag{4}$$

$$\frac{\partial h}{\partial t} = -\nabla \cdot (h\mathbf{v}) + \frac{\gamma - 1}{\gamma} \mathbf{h}^{1-\gamma} \eta \mathbf{j}^2 \tag{5}$$

with $h = (p/2)^{1/\gamma}$, where $\rho, p, \mathbf{v}, \mathbf{B}$ and η denote plasma density, pressure, velocity, magnetic field, and resistivity respectively. All quantities are made dimensionless by a normalization to typical values where times are measured in Alfvén times and length scales are normalized to the width of the current layer which is introduced below. The initial state for the simulation is defined by the following equations.

$$B_x = \begin{cases} 0 \\ 0 \\ 0 \end{cases} \quad B_y = \begin{cases} \cos\phi_0 \\ \cos\phi_0 x \\ \cos\phi_0 \end{cases} \quad B_z = \begin{cases} \sin\phi_0 & x \leq -1 \\ -\sin\phi_0 x & -1 \leq x \leq 1 \\ -\sin\phi_0 & x \geq 1 \end{cases}$$

$$\mathrm{v} = 0 \qquad p = \rho = 1 \qquad \phi_0 = 75^\circ$$

These equations characterize a one-dimensional current sheet, which separates regions of 150° sheared magnetic fields (Fig. 2). This configuration is an equilibrium state which becomes unstable if any non-idealness is available. At time $t = 0$ we apply a localized resistivity of the form

$$\eta = 0.02\cosh^{-2}\frac{x}{2}\cosh^{-2}\frac{y}{3}\cosh^{-3}\frac{z}{3}$$

to the system. At the earth's magnetopause a locally enhanced non-idealness may be produced by appropriate fluctuations in the solar wind of plasma.

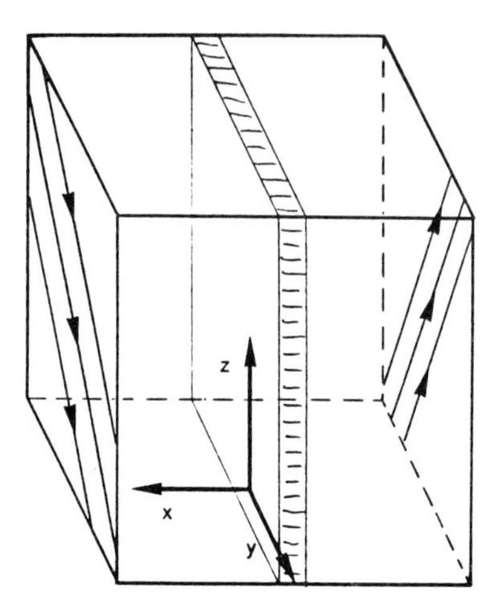

Figure 2: Illustration of the initial state used in the MHD simulation. The lines with arrows on the left and on the right hand side of the box indicate the sheared magnetic field.

It is easy to show that this choice of the initial state and the resistivity ensures some symmetry properties (compare equations (2) to (5)) of the dynamical evolution. This allows that the actual computations may be carried out only in the region with $v_z \geq 0$ in Fig. 2. Considering this symmetry the systems extends from -12 to 12 in x, from -24 to 24 in y and from -80 to 80 in z direction. We remark that for a quantitative application to the magnetopause an asymmetry has to be considered in the equilibrium such that the symmetry with respect to $x = 0$ is lost. However qualitatively the dynamical evolution and the topological aspects of the reconnection process remain unchanged considering the above described initial state.

For the integration of the MHD equation an explicit finite difference method with a leapfrog scheme has been used. From linear theory and two-dimensional simulation it is well known that the resistive tearing instability involves rather small length scales of $O(S^{1/4})$ (S being the magnetic Reynoldsnumber). Therefore a non uniform grid ($51 \times 27 \times 43$ gridpoints in x, y and z-direction respectively) has been considered for all directions where especially for the x-direction the maximum resolution is 0.05 in the center of the current layer.

Let us analyse the dynamical evolution of the system. Initially the localized resistivity leads to slow diffusion of plasma, which provides no efficient transport mechanism. However the perturbation, which is produced by this diffusion, eventually leads to the development of a fast resistive instability. Associated with this instability is an increase of electric field in the region of the localized resistivity. Fig. 3 presents this parallel electric field in a cut through

the three-dimensional system in the center of the current sheet.

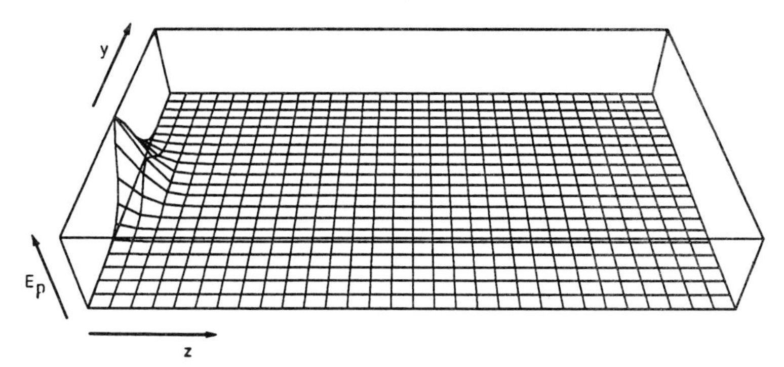

Figure 3: Plot of the parallel electric field E_p in the plane $x = 0$ at $t = 120$. The shown region extends from -20 to $+20$ in the y and from 0 to 75 in the z-direction. The maximum and minimum values of E_p are 0 and 0.05 respectively.

As illustrated in section 2 a parallel electric field is a typical notion of general magnetic reconnection. Hence magnetic line conservation has been violated and magnetic flux must have been newly interconnected. This is indeed the case as shown by the perspective views in Figures 4. These figures illustrate the path of magnetic field lines at $t = 120$ that are defined by the points indicated by the black boxes (with $x = 4$, $y = -8$ and a separation of $\Delta y = 10$). Initially all fieldlines have been straight lines. The fieldlines which are defined by 1 and 4 do not change their basic direction during the evolution. This is different for the field lines that stick to the points 2 and 3. These fieldlines enter the box on the left side with a $-B_z$ component and leave the box on the right side with a $+B_z$ component. Figure 4b shows the same fieldlines for a view into the $-y$-direction. It is obvious that the fieldlines 2 and 3 represent magnetic flux that must have been connected between the regions of $x > 0$ and $x < 0$, such that the magnet topology has changed during the evolution.

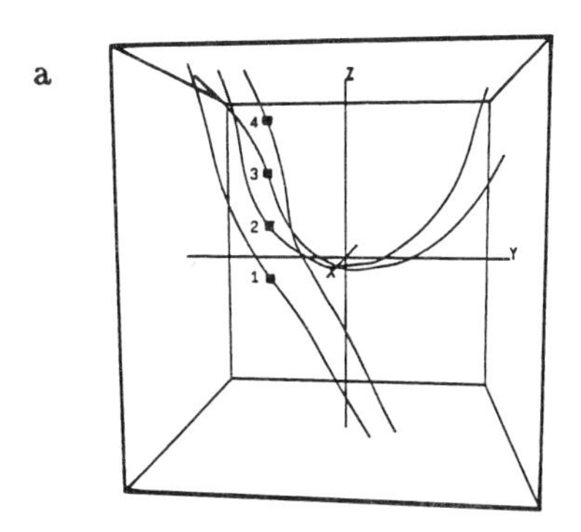

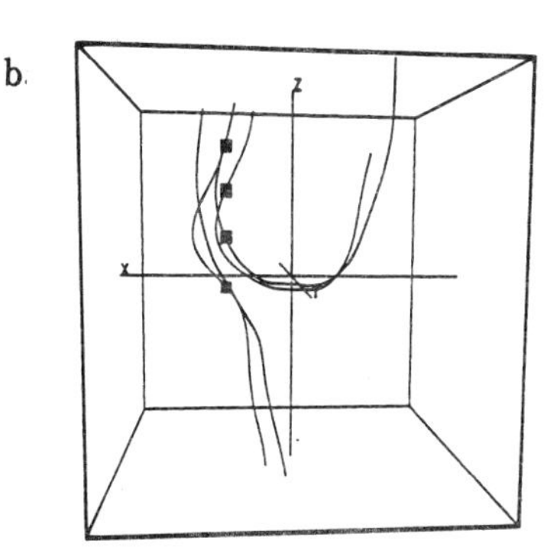

Figure 4: Perspective view of magnetic field lines at $t = 120$. The black boxes indicate the space coordinate which defines the field lines. Figure a represents a view approximately into the $-x$-direction and Figure b a view (of the same field lines) into the $-y$-direction. The size of the shown cube extends from -10 to $+10$ in x, from -20 to $+20$ in y and from 0 to 75 in z-direction.

We remark that up to the end of computation no magnetic nulls occur in this process. Initially the amplitude of magnetic field is one everywhere in the simulation box. If magnetic nulls would be present in the reconnection process (which might not be resolved by the numerical discreteness) we expect that the amplitude of the magnetic field should decrease sufficiently during the dynamical evolution at certain locations. However this is not seen in the simulation. Thus we conclude that a magnetic reconnection process, which is defined by magnetic singularities, can be excluded in this simulation.

The motion of fluid elements in the vicinity of the reconnection region (region of the localized resistivity) is illustrated in Figures 5. The straight line that connects the different fluid trajectories indicates the initial location of the chosen fluid elements (at $t = 0$). This line is chosen parallel to the y-axis with $z = 2$ and $x = 1.2$ and 2.5 for the Fig. 5 a and b respectively. The separation of the initial locations in the y-direction is $\Delta_y = 2$. Hence the fluid elements in Fig. 5a are closer to the reconnection region (the vicinity of $x = 0$). In both cases these fluid particles are in the inflow region of the reconnection zone (compare Fig. 1) which is seen by the fact that the initial motion in both cases is basically in the x-direction. While in Fig. 5b this motion continues for most of the time the fluid elements in Fig. 5a are deviated into the $+z$-direction.

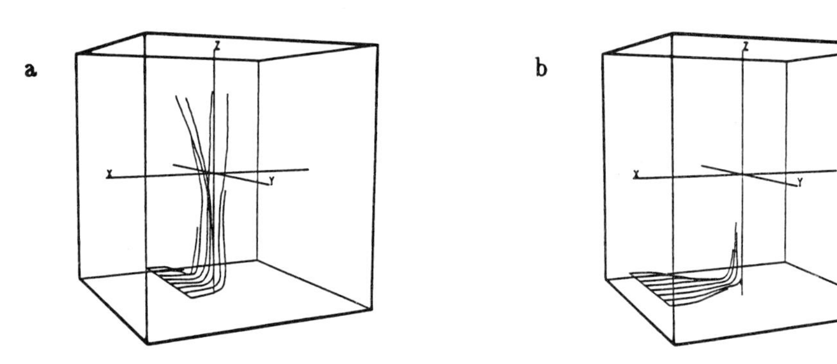

Figure 5: Perspective view of fluid trajectories during the evolution until $t = 120$. The size of the cube is chosen from -3 to $+3$ in x, from -12 to $+12$ in y, and from 0 to 40 in z-direction. Fluid elements start at $x = 1.5$ in Figure a and at $x = 2.5$ in Figure b (indicated by the straight line).

Considering the qualitative sketch of two-dimensional reconnection one would expect fluid trajectories like those shown in Fig. 5. Nevertheless the actual path of a fluid element depends very much on the particular y-coordinate which in addition shows the three-dimensional nature of this reconnection process. We found that all fluid element which have been convected into the outflow region (associated with a dominant v_z-component in the velocity) are bound to reconnected field lines.

Let us now consider the question whether or not a three-dimensional reconnection process of this kind would explain the observed FTE signatures. Fig. 6 shows the normal component of the magnetic field (B_x) in a plane parallel to the current layer. The structure seen in this figure moves basically into the $+z$-direction Hence a satellite at an appropriate location would observe an increase of B_x and a subsequent decrease which leads to the observed bipolar signature of the normal magnetic field component. We have also compared other characteristics of flux transfer events (length and time scales, interior structure) and found the results of the simulations consistent. Therefore we conclude that a general magnetic reconnection process of the kind described in this section presents a reasonable explanation of FTE's.

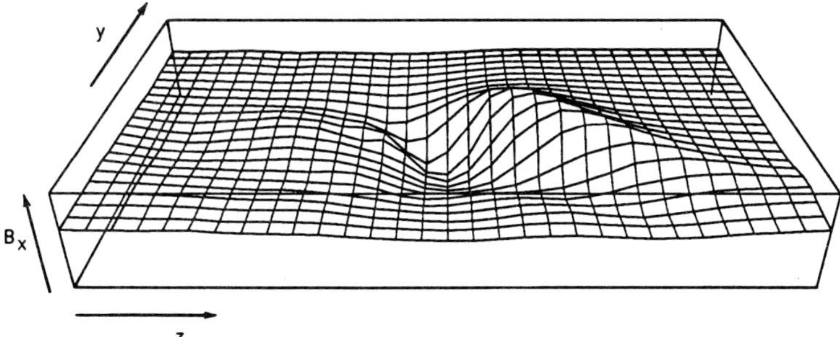

Figure 6: Normal component of the magnetic field B_x at $t = 120$ in a plane parallel to the current sheet ($x = 4.2$). The size of the plane extends from -20 to $+20$ in y and from 0 to 75 in the z-direction. Minimum and maximum values of B_x are roughly -0.3 and $+0.2$.

4. SUMMARY

By a reconfiguration of the magnetic field topology magnetic reconnection leads to a large macroscopic plasma transport, a fast release of stored energy and thus to a relaxation of plasma systems. In the past the two-dimensional topological structure of a reconnection process (neutral lines, separatrices) has been used as a definition of magnetic reconnection. As we have outlined in section 2 these notions of magnetic reconnection are structurally unstable in three dimensions.

In general magnetic reconnection the localized violation of the magnetic line conservation is a basic property. Such a violation of line conservation requires that the integral of the parallel electric field along a magnetic field line is nonvanishing. A concept based on magnetic singularities could be regarded as a special case of GMR. However in section 2 we have shown that the position of magnetic nulls will be altered by a Lorentz transformation. Thus observers in different inertial frames are lead to different interpretations of such a reconnection process, whereas the basic quantity in the GMR concept $(\mathbf{E} \cdot \mathbf{B})$ is Lorentz invariant.

As an illustration of a GMR process we have carried out an MHD simulation of three-dimensional magnetic reconnection. This simulation is also considered as a reasonable explanation for the formation of flux transfer events at the earth's magnetopause. The results of this simulation are consistent with the observed signatures of FTE's.

It is shown that the presence of small region of nonidealness is sufficient to initiate a 3D reconnection process. In the nonideal region the component of the electric field parallel to the magnetic field is non zero. Magnetic nulls do not occur in this simulation. We have illustrated that during the dynamical evolution a reconfiguration of the magnetic field occurs, which leads to a change in the global magnetic field topology. It is also shown that this reconfiguration is associated with a fast plasma transport.

A general magnetic reconnection process will occur at any interface separating plasmas of different origin if a localized nonidealness is available (not necessary resistivity). Satellite observations indeed show that e.g. flux transfer events also occur at the magnetopause of other planets. Thus a GMR process may represent a basic mechanism for the plasma transport at such boundaries.

Acknowledgment. This work was supported by the Deutsche Forschungsgemeinschaft through the Sonderforschungsbereich "Plasmaphysik Bochum/Jülich" and the Schwerpunktsprogramm "Theorie kosmischer Plasmen".

REFERENCES

Axford, W.I., in Magnetic Reconnection in Space and Laboratory Plasmas, E.W. Hones Jr, 25, Geophysical Monographs, AGU, Washington DC, 1984.

Birn, J., M. Hesse and K. Schindler, J. Geophys. Res., **94**, 241, 1989.

Dungey, W. Cosmic Electrodynamics, Cambridge University Press, New York, 1958.

Greene, J.M., J. Geophys. Res., **93**, 8583, 1988.

Hesse, M. and K. Schindler, J. Geophys. Res., **93**, 5559, 1988.

Hones, E.W., Jr., in Dynamics of the magnetosphere, ed. by S.I. Akasofu, D. Reidel, Dordrecht-Holland, 545, 1979.

Lee, L.C. and Z.F. Fu, Geophys. Res. Lett., **12**, 105, 1985.

Ogino, T., R.J. Walker and M. Ashour-Abdalla, Geophys. Res. Lett., **16**, 155, 1989.

Otto, A., in Reconnection in Space Plasmas, Vol II, 223, Conference Proceedings, Potsdam, ESA, ESTEC, Noordwijk, Netherlands, 1989

Paschmann, G., G. Haerendel, I.Papamastorakis, N. Sckopke, S.J. Bame, J.T. Gosling and C.T. Russel, J. Geophys. Res., **87**, 2159, 1982.

Priest, E.R., in Reconnection in Space Plasma, Vol II, 73, Conference Proceedings, Potsdam, ESTEC, Nordwyk, Netherlands, 1989.

Russel, C.T. and R.C. Elphic, Space Sci. Rev., **22**, 681, 1978.

Saunders, M.A., C.T. Russel and Sckopke, Geophys. Res. Lett., 15, 295, 1988.

Schindler, K. , M. Hesse and J. Birn, J. Geophys. Res., **93**, 5547, 1988

Schindler, K. J. Geophys. Res., **79**, 2803, 1974.

Sonnerup, B.U.O., J. Geophys. Res., **92**, 8613, 1987.

Sonnerup, B.U.Ö ., in Magnetic Reconnection in Space and Laboratory Plasmas, E.W. Hones Jr, 25, Geophysical Monographs, AGU,Washington DC, 1984.

Vasyliunas, V.M., in Magnetic Reconnection in Space and Laboratory Plasmas, E.W. Hones Jr, 25, Geophysical Monographs, AGU, Washington DC, 1984.

Vasyliunas, V.M., Rev. Geophys., **13**, 303, 1975.

Wright, A.N. and M.A. Berger, J. Geophys. Res., **94**, 1295, 1989.

Spontaneous Formation of Electric Current Sheets in Astrophysical Magnetic Fields

B.C. LOW

High Altitude Observatory, National Center for Atmospheric Research
Boulder CO 80307, USA

1. INTRODUCTION

The relaxation of a hydromagnetic plasma to an equilibrium state would in general take place with the spontaneous formation of electric current sheets in the limit of letting the electrical conductivity be infinite. In that limit, Faraday's law of induction requires the magnetic field to evolve as though the magnetic flux is frozen into the plasma. The plasma on the other hand can execute discontinuous displacements in response to magnetic stress so that in the relaxation to force equilibrium, the appearance of magnetic discontinuities in the form of electric current sheets are unavoidable. This remarkable property was pointed out by Parker (1972, 1983a, b, 1987, 1989a, b, and in these proceedings) in theoretical studies of astrophysical magnetic fields which generally occur with the magnetic lines of force threading through the boundary of the domain of interest. This property was also pointed out by Arnol'd (1974) and Moffatt (1985) in their considerations of magnetic fields wholly contained in the domain such as encountered in some laboratory plasma devices.

There are two basic processes by which electric current sheets form, namely, the development of an infinite shear in the plasma and the expulsion of magnetic flux from a localized region to bring spatially separate magnetic flux bundles into contact (Parker 1987, 1989, b, Moffatt 1987, Low 1987, 1989, Low and Wolfson 1988, Aly 1989). A pedagogical approach is taken in this paper by illustrating these processes in terms of several explicit examples that have appeared recently in the literature. A point to be made is that the discontinuous plasma displacements capable of creating current sheets arise naturally in three-dimensional systems whereas they can be largely suppressed in geometrically symmetric systems.

2. FORCE-FREE MAGNETIC FIELDS

In many astrophysical situations of interest, the magnetic field may be taken to be embedded in a perfectly conducting plasma so tenuous that the plasma thermal and mechanical energies are negligible compared to the magnetic energy. The equilibrium of such a system is then described to a good approximation by the vanishing of the Lorentz force everywhere. The Lorentz force is proportional to $\mathbf{J} \times \mathbf{B}$, where $\mathbf{J}$ and $\mathbf{B}$ are the electric current density and magnetic field, respectively. Writing $\mathbf{J}$ in terms of $\nabla \times \mathbf{B}$ using Ampere's law, the "force-free" condition for the vanishing of the Lorentz force becomes

$$(\nabla \times \mathbf{B}) \times \mathbf{B} = 0, \tag{1}$$

with $\mathbf{B}$ satisfying Maxwell's equation $\nabla.\mathbf{B} = 0$. A rich variety of smooth mathematical solutions to equation (1) can be found in the literature. The requirement of smoothness in the construction of these solution is usually imposed for mathematical reasons of simplicity and convenience. A proper physical formulation of the problem must, in general, allow for solutions of equation (1) to have a certain type of discontinuity in $\mathbf{B}$. Let us give such a formulation using a variational approach, point out in quite general terms why discontinuities may be unavoidable and then proceed with the illustrative examples.

2.1 A Variational Problem

Consider a magnetic field $\mathbf{B}$ in a given volume of space V with both ends of each line of force anchored at the boundary ∂V. Let B_n be the distribution of the normal field at ∂V which we take to be a rigid perfect conductor. The boundary data B_n of course does not determinine the topology of the lines of force in the interior of V. Magnetic fields of quite different field-line connectivities may have the same boundary distribution B_n. Taking the plasma in V to be a perfectly conducting inertia-less fluid, the magnetic field $\mathbf{B}$ has a fixed field-line topology T which is invariant to hydromagnetic changes in V. Specifically, the topology T is an invariant of the ideal hydromagnetic induction equation

$$\frac{\partial \mathbf{B}}{\partial t} = \nabla \times (\mathbf{v} \times \mathbf{B}), \tag{2}$$

describing the evolution of $\mathbf{B}$ in the plasma moving with velocity $\mathbf{v}$, subject to rigid boundary conditions at ∂V. Any physical specification of this system must include a prescription of that fixed field topology T in the same manner the entropy of an adiabatic gas must be prescribed to define its physical state. The equilibrium of such a system is then described by a solution to equation (1) subject to boundary data B_n and the prescribed field topology T. Since T is an entity quite independent of the force balance equation (1), no smooth solutions of equation (1) can in general be found to be compatible with an arbitrarily prescribed T. This basic point of Parker, Arnol'd, and others can be related to a property regarding the compactness of extremizing sequences of trial functions in variational calculus.

The force-free equation (1) is the Euler Lagrangian equation for the extremization of the total magnetic energy

$$\delta \int_V \frac{B^2}{8\pi}\, dV = 0, \tag{3}$$

subject to boundary data B_n and a prescribed field topology T. If we omit the constraint of a fixed T, equation (3) leads to a potential field with zero currents,

$$\nabla \times \mathbf{B} = 0, \tag{4}$$

as is well known, whereas imposing T in general requires the presence of electric currents (see, e.g., Roberts 1967). It should be pointed out that consideration of stability usually limits our physical interest to those force-free fields for which the total magnetic energy is a minimum. In this case, a useful way of visualizing the minimization of the total magnetic energy is to take the inertia-less plasma to be artificially viscous. A magnetic field in this plasma would evolve to a (stable) minimum energy state by a monotonic decrease of its total energy doing work against the viscous stress.

The reduction of a variational problem to its Euler-Lagrangian equation assumes that the extremum can be attained by a smooth trial function satisfying this partial differential equation, an assumption that cannot be taken for granted in general variational problems (Courant and Hilbert 1963). The problem posed by the force-free magnetic fields is a case in point. In carrying out the variation posed by equation (3) over the set S_T of all smooth fields with the same topology T, we may think of a sequence of fields $\mathbf{B}_i \in S_T$ that take the total energy to an extremum as i runs through the integers. Depending on the form of T, the set S_T may not be compact and we may have the case where all extremizing sequences $\mathbf{B}_i$ are not compact in that the limit $\mathbf{B}_\infty$ does not belong to S_T. In this case, a magnetic field having the prescribed topology can relax to a force-free state only by forming current sheets as represented by the inevitable singularities of $\mathbf{B}_\infty$. A simple illustration of the non-compactness of extremizing sequences can be found in the appendix of Low (1989).

2.2 Two Dimensional Magnetic Fields

It has long been known that a two dimensional magnetic field possessing neutral points readily develops electric current sheets. A well known case in the astrophysical literature is that associated with an X-type magnetic neutral point (e.g., Syrovatskii 1971). When stressed, the four magnetic sectors about the neutral point behave as integral bodies in consequence of flux freezing. To relieve the stress, a tendency is for two opposite sectors to push toward each other and squeeze out the other two sectors. The plasma displacement is discontinuous, creating a gap where opposite magnetic fields of the in-moving sectors come into contact. Thus, the X-type neutral point is stretched into an electric current sheet in the form of a magnetic tangential discontinuity.

It is the interaction of distinct magnetic flux bundles rather than the presence of an X-type neutral point that is crucial in creating the current sheet. Figure 1 shows an example that illustrates the same basic process without a role for the magnetic neutral point (Moffatt 1987, Low 1987, Low and Wolfson 1988, Aly 1989). Consider a two dimensional potential magnetic field $\mathbf{B}^I$ lying in the Cartesian $y-z$ plane. It is easy to construct $\mathbf{B}^I$ such that it has the field topology as sketched in Figure 1a where two adjacent (shaded) flux bundles are separated from a third by a separatrix line. Take the domain of interest V to be the infinite half plane $z > 0$ having a boundary ∂V made up of the line $z = 0$, denoted by ∂V^*, closed by the infinite half circle. Now deform the initially potential field by subjecting the boundary ∂V^* to some compressive displacement along ∂V^* centered, say, at the origin. The displacement is continuous and is continued smoothly into the interior V, deforming the field to some state represented by $\mathbf{B}^D$. This deformed field has a boundary flux distribution $B_z^D(\partial V^*)$ which is different from $B_z^I(\partial V^*)$ as the result of the boundary displacement. The field $\mathbf{B}^D$ will retain the same field topology shown in Figure 1a under the flux-freezing condition and the interesting question we pursue here concerns what happens if the magnetic field is now allowed to relax to a force-free state, with the magnetic footpoints rigidly anchored in their displaced positions on ∂V^*. In general, this magnetic field must develop an electric current sheet in the relaxed state. The proof is simple but we shall state its logic explicitly in anticipation of section 2.4 where the same basic process generalized to three dimensions is posed as a challenging unsolved problem.

The electric current density of a planar two-dimensional magnetic field is necessarily perpendicular to the plane of the magnetic field. The only way for the Lorentz force to vanish everywhere is to have the electric current density vanishing everywhere except, possibly, at singular lines across which tangential magnetic fields reverse directions with no change of magnitude. In other words, the force-free states of a two dimensional magnetic field in a plane are those in which the magnetic field is potential everywhere except possibly at tangential discontinuities. If tangential discontinuities are not allowed, the only force-free states are the classical smooth potential fields. It follows that if the deformed field $\mathbf{B}^D$ can relax to a smooth force-free state, this state is the one given by the potential field $\mathbf{B}^P$ bounded in V and having the same boundary flux $B_z^D(\partial V)$. From classical potential theory, this field $\mathbf{B}^P$ is unique. By direct construction starting with an initial potential field $\mathbf{B}^I$ having the field topology sketched in Figure 1a and deforming it in an arbitrary manner, it is easy to show that the deformed field $\mathbf{B}^D$ in general leads to a potential field $\mathbf{B}^P$ having the field topology sketched in Figure 1c. This topology with an X-type neutral point in V is different from that of $\mathbf{B}^D$, as indicated by the connectivity of individual lines of force identified in Figure 1 with Greek letters. Therefore, in general, the deformed magnetic field $\mathbf{B}^D$ cannot relax into the everywhere potential field $\mathbf{B}^P$ under the flux-freezing condition. Since $\mathbf{B}^P$ is the unique smooth force-free state compatible with the boundary flux of $\mathbf{B}^D$ at ∂V^*, it follows that $\mathbf{B}^D$ must relax to another end state represented by a potential field containing some form of tangential discontinuity. Again by direct construction, it is simple to show that the proper end state is as sketched in Figure 1b where the electric current in the tangential discontinuity preserves the field topology of the field $\mathbf{B}^D$. Note that the current in the sheet has zero intensity at the point of the sheet's intersection with ∂V^* because $B_z^D(\partial V^*)$ is continuous. The formation of the electric current sheet can be understood in terms of the expected dynamical behavior of the three distinct flux bundles in Figure 1a during the relaxation. To relieve the stress of the compressive deformation, the overlying flux bundle is expelled from the region between the other two (shaded) flux bundle resulting in the vertical current sheet indicated in Figure 1b. Only through the introduction of electrical resistive dissipation can the field topology change. In that case, $\mathbf{B}^D$ can relax into the everywhere smooth $\mathbf{B}^P$ by eliminating the current sheet.

2.3 Three Dimensional Fields

In the preceding example, the basic effect that creates the electric current sheet is the expulsion of a magnetic flux bundle from a local region to enable other flux bundles otherwise separated in space to come into contact. The presence of a separatrix line is essential in this two dimensional example. The special role of the separatrix line loses its significance in systems that vary with three dimensions as we now demonstrate. In three dimensional space, it is possible for a stress distribution to compress a set of locally-parallel magnetic flux surfaces to result in the expulsion of magnetic flux from between a pair of these flux surfaces. This is essentially a three dimensional process. In addition to the two dimensional fields in each of the flux surfaces, a strong variation in the third dimension perpendicular to the locally-parallel flux surfaces is required. Depending on the stress distribution, any two flux surfaces may force an expulsion of the flux sandwiched in between and come into contact over a finite area (Low 1989, Parker 1989b). In three dimensions, the field does not have to be antiparallel and the area of contact is in general a rotational discontinuity (Parker 1987). The following three dimensional model provides a simple illustration of this process.

There is a general class of three dimensional force-free magnetic fields with flux surfaces in the form of parallel planes (Chang and Caravillano 1981). Taking these flux surfaces to be planes of constant x in Cartesian coordinates, these force-free fields having no x component are generated by

$$B_y - iB_z = F(w)e^{i\zeta}, \tag{5}$$

where $i = (-1)^{1/2}$, F is any analytic function of $w = y + iz$, and ζ is any real function of x. For a fixed form of F, equation (5) gives a field whose pattern varies across the planes of constant x according to the function ζ. The shear of the field across these flux planes is associated with an electric current density which is everywhere parallel to the magnetic field. The case of a constant ζ has zero shear, or zero electric current, and it is easy to verify that this trivial case corresponds to the two dimensional potential fields invariant in the x direction.

The three-dimensional magnetic fields generated by equation (5) have a particular simplicity that allows us to see how they can develop electric current sheets when subject to appropriate stresses (Low 1989). The situation we are interested in is sketched in Figure 2. Figure 2a represents one of these magnetic fields in terms of a subset of its flux planes evenly spaced. In each of these numbered flux planes is a two dimensional distribution of field generated by some fixed choice of F and ζ. Now subject this initial equilibrium field to a deformation that treats the flux planes as rigid plates displacing them in the x-direction. Consider a displacement in the form of uniform translations in opposite directions in the regions outside planes 3 and 9 such that a uniform expansion of the region between these two planes increases the constant distance between adjacent planes in this region. This displacement does not change the field pattern in each of the constant-x planar flux surfaces. Figure 2b represents the deformed field, no longer in equilibrium, in terms of the flux planes in their displaced positions. Now let the deformed magnetic field relax to a new equilibrium in $z > 0$ under the flux-freezing condition and holding the field anchored rigidly to the lower boundary $z = 0$.

To see physically how the deformed field in Figure 2b would behave, it is important to understand the manner by which force balance is achieved in the initial force-free field in Figure 2a. Equation (1) can be written in the form

$$\frac{1}{4\pi}(\mathbf{B}.\nabla)\,\mathbf{B} - \nabla\left(\frac{B^2}{8\pi}\right) = 0, \tag{6}$$

showing that the force-free field achieves equilibrium by balancing a magnetic tension force against a magnetic pressure force. The initial field in Figure 2a lies in planes of constant x and has no component of its tension force in the x direction. Its equilibrium is maintained with a magnetic pressure uniform in the x direction. The displacement that produces the deformed state in Figure 2b does not change the force balance in the y and z direction. The uniform expansion of the region between planes 3 and 9 results in a lowering of the magnetic pressure relative to that outside of this region. So it is clear that the deformed magnetic field would seek its new equilibrium by a collapse inward into the rarefied region until the magnetic pressure between the two flux surfaces initially represented by planes 3 and 9 is built up sufficiently to halt the collapse. Anchoring the field at $z = 0$ has two effects. The flux surfaces must deform from their initial planar shapes and the build up of magnetic pressure in the region between flux surfaces 3 and 9 requires magnetic flux to be drawn downward as it is squeezed from between the upper reaches of these two collapsing flux surfaces. The second effect is better appreciated if we note that each of the initially planar flux surfaces contains a two dimensional field with lines of force anchored at both ends on $z = 0$. If the rarefaction is sufficiently strong, the downward expulsion of the flux from the upper reaches of the region between the two flux surfaces may be so complete that the two flux surfaces come into contact as sketched in Figure 2c. The field patterns on these two flux surfaces are quite uncorrelated for these patterns are controlled by a free function ζ in the initial state. The contact surface is therefore a rotational discontinuity. It has not been possible to explicitly construct the end state sketched in Figure 2c but it seems clear that electric current sheets can form at ordinary flux surfaces when subject to suitable stresses. Figure 2c represents an extreme case of the end state. To form a rotational discontinuity, flux surfaces 3 and 9 may meet over a finite area, rather than over the infinite area sketched in Figure 2c.

2.4 An Interesting Three-Dimensional Problem

The example in Figure 2 provides a strong plausibility demonstration but is short of a mathematical proof because it is technically formidable to construct the final relaxed state sketched in Figure 2c. The two dimensional example in Figure 1 is rigorous. Given that most three dimensional problems are intractable, a generalization of the problem in Figure 1 to three dimensions is an interesting unsolved problem that merits our attention. Suppose we are given a three dimensional potential field $\mathbf{B}^I$ in V with all lines of force anchored at both ends on the boundary ∂V. Let the footpoints of the magnetic lines of force be displaced in a continuous manner at the boundary ∂V, continuing these displacements smoothly into V. The general displacements would impart shear or even twist to the field so that the deformed magnetic field can only relax into an equilibrium that contains both volume and sheet currents. There is a subset of displacements which imparts no shear to the field so that the deformed field must relax to a state which is almost everywhere potential except for singular surfaces of magnetic discontinuity. If such a subset of displacements can be identified explicitly, it would be interesting to study which of those in this subset require the relaxed state to contain magnetic discontinuities in order to be consistent with the flux-freezing assumption. The study would proceed as in the two dimensional example of section 2.2, where a tactical advantage is exploited, which is that if the end state is free of magnetic discontinuity, it is a unique tractable potential field. The unsolved problem we pose is whether this special subset of footpoint displacements can be constructed in a general way for three dimensional systems.

3. CONCLUSION

Electric current sheets are created from continuous magnetic fields through discontinuous displacements (or tearing) in the plasma medium. A simple way of creating them is to impose such displacements directly in some physical manner. In the model treated in this paper, if the boundary displacement of the magnetic footpoints is discontinuous at some point on the boundary, a current sheet would have been imposed upon the end state extending from that point on the boundary into the interior. The interesting point made with this model in section 2 is that current sheets would form in the interior of the end state even when the boundary displacement of the magnetic footpoints is continuous. It is in this sense the current sheet formation is spontaneous, and therein lies the novelty of this effect first pointed out by Parker.

Since the examples treated in this paper are derived from astrophysics, it seems appropriate to conclude with a brief discussion of an especially interesting astrophysical implication of these examples. The magnetic fields in the outer atmospheres of stars similar to our Sun are believed to evolve in two distinctive phases. There is a quasi-static phase during which the magnetic field is approximately force-free with its field configuration changing slowly as the result of convective transport of the magnetic footpoints anchored at the surface of the star. This quasi-static evolution arises from the convective time scale being much longer then the (Alfvenic) time scale for the field to adjust in the tenuous upper atmosphere from one equilibrium state to another. The high electrical conductivity of the upper atmosphere allows for substantial amount of electric currents to flow and persist in the evolving equilibrium state. In this manner, a considerable amount of energy can be stored in the tenuous atmosphere in the form of volume electric currents. On the other hand, the high electrical conductivity would be an obstacle to releasing that stored energy unless adequately small scale structures can be created so that electrical resistive dissipation has a role to play (over the small scales) despite the high electrical conductivity. The spontaneous formation of electric current sheet is just the mechanism to accomplish that. The dissipation as a quiescent heating has been proposed by Parker (1986) to be a mechanism for heating stellar atmospheres during the quasi-static phase. The catastrophic collapse of magnetic fields to form large scale current sheets like the type envisaged in Figure 2 may initiate an explosive dissipation of currents during which the magnetic field transits into a fully dynamical turbulent state. This second phase of evolution corresponds to flares, outward expulsion of mass and other related astrophysical high-energy phenomena.

The National Center for Atmospheric Research is sponsored by the U.S. National Science Foundation. I thank Annick Pouquet for commenting on this paper.

References

Aly, J.J. 1989, *Astron. Astrophys.* (submitted).

Arnold, V. 1974, in *Proc. Summer School in Differential Equations, The Asymptotic Hopf Invariant and its Applications* (Erevan, Armenian SSR Acad. Sci.).

Chang, H.M. and R.L. Carovillano 1981, *Bull. AAS* **13**, 909.

Courant, R., and D. Hilbert 1963, *Methods of Mathematical Physics*, Vol. 1 (New York: Interscience).

Low, B.C. 1987, *Ap. J.* **323**, 358.

Low, B.C. 1989, *Ap. J.* **340**, 558.

Low, B.C., and R. Wolfson 1988, *Ap. J.* **324**, 574.

Moffatt, H.K. 1985, *J. Fluid Mech.* **159** 359.

Moffatt, H.K. 1987, in *Advances in Turbulence* ed. G. Comte-Bellot and J. Meathieu (New York: Springer), p. 228.

Parker, E.N. 1972, *Ap. J.* **174**, 499.

Parker, E.N. 1983a, *Ap. J.* **264**, 635.

_______ 1983b, *Geophys. Ap. Fluid Dyn.* **23**, 85.

_______ 1986, in *Proc. NASA-SMM Workshop on Coronal and Prominence Plasmas*, ed. A.I. Poland (NASA Pub, CP-2442), p. 9.

_______ 1987, *Ap. J.* **318**, 376.

_______ 1989a, *Geophys. Ap. Fluid Dyn.* **45**, 159.

_______ 1989b, *Geophys. Ap. Fluid Dyn.* **45**, 169.

Roberts, P.H. 1967, *An Introduction to Magnetohydrodynamics* (London: Longmans).

Syrovatskii, S.I. 1971, *Soviet Phys. JETP* **33**, 933.

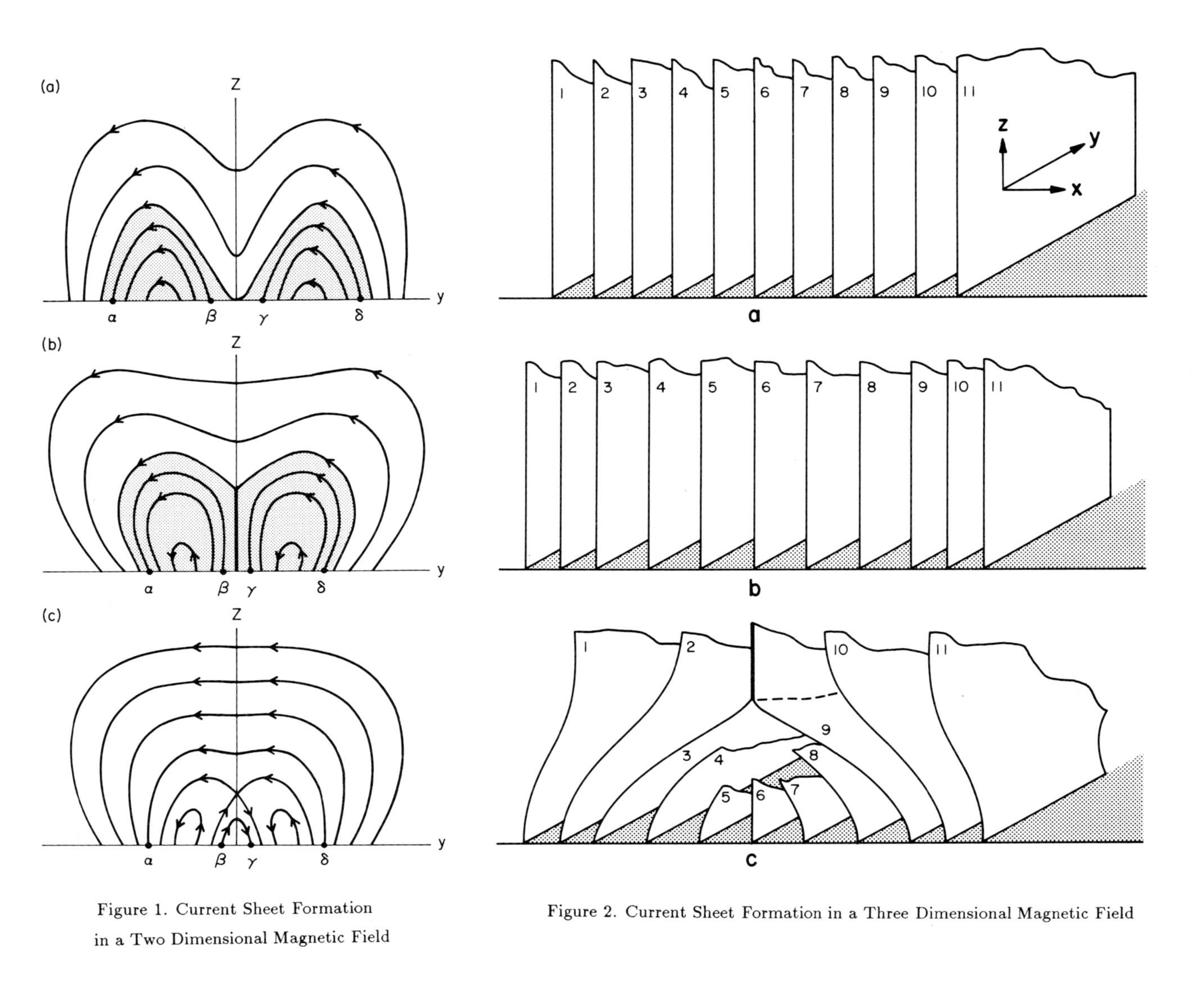

Figure 1. Current Sheet Formation in a Two Dimensional Magnetic Field

Figure 2. Current Sheet Formation in a Three Dimensional Magnetic Field

Current Sheets in Line-tied Magnetic Fields With Pressure

G.B. FIELD

Harvard-Smithsonian Center for Astrophysics
60 Garden St., Cambridge, MA 02138

ABSTRACT

Van Ballegooijen (1985, 1988) argued that if the sun's coronal magnetic field is force free, current sheets do not form in Parker's (1972) model of coronal heating if the velocity field in the solar photosphere is continuous at all times. Here we refine his argument for force-free fields to provide a constraint on current sheets in line-tied magnetic fields. We also extend Van Ballegooijen's argument to current sheets aligned with isobars in a pressurized plasma.

1. INTRODUCTION

Parker (1972) proposed a model for solar coronal heating in which the coronal magnetic field is initially uniform between two plates at $z = 0$ and L, which represent parts of the photosphere with different polarity. The field in the corona ($0 < z < L$) evolves quasistatically as the footpoints of the field lines move with the photospheric plasma. Assuming ideal MHD, Parker argued that as the field becomes more tangled (Figure 1), current sheets would form, and heat would be generated as these sheets dissipate by finite resistivity.

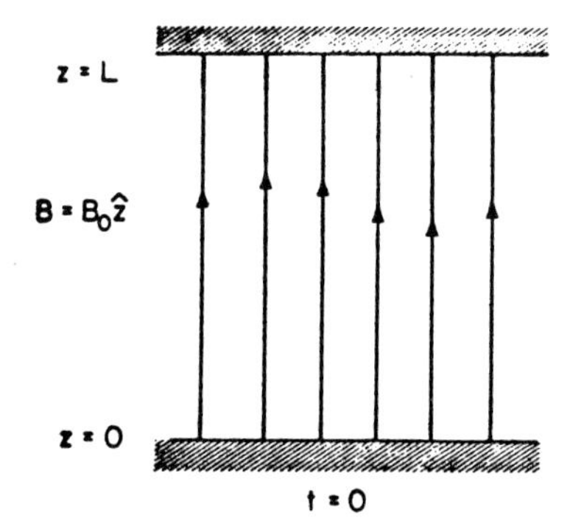

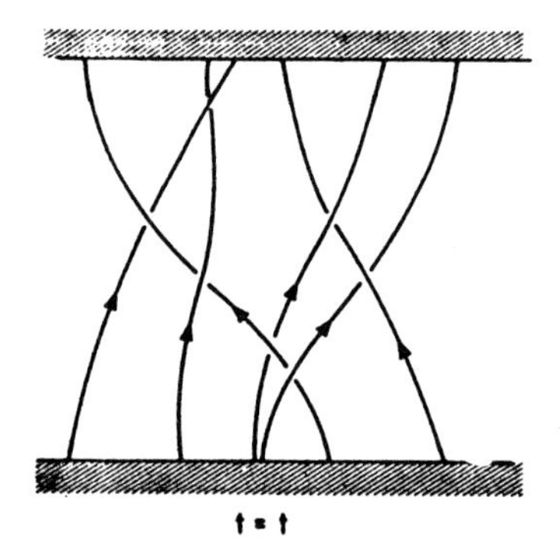

Figure 1. Geometry of the coronal magnetic field **B** as proposed by Parker (1972). After a time t the field lines become tangled because of motion of the footpoints on the boundaries.

In later papers, Parker (1979, 1982, 1983ab, 1986abcd, 1987, 1989abc) and Vainshtein and Parker (1986) elaborated this model, and Sakurai and Levine (1981), Van Ballegooijen (1985, 1988), Seehafer (1986), Aly (1987), Antiochos (1987), Zweibel and Li (1987), and Field (1988) discussed it as well. The original 1972 Parker model involves no separatrices at $t = 0$. Under slow motion of the footpoints, this remains true. As discussed by Aly (1987), current sheets are generic if there are separatrices; recent examples of such models are those of Low (1987) and Low and Wolfson (1988). Here we consider only Parker's 1972 model.

Most discussions of the Parker model have assumed that, as is approximately correct in the solar corona, the pressure p is negligible with respect to $B^2/8\pi$, so that the magnetic field is force free. Exceptions are Parker (1983ab) and Vainshtein and Parker (1986). Van Ballegooijen (1985, 1988) argued that if the magnetic field is force free, postulating the existence of a current sheet in equilibrium appears to lead to a contradiction if the photospheric flow on the boundaries is continuous at all times. He concluded that current sheets do not form under these conditions, attributing this result to the fact that the field lines on opposite sides of a current sheet must have different curvature; hence they exert different magnetic stresses, and that in the case of a force-free field, there is no other force available to balance the difference in magnetic stress. Here we derive Van Ballegooijen's result more formally, and also generalize it to the case in which the current sheet is assumed to be along an isobar in a plasma with finite but continuous pressure.

The equation of magnetohydrostatic equilibrium

$$\mathbf{J} \times \mathbf{B} = c\nabla p \tag{1}$$

implies that

$$\mathbf{B} \cdot \nabla p = 0 \tag{2}$$

so that p is constant on each magnetic field line. Therefore, p is determined throughout the system by its values on either boundary, say $z = 0$. We can characterize $p(x, y)$ by its isobars $p(x, y) = constant$. Here we consider the special case of current sheets whose curves of intersection with the boundary, which we call "traces," coincide with isobars.

We assume that current sheets must extend from one boundary to the other. This has been demonstrated formally by Willette (1988, Proposition III.3.5) for the case of vanishing pressure. Willette's proof relies upon the fact that if a current sheet were to end before reaching a boundary, the current on it must leave along field lines, as the field is force free. This is not possible because the magnetic flux of a set of field lines is finite, whereas that in a current sheet is zero. We shall assume that this is also true if $p \neq 0$. We label one side of the current sheet (1) and the other side (2), and let $\hat{\mathbf{n}}$ be a unit normal to the sheet from side (1) to side (2). If $\mathbf{B}_1$ and $\mathbf{B}_2$ are the local magnetic fields on either side of the sheet, the current density in the sheet is

$$\mathbf{K} = \frac{c}{4\pi}\hat{\mathbf{n}} \times (\mathbf{B}_2 - \mathbf{B}_1) \tag{3}$$

according to Ampere's law. Gauss's law implies that $\mathbf{B}_1 \cdot \hat{\mathbf{n}} = \mathbf{B}_2 \cdot \hat{\mathbf{n}} = B_n$. However, if $B_n \neq 0$, the Lorentz force per unit volume in the sheet, KB_n, would be infinite, so

$$\mathbf{B}_1 \cdot \hat{\mathbf{n}} = \mathbf{B}_2 \cdot \hat{\mathbf{n}} = 0 \tag{4}$$

and the magnetic field must be tangent to the sheet on both sides. Integration of equation (1) through the sheet yields the equation of stress balance across the sheet:

$$\frac{B_1^2}{8\pi} + p_1 = \frac{B_2^2}{8\pi} + p_2 \tag{5}$$

2. THE EFFECTS OF PRESSURE

Since the trace of the current sheet coincides with an isobar by assumption, p_1 and p_2 are both constant on the trace. Further, since p is continuous on the boundary, $p_1 = p_2$ on the trace. Equation (2) shows that both p_1 and p_2 are constant everywhere on the sheet, and since they are equal on the trace, they are equal everywhere ($p_1 = p_2 = p$) on the sheet.

The cross product of $\hat{\mathbf{n}}$ with equation (1) shows that

$$J_{1n}\mathbf{B}_1 = J_{2n}\mathbf{B}_2 = c\nabla p \times \hat{\mathbf{n}} = 0 \tag{6}$$

where we have used equation (4) and the fact that p is constant on the current sheet. It follows from equation (6) that

$$J_{1n} = J_{2n} = 0 \tag{7}$$

everywhere on the sheet. Under these circumstances, Ampere's law implies that for any closed contour C on the sheet

$$\int_C (\mathbf{B}_1 - \mathbf{B}_2) \cdot ds = \frac{4\pi}{c}\int_S (J_{1n} - J_{2n})dS = 0 \tag{8}$$

where S is the surface enclosed by C. Van Ballegooijen (1985, 1988) argued that if the motion of the fluid on both boundaries is continuous at all times any two magnetic field lines which remain close on one boundary remain close on the other boundary. In particular, this is true of two field lines L_1 and L_2 on opposite sides of the current sheet (see Figure 2). Although L_1 and L_2 may be distinct almost everywhere, because of Van Ballegooijen's argument cited above, they must meet somewhere on the bottom trace (P) if they meet on the top trace (P').

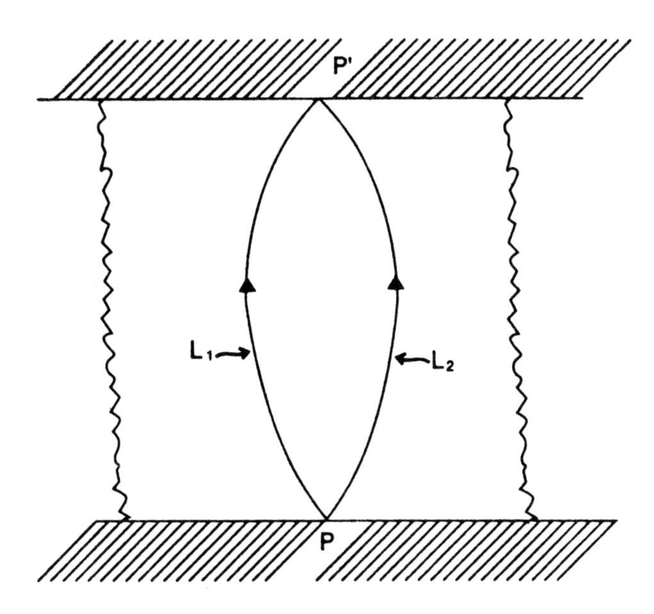

Figure 2. Part of a current sheet that might form in the Parker (1972) model. The line of force L_1 on side 1 of the sheet connects P on the bottom boundary with P' on the top. Line L_2, on side 2, intersects L_1 at P' on the top boundary, and by continuity, must intersect it again on the bottom, at P.

It follows that the contour $C = P - L_2 - P' - L_1 - P$ in Figure 2 is closed, so we can apply equation (8) to it. If ϕ is the local angle between $\mathbf{B}_1$ and $\mathbf{B}_2$, we find that because $ds = \mathbf{B}_2 ds/B_2$ on L_2, and $ds = -\mathbf{B}_1 ds/B_1$ on L_1,

$$\int_C (\mathbf{B}_1 - \mathbf{B}_2)\cdot ds = \int_{L_2:P}^{P'} (B_1 \cos\phi - B_2)ds + \int_{L_1:P'}^{P} (B_2 \cos\phi - B_1)ds \tag{9}$$

Here $\mathbf{B}_1$ and $\mathbf{B}_2$ refer to the magnetic fields on either side of the current sheet at the same location. According to equation (5), $B_1 = B_2 = B$ everywhere on the current sheet because $p_1 = p_2$. Hence equation (9) can be written

$$\begin{aligned} \int_C (\mathbf{B}_1 - \mathbf{B}_2)\cdot ds &= \int_{L_2:P}^{P'} B(\cos\phi - 1)ds + \int_{L_1:P'}^{P} B(\cos\phi - 1)ds \\ &= -2\left[\int_{L_2:P}^{P'} B\sin^2\left(\frac{\phi}{2}\right) ds + \int_{L_1:P'}^{P} B\sin^2\left(\frac{\phi}{2}\right) ds\right] \end{aligned} \tag{10}$$

Equation (10) vanishes as required by equation (8) only if ϕ is zero everywhere, in which case $\mathbf{B}_1$ is parallel to $\mathbf{B}_2$ and, because $B_1 = B_2$, $\mathbf{B}_1 = \mathbf{B}_2$ everywhere on the sheet. But in this case equation (3) shows that $\mathbf{K} = 0$ everywhere on the sheet, an apparent contradiction. This generalizes Van Ballegooijen's (1985, 1988) result for a force-free field, which corresponds to the special case $p = constant$ everywhere.

3. DISCUSSION AND CONCLUSIONS

If the flow on the boundaries is continuous at all times, and the pressure on the boundaries is continuous, the existence of a current sheet in equilibrium along an isobar leads to an apparent contradiction.

If $p \neq 0$, the argument given here fails if the current sheet is not along an isobar. The case in which a current sheet forms across isobars is under study (Field 1989).

If p is constant, so that the field is force-free, it is not obvious how to interpret the apparent contradiction demonstrated here. Van Ballegooijen (1985, 1988) concluded simply that current sheets do not form in this case. However, the examples of current sheets given by Parker seem persuasive, so perhaps one or more of our assumptions is not met in practice. One unproved assumption is that current sheets must extend to the boundaries. At the Symposium, E.N. Parker and B.C. Low pointed out that their calculations indicate that the current sheets that form are usually joined at a three-way vertex, and M.A. Berger pointed out that a current sheet could bifurcate into two sheets as it approaches the boundary. These and other modifications of the assumed topology should be studied further to see if they permit a resolution of the apparent contradiction found here.

4. ACKNOWLEDGEMENTS

Aad Van Ballegooijen has kindly provided much needed criticism and advice, and I am indebted to B.C. Low for criticism of an earlier draft. This work was supported by NASA under Grant NAGW-931.

REFERENCES

Aly, J.J. 1987, in *Proc. Workshop on Interstellar Magnetic Fields, Schloss Ringberg (FRG)*, September 8-12, 1986 (Berlin: Springer Verlag), p. 240.

Antiochos, S.K. 1987, *ApJ.*, **312**, 886.

Field, G.B. 1988, in *Order and Disorder in Planetary Dynamos*, Technical Report WHOI-88-60 (Woods Hole, Massachusetts: Woods Hole Oceanographic Institution), p. 117.

Field, G.B. 1989, in preparation.

Low, B.C. 1987, *ApJ.*, **323**, 358.

Low, B.C. and Wolfson, R. 1988, *ApJ.*, **324**, 574.

Parker, E.N. 1972, *ApJ.*, **174**, 499.

__________ 1979, *Cosmical Magnetic Fields* (Oxford: Clarendon Press).

__________ 1982, *Geophys. Astrophys. Fluid Dynamics*, **22**, 195.

__________ 1983a, *ApJ.*, **264**, 635.

__________ 1983b, *Geophys. Astrophys. Fluid Dynamics*, **23**, 85.

__________ 1986a, *ibid.*, **34**, 243.

__________ 1986b, *ibid.*, **35**, 277.

__________ 1986c, in *Proc. Workshop on Coronal Prominences and Plasmas*, A. Poland, ed. (NASA-GSFC Printing Office), in press.

__________ 1986d, in *Lecture Notes in Physics*, **254**: *Cool Stars, Stellar Systems and the Sun*, M. Zeilik and D.M. Gibson, eds. (Berlin: Springer Verlag), p. 341.

__________ 1987, *ApJ.*, **318**, 876.

__________ 1989a, *Geophys. Astrophys. Fluid Dyn.* **45**, 159.

__________ 1989b, *ibid.*, **45**, 169.

__________ 1989c, *ibid.*, **46** (in press).

Sakurai, T. and Levine, R.H. 1981 *ApJ.*, **248**, 817.

Seehafer, N. 1986, *Solar Phys.*, **107**, 73.

Vainshtein, S.I. and Parker, E.N. 1986, *ApJ.*, **304**, 821.

Van Ballegooijen, A.A. 1985, *ApJ.*, **298**, 421.

__________ 1988, *Geophys. Astrophys. Fluid Dyanmics*, **41**, 181.

Willette, G.T. 1988, *The Formation of Tangential Discontinuities in Line-Tied Force-Free Fields: Topological Heating of the Solar Corona*, Honors Thesis, Department of Astronomy, Harvard University.

Zweibel, E.G., and Li, H.-S. 1987, *ApJ.*, 312, 423.

Flux Constraints on Magnetic Energy Release in a Highly Conducting Plasma

J.J. ALY

Service d'Astrophysique, CEN-Saclay, F-91191 Gif-sur-Yvette Cedex, France.

1 INTRODUCTION

In many astrophysical systems, one has to deal with a region D of space containing a magnetic field $\underset{\sim}{B}$ whose energy density amply dominates that one of the highly conducting plasma in which it is embedded, this field emerging through the boundary ∂D of D from a "plasma-dominated" region D' in which it is passively transported by slow motions of the matter. Under these conditions, the field in D may be generally considered as evolving through a sequence of equilibrium force-free configurations: at each time t, it adjusts quasi-statically to the changes imposed on ∂D by the boundary motions. In some circumstances, however, a highly dynamical behaviour may take place for a short period of time in some part of D: for instance, it may be that the force-free field corresponding to the actual boundary conditions either is not stable, or contains singular layers (current sheets) subject to a non-equilibrium dissipative process. In that case, the field may try to effect an energy-releasing transition to a new force-free state of lower energy. Of course, a complete description of this process may be obtained only by working out the dynamical equations. In some cases, however, it is possible to guess on physical grounds a complete set of physical quantities which are (approximately) conserved during the transition. Making the natural assumption that the dynamic phase allows to reach the minimum energy state among all the states for which the conserved quantities have the values fixed by the initial state, it thus turns out to be possible to predict in a simple way the final configuration and the amount of energy released, without having to deal with the complicated details of the transition. In this Communication, we present several two-dimensional models of astrophysical interest in which we make an explicit use of such a simplified theory.

2 STATEMENT OF THE PROBLEM

2.1 Two-dimensional force-free fields

Let $(0, \hat{x}, \hat{y}, \hat{z})$ be a cartesian frame and Ω be a simply-connected domain of the plane $\{x = 0\}$. We consider in the cylindrical domain $D = R_x \times \Omega$ ($R_x = \{x \mid -\infty < x < \infty\}$) an x-invariant smooth magnetic field

$$\underset{\sim}{B}(y,z) = B_x(y,z)\,\hat{x} + \nabla A(y,z) \times \hat{x}\ ; \tag{1}$$

if Ω is unbounded, we assume that A and $|\underset{\sim}{B}|$ decay to zero when $r = (y^2+z^2)^{1/2} \to +\infty$ $\left(\lim_{r\to+\infty} r^{1+\epsilon}|\underset{\sim}{B}| = 0 \text{ for some } \epsilon > 0, \text{ say}\right)$. Quite generally, the level sets in Ω of the potential A - i.e. the curves $C(a) = \{(y,z) \mid A(y,z) = a\}$ - determine a partition of Ω into "flux cells" Ω_i (open subdomains such that $\bigcup_i \bar{\Omega}_i = \bar{\Omega}$ and $\Omega_i \cap \Omega_j = \emptyset$ if $i \neq j$) which have the following properties:

i) for any value of a, there is at most one connected part $C_i(a)$ of $C(a)$ in Ω_i ;

ii) the C_i belonging to Ω_i either are all closed - Ω_i thus being either disk-shaped or ring-shaped - or they all meet twice the boundary $\partial\Omega$ - $\partial\Omega \cap \partial\Omega_i$ thus having either one or two connected parts;

iii) in Ω_i, A varies monotonically accross the lines from some minimum value $\underline{a}_i$ up to some maximum one $\bar{a}_i$.

The curves of Ω separating the cells Ω_i form the net Γ of the so-called separatrices ; on a connected part Γ_j of Γ, one has $A(y,z) = \gamma_j$ and Γ_j either contains an X-critical point of A (where $\nabla A = 0$), or extends to infinity, or meets $\partial\Omega$ at some bifurcation point (which may be a critical point of A, or a tangential contact of two arcs, ...).

Let us now enlarge a little bit the class of fields we consider, by allowing fields which may possibly have tangential discontinuities -

i.e. current sheets - located on the separatrices Γ, while being smooth in each Ω_i, and let us consider among these fields one which is force-free in Ω. Then we have

$$B_x(y,z) = B_{xi}[A(y,z)] \qquad \text{in } \Omega_i \qquad (2)$$

$$-\Delta A = \frac{d}{dA}\frac{B_{xi}^2}{2} \qquad \text{in } \Omega_i \qquad (3)$$

$$[|\nabla A|^2 + B_x^2] = 0 \text{ and } [\underset{\sim}{B}.\hat{n}] = 0 \qquad \text{on } \Gamma \qquad (4)$$

where $[X]$ denotes the jump of X accross Γ and $\hat{n}$ a normal to that curve. For this $\underset{\sim}{B}$, we associate to each $C_i(a)$ the number

$$X_i(a) = B_{xi}(a) \int_{C_i(a)} \frac{ds}{|\nabla A|} = -\left(B_{xi}\frac{d\mu_i}{da}\right)(a) \qquad (5)$$

where s is an arc length along $C_i(a)$ and $\mu_i(a) = \text{meas}\{(y,z) \in \Omega_i : A(y,z) > a\}$. If $C_i(a)$ is connected to $\partial\Omega$, $X_i(a)$ represents the "shear" of the associated lines of $\underset{\sim}{B}$ (i.e. the difference between the x-positions of their left and right feet on $\partial\Omega$, respectively) ; if $C_i(a)$ is closed, $X_i(a)$ is the pitch of the associated helicoidal lines of $\underset{\sim}{B}$.

2.2 Evolution of a force-free field

Let us assume that we have in D a force-free field $\underset{\sim}{B}_0$ embedded in a perfectly conducting low-β plasma and that we start imposing at some initial time $t = 0$ on ∂D on x-invariant velocity field $\underset{\sim}{v}$ which is "flux-preserving" (no flux emerges or disappears through ∂D) and "slow" (sup $|\underset{\sim}{v}|$ << typical Alfven velocity in D). Thus we may assume that the field evolves quasi-statically through a sequence of force-free configurations $\underset{\sim}{B}_t$. Because of the frozen-in law, the topological structure of the field is conserved during this evolution, as well as the values of $(\underline{a}_i, \bar{a}_i)$ and the functions $X_i(a)$ associated with the "magnetic islands" Ω_i (cells containing closed lines); on the contrary, the functions $X_i(a)$ associated with the Ω_i constituted of lines connected to $\partial\Omega$, change in time ; as A on ∂D, however, these X_i can be easily computed from the knowledge of A_0 (initial potential) and $\underset{\sim}{v}$ on ∂D. Then $\underset{\sim}{B}_t$ may be determined - in principle - by solving at each time t Equ. (2)-(4), with B_{xi} being calculated from the known functions X_i by using Equ. (5), and with A being imposed on $\partial\Omega$.

Actually, such an evolution may lead at some time t_1 to an ideally or resistively unstable state (we now allow for a small resistivity of the plasma), or to a singular state (containing e.g. current sheets on the separatrices). In that case, the system is forced into a fast dynamical evolution (time scale very small compared to the resistive time scale associated with the typical size of the system) during which dissipative effects can no longer be neglected and topology changes and flux rearrangements may occur. In general, however, the frozen-in law is violated only in small parts of space and some set $\mathcal{S}$ of "flux quantities" are approximately conserved during the active phase leading to a new equilibrium. One is thus led quite naturally to the following principle: the final state $\underset{\sim}{B}_1'$ of the transition is an energy minimizer (by energy, we always mean "energy per unit of x-length": $W[\underset{\sim}{B}] = (8\pi)^{-1}\int_\Omega |\underset{\sim}{B}|^2 dydz$) in the set $\mathcal{H}_1$ of all the fields $\underset{\sim}{B}$ having the same values of the quantities in $\mathcal{S}$ as the initial state $\underset{\sim}{B}_1$. Of course, the transition may occur spontaneously only if $\underset{\sim}{B}_1'$ has a lower energy than $\underset{\sim}{B}_1$ (i.e. $W[\underset{\sim}{B}_1'] < W[\underset{\sim}{B}_1]$). Here, we have assumed that $\underset{\sim}{B}_1$ was known a priori to be unstable with respect to an $\mathcal{S}$-preserving process. For an arbitrary $\underset{\sim}{B}_1$, however, we may not be aware if this is the case indeed. But we may still apply the previous procedure and compute the energy minimizer $\underset{\sim}{B}_1'$ in the set $\mathcal{H}_1$ associated to $\underset{\sim}{B}_1$. We can thus conclude at the possibility of a transition as above if $W[\underset{\sim}{B}_1'] < W[\underset{\sim}{B}_1]$. Rather than developing this idea in an abstract way, we prefer to report now three specific examples in which it is used explicitely.

3 CREATION OF MAGNETIC ISLANDS IN A SHEARED ARCADE

3.1 Quasi-static evolution of an arcade

Let us take $\Omega = \{z > 0\}$, and the initial field $\underset{\sim}{B}_0$ be potential in D ($\underset{\sim}{B}_0 = \nabla A_0 \times \hat{x}$; $-\Delta A_0 = 0$), with $g(y) = A_0(y,0)$ satisfying $yg'(y) > 0$ for $y \neq 0$. Then the lines $C(a)$ have a simple topology, being arcades bridging over the origin (there is one single flux cell). Let us now suppose that D is filled up with a perfectly conducting plasma and that we start applying on ∂D at $t = 0$ a stationnary slow velocity field $\underset{\sim}{v}(y) = v(y)\hat{x}$ (with $v(y)$ tending fast to zero when $|y| \to \infty$), what results at t in a shear $X(a,t) = t\zeta(a)$ for any line of $\underset{\sim}{B}$ projecting onto $C(a,t)$. Thus one has the following results (*Aly, 1985, 1988*):

i) the field may evolve quasi-statically through a continuous sequence of 2D-stable force-free configurations up to arbitrarily large times, the associated potential A_t being at each time t a minimizer of the energy

$$C_t[A] = \int_\Omega |\nabla A|^2 dydz + t^2 \int_0^\infty |\zeta[A(\mu)]|^2 \left|\frac{dA(\mu)}{d\mu}\right|^2 d\mu \qquad (6)$$

over the set $\mathcal{H} = \{A | C_1[A] < \infty \ ; \ A(y,0) = g(y) \ ;$ lines of A are arcade-like$\}$ (note that $A_t(y,0)$ is kept invariant by $\underset{\sim}{v}$) ;

ii) there is an initial phase of evolution during which A_t does not change appreciably while $B_{tx}(a) \simeq t\, b(a)$ and $C_t[A_t]-C_0[A_0] \propto t^2$, and an asymptotic phase during which the poloidal structure expands ($\mu(a,t) \propto t^2$) while $B_{tx}(a) \simeq t^{-1}b'(a)$ and $C_t[A_t] \propto \log t$;

iii) when $t \to \infty$, A_t approaches asymptotically an open structure, with all the currents concentrating into a current sheet.

3.2 Stability of an arcade with respect to reconnection

Let us now relax the assumption of perfect conductivity of the plasma and look for the possibility of a reconnection process which, acting at time t_1 on a fast time scale, could change the topology of the lines but would quasi-conserve the magnetic fluxes, thus inducing a transition to a field $\underset{\sim}{B}_1'$ for which (constraints $\mathcal{S}_r$):

i) $A_1'(y,0) = g(y)$ and $0 \leqslant A_1'(y,z) \leqslant A_1(0,0)$;

ii) $X(a,t_1) = t_1\ \zeta(a) = -\sum_i B_{1xi}'\, d\mu_{1i}'/da$ (i.e. the x-flux between the two lines $C(a)$ and $C(a+da)$ is the same for $\underset{\sim}{B}_1$ and $\underset{\sim}{B}_1'$; see also *Grad, Hu and Stevens, 1975*).

Clearly, from the general principle stated in § 2.2, such a reconnection process may happen only if A_{t_1} is not a minimizer of $C_{t_1}[A]$ over the set $\mathcal{H}' = \{A|C_1[A] < \infty ; A(y,0) = g(y)\}$ containing the potentials having the right boundary values, but an arbitrary topology (note that our flux constraint (ii) above is included in the functional form of C_t when we set $B'_{xi} = B'_x$, what may be done because this condition minimizes the energy for a given poloidal structure).

Actually, it may be shown that there is a critical time $t_c[g,\zeta]$ such that:

$$\text{i) for } t < t_c,\ C_t[A_t] = \inf_{\mathcal{H}'} C_t[A] ;$$
$$\text{ii) for } t > t_c,\ C_t[A_t] > \inf_{\mathcal{H}'} C_t[A].$$

Thus reconnection becomes energetically favourable when the shear exceeds some critical value. The reason for this result may be easily understood. For $t = 0$, $C_0[A_0] = \inf_{\mathcal{H}'} C_0[A]$ by an elementary property of harmonic potentials, and a similar result naturally holds for small values of t. On the contrary, for large enough values of t, it is easy to see by using the asymptotic result quoted in § 3.1 that the poloidal energy of $\underset{\sim}{B}_t$ may be decreased if we reconnect that field in some small part of the region where the currents concentrate, while the toroidal energy may be made to change as little as we want (remember that $\lim_{t\to\infty} B_{tx} = 0$).

Thus, at some time $t_r > t_c$, there is a positive amount of energy ΔC_r which may be released by reconnection. However, it must be noted that $\underset{\sim}{B}_t$ is still stable with respect to small perturbations in $\mathcal{H}'$. Therefore, the result above has to be interpreted as meaning that $\underset{\sim}{B}_t$ becomes metastable with respect to reconnection in $\mathcal{H}'$ at t_c. There is an energetic barrier (whose height actually decreases in time) which has to be overpassed (under the effect of a finite perturbation) for initiating the reconnection process, which then occurs in an explosive way, and our model may be relevant to explain flare events taking place in the solar corona or in other astrophysical objects.

4 COALESCENCE OF FLUX TUBES

4.1 Coalescence of flux tubes in a fixed symmetric domain

Let us now take Ω to be a bounded domain partitionned by a force-free field $\underset{\sim}{B}$ into three flux cells Ω_i, all containing closed lines. Ω_1 and Ω_2 are disk-like cells in contact and they are surrounded by the ring-like cell Ω_3. The flux distribution is characterized by the values of $(\underline{a}_i, \bar{a}_i)$, $i = 1,2,3$, with: $\underline{a}_3 = 0$, $\bar{a}_3 = \underline{a}_1 = \underline{a}_2 = \Upsilon > 0$, and $\bar{a}_2 \geqslant \bar{a}_1$; and by the three functions $X_i(a)$ (assumed to vanish only at isolated points). In general, such on equilibrium has current sheets on the separatrix $\Gamma = \partial\Omega_1 \cup \partial\Omega_2$, and any tiny amount of resistivity makes fast reconnection unavoidable. Here, we would like to determine some conditions in which coalescence of the cells into a single cell is energetically favourable, reconnection of the lines occuring in a small dissipative region and thus under flux-conservation conditions quite similar to those $\mathcal{S}_r$ introduced in § 3.2: for the final state $\underset{\sim}{B}'$, one should have:

i) $0 \leqslant A' \leqslant \bar{a}_2$;

ii) $X'(a) = X_3(a)$ for $0 < a < \Upsilon$; $X'(a) = X_1(a) + X_2(a)$ for $\Upsilon < a < \bar{a}_1$; and (if $\bar{a}_1 < \bar{a}_2$) $X'(a) = X_2(a)$ for $\bar{a}_1 < a < \bar{a}_2$.

Thus reconnection may occur spontaneously if the minimizer of

$$C[A] = \int_\Omega |\nabla A|^2 dydz + \int_0^{\text{meas}\,\Omega} |X'|^2 \left|\frac{dA}{d\mu}\right|^2 d\mu \qquad (7)$$

over $\mathcal{H} = \{A|\ A|_{\partial\Omega} = 0;\ \inf A = 0;\ \sup A = \bar{a}_2\}$, has a simple topology.

In general, it seems difficult to answer that question on simple general grounds. This turns out to be possible, however, when Ω is taken to be symmetric and convex with respect to both axis Oy and Oz (Ω is said to be convex w.r.t. Oy, say, if $\partial\Omega$ is not intersected at more than two points by any line parallel to Oy). With such an Ω, the minimum of C[A] defined by Equ. (7) may be taken to be a simple topology field. Suppose indeed that we have a minimizer which has not this property. Then, by applying to it Steiner symmetrizations with respect to both axis (e.g. *Garabedian, 1964, Chap. 11*), we construct a simple topology field belonging to $\mathcal{H}$ and having a not larger energy

- therefore also a minimizer. Thus reconnection to a single cell is possible indeed - although the mathematical theorems available show only the existence of a marginal tendency for this effect to occur. Inversely, it seems likely that a single cell field contained in an Ω constituted of two convex bubbles connected by a sufficiently thin neck, say, would evolve by reconnection towards a complex field similar to the one considered at the begining of this subsection.

4.2 Coalescence of flux tubes confined by a constant pressure

Let us now consider briefly a slightly different situation (which has been also studied independantly by *Choudhuri, 1988*) where we have two flux tubes of cross-section Ω_1 and Ω_2, respectively, which have a simple topology ($\nabla A = 0$ in Ω_i only at a central 0-critical point) and are confined by a constant pressure p_0 (pressure balance implies $8\pi p_0 = |\nabla A|^2 + B_x^2$ on $\partial\Omega_i$). Thus, as discussed in *Aly (1989)* on the basis of some general theorems on non-linear elliptic equations, $\partial\Omega_i$ are necessarily circles. If we now make the two tubes to come into contact, it is easy to see that they will be able to effect an S_r - preserving coalescence process if A has the same sign in Ω_1 and Ω_2 (we choose $A = 0$ on $\partial\Omega_i$), the final state being also a simple topology circular flux tube in which the flux distribution is calculated as in § 4.1 above. Note, however, that the tubes are no longer isolated, being submitted to a constant pressure, and thus the final magnetic field has not to minimize the energy W, but the "enthalpy" $W + 8\pi^2 p_0 R^2$ (R being the radius of the tube).

As an illustration of this general analysis, we have computed explicitely (*Aly, 1989*) the coalescence of two tubes containing each a so-called constant-pitch force-free field (X_i = constant). For instance, in the case of two identical tubes, we find the final state to be also a constant-pitch field, with $X' = 2X$ and a radius R' determined by the equation $f(u) = 2f(u')$, where $f(s) = (1+s^2)^{1/2} \log(1+s^2)$ and $u = 2\pi R/X$, $u' = \pi R'/X$.

5 PROCESSES CONSERVING THE TOTAL HELICITY

5.1 The Heyvaerts-Priest model

Let us take Ω to be the half-strip $\{-L/2 < y < L/2,\ z > 0\}$ and consider a finite energy potential field $\underset{\sim}{B}_0 = \nabla A_0 \times \hat{x}$ (with $A_0(\pm L/2, z) = 0$, $A_0(y,0) = g(y)$ and $\lim_{z\to+\infty} A = 0$) which is sheared by a velocity field $\underset{\sim}{v}(y) = v(y)\hat{x}$. In a model proposed by *Heyvaerts and Priest (1984)*, the boundary motions drive an evolution of the field which is similar to that described in § 3.1 during the interval of time $](n-1)\tau, n\tau[$ (τ being a phenomenological "relaxation" time) ; strong turbulence develops, however, near time $n\tau$, inducing a general reconnection process during which only the total relative helicity H (integral of AB_x over Ω_i ; *Berger and Field, 1984*) is conserved (*Taylor, 1986*); at $n\tau$, the field thus relaxes to the lowest possible energy state in the set $\mathcal{H}[n\tau]$, where $\mathcal{H}(t) = \{\underset{\sim}{B} | A = A_0$ on $\partial\Omega$; $H[\underset{\sim}{B}] = H(t)\}$ and the injected helicity H(t) writes

$$H(t) = t \int_{-L/2}^{+L/2} v(y)\ g(y)\ B_z(y)dy \tag{8}$$

This state is taken to be the unique constant-α force-free field $\underset{\sim}{B}_\alpha(n\tau)$ contained in $\mathcal{H}(n\tau)$; this is justified (at least it $|H|$ is not too large) as it may be proved (*Aly, 1989*) that, for any value of t (t implied hereafter):

a) for $|\alpha| < \pi/L$, $\underset{\sim}{B}_\alpha$ makes indeed the energy an absolute minimum over $\mathcal{H}$; the condition $|\alpha| < \pi/L$ is automatically satisfied (because $\lim_{z\to+\infty} A = 0$) if the Fourier development of g(y) contains a term in $\cos(\pi y/L)$;

b) for $|\alpha| > \pi/L$, there is no longer any global minimizer in $\mathcal{H}$; minimizing sequences in $\mathcal{H}$ converge towards a constant-α force-free field having an helicity $|H'| < |H|$; physically, this state can be reached only if some helicity is ejected at infinity ;

c) if relaxation to 3-D fields is allowed, then the 2-D $\underset{\sim}{B}_\alpha$ stops being a global minimizer at some critical value α_c of $|\alpha|$ with $\alpha_c < \pi/L$, and bifurcation to a 3-D state is energetically favoured.

Actually, this model suffers from a defect: it does not take into account the fact that the total amount of toroidal flux $\Phi(t)$ (integral of B_x over Ω) injected by the boundary motions, is a quantity which is exactly conserved during any dissipative transition taking place in the domain because the assumed boundary conditions on $\partial\Omega$ are of the "perfectly-conducting" type. Unfortunately, this second constraint cannot be incorporated into the model in a straightforward way: a simple variational argument using Lagrange multipliers shows indeed that, for a minimum energy state, one should have: $B_x = \alpha A + \beta$, with α and β being constant but, as in general $\beta \neq 0$, this implies infinite energy, flux, and helicity ! And thus there is the need for ejecting some amount of flux and helicity to reach some "pseudo-minimum energy" state (having $\beta = 0$).

5.2 A modified model

To try to take care of this difficulty, we have proposed (*Aly, 1989*) a modified model which may be described as follows (we present here the simplest situation to keep the presentation clearer). Let $\Omega = \{z>0\}$ and consider an initial potential field with an arcade-like structure (as in § 3.1) and a shearing velocity field on ∂D $\underset{\sim}{v}(y) = v(y)\hat{x}$ which is concentrated near the origin - i.e. at any time the associated displacement $x(a)$ satisfies $x(a) = 0$ for $\lim_{r\to\infty} A = 0 \leqslant a \leqslant a_1 < A(0,0)$. Now it is quite natural to assume that turbulence is confined in the sheared part of the field and therefore that Taylor relaxation occurs at $n\tau$ to the lowest energy state in the set of fields $\mathcal{H}(n\tau)$, with $\mathcal{H}(t) = \{\underset{\sim}{B} | A = A_0$ on $\partial\Omega$; $\Phi[A_1 ; a_1] = \Phi(t)$; $H[A ; a_1] = H(t)\}$ ($\Phi[A; a_1]$ and $H[A; a_1]$ are the integrals over $\Omega[A ; a_1] = \{(y,z) | A(y,z) > a_1\}$ of B_x and AB_x, respectively, and $\Phi(t)$ and $H(t)$ are the amounts of flux and helicity, respectively, injected at t ; both quantities are linear functions of time). Clearly the minimizer $\underset{\sim}{B}_n$ at $n\tau$ - if it exists at all - must be potential in $\Omega/\Omega[A_n ; a_1]$ and force-free in $\Omega[A_n ; a_1]$, with $B_{nx} = \alpha_n A_n + \beta_n$, the constants α_n and β_n being determined by the values of $\Phi(n\tau)$ and $H(n\tau)$; both regions must be separated by a current sheet.

The problem of determining $\underset{\sim}{B}_n$ has not yet been considered in details and we have just developed a preliminary calculation giving an approximate form of this field which should prove to be valid when relaxation occurs at a not too large value of the shear. This simple calculation is based on the remark that, in these conditions, the poloidal structure of the field is not very much perturbed and may therefore be taken to be that of the initial field. This leads to useful estimates for the amount of energy released.

REFERENCES

Aly, J.J. 1985, Astron. Astrophys. 143, 19.

Aly, J.J. 1988, in "Solar and Stellar Flares", IAU Coll. 104, Stanford University.

Aly, J.J. 1989, Preprint, Saclay.

Berger, M.A. and Field, G.B. 1984, J. Fluid Mech. 147, 133.

Choudhuri, A.R. 1988, Geophys. Astrophys. Fluid Dynamics 40, 261.

Garabedian, P.R. 1964, "Partial Differential Equations", John Wiley & Sons, New York.

Grad, H., Hu, P.N., and Stevens, D.C. 1975, Proc. Natl. Acad. Sci. U.S.A. 72, 3789.

Heyvaerts, J., and Priest, E.R. 1984, Astron. Astrophys. 137, 63.

Taylor, J.B. 1986, Rev. Mod. Phys. 58, 741.

IV: Two-dimensional and Quasi-two-dimensional Flows

Coherent Structures, Homoclinic Cycles and Vorticity Explosions in Navier-Stokes Flows

BASIL NICOLAENKO[1] & ZHEN-SU SHE[2]

[1]**Center for Nonlinear Studies, Los Alamos National Laboratory, Los Alamos, NM 87545, USA and Department of Mathematics, Arizona State University, Tempe, AZ 85287, USA**

[2]Applied and Computational Mathematics, Princeton Univeersity, Princeton, NJ 08544, U.S.A.

1. Introduction

Turbulence is basically a spatially extended system which develops complex spatio-temporal behavior. Traditional statistical descriptions of turbulence have been focused on the dynamics of low-order correlation functions, such as the evolution of the energy spectrum and energy transfer spectrum. While statistical closure theories have given a faithful description of energy and enstrophy transfer in two and three dimensional isotropic turbulence (Kraichnan 1966, Yakhot & Orszag 1986), we are still far from understanding the associated physical processes, especially when flows are anisotropic. The dynamical systems approach developed in the last two decades has proved useful in providing new insight into the problem. In addition to the characterization of the deterministic nature of chaotic motions in turbulent flows, the study of the transition to turbulence in spatially extended systems has provided much information on the dynamical mechanisms of the generation of stochastic and coherent structures (Coullet et al. 1987, Chate & Manneville 1987).

Most previous studies on spatially extended chaos have been made upon model systems (one-dimensional partial differential equations, cellular automata, etc.). Often, a regime of coexistence of clusters of laminar regions within a sea of turbulent patches is observed (Chate & Nicolaenko 1989). The same feature appears in time as well: turbulent bursting events occur randomly and intermittently, preceded and followed by much smoother motions. This has been commonly described as spatio-temporal intermittency (Ciliberto & Bigazzi 1988). The present work is aimed at demonstrating the existence of similar phenomena in flow systems governed by the Navier-Stokes equation. In addition, we will develop a dynamical system description of the temporal intermittent bursting events in terms of homoclinic cycles, which have been conjectured for real flow systems by many authors, especially in connection with boundary layer turbulence (Aubry et al. 1987).

The most interesting problem is how to relate a dynamical system description of flow systems to more traditional statistical measures. Turbulence displays such complex spatial and temporal dynamics that a fully deterministic description is unrealistic. The most significant measures of turbulent flows, in both a theoretical and a practical sense, involve some degree of averaging. The spatially averaged second-order correlation functions, which can be related to the energy spectrum, are most commonly used for the characterization of turbulent states. We will show how certain dynamical systems concepts, such as heteroclinic excursion, when combined with statistical measures, can highlight dynamical processes of enhanced turbulent transport.

The paper is organized as follows. In section 2, we will introduce a particular flow system, the so-called two-dimensional Kolmogorov flow (Obukhov 1983), which has been studied in detail in the present work. In particular, we will describe the sequence of bifurcations with increasing Reynolds number which lead to chaotic, and finally to spatio-temporally developed turbulent regimes. In section 3, we present some of the symmetry groups of this flow system, and point out their dynamical significance. These symmetries imply the existence of a series of equivalent states, transitions between which can occur subsequently through heteroclinic excursions; numerical evidence os this process will be reported. In section 4, we attempt to characterize statistically the basic (symmetry) state and the turbulent bursting (excursion) state using correlation functions. Section 5 contains some concluding remarks.

2. The Kolmogorov flow

The two-dimensional Kolmogorov flow is the solution of the 2-D Navier-Stokes equation with a uni-directional force $\mathbf{f} = (\nu \sin y, 0)$. It was introduced by Kolmogorov in the late fifties as an example on which to study transition to turbulence. For large enough viscosity ν, the only stable flow is a plane parallel periodic shear flow $\mathbf{u}_0 = (\sin y, 0)$, usually called the "basic Kolmogorov flow". The Reynolds number of the basic flow is easily found to be $1/\nu$; this will be used later as a free parameter to define the bifurcation sequence. It was shown by Meshalkin and Sinai (1961) that large-scale instabilities are present for Reynolds numbers exceeding a critical value, $\sqrt{2}$. This large-scale instability has been shown by Nepomnyachtchyi (1976) and Sivashinsky (1985) to be of negative-viscosity type, in the sense that the basic anisotropic flow generates a negative viscosity for large scale perturbations. The renormalized viscosity may become negative at large enough Reynolds number, leading to an instability.

The nonlinear regimes of the large scale instability have been studied by She (1988) using direct numerical simulation of the 2-D Navier-Stokes equation. Here we summarize the main results in order to better understand the role of symmetry breaking; details may be found in She (1988). In a 2π-periodic box, the force was set at a small scale: $\mathbf{f} = (\nu k_f^3 \sin k_f y, 0)$, to allow scale separation from large scale perturbations. The basic flow is then $\mathbf{u}_0 = (\nu k_f \sin k_f y, 0)$. When the Reynolds number slightly exceeds $\sqrt{2}$, a supercritical bifurcation occurs, leading to a large scale transverse structure, essentially x-dependent, consistent with the instability analysis. The emergence of large scale modes breaks the translational invariance in the x-direction, so the solution is determined up to an arbitrary phase. The second bifurcation, also steady, is of subcritical type, leading to an excitation of large scale y-dependent modes. This bifurcation shows a tendency for large scale motion to be more isotropic. Note that the new steady solution breaks both the parity and the x-reflection symmetries.

As the Reynolds number continues to increase, a travelling wave bifurcation takes place, through which the whole system propagates in the x-direction. The solution now oscillates in time, due to the periodic boundary condition. The origin of such travelling waves seems to be related to a degeneracy of eigenspace associated with the zero-eigenvalue (Kevrekidis et al. 1989). In the present case, the flow system has both translational and Galilean invariance in the x-direction, which would provide a necessary condition for the formation of waves (Frisch et al. 1986).

A few Hopf bifurcations follow the travelling wave one, leading to more and more complex oscillations, and finally to a weakly chaotic state. This state is characterized by a fairly regular motion of large scale modes with chaotic motions of small scales. Since small scale modes (except the forcing mode) contain little energy, the degree of chaos is very low, distinguished from the turbulent regime discussed below. It is interesting to note that the appearence of a chaotic state seems related to a decrease of the oscillation frequency induced by the travelling wave. We found that this frequency is Reynolds number dependent, and an re-increase of the frequency at higher Reynolds number coincides with a relaminarization of the weakly chaotic state to a quasi-periodic motion.

The most interesting transition occurs at even higher Reynolds number, leading to sparsely distributed bursts in time for a fairly large range of Reynolds number above a certain threshold (see Fig. 3b). The most striking feature of this transition is that the bursts generate substantial spatial disorder and drive developed turbulence. In Fig.1, we show a height plot of the 2-D vorticity distribution, with the z-axis indicating the vorticity amplitude, for a field during the burst (Fig. 1a) and after the burst (Fig. 1b). It is clear that the burst generates a high degree of stochasticity and a large amount of enstrophy, while the flow remains fairly organized in the mean time outside of the bursts.

Note that the temporal intermittent regime extends for a fairly large range of Reynolds number (for $k_f = 8$, the critical Reynolds number for this transition is $R_{cr} \approx 25$, and intermittent bursts are clearly observable until $R = 55$). In the next section, we will discuss in more detail the dynamics associated with the burst events.

3. Intermittent Bursts: Heteroclinic Excursions

The Kolmogorov flow is governed by the 2-D Navier-Stokes equation with a body force:

$$\partial_t(\nabla^2\psi) + \partial_x(\nabla^2\psi)\partial_y\psi - \partial_y(\nabla^2\psi)\partial_x\psi = \nu\nabla^4\psi + \nu k_f^4 \cos(k_f y) \tag{1}$$

where ψ denotes the stream function. It can be easily checked that this flow system has the following important symmetry properties:

$$\begin{aligned}
&y \mapsto y, x \mapsto x + \theta, \qquad \psi \mapsto \psi,\\
&y \mapsto y, x \mapsto -x, \qquad \psi \mapsto -\psi,\\
&x \mapsto x, y \mapsto y + \frac{2n\pi}{k_f}, \qquad \psi \mapsto \psi,\\
&x \mapsto x, y \mapsto -y + \frac{(2n+1)\pi}{k_f}, \qquad \psi \mapsto -\psi,
\end{aligned} \tag{2}$$

where θ is an arbitrary real number and n is an integer; i.e., the group symmetries of the system are $O(2)$ equivariance in x, and discrete D_k equivariance in y.

As we mentioned above, the continuous $O(2)$ group in the x-direction may be deeply related to the travelling wave bifurcation in the Kolmogorov flow. In contrast, the intermittent bursting behavior is more likely to be connected with a subgroup $\Lambda = D_r \times D_s$ of $\Gamma = O(2) \times D_k$; $\Lambda \subset \Gamma$ is discrete in both x and y, its action on any function ψ_0 will introduce a discrete shift in both the x and y directions together with reflections. One important property of Λ is that there exist $\lambda_1, \lambda_2, ...$ such that $\lambda_1 \cdot \lambda_2 \cdot ... = I$. We may now formulate our principal hypothesis:

Hypothesis. There exists a functional subspace Ψ which is the family of equivalent hyperbolic states of the dynamical system (1). For these states, we can construct a discrete group $\Lambda = D_r \times D_s$ and $\Lambda_0 \subset \Lambda$ such that

(1) For $\psi \in \Psi$ and $\lambda_0 \in \Lambda_0$, $\lambda_0\psi \neq \psi$;
(2) For $\psi \in \Psi$ and $\lambda \in \Lambda - \Lambda_0$, $\lambda\psi = \psi$.

The hyperbolic states may be steady or time dependent solutions of the Navier-Stokes equation (1). The key is that each such state should cover a very localized region in phase space during the time evolution. Quasi-periodic or weakly chaotic solutions of (1) do satisfy this constraint. The hypothesis formulated above then allows us to construct explicitly a homoclinic cycle through a series of heteroclinic connections:

$$\psi' \xrightarrow{\lambda_0'} \psi'' \xrightarrow{\lambda_0''} \psi''' \ldots \psi^{(n)} \xrightarrow{\lambda_0^{(n)}} \psi'. \tag{3}$$

The realisability of such heteroclinic connections depends crucially on the local properties of the stable and unstable manifolds around a hyperbolic state $\psi^{(i)}$. It has been shown by Nicolaenko & She (1989) that there exists a submanifold $\mathcal{A}_{i,i+1}$ which passes through $\psi^{(i)}$ and $\psi^{(i+1)}$ such that $\psi^{(i)}$ is a saddle while $\psi^{(i+1)}$ is a sink. Similar heteroclinic connections were established for the Kuramoto-Sivachinsky equation by Kevrekidis et al. (1989) and independently by Armbruster et al. (1989).

In practice, the dynamical trajectories will spend a long transit time in a small neighborhood of $\psi^{(i)}$ before shooting out very rapidly into a region characterized by strongly turbulent behavior, and then connecting to some $\psi^{(i+1)}$. This description is confirmed by direct numerical simulation of the Kolmogorov flow at a Reynolds number corresponding to the temporally intermittent regime. In Fig. 2, we plot the time evolution of the phase for the mode $\mathbf{k} = (1,0)$ and $(0,1)$ during a period which covers three bursts. Here the phase ϕ is defined by $\psi_{\mathbf{k}} = |\psi_{\mathbf{k}}|e^{i\phi}$; it undergoes major change under the action of Γ. It can be seen that in the mean time, the phase of the mode (0,1) remains essentially constant while that of the mode (1,0) has a slow continuous shift reflecting the existence of a travelling wave in x direction. During the bursts, however, the phase makes a finite jump as a result of Γ in both x and y. More interestingly, the second burst (at $t \approx 615$) brings both phases to their values before the first burst (compare the phases at $t = 650$ with $t = 570$). The same happens for the third burst (compare the phases at $t = 610$ with $t = 680$, modulo 2π for the mode (0,1)). We have then clear evidence that two heteroclinic connections form a homoclinic cycle.

4. Statistical Description of Heteroclinic Excursion: Enhanced Transfer

A major characteristic of two-dimensional turbulence is the existence of an inverse energy cascade to large scales and a direct enstrophy cascade to small scales at high Reynolds number. Since the Reynolds number in the present case is not large enough to generate any significant cascade of enstrophy, we will concentrate mainly on the transfer of the kinetic energy. Through the very first bifurcations of the basic flow, large scale structures are developed with an appreciable amount of excitation (see Fig. 5a). This implies that a large scale instability of a negative viscosity type is primarily responsible for "distant" inverse energy transfer from the forcing scales to very large scales (size of the system). At moderately large Reynolds number, however, the interaction between the largest scale modes and the forcing modes excites the whole range of intermediate scales. As we have seen, the most active dynamics occur during temporally intermittent bursts. We will investigate in this section the statistical properties of these bursting events.

In Fig. 3, we show the time evolution of the Taylor microscale length (Fig. 3a) and the total enstrophy (Fig. 3b) during a period covering several bursts. It is clear that there is strong production of vorticity during the burst, as well as a transfer of excitation towards small scales. It may be also seen from Fig. 3a that the variation of the Taylor microscale length is slower during the decaying phase, suggesting more complex transfer. This may be checked in Fig. 4, where we show the energy transfer spectrum at three different times corresponding, respectively, to the regular state, the onset of a burst and the decaying phase of the burst. At the regular stage, the forcing scales receive energy from the external body force, and some amount of energy is transfered into the largest scales. The onset of the burst introduces a transfer into intermediate scales, where the energy will be pushed away, during the decaying phase, to both large and small scales. The energy which continues to be transfered to small scales will dissipate eventually, leading to enhanced dissipation during the whole process. The inverse energy transfer, on the other hand, may be of the same nature as what happens in two-dimensional turbulence; it becomes evident due to enhanced nonlinear interaction between modes of intermediate scales which are fed by large scales during the onset phase of the burst.

One peculiarity of the Kolmogorov flow is that at small scales the flow remains very anisotropic due to the force. In the linear regime, the analysis of large scale instability shows that the large scale transport coefficient, or eddy-viscosity, is anisotropic, leading to an anisotropic energy flow to large scales. We have found that this feature is modified in the inverse transfer due to nonlinear interaction of the intermediate scales. In Fig. 5, we plot both the isotropic energy spectrum (Fig. 5a) and the one-dimensional energy spectrum, averaged respectively in x and y-directions (Fig. 5b and 5c). Fig. 5a gives a clear indication of direct energy transfer during the onset phase (from time t_1 through t_3), and inverse transfer during the later phase (from t_3 to t_4). It may be noticed from Fig. 5b and 5c that the inverse transfer to large scales is more or less isotropic (from t_3 to t_4), except to the largest scale $k = 1$ which is strongly coupled with the forcing mode. This is a cascade very different from an "arithmetic" cascade resulting from successive vortex pairings (She 1987).

5. Concluding Remarks

It is generally recognized that traditional statistical descriptions of turbulence are unable to capture information about structures, information which is crucial for the understanding of the physical mechanisms responsible for dynamical features. The most important dynamical event occuring in such highly nonlinear systems as turbulence is energy transfer. In two dimensional turbulence, energy flows into large scales, leading to formation of large-scale structures. In inhomogeneous situations, turbulent transport is the most remarkable phenomenon; the theoretical description of this process is still quite incomplete. Over the past decades, there has been growing interest in a more geometrical description of turbulence (Moffatt 1985).

We have attempted to take a mixed approach, namely by using modern concepts of dynamical system theory combined with traditional statistical measures to understand certain dynamical aspects of Navier-Stokes turbulence. The case study undertaken here for the two-dimensional Kolmogorov flow has enabled us to describe the bursting events usually associated with strong horizontal shear in terms of symmetry breaking, and also to study more deeply the generation of spatial patterns as a consequence of turbulence production by the bursts.

Boundary layer turbulence is an example of a flow system where strong shear motions dominate the dynamics. One of the important problems there is the turbulence production mechanism. It is believed that bursts near the wall inject energy into the outer regions, and thus are the main source of the turbulence production. It has also been speculated that homoclinic excursions may be the dynamical origin of the bursting events in boundary layer flows. Indeed, Aubry et al. (1987) have found that reduced dynamical systems obtained by projecting the boundary layer turbulence correlation function do contain homoclinic orbits. We have seen that the Kolmogorov flow displays, at high enough Reynolds number, temporally intermittent behavior. In this sense, the Kolmogorov flow provides an explicit example of a Navier-Stokes flow system exhibiting temporal intermittency, without any modelling. Moreover, intermittent bursting events in the Kolmogorov flow are intimately related to the generation of spatially disordered patterns, and thus are directly responsible for the production of developed turbulence.

One characteristic of the Kolmogorov flow is the small scale forcing which introduces a third intrinsic scale into the system, in addition to the dissipation scale and the size of the system. When the Reynolds number is small, the dissipation scale is large compared with the forcing scale, so that the molecular dissipation efficiently removes the energy injected by the force. As the Reynolds number increases, the dissipation scale gradually decreases. When the dissipation scale becomes slightly smaller than the forcing scale, molecular dissipation is no longer able to balance the energy input by the force, so that an accumulation of excitation is expected. The system somehow attains an overheated state, and relaxes the extra excitation through intermittent bursting. From a dynamical systems point of view, this corresponds to the situation where an unstable manifold embedding a homoclinic (or heteroclinic) orbit intersects the weak turbulent state. The system is then allowed to leave the weak turbulent state and to display much higher transfer to small scales for dissipation. We believe that this description is of general interest for the understanding of boundary layer turbulence.

We wish to acknowledge the support and the hospitality of the Center for Nonlinear Studies at Los Alamos National Laboratory. This research is also supported by DARPA under Contract N00014-86-K-0759, and by the Air Force Office for Scientific Research through its URI Grant to the Mathematics Department at Arizona State University. Finally, Dr. Cleve Moler from Ardent Computer Co. has developed the graphics software for the visualization of the Kolmogorov flow on the Titan Graphics Minisupercomputers.

References

Armbruster, D., Guckenheimer, J. & Holmes, Ph., Physica D, (1989).

Aubry, N., Holmes, P., Lumley, J.L. & Stone, E., J. Fluid. Mech. **192**, 112 (1987).

Chaté, H. & Manneville, P., Phys. Rev. Lett. **58**, 112 (1987).

Chaté, H. & Nicolaenko, B., Phase Turbulence, Spatiotemporal Intermittency and Coherent Structures, Proc. Cargese Conf. on Recent Developments in Dynamical Syetems, to appear, Plenum Press (1989).

Ciliberto, S. & Bigazzi, P., Phys. Rev. Lett. **60**, 286 (1988).

Coullet, P., Elphick, C. & Repoux, D., Phys. Rev. Lett. **58**, 431 (1987).

Frisch, U., She., Z.-S. & Thual, O., J. Fluid Mech. **168**, 221 (1986).

Kevrekidis, I., Nicolaenko, B. & Scovel, C., SIAM J. Appl. Math., to appear (1989).

Kraichnan, H.K., Phys. Fluid., **9**(9), 1728 (1966).

Meshalkin, L.D. & Sinai, Ya.G., J. Appl. Math. (PMM), **25**, 1700 (1961).

Moffatt, H.K., J. Fluid Mech., **159**, 359 (1985).

Nepomnyachtchyi, A.A., Prikl. Math. Makh. **40**(5), 886 (1976).

Nicolaenko, B. & She, Z.-S., Phys. Lett. A, submitted (1989).

Obukhov, A.M., Russ. Mach. Survey, **38**, 113 (1983).

She, Z.-S., Proc. on Current trends in turbulence research, AIAA, 374 (1988).

She, Z.-S., Phys. Lett. A **124**, 161 (1987).

Sivashinsky, G.I., Physica **17**D, 243 (1985).

Yakhot, V. & Orszag, S.A., J. Sci. Comp. **1**, 3 (1986).

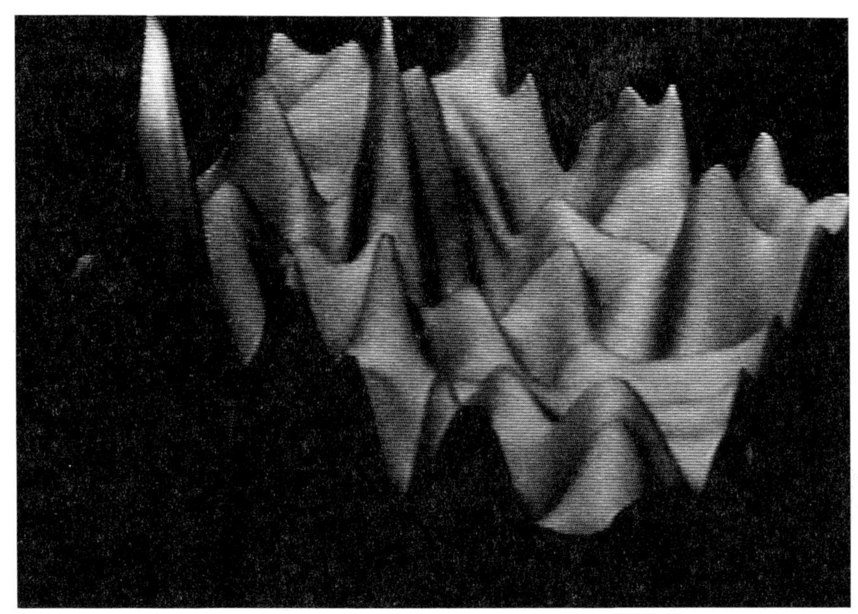

Fig. 1a The 2-D vorticity distribution for a field during the burst, with the z-axis indicating the vorticity amplitude. Here $R = 36$ and $k_f = 8$. The scale of the periodic force controls the typical scale of vorticity structures.

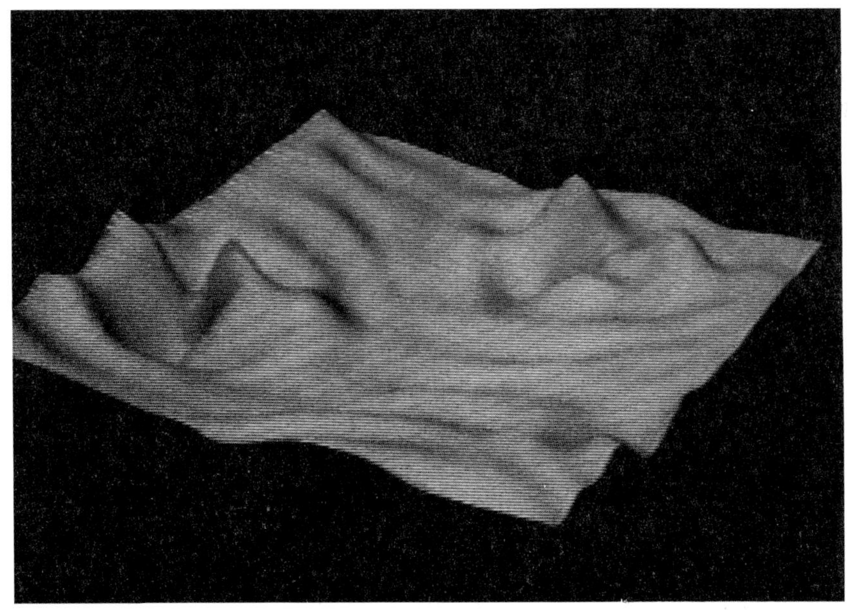

Fig. 1b The 2-D vorticity distribution for a field betwen bursts for the same parameters. Notice the counter-rotating eddies, with tripoles within their core. The scale of the eddies is much larger than the forcing scale. Computations performed on ASU's Titan Graphics Computers.

(See also Colour Plates)

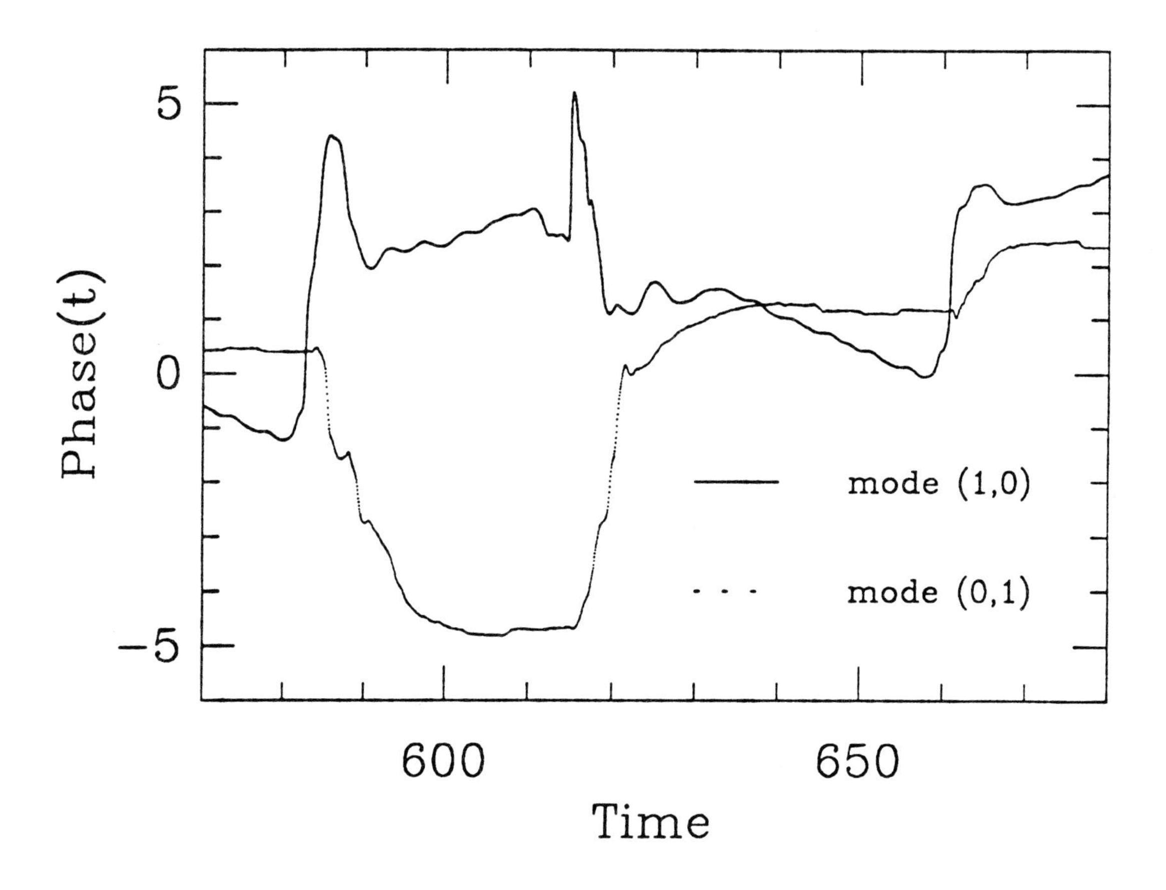

Fig.2 Time evolution of the phase $\phi_{\mathbf{k}}$ for the mode $\mathbf{k} = (1,0)$ (solid line) and the mode $(0,1)$ (dotted line) during a period covering three bursts. Note the jumps of the phase at $t \approx 580$, 615 and 660 reflecting heteroclinic connections. The phase return to its initial value (modulo 2π) after two successive jumps, indicating homoclinic cycles.

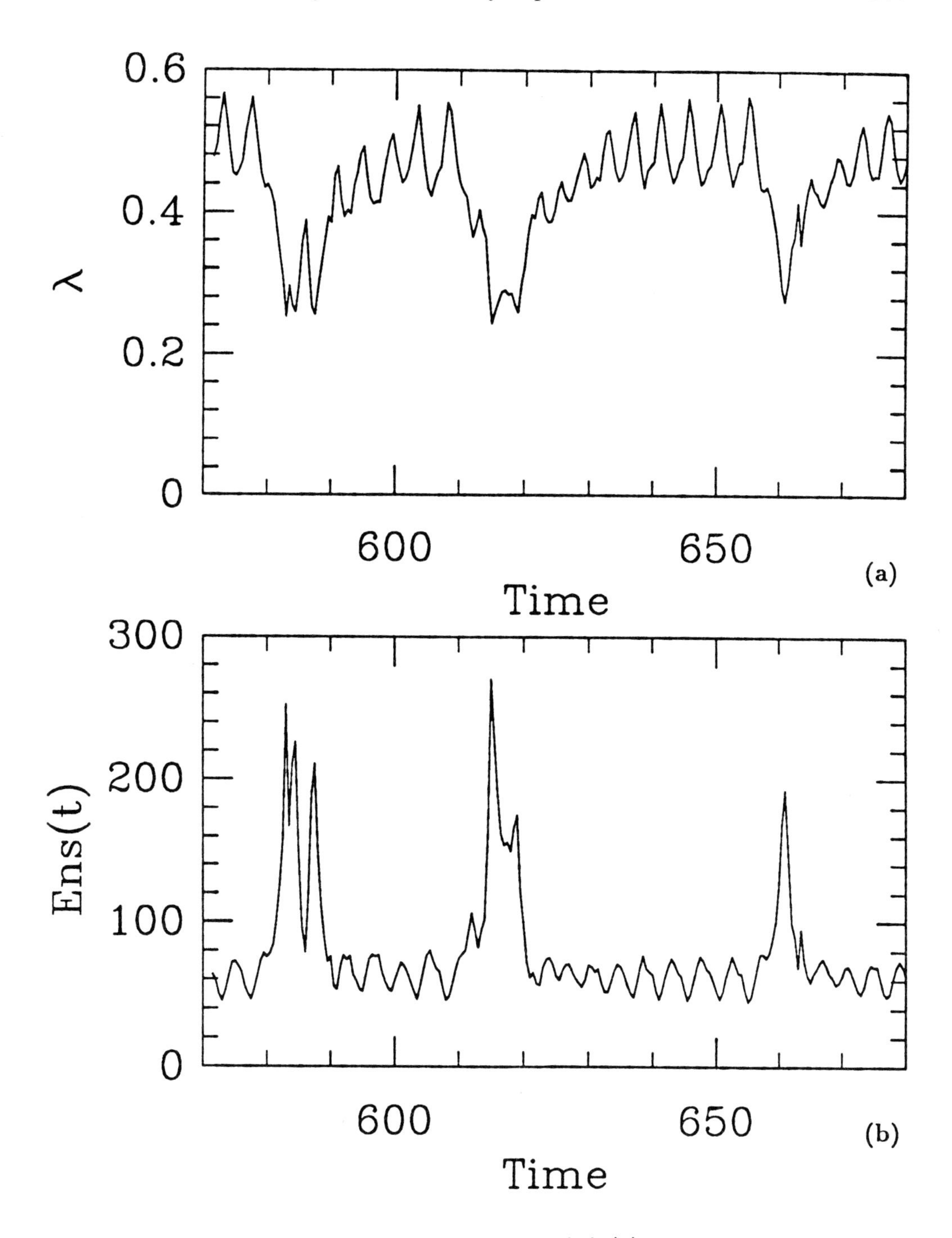

Fig.3 Time evolution of the Taylor microscale length λ (a) and the total enstrophy (b) during the same time interval as in Fig.2. Three intermittent bursts are observed. Note the strong increase of the enstrophy and decrease of the Taylor microscale during bursts, indicating an enhancement of transfer.

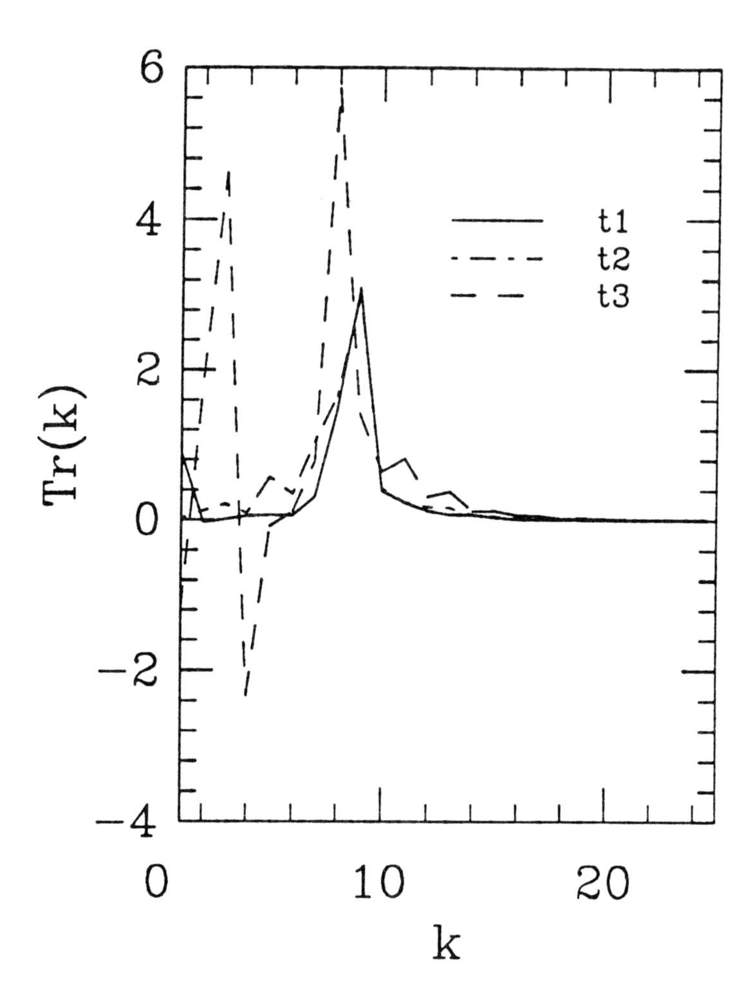

Fig.4 The energy transfer spectrum at three instants corresponding to the pre-burst state (t_1), the onset phase (t_2) and the decaying phase (t_3) of a burst. Note the positive transfer to both large and small scales during the decaying phase of the burst.

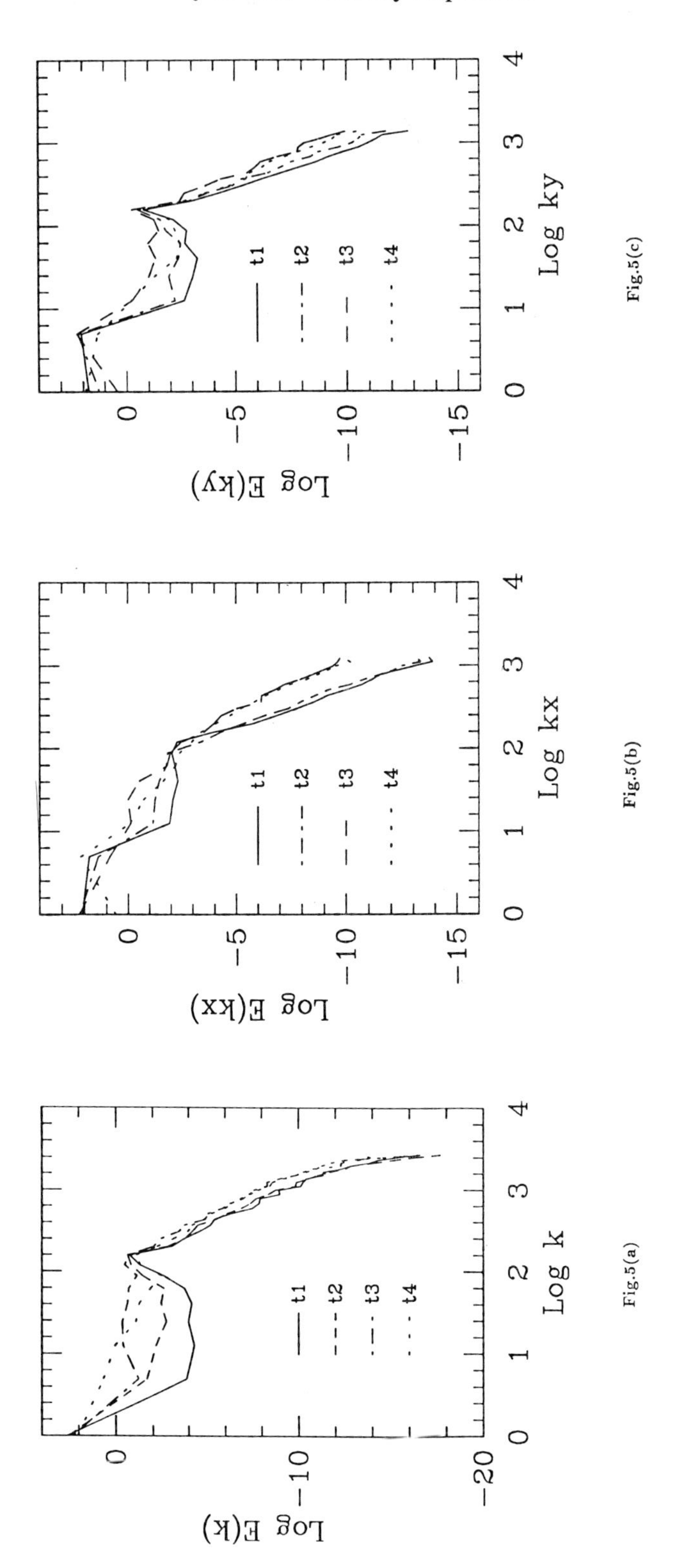

Fig.5 The isotropic energy spectrum (a), one-dimensional energy spectrum averaged in x (b) and in y (c) at four instants. Here t_1 is before a burst, $t_2 - t_3$ is in the onset phase of the burst and t_4 is in the decaying phase.

Isovortical Constraints on the Statistical-Dynamical Behaviour of Strongly Nonlinear Two-Dimensional and Quasi-Geostrophic Flow

THEODORE G. SHEPHERD

Department of Physics, University of Toronto, Toronto M5S 1A7 Canada

1 INTRODUCTION

V.I. Arnol'd's (1966) hydrodynamical stability theorems provide rigorous bounds on the growth of disturbance norms for two-dimensional and (by straightforward extension) quasi-geostrophic flow. This means that phase-space trajectories originating in the neighbourhood of a stable flow must forever remain close to it (in some normed sense). Insofar as the stability theorems depend on the isovortical character of the flow evolution (in the sense that the vorticity is materially conserved), this shows one way in which the isovortical constraints may have a macroscopic effect on the statistical dynamics of the flow.

It does not seem to be widely appreciated that these stability bounds apply to *large*-amplitude disturbances, and are not based in any way on small-amplitude expansions. Because the theory is exact, the choice of the stable basic flow is completely arbitrary, and this arbitrariness may be used to seek the tightest constraint on a given initial condition. Essentially, the idea is to use the stability bounds to constrain the evolution of unstable flows in terms of their proximity to a stable flow.

Two applications of this approach are presented here. The first (§3) is an explicit proof that two-dimensional flow on a beta-plane cannot become isotropic, even in the limit of infinite time, if it is sufficiently zonally anisotropic initially. This contradicts the presumed eventual approach to isotropy that would be expected from statistical theories based on the conservation of energy and enstrophy alone, neglecting the isovortical constraints. New numerical simulations reveal the robust tendency for zonal anisotropy to equilibrate at a non-zero value that is independent of the precise nature of the initial condition.

The second application (§4) is a method for deriving rigorous upper bounds on the nonlinear saturation of instabilities to parallel shear flows. New results are presented concerning bounds on the energy of the non-zonal part of the flow, with application to the point-jet instability.

2 THE FINITE-AMPLITUDE RAYLEIGH STABILITY THEOREM

The system under consideration is taken to be inviscid two-dimensional flow on a beta-plane, which is governed by material conservation of the absolute vorticity $P \equiv \nabla^2\Phi + \beta y$:

$$\frac{DP}{Dt} \equiv P_t + J(\Phi, P) \equiv P_t + \Phi_x P_y - \Phi_y P_x = 0, \tag{1}$$

where Φ is the stream function, t the time, x the zonal coordinate and y the meridional coordinate. The planetary component of vorticity, βy, is meant to mimic the meridional variation of the Coriolis parameter that occurs in spherical geometry, an effect that is fundamental to much of geophysical fluid dynamics insofar as it allows the existence of Rossby waves. The flow is presumed to be zonally-homogeneous, in the sense that the zonal average $\overline{f}$ of any quantity f is well defined. The meridional geometry is taken to be either closed, infinite, or periodic.

Given a flow with stream function Φ and absolute vorticity P, decompose it into a steady, x-invariant "basic state" (Ψ, Q) plus a time-dependent "disturbance" (ψ, q), viz.

$$\Phi = \Psi + \psi, \qquad P = Q + q. \tag{2}$$

Since the Jacobian term $J(\Psi, Q)$ vanishes identically, the basic state is a steady solution to (1) and the disturbance satisfies the exact equation

$$\frac{Dq}{Dt} \equiv q_t + J(\Psi, q) + J(\psi, q) = -J(\psi, Q) = -\psi_x Q_y = -\frac{DQ}{Dt}. \tag{3}$$

It follows from the spatial (zonal) symmetry of both the problem and the basic state that a conservation law exists for the disturbance pseudomomentum (see McIntyre & Shepherd 1987, §7). This conservation law leads to the following finite-amplitude version of Rayleigh's (1880) linearized stability theorem: if the basic-state absolute vorticity $Q(y)$ is a monotonic, piecewise-differentiable function of y then

$$\int \overline{q^2}(x, y, t)dy \le \frac{|Q_y|_{\max}}{|Q_y|_{\min}} \int \overline{q^2}(x, y, 0)dy. \tag{4}$$

This is a statement of Liapunov stability, in the sense that a disturbance norm (here just the r.m.s. vorticity) at any time $t > 0$ is bounded in terms of its value at $t = 0$. The bound is, moreover, valid for disturbances of any magnitude whatever.

The following remarks may be made:

1. The same formalism applies to stratified, quasi-geostrophic flow, in which case P is the potential vorticity.

2. When the basic state is steady but possibly non-parallel there is an analogous conservation law for the disturbance pseudoenergy, arising from the time-invariance

of the problem (see McIntyre & Shepherd 1987, §7). This also leads to a nonlinear stability theorem (Arnol'd 1966), which in the case of parallel flow is the finite-amplitude generalization of Fjørtoft's (1950) linearized result.

3. The stability bound (4) is implicit in the important work of Arnol'd (1966), but appears to have been stated explicitly for the first time by McIntyre & Shepherd (1987, equation (6.28)); a concise derivation is available in Shepherd (1987, §4). The mathematical method leading to the result is not to be confused with Liapunov's "direct method", invoked in Arnol'd's earlier (1965) stability paper, which by itself proves nonlinear stability only for *finite*-dimensional dynamical systems. To establish normed stability in the infinite-dimensional (or continuous) case, additional "convexity" estimates are required.

4. Although the stability bounds discussed above are derived for an inviscid fluid, it turns out that there is at least one important case of a forced-dissipative system for which they apply. When the absolute vorticity is relaxed back to some (possibly unstable) equilibrium profile $P^*(y)$ on some timescale r^{-1}, so that (1) is replaced with

$$\frac{DP}{Dt} = -r(P - P^*), \tag{5}$$

and if the initial zonal-mean profile $\overline{P}(t = 0) = P^*$, then it may be shown that the disturbance pseudomomentum and pseudoenergy are bounded by their initial values (Shepherd 1988*a*, §4). The stability bounds therefore go through, in a certain sense, although they are no longer strictly Liapunov.

5. The system (1) possesses an infinite family of "Casimir invariants" associated with the material conservation of P, namely any integral of the form

$$\int \overline{C(P)}dy \tag{6}$$

where C is some function of P. The time evolution of the system (1) may be considered as a trajectory along an "isovortical sheet" embedded within the phase space of the problem (Arnol'd 1978, Appendix 2). This isovortical sheet consists of all the possible measure-preserving, continuous rearrangements of the given vorticity distribution, and hence corresponds to fixed values of all the Casimir invariants (6). The pseudomomentum invariant which underlies the stability bound (4) is in fact a combination of Kelvin's impulse $\int y\overline{P}dy$ (which is invariant by the zonal symmetry of the problem) together with a particular choice of Casimir invariant (6). (In geometric terms, the stability of the basic state corresponds to the fact that it is an extremum of impulse on the isovortical sheet.) So to the extent that the statistical-dynamical behaviour of the flow is constrained by the stability bound (4), this demonstrates one way in which the isovortical constraints, manifested in the Casimir invariants, can play a significant role in determining the macroscopic character of the flow evolution.

The final remark has important implications concerning statistical theories of turbulence, for example closure theories. Such theories are generally based, *inter alia*, on the assumption that the statistical evolution of the flow is principally constrained by the two quadratic invariants of energy and enstrophy (the mean-square vorticity), with the detailed material-conservation property of vorticity not being essential (e.g. Kraichnan 1975; Thompson 1982). In the presence of flow inhomogeneities with inherent stability properties, however, it is clear from what has already been said that this assumption (which amounts to a kind of ergodic hypothesis – indeed EDQNM closure has been shown by Carnevale, Frisch & Salmon (1981) to satisfy an H-theorem) is unlikely to be true, and that neglect of the so-called "higher-order" invariants (namely the integrals of P^3, P^4 and so on) may give qualitatively incorrect results. The well-known inability of EDQNM closure to represent the statistics of two-dimensional turbulence when dominated by coherent vortices (Herring & McWilliams 1985; Babiano *et al.* 1987) is a case in point. A similar conclusion has also been reached for the case of two-dimensional flow over topography by Carnevale & Frederiksen (1987), who then attempt to outline a framework for building in the higher-order invariants within a statistical theory.

3 ON THE END-STATE OF INVISCID BETA-PLANE TURBULENCE

A case where the influence of the isovortical constraints is brought into sharp relief is that of inviscid two-dimensional turbulence on a beta-plane. As mentioned above, the quadratic invariants of the system are the energy and the enstrophy, viz.

$$E \equiv \int \frac{1}{2}\overline{|\nabla\Phi|^2}dy, \qquad Z \equiv \int \frac{1}{2}\overline{(\nabla^2\Phi)^2}dy. \tag{7}$$

These invariants are independent of β; and they are, moreover, isotropic. So if the flow dynamics is ergodic on the phase-space surface of constant energy and enstrophy, one expects the flow statistics to become isotropic in the limit of long time. (Recall that the inviscid truncated equilibria are isotropic, independently of the value of β (Salmon, Holloway & Hendershott 1976).) On the other hand, numerical simulations (Rhines 1975) reveal the development of strong zonal anisotropy in the form of quasi-zonal jets. This phenomenon is evidently connected with the isovortical constraints: fluid parcels must conserve $\nabla^2\Phi + \beta y$, and for sufficiently strong β this restricts meridional motion while allowing unlimited zonal motion. Closure theory (Holloway & Hendershott 1977) further confirms the inhibition of turbulent interactions by the presence of Rossby waves. The question thus arises: is the timescale characterizing the approach to isotropy simply lengthened, or does β defeat the ergodic hypothesis altogether?

This question has recently been settled (Shepherd 1987) by appealing to the stability bound (4). The idea is quite simple: for given values of E, Z and β, one tries to represent a given flow as a stable basic state plus a disturbance which, by (4), cannot evolve into a state for which the total flow is isotropic. (Here it is absolutely crucial that (4) apply to disturbances of large, and not just small-but-finite, amplitude.) If this can be done for a non-trivial set of initial conditions, then the ergodic hypothesis is disproven, and anisotropy may persist forever. It turns out that such a demonstration is possible whenever the wave steepness $\varepsilon \equiv 2Z/\beta E^{1/2}$ is less than unity. The pivotal role of ε is not surprising in light of the physical processes involved (see Rhines 1975; Holloway & Hendershott 1977).

The above result, while rigorous, is only strictly applicable to the exact (continuous) equations. In a numerical model, although it is possible to exactly conserve the energy and enstrophy, the detailed material-conservation property of vorticity is necessarily lost. It is therefore of some interest to determine whether the isovortical constraints are still felt closely enough in a numerical model that the persistence of anisotropy holds true.

This matter has been addressed using a strictly inviscid, de-aliased spectral model (courtesy of G.K. Vallis), run at a truncation of 32×32 waves. A set of numerical simulations will be presented here, each of which begins with the same initial energy spectrum which is peaked at zonal wavenumber $k = 0$ and meridional wavenumber $\ell = 8$, falling away smoothly in all directions. Dimensions are normalized so that the energy E equals unity and the (doubly-periodic) domain is 2π square; the enstrophy in this case is then 106·4 while the initial anisotropy μ, defined by

$$\mu \equiv \frac{1}{E} \int \frac{1}{2} (\overline{\Phi_y{}^2} - \overline{\Phi_x{}^2}) \, dy, \tag{8}$$

takes the relatively moderate value of 0·67. ($\mu = 1$ corresponds to completely zonal motion, $\mu = -1$ to completely meridional, and $\mu = 0$ to isotropic.) In each simulation the initial phases of the waves are assigned randomly; the stream function and relative vorticity fields for one of the realizations are shown in figure 1, and reveal that although the initial flow is quasi-zonal it certainly is not overly so.

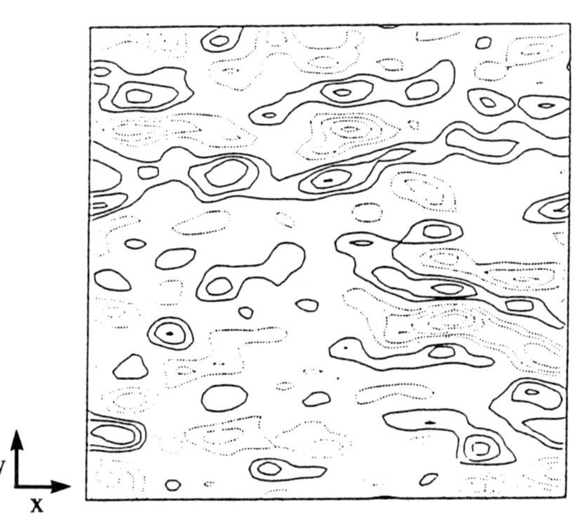

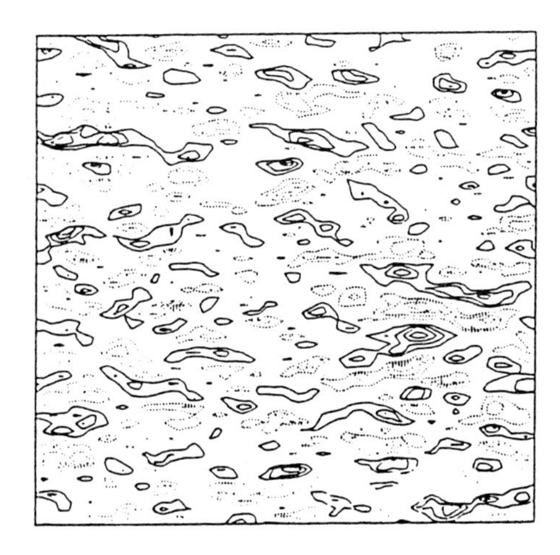

FIGURE 1. Plan view of the stream function Φ (left) and relative vorticity $\nabla^2\Phi$ (right) at $t = 0$ for one of the ensemble of runs discussed in the text.

When $\beta = 0$ one expects the anisotropy to decay rapidly to zero, on the scale of the eddy-turnaround time $Z^{-1/2}$ (Herring 1975). This is confirmed by an ensemble of ten runs (figure 2(a)); here $Z^{-1/2} \approx 0{\cdot}1$. But with $\beta = 100$, so that $\varepsilon \approx 2$, the anisotropy is seen to level out at around $\mu \approx 0{\cdot}3$ (figure 2(b)). This "anti-ergodic" behaviour occurs in spite of the fact that ε is actually greater than unity, and the initial condition is far from satisfying the strict requirement of Shepherd (1987)'s inequality (5.11), suggesting that the persistence of anisotropy is a remarkably robust feature of the dynamics. An example of an ensemble which *does* satisfy the requirements, having a different initial spectrum with $Z \approx 46$, initial anisotropy $\mu \approx 0{\cdot}93$, and $\varepsilon \approx 0{\cdot}9$, is shown in figure 2($c$); there the anisotropy levels out at around $\mu \approx 0{\cdot}75$, and remains close to that level until the end of the run at $t = 300$. (The figure only shows the first third of the time series.) It is noteworthy that in (b) and (c), each realization equilibrates at about the same level of anisotropy.

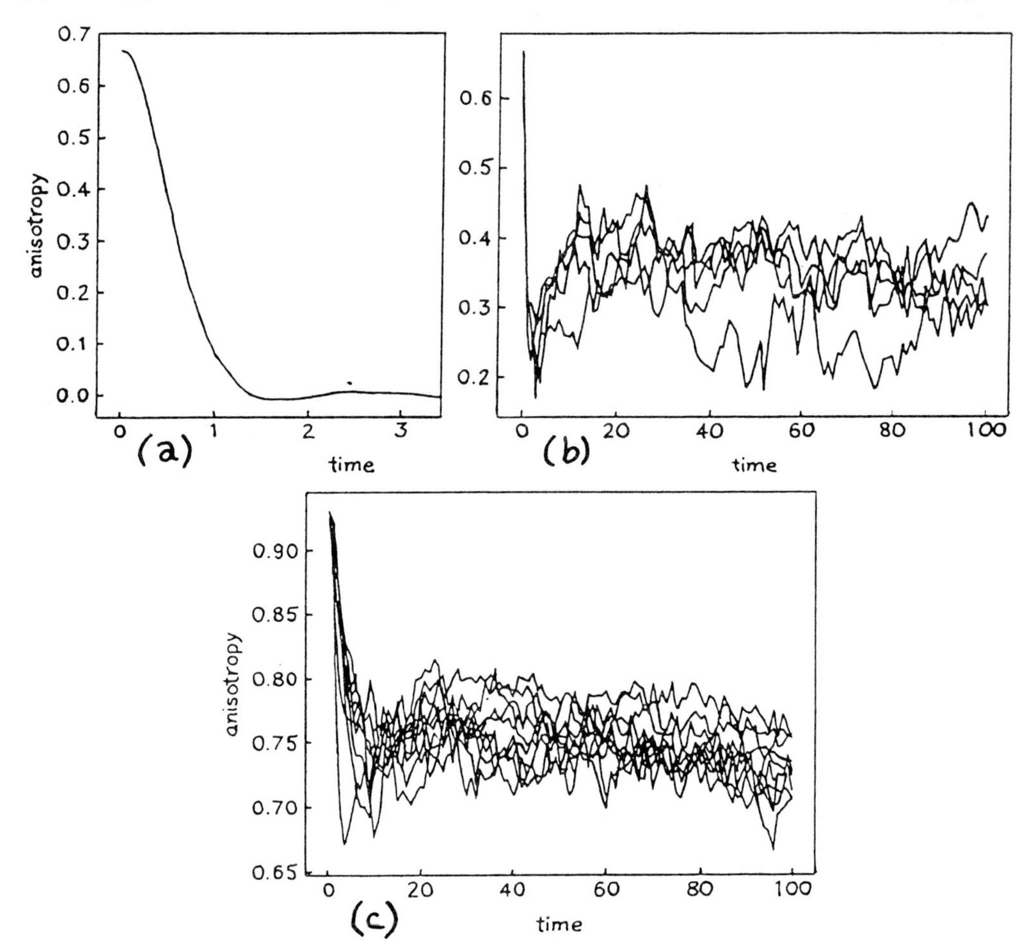

FIGURE 2. (a) Ensemble-average anisotropy for ten runs with $\beta = 0$. (b) Anisotropy for an ensemble of five runs with $\beta = 100$. (c) The same, for an ensemble of ten runs starting from a more anisotropic initial spectrum. The jagged nature of the curves in (b) and (c) is due to the discrete sampling in time.

4 NONLINEAR SATURATION OF SHEAR-FLOW INSTABILITIES

A second example of the power of the isovortical constraints is provided by the saturation of instabilities to parallel (zonal) shear flows on the beta-plane. Suppose that one is given an initial condition $\Phi(t=0)=\Phi_0$ consisting of a zonal-mean component $\overline{\Phi}_0$ that is known to be unstable, plus a wave (i.e. non-zonal) component Φ_0'. The question then arises: can one obtain an *a priori* bound on the maximum amplitude that can be attained by the growing disturbance Φ'?

One method for attacking this problem is provided by weakly-nonlinear theory. Such an approach has the benefit of providing details of the disturbance evolution and mean-flow adjustment, but it suffers from the handicap that the expansions involved are invariably asymptotic, and so the results may be incorrect for finite supercriticalities. A complementary approach will be outlined here that uses the stability bound (4) in a powerful way; it has the twin advantages of simplicity and rigour, but on the other hand provides only crude information on integral quantities.

The idea is once again to bound disturbances in terms of their proximity to a stable flow. To this end, consider the decomposition (2) with $Q(y)$ monotonic but otherwise arbitrary; then (4) follows. Now note that the wave enstrophy at any time t is bounded according to

$$\int \overline{(\nabla^2\Phi')^2}(t)dy = \int \overline{q'^2}(t)dy \leq \int \overline{q^2}(t)dy, \tag{9}$$

where $\nabla^2\Phi' = P' = q'$ since $Q' = 0$, while the initial *disturbance* enstrophy (times two) is given by

$$\int \overline{q^2}(0)dy = \int \overline{\left(\nabla^2\Phi_0'\right)^2}dy + \int \left(\overline{\Phi}_{0yy} - \Psi_{yy}\right)^2 dy. \tag{10}$$

Putting together (4), (9) and (10) then yields the bound

$$\int \overline{(\nabla^2\Phi')^2}(t)dy \leq \frac{|Q_y|_{\max}}{|Q_y|_{\min}} \left\{ z_0 + \int \left(\overline{\Phi}_{0yy} - \Psi_{yy}\right)^2 dy \right\}, \tag{11}$$

where $z_0 \equiv \int \overline{(\nabla^2\Phi_0')^2}dy$ is (twice) the initial wave enstrophy. The right-hand side of (11) is a functional of the (given) initial zonal flow $\overline{\Phi}_0$ and the (arbitrary) basic flow Ψ; one may therefore seek to minimize it by an appropriate choice of Ψ, in order to find the tightest upper bound. It is clear that if one chooses the basic flow to be too close to the initial flow (while yet being stable) in order to minimize the quantity within braces, then $|Q_y|_{\min}$ will be small and the bound will be too large; whilst if the two flows are too far apart from one another then the quantity within braces will be too large. So some optimal intermediate choice is implied. Of course, insofar as one cannot realistically minimize the right-hand side of (11) over *all* possible basic flows Ψ, one never knows whether a given bound could not be improved upon. But in this regard it is useful to note that one is justified in restricting attention to functions Q_y that take their extrema over finite intervals (Haynes 1988), which greatly facilitates the analysis.

It may turn out that the best bound obtainable from (11) will equal or exceed the total amount of enstrophy in the system, in which case the unstable zonal flow cannot be prevented (on these grounds at least) from breaking up into fully-developed turbulence. But whenever the bound (11) is less than the total enstrophy (or, strictly speaking, than the total available to the waves under the given boundary conditions), it may be concluded that the isovortical constraints are playing a role, through (4), in limiting the saturation amplitude of the instability.

Saturation bounds on the wave enstrophy have been worked out for various kinds of two-dimensional (barotropic) instability (Shepherd 1988*a*) and for baroclinic instability (Shepherd 1988*b*, 1989). The details are spelt out in those papers and there is no need to rehearse them here. The saturation bounds are of intrinsic interest insofar as they offer a way of checking the validity of approximate (e.g. weakly-nonlinear or low-order) theories. But experience from the studies mentioned above suggests that the bounds may also frequently be used to estimate saturation amplitudes, not only qualitatively (in terms of parameter dependences) but also quantitatively.

In the remainder of this paper some new results will be presented on saturation bounds for the wave energy, with an application to the point-jet instability. To obtain such bounds, one must employ the pseudoenergy conservation law. In this case of parallel flow the associated stability theorem is the generalized Fjørtoft theorem (see Shepherd 1988*a*, §2.2), from which one can easily derive the bound

$$\int \overline{|\nabla\Phi'|^2}(t)dy \leq \int \overline{|\nabla\psi|^2}(t)dy \leq \int \overline{|\nabla\psi|^2}(0)dy + \left(\frac{-U}{Q_y}\right)_{\max} \int \overline{(\nabla^2\psi)^2}(0)dy \quad (12)$$

for any basic flow with $\Psi'(Q) = -U/Q_y > 0$ everywhere. The right-hand side of (12) involves both the initial wave energy and enstrophy (which will be taken to be negligible in what follows) and the initial zonal-mean disturbance component

$$\int \left(\overline{\Phi}_{0y} - \Psi_y\right)^2 dy + \left(\frac{-U}{Q_y}\right)_{\max} \int \left(\overline{\Phi}_{0yy} - \Psi_{yy}\right)^2 dy. \quad (13)$$

As before, for given $\overline{\Phi}_0$ one may vary Ψ to seek the tightest bound.

In the case of the instability of the point jet $\overline{U}_0(y) = u_0 + \Gamma|y|$, $\overline{P}_0(y) = \beta y - \mathrm{sgn}(y)\Gamma$ in a channel of half-width L, employing the same basic flow that was used to derive the enstrophy bounds in Shepherd (1988*a*, §7.3), namely

$$Q(y) = \left\{ \begin{array}{cc} \beta y + \Gamma & \text{for } y < -y_1 \\ \alpha y & \text{for } -y_1 < y < y_1 \\ \beta y - \Gamma & \text{for } y > y_1 \end{array} \right\} \quad (14)$$

with $0 < \alpha < \beta$ and $y_1 = \Gamma/(\beta - \alpha)$, here yields the saturation bound

$$\int \frac{1}{2}\overline{|\nabla\Phi'|^2}(t)dy \leq \frac{\Gamma^5}{(\beta-\alpha)^3}\left\{\frac{3}{10} + \frac{1}{3\alpha}\left(\frac{L}{\Gamma}(\beta-\alpha)^2 - \frac{1}{2}(\beta-\alpha)\right)\right\}. \tag{15}$$

Of most interest is the limit $\beta L/\Gamma \gg 1$ where the channel walls are unimportant, in which case the right-hand side of (15) takes a relative minimum at $\alpha = \beta/2$ and the bound then equals $B_1 \equiv 4\Gamma^4 L/3\beta^2$. This can be compared with a number of other quantities. First, the wave energy is bounded by $(2L/\pi)^2$ times the wave enstrophy, which from Shepherd (1988*a*)'s equation (7.25) has the rigorous upper bound of $4\Gamma^3/3\beta$; this gives the energy bound $B_2 \equiv 16\Gamma^3 L^2/3\beta\pi^2$. Next, one may make an estimate (as opposed to a bound) of the saturated wave energy by presuming the zonal-mean flow following saturation to be neutrally stable (hence being given by (14) with $\alpha = 0$), and calculating how much energy has been released thereby to the waves. This gives the estimate $E_1 \equiv \Gamma^4 L/3\beta^2$. Finally, the wave energy is evidently bounded by the total amount of energy in the system, giving the bound $B_3 \equiv \Gamma^2 L^3/3$. These various estimates and bounds on the wave energy are ordered asymptotically (for $\beta L/\Gamma \gg 1$) according to

$$E_1 = \tfrac{1}{4}B_1 \ll B_2 \ll B_3. \tag{16}$$

The fact that the "adjustment estimate" E_1 is one-quarter of the rigorous saturation bound B_1 parallels the results for the wave enstrophy, and echoes a familiar pattern in this sort of calculation (cf. Shepherd 1989).

When $\beta L/\Gamma = O(1)$, the four quantities in (16) are all of the same order, and the isovortical constraints are therefore not as important as the geometrical ones. In this case one takes $y_1 = L$ in the construction (14), with the choice $\alpha = \beta(1 - (\Gamma/\beta L))$ (α can now be negative). Since $|Q_y|_{\max} = |Q_y|_{\min} = \alpha$ the saturation bound on the wave enstrophy is just $\Gamma^2 L/3$ (one-third of the total enstrophy), implying an energy bound of $4\Gamma^2 L^3/3\pi^2$ (just under one-half of the total energy). But these bounds could presumably be deduced just as easily from the global properties of the flow, without appeal to stability considerations.

REFERENCES

ARNOL'D, V.I. 1965 Conditions for nonlinear stability of stationary plane curvilinear flows of an ideal fluid. *Dokl.Akad.Nauk.SSSR* **162**, 975-978. (English transl.: *Soviet Math.* **6**, 773-777 (1965).)

ARNOL'D, V.I. 1966 On an a priori estimate in the theory of hydrodynamical stability. *Izv.Vyssh.Uchebn.Zaved.Matematika* **54**, no.5, 3-5. (English transl.: *Amer.Math.Soc.Transl., Series 2* **79**, 267-269 (1969).)

ARNOL'D, V.I. 1978 *Mathematical Methods of Classical Mechanics.* Springer, 462 pp. (Russian original: Nauka, 1974.)

BABIANO, A., BASDEVANT, C., LEGRAS, B. & SADOURNY, R. 1987 Vorticity and passive-tracer dynamics in two-dimensional turbulence. *J.Fluid Mech.* **183**, 379-397.

CARNEVALE, G.F. & FREDERIKSEN, J.S. 1987 Nonlinear stability and statistical mechanics of flow over topography. *J.Fluid Mech.* **175**, 157-181.

CARNEVALE, G.F., FRISCH, U. & SALMON, R. 1981 H-theorems in statistical fluid dynamics. *J.Phys.A* **14**, 1701-1718.

FJØRTOFT, R. 1950 Application of integral theorems in deriving criteria of stability for laminar flows and for the baroclinic circular vortex. *Geofys.Publ.* **17**, no.6, 52 pp.

HAYNES, P.H. 1988 Demonstration that any smooth extremum of Q_y allows an improvement to the saturation bound. *J.Fluid Mech.* **196**, 320-321. (Appendix C of Shepherd 1988*a*.)

HERRING, J.R. 1975 Theory of two-dimensional anisotropic turbulence. *J.Atmos. Sci.* **32**, 2254-2271.

HERRING, J.R. & MCWILLIAMS, J.C. 1985 Comparison of direct numerical simulation of two-dimensional turbulence with two-point closure: the effects of intermittency. *J.Fluid Mech.* **153**, 229-242.

HOLLOWAY, G. & HENDERSHOTT, M.C. 1977 Stochastic closure for nonlinear Rossby waves. *J.Fluid Mech.* **82**, 747-765.

KRAICHNAN, R.H. 1975 Statistical dynamics of two-dimensional flow. *J.Fluid Mech.* **67**, 155-175.

MCINTYRE, M.E. & SHEPHERD, T.G. 1987 An exact local conservation theorem for finite-amplitude disturbances to non-parallel shear flows, with remarks on Hamiltonian structure and on Arnol'd's stability theorems. *J.Fluid Mech.* **181**, 527-565.

RAYLEIGH, LORD 1880 On the stability, or instability, of certain fluid motions. *Proc.Lond.Math.Soc.* **11**, 57-70.

RHINES, P.B. 1975 Waves and turbulence on a beta-plane. *J.Fluid Mech.* **69**, 417-443.

SALMON, R., HOLLOWAY, G. & HENDERSHOTT, M.C. 1976 The equilibrium statistical mechanics of simple quasi-geostrophic models. *J.Fluid Mech.* **75**, 691-703.

SHEPHERD, T.G. 1987 Non-ergodicity of inviscid two-dimensional flow on a beta-plane and on the surface of a rotating sphere. *J.Fluid Mech.* **184**, 289-302.

SHEPHERD, T.G. 1988*a* Rigorous bounds on the nonlinear saturation of instabilities to parallel shear flows. *J.Fluid Mech.* **196**, 291-322.

SHEPHERD, T.G. 1988*b* Nonlinear saturation of baroclinic instability. Part I: The two-layer model. *J.Atmos.Sci.* **45**, 2014-2025.

SHEPHERD, T.G. 1989 Nonlinear saturation of baroclinic instability. Part II: Continuously-stratified fluid. *J.Atmos.Sci.* **46**, 888-907.

THOMPSON, P.D. 1982 On the structure of the hydrodynamical equations for two-dimensional flows of an incompressible fluid: the role of integral invariance. In *Mathematical Methods in Hydrodynamics and Integrability in Dynamical Systems* (ed. M. Tabor & Y.M. Treve), AIP Conf.Proc., vol. 88, pp. 301-317.

The Survivability of Vortices in a Two-Dimensional Fluid at Extremely High Reynolds Numbers

DAVID G. DRITSCHEL

Department of Applied Mathematics and Theoretical Physics,
University of Cambridge, Silver Street, Cambridge CB3 9EW, UK

ABSTRACT

It has now been established that a two-dimensional fluid evolves into a very heterogeneous state characterized by well-separated coherent vortices and a background sea of quasi-passive filamentary debris. The heterogeneity arises from the organization of vorticity that takes place as a result of repeated vortex merging, which, on average, leads to larger and larger vortices. As the average vortex size grows, vortices become more isolated and less frequently encounter one another. The above picture, drawn from a series of numerical studies, implies that a single merging event is efficient, in the sense that more circulation ends up in the final coherent vortex than was in any of the original vortices participating, this despite the inevitable loss of circulation to the filamentary debris expelled during merging. Why this is the case is a theoretical question as yet unanswered.

Herein, some subsidiary problems are examined. These problems address the *generic* structure of vortices in a two-dimensional fluid at Reynolds numbers much higher than considered in the aforementioned numerical studies, but in fact typical of geophysical flows. As a first simplification, a single vortex is modelled in detail, while the remaining flow field is represented by a locally-uniform strain. This model is valid as long as a vortex remains well separated from other vortices in the fluid. It is discovered that the strain field "strips" away weak vorticity from the periphery of the vortex leaving it with an exceedingly steep edge. The steep edge then protects the vortex from further "stripping," and it also enables the vortex to undergo significant shape deformations in response to the external strain without "axisymmetrizing." These results contrast with those of lower Reynolds number numerical studies, in which one sees predominantly smooth-edged, circular vortices.

A second, more elaborate model considers the merging of two vortices in an external strain field. It is found that merging itself can leave the final vortex with much steeper vorticity gradients than had either of the two original vortices. In effect, the straining of one vortex by the other, and *vice-versa*, strips the vortices of their low-lying vorticity. These results are beginning to suggest that two-dimensional turbulence, at sufficiently high Reynolds numbers, is characterized by vortices with exceedingly steep edges capable of significant departures from circular symmetry.

SUMMARY OF RESULTS

Recent computations of two-dimensional vortex dynamics (hereinafter 2VD) at high Reynolds numbers have shown that such flows exhibit a striking tendency toward extreme inhomogeneity (see e.g. Fornberg 1977, Basdevant et al 1981, McWilliams 1984, Legras et al 1988, and many others). What develops are compact, coherent regions of circulation – vortices – within a complex but quasi-passive sea of filamentary vorticity.

Much research, partly analytical and partly numerical, has unravelled various components of this picture. The merging of vortices, essential to the observed growth in size of vortices in 2VD, has received much attention. In one of the first systematic investigations of this process, Zabusky et al (1979) reported that two identical, initially circular patches of uniform vorticity merge (and simultaneously produce filamentary vorticity debris) if their initial centroid separation is less than approximately 3.4 vortex radii. Later investigations suggested that inviscid vortex merger results from an instability mechanism (Dritschel 1985, 1986), and, more recently, general criteria have been developed for the merging of vortices with distributed vorticity (Melander et al 1987a,b).

Further research has examined the behaviour of a single, isolated, distributed vortex, in an effort to explain the observed predominantly circular shapes in numerical computations of 2VD. Such circular shapes are rigorously stable in both a linear and nonlinear sense (see Dritschel 1988a & refs.) when the distribution of vorticity is monotonic. However, noncircular shapes with distributed vorticity have as yet defied proofs of nonlinear stability, suggesting, but only suggesting, that noncircular vortices are less apt to persist than circular ones. In fact, on the basis of numerical computations at finite Reynolds numbers, Melander et al (1987c) have claimed that *all* smooth, nonuniform, noncircular vortices will eventually become circular or "axisymmetrize" by wholly inviscid mechanisms. However, this claim does not appear to be consistent with recent analytical results and numerical computations performed at much higher resolution (Dritschel & Legras 1989) indicating that smooth, compact, noncircular vortices may exist indefinitely.

There have also been studies of the production and stability of filamentary vorticity. Apart from the production of filaments occurring during vortex merging, significant production also occurs in the absence of severe vortex shape deformations. For example, sharp edges of vorticity, a recently recognized feature of 2VD at extreme Reynolds numbers (Legras & Dritschel 1989), succomb to a nonlinear instability called "filamentation," whereby shallow undulating disturbances propagating on a sharp vortex edge repeatedly steepen, break, and extend out into long thin filaments (Dritschel 1988b & refs.). In a coarse-grained view, filamentation diffuses sharp vorticity gradients and as a consequence makes the vortex edge less susceptible to further filamentation. However, other processes discussed below act to re-sharpen vortex edges.

The stability of filaments has been reconsidered recently in order to understand the remarkable persistence or quasi-passive behaviour of filamentary vorticity in numerical calculations. Why do we only seldomly observe the "roll-up" of a vorticity filament into a string of vortices, such as may be expected on the basis of the classic analysis of Rayleigh in the 19th century? The answer is that the conditions for Rayleigh's analysis are almost never satisfied in the numerical calculations. The filaments are constantly being stretched and twisted by strain fields arising principally from the coherent vortices. Under these conditions, the stability problem gives far different results, and one finds that stretching and twisting rates an order of magnitude less than the vorticity anomaly of a filament stabilizes it in most circumstances (Dhanak 1981, Dritschel 1989a,b, Dritschel et al 1989).

The long-range effects of stretching and twisting that arise from a field of coherent vortices have important consequences for the dynamics of the vortices themselves. The very structure and shapes of vortices depend, it seems, crucially on the nature of this external strain flow. At extreme Reynolds numbers, incidentally not uncharacteristic of geophysical flows yet far greater than commonly considered in numerical calculations of 2VD, a surprisingly weak strain flow can strip a vortex edge of practically all of its low-lying vorticity. This can leave a vortex with an exceedingly sharp edge, orders of magnitude thinner than the radius of the vortex itself (Legras & Dritschel 1989).

In many instances, vortices are so extensively stripped by the large-scale straining flow of surrounding vortices that it may be appropriate, for the investigation of further effects of the external straining flow, to idealize a vortex as a patch of uniform vorticity. It is becoming clear from extensive numerical evidence that the vortex patch with its inherent vorticity discontinuity may not be as severe an idealization of the typical vortex structure of 2VD as one might believe. The simplicity of the vortex patch furthermore enables one to investigate, exhaustively, the stability, both linear and nonlinear, of the Moore-Saffman-Kida-Neu elliptical vortices in a general straining flow (Moore & Saffman 1971, Kida 1981, Neu 1984). These elliptical vortices retain their elliptical character, in the absence of disturbances, while generally changing in aspect ratio and orientation. In the presence of disturbances, which are invariably present (e.g. as local variations in the external strain field), the picture changes dramatically (Dritschel 1989b). The pattern of instability behaviour turns out to be surprisingly rich, depending sensitively on the nature of the external straining flow. Furthermore, there appears to be nonlinear modes of instability which are capable of amplifying arbitrarily small disturbances but are not captured by the linear stability (Floquet) analysis. When the external straining flow is very weak, the linear growth rates scale with powers of the strain rate and hence one is not likely to see significant effects of instability before the external strain field itself changes in response to the movement of distant vortices.

It has become apparent that the fundamental problem of vorticity organization in 2VD and the consequent growth of coherent structures boils down to understanding the phenomenon of vortex merging in a general straining flow. It seems that one can largely carry over ideas from previous studies of isolated vortex merging when vortices are close enough together, for then the local straining flow typically dominates the external flow (Dritschel 1989c). However, getting vortices close enough to merge can only be brought about by the external straining flow (i.e. other vortices; see figure 1). The difficulty of the problem is that on the one hand one needs to know the typical external straining flow surrounding a given set of candidate vortices in order to predict, for instance, the likelihood of their merging, but on the other hand one needs to know the statistical properties of vortex merging in order to determine the typical external straining flow. The problem is most acute at extreme Reynolds numbers, for which the average time between successive merging events, itself long compared to the rotation period of a typical vortex, is short compared to the viscous time scale. Direct many-vortex numerical calculations satisfying these constraints do not at present appear possible.

It is perhaps surprising that in a system as idealized as 2VD there can be such complex, as yet unanswered questions. These questions, in fact, go beyond 2VD. For instance, the ability of large-scale straining fields to form and maintain very sharp vorticity gradients is the key ingredient to understanding the observed and numerically simulated structure of the stratospheric polar vortex (see, e.g. Juckes & McIntyre 1987). The somewhat unexpected sharp vortex edge presents problems for conventional models, which must diffuse sharp vorticity gradients in order to maintain numerical stability. Unconventional models of 2VD already have the ability of forming and maintaining vortex edges orders of magnitude thinner than conventional models, and work is in progress to extend this ability to more realistic systems. Finally, the study of 2VD is also providing directions for studying these more realistic systems, and may lead to important insights into tackling some of today's urgent environmental problems.

REFERENCES

Basdevant, C., Legras, B. & Sadourny, R 1981 A study of barotropic model flows: intermittency, waves and predictability. J. Atmos. Sci. 38, 2305-2326.

Dhanak, M.R. 1981 The stability of an expanding circular vortex layer. Proc. R. Soc. Lond. A375, 443-451.

Dritschel, D.G. 1985 The stability and energetics of corotating uniform vortices. J. Fluid Mech. 157, 95-134.

Dritschel, D.G. 1986 The nonlinear evolution of rotating configurations of uniform vorticity. J. Fluid Mech. 172, 157-182.

Dritschel, D.G. 1988a Nonlinear stability bounds for inviscid, two-dimensional, parallel or circular flows with monotonic vorticity, and the analogous three-dimensional quasi-geostrophic flows. J. Fluid Mech. 191, 575-581.

Dritschel, D.G. 1988b The repeated filamentation of two-dimensional vorticity interfaces. J. Fluid Mech. 194, 511-547.

Dritschel, D.G. 1989a On the stabilization of a two-dimensional vortex strip by adverse shear. J. Fluid Mech. 206, 193-221.

Dritschel, D.G. 1989b The stability of elliptical vortices in an external straining flow. J. Fluid Mech. (in press).

Dritschel, D.G. 1989c Contour dynamics and contour surgery: numerical algorithms for extended, high-resolution modelling of vortex dynamics in two-dimensional, inviscid, incompressible flows. Computer Phys. Rep. 10, 77-146.

Dritschel, D.G. and Legras, B. 1989 On the structure of isolated, coherent vortices in a two-dimensional, incompressible fluid at very high Reynolds numbers. J. Fluid Mech. (submitted).

Dritschel, D.G., Haynes, P.H., Juckes, M.N. and Shepherd, T.G. 1989 The stability of a two-dimensional vorticity filament subjected to uniform strain. J. Fluid Mech (submitted).

Fornberg, B. 1977 A numerical study of 2-D turbulence. J. Comput. Phys. 25, 1-31.

Juckes, M.N. & McIntyre M.E. 1987 A high-resolution one-layer model of breaking planetary waves in the stratosphere. Nature 328, 590-596.

Kida, S. 1981 Motion of an elliptic vortex in a uniform shear flow. J. Phys. Soc. Jpn. 50, 3517-3520.

Legras, B. and Dritschel, D.G. 1989 Vortex stripping. J. Fluid Mech. (submitted) See also: Dritschel, D.G. 1989 In: *Mathematical Aspects of Vortex Dynamics* (ed. R.E. Caflisch) Society for Industrial and Applied Mathematics.

Legras, B., Santangelo, P. & Benzi, R. 1988 High resolution numerical experiments for forced two-dimensional turbulence. Europhys. Lett. 5(1), 37-42.

McWilliams, J.C. 1984 The emergence of isolated coherent vortices in turbulent flow. J. Fluid Mech. 146, 21-43.

Melander, M.V., Zabusky, N.J. & McWilliams, J.C. 1987a Symmetric vortex merger in two-dimensions: causes and conditions. J. Fluid Mech. 195, 303-340.

Melander, M.V., Zabusky, N.J. & McWilliams, J.C. 1987a Asymmetric vortex merger in two-dimensions: which vortex is "victorious". Phys. Fluids 30, 2610-2612.

Melander, M.V., McWilliams, J.C. & Zabusky, N.J. 1987c Axisymmetrization and vorticity-gradient intensification of an isolated two-dimensional vortex through filamentation. J. Fluid Mech. 178, 137-159.

Moore, D.W. & Saffman, P.G. 1971 The structure of a line vortex in an imposed strain. In: *Aircraft Wake Turbulence and Its Detection* (eds. J. Olsen, A. Goldberg & N. Rogers) Plenum.

Neu, J.C. 1984 The dynamics of a columnar vortex in an imposed strain. Phys. Fluids 27, 2397-2402.

Rayleigh, Lord 1894. *The Theory of Sound*, 2nd edn. Macmillan. (see also: 3rd edn., 1945).

Zabusky, N.J., Hughes, M.H. & Roberts, K.V. 1979 Contour dynamics for the Euler equations in two dimensions. J. Comput. Phys. 30, 96-106.

Figure 1 (opposite): A numerical calculation of induced vortex merger. The two, largest vortices have identical positive vorticity, while the third, smallest vortex has negative vorticity and half the circulation magnitude of the other vortices. The calculation is performed with contour surgery (see Dritschel 1989c & refs.), in which each contour represents an equal jump in vorticity. Time, in units of 2π divided by the peak vorticity, is indicated in the upper left hand corner of each frame. Note that the negative vortex shows little disruption during the evolution yet induces the two positive vortices to merge. In the process of merging, much stripping of low-lying vorticity occurs, and the edges of the vortices sharpen considerably.

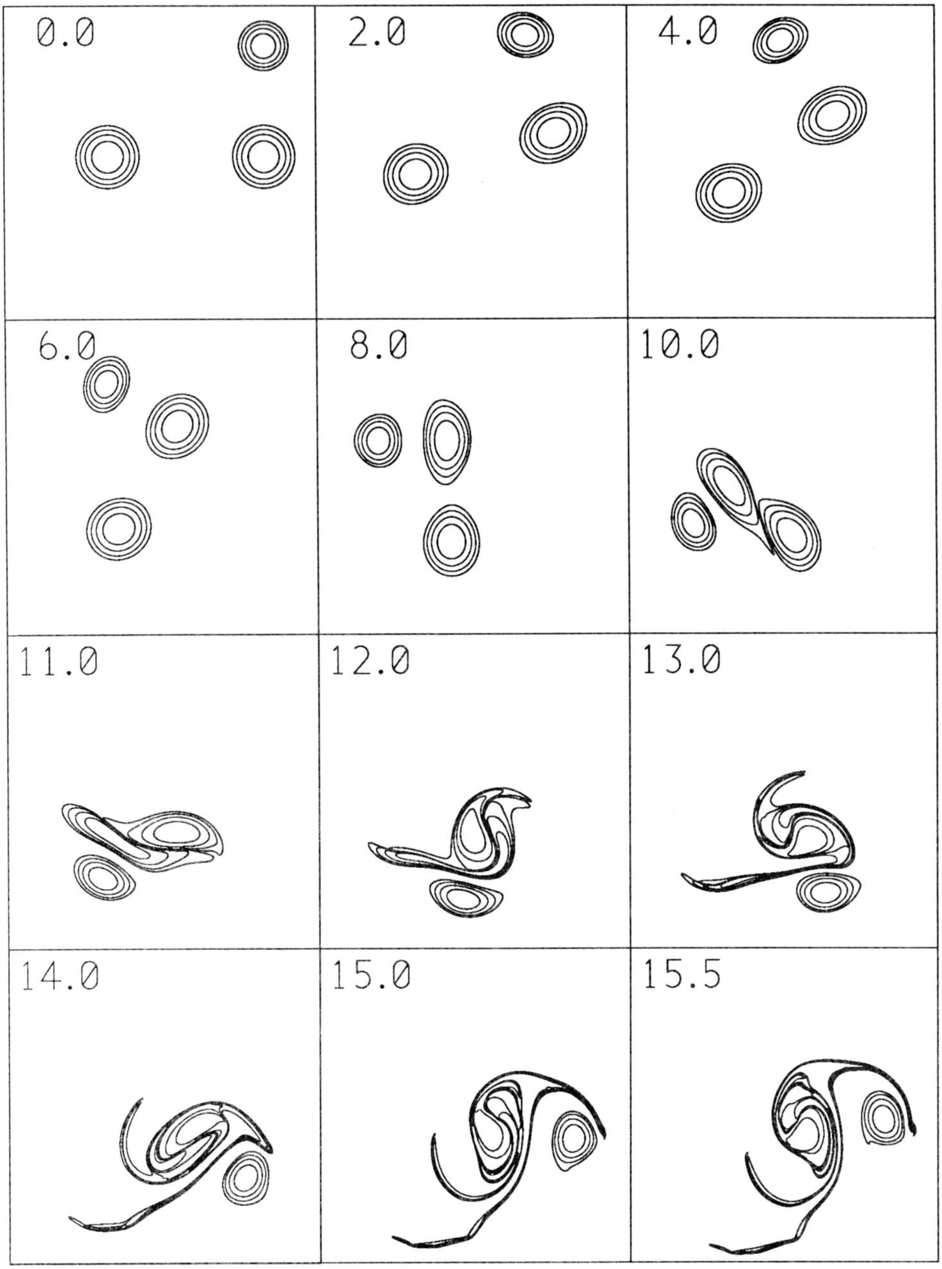
0.0
2.0
4.0
6.0
8.0
10.0
11.0
12.0
13.0
14.0
15.0
15.5

Isovortical Relaxation to Stable Two-Dimensional Flows

G. F. CARNEVALE

University of California, San Diego, Ca., 92093 U.S.A

G. K. VALLIS

University of California, Santa Cruz, Ca., 95064 U.S.A

An artificial modification of inviscid fluid dynamics results in a system which monotonically loses energy while preserving all vorticity invariants. Through numerical simulation of this modified dynamics, the isovortical extraction of energy can be observed as a continuous evolution process. An example of this is provided in the context of quasi-geostrophic flow over topography. It has been found that these simulations can take a nonstationary flow into an Arnol'd stable minimum energy state with a nonlinear relation between streamfunction and vorticity. The minimum energy state constrained only by the value of the enstrophy has a linear streamfunction vorticity relation. Hence, these results demonstrate that simulations of the modified dynamics are able to converge to nontrivial stable stationary states in spite of the lack of strict conservation of nonquadratic vorticity integral invariants due to the inadequacies of finite resolution. This technique can also be used to monotonically increase energy. A simulation of this effect is presented in which a flow is transformed from a stable minimum energy state to a maximum energy state. Furthermore, stable configurations of arbitrary vorticity distributions on a flat plane can also be obtained and an example is provided.

1 INTRODUCTION

Inviscid two-dimensional fluids evolve according to the law of advection of vorticity. This may simply be relative vorticity as in the Euler equation or potential vorticity as in the quasi-geostrophic equation. In any case, there is associated with such motion an infinity of dynamical invariants. That is, in addition to the energy of the flow, the vorticity on all material particles is conserved. This implies, for example, that with suitable boundary conditions, the integral over the domain of any function of the vorticity is invariant in time. Therefore, the evolution of any particular state in the infinite dimensional phase space remains on a hypersurface (an isovortical sheet) defined by the specification of the values of an infinity of vorticity invariants. Isolated points of minimum or maximum energy on these hypersurfaces correspond to nonlinearly stable flows. Here we consider the problem of finding stable stationary points on a given sheet in the vicinity of any given state on that sheet. This requires energy extremization subject to an infinite number of constraints.

The equation of evolution of a two-dimensional flow is simply the advection of potential vorticity:

$$\frac{\partial q}{\partial t} + \mathbf{v} \cdot \nabla q = 0. \tag{1}$$

The velocity field, $\mathbf{v}$, is non-divergent and can be written in terms of a streamfunction, ψ. To be concrete, we take the potential vorticity, q, to be given by $q = \zeta + h$, where $\zeta = \nabla^2 \psi$ is the relative vorticity, and h is a given field independent of time. For quasi-geostrophic flow over topography $-h(x, y)$ is the fractional change of layer depth scaled by the Coriolis parameter. Furthermore, we will assume periodic boundary conditions on all fields throughout this paper.

States evolving according to equation (1) remain on the surfaces of constant energy which intersect the isovortical sheet. If the explicit $\mathbf{v}$ in (1) is replaced by some arbitrary velocity field $\tilde{\mathbf{v}}$, then the evolution would no longer necessarily remain on the constant energy surfaces because the diagnostic relation between the advecting velocity field and q is broken. Nevertheless, the evolution would remain on the same isovortical sheet because the modified dynamics, which is simply advection by $\tilde{\mathbf{v}}$, also conserves q on all particles. In Vallis et al. (1989) and Carnevale and Vallis (1989) we discussed various ways of prescribing $\tilde{\mathbf{v}}$ such that the energy evolves monotonically. Here we define $\tilde{\mathbf{v}}$ through the streamfunction

$$\tilde{\psi} = \psi - \alpha J(\psi, q), \tag{2}$$

where α is an arbitrary constant. That is, the modified dynamics obeys

$$\frac{\partial q}{\partial t} + \tilde{\mathbf{v}} \cdot \nabla q = 0, \tag{3}$$

where $q = \nabla^2 \psi + h$ is defined by ψ and not $\tilde{\psi}$.

By multiplying (3) by ψ, and integrating, we then obtain

$$\frac{dE}{dt} = \alpha \int [J(\psi, q)]^2 dxdy, \tag{4}$$

where $E = \frac{1}{2} \int (\nabla \psi)^2 dxdy$ is the total energy. With α greater (less) than zero the energy increases (decreases) monotonically. We refer to this scheme based on advection with an artificial velocity field as pseudo-advection.

The stationary states of both equations (1) and (3) must have constant potential vorticity on contours of the streamfunction, ψ. Thus any state in which the potential vorticity is a function of the streamfunction (i.e., $q = f(\psi)$) is a stationary state. Arnold (1966) has given sufficient conditions for determining if a given stationary state is nonlinearly stable. If the slope, f', is positive everywhere, then the state is a minimum energy state and Arnold stable. By a second criterion the state is a maximum energy state and Arnold stable if the slope is sufficiently shallow and negative everywhere (cf., Arnold 1966). Note that if a given f satisfies either condition and if there is a solution to $q = f(\psi)$, then that solution is unique (cf. Carnevale and Frederiksen, 1987). However, if there is an Arnold minimum energy state on a sheet, then there can be no other Arnold minimum energy states on the sheet and its energy will be strictly less than that of every other state on the sheet. Similarly, uniqueness also holds for the Arnold maximum energy states (Carnevale and Vallis, 1989). The Arnold conditions are, however, only sufficient and one can anticipate that pseudo-advection may find maximum and minimum energy states that do not satisfy them.

By implementing the pseudo-advection algorithm numerically, one can actually follow the continuous transformation of any flow into a stable stationary flow by monotonic energy extraction or enhancement. Numerical simulation of pseudoadvection is necessarily at finite resolution. This implies that the details of the small-scale motions cannot be accurately reproduced, and the vorticity invariants will in fact change in time. The simulations presented here are spectral. The total potential vorticity is exactly zero for all time. The enstrophy is conserved as accurately as we wish depending on the choice of the time step. Other nonlinear vorticity invariants are more or less conserved depending on the time stepping and, most critically, on the resolution. The degree to which pseudo-advection can be used to explore the structure of the sheets in phase space is determined by the ability of the simulation to preserve the vorticity invariants and thus keep the evolving state near the original sheet. If total vorticity and enstrophy are the only quantities which determine an extremal energy state, then the relationship $q = f(\psi)$ in that state will be linear. Thus, one measure of the fidelity of the simulations of pseudo-advection will be the ability to converge to extremal energy states in which the function f is nonlinear, since this reflects constraints of invariants other than total vorticity and enstrophy in the extremization process. In section 2, we provide examples with resolution as low as 32×32 that demonstrate that by this measure one can actually perform valuable numerical simulations of pseudo-advection. Thus, such simulations can be used as a way of gaining information about the structure of phase space even though, strictly speaking, they cannot remain precisely on a single isovortical sheet.

2 SIMULATIONS

2.1 Energy minimization in flow over topography

Without topography there actually is no nontrivial minimum energy state on any isovortical sheet (assuming periodic boundary conditions). This is because energy can be continually and isovortically decreased under pseudo-advection by a process of filamentation, which is unrestrained in the absence of topography. One can observe this filamentation in simulations (Carnevale and Vallis 1989). Thus to show energy minimization we select an example from flow over topography.

Stationary states with arbitrary relations $q = f(\psi)$ can be constructed if there is freedom to choose the topography. First pick ψ to be any field satisfying the boundary conditions. Then the appropriate h for which this is a stationary state is defined by

$$q \equiv \nabla^2\psi + h = f(\psi). \tag{5}$$

For our first example, we consider a stationary state satisfying

$$q = \psi - 1.6\psi^3 + 1.3\psi^5. \tag{6}$$

We pick the solution ψ to be simply

$$\psi = \sin x \sin y. \tag{7}$$

Then eq. (5) defines the topography. Since the slope of the graph of q versus ψ is everywhere positive, the solution ψ is a nonlinearly stable minimum energy state (Arnold 1966). If we perturb this state isovortically, we can expect the pseudo-advection algorithm to drain the excess energy from the perturbed state, returning it to the unperturbed minimum energy state. For the initial condition we perturb the stationary flow isovortically by shifting the potential vorticity field rigidly in the x-direction by one whole gridpoint interval. Since the resolution in the simulation is only 32×32, this is a substantial perturbation. An alternative method of perturbation is the advection of the unperturbed state for a small period with a randomly generated static velocity field, and this has also been used with no change in the results. The left hand panel in figure (1) shows the $q - \psi$ relation in the perturbed state. The energy in the perturbed state is 1.5% greater than the unperturbed state. The potential enstrophy is of course the same in each state. With this value of enstrophy and the energy in the perturbed state, one can readily calculate the minimum energy for the given enstrophy state. This corresponds to a straight line on the $q - \psi$ scatter plot through the origin and enveloped by the scattered points which represent the perturbed state. As pointed out above, the simulation will accurately conserve only the enstrophy while the energy is being decreased. The interesting question then is whether the relaxation will clearly be toward the nonlinear relation, which represents the unique minimum energy on the initial sheet, or whether the simulation will deviate far from the sheet and reach the even lower energy state with the linear $q - \psi$ relation. In fact, in a reasonably short computation time (roughly 10 box transit times in terms of the time it would take the fastest moving particles to cross a distance equal to the length of a side of the periodic domain) the algorithm does transform the perturbed state back to the original unperturbed state. This is illustrated in the right hand panel in figure (1) where we have drawn the theoretical limit, eq. (6), through the data points. During the period of relaxation the total enstrophy changes by 0.0004%. This can be made smaller by decreasing the time step. Another measure of how closely the simulation remains on a given sheet is the variation of the integral invariants $Q_n = \int q^n dxdy$. The change in the Q_4 and Q_8 for this experiment were 0.5% and 5% respectively. While these numbers imply that one cannot remain precisely on the initial sheet, we interpret the accurate collapse of the scatter diagram onto the nonlinear polynomial relation as evidence that the simulation can indeed be considered as remaining near the original sheet, and is a useful tool for the exploration of the phase space of two dimensional flows.

Similar results have been obtained with other polynomial relationships between ψ and q. Based on tests where the stationary state has a $q = \psi^3$ relationship, one can expect significant improvement in the conservation of the Q_n by doubling the resolution of the simulation. For example, with equal time step, we observed the error in Q_8 to decrease by a factor of 75. The improvement seems to be more dramatic the higher the power n considered.

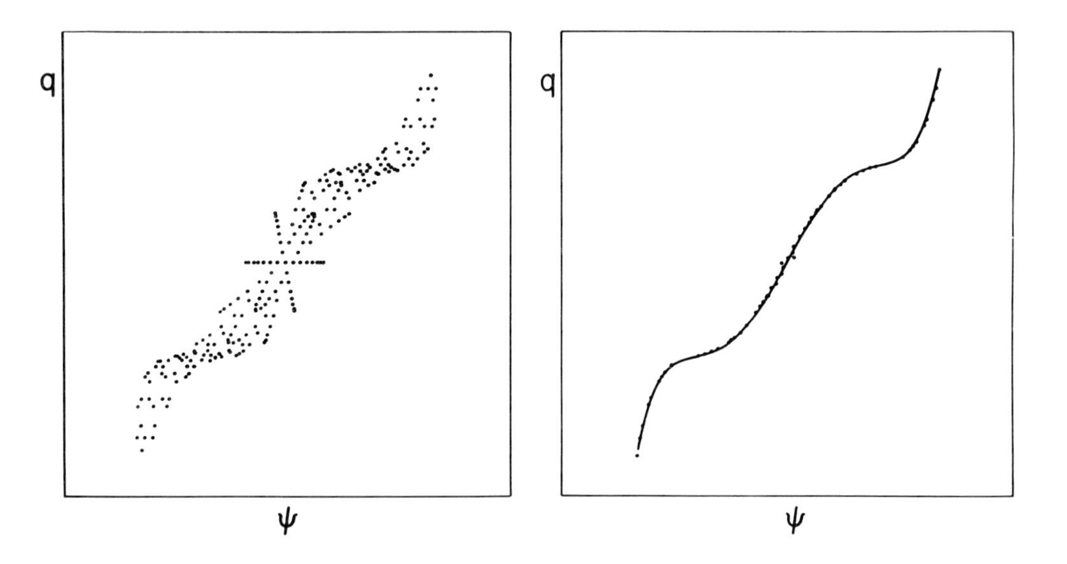

Figure 1. Scatter plots showing the pseudo-advection of the perturbed state back toward the unperturbed state, $q = \psi - 1.6\psi^3 + 1.3\psi^5$.

2.2 Energy maximization in flow over topography

At infinite resolution one could start with a minimum energy state and isovortically increase its energy until a maximum energy state on the same sheet is reached. Here we show such an experiment with a simulation at resolution 32×32 in which we start in the minimum energy state given by $q = \psi^3$ with $\psi = \sin x \sin y$ and topography defined as in eq (5). The energy is continually increased and the state is transformed into a maximum energy state with a $q - \psi$ relation of negative slope. The initial and final potential vorticity fields are 180° out of phase. That is, the initial minimum energy state is one of topography-potential vorticity correlation the final maximum energy state is one of anticorrelation. As one can see in figure (2) the transformation proceeds from an instability which pulls out the potential vorticity contours into small "arms". These elongated pieces then break off and migrate from the hills to the valleys and vice versa. The small scale pieces then reassemble with phase exactly the opposite from the initial condition. In the $q - \psi$

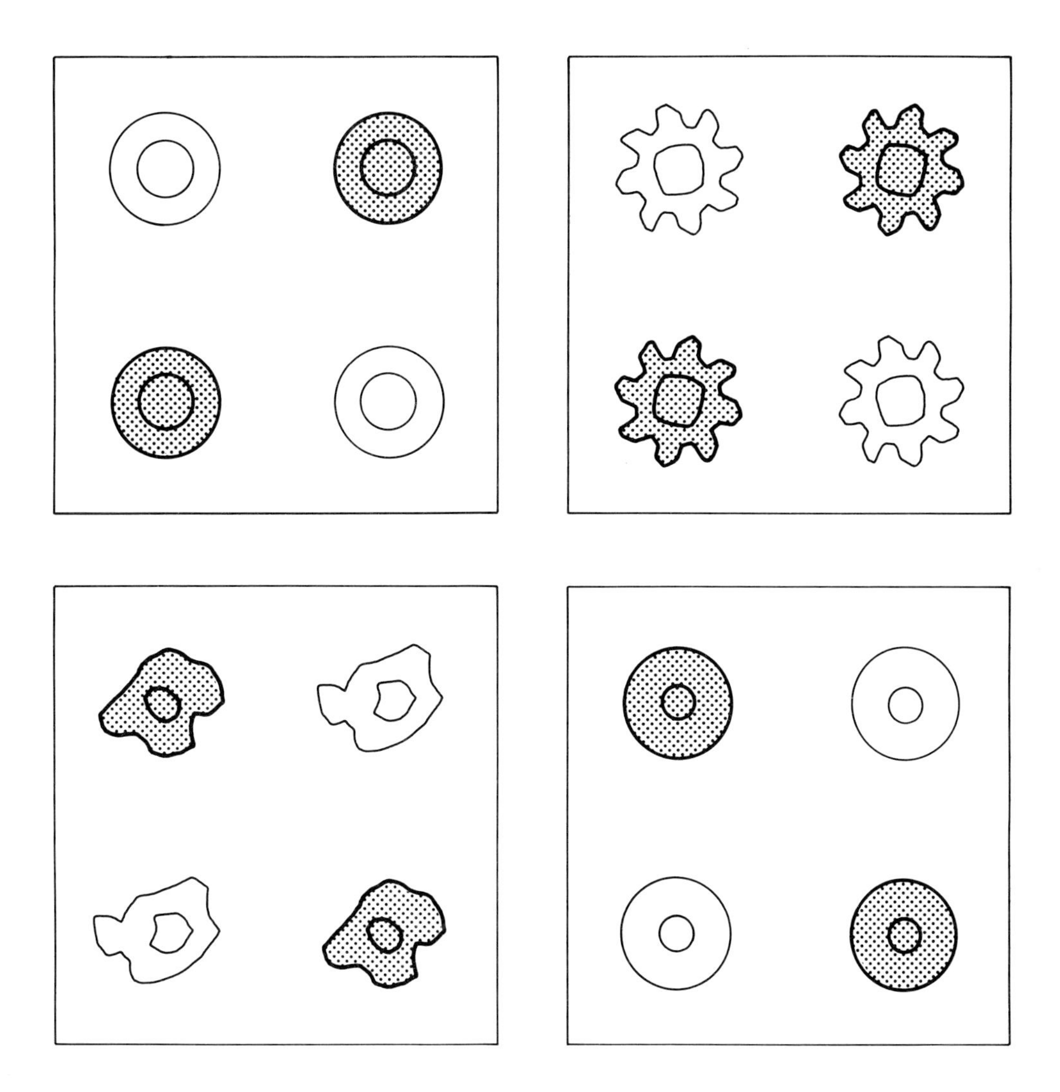

Figure 2. Contours of the potential vorticity field during monotonic increase in energy with $\alpha = 0.1$. Time progresses left to right, top to bottom. The initial minimum energy state has potential vorticity correlated with the topography while final maximum energy state has potential vorticity anticorrelated with topography. Positve potential vorticity contours above value 0.4 are shaded (initially $q_{max} = 1.$).

plots one observes the initial positive functional relation first becomes fuzzy, then the scatter plot becomes completely amorphous (not shown) when the q field has been completely shattered into small-scale pieces, then as the pieces reassemble a functional relationship of negative slope begins to emerge. The scatter diagram approaches closer and closer to a curve as the energy continues to increase. If this were performed at infinite resolution, the breakup of the initial potential vorticity contours could not occur and there would for all time be filaments connecting all of the original contours so that the topology would be conserved. Here, of course these filaments are lost early on due to the coarseness of the grid.

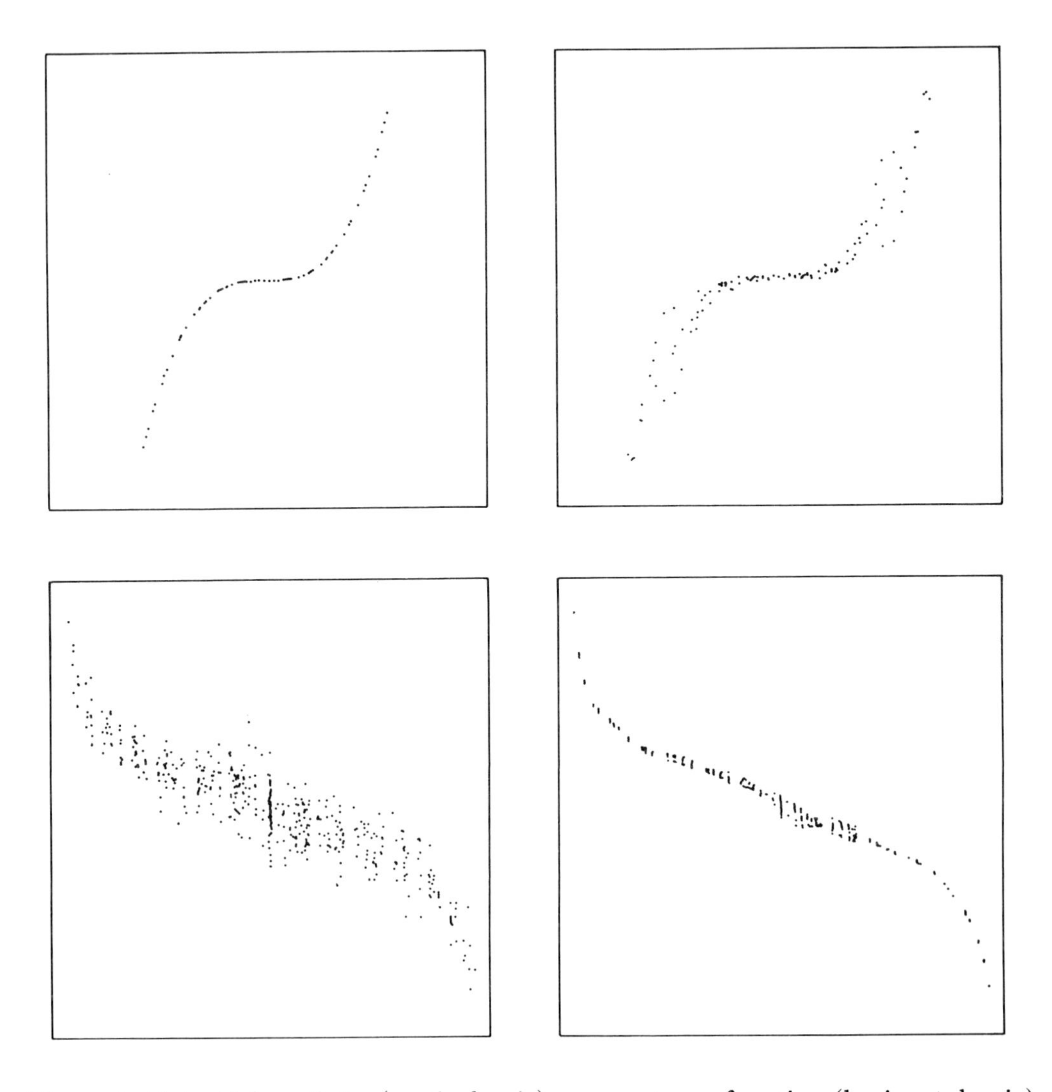

Figure 3. Potential vorticity (vertical axis) versus streamfunction (horizontal axis) scatter plots correspond to the potential vorticity fields shown in the previous figure.

In such a drastic transformation as in this example, the process can only loosely be described as remaining on the same isovortical sheet. In fact, this could be tested by reversing the process and evolving by pseudo-advection from the maximum energy state back toward the minimum energy state. In this process, the potential vorticity field again breaks up into small pieces which reassemble back to a state of positive topography-potential vorticity correlation. The final state however is closer to $q = \psi^3 + \psi$ than to the original $q = \psi^3$ state. These two relations respectively represent the unique minimum energy states on there own isovortical sheets. So we have returned to a different sheet with a different minimum energy state. Presumably the ability to remain near a particular sheet will to some extent rely on the resolution of the simulation; however, for such transformations, that pass through an intermediate state in which substantial energy is transferred to small scale, significant drift between sheets must be expected. However, it is important to note that these simulations still do not relax to a linear $q - \psi$ relation, and so must remember to some extent the potential vorticity distribution of the initial condition.

Note that the maximum energy state toward which the advection tends in this example (lower right panel in fig. 3) does not satisfy the sufficient criteria of Arnold (1966) for nonlinear stability because the slopes near the ends of the graph are too steep. Nevertheless, the state is stable by construction (and tests by simulation show no instabilities). A case in which an Arnold stable maximum energy state is transformed into an Arnold stable minimum energy state is presented in Carnevale and Vallis (1989).

2.3 Energy maximization in flow on a flat bottom

As mentioned above, isovortical energy extraction of a flow on a flat plane tends toward a state of zero energy by a process of filamentation. For isovortical energy increase on a flat plane, one can expect to reach stationary stable rearrangements of the initial vorticity distribution. For example, on an infinite plane with an initial vorticity field of one sign, the corresponding stable distribution has a monotonic relation between vorticity and radius with the peak vorticity at the center (Kelvin 1887). With periodic boundary conditions one can anticipate a similar behavior for an initial condition in which there is a small region of intense vorticity of one sign. The total vorticity must add to zero so the evolution should be toward an axisymmetric state with monotonic decrease in vorticity from the center surrounded by a region of weak vorticity of opposite sign. In the simulation shown here the initial state was created by the perturbation of an initial monopolar vortex with conical distribution of vorticity. The perturbation was a distortion of the vortex by shifting all particles with positive y positions 2 grid points in the x direction while leaving the values on gridpoints with $y \leq 0$ unperturbed. This could be thought of as the effect of a strong static shear over a range of 1 gridpoint spacing. As the energy is isovortically increased by pseudo-advection, the tendency is the expected axisymmetrization.

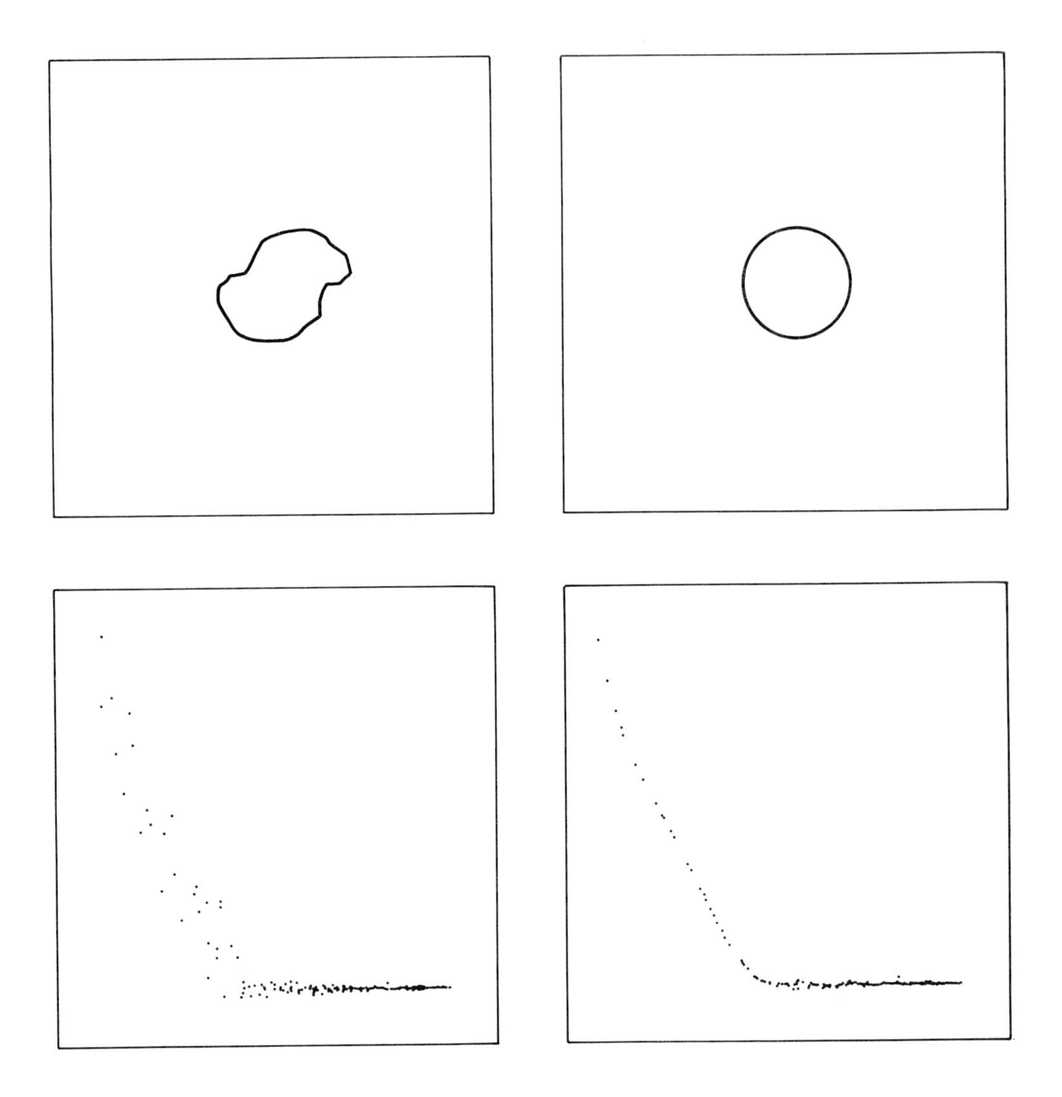

Figure. 4. Top left: Initial relative vorticity contour at value $.05\zeta_{max}$ of the sheared conical distribution. Top right: Shape of the same vorticity contour after pseudo-advection for approximately two circulation times with $\alpha = 0.4$. Bottom: Scatter plots of relative vorticity (vertical axis) vs. streamfunction (horizontal axis) at times corresponding to the upper panels.

3 CONCLUSION

Our experience with simulations of pseudo-advection has shown this to be a useful numerical approach for exploring the phase space of two-dimensional flow. We have demonstrated that the failure of the finite resolution representation of the pseudo-advection process to conserve accurately all vorticity invariants is not so severe as to prevent very precise convergence to a nearby nonlinearly stable state from a given initial condition. Investigations into the feasibility of using these methods for more complicated two-dimensional flows (e.g., β-plane flow) are now underway. For three-dimensional flow similar prescriptions can also be made yielding equations which decay or enhance energy while preserving the topological invariants (e.g., helicity) and research in this area is also proceeding.

We gratefully acknowledge valuable discussions with W.R. Young and T.G. Shepherd. This research has been supported by the National Science Foundation (grant OCE 86-00500) and the Office of Naval Research (grant N00014-85-C-0104). Additional support was provided by DARPA and ONR through University Research Initiatives N0014-86-K-0752 and N0014-86-K-0758, the European Center for Scientific and Engineering Computing (IBM), the Istituto Fisica della Atmosfera (CNR), and the University of Rome. The numerical simulations were performed at the San Diego Super Computer Center.

REFERENCES

ARNOL'D, V.I., Izv. Vyssh. Uchebn. Zaved. Matematika **54**, 3 (1966) (English transl.: Am. Math. Soc. Transl., **79**, 267 (1969)).

CARNEVALE, G.F., & FREDERIKSEN J.S., 1987: Nonlinear stability and statistical mechanics of flow over topography. *J. Fluid Mech.* **175,** 157-181.

CARNEVALE G.F., & VALLIS G.K., 1989: Pseudo-advective relaxation to stable states of inviscid two dimensional fluids (submitted to *J. Fluid Mech.*)

KELVIN, LORD, 1887: On the stability of steady and of periodic fluid motion. *Phil. Mag.* **23,** 459-464.

VALLIS G.K., CARNEVALE G.F., & YOUNG W.R. 1988: Extremal energy properties and construction of stable solutions of the Euler equations. *J. Fluid Mech.* (in press).

A Simple Hydrodynamic System with Oscillating Topology

A.M. OBOUKHOV, F.V. DOLZHANSKII & A.M. BATCHAEV

Institute of Atmospheric Physics, USSR Academy of Science, Pyzhevsky 3, 109017 Moscow Zh-17, USSR

At present, great interest is shown in hydrodynamic systems having a strange attractor behaviour. At the same time, systems behaviour that is regular even at substantial supercriticality are of equal interest. An example of this kind of system is the 'hydrodynamic clock' realized in a rectangular dish excited by constant hydromagnetic forcing. During the cycle the streamline topology changes, as shown by the results of a laboratory experiment; the changes can be described analytically by an eight-mode model.

The oscillatory changing of the two-dimensional flow topology associated with reconnection of streamlines has been observed in the Hele-Shaw convective cell (Liubimov, Putin & Chernotynskii 1977), the self-oscillation process in the supercritical domain of the external parameter space being as a rule stochastic, except for a narrow interval of Rayleigh numbers. This phenomenon can be simulated more simply in a system with regular oscillations of four quasi 2D vortices excited in a homogeneous (electrically conducting) fluid by hydromagnetic forcing. Strictly periodical oscillations observed in a wide range of supercritical exte-nal parameters make it possible to call such system the *hydrodynamic clock.* The scheme of the apparatus is given in figure 1.

A rectangular dish with dimensions $240 \times 120 \times 30\text{mm}^3$ containing a thin layer of weakly conducting fluid (a solution of copper sulphate) rests on a system of heteropolar permanent magnets, magnetic force lines piercing the fluid layer normally to its surface. The arrangement of the magnets is chosen so that when the electric current runs through the fluid parallel to the short side of the rectangle, a system of four vortices is formed under the influence of the Ampere force, $\mathbf{f}_A = [\mathbf{j}, \mathbf{B}]$, as shown in figures 1d and 2a. ($\mathbf{B}$ denotes the magnetic flux, $\mathbf{j}$ the electric currrent density.) It should be emphasized that the parameters of the arrangement considered are chosen so as to prevent the magnetic Reynolds number being too large (typical value of B is of the order of several hundred Gauss, of j – several tens of mA). Thus, the effect of the velocity field on the magnetic field is practically absent here (the so-called weak conductivity approximation is valid). Ampere's force imitates in this case the baroclinic force in the convective Hele-Shaw cell, which creates the four-vortex circulation pattern. When the current reaches a certain critical value, self-oscillations are weakly excited in the form of strictly periodical merging and splitting of the vortices situated in diagonally opposite corners of the rectangle. This is shown in figures 2b and 3 for different values of the Reynolds number (defined in terms of the force). Note that the hydrodynamic clock was first produced in an oval dish (Gak 1981). However, for the theoretical treatment of the experimental data it is convenient to use a rectangular dish (Pleshanova 1982). A detailed study of the oscillation characteristics for a wide range of the external parameters was performed by Batchaev (1989). It should be noted that the process takes place in a fluid layer of finite thickness, $h \ll L$ (L is the typical horizontal size of the dish), so that the flow is quasi-two-dimensional rather than strictly two-dimensional. Thus, the parameter space of the system can be represented by the Reynolds number $Re = \rho^{-1}\,(L/2\pi)^3\,(jB_0/\nu^2)$ (in terms of the force), the geometric parameter, $\epsilon = L_1/L_2$, and $\sigma = L^2/h^2$, σ defining the influence of the bottom friction. Here ρ is the fluid density, ν the viscosity, B_0 the typical value of the magnetic flux, and L_1 and L_2 the length and width of the dish. It should be noted that the bottom friction is of principal importance for the stability characteristics and physical analysis of quasi 2D flows (Oboukhov 1983, Dolzhansky 1987, 1988).

The form and dimensionless period of the oscillations at different values of Re and σ and fixed ϵ are given in figures 4 and 5. The principal conclusion is that with increasing supercriticality, the spectrum is substantially enriched with the harmonics of the main frequency (defined from linear stability theory of the primary flow). But even at quite high supercriticality ($Re/Re_* \sim 4$), transition to irregularity was not observed. With Re increasing, a self-synchronisation takes place, which results in the system functioning like a clock, but with a shorter oscillation period (figure 5). It is interesting to note that the rate of the period decrease with the increase of Re/Re_* reduces noticeably when σ increases (i.e. when the layer thickness decreases). This can be connected to the competitive influence of the Van-der Pol effect, which implies growth of oscillation period with increase of number of overtones observed when σ increases (figure 4).

The experiments carried out lead to another important conclusion, namely that with increasing σ, the Re-dependence of oscillation characteristics weakens. Beyond a value $\sigma \sim 800$, the system becomes effectively independent of Re. In this case, at fixed ϵ the system is governed by a single external parameter, which can be taken to be $R_\lambda = \sigma^{-2} Re = U_0 h^2/L\nu$ (U_0 is the typical primary flow velocity) (see Dolzhanskii 1987). This is a Reynolds number based on bottom friction (Rayleigh friction) $\lambda = \nu/h^2$.

Specific features of the process observed in the experiment can be described by the eight-mode model (Pleshanova 1982) constructed by Galerkin's method from the original hydrodynamic equations, which in this case can be written in terms of dimensionless streamfunction:

$$\frac{\partial \triangle \Psi}{\partial t} + \epsilon^{-1}[\triangle\Psi, \Psi] = \sigma^{-1}\triangle^2\Psi - \triangle\Psi - R_\lambda \sin 2\epsilon y \sin 2x, \tag{1}$$

with boundary conditions

$$\Psi|_\Gamma = 0, \ \Psi(x,y) = \Psi(x+2\pi, y) = \Psi(x, y+2\pi). \tag{2}$$

The periodicity on spatial variables means that we may neglect side-wall friction. $L_1/2\pi$ and $h^2/2\nu = (2\lambda)^{-1}$ were taken as length and time scales. The second term on the RHS of (1) corresponds to the Rayleigh friction $\sim \lambda \mathbf{u}$, and the last term approximates the field of the external force which simulates the four-vortex pattern. The solution is sought in the form

$$\Psi = \sum_{k,l=1}^{\infty} \varphi_{kl}(t) \sin kx \sin \epsilon l y, \tag{3}$$

and we truncate this expansion at $k, l = 3$. As a result we have the following dynamic system of order 8 (the mode φ_{33} is practically not excited):

$$\begin{aligned}
\dot{\varphi}_{11} = & -\frac{9(1-\epsilon^2)}{4(1+\epsilon^2)}\varphi_{12}\varphi_{21} - \frac{5-3\epsilon^2}{1+\epsilon^2}\varphi_{31}\varphi_{22} - \frac{3-5\epsilon^2}{1+\epsilon^2}\varphi_{13}\varphi_{22} + \\
& \frac{3+5\epsilon^2}{4(1+\epsilon^2)}\varphi_{12}\varphi_{23} - \frac{5+3\epsilon^2}{4(1+\epsilon^2)}\varphi_{21}\varphi_{12} + \frac{25(\epsilon^2-1)}{4(\epsilon^2+1)}\varphi_{23}\varphi_{32} - \\
& \frac{\sigma+\epsilon^2+1}{\sigma}\varphi_{11}, \\
\dot{\varphi}_{21} = & -\frac{9\epsilon^2}{4(4+\epsilon^2)}\varphi_{12}\varphi_{11} - \frac{5(8-\epsilon^2)}{4(4+\epsilon^2)}\varphi_{31}\varphi_{12} - \frac{25\epsilon^2}{4(4+\epsilon^2)}\varphi_{13}\varphi_{12} + \\
& \frac{8+3\epsilon^2}{4(4+\epsilon^2)}\varphi_{11}\varphi_{32} + \frac{7(5\epsilon^2-8)}{4(4+\epsilon^2)}\varphi_{13}\varphi_{32} - \frac{\sigma+4+\epsilon^2}{\sigma}\varphi_{21}, \\
\dot{\varphi}_{12} = & \frac{9}{4(1+4\epsilon^2)}\varphi_{21}\varphi_{11} - \frac{5(3-8\epsilon^2)}{4(1+4\epsilon^2)}\varphi_{13}\varphi_{21} + \frac{25}{4(1+4\epsilon^2)}\varphi_{21}\varphi_{31} -
\end{aligned}$$

$$\frac{3+8\epsilon^2}{4\left(1+4\epsilon^2\right)}\varphi_{11}\varphi_{23}+\frac{7\left(8\epsilon^2-5\right)}{4\left(1+4\epsilon^2\right)}\varphi_{31}\varphi_{23}-\frac{\sigma+1+4\epsilon^2}{\sigma}\varphi_{12},$$

$$\dot{\varphi}_{22}=\frac{2}{1+\epsilon^2}\varphi_{11}\varphi_{13}-\frac{2\epsilon^2}{1+\epsilon^2}\varphi_{11}\varphi_{13}+\frac{4\left(\epsilon^2-1\right)}{1+\epsilon^2}\varphi_{31}\varphi_{13}-$$

$$\frac{\sigma+4+4\epsilon^2}{\sigma}\varphi_{22}+R_\lambda/4\left(1+\epsilon^2\right),$$

$$\dot{\varphi}_{31}=-\frac{3\left(1+\epsilon^2\right)}{9+\epsilon^2}\varphi_{11}\varphi_{22}+\frac{15\left(1-\epsilon^2\right)}{4\left(9+\epsilon^2\right)}\varphi_{12}\varphi_{21}-\frac{2\left(5\epsilon^2-3\right)}{9+\epsilon^2}\varphi_{13}\varphi_{22}-$$

$$\frac{7\left(3+5\epsilon^2\right)}{4\left(9+\epsilon^2\right)}\varphi_{12}\varphi_{23}-\frac{\sigma+9+\epsilon^2}{\sigma}\varphi_{31},$$

$$\dot{\varphi}_{13}=\frac{3\left(1+\epsilon^2\right)}{1+9\epsilon^2}\varphi_{11}\varphi_{22}+\frac{15\left(1-\epsilon^2\right)}{4\left(1+9\epsilon^2\right)}\varphi_{12}\varphi_{21}-\frac{2\left(3\epsilon^2-5\right)}{1+9\epsilon^2}\varphi_{31}\varphi_{22}+$$

$$\frac{7\left(5+3\epsilon^2\right)}{4\left(1+9\epsilon^2\right)}\varphi_{21}\varphi_{32}-\frac{\sigma+1+9\epsilon^2}{\sigma}\varphi_{13},$$

$$\dot{\varphi}_{32}=-\frac{3}{4\left(9+4\epsilon^2\right)}\varphi_{11}\varphi_{21}-\frac{7\left(8\epsilon^2-3\right)}{4\left(9+4\epsilon^2\right)}\varphi_{13}\varphi_{21}-\frac{5\left(3+8\epsilon^2\right)}{4\left(9+4\epsilon^2\right)}\varphi_{11}\varphi_{23}-$$

$$\frac{\sigma+9+4\epsilon^2}{\sigma}\varphi_{32},$$

$$\dot{\varphi}_{23}=\frac{3\epsilon^2}{4\left(4+9\epsilon^2\right)}\varphi_{11}\varphi_{12}-\frac{7\left(3\epsilon^2-8\right)}{4\left(4+9\epsilon^2\right)}\varphi_{31}\varphi_{12}+\frac{5\left(8+3\epsilon^2\right)}{4\left(4+9\epsilon^2\right)}\varphi_{11}\varphi_{32}-$$

$$\frac{\sigma+4+9\epsilon^2}{\sigma}\varphi_{23}, \tag{4}$$

The structure of the system (4) can be easier understood in the variables $\mathbf{x} \equiv \left(x^1,x^2,x^3,x^4\right) \equiv \left(\varphi_{11},\varphi_{31},\varphi_{13},\varphi_{22}\right)$ and $\mathbf{y} \equiv \left(y^1,y^2,y^3,y^4\right) \equiv \left(\varphi_{21},\varphi_{12},\varphi_{32},\varphi_{23}\right)$, in terms of which it is written as

$$\dot{x}^i = \mathbf{x}^\tau A^i\mathbf{x}+\mathbf{y}^\tau B^i\mathbf{y}-\alpha^i_j x^j+f^i,$$
$$\dot{y}^i = \mathbf{x}^\tau C^i\mathbf{y}-\beta^i_j y^j, \quad i,\ \ j=1,2,3,4,$$

where $A^i = \|A^i_{jk}\|$, $B^i = \|B^i_{jk}\|$, $C^i = \|C^i_{jk}\|$, are the coefficients of the i-th line of the system (4). The superfix τ indicates the transpose. The repeating indices mean summation. So the system (4) divides into two interacting subsystems X and Y, the former acting upon the latter parametrically, and the latter upon the former as an external force. Note that the system (4) is invariant with respect to the substitutions $\epsilon \to \epsilon^{-1}$, $\sigma \to \epsilon^2\sigma$, $\varphi_{ij} \to -\varphi_{ji}$, and has two quadratic positive inviscid invariants

$$E=\sum_{i,j}\varphi^2_{ij}\left(i^2+\epsilon^2 j^2\right) \quad , \quad I=\sum_{i,j}\varphi^2_{ij}\left(i^2+\epsilon^2 j^2\right)^2,$$

which correspond to the principles of conservation of energy and enstrophy in the absence of dissipation. The general principles of construction of dynamic systems which have fundamental hydrodynamic invariants are discussed in the book by Gledzer, Dolzhanskii & Oboukhov (1981).

Analysis shows (Pleshanova 1982) that the loss of stability of the primary flow in a certain range of external parameters does not result in excitation of oscillations. Instead, a secondary stationary regime is established which has the form of merging of a pair of vortices situated along one of the rectangle's diagonals. The system can stay in this state for an indefinitely long time. In particular, this happens when $\epsilon_0^{-1} \leq \epsilon \leq \epsilon_0 = 1.58$, i.e. for rectangles close to a square. This was experimentally confirmed in modelling flows in a square dish, in which both the primary and secondary regimes were observed (figure 6), while oscillations were not observed. A similar result was obtained experimentally by Verron & Sommeria (1987). In the range $\epsilon_0 < \epsilon < \epsilon_{max} = 4.4$ the system behaviour depends significantly on the bottom friction. Thus, at moderate σ (the upper limit depends on the value of ϵ) the primary regime with growing R_λ loses stability and oscillations occur, with characteristics in agreement with the experiments. With increasing σ, theory and experiment do not agree. The large σ range (when effects of bottom friction are dominant) requires further research.

REFERENCES

Batchaev A.M. *Isv. Acad. Nauk SSSR, Atmos. and Oceanic Phys.*, 1989, vol. 25, No. 4, p. 434-439

Batchaev A.M. *Prikl. Mat. Fiz.* (in press).

Dolzhanskii F.V. *Izv. Akad. Nauk SSSR, Atmos. and Oceanic Phys.*, 1987, vol. 23, No. 4, p. 348-356.

Dolzhanskii F.V. In: Nonlinear and turbulent processes, vol. 2 (Proceedings of the III International Workshop). Kiev, Naukova Dumka, 1988, p. 25-28.

Gak M.Z. *Izv. Akad. Nauk SSSR, Atmos. and Oceanic Phys.*, 1981, vol. 17, No. 2, p. 201-205.

Gledzer E.B., Dolzhanskii & F.V., Oboukhov A.M. The systems of hydrodynamic type and their applications. 1981, Moscow, Nauka Publ.

Liubimov D.V., Putin G.f. & Chernotynskii V.I. *Dokl. Akad. Nauk SSSR*, 1977, vol. 235, p. 554-556.

Pleshanova L.A. *Isv. Acad. Nauk SSSR, Atmos. and Oceanic Phys.*, 1982, vol. 18, No. 4, p. 339-348.

Verron J. & Sommeria J. *Phys. Fluids*, 1987, vol. 30, No. 3, p. 732-739.

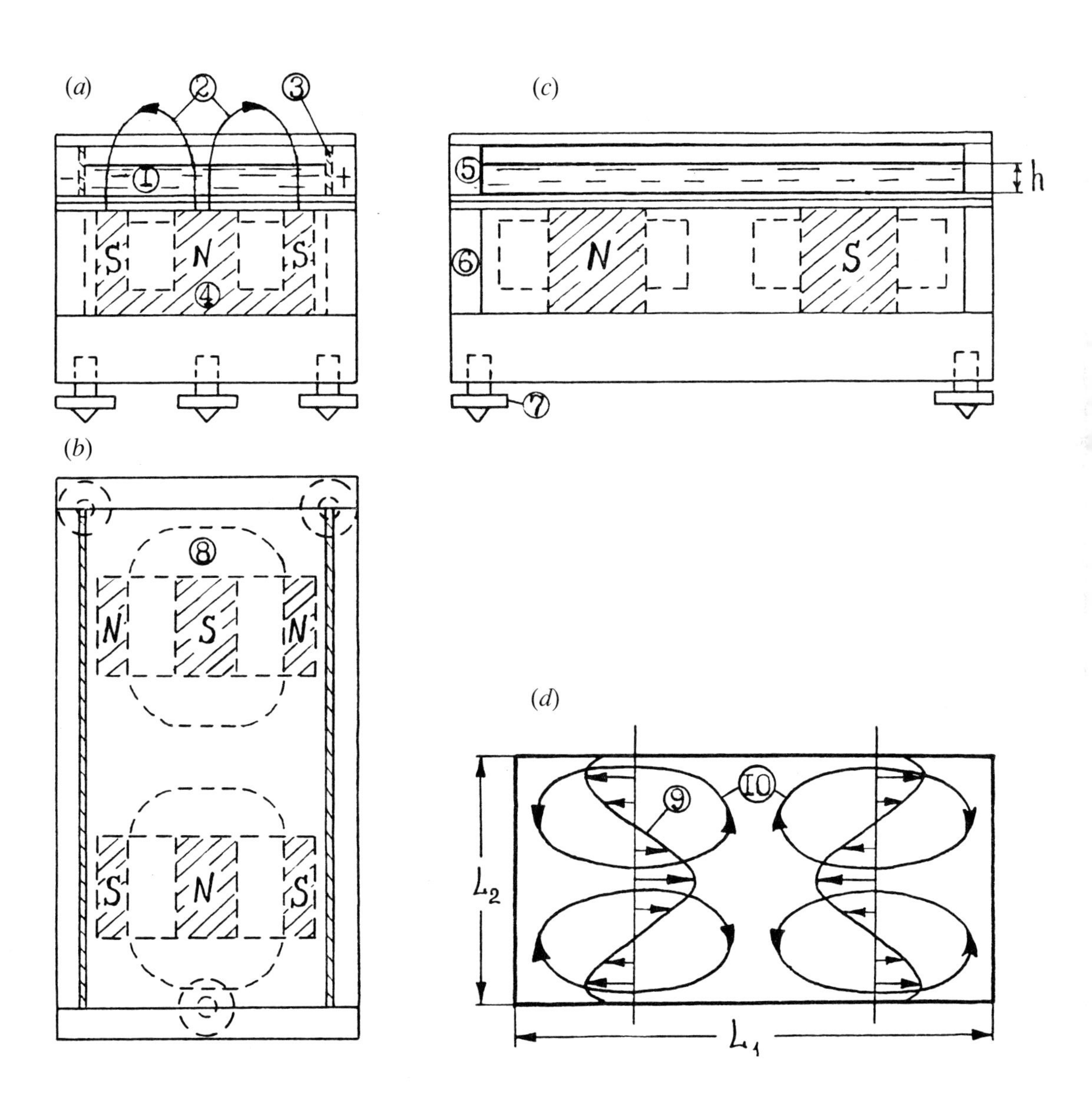

FIGURE 1. The experimental apparatus (a,b,c).
Profile of MHD force and streamlines (d).

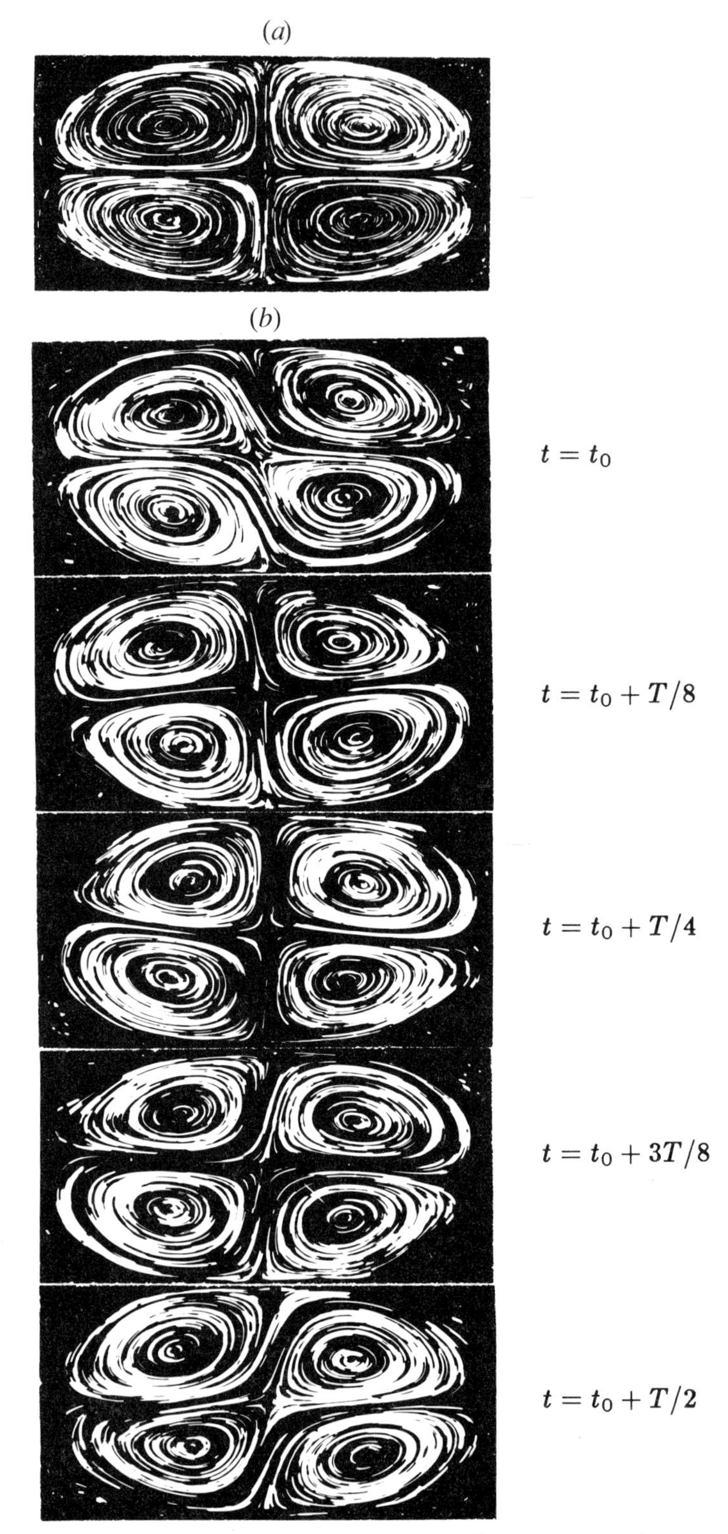

FIGURE 2. *a*) Primary flow at $Re = 46000 (Re^* = 65000)$
b) Self-oscillation at $Re/Re^* = 1.2$ and $\sigma = 843$.

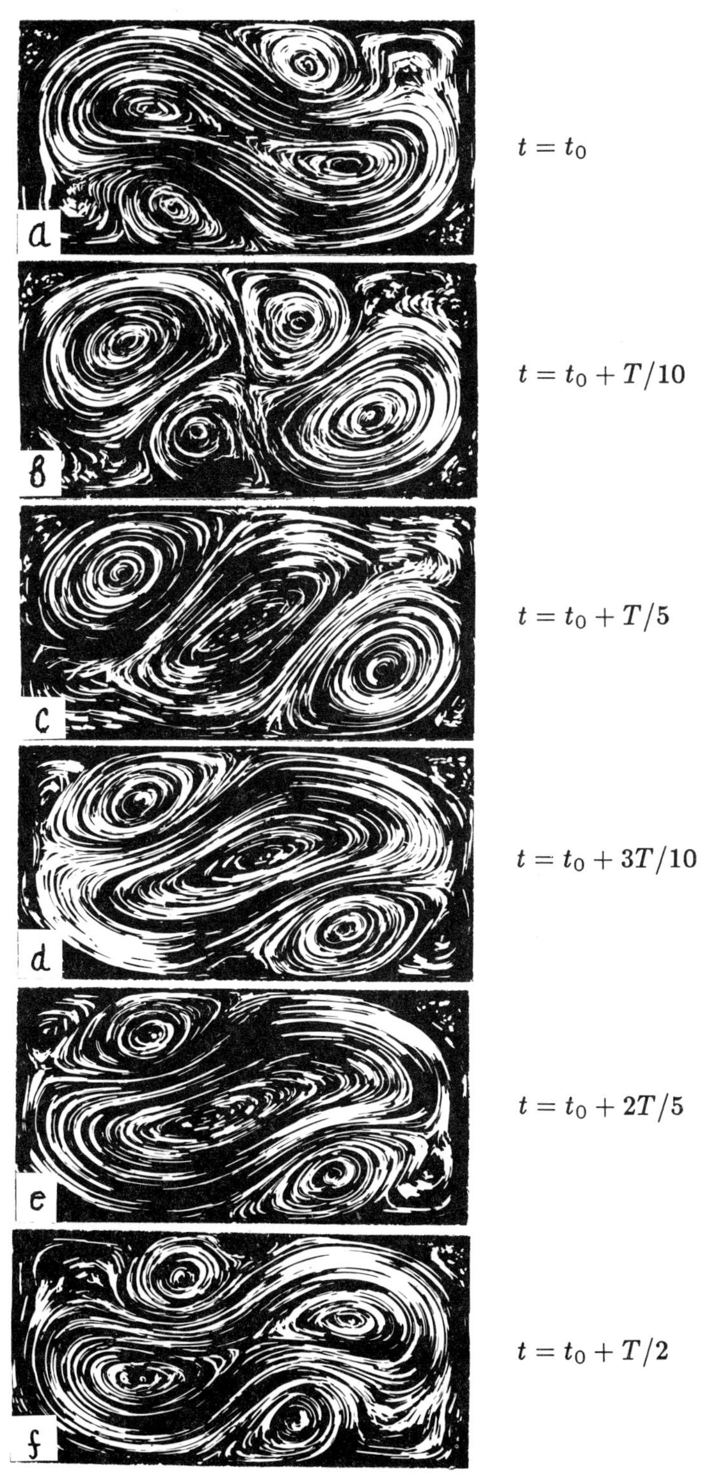

FIGURE 3. Self-oscillations at $Re/Re^* = 3.1$ and $\sigma = 843$.

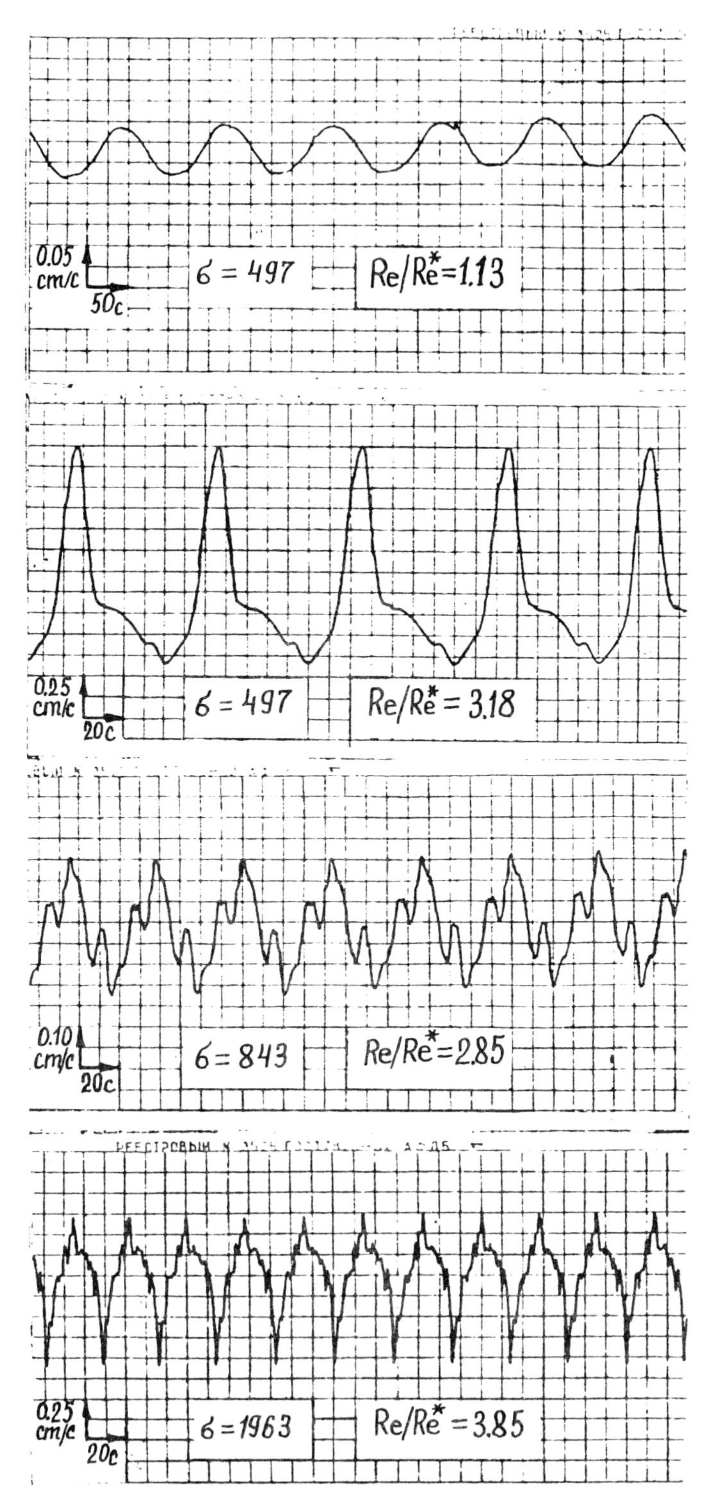

FIGURE 4. Velocity record of the self-oscillation for different Re/Re^* and σ.

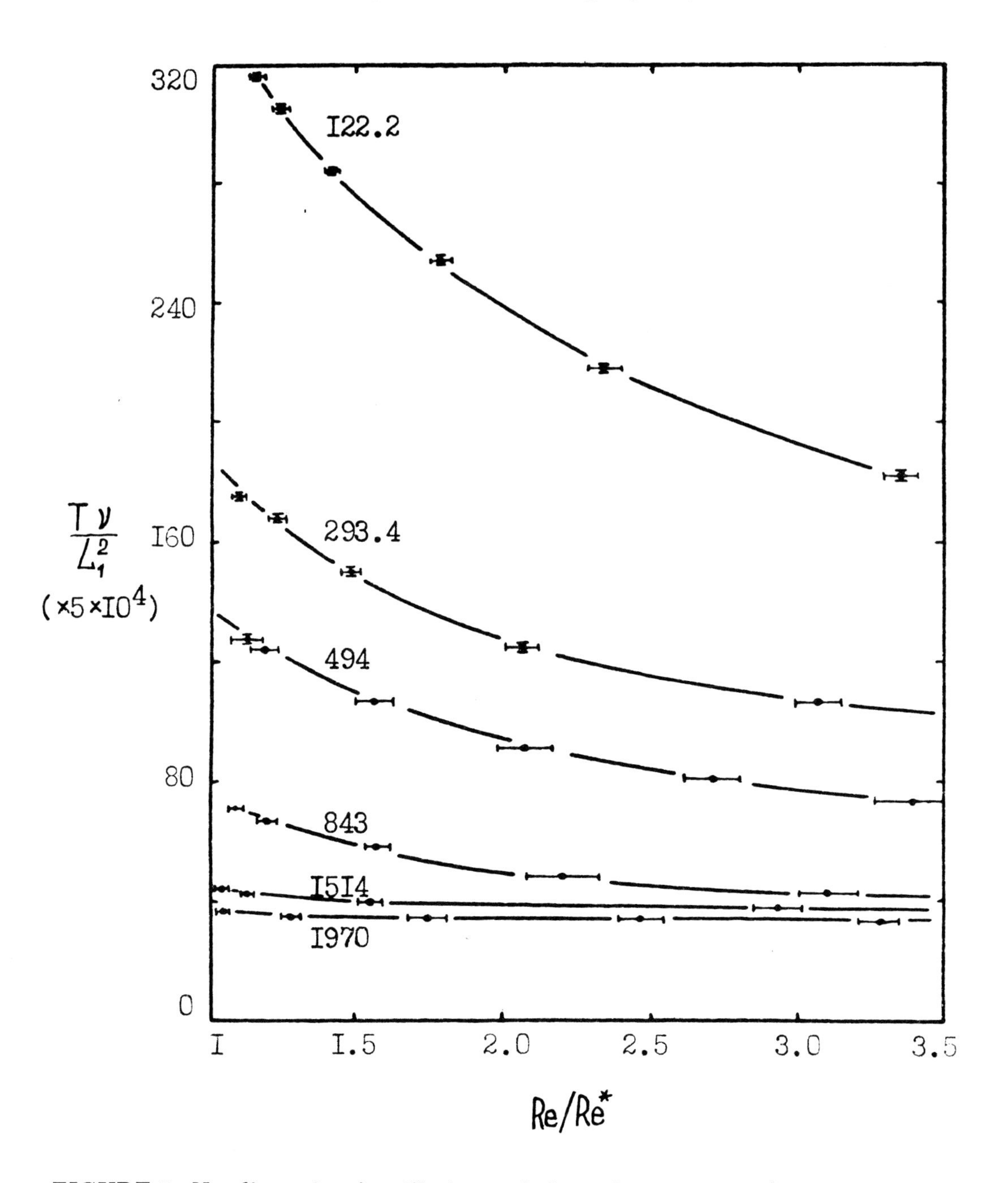

FIGURE 5. Nondimensional oscillation period as a function of Re/Re^* for different σ.

FIGURE 6. Primary (a) and supercritical stationary regimes in a square cuvette ($\epsilon = 1$).

Small-Scale Structures of Two-Dimensional MHD Turbulence

H. Politano[1], P.L. Sulem[1,2], A. Pouquet[1,3]

[1] - CNRS, Observatoire de Nice, BP 139, 06003 Nice-Cedex, France

[2] - School of Mathematical Sciences, Tel Aviv University, Israel

[3] - High Altitude Observatory, NCAR, Boulder, Col. 80307, USA

1. INTRODUCTION

In MHD flows, the dissipation range does not reduce to a simple sink which eliminates the energy cascading along the inertial range. Indeed, current sheets formed near neutral X-points may be destabilized by resistive instabilities, which leads to magnetic reconnection and changes in the flow topology. This effect is known to play an important role in cosmic plasmas as well as in laboratory tokamacs.

Most analyses of resistive instabilities are restricted to weak perturbations of strongly sheared magnetic fields (Furth, Killeen and Rosenbluth 1963, Matthaeus and Lamkin 1986). These basic structures idealize the small-scale configuration obtained by free evolution of a high Reynolds number two-dimensional MHD flow with large scale initial conditions. A complete description of the evolution of all the scales of the flow, including inertial dynamics and resistive instabilities, is in fact accessible by direct numerical integrations of the MHD equations (Politano, Pouquet and Sulem 1989). It requires a number of collocation points sufficiently large to resolve precisely the inner dynamics of vorticity and current sheets.

2. TWO-DIMENSIONAL TURBULENCE

The two-dimensional MHD equations

$$\begin{aligned} &\partial_t \mathbf{v} + \mathbf{v}.\nabla\mathbf{v} = -\nabla p + \nu\nabla^2\mathbf{v} + \mathbf{b}.\nabla\mathbf{b}, \\ &\partial_t \mathbf{b} + \mathbf{v}.\nabla\mathbf{b} = \mathbf{b}.\nabla\mathbf{v} + \eta\nabla^2\mathbf{b}, \\ &\nabla.\mathbf{v} = 0, \ \nabla.\mathbf{b} = 0, \end{aligned} \qquad (1)$$

with a unit Prandtl number ($\nu = \eta$) are integrated with a (pseudo-)spectral method in space and a second order finite difference scheme in time.

A prototype of two-dimensional MHD flows is the Orszag-Tang (OT) vortex (Orszag and Tang 1979) defined by the initial stream function $\psi_0(x,y) = 2(\cos x + \cos y)$ and magnetic potential $a_0(x,y) = 2\cos x + \cos 2y$. These initial conditions correspond to a kinetic vortex with a central stagnation point and two magnetic vortices with a central neutral X-point. The viscosity and magnetic diffusivity are $\nu = \eta = 10^{-3}$. The initial kinetic and magnetic Reynolds numbers are thus of order 10^4. A resolution of 1024^2 collocation points and a time step $dt = 2.5\ 10^{-4}$ were used.

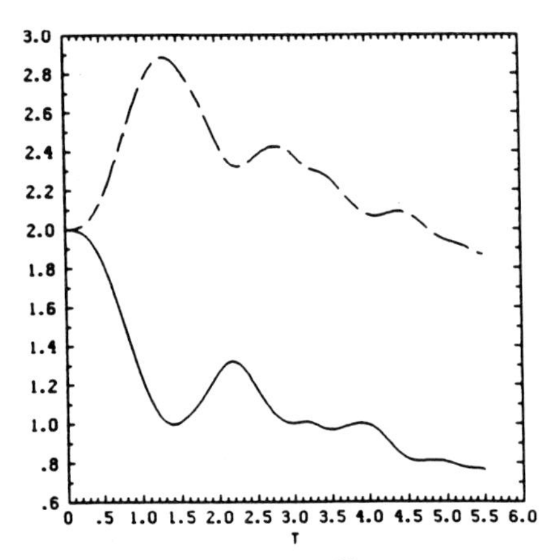

Fig.1 : Kinetic energy $E^V(t)$ (solid line) and magnetic energy $E^M(t)$ (dashed line) vs time for the OT vortex.

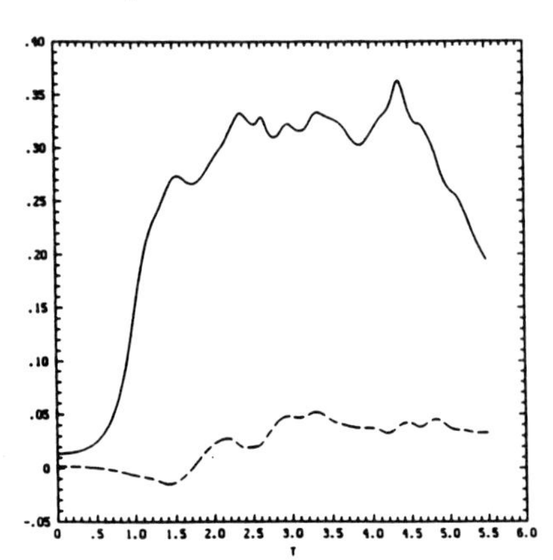

Fig.2 : Temporal evolution of energy dissipation $\epsilon^T(t)$ (solid line) and correlation dissipation $\epsilon^C(t)$ (dashed line) for the OT vortex.

We begin by showing the time evolution of a few global quantities characteristic of the flow. Fig.1 displays the exchanges of energy between the kinetic and magnetic forms, $E^V(t) = \frac{1}{2}\int |\mathbf{v}|^2 d\mathbf{x}$ and $E^M(t) = \frac{1}{2}\int |\mathbf{b}|^2 d\mathbf{x}$. Starting with the same value, the magnetic energy dominates the kinetic energy after a short transient. The total energy $E^T(t) = E^V(t) + E^M(t)$ is conserved by the non-linear interactions, but decays in time because of viscosity and magnetic diffusivity. This is also the case for the cross-correlation $E^C(t) = \int \mathbf{v}.\mathbf{b}\, d\mathbf{x}$. We see on Fig.2 that the energy dissipation $\epsilon^T = -\partial_t E^T$ is significantly stronger than the correlation dissipation $\epsilon^C = -\partial_t E^C$. As a consequence, the correlation coefficient $\rho(t) = E^C(t)/E^T(t)$, which measures the degree of correlation between velocity and magnetic fields, grows monotonically in time (Orszag et al 1979). It was 50% initially, it reaches 66.4% at $t = 5.5$. We also observe that the dissipations are quasi-stationary for $2.3 < t < 4.7$.

Fig.3a shows the energy flux $\Pi^T(k)$ versus the wavenumber k, at time $t = 3.85$. This flux is almost constant in an "inertial range" extending approximately on $5 < k < 25$. The kinetic and magnetic energy spectra E_k^V and E_k^M are presented in Fig.3b. They display an (exponential) dissipation range whose logarithmic decrement agrees with the dissipation wavenumber $k_d \approx 25$, defined from the energy flux. In the inertial range, power laws are obtained for the spectra $E_k^{\pm}$ of the Elsässer

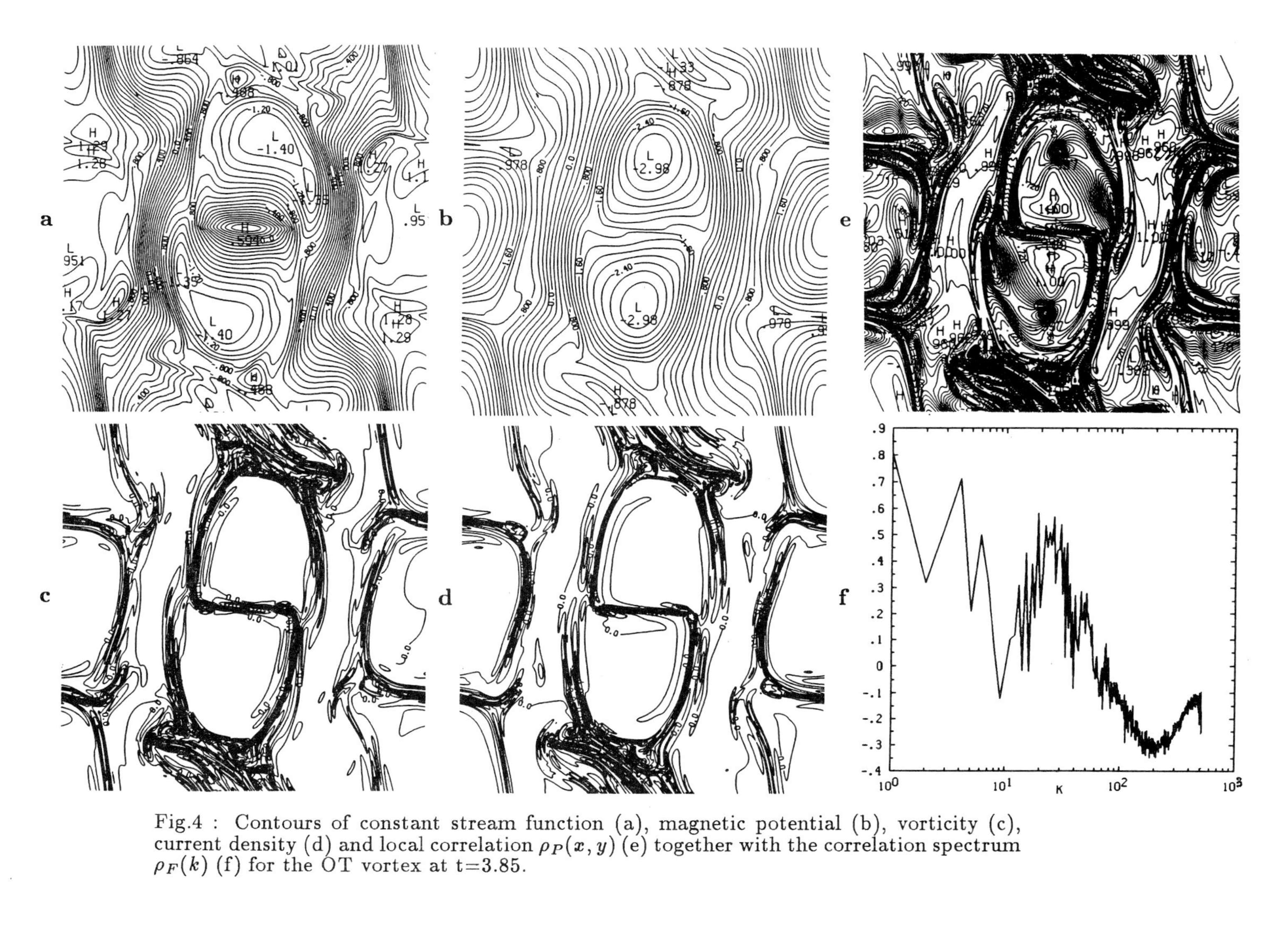

Fig.4 : Contours of constant stream function (a), magnetic potential (b), vorticity (c), current density (d) and local correlation $\rho_P(x,y)$ (e) together with the correlation spectrum $\rho_F(k)$ (f) for the OT vortex at t=3.85.

variables $z^{\pm} = v \pm b$. In agreement with phenomenology predictions (Iroshnikov 1963, Kraichnan 1965, Grappin, Pouquet and Léorat 1983), the sum of these power law exponents is found to be close to -3, their individual values depending on the correlation coefficient (Politano et al. 1989, Pouquet, Sulem and Meneguzzi 1988).

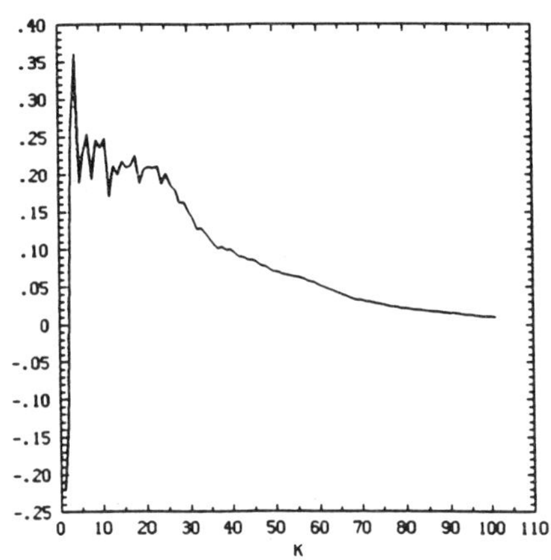

Fig.3a : Energy flux $\Pi^T(k)$ in the range $0 < k < 110$ at t=3.85, for the OT vortex.

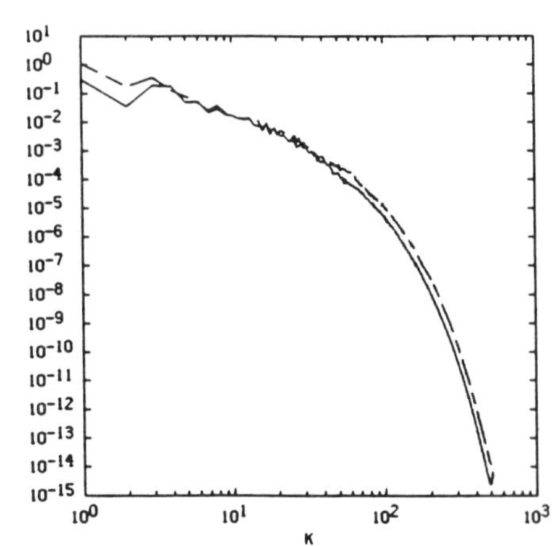

Fig.3b : Kinetic and magnetic energy spectra E_k^V (solid line) and E_k^M (dashed line), at t=3.85 for the OT vortex.

The tendency of the velocity and magnetic fields to become correlated is visible when comparing the initial conditions of the OT vortex, with Figs.4. The stream function and magnetic potential contours suggest that the large scales are strongly correlated (Figs.4a and 4b). The vorticity and current contours show that this is also the case for the small scales (Figs.4c and 4d). This observation is made conspicuous on the contours of the local correlation coefficient $\rho_P(\mathbf{x}) = 2\mathbf{v}(\mathbf{x}).\mathbf{b}(\mathbf{x})/(|\mathbf{v}(\mathbf{x})|^2 + |\mathbf{b}(\mathbf{x})|^2)$ (Fig.4e). They are positive at large scales and negative at small scales. Furthermore the strong gradients of $\rho_P(\mathbf{x})$ are localized near the vorticity and current sheets. The distribution of the correlation among the various scales is provided by the normalized correlation spectrum $\rho_F(k) = E_k^C/E_k^T$, defined as the ratio of the correlation and energy spectra (Fig.4f). It displays a positive part at small wavenumbers and a (somehow weaker) negative part at large wavenumbers, in agreement with the observations in physical space.

3. TEARING INSTABILITIES

The destabilization of the current and vorticity sheets is visualized in Figs.5 which display local snapshots of the magnetic potential, stream function, vorticity and current near the center of the central sheet of the OT vortex. We notice that the central magnetic current sheet is located on a kinetic vortex which has been generated during the early time evolution. This eddy distorts the magnetic sheet and induces a tearing instability. Reconnection takes place, leading to the formation of two and then four magnetic islands in a configuration consistent with the global flow symetries. Simultaneously, the kinetic vortex is elongated under the action

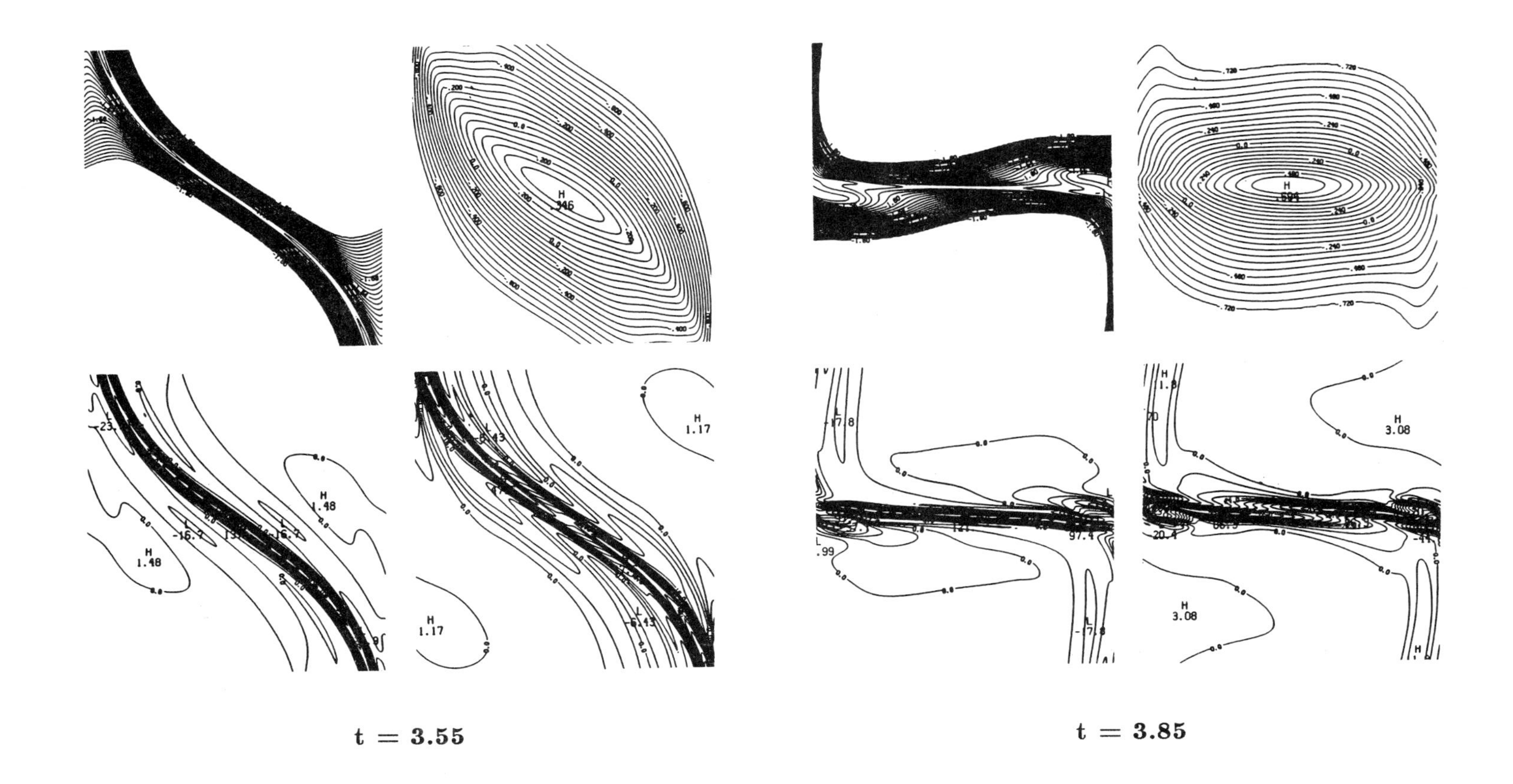

Fig.5 : Local snapshots of the magnetic potential (top left), stream function (top right), current (bottom left) and vorticity (bottom right) near the central sheet of the OT vortex.

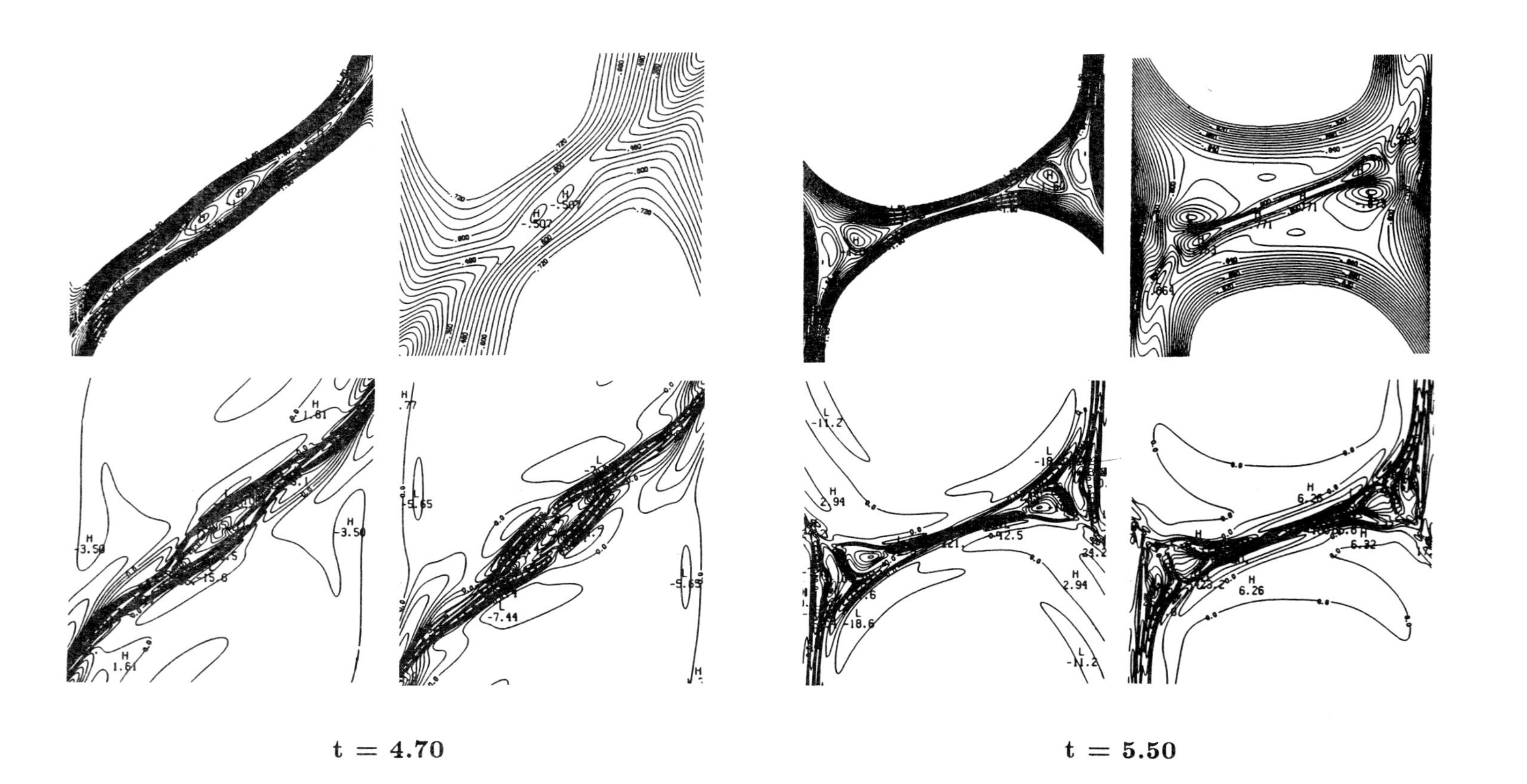

Fig.5 (continuated)

of the dynamics in the outer domain. Note also the development of some kind of cat's-eyes for the stream function, rather correlated to the magnetic islands.

As a result of the formation of the magnetic islands and kinetic cat's eyes, the contours of vorticity and current become significantly more complex. Ultimately, however, after the islands have been ejected to the edges of the sheet, the current tends to recover its simple geometry, with a quadrupole structure for the vorticity. This suggests that a recursive process could take place in very high Reynolds number flows.

In order to test the genericity of the observation made on the OT vortex, we also consider two random initial conditions. They both corresponding to initial kinetic and magnetic energy spectra proportional to $kexp[-(k/k_0)]$ with $k_0 = 1.3$ and kinetic and magnetic energies of about 1.5. They only differ by the correlation coefficient which is close to 10% for one run and to 80% for the other. The viscosity is again $\nu = 10^{-3}$. When developed, the flows display a multiplicity of small scale activity regions in the vicinity of current sheets where reconnection takes place. Figs.6 show for the 10% correlated run, a dynamics strongly evocative of an "impulsive bursty reconnection" (Priest 1985). The region represented has linear dimensions corresponding to a half of the 2π-period. The magnetic islands reach a size which at its maximum dimension is about a tenth of the period and still survive at the end of the run. Figs.7 correspond to the 80%-correlated flow. The linear size of the picture is a quarter of the period. We see the appearance and growth of a magnetic island. Similar structures develop at several places in both random flows. We also observe the formation of small magnetic bubbles with short life-times.

4. CONCLUDING REMARKS

We have integrated the two-dimensional MHD equations without any subgrid-scale modeling at a resolution of 1024^2, which enables us to simulate turbulent regimes displaying both an inertial range extending over about one decade of wavenumbers and a dissipation range where resistive instabilities take place. We were able to visualize tearing instabilities in current sheets that result from the dynamical evolution of initially large scale velocity and magnetic fields. We observe the formation of magnetic islands whose size can grow to about one tenth of the energy containing scales and survive a very long time. Our simulations contrast with previous investigations where a sheared magnetic field is prescribed in a generally flowless initial plasma. They also differ from numerical computations using hyper-viscosities and magnetic diffusivities (Biskamp and Welter 1989). At a given resolution, these models reduce the spectral extension of the dissipation range, thus making the inertial range longer. Their effects on the inner dynamics of the current sheets and on the tearing instabilities are however uncontroled.

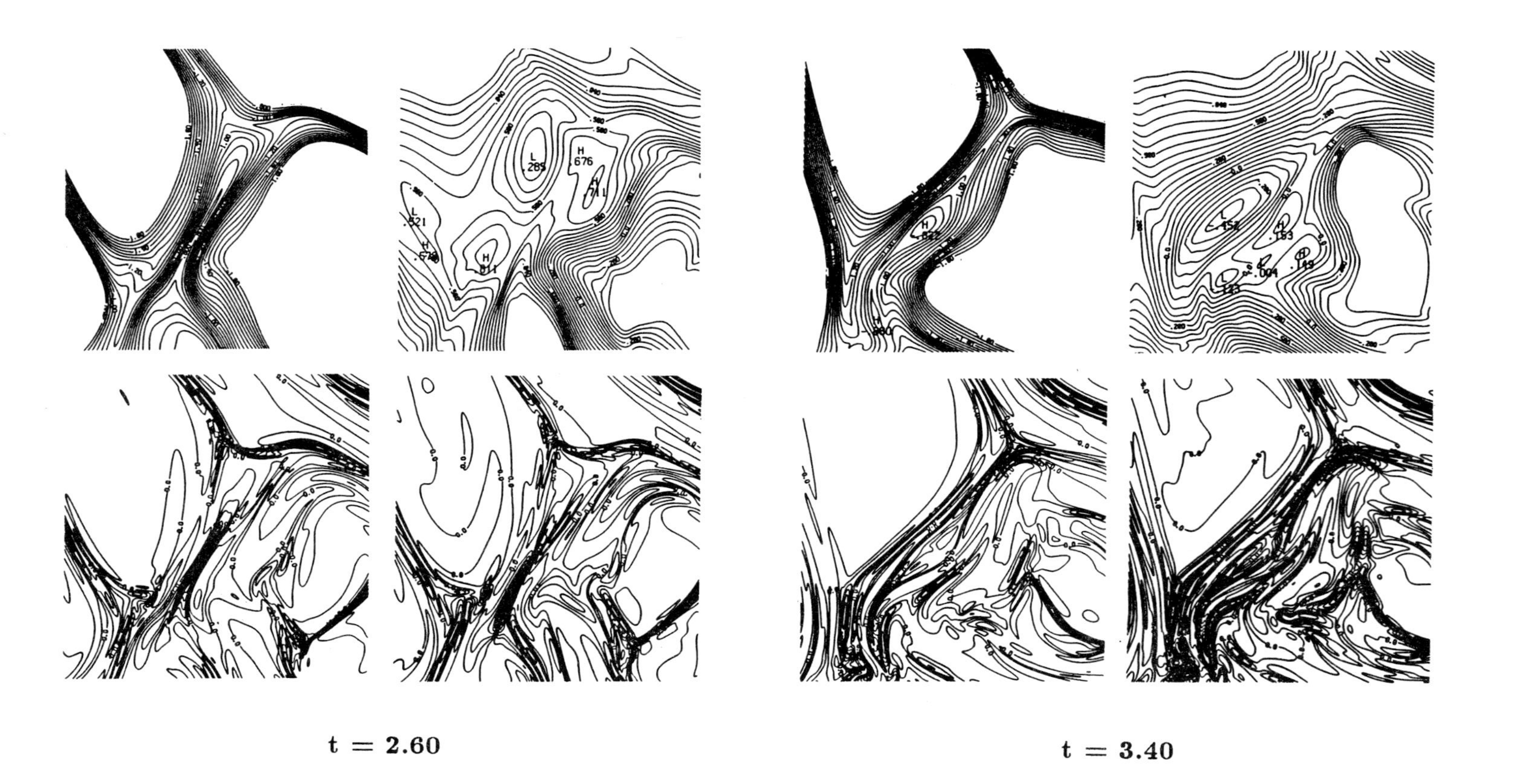

Fig.6 : Local snapshots of the magnetic potential (top left), stream function (top right), current (bottom left) and vorticity (bottom right) during reconnection event, for the 10% initially correlated run.

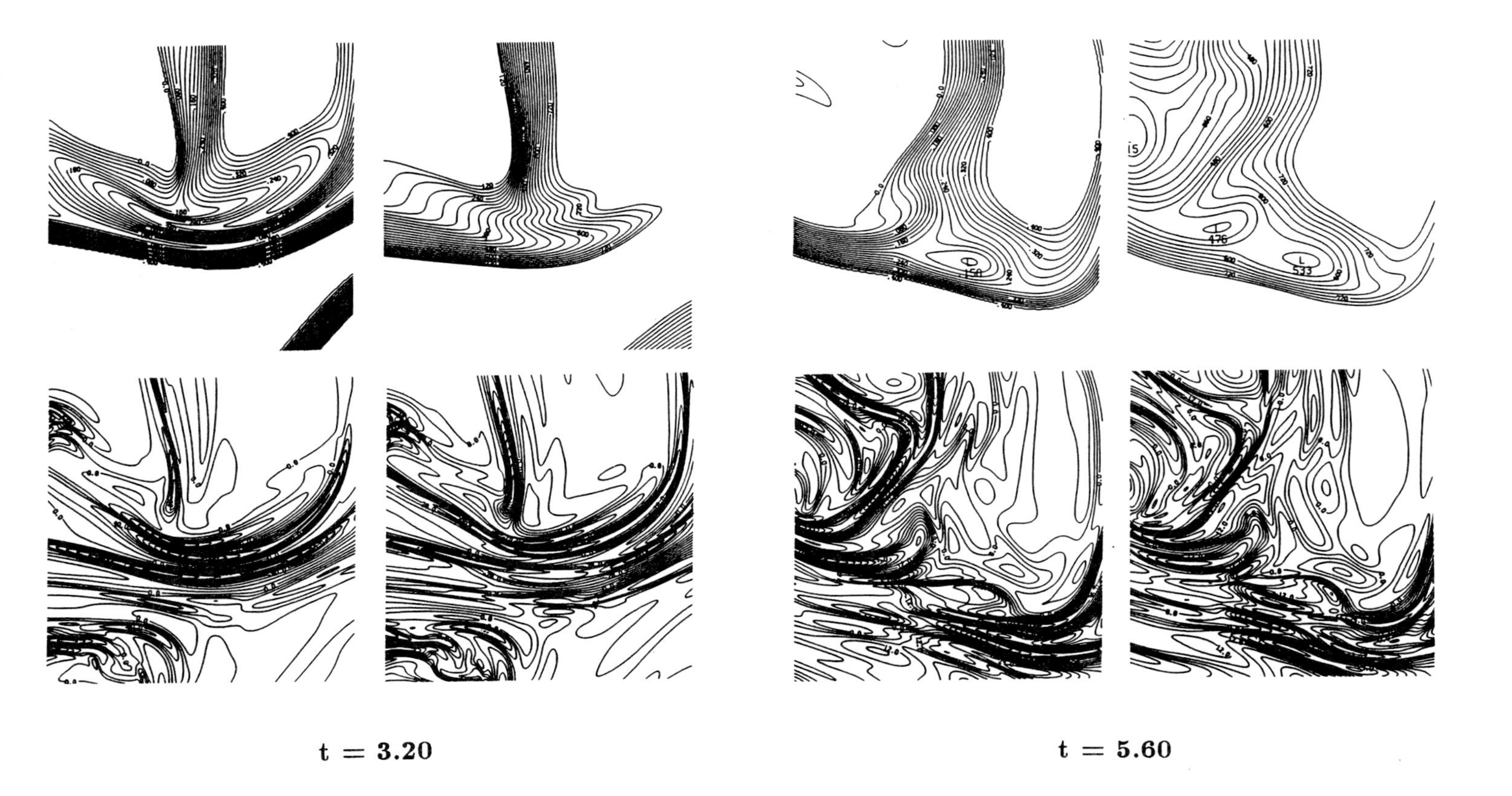

Fig.7 : Same as figure 6, for the 80% initially correlated run.

ACKNOWLEDGMENTS

The computations reported in this paper were performed on the Cray–2 of the Centre de Calcul Vectoriel pour la Recherche (Palaiseau, France), and on the Cray–XMP of the National Center for Atmospheric Research (Boulder, Colorado), which is thanked for its hospitality. The National Center for Atmosphere Research is supported by the National Science Foundation.

REFERENCES

H.P. Furth, J. Killeen and M.N. Rosenbluth 1963, *Phys. Fluids*, **6**, 459.

W.H. Matthaeus and S. Lamkin 1986, *Phys. Fluids*, **29**, 2513.

H. Politano, A. Pouquet, P.L. Sulem 1989, *Inertial ranges and resistive instabilities in two-dimensional MHD turbulence*, *Phys. Fluids B*, (to appear).

S.A. Orszag and C.M. Tang 1979, *J. Fluid Mech.*, **90**, 129.

M. Dobrowolny, A. Mangeney and P. Veltri 1980, *Phys. Rev. Lett.***45**, 144.

P.S. Iroshnikov 1963, *Astron. J. SSSR*, **40**, 742; *Soviet Astron.*, **7**, 566.

R.H. Kraichnan 1965, *Phys. Fluids*, **8**, 1385.

R. Grappin, A. Pouquet and J. Léorat 1983, *Astron. Astrophys.*, **126**, 51.

A. Pouquet, P.L. Sulem and M. Meneguzzi 1988, *Phys. Fluids* **31**, 2635.

E.R. Priest 1985, *Report Prog. Phys.*, **48**, 955.

D. Biskamp and H. Welter 1989, *Dynamics of Decaying Two-Dimensional Magnetohydrodynamics Turbulence*, preprint Max Planck Institute, Garching.

The Finite-Time Formation of Singularities on the Boundaries of Patches of Constant Vorticity

Thomas F. Buttke

Program in Applied and Computational Mathematics
Princeton University
Princeton, New Jersey 08544

Abstract

Euler's equation is solved numerically in two dimensions for the case where the flow can be described by patches of constant vorticity. For the case where the vorticity is described initially by two circular patches, it is found that when the minimum distance between the two patches is initially less than the radius of the patches a singularity forms in finite time on the boundary of the patches. The singularity is a jump discontinuity in the tangent vector of the boundary curve.

Introduction

In two dimensions it is well known that weak solutions of Euler's equation exist for all times [1]. This does not preclude the formation of singularities in the boundaries of patches of constant vorticity, since the theory only requires that the velocity remains continuous; this requirement is unaffected by the formation of singularities on the boundary. Since the equation of motion for the tangent vector to the boundary in two dimensions is similar to the vorticity evolution equation in three dimensions it has been conjectured that singularities form in finite time along the boundary of patches of constant vorticity in two dimensions [2]. If singularities are found in two dimensions it indicates that singularities may also form in the vorticity in three dimensions.

In the first part of the paper we investigate the curvature of a patch as it evolves in time. By studying the Fourier spectrum of the curve, we find that the curve remains smooth for all time. In the second part of the paper we investigate the evolution of two circular patches of constant vorticity as a function of their initial separation. We find that for initial separations less than 1.0 a jump discontinuity forms in finite time in the tangent vector to the boundary curve; for larger initial separations it is not known if a singularity forms.

The calculations in this paper were made possible by the development of a new numerical method for solving the equation of motion for patches of constant vorticity when they are governed by Euler's equation [5]. The method is a vortex blob type method with the size of the blobs chosen adaptively over a large range of sizes. The adaptive choice of the blobs is done in such a way that the number of blobs approximating a patch of vorticity is inversely proportional to the smallest blob size; whereas in conventional vortex blob methods the number of blobs approximating a patch of vorticity is inversely proportional to the smallest blob size squared. This adaptive method of choosing blobs allows us to calculate the evolution of the patches of vorticity to a much higher accuracy than was previously possible. The accuracy of the method is unaffected by the formation of long thin filaments and by the formation of singularities on the boundary curve of the patch.

The Formation of High Curvature Regions

In this section we investigate the evolution of three circular patches as shown in Fig. 1. The central patch has vorticity $\omega = 2.0$ and the upper and lower patches have $\omega = -1.0$. As can be seen in Fig. 1, a region of high curvature forms on the patches with negative vorticity. We look at the rate at which the high curvature region forms and conclude that the curvature remains finite for finite times.

A closed curve C in two dimensions can be considered to be a complex function which is periodic in the arclength s by identifying the complex variable $z(s) = x(s) + iy(s)$ with the position vector of the curve $\mathbf{r} = x(s)\hat{x} + y(s)\hat{y}$, where $\hat{x}$ and $\hat{y}$ are unit vectors in the x and y directions respectively. Expressing $z(s)$ in terms of its Fourier transform we have that

$$z(s) = \sum_{n=-\infty}^{\infty} a_n e^{i2\pi ns/L}$$

where

$$a_n = \frac{1}{L}\int_0^L z(s)\, e^{-i2\pi ns/L}\, ds \;,$$

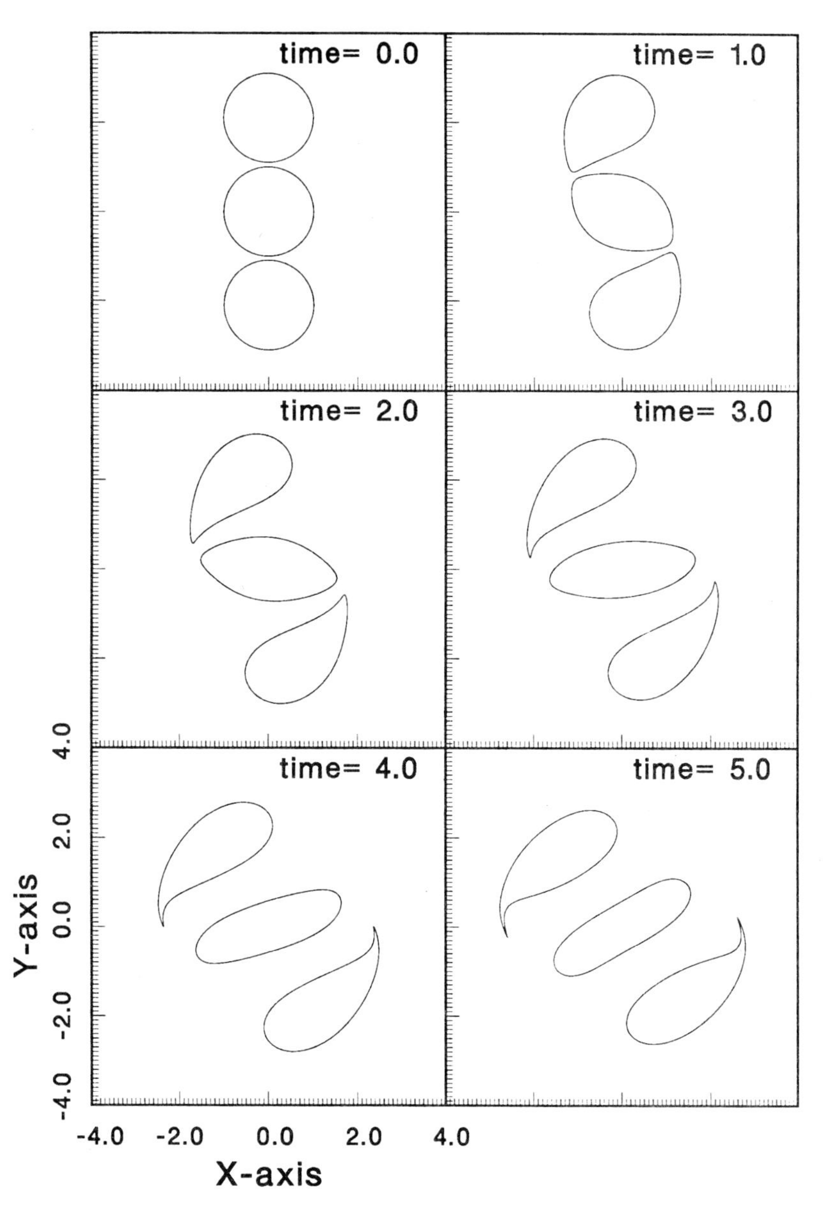

Figure 1; The evolution of three circular patches of constant vorticity. The middle patch has vorticity $\omega = 2.0$ and the top and bottom patches have vorticity $\omega = -1.0$. The radius of the patches is 1.0. The distance between the centers of the two closer patches is 1.1.

and L is the perimeter of the closed curve.

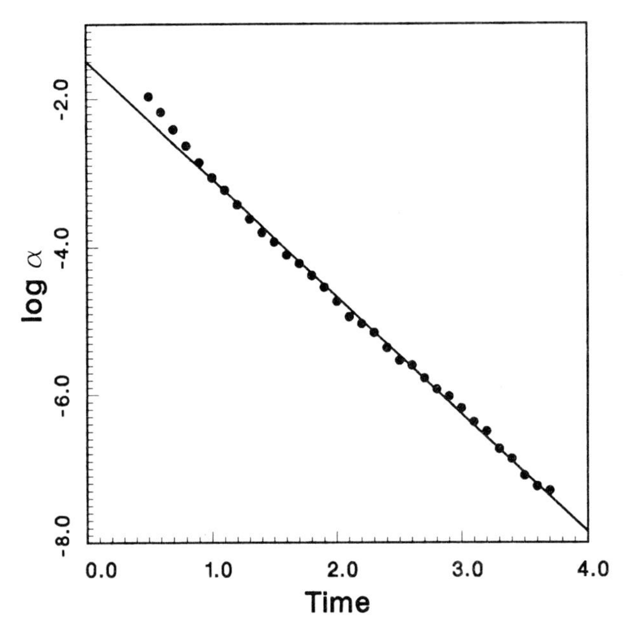

Figure 2; The logarithm of α as a function of time. The solid dots represent the calculated values of α at various times for the top and bottom curves shown in Fig. 1. The line is a linear least-squares fit of the data.

Curves which are continuously differentiable k times have spectra whose magnitude behaves asymptotically as $1/n^{k+2}$. A class of curves which are infinitely differentiable have spectra whose magnitude behaves asymptotically as $e^{-\alpha|n|}/|n|^{\beta}$. We calculate the fourier coefficients of the curves shown in Fig. 1. We assume that the spectra behave asymptotically as

$$|a_n| \approx \gamma \frac{e^{-\alpha|n|}}{|n|^{\beta}}, \tag{1}$$

where a_n is the n-th fourier coefficient. The constants α, β, and γ are different for positve and negative n. We look at the evolution of α, β and γ as a function of time. We conclude that $\alpha > 0$ for all times and thus that the curves remain infinitely differentiable for all times.

We calculate the exact fourier coefficients of the polygons which approximate the exact curves. The contribution to the fourier coefficient made by a line segment between the points $\mathbf{r}_j$ and $\mathbf{r}_{j+1}$ is

$$\frac{1}{L} \int_{s_j}^{s_{j+1}} (z_j + s\,(z_{j+1} - z_j))e^{-i2\pi n s/L}\, ds$$

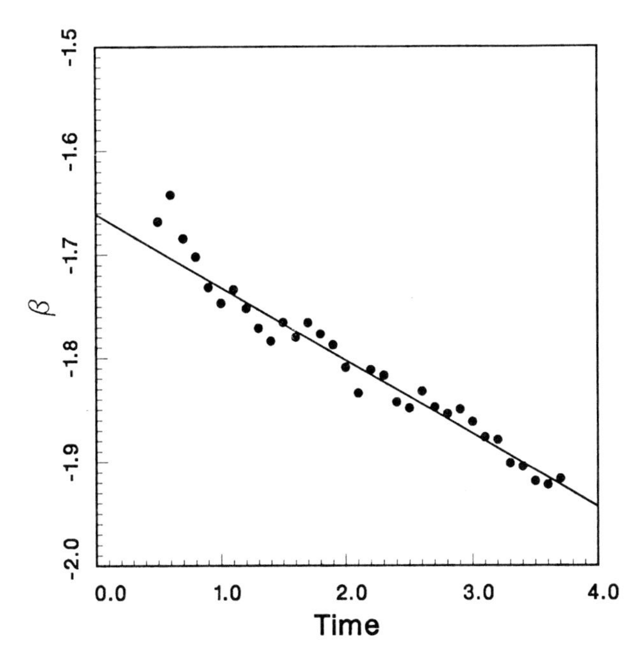

Figure 3. β as a function of time. The solid dots represent the calculated values of β at various times for the top and bottom curves shown in Fig. 1. The line is drawn only for comparison.

$$= \frac{\Delta s_{j+1/2}}{L} e^{-i2\pi n s_{j+1/2}/L} \left[\bar{z}_{j+1/2} \frac{\sin(\chi)}{\chi} + i \Delta z_{j+1/2} \frac{\chi\cos(\chi) - \sin(\chi)}{\chi^2} \right]$$

where $\bar{z}_{j+1/2} = (z_j + z_{j+1})/2$, $\Delta z_{j+1/2} = (z_{j+1} - z_j)/2$, $\Delta s_{j+1/2} = |z_{j+1} - z_j|$, $\chi = \Delta s_{j+1/2} n\pi/L$, s_j is the arclength measured along the polygon at the point z_j, L is the total arclength of the polygon, and z_j is to be identified with $\mathbf{r}_j$ as mentioned previously.

In order to determine the functional form of the spectra we fit the spectral data to the functional form given in eq (1). We minimize the error E as defined by

$$E = \sum_{n=n_0}^{n_1} \left[\log|a_n| - \log\gamma + \alpha|n| + \beta\log|n| \right]^2 ,$$

where n_0 and n_1 determine which modes are included in the least-squares fit. n_0 was chosen between 5 and 10 and n_1 was chosen as the largest value of n where the value of $|a_n|$ was approximately equal to the error between the calculated and exact value of $|a_n|$. Initial guesses of n_0 and n_1 were determined from the plot of $|a_n|$ as a function of $|n|$ and then these initial guesses were varied slightly until E was minimized.

In Fig. 2 and 3 we show α and β, respectively, as a function of time as determined from the least-squares fits. We see that $\alpha(t)$, where t is time, appears to be of the form $\alpha(t) \approx C_0 e^{-C_1 t}$ for the largest times observed. This indicates that $\alpha \rightarrow 0$ only in the limit as $t \rightarrow \infty$ and thus the curves remain infinitely differentiable. This indicates that in other similar situations where the curvature at a point on the boundary of a patch is rapidly increasing, the curvature does not become infinite in finite time and a singularity does not form at these points.

Singularity Formation

We show the evolution of two circular patches for the case when the patches are separated by a distance d_0 of 0.5, where we measure distance in units of the radius of the initial patches. (See Fig. 4) The initial separation d_0 is defined as the minimum distance between the two patches. The vorticity ω=1.0 for the two patches. We have investigated the evolution of patches with initial separations ranging from 0.0 to 1.5 . For patches with initial separations greater than 1.5 the patches remain approximately circular and orbit each other as in the point vortex case [6].

We have observed the formation of singularities in patches with separations $d_0 = 0.25, 0.50, 0.75$. The singularity forms at approximately the same time in the three observed cases, with the singularity time increasing slightly as the separation increases. The singularity forms at time $t \approx 16$ when $d_0 = 0.25$; at $t \approx 17$ when $d_0 = 0.50$; and at $t \approx 19$ when $d_0 = 0.75$. It appears that singularities also form for separations $d_0 \geq 1.0$; however, we have not yet confirmed this.

The fact that the positive singularity forms in a region of negative curvature provides strong evidence that the singularity appears in finite time. The singularity appears instantaneously with a strength equal to zero at the singularity time t_c. The strength of the singularity then grows monotonically (See Fig. 5). $\Theta(s)$ is defined as the angle that the tangent vector to the boundary curve makes with the x-axis, where s is the arclength of the boundary curve. In two dimensions Θ is related to the curvature $\kappa(s)$ by $d\Theta/ds = \kappa(s)$. In Fig. 5 we see that the curvature remains negative at all times in the regions neighboring the singularity and therefore it is impossible that the singularity appears as the asymptotic limit of a region of large positive curvature as the curvature tends to infinity.

As seen in Fig. 5 the jump in $\Theta(s)$ appears to occur between a single mesh point. The reason that we were able to observe the singularity formation initially is that we use Lagrangian points to mark the boundary of the patch. Before the singularity forms, the points near the singularity flow towards the singularity point and

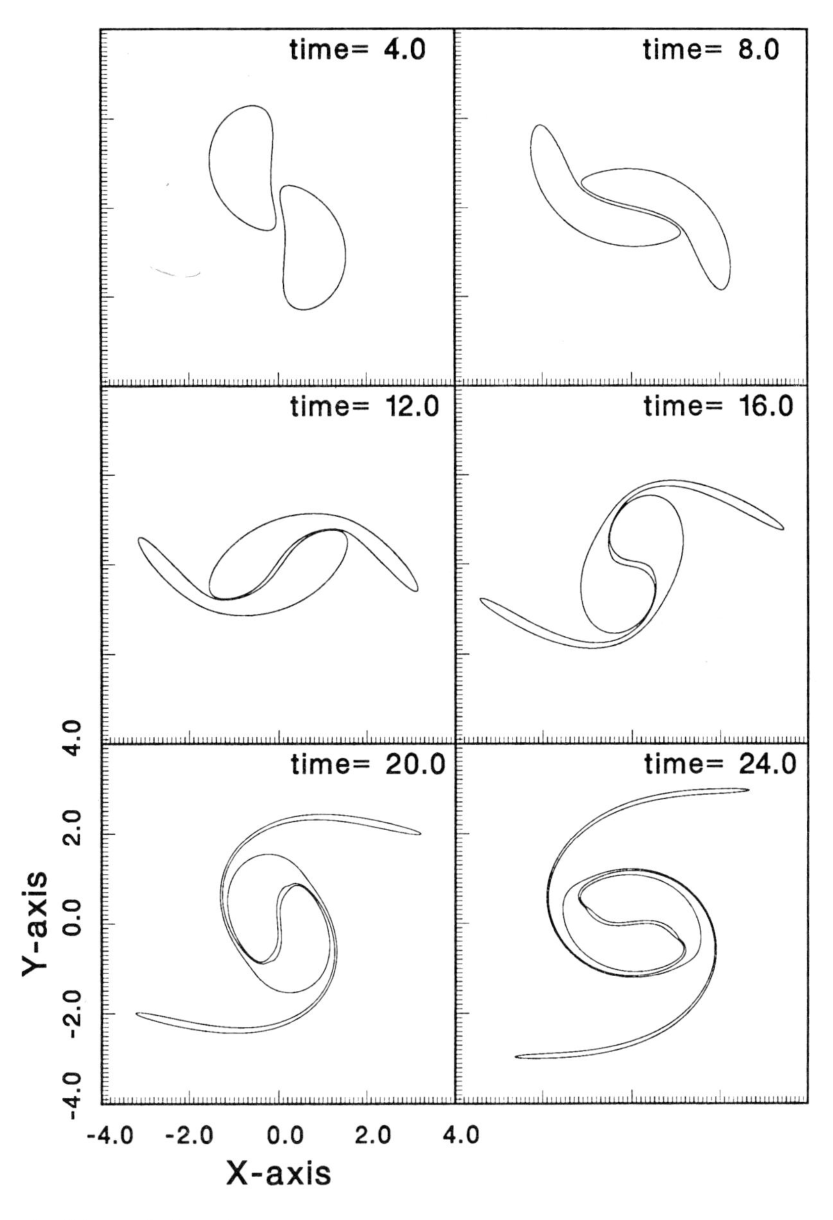

Figure 4a. The evolution of two patches of constant vorticity. The evolution of two patches of constant vorticity $\omega = 1.0$, separated initially by a distance of $d_0 = 0.5$, is shown at various times. More points are added as the contours stretch; at time t=24.0 there are 585 points approximating each contour.

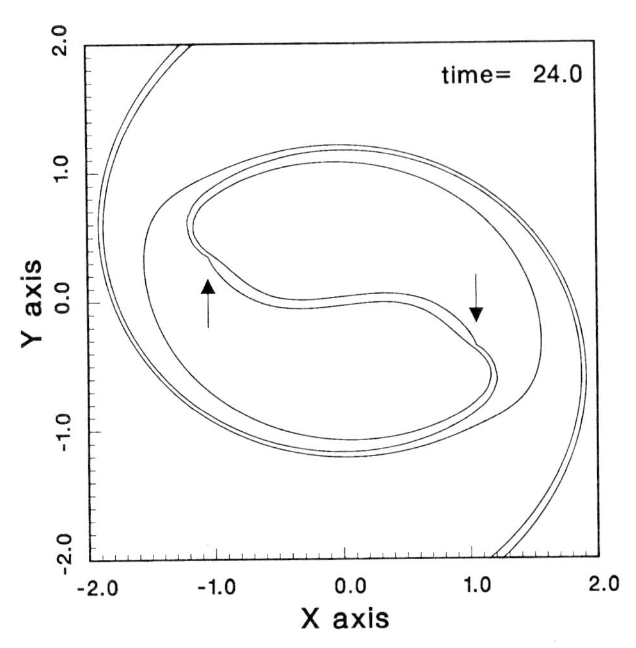

Figure 4b. Two patches of constant vorticity. An enlargement of the curves shown in Fig. 4a is shown at time t=24.0. The arrows in Fig. 4b point to the location of the singularity.

automatically provide a highly refined numerical mesh at the singularity point. Near the singularity the Lagrangian points are spaced a distance of approximately 1×10^{-3} whereas over the rest of the curve the spacing is approximately 4×10^{-2}. Thus there is an automatic refinement by a factor of 40 near the singularity which accounts for the noisy curve near the singularity in Fig. 5; we are sampling the numerical error in solving the equations of motion for the boundary points. This error is more apparent in Fig. 5 than in Fig. 4 because we are looking at the first derivative of the boundary curve. For the calculation shown in Fig. 5 there were a total of 1212 points marking the two boundary curves at t = 25.0; if the finest spacing were used uniformly throughout the curve it would require 57000 points.

In order to show that the singularity formation was not an artifact of the closely spaced mesh near the singularity point, we repeated the calculation with a Lagrangian mesh which was uniformly spaced at a time just prior to t_c. We did this by calculating the Fourier transform (viewing the boundary curve to be a periodic complex function in the arclength) of the curve at time t =16.0 and then keeping only the first 80 modes we reconstruct the curve with a Lagrangian mesh consisting of 400 points on each curve which are uniformly spaced in arclength. With this procedure we eliminate any effects of the nonuniform mesh and actually smooth the boundary curve. When we continue the calculation with the smoothed

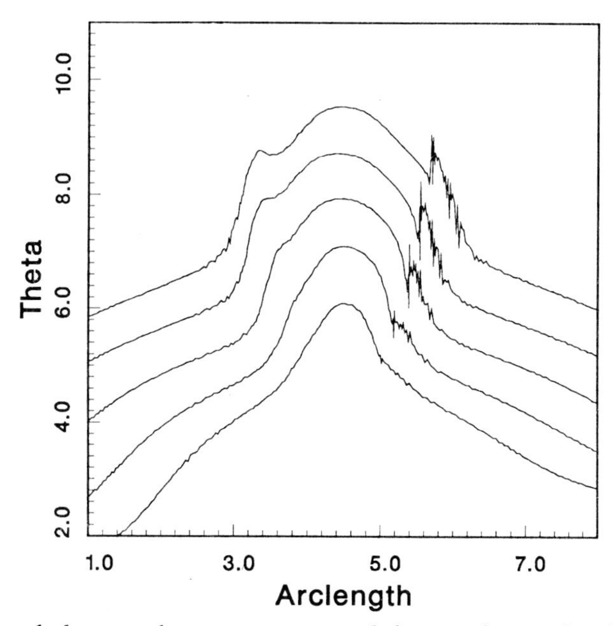

Figure 5. The angle between the tangent vector and the x-axis as a function of the arclength. The angle Θ, in radians, is plotted as a function of arclength for the times near the singularity time for values of the arclength near the singularity point. The data is taken from the calculation shown in Fig. 1. The curves correspond to the times t = 16.0, 18.0, 20.0, 22.0, and 24.0. Time increases from bottom to top.

mesh we find that the singularity still forms, as shown in Fig. 6.

We repeated the calculation after we had refined all of the numerical parameters which govern the accuracy of the numerical method and we find that the singularity appears at the same place on the curve and at the same critical time. The results shown in Fig. 5 were obtained with a mesh which initially has 160 points uniformly spaced along each circle and a vortex blob size equal to 1/512. The calculation was repeated using mesh spacings where 80 and 320 points were placed along each circle and using vortex blob sizes of 1/256, 1/512, 1/1024, and 1/2048; all calculations show the same singularity formation. An additional check on the validity of the numerics is the symmetry of the problem; the patches should be mirror images of each other. No such symmetry was imposed on the numerics, and yet in all cases the singularities appear at the analogous points on the two curves. We also repeated the calculation with the patches oriented along the line $y=x$ and found the same results.

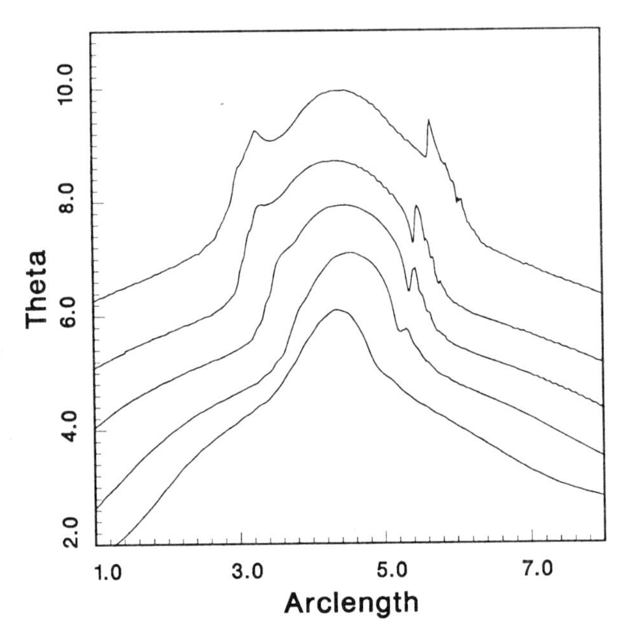

Figure 6. The angle between the tangent vector and the x−axis as a function of arclength. The calculation shown in Fig. 4 is repeated except at time $t = 16.0$ a smoothed and equal-spaced Lagrangian mesh is used. The times shown are the same as those appearing in Fig. 5. The singularity appears as in Fig. 5 except that the leading edge of the singularity is slightly rounded due to the reduced resolution caused by the equal-spaced mesh.

This work was performed while the author was a National Science Foundation Mathematical Sciences Postdoctoral Fellow at Princeton University.

Bibliography

[1] V. I. Yudovitch, Zh. Vych. Mat. 3, 1032-1066 (1963). (in Russian)

[2] A. Majda, Comm. Pure Appl. Math. 39, s187-s220 (1986).

[3] M.J. Shelley and G.R. Baker, *On the connection between thin vortex layers and vortex sheets. Numerical Study,* to be published J. Fluid. Mech.

[4] T.F. Buttke, *The finite time formation of singularities on the boundaries of patches of constant vorticity.* to be published.

[5] T.F. Buttke, *A fast adaptive vortex method for patches of constant vorticity in two dimensions*, to be published J. Comput. Physics.

[6] N.J. Zabusky, M.H. Hughes, and K.V. Roberts, J. Comp. Phys., 30, 96-106 (1979).

Steadily Rotating Quasigeostrophic Vortices in a Stratified Fluid

STEPHEN P. MEACHAM

Massachusetts Institute of Technology, Cambridge, MA 02139, USA

In a rotating reference frame (f-plane), we consider compact volumes of uniform potential vorticity embedded in an otherwise quiescent, uniformly stratified fluid of infinite, three-dimensional extent. We show that, under the quasigeostrophic approximation, there exist families of steadily rotating equilibria each with a different configuration for the bounding surface of the vortex. The shapes, rotation rates and stability properties of these solutions are explored.

1 INTRODUCTION

Observations indicate the presence of persistent, isolated, three-dimensional, monopolar vortex structures within the ocean at horizontal scales comparable to the meso-scale (McDowell and Rossby, 1978, Armi et al. 1989, Manley and Hunkins, 1985). Simulations of decaying, geostrophic turbulence (McWilliams, 1989) have shown that, under some circumstances, the decay is dominated by the emergence of, and interactions between, monopolar, coherent vortices.

We consider a simple paradigm for quasigeostrophic, quasi-isolated vortices in a rotating, stratified fluid which contains the following ingredients: a uniformly stratified fluid at rest at infinity with respect to a frame of reference that is rotating steadily at a rate $f/2$ about the vertical axis; a blob of fluid about the coordinate origin that has uniform potential vorticity of unity and whose bounding surface is such that the blob is a steadily rotating figure of equilibrium; and zero potential vorticity outside the blob.

The simplest such equilibrium is a sphere of uniform potential vorticity. This is stable in a Lyapunov sense. However, a self-contained vortex drifting through the ocean will encounter shear fields which will distort the shape of the vortex. Between encounters with localised currents, the vortex will be in some time-dependent state representing a fluctuation about an equilibrium state that need not be spherical. This motivates us to examine some of the possible rotating equilibria of a simple uniform potential vortex.

2 NORMAL MODES ON THE SPHERE

We consider an unbounded, incompressible fluid that is vertically stratified with a uniform buoyancy frequency, N. Our reference frame is assumed to be rotating with an angular speed $\frac{1}{2}f$ about the vertical axis of some inertial coordinate system. The undisturbed fluid is taken to be at rest in the rotating frame and our spatial coordinate system is embedded in the rotating frame. We will perturb this equilibrium by introducing a compact, simply connected region, D, of uniform, unit potential vorticity in the neighborhood of the origin. We further assume that the velocities and length scales of the relative motion induced by this distribution satisfy the quasigeostrophic approximation. The fluid velocities are quasi-horizontal and can be described by a streamfunction $\tilde{\psi}$. The potential vorticity, Q, is given by

$$Q = \left(\partial^2_{x^*}\tilde{\psi} + \partial^2_{y^*}\tilde{\psi} + \partial_{z^*}\left(\frac{f^2}{N^2}\partial_{z^*}\tilde{\psi}\right)\right)$$

Since f and N are independent of position, it is convenient to rescale the vertical coordinate z^*; we will therefore adopt coordinates defined by $(\tilde{x}, \tilde{y}, \tilde{z}) = (x^*, y^*, \frac{N}{f}z^*)$ in which the potential vorticity simply becomes $\tilde{\nabla}^2\tilde{\psi}$. This quantity is conserved following the motion and so we have the following equation of motion,

$$\partial_t\tilde{\nabla}^2\tilde{\psi} + \tilde{J}(\tilde{\psi}, \tilde{\nabla}^2\tilde{\psi}) = 0$$

We will usually be considering a rotating equilibrium with angular velocity Ω. In the co-rotating frame, having Cartesian coordinates x, y, z, the streamfunction becomes $\psi = \tilde{\psi} - \frac{1}{2}\Omega(x^2 + y^2)$.

We begin by briefly considering the case in which D is a unit sphere. Using radial coordinates (r, θ, ϕ), we find that normal modes have the structure

$$(\phi_i, \phi_e, \eta) = (r^n, r^{-(n+1)}, (\Omega - \frac{1}{3})^{-1})e^{im(\phi-\Omega t)}L_n^{(m)}(\theta)$$

where ϕ_i and ϕ_e are the perturbation streamfunction within and outside the sphere, while η is the normal displacement of the surface of the sphere, $L_n^{(m)}$ is an associated Legendre function and $0 \leq n, \quad 0 \leq m \leq n$. The rotation rate Ω of the modes with non-trivial azimuthal structure is given by $\Omega = \{1/3 - 1/(2n+1)\}$

Each of these normal modes represents the beginning of a separate family of rotating equilibria that bifurcates directly from the spherical solution. Each family is characterised by a different azimuthal and vertical structure. In particular, for each n there is a collection of n modes that have different azimuthal structures but a common rotation rate Ω, and one stationary mode ($m = 0$). The stationary, $m = 0$ modes are azimuthally symmetric.

All of the linear eigenmodes are neutral. An argument due to Dritschel (1988) shows that azimuthally symmetric blobs of uniform potential vorticity are stable in a finite amplitude sense. If $a^2(z)$ is the squared, horizontal radius of the undisturbed symmetric vortex and this is perturbed to $a^2(z) + \eta(z, \phi, t)$, one can show that the quantity

$$I = \frac{1}{8} \int dz \int d\phi \eta^2(z, \phi, t)$$

is invariant. For any initial perturbation $\eta(z, \phi, 0)$, this constrains the degree to which the subsequent evolution of the vortex can depart from the basic configuration, $a(z)$. In the sense of this norm, the sphere is therefore stable to disturbances of small but finite amplitude. The same is also true for the family of spheroids.

3 TWO AND THREE-MODE INTERACTIONS ON A SPHERE

We consider interactions between pairs and triplets of normal modes at small but finite amplitudes, taking streamfunctions of the form.

$$\tilde{\psi}_e = -\frac{1}{3r} + \epsilon \sum_{j=1}^{3} a_j(\epsilon t) e^{\imath m_j(\phi - c_j t)} L_{n_j}^{m_j}(\theta) r^{-(n_j+1)} + O(\epsilon^2)$$

$$\tilde{\psi}_i = \frac{1}{6}(r^2 - 3) + \epsilon \sum_{j=1}^{3} a_j(\epsilon t) e^{\imath m_j(\phi - c_j t)} L_{n_j}^{m_j}(\theta) r^{n_j} + O(\epsilon^2)$$

where $\epsilon \ll 1$ and $0 < m_j \leq n_j$. Setting $T = \epsilon t$, one can obtain evolution equations for the complex amplitude coefficients a_j.

2-mode interactions

Setting $a_3 = 0$, we find that pairs of modes can interact when $m_2 = 2m_1 = 2m$ and $n_2 = n_1 = n$ ($m \leq n/2$). Thus

$$\frac{d}{dT} a_1 = \imath N_1 a_1^* a_2 \qquad \frac{d}{dT} a_2 = \imath N_2 a_1^2$$

When $|a_1| \ll |a_2|$, these equations show that the combination of the sphere and a small amplitude L_n^{2m} mode is linearly unstable to the L_n^m mode. In the more general case, when a_1 and a_2 have comparable magnitudes, there is a periodic exchange of energy between the two modes. A particular example of this is the case

$n = 2$, $m = 1$. The L_n^{2m} mode is then an ellipsoidal perturbation while the L_n^m mode is a vertically antisymmetric mode with azimuthal wavenumber one.

3 mode interactions

When the waves satisfy certain resonance conditions, their amplitudes evolve according to

$$\frac{d}{dT}a_j = \imath M_j a_k^* a_l^*$$

where (j,k,l) is a cyclic permutation of (1,2,3). The constants M_j are real. When $|a_1|$, $|a_3| \ll |a_2|$, $|a_2|$ remains aproximately constant while

$$\frac{d^2}{dT^2}\{a_1, a_3\} = M_1 M_3 |a_2|^2 \{a_1, a_3\}$$

A (n_2, m_2) mode will be unstable to a pair of sidebands, (n_1, m_1) and (n_3, m_3), if $M_1 M_3 > 0$.

The resonance can be satisfied most simply when $n_1 = n_2 = n_3, = n^*$ say. The lowest n^* for which distinct 3-mode resonances of this type can occur is $n^* = 4$. Resonances can also occur when n_1, n_2 and n_3 are different. The lowest value of the maximum of $\{n_1, n_2, n_3\}$ for which such a resonance occurs is 7.

4 THE ELLIPSOIDAL FAMILY

The continuation of the ($n = 2$, $m = 2$) and ($n = 2$, $m = 0$) families of solutions away from their points of bifurcation from the sphere is relatively simple. Together they span the family of ellipsoidal figures of equilibrium. At small amplitude, the (2, 0) mode corresponds to a spheroidal perturbation while the (2, 2) mode is an ellipsoidal perturbation that conserves the area of the horizontal cross-sections. The rotation rate of a general ellipsoid can be derived analytically in terms of elliptic functions (Meacham, 1989).

The rotation rate $\Omega = \Omega(\alpha/\gamma, \beta/\gamma)$, a function only of the ratios of the lengths of the principal axes, is contoured in Figure 1 for $\gamma = 1$. This rate is largest when the ellipsoid is a spheroid, $\beta = \alpha$, and, for a given elliptic cross-section of aspect ratio λ, $(\lambda = \beta/\alpha)$, yields the appropriate 2-D Euler limit as $\alpha \to 0$.

5 NORMAL MODES ON AN ELLIPSOID

Following methods similar to those used by Love (1893) for the Kirchhoff ellipse, we can obtain dispersion relations for normal modes on a general ellipsoid. In the co-rotating frame, we expand the surface displacement Λl in surface harmonics, $L_n^{(j)}(q_2, q_3)$,

$$\Lambda l = \sum_{n=0}^{\infty} \sum_{j=1}^{2n+1} T_n^{(j)}(t) L_n^{(j)}(q_2, q_3)$$

One obtains a simple set of linear ODE's coupling together the evolution of the $2n+1$ harmonics of order n:

$$\frac{d}{dt} T_n^{(i)} = - \sum_{j=1}^{2n+1} T_n^{(j)} \Gamma_{nj}^{i} \tag{1}$$

In this equation, the $\{\Gamma_{nj}^{i} : n = 0, 1, 2, \dots \; ; \; 1 \leq i, j \leq 2n+1\}$ are just constants that depend on i, j, n and the geometry of the ellipsoid.

At each order n (which corresponds roughly to an index of the complexity of the vertical structure of the perturbation mode) we therefore obtain from (1) $2n+1$ eigenmodes with eigenfrequencies $\{0, \pm\omega_j : 1 \leq j \leq n\}$.

We note that for each n there is a mode that is stationary for all values of $(\alpha/\gamma,\ \beta/\gamma)$.

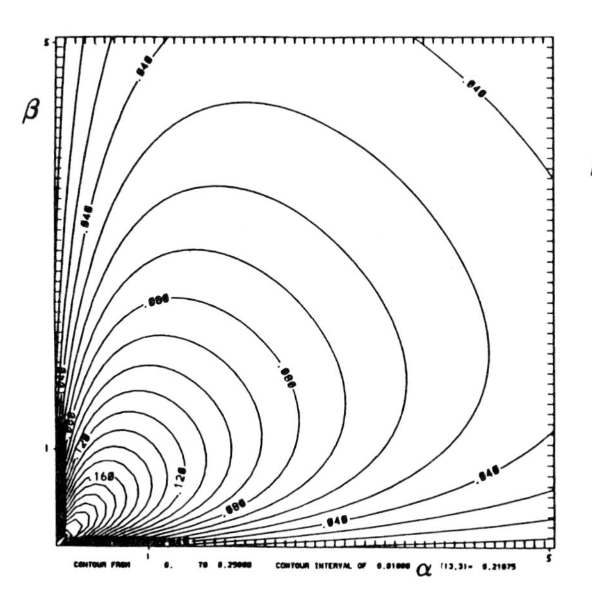

Fig. 1: Contours of the rotation rate, Ω, of an ellipsoid as a function of α and β. ($\gamma = 1.0$)

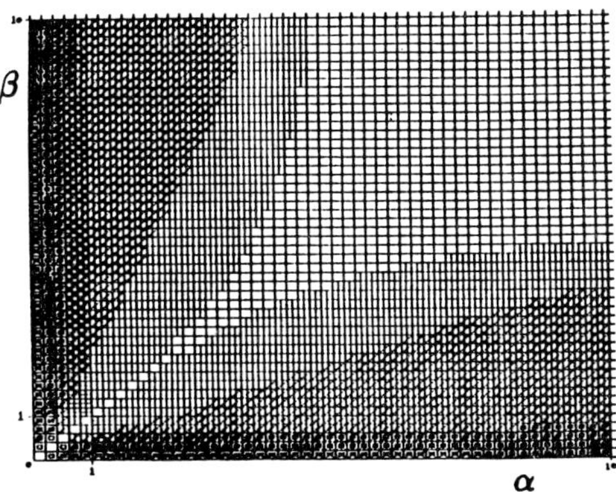

Fig. 2: The parts of the (α, β) plane that are unstable to $n = 2$ and $n = 3$ modes. The hatched areas are regions of instability for the several modes.

The single $n = 0$ mode corresponds to a self-similar stretching of the ellipsoid which preserves its rotation rate. The three $n = 1$ modes correspond to horizontal and vertical displacements of the ellipsoid. For $n \geq 2$ the modes become more interesting. All of the time-dependent $n = 2$ and $n = 3$ modes are unstable for sufficiently anisotropic ellipsoids. In Figure 2 we present a composite of the regions of the (α, β) plane in which the various $n = 2$ and $n = 3$ modes are unstable. It can be seen that only a zone near the line $\alpha = \beta$ is neutral to all of these low index modes. This zone is an increasingly narrow strip in the vicinity of $\alpha/\gamma = \beta/\gamma = 1$ and only broadens out for values of α/γ and β/γ greater than about 4.

The unstable domain of one of the $n = 2$ modes extends all the way to the spherical limit. If $\alpha = 1 - \epsilon$, $\gamma = 1$ and $\beta = 1 + \epsilon$, the growth rate of this mode $\sim O(\epsilon)$ as $\epsilon \to 0$. The earlier two mode analysis demonstrates why. The family of ellipsoids is a two parameter family (we parameterised it with the ratios α/γ and β/γ above) and close to its bifurcation point, i.e. when the ellipsoids are almost spherical, this family can be parameterised as a sum of two of the normal modes of the sphere, the (2, 2) mode and the spheroidal (2, 0) mode. The pure (2, 2) mode corresponds locally to the axis $\alpha/\gamma = 2 - \beta/\gamma$ whilst the (2, 0) mode corresponds to the axis $\alpha/\gamma = \beta/\gamma$. Since the former mode is shown by the analysis above to be unstable to the (2, 1) mode, ellipsoids near the former axis are unstable. Those that lie close to the spheroidal axis are not.

The small amplitude analysis suggests that a similar pattern of instability is likely to befall families of equilibria bifurcating from the sphere that have an even number of azimuthal lobes. Those families having an odd number of azimuthal lobes may be stable in the neighbourhood of the sphere.

6 SECONDARY BIFURCATIONS FROM AN ELLIPSOID

We noted that for each $n \geq 0$, a general ellipsoid possesses a mode that is stationary in the co-rotating frame, regardless of the values α/γ, β/γ. These modes represent secondary bifurcations for the ellipsoidal family of equilibria to steadily rotating equilibria with more complicated azimuthal and vertical structures. When the departure from an ellipsoidal state is small, the form of these structures and their stability could be studied asymptotically using the known analytical form of the stationary eigenmodes, however, to continue these solution branches will probably require a numerical approach.

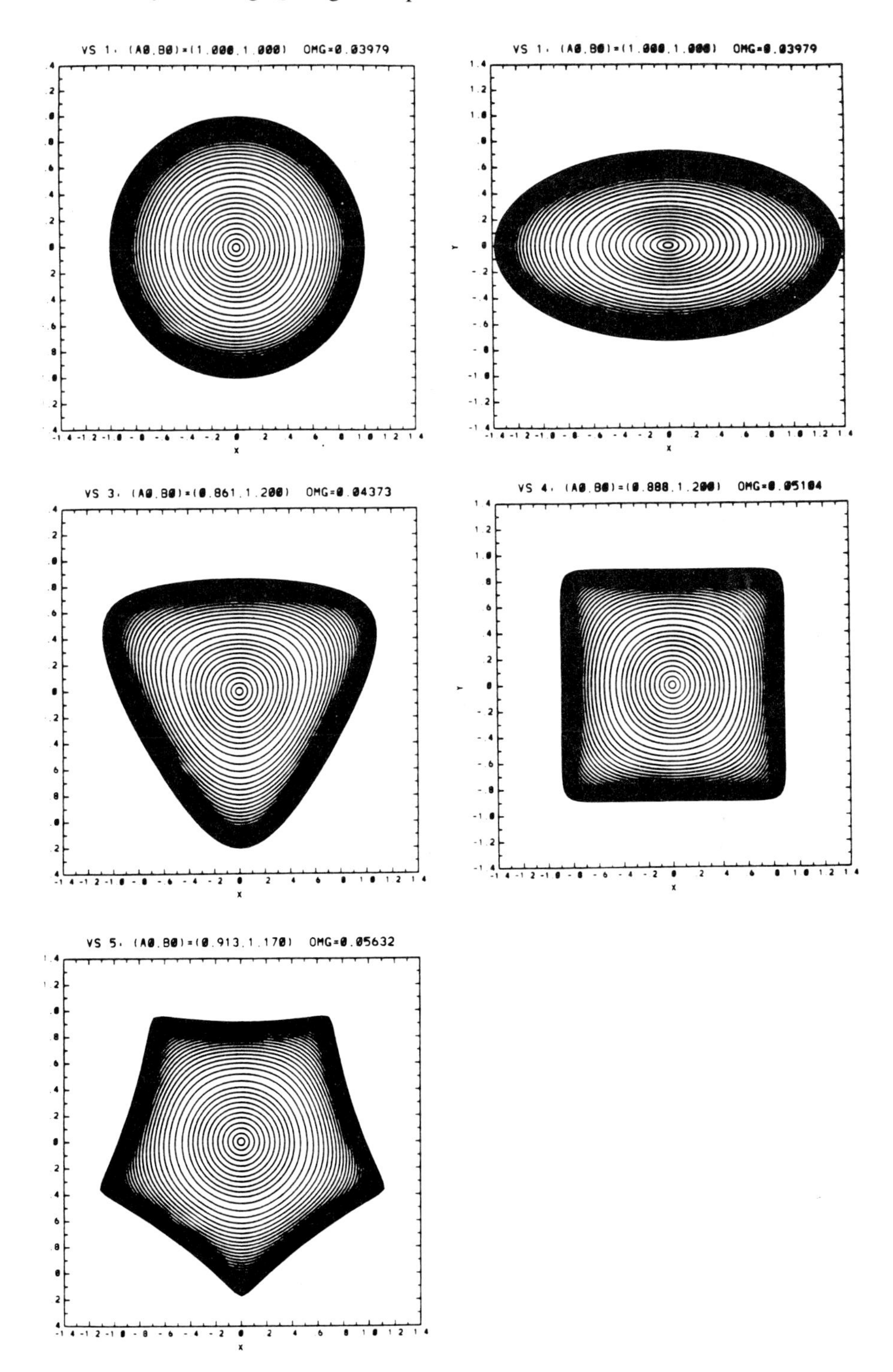

Fig. 3: Examples of rotating equilibria with azimuthal symmetries of ∞,2,3,4 and 5. Contours are isolines of surface elevation as seen from above.

7 OTHER FAMILIES BIFURCATING FROM THE SPHERE

As pointed out in §2, an infinite set of families of steadily rotating equilibria bifurcate from the spherical solution. The continuation of $n = 2, m = 2$ and $n = 2, m = 0$ can be obtained analytically as described in §4. The continuation of other families can be obtained numerically. We have begun to do this for some of the simpler families. In Figure 3, we present examples of members of families of solutions with $m = 3$, $m = 4$ or $m = 5$ azimuthal symmetry. We have also begun to numerically determine the stability of some of the members of these families.

We have examined a series of threefold symmetric solutions which have a vertical axis whose endpoints are fixed at $z = +1$ and $z = -1$ and are chosen to be vertically symmetric about the plane $z = 0$. The solutions are constrained in such a way that the area of each horizontal section is equal to that of the similar section in the undisturbed sphere; this filters out all of the $m = 0$ modes. At small amplitudes this series is almost pure (3,3) mode. This family can be parameterised by the maximum radius of the cross-section at $z = 0$, r_m. We have continued this series up to $r_m = 1.23$. We are using a model resolution of 7776 points. The family of solutions continues beyond $r_m = 1.23$ but to accurately resolve these cases will require higher resolution. We have computed the eigenfrequencies for the first fifteen normal modes of each of the triangular figures in the series. Growth rates of the unstable modes are shown in Figure 4. Unlike the ellipsoidal family, this series of solutions is numerically stable near its bifurcation from the sphere, at least to the low order modes that we have examined. This is consistent with our two-mode analysis above. For the limited number of modes that we have examined numerically, instability first sets in between $r_m = 1.05$ and $r_m = 1.08$. There are two unstable modes with the same growth rates and equal but opposite propagation speeds. Each is due to a resonance between two modes, one, the continuation of a $(2,2)$ mode and one, the continuation of a $(4,4)$ mode. We identify this as *de*. As one might expect from the resonant nature of this instability, this instability exists only over a finite range of r_m, ceasing somewhere between $r_m = 1.14$ and $r_m = 1.17$. Two other resonant instabilities, (*km* and *il*) can be seen at larger r_m.

We have also examined families of $m = 4$ and $m = 5$ equilibria. The growth rates of those of the first fifteen eigenmodes of the $m = 4$ equilibria that are unstable are plotted in Figure 4. Near the spherical limit, the instability of the $(4, 2)$ mode, predicted by the two-mode theory, can be seen together with a resonant instability, suggested by the three-mode analysis.

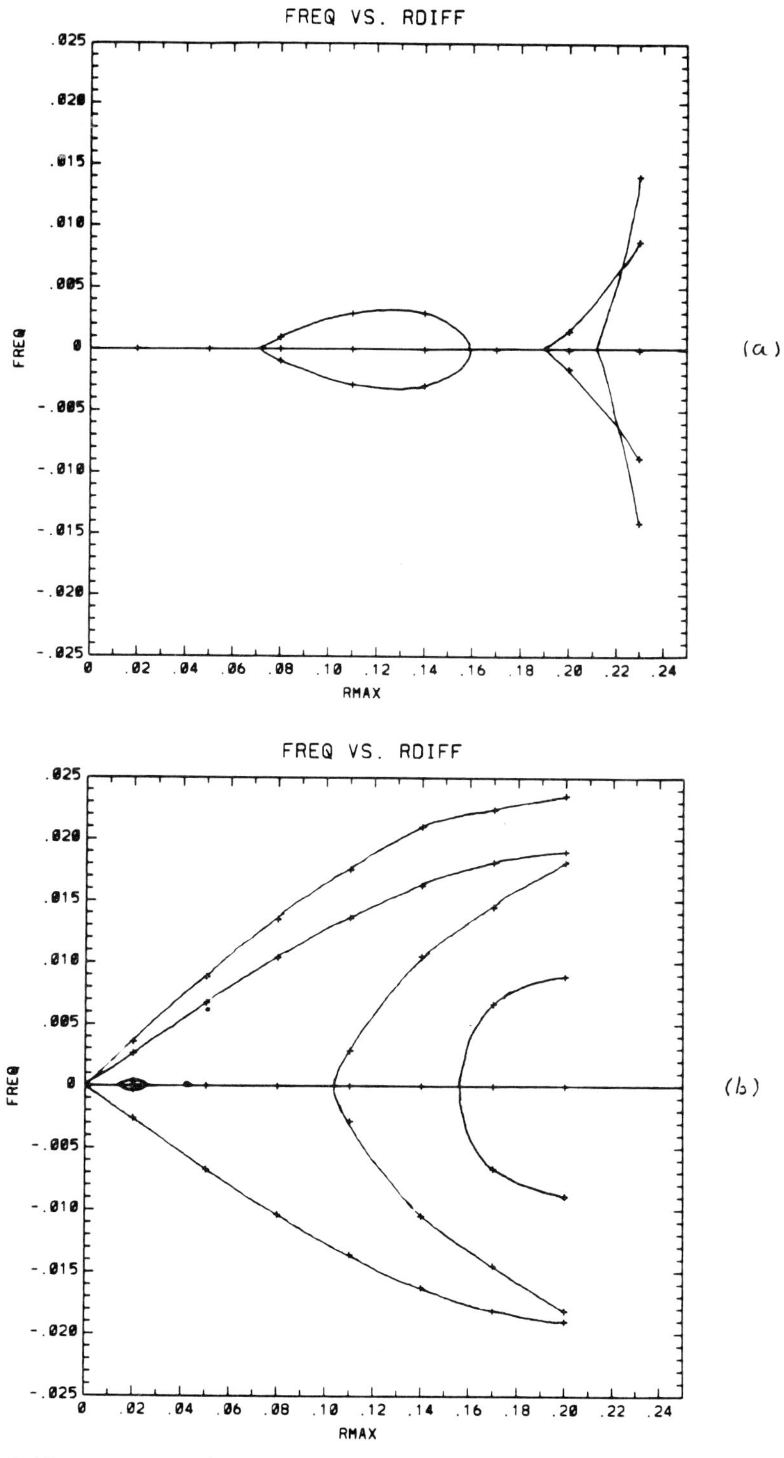

Fig. 4: Growth/decay rates of complex eigenmodes on rotating equilibria. (a) – threefold symmetric, (b) – fourfold symmetric.

8 REMARKS

In comparison to the theory of two-dimensional quasigeostrophic vortices, a three-dimensional quasigeostrophic vortex has a much richer variety of possible rotating and periodic equilibrium shapes whilst retaining the simple dynamics of quasigeostrophic flows. We have only touched upon some of the possibilities here.

Unlike the two-dimensional case, most three-dimensional vortices with an elliptical, horizontal cross-section and a moderate vertical aspect ratio are unstable. This does not necessarily mean that a disturbed vortex will quickly relax to a spheroidal state. The coupling between the ($n = 2$, $m = 2$) and the ($n = 2$, $m = 1$) modes suggest that a periodic vacillation between an elliptically deformed state and a vertically tilted state is possible.

The relative stability of the threefold symmetric state, when compared to a vortex with an elliptic cross-section, suggests that some distortions might lead to vortices having roughly triangular cross-sections.

REFERENCES

Armi L., D. Hebert, N. Oakey, J. Price, P. Richardson, T. Rossby and B. Ruddick. (1989) Two years in the life of a Mediterranean salt lens. *J. Phys. Oceanog.* **19**, 354-370.

Dritschel, D. (1988) Nonlinear stability bounds for inviscid, 2-D, parallel or circular flows with monotonic vorticity, and the analagous 3-D quasigeostrophic flows. *J. Fluid Mech.* **191**, 575-581.

Love, A.E.H. (1893) On the stability of certain vortex motions. *Proc. Lond. Math. Soc.* **25**, 18-42.

Manley, T.O. and K. Hunkins. (1985) Mesoscale eddies of the Arctic Ocean. *J. Geophys. Res.* **90C3**, 4911-4930.

McDowell, S.E. and H.T. Rossby. (1978) Mediterranean Water: An intense mesoscale eddy off the Bahamas. *Science* **202**, 1085-1087.

McWilliams, J.C. (1989) Statistical properties of decaying geostrophic turbulence. *J. Fluid Mech.* **198**, 199-230.

Meacham, S.P. (1989) Quasigeostrophic, ellipsoidal vortices in a stratified fluid. Submitted to *Dyn. Atmos. and Oc.*

High-resolution Three-dimensional Modelling of Stratospheric Flows: Quasi-two Dimensional Turbulence Dominated by a Single Vortex

PETER H. HAYNES

Department of Applied Mathematics and Theoretical Physics,
University of Cambridge, Silver Street, Cambridge CB3 9EW, UK

1 INTRODUCTION

The winter stratosphere is dominated by a strong westerly circumpolar vortex, associated with strong horizontal temperature gradients at the edge of the polar night. The vortex is disturbed through the upward propagation of quasi-stationary planetary waves from the troposphere. The latitudinal and vertical variation of the background flow, together with the decrease in density with height, leads to marked variation in space in the character of the disturbances. This has been clearly revealed by maps of potential vorticity on isentropic surfaces derived from satellite observations, albeit at rather poor spatial resolution (McIntyre and Palmer 1984). Such maps show the vortex as a material entity, the planetary waves manifested through changes in its shape or position. Surrounding the vortex is a region in which potential vorticity contours appear to be strongly deformed and often fragmented. The motion here is clearly not wave-like, although it is still fair to say that it has arisen through the upward propagation of planetary waves, and it has been argued that it is appropriate to regard the motion as a manifestation of planetary-wave breaking (McIntyre and Palmer 1985 and refs.).

Recent numerical simulations based on integration of the barotropic vorticity equation on the sphere (Juckes and McIntyre, 1987) have succeeded in reproducing flows which are qualitatively similar to those observed. In a typical experiment a pre-existing vortex was disturbed by a vorticity forcing, equivalent to shallow topography which was purely wave-1 in longitude. The following qualitative features were of note. As the disturbance grew, tongues of high-vorticity fluid were pulled out of the vortex, and subsequently stirred into the region outside. This process lead to the formation of sharp gradients at the edge of the vortex. Once the strong gradients had formed the decrease in size of the vortex slowed, although intermittently further tongues were pulled out. Some of these tongues were stretched out quasi-passively by the strong strain fields associated with the distorted main vortex. Others would roll up into vortices, which would then persist for a few days.

These features of the flow have important implications for transport of chemical tracers and it is therefore important to establish through numerical simulation that similar features are possible in fully three-dimensional flows. Some prelimary results of such simulations, which seem of interest not only for their implications for the stratosphere, but also from a purely fluid-dynamical viewpoint, are summarised in this paper.

2 NUMERICAL EXPERIMENTS AND FLOW DIAGNOSTICS

The numerical model is based on that described by Hoskins and Simmons (1975). The only major modification is that a semi-implicit time scheme for the vorticity equation has been included (allowing larger timesteps when the component of the flow along latitude circles is large). The model uses finite-differenced sigma-coordinates in the vertical and represents horizontal variation using spherical harmonics. For reasons of computational economy hemispherical symmetry was assumed. In order to simulate a notional stratospheric circulation the lower boundary of the model, which in sigma coordinates is a material surface, was taken at 10km. Disturbances were forced by an asymetrical deformation of this surface. This might be regarded as a crude representation of the tropospheric forcing. The initial flow was taken to be characteristic of that observed in the winter stratosphere, with a jet increasing in strength with height, and centred around 55N. A cross-section of a typical initial flow is shown in Figure 1.

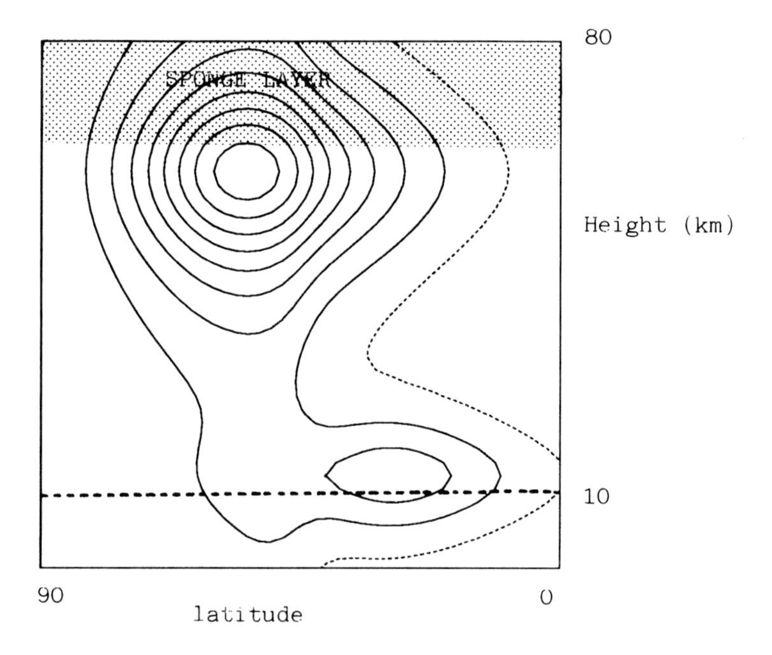

Figure 1: Cross-section of a typical initial flow, showing contours of flow speed at intervals of $10ms^{-1}$. The zero contour is finely dashed. The artificial lower boundary is at 10km and there is a sponge layer extending from 65km to 80km.

A convenient way to examine the structure of the flow is through maps of potential vorticity on isentropic surfaces. It is also interesting to follow McWilliams (1984) and Brachet et al. (1988) who, for two-dimensional flows, have examined the spatial variation of the quantity Q, defined by

$$Q = \frac{1}{2}\underline{\underline{e}} : \underline{\underline{e}} - \frac{1}{4}\zeta^2, \tag{1}$$

where $\underline{\underline{e}}$ is the rate of strain tensor and ζ the relative vorticity. It is well known that, in a flow where the velocity gradient is spatially constant, line elements ultimately increase in length as $\exp(Q^{\frac{1}{2}}t)$ if $Q > 0$, and do not systematically increase in length if $Q < 0$. Providing that spatial gradients are not too large, the distribution of Q therefore shows where material lines are likely to be predominantly stretched or rotated. In the three-dimensional case under consideration here, which is envisaged to be dominated by quasi-horizontal advection, the relevant definition of Q is

$$Q = \frac{1}{2}(\frac{1}{cos\theta}\frac{\partial u}{\partial \phi} - vtan\theta)^2 + \frac{1}{2}(\frac{\partial v}{\partial \theta})^2$$

$$+\frac{\partial u}{\partial \theta}(\frac{1}{cos\theta}\frac{\partial v}{\partial \phi} + utan\theta), \tag{2}$$

where θ and ϕ are latitude and longitude respectively, u the velocity in the ϕ direction, and horizontal derivatives are taken along isentropic surfaces.

3 RESULTS

3.1 Strongly-forced vortex

In this case the model extended to 80km in the vertical. The considerable increase in wave amplitude with height lead to the vortex being strongly disturbed at upper levels. The horizontal resolution of the model was taken as T63 (i.e. triangular truncation at total wavenumber 63) and the vertical resolution was 2km. The forcing at the lower boundary was switched on between days 0 and 4, sustained, and then switched off between days 16 and 20. In the early part of the run the PV contours deform weakly, consistent with the theory of linear Rossby waves. In part of the flow this deformation persists and contour displacements become large. The spatial distribution of the strain on each isentropic surface tends to increase isentropic PV gradients in certain locations, particularly around the edge of the main vortex. This gradient intensification is apparent in Figure 2, and regions of positive Q are seen to extend around the vortex and away from it in a spiral structure. Similar features are seen at other levels.

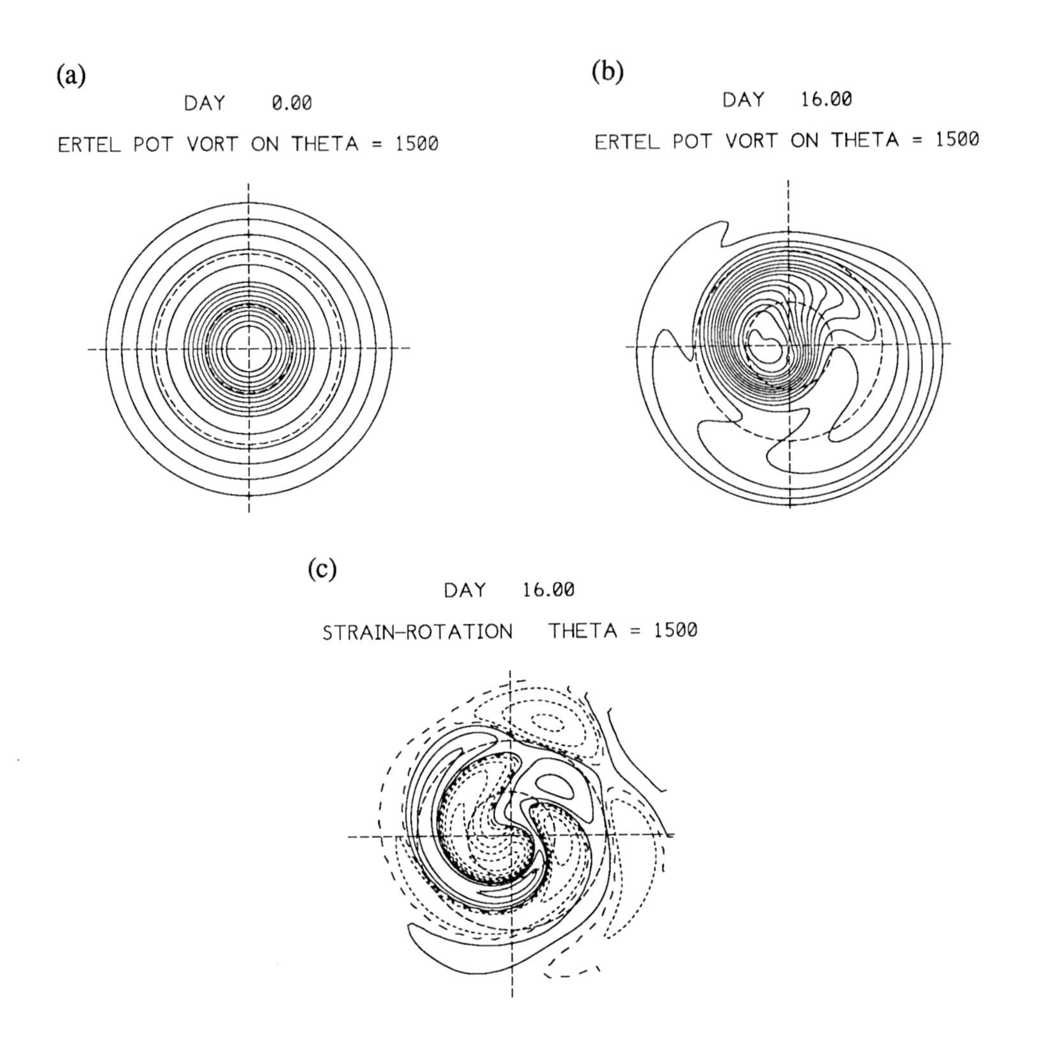

Figure 2: (a) Contours of PV on 1500K surface at days (a) 0 and (b) 16. (c) Contours of $|Q|^{\frac{1}{2}}$ on 1500K surface at day 16. Positive values of Q have solid contours, and negative values finely dashed contours. The zero contour is coarsely dashed. The contour interval is 0.1 x 2πdays^{-1}.

Subsequently, around 22 days, a large tongue of higher PV fluid begins to be pulled away. This tongue is a deep feature, which tilts gently poleward with height. As the tongue lengthens, around 23 days, its tip begins to roll up into a cyclonic vortex. The resulting strain field thins the middle of the tongue and around day 25 the vortex effectively detaches. The PV fields on the 1650K surface for days 23 and 25 are shown in Figure 3.

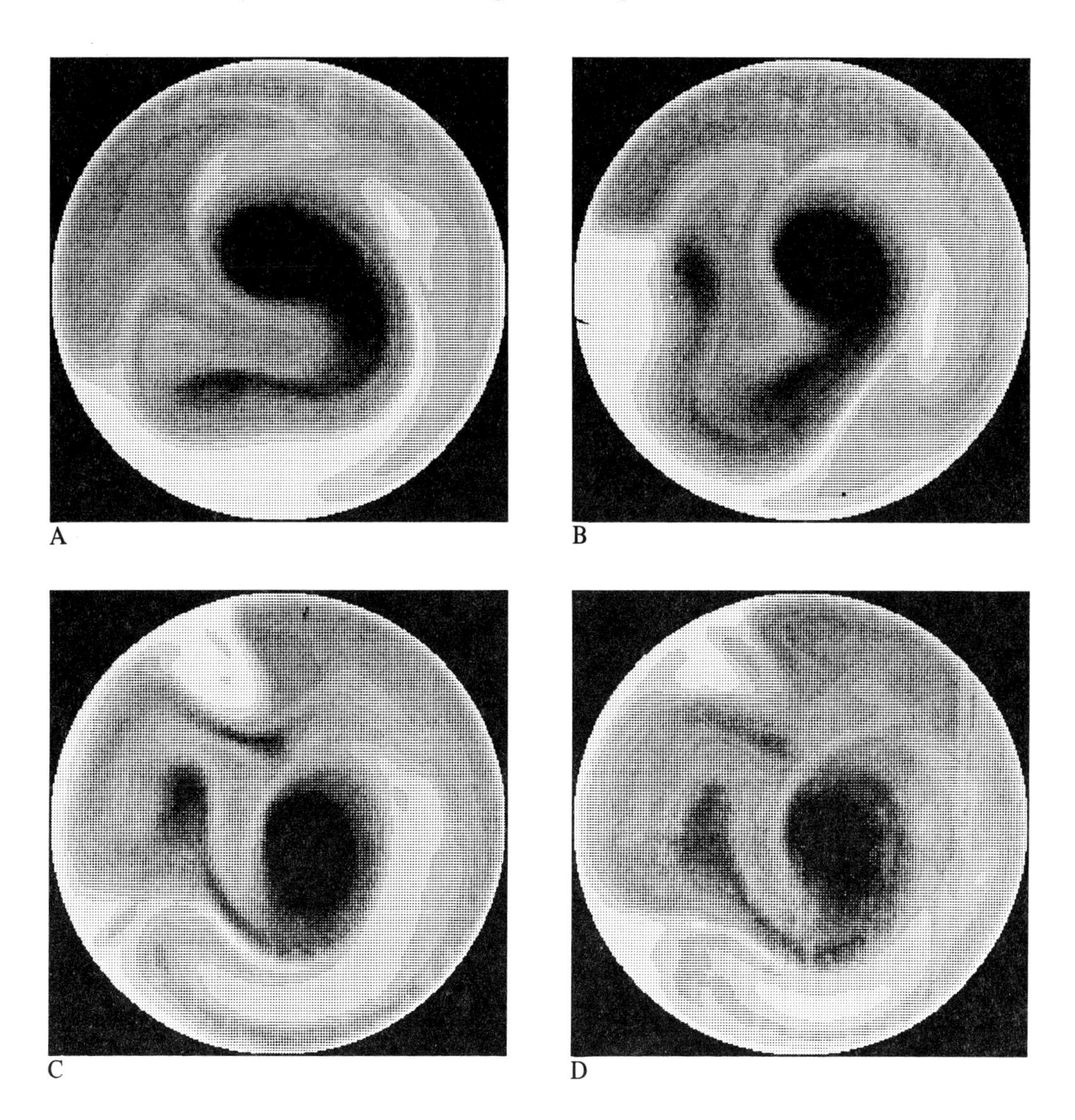

Figure 3: Grey-scale pictures of the PV distribution on the 1650K surface at days (a) 23, (b) 25 and (c) 27, and at day 27 on the (d) 1500K surface.

The part of the tongue left attached to the main vortex continues to extrude and by day 27 forms a second new vortex. The first vortex has now been advected around the anticyclone and entered a region where the external strain field is large enough stretch to it out and destroy it. A similar process is taking place through a deep layer, as may be seen from the PV fields for day 27 on the 1650K, 1500K, 1350K and 1200K isentropic surfaces, also shown in Fig. 3. However, note the tilt of the second detached vortex with height and the fact that it has all but disappeared at the 1200K surface.

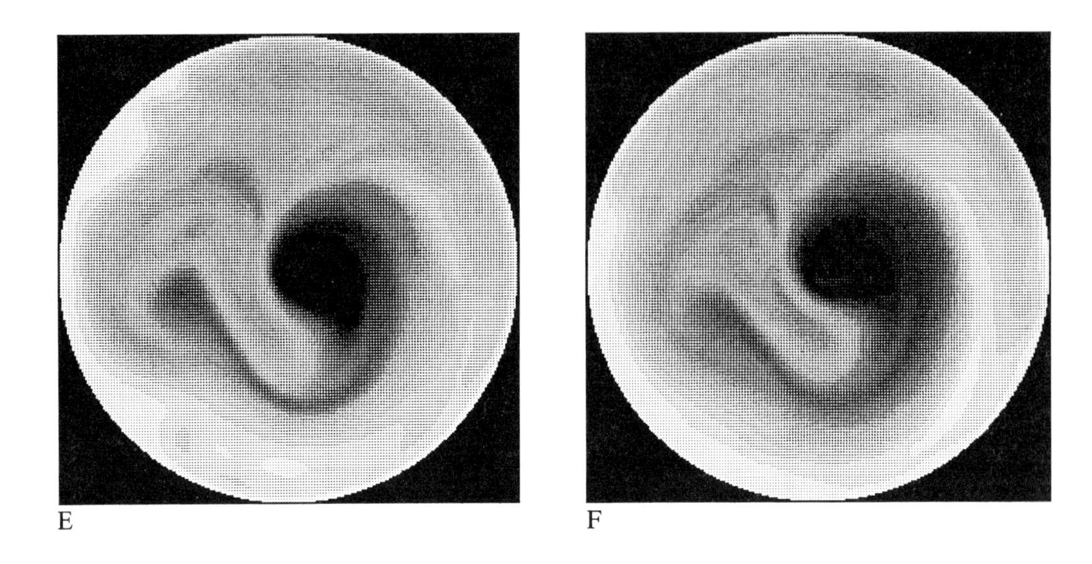

Figure 3: (continued) Grey-scale pictures of the PV distribution at day 27 on the (e) 1350K and (f) 1200K surfaces.

Examination of relevant Q fields, shown in Fig. 4 for the 1650K surface shows the second vortex still associated with a region of locally negative Q, whilst the first vortex has entered a region of strongly positive Q.

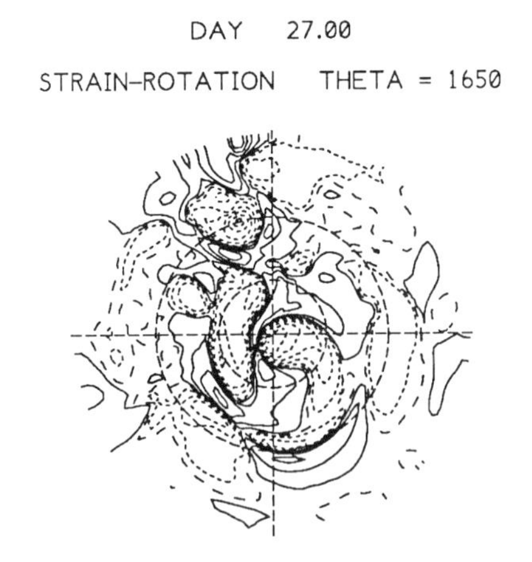

Figure 4: As Figure 2(c) but on 1650K surface at day 27.

On some surfaces the area of the main vortex reduces further through a repeat of the process described above in which the size of the (third) new vortex is almost as large as that of the remaining main vortex.

3.2 Weakly-forced vortex

In another case the model extended to only 45km. The vertical grid size was 1.75km the horizontal resolution was increased to T106. After the initial switch-on over 4 days the forcing was sustained through the duration of the experiment. At lower levels than those considered in §3.1 the wave amplitude is smaller. The centre of the main vortex never displaces far from the pole and a large body of air at the centre of the vortex remains relatively undisturbed. This may be seen from an example PV field, on the 660k surface on day 37, shown in Fig. 5(a). The characteristic contrast between the main vortex and its exterior is clear. To highlight features in the exterior Fig. 5(a) has been repeated, with different grey-scale, and with the main vortex blacked out, in Fig. 5(b). The streakiness of the PV field in much of this region is evidence of passive straining. However there is also evidence of dynamically active features. A small anticyclonic vortex that has formed through the roll-up of a tongue of low-latitude fluid is travelling around the boundary of the main vortex. There is evidence of another coherent anticyclonic circulation, seen in Fig. 5 to the right of the main vortex. It also appears as if the streak of high vorticity which is seen on the left-hand side of the Fig. 5 and lies closest to the equatorial boundary is beginning to form closed cyclonic vortices, perhaps through a barotropic-instability mechanism. The relevant Q field is full of small-scale features and is not shown, but not surprisingly, the anticyclonic vortex is associated with a region of negative Q, and surrounded by a region of positive Q values far higher than those typical of much of the region exterior to the main vortex.

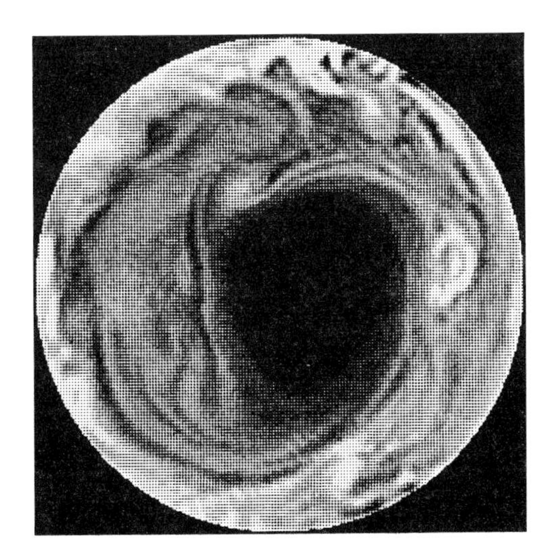

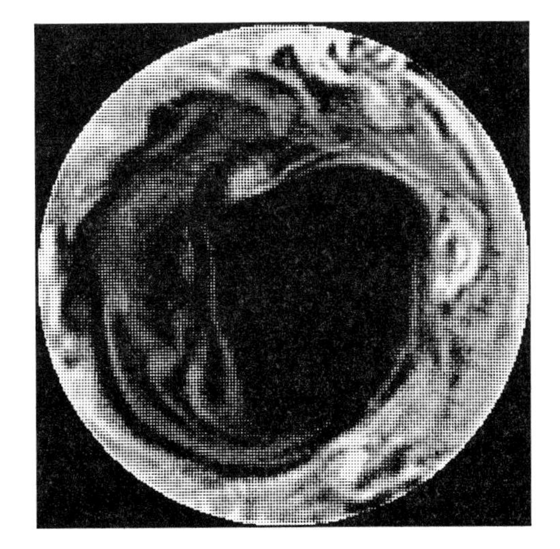

Figure 5: (a) Grey-scale picture of the PV distribution on the 660K surface in the run described in §3.2 at day 37 and (b) the same with the contrast in the region outside the vortex enhanced, and with the interior of the vortex blacked out.

3.3 Gradient intensification

There is much evidence in the simulations performed so far of the intensification of PV gradients through straining and that the resulting sharp-gradient features are relatively long-lived. The intensification of gradients at the edge of the main vortex was identified in two-dimensional simulations by Juckes and McIntyre (1987), and there is evidence of the same process in the results of very high resolution simulations presented by Brachet et al. (1988).

Here the sharpening of the vorticity gradients at the edge of the main vortex is made particularly clear in an experiment where the initial vortex is rather broad. This experiment uses at T63 horizontal resolution. PV fields on the 660K and 850K surfaces are shown in Figure 6, initially and after 100 days, during which time the wave forcing at the lower boundary was repeatedly applied and removed.

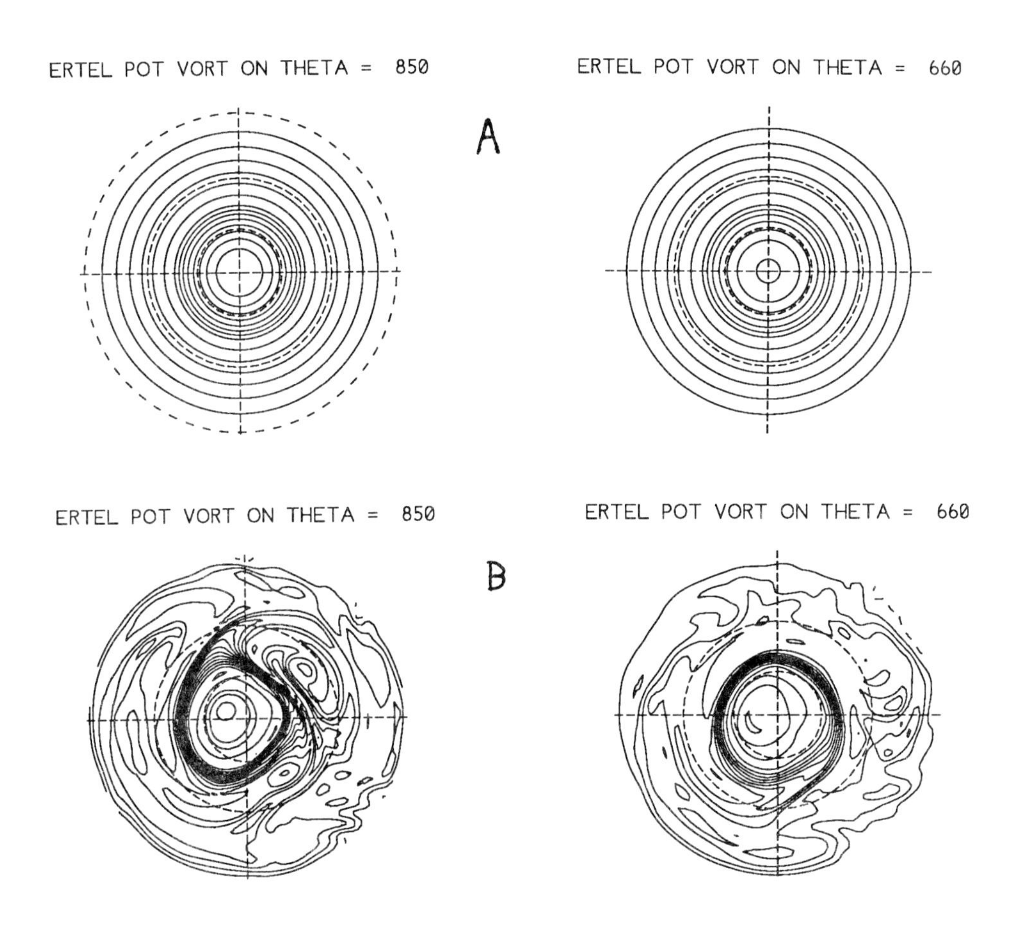

Figure 6: Contours of PV on the 850K and 660K surfaces at days (a) 0 and (b) 100.

Experiments at different horizontal resolution suggest that the strength of the gradients is limited only by the hyperdiffusion applied to inhibit features on the smallest scales resolved by the model. This is consistent with experience of two-dimensional simulations (Juckes and McIntyre, 1987) and suggests that there is no three-dimensional, purely dynamical mechanism that can inhibit gradient intensification. The reason why sharp edges have not been clearly observed in more conventional two-dimensional simulations (such as that of McWilliams (1984), who presents an example vortex cross-section which is relatively smooth) is presumably that individual vortices are relatively badly resolved in such simulations.

At 100 days the vortex seems to have reached an equilibrium size and one is tempted to speculate that further application of forcing of the same amplitude will not erode it further. This is consistent with the result that, in two dimensions, sharp-edged vortices survive applied external strain rates up to a finite proportion of the vorticity jump at their edges (Moore and Saffman, 1971). In an experiment where the applied forcing was 50% stronger the vortex was almost entirely eroded after 100 days.

4 DISCUSSION

The features seen in these simulations seem entirely consistent with the notion that two-dimensional vortex dynamics can provide considerable insight into the behaviour of three-dimensional stratified and rapidly rotating flow. In particular, these simulations have shown the distinct two-zone structure also observed in high-resolution simulations of decaying two-dimensional turbulence, namely coherent vortices, inside which scale-cascading processes are relatively slow, surrounded by a region in which the scale cascade in the vorticity field is much more rapid. In the stratospheric simulations, where PV plays the same role as the vorticity in the two-dimensional simulations, the flow is dominated much of the time by a single pre-existing vortex. This is perturbed, not so much by interactions with neighbouring vortices, but by upward propagating planetary waves. Nonetheless, the dominant fluid-dynamical processes are fundamentally the same. Indeed it is possible that stratospheric, single-vortex, simulations, both two- and three-dimensional can give insight into the structure of the vortices which emerge in simulations of decaying two-dimensional, and three-dimensional stratified and rotating, turbulence.

REFERENCES

BRACHET, M.E., MENEGUZZI, M., POLITANO, H. & SULEM, P.L. 1988 The dynamics of freely decaying two-dimensional turbulence. *J. Fluid Mech.* **184**, 399-422.

HOSKINS, B.J. & SIMMONS, A.J. 1975 A multi-layer spectral model and the semi-implicit method *Quart. J. Roy. Met. Soc.* **101**, 637-655.

JUCKES, M.N. & MCINTYRE, M.E. 1987 A high-resolution one layer model of breaking planetary waves in the stratosphere. *Nature* **328**, 590-596.

MCINTYRE, M.E. & PALMER, T.N. 1984 The stratospheric 'surf zone'. *J. atmos. terr. Phys.* **46**, 825-849.

MCINTYRE, M.E. and PALMER, T.N. 1985 A note on the general concept of wave breaking for Rossby and gravity waves. *Pure appl. Geophys.* **123**,964-975.

MCWILLIAMS, J.C. 1984 The emergence of isolated coherent vortices in turbulent flow. *J. Fluid Mech* **146**, 21-43.

MOORE, D.R. and SAFFMAN, P.G. 1971 Structure of a line vortex in an imposed strain. In *Aircraft Wake Turbulence and its Detection* (ed. J.H. Olsen, A. Goldburg & M. Rogers), pp. 339-354. New York: Plenum.

Nonlinear Vorticity or Potential Vorticity Inversion

M.E. McINTYRE and W.A. NORTON

Department of Applied Mathematics and Theoretical Physics,
Cambridge CB3 9EW, U.K.

1 INTRODUCTION

The notion of vorticity or potential vorticity (PV) inversion is related to the notions of balanced flow and nonlinear vortical mode decomposition. Balanced flows are just those fluid flows that are controlled by vorticity or PV evolution. In other words, balanced flows are those fluid motions to which an 'invertibility principle' for vorticity or PV applies, in the sense that the vorticity or PV field can be inverted to yield the velocity field and any other relevant dynamical information. The usual incompressible case (unstratified vortical flows in the zero Mach number limit), is merely the most familiar special case, the inversion then being given by the Biot-Savart or inverse Laplacian type of integral. Other cases are discussed in the review by Hoskins *et al.* (1985), and in a forthcoming paper by the present authors (1989b, hereafter MN).

The general notion of vorticity or PV invertibility is by no means a trivial one, if only because the most accurate inversion operators are nonlinear at finite Froude, Mach or Rossby numbers, and also because inversion seems to be an inherently approximate process, as discussed briefly in Hoskins *et al.* and more extensively in MN. In practice, however, the approximations involved can be remarkably good, in comparison with what one might guess from a *prima facie* consideration of Froude or Mach number values.

2 EXAMPLE: 2D FLOW AT MACH NUMBER $\lesssim 0.5$

Figure 1 shows an example. The system is a shallow, hydrostatic layer of fluid with a free upper surface, on a hemisphere viewed in polar stereographic projection in a reference frame rotating with angular velocity $\Omega = 2\pi/1\,\text{day}$. It was originally studied as a simplified model atmosphere, but can also be thought of, in the usual way, as a two-dimensional acoustic system, with sound waves in place of gravity waves, in a hypothetical 'perfect gas' with ratio of specific heats $\gamma = 2$, and with density ρ replacing the layer depth h. The system has an invariant, Q, that is materially conserved in the absence of viscosity:

$$Q = \zeta^{\mathrm{a}}/h \quad \text{or} \quad \zeta^{\mathrm{a}}/\rho\,, \qquad DQ/Dt = 0\,, \tag{1}$$

where ζ^{a} is the radial component of absolute vorticity. For want of any other name Q will be referred to indiscriminately as the PV both in the free-surface and in the compressible interpretation.

Figure 1a (top left) shows the velocity field and the h or ρ field taken from a high resolution numerical simulation of nearly-inviscid evolution of this model fluid system. The local Froude or Mach number M, defined as the ratio of the local flow speed $|\mathbf{u}|$ to the gravity or sound wave speed, takes values up to about 0.5. The local Rossby number R, defined as $|\mathbf{u}|/2L\Omega\sin\phi$, where ϕ is latitude and L is a length scale of the flow, is of course infinite at the equator. It reaches a local maximum of the same order as M, 0.5, in the region of strong subtropical winds. The flow was set up by means of an artificial forcing, in a manner that need not concern us here except to say that care was taken to apply the forcing smoothly, so that minimal gravity or acoustic wave activity was introduced as a direct result. Figure 1b (top right) shows the two-dimensional divergence field. Figures 1c,d show the associated distribution of Q, both in conventional contouring, and in grayscale for better visibility. The smallest features are well resolved numerically; a pseudospectral method is used, isotropically representing the fields using total spherical harmonic wavenumbers up to 106, corresponding to a mesh size of the order of a degree of latitude.

Figures 1e,f show the results of a nonlinear inversion. The velocity, density and divergence fields were reconstructed, from the Q field alone, using one of the accurate PV inversion algorithm described in MN. Comparison with figures 1a,b shows that the invertibility principle applies with remarkable accuracy to this particular flow. Despite the not very small values of the local Froude, Mach and Rossby numbers, and the large departure from incompressibility (the density can be seen from figures 1a and 1e to vary by over 40% away from its area-mean), almost all the detail is recovered by the inversion, even in the divergence field. It follows that the flow is accurately balanced, in the sense proposed above, even though it is clear that a simple incompressible or Biot-Savart or inverse Laplacian type of vorticity inversion would be grossly inaccurate.

This particular example supplements a more extensive set of examples given in MN. The qualitative impression given by figure 1 is typical of all the other examples, which include cases with local Froude or Mach numbers up to 0.7 albeit not simulated at such high spatial resolution. The discussion in MN also describes some tests of *cumulative* accuracy, in which the exact evolution over many eddy turnaround times was compared with the evolution of a corresponding balanced model, defined by alternately inverting and time-stepping Q. Even with maximum local Froude, Mach and Rossby numbers in excess of 0.7, the evolution was reproduced with an accuracy not far short of that in figure 1. The fact that the balance and inversion seem so remarkably accurate is presumably connected with another, better known fact, namely the equally remarkable weakness of spontaneous emission of acoustic or gravity waves predicted by the Lighthill theory of aerodynamic sound generation (e.g. Crighton 1975, 1981).

Our results to date, then, in combination with the Lighthill theory, strongly encourage the belief that the concepts of balance, invertibility, etc., as defined above, are far more widely applicable than one might imagine from approximate theories that restrict attention to Mach, Froude and/or Rossby numbers much less than unity. This would appear to have far-reaching consequences – e.g. leading to a fresh view of wave-mean interaction theory (McIntyre and Norton 1989a). But the invertibility principle has yet to be fully tested, as far as we know, in the case of three-dimensional homentropic (unstratified) flows at substantial Mach numbers. We would be interested to hear of related work.

REFERENCES

Crighton, D.G., 1975: Basic principles of aerodynamic noise generation. *Prog. Aerospace Sci.*, **16**, 31-96.

Crighton, D.G., 1981: Acoustics as a branch of fluid mechanics. *J. Fluid Mech.*, **106**, 261-298.

Hoskins, B.J., McIntyre, M.E. and Robertson, A.W., 1985: On the use and significance of isentropic potential-vorticity maps. *Quart. J. Roy. Meteorol. Soc.*, **111**, 877-946. Also **113**, 402-404.

McIntyre, M.E., and Norton, W.A., 1989a: Dissipative wave-mean interactions and the transport of vorticity or potential vorticity. *J. Fluid Mech.*, (G.K. Batchelor Festschrift Issue), to appear.

McIntyre, M.E. and Norton, W.A., 1989b: Potential vorticity inversion on a hemisphere. In preparation.

Figure 1. Demonstration of balance and invertibility in a compressible two dimensional flow with substantial density variations, from a numerical experiment of flow on a hemisphere. This was motivated as an atmospheric model but is equally well interpretable as a compressible two-dimensional flow in a hemispherical shell. The system is a shallow water free-surface model with area-mean depth 2 km and corresponding gavity wave speed 140 ms^{-1}, or equivalently a fictitious 'perfect gas' with ratio of specific heats $\gamma = 2$ and sound speed $140\mathrm{ms}^{-1}$ at area-mean density. Solid contours show positive values, long dashed contours negative values, and dotted contours zero. Projection is polar stereographic; the radius of the hemisphere is 6371 km. Panel (a): arrows show the velocity field on the scale indicated; contours show departures of density or layer depth from the area-mean value. The contour interval is one twentieth of the mean; in the two-dimensional compressible system it can also be regarded as the anomaly in the square root of the pressure. Panel (b): divergence field contoured at intervals of 0.6×10^{-6} s^{-1}. Panels (c), (d): the quantity Q defined in equation (1). The right hand plot is contoured at interval 1×10^{-8} m^{-1} s^{-1} in units appropriate to the first formula in (1). The shading in the contour plot highlights values lying between 4 and 6 of these units. The grayscale representation of the same information on the left is monotonic from light to dark. Panels (e), (f): as at the top, but reconstructed from Q alone using an accurate nonlinear PV inversion algorithm.

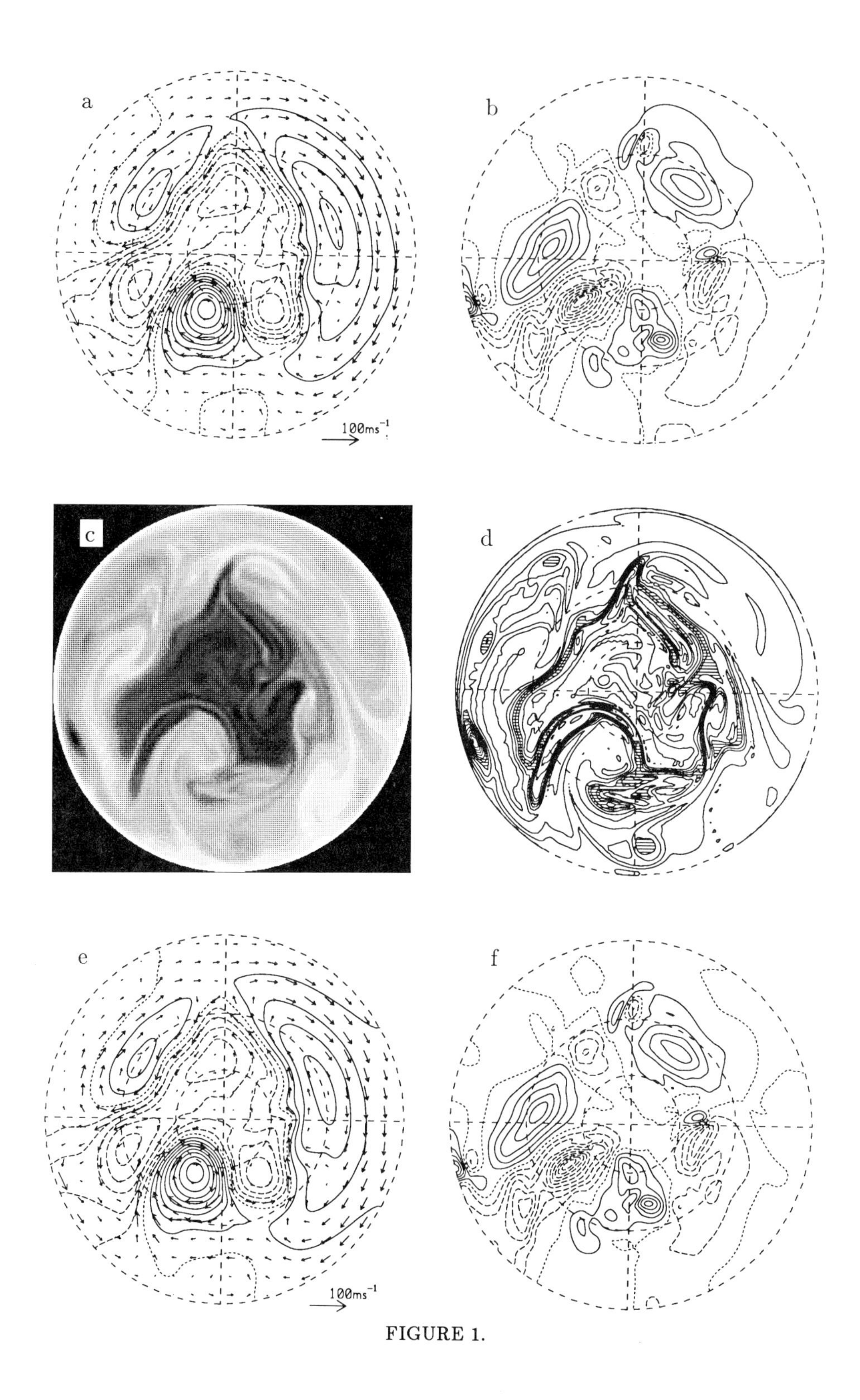

FIGURE 1.

V: Topology of Three-dimensional Flows

Analysis and Visualization of Flow Topology in Numerical Data Sets

JAMES HELMAN
Department of Applied Physics, Stanford University, Stanford, CA 94305-4035, USA

LAMBERTUS HESSELINK
Departments of Aeronautics & Astronautics and Electrical Engineering, Stanford University, Stanford, CA 94305-4035, USA

1 INTRODUCTION

Visualization has always played an important role in the study of fluid dynamics. Traditionally, flow visualization has been accomplished experimentally as part of the measurement process. For example, by seeding the flow with smoke or by photographing patterns of oil streaks on the surface of a body in the flow, a wealth of information concerning global flow structures has been obtained [Hesselink, 1988]. For example, the discovery of large scale eddies by Brown and Roshko[1974] has had a profound influence on flow modeling studies.

With the advent of supercomputers, substantial progress has been made towards simulating many of these flows numerically. The solutions generated by these calculations must be compared with experimental results to verify the accuracy of the simulation methods. These data sets are usually much larger in size than those found in experimental work and present a formidable challenge for evaluation efforts. Most approaches of data interpretation involve computer aided visualization methods, many of which try to mimic experimental results by generating images of skin friction lines or particle trajectories [Buning and Steger, 1985]. These visualization methods have been very successful in rapidly conveying large amounts of information to computational fluid dynamicists.

Although these solution sets usually contain velocity field information, most visualization techniques only make partial use of it through the display of skin friction lines (figure 1) or groups of streamlines (figure 2). Though these curves are extremely valuable, when more than a few well grouped curves are displayed, the graphic representation becomes confusing and difficult to understand because the isolated curves are poorly suited to depth cuing and visual interpretation. Animation has been used to reduce the problem by only displaying a subset of patterns at any one time[Lasinski *et al.*, 1987], but there remains the difficulty of manually selecting and refining starting points for the curves in order to locate structures in the flow.

Even though computational simulations provide more quantities and more information than most experiments, when a researcher finally tries to deduce the topology of a fluid flow from the solution set, he or she frequently ends up following a course similar to that applied to laboratory visualizations, i.e. inferring the types, locations and connections of structures from pictures of skin friction lines and particle paths. This is a very time consuming process, and in the end, diagrams and surfaces representing the inferred topology are drawn by hand as in Ying et al.[1987]. These figures and the complex structures drawn by Dallmann[1983] provide convincing evidence that topological information is best conveyed visually. In the case of numerical data, these figures would be even more useful if they were not only qualitatively correct, but accurate about the positions and windings of the surfaces.

In order to address this need, we have been developing methods for automating the interpretation and display of flow topology in numerical data sets. In general, topology provides a very effective way of summarizing the properties of any continuous vector field, whether or not it is related to a fluid flow. For example, in solid mechanics studies one usually computes vector and tensor quantities related to stress and strain fields that can be equally well displayed and interpreted using our topological approach.

Our aim is to allow the flow topology to be directly visualized for data understanding purposes and to provide a representation of the topology, in the form of a graph, which allows comparisons and characterizations of topological transitions to be made more easily.

2 ANALYSIS AND REPRESENTATION

We wish to represent the flow topology by means of surfaces in three-dimensional domains or curves in two-dimensional domains which divide the flow into separate regions. Two sets of surfaces or curves are of particular interest[Hesselink and Helman, 1987],[Helman and Hesselink, 1989]. The first set is that along which the flow close to the wall of a body attaches to or separates from that wall, i.e. those tangent curves which stop on the surface rather than moving along it. The second set contains lines emanating from the principal directions of a critical point. Of these the most important curves come from saddle points because these lines denote the boundaries of regions which can divert or bring together tangent curves.

3 TWO-DIMENSIONAL ANALYSIS

Starting from a two-dimensional vector field defined on a regular grid of points, which is specified by a separate array, we locate and characterize the critical and wall points. Each of these points is represented by a data object which has slots containing intrinsic information about the point, including position, the Jacobian matrix, eigenvalues, and eigenvectors [Perry and Chong, 1987]. For each point, these quantities are computed and stored both relative to the grid and relative to physical space. Slots are also provided, as appropriate to the class of the point, to link it to incoming and outgoing tangent curves.

The first step is to locate those points in the flow where the magnitude of the vector vanishes. The Jacobian matrix of the vector (u, v) with respect to position at the critical point $(x0, y0)$,

$$\left.\frac{\partial(u,v)}{\partial(x,y)}\right|_{x0,y0} = \left.\begin{bmatrix} \frac{\partial u}{\partial x} & \frac{\partial u}{\partial y} \\ \frac{\partial v}{\partial x} & \frac{\partial v}{\partial y} \end{bmatrix}\right|_{x0,y0}$$

is calculated and the eigenvalues and eigenvectors computed. According to the eigenvalues, each critical point is classified to first order as an *attracting node (an)*, a *repelling node (rn)*, an *attracting focus (af)*, a *repelling focus (rf)*, a *center (ce)*, or a *saddle point (sp)*.

On walls where the velocity is constrained to be zero, tangent curves usually run parallel to the surface, but at certain points they terminate on the surface either attaching to or detaching (separating) from it. These points are classified as *attachment points (at)* and *detachment points (de)*.

The saddle points, attachment and detachment points are the only points at which there are a small number of curves which end on the point itself. These curves are tangent to the eigenvectors of the Jacobian matrix at the point. Numerical integrations are started, forwards or backwards depending on the sign of the corresponding eigenvalue, in the directions of the eigenvectors near these three types of critical points. These curves are used to link the points into a partially connected graph. The order in which the linking is carried out is important to avoid topologically impossible combinations which can result from numerical integration errors.

Some curves may leave the domain of available data. To complete the representation, we call the points at which these tangent curves leave the domain *boundary incoming (bi)* and *boundary outgoing (bo)* points.

Given a two-dimensional vector field, the algorithm is as follows:

I. Locate, characterize and classify all critical and wall points.

II. Integrate tangent curves out of the originators. This generates one curve for each attachment or detachment point and four curves for each saddle point. The beginning of each curve is linked to its originator.

III. Identify the end point of each tangent curve and link the curve to it.

  A. If the curve ends on the boundary, a new boundary point is created to represent it, and the curve is linked to it.

C. If the curve ends at a saddle, detachment or attachment point, then the corresponding tangent curve for that point is replaced by the new curve.

B. Otherwise, the curve ends a critical point, and the end of the curve is linked to it.

The representation resulting from the application of the above steps to the instantaneous velocity field of a computed two-dimensional flow around an airfoil is shown in figure 3a. The data used here represent artificial results from a numerical simulation code[Rogers and Kwak, 1988] and do not correspond to a physical flow. We use it here to demonstrate the method as applied to a vector field with non-zero divergence. A graph representation of the topology is shown in figure 3b.

When the method described above is applied to vector fields with zero or very small divergence, the integration may or may not yield the closed streamlines desired for interpretation. This problem arises due to errors associated with the differencing scheme used for computing the divergence and the integration method. For this reason, we provide for the display of the topology as computed or with streamlines which have been closed by connecting the appropriate streamlines at their point of closest approach.

Figure 4 shows the instantaneous topology at two time steps in the two-dimensional flow around a circular cylinder computed by Rogers and Kwak[1988].

4 ANALYSIS OF TWO-DIMENSIONAL PARAMETER-DEPENDENT FLOWS

When a two-dimensional vector field depends on time or another parameter, the representations of instantaneous topologies can be linked together to denote the time evolution of the flow. The adjacent representations are joined by linking their corresponding points and tangent curves.

The order in which the points and curves in different slices are linked is important because the distance in the plane between two points of the same class is not sufficient to indicate that they are corresponding points in the flow topology. This is particularly true of the *boundary incoming* and *boundary outgoing* points which are often very close together. For this reason, the saddle, detachment and attachment points are used as a basis for the linking, and points that are connected to corresponding curves from these points are linked recursively to ensure consistency.

The set of stacked topological representations can be displayed as a set of surfaces with the third dimension corresponding to time. The surfaces are created by filling in strips between corresponding tangent curves in adjacent slices of the representation. These strips are drawn only when the start and end points of the curve in one slice are linked to the start and end points of the curve in the next slice. If a topological transition has occurred, the surface between the time steps cannot be drawn without knowledge of the intermediate topology. Currently, these surfaces are omitted from the display.

Figure 5 shows the surfaces in the periodic flow around a two-dimensional circular cylinder. Time increases from back to front along the cylinder. The display utilizes several cues to aid visualization. The surfaces are lighted and shaded, and they are colored according to their type. Surfaces corresponding to the incoming separatrix of a saddle point are colored yellow. Those surfaces corresponding to the outgoing direction are colored blue. Surfaces from attachment points are colored orange, and those from detachment points are colored purple. The periodic vortex shedding can be seen in the repeated development and movement downstream of a saddle-center pair.

5 THREE-DIMENSIONAL SEPARATED FLOWS

The two-dimensional method described above carries over almost directly to the surface of a body in a three-dimensional flow. The only differences are that the two-dimensional vector field being considered is the tangential velocity field at the first grid point off of the body surface and that the normal velocity component is used to further subdivide the classification of each point into two subclasses depending on the sign of the normal component. For example, a saddle point may now either be a saddle of attachment or a saddle of separation.

Figure 6 shows the representation of the surface topology for the flow past a hemisphere cylinder at 19 degrees angle of attack. This computed flow, which is the same as shown in figures 1 and 2, was generated by Ying et al[1987]. The vertical and horizontal axes correspond to the radial and longitudinal grid indices, respectively. *Sa* and *Ss* denote a saddles of attachment and separation, respectively. *Na* and *Ns* represent nodes (both star nodes and foci) of attachment and separation, respectively. The topology of hemisphere cylinder flows, both real and simulated, have been studied extensively by Tobak and Peake [1979][1982] [1982].

This skin friction skeleton matches that inferred by Ying et al., except for the position and type of the focus paired with the saddle of separation on the far right. The nodes denoted *rn-1* and *rn-4* are repelling nodes on the surface which are also repelling in the normal direction and indicate an anomaly in the data set which was not seen until this method was applied to it. Given the extreme coarseness of the grid in the aft region of the body, the anomaly is not surprising, or particularly significant. And the interpretation of Ying et al. is born out by the structures of the surface of secondary separation. But this does demonstrate the utility of the methods we have presented for interpreting data sets and their ability to elucidate features which otherwise might be missed.

Figure 7 shows the skin friction topology on the body of the hemisphere cylinder. Figure 8 shows portions of the surfaces of separation at the nose and along the line of primary separation. The latter can be seen wrapping up into two vortices.

6 FUTURE WORK

Our work on developing methods for constructing representations of the topology of three-dimensional separated flows is far from complete. Eventually we hope to be able to fully characterize the topology and to generate surface representations of the complex structures that have traditionally been hand drawn. These representations rely on both graphics approaches and graph theory. For example, the graph representation of the topology can be used to infer the intermediate topology from graph representations before and after a transition.

7 CONCLUSIONS

By reducing the original vector field to a set of critical points and their connections, we have arrived at a representation of the topology of a two-dimensional vector field, which is much smaller than the original data set but retains with full precision the information pertinent to the flow topology. This representation can be displayed as a set of points and tangent curves or as a graph, which is especially useful for comparing data sets and detecting topological transitions. When time defines a third dimension, the representation can be readily displayed as surfaces.

As an intermediate step towards the development of a complete representation of three-dimensional flow topology, we have generalized the aforementioned two-dimensional methods to generate a representation of the topology on the surface of a body in a three-dimensional flow.

ACKNOWLEDGEMENTS

We are indebted to Drs. Lewis Schiff and Stuart Rogers of NASA Ames for providing us with interesting data sets to visualize and especially for sharing unpublished results from debugging runs. We also wish to thank Prof. Brian Cantwell of Stanford for sharing his ideas and experimental data and Dr. Murray Tobak of NASA Ames for elucidating comments and discussions.

This work is supported by NASA under contract NAG-2-489-S1, including support from the NASA Ames Numerical Aerodynamics Simulation Program and the NASA Ames Fluid Dynamics Division.

REFERENCES

G. L. BROWN and A. ROSHKO. On density effect and large structure in turbulent mixing layers. *Journal of Fluid Mechanics*, 64:775–816, 1974.

P. BUNING and J. STEGER. Graphics and flow visualization in computational fluid dynamics. *AIAA 7th Computational Fluid Dynamics Conference: A Collection of Technical Papers*, American Institute of Aeronautics and Astronautics, June 1985. Paper 85-1507-CP.

U. DALLMANN. *Topological Structures of Three-Dimensional Flow Separation.* Technical Report DFVLR-IB 221-82 A 07, Deutsche Forschungs- und Versuchsanstalt für Luft- und Raumfahrt, April 1983.

J. L. HELMAN and L. HESSELINK. Representation and display of vector field topology in fluid flow data sets. *IEEE Computer*, :, August 1989.

L. HESSELINK. Digital image processing in flow visualization. *Annual Review of Fluid Mechanics*, 20:421–485, 1988.

L. HESSELINK and J. L. HELMAN. Evaluation of flow topology from numerical data. *AIAA 8th Computational Fluid Dynamics Conference: A Collection of Technical Papers*, pages 825–831, American Institute of Aeronautics and Astronautics, June 1987. Paper 87-1181-CP.

T. LASINSKI, P. BUNING, D. CHOI, S. ROGERS, G. BANCROFT, and F. MERRIT. Flow visualization of cfd using graphics workstations. *AIAA 8th Computational Fluid Dynamics Conference: A Collection of Technical Papers*, pages 814–824, American Institute of Aeronautics and Astronautics, June 1987. Paper 87-1180.

D. J. PEAKE and M. TOBAK. *Three-Dimensional Flows About Simple Components at Angle of Attack.* Technical Report, NASA Ames Research Center, April 1982.

A. E. PERRY and M. S. CHONG. *Annual Review of Fluid Mechanics*, chapter A description of eddying motions and flow patterns using critical point concepts, pages 125–156. Annual Reviews Inc., 1987.

S. ROGERS and D. KWAK. An upwind differencing scheme for time accurate incompressible navier-stokes equations. *AIAA 6th Applied Aerodynamics Conference: A Collection of Technical Papers*, American Institute of Aeronautics and Astronautics, June 1988. Paper 88-2583.

M. TOBAK and D. J. PEAKE. Topology fo two-dimensional and three-dimensional separated flows. *AIAA 12th Fluid and Plasma Dynamics Conference*, July 1979.

M. TOBAK and D. J. PEAKE. Topology of three-dimensional separated flows. *Annual Review of Fluid Mechanics*, 14:61–85, 1982.

S. X. YING, L. B. SCHIFF, and J. L. STEGER. A numerical study of three-dimensional separated flow past a hemisphere cylinder. *AIAA 19 Fluid Dynamics, Plasma Dynamics and Lasers Conference*, American Institute of Aeronautics and Astronautics, June 1987. Paper 87-1207.

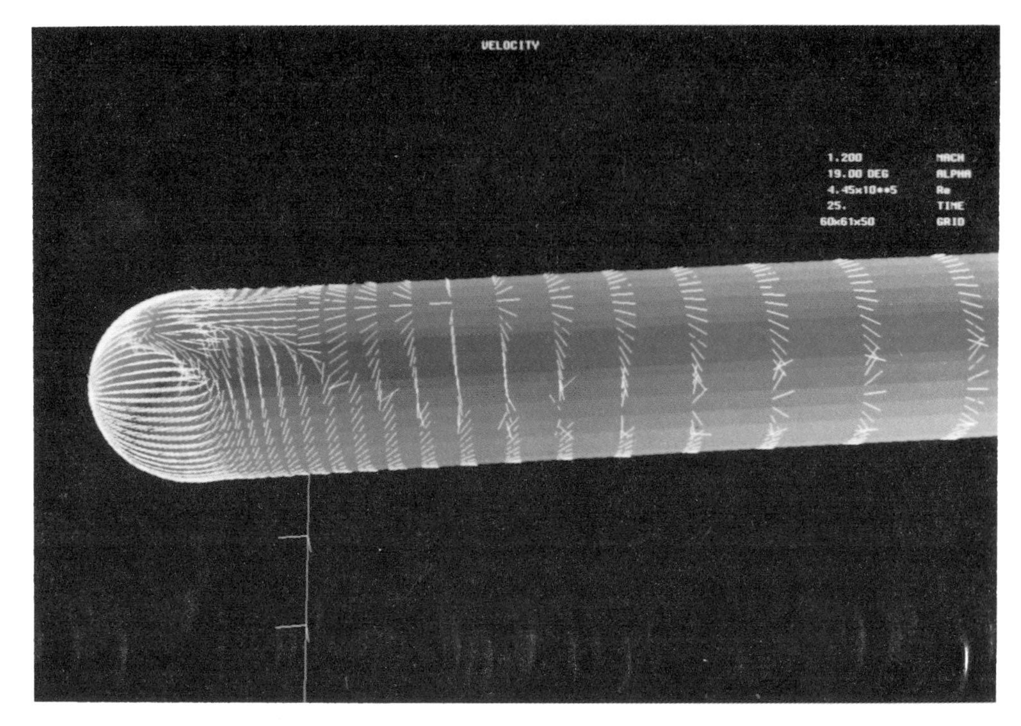

Figure 1: Skin friction vector field on the surface of a hemisphere cylinder.

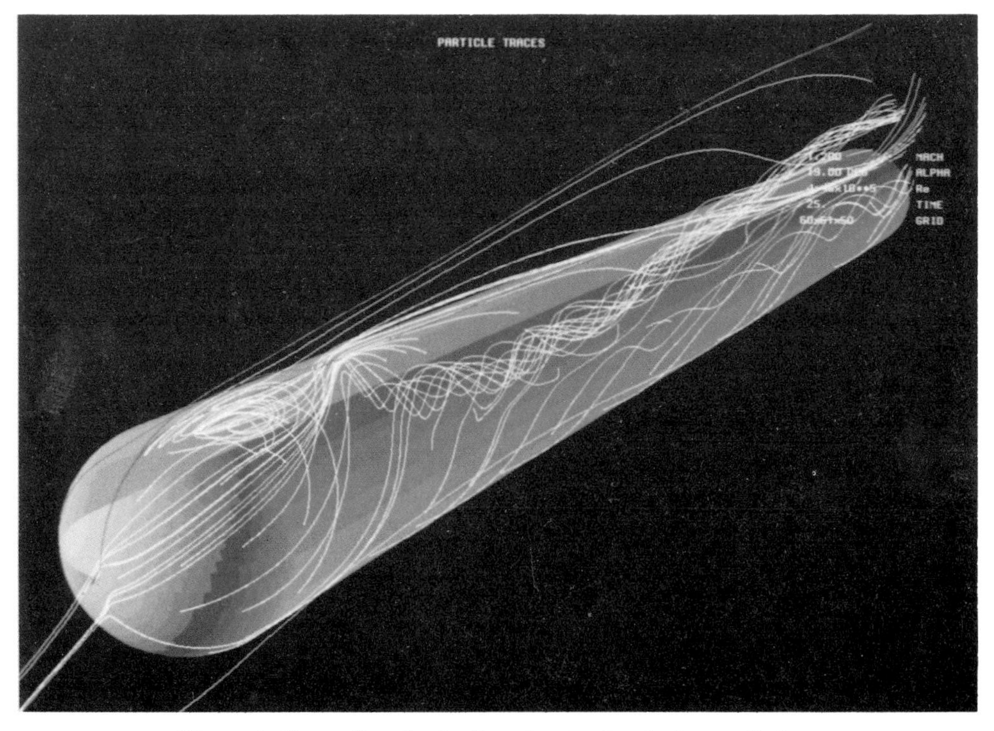

Figure 2: Streamlines in the flow about a hemisphere cylinder.

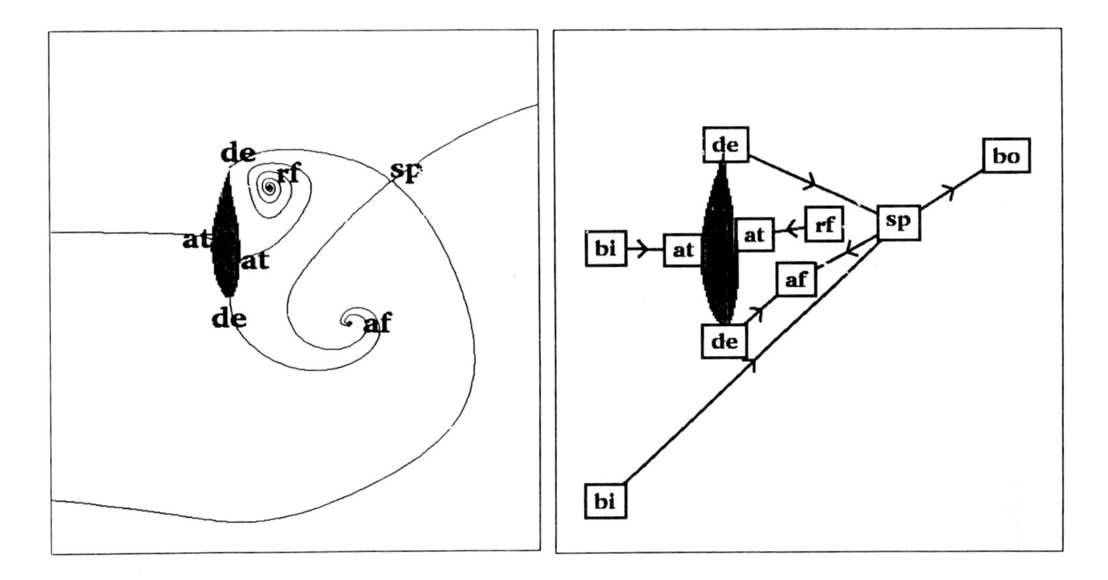

Figure 3: a. Critical points and connecting instantaneous streamlines in a two-dimensional flow around an airfoil. The data set was generated during the development of a simulation program and does not represent a physical flow. b. Corresponding graph of flow topology.

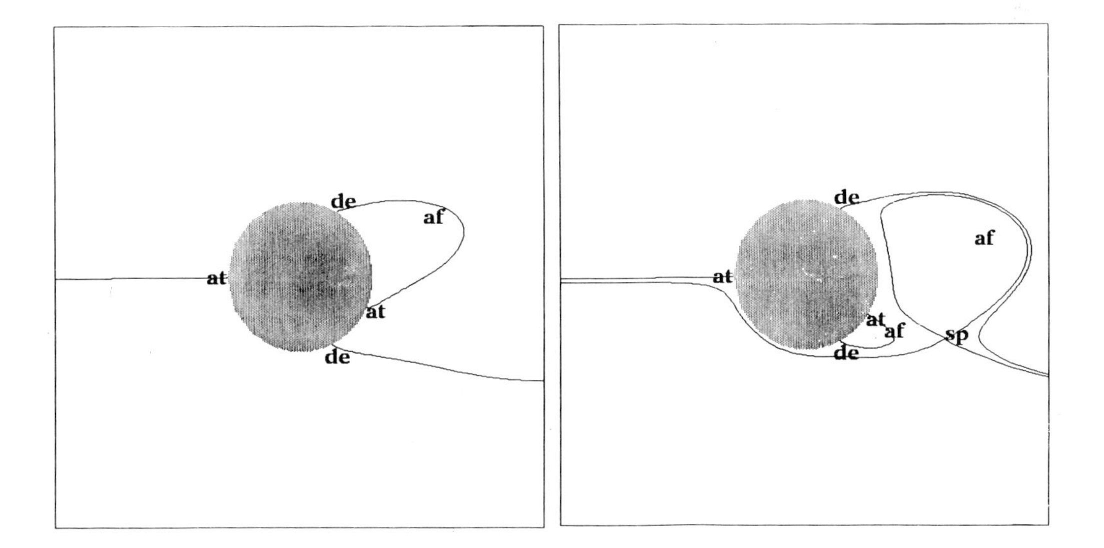

Figure 4: a. Critical points and connecting instantaneous streamlines in a computed flow around a circular cylinder. b. Same, at a later time step.

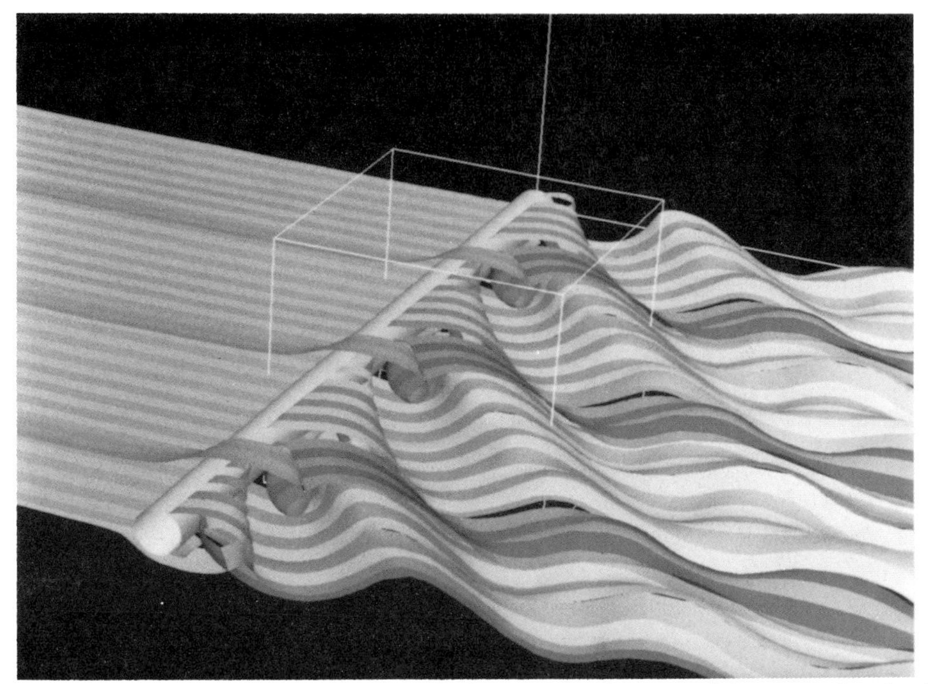

Figure 5: Topological surfaces showing time evolution of a two-dimensional flow around a circular cylinder.

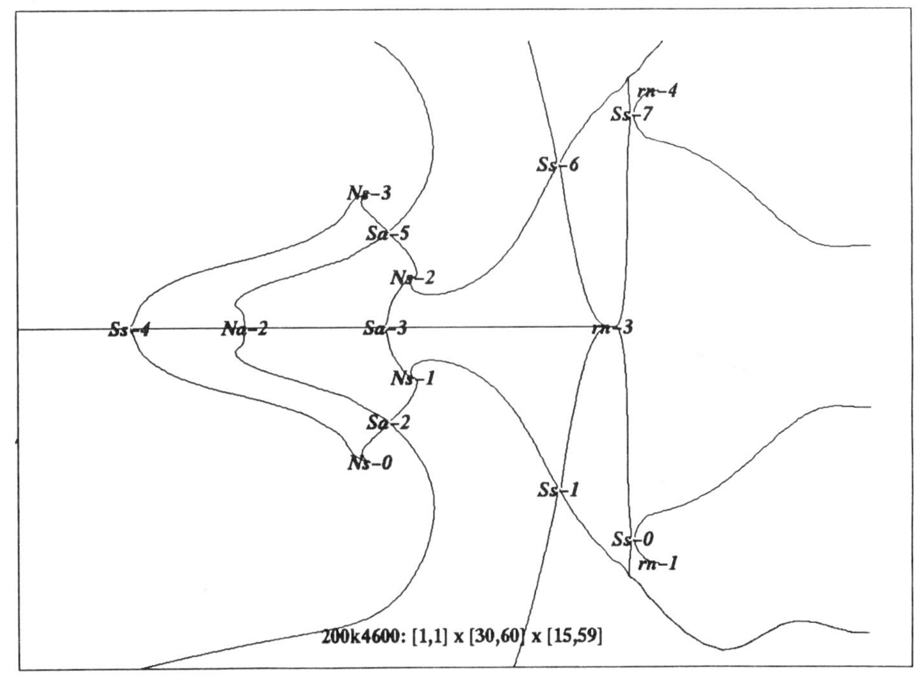

Figure 6: Skin friction curves and critical points from the analysis of the surface topology of a hemisphere cylinder.
(See also Colour Plates)

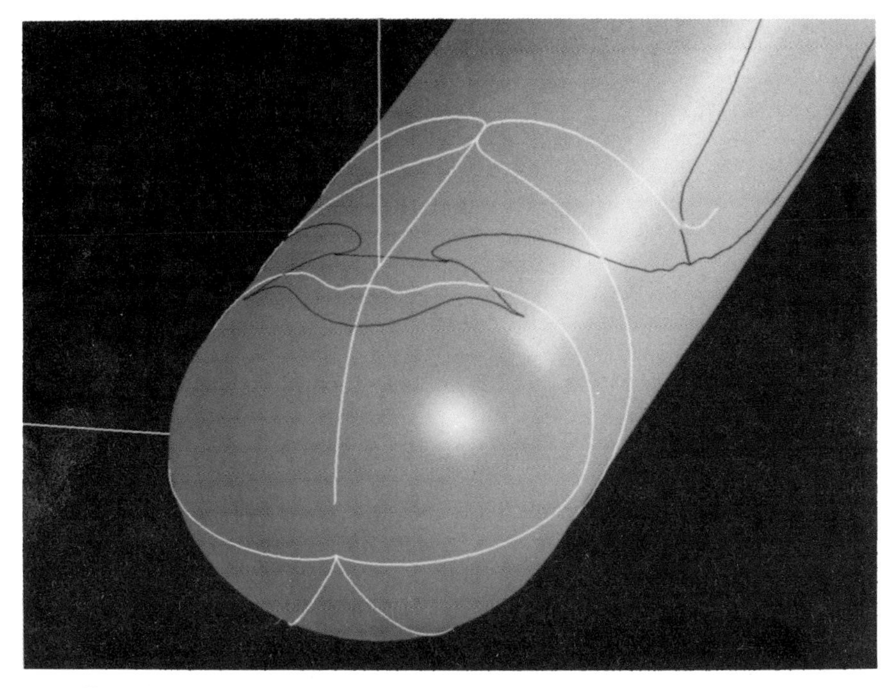

Figure 7: Skin friction curves representing topology on the surface of a hemisphere cylinder.

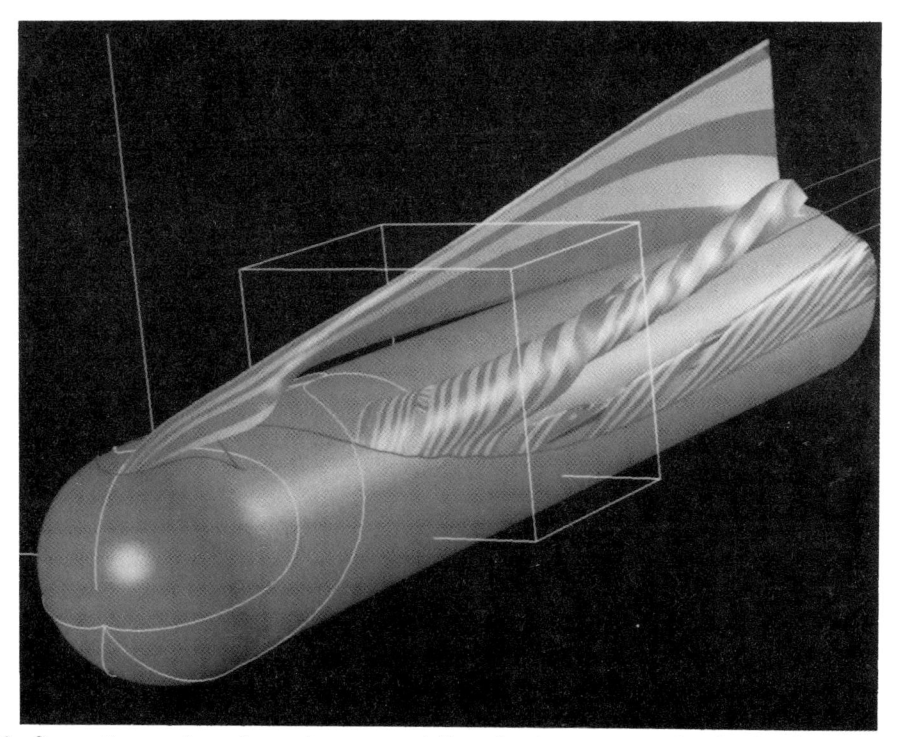

Figure 8: Separation surfaces from the nose and line of primary separation in the flow around a hemisphere cylinder.

(See also Colour Plates)

Topological Changes of Axisymmetric and Non-axisymmetric Vortex Flows

U. DALLMANN, & B. SCHULTE-WERNING

Institute for Theoretical Fluid Mechanics, DLr, D-3400 Göttingen, FRG

1. INTRODUCTION

Separated flows around simple geometric bodies may exhibit a variety of complex flow structures. The flow around a sphere is one such example. In order to validate any numerical simulation or to compare a calculated flow field with an experimental simulation we are forced to reduce the informations contained in huge data fields. In addition we have to face the problem that small scale perturbations are present in every simulation. The most likely unknown initial conditions in experiments, or the various differencies in numerical grids or algorithms, or different boundary conditions, etc., may cause considerable quantitative and qualitative differences in the simulations we compare, expecially when the flow exhibits separated flow regions. Furthermore, small variations or differences in parameters like Reynolds and Mach number can lead to flow bifurcations where changes of flow field symmetries occur. Hence, in order to validate any two simulations it is necessary that the *sequences of topological changes* that occur in the fields are identical in both experimental and/or numerical simulations. It is the aim of this paper to show by a topological analysis of a numerical simulation of the flow around a sphere at a low Mach number what kind of, either simulation dependent or independent ("physical") topological changes one has to face and what kind of technical problems and requirements arise in this context when a computer-aided graphical topological analysis is necessarily performed. We do not intend at this stage that the simulation results presented below are in perfect agreement with any other experimental or numerical results, we only emphasize the necessity of analyzing *sequences of topological changes* that occur in flow fields before any further (for instance quantitative) comparison is of any value.

Upon analyzing the topological structure of any flow field the following question arises: What are the necessary and sufficient topological (rather than analytical) informations about a flow that have to be compared and have to be equivalent in any two simulations such that any further topological properties of any other quantity (variables or invariants) will be preserved in these two simulations? In the present paper we only analyze the topological structure of the velocity fields and vorticity fields. However, it is at least necessary that all the topological properties of all the dependent variables of the underlying Navier-Stokes equations will be discussed in a further study.

2. ON THE TOPOLOGICAL ANALYSIS OF COMPLEX FLOW FIELDS

The formation of three-dimensional, non-axisymmetric, steady or unsteady flows under axisymmetric boundary conditions is of fundamental interest in fluid mechanics. Flow separation and vortex breakdown are phenomena where such topological changes are of importance. In our contribution we investigate the formation of a three-dimensional near-wake behind a sphere by a topological analysis of a numerical simulation.

Numerical simulations as well as experimental investigations of three-dimensional flows provide a tremendous amount of data. In order to analyze these data for the purpose of physical modelling of local or global flow properties a procedure is necessary that reduces the information contained in the data fields in a systematic way. The description of topological properties of the flow is of great value for that purpose. Investigations of the topological structure of complex flow fields, namely, of the separated transonic flow around a hemisphere-cylinder combination (Kordulla et al. 1986), of the three-dimensional convection flow in a rectilinear container (Dallmann 1985b) and of the topological changes of separating flows around cylinders and spheres at transition Reynolds numbers (Dallmann et al. 1987) provided new insights into the complex structure of three-dimensional vortex flows. Three-dimensional, elementary vortex structures, which may form at a wall, were derived by Dallmann (1988).

In order to define the topological structure of any vector field one has to locate, at first, all the so called critical points in the field. Such critical points, where the direction field is non-unique, may appear on rigid boundaries as well as in mid-air. It is this last feature that may cause insuperable technical difficulties for a topological analysis, since the resulting, graphically displayed flow structure may be highly dependent on the "proper" choice of initial conditions for instance in streamline and streamsurface integration (Kordulla et al. 1986, Dallmann 1985b). Once the critical points are identified in the flow field the eigendirections of the streamline trajectories at these critical points provide the initial directions along which the integration of the so called singular streamsurfaces (Dallmann 1982, 1985a, 1988) has to be performed. The topology of the instantaneous velocity field is then given by the complete set of singular streamsurfaces which are determined by the set of all of those streamlines which connect, reach or leave the critical points on the flow boundaries (fig. 1) and in mid-air. For instance, the boundary of a body submerged in a flow always defines one such singular streamsurface where the "streamlines" on the wall, the skin-friction trajectories, run out of the front stagnation point, cover the whole surface of the body and finally run either into the rear separation point or into well-defined ensembles of critical points. Since streamline as well as any other trajectories' integration can, at best, be started in the vicinity of critical points considerable sensitivity on these initial conditions may influence the resulting graphic display.

Another problem has to be mentioned. Although sectional streamlines are widely used to identify vortices the result is in general strongly dependent on the selected sequence of sections. Only the sequence of "Poincaré-sections", derived in the well-known way, sketched in fig. 1, provide us with the correct information. They visualize the intersections of the singular stream- (vorticity-, etc.) surfaces with the selected sections.

Although the formation of a vortex on a boundary can be sufficiently identified by the formation of foci on these walls there is the additional possibility that vortices form or annihilate in mid-air. In fig. 2 we show examples of "annihilating" vortices (decay of swirling motions) in local solutions of the Navier-Stokes equation at a rigid wall. Vollmers (1983) and Dallmann (1982) have formulated criteria, based on the invariants of the Jacobian matrix of three-dimensional vector fields, that allow to locate regions in a flow field where locally swirling, vortical flow appears relative to an observer who moves with the local velocity through the fluid. The "vortices" which can be identified by this local investigation of the flow field may indeed form in mid-air. One such example is provided by the separating leeside vortices behind an ellipsoid at angle of attack. The surface flow pattern only shows local convergence of neighbouring skin friction lines towards a so called open separation line. Apart from the front and rear stagnation points no other critical point which could be identified as the apex of a separating vortex appears on the boundary. Hence we conclude that singular streamsurfaces do not provide the complete information about the structure of a flow field. The topology of the vorticity field provides additional information.

Elementary topological structures of vorticity fields have been described by Dallmann (1988). On a rigid wall, where the no-slip boundary condition holds, the skin friction lines are always orthogonal to the wall vortex lines. It follows, that the wall critical points of one set of trajectories are also critical points of the other set of trajectories and, therefore, singular vortex surface integration follows the same rules and faces the same sensitivity on initial conditions as singular streamsurface integration.

A complete topological analysis of a flow requires the description of the topological structures and properties of all the dependent flow variables. The analysis of instantaneous velocity and vorticity fields is only part of this task.

3. NUMERICAL SIMULATION OF FLOW SEPARATION AND VORTEX SHEDDING FROM A SPHERE

As an example we analyze a numerical simulation of the formation of three-dimensional separating flow around a sphere. In order to validate the physical relevance of such a numerical simulation a topological reduction of the information contained in the data is extremely helpfull. For a given simulation we should describe the topological structures of the dependent variables or of an associated set of invariant vector fields. Any changes in the calculation procedure, changes of grids, algorithms, numerical damping terms imposed, etc. may then be identified in its effects on the topological structure of these variables. In the following we give such a description for one such simulation in terms of the instantaneous velocity and vorticity field.

A survey of the physical properties and the evolution of the sphere wake as a function of the *Re*-number based on experimental investigations can be found in Clift et al. (1978) and Möller (1938).

The simulation was performed by calculating the unsteady, compressible equations of motion using finite-difference solvers of the Beam-Warming type. The axisymmetric flow was computed by a fully implicit scheme with second order accuracy both in time and space (Müller 1985), the three-dimensional flow was computed by a hybrid implicit/explicit method with first order accucary in time and second order accuracy in space, which was initially developed by Riedelbauch & Müller (1987) and extended by Schulte-Werning (1990). The results presented below are considered to be preliminary, since no attempt is made in the present paper to discuss qualitative and quantitative changes of the calculated flow field due to changes of mesh, numerical damping terms etc. Nevertheless, we have to mention that no perturbations are introduced by the numerical algorithm itself apart from round-off and truncation errors, only the explicit numerical damping terms could be a source of perturbations, which may cause bifurcation from an axisymmetric to a three-dimensional flow in a similar way as perturbations do in experiments. Concerning the effect of boundary and initial conditions, *Re*- and *Ma*-number variation and variation of numerical damping we refer to Schulte-Werning et al. (1989). The calculation of the three-dimensional flow around a sphere is performed on a spherical grid with equidistant spacing in circumferential direction and without imposing any symmetry condition onto the evolving flow.

The freestream Mach number for all calculations was $Ma = 0.40$, thus we are away from transonic flow phenomena which may considerably influence the flow behaviour. This study will be part of an investigation of compressibility effects on flow separation at low *Ma*-numbers.

In fig. 3 we show the simulation of axisymmetric flows for $Re = 10$ and 1000. Flow separation at the rear stagnation point starts at $Re \geq 10$ and the now appearing closed axisymmetric separation line moves upstream. Above $Re = 500$ secondary separation takes place which leads, due to another bifurcation in the flow field, into multiply structured recirculation zones. Fig. 4 shows the development of the recirculation zones in a plane of axial symmetry for $Re = 300/\ 600/\ 1000$.

Fig. 5 shows, for $Re = 100$, the onset of three-dimensional flow separation. A slight swirl appears near the separation line indicating the bifurcation from the (structurally unstable) axisymmetric flow without swirl to one of the (structurally stable) three-dimensional types of separations. An axially swirling, separated near wake arises first. The apparent swirl in the flow indicates a topological bifurcation which clearly stimulates further investigations on the dependence of this topological change upon the numerical algorithm used in the simulation. We can demonstrate that the direction of the resulting swirl is initiated by a non-axisymmetric influence of the numerical damping terms. It is of interest that such a "fictitious" or "physical" swirl (for instance in experiments) is only apparent above a critical Reynolds number and can be recognized only in association with the separated flow region. The attached flow region remains axially symmetric without any, measurable swirl. In addition, fig. 5 indicates that care has to be taken in numerical streamline integration. One recognizes convergence of streamlines towards a closed vortical loop in the center of the near wake. This phenomenon cannot appear in an axially symmetric swirling wake flow but will result from insufficient accuracy in the streamline integration.

For $Re = 500$ fig. 6 indicates that the surface flow trajectories in the attached flow region upstream of the separation line remain fairly axisymmetric. As the flow approaches the separation line it becomes three-dimensional and the skin-friction trajectories as well as the streamlines in the separated flow exhibit a plane of flow symmetry which has not been imposed explicitely but which is a large scale response of the non linear dynamics of the vorticity field to small-scale three-dimensional perturbations introduced via numerical errors. The topological structure of the skin-friction lines is all of a sudden characterized by a sequence of critical points, saddles and nodal points. The apparent change in surface flow symmetry is sketched in fig. 7. All skin friction lines leaving the front stagnation point (node) run into the node on top of the sphere except the one line running into the saddle point just in the opposite of the aforementioned node. The skin friction lines emanating from the rear stagnation point (node) exhibit the same behaviour. It is worthwhile noting that simultaneously with the development of three-dimensionality at the separation line the rear stagnation point starts moving off the geometrical axis.

At $Re = 2000$ the flow has become time-periodic such that the surface flow patterns shown in fig. 8 change periodically in time. One recognizes the formation of nodal points and especially of foci on both sides of a slowly rotating plane of flow symmetry (see also fig. 7). (We remind the reader, that the apparent global swirl of the separated flow region has been identified as a topological change at a lower Reynolds number and its cause has been discussed above). Vorticity will be advected towards or away from these nodal points and will leave or reach the vicinity of the sphere under the formation of streamwise vorticity and vortical structures in the velocity field. Although the surface flow is rather simple to be described it is very hard to describe the other complex convoluted singular streamsurfaces in the separated flow region of the near wake. Fig. 9 shows an attempt for extracting information from the calculated flow on the sphere surface in order to "unfold" the flow complexity in the separated flow region. Since the three-dimensional wake has been formed by bifurcations from a steady, axisymmetric closed near wake bubble and has preserved one plane of flow symmetry we may sketch the plane-of-symmetry streamlines as shown in fig. 9; but we have to consider various possible reconnections of neighbouring streamlines (indicated by $\ominus$). These reconnections immediately lead to complex three-dimensional vector fields (fig. 9c shows one simplifying example) and singular streamsurfaces become extremly convoluted. If we simulate an experimental smoke/dye-injection by placing the starting points of streamline (or streakline) integration near to the stagnation line in front of the sphere a fairly chaotic behaviour of the streamlines in the wake results, only showing a certain spatial symmetry (fig. 10). Streamlines starting at the foci in the rear of the sphere exhibit a considerably more regular behaviour, but even in this case the formation of a "vortex configuration" cannot be identified. Hence, we investigate the vorticity field (fig. 10).

Vortex lines will form either closed loops or will leave and reach the surface of the sphere at critical points. Hence it is of interest to follow those vortex lines which leave or reach the vicinity of the nodal points and especially those which leave the foci seen in the wall flow patterns. These vortex lines and the associated singular vorticity surfaces exhibit a "vortex configuration" in the wake of the sphere. A similar structure has been sketched by Achenbach (1974) and Cometta (1957). However, their sketches do not precisely define what is meant by "vortex". Kuwahara et al. (1987) have obtained a similar vortex line behaviour in their incompressible flow simulation. Although there are considerable similarities in the sequence of topological changes of the surface flows in their simulations, their incompressible flow calculations do not show all the topological changes that occur in our compressible low-Mach number simulations.

4. CONCLUDING REMARKS

We have chosen results of a numerical simulation of the separated flows around a sphere to indicate the needs and requirement for computer-aided analysis of topological flow changes. Here, only a few basic considerations of the velocity and vorticity fields could be presented. Various other quantities have to be analyzed in order to obtain a more complete picture of this complex flow. The topological analysis of compressibility effects, the redistribution and reconnection of the vorticity field in the wake, the spatial distribution of viscous forces and of dissipation is presently being analyzed in order to derive simplifying physical and mathematical models for better understanding of vortex separation and shedding from a sphere.

ACKNOWLEDGEMENT

The study has been partly supported by Deutsche Forschungsgemeinschaft DFG under *Da 183/1-5*. The authors highly appreciate the support of H. Vollmers providing computer graphic codes.

REFERENCES

ACHENBACH, E. (1974) Vortex shedding from spheres. *J. Fluid Mech. 62.*

CLIFT, R., GRACE, J.R. & WEBER, M.E. (1978) Bubbles, Drops and Particles. *Academic.*

COMETTA, C. (1957) An Investigation of the Unsteady Flow Pattern in the Wake of Cylinders and Spheres using a Hot Wire Probe. *Div. Engng., Brown Univ. Techn. Rep. WT-21.*

DALLMANN, U. (1982) Topological Structures of Three-Dimensional Flow Separations. *DFVLR - IB 221-82 A 07.*

DALLMANN, U. (1985a) On the Formation of Three-Dimensional Vortex Flow Structures. *DFVLR - IB 221-85 A 13.*

DALLMANN, U. (1985b) Structural Stability of Three-Dimensional Vortex Flows. *In: Nonlinear Dynamics in Transcritical Flows, (H.L. Jordan, H. Oertel & K. Robert, eds.), Springer Lecture Notes in Engng.*

DALLMANN, U. (1988) Three-Dimensional Vortex Structures and Vorticity Topology. *Proc. IUTAM Symp. on Fundamental Aspects of Vortex Motion, Tokyo, Sept. 1987, (H. Hasimoto & T. Kambe, eds.), North-Holland, Amsterdam, pp.183-189.*

DALLMANN, U., SCHEWE, G. (1987) On Topological Changes of Separating Flow Structures at Transitions Reynolds Numbers. *AIAA 19th Fluid Dynamics, Plasma Dynamics and Lasers Conference, Honolulu, Hawaii, AIAA-87-1266.*

KORDULLA, W., VOLLMERS, H., DALLMANN, U. (1986) Simulation of Three-Dimensional Transonic Flow with Separation Past a Hemisphere-Cylinder Configuration. *AGARD Meeting, Aix-en-Provence; AGARD-CP 412, pp. 31-1 – 31-15.*

KUWAHARA, K. & SHIRAYAMA, S. (1987) Simulation of Unsteady Flow Separation. *Forum on Unsteady Flow Separation. ASME, Cincinnati, Ohio.*

MÖLLER, W. (1938) Experimentelle Untersuchungen zur Hydromechanik der Kugel. *Physikalische Zeitschrift, Vol.39, 57-80.*

MÜLLER, B. (1985) Navier-Stokes Solution for Hypersonic Flow over an Intended Nose Tip. *AIAA Paper 85-1504.*

RIEDELBAUCH, S. & MÜLLER, B. (1987) The Simulation of Three-Dimensional Viscous Supersonic Flow past Blunt Bodies with a Hybrid Implicit/Explicit Finite-Difference Method. *DFVLR - FB 87-32.*

SCHULTE-WERNING, B. (1990) PhD-thesis, *in preparation.*

SCHULTE-WERNING, B., DALLMANN, U., MÜLLER, B. (1989) Zur Numerischen Lösung der Zweidimensionalen Symmetrischen Zylinder- und Kugelumströmung. *DLR - IB 221-89 A 14.*

VOLLMERS, H. (1983) Separation and Vortical-Type Flow around a Prolate Spheroid. Evaluation of Relevant Parameters. *AGARD Symp. on Aerodyn. of Vortical Type Flow in Three Dimensions, Rotterdam, AGARD-CP 342, pp. 14-1 – 14-14.*

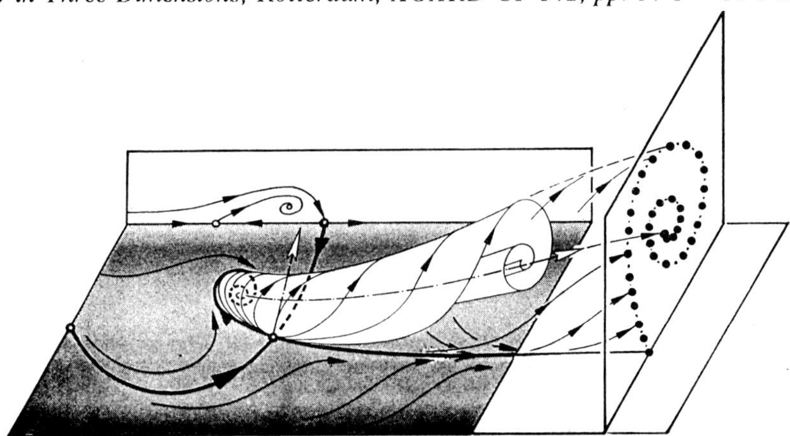

Fig. 1. Principle sketch indicating singular streamsurfaces (wall W, "vortex" V, possible symmetry plane S) and "Poincaré section" Σ. A sectional streamline pattern in Σ would provide a very different information.

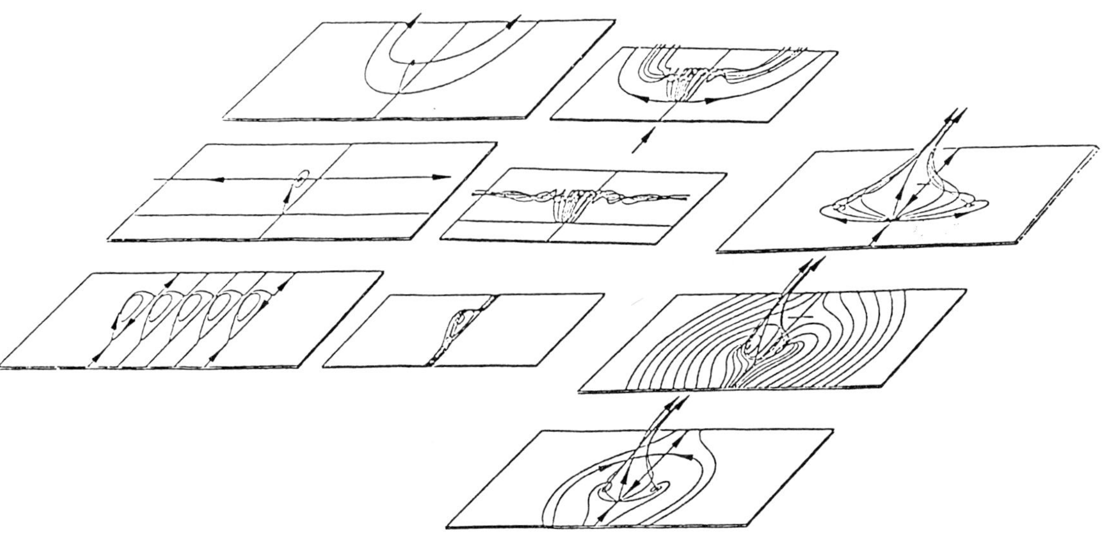

Fig. 2. Simulation of elementary flow structures with vortices created at a wall indicating the possible formation and annihilation of swirling motions ("vortices").

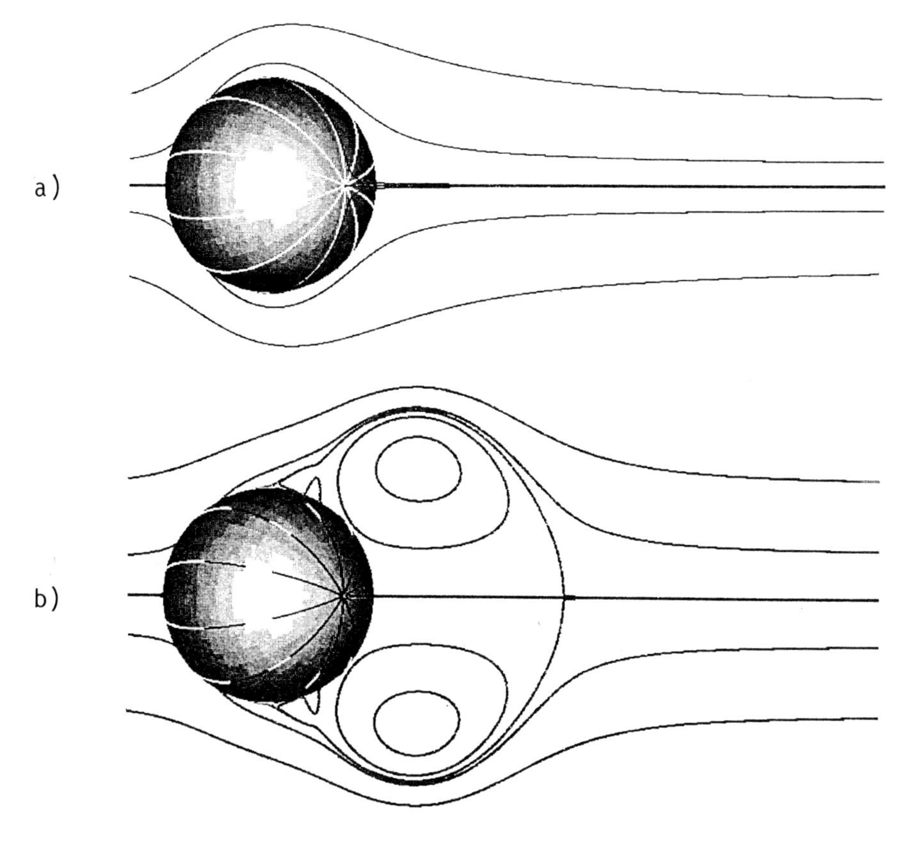

Fig. 3. Streamlines of axisymmetric compressible flow around a sphere at $Ma = 0.4$ a) $Re = 10$, attached flow b) $Re = 1000$, recirculation zones with secondary separation.

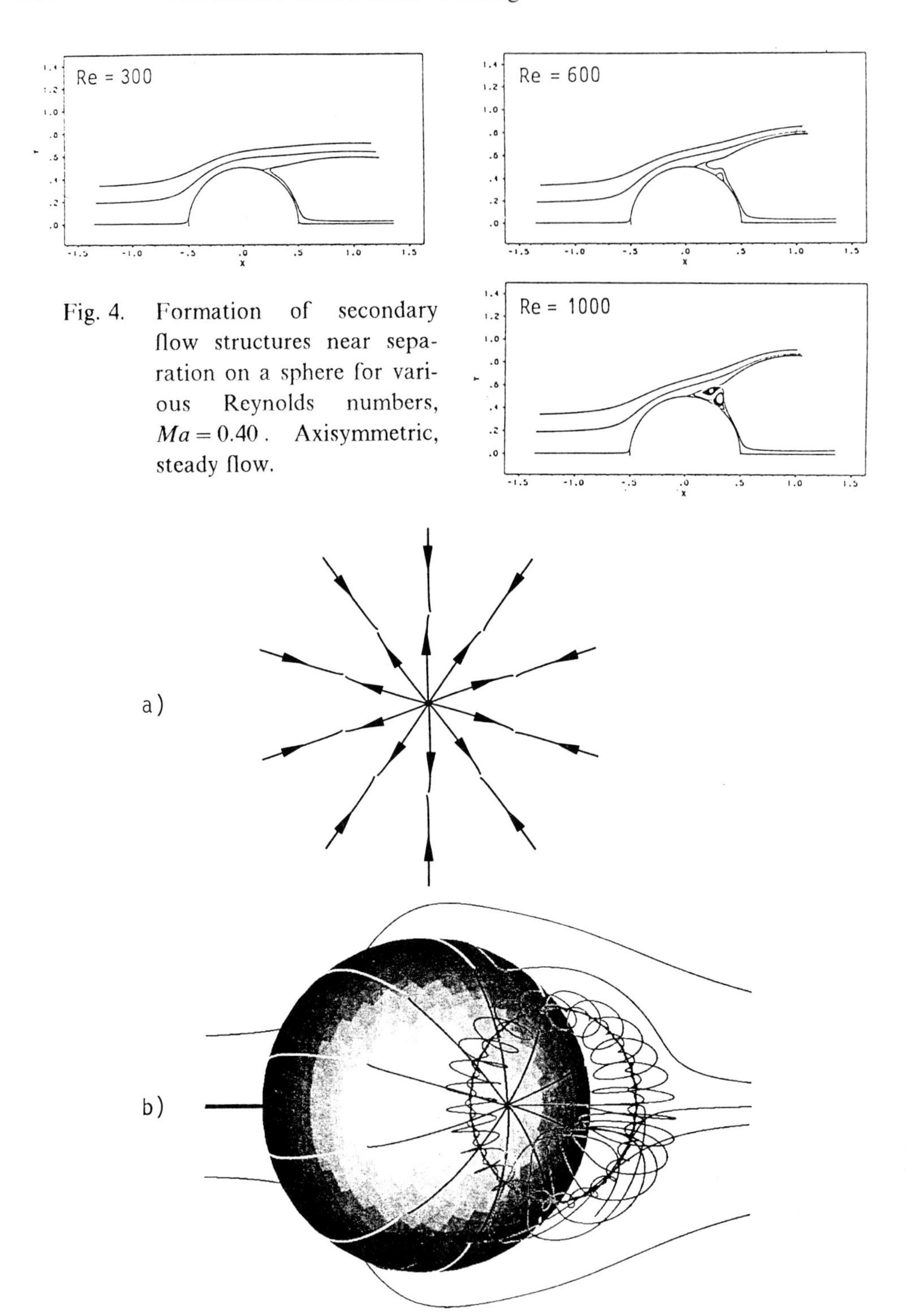

Fig. 4. Formation of secondary flow structures near separation on a sphere for various Reynolds numbers, $Ma = 0.40$. Axisymmetric, steady flow.

Fig. 5. Skin-friction lines and streamlines of 3-D flow around a sphere at $Re = 100$ indicating swirl flow at separation line, a) rearview (conformal mapping of spherical surface streamlines onto a plane) and b) inside the wake.

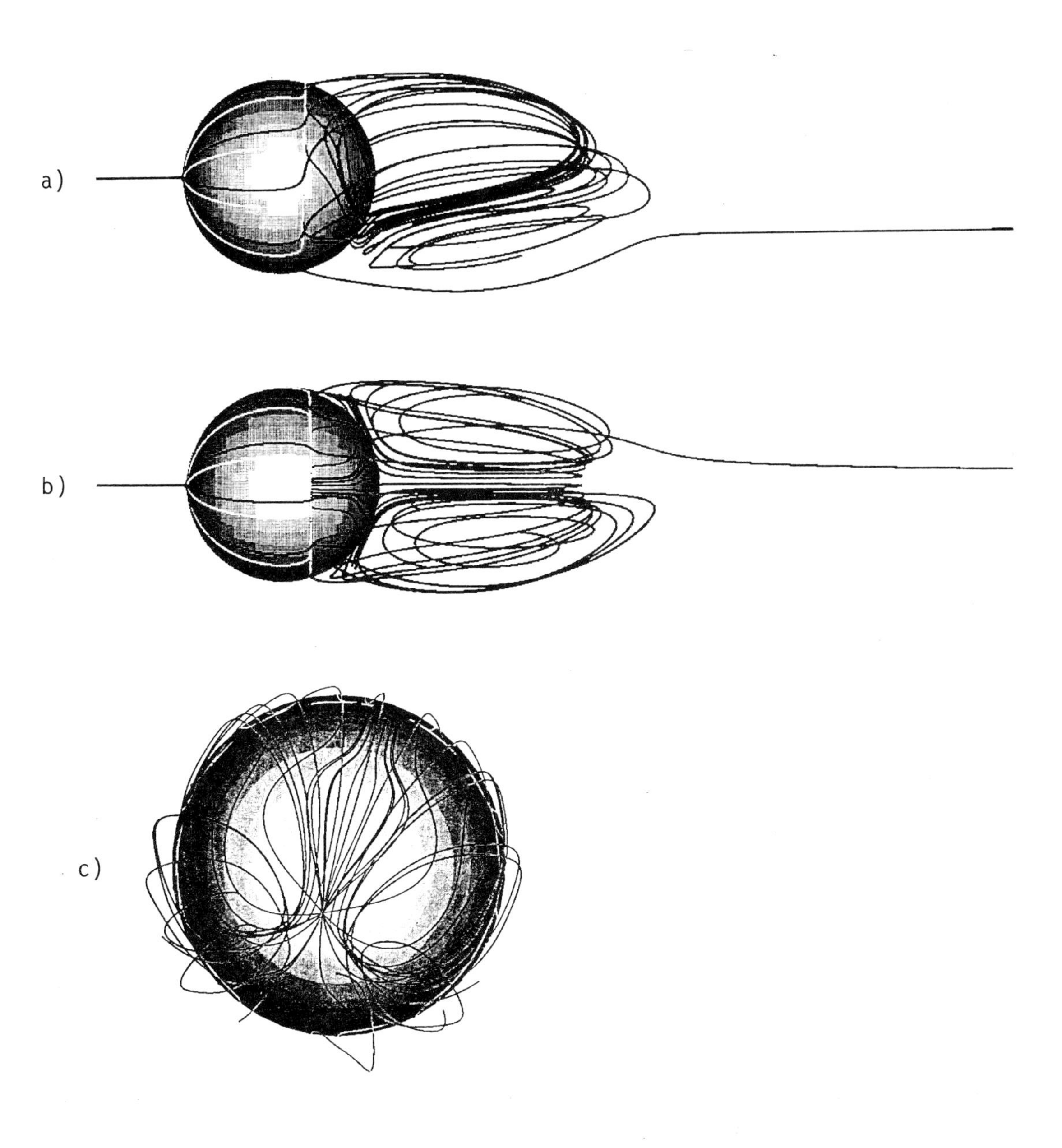

Fig. 6. Skin friction and streamlines of 3-D flow around a sphere at $Re = 500$, a) side view, b) top view, c) rear view, indicating the formation of a plane of flow symmetry.

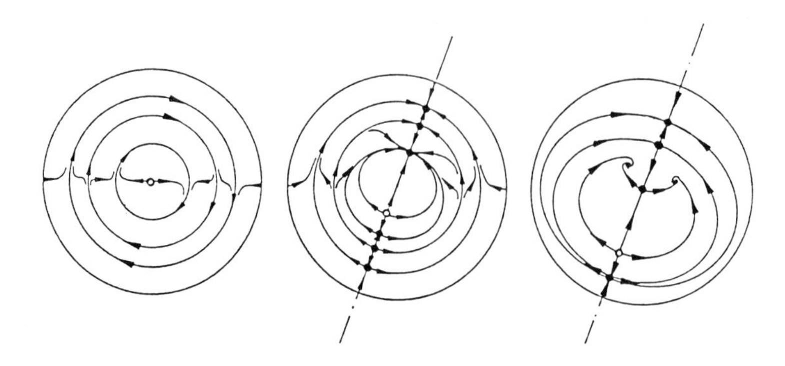

Fig. 7. Principle sketch of symmetry and topological changes in surface flow patterns.

Fig. 8. Time-sequence of skin friction lines on a sphere. Numerical simulation $Re = 2000$, $Ma = 0.40$. Upstream view onto the separated flow region shows the periodic formation of foci where streamwise vortices form on the sphere.

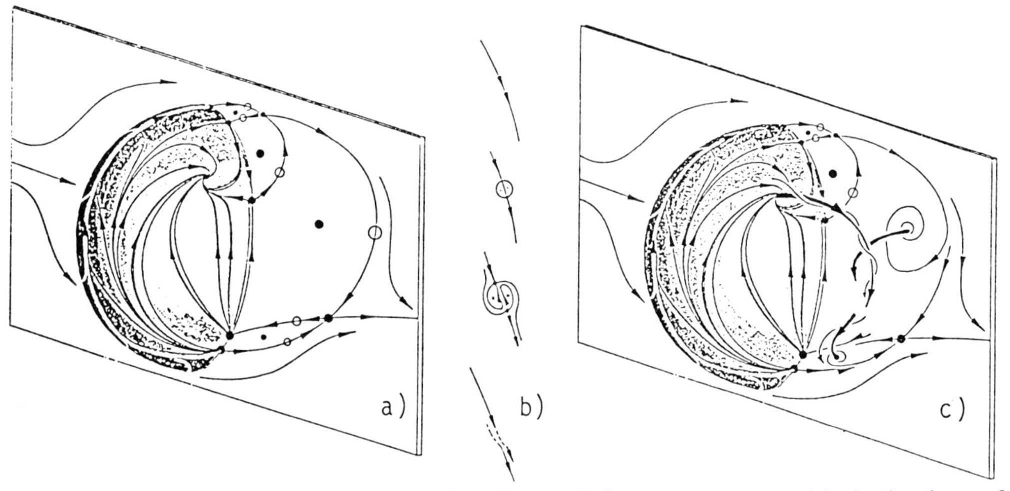

Fig. 9. Sketch of conjectured flow in plane of flow symmetry with indication of reconnections (b) of neighbouring streamlines due to three-dimensional perturbations leading to complex large-scale flow three-dimensionality. (c, simplified sketch of one such flow).

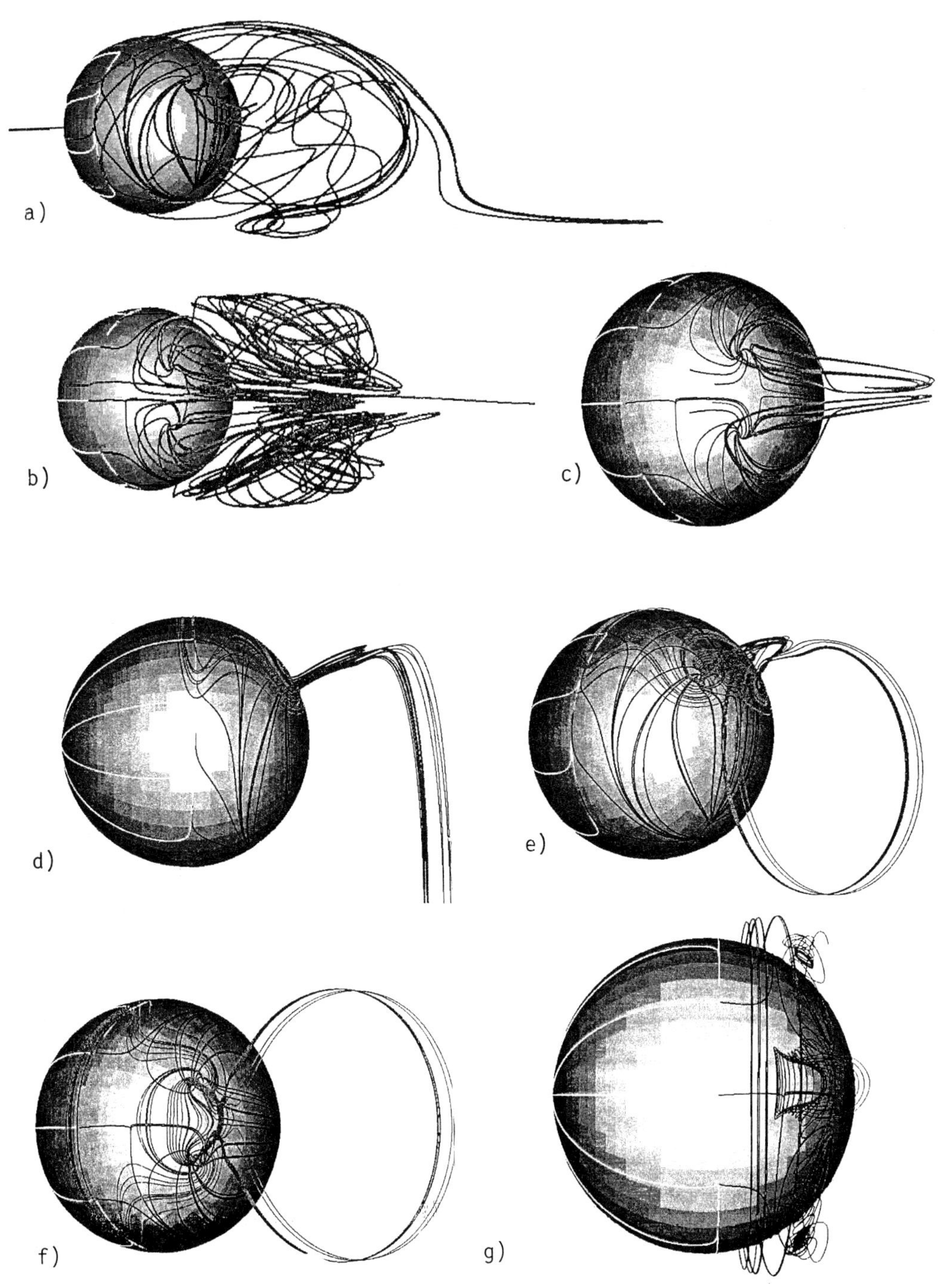

Fig. 10. Skin friction lines and vorticity lines of 3-D flow around a sphere at $Re = 2000$, streamlines starting at the front stagnation line (a), selected streamlines (b), streamlines starting at the foci (c), vorticity lines starting at the foci (d, e, f) and closed vorticity loops (g) which are getting folded in the separated flow region. "See also colour plates".

On the Topology of Three-Dimensional Separated Flow Structures and Local Solutions of the Navier-Stokes Equations

P.G. Bakker and M.E.M. de Winkel.

Department of Aerospace Engineering, Delft University of Technology, The Netherlands.

1 Introduction.

The topology of three-dimensional (3D) separated flow patterns is a subject of several contemporary studies in aerodynamics. Interesting results are performed by Hornung (1983), Dallmann (1983) and Perry & Chong (1986). Hornung introduces the vortex-skeleton model in order to describe unambiguously the qualitative features of the structure in a steady 3D flow. Dallmann analyses a special set of local solutions of the Navier-Stokes (NS) equations describing the flow patterns with some special symmetry conditions. Later on, Perry & Chong have developed an algorithm that enables them to generate local solutions of the NS-equations numerically. Recently we have developed a classification strategy along which viscous flow patterns with complicated topology can be analyzed in a systematic way. The method relies on the qualitative theory of differential equations and the theory of bifurcations. For 2D flows the method is outlined by Bakker (1988), the investigation of 3D flows is taken up by de Winkel (1988) and Kooij (1989). Some results will be presented in this paper.

2 Local solutions of the Navier-Stokes equations, general theory.

The flow is assumed to be steady, viscous, incompressible and it satisfies no-slip boundary conditions on the wall. The topology of such a flow is studied on the basis of local solutions of the NS-equations. The streamline pattern is represented by the trajectories of a third-order dynamical system $\dot{x} = u(x,y,z)$, $\dot{y} = v(x,y,z)$, $\dot{z} = w(x,y,z)$ with u, v, w denoting the velocity components in an orthogonal cartesian reference system x, y, z; $(\dot{\ })$ denotes $\frac{d}{dt}$ and t is real time. The wall is represented by $y = 0$. The local solutions are obtained by performing Taylor expansions of the velocity vector field.

Consider the flow pattern that results if the velocity vector field is expanded up to the Nth order near an arbitrary point P, then the trajectory pattern near P is governed by the system (S):

$$\dot{x} = u = \sum_{i=0}^{N} \sum_{j=0}^{N-i} \sum_{k=0}^{N-i-j} U_{i,j,k} x^i y^j z^k + O(N+1)$$

$$\dot{y} = v = \sum_{i=0}^{N} \sum_{j=0}^{N-i} \sum_{k=0}^{N-i-j} V_{i,j,k} x^i y^j z^k + O(N+1) \qquad (S)$$

$$\dot{z} = w = \sum_{i=0}^{N} \sum_{j=0}^{N-i} \sum_{k=0}^{N-i-j} W_{i,j,k} x^i y^j z^k + O(N+1)$$

where O(N+1) denote higher-order terms of at least order N+1, composed by powers of x, y and z. The coefficients $U_{i,j,k}$, $V_{i,j,k}$ and $W_{i,j,k}$ are unknown constants. Due to the flow equations and the boundary conditions some relationships between $U_{i,j,k}$, $V_{i,j,k}$ and $W_{i,j,k}$ exist, reducing the number of coefficients that can be chosen independently. Our main objective will be to give for finite N an unified description of all topologically different flow patterns that will arise near P if the remaining coefficients are varied independently.

Instead of the phase portraits of system (S) we study those of the truncated system (S_N), which is system (S) omitting the O(N+1) terms. Since (S_N) contains only a finite number of terms the question arises whether it gives a proper description of the local flow topology in flows governed by the full NS-equations. With regard to this question we know that phase portraits are characterized by the number and position of singular points, the local trajectory pattern near these points and the position of separatrices. If the topological structure of singular points of system (S_N) can not be disturbed if higher order terms are added, then system (S_N) suffices to obtain a qualitative description of the local flow topology of system (S). For isolated hyperbolic singularities a linear expansion (N=1) is already sufficient for determining the local topology. For isolated non-hyperbolic singularities also a truncated system (S_N) can be found provided that N is chosen sufficiently large so that O(N+1) terms do not disturb the topological character of the degenerate singularities. Normal form theory is used to examine whether higher order terms play a role for the local flow topology. This implies that for a given N a classification of degenerate singularities can be established, and each of them represents a local flow topology which can occur in flows governed by the full NS-equations.

Moreover system (S_N) enables us also to encounter local flow patterns consisting of a cluster of singular points forming a coherent flow structure. In order to guarantee that such a cluster is a local solution of the NS-equations it will be sought as an unfolding of a non-hyperbolic singularity of system (S_N).

3 Mathematical formulation.

Consider the incompressible steady laminar flow in the neighbourhood of a plane wall at rest. The flow near the wall occurs in the upper halfplane $y \geq 0$. It satisfies:

the continuity equation: $\operatorname{div} \underline{V} = 0$

the NS-equations conserving momentum: $(\underline{V} \cdot \nabla)\underline{V} + \nabla p^* = \nu \Delta \underline{V}$

the no-slip boundary conditions: $u(x,0,z) = v(x,0,z) = w(x,0,z) = 0 \quad \forall x,z \in R$

with $p^* = \frac{p}{\rho}$ the kinematic pressure, ν the kinematic viscosity, p the pressure and ρ the mass density. Applying the continuity equation and the no-slip condition system (S) takes the form

$$\dot{x} = y u'(x,y,z),\ \dot{y} = y^2 v'(x,y,z),\ \dot{z} = y w'(x,y,z) \tag{1}$$

The plane y = 0 is a plane with singular points (u=v=w=0) implying a quasi non-hyperbolic singular character at the wall surface. Since the trajectories of system (1) at $y \neq 0$ are identical with those of the equivalent system:

$$\dot{x} = y^{-1}u(x,y,z),\ \dot{y} = y^{-1}v(x,y,z),\ \dot{z} = y^{-1}w(x,y,z) \tag{2}$$

the singular character of the plane $y = 0$ can be removed by investigating (2) instead of (1). The equivalent system which governs the flow above the wall becomes:

$$\begin{aligned}
\dot{x} &= a_1 + a_2x + a_3y + a_4z + a_5x^2 + a_6xy + a_7xz + a_8y^2 + a_9yz + a_{10}z^2 + O(3)\\
\dot{y} &= y\{b_1 + b_2x + b_3y + b_4z\} + O(3)\\
\dot{z} &= c_1 + c_2x + c_3y + c_4z + c_5x^2 + c_6xy + c_7xz + c_8y^2 + c_9yz + c_{10}z^2 + O(3)
\end{aligned} \tag{3}$$

where $a_i, b_j \,\&\, c_k \in R$. To fulfil the continuity equation and the NS-equations relations between the coefficients $a_i, b_j \,\&\, c_k$ exist. The one which will be often used reads $a_2 + 2b_1 + c_4 = 0$. The coefficients $a_i, b_j \,\&\, c_k$ can also be expressed in the physical quantities p^* and τ, σ which are the components of the shear stress in resp. x- and z-direction defined by $\tau = \mu(u_y)_{y=0}$ & $\sigma = \mu(w_y)_{y=0}$. Considering system (3) note that the plane $y = 0$ is now filled with solution curves $\dot{x} = u'(x,0,z),\ \dot{z} = w'(x,0,z)$. Since $(u_y)_{y=0} = u'(x,0,z)$ and $(w_y)_{y=0} = w'(x,0,z)$ these solution curves are identical with the skin friction lines defined by $\dot{x} = \tau,\ \dot{z} = \sigma$.

4 Singular points on the wall, Jordan normal forms.

Singularities of system (3) can be interpreted in two ways, those lying above the wall represent stagnation points in the flow, those located at the wall have a vanishing shear stress vector indicating either flow separation from the wall or flow attachment to the wall. Since we are particularly interested in the topology of separation and attachment structures the origin of the reference system is taken in a wall singularity. Then $a_1 = c_1 = 0$ and in shorthand notation we may write with $\underline{x} = (x,y,z)^T$

$$\dot{\underline{x}} = A\underline{x} + f(\underline{x}) + O(3) \qquad \text{where } A = \frac{1}{\mu}\begin{pmatrix} \tau_x & \frac{1}{2}p_x & \tau_z \\ 0 & \frac{1}{2}p_y & 0 \\ \sigma_x & \frac{1}{2}p_z & \sigma_z \end{pmatrix} = \begin{pmatrix} a_2 & a_3 & a_4 \\ 0 & b_1 & 0 \\ c_2 & c_3 & c_4 \end{pmatrix} \tag{4}$$

and f(x) is a vector polynomial containing the second and third order terms of system (3). To analyse system (4) we may apply Hartman-Grobman's theorem which states that the character of a singularity is determined by the eigenvalues of the linear part of the system, unless there are eigenvalues having real parts equal to zero, Re $(\lambda) = 0$; for details see Guckenheimer & Holmes (1983). Points having all Re $(\lambda) \neq 0$ are called <u>hyperbolic</u>. Otherwise they are called <u>non-hyperbolic</u>, degenerate or higher-order points. The linear part $\dot{\underline{x}} = A\underline{x}$ yields the eigenvalues $\lambda_{1,3} = -b_1 \mp \sqrt{(a_2+b_1)^2 + a_4c_2},\ \lambda_2 = b_1$. The continuity equation gives the relation $\lambda_1 + 2\lambda_2 + \lambda_3 = 0$ reflecting the fundamental property that parts of the flow will enter and other parts of the flow will leave the neighbourhood of the singular point.

A suitable linear transformation $\underline{x} = T\underline{u}$ which does not affect the plane $y = 0$ may be applied and bring the non-linear system (4) into the equivalent form

$$\dot{\underline{u}} = J\underline{u} + g(\underline{u}) \tag{5}$$

where $J = T^{-1}AT$ is one of the Jordan normal forms listed in Table 1 and $g(\underline{u}) = T^{-1}f(T\underline{u})$.

Hyperbolic	Non-hyperbolic
1) all eigenvalues are real and distinct $\begin{pmatrix} \lambda_1 & 0 & 0 \\ 0 & \lambda_2 & 0 \\ 0 & 0 & -(\lambda_1+2\lambda_2) \end{pmatrix}$ (1a)	1) all eigenvalues are real and distinct $\begin{pmatrix} 0 & 0 & 0 \\ 0 & \lambda_2 & 0 \\ 0 & 0 & -2\lambda_2 \end{pmatrix}$ (1b) $\begin{pmatrix} \lambda_1 & 0 & 0 \\ 0 & 0 & 0 \\ 0 & 0 & -\lambda_1 \end{pmatrix}$ (1c)
2) all of them are real and two are equal $\begin{pmatrix} \lambda_1 & 0 & 0 \\ 0 & \lambda_1 & 0 \\ 0 & 0 & -3\lambda_1 \end{pmatrix}$ (2a) $\begin{pmatrix} \lambda_1 & 0 & 0 \\ 0 & -\lambda_1 & 0 \\ 0 & 0 & \lambda_1 \end{pmatrix}$ (2b) $\begin{pmatrix} \lambda_1 & 1 & 0 \\ 0 & \lambda_1 & 0 \\ 0 & 0 & -3\lambda_3 \end{pmatrix}$ (2c) $\begin{pmatrix} \lambda_1 & 0 & 1 \\ 0 & -\lambda_1 & 0 \\ 0 & 0 & \lambda_1 \end{pmatrix}$ (2d)	2) all of them are real and two are equal ----------
3) all of them are real and equal ----------	3) all of them are real and equal $\begin{pmatrix} 0 & 0 & 1 \\ 0 & 0 & 0 \\ 0 & 0 & 0 \end{pmatrix}$ (3a) $\begin{pmatrix} 0 & 1 & 0 \\ 0 & 0 & 0 \\ 0 & 0 & 0 \end{pmatrix}$ (3b) $\begin{pmatrix} 0 & 0 & 1 \\ 0 & 0 & 0 \\ 0 & 1 & 0 \end{pmatrix}$ (3c) $\begin{pmatrix} 0 & 0 & 0 \\ 0 & 0 & 0 \\ 0 & 0 & 0 \end{pmatrix}$ (3d)
4) one eigenvalue is real and two are complex conjugated $\begin{pmatrix} -\lambda_2 & 0 & \mathrm{Im}\lambda_1 \\ 0 & \lambda_2 & 0 \\ -\mathrm{Im}\lambda_1 & 0 & -\lambda_2 \end{pmatrix}$ (4a)	4) one eigenvalue is real and two are complex conjugated $\begin{pmatrix} 0 & 0 & \mathrm{Im}\lambda_1 \\ 0 & 0 & 0 \\ -\mathrm{Im}\lambda_1 & 0 & 0 \end{pmatrix}$ (4b)

Table 1. Jordan normal forms of the linear part of system (4).

5 Elementary singularities in skin friction patterns.

The qualitative behaviour of trajectories around a hyperbolic point informs us about the possible types of elementary singularities that can appear in a skin friction field. To classify them we use the Jordan normal forms as listed in Table 1. A singularity is called an attachment point ($\lambda_2 < 0$) or a separation point ($\lambda_2 > 0$) if a particular streamline above the wall can be found which enters or leaves the singular point on the wall. Since $\lambda_2 = b_1 = \frac{1}{2}p_y$ (Table 1) attachment (separation) points indicate a pressure decrease (increase) when going in off wall direction. We find three basically different skin friction patterns near an attachment point; the singularity is either a saddle (τ_x and σ_z have different sign), an unstable node (general, starlike or inflected, τ_x and σ_z have the same sign) or an unstable focus ($(\tau_x - \sigma_z)^2 + 4\tau_z\sigma_x < 0$). Very similar skin friction patterns arise near a separation point, however the node and the focus must be stable in that case. The possible flow structures near an attachment point are shown in Fig. 1.

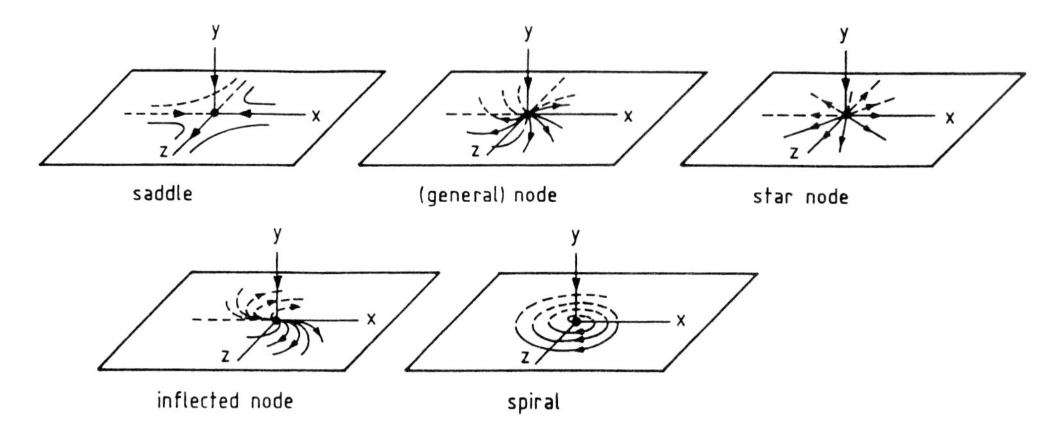

Fig. 1. The possible flow structures near attachment points.

Some of these local flow structures can be recognized in Aerodynamics. The (general) node appears for instance on the surface of a body at the attachment point of the free stream. In case of a symmetrical body (like a sphere) one might expect the star node as an attachment point. The focus corresponds to a vortex standing on the surface of the body and saddles occur for example in flows with closed separation bubbles. The inflected node is not much familiar in aerodynamic applications because it may appear as a transition pattern between node and focus.

6 Non-hyperbolic singularities and bifurcations.

The proper determination of the local flow topology near a non-hyperbolic singularity requires several dominant non-linear terms to be retained in the series expansion. These terms can be found by applying a near-identity transformation $\underline{u} = \underline{y} + P(\underline{y})$ yielding a normal form of system (5); $P(\underline{y})$ being a vector polynomial (at least of degree 3) mapping the wall onto $y_2 = 0$. Then a blow down/up technique, see Andronov (1973), is used to analyze the structurally unstable flow patterns. They can fall apart when subjecting them to small disturbances. Bifurcation theory is applied to find the resulting structurally stable flow structures. To illustrate the procedure two cases of Table 1 namely (4b) and (3c) will be treated in the next paragraphs.

6.1 Flow structures associated with the Jordan normal form (4b).

Consider system (4) with Jordan form (4b). Using normal form theory, applying the continuity equation and using cylindrical coordinates system (4) is equivalent to

$$\dot{r} = r\{cy + er^2 + fy^2 + O(3)\}$$
$$\dot{y} = y\left\{-\frac{2}{3}cy - 2er^2 - \frac{1}{2}fy^2 + O(3)\right\} \qquad (6)$$
$$\dot{\phi} = -\omega + by + gy^2 + jr^2 + O(3)$$

In (6) the ϕ-equation is uncoupled from the $\dot{r}$- and $\dot{y}$-equation indicating that the local flow can be investigated in a meridional plane ϕ = constant. It can be proven that the uncoupled behaviour remains if higher order terms are added. Fig. 2. shows typical meridional structures of degenerate 3D separations. For sake of convenience these structures are referred to as a bowl- (Fig. 2a), a saddle- (Fig. 2b) and a cone type separation.

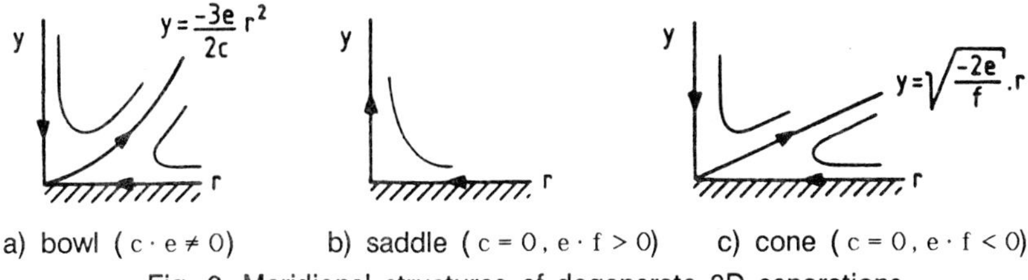

a) bowl $(c \cdot e \neq 0)$ b) saddle $(c = 0, e \cdot f > 0)$ c) cone $(c = 0, e \cdot f < 0)$

Fig. 2. Meridional structures of degenerate 3D separations.

6.2 Burst of a closed separation bubble, bifurcation of system (6).

The unfolding of the structurally unstable cone type separation (Fig. 2c) will now be analysed. It can be obtained by using Mather's technique, as explained by Shirer & Wells (1983). Then the leading terms of (6) are supplemented with lower order terms which contain the bifurcation parameters μ_1 and μ_2 and satisfy the governing flow equations. The unfolding in cylindrical coordinates reads

$$\begin{array}{lll} & \text{supplementary terms} & \text{degenerate singularity} \\ \dot{r} = & \mu_1 r + \mu_2 r y & + r\{e r^2 + f y^2\} \\ \dot{y} = & -\mu_1 y - \frac{2}{3}\mu_2 y^2 & + y\left\{-2 e r^2 - \frac{1}{2} f y^2\right\} \\ \dot{\phi} = & & -\omega + b y \end{array} \qquad (7)$$

Since the ϕ-equation is uncoupled and the bifurcation parameters do not appear in this equation the bifurcation can be analysed in a plane ϕ = constant. The following bifurcation sets can be identified (Fig. 3):

Higher order Saddle (HS)	$\mu_1 = 0 \ \& \ \mu_2 \neq 0$
Third order Saddle (TS)	$\mu_1 = \frac{2}{9f}\mu_2^2 \ \& \ \mu_2 < 0$
Cusp (C)	$\mu_1 = \frac{8}{27f}\mu_2^2 \ \& \ \mu_2 < 0$

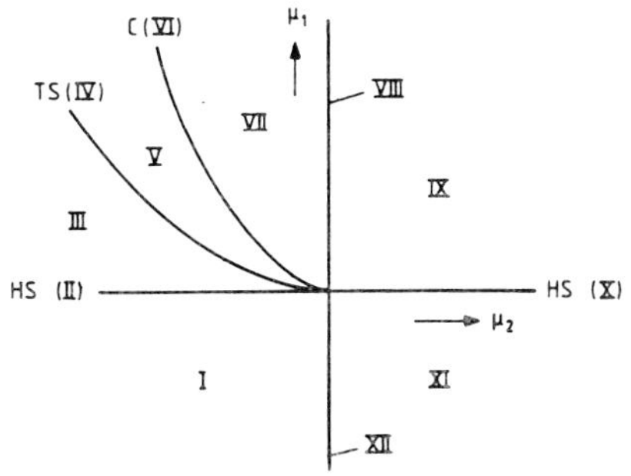

Fig. 3. Bifurcation sets.

The various flow patterns that result from bifurcation are shown in Fig. 4. In region I we observe a flow pattern having two saddle points on the y-axis. The 3D flow reveals two separate flow domains which interact together in a rotational movement. There is a part A which moves on to the wall more or less in perpendicular direction. Next to it a part B exists which flows along the wall towards the axis of rotation. Both flows are separated by a 3D bowl-shaped separation surface fully embedded in the flow above the wall. No separation from the wall is observed. Going from region I towards III bifurcation of a higher order saddle at $r = y = 0$ occurs. The local flow structure near this higher order singularity is already depicted in Fig. 2, case $c \neq 0$. The higher order saddle bifurcates into a closed bubble fitted to the wall (region III). Due to the formation of the bubble flow separation in part B occurs along a closed separation line. This closed separation line appears as a stable limit cycle in the skin friction field on the wall. Inside the bubble particle paths lie on tori and they form a center in the meridional plane. Since the flow in the r-y-plane is governed by a hamiltonian system these centers remain also if higher order terms should be taken into account. This implies that mass is conserved inside the bubble.

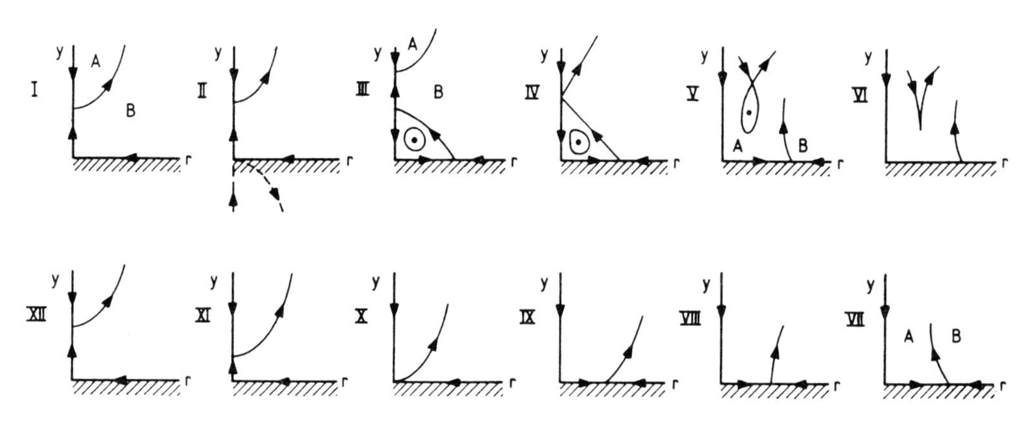

Fig. 4. The bifurcated flow structures in the r-y-plane.

Moving from region III to region V a third order saddle point appears on the y-axis above the wall. The closed bubble interferes with the bowl-shaped separation surface. Then the flow in part B is pushed away from the axis of rotation and an open separation surface standing on the wall results. The flow in part A, forces the bubble to leave the wall and generates an annulus of recirculating flow (with probably two commensurate frequencies) inside the flow A. The observed phenomenon may be referred to as bursting of the closed separation bubble. Moving to VII the annular region of circulating flow shrinks until we pass region VI where cusp bifurcation appears in the meridional plane. The cusp disappears and we end up with a flow having a 3D open separation surface originating from the wall (region VII-IX). Going from IX to XI the separation line on the wall moves to the origin (region X) where bifurcation causes lift off of the separation surface (region XI, XII, I).

6.3 Flow structures associated with the Jordan normal form (3c).

Consider system (4) with Jordan normal form (3c). Using normal form theory and applying the continuity equation system (4) is equivalent to:

$$\dot{x} = z + x^2 - \frac{2}{3}y^2 + O(3)$$

$$\dot{y} = -xy + O(3) \tag{8}$$

$$\dot{z} = y + cx^2 - \frac{1}{3}cy^2 + O(3)$$

The local flow structures of this higher order singularity are for the first time discussed by Kooij (1989) for the case $c \neq 0$. The coefficient c is related to the local value of σ_{xx} in the singularity; for $c > 0$ and $c < 0$ the 3D flow patterns are shown in Fig. 5.

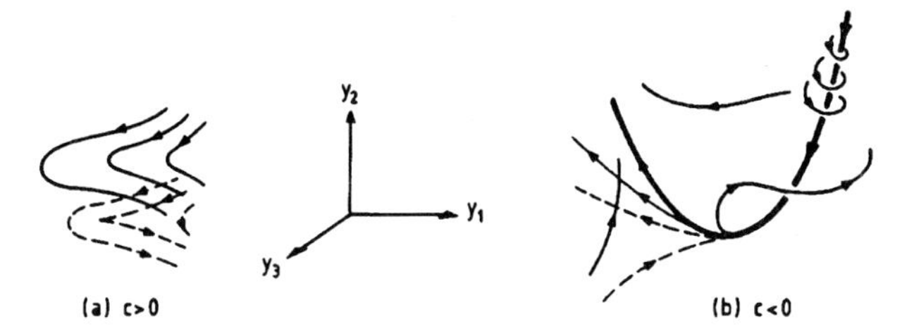

Fig. 5. The local flow structures of system (8) near the origin.

The skin friction pattern as resembled by the dotted lines has a cusp singularity which points in x-direction.

6.4 Development of an open separation surface (case $c < 0$),bifurcation of system (8).

The unfolding of the case $c < 0$ will now be analysed. Applying Mather's technique the unfolding of system (11) becomes

$$\dot{x} = \mu x + z + x^2 - \frac{2}{3}y^2 + O(3)$$

$$\dot{y} = y\left\{-\frac{1}{2}\mu - x + O(2)\right\} \tag{9}$$

$$\dot{z} = \lambda + y + cx^2 - \frac{1}{3}cy^2 + O(3)$$

where λ and μ represent small perturbations of the shear stress quantities σ and τ_x respectively. The various bifurcations can be discussed, let us elaborate the case $c < 0$ in more detail. The bifurcation sets and corresponding flow patterns are shown in Fig. 6 and 7, respectively. The bifurcations that can occur are

Hopf ($p^+ = 0$):

$\lambda = \frac{1}{4}\mu^2 \quad (\mu < 0)$

Saddle-Node ($D = 0$ and $p^- = 0$):

$\lambda = 0 \ \& \ \lambda = \frac{1}{4}\mu^2 \quad (\mu > 0)$

Saddle-loop (S):

$\lambda = \left(\frac{7}{10}\right)^2\mu^2 \quad (\mu < 0)$

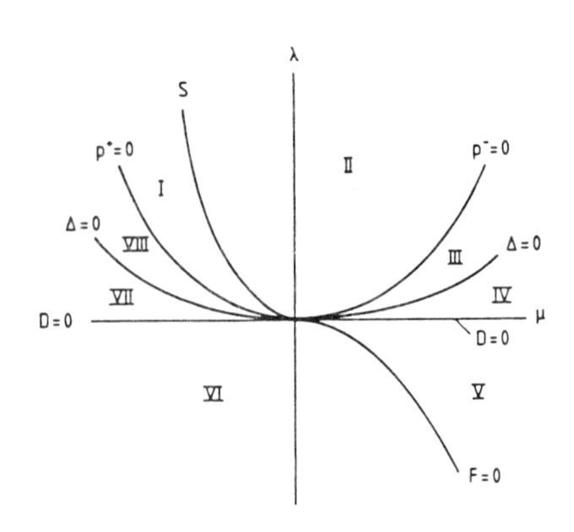

Fig. 6. The bifurcation sets of system (9) for $c < 0$.

Along $\lambda = \frac{1}{64}\mu^4\,(\Delta = 0)$ and $\lambda = -\frac{2}{9}\sqrt{3}\mu^{\frac{3}{2}}\,(F = 0)$ foci and nodal points transfer into each other by passing an improper node.

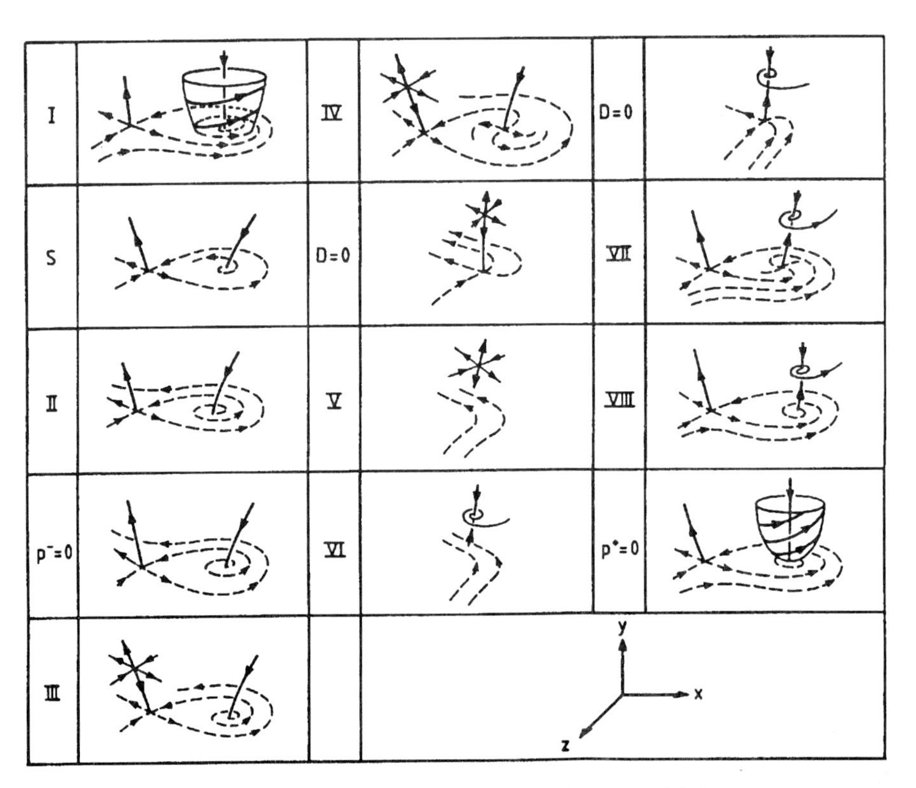

Fig. 7. The bifurcated flow structures of system (9) for $c < 0$

Moving in the parameter plane onto various domains, a sequence of complicated flow structures in R^3 is encountered. The most salient features of them will be revealed briefly. If the degenerate state $\mu = 0, \lambda = 0$ is perturbed by taking $\lambda < 0$ a regular skin friction field results (VI). Additionally there appears a stagnation point in the flow where two vortex tubes seem to terminate in a focal singularity which carries the fluid away from the tube axis. No separation from nor attachment to the wall surface is observed. The focal singularity is replaced by a singularity having a 1D stable eigenspace and 2D unstable eigenspace when moving to domain V. A perturbation to $\lambda > 0$ can result into flow patterns having separated flow regions. Two distinct singularities appear in the skin friction field. If separation is present then there is at least one singularity with an unstable manifold which carries the fluid away from the wall. The dimension of the unstable manifold is one or two. The first case corresponds with a stable focus (domain VIII) or with a saddle point (domains I,II,VII,VIII) in the skin friction field. If both singularities are present (domain VIII) then the well-known saddle-focus pattern as observed by Legendre (1965) results. A 2D unstable manifold occurs in correspondance with a higher-order stable fine focus ($p^* = 0$) in the skin friction pattern. In this case the manifold is more or less bowl-shaped, it originates from the fine focus on the wall and it forms a separation surface in the flow. The flow pattern is structurally unstable. A perturbation via a Hopf bifurcation creates a structurally stable flow pattern containing a separation surface forming a closed separation line on the wall (domain I). This separation line appears as a limit cycle in the skin friction field and encloses an unstable focus. Such a limit cycle is a rather new phenomenon in skin friction patterns. Just as the open separation phenomenon as observed by K.C. Wang (1983) it is an example where a separation line has no terminating singular points.

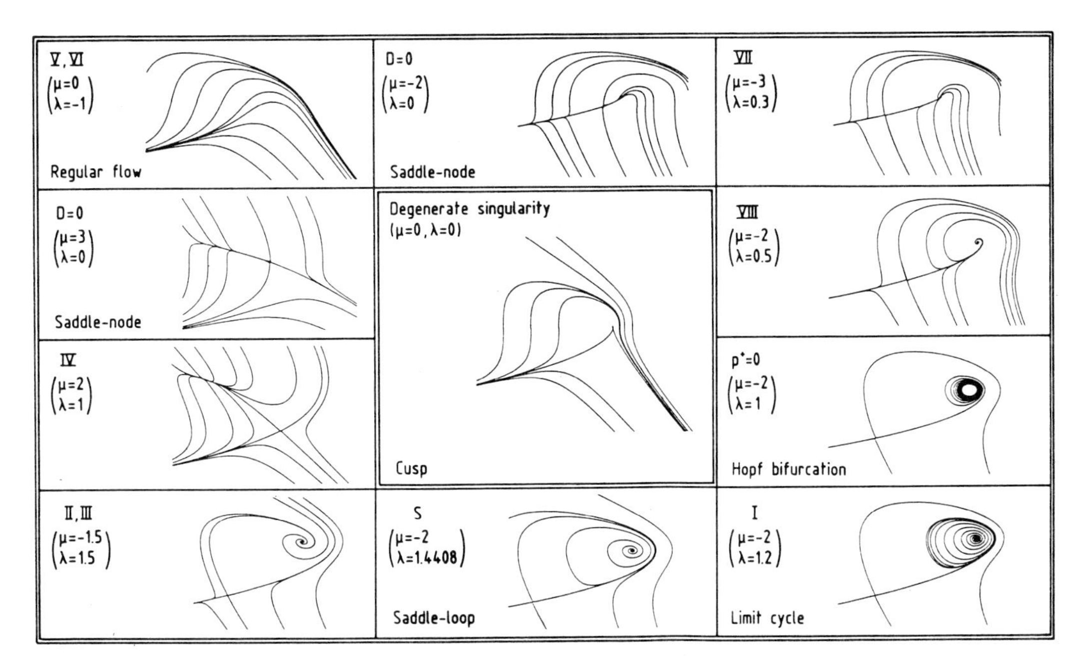

Fig. 8. Skin friction patterns generated from bifurcation of a cusp singularity.

Results of a numerical calculation of the various skin friction patterns in the parameter plane, which are described by $\dot{x} = z + \mu x + x^2$ & $\dot{z} = \lambda - x^2$ so $c = -1$, are depicted in Fig. 8.
The limit cycle (closed separation line) can grow onto a structurally unstable saddle-loop which is the union of a homoclinic cycle and a saddle point singularity. The saddle-loop can be perturbed by a global bifurcation (S) with the effect that the closed separation line vanishes.

7 Conclusions and Prospectives.

The method described in this paper yields local solutions of the NS-equations by considering bifurcations of degenerate singularities in the velocity vector field of a 3D flow. These solutions are of particular interest as they provide insight into complicated 3D separated flow structures. Several cases are still unresolved, especially the Jordan forms (3a) & (3d) have to be analyzed. At the moment case (3b) with additional symmetries is under consideration. Furthermore some basic experiments are planned in order to verify the theoretical results. Finally it seems possible to extend the method so that curved walls, moving walls and unsteady flow problems can be treated as well.

8 References.

Andronov, A.A. et.al. (1973) Qualitative theory of second order dynamical systems. Wiley, New York.

Bakker, P.G. (1988) Bifurcations in flow patterns. PhD thesis, Delft University of Technology.

Dallmann, U. (1983) Topological structures of three-dimensional vortex flow separation, AIAA-paper 1735.

Guckenheimer, J. (1983) & Holmes, P.J. Nonlinear oscillations, dynamical systems and bifurcation of vector fields. Springer, New York, Berlin.

Hornung, H.G. (1983) The vortex-skeleton model for three-dimensional steady flows. AGARD Conference Proc. 342.

Kooij, R.E. (1989) & Bakker, P.G. Three-dimensional viscous flow structures from bifurcation of a degenerate singularity with three zero eigenvalues, Report LR-572, Delft University of Technology.

Legendre, R. (1965) Lignes de courant d'un ecoulement continu, La Recherche Aerospatiale, 105.

Perry, A.E. (1986) & Chong, M.S. A series-expansion study of the Navier-Stokes equations with three-dimensional separation points, J. Fluid Mech., vol. 173.

Shirer, H.N. (1983) & Wells, R. Mathematical structures of the singularities at the transitions between steady states in hydrodynamic systems, Lecture notes in physics, vol. 185, Springer, Berlin, New York.

Wang, K.C. (1983) On the disputes about open separation, AIAA-paper 83-0296.

Winkel, M.E.M. de (1988) & Bakker, P.G. On the topology of three-dimensional viscous flow structures near a plane wall. A classification of hyperbolic and non-hyperbolic singularities on the wall, Report LR-541, Delft University of Technology.

Topological Structures of Separated Flows About a Series of Sharp-Edged Delta Wings at Angles of Attack up to 90°

WENHAN SU, MOUJI LIU & ZHIZHONG LIU

Institute of Fluid Mechanics, Beijing University of Aeronautics and Astronautics, Beijing, China

1. INTRODUCTION

Topological analysis has begun to play an important role in the study of 3D separated flows in recent years. The study presented in this paper consists of some topological analyses of the separated flows about delta wings, based on a large amount of experimental data from low speed wind tunnel tests (figure 1). There are two kinds of relevant references in previous works. Papers about topological analysis of 3D separated flows are associated with the names of Legendre, Oswatitsch, Lighthill, Smith, Andronov et al, Perry & Chong, Hunt et al, Tobak & Peake, Dallmann, Liu & Su et al, Hornung, and Chapman. Papers on the experimental study of separated flows about delta wings are associated with Squire et al, Werle, Earnshaw & Lawford, Wentz et al, Hummel, Erickson, Liu et al, Su et al, Liu & Su, and Brennenstuhl & Hummel (see references). Much work has been done in this field, but more work on the global behaviour of 3D separated flow and the evolution of topological structure with parameter variation is still needed. In particular, more work on separated flow about a delta wing at high angle-of-attack is needed. We present in this paper some results and discussions concerning this problem.

2. TOPOLOGICAL ANALYSIS OF 3-D SEPARATED FLOWS

The task of topological analysis is to determine the topological structure of a flow pattern and the evolution of the topological structure with relevant parameters. The set of all topological properties of the flow pattern describes the topological structure. The number and types of singular points of a flow pattern and the structure of paths connecting the singular points are the major topological properties for 3D separated flow. The 3D streamline equation, skin-friction line equation and the equation for streamlines in 2D sections of 3D flow describe the flow pattern of a 3D steady flow. These equations correspond to their nonlinear dynamical systems respectively. The point at which all of the three velocity components are equal to zero is a 3D singular point. The point at which both of the two surface-shear-stress components are equal to zero is a 2D singular point for the skin-friction pattern. The 2D singular point for skin-friction is the representation of a 3D singular point on the body surface. The major singular points concerned with this study are shown in figure 2.

Based on two asSigmaptions (i) the vector fields of velocity and surface shear stress are continuous and (ii) the body is simply connected, we have the following two topological rules connecting the number of nodes N, saddles S, half-nodes N' and half-saddles S':

a. $\Sigma N - \Sigma S = 2$ (Lighthill 1963)

for a skin-friction line on a 3D body.

b. $(\Sigma N + \frac{1}{2}\Sigma N') - (\Sigma S + \frac{1}{2}\Sigma S') = -1$ (Hunt et al 1978)

for a streamline on a 2D plane cutting a 3D body.

In a study of the evolution of topological structure with changing parameter, we need some concepts of bifurcation theory. The evolution equation of the problem is of the form $\dot{X} = F(X, \lambda)$, where X, F, for example, may be $(x, y, z), (u, v, w)$, and λ is a parameter (Λ or α in this study). Where the parameter is varied and approaches a critical value (the bifurcation point) a topological change of the flow pattern occurs. As the parameter crosses the critical value, the topological structure of the flow pattern changes. The occurrence and bifurcation of a high order singular point (HOSP) play important roles for this transition. Here are two examples of the bifurcation of a HOSP:

$$Ns - S \rightleftharpoons NsS \rightleftharpoons \text{regular point} \; ; \quad Na - Ss - Na \rightleftharpoons NaSsNa \rightleftharpoons Na$$

Here Na means a nodal point of attachment, Ns a nodal point of separation, and so on. In a structurally stable flow pattern the singular points are all elementary singular points (ESP). However, in this flow pattern there is still a kind of special point which is very similar to a HOSP but it is not a HOSP theoretically. Two or more ESP get together at this point very close to each other. This special point is called a group of elementary singular points (GESP). It is very important to determine the topological structure of a GESP in the analysis.

3. THE TYPES OF SEPARATED FLOWS ABOUT DELTA WINGS AND THE BOUNDARIES BETWEEN THEM

3.1 *The nine types of separated flows about delta wings*

By means of the smoke-wire and oil-flow visualizations in low-speed wind tunnels, data on separated flow patterns were obtained. Based on these experimental results and according to the physical character of the flow patterns, a classification of separated flows about delta wings is suggested: In the (Λ, α)-region studied (see figure 1 for notation) there are nine types of separated flow, as shown in the photographs of figure 3.

At low α there is a 3D bubble near the leading-edge of the wing. It is an open viscous region with some helical flow and a spiral point on the cross-flow plane, but no free shear layer rolls up into a vortex. As α increases, a bubble vortex appears above the wing with low swept back angle (Λ). In this pattern the free shear layer has rolled up but no concentrated vortex core is formed. For a wing with high Λ, there are two arrays of streamwise vortices with counter-rotation above the left panel and right panel of the wing. They appear to be larger than the thickness of the wing boundary layer. As α is increased further, the streamwise vortices on a half wing interact with each other and coalesce into a concentrated vortex. The concentrated vortex has a core with highly concentrated vorticity, but there is no vortex bursting above the wing.

A vortex with bursting point between the trailing-edge and the apex of the wing is called a bursting vortex. There are two types of skin-friction line pattern for this vortex: a kink point is on the secondary separation line of a high Λ wing; a spiral point is at the end of the secondary separation line of a low Λ wing. For a wing with very high Λ, at high enough α an asymmetric vortex appears. The bursting point of an asymmetric bursting vortex is often unstable. After the bursting point has moved forward to the apex, spiral flow occurs. In this flow there are two spiral points on the wing and the vorticity sheds from these points unsteadily. The reversed flow is characterized by reversal of the flow direction near the upper surface. The completely separated flow is characterized by existence of a nodal point of attachment on the upper surface; this means that there is a strong influence of the unsteady vortices shedding from the trailing-edge on the upper surface flow pattern. The last three separated flows are unsteady flows and the oil-flow patterns are then just the time-averaged flow patterns.

In the whole (Λ, α)-plane there is only one singular point, the first attachment point Na_1, on the lower surface, and the flow pattern is not conical.

3.2 The flow boundary chart

The flow boundary chart in figure 4 shows how Λ and α influence flows. This is useful for the study of control of vortex and separated flow. Every point on the boundary lines is a bifurcation point of the flow pattern.

4. TOPOLOGICAL STRUCTURES OF THE NINE SEPARATED FLOWS AND THEIR PARAMETRIC EVOLUTION

4.1 Topological structures of nine separated flows

The topological structures of the concentrated vortex are shown in figure 5, and the rest are shown in figure 6. For the concentrated vortex there is a GESP (see section 2) at the apex of the wing. In the analysis, we treat the sharp apex as a rounded one so that, as pointed first by Lighthill, the GESP $\underline{Na_1SNa_2}$ would be split in three $Na_1 - S - Na_2$ ((for cases (1), (2)) and the GESP $\underline{Na_1S_1Na_3S_2Na_2}$ in five $Na_1 - S_1 - Na_3 - S_2 - Na_2$ (for case (3)). We use a line segment near the singular points concerned to denote a GESP in this paper. We also treat the sharp edges (leading-edge and trailing-edge) as rounded so that there are just a finite number of isolated singular points on the edge. It seems that these treatments are reasonable. For the cases of figure 6, the application of the appropriate topological rule are as follows:

	a	b	e_1	e_2	g_1	g_2	g_3
$\Sigma N - \Sigma S =$	5-3	7-5	14-12	10-8	13-11	14-12	10-8
	h_1	h_2	i_1	i_2	c	f	
$\Sigma N - \Sigma S =$	10-8	5-3	10-8	5-3	16-14*	8-6	

* for the case with two pairs of streamwise vortices.

4.2 The parametric evolution of topological structures

From figures 4-6, it is not difficult to see how the topological structures evolve with parameters Λ and α. As an example, we discuss the evolution of the topological structure from spiral flow to reversed flow. For the low Λ case, this is the evolution from g_1 to h_1 in figure 6. The topological changes are as follows:

$$Ns{-}S_1 - Ns \rightarrow NsS_1Ns \rightarrow Ns;$$
$$S{-}Na_3 - S \rightarrow SNa_3S \rightarrow S;$$
$$N_F{-}S_2 - N_F \rightarrow Ns - S_2 - Ns \rightarrow NsS_2Ns \rightarrow Ns.$$

For the high Λ case, it is from g_3 to h_2. The changes are:

$$\underline{NsS} \rightarrow NsS \rightarrow \text{regular point.}$$
$$\underline{S_1Na_3S_2Na_2} \rightarrow S_1Na_3S_2Na_2 \rightarrow \text{regular point.}$$
$$N_F - S - N_F \rightarrow Ns \quad \text{as in the low } \Lambda \text{ case .}$$

5. APPLICATION TO OTHER CONFIGURATIONS

The method of topological analysis presented here can be applied to other delta-like configurations. As an example, the topological analyses for two typical flow patterns about double-delta wing are shown in figure 7. For these cases, we have $\Sigma N - \Sigma S = 16 - 14$ (case a), $\Sigma N - \Sigma S = 14 - 12$ (case b).

6. CONCLUDING REMARKS

In the (Λ, α)-plane studied, there are nine types of separated flows about the wings, having different physical characters and different topological structures. A method has been presented, which is useful in analysing the topological structures of 3D separated flow about delta-like configurations.

ACKNOWLEDGEMENT

The first author (Wenhan Su) expresses his warm gratitude to his teachers the late Prof. Shi-Jia Lu and Prof. Li-Yee Wu for their instruction and encouragement.

This study is supported by the Aeronautical Science Foundation.

REFERENCES

Andronov A.A. et al 1973 "Theory of Oscillation".

Brennenstuhl U. & Hummel D. 1987 *Z. Flugwiss. W.* **11**, 37-49.

Chapman G.T. 1986 AIAA Paper 86-0485.

Dallmann U. 1983 AIAA Paper 83-1735.

Earnshaw P.B. & Lawford J.A. 1964 ARC R & M No. 3424.

Erickson G.E. 1982 ICAS Paper 82-661.

Hornung H.G. & Perry A.E. 1984 *Z. Flugwiss. W.* **8**, 155-160.

Hummel D. 1978 AGARD Cp-247, No. 15.

Hunt J.C.R. et al 1978 *J. Fluid Mech.* **86**, 179-200.

Legendre R. 1977 *Recherche Aerospatiale* **6**, 327-355.

Lighthill M.J. 1963 "Laminar Boundary Layers" 72-82.

Liu M.J. et al 1980 *J. Aircraft* **17**, 332-338.

Liu M., Su W. et al. 1983 Research Paper of BUAA. Bh-B1104.

Liu M. & Su W. 1985 *ACTA Aero. et Astro. Sinica* **6**, 1-12.

Oswatitsch K. 1957 "Boundary Layer Research" 357-367.

Perry A.E. & Chong M.S. 1987 *Ann. Rev. Fluid Mech.* **19**, 125-155.

Smith J.H.B. 1972 *Progress in Aero. Sciences* **12**, 241-272.

Squire L.C. et al 1961 ARC R & M No. 3305.

Su W. et al 1982 ICAS Paper 82-662.

Tobak M. & Peake D.J. 1982 *Ann. Rev. Fluid Mech.* **14**, 61-85.

Wentz W.H., Jr. et al 1968 NASA CR-98737.

Werle H. 1982 ONERA TP No. 1982-5.

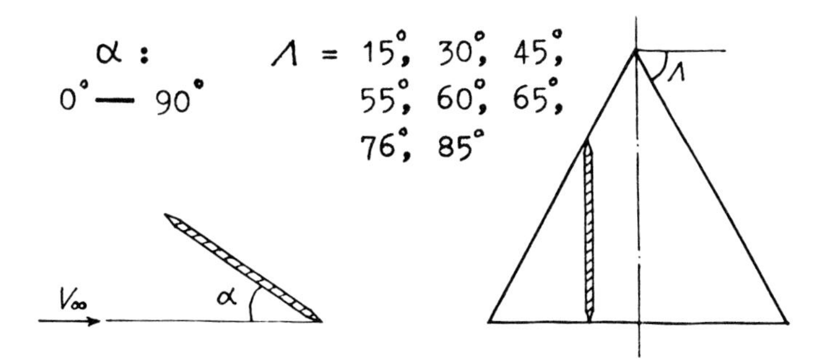

FIGURE 1. The test models and the α-region studied.

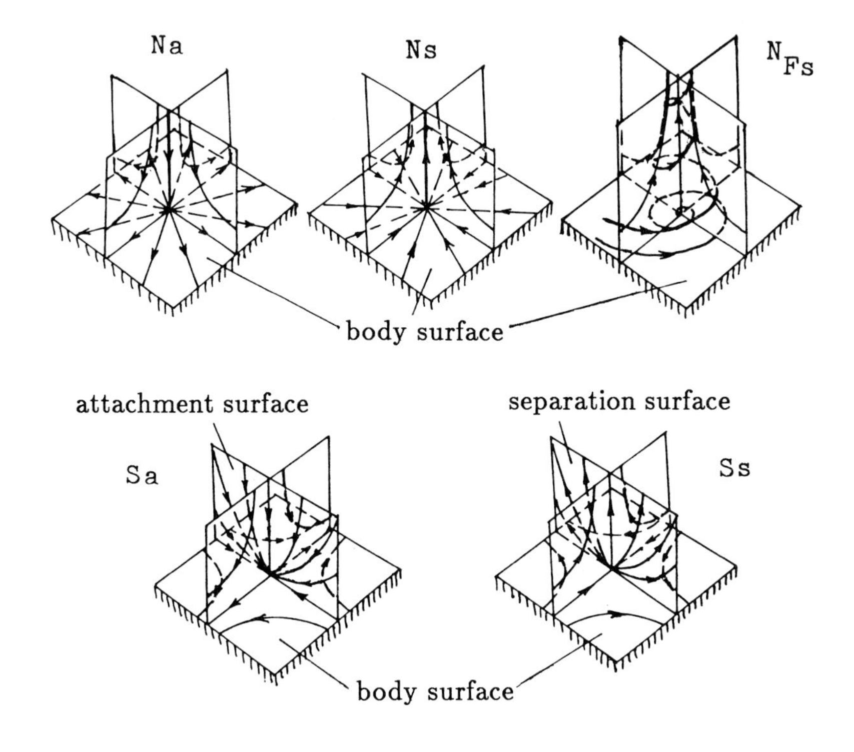

Na – nodal point of attachment
Ns – nodal point of separation
N_{Fs} – focus of separation (spiral point of separation)
Sa – saddle point of attachment
Ss – saddle point of separation

FIGURE 2. Major singular points concerned with this study. The lines on the body surface are skin-friction lines. The lines above the body surface are 3D streamlines.

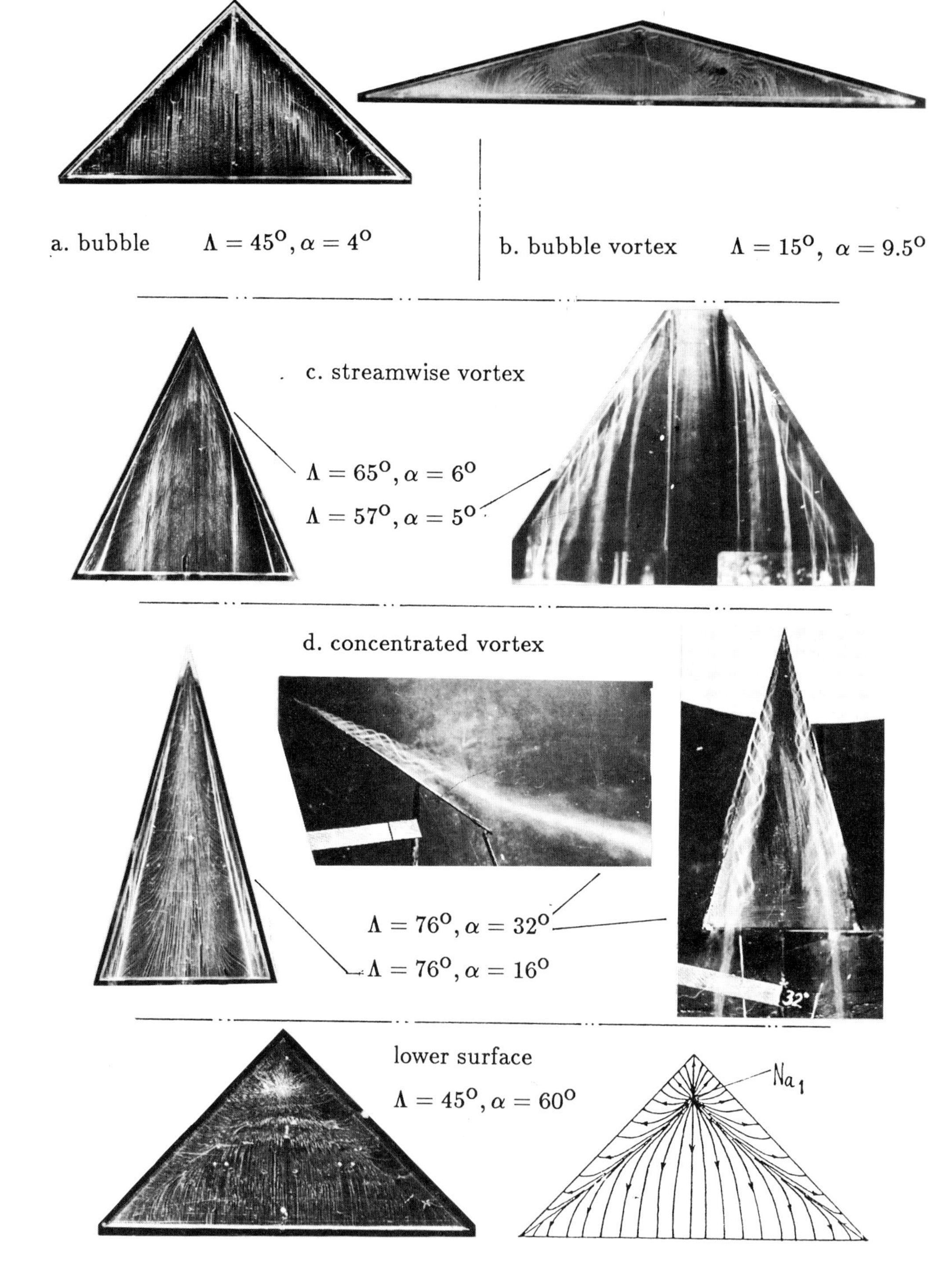

FIGURE 3. Nine types of separated flows about delta wings.

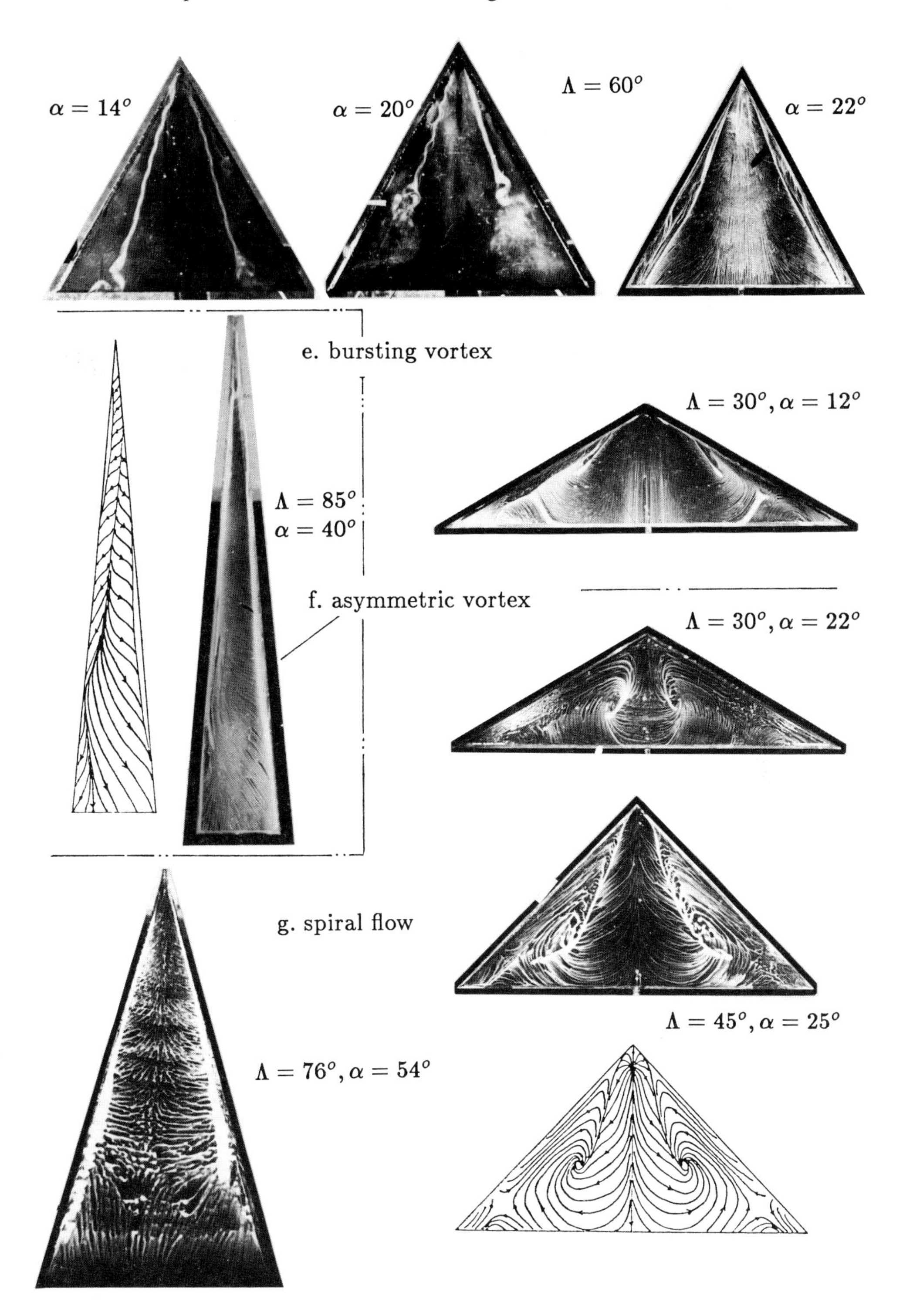

Figure 3 (continued)

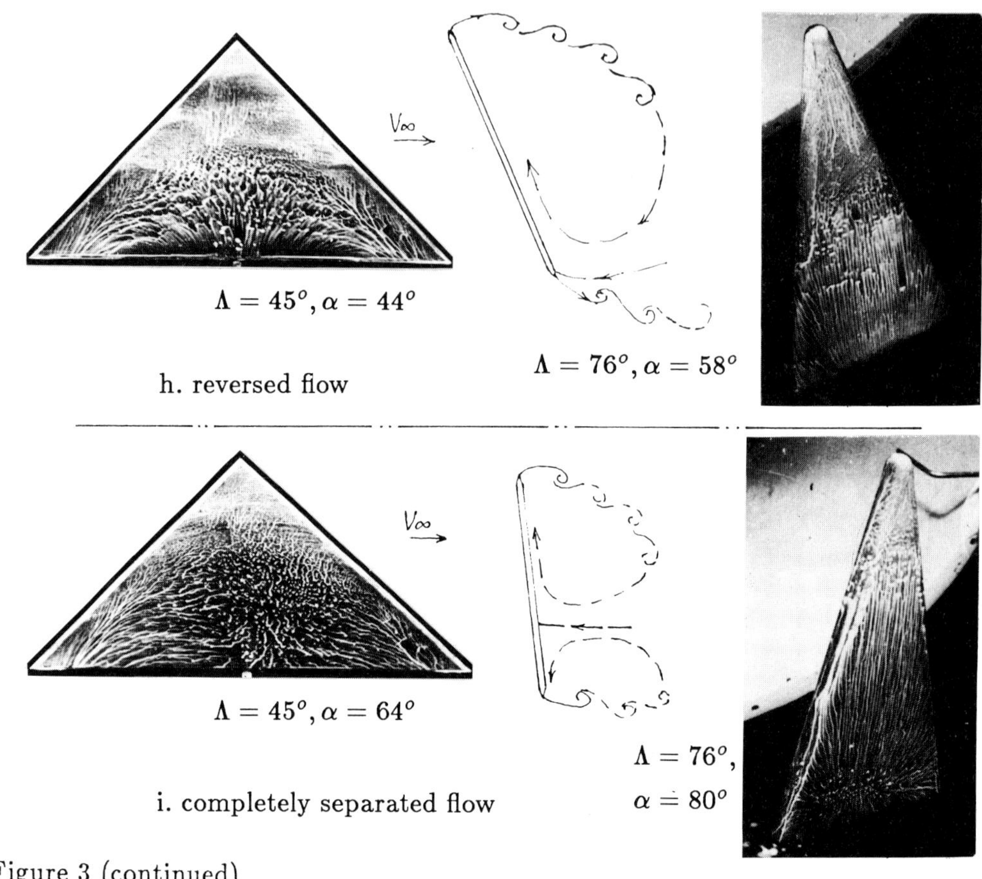

Figure 3 (continued)

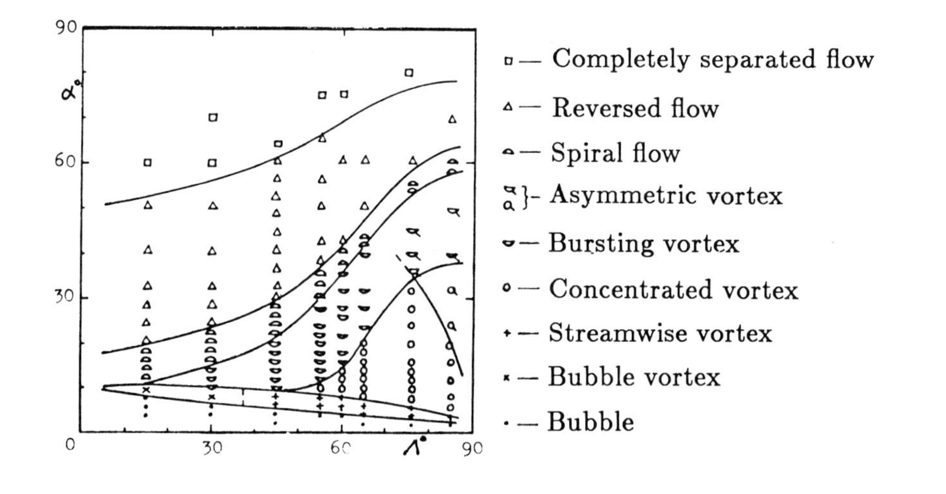

FIGURE 4. The types of separated flows about delta wings and the boundaries between them.

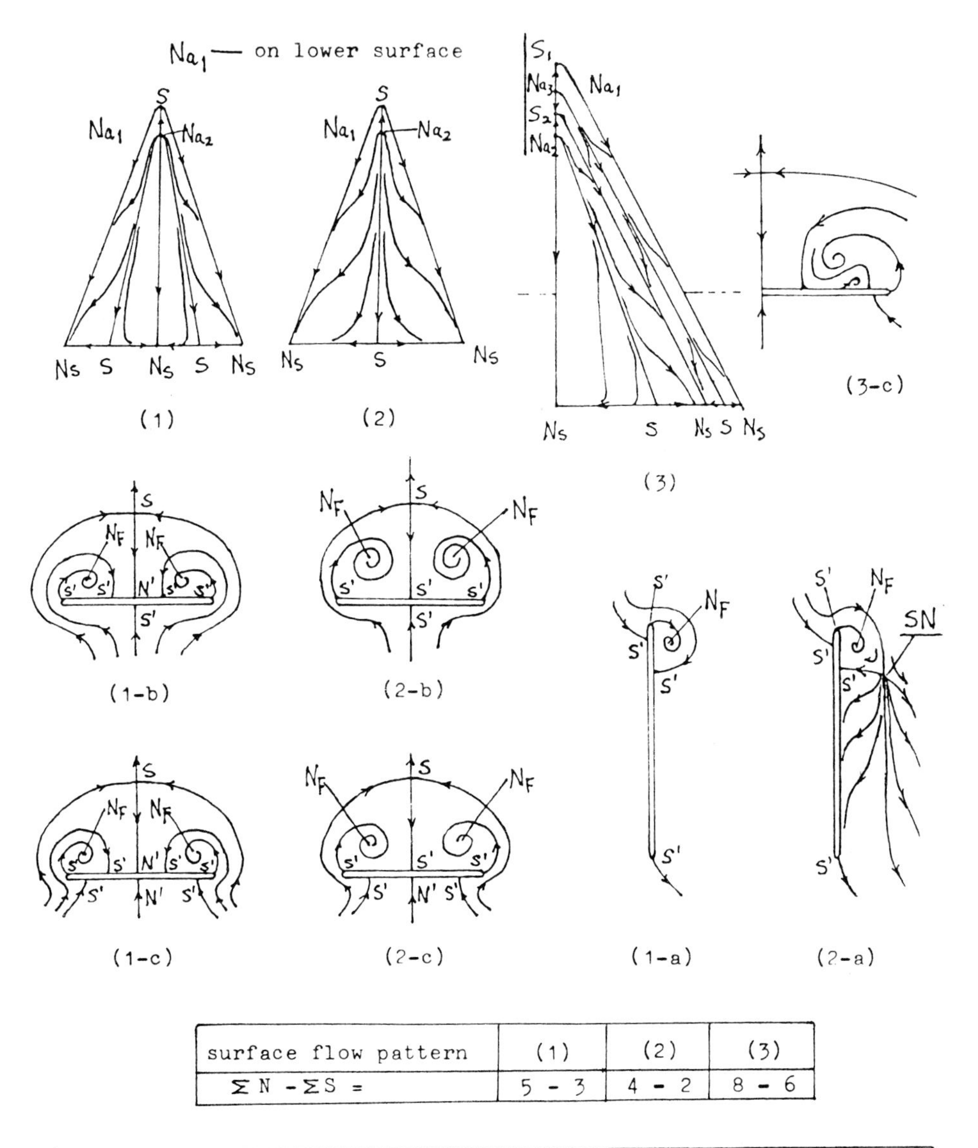

surface flow pattern	(1)	(2)	(3)
$\Sigma N - \Sigma S =$	5 - 3	4 - 2	8 - 6

	cross section flow pattern					symmetric plane flow pattern	
	(1-b)	(1-c)	(2-b)	(2-c)	(3-c)	(1-a)	(2-a)
$(\Sigma N+\frac{1}{2}\Sigma N') - (\Sigma S+\frac{1}{2}\Sigma S') =$	2.5-3.5	3-4	2-3	2.5-3.5	5-6	1-2	2-3

FIGURE 5. Topological structures of concentrated vortices.

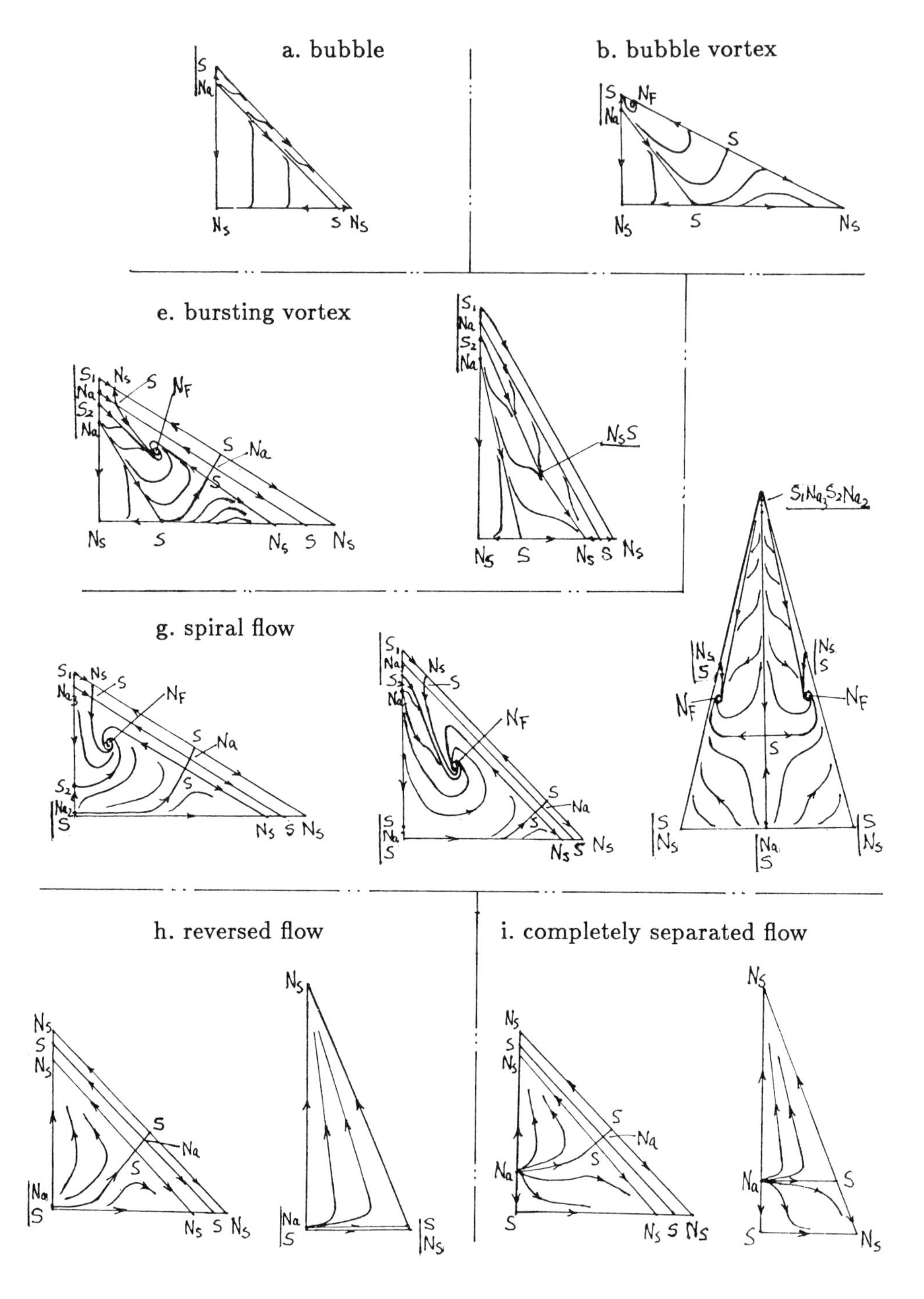

FIGURE 6. Topological structures of the separated flows concerned.

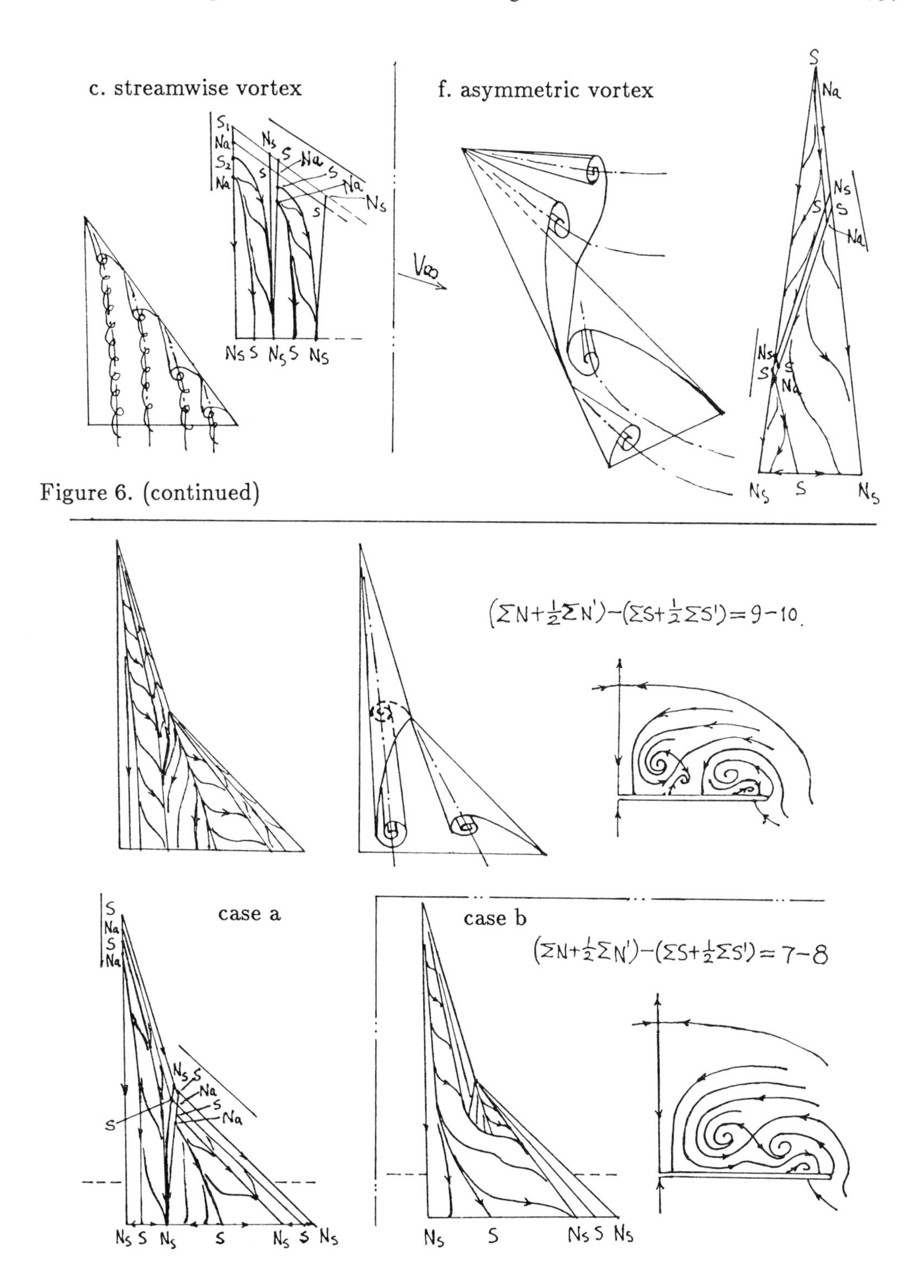

Figure 6. (continued)

FIGURE 7. Topological structures of the separated flow about double-delta wing.

A General Classification of Three-dimensional Flow Fields

M.S. CHONG[1], A.E. PERRY[1] & B.J. CANTWELL[2]

[1]Mechanical Engineering Department, University of Melbourne, Victoria, Australia
[2]Department of Aeronautics and Astronautics, Stanford University, Stanford, California, USA

1. SUMMARY

The most general classification of linearized critical points in flow fields is presented in terms of three invariants of the Jacobian matrix. The use of the properties of such elementary patterns is reviewed and it is shown how complex three-dimensional time dependent flow fields can be interpreted and described in some manageable way.

2. THE TOPOLOGICAL CLASSIFICATION OF BASIC FLUID MOTIONS

2.1 The Rate of Deformation Tensor

Consider a non-rotating observer moving with a fluid particle. He observes in his immediate neighbourhood the instantaneous velocity vector field and his region of consideration is sufficiently local for a linearized approximation of the field to be adequate. Relative to this observer, the velocity field can be expressed as:

$$\dot{\mathbf{x}} = \mathbf{A} \, . \, \mathrm{x} \tag{1}$$

$$\text{where } \dot{\mathbf{x}} = d\mathbf{x}/dt$$

This equation is three coupled linear differential equations and $\mathbf{A}$ is a 3 x 3 Jacobian matrix. The elements of $\mathbf{A}$ are given by $a_{ij} = \partial \dot{x}_i / \partial x_j$. This is the rate of deformation tensor and can be split into a symmetric and anti-symmetric part i.e.

$$a_{ij} = S_{ij} + R_{ij} \tag{2}$$

where $$S_{ij} = (\partial x_i/\partial xj + \partial x_j/\partial x_i)/2 \tag{3}$$

and $$R_{ij} = (\partial x_i/\partial x_j - \partial x_j/\partial x_i)/2 \tag{4}$$

S_{ij} is the rate of strain tensor and R_{ij} is the rotation or spin tensor. This latter quantity has only three independent components and these are the three components of vorticity.

The simple linear set of equations given by (1) and the coefficients given by (2), (3) and (4) represents the most basic fluid motions. All fluid motions, no matter how complex (even in turbulence) must reduce to the above set of motions locally relative to an observer moving with a fluid particle. In spite of the elementary basic formulations, the properties of the flow field represented by the matrix **A** can be surprisingly complex and the topological classification of these local patterns is far from trivial. The properties of these patterns are completely described in terms of the eigenvalues and eigenvectors of the matrix **A**. The eigenvalues are given by the solution to the cubic

$$\lambda^3 + P\lambda^2 + Q\lambda + R = 0 \tag{5}$$

$$\text{where } P = -(a_{11} + a_{22} + a_{33}) = -\text{ trace } [\mathbf{A}] = -S_{ii} \tag{6}$$

$$Q = \begin{vmatrix} a_{11} & a_{12} \\ a_{21} & a_{22} \end{vmatrix} + \begin{vmatrix} a_{11} & a_{13} \\ a_{31} & a_{33} \end{vmatrix} + \begin{vmatrix} a_{22} & a_{23} \\ a_{32} & a_{33} \end{vmatrix} \tag{7}$$

$$= (1/2)\,(P^2 - \text{trace } [\mathbf{A}^2]) = (1/2)\,(P^2 - S_{ij}S_{ji} - R_{ij}R_{ji})$$

$$R = -\begin{vmatrix} a_{11} & a_{12} & a_{13} \\ a_{21} & a_{22} & a_{23} \\ a_{31} & a_{32} & a_{33} \end{vmatrix} = -\det[\mathbf{A}] \tag{8}$$

$$= (1/3)\,(-P^3 + 3PQ - \text{trace } [A^3])$$

$$= (1/3)\,(-P^3 + 3PQ - S_{ij}S_{jk}S_{ki} - 3R_{ij}R_{jk}S_{ki})$$

The characteristic equation (5) can have (1) all real roots which are distinct, (2) all real roots where at least two roots are equal, or (3) one real root and a conjugate pair of complex roots.

The classification of all possible patterns can be made with the aid of a space where the coordinates are the three matrix invariants P, Q and R. Various regions of P-Q-R-space correspond to the various possible topologies of the flow patterns. The first to recognize the significance of this space was Reyn (1964) and Blaquire (1966) but as far as the authors are aware it is only recently that the geometrical features of the various critical boundaries within this space have been understood and mapped out. In fact this study is still undergoing development by the authors. Figure 1 shows a very important surface S_1 in the P-Q-R space by showing how this surface intersects with planes of constant P. All points above this surface give complex eigenvalues for **A**. All points below this surface S_1, give only real eigenvalues for **A**.

The surface S_1, is made up of two surfaces S_{1a} and S_{1b} which are respectively given by

$$\frac{1}{3}P(Q-\frac{2}{9}P^2)-\frac{2}{27}(-3Q+P^2)^{3/2}-R=0$$
$$\frac{1}{3}P(Q-\frac{2}{9}P^2)+\frac{2}{27}(-3Q+P^2)^{3/2}-R=0 \qquad (9)$$

Surface S_{1a} osculates (or kisses) surface S_{1b} to form a cusp. Note that S_1 is anti-symmetrical with respect to P.

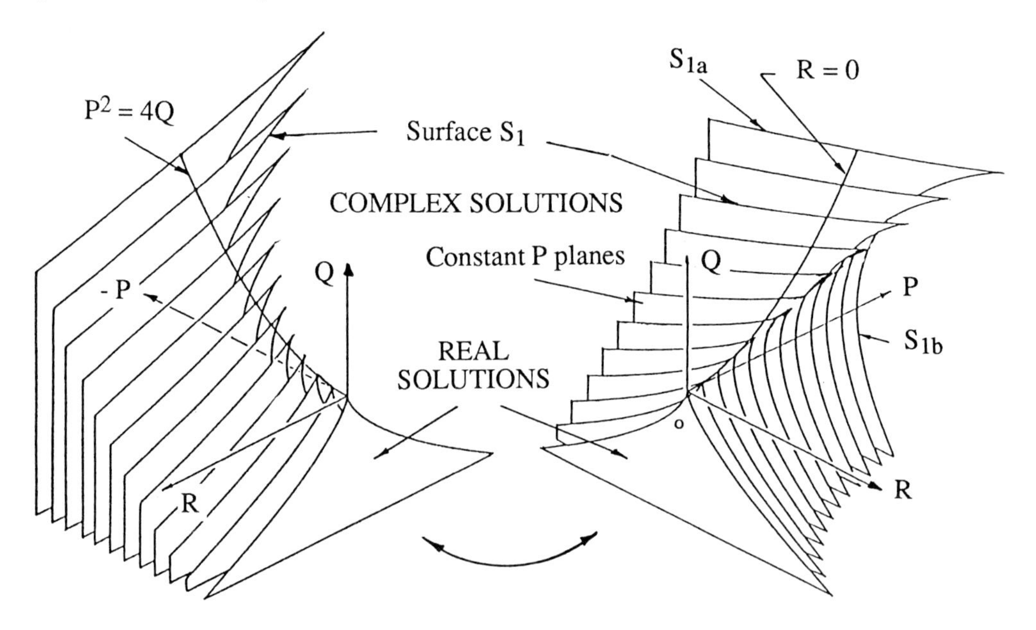

FIGURE 1. P-Q-R space showing surface S_1 which divides real solutions from complex solutions.

It can be seen from equations (6), (7) and (8) that the three matrix invariants P,Q and R are related to the physically indentifiable quantities S_{ij} and R_{ij} and it is felt that further study and development of these relationships will yield deeper insights into the physical understanding of these basic fluid motions.

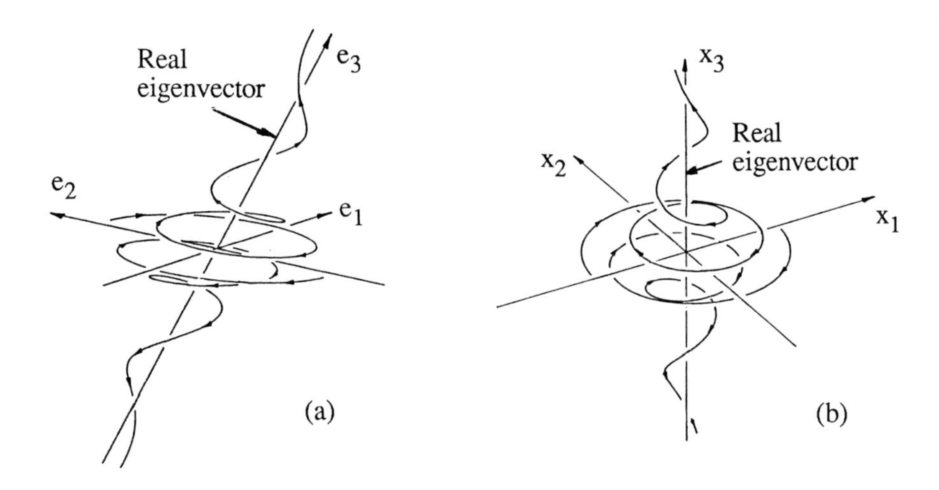

FIGURE 2. Critical point with complex eigenvalues.
(a) Noncanonical case. (b) Canonical case.

When the eigenvalues are complex there is only one plane which contains solution trajectories and the pattern within that plane is a focus. A typical pattern is shown in figures 2(a) and (b)

The solution trajectories (instantaneous streamlines) are obtained by integrating equations (1) in pseudo time i.e. the velocity field is assumed frozen in time. Trajectories close to the plane indicated in the figure ultimately move away and wrap around the real eigenvector. In figure 2(a) the vorticity field is not aligned with the principle axes of the rate of strain field (i.e. the eigenvectors of S_{ij}) and the pattern is in noncanonical form. In figure 2(b), the vorticity vectors are aligned with the eigenvectors of S_{ij} to give a canonical form.

One can convert the pattern in noncanonical form to canonical form by an affine transformation which consists of differential rotation and stretching of the three coordinates e_1, e_2 and e_3 to correspond to our coordinate system x_1, x_2 and x_3.

If the eigenvalues are real, then there are three planes which contain solution trajectories and these are defined by the three real eigenvectors of **A**. A typical pattern is shown in figure 3 where in the e_1, e_2 plane we have a saddle,

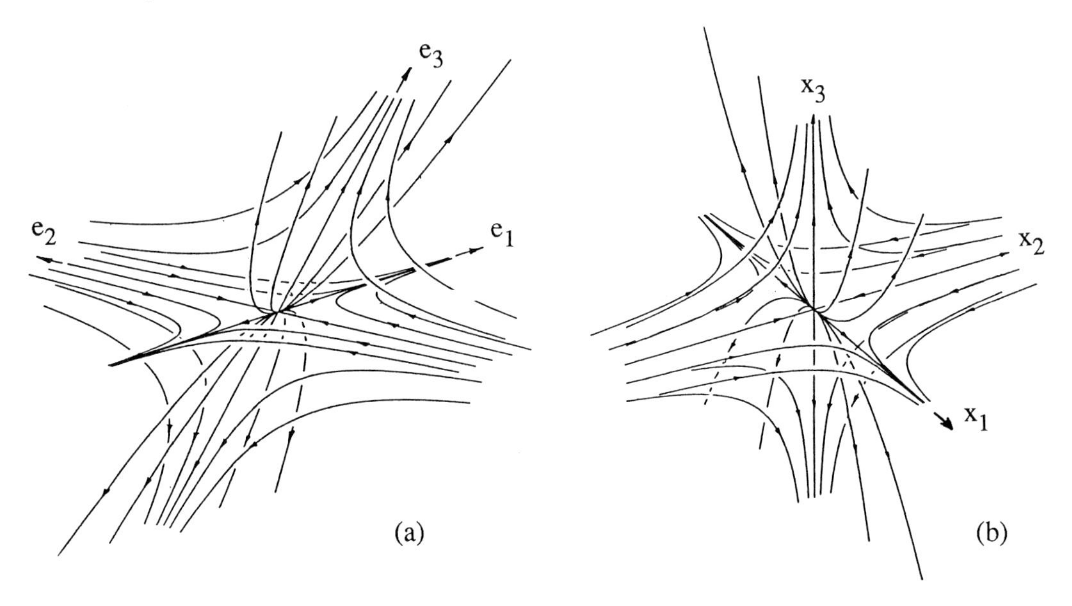

FIGURE 3. Critical point with real eigenvalues. (a) Noncanonical case. (b) Canonical case.

In the e_1- e_3 plane we have a node and in the e_2-e_3 plane and e_1-e_2 plane we have a saddle. Figure 3(a) shows a noncanonical form whereas a canonical form is shown in figure 3(b). In figure 3(a) the flow has vorticity while in figure 3(b), the flow is irrotational. Whenever we have irrotational flow all eigenvalues are real and in addition all eigenvectors are orthogonal.

If we confine ourselves to the planes which contain solution trajectories we can use simple "phase-plane" analysis for the topological classifications. For instance, if in a plane we have $\dot{\mathbf{y}} = \mathbf{C} \cdot \mathbf{y}$, where $\mathbf{y}$ is a position vector in the plane, then $\mathbf{C}$ is a 2 x 2 matrix and we can use the familiar p-q chart for classifications where $p = -\operatorname{trace}[\mathbf{C}]$ and $q = \det[\mathbf{C}]$. This chart is described in many elementary textbook on differential equations and is used for determining whether we have nodes, foci or saddles. Perry and Fairlie (1974) applied the phase-plane method to three-dimensional patterns by examining solutions on the various eigenvector planes.

As we 'tour' the P-Q-R space the pattern undergoes topological changes as we cross the various critical boundaries which divides the space into various regions. Degenerate cases occur on all the P-Q-R space boundaries. A complete description of all boundaries and cases is not possible here and the reader is referred to the Stanford report by Chong et al (1988).

2.2 Critical Points in Flow Fields

The patterns just discussed are equally valid about certain points in a flow field where the velocity of the fluid is zero and the streamline slope is indeterminate. Such critical points occur in complex three-dimensional separation patterns as shown in figure 4. The critical points are labelled and by showing the location and classification of these points a quantitative understanding of the entire pattern can be readily grasped.

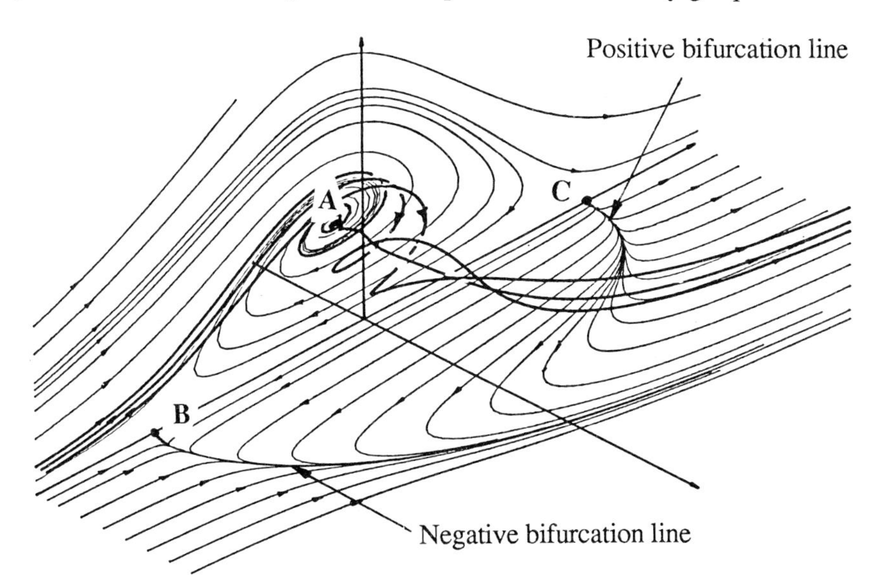

FIGURE 4. Critical points in a three-dimensional separation.
A is a free-slip critical critical point. **B** and **C** are no-slip critical points.

The pattern shown is on a no slip boundary. There are two types of critical points. There are those located above the boundary (free-slip critical points) and those on the boundary (the no-slip critical points). In the former, the formulation given by equation (1) is appropriate but in the latter, the formulation must be written as

$$\overset{\circ}{\mathbf{x}} = \mathbf{A}\ \mathbf{x} \tag{10}$$

where $\overset{\circ}{\mathbf{x}} = \mathbf{x}/x_3$ or $d\mathbf{x}/d\tau$ where $d\tau = x_3 dt$ and τ can be thought of as a transformed time. The inclusion of x_3 as given above is necessary to ensure we satisfy the no-slip boundary condition at $x_3 = 0$. The matrix **A** is no longer the rate of deformation tensor but is made up of elements related to the local pressure gradients and gradients of vorticity (see Perry & Fairlie, 1974). Nevertheless the classifications are the same as given earlier and the properties of (10) were first examined in detail by Oswatitsch (1958) and are restricted to certain surfaces in the P-Q-R space which restricts the topological possibilities.

In the case of free-slip critical points, incompressible flow is confined to the Q-R plane and this restricts the combination of saddles and nodes in the various eigenvector planes (i.e. one node and two saddles is the only combination possible for the case of three real eigenvalues). With compressible flow, nodes in all three planes are possible e.g. a three dimensional star node which occurs in isotopic expansion or compression.

2.3 Bifurcation Lines

Besides critical points, there are other related flow pattern features called bifurcation lines as discussed by Hornung & Perry (1984) and Perry & Hornung (1984). The addition of a zeroth other term to the formula given by (1) or (10) quite often produces a trajectory to which neighbouring trajectories asymptote exponentially if we are sufficiently close. Such bifurcation lines or "asymptote trajectories" are shown heavy in figure 5. The mathematical properties of these lines need to be investigated further.

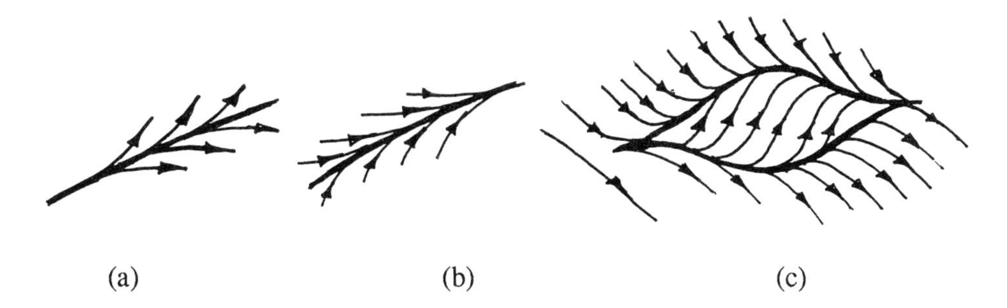

FIGURE 5. Bifurcation lines. (a) Positive. (b) Negative.
(c) Combination produced by a vortex close to a surface.
Perry & Hornung (1984).

3. APPLICATION OF CRITICAL POINT THEORY

3.1 Three-Dimensional Separation and Wall Turbulence

What we have been discussing in Section 2 is critical point theory. One may ask what use is this theory and why do we bother with topological classifications and the geometrical properties of such critical point patterns. Quite apart from the intrinsic beauty and the fascination it generates does it have a "use"? The answer to this question is becoming clearer as we move further into the computer and supercomputer age. It is now possible to produce large amounts of experimental data from modern high speed data acquisition systems. Also with supercomputers enormous amounts of computational "data" are being produced by direct numerical simulations of the Navier-Stokes equations. What are we going to do with all this data and how are we going to display, describe and interprete it? Complex three dimensional time dependent flow fields need to be described and understood from two dimensional images and critical point theory appears to be one of the few tools we have available for describing and interpreting this data. Critical points and bifurcation lines are the salient features of a flow pattern. In fact, they are probably the only identifiable features of flow patterns.

Given an arrangement of such points and lines and a knowledge of their type, much of the remaining flow field and its gross transport properties can be grasped. There is also a much more direct and analytical use of critical point theory. For instance, Perry and Chong (1986) have been studying local Taylor series expansion solutions of the time dependent Navier-Stokes equations for three dimensional unsteady separation patterns at low Reynolds numbers. The strategies and techniques for developing and testing the various algorithms required the generation of known solutions with certain geometrical properties and knowledge of critical point theory, in particular the equations of Oswatitsch (1958) and of Perry & Fairlie (1974), were found to be indispensable for mathematically synthesizing complex three dimensional separation bubbles with the required features. See also Perry and Chong (1987).

An better understanding of three-dimensional flow field may come from a study of other quantities besides the velocity field. The vorticity field is a natural candidate since vorticity is independent of the velocity of the observer. The vorticity field has its own set of critical points (points where the vorticity is zero and the slope of the vortex lines is indeterminate). Lines of vorticity can be obtained using equation (1) by treating the 3 components of vorticity ω_i as velocity $\dot{x}_i$. Hence matrix **A** has elements $a_{ij} = \partial\omega_i/\partial x_j$. [In fact, this procedure can be applied to any other smooth vector field]. The topological classifications are the same as the velocity field. However, because of the solenoidal condition for vorticity all critical points will be confined to the R-Q plane of P-Q-R space as were the free-slip incompressible velocity fields.

Hornung & Perry (1984) and Perry and Hornung (1984) discuss the possibility of describing complex three-dimensional flow patterns with the aid of "vortex skeletons" which are much simpler than the velocity field. Perry & Tan (1984) have described the periodic array of eddies in jets and wakes successfully in terms of simple vortex skeletons and Perry & Hornung have shown that a pattern topologically similar to that shown in figure 4 could be generated by a "U"-shaped vortex as shown in figure 6(a) using the Biot-Savart law together with an image vortex under the plane with a uniform flow superimposed. Such a simple "vortex skeleton" description might be appropriate at high Reynolds numbers but at the Reynolds numbers of order 10, the vorticity field is quite different from expectations based on high Reynolds numbers. Figure 6(b) shows such a vorticity field. On the no-slip boundary, the lines of vorticity are othogonal to the velocity-field limit-trajectories (Lighthill, 1963) but away from the boundary there are no simple rules.

With the advent of supercomputer computations which are full-direct numerical simulations of the Navier-Stokes equations, it is now possible to look at instantaneous velocity fields and instantaneous fields of vorticity, e.g. Spalart (1988) and Moin & Spalart (1987). Simple vortex skeleton models have been proposed for wall turbulence and supercomputer data is now available for testing such models although so far only low Reynolds number turbulence ($R_\theta < 1410$) has been computed. In these models, wall turbulence is thought to consist of "$\cap$"-shaped vortex loops attached to the boundary and inclined at 45° to the downstream direction e.g. see Perry et al (1987). Instantaneous vorticity fields shows this to some extent, e.g. see Moin & Spalart (1987), but it is obvious that there are other motions present (perhaps detached Kolmogorov eddying motions) and so these vortex loops do not show up clearly unless some conditional averaging is carried out.

As was mentioned earlier, critical points and bifurcation lines (or asymptote trajectories) are salient features of flow patterns which are readily recognizable and once classified can be interpreted physically. The instantaneous skinfriction lines from a supercomputer calculation of wall turbulence is shown in figure 7. The network of interconnecting bifurcation lines look suspiciously like the footprints of vortices above the surface with were observed by Perry & Hornung (1986) with their electromagnetic simulations, an example of which was shown in figure 5(c) .

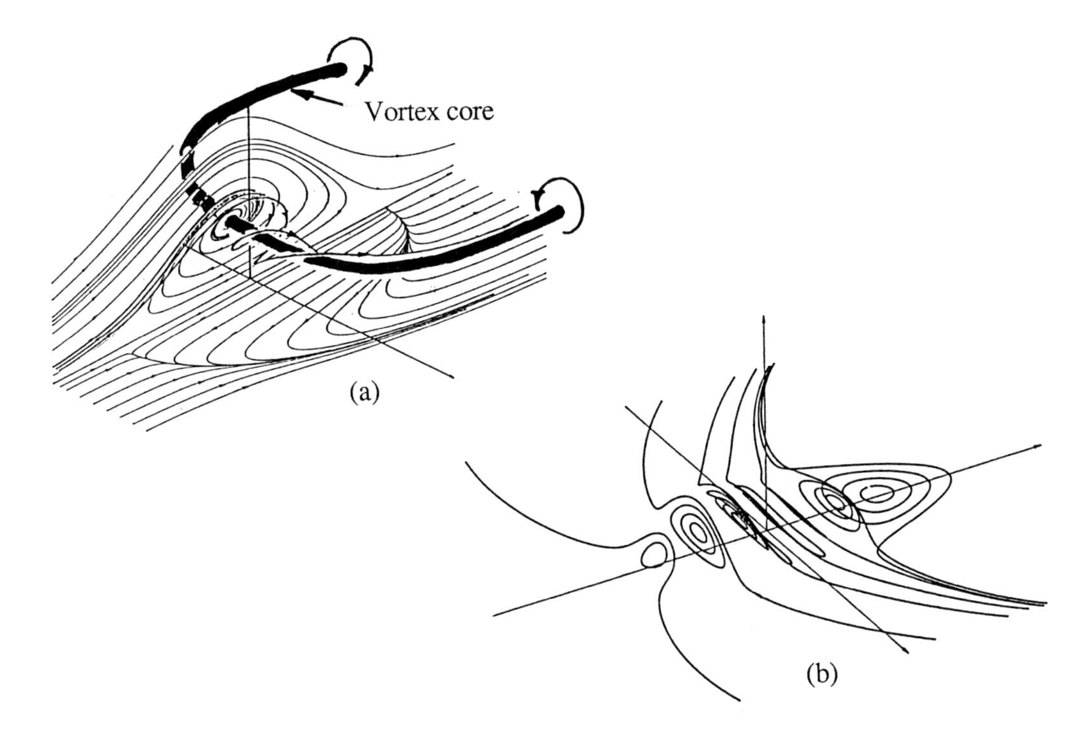

FIGURE 6. "U"-shaped separation. (a) Conjectured high Reynolds number vortex skeleton. (b) Actual vorticity field at Reynolds number of order 10.

Another possible flow field quantity to use for gaining possible insights into the physical processes occurring is the vector field of the pressure gradient ∇P. Chen, Cantwell & Mansour (1989) make use of this quantity which has its own set of critical points. One feature of this vector field is that its curl must be zero and so the asociated Jacobian matrix **A** must have only real eigenvalues and all eigenvectors must be orthogonal. The P-Q-R classification are the same as for other quantities. ∇P is an extremely attractive quantity to calculate and examine since it is directly related to a force field.

3.2 Definition of a Vortex

Recently Robinson, Kline & Spalart (1988) have studied many graphical images of the various quantities in the turbulent boundary layer as generated by supercomputer calculations. Quantities such as pressure (but not ∇P), stress fields, velocity vector fields

FIGURE 7. Instantaneous limiting streamlines or skinfriction lines from Moin & Spalart (1987).

and vorticity fields have been examined. From this work there is currently occurring a debate as to how a vortex should be defined. This is related to the conjectured counter rotating streamwise vortices just above the sublayer and the attached "∩"-shaped vortices. Statements such as: "there may be a region of vorticity represented by a bundle of lines of vorticity but this does not neccessarily mean that there is a vortex" or "we do not often see the "∩"-shaped vortices and if we do, they have only one leg" have been made. Such statements have a implied definition of a vortex and as far as the authors can make out, this definition is as follows: If there is a region of space where the rate of the deformation tensor has complex eigenvalues, then that region belongs to a vortex core. The instantaneous streamlines generated by such local solutions will orbit or wrap about an axis through the point at rest relative to the observer as shown in figure 2(a). Most workers would certainly deem such a pattern to be a vortex. The first to suggest such a formal definition was Volmers (1983) and Dallmann (1983). However, suppose we impose on our bundle of vortex lines an irrotational rate of strain induced by vorticity remote from our point of interest. If the stretching is mainly orthogonal to the vortex lines, then as the strength of the rate of strain field is increased while the vorticity is kept fixed, we will move in the P-Q-R space towards the surface S_1, cross it and then be in the region of real eigenvalues like the pattern shown in figure 3(a). This would have a crucial effect on the local mixing processes but as far as the Biot-Savart law is concerned the contribution the region of vorticity makes to the velocity field is unchanged. In summary, according to the definition given here, whether a region of vorticity constitutes a vortex, depends on the environment in which it is placed.

4. CONCLUSION

It can be seen that a study of the topology of the elementary local solutions for fluid motions provides a framework for describing the geometry and topology of more complex flow patterns and for gaining some physical insights into processes described by supercomputer data. Which flow field quantity or combination of quantities turn out to be the most useful still needs to be discovered and much of this work is still in its infancy. Some simple case studies have been given here to illustrate potential uses. The authors anticipate new and exciting developments in this young and rapidly growing field.

Support from the Australian Research Council and AFSOR Grant 84-0373D is gratefully acknowledged.

REFERENCES

BLAQUIRE 1966 *Nonlinear system analysis*, Academic Press: New York and London.

CHEN, J.H., CANTWELL, B.J. and MANSOUR, N.N. 1989 The effect of Mach number on the stability of a plane supersonic wake. *AIAA* paper 89-0285, Reno.

CHONG, M.S., PERRY, A.E. & CANTWELL, B.J. 1988 A general classification of three-dimensional flow patterns. Report No. SUDAAR 572, Dept. of Aeronautics and Astronautics, Stanford University.

DALLMANN, U. 1983 Topological structures of three-dimensional flow separations. DFVLR Rep. IB 221-82-A07. Gottingen. West Germany.

HORNUNG, H.G. and PERRY, A.E. 1984 Some aspects of three-dimensional separation. Part I. Streamsurface bifurcation. *Z. Flugwiss. Weltraumforsch.* **8**, pp.155-60.

LIGHTHILL, M.J. 1963 *In Laminar Boundary Layers* (ed. L. Rosenhead), pp. 48-88. Oxford University Press.

MOIN, P and SPALART, P.R. 1987. Contributions of numerical simulation data bases to the physics, modeling, and measurement of turbulence. *NASA* Technical Mem. 100022.

OSWATITSCH, K. 1958 *Die Ablosungsbedingung von Grenzshichten.* In K Oswatitsch: Contributions to the Development of Gasdynamics - Selected Papers translated (to English) on the occasion of K. Oswatitsch's 70th Birthday (ed. W. Schneider & M. Platzer), pp. 6-18. Braunschweig. Wiesbaded: Vieweg 1980.

PERRY, A.E. and FAIRLIE, B.D. 1974 Critical points in flow patterns. *Adv. Geophys.* **18B**, pp.299-315

PERRY, A.E. and TAN, D.K.M.1984 Simple three-dimensional motions in coflowing jets and wakes. *J. Fluid Mech.* **141**, pp.197-231.

PERRY, A.E. and HORNUNG, H.G. 1984 Some aspects of three-dimensional separation. Part II. Vortex skeletons *Z. Flugwiss. Weltraumforsch.* **8**, pp.155-60.

PERRY, A.E. and CHONG, M.S. 1986 A series expansion study of the Navier-Stokes equations with applications to three-dimensional separation patterns. *J. Fluid Mech.* **173**, pp.207-223.

PERRY, A.E. and CHONG, M.S. 1987 A description of eddying motions and flow patterns using critical point concepts. *Ann. Rev. Fluid Mech.* **19**, pp. 125-55.

REYN 1964 Classification and description of the singular points of a system of three linear differential equation. *ZAMP* **15**, pp.540-577.

ROBINSON, S.K., KLINE, S.J. and SPALART P.R. 1988 Quasi-coherent structures in the turbulent boundary layer. Part II. Verification and new information from a numerically simulated flat-plate layer. *Zorin Zaric Memorial International Seminar on Near-Wall Turbulence*, Dubrovnik, Yugoslavia 16-20 May, Hemisphere Publishing.

SPALART, P.R. 1988 Direct simulation of a turbulent boundary layer up to Rq = 1410 *J. Fluid Mech.* **187**, pp.61-98. Also NASA Technical Memorandum 89407 (1986).

VOLLMERS, H. 1983 Separation and vortical-type flow around a prolate spehroid. Evaluation of relevant parameters. *AGARD Symposium on Aerodyn. of Vortical Type Flow in Three-Dimensions.* Rotterdam. AGARD-CP 342 pp. 14-1 to 14.14.

Construction of Frozen-in Integrals, Lagrangian and Topological Invariants in Hydrodynamical Models

R.Z. SAGDEEV, A.V. TUR & V.V. YANOVSKY

Space Research Institute, Academy of Sciences
Profsoyuznaya 94/72, Moscow, USSR

Studies of turbulence in strongly interacting systems meet a number of generally known difficulties. Therefore it is important to find ways to overcome some of these difficulties. The analysis of the invariant properties of hydrodynamic models may, on the one hand, assist in advancing in this direction and on the other hand, indicate some finer properties of turbulent motion.

Let us consider the principal invariants of hydrodynamic models and obtain general relations between them. Lagrangian invariants are the quantities that satisfy, in terms of the Eulerian variables, the equation

$$\frac{\partial I}{\partial t} + (\mathbf{V} \cdot \nabla)\, I = 0, \tag{1}$$

that describes the conservation of I when it is convected by the fluid. Another type of invariant is represented by the freezing-in integrals described by the equation

$$\frac{\partial \mathbf{J}}{\partial t} + (\mathbf{V} \cdot \nabla)\, \mathbf{J} = (\mathbf{J} \cdot \nabla)\, \mathbf{V}. \tag{2}$$

This equation describes the freezing of the lines of force of the field $\mathbf{J}$ into a fluid.

Let us consider some relations that are independent of the specific form of the equations describing the velocity field. These were derived using only the continuity equation for the density ρ (or some other quantity). Hence the relations derived below are valid for any dissipationless hydrodynamic model. Suppose that, in addition to the Lagrangian invariants x_O (the initial coordinate of Lagrangian particles), we have a number of other Lagrangian invariants I_1, I_2, I_3. Choosing these as the basis of a new Lagrangian coordinate system we change from x_1^0, x_2^0, x_3^0 to $I_1, I_2, I_3 (x^0 \to x \to I)$. Clearly because the volume elements in terms of both the old and the new coordinates are Lagrangian invariants, the transition Jacobian is also a Lagrangian invariant. Thus we obtain the Lagrangian invariant

$$I_4 = \frac{1}{\rho} \frac{\partial (I_1, I_2, I_3)}{\partial (x_1, x_2, x_3)} \tag{3}$$

The derived relationship can be used in constructing Lagrangian conservation laws in terms of three or more known Lagrangian invariants by means of repeated consecutive use of (3).

We may also construct Lagrangian invariants in terms of known freezing-in integrals by using the following (1):

$$I' = \mathbf{J} \cdot \nabla I. \tag{4}$$

Indeed, direct calculation of dI'/dt, using (1) and (2), shows that I' is a Lagrangian invariant. This can be also easily verified by simple physical arguments. To do this, let us consider a surface $I =$ cst. and choose an elementary contour lying on this surface. It is evident that for any motions the contour will remain on this surface due to the conservation of the Lagrangian invariant I. The orientation of the element dS of the elementary contour coincides with the orientation of ∇I. Moreover, noting that the equation for ∇I coincides with that for $\rho d\mathbf{S}$, one can select the constant so that $\nabla I =$ const $\times\ \rho d\mathbf{S}$. The relation (4) may be then interpreted as the conservation of the flux of a frozen-in quantity, thus proving that the quantity I' is a Lagrangian invariant. The relation (4) admits a certain generalization because in the above argument the only important fact is that the quantity ∇I is a vector field satisfying an equation whose form coincides with that of the equation for $\rho d\mathbf{S}$. Thus, introducing fields $\mathbf{S}$ which satisfy the equation

$$\frac{\partial S_i}{\partial t} + (\mathbf{V} \cdot \nabla)\, S_i = -S_j \frac{\partial V_j}{\partial x_i} \tag{5}$$

we obtain a generalization of (4), namely

$$I' = \mathbf{J} \cdot \mathbf{S}. \tag{6}$$

For an incompressible fluid this relation is given in Kuzmin (1983). The introduction of the field $\mathbf{S}$ is motivated by physical arguments. In particular, it emerges as the momentum of a small annular vortex (Roberts 1972).

Now let us construct the frozen-in quantities from Lagrangian invariants. Consider in a Lagrangian coordinate space the surfaces $I_1(X_0) = I_1^0$ and $I_2(x_0) = I_2^0$ on which two Lagrangian invariants have fixed values. In the general case these surfaces intersect along a curve. Let us choose an element $d\mathbf{l}$ along this curve, this being the element of length of a fluid line specified by the invariants I_1 and I_2. It is clear that the vector $[\nabla I_1, \nabla I_2]$ is tangential to $d\mathbf{l}$, i.e. it always coincides with the latter's orientation. This makes it possible, by passing to the Eulerian coordinate system, to derive the expression for the frozen field (Moiseev et al 1982).

$$\mathbf{J} = \frac{1}{\rho}[\nabla I_1, \nabla I_2]. \tag{7}$$

Thus, the Lagrangian integrals enable us to construct vector quantities frozen into the medium and satisfying (7). Of course, this can be also verified directly, by differentiating (7) with respect to time. The linearity of equation (2) in $\mathbf{J}$ implies that the sum of frozen-in quantities will be again frozen-in. Moreover, multiplying $\mathbf{J}$ by a Lagrangian invariant does not violate its frozen-in property. Proceeding from this simple reasoning one can easily see that the quantity $\mathbf{J}$ constructed as follows from three Lagrangian invariants I_1, I_2, I_3 is frozen into the medium

$$J_\alpha = \frac{1}{\rho}\epsilon_{\alpha\beta\gamma}\ \epsilon_{ikl}\ I_i \frac{\partial I_k}{\partial x_\beta}\frac{\partial I_l}{\partial x_\gamma}, \tag{8}$$

where the subscripts take the values 1, 2, 3. This quantity has an independent sense as well, e.g. in terms of n-fields (Moiseev et al 1981).

It should be noted that the relation (7) can also be generalised (by reasoning similar to the above) to s-fields. Thus, (7) transforms to the following relation (Kuzmin 1983)

$$\mathbf{J} = \frac{1}{\rho}[\mathbf{S}_1, \mathbf{S}_2].$$

Let us stress one specific feature of (7). All known fields frozen into a medium (e.g. $\mathbf{J} = \frac{\mathbf{H}}{\rho}, \frac{1}{\rho}$ curl $\mathbf{V}$, etc), like the quantities in (7), possess the property of div $(\mathbf{J}\rho) = 0$. In physical terms this means that the "charges" of the field $\rho\mathbf{J}$ are absent. In other words, the lines of force of these fields can be either closed lines, or may go to infinity, or may everywhere densely fill a manifold of dimension higher than 1. Generically, the constructed fields (8) do not have such a property and exemplify frozen-in quantities of a new type. Moreover, the two forms of the differential relations for a frozen-in state, i.e. (2) and

$$\frac{\partial \rho\mathbf{J}}{\partial t} = -\text{curl}\ [\rho\mathbf{J}, \mathbf{V}] \tag{9}$$

are not equivalent since (2) implies the following equation for $\rho\mathbf{J}$

$$\frac{\partial \rho\mathbf{J}}{\partial t} = -\text{curl}\ [\rho\mathbf{J}, \mathbf{V}] - \mathbf{V}\ \text{div}\ \rho\mathbf{J},$$

which, when div $\rho\mathbf{J} \neq 0$, does not coincide with (9). As a consequence of nonvanishing div $\rho\mathbf{J}$ we have the nonzero flux of a frozen-in quantity through a closed surface and, thus, the appearance of an analogue of charges for these fields. If the Lagrangian invariants in (8) are interdependent (i.e. there exists a relation $\phi(I_1, I_2, I_3) = 0$), then div ρJ vanishes and we return to conventional frozen-in quantities.

Besides the above difference, equations (7) and (8) differ also topologically within the same class. However, these differences will be discussed after introducing the topological characteristics of frozen-in integrals. To obtain these integrals we consider a purely vortical field of the quantities $\rho\mathbf{J}$ (i.e. $I_3 = f(I_1, I_2)$). In this case the vector potential $\mathbf{A}$ ($\rho\mathbf{J} =$ curl $\mathbf{A}$) can be introduced. Consider the quantity (Moiseev et al 1982)

$$I^{\tau} = \int_{v_0} \rho\mathbf{J}\cdot\mathbf{A}v. \tag{10}$$

where the integration is performed over the volume bounded by the surface S_0 whose normal is always orthogonal to $\mathbf{J}$, i.e. $(\mathbf{n}\cdot\mathbf{J}) = 0$.

The quantity introduced describes the entanglement of the lines of force of the frozen-in quantities and is conserved. To prove this, we use the equation for the vector potential $\mathbf{A}$ that follows from the specific form of $\mathbf{J}$, namely

$$\frac{\partial\mathbf{A}}{\partial t} = [\mathbf{V},\ \text{curl}\ \mathbf{A}] + \nabla\Psi. \tag{11}$$

The form of the function Ψ is not important here. Using (11) and (2) we obtain

$$\frac{\partial\mathbf{J}\cdot\mathbf{A}}{\partial t} = (\mathbf{J}\cdot\nabla)(\Psi + \mathbf{A}\cdot\mathbf{V}).$$

Taking into account this equation one can easily show that

$$\frac{dI^{\tau}}{dt} = \int_{S_0} \rho(\mathbf{J}\cdot\mathbf{n})(\Psi + \mathbf{A}\cdot\mathbf{V})\,dS = 0$$

since $\mathbf{J}\cdot\mathbf{n} = 0$, by definition; hence I^{τ} is conserved. The physical interpretation of the Lagrangian invariants (10) is completely analogous to the interpretation of their particular cases for a compressible fluid and MHD (Moffatt 1969, Arnold 1986).

As another example, consider invariants arising in two-fluid plasma hydrodynamics. Let us illustrate another possibility of applying the above relations without use of canonical variables. To derive the initial set, we use the known system of equations of the two-fluid plasma hydrodynamics without dissipation in the adiabatic case. Evaluating the curl of the equations of motion one can easily obtain the frozen-in integral (Moiseev et al 1981)

$$\frac{d}{d}\left\{n_\alpha^{-1}\left(\text{ curl } v_\alpha+\frac{e_\alpha \mathbf{H}}{m_\alpha c}+\boldsymbol{\Omega}_\alpha\right)\right\}=\left\{n_\alpha^{-1}\left(\text{ curl } \mathbf{v}_\alpha+\frac{e_\alpha \mathbf{H}}{m_\alpha c}+\boldsymbol{\Omega}_\alpha\right)\cdot\nabla\right\} v_\alpha . \tag{12}$$

where $\boldsymbol{\Omega}_\alpha$ is the rotation velocity of the α-component as a whole (the remaining notation is conventional).

Thus, we have a frozen-in integral and a Lagrangian invariant S_α (entropy) for each component. By (4), we obtain another Lagrangian invariant

$$I_1^\alpha=\nabla S_\alpha\cdot\left(\text{ curl } \mathbf{v}_\alpha+\frac{e_\alpha \mathbf{H}}{m_\alpha c}+\boldsymbol{\Omega}_\alpha\right) n_\alpha^{-1} \tag{13}$$

that is analogous, to some extent, to the Ertel (1942) integral for a compressible fluid. Using consecutively the relations (4) and (7), one can construct the frozen-in integrals and the Lagrangian invariants of higher order with respect to derivatives, e.g.

$$J_1^\alpha=\frac{1}{n_\alpha}[\nabla S_\alpha,\ \nabla(\nabla S_\alpha\cdot(\text{ curl } \mathbf{v}_\alpha+\frac{e_\alpha \mathbf{H}}{m_\alpha c}+\boldsymbol{\Omega}_\alpha)n_\alpha^{-1}] \tag{14}$$

etc. Finally, by using (10), we obtain the topological invariants

$$I^\tau=\int_{v_0}\left(\text{ curl } \mathbf{v}_\alpha+\frac{e_\alpha \mathbf{H}}{m_\alpha c}+\boldsymbol{\Omega}_\alpha\right)\cdot\left(\mathbf{v}_\alpha+\frac{c_\alpha \mathbf{A}}{m_\alpha c}+\tfrac{1}{2}[\boldsymbol{\Omega}_\alpha,\mathbf{r}]dv_0\right. \tag{15}$$

where the integration is performed over the volume whose surface has a normal vector that is orthogonal to the vector $\left(\text{ curl } \mathbf{v}_\alpha+\frac{e_\alpha \mathbf{H}}{m_\alpha c}+\boldsymbol{\Omega}_\alpha\right)$. The invariant (15) reflects the entanglement of the lines of force of the curl of the generalized momentum field for each plasma component.

The invariants for the hydrodynamic models considered here play an important part in studies of dynamic processes in continuous media. They facilitate the use of graphic geometric properties of the hydrodynamic models for a qualitative analysis of specific situations (e.g. the change in the 'minimum B' principle for plasmas, with vortical motions present, due to (12)). Their use substantially facilitates both the search for exact solutions of dynamic equations (Moiseev et al 1982) and the construction of various approximate computational schemes. Moreover, the dynamic invariant features manifest themselves also in turbulent processes, and, in principle, allowance for them must result in a finer classification of turbulence patterns than accepted at present. Let us consider how these properties are related to the invariant properties of strong homogeneous turbulence (Sagdeev et al 1986).

As before, we shall not specify a hydrodynamic model since the resulting relations will be valid in all dissipationless media. (Moreover, in the relations described below dissipation is of minor importance.) Consider an arbitrary hydrodynamic model with Lagrangian invariants $I^\alpha(\mathbf{x},t)$ $(\alpha = 1,2,\ldots)$. Let us introduce the correlator

$$K^{\alpha\beta}(r,t) = < I^\alpha(\mathbf{x}_1,t)\, I^\beta(\mathbf{x}_2,t)\, \rho(\mathbf{x}_1,t)\, \rho(\mathbf{x}_2,t) > .$$

By virtue of homogeneity this depends only on $\mathbf{r} = \mathbf{x}_1 - \mathbf{x}_2$. For simplicity, we assume that $< I^\alpha \rho >= 0$ (this is not essential since otherwise one can consider $K^{\alpha\beta} - < I^\alpha\rho >< I^\beta\rho >$). The equation describing the temporal evolution of the correlator can be easily derived using the continuity equation and Eq. (1):

$$\frac{\partial K^{\alpha\beta}}{\partial t} + \operatorname{div} \mathbf{P}^{\alpha\beta} = 0$$

where

$$P^{\alpha\beta} = <I^\alpha(\mathbf{x}_1,t)\, I^\beta(\mathbf{x}_2,t)\, \rho(\mathbf{x}_1,t)\, \rho(\mathbf{x}_2,t)\, \mathbf{V}(\mathbf{x}_1) > - \\ - <I^\alpha(\mathbf{x}_1,t)\, I^\beta(\mathbf{x}_2,t)\, \rho(\mathbf{x}_1,t)\, \rho(\mathbf{x}_2,t)\, \mathbf{V}(\mathbf{x}_2) > .$$

This equation implies that, taking into account $\mathbf{P}^{\alpha\beta}\|_{\mathbf{r}\to\infty} \longrightarrow 0$, the quantity

$$I_\tau^{\alpha\beta} = \int_{-\infty}^{+\infty} K^{\alpha\beta}(\mathbf{r},t)\, d^3\mathbf{r} \tag{16}$$

is conserved, i.e.

$$\partial_\tau I_\tau^{\alpha\beta} = 0.$$

Thus, in the case of homogeneous turbulence, the presence of Lagrangian invariants gives rise to conservation laws that characterize its large-scale properties. It can be easily understood that using the relations (4) generates further invariants of homogeneous turbulence that include the frozen-in integrals

$$I_\tau^{\alpha\beta} = \int < \nabla I^\alpha(x_1 t) \cdot \mathbf{J}(x_1 t)\, \nabla I^\beta(x_2 t) \cdot \mathbf{J}(x_2 t)\, \rho(x_1 t)\, \rho(x_2 t) > d^3\mathbf{r}. \tag{17}$$

The relations (16) and (17) enable us to construct statistical invariants in specific hydrodynamic models. This, quite naturally, leads to different kinds of homogeneous turbulence which are characterized by nonvanishing statistical invariants. Such a classification of turbulent models may be justified since the presence of the invariants affects the basic turbulence properties (such as spectral parameters, transfer coefficients, etc.). To see that this is so, note that the invariants if they exist enter into the set of arguments of the characteristic functional. This, in turn, modifies the scaling theorem and results in the emergence of spectral regions of a non-Kolmogorov form. The effect of the additional invariants on spectral characteristics can be illustrated by helical or two-dimensional turbulence.

One can reformulate the proposed scheme of the derivation of frozen-in integrals by means of external differential forms. This allows one to obtain also some new integral and topological invariants (Tur et al 1989). It is also possible to prove that the frozen-in integrals form a Lie algebra. With the use of frozen-in integrals one can construct a number of solutions in the usual and magnetic hydrodynamics, which have nontrivial streamline topology. These solutions can be considered in terms of topological charge (see, for example, Sagdeev et al 1986, Tur et al 1984).

As shown in Moiseev et al (1983, 1985), Frisch et al (1989), topological solitons can be generated in hydrodynamic helical turbulence, that is when the mean helicity of the turbulence $< \mathbf{v} \cdot \text{rot}\, \mathbf{v} >$ does not vanish.

The helicity, as is well known, is a topological characteristic of the turbulence itself. In Rutkevich et al (1988) it was shown that topological solitons may be generated in the fluid also by means of an external driving force of Arnold-Beltrami-Childress type. This force also has motivated topological characteristics.

REFERENCES

ARNOLD V.I. (1986) *Selecta Mathematica Sovietica* Vol. 5, **4**, 327.

ERTEL H. (1942) *Meteorol. Zeitschr.* B**159**, 277.

FRISCH U., SHE Z.S., SULEM P.L. (1987) *Physica* **28D**, p. 283.

KUZMIN G.A. (1983) Preprint IYaF SO AN SSSR, No. 87-83, Novosibirsk.

MOFFATT H.K. (1969) *J. Fluid Mech.* **35**, 117.

MOISEEV S.S., SAGDEEV R.Z., TUR A.V. & YANOVSKY V.V. (1986) in "Nonlinear Phenomena in Plasma Physics and Hydrodynamics", Ed. R.Z. Sagdeev. Mir, Publ. Moscow, pp. 137-180.

MOISEEV S.S., SAGDEEV R.Z., TUR A.V. & YANOVSKY V.V. (1982), *Eksp. Teor. Fiz.* **83**, 1 (7), 215.

MOISEEV S.S., SAGDEEV R.Z., TUR A.V. & YANOVSKY V.V. (1983) *Sov. Phys. JETP*, **58**, 1144.

MOISEEV S.S., RUTKEVICH P.B., TUR A.V. & YANOVSKY V.V. (1988), *Sov. Phys. JETP*, **67** (2), p. 294.

PENFOLD P. (1966) *Phys. Fluids* **9**, 1184.

ROBERTS P.H. (1972) *Mathematics*, **19**, 169.

RUTKEVICH P.B., SAGDEEV R.Z., TUR A.V. & YANOVSKY V.V. (1988) Preprint IKI AN USSR, No. 1987.

SAGDEEV R.Z., MOISEEV S.S., TUR A.V. & YANOVSKY V.V. (1986), in "Nonlinear Phenomena in Plasma Physics and Hydrodynamics", Ed. R.Z. Sagdeev. Mir. Publ. Moscow, p. 137-180.

TUR A.V. & YANOVSKY V.V. (1984) in "Nonlinear and Turbulent Processes" Harwood Acad. Publ. N.Y., Vol. 2., p. 1079.

TUR A.V. & YANOVSK V.V. (1988) Preprint IKI AN USSR.

A Natural Method for Finding Stable States of Hamiltonian Systems

G. K. VALLIS

Division of Natural Sciences, University of California, Santa Cruz

G. F. CARNEVALE

Scripps Institution of Oceanography, University of California, San Diego

T. G. SHEPHERD

Department of Physics, University of Toronto

It is shown that associated with any dynamical system that can be expressed in Hamiltonian form there exists a modified system in which the energy changes monotonically, yet in which all the Casimirs, or distinguished functions, of the original system are preserved. The steady solutions of the modified dynamics and the original system are identical. Thus the modified system evolves toward an extremal energy state of the original system, constrained by the initial values of all the Casimir invariants. Since such states are generally nonlinearly stable, the method provides a means to find stable states of Hamiltonian dynamical systems.

When applied to the Euler equations of fluid motion the procedure yields a system of dynamics in which the energy grows or decays monotonically yet which stays on the same isovortical surface, thereby preserving all the vortical or topological invariants, namely helicity, circulation and potential vorticity. The modified system thus relaxes toward a stable solution of the Euler equations. The application to magnetohydrodynamics is also possible.

---

1. PREAMBLE

Recently, a natural method was presented for searching for stable solutions of the Euler equations of fluid dynamics and various other fluid systems (Vallis, Carnevale and Young, 1989). The method is 'natural' because it utilizes the structure of the evolution equations themselves, rather than using some generic algorithm, like steepest descents, for finding extremal energy states. It involves modifying the equations of motion in such a way that all the topological invariants of the original system associated with the advection of vorticity (namely circulation around marked fluid parcels, helicity, and potential vorticity) are preserved in the new system yet the energy monotonically changes. Further, the steady states of the original system and modified one are identical. Thus, the modified system evolves toward an energy extremum of the original system, on the isovortical sheet on which it began. Arnol'd (1966) had previously shown that such energy extrema, where they exist, would generally be stable (see also Holm et al., 1985, and McIntyre and Shepherd, 1987). This result can be seen heuristically as follows. The phase space of the fluid may be considered to comprise multi-dimensional isovortical surfaces; the evolution of the fluid is constrained to lie on the intersection of these surfaces and surfaces of constant energy. An isolated energy extremum on an isovortical sheet will therefore be at least linearly stable, and may be Lyapunov stable, because a state perturbed from such an extrema must stay close to its parent since it must follow the energy contours on the isovortical sheet. Thus, the method provides a potentially powerful tool for finding stable flow configurations of potentially arbitrary vortex topology, although there is no guarantee of convergence from all initial conditions. Carnevale and Vallis (1990) present a number of numerical examples demonstrating the feasibility of the method.

In this paper we show that the method may be applied quite generally to both discrete and continuous, generally non-canonical, Hamiltonian dynamical systems. Typically, in a Hamiltonian dynamical system the Hamiltonian itself is an invariant. Additionally there may exist Casimir invariants. (Impulse invariants associated with a translation invariance may also exist (Benjamin, 1984), but we shall say nothing about them here.) Casimirs are independent of the Hamiltonian, being functions of the evolution operator, or the bracket, governing the system and depend on the non-canonical nature of the bracket (e.g. Littlejohn, 1982). They play the same role as the vortical or 'topological' invariants do in fluid mechanics—indeed the helicity of a fluid *is* a Casimir. Our method consists of modifying the original dynamical system in such a way that in the new system all the Casimirs remain invariant while its energy monotonically changes. The steady state of the modified system are identical to those of the original system. Thus, the system evolves toward an extremal energy state on an 'iso-Casimir' surface, or distinguished surface. The potential importance of the method lies in the observation that isolated extremal energy states, where they exist, are stable configurations of the system. In this paper we shall begin abstractly and then show how the fluid dynamical forms follow as special cases.

2. MODIFIED HAMILTONIAN DYNAMICS

We first restrict attention to finite dimensional Hamiltonian systems, although the extension to continuous systems will be straightforward. Consider the system

$$\dot{u} = J\nabla_u H(u). \tag{2.1}$$

The dependent variable u is the state vector of the system and is a function of time and $H(u)$ is the Hamiltonian. (The subscript u on ∇ will subsequently be dropped.) J is a skew-symmetric operator (which may be a matrix) satisfying $(a, Jb) = -(Ja, b)$, where $(*, *)$ is the appropriate inner product. Specification of J and the Hamiltonian determines the system evolution. The associated (possibly non-canonical) Poisson bracket $\{*, *\}$ may be determined from J by noting that

$$\frac{dF}{dt} = \{F, H\},$$

but from (2.1)

$$\dot{F} = (\nabla F, u_t) = (\nabla F, J\nabla H),$$

implying the bracket:

$$\{F, G\} = (\nabla F(u), J\nabla G(u)) \tag{2.2}$$

where F and G are arbitrary functions of u and the right hand side is the inner product of the two functions. By the skew-symmetry of J,

$$\{F, G\} = (\nabla F, J\nabla G) = -(J\nabla F, \nabla G) = -(\nabla G, J\nabla F) = -\{G, F\}. \tag{2.3}$$

An evolution equation entirely equivalent to (2.1) is thus:

$$\dot{F} = \{F, H\}, \tag{2.4}$$

if the bracket is determined from J by (2.2), or possibly *vice versa*. These aspects are discussed further in Olver (1986, chapter 6), Salmon (1988) and elsewhere. The Hamiltonian itself is obviously an invariant, since

$$\frac{dH}{dt} = (\nabla H, \dot{u}) = (\nabla H, J\nabla H) = 0.$$

Another class of invariants are the Casimirs $C(u)$, or distinguished functions. These are invariants not because of the form of the Hamiltonian, but because given the form of the bracket, or of J, they, by definition, commute with everything. That is

$$\{G, C\} = (\nabla G, J\nabla C) = 0 \tag{2.5}$$

for all admissable G, implying

$$J\nabla C = 0, \tag{2.6}$$

which may also be taken as their defining property. From (2.4) they are therefore invariants. Littlejohn (1982) discusses their properties and existence in more detail.

Now, we seek a dynamical system in which energy will monotonically change yet for which the Casimirs remain invariant. To this end, consider the system

$$\dot{u} = J\nabla H + J\alpha J\nabla H \tag{2.7}$$

where α is a positive or negative definite matrix. An equivalent 'Poisson bracket' form is

$$\dot{F} = \{F, H\} + (\nabla F, J\alpha J\nabla H). \tag{2.8}$$

The second term on the right hand side of (2.8) cannot generally be put into a true Poisson bracket form because the operator $J\alpha J$ is symmetric whereas J is skew symmetric. This precludes the existence of any quantity H' such that

$$\{F, H'\} = (\nabla F, J\nabla H') = (\nabla F, J\alpha J\nabla H),$$

or

$$\alpha J\nabla H = \nabla H'.$$

For example, we see that in canonical systems the existence of an H' requires (with α having unit entries), $\partial H'/\partial q = -\partial H/\partial p$ and $\partial H'/\partial p = \partial H/\partial q$, or $\nabla^2 H = 0$.

The energy budget follows easily from (2.8) by putting $F = H$. The first term vanishes leaving

$$\dot{H} = (\nabla H, J\alpha J\nabla H) = -(J\nabla H, \alpha J\nabla H), \tag{2.9}$$

using the skew symmetry of J. The right hand side is positive or negative definite, according as α is chosen.

Casimir invariance follows by letting $F = C$ in (2.8). The first term immediately vanishes (by definition of the Casimir). The second term is also zero by (2.6).

The modified system will thus evolve to a new state with higher or lower energy whilst staying on the same distinguished surface. The energy may or may not be bounded from above or below, depending, for example, on whether the Casimirs provide a bound on the energy through some integral inequality. If the modified system does reach a steady state then, from (2.6), $J\nabla H = -J\alpha J\nabla H$. However, each term must individually vanish, since from (2.9) the energy only ceases to evolve when $J\nabla H = 0$. But this is precisely the condition that the *original* system be in a steady state. Thus the steady states of the original and modifed system are identical.

The application to continuous systems (through the use of non-canonical formulations, cf. Morrison and Greene, 1980) is relatively straightforward and indeed may be the area where the method is of most use. The equations of motion corresponding to (2.1) are

$$u_t = J\frac{\delta \mathcal{H}}{\delta u}, \tag{2.10}$$

which has the equivalent Poisson bracket formulation

$$\frac{d\mathcal{F}}{dt} = \{\mathcal{F}, \mathcal{H}\}, \tag{2.11}$$

where

$$\{\mathcal{F}, \mathcal{H}\} = \left(\frac{\delta \mathcal{F}}{\delta u}, J\frac{\delta \mathcal{H}}{\delta u}\right). \tag{2.12}$$

In (2.10) through (2.12) the script symbols are functionals; the Hamiltonian $\mathcal{H}$, for example, is obtained as an integral of a Hamiltonian density $H(x, y, z)$ thus

$$\mathcal{H} = \int H(x, y, z)dV.$$

The variational or functional derivative $\delta\mathcal{F}/\delta u$ is defined in the usual way by

$$\delta\mathcal{F} = \mathcal{F}(u + \delta u) - \mathcal{F}(u) = \left(\frac{\delta \mathcal{F}}{\delta u}, \delta u\right) + O(\delta u^2) \tag{2.13}$$

where the arbitrary variation δu vanishes at the domain boundary. (2.13) serves to define $\delta\mathcal{F}/\delta u$.

The associated modified system is given by

$$u_t = J\frac{\delta \mathcal{H}}{\delta u} + J\alpha J\frac{\delta \mathcal{H}}{\delta u} \tag{2.14}$$

or

$$\frac{d\mathcal{F}}{dt} = \{\mathcal{F}, \mathcal{H}\} + \left(\frac{\delta \mathcal{F}}{\delta u}, J\alpha J\frac{\delta \mathcal{H}}{\delta u}\right). \tag{2.15}$$

The invariance of the Casimirs (which satisfy $\{\mathcal{C}, \mathcal{G}\} = 0$ or $J\delta\mathcal{C}/\delta u = 0$) follows immediately from (2.15). The energy equation is

$$\frac{d\mathcal{H}}{dt} = \left(\frac{\delta \mathcal{H}}{\delta u}, J\alpha J\frac{\delta \mathcal{H}}{\delta u}\right) = -\left(J\frac{\delta \mathcal{H}}{\delta u}, \alpha J\frac{\delta \mathcal{H}}{\delta u}\right) \tag{2.16}$$

which again is of definite sign, and is zero in a steady state.

3. AN ELEMENTARY EXAMPLE

We illustrate the above abstractions with perhaps the simplest example of a Hamiltonian system for which Casimirs exist, namely a rigid body. It is useful example because the system is discrete, with only three degrees of freedom, so the construction is easily made. However, the final form is not one easily guessed or arrived at by other methods. The structure matrix is

$$J(\mathbf{m}) = \begin{pmatrix} 0 & -m_3 & m_2 \\ m_3 & 0 & -m_1 \\ -m_2 & m_1 & 0 \end{pmatrix}, \tag{3.1}$$

where m_i, $i = 1, 2, 3$, or $\mathbf{m}$, are the angular momenta. Using (2.1)

$$\dot{\mathbf{m}} = \mathbf{m} \times \nabla_{\mathbf{m}} H(\mathbf{m}). \tag{3.2}$$

The corresponding Poisson bracket is

$$\{F, G\} = (\nabla F, \mathbf{m} \times \nabla G) = -\mathbf{m} \cdot (\nabla F \times \nabla G). \tag{3.3}$$

The inner product in this case is just the dot product in *m*-space. From the form (3.3) of the bracket it is clear that $[\mathbf{m}^2, G] = 0$ for any G, so the magnitude squared of the angular momentum, $A = \mathbf{m}^2$ is a Casimir. Indeed without specifying the Hamiltonian, it is apparent by taking the dot product of (3.2) with $\mathbf{m}$ that A is conserved.

The Hamiltonian is in fact

$$H = \frac{1}{2}\left(\frac{m_1^2}{I_1} + \frac{m_2^2}{I_2} + \frac{m_3^2}{I_3}\right), \tag{3.4}$$

which, using (3.2), leads to the equations of motion

$$\dot{m}_1 = \frac{I_2 - I_3}{I_2 I_3} m_2 m_3, \qquad \dot{m}_2 = \frac{I_3 - I_1}{I_1 I_3} m_1 m_3, \qquad \dot{m}_3 = \frac{I_1 - I_2}{I_2 I_1} m_2 m_1. \tag{3.5}$$

Using the recipe of §2 the modified dynamics takes the form

$$\dot{\mathbf{m}} = (\mathbf{m} \times \nabla H) + \alpha(\mathbf{m} \times (\mathbf{m} \times \nabla H)) \tag{3.6}$$

where we may simply take α to be a scalar. Explicitly evaluating the extra terms, the modified equations of motion become

$$\begin{aligned} \frac{dm_1}{dt} &= \frac{I_2 - I_3}{I_2 I_3} m_2 m_3 + \alpha\left(\frac{I_1 - I_2}{I_2 I_1} m_2^2 m_1 - \frac{I_3 - I_1}{I_1 I_3} m_1 m_3^2\right) \\ \frac{dm_2}{dt} &= \frac{I_3 - I_1}{I_1 I_3} m_1 m_3 + \alpha\left(\frac{I_2 - I_3}{I_2 I_3} m_2 m_3^2 - \frac{I_1 - I_2}{I_2 I_1} m_2 m_1^2\right) \\ \frac{dm_3}{dt} &= \frac{I_1 - I_2}{I_2 I_1} m_2 m_1 + \alpha\left(\frac{I_3 - I_1}{I_1 I_3} m_1^2 m_3 - \frac{I_2 - I_3}{I_2 I_3} m_2^2 m_3\right) \end{aligned} \tag{3.7}$$

The first term on each right hand side is the unmodified term of the true dynamics. From inspection of (3.7) it is clear that angular momentum is still conserved. The energy evolution is given by

$$\frac{dE}{dt} = -\alpha\Big\{\Big(\frac{m_1m_2(I_1-I_2)}{I_1I_2}\Big)^2 + \Big(\frac{m_2m_3(I_2-I_3)}{I_2I_3}\Big)^2 + \Big(\frac{m_1m_3(I_1-I_3)}{I_1I_3}\Big)^2\Big\}. \quad (3.8)$$

The system above is equivalent to a triad interaction in incompressible two-dimensional fluid mechanics. The associations are $I_i \Leftrightarrow k_i^2$ and $m_i \Leftrightarrow \zeta_i$, where ζ_i is the vorticity of the i'th mode with wavenumber k_i. Angular momentum corresponds to enstrophy, and energy to energy.

Thus, the modified rigid body will seek an energy minimum or maximum subject to the constraint that angular momentum is conserved. This is just motion about, respectively, the axis of largest or smallest moment of inertia, known from elementary considerations to be stable. It is interesting to speculate whether the above modified system has any physical significance other than being a means to evolve to a stable solution. If a spinning rigid body is released into superficially free motion, there are in practice likely to be dissipative mechanisms acting on the body, due for example to straining and flexing of its parts. Hence its kinetic energy may fall (the energy being converted to heat and radiated away), although its angular momentum is conserved, so that the body evolves to a state of minimum possible kinetic energy. Similar effects may manifest themelves in many-body problems with non-elastic collisions, for example some astrophysical situations.

4. FLUID DYNAMICAL SYSTEMS

In this section we show how application of the general method to fluid dynamical systems indeed reproduces the forms previously found by Vallis *et al.* (1989). A number of other examples, as well as further discussion of the general method, are given by Shepherd (1990).

Consider three-dimensional incompressible flow,

$$\frac{\partial \mathbf{u}}{\partial t} - \mathbf{u}\times\boldsymbol{\omega} = -\nabla b \quad (4.1)$$

with

$$\nabla\cdot\mathbf{u} = 0, \qquad \boldsymbol{\omega} = \nabla\times\mathbf{u}, \qquad b = p + \frac{1}{2}\mathbf{u}^2. \quad (4.2)$$

To put these equations in Hamiltonian form first take the curl of (4.1) to give:

$$\frac{\partial \boldsymbol{\omega}}{\partial t} - \nabla\times(\mathbf{u}\times\boldsymbol{\omega}) = 0, \quad (4.3)$$

If boundary conditions are such that $\mathbf{u}$ may be determined uniquely for given $\boldsymbol{\omega}$ then (4.3), with $\nabla \cdot \mathbf{u} = 0$ and $\boldsymbol{\omega} = \nabla \times \mathbf{u}$, suffices to describe the system. Following Benjamin (1984) or Olver (1986) the system may be expressed in the manifestly Hamiltonian form (2.10). The state variable is $\boldsymbol{\omega}$, the Hamiltonian is given by

$$\mathcal{H} = -\frac{1}{2} \int \boldsymbol{\psi} \cdot \boldsymbol{\omega} dV, \tag{4.4}$$

and the skew symmetric evolution operator is

$$J = -\nabla \times (\boldsymbol{\omega} \times \nabla \times (\cdot)). \tag{4.5}$$

Now, $\delta\mathcal{H}/\delta\boldsymbol{\omega} = \boldsymbol{\psi}$, and since $\nabla \times \boldsymbol{\psi} = \mathbf{u}$ and $\nabla \cdot \boldsymbol{\psi} = 0$, substitution of (4.5) into (2.10) indeed gives (4.3). The modified dynamics are obtained from (2.14), giving

$$\frac{\partial \boldsymbol{\omega}}{\partial t} - \nabla \times (\tilde{\mathbf{u}} \times \boldsymbol{\omega}) = 0, \tag{4.6}$$

where

$$\tilde{\mathbf{u}} = \mathbf{u} + \alpha \nabla \times \nabla \times (\boldsymbol{\omega} \times \mathbf{u}) \tag{4.7}$$

The energy equation of the modified system is

$$E = \frac{1}{2} \int \mathbf{u}^2 dV, \qquad \frac{dE}{dt} = -\alpha \int [\nabla \times (\mathbf{u} \times \boldsymbol{\omega})]^2 dV \tag{4.8}$$

so there is indeed a monotonic energy change until the steady state, with $\nabla \times (\mathbf{u} \times \boldsymbol{\omega}) = 0$, is reached.

Note the form of the modified dynamics: in the vorticity equation the vorticity is evolved, not by advection by the true velocity (by true we mean such that $\boldsymbol{\omega} = \nabla \times \mathbf{u}$) but by a modified velocity given by (4.7). Because this is the only change to the equations, the technique may be labelled 'pseudo-advection'. Now, the evolution of the vorticity field is the same as that of an incompressible passive vector field, say $\mathbf{B}$, which obeys the equation $\partial \mathbf{B}/\partial t - \nabla \times (\mathbf{u} \times \mathbf{B}) = 0$. It is common to say that both $\boldsymbol{\omega}$ and $\mathbf{B}$ are frozen to the flow, meaning that the material curve coincident with a $\mathbf{B}$–line at $t = 0$ remains coincident for all future time. This topological property leads directly to the conservation of circulation around each material line and to the conservation of helicity, $\mathcal{I}$, where $\mathcal{I} = \int \boldsymbol{\omega} \cdot \mathbf{u}\, dV$. It is clear that under the modified dynamics we may also say that vorticity is frozen to the fluid, moving with velocity $\tilde{\mathbf{u}}$. Thus, the helicity and circulation invariants are preserved in the modified dynamics. This is demonstrated explicitly in Vallis *et al.* (1989). The conservation of helicity is expected since helicity is a Casimir, indeed according to Olver (1986) the *only* non-trivial Casimir.

For two-dimensional incompressible flow the equations of motion are

$$\frac{D\zeta}{Dt} = \frac{\partial \zeta}{\partial t} + \partial(\psi, \zeta) = 0 \tag{4.9}$$

where $\partial(\psi, \zeta) = \psi_x \zeta_y - \psi_y \zeta_x$ is the Jacobian. These equations may be directly derived from the three-dimensional equation (4.3) by specializing to two dimensions and setting $\zeta = \boldsymbol{\omega} \cdot \hat{\mathbf{z}}$ and $\mathbf{u} = \hat{\mathbf{z}} \times \nabla\psi$, so $\zeta = \nabla^2\psi$. In Hamiltonian form the vorticity is again the state variable, and we have

$$J = -\partial(\zeta, \cdot) \tag{4.10}$$

and

$$\mathcal{H} = -\frac{1}{2}\int \psi\zeta \, dA. \tag{4.11}$$

Since $\delta\mathcal{H}/\delta\zeta = -\psi$, substitution of (4.10) into (2.10) gives (4.9). The modified dynamics are obtained from (2.14), giving

$$\frac{\tilde{D}\zeta}{Dt} = \frac{\partial\zeta}{\partial t} + \partial(\tilde{\psi}, \zeta) = 0 \tag{4.12}$$

where

$$\tilde{\psi} = \psi + \alpha\partial(\psi, \zeta) = 0. \tag{4.12b}$$

The energy equation is given by

$$E = \mathcal{H} = -\frac{1}{2}\int \psi\zeta \, dA = \frac{1}{2}\int (\nabla\psi)^2 dA, \qquad \frac{dE}{dt} = -\alpha\int [\partial(\psi, \zeta)]^2 dA.$$

The Casimirs for two-dimensional flow are determined by inspection of (4.10) or the corresponding bracket:

$$\{F, G\} = \int \frac{\delta\mathcal{G}}{\delta\zeta}\partial(\zeta, \frac{\delta\mathcal{F}}{\delta\zeta})dA.$$

A functional $\mathcal{C}$ will evidently be a Casimir if $\partial(\zeta, \delta\mathcal{C}/\delta\zeta) = 0$. If $\mathcal{C} = \int C(\zeta)dA$ then

$$\partial(\zeta, \frac{\delta\mathcal{C}}{\delta\zeta}) = \partial(\zeta, C'(\zeta)) = 0.$$

Thus any integral functions of vorticity alone are Casimirs, and are conserved under the modified dynamics. This corresponds precisely to the conservation of any integral function of ζ by the modified dynamics (4.12). That is,

$$\frac{d}{dt}\int F(\zeta)dA = 0,$$

which follows directly from (4.12), no matter how $\tilde{\psi}$ is defined, because the vorticity on material parcels is conserved and the evolution effects only a rearrangement of vorticity.

4. DISCUSSION

We have seen that an algorithm exists which will transform an energy conserving Hamiltonian dynamical system into a system in which the energy monotonically changes, yet which preserves all the Casimir invariants of the original system. Inviscid fluid dynamical systems may be cast in Hamiltonian form, and the algorithm then reproduces the recipes given by Vallis et al. (1989). The Casimir invariants then correspond to the vortical invariants (potential vorticity, helicity) which determine the topology of the flow, preserved by the modified evolution operator. In this paper we have provided a general prescription which may be applied in more complex settings where the form of the modification to the original system cannot be guessed; Shepherd (1990) provides a number of fluid dynamical examples where this is so. The magnetohydrodynamic case may be similarly treated, and this will be given in a subsequent paper.

The modified systems will therefore seek extremal energy states, constrained by the initial values of the Casimir invariants. The steady states of the modified dynamics and the original evolution equations are identical, and at a steady state construction of a stable extremal energy state of the dynamical system has therefore been effected.

A number of caveats concerning the method may be appropriate. First, there is of course no guarantee that extremal energy states on a given surface of constant Casimirs (a 'distinguished surface') exist. Whether they do will depend on the nature of the system at hand. For example, in the flow of an ideal two-dimensional fluid any integral function of the vorticity, for example the enstrophy $\int \zeta^2 dA$, is a Casimir. In a finite domain we may use Poincaré's inequality in the form $\int (\nabla^2 \psi)^2 dA > C \int (\nabla \psi)^2 dA$ to show that there always *exists* an energy maximum on an isovortical sheet: since the left hand side is constant the right hand side is bounded. In three dimensions enstrophy is not bounded and it seems unlikely that there generally exist energy extrema on isovortical surfaces. Second, the modified dynamics may find only a *local* maximum, not a global maximum. Thus, any steady state found may not satisfy Arnol'd's (1966) criterion for Lyapunov stability. However, the state may still be nonlinearly stable, if not provably so by Arnol'd's theorems. Further discussion is to be found in Carnevale and Vallis (1990). Third, even when energy is bounded from above, convergence is certainly not assured. An asymptotic approach to a steady state may ensue, but the grail may not be found. This is likely when the energy extrema is to be found at the 'edge' of the distinguished surface; the modified dynamics may then take an infinite time to reach it. In this case, there will in general be no guarantee that the topology of the original flow configuration be preserved. In fluid dynamics this will be manifested by the reconnection of vortex lines. Consider, for example, the case of a circular patch of constant vorticity ζ_1 within a larger circular patch ζ_2, beyond which $\zeta = 0$. Suppose $\zeta_2 > 0 > \zeta_1$. Then the initial configuration is unstable, and clearly no stable configuration exists with this topology. Under the modified evolution the configuration will 'explode' and can only achieve a stable steady state if vortex line reconnection occurs. These caveats aside, it seems that the method may be a useful tool for constructing new stable solutions to a variety of dynamical systems.

We are grateful to Rick Salmon for some some useful remarks. This work was funded by NSF under grant ATM-8914004 and by the ONR.

REFERENCES

ARNOL'D, V.I. 1966: On an a priori estimate in the theory of hydrodynamic stability. *Izv. Vyssh. Uchebn. Zaved. Matematika,* **54,** no. 5, 3-5. (English transl.: *Am. Math. Soc. Transl., Series 2,* **79**, 267-269 (1969).)

BENJAMIN, T.B. 1984: Impulse, flow force and variational principles. *IMA J. Appl. Math.,* **32,** 3-68.

CARNEVALE, G.F. & VALLIS, G.K. 1990: Pseudo-advective relaxation to stable states of inviscid two-dimensional fluids. *J. Fluid Mech.* (to appear)

HOLM, D.D., MARSDEN, J.E., RATIU, T. & WEINSTEIN, A. 1985: Nonlinear stability of fluid and plasma equilibria. *Phys. Rep.* **123,** 1-116.

LITTLEJOHN, J. 1982: Singular Poisson tensors. In *Mathematical Methods in Hydrodynamics and Integrability in Dynamical Systems* (ed. M. Tabor & Y.M. Treve), Am. Inst. Phys. Conf. Proc. **88,** 47-66.

MCINTYRE, M.E. & SHEPHERD, T.G. 1987: An exact conservation theorem for finite amplitude disturbabces to non-parallel shear flows, with remarks on Hamiltonian structure and on Arnol'd's stability theorems. *J. Fluid Mech.* **181,** 527-565.

MORRISON, P.J. & GREENE, J. 1980: Non-canonical Hamiltonian density formulation of hydrodynamics and ideal magnetohydrodynamics. *Phys. Rev. Lett.* **45,** 790-794.

OLVER, P.J. 1986: *Applications of Lie Groups to Differential Equations.* Springer-Verlag.

SALMON, R. 1988: Hamiltonian fluid mechanics. *Ann. Rev. Fluid. Mech.* **20,** 225-256.

SHEPHERD, T.G. 1990: A general method for finding extremal energy states of Hamiltonian dynamical systems, with applications to perfect fluids. *J. Fluid Mech.* (to appear)

VALLIS, G.K., CARNEVALE, G.F. & YOUNG, W.R. 1989: Extremal energy properties and construction of stable solutions of the Euler equations. *J. Fluid Mech.* **207,** 133-152.

Third Order Invariants of Randomly Braided Curves

MITCHELL A. BERGER

Department of Mathematical Sciences, University of St. Andrews, Scotland, UK[1]

ABSTRACT

Some aspects of topological structure in a set of curves or a vector field can be quantified by linking numbers, winding numbers, or helicity integrals. These quantities involve relations between *pairs* of curves or field lines. They do not completely describe the topology, however. For example, fields with differing topological structures can have the same helicity, and a topologically nontrivial field can have zero helicity.

This report presents a topological invariant $\mathcal{T}$ involving *triplets* of braided curves. This invariant measures the total amount of information needed to describe the topology of the curves. A statistical ensemble of randomly braided curves can be studied by finding probability distributions of $\mathcal{T}$. We consider sets of three curves confined inside a cylinder, where the endpoints of the curves random walk about each other. Simple theoretical arguments suggest that $\mathcal{T}$ grows in time like a one-dimensional random walk with a probability of 2/3 for travelling to the right. A numerical simulation verifies this behavior.

1. INTRODUCTION

Topological invariants related to knots, links, and braids are employed in several areas of physics. For example, an important quantity in fluid mechanics, the helicity integral, measures the net linking of vortex lines or magnetic field lines [Moffatt 1969; Berger and Field 1984]. Braid and knot polynomials aid the study of solvable models in statistical mechanics and conformal field theory [Akutsu and Wadati 1987]. In polymer physics, the entangling of polymer filaments reduces the mobility of the filaments. A topological description of the entangling involves quantities similar to helicity integrals such as linking numbers and winding numbers [Prager and Frisch 1967; Edwards 1967].

The linking number is the simplest topological relation between two *closed* curves (e.g. if they are unlinked the linking number is zero). Winding numbers are the analogous quantity for *open* curves. For example, consider two curves, each having endpoints on the planes

[1] Present address: Department of Mathematics, University College London

$z = 0$ and $z = 1$. As these curves rise from $z = 0$ to $z = 1$, they twist about each other through a net angle $\theta(1) - \theta(0)$ called the winding number ($\theta(z)$ equals the orientation of a line segment drawn between the two curves in the plane z = const.). Helicity integrals sum the linking or winding numbers of all pairs of fieldlines within a volume. For example, the fluid helicity in a volume where no vortex line crosses the boundary, $\omega \cdot \hat{n} = 0$, equals the sum of all the linking numbers of pairs of vortex lines [Moffatt 1969]. This remains true in an asymptotic sense if the vortex lines are ergodic [Arnold 1974]. Similarly, the *open helicity* (or relative helicity) defined in Berger and Field [1984] equals the sum of all the winding numbers between pairs of vortex lines [Berger 1986]. (Assuming all lines cross the boundary. More generally both open and closed lines exist within a volume; the open helicity is then a sum of both winding numbers and linking numbers).

These numbers only provide partial information about a configuration. The Borromean rings (figure 1a) are a famous example – no two of the three rings link each other, and yet the rings cannot be pulled apart. A similar situation occurs for braids (figure 1b). A braid [Artin 1950; Birman 1974] can be visualised as a collection of strings extending between two parallel planes. In figure 1b, all winding numbers are zero, but the strings cannot be deformed into three straight lines. If we desire a more complete description of the topology of a set of curves, we must go beyond second order invariants such as helicity, linkage, and winding.

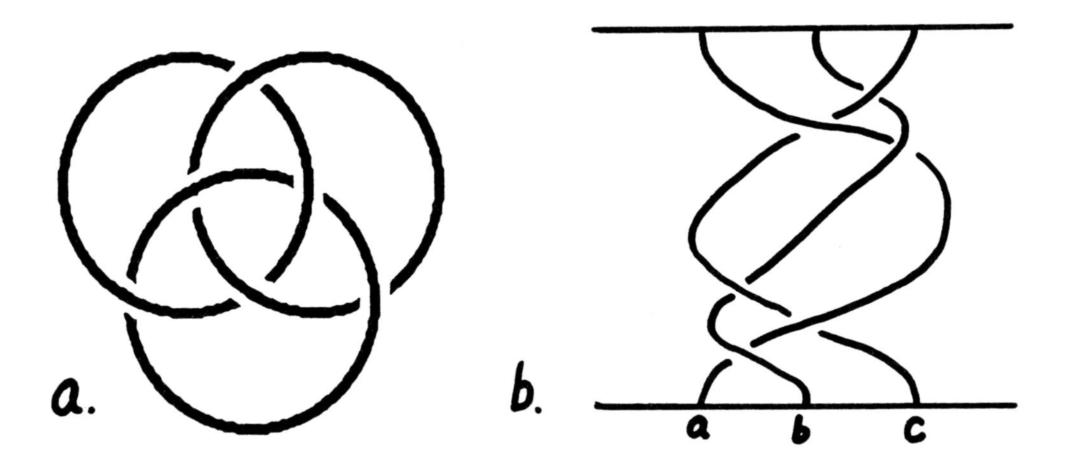

Figure 1. a) Borromean rings. b) Corresponding braid. The Borromean rings can be constructed by identifying top and bottom endpoints.

In §2 of this paper, we present a procedure for calculating third-order invariants for braids, which only requires simple geometrical concepts. In particular we will define a *tangle number* $\mathcal{T}$, which measures the net topological complexity of the braid. This number equals

6 for the pigtail braid in figure 1b, but 0 for three unbraided curves ($\mathcal{T}$ is related to the minimum number of crossovers seen in a 2-dimensional projection of the curves).

Many physical processes exhibit random behaviour. Can statistical methods be used to study topological aspects of random processes? To give a few recent examples, Levich and Shtilman [1988] have studied fluctuations in fluid helicity for random flows. Winding number distributions for random walks or Brownian motions have been found using several different methods [Pitman and Yor 1986; Berger 1987; Rudnik and Hu 1988; Berger and Roberts 1988; DuPlantier and Saleur 1988] (see chapter 4 of Wiegel 1986 for more examples of the statistical physics of topological quantities).

Winding numbers and tangle numbers can be readily calculated for a statistical ensemble of braids. If the endpoints at one plane are moved about each other, then the strings will become braided. In §3 we consider what happens when the endpoints random walk about each other. So that the density of strings remains constant, the endpoints are confined to a disk of radius R. We employ numerical methods to generate distributions of winding numbers and tangle numbers, and compare with our theoretical predictions.

2. THIRD ORDER BRAID INVARIANTS

2.1 Phase space description of the braid

For definiteness, visualize a braid with three strings as follows: The three strings are labelled a, b and c. They stretch from the plane $z = 0$ to $z = 1$. They always move upwards in z; i.e. if s is arc-length along a string, then $dz/ds \neq 0$.

The braid can be described by the six-dimensional vector

$$\vec{\gamma}(z) = (x_a(z), y_a(z), x_b(z), y_b(z), x_c(z), y_c(z)).$$

Here $(x_a(z), y_a(z))$ specifies the position of string a at height z. This is the *geometrical braid.* The *topological braid* is an equivalence class of geometrical braids. If one geometrical braid can be distorted into another without having strings break then they are equivalent.

The phase space for the geometrical braid is six dimensional: the braid follows a curve $\vec{\gamma}(z)$ in the space $(x_a, y_a, x_b, y_b, x_c, y_c)$. To determine the topological braid, much of the information carried by this curve is unnecessary. A uniform translation can always be added - this shifts each plane $z =$ const. by a vector $(S_x(z), S_y(z))$ leaving the bottom and top plane fixed:

$$\vec{\gamma}(z) \rightarrow \vec{\gamma}(z) + \big(S_x(z), S_y(z), S_x(z), S_y(z), S_x(z), S_y(z)\big)$$

where

$$(S_x(0), S_y(0)) = (S_x(1), S_y(1)) = 0.$$

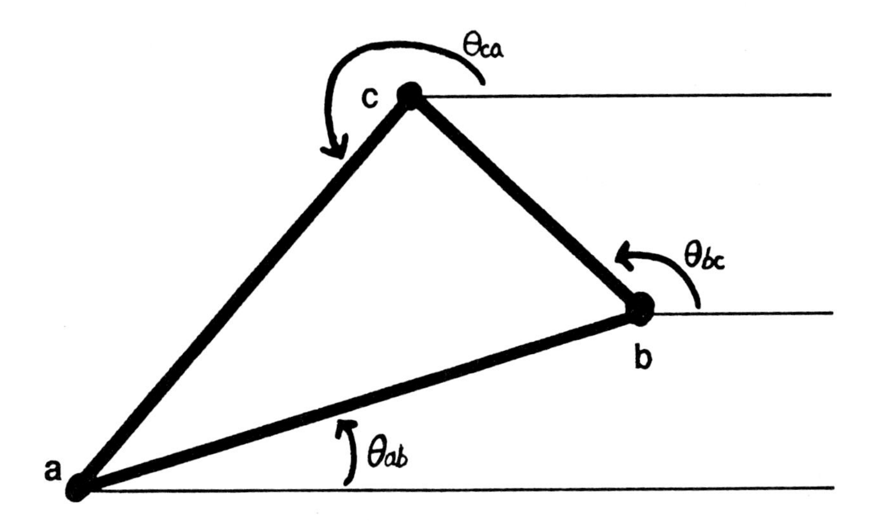

Figure 2. Winding functions θ_{ab}, θ_{bc}, and θ_{ca} in a plane z = constant.

Also we could expand each plane by a factor $p(z)$ without changing the equivalence class:

$$\vec{\gamma}(z) \to p(z)\vec{\gamma}(z) \qquad p(0) = p(1) = 1; \qquad p(z) > 0 .$$

The variables $S_x(z)$, $S_y(z)$ and $p(z)$ employ three unnecessary degrees of freedom specifying the size and position of the triangle $abc(z)$ with coordinates $\vec{\gamma}(z)$. By removing these degrees of freedom we can reduce the phase space to three dimensions without losing essential topological information.

The three remaining degrees of freedom specify the shape and orientation of the triangle $abc(z)$. These three degrees of freedom can always be specified by three functions of z, the winding angles $\theta_{ab}(z)$, $\theta_{bc}(z)$ and $\theta_{ca}(z)$ (see figure 2) where, e.g.,

$$\tan\ \theta_{ab}(z) = \frac{y_b(z) - y_a(z)}{x_b(z) - x_a(z)} .$$

We must distinguish between two strings winding about each other by 2π from two parallel strings. Thus the winding functions have range $(-\infty, \infty)$.

Let us now think about phase curves in the space $(\theta_{ab}, \theta_{bc}, \theta_{ca})$. For the braid in figure 1b, the phase curve forms a closed loop, since $\theta_{ab}(1) = \theta_{ab}(0)$, etc. If we deform the strings shown in 1b, then the phase curve should also deform (unless the deformation involves only uniform translations and expansions). Why can we not deform the strings into three vertical lines? This would correspond to shrinking the phase space curve to a point. Evidently there is some obstruction in the phase space which prevents such a shrinkage. Some points in phase space are forbidden – for example no triangle can have the angles

$(\theta_{ab}, \theta_{bc}, \theta_{ca}) = (0, 0, \pi/2)$. As we will show, the phase curve for figure 1b forms a loop about a forbidden region consisting of such points. First let us reduce our phase space by one further dimension.

The average winding function $\Theta(z)$ is

$$\Theta(z) = 1/3(\theta_{ab}(z) + \theta_{bc}(z) + \theta_{ca}(z)).$$

If we apply a uniform rotation to each plane z by an angle $\mu(z)$, where $\mu(0) = 0$ and $\mu(z)$ is continuous $(-\infty < \mu < \infty)$ then

$$\Theta(z) \to \Theta(z) + \mu(z) \tag{1}$$

Note that the bottom and top planes are fixed so that $\mu(0) = \mu(1) = 0$. Thus the mean winding number $\mathcal{W} \equiv \Theta(1) - \Theta(0)$ is an invariant. There is a simple correspondence between the mean winding number and helicity integrals. Suppose each string were a vortex tube (or magnetic flux tube) with flux Φ. Then the gauge-invariant helicity of the tubes [Berger and Field 1984] would be $H = 3\mathcal{W}\Phi^2/\pi$, plus a term arising from the twisting of vortex lines within the tubes.

It will be convenient to choose coordinates so that $\Theta(0) = 0$. Because we can always make the transformation in equation (1), for any topological braid we can always find a geometrical braid where, say

$$\Theta(z) = 0 \qquad \text{for } 0 \leq z \leq 0.5 \tag{2}$$

$$\theta_{\alpha\beta}(z) = \theta_{\alpha\beta}(0.5) + 2(z - 0.5)\mathcal{W} \qquad \text{for } 0.5 \leq z \leq 1 \tag{3}$$

Between $0 \leq z \leq 0.5$, the total winding is zero. What is left I will call *tangling*. Between $0.5 < z \leq 1$ there is only a uniform rotation, without tangling. (In the language of group theory, uniform rotations commute with all other braid elements; they comprise the center of the braid group. The 'tangled' part of the braid corresponds to the braid group modulo its center.)

2.2 Tangles

By ignoring the uniform rotation part of the braid, we reduce the phase space of the braid to two dimensions. Let us look at what is left of phase space for the tangled part of the braid. Consider the plane $\Theta = 0$ in the three-space $(\theta_{ab}, \theta_{bc}, \theta_{ca})$ (see figure 3). The phase curve for figure 1b is shown; it neatly makes one loop about a forbidden region.

The forbidden regions have the shape of regular hexagons, and the allowed regions are equilateral triangles. The point at the center of each phase space triangle corresponds to the physical triangle $abc(z)$ being equilateral. The sides (which belong to the forbidden

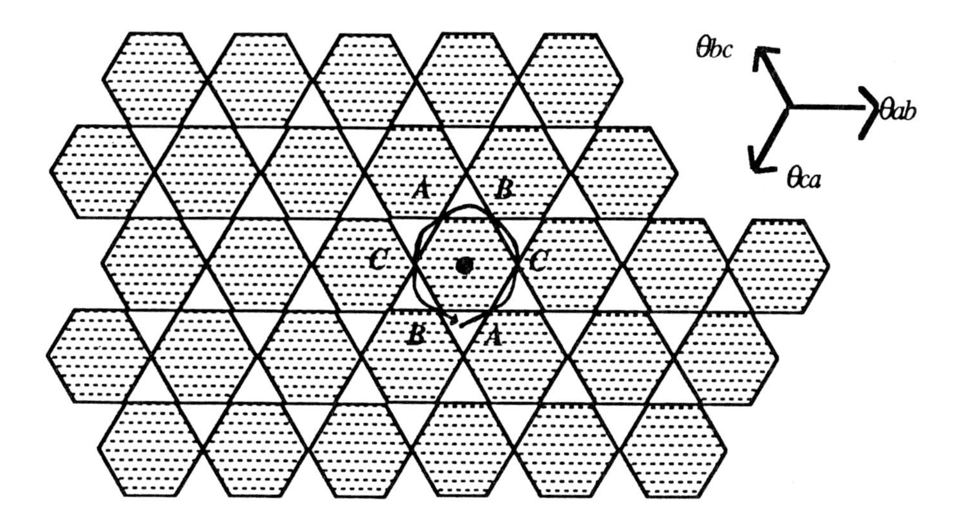

Figure 3. The plane $\theta_{ab} + \theta_{bc} + \theta_{ca} = 0$ in the space $(\theta_{ab}, \theta_{bc}, \theta_{ca})$. The crossed regions are forbidden; they represent impossible combinations of θ_{ab}, θ_{bc}, and θ_{ca}. The path for figure 1b is through the sequence of vertices $ACBACB$. This path encircles the point $(0,0,0)$. The arrows show directions of increasing θ_{ab}, θ_{bc}, or θ_{ca} (these arrows lie along projections of the three coordinate axes).

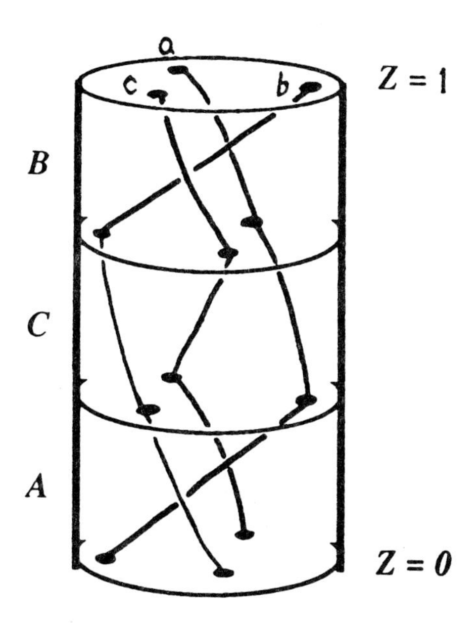

Figure 4. Example of a braid with tangle sequence ACB. In each section one string moves between the other two.

regions) correspond to one of the three points $a(z)$, $b(z)$, $c(z)$ being at infinity in the $x-y$ plane.

Most importantly, the vertices of the phase triangles correspond to the three points being colinear. We can label each vertex A, B, or C depending on which of the colinear points $a(z)$, $b(z)$, or $c(z)$ is in the middle (see figure 4). The phase curve passes through a sequence of vertices. This sequence can be notated as a sequence of letters, like $ABACBACB$. Repeated letters (for example AA) can be removed from the sequence; they correspond to the phase curve passing through a vertex and immediately coming back. A deformation of the strings can remove such a trivial path. Once repeated letters are removed, the sequence is invariant to deformations.

The topology of a three-braid can now be completely specified by giving the winding number $\mathcal{W}$ and the sequence of vertices. Let us call the length of the sequence (number of letters) the *tangle number* $\mathcal{T}$. The tangle number provides a measure of topological complexity which can be non-zero even when the winding numbers are zero.

3. RANDOM BRAIDING

How much topological structure is present in a set of randomly braided curves? Recall from the previous section that a three-braid is completely specified by the tangle sequence and the winding number $\mathcal{W}$. Thus the length $\mathcal{T}$ of the tangle sequence tells us how much information the braid contains. For this reason $\mathcal{T}$ provides a good measure of net topological structure.

Randomly braided curves can be generated by random motions of points in a plane. Start out with a trivial three-braid, i.e. three parallel vertical strings extending from $z = 0$ to $z = 1$. Then have the three lower endpoints of the strings random walk about each other. As they do so, the strings themselves become more and more braided. Winding numbers and tangle numbers can then be found as a function of number of steps.

We have done this in a Monte Carlo simulation with 2500 sets of curves. Each endpoint makes $N = 200$ steps of constant step size s. In the present simulation, the endpoints are confined within a disk of radius $r = R$. If an endpoint attempts to cross $r = R$, it bounces elastically back into the interior of the disk (equal angles; total step size = s). Without confinement, $\mathcal{W}$ and $\mathcal{T}$ would grow very slowly because the endpoints would wander far away from each other (for example, the rms $\mathcal{W}$ grows as $\log N$ for unconfined motion as opposed to $N^{1/2}$ for confined motion [Pitman and Yor 1986; Berger 1987]). By keeping the typical distance between endpoints constant, confinement simplifies the analysis and description of the problem. Furthermore, one would eventually like to analyse systems with a large number of braided curves, where the density of curves is independent of z. What we learn from three-braids may be more applicable to the general case if the three curves are artificially confined.

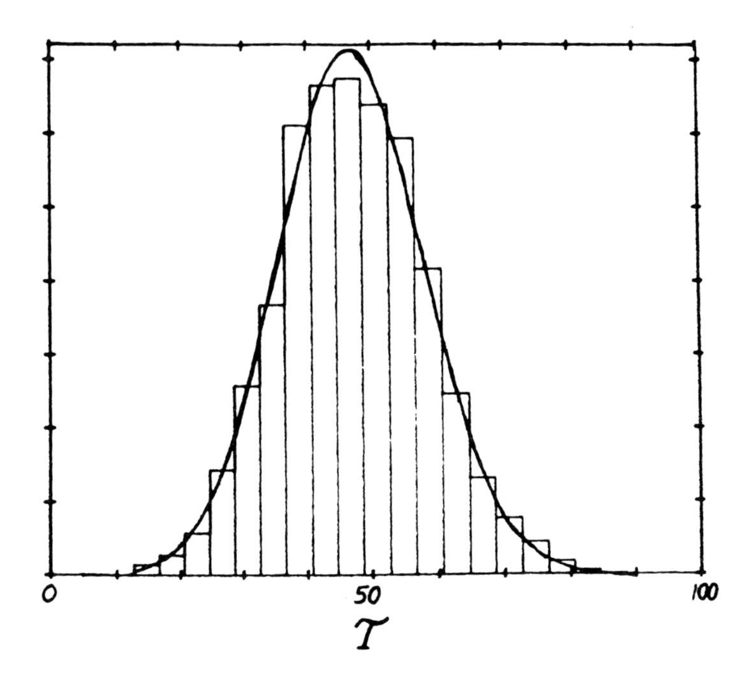

Figure 5. Distribution of $\mathcal{T}$ after 200 steps. The mean $\overline{\mathcal{T}} = 46.8$. The Gaussian fit has $\overline{\mathcal{T}^2} = \left(\overline{\mathcal{T}}\right)^2 + 8/3\overline{\mathcal{T}}$.

As the endpoints random walk, the phase curve passes through a random sequence of vertices as in figure 3. For small s, one might guess that $\overline{\mathcal{T}} \sim (s^2/R^2)N$. The factor (s^2/R^2) allows for the number of steps it takes for one randomly moving endpoint to go around another endpoint (or for the phase curve to travel between vertices). This dependence was verified in the simulation. For $s = R$ and $N = 200$ the distribution of $\mathcal{T}$ is shown in figure 5. The rms winding number for this simulation was $(\overline{\mathcal{W}^2})^{1/2} = 5.1\pi$.

A Gaussian with $\overline{\mathcal{T}^2} = \left(\overline{\mathcal{T}}\right)^2 + 8/3\overline{\mathcal{T}}$ fits the distribution extremely well. Such a Gaussian would be generated (approximately) by a one-dimensional random walk with a probability of 1/3 for travelling one step to the left, and 2/3 for travelling one step to the right. In terms of the tangle sequence, this corresponds to randomly generating a sequence of the letters A, B, and C (e.g. $ABCAACA = ABA$). There is a probability of 1/3 for repeating a letter, thereby decreasing $\mathcal{T}$ by one.

An application for this calculation to the theory of solar coronal heating will be described elsewhere. Very briefly, the magnetic flux at the solar photosphere is highly localized into intense tubes. Most of these tubes connect to coronal arches – the field lines from the tubes rise into the atmosphere before plunging down into tubes of the opposite polarity. These arches can become braided if the photospheric tubes random walk owing to the turbulent convection below the photosphere. The magnetic energy stored in the braiding structure, when released by reconnection, may be a significant source of heat [Parker 1983]. For three braided flux tubes, the energy stored is proportional to $\mathcal{T}^2$. Calculations of $\mathcal{T}$ distributions can thus provide an indication of the efficiency of the heating process.

AKUTSU Y and WADATI M 1987 Exactly solvable models and new link polynomials. I. N-state vertex models *J. Phys. Soc. Japan* **56** 3039

ARNOL'D VI 1974 The asymptotic Hopf invariant and its applications, in *Summer School in Differential Equations* Erevan; Armenian SSR Acad. of Sciences [English transl.: *Sel. Math. Sov.* **5** 327 (1986)]

ARTIN E 1950 The theory of braids *American Scientist* **38** 112

BERGER MA 1986 Topological invariants of field lines rooted to planes *Geophysical and Astrophysical Fluid Dynamics* **34** 265

BERGER MA 1987 The random walk winding number problem: convergence to a diffusion process with excluded area *J. Physics A: Mathematical and General* **20** 5949

BERGER MA and FIELD GB 1984 The topological properties of magnetic helicity *J. Fluid Mechanics* **147** 133

BERGER MA and ROBERTS PH 1988 On the finite step winding number problem *Adv. Applied Probability* **20** 261

BIRMAN JS 1974 *Braids, Links, and Mapping Class Groups* Princeton U. Press

DUPLANTIER B and SALEUR H 1988 Winding angle distributions of two-dimensional self-avoiding walks from conformal invariance *Physical Review Letters* **60** 2343

EDWARDS SF 1967 Statistical mechanics with topological constraints: I. *Proc. Phys. Soc. London* **91** 513

LEVICH E and SHTILMAN L 1988 Coherence and Large fluctuations of helicity in homogeneous turbulence *Phys. Letters A* **126** 243

MOFFATT HK 1969 The degree of knottedness of tangled vortex lines *J. Fluid Mechanics* **35** 117

PARKER EN 1983 Magnetic neutral sheets in evolving fields. II. Formation of the solar corona *Astrophysical J.* **264** 642

PITMAN JW and YOR M 1986 Asymptotic laws of planar Brownian motion *Annals of Probability* **14** 733

PRAGER S and FRISCH HL 1967 Statistical mechanics of a simple entanglement *J. Chemical Phys.* **46** 1475

RUDNICK J and HU Y 1987 The winding angle distribution for an ordinary random walk *J. Physics A: Mathematical and General* **20** 4419

WIEGEL FW 1986 *Introduction to Path-Integral Methods in Physics and Polymer Science* World Scientific

Topological Torsion, Pfaff Dimension and Coherent Structures

R.M. KIEHN

Physics Department, University of Houston, Houston TX 77004, USA

ABSTRACT

A coherent structure is viewed as a deformable connected domain of velocity space with certain similar topological properties. The topology of interest is the topology induced by the constraint on the variety {x,y,z,t} such that the vector field of flow satisfies a kinematic system of ordinary differential equations, as well as a dynamical partial differential equation of evolution. The Pfaff dimension or class of a domain is a topological property whose evolution may be computed. Of the four Pfaff classes of coherent structures admitted over space time, potential flow and integrable vorticial flows make up the first two Pfaff classes. The third and fourth Pfaff classes lead to the notion of Topological Torsion and Topological Parity. A concept of Helicity current is introduced, a current whose non-zero divergence signals the production of 3-dimensional defects internal to the flow field. Invariant surfaces of separation associated with the Jacobian of the unit tangent field may be used to define topological domains. An example is given demonstrating that these domains, as deformable domains of similar topology in the flow velocity field, may be put into correspondence with coherent thermodynamic phases.

1 INTRODUCTION

Topological properties of an evolutionary process are properties that stay the same under continuous and reversible transformations . From this viewpoint, a coherent structure is defined to be a connected, convectively deformable domain with certain topological properties that remain dynamically invariant.

Examples of topological properties are: Connectivity, Boundary, Orientation, Pfaff Dimension, and Limit points, which are to be distinguished from geometrical properties of Size and Shape.

The fundamental problem is how to extract, both theoretically and experimentally, topological information from a hydrodynamic flow. The kinematic velocity field is viewed as a set of differential constraints on R4, prescribing a topology on the variety {x,y,z,t}.

Herein, two related - but distinct - methods are suggested for extracting such topological information. Both methods have the same point of departure: both use the concept of a normalized, or unit speed, tangent velocity field:

$$\mathbf{t} = \mathbf{V}(x,y,z,t)/(\mathbf{V}.\mathbf{V})^{1/2}\,, \tag{1}$$

which to a differential geometer is a conformal Gauss map of the velocity space.

The first method utilizes the Cartan calculus based on the 1 form of action, or unit arclength, $A = ds = t_\mu dx^\mu$ (an imperfect differential constructed from the unit tangent field) to define the topological property of Pfaff dimension and Topological Torsion.

The second method of extracting topological information considers the null sets of the six invariant scalar functions associated with the Jacobian of the unit tangent vector field. As deformation invariants, these surfaces separate domains of different topological properties. It will be demonstrated that these domains can be put into correspondence with topological or thermodynamic phases.

2 TOPOLOGICAL TORSION

The traditional hydrodynamic vectors of velocity and vorticity are tensors of the first and second rank. The Cartan analysis focuses attention on the utility of a third rank tensor field, called Topological Torsion, and its divergence, which is a fourth rank tensor field, called Topological Parity. Although these fields are of third and fourth rank, they are set intersections of the velocity and the vorticity, and require no more than first derivatives of the velocity for their construction. Therefore, these fields are as amenable to measurement as are the traditional fields of velocity and vorticity, but they carry different kinds of information. Their components, in principle, can be measured directly with a 9-wire probe.

Following the methods of Cartan, a certain amount of topological information can be obtained by the construction of the Pfaff sequence based on the 1-form of Action, A . (Note that the closed integral of the action is related to the hydrodynamic concept of circulation.)

TOPOLOGICAL ACTION	A	$= t_\mu dx^\nu$
TOPOLOGICAL VORTICITY	$F = dA$	$= F_{\mu\nu} dx^\mu \wedge dx^\nu$
TOPOLOGICAL TORSION	$H = A \wedge dA$	$= H_{\mu\nu\rho} dx^\mu \wedge dx^\nu \wedge dx^\rho$
TOPOLOGICAL PARITY	$K = dA \wedge dA$	$= K_{\mu\nu\rho\sigma} dx^\mu \wedge dx^\nu \wedge dx^\rho \wedge dx^\sigma$

The Pfaff dimension of a domain is the rank of the largest non-zero element of the above sequence. The topological torsion tensor of an arbitrary 1-form of Action has four components in space-time, the first three of which form a Helicity or Torsion current, **T**. The fourth component, Ω, is related to the helicity density, h, introduced to hydrodynamics by Moffatt [1969]. The differential topology of the "torsionlines" of the Helicity current, $dx/\mathbf{T}_x = dy/\mathbf{T}_y = dz/\mathbf{T}_z = dt/\Omega$, can be evaluated and studied in a manner similar to the non-linear analysis applied to the "flowlines".

A first topological result is that if the four components of this third rank tensor of Helicity current and helicity density vanish over a domain, then the hydrodynamic flow satisfies the Frobenius complete integrability theorem [Flanders 1963], and the flow is never chaotic, nor braided, nor knotted over the domain. It follows that the concept of Topological Torsion is necessary for the understanding of turbulent and chaotic flow.

The Pfaff sequences, and the Topological Torsion tensor, can be constructed for arbitrary fields, but the importance of the Gauss map and the unit tangent field to global issues cannot be overemphasized. Sometimes, the technique is surprising. For example, rigid body rotation and the potential vortex have the same Gauss map! The two flows are Cremona duals of one another, being projections exterior to and interior to the unit sphere, respectively. In terms of the unit tangent field, the four components of topological torsion may be written as:

$$\mathbf{T} = - (\partial \mathbf{t}/\partial t) \times \mathbf{t}, \quad \text{and} \quad \Omega = \mathbf{t} \,.\, \text{curl}\, \mathbf{t} = \mathbf{v}.\text{curl}\, \mathbf{v} \,/\, (\mathbf{v}.\mathbf{v}) = h \,/\mathbf{v}.\mathbf{v} \qquad (2)$$

An important result is the fact that the divergence of the Topological Torsion tensor is not necessarily zero, and does not depend explicitly upon viscosity:

$$\text{div}\, \mathbf{T} + \partial\Omega/\partial t = 2 \ (\partial \mathbf{t}/\partial t) \times \mathbf{t} \,.\, \mathbf{curl}\ \mathbf{t} = K. \qquad (3)$$

The implication is that 3 dimensional defects of Helicity current can be created or destroyed spontaneously within the bulk medium of a fluid. The production of defects is an entropy increasing process, and therefore these processes, if continuous, must be irreversible. It is conjectured that these processes are part of the turbulent process. A visual signature of such hydrodynamic defects will be irreducibly 3 dimensional helical contrails and scroll waves.

The vector, curl **t,** is related to the Darboux vector, **D** , of the system of stream-lines associated with **t**. It may be shown that the norm of **D** is exactly equal to the sum of the squares of the Frenet torsion and the Frenet curvature of these streamlines. When the unit normal field is integrable, the Frenet torsion, τ, is precisely the 4th component, Ω, of the Topological Torsion tensor of the unit tangent field! In this sense, the helicity density divided by the square of the velocity, **v**.curl**v**/(**v**.**v**), is an intrinsic property of the flow, independent from the observer's frame of reference. Similarly, the intrinsic property of the square of the Frenet curvature may be evaluated as

$$\kappa^2 = (\text{curl}\,\mathbf{t} \times \mathbf{t})^2, \text{ with } \quad \tau^2 = \Omega^2 = (\text{curl}\,\mathbf{t} \cdot \mathbf{t})^2 . \tag{4}$$

It is important to recall that the topology of interest is the topology induced on the underlying 4 dimensional variety by a system of kinematic constraints. These constraints are specified by the condition that the velocity field, which in this case is a solution of a partial differential system (the Navier-Stokes equations), form a kinematical system of ordinary differential equations

$$dx - u\,dt = 0,\ dy - v\,dt = 0,\ dz - w\,dt = 0. \tag{5}$$

3 PFAFF DIMENSION

The Pfaff dimension of the unit tangent field is an invariant of a continuous deformation of the domain. The primitive idea is: what is the minimum number of functions that are required to describe a vector field over a domain? Gradient fields are adequately described by 1 function, hence a potential flow is of Pfaff dimension 1. Non-chaotic, or completely integrable vorticial flows are of Pfaff Dimension 2. Flows with domains of Helicity current and/or helicity density are of Pfaff dimension 3. The point of departure beyond Clebsch is that in space time the flows may be of Pfaff dimension 4!

Connected domains of the same Pfaff dimension can serve as candidates for coherent structures. In this sense, a vortex ring or a vortex tube is a coherent structure of Pfaff dimension 2 embedded in a larger domain of Pfaff dimension 1. Connected toroidal tubes such as those found in the ABC flow would be candidates for a coherent structure of Pfaff dimension 3. These domains or struc tures of different Pfaff dimension can evolve with deformation, but their connectivity and the sense of coherence instilled by their topological sameness, will be preserved with respect to homeomorphic flows. Domains of Pfaff dimension 4 form coherent structures that are irreducibly space-time dependent. The domains of Pfaff dimension 3 or greater may or may not be chaotic. For example, the toroidal bundles of ergodic trajectories suggested by Moffatt as candidates for coherent structures would fall into domains of Pfaff dimension 3.

4 DETERMINISTIC CHAOS vs. NON-DETERMINISTIC VELOCITY FIELDS

Vector fields that define a hydrodynamic flow may be classified into flows that are deterministic and separable, and those that are not. Separability implies that the time dependent flow vector field factors into a vector function of coordinates and a common factor of time and space: $V^\mu = \mathbf{v}(x,y,z)\ T(x,y,z,t)$.

According to Eisenhart [1963], the separability condition is necessary and sufficient for the vector field to be the generator of a single parameter semi-group of transformations. The physical interpretation of the single parameter is that it represents time. The remarkable feature of the separable flows is that the associated unit tangent field, **t**, is explicit *independent* from time! The Helicity current, **T**, of the unit tangent field vanishes, and the Pfaff sequence terminates at dimension 3. The solution to the system of unit tangent ODE's form a set of streamlines! The concept of Frenet torsion and Topological Torsion are, in this case, the same. The Rott solution [Rott ,1958] is an example of a non-separable solution to the Navier-Stokes equations that has a non-null Helicity current. The integrable flows (Pfaff dimension 2 or less) demonstrate a 1-1 correspondence between an increment of arclength and an increment of time; the non-integrable flows do not.

It is suggested that turbulent flows are non-separable flows of irreducible Pfaff dimension 4, and are to be distinguished from deterministic chaotic flows which are of Pfaff dimension 3. The remarkable feature of the concept of Pfaff dimension is that it simultaneously offers a precise definition of a coherent structure (a domain of the same Pfaff dimension), and distinguishes between a chaotic (Pfaff dimension 3) and a turbulent state (Pfaff dimension 4).

5 TOPOLOGICAL SURFACES AND THERMODYNAMIC PHASES

Any vector field, **V**(x,y,z,t), including those which are solutions to the Navier Stokes equations, may be normalized to 1 everywhere except at its fixed points, by the Gauss map, given by Eq. 1. This unit tangent field may be used to construct the Jacobian matrix, [**J**],

$$[\mathbf{J}] = [\ \partial \mathbf{t}_{\mu}/\partial x^{\nu}\ ], \tag{6}$$

as well as the one form of action, A = ds, based on the unit arc length.

There are six primary invariant functions of [**J**], which are to be associated with its complex eigenspectra. The Jacobian is to be viewed as a dyadic of functions rather than as a matrix of values. No linearization process is subsumed.

The six invariant functions are defined as:

$P(x,y,z) = \text{trace}\,[\mathbf{J}] = \text{div}\,\mathbf{t}$, $\quad L(x,y,z) = \langle \mathbf{D.D} \rangle$ ~ enstrophy, $\mathbf{D} = \text{curl}\,\mathbf{t}$

$Q(x,y,z) = \text{trace}\,[\text{co}\mathbf{J}]$ $\quad M(x,y,z) = \langle \mathbf{D[J]D} \rangle$ ~ vortex stretching

$R(x,y,z) = \det\,[\mathbf{J}]$ $\quad N(x,y,z) = \langle \mathbf{D[J][J]D} \rangle$.

The null sets of these functions, or combinations of these functions, form invariant surfaces of separation, which globally separate domains of different topology. The idea to be developed is that these surfaces of separation (in velocity space) define domains of thermodynamic phase! What better way to think of a coherent structure, than as domains of pure or mixed thermodynamic phase.

The invariant surfaces may also be viewed as sets of partial differential equations whose solutions have interesting properties:

$P(x,y,z) = 0$ defines a minimal surface (soap film of zero mean curvature)
$L(x,y,z) = 0$ defines an asymptotic surface.
$Q(x,y,z) = 0$ defines a surface of zero Gauss sectional curvature
$M(x,y,z) = 0$ defines a surface of null vortex stretching rate

The three dimensionality of the Jacobian immediately focuses attention on the Cardano function, C(x,y,z), whose null set generates a surface that separates domains of complex versus real cubic roots. In terms of the invariant scalars, the Cardano function may be constructed as $C = -27R^2 - P(P^2 - 18Q)R - (4Q-P^2)Q^2$.

When $C > 0$ there are 3 distinct real solutions to the Hamilton characteristic equation for [**J**]. As shown in appendix A, this condition corresponds to the mixed phase region of a Van der Waals gas.

When $C = 0$ the eigen values of the Jacobian are real, but at least 2 eigenvalues are equal. If, in addition, the rank of the Jacobian is less than 3, then the Cardano condition corresponds to the binodal and spinodal lines on the equilibrium surface of a Van der Waals gas. The criteria of rank less than 3 implies that $R = 0$, and indicates that the solution set is satisfied by the spinodal constraint $Q = 0$, or the binodal constraint, $4Q - P^2 = 0$.

When $R = P^3/27$ and $Q = P^2/3$, all three eigen values are real and equal. For $R = 0$, then $P = 0$ and $Q = 0$. This condition corresponds to the critical point of a Van der Waals gas. See Appendix A. When $C < 0$ there is 1 real eigen value and two conjugate complex solutions. This condition corresponds to the pure phase domains of the Van der Waals gas.

Define two functions, A and B, such that $3B = Q - P^2/3$ and $2A = -R + QP/3 -2P^3/27$. Use A and B as the coordinates of the bifurcation space. Then the functions P, Q, or R can be chosen as the bifurcation parameter, for constant values of the other scalar functions. This third order system fits a GLOBAL bifurcation scheme which is : Hysteretic in the variable R; a Pitchfork in the variable Q, and a winged cusp in the variable P [Golubitsky 1985].

When $R = 0$ the system can be put into correspondence with an equilibrium thermodynamic system, which is not hysteretic. When R # 0, the system is irreducibly 3-dimensional, and the non-equilibrium thermodynamic system admits irreversible hysteresis as R varies. It may be shown that tertiary bifurcations, such as the Hysteretic Hopf bifurcation [Langford 1983] can lead to intermittency. Such flows have non-zero Topological Torsion. An example of an exact solution to the Navier-Stokes equations that yields an intermittent transverse torsion wave packet is given in Appendix B.

The above analysis indicates that kinematic domains have properties that are topologically equivalent to thermodynamic phases. Borrowing from thermodynamic experience, cooperative and coherent behavior is to be expected in complex kinematic flows, along with kinematic phase transitions, depending on initial conditions and parameters.

6 TOPOLOGICAL EVOLUTION AND PFAFF DIMENSION

A fundamental question of turbulence is: How do systems originally of Pfaff dimension 3 or less evolve into systems of Pfaff dimension 4? Such evolutionary flows involve changing topology, and are either irreversible, discontinuous, or both.

The decomposition theorems of deRham [Goldberg 1982] may be used to form 4 equivalence classes of flows, **V**, relative to the 1-form of Action, A. Consider all flows that satisfy either:

	1. $i(\mathbf{V})\, dA = 0$	Hamiltonian flows
	2. $i(\mathbf{V})\, dA = dP$	Eulerian flows
	3. $i(\mathbf{V})\, dA = dP + G$	Stokes flows
or	4. $i(\mathbf{V})\, dA = dP + G + {*\partial*Z}$	Open flows

where G is harmonic ($dG = 0$, ${*\partial*G} = 0$). The first three equivalence classes are closed with respect to exterior differentiation, $d(i(\mathbf{V})\, dA = 0$, while the 4th is not.

To test for topological evolution, form the integral over a closed domain of one of the 4 Pfaff classes. Then, construct the Lie derivative with respect to **V** of the integral. If the Lie (convective) derivative vanishes, then the topological property represented by the integral is an invariant of the evolution. Conversely, a non-null value of the Lie derivative of the integral indicates topological change.

An extraordinary result is that for all closed flows of the deRham categories (1,2,3 above), the Lie variation of all even dimensional Pfaff sequences vanishes,

$$L(\mathbf{V})\, dA\wedge dA\wedge ... = 0.$$

However, it may be demonstrated that relative to closed flows, the odd dimensional sets, that is, the 1 and 3 dimensional Pfaff structures of Circulation and Helicity, may undergo topological evolution with the production of defects, while the even dimensional Pfaff structures of Vorticity and Parity remain invariant. These defects may be associated with the production of topological "holes and handles". Moreover, according to the Brouwer theorem, the evolutionary change of topology is quantized to the integers (you cannot create half a hole).

If the view is taken that the limit points of the Pfaff topology are given in terms of the exterior derivative operator, then the invariance of the even dimensional sets relative to closed flows implies invariance of the limit points, and it follows that such closed flows are continuous relative to the Pfaff topology! Hence, when topology changes by variation of the odd dimensional sets, the closed process must be irreversible.

The topological evolution of the 1 dimensional Pfaff structures may be put into 1 to 1 correspondence with Newton's laws for the evolution of energy,

$$L(\mathbf{V})\,A = W \quad \sim \quad \mathbf{F}\,.\,d\mathbf{r} \quad = \text{Work}.$$

The corresponding evolutionary law for the third rank field of topological torsion corresponds to the evolution of entropy:

$$L(\mathbf{V})\,H = S.$$

APPENDIX A : TOPOLOGICAL AND THERMODYNAMIC PHASES

Consider the non-linear flow:

$$u = -(3x/8)^2/(3y-1),\quad v = +3\,x/(3y-1)^2 - 3/y^2,\quad w = 1.$$

Transform the variables by means of the equations: $x = \exp 3/8\, s$, $y = V$, $z = U$, redefining $ds/dt = 8/3\, d\ln x/dt$ as $-T$, dy/dt as p, and $\rho = 1/V$. The solution to the system is given by the expression :

$$\Psi = U - \rho \exp(3s/8)/(3-\rho) + 3\rho\,.$$

The unit tangent field may be constructed as:

$$t_s = T/(1+T^2+p^2)^{1/2},\quad t_v = -p/(1+T^2+p^2)^{1/2},\quad t_u = -1/(1+T^2+p^2)^{1/2}$$

from which it is possible to compute the Jacobian Dyadic. As the system depends only on two variables, it is possible to show that $R = 0$ identically. A plot of the functions p vs ρ for various T yields a pressure density relationship typical of Van der Waals gas (See Figure 1). The intersection of this surface and the surface $Q = 0$ defines the Spinodal line of absolute phase instability; it is the line of zero Gauss curvature for the fundamental surface, $\Psi = 0$.

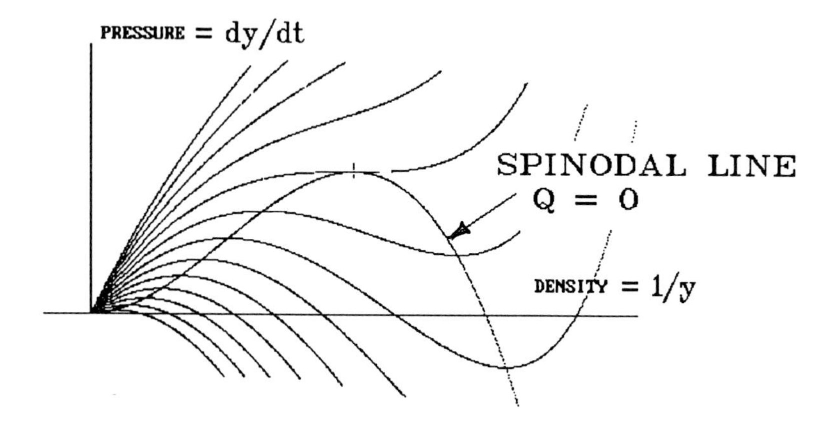

Figure 1 The Kinematic Surface Topology is similar to a Van der Waals Gas.

A more dramatic result is given for the Gibbs surface $g = u\text{-}Ts+pV$ which is plotted in Figure 2. The intersection of the Gibbs surface, $g(T,\rho) = 0$ and the surface $Q = 0$ of zero Gauss sectional curvature forms a line in space which is the envelope, or tac locus, of the cuspoids of the Gibbs surface. It is apparent that the cuspoidal edge corresponds to the Cardano condition $C = 0$ where the three eigenfunctions are real, but two are degenerate. Moreover, the point $R = 0$, $Q = 0$, and $P = 0$ defines the thermodynamic critical point.

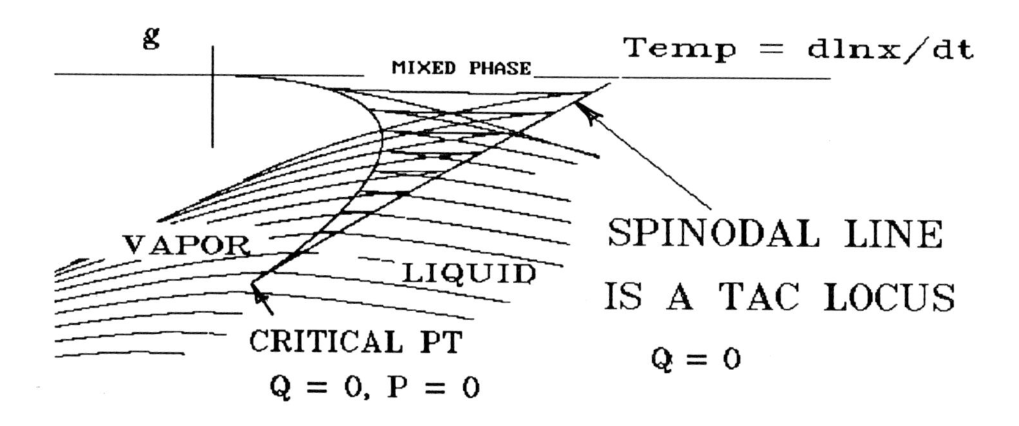

Figure 2 The Kinematic Gibbs Surface Topology of a non-linear flow .

APPENDIX B : INTERMITTENT TORSION BURSTS

Consider the non-linear flow:

$$dx/dt = -\Omega y + xz$$
$$dy/dt = +\Omega x + yz$$
$$dz/dt = F + Az - Dz^3 - B(x^2+y^2)$$

which as a Velocity field, **V**, is not only an exact solution of the Navier-Stokes equations in a rotating frame of reference, but also is a model of the tertiary hysteretic bifurcation studied by Langford. A plot of the transverse torsion wave packet burst generated by such a solution is presented in Figure 3. The time between bursts is not periodic.

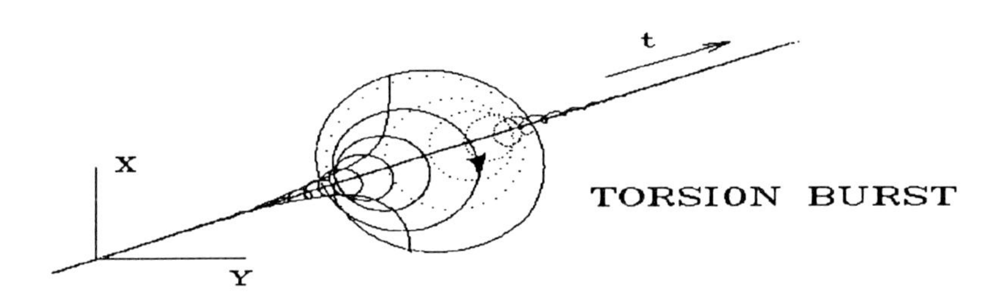

Figure 3. An intermittent torsion wave packet in a Navier-Stokes fluid.

ACKNOWLEDGMENTS

The author would like to acknowledge the continuing interest of F. Hussain in this abstract topological approach to the problems of hydrodynamics. M. Gorman's encouragement and computational support stimulated the development of the basic ideas. This work was supported in part by DOE grant DE-FG05-88ER13839.

REFERENCES

L. P. Eisenhart, "Continuous Groups" (Dover, N. Y. 1963)

H. Flanders, "Differential Forms", (Academic Press, New York,1963)

S. I. Goldberg, "Curvature and Homology Theory" (Dover, N.Y., 1982)

M. Golubitsky and D. G. Shaeffer, "Singularities and Groups in Bifurcation Theory" Springer Verlag, N.Y. (1985)

W. F. Langford, in "Non-linear Dynamics and Turbulence" Edited by G.I. Barrenblatt, et. al. Pitman, London (1983)

H. K. Moffatt, J. Fluid Mech. **35**,117 (1969)

N. Rott, ZAMP **9** (1958),543; **10** (1959),73

Steady Confined Stokes Flows with Chaotic Streamlines

K. BAJER*, H.K. MOFFATT & FRANCES H. NEX

Department of Applied Mathematics and Theoretical Physics,
University of Cambridge, Silver Street, Cambridge CB3 9EW, UK

A family of incompressible steady flows, called STF flows, is considered. These are Stokes flows whose velocity is a quadratic function of cartesian coordinates satisfying $\mathbf{u}^{STF}\cdot\mathbf{n} = 0$ on a sphere $r = 1$. When the flow is an $O(\epsilon)$ perturbation of a flow $\mathbf{u}_0^{STF}$ with closed streamlines the particle paths are, for a long time, constrained near some two-dimensional surfaces (adiabatic drift) but they jump randomly from one such surface to another when they approach a stagnation point. Because of these random jumps, which we call 'super-adiabatic drift', streamlines are apparently space-filling throughout entire domain. The Lyapunov exponent λ is computed and it is found to be positive on the space-filling streamlines. There is some evidence that λ remains bounded away from zero when $\epsilon \to 0$. The separation of two distinct but initially close fluid particles is also computed. It indicates efficient mixing of a passive scalar in the STF flows.

1. INTRODUCTION

It has been known for some years that even simple fluid flows may lead to exceedingly complicated particle paths. Three-dimensional steady flows can have chaotic streamlines which densely fill some regions of space – a property relevant, for example, to the kinematic dynamo problem. In two-dimensional flows even weak time dependence may result in the chaotic motion of fluid particles, which can be useful for stirring and mixing.

Several time-dependent, two-dimensional flows have been investigated so far: blinking point vortices (Aref 1984), Stokes flow between the alternately rotating, excentric circular cylinders (Aref & Balachandar 1986; Chaiken *et al.* 1986,1987; Ottino *et al.* 1988); and also simple models of such physical flows as: weak waves in Taylor vortices (Broomhead & Ryrie 1988) or tidal flows in shallow seas (Pasmanter 1988).

---

* On leave of absence from the Institute of Geophysics, University of Warsaw, Poland

The structure of streamlines has been studied in some space-periodic steady flows.¶ The best known are the so-called ABC flows (Henon 1966; Dombre *et al.* 1986) and Q-flows (Beloshapkin *et al.* 1989). In this paper we introduce a class of three-dimensional incompressible steady flows in a sphere which have a particularly rich Lagrangian structure. These flows have a very simple Eulerian representation: the cartesian components of the velocity are quadratic functions of the coordinates. Similar flows were considered before (Moffatt & Proctor 1985) in connection with Stretch-Twist-Fold fast dynamo processes and for this reason we adopt the name STF flows for their modified version presented below.

2. STF FLOWS

We consider a two-parameter family of quadratic flows:

$$\mathbf{u}^{STF} = \left(\alpha z - 8xy,\ 11x^2 + 3y^2 + z^2 + \beta xz - 3,\ -\alpha x + 2yz - \beta xy\right) \quad , \tag{1}$$

where α, β are positive real parameters. Clearly $\mathbf{u}^{STF}$ satisfies: $\nabla \cdot \mathbf{u}^{TSF} = 0$; $\mathbf{u}^{STF} \cdot \mathbf{n}\big|_S = 0$ where S is the surface of a unit sphere.*

The flow (1) is a sum of the rigid rotation $\mathbf{R}$ around the y-axis with angular velocity α and the remaining part $\mathbf{u}_0^{STF}$. Streamlines of $\mathbf{u}_0^{STF}$ are mostly closed loops with the exception of some which are heteroclinic lines linking the hyperbolic stagnation points at $(x, y, z) = (0, \pm 1, 0)$. Hence, for $\alpha \ll 1$, $\mathbf{u}^{STF}$ is a weakly perturbed integrable flow with closed streamlines.

The STF flow, being quadratic in space coordinates, satisfies the Stokes equation $\nabla^2 \mathbf{u} = \nabla p$. Stokes flows are uniquely determined provided the velocity is prescribed on the boundary. Hence they can be realised in a viscous fluid by suitable motions of the boundaries.

3. THE STRUCTURE OF STREAMLINES

In fig. 1*a*) we show a Poincaré section of a single streamline of $\mathbf{u}^{STF}$ with $\alpha = 0.1$ and in fig. 1*b*) a projective view of the same streamline. Over 8000 points of section are plotted. The streamline apparently penetrates most of the sphere but in a

¶ This problem is related to the structure of the magnetic field lines (Moffatt 1985,1986). Despite the formal equivalence between steady incompressible flows and magnetic fields the ways of inducing them are very different. Therefore solenoidal fields regarded as physically realistic will be very different in the two cases.

* A general quadratic flow satisfying these conditions has seven independent parameters (Bajer & Moffatt 1989).

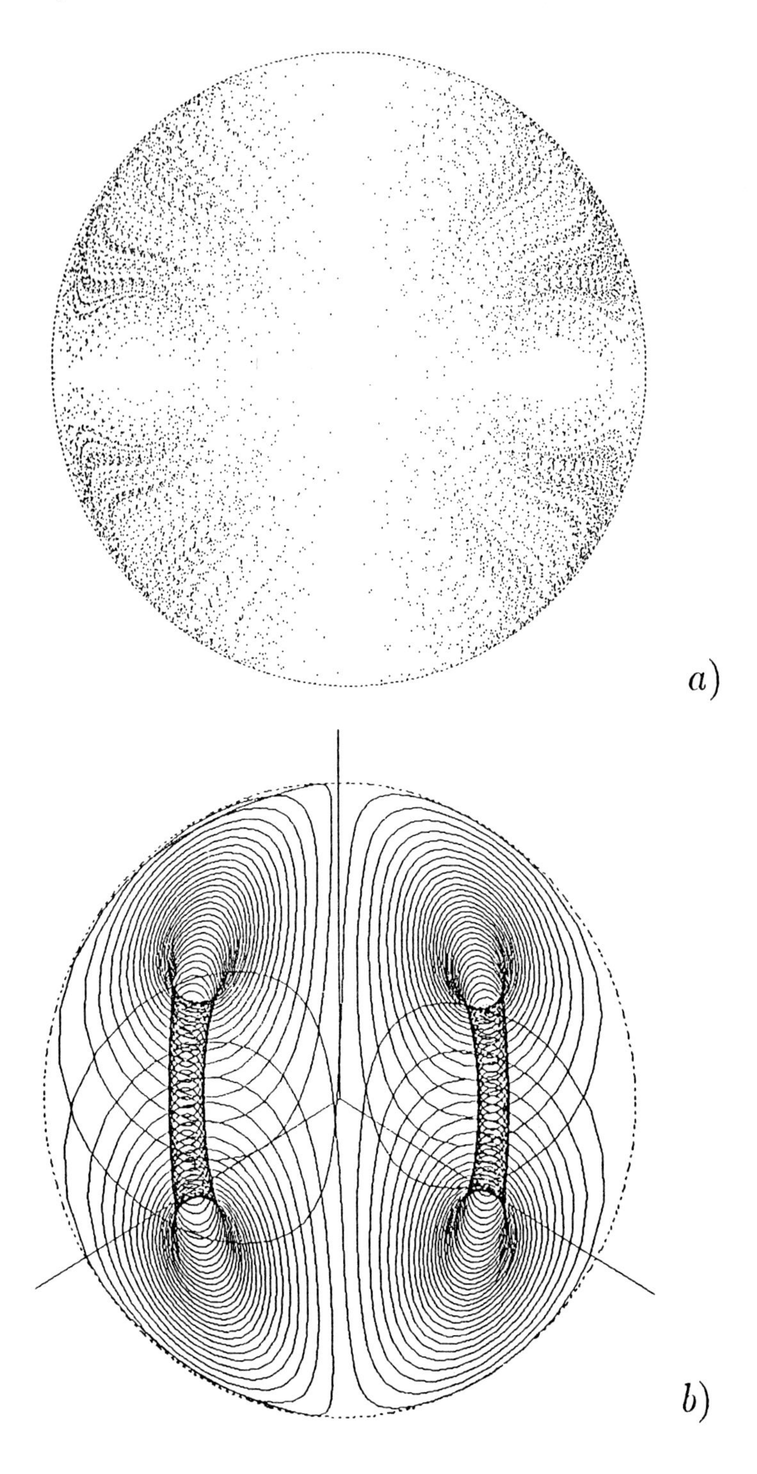

FIGURE 1. A single chaotic streamline of the STF flow: *a*) The Poincaré section; *b*) The projective view.

rather orderly manner. Watching the Poincaré section being plotted one may notice that for a long time points closely follow curves. These curves are sections of two-dimensional surfaces which meet at the stagnation points on the surface of a unit sphere and the streamline jumps from one such surface to another every time it comes near the stagnation point. The set of these curves is traced out with apparently random spacing. It forms the distinct layered pattern seen in fig. 1*a*).

The motion along the curves can be explained in terms of an adiabatic invariant (Bajer 1989; Bajer & Moffatt 1989) and therefore we should call it adiabatic motion (or drift). The relatively large jumps mentioned above form what we term as 'super-adiabatic drift'. These two kinds of drift persist as we decrease α. For $\alpha = 0$ streamlines are closed loops, i.e. the Poincaré section of an orbit consists of only two points. For an arbitrarily small but finite α orbits spread apparently over the entire domain and there are no visible islands of regularity. This is a very different picture from, for example, the ABC flow where, in the integrable case, streamlines fill toroidal surfaces, most of them densely. Then the KAM tori surviving a small perturbation ensure confinement of the chaotic streamlines. The KAM theory does not apply to flows with closed streamlines and hence the global chaos is possible for arbitrarily small perturbations.

4. LYAPUNOV EXPONENT AND THE SEPARATION OF PARTICLES

In order to substantiate the above claim we compute the Lyapunov exponent associated with the STF flow. The Lyapunov exponent λ is the asymptotic growth-rate of the length of an infinitesimal material line element $\boldsymbol{\xi}$ advected by the flow:

$$\lambda = \lim_{t \to \infty} \frac{1}{t} \log |\boldsymbol{\xi}| \quad . \tag{2}$$

Positive λ means exponential stretching – the signature of a chaotic flow. We solve numerically the sixth order system of ODE's for $\boldsymbol{\xi}$:

$$\begin{aligned} \frac{d\mathbf{x}}{dt} &= \mathbf{u}^{STF} \\ \frac{d\boldsymbol{\xi}}{dt} &= (\boldsymbol{\xi} \cdot \nabla)\, \mathbf{u}^{STF} \end{aligned} \tag{3}$$

We integrate (3) with $\alpha = 0.01$ and plot in fig. 2*a*) the function $\lambda(t) = \frac{1}{t} \log |\boldsymbol{\xi}(t)|$. The corresponding Poincaré section of $\mathbf{u}^{STF}$ is shown in fig. 2*b*). The value of the function $\lambda(t)$ was plotted (marked with a single dot in fig. 2*a*) every time the streamline crosses the Poincaré plane of section. Gaps in the plot of $\lambda(t)$ correspond to the long time that a particle spends near the stagnation points.

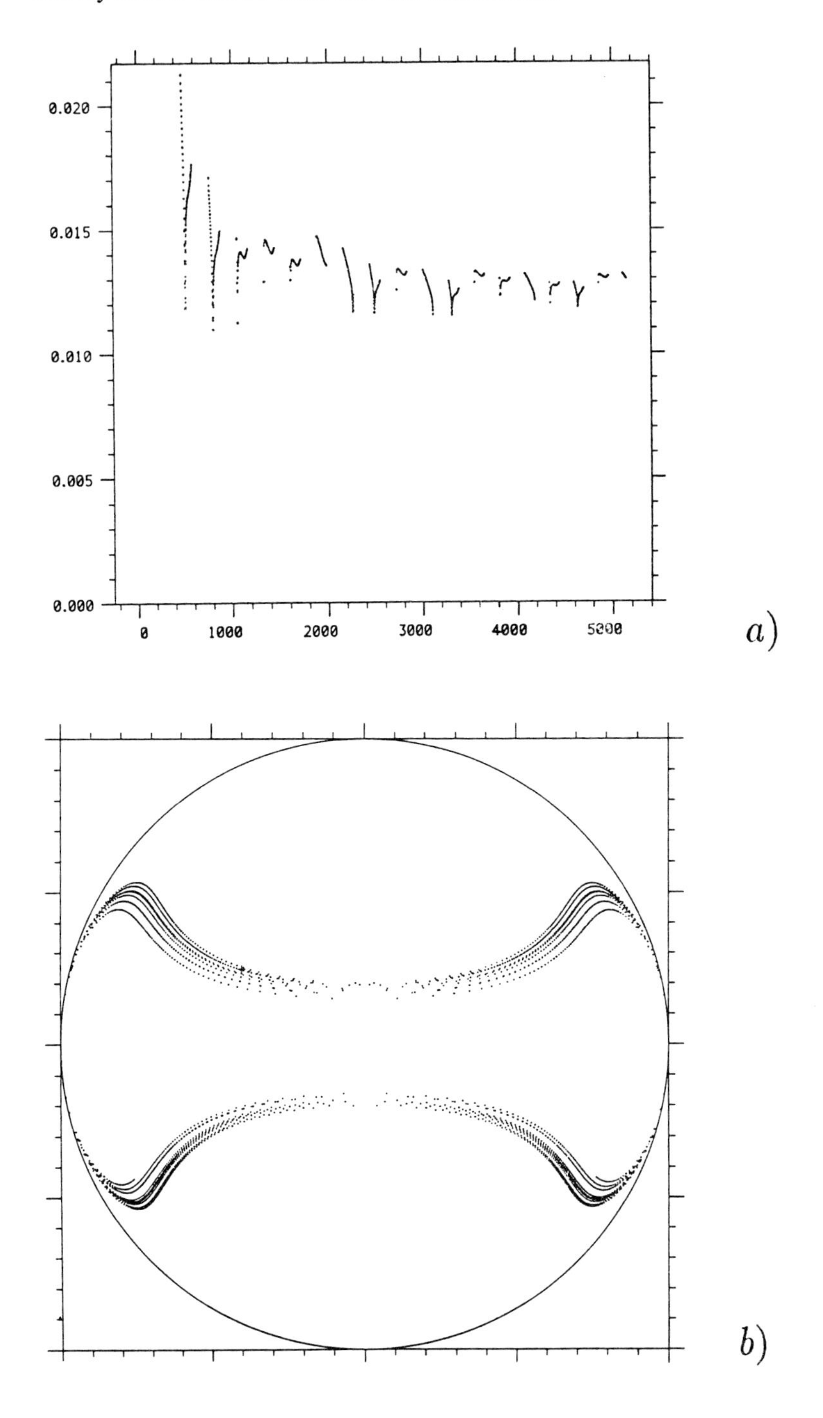

FIGURE 2. $a)$ The function $\lambda(t)$. Its asymptotic value for $t \to \infty$ is equal to the Lyapunov exponent of a streamline; $b)$ Corresponding Poincaré section.

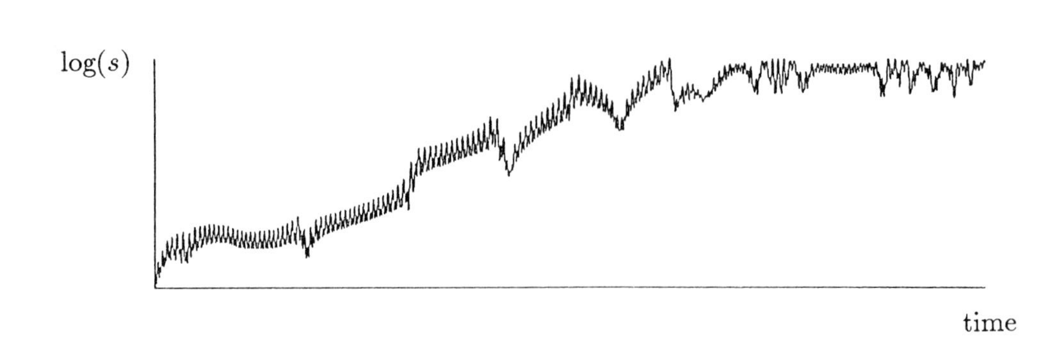

FIGURE 3. Logarithm of the separation between two close particles as a function of time.

The Lyapunov exponent has the dimension of $(\text{time})^{-1}$, the same as the small parameter α. The time-scale of the unperturbed flow $\mathbf{u}_0^{STF}$ is not well defined. For the streamlines close to heteroclinic lines the turnover time is arbitrarily long. However, the perturbation $\mathbf{R}$ has a unique time-scale $\frac{1}{\alpha}$. In these units the value of the Lyapunov exponent, as read from fig. 2*a*), is $\lambda \approx 1.2$. Further computations indicate that λ remains $O(1)$ as we increase α (up to $\alpha = 0.1$).

Finally in fig. 3 we plot the separation s between two distinct fluid particles which initially were very close (10^{-6} compared to the unit radius of the sphere). Here $\log(s)$ is plotted as a function of time for $\alpha = 0.1$. Jumps in $\log(s)$ correspond to the jumps between different adiabatic surfaces.

In a finite domain the separation, unlike the length of the infinitesimal line element, is bounded. Therefore a systematic increase in s saturates, and then s varies randomly between $s = 0$ and its maximal value $s = 2$. The time needed for the systematic increase to saturate could be regarded as the characteristic time-scale for mixing of the spatial scales of the size of the initial separation (10^{-6} in this case).

5. CONCLUSIONS

We have shown that STF flows (1) in the limit of small α exhibit Lagrangian chaos of a kind so far unknown in the context of flow kinematics. Because of the degenerate character of the unperturbed flow $\mathbf{u}_0^{STF}$ which has closed streamlines there are no KAM tori, and for an arbitrarily small perturbation the super-adiabatic drift spreads the streamlines over the entire domain. The Lyapunov exponent is positive, and when measured on the time-scale of the slow perturbation it appears to be bounded away from zero when $\epsilon(=\alpha) \to 0$.

This new type of chaos is likely to be typical for perturbed flows with closed streamlines. Its transport properties such as the chaotic advection of a passive scalar or the possibility of a fast dynamo deserve further investigation.

REFERENCES

AREF, H. (1984) Stirring by chaotic advection. *J. Fluid Mech.* **143**, 1-21.

AREF, H., & BALACHANDAR, S. (1986) Chaotic advection in a Stokes flow. *Phys. Fluids* **29**, 3515-3521.

BAJER, K. (1989) Flow kinematics and magnetostatic equilibria. Ph.D. Thesis. Cambridge University.

BAJER, K., MOFFATT, H.K. (1989) On a class of steady confined Stokes flows with chaotic streamlines. *J. Fluid Mech.* (submitted).

BELOSHAPKIN, V.V., CHERNIKOV, A.A., NATENZON, M.Ya., PETROVICHEV, B.A., SAGDEEV, R.Z., & ZASLAVSKY, G.M. (1989) Chaotic streamlines in pre-turbulent states. *Nature* **337**, 133-137, Jan 1989.

BROOMHEAD, D.S., & RYRIE, S.C. (1988) Particle paths in wavy vortices. *Nonlinearity* **1**, 409-434.

CHAIKEN, J., CHEVRAY, R., TABOR, M., & TAN, Q.M. (1986) Experimental study of Lagrangian turbulence in a Stokes flow. *Proc. Roy. Soc.* **A 408**, 165-174.

CHAIKEN, J., CHU, C.K., TABOR, M., & TAN, Q.M. (1987) Lagrangian turbulence and spatial complexity in Stokes flow. *Phys. Fluids* **30**, 3, 687-694.

DOMBRE, T., FRISCH, U., GREENE, J.M., HÉNON, M., MEHR, A., & SOWARD, A.M. (1986) Chaotic streamlines in the ABC flow. *J. Fluid Mech.* **167**, 353-391.

HÉNON, M. (1966) Sur la topologie des lignes de courant dans un cas particulier. *C. R. Acad. Sci.* **262**, 312-314.

MOFFATT, H.K., & PROCTOR, M.R.E. (1985) Topological constraints associated with fast dynamo action. *J. Fluid Mech.* **154**, 493-507.

MOFFATT, H.K. (1985) Magnetostatic equilibria and analogous Euler flows of arbitrarily complex topology. Part 1. Fundamentals. *J. Fluid Mech.* **159**, 359-378.

MOFFATT, H.K. (1986) Magnetostatic equilibria and analogous Euler flows of arbitrarily complex topology. Part 2. Stability considerations. *J. Fluid Mech.* **166**, 359-378.

OTTINO, J.M., LEONG, C.W., RISING, H., & SWANSON, P.D. (1988) Morphological structures produced by mixing in chaotic flows. *Nature* **333**, 419-425, June 1988.

PASMANTER, R.A. (1988) Anomalous diffusion and anomalous stretching in vortical flows. *Fluid Dyn. Res.* **3**, 320-326.

VI: Vortex Interaction and Reconnection

Collapsing Solutions in the 3-D Euler Equations

ALAIN PUMIR[1,2] & ERIC D. SIGGIA[1,3]

[1]LPS, ENS 24 rue Lhomond, 75231 Paris Cedex, France
[2]SPT, CEN Saclay, 91191 Gif sur Yvette, France
[3]LASSP, Cornell University, Ithaca, NY 14853, USA

ABSTRACT
Extensive results of numerical simulations of vortex tubes are presented. The mechanism of vortex stretching, as well as the core deformations can be understood at early times by simple considerations, based on the Biot-Savart model. At later times, the vorticity tends to develop sharp jumps in the tubes and the rate of amplification seems to saturate, thus resulting in an exponential growth of vorticity.

Many outstanding questions of both fundamental and practical importance in 3-d hydrodynamics of incompressible flows remain open. The structure of turbulent flows, which are of obvious importance in many natural and industrial phenomena is certainly the best example. Decades of experimental work on grid turbulence, or boundary layer turbulence have uncovered new and interesting phenomena which remain unexplained theoretically.

The well established intermittent nature of turbulent fluctuations (Townsend 1951), observed in very different flow configurations is closely associated with the spontaneous formation of very large gradients of the velocity field. This occurence of large gradients is by no means a simple curiosity. Indeed, careful measurments have shown that about half of the Reynolds stress in boundary layer turbulence is produced during the bursting events (Willmarth and Lu 1973), that are precisely associated with variations of the velocity field over extremely small lengths (Willmarth and Bogar (1977) have reported variations on lengths smaller than the Kolmagarov length).

Theoretically one has to explain how these large gradients are generated by the flow. Also, it is obviously of fundamental importance to understand the nonlinear mechanisms involved in the 3-d Euler and Navier-Stokes equations. In particular, vortex stretching could induce large values of the vorticity, possibly infinite in a finite time. Singularities in the fluid equations may provide a paradigm for many observed properties of turbulent flows (Pumir and Siggia 1989a).

Although the Kolmogorov analysis of turbulent flows suggests that in the inviscid limit, singularities should exist, the question has not been settled mathematically. In fact very little is known concerning the Euler equations. It has been proven by Beale et al. (1984) that a solution that diverges in a finite time must lead to unbounded values of vorticity, hence to infinite gradients. One of the main difficulties come from both the nonlinear and non local character of the 3d fluid equations. Numerical analysis can provide new insight on these questions, like in many other fields in nonlinear sciences. In this paper we present some results of a numerical attempt to simulate collapsing solutions of the fluid equations.

Previous attempts with a fixed mesh code have been unsuccessful in tracking reliably such solutions (Pumir and Kerr 1987, Kerr and Hussain 1989). Our overriding concern is to simulate a large range of scales. The number of modes should not increase significantly as the collapse proceeds. In addition an accurate way is needed to reallocate grid points so as to always have the flow fully resolved. Essentially we require that the dynamics is sufficiently local. For a singularity to occur requires real nonlinearity : if the small scale vorticity containing modes are acted upon by a strain field which is dominated by the large scales, only exponential growth will result. One could however imagine weak singularities which arise when octave of scales, $2^{-n-1} < \lambda < 2^{-n}$ make a constant contribution to the strain rate. Examination of our flow field suggests that this is not occuring, but to simulate more scales than is possible with a fixed mesh code, some approximation has to be made.

The above considerations led us to write the following code. Separable rectangular coordinates are emloyed. The real line is mapped onto the interval $\xi \in (-1, 1)$ with a suitable function and a uniformly spaced grid is laid down in ξ. Derivatives are computed with centered differences which are so arranged that the discrete equations of motion explicitly respect incompressibility and conserve energy and momentum. Since we are ultimately interested only in small scales we chose boundary conditions with velocity and vorticity zero at infinity. The time stepping is also second order accurate.

When gradients became too large we remeshed the flow by adjusting the scales in the coordinate transformation, which can be done separately in the x,y and z direction. Interpolation from one mesh to another is done with cubic splines. Vorticity, strain rate and velocity are monitored before and after to assess errors. For the Euler equations, we had no difficulty in maintaining good resolution around the points of maximum vorticity.

The simplest non-trivial flow configuration one can imagine is just a vortex filament, in an otherwise irrotational flow. Equations of motion for the vortex filament have been derived by Moore and Saffman (1972) and implemented numerically by Siggia (1985). The validity of the model depends in a very crucial way on the absence of core distortion. Numerical analysis shown that two antiparallel vortex filaments spontaneously pair, and then stretch very efficiently. In fact, a systematic study of the Biot-Savart model for one or several vortex filaments (Siggia and Pumir 1985 and 1987) showed that a finite time singularity resulted, the vorticity blowing up as $1/(t^*-t)$ up to logarithmic corrections. The collapse that followed was universal i.e., independent of initial conditions. We therefore initialized our Euler code with paired vortex filaments. We show our initial condition in Fig. 1. The tubes contain vorticity distributed according to a Gaussian distribution. The vorticity field is parallel to the tubes, and is pointing upwards for $(x>0)$, and downwards for $(x<0)$; the $z = 0$ plane is also a plane of symmetry.

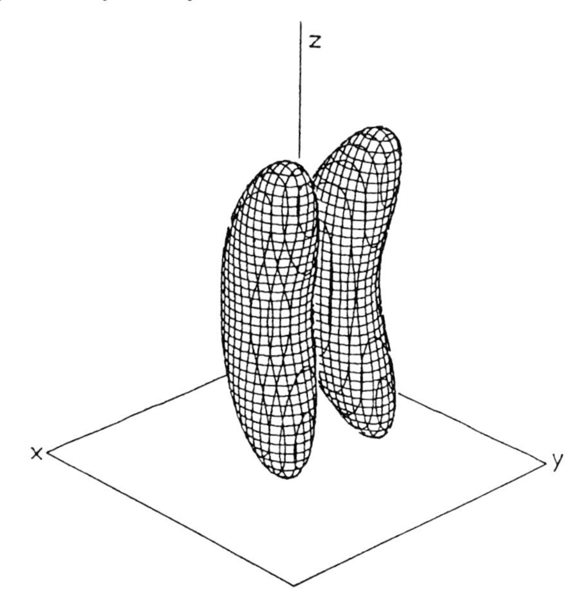

Figure 1. Surface plots of regions of intense vorticity at initial time. Space coordinates are limited by $(-2 < x,y,z < 2)$.

Since our initial conditions can be reasonably approximated as vortex tubes, one would expect that the Biot-Savart formula would give an approximate account of the velocity field they induce and hence their initial evolution. However, it has been shown that the cores undergo severe flattening when they evolve (Pumir and Kerr (1987), Meiron et al. (1989)). The initial core distortion can be explained as essentially due to the non-uniformity in the vortex stretching which for $z = 0$, which is just $\partial_z v_z$ by symmetry. The other components of strain are of course present and have been examined but they are dominated by purely two dimensional effects which for early times will slightly distort the cores but basically produce a stable propagating dipole similar to the Batchelor dipole (Batchelor 1967). The existence of many stable vortex dipole solutions in two dimensions argues against low pressure around the center of the pair as a cause of the flattening, since it is predominately of 2-D origin.

It will be instructive to explicitly display the expression for $\partial_z v_z$ for a single vortex tube lying in the plane $x = 0$, symmetric about $z = 0$ and given parametrically by $y = \rho\xi(\theta)$, $z = \rho\eta(\theta)$ where ρ is the radius of curvature at $z = 0$ and $\theta \in [-1, 1]$ viz.

$$\partial_z v_z(x,y,0) = \frac{-3\,\Gamma}{4\pi\rho^2}\left(\frac{x}{\rho}\right)\int_{-1}^{1} \frac{\eta\,(d\xi/d\theta)\,d\theta}{((x/\rho)^2 + (y/\rho - \xi)^2 + \eta^2 + (\sigma/\rho)^2)^{5/2}} \tag{1}$$

where Γ is the circulation and σ the (uniform) core size. Therefore $\partial_z v_z$ as a function of x,y acquires the spatial scale of the curve in the perpendicular direction, which will be important in the following. It results immediately from the Biot Savart formula that $\partial_z v_z(x,y,0)$ depends only on ω_x and ω_y and therefore knows about ω_z only implicitly through $\nabla.\omega = 0$.

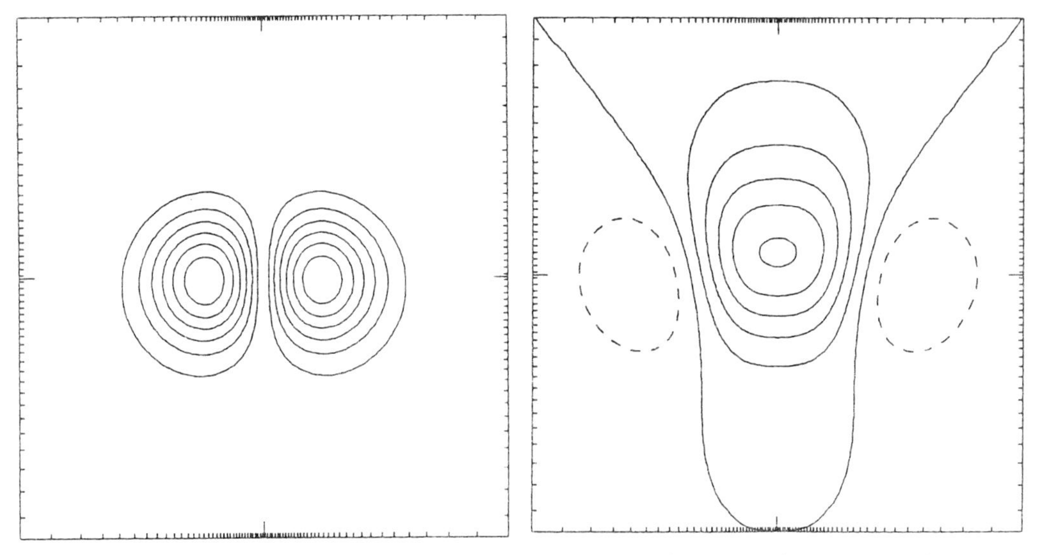

Figure 2. The vorticity distribution (left) and the rate of strain (right) in the $z = 0$ (symmetry) plane at initial time. Coordinates are defined by $(-2 < x,y < 2)$

In Fig. 2a,b we show the actual vorticity profile and strain rate in the plane z = 0 for our initial conditions. Evidently the vorticity will grow most rapidly for $|x| < c/2$, where c is the distance between the two centers of the tubes in the plane (z=0).

The maximum will increase only slightly for $|x| \leq c/2$, and the vorticity will actually decrease for $|x| \geq c/2$ where $\partial_z v_z$ is negative. Hence there will be a buildup of ω_z in the center without much change in the y extent of the tubes.

Figure 2b can be understood intuitively by comparing it with $\partial_z v_z$ in Fig. 3a computed from the Biot-Savart formula for the two centers of the tubes. A core size of 0.5 was used but in principal this could be adjusted to improve the fit. In Fig. 3b we show $\partial_z v_z$ for a single curve which makes it evident (cf. Eq. 1) that the sign change in $\partial_z v_z$ and therefore the principal core distortion is due to the stretching action of each tube on itself. A second tube placed at $x \approx -c/2$ in Fig. 3b would experience a net amplification. Only this second effect, the interfilament strain, was included in the Biot-Savart model solved in Siggia and Pumir (1985, 1987). There it was interpreted as the result of the difference in radial, (i.e., normal to the pair of tubes, in the direction $(\hat{e}_y + \varepsilon\,\hat{e}_z)$, velocities which leads to an increase in arc length of the filament pair and hence vortex stretching.

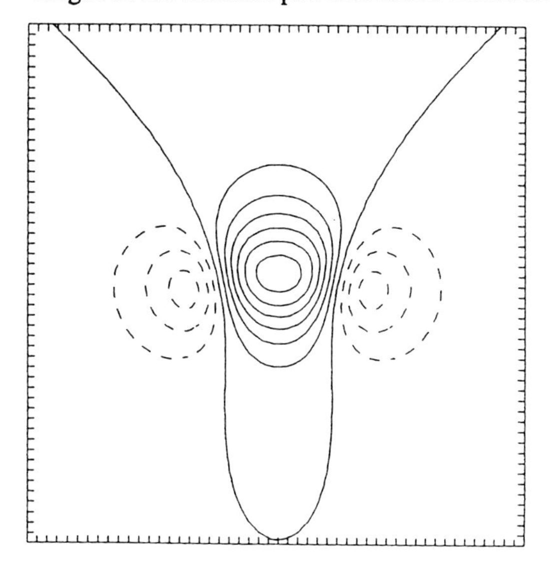

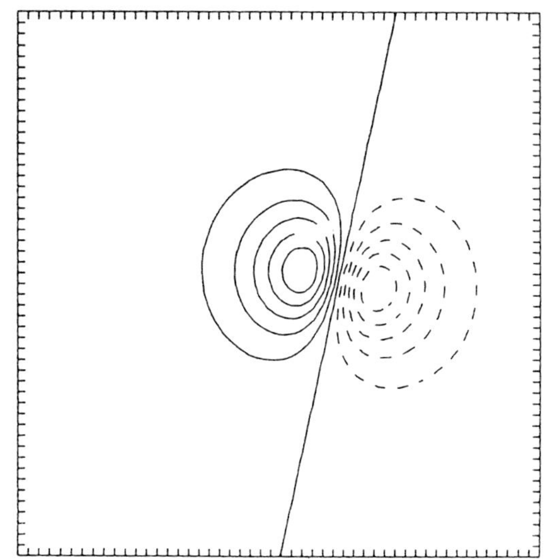

Figure 3. The rate of strain computed from the Biot Savart integral for two vortex filaments (left) and for only one single filament (right). Core deformation comes from inhomogeneities of the amplification rate.

Obviously, once the core has significantly flattened, important differences with the predictions of the Biot Savart model are to be expected. In fact, although we did see a huge contraction of scales, as documented below, we did not find any evidence of a 1/(t*-t) blow up of vorticity in what we consider to be the asymptotic regime in our simulation.

The range of parameters over which we were able to follow the flow is shown in table 1.

	T = 0.	T = 8.
x-scale	0.5	$1.\ 10^{-3}$
y,z-scale	1.0	1.10^{-2}
max vel.	0.5	1.0
$\lvert\omega\rvert_{max}$	3.	100.
$\omega.e.\omega_{max}$	2.5	$6.\ 10^{+3}$

Table 1. Ranges of values obtained in our simulation

The time dependence of the maximum vorticity and strain rate along the vorticity are shown in Fig.4.

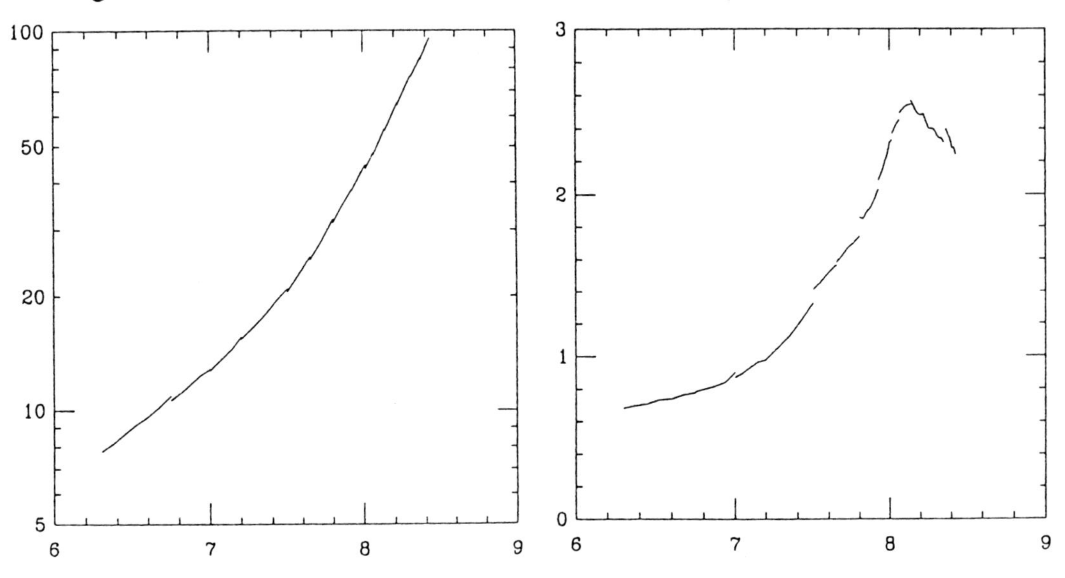

Figure 4. The maximum of vorticity (left) and the ratio $\omega e.\omega_{max}/\omega^2{}_{max}$ (right). The behavior turns exponential at t ~ 7.5

Several comments are in order. Firstly it is apparent that the growth rate of vorticity becomes exponential. This is perhaps most obvious in the normalized rate of strain plot which must diverge if there is to be a finite time singularity. Although we find no evidence of a singularity here, the range of scales explored is very large. If an adaptive mesh scheme was not used, a fixed mesh of order $6.10^4 \cdot 5.10^3 \cdot 5.10^3$ would have been necessary, which of course is absurdly large. In practice we were able to do the entire run with a 100^3 mesh.

The limited growth in velocity strongly suggests that the solution would not lead to singularities in the Navier-Stokes equations: indeed, it has been shown by Leray (1934) that if a solution of the 3d Navier-Stokes equations looses regularity at time t*, then,

$$\max_x \left|\vec{v}(x,t)\right| \geq A\sqrt{\nu/(t^*-t)}$$

where A is a universal constant. This has to be contrasted with the results of Siggia and Pumir (1985, 1987) in the case of the Biot-Savart model, where it was found that the velocity was also blowing up as $1/(t^*-t)^{1/2}$.

We have found that the eigenvalues of the strain matrix are proportional to $(1, \varepsilon, -1-\varepsilon)$, $0 < \varepsilon << 1$ and that the vorticity is parallel to the direction of weak stretching. The strongest stretching is thus perpendicular to the vorticity. Ashurst et al. (1987) and Kerr (1987) found similar results in simulations of turbulent channel flows and homogeneous isotropic turbulence.

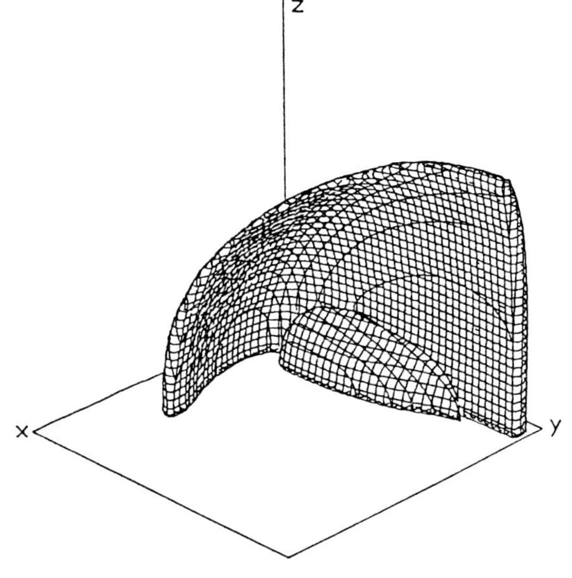

Figure 5. Surface plots of regions of intense vorticity at time $t = 8.3$ ($0 \leq x \leq .025$; $-.03 \leq y \leq .07$; $0. \leq z \leq .175$) clearly demonstrating the formation of 'shells' of vorticity.

In figure 5, we show 3 dimensional surface plots of the regions within which the vorticity is larger than some threshold. The vortex tubes have become paired vortex sheets. The crossover to exponential growth can be understood fluid mechanically as the concentration of the region of maximum vorticity into a thin sheet which contains negligible circulation. The stretching comes from the modes with the dominant circulation which also form paired vortex sheets but on a larger scale and with a maximum vorticity which is significantly less. Hence, the maximum vorticity undergoes passive stretching and whether the amplification in the region containing most of the circulation itself eventually diverges is unclear.

The paired sheets never become unstable because there is always some self induced stretching present. The time of integration is many times the spacing divided by velocity so any instabilities present would certainly have time to develop.

Although we have succeeded in simulating very large gradients and a very dramatic contraction of spatial scales, we have not found any evidence that our solution must collapse in a finite time. Of course, our calculation does not disprove the existence of such solutions. There may very well be unstable collapsing solutions (as is suggested by the initial rapid growth of vorticity, followed by the exponential amplification). Tracking such solutions would require a better understanding of the phase space. The weakening of the effective threedimensionality in our flow field is itself of great interest. It is due to the fact that the stretching when amplifying the vorticity field, also concentrates the circulation on (locally) 2-dimensional sheets, that do not generate self stretching. Details will be published elsewhere (Pumir and Siggia 1989b).

References

Ashurst W., A. Kerstein, R. Kerr and C. Gibson, 1987, Phys. Fluids **30**, 2343

Batchelor G.K. 1967, 'An introduction to Fluid Dynamics', Cambridge University Press

Beale J.T., T. Kato and A. Majda, 1984, Commun. Math. Phys. **94**, 61.

Kerr R.M. 1987, Phys. Rev. Lett. **59**, 783

Kerr R.M. and F. Hussain, 1989, Physica D (in press)

Leray J. 1934, Acta Math. **63**, 193

Meiron D., M. Shellye, W. Ashurst and S. Orszag, 1989, "Mathematical Aspects of Vortex Dynamics", SIAM Proceedings

Moore D. and P. Saffman 1972, Philos. Trans. R. Soc. London Ser. **A272**, 403

Pumir A. and R.M. Kerr, 1987, Phys. Rev. Lett. **58**, 1963

Pumir A. and E.D. Siggia, 1987, Phys. Fluids **30**, 1606

Pumir A. and E.D. Siggia, 1989a, Physica D (in press)

Pumir A. and E.D. Siggia, 1989b, preprint

Siggia E.D. 1985, Phys. Fluids **28**, 794

Siggia E.D. and A. Pumir, 1985, Phys. Rev. Lett. **55**, 1749

Townsend A.A., 1951, Proc. Roy. Soc. **A208**, 534

Willmarth W. and S. Lu , 1973, J. Fluid Mech. **48**, 775

Willmarth W. and T. Bogar, 1977, Phys. Fluids. **20**, S9

Vortex Nulls and Magnetic Nulls

John M. Greene

General Atomics, San Diego, CA 92138-5608, U.S.A.

1. INTRODUCTION

In this paper vortex reconnection will be compared with the perhaps slightly better understood magnetic reconnection. The latter phenomenon is usually, but not always, treated in terms of separatrices. Distinct flux bundles, sometimes called "islands," are identified together with the separatrices that divide them. The exchange of flux across these separatrices is reconnection. The elements of a similar theory are outlined here.

In magnetic reconnection, breaking of the condition of frozen magnetic field lines is associated with a component of the electric field parallel to the magnetic field. Here the equivalent quantity for vortex reconnection is identified as the component of the viscous force parallel to the vorticity.

The role of the viscous force in reconnection is discussed in Sec. 2. In this section there is a careful distinction between between breaking the Helmholtz invariant by shifting the vortex structure through the fluid, and genuine topological change of the vortex structure. Section 3 is concerned with a description of the role of vorticity nulls in producing separatrices. A beginning at applying these ideas to existing data is made in Sec. 4, and Sec. 5 has some conclusions.

2. EVOLUTION OF VORTEX LINES

It is well-known that vortex lines are convected with the fluid in inviscid flow. Following Newcomb's (1958) analysis of the analogous flow of magnetic field lines, this will be separated into two distinct ideas. The first to be discussed will be the conditions on a velocity field that must be satisfied if it is to convect vortex lines. Then the relation between these velocity fields and the actual flow velocities will be discussed. Of particular interest is the fact that in many circumstances there is no velocity field that simply convects vortex lines. In that case, vortex topology must be altered by the evolution. In other words, there must be reconnection.

We take incompressible flow, with unit density,

$$\frac{\partial}{\partial v} \underline{v} + \underline{\omega} \times \underline{v} + \underline{\nabla}\left(P + \frac{1}{2} v^2\right) = \underline{F}_\nu \quad . \tag{1}$$

Here the viscous force is denoted $\underline{F}_\nu$. Then the evolution of the vorticity, $\underline{\omega} = \underline{\nabla} \times \underline{v}$, is given by

$$\frac{\partial}{\partial t} \underline{\omega} + \underline{\nabla} \times (\underline{\omega} \times \underline{v}) = \underline{\nabla} \times \underline{F}_\nu \quad . \tag{2}$$

Following Newcomb, if the evolution of the vorticity can be written in the form

$$\frac{\partial}{\partial t} \underline{\omega} = \underline{\nabla} \times (\underline{\widehat{v}} \times \underline{\omega}) \quad , \tag{3}$$

then vorticity lines are convected with the velocity $\underline{\widehat{v}}$. Note that this velocity is not unique. As the vorticity field evolves, there is some arbitrariness about the identification of a vortex line from one instant to the next, and this is reflected in the non-uniqueness of $\widehat{v}$.

Upon adding and subtracting $\underline{\nabla}\chi$, where χ is arbitrary, the evolution of the vorticity can be expressed

$$\frac{\partial}{\partial t} \underline{\omega} = \underline{\nabla} \times \left[\underline{v} \times \underline{\omega} + \underline{F}_\nu - \underline{\omega}\,(\underline{\omega} \cdot \underline{\nabla}\chi)/\omega^2 \right.$$
$$\left. - (\underline{\omega} \times \underline{\nabla}\chi) \times \underline{\omega}/\omega^2 + \underline{\nabla}\chi\right] \quad , \tag{4}$$

where the last term clearly vanishes. Places where the vorticity vanishes will be treated separately. Now, if a decent function χ can be found that satisfies

$$\underline{\omega} \cdot \underline{\nabla}\chi = \underline{\omega} \cdot \underline{F}_\nu \quad , \tag{5}$$

then

$$\underline{\widehat{v}} = \underline{v} + \underline{\omega} \times (\underline{F}_\nu - \underline{\nabla}\chi)/\omega^2 \quad , \tag{6}$$

satisfies the condition, Eq. (3), that the vorticity is convected with the velocity $\underline{\widehat{v}}$.

Analysis of the helicity, $\underline{\omega} \cdot \underline{v}$, follows a similar pattern. Its evolution is given by

$$\frac{\partial}{\partial t}\left(\underline{\omega}\cdot\underline{v}\right)+\underline{\nabla}\cdot\left[\underline{v}\left(\underline{\omega}\cdot\underline{v}\right)+\underline{\omega}\left(P-\frac{1}{2}\,v^2\right)+\underline{v}\times\underline{F}_\nu\right]=2\,\underline{\omega}\cdot\underline{F}_\nu \quad . \tag{7}$$

Again, introducing an arbitrary χ in a way that does not affect the dynamics, we find

$$\frac{\partial}{\partial t}\left(\underline{\omega}\cdot\underline{v}\right)+\underline{\nabla}\cdot\left[\underline{v}\left(\underline{\omega}\cdot\underline{v}\right)+\underline{\omega}\left(P-\frac{1}{2}\,v^2\right)+\underline{v}\times\left(\underline{F}_\nu-2\,\underline{\nabla}\chi\right)\right]$$
$$=2\left(\underline{\omega}\cdot\underline{F}_\nu-\underline{\omega}\cdot\underline{\nabla}\chi\right) \quad . \tag{8}$$

When Eq. (5) is satisfied, helicity is conserved. A slightly different point of view is adopted here, in that a quantity is considered to be conserved if the evolution of its integrated value can be calculated from a well-defined flux on the boundary of the region.

The result is that if Eq. (5) has a decent solution, helicity is conserved. A simple case is that in which the viscous force, $\underline{F}_\nu$ vanishes. Another case is that in which the topology of the vortex lines is trivial. If the lines all run together through the region of interest, then Eq. (5) is a simple first order equation. With appropriate boundary conditions, a solution can be easily constructed. In such a circumstance, vortex lines are not convected with the fluid. They slip with respect to the fluid while maintaining their topological relations.

A situation in which Eq. (5) does not have a decent solution is that in which a vortex line forms a closed loop, and the integral of the tangential viscous force around the loop is nonvanishing. Then the straightforward solution for χ is not periodic around the vortex line, and thus is not a decent function. In this case, the vorticity flux passing through the loop is changing in time. The vortex lines that crowd in to make up the increase in flux must break and reform as they pass through the closed vortex loop.

Thus, there is a very distinct difference in the nature of flow that depends on whether Eq. (5) has a decent solution or not. The latter case will be defined to be reconnection.

Closed vortex lines are not the only configuration that leads to singularities in Eq. (5). In this note, the role of vortex nulls is emphasized. Details of the topology generated by nulls is given in the next section.

3. VORTICITY NULLS

This section contains a brief discussion of the neighborhood of a point where the vorticity vanishes. Fuller discussions can be found in papers by Greene (1988) and Lau and Finn (1989).

The first point to be made is that configurations with isolated vortex nulls are topologically stable, but configurations in which the vorticity vanishes on a line, on a surface, or throughout a region are not. Small perturbations of the flow can induce pervading, nonvanishing vorticity, but they can only move isolated null points. Thus, here only the general case of isolated nulls is considered. Limiting situations in which the vorticity vanishes more extensively frequently behave similarly to those discussed here, but the nature of these limits are outside the scope of this paper.

Consider the vorticity in the neighborhood of a null. Then, in the linear approximation, and in some Cartesian coordinates centered on the null, the vorticity can be written

$$\underline{\omega} = \delta\omega \begin{bmatrix} x \\ y \\ z \end{bmatrix} \quad , \tag{9}$$

where $\delta\omega$ is a 3×3 matrix of constants. The divergence of the vorticity is trace $(\delta\omega)$, so that the latter must vanish. Since we take the null to be isolated, none of the eigenvalues of $\delta\omega$ vanishes. Thus, there are two cases to be treated, that in which two eigenvalues have negative real part, and that in which only one eigenvalue has negative real part. The former will be called A nulls, and the latter will be called B nulls.

Consider the vicinity of an A null. Vortex lines lying in the stable plane, which is defined to contain the two eigenvectors whose eigenvalues have negative real parts, all converge on the null. Here, convergence or divergence is relative to the direction of the vorticity vector. The line that lies along the third eigenvector diverges from the null, either along the eigenvector direction or against it. They will be called the unstable spindle lines. If the vortex lines in the stable plane are followed back away from the null, they form a sheet. This sheet may be fluted or twisted, but it forms a separatrix that divides vortex lines into two classes. Lines on one side of the sheet converge toward the null and then diverge along one of the spindle lines, while those on the other side of the sheet diverge along the other side. Thus, there is an absolute separation of vortex lines, depending on how they leave.

Around a B null everything is reversed. It produces an unstable sheet that divides vortex lines according to their route of entry. This entry tends to be aligned along stable spindle lines.

The stable sheets centered on A nulls, and the unstable sheets centered on B nulls frequently, or perhaps generally, intersect. The intersection is a vortex line that connects the two nulls. In magnetic terminology, the intersection of two separatrices is called a separator. Such a separator sits at the junction of four vortex bundles, or vortices, that are identified according to their entrance and departure from the region of interest.

Now return briefly to the considerations of Sec. 2. The particular problem that arises when trying to solve Eq. (5) in the vicinity of a null point is that vortex lines come in from all directions, having experienced different boundary conditions and sampled different viscous forces. Thus, in general, χ will be discontinuous across a separatrix. This fits with an intuition that reconnection should be associated with separatrices.

The conclusion of this section is that vorticity nulls are associated with separatrices and can force reconnection as defined in Sec. 2. In the next section, some previous results will be examined to show how these ideas can be applied.

4. EXAMPLES

The configurations studied by Hussain and coworkers provide a simple first example [Kida *et al.* (1989), Melander and Hussain (1988), Melander and Hussain (1989)]. Taking a convenient coordinate system and with some simplification, these papers display antiparallel flux bundles, with vorticity primarily in the x-direction, lying side by side in the y direction. From symmetry, the vorticity vanishes on the axis $x = y = 0$ at all times. This is a special case, as discussed in the previous section, but as will be discussed below, it can be treated as a limit.

When the vortices are interacting, the vorticity does not vanish in the neighborhood of this line, so an analysis paralleling that of Eq. (9) can be performed for each point on the line. Clearly, $\delta\omega$ will have one vanishing eigenvalue, with eigenvector in the z-direction. In the perpendicular direction, there will be a stable and an unstable eigenvalue, with eigenvectors making some angle with respect to each other. The angle will depend on time and position in the z-direction. Thus, the set of stable eigenvectors arising from different points on the z axis will form a twisted sheet, and the unstable eigenvectors will form another, intersecting sheet. These sheets can be extended and will form separatrices.

It is clear that a small perturbation that introduces a z-directed vorticity along the $x = y = 0$ axis will have a small effect on the separatrix structure. In fact, this structure is probably quite rugged. What is gained by treating the general case of vorticity nulls is a way of identifying the separatrix structure in the absence of a symmetry that makes it obvious.

The configuration described in these papers consists initially of two oppositely directed vortices. Later, there are a pair of bridging vortices connect the original pair and that lie on opposite sides of the $x = y = 0$ axis. Thus, there are four very distinct vortices. It is clear from the discussion above that they are divided by the separatrices that emanate from the z axis.

A more complex case has been considered by Melander and Zabusky (1988). They have two vortex cores displaced in the z-direction, one with vorticity in the x-direction and the other with vorticity in the y-direction. Their cores are bathed in a screening vorticity background. Here the symmetry leads to the line $x = y$, $z = 0$ being a vortex line. The screening vorticity causes this vortex line to end at null points. The two-dimensional stable and unstable manifolds of these null points intersect on vortex lines. Results show that these manifolds evolve in a complex way.

5. CONCLUSIONS

Vortex topology and reconnection are closely associated with separatrices. Nulls are probably a copious source of separatrices, and so should be studied.

One recurrent question in magnetic reconnection concerns the timescale for reconnection. The dynamics tends to create singularities along separators, and thus greatly accelerates the process of reconnection. It would be interesting to understand if something similar happens in vortex reconnection. The crucial quantity would be the scaling of gradients with Reynolds number, and hence of the viscous force, at separators.

REFERENCES

GREENE, J.M. 1988 *J. Geophys. Res.*, **93** 8583-8590.

KIDA, S, TAKAOKA, M. & HUSSAIN, F. 1989 *Phys. Fluids A*, **1**, 630-632.

LAU, Y.-T. & FINN, J.M. 1989 submitted to *J. Geophys. Res.*

MELANDER, M.V. & HUSSAIN, F. 1989 *Phys. Fluids A*, **1**, 633-636.

MELANDER, M.V. & HUSSAIN, F. 1988 *CTR Report No. CTR-S88*, 257-286.

MELANDER, M.V. & ZABUSKY, N. 1988 submitted to *Phys. Rev. Lett.*

NEWCOMB, W.A. 1958 *Annals of Physics NY*, **3**, 347-385.

Topological Aspects of Vortex Reconnection

MOGENS V. MELANDER* AND FAZLE HUSSAIN

*Department of Mathematics, Southern Methodist University, Dallas, Texas
Department of Mechanical Engineering, University of Houston
Houston, TX 77204-4792, USA

1. INTRODUCTION

This work is motivated by our speculation that vortex reconnection is a frequent event in turbulent flow and that it plays a key role in mixing, and generation of helicity, enstrophy and aerodynamic noise (Hussain 1983, 1986). Also called cut-and-connect or cross-linking, this phenomenon has recently received considerable interest because it is an example of nonpreserving topology. It has been argued that vortex reconnection is a prime candidate for a finite-time-singularity in the Navier-Stokes equation (Kerr *et al.* 1989, Siggia & Pumir 1985). This study is also triggered by our long-standing warning against relying too heavily on flow visualization for understanding vortex dynamics or coherent structures in turbulent flows because the marker boundary coincides with the structure boundary, even at unity Schmidt number, *only* in 2D flows; as we also show here, they can be quite different in 3D flows. Thus, flow visualization can be confusing, and even grossly misleading, in studies of vortex dynamics and mature turbulence. Our interest was also piqued by the observation that while viscosity is crucial for reconnection (Fohl & Turner 1975), the process seems to occur on a convective timescale. For a review of these matters and other simulation details, see Melander & Hussain (1988, 1989); some related matters are discussed by Kida *et al.* (1989). We are also interested in exploring flow topology, in particular the connection between helicity and dissipation (Tsinober & Levich 1983; Moffatt 1985).

2. SIMULATION METHOD AND INITIAL CONDITIONS

The flow is simulated using a spectral method with a fourth-order predictor-corrector for time stepping, dealiased by a 2/3 k-space truncation. Periodic boundary conditions are appropriate because reconnection is a strong local vortex interaction, insensitive to nonlocal effects such as the influence of adjacent boxes. The initial condition consists of two antiparallel vortices with mutually inclined symmetric sinusoidal perturbations (figure 1a). Initially, each vortex has a circular core with compact support. We also studied an asymmetric initial condition as a perturbation of figure 1a. The symmetric initial condition is particularly useful for clearly identifying the effects of viscosity because the two symmetry planes are preserved in time and are material planes in the absence of viscosity. For clarity, the xy-plane is called the 'symmetric plane' or π_s, while the yz-plane is the 'dividing plane' or π_d. In both planes, the normal velocity and the tangential vorticity vanish at all times; furthermore, $x\omega_3 \leq 0$ everywhere in π_s; this eliminates the problem of self-annihilation within each vortex which occurs in the simulation of Meiron *et al.* (1989). Consequently, the circulations Γ_s and Γ_d in the half planes π_{s+} (x > 0) and π_{d^-} (z < 0) change exclusively by viscous vorticity annihilation. We also know that $d\Gamma_d/dt = -d\Gamma_s/dt$, since an arbitrary vortex line must intercept either π_s or π_d.

3. THREE PHASES OF RECONNECTION

The evolution of the vortices as well as of the vorticity distributions in the symmetry planes is shown in figure 1 for the Reynolds number Re = Γ/ν = 1000. Figure 1 shows wire-frame plots of the vorticity norm $|\omega|$ at t = 0, 2.25, 3.5, 4.75 and 6; the level surface here represents 30% of the initial peak vorticity $\omega_{max}(0)$, and t is time t^* nondimensionalized by $\omega_{max}(0)$ such that $t = t^*\omega_{max}(0)/20$. When we compare figure 1 with figure 5i it is clear how the reconnection evolves through *three phases*.

The *first* phase is governed by inviscid vortex dynamics as almost no circulation is transferred from π_s to π_d. During this phase the two vortices move upward by mutual induction, while self-induction presses the vortices against each other, flattening them and forming a contact zone with steep vorticity gradients. In the *second* phase a dramatic change in topology takes place as a rapid transfer of circulation from π_s to π_d (figures 3,4) occurs due to viscous vorticity annihilation (by cross-diffusion) in the contact zone. This phase, marked by 'bridging' (growing humps in figures 1(c,d)), is the heart of the reconnection process (discussed later). The growing bridges acquire a curved horseshoe shape, causing them to pull apart by self-induction as they stretch the dipole and sustain annihilation. The *third* phase is the evolution of the remnants of the dipole after the rapid circulation transfer. The slender vortices, which we call 'threads', are well defined because of the stretching by the bridges. These threads arise as a consequence of the incompleteness of the reconnection and are the longest lasting features of the reconnection.

The simulation with initial asymmetries produced virtually the same wireframe plots as in figures 1(a-e) even at t = 6, thus vindicating both the analysis (Takaki & Hussain 1985) and the high-resolution simulation (Kerr & Hussain 1989) performed by invoking symmetry. The Re dependence of the crosslinking process was tested and found to consist of the same sequence of stages with identical successive spatial patterns, except that the corresponding timescales increase with decreasing Re.

4. THE BRIDGING MECHANISM

The rapid circulation transfer from π_s to π_d begins when the vortices come in contact near $\pi_s \cap \pi_d$ (figures 1b,3b) and form a contact zone C(t) with steep vorticity gradients. As the flow is highly viscous around C and nearly inviscid elsewhere, we may think of the vortexlines as material lines far away from C. We use this idea schematically in figures 5a-f, where each vortex is represented by three vortexlines; one vortexline from each vortex intercepts C (figure 5a), and shortly afterwards these vortexlines annihilate each other at x = z = 0 by cross-diffusion (figure 5b). This produces two reconnected lines intercepting π_d at

$z = \pm z_o$ (Figure 5c). The π_d-intercept z_o quickly retreats away from $x = z = 0$, because of strong diffusion and nearly antiparallel vorticity vectors.

A cross section AB (figure 5e) at the edge of *C* reveals the two-dimensional dipole structure of the original vortices. In a co-translating frame the streamline pattern becomes the well known recirculation bubble of a vortex dipole (figures 5g,6a). Although this frame is not precisely defined for an unsteady dipole, the existence of two stagnation points - at the front and back of the dipole - S_F and S_B is clear. The large velocity between S_F and S_B advects z_o towards S_F, ahead of the dipole (figure 5d). At S_F the relative upward advection of z_o stops, and our reconnected vortexline unfolds along the diverging separatrix a-a (figure 5g). Here, the vortexline continuously stretches and begins to wrap around the dipolar vortex pair (figure 5f). Meanwhile, the circulatory motion in the dipole continuously pumps fresh non-reconnected lines into *C* near S_B (figure 5g). Due to this mechanism, reconnected vortexlines accumulate ahead of the antiparallel vortices and align orthogonal to the latter. This schematic scenario is highlighted by the actual vortex line tracing shown in figures 11(a-c).

During bridging, a complex interaction of competing effects governs the rate of annihilation by changing the vorticity gradients in *C*. These effects are diffusion, stretching in the z-direction, and local 'self-induction'. Diffusion counteracts the steepening of gradients and can only decrease the peak vorticity in the symmetry plane ω_m, while any increase in ω_m is due to stretching (normal to π_s). This 'axial' stretching is generated mainly by the vorticity distribution $\varepsilon(t)$ away from *C* (ε is shown in Figure 5a; note that ε includes the bridges), but also in part by self-induced lengthening of vortex lines near *C* (analogous to the axisymmetric collision of vortex rings). Depending on the relative positions of ε and the dipole in π_s, the external stretching is either positive (vortex stretching) or negative (vortex compression): positive if the dipole has fallen behind ε (figure 5h), and negative if the dipole has advanced ahead of ε. The self-induction on each side of the sheet-like region *C* is in π_s, and is directed toward π_d whenever the non-reconnected vortex lines in *C* curve upward, and away from π_d when they curve downward. The motion toward π_d increases gradients across the y-axis by compressing the two vortex cores into a characteristic head-tail structure (figure 3d), while motion away from π_d decreases the gradients by separating the vortex cores in π_S. The above three effects – diffusion, stretching and self-induction – dominate the evolution in the following order. At first the vortex lines in *C* curve upward, and the gradients across π_d increase as self-induction toward π_d presses the dipole cores into a head-tail shape; meanwhile, the external stretching being negative, the increase in ω_m is small. However, as the dipole in π_s falls behind ε (discussed later) such that the vortex lines in *C* curve downward, the axial stretching picks up and produces a large increase in ω_m, and consequently large gradients. Finally, toward the end of the rapid circulation transfer from π_s to π_d (figure 5i), the local self-induction away from π_d arrests the cross-diffusion by producing a near balance between axial stretching, diffusion and separation of the dipole cores in π_s, causing the threads to linger.

The curvature reversal, which causes the incomplete reconnection, is a characteristic and crucial event. It occurs because the upward motion of the dipole in π_s slows down relative to that of ε (figures 3,4; note that the y-position can be compared using the numbers on the right side of each panel); but why does the dipole in π_s slow down? A downward velocity field in C – downwash – opposes the dipole's own upward motion in π_s (figure 5h). Furthermore, the dipole loses circulation through vorticity annihilation; its upward motion weakens and it also relaxes into a head-tail shape. The downwash is induced by the growing bridges as well as by non-reconnected vortex lines in ε, so downwash will be present even at $Re = \infty$. The relaxation into a head-tail shape, an inviscid process, virtually decreases the dipole circulation as the dipole propagation velocity is determined almost exclusively by the head. Although this inviscid effect is weak at $Re = 1000$, it may well be the primary mechanism for curvature reversal at higher Re. An indication of this would be that the area A of the nearly self-similar head decreases with higher Re, such that $\omega_m A^{1/2}$ (propagation speed of the head to leading order) is bounded as $Re \rightarrow \infty$. This indicates the possibility that the amount of circulation transferred from π_s to π_d tends to zero as Re tends to infinity while the transfer is still occurring within a convective timescale. In fact, a cascade of such short bursts of circulation transfers may occur at high Reynolds numbers.

5. TOPOLOGICAL PROPERTIES

The direct simulation provides considerable amounts of spatial data inaccessible experimentally with current measurement technology. These provide valuable information on the topology and dynamics of interacting vortices. While post-processing and visual examination on a color graphics workstation proves extremely illuminating, this visual information is typically overwhelming and must be properly synthesized and interpreted. The approach undertaken here evolved from that employed to study coherent structures in turbulent shear flows (Hussain 1980, 1983). Of the variety of properties that can be useful in studying the topology and dynamics, we limit our attention to scalar intensity, enstrophy production, dissipation and helicity density. We find that the spatial distribution of a scalar in a vortical structure is best examined when bundles of vortex lines are colored by the magnitude of the scalar quantity. This numerical visualization technique emphasizes the true geometry and topology of the structure (as given by the vorticity field) while revealing all details of the spatial distribution. (Note that this technique is useful only when vortexlines at the noise level are deleted). Based on such visualization we have selected a few planar cross-sections which capture striking new dynamical and topological aspects of the reconnection process. Figures 8-10 show planar cross sections of vorticity norm $|\omega|$, scalar s, dissipation $\varepsilon = 2\nu s_{ij}s_{ij}$, helicity $h = u_i\omega_i$, enstrophy production $P_\omega = \omega_i s_{ij}\omega_j$ and relative helicity $u_i\omega_i/(|u||\omega|)$ in a plane parallel to the dividing plane, but at a distance 0.196 from it (this plane essentially goes through the threads). Three instants were selected: $t = 3.5$ at the middle of the circulation transfer, $t = 4.75$ the time of peak vorticity in the bridges and $t = 6$ representing the slowly evolving post-reconnection vortical structure.

In this simulation, the Schmidt number is unity, and initially the scalar distribution exactly marks only *one vortex*. Even under such ideal conditions the scalar does not continue to faithfully mark the vortical fluid (compare figures 1 and 2). At t = 4.75, vortex stretching has diluted the scalar in the contact zone in A (figures 2d,9b) while the vorticity, being accentuated, remains large, with the result that a key feature of the vortical structure, the threads, is invisible in experimental flow visualization (figures 2e,10b); see Schatzle (1987). Another interesting feature of the scalar distribution is seen at t = 3.5, where strong diffusion across π_d in conjunction with dipole pumping causes scalar to accumulate in the bridge across the dividing plane A-B (figure 6b), with the surprising result that the bridge consists of mixed fluid, thereby underlining the basic idea in the bridging mechanism.

Throughout the first phase (the inviscid advection), the flow has little or no helicity and the vortex lines inside each vortex are almost parallel. During and especially towards the end of the bridging phase, high helicity develops locally on each side of the dividing plane (figures 7a,b). Far away from π_d the helicity remains low and the vortexlines are untwisted. However, in the threading phase the peak helicity decays and the helicity distribution gradually spreads outward from π_d. The helicity peaks remain in the threads after the bridging phase and are the most dominant feature of the helicity distribution (figure 7b). The reason is that while the bridges build up circulation the weaker threads are being stretched and aligned orthogonal to the bridges (figure 1e). The high helicity in the threads is a consequence of this near orthogonality of threads and bridges -- the strong swirling velocity of the bridges nearly aligning with the threads.

The helicity distribution after the reconnection (*e.g.* t = 4.75 and t = 6) is particularly interesting, as it consists of intertwined regions of positive and negative helicity. Apart from the threads, where helicity stands out clearly (figures 7a,b), the structure is an enigma until one examines bundles of vortexlines. Figures 11(b,c) show that vortexlines emerging from the threads do not wrap all the way around the bridges as one would expect from $|\omega|$ surfaces, *e.g.* expanded views of figures 1(d,e). Instead, they form asymmetric hairpin-like structures (see A and B, figure 11b). The vorticity is high in one hairpin leg and low in the other. Vortical structures of this kind are unfortunately not faithfully reflected in the surface plots of $|\omega|$. The intense leg (A) is the extension of the thread (C) and its high vorticity makes its helicity stand out in figure 7. The diffuse leg is difficult to identify in the helicity distribution as the $|\omega|$ there is low and the angle between vorticity and velocity is larger. The intense leg induces a flow away from π_d on the outside of the bridges causing vortex stretching in the outer part of the bridge cores (see point A in figures 9e,10e). The relative helicity also reflects this axial flow away from π_d (see point b in figures 9f,10f).

Inside the bridge the vortex lines have a slow helical twist near π_d (barely discernible in figure 11c). Contrary to the intense hairpin leg, the twist produces an axial flow toward π_d at the center of the bridge. Since this axial flow is toward π_d from both sides, it results in a negative enstrophy production (dotted lines in figures 9e,10e). The twisting of vortexlines is the result of a skewed (nonconcentric) vorticity distribution in the bridges (figures 4b-d). The peak vorticity in the bridge is not at the location of the vorticity centroid, but it is far away from the contact zone, where vorticity is lower than in bridges -- a consequence of earlier stretching by the dipole during bridging. Figures 11(b,c) show that a vortexline which neatly follows the centerline of the vortex far away from π_d ceases to do so in the bridge. This effect clearly results in vortexline twisting and thus induced axial flow (figure 11d). The direction of the axial flow is determined by the orientation of the twist, which in turn is determined by the fastest swirling velocity along the vortex. The bridges, having the highest peak vorticity, also have the highest swirling velocity. Hence the axial flow is toward π_d from both sides, resulting in vortex compression (*i.e.* negative P_ω in figures 8 - 10) which decreases the peak vorticity and thereby also the twisting rate. This constitutes a new inviscid mechanism, clearly distinct from diffusion, for smoothing the vorticity intensity along the reconnected vortex lines.

The relative helicity is sensitive to noise level amplitudes and must be treated with care. Interestingly enough, numerical noise in almost all irrotational regions exhibits large relative helicity. Therefore it is essential to condition the relative helicity by both vorticity and velocity amplitude thresholds. One thus has a physically meaningful quantity as a measure for the angle between velocity and vorticity vectors. In studying the present simulation we found the *conditional relative helicity* to be useful for identifying sharp turns in vortex lines, such as those occurring at the tip of hairpin-like structures. Near such structures the relative helicity exhibits a characteristic sharp variation from positive to negative values. The conditional relative helicity thereby highlights asymmetric hairpin structures which elude other diagnostics. Furthermore, the relative helicity clearly marks the centerline of a vortex. This is a simple consequence of the fact that the swirling velocity vanishes at the centerline of a vortex whereas other components of the velocity such as axial and nonlocal components in general do not; figure 11e illustrates that the angle Ψ_0 on the vortex axis is lower than the angle Ψ away from the axis. In figures 8f, 9f and 10f one therefore observes a sharp peak in relative helicity near the vortex center.

Until the threading is well under way, ε is high in the center of the contact zone and the peak is located in π_s. Since, as a simple consequence of the symmetry, neither of the symmetry planes have helicity h, the peaks of h and ε remain spatially exclusive during the bridging phase. However, during the threading phase the peak of the ε bifurcates into two peaks, which move away from each other in π_d. Thereby, the peaks of ε locate themselves between two peaks of h, one on each side of the dividing plane. The spatial exclusiveness remains, but is not as clear as at earlier times; for example in the plane cross-sections (figures 10c,d) the peaks almost coincide. We have found no unique correspondence between dissipation and helicity.

This work is supported by Office of Naval Research Grant N00014-87-K-0126 and Department of Energy Grant DE-FG05-88ER13839. The computations were performed at NASA-Ames.

6. REFERENCES

Fohl, T. & Turner, J.S. 1975 *Phys. Fluids* **18**, 434.

Hussain, A.K.M.F. 1980 *Lect. Notes Phys.* **136**, 252.

Hussain, A.K.M.F. 1983 *Phys. Fluids* **26**, 2816.

Hussain, A.K.M.F. 1986 *J. Fluid Mech.* **173**, 303.

Kerr, R.M. & Hussain, F. 1989 *Physica D* (in press).

Kerr, R.M., Virk, D. & Hussain, F. 1989 (in this Proceedings).

Kida, S., Takaoka, M. & Hussain, F. 1989 *Phys. Fluids A* **1**, 630.

Kida, S., Takaoka, M. & Hussain, F. 1989 (in this Proceedings).

Melander, M. V. & Hussain, F. 1988, NASA report CTR-S88, 257.

Melander, M. V. & Hussain, F. 1989 *Phys. Fluids A* **1**, 633.

Meiron, D.I., Shelley, M.J., Ashurst, W.T. & Orszag, S.A. 1989, in *Mathematical Aspects of Vortex Dynamics* , SIAM, edited by R. Caflisch, p. 183.

Moffat, H.K. 1985 *J. Fluid Mech.* **159**, 359.

Schatzle, P.R. 1987, PhD thesis, California Institute of Technology.

Siggia, E.D. & Pumir, A. 1985 *Phys. Rev. Lett.* **55**, 1749.

Takaki, R. & Hussain, A.K.M.F. 1985 *Turbulent Shear Flows V*, Cornell U., 3.19.

Tsinober, A. & Levich, E. 1983 *Phys. Lett. A* **99**, 321.

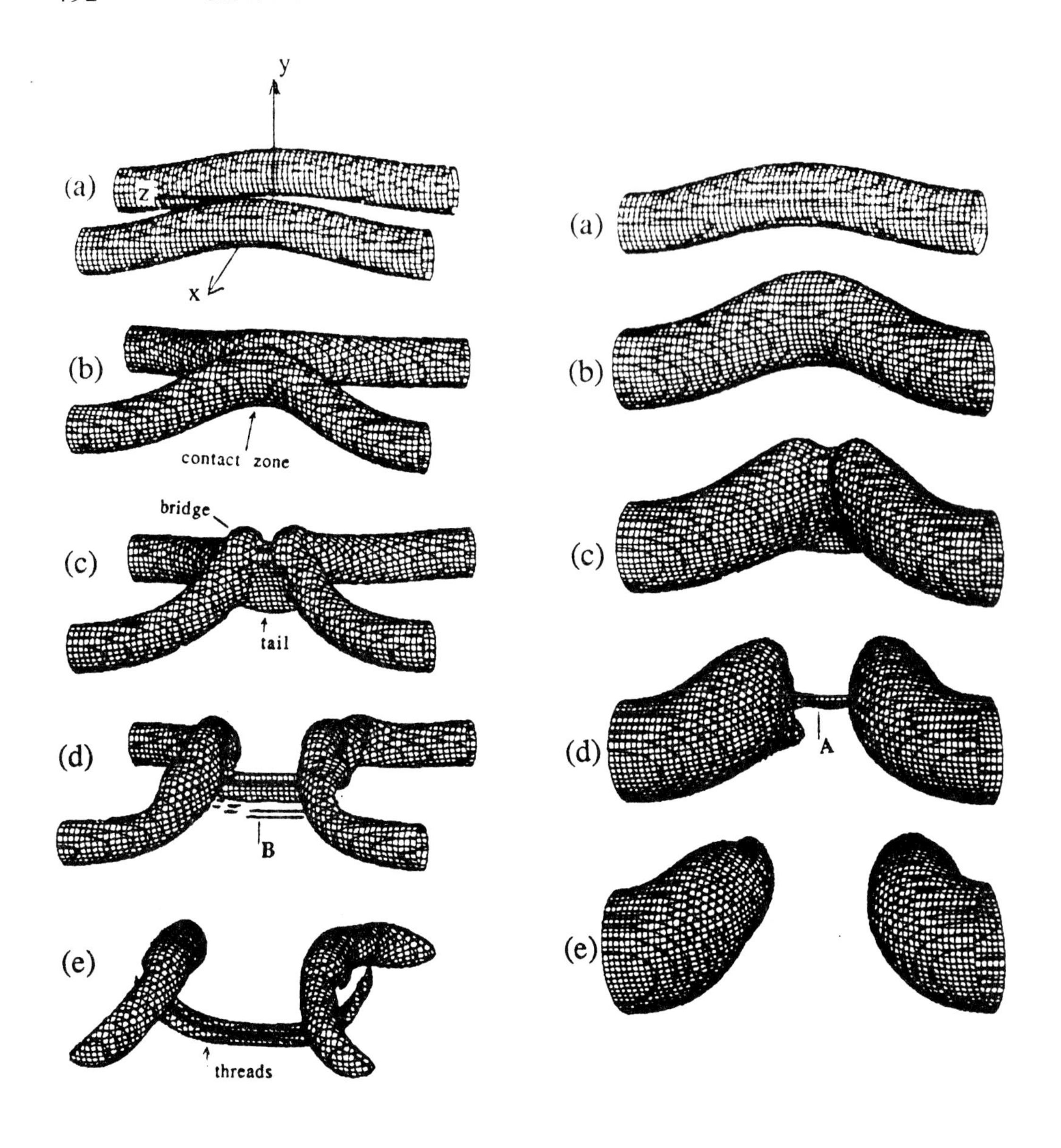

Figure 1.

$|\omega|$ surface at 30% of the initial peak. (a) t=0; (b) t=2.25; (c) t=3.5; (d) t=4.75; (e) t=6.0

Figure 2.

Scalar at 5% of the initial peak. (a) t=0; (b) t=2.25; (c) t=3.5; (d) t=4.75; (e) t=6.0

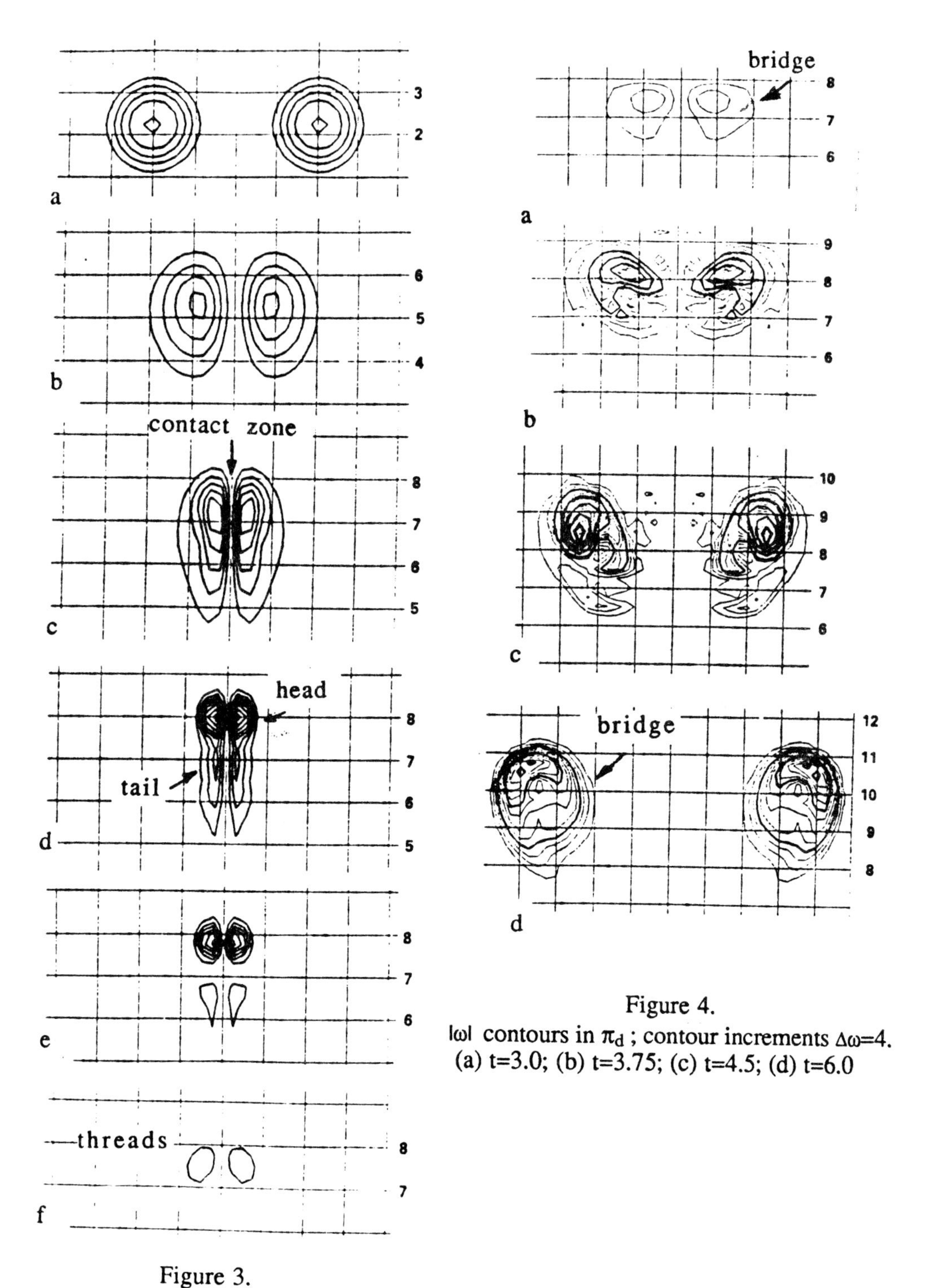

Figure 4.
$|\omega|$ contours in π_d ; contour increments $\Delta\omega=4$.
(a) t=3.0; (b) t=3.75; (c) t=4.5; (d) t=6.0

Figure 3.
$|\omega|$ contours in π_s ; contour increments $\Delta\omega=4$.
(a) t=0; (b) t=2.0; (c) t=3.0; (d) t=3.75;
(e) z=4.5; (f) t=6.0

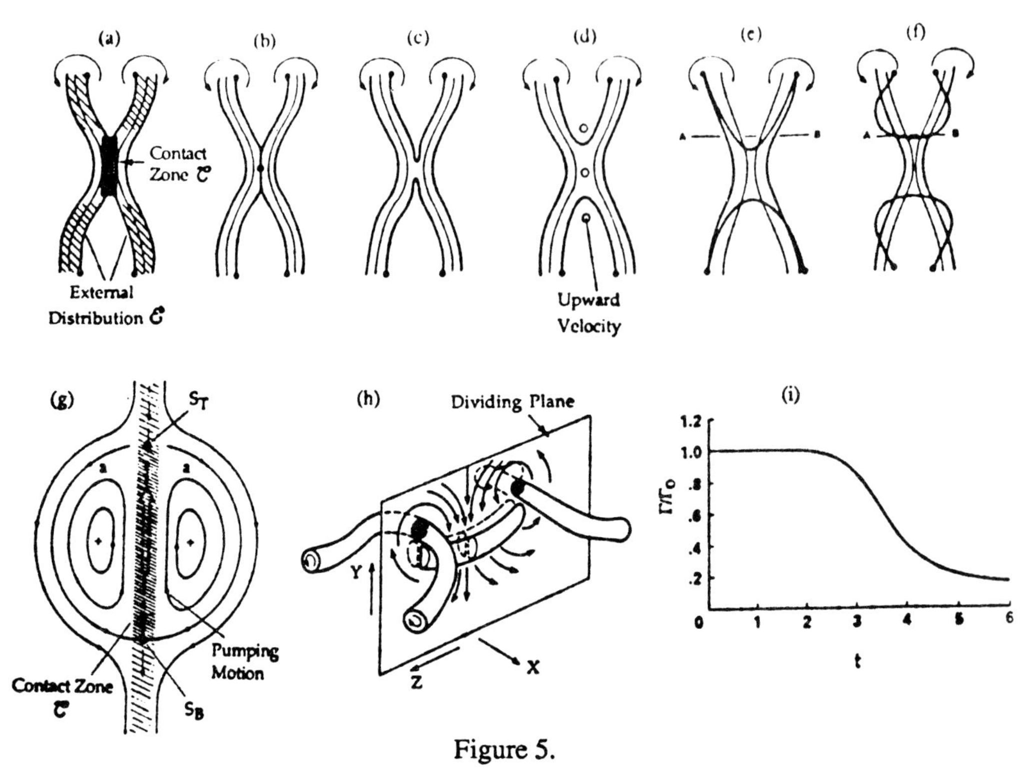

Figure 5.

(a-f) Explanation of bridging mechanism; (g) streamwise pattern in π_s ; (h) downwash; (i) circulation in π_s^+ $(x > 0)$.

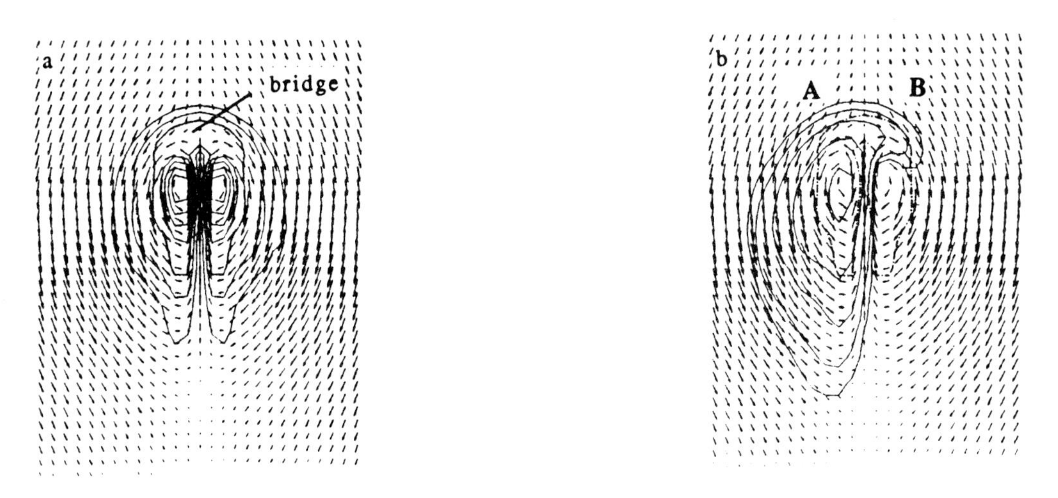

Figure 6.

Cross-section parallel to π_s in the bridges at t=3.5 with velocity vectors overlaid.
(a) Vorticity norm $\Delta\omega$=5; (b) passive scalar, contour levels 2.5%, 5%, 10%, 20%, 40% of initial peak concentration.

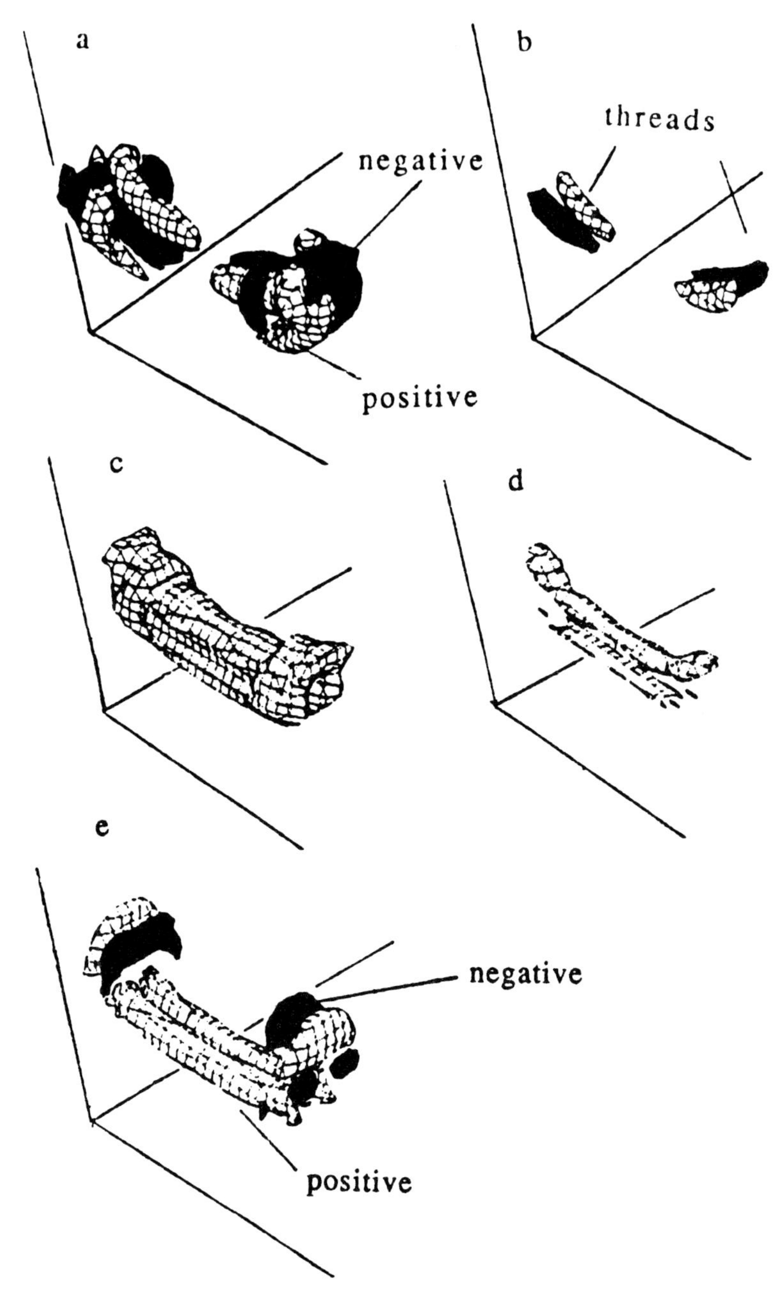

Figure 7.

(a) h at z=4.5, 20% of peak value; (b) h at 50% level; (c) ε at 20% level; (d) ε at 50% level; (e) P_ω at 20% of positive and negative peak values.

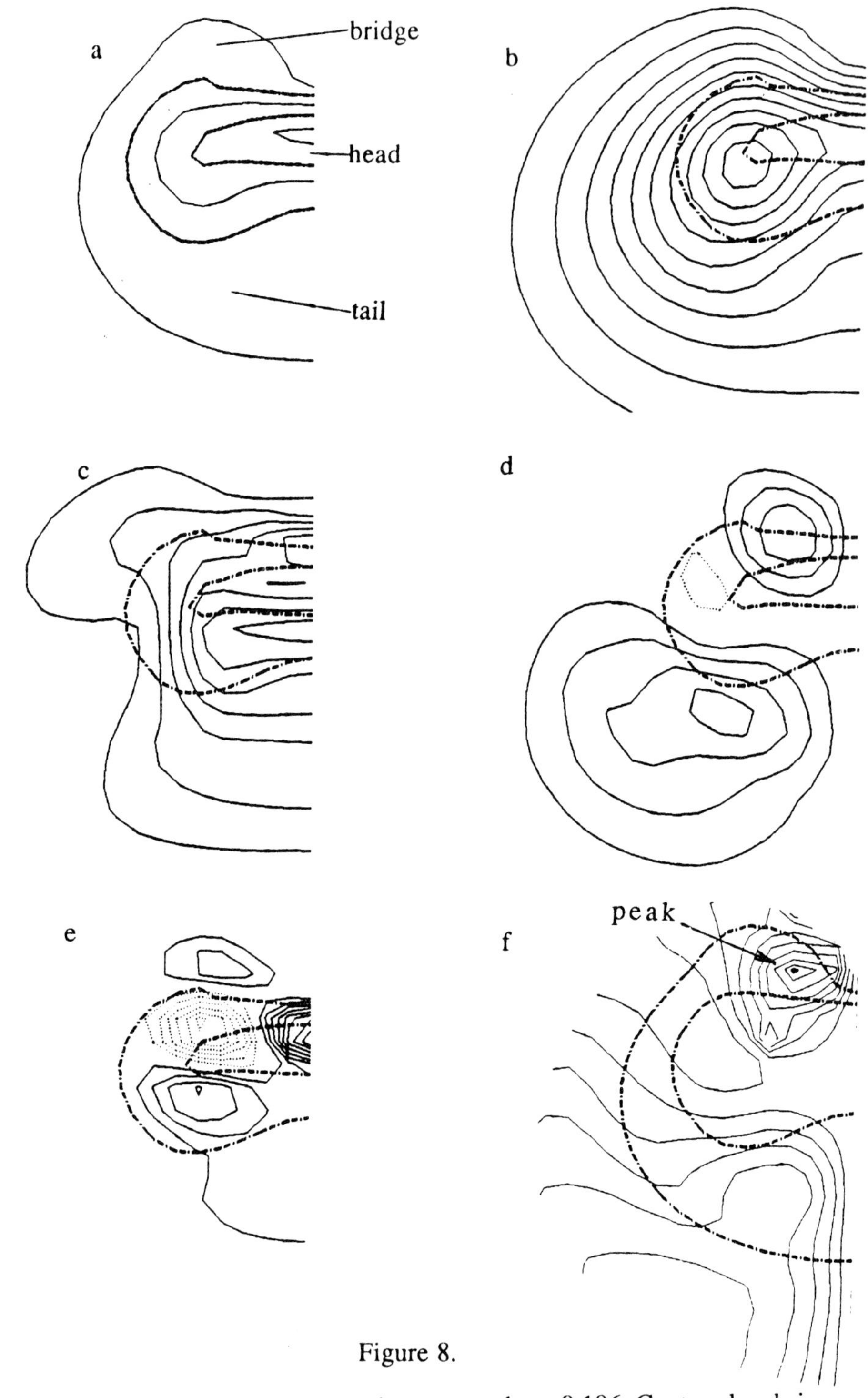

Figure 8.

Plane cross-section at t=3.5 parallel to π_d but removed Δx=0.196. Contour levels in panels (a,b) are scaled by the peak values in the plane. (a) $|\omega|$; (b) s; (c) ε; (d) h; (e) P_ω; (f) h_r contours giving the angle between vorticity and velocity vectors in increments $\Delta\theta$=10°. Solid lines represent positive values, dotted lines are negative values.

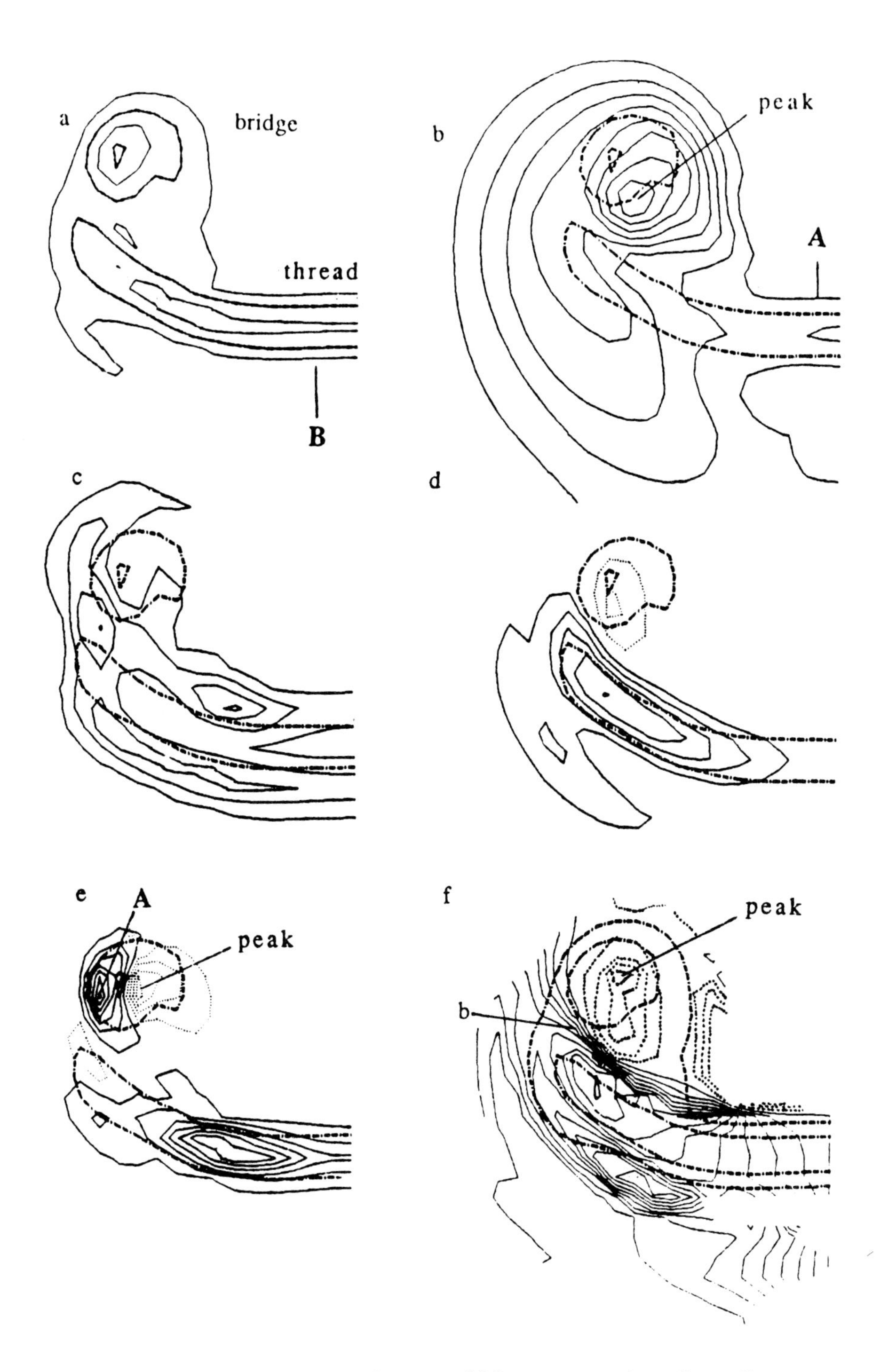

Figure 9. Plane cross-sections at t=4.75 (see caption from figure 8).

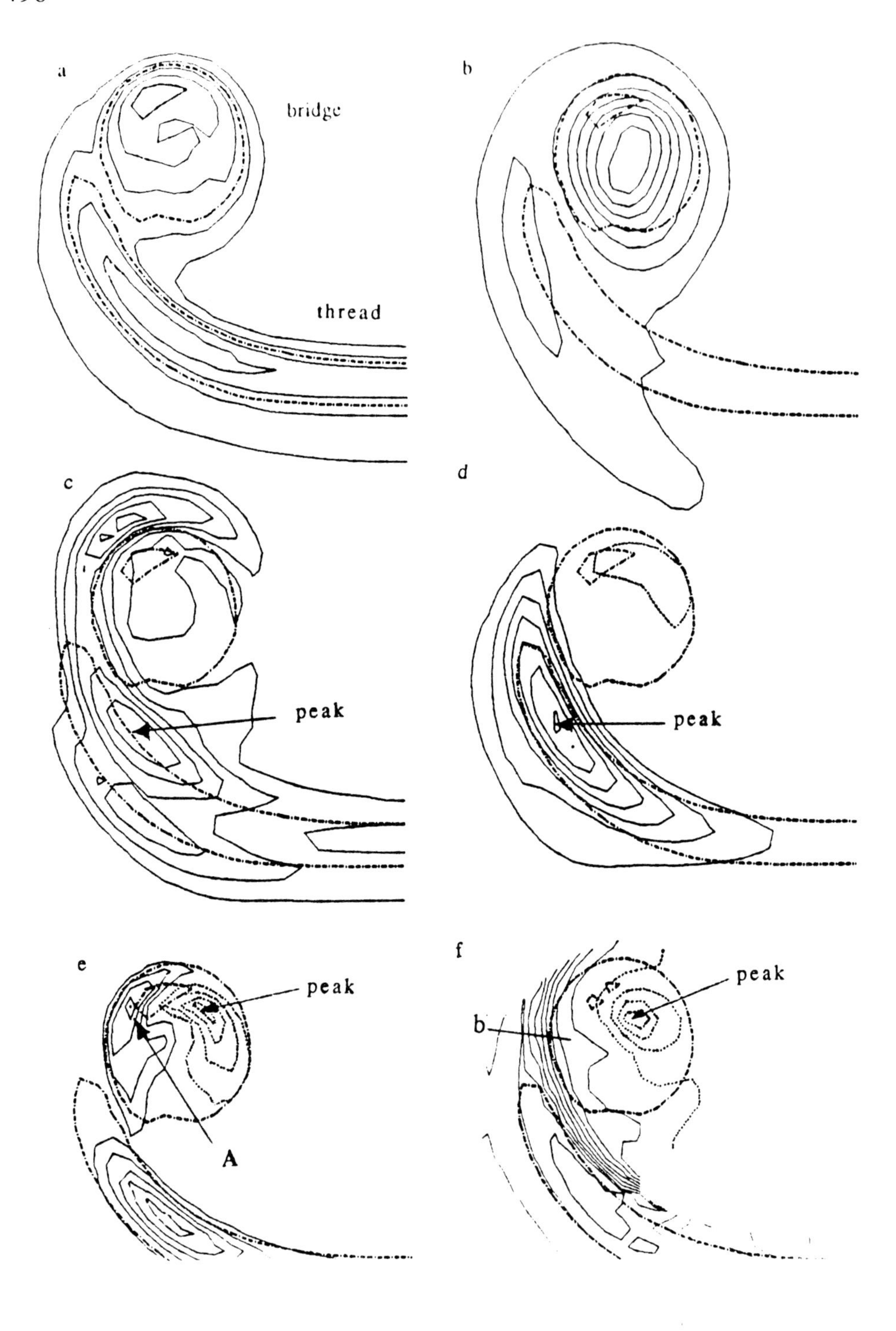

Figure 10. Plane cross-sections at t=6.0 (see caption from figure 8).

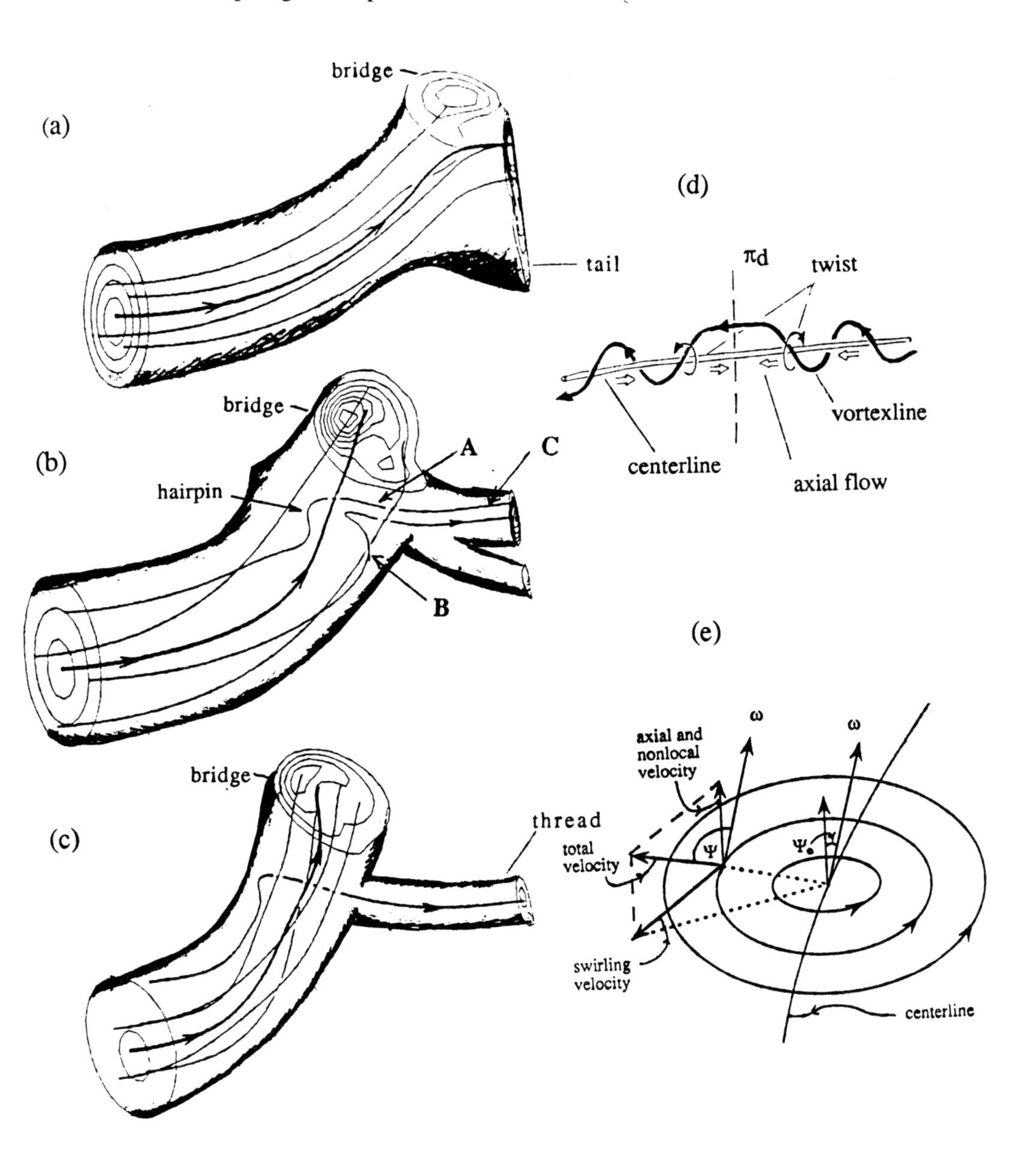

Figure 11.

Vortex lines in a quarter of the computational domain. Also shown are the plane cross-sections of the vorticity norms in π_s, π_d and in the box side. A sketch of the vortex has been overlayed in order to orient the reader. (a) t=3.5; (b) t=4.75; (c) t=6.0. (d) axial flow due to vortexline twist; (e) explanation for higher h_r on vortex centerline.

Effects of Incompressible and Compressible Vortex Reconnection

ROBERT M. KERR[1], DAVINDER VIRK[1,2] & FAZLE HUSSAIN[2]

[1]Geophysical Turbulence Program, National Center for Atmospheric Research, PO Box 3000, Boulder CO 80307-3000, USA
[2]Department of Mechanical Engineering, University of Houston, Houston, TX 77004, USA

ABSTRACT

The reconnection of two anti-parallel viscous vortices is studied by both incompressible and compressible simulations in a periodic domain. Using symmetries along with increased resolution in the direction normal to the dividing plane allows Reynolds numbers (Re) based on circulation divided by viscosity range up to 3200 for the incompressible case. As Re is increased there is a significant increase in the peak vorticity that is consistent with a singularity of the three-dimensional, incompressible Euler equations in a finite time. However, a significant increase in the volume integral of enstrophy is not observed. After reconnection the energy spectrum obeys k^{-2}, consistent with vortex sheet formation in the reconnection region. In the compressible calculations weak shocks are formed. The vortex-shock interaction causes the vortices to be flattened. The peak vorticity in the symmetry planes is much lower for the high Mach number (M = 2) case than that in the low M and incompressible calculations.

1 INTRODUCTION

Despite evidence that flow structures are dynamically significant at both the large and small scales of turbulent motion, it is still not understood what role these structures play in turbulent cascade, transport and noise production. Some experimental examples of structures in shear flows are horseshoe vortices in boundary layers and ribs that form the three-dimensional structure of a mixing layer (For reviews of experimental and numerical studies of coherent structures in turbulent shear flows see Metcalfe et al. (1985) and Hussain (1986).) Attempts to understand these structures using idealized vortex elements are limited by our ability to represent the Euler and Navier-Stokes equations with these methods. Attempts to understand their structure with viscous calculations of homogeneous turbulence are limited by resolution. For example, visualization of vorticity in viscous calculations suggests that vortex filaments are the dominant structures (Rogers & Moin, 1987). But analysis of strain rates in simulations of isotropic, forced turbulence and the homogeneous shear calculation of Rogers & Moin (1987) suggest that the smallest resolved structures are sheets (Ashurst et al., 1987).

A related question is whether there is a singularity of the three-dimensional, incompressible Euler equations. A numerical study of vortex interactions by Kerr & Hussain (1989) (to be referred to as KH), where symmetries allow sufficient resolution to clearly identify vortex sheets, suggests that there is at least a singularity of the peak vorticity. KH reached their conclusions by using direct simulation of the Navier-Stokes equations with small viscosities to study vortex reconnection, or cut-and-connect. After reviewing those results and discussing their limitations, the role vortex reconnection might play in the turbulent cascade will be discussed and some new calculations exploring reconnection in compressible flows will be discussed.

2 INCOMPRESSIBLE RECONNECTION

All of the vortex reconnection calculations to be discussed are to some degree inspired by the vortex filament calculation of Siggia (1984). He emphasized two dynamic processes, first that vortex filaments tend to be anti-parallel at the point of closest approach and second that through self-induction this anti-parallel pair collapses in a finite time. Unfortunately Siggia's study fails as a model for the Euler equations because it does not address core deformation as the vortex filaments collapse and cannot determine the importance of viscosity, which is crucial to the reconnection process. The most extensive test of the validity of the late stages of Siggia's calculation are the three-dimensional viscous calculations of KH using spectral methods. In those calculations symmetries are assumed at planes between the vortices and along the vortices, to be called the dividing and symmetry planes respectively, and are used along with a fine mesh perpendicular to the dividing plane to provide enough resolution to simulate Reynolds numbers based on circulation divided by viscosity (σ/ν) up to 3200.

In agreement with the calculation of Pumir & Kerr (1987) KH find that the reconnection region is dominated by vortex sheets. Figure 1 shows contours of local vorticity in the symmetry plane of the two original vortex filaments of KH for Re$=$3200. Note how at a late time (figure 1c) the original vortex breaks into a head and what Melander & Hussain (1989) describe as threads. The second conclusion of KH is that while the vortex sheets inhibit singularity formation, figure 2 shows that the peak vorticity appears to increase rapidly (up to a factor of 8) as the viscosity approaches zero. They also find that the enstrophy increases slowly without any indication of singular behavior, suggesting that if there is a singularity in the peak vorticity that it occurs on a set of measure zero. At roughly the time of peak vorticity KH show that the drop in circulation in the plane shown in figure 1 becomes steeper as Re increases and the vorticity in the dividing plane between the two vortices increases significantly. Additional results not reported by KH are that the enstrophy production rate increases rapidly up to this time, then decreases rapidly, and that the peak rate-of-strain in the symmetry plane, defined by the enstrophy production divided by the enstrophy, increases at a slightly faster rate than the peak vorticity (that is greater than a factor of 8) and does not show any signs of saturation.

The suggestion by KH that there is only a singularity in the peak vorticity, if it is generalizable to all fluid turbulence, is not consistent with accepted dogma, which states that the volume integral of enstrophy has a singularity. This belief stems primarily from spectral closure analysis starting with Proudman & Reid (1954), but it is supported by the experimental observation that the velocity derivative skewness, which is related to the normalized enstrophy production, is independent of Re. In analogy with the two-dimensional magneto-hydrodynamic equations where a spectral closure (Pouquet, 1978) predicts a singularity, in a direct calculation (Frisch et al., 1983) current sheets inhibit the formation of a singularity. For the reconnection problem KH shows that vortex sheets inhibit singularity formation in a similar manner, but the analogy cannot necessarily be extended to turbulence in general unless the Re independence of the experimentally observed skewness can be accounted for. Another piece that must fit into this puzzle is the observation by Kerr (1987) that strain statistics support the appearance of sheet-like structures at small scales and that the dominant orientation of the strain is related to the skewness. Some aspects of this orientation are reported experimentally by Tsinober in these proceedings.

Energy spectra near the time of peak vorticity have been analyzed by fitting a $k^{-\gamma}\exp(-\delta k)$ form to the energy spectrum. Ideally, δ should be small enough that the simplier form $k^{-\gamma}$ could be fit by hand over some range, but in practice it was found that there was always a significant exponential component to the spectrum and that simply fitting $k^{-\gamma}$ overestimated γ. For the Re=3200 calculation of KH $k^{-\gamma}\exp(-\delta k)$ was fit between wavenumbers 10 and 160 in order to capture both inertial subrange effects and dissipation range effects. Figure 3 shows that for Re=3200 γ decreases up to the time of the peak vorticity, then appears to settle at a value of -2, which is consistent with vortex sheets. This result suggests that some of the structures observed in more viscous calculations of reconnection by Melander & Hussain (1989) are dominated by vortex sheets at higher Re. The length scale δ also decreases up to the time of the peak vorticity then is relatively constant. Because reconnection is a self-induction mechanism for producing small-scale structures, it had been hoped that it might provide a means for energy to be transferred to small scales and producing the experimental $k^{-5/3}$ spectrum without invoking classical cascade arguments. Clearly this goal was not reached since most of the energy stays in the large scales after the time of peak vorticity and the spectrum is not $k^{-5/3}$. Since a k^{-2} energy spectrum would indicate a logarithmic divergence in the integral enstrophy, determining how the energy spectrum reaches k^{-2} might shed some light on the singularity issues discussed above.

Before the time of peak vorticity vortex sheets can be identified only in the plane of symmetry, as in figure 1a. As shown in figure 7c after that time the flow can be divided into two regions. First a region of paired vortex sheets, that grow normal to the symmetry plane. We expect that a viscous timescale is required to fully diffuse these sheets. Second, there is a region of reconnected vortices crossing the dividing plane between the two original vortices. In the vortex sheet region there is a self-induced velocity in the direction of the original vortices that pulls the reconnected vortices outward. This flow is very similar to the predictions for X-type magnetic field line reconnection by Parker (1957), although whether the MHD equations have singular behavior has not been shown. A singularity of the inviscid MHD equations might explain nanoflares in the solar corona (Parker, private communication).

3 RELATION TO TURBULENT CASCADE

The inconsistency between the k^{-2} energy spectrum for reconnection and the observation of a $k^{-5/3}$ is similar to an inconsistency in two-dimensional turbulence that was recently resolved. It was suggested by Saffman (1971) that in analogy with the one-dimensional Burger's equation, where shocks yield a k^{-2} energy spectrum, that vorticity jumps in two dimensions should yield a k^{-4} energy spectrum. This was inconsistent with thermodynamic arguments by Batchelor (1969) and Kraichnan (1967) and atmospheric observations of large-scale quasi-two dimensional motion that suggested a k^{-3} energy spectrum. Recent calculations by Brachet et al. (1986) and analysis by Gilbert (1988) (and a mid-1970s calculation by Rogallo, private communication) find that initially there is a k^{-4} energy spectrum as vorticity jumps form from random initial conditions, then as these jumps are piled up and wrap around vortex cores a k^{-3} spectrum can form. Could an analogous process happen in three dimensions, converting vortex sheets with a k^{-2} spectrum into a $k^{-5/3}$ spectrum?

A process like this has been suggested in two theoretical papers by Parker (1969) and Lundgren (1982). With certain assumptions each paper shows how by piling up vortex sheets a $k^{-5/3}$ spectrum can be obtained. But several secondary predictions from these studies are not consistent. Parker makes a closure type of assumption and as a result predicts a singularity in the enstrophy. The closure assumption used assumes local wavenumber interactions, which is inconsistent with a calculation of low Re turbulence by Domaradzski (1988). Lundgren assumes a large-scale strain that converts a two-dimensional analysis similar to Gilbert (1988) into a three-dimensional spectrum and finds only an exponential increase in the enstrophy, which might be consistent with KH. The major drawbacks to Lundgren's analysis are that he does not provide a mechanism for producing the vortex sheets and his analysis assumes a single sheet wrapped around a core, whereas the calculation of Siggia (1984) shows that anti-parallel structures dominate.

What type of vortex structures appear in a calculation without the symmetries imposed by KH, but remaining more faithful to the Euler equations than the calculation of Siggia (1984)? The most relevant calculations of this type are by Melander & Zabusky (1986) and Chua & Leonard (1987), where vortices seem to pair, form sheets, then loop around the original filaments. Could this looping process produce the vortex piling we suspect is necessary to obtain the $k^{-5/3}$ spectrum? If so, how can we model this complex process? Could helicity constrain or enhance this process?

4 COMPRESSIBILITY

While reconnection has not answered all our questions about the cascade process, it still provides a unique mechanism for producing small-scale motion and the effects of this need to be investigated. Introducing compressibility effects provides more mechanisms to alter enstrophy and vorticity, e.g. non-barotropic production, viscous production. Also, shocks can form and interact with the vortices in the compressible simulations. To investigate the influence of these mechanisms on vortex reconnection we have simulated a case with the same symmetries as the incompressible case of KH. However, due to higher computational requirements, for the compressible case Re is lower and equal to 700. We used divergence-free initial velocity fields with the Mach number (M) of 0.5 and 2.0 and a Prandtl number of 0.67.

Figures 4a-d show the circulation and the peak vorticities in the symmetry planes and dividing planes for both the M=0.5 and M=2.0 calculations. The M=0.5 calculation shows behavior that is very similar to the incompressible calculations, even though density plots show the formation of weak shocks on both sides of the vortices. In figure 4a the sum of the circulations in the symmetry and dividing planes is nearly constant, with the curves crossing near the time of peak vorticity in the symmetry plane in figure 4b. The peak vorticity in the symmetry plane in figure 4b drops a small amount initially, then increases by a factor 1.4, which is consistent with an incompressible calculation at Re = 800. Vorticity plots show the same three regimes in time noted by Melander & Hussain: approach and deformation, bridging, and finally threading and dissipation. However, for M=2.0 the compressibility effects significantly change the scenario. While figure 4c shows that the sum of the circulations is still constant, the time at which the curves in figure 4c cross has changed and the behavior of the peak vorticity in the symmetry plane is significantly different. First there is a sharp drop in the peak vorticity, then a series of peaks due to the combined effects of periodicity, shocks, and non-barotropic vorticity production.

These effects can be partially explained by referring to contour plots. Figures 5a-e and 6a-e show the vorticity and density for a series of times from the M=2 calculation. In figure 5a at t = 1.8 the initial Gaussian vortex has begun to flatten in a manner reminiscent of the incompressible cases, but in figure 5b note that the vortex has not formed as close to the lower dividing plane as it did for the incompressible case in figure 1. This will be shown to be due to negative non-barotropic production of vorticity that is of the same magnitude as vorticity production due to vortex stretching. Only in figure 5e at t=36.5, by which time dissipation has made the flow relatively incompressible, are the vorticity profiles similiar to the incompressible case with a characteristic head and following threads.

The strong non-barotropic production is associated with shock formation. The initial density in these calculations was uniform, but by t = 1.8 in figure 6a the density in the vortex core is half that in the surrounding fluid. Figure 6a also shows two shocks forming in front of and behind the vortex core. In figure 6b these shocks have moved away from the vortex core, with the trailing vortex beginning to cross the periodic boundary on the left. By t = 14.8 the shocks have crossed one another and the trailing shock is interacting with the leading edge of the vortex core in the center of the figure. The numerous minor maxima of the peak vorticity in figure 4d are associated with positive non-barotropic production as the shocks cross the periodic boundaries and interact with the original vortices.

Despite the distinctly different behavior of the peak vorticity and circulation for the M=2.0 calculation the overall aspects of the enstrophy are very similar to the incompressible calculations. The plot of enstrophy at t=45.7 (figure 7b), slightly before the crossover of the circulation, shows bridges perpendicular to the original vortex tubes similiar to an incompressible calculation at Re = 1600 (figure 7c). Note that only one-fourth of the two filaments is shown because of the symmetries highlighted in figure 7a.

The plots of ω_z in the xy dividing plane, for the M = 2.0 and M = 0.5 calculations are shown in figures 8a and 8b at the times when the circulation in xz symmetry plane has decayed to 20% of initial value. Note that the bridges are more flattened in the M = 2 calculation. This might be due to the shock interaction discussed. A density plot for M = 2 at t = 76.2 is shown in figure 9.

A comparison of the enstrophy production terms shows that the stretching production $\omega_i e_{ij} \omega_j$ (b) is dominant in the M = 0.5 simulation at all times, thus the results are similar to the incompressible case. However, in the M = 2.0 case the non-barotropic term $(\omega.(\nabla P \times \nabla(1/\rho))$ (c), the divergence term $(\omega.\omega(\nabla.\vec{u}))$ (d) (even when $(\nabla.\vec{u})$ is zero initially), and the viscous production term $\mu\omega.(\nabla(1/\rho) \times (\nabla \times \omega))$ (e) are of the same order of magnitude. Three-dimensional iso-surface plots of these terms show that they are most significant in the bridges and near the symmetry plane. The contour plots of these terms in a plane passing through the bridges, for the M = 2.0 calculation, are shown in figure 10(a-e). In the enstrophy equation, the terms b and c are added while d and e are subtracted. As expected the viscous term causes dissipation in both the bridges and the threads. The non-barotropic and divergence terms cause reduction of vorticity at the symmetry plane and production away from it. This would explain the flattening of the vortices away from the symmetry plane observed in figure 5b.

A comparison of the spectra of the solenoidal and compressible parts of the velocity field in the M = 0.5 and M = 2.0 cases was also done. We find that the energy is higher in the solenoidal part for the M = 0.5 case at all times, consistent with the predominantly incompressible behavior of this simulation. In the M = 2.0 case, however, at later times the energy is higher in the solenoidal part for low wavenumbers, but at higher wavenumbers the compressible part of the velocity has higher energy. This implies that the M = 2.0 calculation does have significant compressibility. This behavior of the energy is consistent with the simulations of isotropic turbulence by Passot & Pouquet (1987). Figures 11a,b show the solenoidal and compressible parts of the velocity field for M = 2.0 at t = 42.7, which is the time of maximum peak vorticity, to demonstrate this. The peak in the compressible part near k = 10 was not seen by Passot & Pouquet (1987) and might be related to acoustic noise generation.

In summary, it is increasingly becoming apparent that understanding turbulent structures will be an important part of any theory of turbulence. KH has shown how vortex sheets form and could be related to singularity formation. After the time of peak vorticity they observe vortex bridging across a plane of symmetry. These are only some of the structures that have been observed in direct simulations of turbulence. But great care must be taken even in calculations where bridging similar to that described by KH and Melander & Hussain (1989) does appear. For example, in calculations by Kida & Takaoka (1987) and Melander & Zabusky (1986) another type of bridging forms parallel to the vortex sheets, but is not observed by KH and might be a viscous structure that is unrelated to the underlying inviscid phenomena. More comparisons with observed structures in incompressible flows, such as horseshoe vortices in boundary layers and ribs in mixing layers, need to be done to determine how important reconnection and vortex sheet formation are in observed turbulent flows. The compressible calculations demonstrate that some of the incompressible conclusions might not apply to the interaction of supersonic vortices. In particular, the evidence suggests that in the inviscid limit the peak vorticity is not singular. This could be because shock formation, which in the inviscid limit is singular, occurs first and introduces dissipative effects at an early time.

ACKNOWLEDGEMENTS

RMK acknowledges support of ARO MIPR No. 103-89 at the National Center for Atmospheric Research. DV acknowledges support of the Advanced Study Program at NCAR. NCAR is funded by the National Science Foundation. FH acknowledges support of DOE Grant No. DE-FG05-88ER13839.

REFERENCES

Ashurst, W. T., Kerstein, A. R., Kerr, R. M. & Gibson, C. H. 1987 *Phys. Fluids*, **30**, 2343.

Batchelor, G.K. 1969 *Phys. Fluids Suppl. II*, 233.

Brachet, M.E., Meneguzzi, M. & Sulem, P.-L. 1986 *Phys. Rev. Lett.* **57**, 683.

Chua, K. & Leonard, A. 1987 *Bull. Amer. Phys. Soc.* **32**, 2098.

Domaradzki, J. A., 1988 *Phys. Fluids*, **31**, 2747.

Frisch, U., Pouquet, A., Sulem, P.-L. & Meneguzzi, M. 1983 *J. Mécanique Théor. Appliquée* **2D**, 191.

Gilbert, A.D. 1988 *J. Fluid Mech.* **193**, 475.

Herring, J. R. & Kerr, R. M. 1982 *J. Fluid Mech.* **118**, 205.

Hussain, F. 1986 *J. Fluid Mech.* **173**, 303.

Kerr, R. M. 1987 *Phys. Rev. Lett.* **59**, 783.

Kerr, R. M. & Hussain, F. 1989 *Physica D.*

Kida, S. & Takaoka, M. 1987 *Phys. Fluids* **30**, 2911.

Kraichnan, R.H. 1967 *Phys. Fluids* **10**, 1417.

Lesieur, M. & Rogallo R. 1989 *Phys. Fluids A***1**, 718–722.

Lundgren, T.S. 1982 *Phys. Fluids* **25**, 2193.

Melander, M. & Hussain, F. 1989 *Phys. Fluids A***1**, 633.

Melander, M. & Zabusky, N. 1986 *IUTAM symposium on fundamental aspects of vortex motion*, Tokyo 116.

Metcalfe, R., Hussain, F. & Menon, S. 1985 *Turb. Shear Flows* **V** Cornell U., 4.13.

Parker, E. 1957 *J. Geophys. Res.* **62**, 509.

Parker, E. 1969 *Phys. Fluids* **12**, 1592.

Pouquet, A. 1978 *J. Fluid Mech.* **8**, 1.

Proudman, I. & Reid, W. H. 1954 *Phil. Trans. Roy. Soc. (London)* **247**, 163.
Pumir, A. & Kerr, R. M. 1987 *Phys. Fluids* **58**, 1636.
Pumir, A. & Siggia, E. D. 1987 *Phys. Fluids* **30**, 1606.
Saffman, P.G. 1971 *Stud. Appl. Maths* **50**,377.
Rogers, M. M., & P. Moin, 1987 *J. Fluid Mech.*, **176**, 33.
Siggia, E. D. 1984 *Phys. Fluids* **28**, 794.

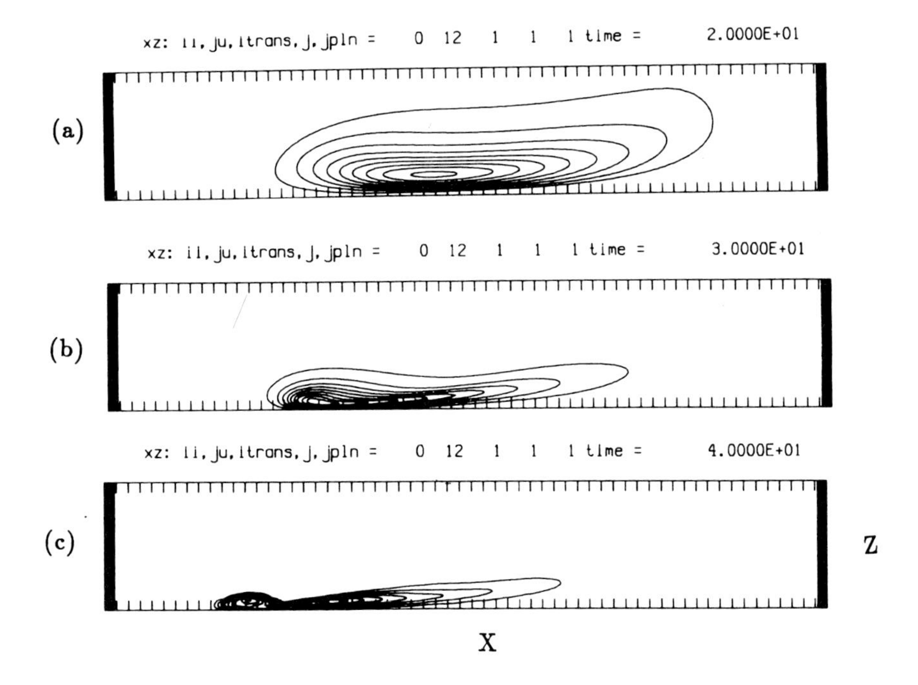

Figure 1. Contours of ω_y in xz symmetry plane at three times for Re = 3200, a: t = 20, b: t = 30, and c: t = 40.

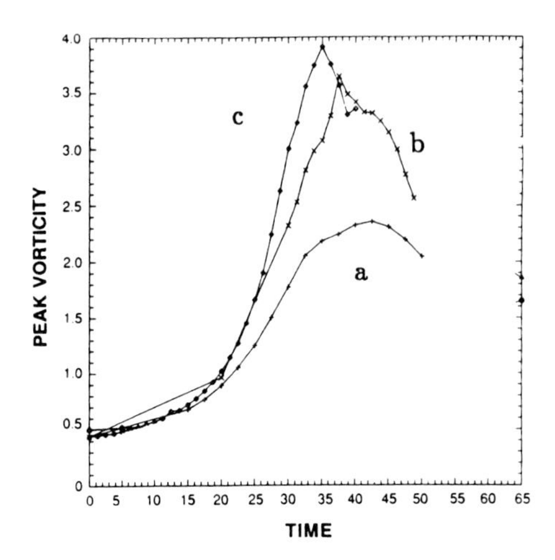

Figure 2. Peak vorticity as a function of time for three calculations. a: Re = 1600, b: Re = 2300, and c: Re = 3200.

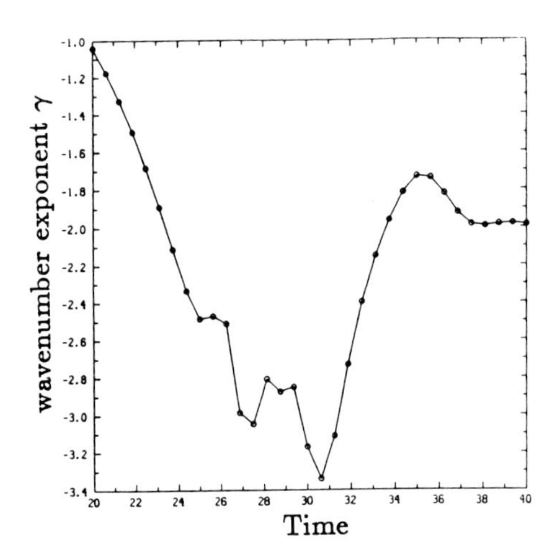

Figure 3. Wavenumber exponent γ as a function of time for Re = 3200.

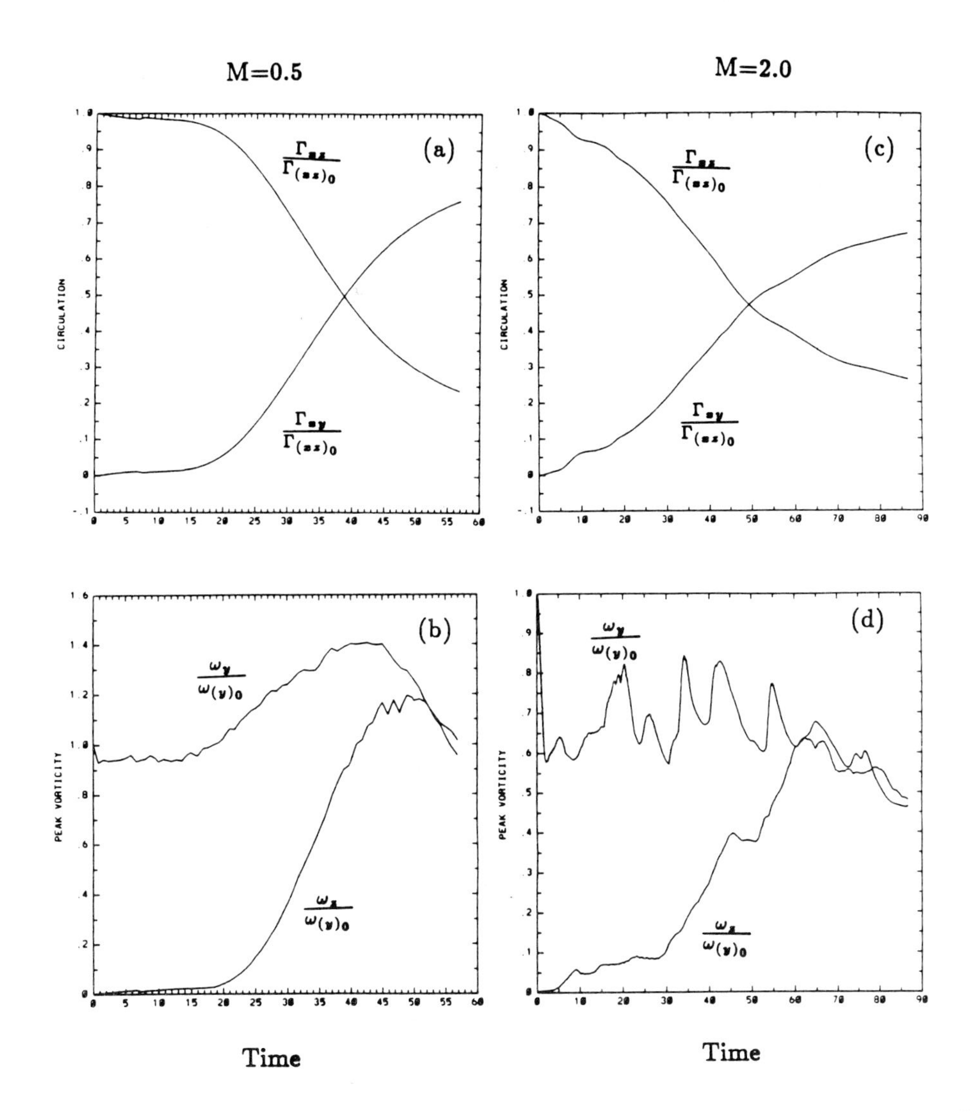

Figure 4. Circulation and peak vorticity in xz symmetry plane (Γ_{xz}, ω_y) and xy dividing plane (Γ_{xy}, ω_z), normalized by initial circulation ($\Gamma_{(xz)_0}$) and peak vorticity ($\omega_{(y)_0}$) in the xz symmetry plane.

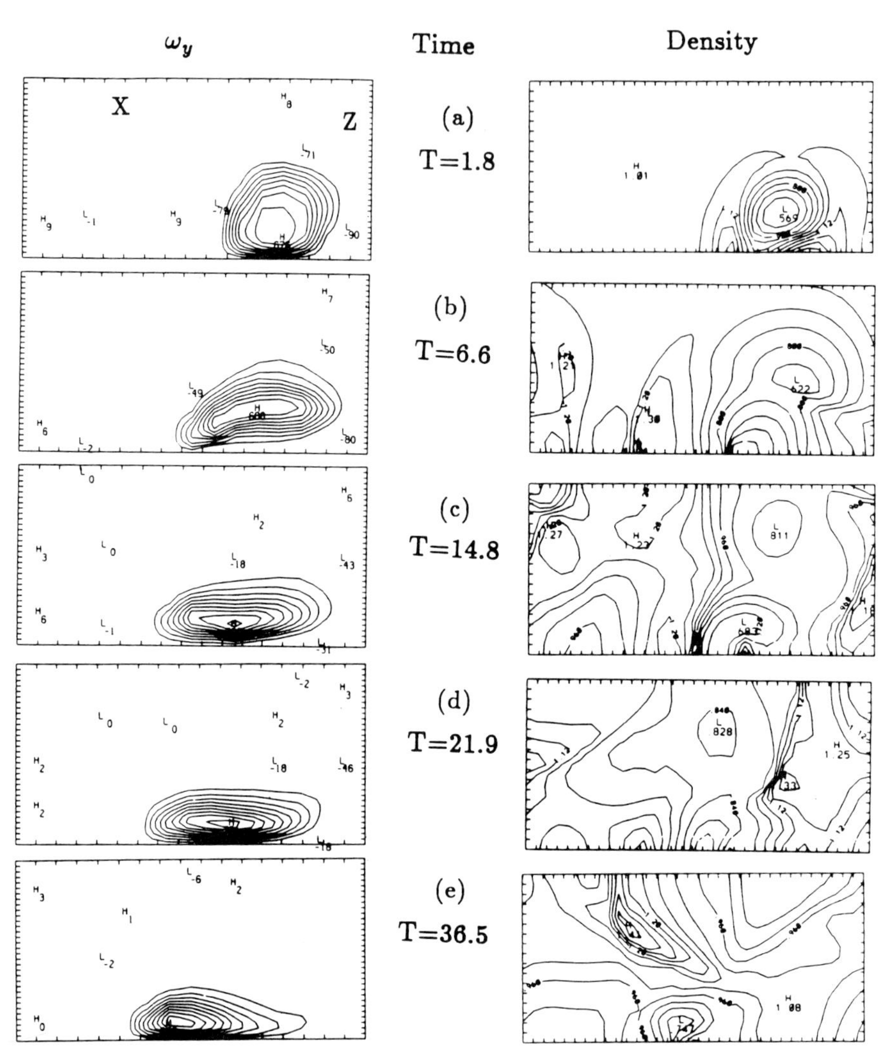

Figure 5. ω_y in xz symmetry plane at t = 8.6, for compressible case with M = 2.0.

Figure 6. ρ in xz symmetry plane at t = 8.6, for compressible case with M = 2.0.

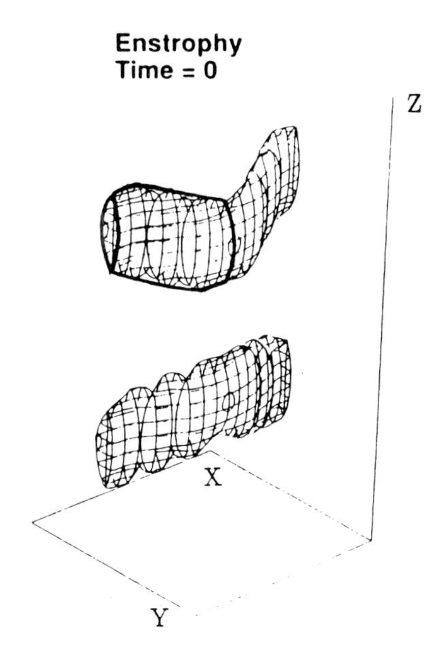

Figure 7a. Coordinate system and initial enstrophy field in a fully periodic domain. The quadrant that is actually calculated is highlighted.

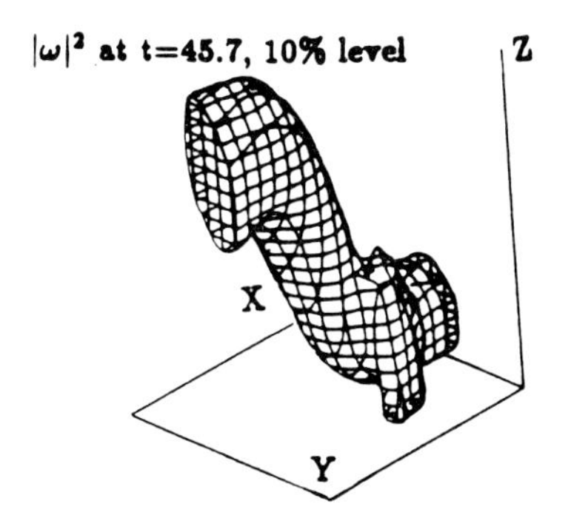

Figure 7b. Enstrophy at $t = 45.7$ for the compressible case with $\mathbf{M} = 2.0$.

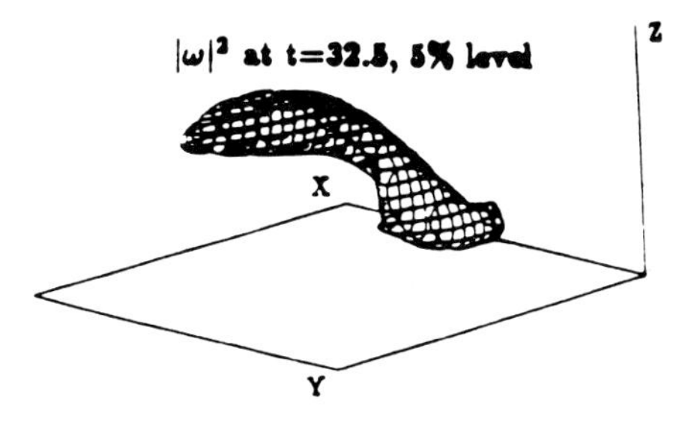

Figure 7c. Enstrophy at $t = 32.5$ (time of peak vorticity) from an incompressible calculation at $Re = 1600$.

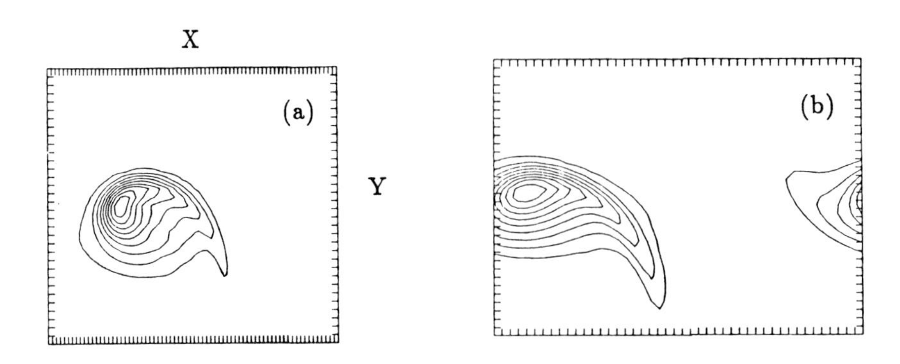

Figure 8. ω_z in xy dividing plane for compressible simulations. (a) M = 0.5, t = 56.8, right-hand domain is not shown. (b) M = 2.0, t = 76.2, central domain in x is not shown.

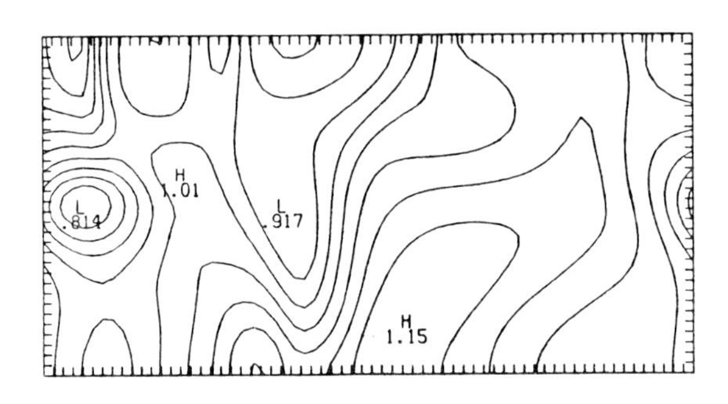

Figure 9. ρ in xy dividing plane for M = 2.0 compressible case at t = 76.2.

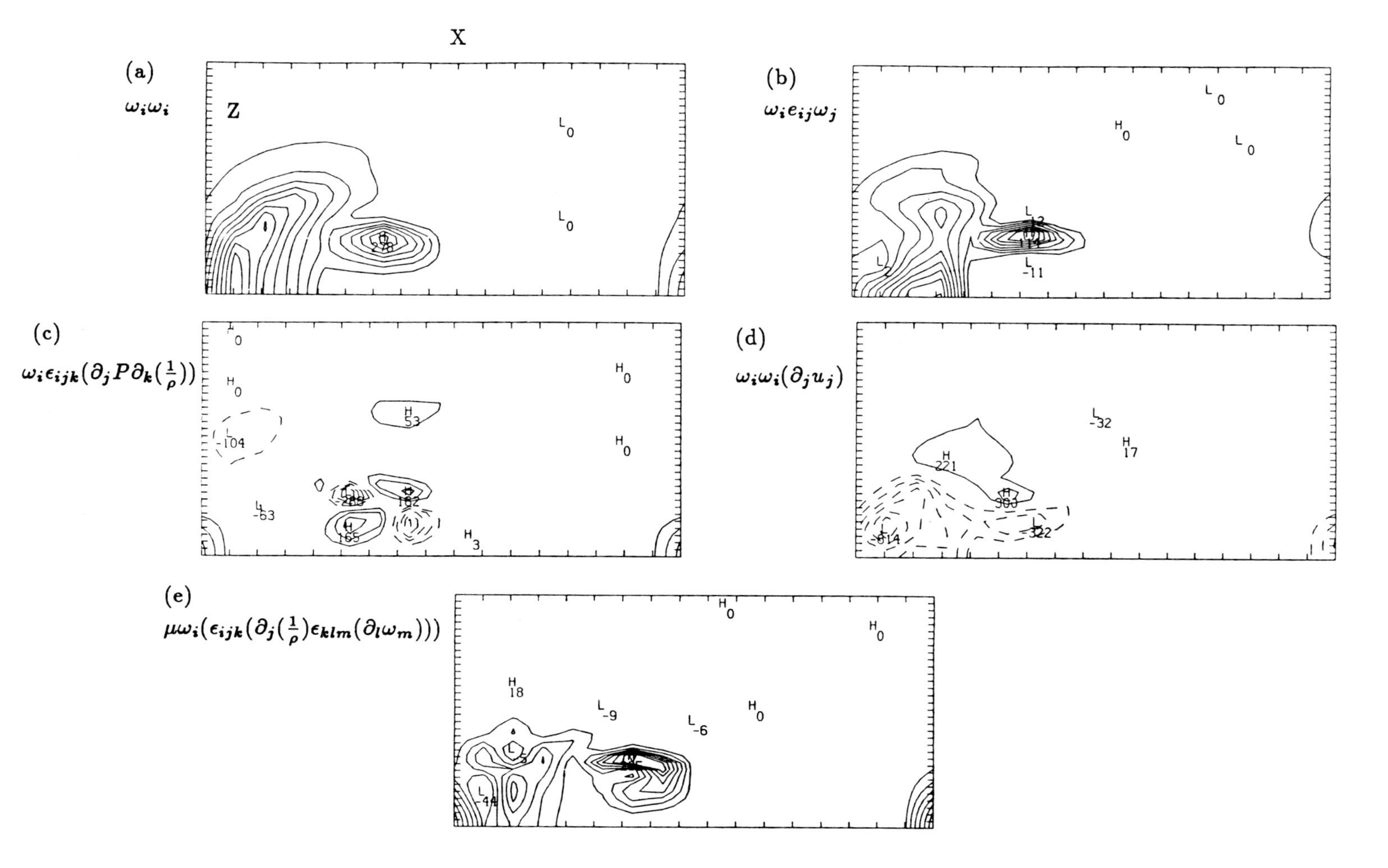

Figure 10. Contours plot in a xz plane passing through the bridges at t = 76.2, for compressible case with M = 2.0.

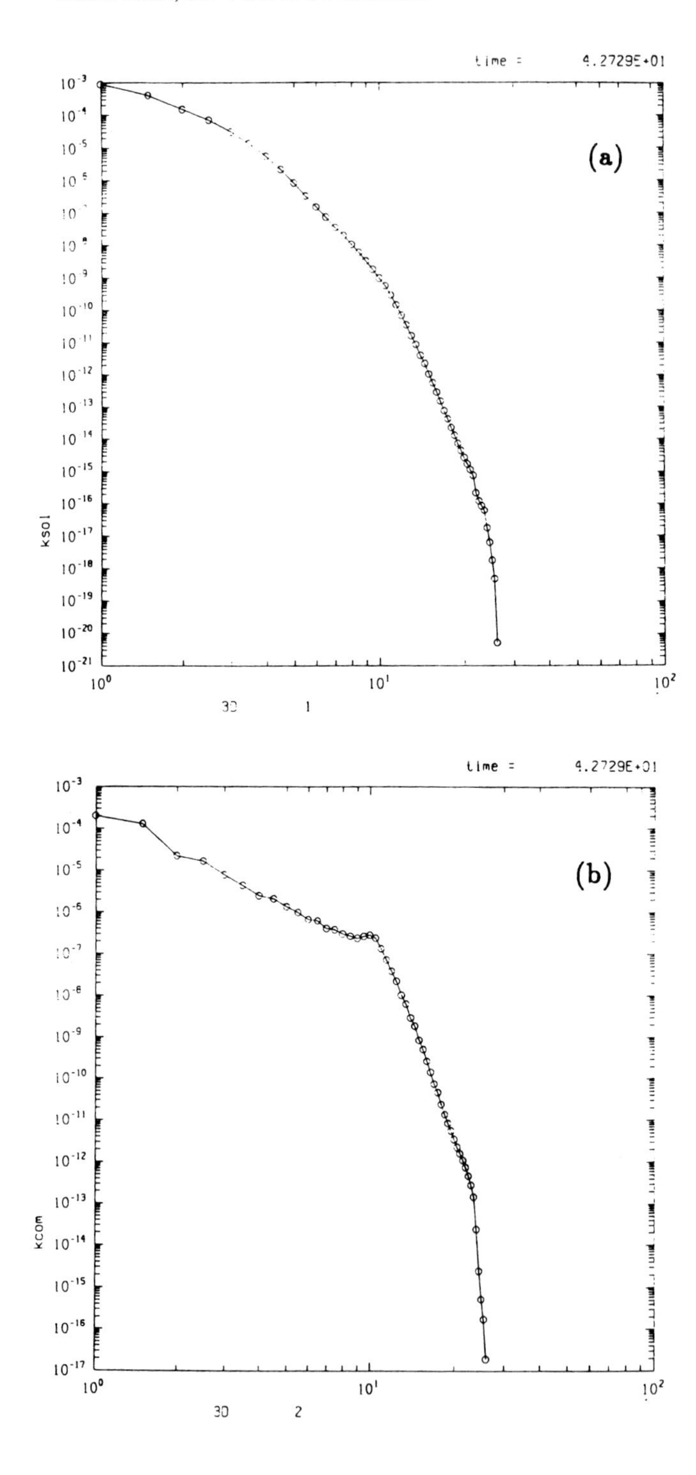

Figure 11. Three-dimensional kinetic-energy spectra for M = 2.0 at t = 42.7. (a): solenoidal part; (b): compressible part.

Oblique Collision of Two Vortex Rings and its Acoustic Emission

T. KAMBE[1], T. MINOTA[2], T. MURAKAMI[2] & M. TAKAOKA[3]

[1]Dept. of Physics, Univ. of Tokyo, Tokyo 113, Japan
[2]Dept. of Applied Science, Kyushu Univ., Fukuoka 812, Japan
[3]Dept. of Physics, Kykto Univ., Kyoto 606, Japan

1 INTRODUCTION

Collision of two vortex rings results in reconnection of vortex lines due to the viscosity, which is immediately followed by a violent motion of vortex lines. This event excites an acoustic wave. The process of sound generation can be formulated in the framework of the theory of vortex sound, giving explicit representation of wave profiles in terms of the time dependent vorticity field $\boldsymbol{\omega}(\boldsymbol{x},t)$. Experimental detection of the vortex sound with using vortex rings, helped by theoretical analysis, has been made in the past decade, and the results are reported in Kambe & Minota [1] and Kambe [2] for three typical problems: (i) acoustic emission by head-on collision of two vortex rings, and emission by a vortex ring passing by (ii) a circular cylinder and (iii) a sharp edge (a half plane). Studies are also made for a sphere [3] and for a finite plate [4]. The vortex sound theory is also reviewed by Müller & Obermeier [5].

The geometry of the vortex motion studied in this paper is as follows. Initial state is given in such a way that two vortex rings are set to move along the paths intersecting at right angles and collide with each other in due course.

2 MATHEMATICAL FORMULATION

We present here a general formulation, in which multipole expansion is carried out to any order n. Suppose that there exists an unsteady fluid motion characterized with a localized vorticity $\omega(\mathbf{x},t)$, where the term *localized* means that ω becomes exponentially small as the distance $|\mathbf{x}| \to \infty$. Pressure fluctuations will be generated at large distances, which drive acoustic waves. If characteristic scales of the velocity and length are denoted by u and l, a time scale of the vortex motion will be $\tau = l/u$ and a length of the wave generated will be $\lambda = c\tau = l/M$, where c is the sound speed and $M = u/c$ the Mach number. It is assumed that the

Mach number M is much less than unity. This leads to formation of two regions scaled on l and λ respectively since $\lambda \gg l$. The vortex motion scaled on l is called an *inner* region, while the wave field scaled on λ is called an *outer* region (Kambe 1986).

2.1 Governing equations

The basic equation of the aerodynamic sound generation is given by the inhomogeneous wave equation for the pressure p:

$$\frac{1}{c^2}\frac{\partial^2 p}{\partial t^2} - \nabla^2 p = \frac{\partial^2}{\partial x_i \partial x_j}(\rho v_i v_j - \tau_{ij}) \tag{1}$$

[6], where τ_{ij} is the viscous stress tensor and ρ the density.

In the inner region, an appropriate scaling lead to the following equation,

$$-\nabla^2 p = \rho_0 \partial_i \partial_j (v_i v_j) , \tag{2}$$

neglecting $O(M^2)$ terms, where $\partial_i = \partial / \partial x_i$ and ρ_0 is the undisturbed density. This is equivalent to taking as $c = \infty$. Therefore this equation is nothing but the one obtained by taking the divergence of the Navier-Stokes equation under the solenoidal condition:

$$\text{div } \mathbf{v} = 0 , \tag{3}$$

for which $\partial_i \partial_j \tau_{ij} = 0$.

On the other hand, in the outer wave region where the fluid motion decays, the equation (1) is linearized and reduced to the equation of sound wave:

$$\frac{1}{c^2}\frac{\partial^2 p}{\partial t^2} - \nabla^2 p = 0 , \tag{4}$$

which is valid in a region where the outer variable $\hat{\mathbf{x}} \equiv \mathbf{x}/\lambda$ is of order unity and viscous damping is neglected. First, the inner equation is solved to represent a source flow of a viscous vortex motion. Then an outer solution is sought so as to match the inner solution just obtained. This solution represents an acoustic wave emitted by the vortex motion.

2.2 Asymptotic formula of the vortex sound

Suppose that the vorticity field $\omega(\mathbf{x}, t)$ and its associated velocity $\mathbf{v}(\mathbf{x}, t)$ satisfy the equations (2) and (3) for a viscous incompressible fluid. A vector potential $\mathbf{A}$ of the velocity is introduced by $\mathbf{v} = \nabla \times \mathbf{A}$. Since $\omega = \nabla \times (\nabla \times \mathbf{A}) = -\nabla^2 \mathbf{A}$ with a subsidiary condition $\nabla \cdot \mathbf{A} = 0$, the vector potential is given by

$$\mathbf{A}(\mathbf{x}, t) = \frac{1}{4\pi} \int \frac{\omega(\mathbf{y}, t)}{|\mathbf{x} - \mathbf{y}|} d^3\mathbf{y} \tag{5}$$

This can be expanded as follows:

$$A_k(\mathbf{x},t) = \frac{1}{4\pi}\sum_{n=1}^{\infty}\frac{(-1)^n}{n!}\sum_{p_1}\cdots\sum_{p_n} W^k_{p_1\cdots p_n}\partial_{p_1}\cdots\partial_{p_n}\frac{1}{r}$$

where $r = |\,\mathbf{x}\,|$ and

$$W^k_{p_1\cdots p_n}(t) = \int \omega_k(\mathbf{y},t)\; y_{p_1}\cdots y_{p_n} d^3\mathbf{y}.$$

The velocity is expressed by $v_i = \varepsilon_{ijk}\partial_j A_k$, where ε_{ijk} is the third-order skew symmeric tensor.

After some substantial manipulation, an asymptotic expansion of the velocity is given in the form:

$$v_i(\mathbf{x},t) = \partial_i\Phi = \partial_i\Phi_1 + \partial_i\Phi_2 + \partial_i\Phi_3 + \cdots$$

where

$$\begin{aligned}\Phi &= \sum_{n=1}^{\infty}\Phi_n \\ &= \sum_{n=1}^{\infty}\left[\sum_{p_1}\cdots\sum_{p_n} Q_{p_1\cdots p_n}(t)\;\partial_{p_1}\cdots\partial_{p_n}\frac{1}{r}\right] \qquad (6)\\ Q_{p_1\cdots p_n}(t) &= \frac{(-1)^{n+1}n}{4\pi(n+1)!}\int(\mathbf{y}\times\omega)_{p_1}y_{p_2}\cdots y_{p_n}d^3\mathbf{y}\;. \qquad (7)\end{aligned}$$

[7]. The first three terms are

$$\Phi_1 = \sum_{i=1}^{3} Q_i(t)\frac{\partial}{\partial x_i}\frac{1}{r}\;, \qquad Q_i(t) = \frac{1}{8\pi}\int(\mathbf{y}\times\omega)_i d^3\mathbf{y}\;, \qquad (8)$$

$$\Phi_2 = \sum_i\sum_j Q_{ij}(t)\frac{\partial^2}{\partial x_i\partial x_j}\frac{1}{r}\;, \qquad Q_{ij}(t) = -\frac{1}{12\pi}\int(\mathbf{y}\times\omega)_i y_j d^3\mathbf{y}\;, \qquad (9)$$

$$\Phi_3 = \sum_i\sum_j\sum_k Q_{ijk}(t)\frac{\partial^3}{\partial x_i\partial x_j\partial x_k}\frac{1}{r}\;, \quad Q_{ijk}(t) = \frac{1}{32\pi}\int(\mathbf{y}\times\omega)_i y_j y_k d^3\mathbf{y}\;, \qquad (10)$$

the first two being already given in Kambe & Minota [8]. At large distances from the region of vortex motion, the equation of motion will be linearized with respect to the velocity, and the asymptotic form of the inner pressure is given by

$$\begin{aligned}p_{\mathrm{I}} &= -\rho_0\frac{\partial}{\partial t}\Phi \\ &= -\rho_0\dot{Q}_i\,\partial_i\frac{1}{r} - \rho_0\dot{Q}_{ij}\,\partial_i\partial_j\frac{1}{r} - \rho_0\dot{Q}_{ijk}\,\partial_i\partial_j\partial_k\frac{1}{r} + O(r^{-5}) \qquad (11)\end{aligned}$$

where a dot denotes a time differentiation and the summation convention is used.

A general solution to the wave equation (4) is represented in the form of multipole expansion which should be valid in the outer wave region for $|\hat{\mathbf{x}}| \geq O(1)$. The wave solution matching to the inner solution (11) as $|\hat{\mathbf{x}}| \to 0$ is given by

$$p_W(\mathbf{x},t) = -\rho_0\,\partial_i\left[\frac{\dot{Q}_i(t_r)}{r}\right] - \rho_0\,\partial_i\partial_j\left[\frac{\dot{Q}_{ij}(t_r)+Q_0(t_r)\delta_{ij}}{r}\right] - \rho_0\,\partial_i\partial_j\partial_k\left[\frac{\dot{Q}_{ijk}(t_r)}{r}\right] + \cdots \tag{12}$$

where $t_r = t - r/c$ is the retarded time. Applying the space derivatives to the factor $1/r$ and retaining only those terms, we obtain the expression (11) since $Q_0\delta_{ij}\partial_i\partial_j(1/r) = Q_0\nabla^2(1/r) = 0$ $(r \neq 0)$ and the other terms are higher order in an intermediate region as as $|\mathbf{x}|/\lambda \to 0$ but $O(|\mathbf{x}|/l)$ being sufficiently large (existence of such a region is assumed). The origin of the term $Q_0(t)\delta_{ij}$ in the second parenthesis is considered in [9, 1]. From the dynamical equation of motion (Navier-Stokes equation), they found

$$Q_0(t) = -\frac{5-3\gamma}{12\pi}\,K(t)\,, \quad K(t) = \frac{1}{2}\int v^2(\mathbf{y},t)\,d^3\mathbf{y} \tag{13}$$

where K is the total kinetic energy and γ the ratio of the specific heats ($\gamma = 7/5$ for the air).

The expression (12) is the pressure of the acoustic wave generated by the vortex motion described by $\omega(\mathbf{x},t)$. The first term represents a *dipole* component which vanishes in the present case without an external body and free from an external force, since the dipole coefficient Q_i is related to the force acting on the system (Kambe 1986). The second Q_{ij} terms are usually called *quadrupole*.

In the far-field where observations are often made, the expansion becomes simpler. As $|\hat{\mathbf{x}}|$ becomes large, the space derivatives applied to r^{-1} become higher order of smallness there. Neglecting those terms and retaining only the terms of $O(r^{-1})$, we obtain the far-field expression:

$$p_F = -\frac{\rho_0}{c^2}\,Q_0^{(2)}(t_r)\frac{1}{r} - \frac{\rho_0}{c^2}\,Q_{ij}^{(3)}(t_r)\frac{x_i x_j}{r^3} + \frac{\rho_0}{c^3}\,Q_{ijk}^{(4)}(t_r)\frac{x_i x_j x_k}{r^4} + \cdots \tag{14}$$

where superscript (n) denotes the n-th time derivative. The first term is isotropic, whence it may be termed a *monopole* which is related to the rate of energy dissipation $\varepsilon = -\dot{K}$, and written as (for the air),

$$-\frac{\rho_0}{c^2 r}Q_0^{(2)} = \frac{\ddot{K}}{15\pi c^2 r} = -\frac{\dot{\varepsilon}}{15\pi c^2 r}\;. \tag{15}$$

Here we have retained up to the Q_{ijk} terms, the third order moments of the vorticity ω (see (10)). This is because in the experimental observation these components have been detected, as described below.

We consider the far-field pressure p_F in the frame of spherical coordinates (r, θ, ϕ):

$$x_1 = r \sin\theta \cos\phi \ , \quad x_2 = r \sin\theta \sin\phi \ , \quad x_3 = r \cos\theta \ .$$

Then the n-th order form, like the expressions $C_{ij}\, x_i x_j / r^2$ or $C_{ijk}\, x_i x_j x_k / r^3$, can be represented in terms of the n-th order spherical harmonics:

$$P_n^0(\zeta) \ , \quad P_n^1(\zeta)(\cos\phi, \ \sin\phi) \ , \cdots, \ P_n^n(\zeta)(\cos n\phi, \ \sin n\phi) \tag{16}$$

where $\zeta = \cos\theta$, and $P_n^0(\zeta)$ and $P_n^k(\zeta)$, $(k = 1, \cdots, n)$, are the Legendre polynomial and its associated polynomials, respectively.

3 EXPERIMENT

A vortex ring is made by using a shock impulse emerging from a circular nozzle which is connected to a shock tube (see [1] for the details). Two straight nozzles are set at right angles. Two vortices, formed simultaneously at each nozzle exit, approach and make almost 90° collision. This collision process was observed by means of a photo-sensor and also visualized by Schlieren photographs with seeding CO_2 gas to the vortex rings. The acoustic waves generated by the vortex collision were detected by four microphones placed in the far-field. The detected signals are processed by digital methods. The test procedure and extraction method of the wave profiles imbedded in the original noisy signals are described in [10]. The extracted signals of the wave profiles are averaged over ten data sets.

In an initial stage, two vortex rings of diameter $D = 9.4$mm start to move along two paths intersecting at right angles with the velocity $U = 27$m/s (Reynolds number $UD/\nu = 1.7 \times 10^4$). The bisecting straight line between the two paths is taken as the polar axis (also x_3 axis) of the spherical polar coordinates and the symmetry plane bisecting the paths are taken as the plane $\phi = 0$ and π (also (x_1, x_3) plane), the origin being chosen at the point of intersection (figure 1). The velocity field $\mathbf{v}$ has a mirror symmetry (right-left symmetry) with respect to this plane, and also the plane $\phi = \pi/2$ and $3\pi/2$ ((x_2, x_3) plane) is a mirror plane with respect to $\mathbf{v}$ (top-bottom symmetry). The initial directions of the vortex ring centres are given by the angles $(\theta, \phi) = (\pi/4, \pi/2)$ and $(\pi/4, 3\pi/2)$.The observed loci of the vortex cores in the plane (x_2, x_3) are shown in figure 2.

In this geometry, the acoustic pressure $p(\theta, \phi, t)$ observed on a sphere S in the far-field is characterized by the symmetries:

$$p(\theta, -\phi, t) = p(\theta, \phi, t) \ , \tag{17}$$

$$p(\theta, \pi - \phi, t) = p(\theta, \phi, t) \ . \tag{18}$$

The far-field pressure (14) on the sphere S can be represented by the spherical harmonics (16). The acoustic pressure in the present case contains only those terms satisfying the symmetries (17) and (18), and is expressed as

$$\begin{aligned} p(\theta,\phi,t) &= A_0(t) + A_1(t)\,P_2^0(\cos\theta) + A_2(t)\,P_2^2(\cos\theta)\,\cos 2\phi \\ &\quad + B_1(t)\,P_3^0(\cos\theta) + B_2(t)\,P_3^2(\cos\theta)\,\cos 2\phi \end{aligned} \tag{19}$$

up to the 3rd-order, where higher order terms are omitted since the experimental signals were not significant enough for those terms.

The wave pressures were detected on the large sphere S of radius $r = 620$mm at every 10° positions lying on the three orthogonal great circles, obtained as intersections with the three planes: (i) $\theta = \pi/2$, (ii) $\phi = 0, \pi$, and (iii) $\phi = \pi/2, 3\pi/2$ (about hundrred positions altogether). Figure 3 shows the average profiles (bold curves) of the acoustic pressure at six angular positions in the plane $\phi = \pi/2$ (the light curves denote r.m.s. error of the average curves at each time point). On each great circle the detected signals are represented by Fourier series with respect to the angle at each instant. From these data, the main mode amplitudes $\bar{\mathbf{A}} = [A_0, A_1, A_2, B_1, B_2]$ of (19) have been determined as a function of time t and are shown in figure 4. Actually they can be determined from the data on two orthogonal circles. Therefore we obtain two sets of the values for $\bar{\mathbf{A}}$, and fair agreement between them has been found. This confirms the credibility of the signal detection. It is interesting to find a substantial component of the 3rd-order mode $P_3^0(\cos\theta)$. The 3rd-order modes are not usually discussed in the literatures.

4 NUMERICAL SIMULATION

Numerical simulation of vortex collision were carried out by Kida, Takaoka & Hussain [11] for the parallel initial start and also for several other angles. Collision of two vortex rings at 90° initial start (corresponding to the experiment) has been simulated in detail by Takaoka, with the same computational methods in order to estimate the acoustic emission (figure 5).

The wave profiles are calculated from the field variables of the vortex motion. The isotropic component is proportional to the time derivative of the energy dissipation rate ε (see (15)), and the quadrupole components and higher modes are related to the change of moments of vorticity distribution. Using the data from the simulation, we can obtain the tensor coefficients $Q_{ij}(t)$ and $Q_{ijk}(t)$ of (9) and (10). These give rise to an acoustic wave, and the main-mode coefficients $[A_0, \cdots, B_2]$ of the acoustic pressure (19) are readily computed.

The mode amplitudes $[A_0, \cdots, B_2]$ obtained from the numerical simulation are shown in figure 6 for comparison with the experimental amplitudes of figure 4. Figure 7 compares the rate of energy dissipation $\varepsilon(t)$ for the present case with twice that of a single isolated vortex ring $\varepsilon_0(t)$. The difference $\triangle\varepsilon = \varepsilon(t) - 2\varepsilon_0(t)$ is considered to be the rate enhanced by the vortex collision.

5 DISCUSSION AND CONCLUSION

Comparison of the mode amplitudes has been made between the experiment and the numerical simulation (figures 4 and 6). Concerning the general appearance, agreement is satisfactory, although the details look differently.

Figure 8 shows the directivity plot of the acoustic pressure with respect to the angle ϕ in the plane $\theta = \pi/2$. The upper diagram (a) is obtained from the experimental mode amplidutes A_0, A_1, A_2 (monopole and quadrupole components of figure 4), while the lower one (b) is obtained from those of figure 6. These illustrate time variations of the wave pattern, which reflects the change of the vortex motion, the wave source.

It is found that the quadrupole component changes its nature at the instant of reconnection of vortex-lines when the rate of energy dissipation is enhanced, and that the computation can reproduce the main components (especially the quadrupoles) observed in the experiment, to a reasonable degree.

References

[1] Kambe, T. & Minota, T. 1983 *Proc. R. Soc. Lond.* A **386**, 277–308.
[2] Kambe, T. 1986 *J. Fluid Mech.* **173**, 643–666.
[3] Minota, T., Kambe, T. & Murakami, T. 1988 *Fluid Dyn. Res.* **3**, 357–362.
[4] Minota, T., Murakami, T. & Kambe, T. 1988 *Fluid Dyn. Res.* **4**, 57–71.
[5] Müller, E.-A. & Obermeier, F. 1988 *Fluid Dyn. Res.* **3**, 43–51.
[6] Lighthill, M.J. 1952 *Proc. R. Soc. Lond.* A **211**, 564–587.
[7] Hasimoto, H. 1985 "Dynamics of Vortex Motion". (in Japanese).
[8] Kambe, T. & Minota, T. 1981 *J. Sound Vib.* **74**, 61–72.
[9] Kambe, T. 1984 *J. Sound Vib.* **95**, 351–360.
[10] Kambe, T., Minota, T. & Ikushima, Y. 1985 *J. Fluid Mech.* **155**, 77–103.
[11] Kida, S., Takaoka, M. & Hussain, F. (the article in this Proceedings).

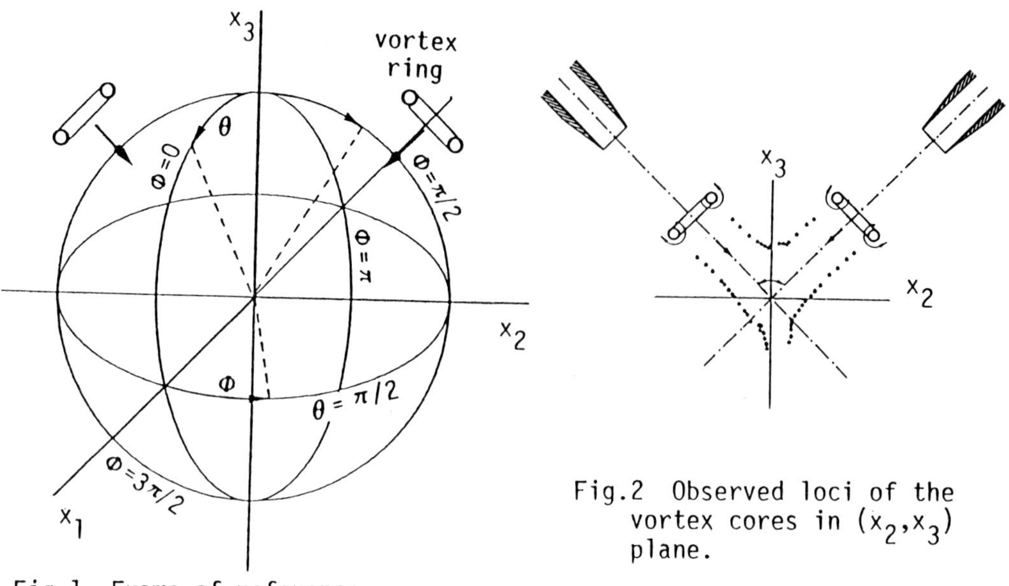

Fig.1 Frame of reference

Fig.2 Observed loci of the vortex cores in (x_2,x_3) plane.

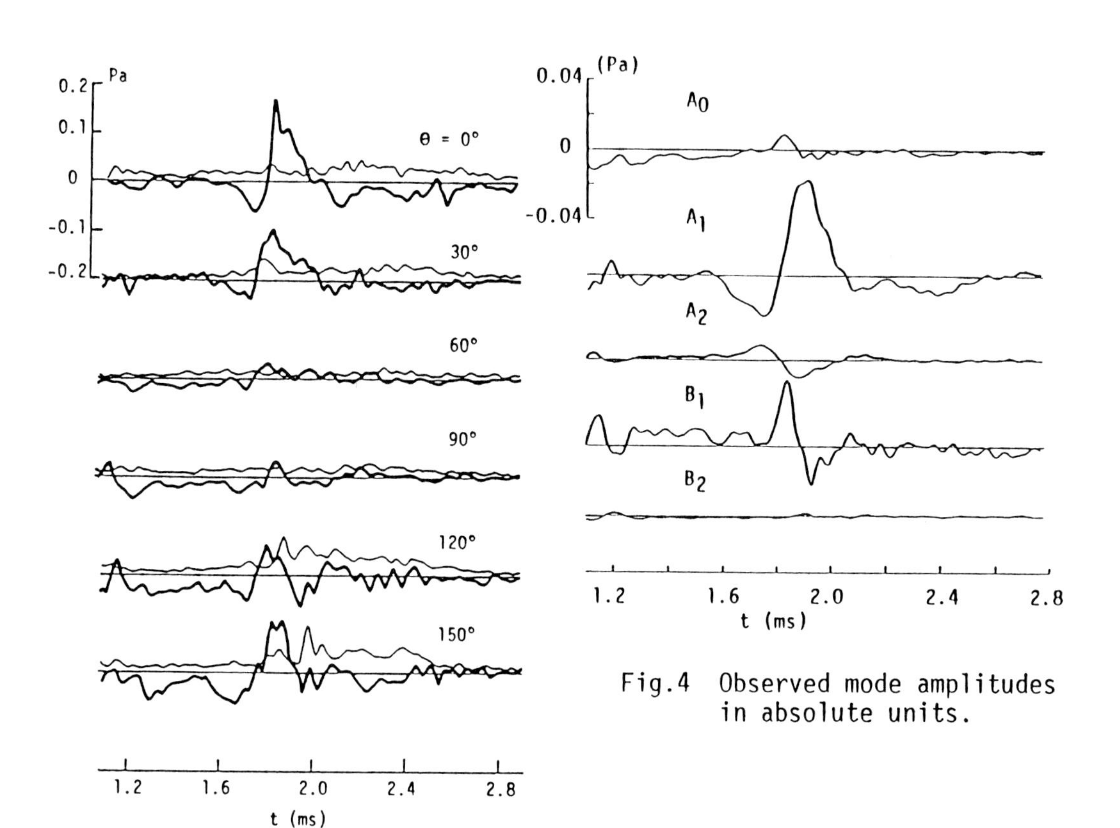

Fig.3 Average profiles (bold lines) of the acoustic pressure measured at six positions in the plane (x_2,x_3).

Fig.4 Observed mode amplitudes in absolute units.

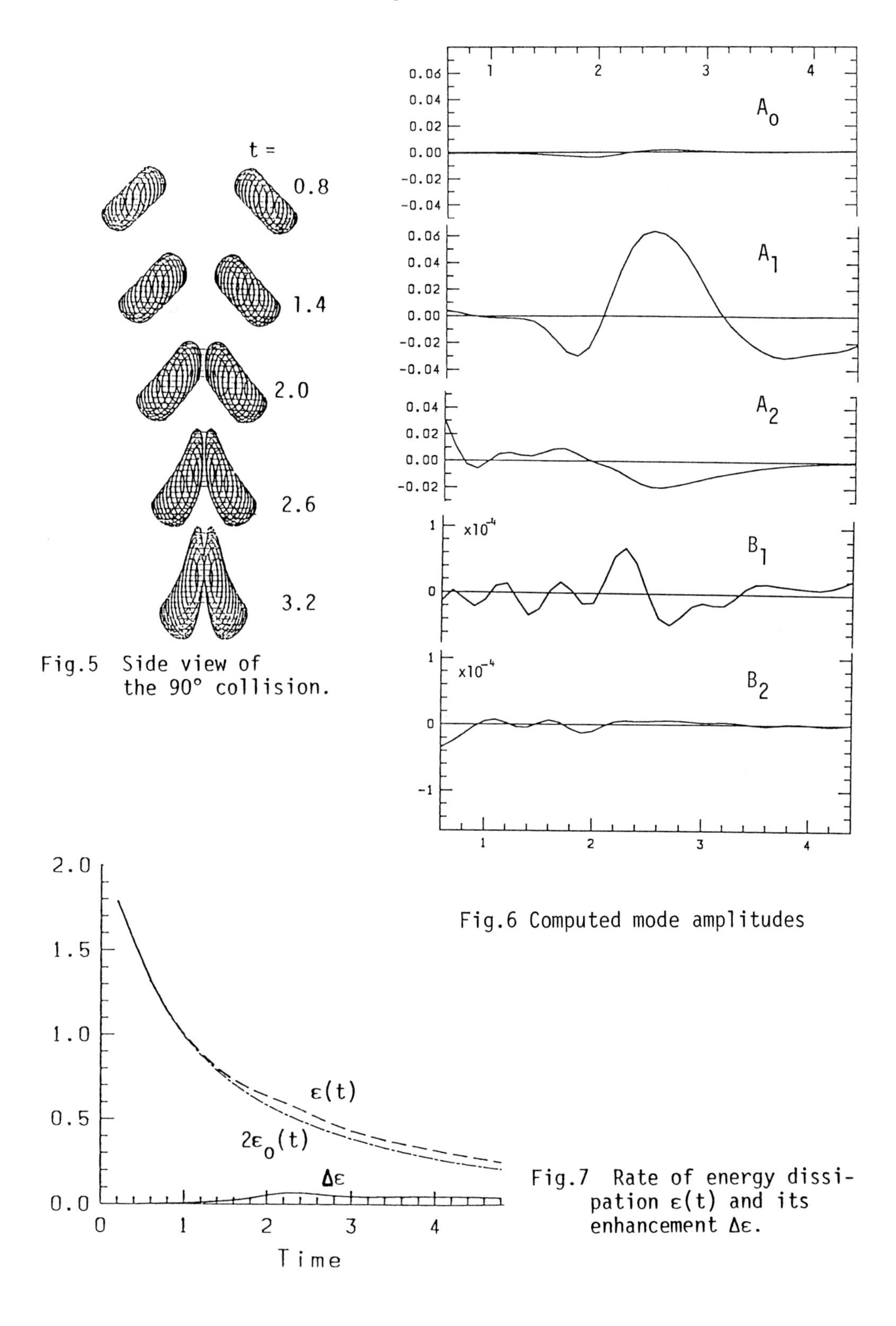

Fig.5 Side view of the 90° collision.

Fig.6 Computed mode amplitudes

Fig.7 Rate of energy dissipation $\varepsilon(t)$ and its enhancement $\Delta\varepsilon$.

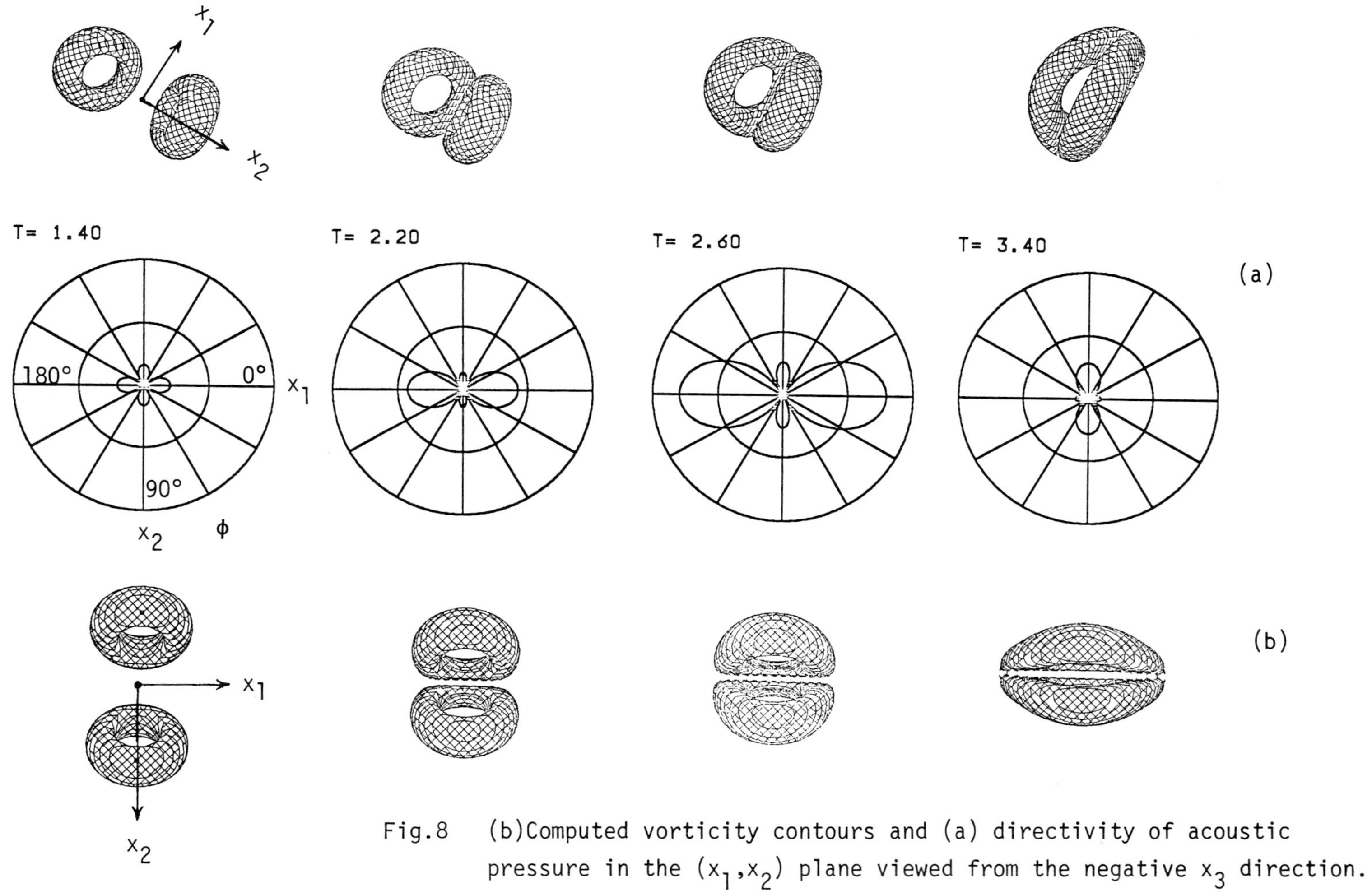

Fig.8 (b)Computed vorticity contours and (a) directivity of acoustic pressure in the (x_1, x_2) plane viewed from the negative x_3 direction.

Reconnection of Two Vortex Rings

S. KIDA[1], M. TAKAOKA[2], & F. HUSSAIN[3]

[1]Applied and Computational Mathematics, Princeton University, Princeton, NJ 08544, U.S.A.
[2]Department of Physics, Faculty of Science, Kyoto University, Kyoto 606, Japan
[3]Department of Mechanical Engineering, University of Houston, Houston, TX 77004, USA

1 INTRODUCTION

In this paper we study numerically the interaction of two circular vortex rings which evolve from a side-by-side configuration. The primary purpose of this study is to (1) realize on a computer *two* successive vortex reconnection processes which were visually observed in laboratory (Oshima and Asaka 1977; Schatzle 1987) but have not been realized by numerical simulation yet, (2) explain the detailed mechanism of the vortex reconnection process by examining 3D vorticity fields not available experimentally, (3) examine the dependence of the reconnection process on the initial condition as well as on viscosity, (4) compare the spatial structure of enstrophy, energy-dissipation, enstrophy-production and helicity density fields and explore a possible connection of these quantities with coherent structures, (5) discuss similarity and difference between vorticity and passive scalar fields, and (6) explain timescale of this process and its role in enstrophy cascade and noise generation.

We present here, however, only items (1), (2) and (4) briefly. A detailed discussion of them and other related matters will be made elsewhere (Kida *et al.* 1989a, b).

2 NUMERICAL METHOD

We trace the motion of two vortex rings by solving the incompressible vorticity equation by the spectral method on $N^3(= 64^3)$ grid points (Orszag 1971). The velocity field is assumed to be periodic with period 2π in three orthogonal directions. For convenience of description, we take the Cartesian coordinate system (x_1, x_2, x_3)

in such a way that the origin is located at the center of the box, the x_3 axis is vertical and the x_1 and the x_2 axes are on diagonal planes. The Runge-Kutta-Gill scheme with time increment Δt $(= 0.02)$ is used for time marching.

Initially, two identical circular vortex rings are set up as shown in figure 1(a). The centers of the vortex rings, both on the x_1-axis, are separated by $= 37.2\Delta x$, Δx being the mesh size $2\pi/N$ $(= 0.098)$. The core centerlines of the rings are on the (x_1, x_2) plane. The radius of the centerlines is $10\Delta x$. Thus, the velocity field is symmetric with respect to the (x_1, x_3) and the (x_2, x_3) planes. These symmetries are preserved very well over the whole period calculated in the present numerical simulation though we do not impose them in our numerical code.

The initial vorticity is distributed parallel to the centerlines with Gaussian core $\omega(r) = \omega_0 \exp\left[-(r/a)^2\right]$. Here r is the distance from the centerlines, ω_0 $(= 23.8)$ is the maximum vorticity at the center and a $(= 4\Delta x)$ is the e^{-1}-fold radius of the core. The circulation of the vortex ring $\Gamma = 2\pi \int_0^\infty r\omega(r)\,dr$ is then $\pi\omega_0 a^2$ $(= 11.53)$. The tails of the core of the vortex rings are terminated at $0.01\omega_0$ so that the vorticity of the two vortex rings are not overlapped at the start and reconnection commences only on contact. The vorticity field is made to be exactly divegence-free by orthogonalizing its Fourier components with wavenumber. The viscosity ν is set at 0.01 so that the initial Reynolds number Γ/ν is 1153.

3 RECONNECTION OF VORTEX RINGS

The global movement of the two vortex rings is conveniently represented by the iso-surfaces of vorticity norm $|\omega|$. They are plotted in figures 1 at several representative stages of evolution. They are viewed from the (2,1,5) direction. The levels of the iso-surfaces are 25% of the instantaneous maximum of vorticity norm.

The self and the mutual induction velocities cause the two vortex rings to travel to the x_3 direction and to turn toward the (x_2, x_3) plane. They collide on the (x_2, x_3) plane at around $t = 3$. The opposite-signed vorticities are cancelled on the symmetry plane by cross-diffusion. The remainder of the cancelled vortex line in one ring is connected with the corresponding vortex lines in the other ring. Therefore, there appears in the (x_2, x_3) plane an amount of circulation, which exactly equals that of the x_2 component lost on the (x_1, x_3) plane. This annihilation continues because the nonlocal interaction keeps the two rings pressed against each other, and as a result a single big distorted ring is formed. This is the first reconnection. (The mechanism of cross-linking called *bridging* will be explained in §4.) The cancellation of the vorticity is incomplete, and the uncancelled parts of the dipole vortices, or *threads*, are pulled around the main tubes advected by a strong swirling motion induced by the latter [see figures 1(e-g)]. The intensity of the threads is weakened by viscous diffusion between the threads and the main tubes.

The new distorted ring moves farther in the x_3 direction. The velocity induced by its own vorticity distribution causes stretching of the tube in the x_2 direction and shrinking in the x_1 direction [figures 1(e-g)], taking a gourd-shape at around $t = 7$. The curvature at the two round ends of the ring becomes larger and induces a higher self-induction velocity there to make these parts travel faster. This makes the distorted ring somewhat planar. The neck of the gourd-shaped vortex, on the other hand, becomes narrower as its facing parts touch each other at around $t = 9$. Then, the second reconnection occurs. Near the closest point, the facing parts of the tube are linked by *bridges* which start formation of two new circular rings. The mechanism of the reconnection is similar to that at the first reconnection process. Two humps appear in front of the vortex ring [figure 1(i)]. Again the cancellation of opposite signed vorticity is only partial and two remnant vortex tubes, called *legs* (Kida *et al.* 1989a), are left behind the bridges. The global shape of the vortex tube at $t = 11$ resembles eyeglasses with a nose-rest.

The time development of circulation of one of the interacting vortices in the (x_1, x_3) plane is plotted as a solid line in figure 2. A broken line represents the circulation of a vortex core in the (x_2, x_3) plane and a dash-dot line represents the sum of these two. The x_2-component of circulation decreases slowly at early times ($t \lesssim 2$), which results from the annihilation at the edges of the cores which are fattened by viscous diffusion. At around $t = 2.5$, just when the first reconnection starts, circulation begins to drop appreciably, losing 80% of its value by $t = 4.5$. Simultaneously the x_1-component of circulation appears and increases rapidly to account for the decrease of the x_2-component. During the reconnection, say $2.5 \leq t \leq 4.5$, 70% of the initial circulation is lost, but the sum of the x_1- and x_2-components is almost constant in time. Afterwards ($t \gtrsim 4.5$), the x_2 component decays exponentially in time. The circulation, therefore, decreases to zero only asymptotically for large t. This is due to the incompleteness of the vortex reconnection; the threads undergo uneventful viscous decay and then collapse on the main ring causes the final decay of the circulation around the thread [figures 1(f, g)].

4 BRIDGING

The cross-linking of vortex lines in the present reconnection process can almost (but not completely) be explained by the *bridging* mechanism observed in an interaction of a pair of anti-parallel vortex tubes with sinusoidal disturbance (Melander and Hussain 1988).

The mechanism of bridging is illustrated in figure 3. The directions of the vorticity are shown by double arrows. The vortex rings rotate in the directions indicated by curved arrows so that they move out of the paper. The two vortex rings are pressed against each other from left and right by the self-induction of the vortex rings themselves.

When two vortex rings come into contact, the outermost vortex lines, the directions of which are opposite, are cancelled by viscous cross-diffusion in the interaction zone [the hatched areas in figure 3(a)]. At the same time, these lines are connected with the counterpart in the other vortex rings at the ends of the interaction zone [figure 3(a)]. We denote, in figures 3, by a the typical vortex lines which have been cut-and-connected during this process, and by b those lines which have not been reconnected yet. Since the two vortex rings are pushed against each other all the time by the converging flow, the number of the cut-and-connected vortex lines increases in time. As the vortex lines are rotating around each vortex core, lines a and b must be tangled at the ends of the interaction zone [figures 3(b) and (c)]. [In this argument the viscous effect is neglected, which acts against the entanglement (Kida *et al.* 1989b)] The vortex lines are strongly twisted especially in the regions denoted by rectangles in figure 3(c). They are twisted like right-handed and left-handed screws in shaded and blank rectangles, respectively. This means that the super-helicity density $\boldsymbol{\omega} \cdot \boldsymbol{\chi}$, where $\boldsymbol{\chi} = \nabla \times \boldsymbol{\omega}$, takes positive and negative values in the respective regions. It is observed that the helicity density $\boldsymbol{u} \cdot \boldsymbol{\omega}$ also takes the same signs as the super-helicity density, so the streamlines wind up in the same sense as the vorticity lines in these regions.

The portions of line a that links the two rings are called *bridges*. A remarkable feature of the reconnection process is that the reconnected vortex lines locate themselves ahead of the vortex dipole. As will be discussed below, bridges are actually created at the position of the maximum strain rate.

In the meanwhile the canceling parts [the two vertical lines at the center in figures 3(b) and (c)] of the vortex rings are bent by the velocity field induced by the upper and lower parts of the vortex rings (not necessarily by the bridges) in such a way that the center of the canceling part comes out of the paper. Thus the curvature of the dipole is reversed. This in turn causes them to move apart from each other due to the reversed self-induction and prevents further cancellation.

The streamlines in planes parallel and perpendicular to the interacting vortex dipole are helpful for understanding of the formation of the bridges. Figure 4(a) represents *actual* streamlines in the (x_1, x_2) plane denoted by line AA' in figure 3(b), in a frame moving with the dipole at $t = 3$. The velocity component normal to this plane is identically zero because of symmetry. The contour of vorticity ω_2 is superimposed to show the position of the dipole. The contour levels are 10 and 40% of the maximum of $|\boldsymbol{\omega}|$ at this time. A fluid particle in the right half of the dipole rotates counterclockwise and the left half clockwise. There are two stagnation points on the centerline above and below the dipole. The directions of the flow around the stagnation points are indicated by arrows.

Advected by this velocity field, the vortex lines, which are almost perpendicular to the plane and which come close to the centerline between the two stagnation points are transported along the centerline to the lower stagnation point. It was argued that such vortex lines are advected and accumulated at the lower stagnation point to make a bridge [see figure 7(g) and related comments in Melander and Hussain 1988], but actually the bridge is created at the point of largest strain rate as will be described below.

Figure 4(b) shows the streamlines and the contours of the normal component of vorticity ω_1 on the (x_2, x_3) plane [BB' in figure 3(b)] on the same moving frame as figures 4(a). The contour level of the vorticity is 2% of the maximum of $|\omega|$ at the time. The two islands represent the embryos of the bridges. The bridges are close to but do not coincide with the lower stagnation point. [Notice that the locations of the stagnation points in planes parallel to the (x_1, x_3) plane are almost independent of x_2.] The x_3 coordinate of the center of the bridges is about $40\Delta x$, whereas that of the stagnation point is about $37\Delta x$.

The rate of strain tensor $e_{ij} = \frac{1}{2}\left(\partial u_i/\partial x_j + \partial u_j/\partial x_i\right), i, j = 1, 2, 3$ is a measure of the deformation of fluid element in a flow. This is useful as an indicator of the vortex stretching.

In figure 5 are shown the angular dependence of the stretching rate on planes parallel to (a) the (x_1, x_2) and (b) the (x_1, x_3) planes at $t = 3$. The 10 and 40% levels of contours of vorticity norm are superimposed for reference. The distance from the centre of a butterfly or a dumbbell shape to a point of its periphery is proportional to the stretching rate in the direction of the point seen from the centre. Positive part is blacked out. The x_3 coordinates of planes parallel to the (x_1, x_2) plane and the x_2 coordinates of planes parallel to the (x_1, x_3) plane are chosen so that these planes pass through bridges.

It is seen in figure 5(a) that fluid particles on the x_1 axis are being stretched in parallel with the x_2 axis. The magnitude of the stretching rate changes gradually for $|x_2| \lesssim 8\Delta x$, but a closer inspection reveals that it has peaks at $|x_2| \approx \pm 4\Delta x$. These positions of the largest stretching rate coincide with the x_2-coordinates of the initial bridges [see the x_2 coordinates of bridges in figure 4(b)].

At the lower half of the x_3 axis in figure 5(b) we see that fluid particles are being stretched in the x_1 direction. The x_3 coordinate of the largest strain rate is about $40\Delta x$, which agrees with the x_3 coordinate of the position of the center of a bridge [see figure 4(b)].

5 SUMMARY AND DISCUSSION

The interaction due to the collision of two identical vortex rings has been investigated by solving the Navier-Stokes equation numerically. The vortices undergo two successive reconnections, fusion and fission, as have been visualized before, but the structure of reconnected vortex rings in the second reconnection are quite different from visualization. The mechanism of the reconnections is explained by *bridging* discovered by Melander and Hussain (1988) except for the position of the bridges. It was shown that a bridge is created not at the stagnation point but at the position of the maximum rate of strain.

The bridging mechanism manifests itself very clearly in the present simple geometry of an interacting vortex pair which are anti-parallel, of equal strength and in symmetry configuration. The *bridges* observed in a trefoiled vortex tube (Kida and Takaoka 1987; 1988) and the *fingers* in two orthogonal straight vortex tubes (Melander and Zabusky 1988) may have the same characteristic feature as that of the bridges mentioned above. The viscosity takes an essential role in bridging. It is therefore anticipated that the bridging may not occur in the inviscid limit, or the total helicity may be invariant in this limit.

The characteristic time of the bridging depends upon not only the vorticity distribution in the interacting vortex cores but also the strength of the converging flow induced at the interaction zone. Because of the rather complicated mechanism of the vortex reconnection and long-range interaction, it is hard to believe that the interaction time may be expressed in terms of local quantities.

We compared various field quantities. It is found that the regions in which the energy dissipation is large is highly localized in space compared with the enstrophy (or vorticity) concentrated regions. The energy dissipation is dominant in the interaction zone before the reconnections, and in bridges and threads (or legs) after the reconnections.

The high enstrophy production ($\boldsymbol{\omega} \cdot \boldsymbol{e} \cdot \boldsymbol{\omega}$) and the high energy dissipation ($2\nu \boldsymbol{e} : \boldsymbol{e}$) regions are nearly the same before the reconnections but not after the reconnections. In the latter stages the peaks of energy dissipation rate intervene those of the enstrophy production rate. This suggests that these two quantities, or $\boldsymbol{\omega}$ and $\boldsymbol{e}$, are quite independnt of each other.

The helicity and super-helicity densities represent the skewed structure of the velocity and the vorticity fields, respectively. The vorticity is large in the main vortex tubes, the form of which is preserved for a relatively long time. On the other hand, it is the helicity density that takes large values in bridges and threads. These shapes change rapidly in time. We therefore conclude that the long-lived structure may carry strong vorticity rather than high helicity density.

The decay law of the energy is examined. It obeys the same power law observed in turbulent flows (t^{-p} where $1.2 \lesssim p \lesssim 1.4$).

A passive scalar is also included in order to compare the vorticity and scalar fields. Scalars such as dye and smoke have been frequently used in laboratory experiments to trace the motion of vortices or coherent structures. A stretching of fluid element dilutes the density of a passive scalar, whereas it intensifies the magnitude of vorticity. The concentration of a passive scalar does not always simulate the magnitude of vorticity and the departure increases with time/space from the point of injection of markers, confirming our longstanding warning against relying heavily on flow visualization (Hussain 1983, 1986).

The initial condition dependence of the reconnection characteristics has been examined by inclining the initial vortex rings toward the (x_2, x_3) plane. It was found that the reconnections are more complete for initial inclination angles 15° and 30° and less complete for 0° and 45°. The shapes of the vortex tubes are different for different initial conditions, but the mechanism of the reconnections is explained by *bridging*.

A detailed discussion of these subjects will appear in a forthcoming paper (Kida *et al.* 1989b).

This work was supported by DOE Grant DE-FGO5-88ER13839 and by the Grant-in-Aid for Scientific Research from the Ministry of Education, Science and Culture of Japan.

REFERENCES

Hussain, F. 1983 Phys. Fluids **26**, 2816.
Hussain, F. 1986 J. Fluid Mech. **173**, 303.
Kida, S. and M. Takaoka 1987 Phys. Fluids **30**, 2911.
Kida, S. and M. Takaoka 1988 Fluid Dyn. Res. **3**, 257.
Kida, S., M. Takaoka and F. Hussain 1989a Phys. Fluids **A1**, 630.
Kida, S., M. Takaoka and F. Hussain 1989b (in preparation)
Melander, M.V. and F. Hussain 1988 in *Center for Turbulence Research Proceedings of the Summer Program 1988* p. 257.
Melander, M.V. and N.J. Zabusky 1988 Fluid Dyn. Res. **3**, 247.
Orszag, S.A. 1971 Stud. Appl. Math. **50**, 293.
Oshima, Y. and S. Asaka 1977 J. Phys. Soc. Japan **42**, 708.
Schatzle, P. 1987 *Ph.D. Thesis, California Institute of Technology.*

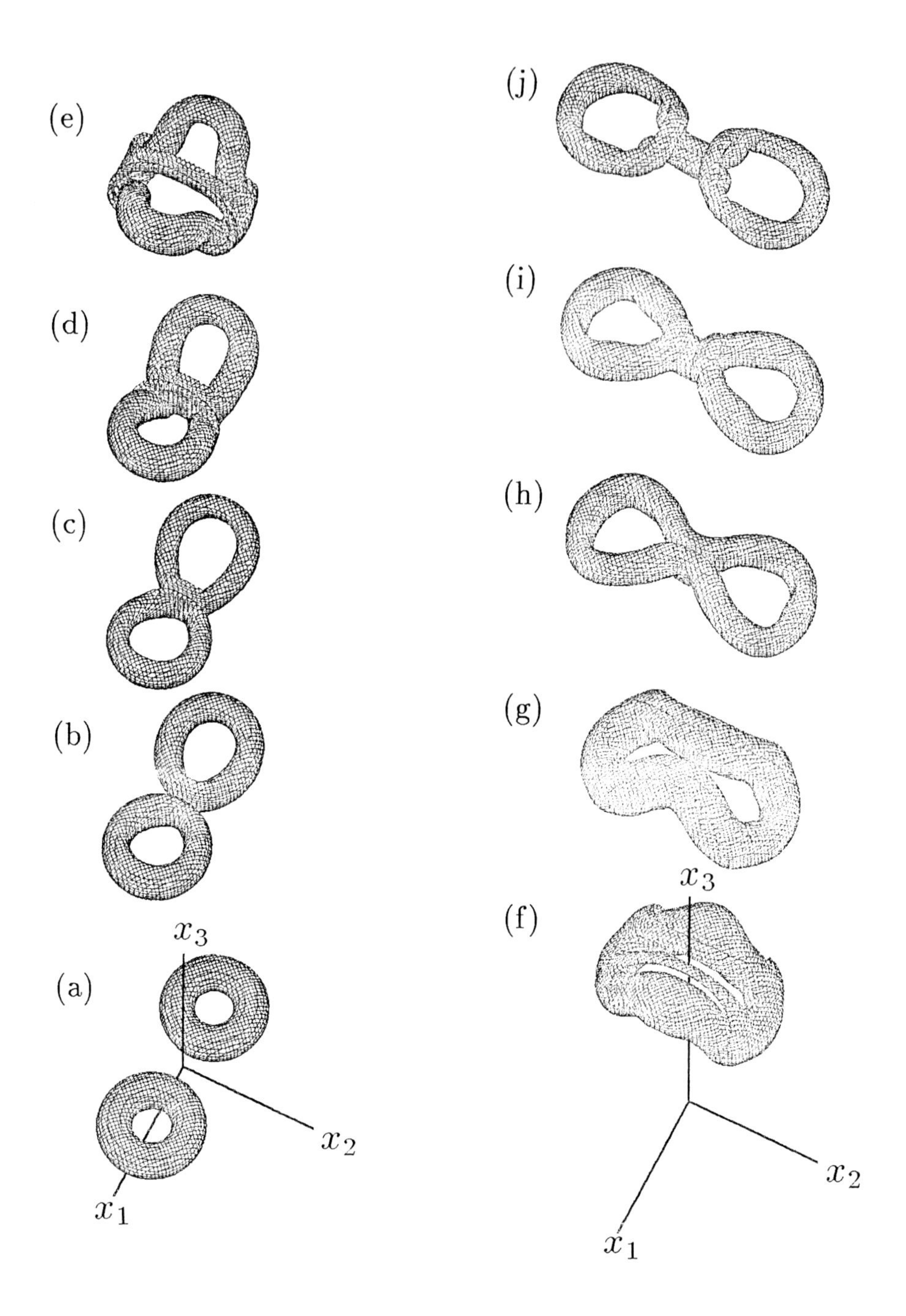

Figure 1. Perspective view of the iso-surface of vorticity norm which is seen from the (2,1,5) direction. $t = 0$ (a), 3 (b), 3.5 (c), 4 (d), 5 (e), 6 (f), 7 (g), 9 (h), 10 (i), 11 (j). The surface level is 25% of the instantaneous maximum of the vorticity norm.

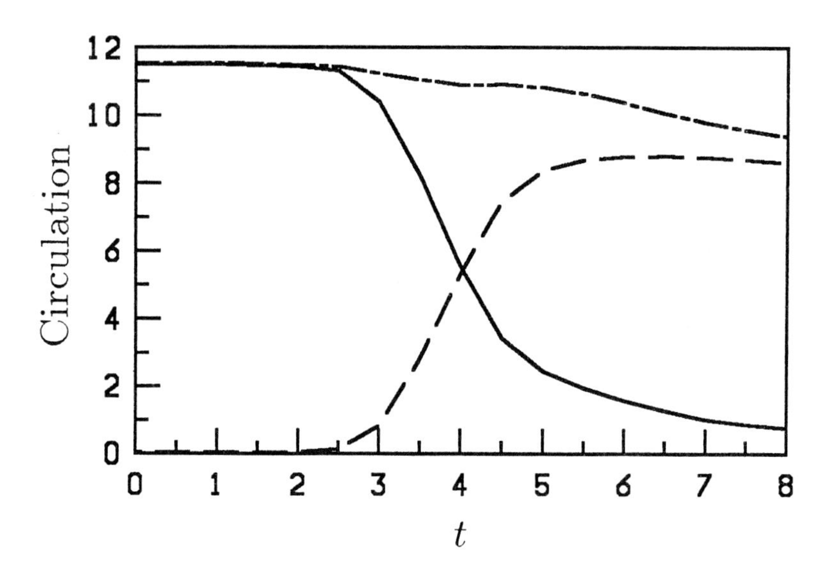

Figure 2. Time-developments of circulations around interacting vortex rings during the first reconnection. ———, circulation around a cross-section of an inner core on the (x_1, x_3) plane; — — —, circulation around a cross-section of a bridge on the (x_2, x_3) plane; - - - - -, sum of the two circulations.

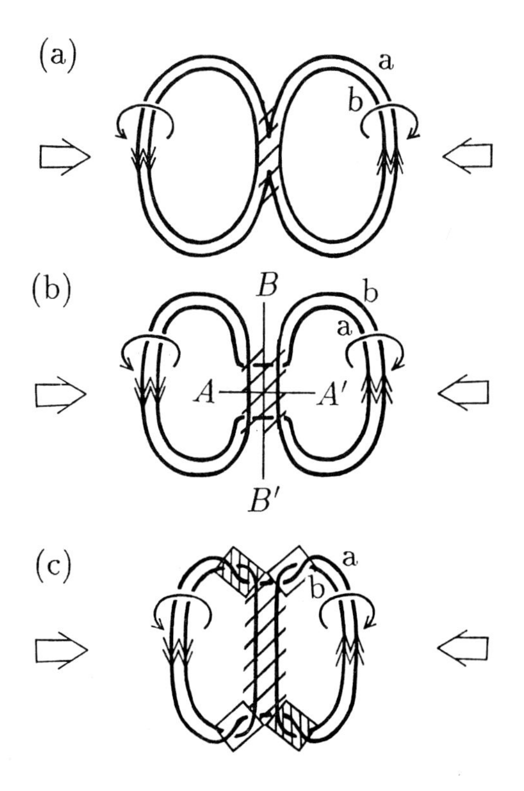

Figure 3. Illustration of bridging. Two vortex rings are interacting at the center. They are pushed from left and right by an converging flow designated by big arrows. Double and curved arrows indicate the directions of the vorticity and the rotation of the vortex lines, respectively. Letters a and b represent cut-and-connected and still-unreconnected vortex lines. The number of reconnected lines increases as time goes on from (a) to (c).

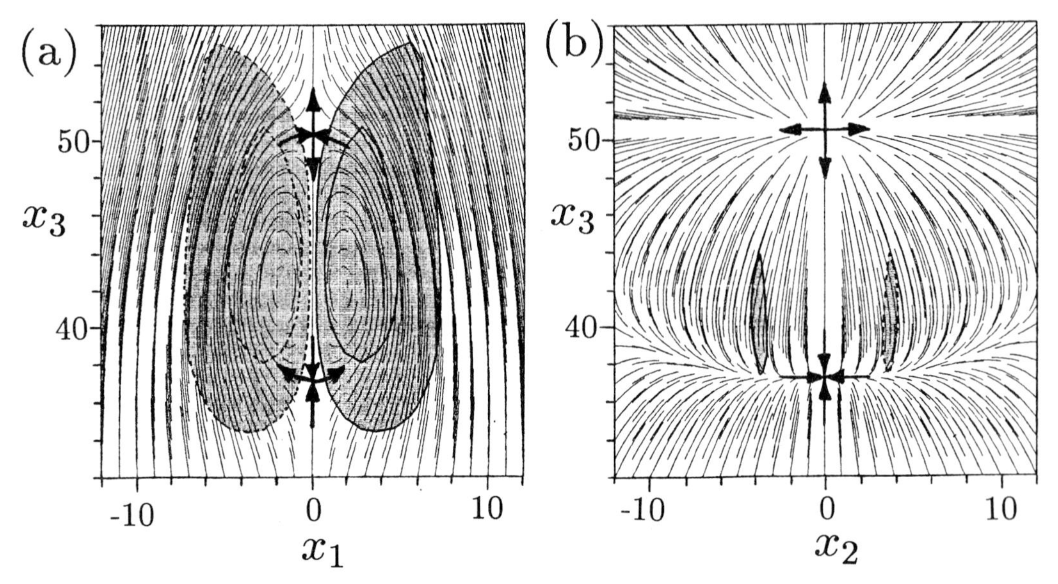

Figure 4. The streamlines and the contours of vorticity norm $|\omega|$ in a frame moving with the interacting vortex dipole at $t = 3$. (a) (x_1, x_3) plane. (b) (x_2, x_3) plane. Thin curves are parallel to the velocity vectors on the planes. The contour levels drawn are 10% and 40% of the maximum at this time for (a) and 2% for (b). Arrows on the centerlines in (a) and (b) indicate the directions of the flow around stagnation points.

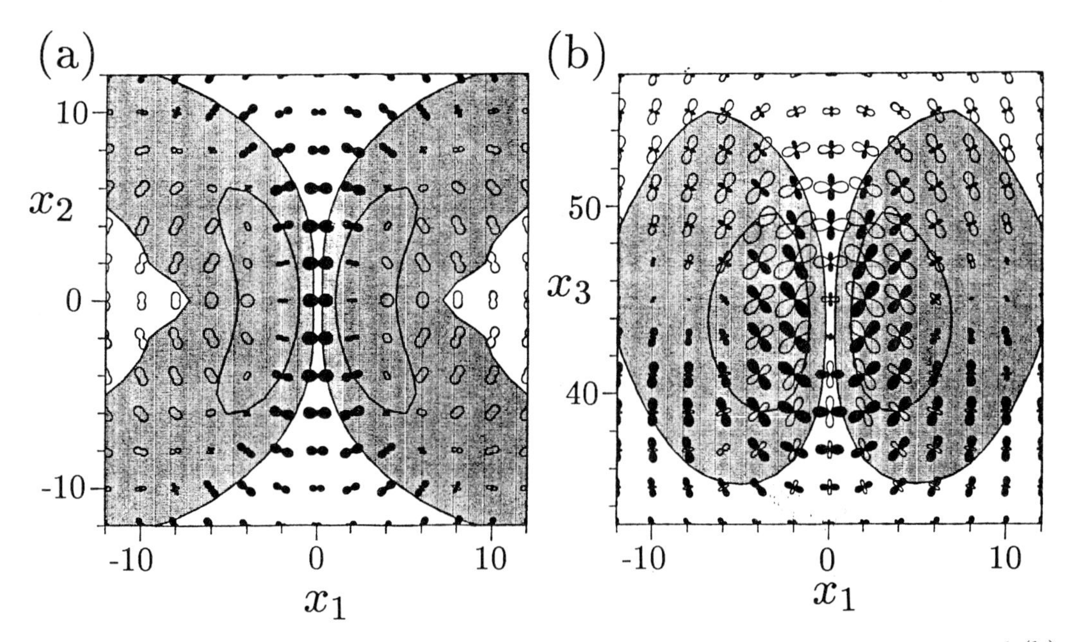

Figure 5. Angular dependence of stretching rate on planes (a) $x_3 = 40\Delta x$ and (b) $x_2 = 4\Delta x$ at $t = 3$. These planes pass through the bridges.

Comment on Vortex Ring Reconnections

HASSAN AREF*† AND IRENEUSZ ZAWADZKI *

University of California, San Diego
La Jolla, CA 92093, USA

1. INTRODUCTION

Using a "vortex method" we have performed computations of vortex ring reconnections very similar to those reported in the papers by Melander & Hussain and Kida, Takaoka & Hussain presented at the Symposium. Our results, however, appear to differ. Since the basic representation of the flow field in a Lagrangian vortex method is quite different from that in an Eulerian spectral code, as used by the authors mentioned, the discrepancies could have many origins in either numerics or physics.

The code that we have developed is a three-dimensional version of the so-called "vortex-in-cell" method, initially described by Christiansen (1973) for two-dimensional flow. The main function of the grid, that is used along with the Lagrangian vortex particles in this method, is to allow fast transform inversions of Poisson's equation. In two dimensions the basic equation to be inverted is the one relating stream-function to vorticity:

$$\nabla^2\psi = -\zeta \tag{1}$$

In three dimensions the velocity potential for incompressible flow, defined by

$$\mathbf{V} = \nabla\times\mathbf{A}, \tag{2}$$

leads to

$$\omega = \nabla\times\mathbf{V} = \nabla(\nabla\cdot\mathbf{A}) - \nabla^2\mathbf{A}. \tag{3}$$

Thus, if a gauge is chosen in which **A** is divergence-free, we are led to three Poisson equations relating the components of the vector potential to those of the vorticity. The grid is useful for reducing the operation count for the inversion of these. Equation (3) plays the same role for our three-dimensional method as Eq.(1) does for the two-dimensional version.

A detailed exposition of the method along with several computed examples appears in Zawadzki & Aref (1989).

---

*) Affiliated with Department of Applied Mechanics and Engineering Science
†) Affiliated with Institute of Geophysics and Planetary Physics, and San Diego Supercomputer Center.

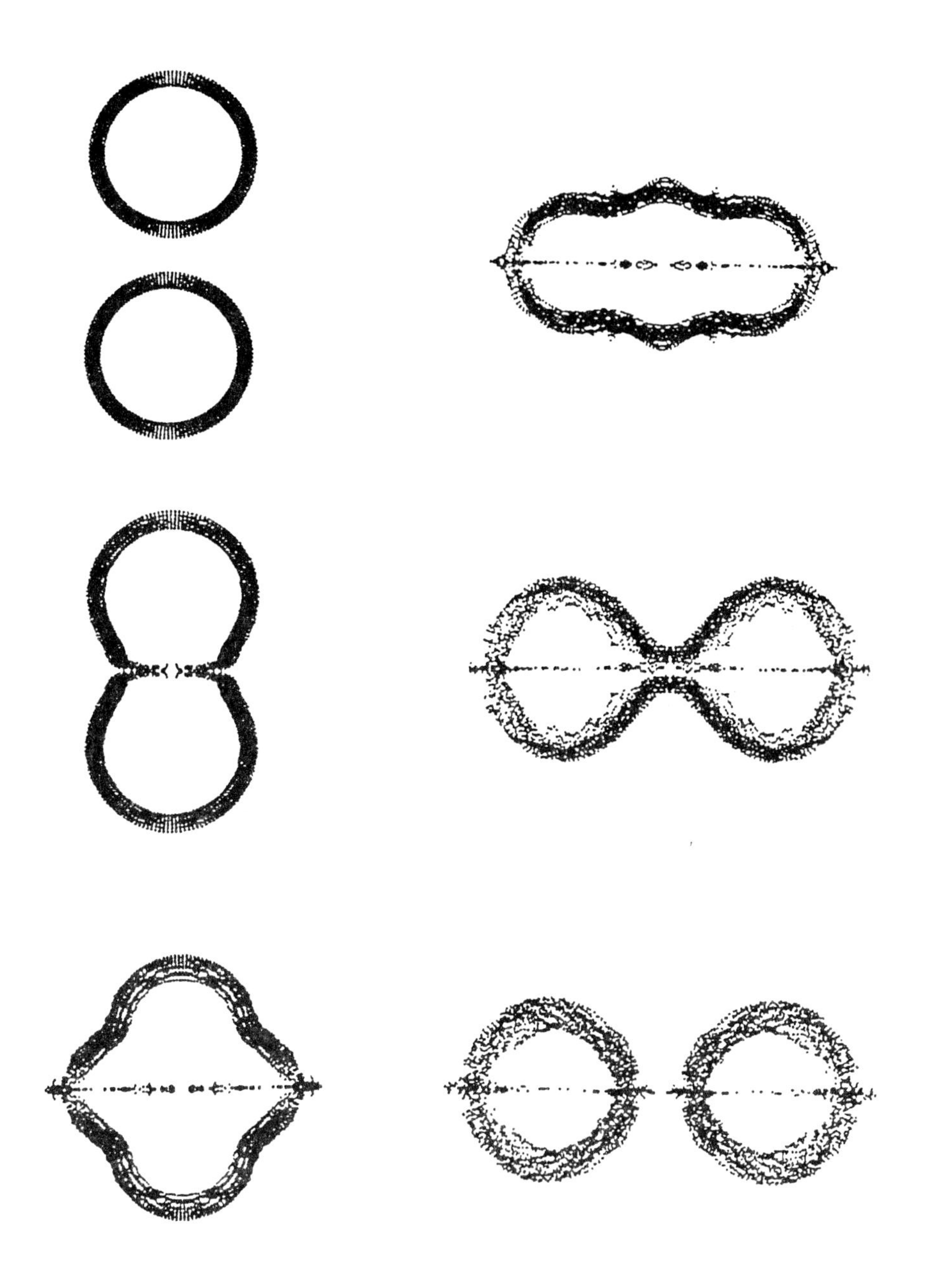

Figure 1: Vortex ring collision, reconnection and emergence of two "new" rings each consisting of half of either "old" ring. The rings initially are inclined 20° from the vertical. Some 7,500 Lagrangian vortex "arrows" make up each ring. The underlying grid (not shown) is 64^3. In the plot a point is shown for each of about 30% of the vortex points. This view of the process is the one shown by Oshima & Asaka (1975). Images proceed from top to bottom left to right. Note the "debris" left along the symmetry plane.

2. VORTEX RING COLLISIONS

Two vortex rings propagating along slightly inclined directions has become something of a paradigm for the reconnection problem. This case is particularly interesting because it presents an example in which viscous effects play an essential role away from rigid surfaces or two-fluid interfaces. In an inviscid fluid vortex reconnection cannot take place. In the limit of small viscosity reconnection takes place at any finite value of the viscosity.

The main point at issue is the morphology of the states that ensue as the two vortex rings reconnect and then attempt to separate. Kida et al. discuss at length the appearance of two essentially parallel strands of vorticity connecting the new rings that form, or try to form, in the final state. In our calculations, shown in Figs.1-2 in three perpendicular views, we see much less of these strands. We may note that the strands seen by Kida et al. are not revealed by flow visualization photographs of Coles & Schatzle, who studied this process in carefully controlled laboratory experiments. Earlier experiments by Oshima & Asaka (1975) did not show the strands either. The explanation given is that the distributions of vorticity and of an advected scalar differ in the computations. The scalar is diluted in the region of the strands because substantial stretching is taking place there. Our calculations, on the other hand, agree extremely well with the images of Coles & Schatzle.

In our Lagrangian calculations the key difference is that no particles seem to accumulate in the regions between the emerging rings where Kida et al. see the strands. At the Symposium the suggestion was made that this might be due to an insufficient number of particles having been used (15,000 in Figs.1-2). We have repeated the calculations with 50,000 particles and we still see a very different picture from that of Kida et al. To the extent that strands appear at all, we tend to see the two parallel strands merge to one, in which opposite signed vorticity gradually annihilates. The particles that are lost from the rings themselves, and then migrate in the symmetry plane of the flow, carry very little vorticity. Faint traces of this "debris" is also seen in the Coles & Schatzle photographs.

Our present interpretation of these results is rather pedestrian. We tend not to think that the existence of two persistent parallel strands as reported by Kida et al. is generic, but that it may occur for a particular initial angle of inclination and separation of the rings. Basically, the two strands survive for a long time if they come out to be almost exactly straight and parallel. That is unlikely to happen except for very special initial conditions. If the two strands are not straight as they form, self-induction will move them together and via viscous dissipation lead to their annihilation. This we believe to be the generic scenario.

There seem to be several items worthy of exploration in this problem. We are particularly interested in using our numerical Lagrangian representation to study details of particle motion during the reconnection process. The vortex-in-cell code promises to be an extremely valuable tool for such an investigation. We note that the vortex-in-cell computations also appear to be competitive in terms of computer resources. The run shown in Figs.1-2 consumed less that 30 minutes of Cray XMP time.

This work is supported by NSF/PYI award MSM84-51107, and by DARPA/URI grant N00014-86-K-0758. The computations were performed at the San Diego Supercomputer Center.

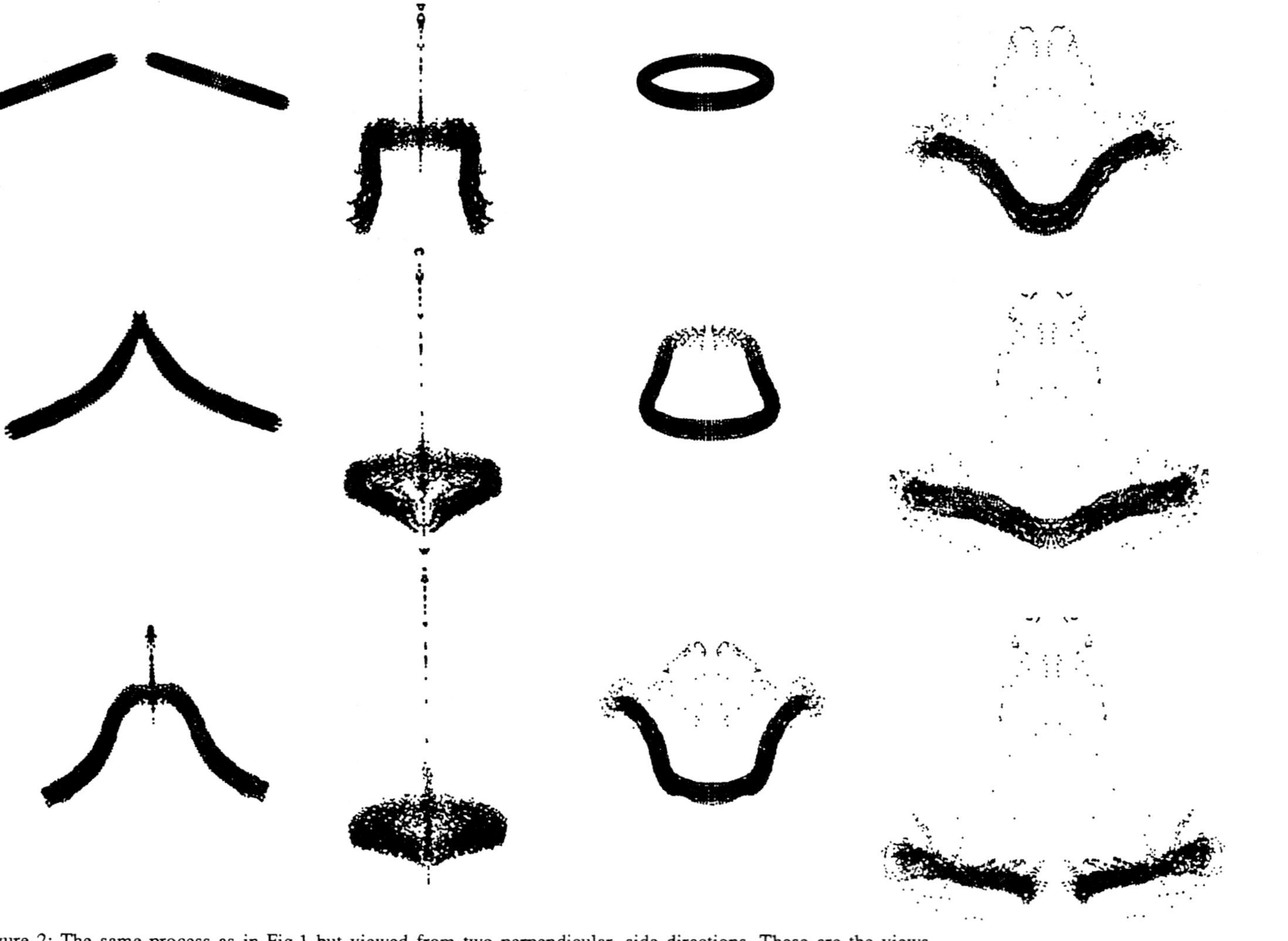

Figure 2: The same process as in Fig.1 but viewed from two perpendicular, side directions. These are the views presented by Coles & Schatzle. Panels correspond exactly in each of the three views shown. Note the approximate reversibility symmetry between the sequence on the left followed from start to finish, and the one on the right traced backwards from finish to start.

REFERENCES

Christiansen, J.P. 1973 Numerical simulation of hydrodynamics by the method of point vortices. *J. Comp. Phys.* **13**, 262-379.

Oshima, Y. & Asaka, S. 1975 Interaction of two vortex rings moving side by side. *Nat. Sci. Rep. Ochanomizu Univ.* **26**, 31- 37.

Zawadzki, I. & Aref, H. 1989 Numerical experiments using a three-dimensional vortex-in-cell method. *J. Comp. Phys.* (submitted).

Two Bifurcation Phenomena for Inviscid Fluid Flow

M.S. BERGER*

Center for Applied Mathematics University of Massachusetts, Amherst, USA

In this article I consider two well-known problems for inviscid fluid flow involving vortex motion. I show circumstances under which each fluid motion can be regarded as a bifurcation process. This bifurcation process means that a "topological change" in the qualitative fluid flow takes place. In fact, such processes are observed and are usually referred to as "fluid instabilities". Such processes, for example the famous occurrence of Taylor vortices between concentric rotating cylinders, are well-known for viscous fluid flow governed by the Navier-Stokes equation. New examples of these processes were also discussed, for example, recently by Benjamin (1962). However, for inviscid fluid motion a more subtle analysis is required.

A quantitative understanding of such viscous and inviscid processes generally requires a mathematical discussion. This fact and its converse define the point of view taken here. The converse statement implies that by looking carefully at the nonlinear mathematical governing equations of a fluid motion a bifurcation process can be seen to lie behind the observed phenomenon. This linking of a change in topological behavior and nonlinear mathematical equations plays a key role in the subsequent discussion. Moreover, in the subsequent discussion I shall focus attention on the nonlinear qualititative processes involved and not the mathematical technicalities, which are covered in the references cited.

1. Leapfrogging of vortex filaments of different strength

The leapfrogging of vortex rings in an ideal fluid in three dimensions was first described by Helmholtz in his famous paper (1889) on vortex motion. To describe this interaction of two thin vortex rings of equal strength propagating along a common axis, Love (1894), found a quantitative analysis by assuming the two thin vortex rings could be represented by two pairs of two point vortices of equal strength placed symmetrically along a common axis. Leapfrogging in this case means that the relative motion of the two pairs of point vortices is periodic. Recently, Aref & Pomphrey (1982) showed that the motion of 4 point vortices could be chaotic if the

*Research partially supported by an AFOSR grant

point vortices are not placed symmetrically. Here, in this paper, I demonstrate the leapfrogging phenomenon for two pairs of point vortices of different strengths but symmetrically placed, as a bifurcation process.

We analyze this problem as a bifurcation process from an equilibrium for a Hamiltonian system. The usual approach of analyzing the geometry of level surfaces of the Hamiltonian seems difficult in this case.

1.1 The physical problem

Consider two pairs of two point vortices propagating along the x-axis (as common axis) initially symmetrically placed in R^2 with Cartesian coordinates $P_1 = (x_1, y_1)$, $P_1' = (x_1, -y_1)$, $P_2 = (x_2, y_2)$ and $P_2' = (x_2, -y_2)$. These point vortices in R^2 are to represent two pairs of cylindrical vortices of infinitely thin cross-section of different strengths. Here the circulation about the two point vortices of each pair are chosen equal and of opposite sign. The line of symmetry coincides for each pair. Thus we choose the strength of $P_1 = \omega_1, P_1' = -\omega_1, P_2 = \omega_2, P_2' = -\omega_2$.

We set the ratio of vortex strengths $\omega_1/\omega_2 = \lambda$. The case $\lambda = 1$ is classical as mentioned above. We show that if λ is any number between 1 and λ_0 (a critical number approximately equal to 0.61623), the same leapfrogging phenomenon occurs as in the classic case.

In general for n point vortices with coordinates $(x_i, y_i)(i = 1, 2, \dots n)$ it is known that the motion of these n point vortices is governed by a Hamiltonian dynamical system of ordinary differential equations. Thus, in the special case of the 4 symmetrically placed point vortices with strength $\alpha_1, -\alpha_1, \alpha_2, -\alpha_2$ we may suppose above that $\alpha_1 = \lambda\alpha_1$, so that λ is the bifurcation parameter and without loss of generality we suppose $\alpha_2 = 1$. Thus, to study relative motion we set

$$x = x_1 - x_2, \qquad y = y_1 - y_2.$$

Thus the relative motion is described by a Hamiltonian system with the Hamiltonian H_1 depending only on x, y and λ. In the sequel we shall demonstrate the leapfrogging of the two vortex pairs P_1P_1' and P_2P_2' by proving

Theorem The reduced Hamiltonian system has a smooth nonconstant periodic solution for each fixed λ in the open interval $(\lambda_0, 1)$, i.e. the relative motion of the vortex pairs is periodic. Here λ_0 is a critical parameter described in Section 2, and is approximately equal to 0.61623.

1.2 Analysis of the Hamiltonian system

We analyze the system with Hamiltonian H_1 by first showing that for λ in a large open interval $(\lambda_0, 1)$ the system has a unique equilibrium point $(0, y_\lambda)$ where y_λ is a real number implicitly defined by the unique real root of a cubic equation. Notice this fact does not extend to the case of vortices with equal strength when $\lambda = 1$.

Then one writes the Hamiltonian system for each λ in the open interval $(\lambda_0, 1)$ in terms of displacement from $(0, y_\lambda)$. The relevant linear term involves the Hessian of the Hamiltonian, near the singular point. One proceeds to show that the desired periodic solution bifurcates from this stationary point for each λ in the open interval $(\lambda_0, 1)$. To achieve this result we proceed by introducing a new parameter ω into (3) by changing the time scale $t = \omega s$. This allows the study of solutions of period ω in t to be reduced to the study of 1-periodic solutions in s. The equation (3) becomes a nonlinear eigenvalue problem.

To analyze the 1-periodic solutions of this nonlinear eigenvalue problem it will suffice to consider the linearized problem about the stationary point $(0, y_\lambda)$, and then utilize standard nonlinear results.

1.3 Analysis of the linearized problem and the nonlinear problem

We consider now the periodic solutions of this linearized system, which can be reduced to a Sturm Liouville system of the form

$$v_{ss} + \omega^2 a(\lambda) b(\lambda) v = 0$$

where a and b are constants depending on the parameter λ. Clearly we have periodic solutions for positive ω provided the product ab is positive where $v(s)$ denotes the relevant displacement. So $v(s)$ will be periodic provided the product $a(\lambda)b(\lambda)$ is positive. Using numerical analysis we find that the product $a(\lambda)b(\lambda) > 0$ provided λ is in the open interval $(0.61623, 1)$. This determines precisely the relevant strength ratio.

The nonlinear result will follow immediately from the standard Liapunov-Schmidt method for bifurcation theory, cf. Berger (1977), once one restricts attention to periodic solutions of the nonlinear eigenvalue problem that are odd in $x(s)$ and even in $y(s)$. Here we use the symmetry of $H_1, H_1(x, y) = H_1(-x, y)$. This yields the fact that 1-periodic solutions with the stated symmetry are one dimensional, a required prerequisite for the applicability of the standard theory. In order to utilize the Liapunov-Schmidt technique we simply note that there is a mapping A defined by the nonlinear eigenvalue problem between the Hilbert spaces of periodic functions X and Y with X a Sobolev space of 2-vector symmetric functions and Y an L_2 space of vector functions with reverse symmetry. To this mapping A the Liapunov-Schmidt method is applicable. Indeed A is a C^1 Fredholm mapping of index zero between the Banach spaces X and Y that can be used for the standard Liapunov-Schmidt theory.

The theory results in a nonzero-periodic solution $z_\omega(s)$ bifurcating from the equilibrium state $(0, y_\lambda)$ for each λ in the open interval $(\lambda_0, 1)$. In fact these periodic solutions correspond to the nonconstant ω-periodic solutions of (2) as stated in the Theorem of 1.1.

For full details of the discussion of this section we refer the reader to Berger & Nee (1989).

2. On vortex breakdown for inviscid flow as a bifucation process

The second fluid flow problem to be discussed as a bifurcation process is vortex breakdown of bubble type in a steady inviscid approximation. Vortex breakdown involves the fact that, under certain conditions, a vortex possessing axial flow and swirl in an incompressible fluid exhibits a sudden finite transition from one state to another. Here I consider the governing Euler equation in a stream function formulation with axial symmetry and parameter dependence, the so-called "Squire-Long" equation. I shall discuss briefly new mathematical methods for obtaining multiple solutions for this equation subject to appropriate boundary conditions. The resulting nonlinear vortex structures involving separated flow arise as a bifurcation process.

2.1 Special solutions of the Squire-Long equation

(a) Classical vortex ring solutions

Fraenkel & Berger (1974) established mathematically a family of axially symmetric vortex ring solutions of varying cross section interpolating between the explicit Hill's spherical vortex and Helmholtz's circular vortex. This family was characterized by an isoperimetric variational principle and could be calculated by optimization techniques. The family had been conjectured by Batchelor (1967) and were without swirl and exhibited no bifurcation. Recently, Moffatt (1988) found these solutions by the method of magnetic relaxation as described in his article of these proceedings.

(b) Vortex ring solutions with swirl

When a swirl component is added to the Squire-Long equation, the vortex ring solutions described in (a) above continue to exist. Such solutions have also been found by Moffatt by magnetic relaxation techniques and by Turkington (1989) using integral equations coupled with a calculus of variations technique. Such solutions in constrast to (a) are new and were not conjectured in classic texts. The helicity of these new vortex ring solutions differs from zero. In the classical case of (a) above, the helicity of the vortex rings vanishes.

(c) Bifurcated columnar vortex solutions

Liebowitz (1987) found a bifurcated family of axisymmetric solutions depending on a swirl magnitude parameter. These solutions bifurcated from a columnar flow but did not exhibit the supercritical-subcritical transition necessary for Benjamin's vortex breakdown theory.

(d) Combined force and free vortex

Benjamin (1962) found a "transition" from a vortex-breakdown model equation whose primary flow A consisted of two parts (i) a core with solid body rotation and (ii) a surrounding annular region of irrotational vortex motion. The secondary flow B also consisted of two regions with the boundary between the regions displaced to preserve continuity of the relevant physical quantities. This model problem did exhibit the desired supercritical-subcritical transition. However, to date the problem has not been extended to the full Squire-Long equation.

2.2 Bifurcation for vortex solutions of the Squire-Long equations

To analyze bifurcation phenomena for the Squire-Long equations, one first introduces a dimensionless parameter whose magnitude serves as a "nonlinear eigenvalue".

In an invited address to the American Physical Society in 1987, Professor J.T. Stuart showed there is only one such parameter which he called the "Squire number". The major effect of swirling flow for the Squire-Long equation at large Squire number is that bifurcation occurs in the associated free boundary problem. Thus the primary state (a vortex ring with swirl described in (b) above) for sufficiently large parameter value bifurcates and a new secondary flow, also axisymmetry appears. To date the only way known to analyze such phenomena is the use of mathematics. The full details of the work will appear in a forthcoming paper.

The key mathematical idea involved is the use of variational methods which retain their validity for inviscid flow. Indeed, the vortex rings with swirl mentioned in (b) are determined by use of variational principles. To demonstrate bifurcation, one shows that such variational principles can be modified to detect secondary flows, (cf. Berger (1989)).

REFERENCES

Aref H. & Pomphrey S. (1982) "Integrable and chaotic motions of four vortices". *Proc. Royal Soc. London A*, pp. 359-387.

Batchelor G.K. (1967) Introduction to Fluid Mechanics, Cambridge University Press.

Benjamin T.B. (1962) "Theory of vortex breakdown phenonena". *Journal of Fluid Mechanics* pp. 593-629.

Berger M.S. (1977) Nonlinearity and functional analysis, Academic Press, New York, N.Y.

Berger M.S. (1989) "Remarks on vortex breakdown", AMS-SIAM Symposium on Vortex Dynamics (ed. R. Caflisch), pp. 171-181.

Berger M.S. & Nee J. (1989) "Leapfrogging of vortex filaments in an ideal fluid" (to appear in Contemporary Mathematics).

Fraenkel L.E. & Berger M.S. (1974) "A global theory of vortex rings", *Acta Mathematica.*

Helmholtz H. (1887) "On the integrals of the hydrodynamical equations that express vortex motion". *Phil. Mag.* (**4**) 33, pp 485-512. English translation of 1858 German original.

Liebowitz S. (1987) "Wave and bifurcations in vortex filaments", Studies of vortex dominated flows, pp. 3-15, Springer Publishers.

Love E.H. (1894) "On the motion of paired vortices with a common axis". *Proc. London Math. Soc.*, Vol. **25**, pp. 185-195.

Moffatt H.K. (1988) Generalised vortex rings with and without swirl. *Fluid Dynamics Research* **3**, 22-30.

Turkington B. (1989) Vortex rings with swirl: axisymmetric solutions of the Euler equations with nonzero helicity *SIAM J. Math. Anal.* **20**, 57-73.

VII: Homogeneous Turbulence

Anomalous Helicity Fluctuations and Coherence in Turbulence: Theory and Experimental Evidence

E. LEVICH

Faculty of Engineering, Tel-Aviv University, Tel-Aviv, Israel

ABSTRACT

In his seminal book, "The Structure of Turbulent Shear Flow", published in 1956, Townsend conjectured and phenomenologically justified the imperative for the existence of Big Eddies in turbulent flows. The big eddies were envisaged as correlated regions of turbulence united by a large scale coherent circulation on the scale $\xi_c >> L_E$, L_E being the energy containing or integral scale of the prime turbulent structure. Although it was pointed out by Townsend that the size ξ_c is eventually determined in conjunction with the large scale properties of turbulent shear flow, the underlying reason for the existence of Big Eddies was asserted to be fluctuations of circulation in primary small scale isotropic and homogeneous turbulence.

The purpose of this work is to show that in fact the existence of large scale correlated regions of turbulence is intrinsically induced by the dynamics of dissipation of small scale turbulence. Moreover, this induced organization is caused by topological properties of vorticity ω = curl v field lines subsequent to the Kelvin's theorem of circulation. Expanding on the conjectures made earlier by (Levich, Tsinober 1983; Moffatt 1984; Moffatt 1985; Levich 1987; Levich, Shtilman 1988), it will be demonstrated below that turbulence is a union of primary fluctuations/structures each with topologically complicated entangled vorticity lines. It will be claimed that these primary fluctuations/structures of the size of the order of L_E strongly interact with each other, forming coherent clusters on the scale $\xi_c >> L_E$. Numerical evidence indicates that the nature of the interaction between helical regions is preference of helical motions of opposite orientation to screen each other, so that the resulting global helicity would tend to zero.

1.1. Definition of Normal and Anomalous Helicity Fluctuations

There is only one known topological invariant for inviscid flows of fluids resulting from the Kelvin's theorem of circulation and expressed through the volume integral. This is helicity, a measure of entangleness of vorticity lines (Moffatt 1969).

$$H_V = \int_D \mathbf{v}\boldsymbol{\omega}\, dV = \text{invar} \tag{1}$$

for any compact domain D, e.g., $\boldsymbol{\omega}\cdot\mathbf{n}\Big|_{\partial D} = 0$. In turbulent flows the appropriate modification of (1) is the mean helicity density $H = \langle\mathbf{v}\cdot\boldsymbol{\omega}\rangle$. However in isotropic homogeneous turbulence H=0, due to the reflexional symmetry. The same is true for the mean helicity density, $H(k) = 4\pi(2\pi)^{-3}k^2\int\langle\mathbf{v}(0)\cdot\boldsymbol{\omega}(\mathbf{r})\rangle\exp(i\mathbf{kr})dV = 0$. The H(k) spectrum is defined similarly to the mean energy spectral density in isotropic homogeneous turbulence, $E(k) = 4\pi(2\pi)^{-3}k^2 \int\langle\mathbf{v}(0)\cdot\mathbf{v}(\mathbf{r})\rangle \exp(i\mathbf{kr})dV$, and as $E(\mathbf{k})$ is related to the symmetric part of the velocity two point correlation tensor, so $H(\mathbf{k})$ is to the antysymmetric one. When mirror symmetry is imposed the latter is zero, $H(\mathbf{k}) = 0$.

As any other quantity helicity should fluctuate in turbulence. The global measure of fluctuations is the variance per unit volume

$$I = \lim_{V\to\infty} V^{-1} \langle H_V^2\rangle = \int \langle\gamma(0)\gamma(\mathbf{r})\rangle dV = \int I(k)dk \tag{2}$$

$$I(k) = \lim_{V\to\infty} V^{-1} \int \langle H(\mathbf{k})H(\mathbf{k}')\rangle d\mathbf{k}$$

where $\gamma(\mathbf{r}) = \mathbf{v}\cdot\boldsymbol{\omega}$, $H(\mathbf{k}) = \mathbf{v}(\mathbf{k})\cdot\boldsymbol{\omega}(-\mathbf{k})$. For different definitions of I in (2) we have used homogeneity in physical space and the Parceval theorem $\int\mathbf{v}\cdot\boldsymbol{\omega}dV = \int\mathbf{v}(\mathbf{k})\cdot\boldsymbol{\omega}(-\mathbf{k})d\mathbf{k}$. The fluctuations are defined as normal if I is finite, and not zero, in the limit $V\to\infty$. On the contrary the anomalous fluctuations are defined as such, if in this limit I is infinite, or zero, signifying respectively acute correlations or antycorrelations. When I is infinite this simply means that due to enhanced correlations between fluctuations with the same sign of helicity the space average $\bar{H}^V = \langle H_V^2\rangle^{1/2} V^{-1} = 0(V^{-1/2} + |\delta|)$. On the contrary I = 0, means that due to antycorrelations of such fluctuations $\bar{H}^V = 0(V^{-1/2} + |\delta|)$. For normal fluctuations δ=0, i.e., in any volume V large enough one finds a large number of statistically independent fluctuations.

The above definitions are quite general and do not depend on a specific nature of fluctuating fields. The anomalous fluctuations in statistical physics are best learned in the theory of second order phase transitions. For instance ferromagnetics at the Curie temperature have infinite variance of the total magnetization. On the other hand antyferromagnets are totally coherent with zero variance of the same for zero temperature. Mixed situations are also known. In nonequilibrium systems fluctuations may be driven anomalous by specific constraints on the dynamics. This is what happens in turbulence with regard to helicity fluctuations. It should be noted that topological transformations require only small scale fluctuations of the vector of normalized vorticity $\omega/|\omega|$. Such fluctuations can leave velocity field $\mathbf{v}$ almost unaltered, and hence require infinitesimal work to be done by the viscous forces in the Navier-Stokes (NS) equation. Nevertheless, the large scale topology of ω - field is unlikely to fluctuate strongly since as will be explained below this would require fast coherent response by large scales to transformations occurring at the small ones.

1.2 Lack of Dissipation of Helicity Fluctuations

Helicity is conserved in inviscid flows, i.e., it is conserved by the nonlinear terms in the NS equation, as energy is. In homogeneous turbulence this is true for the mean helicity density, on a par with the mean energy density. This is expressed through the following balance equations

$$\frac{d\langle H\rangle}{dt} = \frac{d}{dt}\int H(k)dk = -\,2\nu\int k^2H(k)dk \tag{3}$$

$$\frac{d\langle E\rangle}{dt} = \frac{d}{dt}\int E(k)dk = -\,2\nu\int k^2E(k)dk \tag{4}$$

Although Eqs. (3) and (4) look similar there is a difference. In (4) the viscous term is purely dissipative. On the contrary in (3) the viscous term can both dissipate and generate helicity. This is because helicity is pseudo-scalar, and hence H(k) can be either positive or negative. Subsequently the r.h.s. of (3) can be either positive or negative dependent on a particular H(k). Further, since the total H is conserved by the nonlinear terms in the NS equation, so is helicity variance (if it is finite). Therefore the balance equation for I follows

$$\frac{dI}{dt} = -\,2\nu\int\langle[\omega(0)\cdot\mathbf{v}(0)\omega(\mathbf{r})\cdot\mathrm{curl}\,\omega(\mathbf{r})]\rangle dV = -\lim_{V\to\infty}$$

$$V^{-1}2\nu\int k^2\langle H(\mathbf{k})H(\mathbf{k}')\rangle d\mathbf{k}d\mathbf{k}' \tag{5}$$

To calculate I and dI/dt requires the knowledge of the corresponding fourth order velocity-vorticity correlation functions. These cannot be calculated theoretically. Still the fourth order correlation functions can be estimated, if we assume that **v**(r,t) is quasi-Gaussian statistical field, the assumption routinely used in closure schemes, (Monin, Yaglom 1975). With this assumption fourth order correlation functions are factorized as products of pair correlation functions. Using isotropy and collecting all the remaining terms yield the Gaussian value of helicity variance corresponding to incoherent fluctuations (Levich 1987)

$$I_G = 8\pi^2 \int E(k)^2 dk \tag{6}$$

From (6) we see that the dominant contribution to I_G comes from the energy containing wave numbers $|\mathbf{k}| \simeq L_E^{-1}$. Even smaller wave numbers are not contributing, since $E(k \to 0) = 0(k^2)$, the equipartition law. If we assume the Kolmogorov law in the inertial range $k_E = << k << k_d$, $E(k) \propto \langle\epsilon\rangle^{2/3} k^{-5/3}$ where $\langle\epsilon\rangle$ is the mean rate of energy density dissipation/injection, it becomes obvious that the values of $|\mathbf{k}|$ from the inertial range and larger are not essential for the value of I_G. Thus in the Gaussian approximation helicity variance is necessarily finite and not zero describing uncorrelated, statistically independent fluctuations of helicity, with the size of the order of L_E. Let us compare now the balance equations (3), (4) with $E(k) \propto \langle\epsilon\rangle^{2/3} k^{-5/3}$. We have evidently

$$\frac{d<E>}{dt} = - <\epsilon> = - 2\nu \int k^2 E(k) dk \propto - \nu <\epsilon>^{2/3} k_d^{4/3} \tag{7}$$

$$\frac{dI_G}{dt} = - 2\nu(2\pi)^3 \int k^2 E(k)^2 dk \propto - \nu <\epsilon>^{4/3} L_E^{1/3} \tag{8}$$

In the limit $\nu \to 0$ (or equivalently in dimensionless form $Re \to \infty$) for the total rate of energy density dissipation to stay finite implies $k_d \propto <\epsilon>^{1/4} \nu^{-3/4} = k_E Re^{3/4} \to \infty$. At the same time the integral in (8) is convergent for large $|\mathbf{k}|$, k_d independent, and hence is determined by the low values of $|\mathbf{k}| \simeq k_E$. Then in the limit $\nu \to 0$, $dI_G/dt \to 0(\nu) \to 0$. Thus in this limit I_G stays adiabatically invariant. Consider the limit of large times $t \to \infty$, when $E(k,t) \to 0$, but in the same time still $Re \to \infty$. This limit can be easily achieved for the decay turbulence in a thought experiment, by manipulating with L_E and ν. Further the self similarity assures independence of all conclusions from the particular values of these parameters, as long as $Re \to \infty$. However, the limit $E(k,t) \to 0$, is contradictory with having I_G = const. This latter inevitably imposes a constraint on the decay of energy containing eddies, resolved only if one suggests that a significant part of energy propagates upscale in the space of scales (Levich, Tzvetkov 1985; Levich, Tsinober 1983). The nature of the above constraint is easy to understand. In accordance with (6),turbulence consists of uncorrelated helicity fluctuations/structures, i.e., lumps of entangled vorticity lines of opposite signs of helicity.

The structures typical scale is L_E and the space averaged helicity of each of them $E\ L^{-1}_E$. The energy transfer to small scales is intrinsically accompanied by the growth of ensotropy $<\omega^2>$. The mechanism of this growth is the vorticity lines stretching. Since in the Gaussian approximation the structures are statistically independent, each of them can be considered separately. The vorticity lines stretch by folding and wrinkling at progressively smaller scales. If we assume a trivial helical configurations, that is of linked vorticity tubes of the scale L_E, it is clear that such folding and wrinkling will inevitably cause the linkages to impinge on each others. Generally we would expect the vorticity lines to break and reconnect, due to viscous forces enhanced at the locations of nonanalyticity of ω. The break and reconnection of the vorticity lines would change the topology, and ease the constraint imposed by (6). In the Gaussian approximation, however, this does not happen. Indeed, the vorticity lines reconnections occurring at the smallest scales of the order of $\ell_d = k^{-1}_d$, imply simultaneous change of topology determined however by the scales $\simeq L_E$. This would imply a rapid response of large scales to perturbations of the small ones, i.e., coherence between the two. <u>The topology dynamics is inherently coherent.</u> The Gaussian approximation on the other hand precludes such coherence. This is why I_G would have become adiabatic invariant. As a result, either a substantial part of energy should start flowing upscale, as was mentioned as a possibility above, or the nature of nonanalyticity of ω is such that the constraint imposed by the lack of dissipation of I slackens. The former possibility, (in forced turbulence taking the form of genuine inverse cascade of energy), at least for isotropic homogeneous turbulence is not supported by experiment. The latter one allows several alternatives. The simplest is that the small scale intermittency of ω^2, e.g., concentration of dissipation ϵ and ω^2 on a fractal in the limit limit $Re \to \infty$, while leaving almost intact $I \simeq I_G$, in the same time enhances dI/dt, so that Eqs. (3) and (4) would become compatible. We estimate $\sup|<\gamma(0)\gamma(\mathbf{r})>|$ at $|\mathbf{r}| \geq k_d^{-1}$. It is sufficient to consider $<\gamma(0)^2>$. Applying Shwartz inequality yields $<\gamma(0)^2> \leq <v^2\omega^2> \leq <v^4>^{1/2}<\omega^4>^{1/2}$. With a good accuracy $<v^4> = 3<v^2>^2 \propto L_E^{2/3}$. Assuming that ω^2 and ϵ have the same intermittency we obtain $<\omega^4>/<\omega^2>^2 = <\epsilon^2>/<\epsilon>^2 \propto (L_E k_d)^\mu = Re^{3\mu/4}$. The parameter of intermittency $\mu < 0.5$ according to all available experimental data (Monin, Yaglom, 1975; Meneveau, Sreenivasan 1987). Using this empirical fact we evaluate the maximal possible contribution of small scales of the order of ℓ_d to I. Disregarding all k_d independent factors

$$\sup\left\{\int_0^{\ell_d} <\gamma(0)\gamma(\mathbf{r})> r^2 dr\right\} \propto k_d^{1/4} k_d^{4/3} k_d^{-3} = k_d^{-17/12} \propto Re^{-17/16} \to 0 \tag{9}$$

where $<\omega^2> \propto k_d^{4/3}$, $k_d^{-3} = \ell_d^3$ is the volume factor, $k_d^{1/4}$ is the intermittency factor with $\mu = \mu_{max} = 0.5$. Hence the contribution of small separation scales to I is negligible. Consider now dI/dt. This necessitates more assumptions. If I is finite in the limit $Re \to \infty$, (possibility of nonanalytical dependence on Re is disregarded), and scaling in the inertial range is assumed, it follows that I(k), defined in (2), is a power law $I(k) \propto k^{-x}$ in the inertial range and Re independent (although generally L_E dependent). This is enough to determine the lower bound value of x from (9), $x \geq 29/12$, so that $I_{max} \propto k_d^{-17/12}$. Hence the estimate

$$\left|\frac{dI}{dt}\right|_{max} = \left|2\nu \int_{k_d}^{\infty} k^2 I(k) dk\right| \propto k_d^{-4/3} k_d^{7/12} = k_d^{-3/4} \propto Re^{-9/16} \to 0 \tag{10}$$

and the conclusion that intermittency is unlikely to enhance dI/dt .

It should be pointed out that if turbulence lacks global reflexional symmetry for instance under the action of external forces than instead of I one rather considers

$$I' = \lim_{V \to \infty} V^{-1} <\Delta H^2_V> \tag{11}$$

where $\Delta H_V = H_V - <H_V>$. It is not difficult to show that in the Gaussian approximation $I'_G \equiv I_G$, and hence helicity fluctuations in this approximation impose exactly the same constraints.

The above considerations lead one to suggest another alternative. Namely that the helical structures instead of being statistically independent, as the Gaussian approximation implies, in fact strongly interact in such a way that the resulting I and I(k) are Reynolds number dependent. The Reynolds number dependence in practice should mean that in the double limit $V \to \infty$, $Re \to \infty$, both I and I(k) should tend either to infinity or to zero, as a function of Re, the only parameter in the problem. The former case corresponds to long range correlations between the prime structures of the same helicity sign, and the latter to anty-correlations of structures with same helicity sign. Note, that self-similarity in the inertial range is not violated if I(k) is Reynolds dependent, since it is defined in (2) as the integral over all values of $|\mathbf{k}'|$, including $|\mathbf{k}'|$ outside the inertial range. Since the small separation scales are weakly contributing to the values of I and dI/dt, it should be that the large scales of separation $|\mathbf{r}| > L_E$ contribute anomalously to these quantities. Importantly we should not expect anomalies for the separation scales within the interval $L_E < |\mathbf{r}| < \ell_d$. Due to self similarity in the inertial range its contribution to I, should be of the same order as in the Gaussian approximation, e.g. $0(L^{7/3}{}_E)$. This is simply proved based on the dimensional considerations, the fact that L_E is the only external scale in the problem and positivity of $\langle\gamma(0)\gamma(\mathbf{r})\rangle$. Also, one should disregard the possibility that I may be zero, because of a trivial reason such as that the values of $\gamma = \mathbf{v}\cdot\boldsymbol{\omega}$ are uncorrelated on the scales $|\mathbf{r}| > \ell_d$. This would obviously contradict the concept of self-similarity in the inertial range, since both velocity and vorticity fields correlate at all separation scales from the inertial range in accordance with simple scaling laws (and hence $\langle\gamma(0)\gamma(\mathbf{r})\rangle$ is positive everywhere in this range).

A note of caution should be introduced. By definition helicity is a topological quantity only for compact flow domains,, i.e., in our case bounded by vorticity surfaces. It is unlikely that turbulent flow consists of such ideal compact structures, unless they are statistically independent. If vorticity lines stick out as they would for strongly interacting flow regions then helicity as a measure of entangleness should be redefined in order to subtract spurious surface terms. Such procedure is well defined, (Berger, Field 1984; Levich 1987). For an arbitrary domain D_λ one introduces $\boldsymbol{\omega}^\lambda = \boldsymbol{\omega} - \boldsymbol{\omega}'$, where $\boldsymbol{\omega}' = \boldsymbol{\omega}$ outside D_λ, $\omega' = \nabla\xi$ inside D_λ and ξ is harmonic function $\nabla^2\xi = 0$ uniquely determined by the boundary condition $\nabla\xi\cdot\mathbf{n}\Big|_{\partial D_\lambda} = \boldsymbol{\omega}\cdot\mathbf{n}\Big|_{\partial D_\lambda}$. The redefined helicity is

$$\tilde{H}_V = \int_\infty \mathbf{v}\cdot\boldsymbol{\omega}^\lambda dV \equiv \int_{D_\lambda} \mathbf{v}\cdot(\boldsymbol{\omega} - \nabla\xi)dV \tag{1a}$$

It is clear that $\tilde{H}$ is a measure of entangleness of the field $\boldsymbol{\omega}^\lambda$ for which D_λ is a compact domain. The subtraction of $\nabla\xi$ has the meaning of defining a reference frame for comparing the entangleness of $\boldsymbol{\omega}$ - lines and $\nabla\xi$ - lines. The entangleness of $\nabla\xi$ - lines is assumed in a certain sense as a minimal one. It is easy to show that $\boldsymbol{\omega}' = \nabla\xi$ minimises the global enstropy in D_λ. For compact D_λ obviously $\tilde{H}_V \equiv H_V$. We should have used this redefined quasi-helicity for calculation of the variance. Happily, I stays invariant for the infinite (or periodic) flow volume. Therefore, for the sake of clarity, we shall use as a rule the terminology of helicity for parts of the flow with understanding that it is in fact quasi-helicity.

We conclude that extremely general assumptions such as self-similarity of turbulence in the inertial range are sufficient to imply intrinsic fluctuations of helicity of the size L_E and opposite helicity signs. This result does not dependent on the statistical properties of fluctuations, and would be true even with the assumption of Gaussian statistics. Nevertheless, properties of the energy cascade strongly suggest that statistical independence is not possible. The helicity lumps instead interact strongly so that structures with opposite helicity sign either amplify or on contrary screen each other. This strong interaction between primary helical fluctuations constitutes the nature of intrinsic coherence in turbulence. The larger Re is, the more extended are the clusters of corrrelated helicity regions. In the limit $Re \to \infty$, $\xi_c \to \infty$. The above theory however cannot answer the question which of the two alternatives, the amplification or screening, is more likely.

1.3 Analysis of the BigBox Turbulence

The analysis is based on the data sets developed in (Shtilman, Polifke, 1989; Polifke, Shtilman 1989) for $(64)^3$ modes resolution, and also (Levich, Shtilman, 1988), $(128)^3$ modes resolution. First we formulate those quantities in the periodic box with elementary cell volume $V_c = (2\pi)^3$ which are equivalent to I and dI/dt in the continuous spectrum turbulence. Evidently we obtain

$$I = \left(\frac{V_c}{N} \right)^2 V_c^{-1} < (\Sigma\, \mathbf{v} \cdot \boldsymbol{\omega})^2 > = (2\pi)^3 < (\overline{H}^N)^2 > \tag{12}$$

where by comparison with Eq. (2) the volume element $dV \to V_c/N$, integration is substituted by summation over all grid points in the periodic box, and $\overline{H}^N = N^{-1}\, \Sigma\, \mathbf{v} \cdot \boldsymbol{\omega}$, is the space averaged helicity density. Also it is easy to derive

$$I(k) = (2\pi)^3 < H^N\, H(k)^s > \tag{13}$$

$$\frac{dI}{dt} = -\, 2\nu\, \Sigma\, k^2 I(k) \tag{14}$$

where $H(k)^s$ is the shell averaged helicity spectral density, so that $\overline{H}^N = \Sigma\, H(k)^s$. If we accept the Gaussian approximation, then by its very meaning of describing statistically independent fluctuations, it follows that I_G and dI_G/dt should be the same as for the continuous spectrum and infinite flow domain turbulence. No dependence on V_c or N may be relevant. Hence, $\int dk \rightarrow \Sigma_k$, and

$$I_G = 8\pi^2 \Sigma < [E(k)^s]^2 > ; \; I_G(k) = 8\pi^2 <[E(k)^s]^2> ;$$
$$\frac{dI_G}{dt} = - 16\pi^2\nu \Sigma k^2<[E(k)^s]^2> \qquad (15)$$

In Eqs. (13)-(15), <> is the averaging over realizations of the BigBox decay turbulence. It is usual, to assume that the number of grid points in numerical turbulence is large enough to consider single realization as a typical one, in a sense that $E^s(k) \simeq <E(k)^s>$, etc. Also the values of I from realizations should not greatly deviate from the ensemble average I. The situation is quite different with $H(k)^s$. Indeed it should be for statistically reflexionally symmetric turbulence, $<H(k)^s> = 0$. On the other hand, since fluctuations are inevitable, this implies $I \neq 0$ for a finite flow volume for all theoretical alternatives. Hence, trivial reasoning leads us to a non-trivial conclusion, that whatever the initial conditions are, including $H(\mathbf{k},t=0) \equiv 0$, it must be $\overline{H}^N(t>0) \neq 0$. In particular the Gaussian assumption yields $\overline{H}^N_G \simeq \pi^{-1/2} \{\Sigma[E(k)^s]^2\}^{1/2}$. It should be pointed out, however, that this is a consequence of the finiteness of the flow volume $V_c = (2\pi)^3$. For unbounded flows, of course, $\overline{H}^N = \{\Sigma[E(k)^s]^2\}^{1/2} (8\pi^2V^{-1})^{1/2} = 0(V^{-1/2}) \rightarrow 0$, as was explained above (see section 1.1). Dependent on the nature of interaction between the fluctuations, it should be $\overline{H}^N << \overline{H}^N_G$, or $\overline{H}^N >> \overline{H}^N_G$, <u>provided that the ratio $\alpha = V^{1/3} L^{-1}_E >> 1$.</u> When this condition is violated, peculiarities connected with dependence on α, are expected. These should be absent only in the case of normal fluctuations of helicity. Similar comparison can be made with regard to dI_G/dt versus dI/dt, etc.

We consider now a typical run of $N = (64)^3$ modes resolution turbulence with $H(\mathbf{k},t=0) < 10^{-12}$ in dimensionless units, $L_E(t=0) \simeq (2\text{-}3)^{-1}2\pi$, and the Taylor microscale Reynolds number $Re_\lambda(t=0) = 48$. The evolution of I is as follows. It grows with time from zero dramatically, in fact faster than exponentially till the time t_{max} at which enstrophy reaches its highest value. It is customary assumed in numerical simulations of turbulence that the time interval around $t=t_{max}$ corresponds to the most developed stage. For $t > t_{max}$ turbulence decays very rapidly, so that $\overline{E}^N$ and ω^{2N} fall off exponentially with time. Therefore it is most appropriate to compare I and I_G at $t \geq t_{max}$. It follows that $I\, I^{-1}_G \simeq 0.035$. In more detail we consider the spectrum $H(k,t_{max})^s$. It follows that for low values, $|\mathbf{k}| < 10$, $H(k)^s$ is of alternating signs. Thus although the amplitudes are quite significant, the overall value of $|\overline{H}^N| << |\overline{H}^N_G|$. In other words, at low values of $|\mathbf{k}|$, the spectrum shows significant anty-correlations between $H(k)^s$ of opposite signs. In the same time for $|\mathbf{k}| \geq 10$, $H(k)^s$ is almost everywhere of the same sign and for these modes $[\Sigma H(k)^s]^2 \simeq 3\pi^{-1}\{\Sigma[E(k)^s]^2\}$, signifying significant coherence (note the factor 3).

At all times the dynamics is obviously very different from what this would have been in the Gaussian approximation. For instance $dI/dt > 0$, very persistently, whereas from its definition $dI_G/dt < 0$. It may occur that we started from a strongly non-Gaussian constraint upon $H(k)^s$, and turbulence did not adjust rapidly to have $I \simeq I_G$, $dI/dt \simeq dI_G/dt$. However, starting from $I(t=0) \simeq I_G(t=0)$ does not help, at all. Indeed, again at $t \simeq t_{max}$ we obtain $I << I_G$. The above results repeat at all runs in their main qualitative features, including $(128)^3$ modes resolution runs (Levich, Shtilman, 1988), and therefore hold in the sense of ensemble average.

It is reported (Polifke 1989) that for very advanced times, $t >> t_{max}$ the continuous growth of I persists, so that for these times $I \simeq I_G$, contrary to what takes place at $t \geq t_{max}$. Although as was mentioned at these times numerical turbulence is a suspect in terms of its relevance to developed turbulence, still it seems that the reason for such growth of I beyond I_G is the growth of scale of primary structures, so that fewer of them would fit inside the box, probably accompanied by a loss of correlation and reduced screening. Importantly, it also appears that in the runs with larger α, the ratio $I\, I^{-1}_G$ is further reduced, and the quantities $I(k)$, dI/dt are sensitive to the value of α in a predictable way.

Numerical evidence supports the one of the alternatives of coherent fluctuations in which primary fluctuations of opposite sign of helicity screen each other with subsequent value of $I \rightarrow 0$ in the limit $V \rightarrow \infty$, $Re \rightarrow \infty$. This result, although mentioned by the author as a possibility in the past still has been systematically neglected, with expectation of the opposite case of $I \rightarrow \infty$. Perhaps the inner reason for this has been the fear of acute difficulty, both for understanding and experimental ellucidation of such virtual helicity fluctuations. (Levich 1987).

1.4 Laboratory Evidence

The experimental details are in (Dracos, et al 1989). Here an interpretation of relevant experiments is briefly discussed. This is the direct measurement of the time averaged helicity $\overline{H}^T$ in grid generated turbulence. This quantity plays in experiment the role similar to

$\overline{H}^N$ in numerical experiment. In all experiments with both helicity enhanced at the grid and not the following basic features were observed: a) at moderate distances from the grid $\overline{H}^T$ does not exceed experimental accuracy and in this sense can be considered zero (even when helicity is injected at the grid); b) at larger distances $\overline{H}^T$ shows significant increase reaching (4-6)% of its maximal possible value; c) the sign of $\overline{H}^T$ at these large distances is controlled by the spiral orientation induced by external perturbations at the grid.

The above properties seem nontrivial. In particular they indicate that the growth of $\overline{H}^T$, far away from the grid, except for the sign, does not critically depend on whether helicity is injected extrinisically or not. The observations can be interpreted with the help of the above concept. Indeed, the grid turbulence is a finite volume follow, with its dimension approximately determined by the transverse cross section. At moderate distances from the grid L_E is small by comparison with this dimension, $L_E << D$. Hence, helicity structures of opposite sign can screen each other well. Far from the grid the energy containing scale grows, so that $L_E \lesssim D$. Then no compensation is possible, similarly to what takes place in numerical turbulence for $t >> t_{max}$, and as a result $\overline{H}^T$ grows. In all numerical runs the sign $\overline{H}^N$ even at very advanced times is determined by that at the earliest stage $t << t_{max}$, even though $\overline{H}^N$ is very small at these times. The same is likely to be observed in grid turbulence when the initially induced reflexional asymmetry controls the sign of $\overline{H}^T$ at the late stage of turbulence decay. Clearly much more convincing evidence would have been the direct measurement of $I = \int <\gamma(0)\gamma(\mathbf{r})> dV$ which probably can be done in online experiments. Qualitatively, it is clear that for I to be small by comparison wth I_G implies that for certain $|\mathbf{r}| > L_E$ $<\gamma(0)\gamma(\mathbf{r})>$ becomes distinctly negative. Analysis of the data from (Dracos et al. 1989) shows that the correlator $<\gamma(0)\gamma(\mathbf{r})>$ is almost indistinguishable from $<\omega(0)\cdot\omega(\mathbf{r})>$, and indeed both become negative for $|\mathbf{r}| \geq L_E$, and stay negative henceforth. While precise determination of I is not possible, since it becomes unphysically negative in experiment, indicating insufficient accuracy for $|\mathbf{r}| >> L_E$, still the data allows to deduce the upper bound for I. It appears that $I\, I^{-1}_G < (\ 1/6\ -\ 1/3\ )$ dependent on the distance from the grid. In case the result is supported by further experiments this would be a remarkably clear evidence in support of the above theory.

Conclusions

The problem of understanding helical fluctuations and their observation is of challenging complexity both from theoretical and experimental point of view. Theoretically we are yet to relate the concept of fluctuations with the dynamics of energy transfer, etc. The driving idea is that coherence in the regions of enhanced quasi-helicity of the size $\simeq L_E$, i.e., in primary structures/fluctuations, reduces the nonlinear interaction by comparison with its incoherent value, while the interaction betweem them upholds their coherence (Levich, 1987). It should be pointed out that up to now there is no available quantitative data concerning the magnitude of quasi-helicity in the primary structures. The value of $\overline{H}^N$ (or $\overline{H}^T$) is not at all useful in this regard. The indirect qualitative evidence of induced coherence in the structures by comparison with the abstract situation of statistically uncorrelated structures (modelled for instance by initial Gaussian conditions in numerical simulations) is the enhanced allignment of $\mathbf{v}$ and $\boldsymbol{\omega}$ in physical space, the well established fact by now (Pelz et al 1985; Rogers, Moin 1987). Incidentally this allignment reduces for the times $t \gg t_{max}$, apparently indicating a loss of correlations (Polifke, 1989). On the level of interpretation the reduction of nonlinear coupling observed in (Panda, Kraichman 1988; Shtilman, Polifke 1989) can be seen as giving support to the above concept.

On experimental front numerical simulations so far have been the most useful. However, they are insufficient because of severe limitations on resolution, and subsequently on the value of Reynolds number, $N \propto Re^{9/4}$. No real inertial range can be achieved in numerical turbulence in the forseeable future.

The existing laboratory experiment is intrinsically badly suited for the ellucidation of helical structures, because it does not provide the data on 3-D vorticity field. Still very precise 1-D measurements seem to capture residual effects. Also visualization and other methods employed in studying coherent structure in turbulence may be able to establish the double helix or braid-like patterns as probably universal in turbulence. These seem to be the simplest ones, characterstic for mutual screening of the prime structures with opposite orientations of spiral motion.

Acknowledgements

The author is indebted to M. Azbel, E. Kit, D. Lilly, A. Migdal, B. Moishezon and A. Tsinober for illuminating discussions during various stages of this work. A special gratitude is extended to W. Polifke and L. Shtilman for the development of numerical runs and indispensable criticism which significantly contributed to the authors understanding of the relation between periodic box turbulence and turbulence in infinite flow domains. The work was supported in part by the USA Department of Energy Grant, DE-FG02-88ER 1383J.

References

1) V. Arnold, The asymptotic Hopf invariant and its application (in Russian), 1974, Proc. Summer School in Differential Equations, Erevan, Armenian S.S.R. Acad. Sci.
2) M.A. Berger, G.B. Field, 1984, JFM, 147, 133.
3) T. Dracos, M. Kholmyansky, E. Kit, A. Tsinober, 1989, Topological Fluid Dynamics, Cambridge, England.
4) E. Levich, L. Shtilman, 1987, Phys. Lett. A.
5) E. Levich, A. Tsinober, 1983, Phys. Lett. A, 96, 292.
6) E. Levich, E. Tzvetkov, 1985, Phys. Rep. 120, 1.
7) E. Levich, 1987, Phys. Rep. 151, 131.
8) C. Meneveau, K.R. Sreenivasan 1987, Phys. Rev. Lett. 59, 1424.
9) A.S. Monin, A.M. Yaglom 1975, Statistical Fluid Mechanics, Vol. 2 (MIT Press).
10) H.K. Moffatt, 1969, JFM 35, 117.
11) H.K. Moffatt 1984, Proc. IUTAM Symp. Turbulence and Chaotic Phenomena in Fluids, Kyoto 1983 (North Holland).
12) H.K. Moffatt, 1985, JFM 159, 359.
13) R. Pelz, L. Shtilman and A. Tsinober 1986, Phys. Fluids 29, 3506.
14) W. Polifke, 1989, Private communication.
15) W. Polifke, L. Shtilman 1989, submitted to Physics of Fluids.
16) M. Rogers, P. Moin, 1987, Phys. Fluids 30, 2662.
17) L. Shtilman, W. Polifke 1989, accepted for publication as a letter to Phys. Fluids.
18) K.R. Sreenivasan, C. Meneveau, Singularities of the equations of fluid motion, 1988, preprint to appear in Phys. Rev. A.

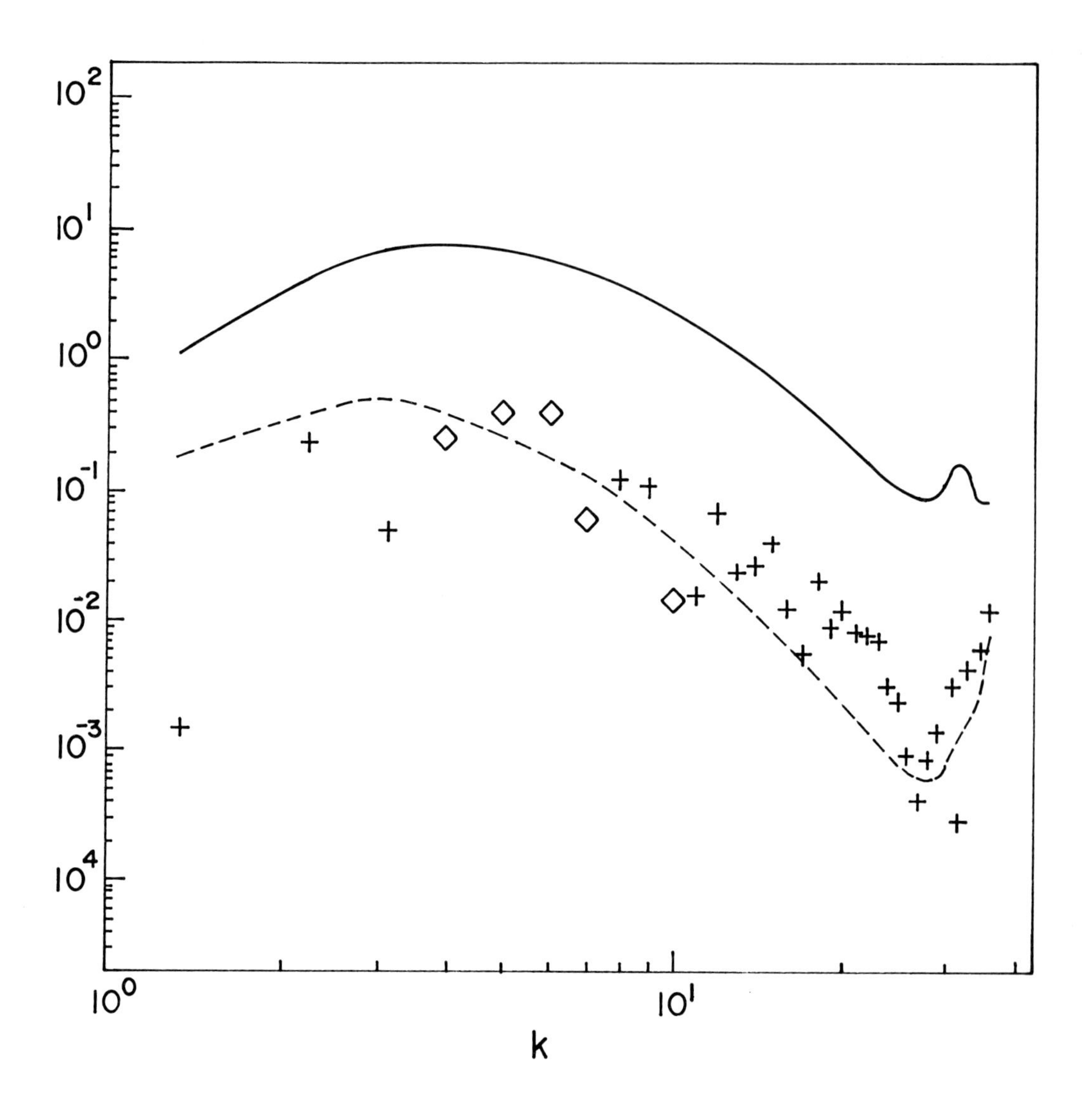

Fig. 1 The spectrum $H(k)^s$ for $t = t_{max}$, is marked by squares, negative values and crosses, positive values; the dash line is the standard Gaussian deviation for the shell averaging (Polifke 1989) $|H_G^s(k)| = \pi^{-1/2} E(k)^s$, the bold line is $2kE(k)^s$ for comparison, $\alpha = 2$ at $t = 0$.

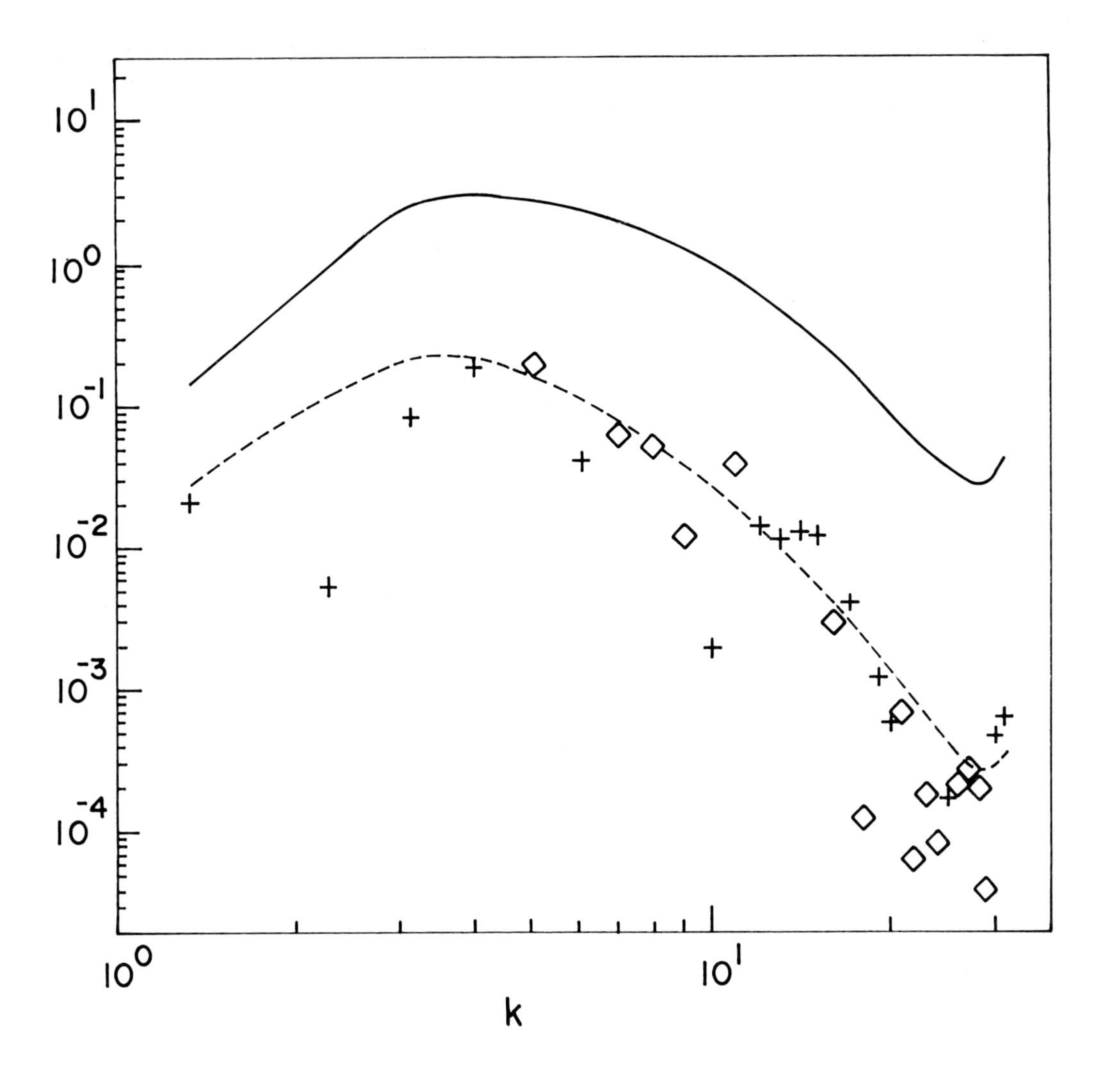

Fig. 2 The same as Fig. 1, but for a different run with $\alpha = 4$, at $t = 0$, but the same $Re_\lambda(t{=}0)$; the corresponding values of $I\, I_G^{-1}$, $|I(k)|_G^{-1}\, I(k)$ and $dI/dt\, (dI_G/dt)^{-1}$ are distinctly smaller than in the first run discussed in the text.

Some Experimental Results on Velocity-Velocity Gradients Measurements in Turbulent Grid Flows

T. DRACOS[1], M. KHOLMYANSKY[2], E. KIT[2], & A. TSINOBER[2]

[1]Institut für Hydromechanik und Wasserwirtschaft, ETH – Hönggerberg, CH-8093, Zurich, Switzerland
[2]Faculty of Engineering, Tel Aviv University, Tel Aviv 69978, Israel

ABSTRACT

We present, in this paper, results of two kinds of experiments on turbulent grid flow: flow in the wind tunnel of the Institute for Hydromechanics, ETH, Zurich and salted water flow in the water tunnel of the Laboratory for Vorticity-Helicity studies, Faculty of Engineering, Tel Aviv University.

The air flow measurements included three velocity components and their nine gradients. This was done by a 12 hot-wire probe (3 arrays x 4 wires), produced for this purpose using specially made equipment (micromanipulators, etc.) calibration unit and calibration procedure. The probe had no common prongs and the calibration procedure was based on constructing a calibration function for each combination of three wires in each aray (total 12) as a three-dimensional Chebishev polynomial of fourth order. A variety of checks were made in order to estimate the reliability of the results.

The water flow experiments included measurements of the longitudinal components of velocity and vorticity. This was made by a probe consisting of a seven electrode probe, allowing to perform absolute measurements of ω_1, assembled together with a commercial hot film probe. The main bulk of measurements was made past a grid with circular mesh. This was necessary to perform experiments with controlled sign of mean helicity of the turbulent fluctuations which was achieved by installing various propellers in the grid holes.

The results with air flow include, among others, comparison of three-dimensional spectra of vorticity obtained from velocity spectrum and directly, checks of the relation $<s_{ij} s_{jk} s_{ki}> = - 3/4 (\omega_i \omega_j s_{ij})$; $s_{ij} = \frac{1}{2} (\partial u_i/\partial x_j + \partial u_j/\partial x_i)$, $\omega_i = \epsilon_{ijk} \partial u_j/\partial x_k$, comparison of the enstrophy generating term $<\omega_i \omega_j s_{ij}>$ with skewness of velocity derivatives, statistical properties of the Lamb vector $\lambda_k = \epsilon_{ijk} \omega_i u_j$, alignment between the vorticity vector and eigenvectors of the rate of strain tensor, alignment between the velocity and vorticity vectors, helicity, etc.

The results with salted water flow include mainly various helicity related quantities (mean, spectra, etc.) for grids with propellers producing perturbation clockwise, anticlockwise or both as well as for grid with empty holes. Other results like the ratio $<\omega_1^2(\partial u_1/\partial x_1)>/<(\partial u_1/\partial x_1)^3>$ both for air and water flow are reported.
Apart from confirming the phenomenon of symmetry breaking (Kit et al (1987, 1988), Tsinober et al (1988)) new results about properties of the vorticity field are included. Finally, we present a few preliminary results on measurements in the outer region of a boundary layer over a smooth plate and compare them with those for the flow past a grid.

1. INTRODUCTION

Obtaining experimental information on the field of velocity derivatives in turbulent flows was motivated mostly by the desire to get information on their small scale structure and dynamics and in the first place vorticity (for an updated review see Foss and Wallace (1989)). Other quantities of importance are the rate of strain tensor, dissipation, etc. (e.g. see Antonia et al (1988) and references therein). In spite of a great number and variety of efforts as described by Foss and Wallace (1989), there are still many doubts as to the reliability of measurement methods of velocity fluctuations derivatives (e.g. Hussain (1986), Aref and Kambe (1988)).

For this reason (among others) we have chosen to concentrate our efforts on the grid turbulent flow as the most appropriate allowing to carry out checks of reliability of the measurements especially of velocity derivatives.

On the other hand, the particular significance of this kind of flows follows from the fact that homogeneous (and quasi-isotropic) flow is free from external influences like mean shear, centrifugal forces (rotation), buoyancy, magnetic field, etc., which usually act as an organizing factor, favouring the formation of coherent structures of different kinds (quasi-two-dimensional, helical, etc.). We therefore believe that the turbulent grid flow still is one of the most suitable for studying the universal properties of turbulent flows and especially their small scale structure.

This paper contains a report on attempts to measure some characteristics of the field of velocity derivatives along with conventional properties of the velocity fluctuations.

2. EXPERIMENTAL FACILITIES INSTRUMENTATION, DATA ACQUISITION AND PROCESSING

The experiments were performed in two experimental facilities:

- wind tunnel of the Institute for Hydromechanics, ETH, Zurich and
- water tunnel of the laboratory for vorticity-helicity studies, Faculty of Engineering, Tel Aviv University.

2.1. Air Flow Experiments

The air flow experiments were performed in the open-circuit wind tunnel of the Institute for Hydromechanics, ETH, Zurich with a rectangular cross section 1.4 x 1.2m². The grid was made of wooden rods 1.5 cm in diameter with square mesh of size 6 cm (solidity of 0.44). The measurements were made at the distances x/M = 8; 17; 30; 38; 64; 90 from the grid and without grid. Most of the measurements were made at the mean velocity ~ 7m/s.

We used a multihotwire technique similar to that suggested by Balint et al (1987, 1988) which is based on use of arrays consisting of several hot wires (usually three). The voltages obtained from each wire are produced by all the three components of velocity (fluctuations). These latter are separated by means of some calibration procedure. This is an intrinsic feature and shortcoming of the method since any imperfection in the measuring system (mechanical, electronic, etc.) and/or experimental facility leading to an error in the signals from the hot wires gives rise to errors in all the three velocity components. The important point is that these errors are (spuriously) correlated. Consequently, so will be many quanitaties obtained from these velocity components. Errors originating in this way may become dominating when measuring quantities like $\langle u_i u_j \rangle$, $i \neq j$ (in grid turbulence they should vanish), $u_i \omega_i$ etc.

It is therefore of special importance to make any possible improvement of the measuring system, calbriation procedure, etc. At the present stage we made the following improvements.

One of the possible error sources of the conventional 9 hot wire probes is the presence of a common prong in every array of three hot wires. This may lead to electronic cross-talking between these hot wires and consequently to unpredictable errors which may significantly contaminate the data. Thus, we have developed a multihotwire probe without common prongs. The problem was solved by producing a compound, splitted prong (instead of the common one), consisting of several (three or four) tungsten wires coated by teflon and glued together. It is noteworthy that glueing requires a specially developed teflon etching procedure. This, along with other reasons, required to give up the conventional sharpening of the tungsten prongs thermally and mechanically. Instead, an electrochemical erosion method was used, allowing to achieve thickness of the tips of the prongs less than 20 microns. Getting rid of common prongs is also important for reducing the length of individual wires, since with common prongs one has to keep their resistance very large comparing to the common resistance.

Another problem we encountered is that the single wire (i.e., the one not having a partner) is incapable of catching sufficient information on the corresponding velocity component even in a probe without common prongs. Tests of four wire arrays showed that they are capable of properly measuring all the three velocity components. We therefore built a probe consisting of 12 hot wires (3 arrays x 4 wires). It's tip is shown in Fig. 1 together with a scheme of the whole probe.

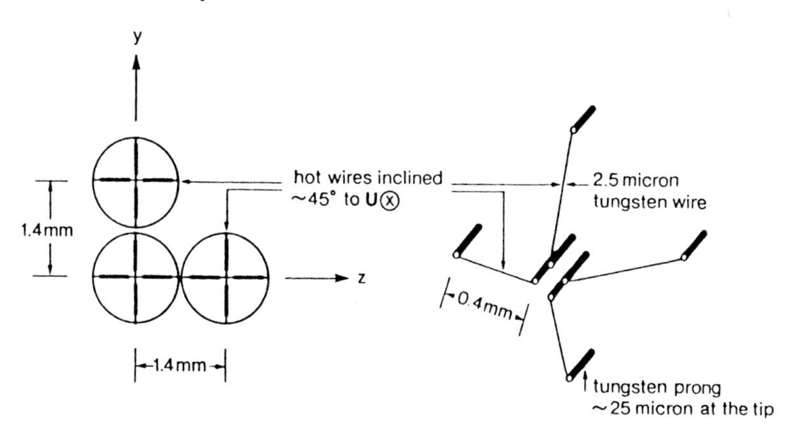

Fig.1. Schematic of a twelve hot wire probe and a 4 wire array.

Manufacturing the above probe required to develop a sophisticated manipulator consisting of three manipulating units each having six degrees of freedom (three translational and three rotational), a special unit for manipulating of 2.5 micron wires and a unit for microwelding. Three manipulating units were necessary to manipulate in an independent way the probe, the welding electrode and the rod carrying a small piece (2-3 mm long) of 2.5 micron wire glued to its tip. The special unit mentioned above in conjunction with one of the manipulating units with six degrees of freedom allowed to glue small pieces of 2.5 micron tungsten wire at the tip of the steel rods.

The third problem is that the existing calibration procedure had the following shortcomings:

i - It employs restrictive assumptions about the symmetry of each array and the flow around the probe. ii - King's law is used for every wire. iii - On the basis of i and ii very limited information is being used to obtain the calibration.

For the above reasons, a new calibration procedure was developed. The main feature of it is the following:

i - For each combination of three wires in each array (i.e. total 4x3=12) a calibrating function is produced which gives the relation between the three velocity components and three voltages obtained from each wire in the array.

ii - The calibration functions for every velocity component in each array was construced as three-dimensional polynomials of the fourth order using Chebishev orthogonal polynomials. Each calibration function requires determination of 35 coefficients, which were found from measurements at 81 space points (9 yaw and 9 pitch angle positions) and 7 velocities.

To produce this amount of information, it was necessary to develop an automatic calibration unit. This consisted of two miniature stepping motors and mechanical arrangement producing yawing and pitching in respect to the tip of the probe remained in a fixed spatial position. This allowed to avoid errors due to flow inhomogeneities. The calibration unit was driven by a controller specially produced for this purpose.

The above modifications led to a considerable reduction of errors and improvement of results. For example, the isotropy relations between the terms of the matrix $<(\partial u_i/\partial x_j)^2>$ were satisfied within 15% of error (previously 30%, see Kit et al (1987, 1988)) as can be seen from Table 1

Table 1. Values of the matrix $<(\partial u_i/\partial x_j)^2>^{1/2}$ ·, s^{-1}

X/M	8	17	30	38	64	90	B. Layer y/δ 0.7	0.2
$<(\partial u_1/\partial x_1)^2>^{1/2}$	150	64.4	26.5	20.1	15.0	7.95	98.9	114
$<(\partial u_1/\partial x_2)^2>^{1/2}$	207	94.9	41.2	31.0	23.8	12.4	150	184
$<(\partial u_1/\partial x_3)^2>^{1/2}$	219	99.6	42.6	28.6	24.9	13.2	136	154
$<(\partial u_2/\partial x_1)^2>^{1/2}$	175	72.8	32.4	23.7	14.7	9.6	122	139
$<(\partial u_2/\partial x_2)^2>^{1/2}$	143	58.0	24.2	19.6	11.1	7.0	106	129
$<(\partial u_2/\partial x_3)^2>^{1/2}$	184	79.3	35.2	26.3	16.4	10.8	129	146
$<(\partial u_3/\partial x_1)^2>^{1/2}$	175	73.9	32.1	24.5	15.0	9.5	122	144
$<(\partial u_3/\partial x_2)^2>^{1/2}$	186	81.4	35.8	27.0	17.7	12.1	136	186
$<(\partial u_3/\partial x_3)^2>^{1/2}$	152	66.3	28.9	18.4	13.9	9.1	106	131

Table 1a. Values of rms of velocity and vorticity fluctuations

x/M	8	17	30	38	64	90	B. Layer y/δ 0.7	0.2
$<u_1^2>^{1/2}$, m/s·10²	52.6	19.7	17.5	14.4	13.7	8.7	80	72
$<u_2^2>^{1/2}$	43.8	23.5	14.3	11.7	9.0	7.0	55	38
$<u_3^2>^{1/2}$	41.3	23.4	13.8	11.9	8.8	6.7	72	72
$<\omega_1^2>^{1/2}$, s^{-1}	275	118	52.5	30.6	24.9	16.8	201	252
$<\omega_2^2>^{1/2}$	303	134	57.5	40.9	31.1	17.5	196	223
$<\omega_3^2>^{1/2}$	292	131	57.0	42.6	30.6	16.9	209	248
$Re_\lambda = \frac{u_1\lambda_1}{\nu}$	96	88	74	67	82	63	415	290

and the incompressibility - Taylor hypothesis test (correlation coefficient C_T between $\partial u_1/\partial x_1 = -U^{-1}\, \partial u/\partial t$ and $-\, \partial u_2/\partial x_2 - \partial u_3/\partial x_3$) produced a value of about 0.7 without any smoothing or other similar processing of the raw data (previously ~0.35). It is noteworthy that if the derivatives $\partial u_2/\partial x_2$ and $\partial u_3/\partial x_3$ are obtained *as one sided* (as in Fig. 1), C_T is limited from above by a value less than one, *independently of how small the probe is.* This can be shown assuming isotropy in small scales and obtaining that $C_T \leq (2/3)^{1/2} \approx 0.82$. When the above derivatives are calculated as central ones then the upper limit for C_T becomes equal to 1, i.e. $C_T \leq 1$. This shows why it is strongly desirable to use a five array probe (i.e. 20 hot wires total) which will enable proper computation of any other quantities involving velocity derivatives like vorticity, helicity density, enstropy production term etc. An additional important benefit from a 5 array probe would be that also second derivatives of the velocity field could be evaluated at the same point.

Along with the above mentioned improvements some shortcomings still remained. The main of them is that one point correlations $<u_1u_2>/u_1'u_2'$ and $<u_1u_3>/u_1'u_3'$ were about 0.1 ÷ 0.15, i.e. were not small enough as they should be in grid turbulence (here u_1', etc. are RMS value of u_1, etc.). As mentioned above this happens due to intrinsic imperfections of the method leading to erros in u_1, u_2, u_3 which are correlated. Our belief is that these can be reduced further by improving the mechanical precision of the calibration unit, using smaller steps in pitch and yaw in the vicinity of zero instead of uniform stepping, etc.

2.2. Water Flow Experiments

The water flow experiments were performed in the big salted (~1%) water tunnel of the Laboratory of Vorticity-Helicity studies, Faculty of Engineering, Tel Aviv University.

This tunnel was built after a series of first exploratory experiments on vorticity measurements, based on a principle described below, were performed on a small scale facility. The big tunnel was built to improve the scale resolution by increasing the scale of the flow. The test section was 290mm in diameter. The magnetic field was about 1 Tesla and was produced by a superconducting solenoid cooled by liquid helium. Two kinds of grid were used; one with square mesh of size 32mm and solidity 0.44 and the other with circular mesh 32mm in diameter and solidity 0.55. The last one was necessary to perform experiments with *controlled sign of mean helicity of the turbulent fluctuations*. This was achieved by installing various propellers in the grid holes. These propellers were able to produce perturbations which were clockwise, anticlockwise or both.
Experiments were performed with either freely rotating or with fixed propellers. In the latter case, larger helicity perturbation were produced. The measurements were made at the distances x/M = 15; 25; 40; 50 from the grid and without grid. Most of the measurements were taken at mean velocity 0.6m/sec.

Simultaneous measurements of longitudinal components of velocity u_1 and vorticity ω_1 were performed. The method of vorticity measurements is based on the equation div $\mathbf{j}=0$ and the Ohm's law $\mathbf{j}=\sigma(\nabla\phi+\mathbf{u}\times\mathbf{B})$, where $\mathbf{j}$ - is the electric current, ϕ - the potential of the electrical field, $\mathbf{u}$ - velocity, $\mathbf{B}$ - external magnetic field intensity. The consequence is $\nabla^2\phi=\mathbf{B}\cdot\mathrm{rot}\mathbf{u}$ (the term $\mathbf{v}\cdot\mathrm{rot}\mathbf{B}$ is several orders smaller). This enables one to measure the component of the vorticity ω_B parallel to the magnetic field, by measuring the $\nabla^2\phi$ which in turn could be measured using a 7 electrodes probe realizing a central difference approximation of the Laplacian. This idea was suggested by Grossman et al (1957) and was implemented in turbulent grid flow of salted water on a small scale facility (with test section of 5x5 cm^2 in cross section) by Tsinober et al (1987). The method has the *principal advantage to be an absolute one*, i.e. *it does not require calibration at all.*

The results to be reported were obtained by a probe consisting of a seven electrode probe as described in Tsinober et al (1987) assembled together with a commercial 55RM DANTEC hot film probe. This allowed to measure simultaneously and at the same location the longitudinal components of velocity and vorticity. In these experiments the scale of the probe was kept not too small (~2mm) to avoid mechanical contamination of the signal. The tip of the probe is shown in Fig.2.
Building of such a probe is quite a delicate procedure too. As with the multihotwire probe it requires to use a micromanipulator, microwelding, micropositioning of the electrode tips etc. A special procedure is electrolytical coating of the silver electrode tips (~0.1mm in diameter) with silver chloride which is vital to reduce the so called electrochemical offset

potential to a level ~3.10^{-7}V to the input in the frequency range up to 200Hz, which is several orders (!) of magnitude smaller than that without the AgCℓ coating.

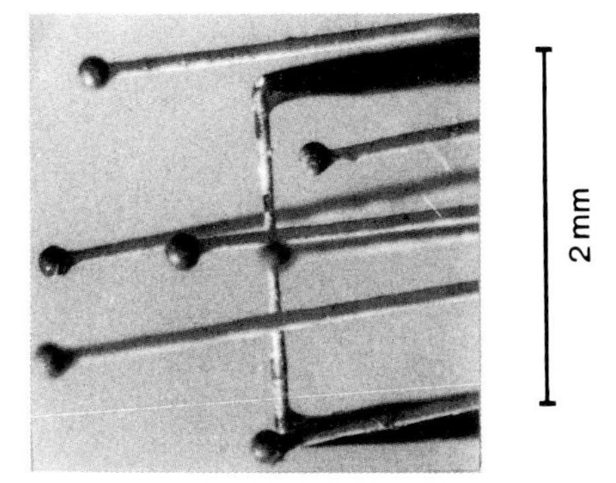

Fig.2. Vorticity-velocity probe consisting of a seven-electrode vorticity probe and a commercial hot-film probe.

3. RESULTS

3.1. Air Flow Experiments

Velocity fluctuations. The RMS values of velocity fluctuations exhibit some degree of anisotropy in that the RMS of the longitudinal velocity fluctuations u_1 is about 10%-20% higher than that of u_2 and u_3. The RMS values of u_2 and u_3 are equal within ~1%. It is noteworthy that all the three arrays produced essentially the same results. It is also of interest that the same is true in respect with the one-dimensional spectra of u_1,u_2,u_3 which are shown in Fig.3.

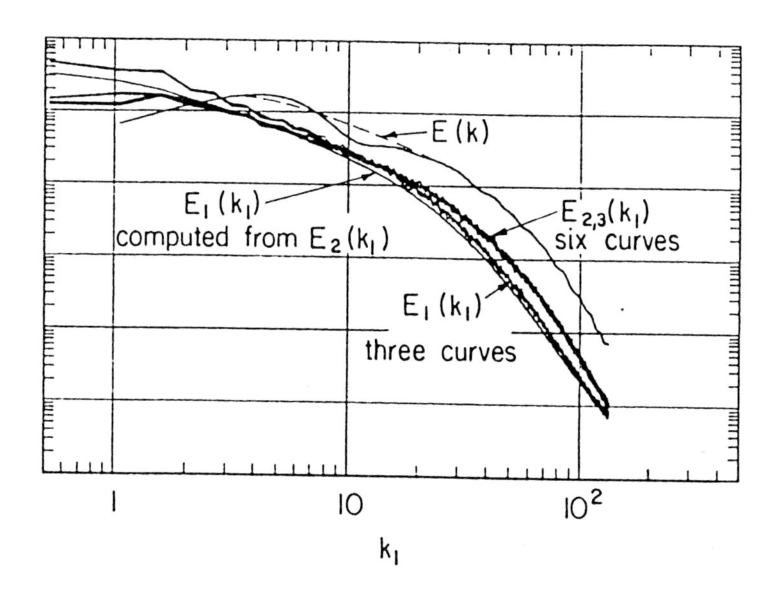

Fig.3. Spectra of velocity fluctuations.

Again, all the six spectra of u_2 and u_3 (from the three arrays) are essentially the same indicating a high degree of isotropy in the plane x_2,x_3; $k_1 = 2\pi fM/U$. The anisotropy of the velocity field is located mostly in the large scales which is seen from comparison (see Fig. 3) of the energy spectrum $E_1(k_1)$ of u_1 calculated from the isotropy relation $E_1(k_1) = 0.5\, k_1 \int_0^{k_1} t^{-2} E_2(t)\, dt$, where $E_2(k_1)$ is the energy spectrum of u_2.

It is of interest to also look at the three-dimensional spectrum E(k), computed from the relation $E(k) = -\frac{k}{2}\frac{dE_1(k)}{dk} - k\frac{dE_2(k)}{dk}$ using an appropriate smoothing procedure to $E_1(k)$ and $E_2(k)$ before taking the first derivatives (see Fig.4). A comparison between $k^2E(k)$ and $\Omega(k)$ computed in a similar way from the field of vorticity fluctuations is shown in Fig. 4 and, the agreement between the two is quite reasonable, especially at large k as should be if isotropy does really exist in the small scales.

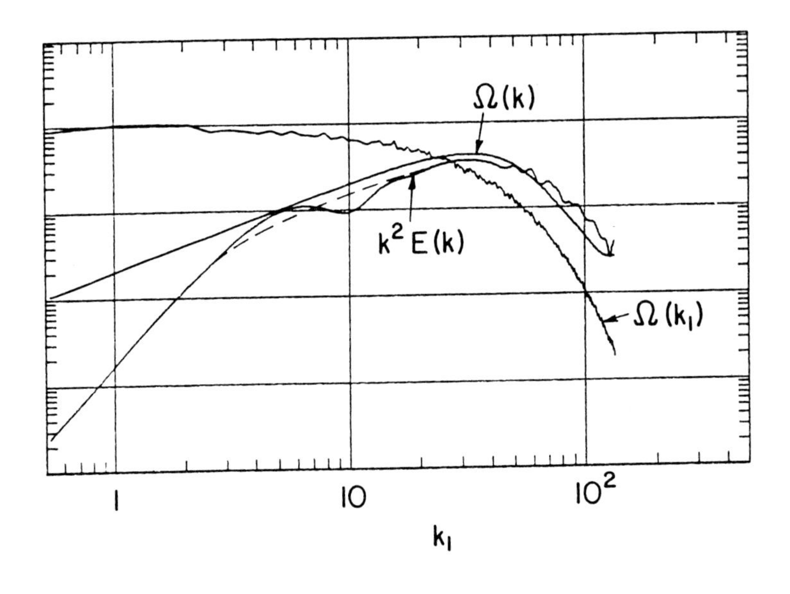

Fig.4. Comparison of $k^2E(k)$ and $\Omega(k)$.

As mentioned above, the one point correlation coefficients $\langle u_1u_2\rangle/u_1'u_2'$ and $\langle u_1u_3\rangle/u_1'u_3'$ were about 0.1÷0.15, i.e. were not small enough as they should be in grid turbulence. In spite of this the triple correlations behaviour was rather satisfactory. Two of them $\langle u_1^2(x)u_1(x+r)\rangle$ and $-2\langle u_2^2(x)u_1(x+r)\rangle$ shown in Fig.5 should be equal if isotropy is assumed. It is seen that they do not differ much. The results shown in Fig.5 are in good agreement with those obtained by previous authors (e.g. Frenkiel et al (1979)).

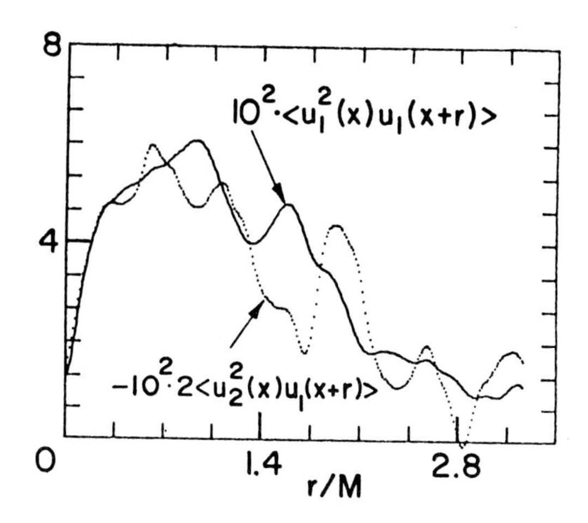

Fig.5. Triple correlations $\langle u_1^2(x)u_1(x+r)\rangle$ and -2 $\langle u_2^2(x)u_1(x+r)\rangle$.

Derivatives of velocity fluctuations

Along with calculations of the matrix $<(\partial u_i/\partial x_j)^2>$ and the incompressibility - Taylor hypothesis test mentioned above a more delicate test of velocity derivatives measurements was carried out. It is based on the relation imposed by the condition of *homogeneity only* on the one point triple correlations of the velocity derivatives. This condition was obtained by Townsend (1951) and Betchov (1956) and is as follows

$$<s_{ij}s_{jk}s_{ki}> = -3/4 <\omega_i\omega_j s_{ij}>$$

where $s_{ij} = \frac{1}{2}(\partial u_i/\partial x_j + \partial u_j/\partial x_i)$ is the rate of strain tensor and $\omega_k = \epsilon_{ijk}\, \partial u_j/\partial x_i$ is the vorticity vector.

The value of the ratio $-3/4<\omega_i\omega_j s_{ij}>/<s_{ij}s_{jk}s_{ki}>$ obtained from experimental data at different values of x/M is given in the Table 2. It is seen that the experimental values for the grid flow are quite close to unity, except for the closest to the grid cross section, while for the boundary layer this ratio is essentially different from unity as can be expected for a non-homogeneous flow.

Many more relations between triple correlations of velocity derivatives are imposed by isotropy (e.g. Townsend (1951)), Champagne (1978), p.106). The one which is generally used relates to the enstrophy generation term with $<(\partial u_1/\partial x_1)^3>$

$$<\omega_i\omega_j s_{ij}> = -35/2 <(\partial u_1/\partial x_1)^3>$$

Table 2. Values of $I_1 = 2<\omega^2>/<s_{ij}s_{ij}>$; $I_2 = -(3/4)<\omega_i\omega_i s_{ij}>/<s_{ij}s_{jk}s_{ki}>$

X/M	8	17	30	38	64	90	B. Layer y/δ 0.7	0.2
I_1	0.80	0.82	0.85	0.91	0.84	0.84	0.81	0.78
I_2	0.78	0.85	0.91	0.89	0.93	1.06	0.59	0.32

The values of the ratio $<\omega_i\omega_j s_{ij}>/<(\partial u_\alpha/\partial x_\alpha)^3>$ (no summation over greek indices, α = 1,2,3) are shown in Table 3.

It is seen that these values are somewhat smaller than 17.5 most probably due to insufficient resolution of the small scales.

Table 3. Values of $J_\alpha = <s_{ij}s_{ij}> / <(\partial u_\alpha/\partial x_\alpha)^2>$ and $K_\alpha = -<\omega_i\omega_j s_{ij}> / <(\partial u_\alpha/\partial x_\alpha)^3>$, $\alpha = 1,2,3$

x/M	8	17	30	38	64	90	B. Layer y/δ 0.7	0.2
J_1	7.1	7.3	7.8	7.0	6.6	8.2	7.8	8.6
J_2	7.7	7.2	9.4	7.0	12.0	10.5	6.8	6.8
J_3	6.8	8.8	6.6	7.4	7.7	6.3	6.8	6.5
K_1	8.5	8.8	11.6	11.3	7.8	12.2	7.2	6.7
K_2	12.3	13.8	17.3	11.0	24.0	24.1	10.2	36.0
K_3	10.5	10.9	11.8	22.2	24.5	29.0	4.8	1.00

The skewnesses for the three velocity derivatives are shown in Table 4 along with the quantity $<\omega_i\omega_j s_{ij}>/<\omega^2><s_{ij}s_{ij}>^{1/2}$.

Table 4. Values of $s_\alpha = <(\partial u_\alpha/\partial x_\alpha)^3>/<(\partial u_\alpha/\partial x_\alpha)^2>^{3/2}$ and $s = -<\omega_i\omega_j s_{ij}>/<\omega^2>\cdot<(s_{ij}s_{ij}>^{1/2}$

x/M	8	17	30	38	64	90	B. Layer y/δ 0.7	0.2
S_1	0.41	0.46	0.50	0.50	0.50	0.50	0.56	0.35
S_2	0.32	0.41	0.44	0.55	0.40	0.37	0.32	0.045
S_3	0.31	0.34	0.38	0.33	0.20	0.14	0.68	-1.61
S	0.36	0.43	0.52	0.6	0.46	0.52	0.37	0.19

We would like to emphasize that (see Table 4) *measurements of the enstrophy generating term show that its mean is essentially a positive quantity.* It was obtained as an average from 200 events each 1024 points long with 10^{-4} sec. interval between two adjacent points. For each such event the enstrophy generating term was positive.

It is of special interest to look at some properties of the Lamb vector $\lambda_k = \epsilon_{ijk} \omega_j u_i$ ($\boldsymbol{\lambda}=\boldsymbol{\omega}\mathrm{x}\mathbf{u}$). Shtilman and Polifke (1989) have found in their direct numerical simulations of decaying box turbulence that the Lamb vector contains a substantial potential part. They calculated the probability density function of the angle between the Fourier image of the Lamb vector and the wave vector k and found that there was a high probability that they were aligned. The implication is that in physical space the Lamb vector should consist mostly of a potential part . Experimentally one can have a look at correlations of the type $\langle\lambda_\alpha(x)\ \lambda_\alpha(x+r)\rangle$, $\alpha = 1,2,3$ and corresponding spectra of λ_α. There exist simple relations between one-dimensional spectra and correlation functions for pure potential vectors assuming isotropy. These are $E^{(p)}_1 = - k_1\ dE^{(p)}_{2,3}/dk_1$. One-dimensional spectra $E^\lambda_1(k_1)$ and $E^\lambda_{2,3}(k_1)$ of λ_1 and $\lambda_{2,3}$ correspondingly are shown in Fig. 6.

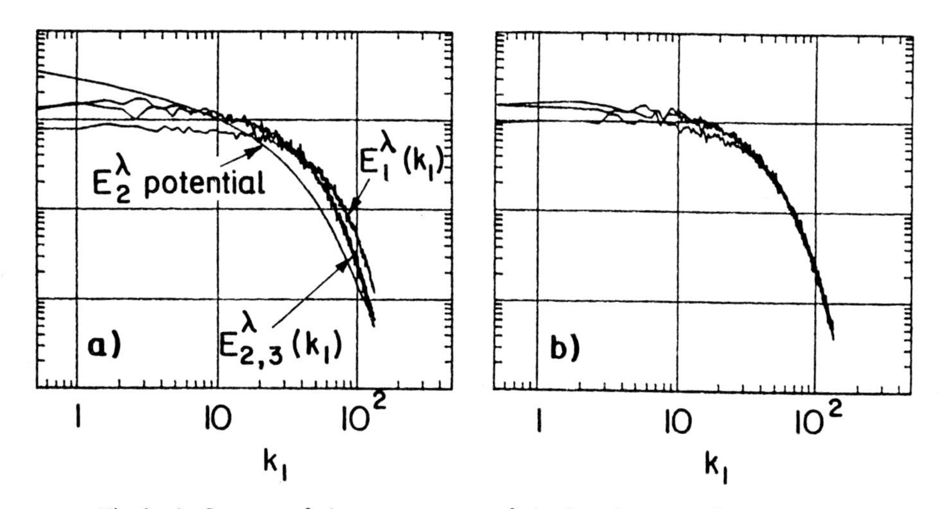

Fig.6. a) Spectra of the components of the Lamb vector $\boldsymbol{\lambda} = \boldsymbol{\omega}\mathrm{x}\mathbf{u}$.
b) Same for $\boldsymbol{\omega}$ and $\mathbf{u}$ independent.

together with the curve $E^\lambda_{2\,\mathrm{potential}}$ computed from the $E^\lambda_1(k_1)$ using the above relation for a potential vector field. The qualitative indication is that the Lamb vector contains a substantial potential part. This is seen from the comparison of the spectra obtained for the real Lamb vector (Fig. 6a) and those for the Lamb vector obtained as a vector product of two independent vectors (Fig. 6b). These vectors were chosen as velocity and vorticity at large time separation. In the latter case, the spectra for all components are identical in the high k region, while for the real Lamb vector the spectrum of λ_1 is higher, which corresponds to the tendency of having a substantial potential part. Still the spectrum $E^{\lambda}_{2,3\,potential}$ computed from isotropy relation is different from the real ones which indicates that the solenoidal part of the Lamb vector is far from being negligible.

Among the most basic questions about the properties and dynamics of the vorticity field in turbulent flows are the relations between the vorticity field and the field of the rate of strain tensor. Ashust et al (1987) discovered in their computations quite a peculiar relation between these fields. They found that the vorticity tends to be aligned with the intermediate eigenvector of the rate of strain tensor and that the strain in this direction is mostly positive (80%). Our experimental data clearly show the same tendencies (see Fig.7), though the alignment is not that strong as in the numerical simulations.

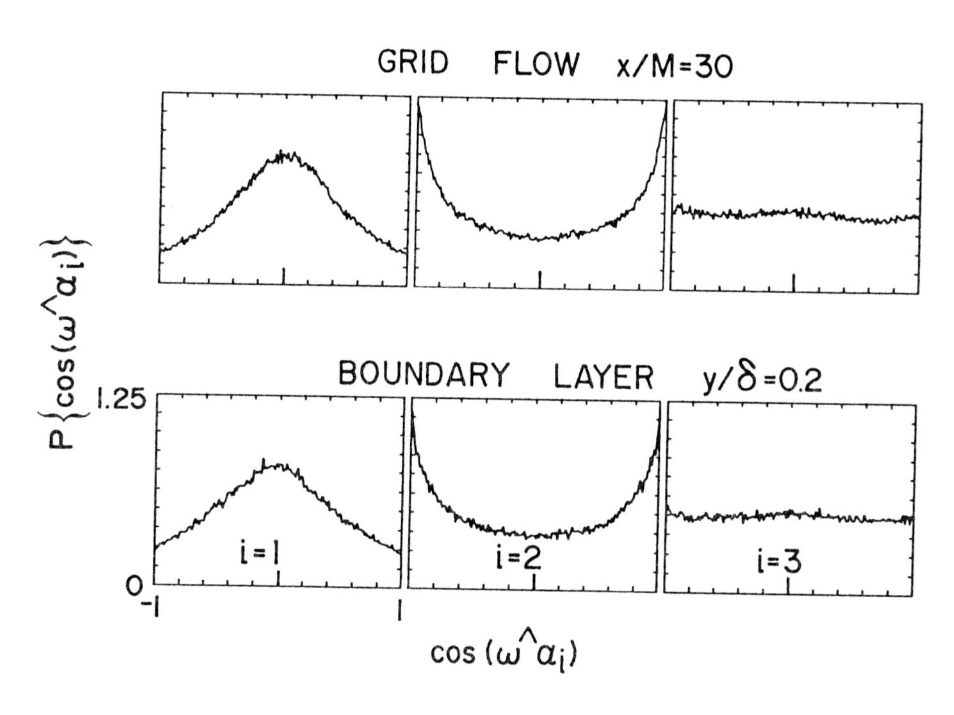

Fig.7. Probability density functions of cosine of the angle between vorticity vector and eigenvectors of the rate of strain tensor. $\alpha_1<\alpha_2<\alpha_3$.

On the other hand, we found in our experiments that in 65% of the sample points the intermediate principal rate of strain was positive in agreement with Betchovs' (1956) conclusions.

The alignment between velocity and vorticity is exhibited less than in previous experiments (Fig. 8).

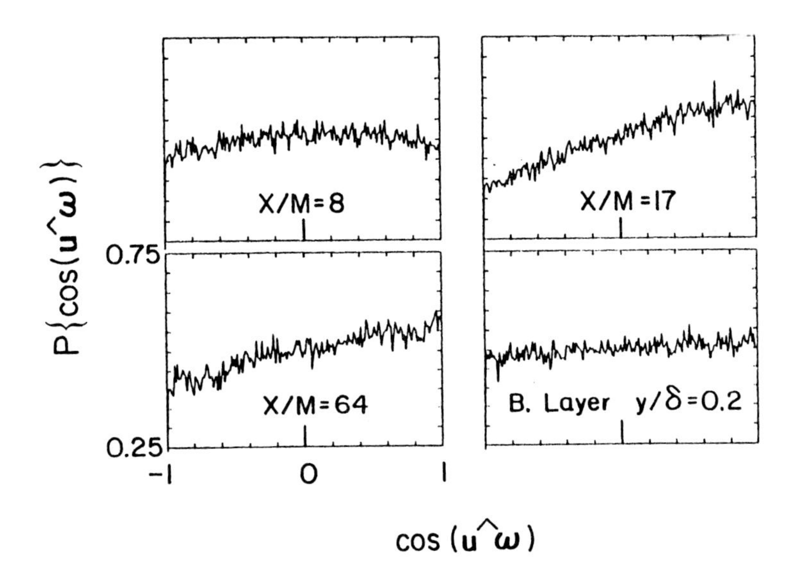

Fig.8. Probability density functions of cosine of the angle between vorticity and velocity vectors.

However, an important feature is that starting from X/M = 17, the probability density function of cosine between **u** and ω becomes asymmetric clearly indicating lack of reflexional symmetry in the flow. The quantity $<(\mathbf{u}\cdot\omega)^2>/<u^2><\omega^2>=$ 0.4 in all the experiments and distances from the grid as in our previous experiments. Although the mean helicity $<\mathbf{u}\cdot\omega>$ was positive, almost in all experiments it's value was within the experimental error.

3.2. Water flow experiments

To verify the question of lack of reflexional symmetry we turned to experiments with salted water in which - unlike the air flow - the vorticity and velocity have been measured in an independent way and the vorticity was measured by an absolute method (see 2.2). Moreover, it was possible to control the sign of the mean helicity produced by the grid (see 2.2) thereby allowing to increase the reliability and our confidence of the results.

In *all* cases when a particular sign of mean helicity (more precisely $<u_X \omega_X>$ was expected (corresponding propellers (screws) installed) it was indeed observed. Some results are given in Table 5.

Table 5. Mean normalized helicity $10^2 <\omega_x u_x>/<\omega_x^2>^{1/2}, <u_x^2>^{1/2}$ in water experiments with various grids

x/M	Empty	RH screws rotating	LH screws rotating	RH+LH screws steady	RH screws steady	LH screws steady
15	0.0	2.4	-1.1	0.6	6.4	-0.8
25	-0.7	2.5	-2.2	-0.9	3.0	-1.7
40	-1.3	2.3	-3.0	-2.3	1.9	-3.4
50	-4.4	2.2	-6.6	-3.5	1.6	-6.6

It is noteworthy that the normalized mean helicity (i.e. $<u_x \omega_x>/<u_x^2>^{1/2}<\omega_x^2>^{1/2}$) is *increasing downstream from* the grid in most of the experiments. It is of special importance that the normalized **mean helicity is generated past a grid with empty holes too** and is increasing downstream from the grid (see Table 5). The same flow behaviour was observed for the grid with an equal number of left and right-handed propellers. In both latter cases the sign of the generated mean helicity was the same as for the grid with the left-handed propellers. An important point is that without grid no measurable signal above the noise of the electronic equipment was observed.

In all cases the main contribution to mean helicity comes from *largest scales* as can be seen from the one-dimensional spectra for different cases shown in Fig. 9. These scales are typically of the order of the diameter (~30cm) of the test section. Though, at present, we do not understand why most of the mean helicity generated past the grid is in the large scales, there is an indication that a small disturbance in helicity is picked up and amplified by the turbulent flow past the grid.

It is of interest to compare the ratio $-<\omega_1^2(\partial u_1/\partial x_1)>/<(\partial u_1/\partial x_1)^3>$ obtained in water experiments with those from experiments with air flow. The results are shown in Table 6. One can see that this ratio is more or less the same in both experiments and smaller than the theoretical value 7/3 following from isotropy.

It is noteworthy that the quantity $<\omega_1^2(\partial u_1/\partial x_1)>$ computed for each event was negative for about 10% of the events each 1.2 sec duration (i.e. ~ 70 cm long!). Each event in water flow experiments consisted of 512 sample points with sampling frequency 400 Hz. The same behaviour was observed in air flow experiments for every quantity $<\omega_\alpha^2(\partial u_\alpha/\partial x_\alpha)>$ separately for events of duration 0.1 sec (i.e. ~ 70 cm long), while the total enstrophy generation term was almost always positive. The above observation may be

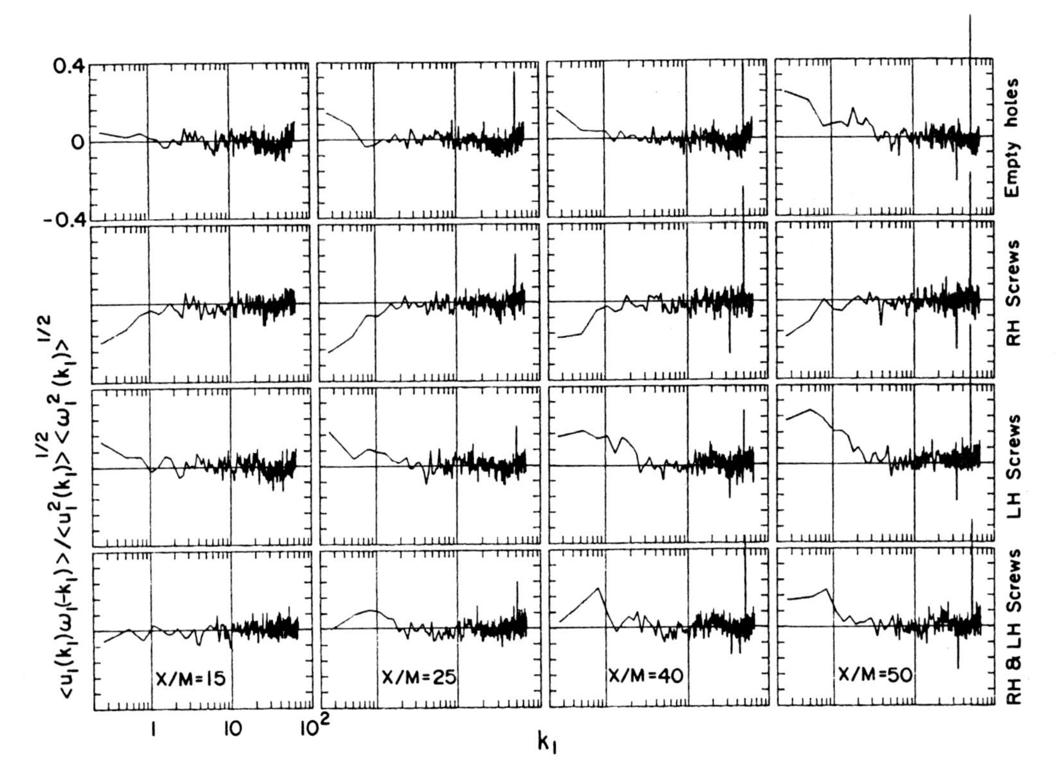

Fig.9. One-dimensional normalized helicity spectra $<u_1(k_1)\omega(-k_1)>/<u_1^2(k_1)>^{1/2}$ $<\omega_1^2(k_1)>^{1/2}$ in water flow experiments.

Table 6. Values of the quantities $I_X = -<\omega_1^2 \partial u_1/\partial x_1>/<(\partial u_1/\partial x_1)^3>$ and $K_X = <\omega_1^2 \partial u_1/\partial x_1>/<\omega_1^2><(\partial u_1/\partial x_1)^2>^{1/2}$ for air and water flow

$$S = -<(\partial u_1/\partial x_1)^3>/<(\partial u_1/\partial x_1)^2>^{3/2}$$

	AIR FLOW						WATER FLOW			
x/M	8	17	30	38	64	90	15	25	40	50
I_X	0.81	0.85	1.3	1.8	0.95	1.7	0.95	1.1	1.6	1.01
K_X	0.10	0.12	0.18	0.20	0.16	0.18	0.13	0.15	0.16	0.17
S	0.40	0.46	0.52	0.5	0.5	0.5	0.33	0.34	0.3	0.34

useful in the context of detecting coherent structures.

Finally in Fig. 10 we show the time series of various quantities during one event (i.e. 70cm long). Perhaps the most spectacular is the behaviour of the enstrophy generating term which exhibit bursts of extremely large amplitude (from 1 to 5 such bursts in one event): quite an intermittency. This amplitude is up 50 (!) times larger than the average and more than 60% of contribution to the average of the enstrophy generating term comes from the fluctuations with amplitude larger than three averages. The scale of these extremely violent events (or if one likes very active structures) is ~0.5 ÷ 1 cm and in between there exist extremely inactive regions (strucutres). The scale of the latter is an order of magnitude larger than that of the active events. It is noteworthy that the same kind of behaviour is observed in the boundary layer (in its outer part). It is easy to guess that we have much more to say about all the material presented above and we intend to do this in subsequent papers. At this stage we venture to suggest that the enstrophy generating term, being an invariant (i.e. indpendent of the system of coordinates and in Galilean sense), seems to be one of the most appropriate quantities to characterize the structural nature of the grid turbulence as well as shear turbulence.

Acknowledgements

Our special acknowledgement is to Professor J.M. Wallace and Dr. J.L. Balint for their kind hospitality during the summer of 1986 and for introducing two of us (EK and AT) to all the secrets of their multihotwire technology etc.

Part of this work was made in the frame of collaborative program with the B. Levich Institute for PCH, City College, CUNY.

We are grateful to Professor E. Levich and Dr. L. Shtilman for many useful discussions.

This research was supported, in part, by the Eidgenossische Technische Hochschule, Zürich; by grant DE-FGOZ-88ER 13837 from the U.S. Department of Energy and by grant No. 85-00347/1 from the U.S.-Israeli Binational Scientific Foundation.

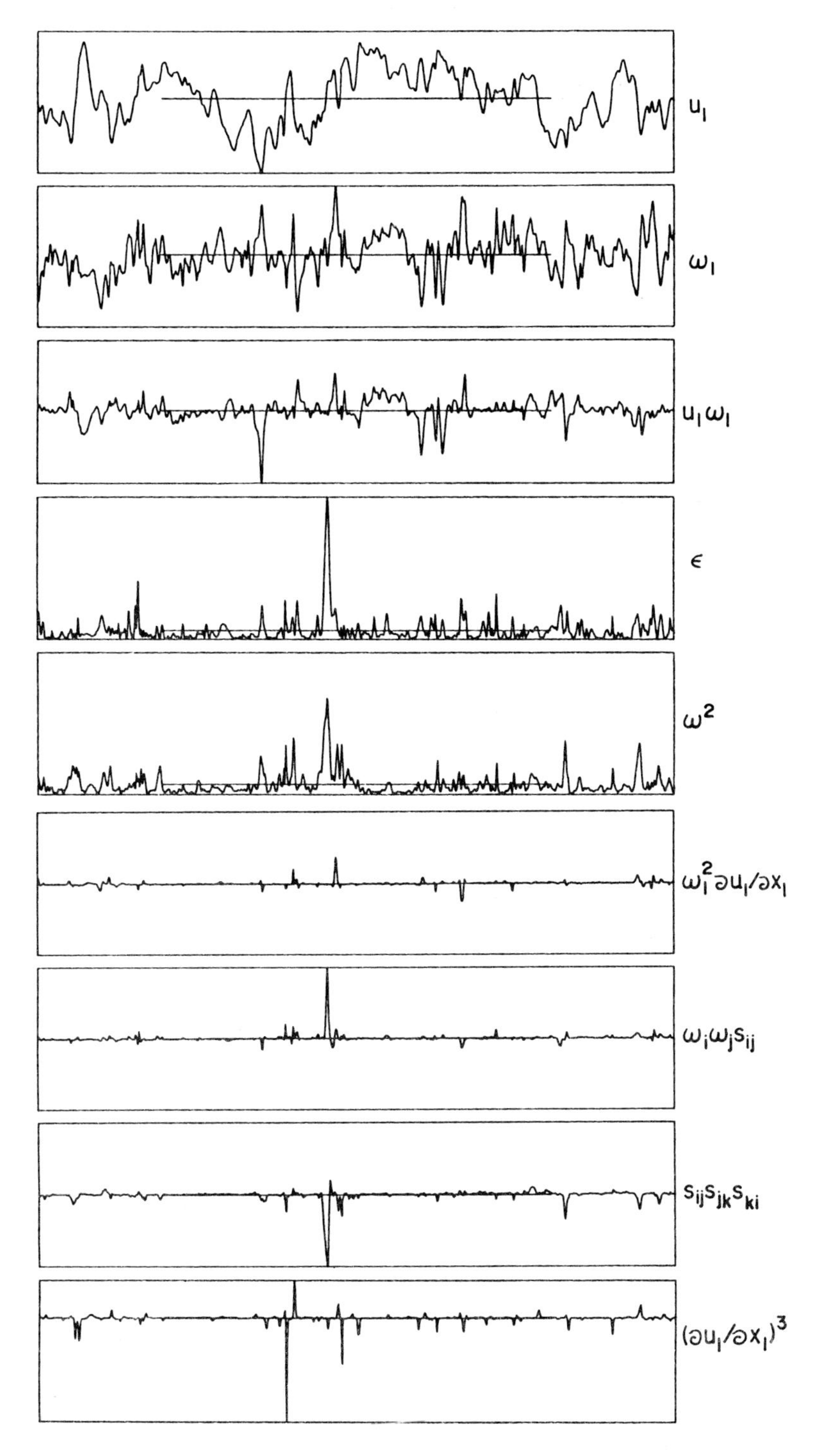

Fig. 10a. Time series of various quantities for one event (70 cm long) in air flow.

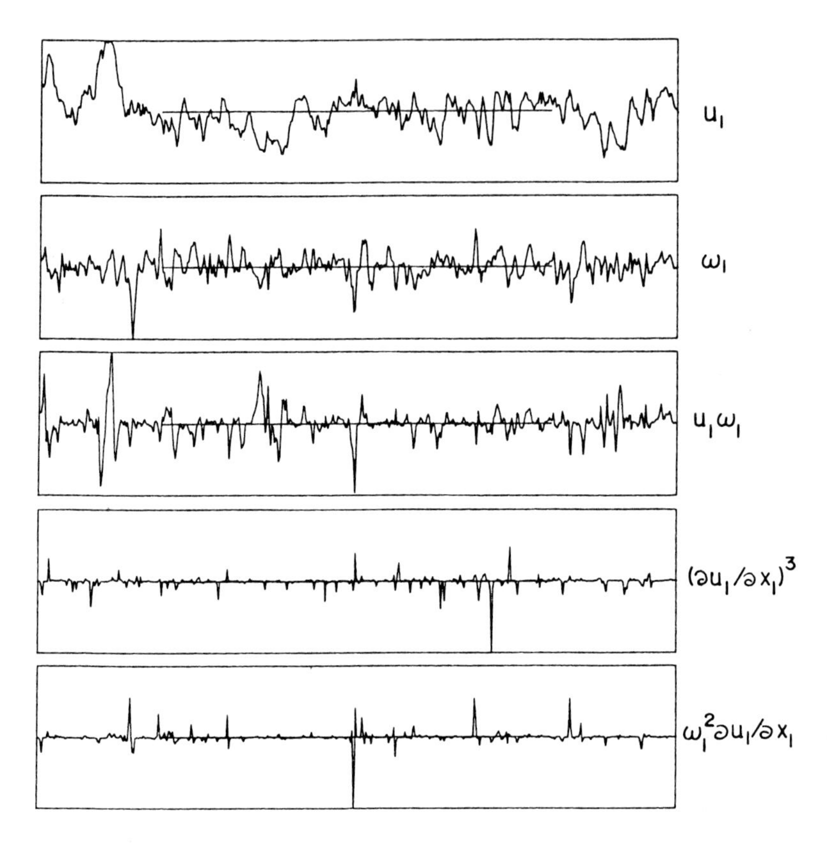

Fig. 10b. Time series of various quantities for one event (70 cm long) in water flow.

References

Antonia, R.A., D.A. Shah and L.W.B Browne, 1988. Dissipation and vorticity spectra in a turbulent wake, *Phys. Fluids*, **31**, 1805-1807.

Aref, H. and T. Kambe, 1988. Report on the IUTAM Symposium: fundamental aspects of vortex motion, *J. Fluid Mech.*, **190**, 571-595.

Ashust, Wm. T., A.R. Kerstein, R.A. Kerr and C.H. Gibson, 1987. Alignment of vorticity and scalar gradient with strain rate in simulated Navier-Stokes turbulence, *Phys. Fluids*, **30**, 2343-2353.

Balint, J.L., P. Vukoslavcevic and J.M. Wallace, 1987. A study of the vortical structure of the turbulent boundary layer, in: *Advances in Turbulence*, eds. G. Compte-Bellot and J. Mathieu (Springer, Berlin), pp. 456-464.

Balint, J.L., P. Vukoslavcevic and J.M. Wallace, 1988. The transport of enstrophy in a turbulent boundary layer, *Proc. Zaric Mem. Inst. Sem. on Wall Turb.* (Dubrovnik, May 1988).

Betchov, R. 1956. An inequality concerning the production of vorticity in isotropy turbulence, *J. Fluid Mech.*, **1**, 497-504.

Champagne, F., 1978. On the fine-scale structure of the turbulent velocity field, *J. Fluid Mech.*, **86**, 67-108.

Foss, J.F. and J.M. Wallace, 1989. The measurement of vorticity in transitional and fully developed turbulent flows, To appear in: Frontiers in Experimental Fluid Mechanics, ed. Gad-el-Hak, *Lect. Notes in Engineering*, Springer.

Frenkiel, F.N., P.S. Klebanoff and T.T. Huang, 1979. Grid turbulence in air and water, *Phys. Fluids*, 22, 1606-1617.

Grossman, L.M., H. Li and H. Einstein, 1958. Turbulence in civil engineering, *J. Hydraul. Piv. ASCE*, **83**, 1-15.

Hussain, A.K.M.F., 1986. Coherent structures and turbulence, *J. Fluid Mech.*, **173**, 303-356.

Kit. E., A. Tsinober, J.L. Balint, J.M. Wallace and E. Levich, 1987. An experimental study of helicity related properties of a turbulent flow past a grid, *Phys. Fluids*, **30**, 3323-3325.

Kit, E., A. Tsinober, M. Teitel, J.L. Balint, J.M. Wallace and E. Levich, 1988. Vorticity measurements in turbulent grid flows, *Fluid Dyn. Res.*, **3**, pp. 289-294.

Shtilman, L. and Polifke, W., (1989). On the mechanism of the reduction of nonlinearity in the incompressible Navier-Stokes equation, *Phys. Fluids*, **32**, in press.

Tsinober, A., E. Kit, and M. Teitel, 1987. On the relevance of the potential difference method for turbulence measurements, *J. Fluid Mech.*, **175**, pp. 477-461.

Tsinober, A., E. Kit and M. Teitel, 1988a. Spontaneous symmetry breaking in turbulent grid flow, Presentation at the 17th IUTAM Congress, Grenoble, (August 21-27, 1988).

Townsend, A.H. 1951. On the fine-scale structure of turbulence, *Proc. Roy. Soc. London*, **A208**, 534-542.

An Experimental Study of Helicity and Related Properties in Turbulent Flows

JAMES M. WALLACE & JEAN-LOUIS BALINT

The University of Maryland, College Park, MD 20742, USA

1 INTRODUCTION

For a barotropic inviscid flow of infinite extent or for a finite volume with zero vorticity normal to the surface and with conservative body forces, Moffatt (1969) has shown that the helicity

$$\mathcal{H} = \int_{\forall} \mathbf{U} \cdot \mathbf{\Omega} \, d\forall \tag{1}$$

is an invariant which is equal to the degree of linkage of the vortex lines. For bounded viscous flows, boundary layers will develop on the solid surfaces, and the time rate of change of helicity is

$$\frac{d\mathcal{H}}{dt} = -2\nu \int_{\forall} \mathbf{\Omega} \cdot (\nabla \times \mathbf{\Omega}) \, d\forall. \tag{2}$$

Based on an analysis of two helical waves, Kraichnan (1973) has shown that, for a given energy spectrum shape, overall energy transfer should be less for turbulence with maximal helicity; the flow of energy to higher wave-numbers can be expected to be inhibited by strong helicity. In a study of three-dimensional isotropic turbulence at very high Reynolds number, André and Lesieur (1977) have shown that the presence of helicity in the flow inhibits the energy transfer toward large wave-numbers in agreement with Kraichnan (1973). Utilizing the analogy between the transport equations for the magnetic and vorticity fields Moffatt (1985) speculated that turbulence might be described as steady solutions to the Euler equation about which unsteady solutions evolve. In the subdomains where the steady Euler solutions exist, the relative helicity density

$$h = \frac{\mathbf{U} \cdot \mathbf{\Omega}}{|\mathbf{U}||\mathbf{\Omega}|} = \cos\theta, \tag{3}$$

where θ is the angle between $\mathbf{U}$ and $\mathbf{\Omega}$, should be maximal at ± 1. He suggested that these subdomains could be considered coherent structures, and that the regions between these subdomains should be vortex sheets which are likely to be the principal locus of viscous dissipation. If this description is valid Moffatt (1985) argued, then high relative helicity density should be correlated with low energy dissipation and vice versa.

Levich and Tsinober (1983) speculated that organized large scale coherent motions in turbulent flows may be a universal and intrinsic property of turbulent flows, even homogeneous ones, and not necessarily the result of shear instabilities. Tsinober and Levich (1983) further conjectured that, in fully developed turbulent flows, all three-dimensional, organized, coherent motions should be helical. Shtilman (1985) found from a numerical simulation of the Taylor-Green vortex that, when the probability of the relative helicity density h was conditioned on low dissipation, a higher probability occurred that the vorticity and velocity vectors would be aligned. Pelz et al. (1985) extended this study adding a numerical simulation of the turbulent channel flow. The PDDs of h and of $h' = \cos\theta'$ were studied, where θ' is the angle between only the fluctuating velocity and vorticity fields. For both the Taylor-Green vortex flow and for the channel flow, the tendency of the velocity and vorticity vectors to be aligned was pronounced for realizations where the instantaneous dissipation is low; where the dissipation is high the distributions were either flat or showed a higher probability of the vectors to be orthogonal as Moffatt (1985) had conjectured.

In an analysis of isotropic turbulence where energy is forced into the flow at high wave-numbers, Kerr and Gibson (1985) observed PDDs which show some alignment of the vectors, but they were also quite asymmetric; they attributed this asymmetry possibly to the mean helicity injected into the flow by the forcing. They found little change in the shapes of the PDDs conditioned on high dissipation. Pelz et al. (1986) studied the effects of the initial conditions, with and without mean helicity, on nearly isotropic turbulence simulations. They found that, irrespective of the initial conditions, the isotropic turbulence decays with time to PDDs with distributions indicating the preferred alignment of the velocity and vorticity vectors. It was argued by Pelz et al. (1985) that, as a consequence of the vector identity

$$|\mathbf{U} \cdot \mathbf{\Omega}|^2 + |\mathbf{U} \times \mathbf{\Omega}|^2 = |\mathbf{U}|^2 |\mathbf{\Omega}|^2, \tag{4}$$

that the cascade of turbulent kinetic energy, represented by the acceleration term in the Navier-Stokes equation, or by the so-called Lamb vector in the equation in rotational form

$$\frac{\partial \mathbf{U}}{\partial t} - \mathbf{U} \times \mathbf{\Omega} = -\nabla \left(\frac{P}{\rho} + \frac{\mathbf{U} \cdot \mathbf{U}}{2} \right) + \nu \nabla^2 \mathbf{U}, \tag{5}$$

would be inhibited in regions of the flow where the helicity density $H = \mathbf{U} \cdot \mathbf{\Omega}$ is large because the Lamb vector must necessarily be small. Speziale (1987) has raised questions about this argument based on a Helmholtz decomposition of the Lamb vector into its irrotational and solenoidal parts. Hussain (1986) discusses the role of helicity in turbulent flows and presents contours of vorticity, relative helicity density and dissipation for a numerical simulation of a two-stream mixing layer; these show that, although there are overlapping regions of relative helicity density and dissipation, the peak regions do not spacially overlap. He points out that this flow is still in a transitional stage, and that in fully developed turbulent flow dissipation will occur within the coherent structures which are not exclusively helical.

Rogers and Moin (1987) examined several simulations of homogeneous and inhomogeneous turbulent flows. They point out that the mean square values of either h or h' are equal to 0.333 for two isotropic vector fields which are uncorrelated, and that the PDDs are flat in this case for all values of $h = \cos\theta$. Thus mean square values of h or h' are a measure of the tendency of the velocity and vorticity vectors to align, with small values of $\overline{h^2}$ indicating nearly orthogonal conditions and large values indicating near alignment.

Kit et al. (1987) made measurements at two locations $X/M = 17$ and 30 in a turbulent grid flow. Analysis of the PDDs of h' indicated a slight tendency for the vectors to align, but the data showed considerable mean helicity which was reflected in the pronounced asymmetry of the PDDs. The flow also only roughly approximated isotropic conditions; some of the velocity gradients signifigantly deviated from the ideal. The mean helicity and anisotropies were thought to be due to a combination of measurement errors and possibly to being too close to the grid for the $X/M = 17$ case. Kit et al. (1988) also found asymmetries for the measurements of one component of helicity in a conducting water flow with a probe consisting of seven electrodes.

The disagreement between the results from the numerical simulations mentioned above and the anomalies and asymmetries in the sparse data taken in laboratories motivated us to carefully examine data sets obtained in three turbulent flows with our miniature nine-sensor hot-wire probe. We hoped to help resolve some of the still open questions about the role of helicity and related properties in turbulence.

2 MEASUREMENT OF THE VELOCITY AND VORTICITY VECTORS IN TURBULENT FLOWS

A miniature multi-sensor hot-wire probe has been developed to measure these vectors with a spacial resolution of only a few Kolmogorov dissipation lengths. The geometrical configuration of the probe and its operating properties have been described in detail in Balint (1986); a summary is given in Vukoslavčević et al. (1989). The three types of turbulent flows described below were investigated.

2.1 Turbulent Boundary Layer

This is a nominally zero pressure gradient boundary layer which developed over 8 m downstream of a 5 mm diameter trip wire on the lower wall of an open return wind-tunnel (Balint (1987)). The velocity and vorticity vector components were determined at locations in the buffer layer, the logarithmic region, and the wake region. The principal characteristics of the boundary layer investigated are

U_∞ (m/s)	δ (m)	θ (m)	Re_θ	Re_x	u_τ/U_∞
3.513	0.1245	0.0121	2688	1.51×10^6	0.040

where U_∞ is the free stream velocity, δ is the boundary layer thickness, θ is the momentum thickness, Re_θ is the Reynolds number based on θ, Re_x is the Reynolds number based on the streamwise measurement location x, and u_τ is the friction velocity.

2.2 Two-Stream Turbulent Mixing Layer

Data at a streamwise measurement station of 109 cm downstream of the splitter plate of the University of Illinois wind-tunnel (Balint (1988)) have been analysed for helicity properties. At this station the mean characteristics of the mixing layer are

U_1 (m/s)	U_2 (m/s)	θ (m)	$d\theta/dx$	δ_ω (m)	$Re = U_{.5}\theta/\nu$	$Re = U_{.5}\delta_\omega/\nu$
9.959	4.431	0.0138	0.0105	0.072	6406	33422

where θ is the momentum thickness of the flow, U_1 and U_2 are the mean velocities on either side of the splitter plate, δ_ω is the vorticity thickness $\Delta U/(dU/dy)_{max}$, $U_{.5} = 0.5(U_1 + U_2)$ and $\Delta U = U_1 - U_2$.

2.3 Turbulent Grid Flow

These data were obtained in the Johns Hopkins University wind-tunnel. The properties of the grid flow in this tunnel are described in Comte-Bellot and Corrsin (1966) who installed a secondary 1.27 : 1 contraction to accelerate the flow downstream of the grid location in order to improve the isotropy. We used a $M = 5.08cm$ square mesh of round rods with a solidity of 0.44 located at $40M$ upstream of the measuring station. At this station the turbulence intensity of the flow without the grid in place was about 0.06%. The mean velocity was $\overline{U} = 9.216$ m/s. Some of the statistical properties of this flow are

Velocity and Vorticity Fluctuations

	u(m/s)	v(m/s)	w(m/s)	$\omega_x(s^{-1})$	$\omega_y(s^{-1})$	$\omega_z(s^{-1})$
rms	0.183	0.159	0.152	64.05	54.75	61.53
skewness	−0.031	0.011	−0.020	0.041	−0.032	0.063
flatness	2.989	2.930	2.907	4.703	4.638	4.735

Variance of Fluctuating Velocity Gradients (s^{-2})

$\overline{u_{,x}{}^2}$	$\overline{u_{,y}{}^2}$	$\overline{u_{,z}{}^2}$	$\overline{v_{,x}{}^2}$	$\overline{v_{,y}{}^2}$	$\overline{v_{,z}{}^2}$	$\overline{w_{,x}{}^2}$	$\overline{w_{,y}{}^2}$	$\overline{w_{,z}{}^2}$
565.0	2483.0	2042.1	690.6	1154.6	2006.1	565.5	1441.7	694.3

Dissipation (m^2/s^3) Helicity

$\overline{\epsilon}$	$\nu\overline{\omega_i{}^2}$	$\overline{\cos\theta}$	$\overline{\cos\theta'}$	$\overline{h^2}$	$\overline{h'^2}$
0.19226	0.1688	0.4739	−0.0885	0.4713	0.3439

3 EXPERIMENTAL RESULTS

3.1 Boundary Layer

The PDDs of h and h' are shown in figure 1 for data taken at $y^+ = 10.1$, 27, 72, and 900 ($y/\delta = 0.8$). At each location the analysed data showed small mean values of the transverse velocity components $\overline{V}$ and $\overline{W}$ which should be nearly and exactly zero respectively. Likewise the analysis yielded small mean values of the streamwise and normal vorticity components $\overline{\Omega}_x$ and $\overline{\Omega}_y$ which should be identically zero in this flow because of symmetry. These small mean values result from an accumulation of slight probe misalignment and from an assortment of other small measurement errors. These errors give rise to mean relative helicity densities which cause asymmetries in the PDDs at each of the positions in the flow similar to those found by Kit et al. (1987), as seen from the dotted curves in figure 1. By assuming a small misalignment with the mean flow of the axis of each array (angles $< 2.5^o$) and redefining the cartesian axis of the flow with respect to the array direction, the mean values $\overline{V}$ and $\overline{W}$ were brought close to zero. This correction also reduced $\overline{\Omega}_x$ and $\overline{\Omega}_y$ and virtually eliminated the mean helicity making the PDDs much more symmetric at almost all positions. These corrected distributions are shown as solid lines in the the figures. We believe the mean helicity in the experiment of Kit et al. (1987) was also due to these small residual mean velocities and vorticities and not to the proximity of the grid or the "spontaneous breaking of symmetry" as conjectured in that paper.

Also shown in the figure are the values of the uncorrected and corrected mean square values $\overline{h^2}$ and $\overline{h_c^2}$ for each location. It is quite evident that these increase with y^+ from nearly zero to substantially above the value of 0.333 characteristic of two uncorrelated isotropic vector fields. Thus, the velocity and vorticity vectors have an increasing probability to be aligned with distance from the wall, as is also seen from the shapes of the PDDs. Just above the sublayer at $y^+ = 10.1$, where the vortex lines are predominantly spanwise and the streamlines are predominantly streamwise, $P(h)$ is concentrated around $\cos\theta = 0$. At the outer edge of the buffer layer, at $y^+ = 27$, the distribution has become much wider and even has begun to show a slight double peaked shape at $h = 0$ and ± 1. In the logarithmic layer, at $y^+ = 72$, the PDD of h clearly shows a higher probability for the vectors to be aligned, a trend which is further accentuated in the wake region at $y^+ = 900$. These results are physically plausible because, in the outer part of the boundary layer, the rms streamwise vorticity component becomes signifigantly larger than the rms spanwise and normal vorticity components (see Balint et al. (1987)), and the mean (spanwise) vorticity is very small. Thus the vorticity vector is predominantly oriented in the same direction as the mean velocity. This trend toward greater alignment was not seen in the logarithmic region in the channel flow simulation of Rogers and Moin (1987) however. The PDDs of h' in figure 1 are much broader than those for h, as was found by Pelz et al. (1985) and by Rogers and Moin (1987), but do not show the higher probability of the velocity and vorticity vectors to align, in agreement with

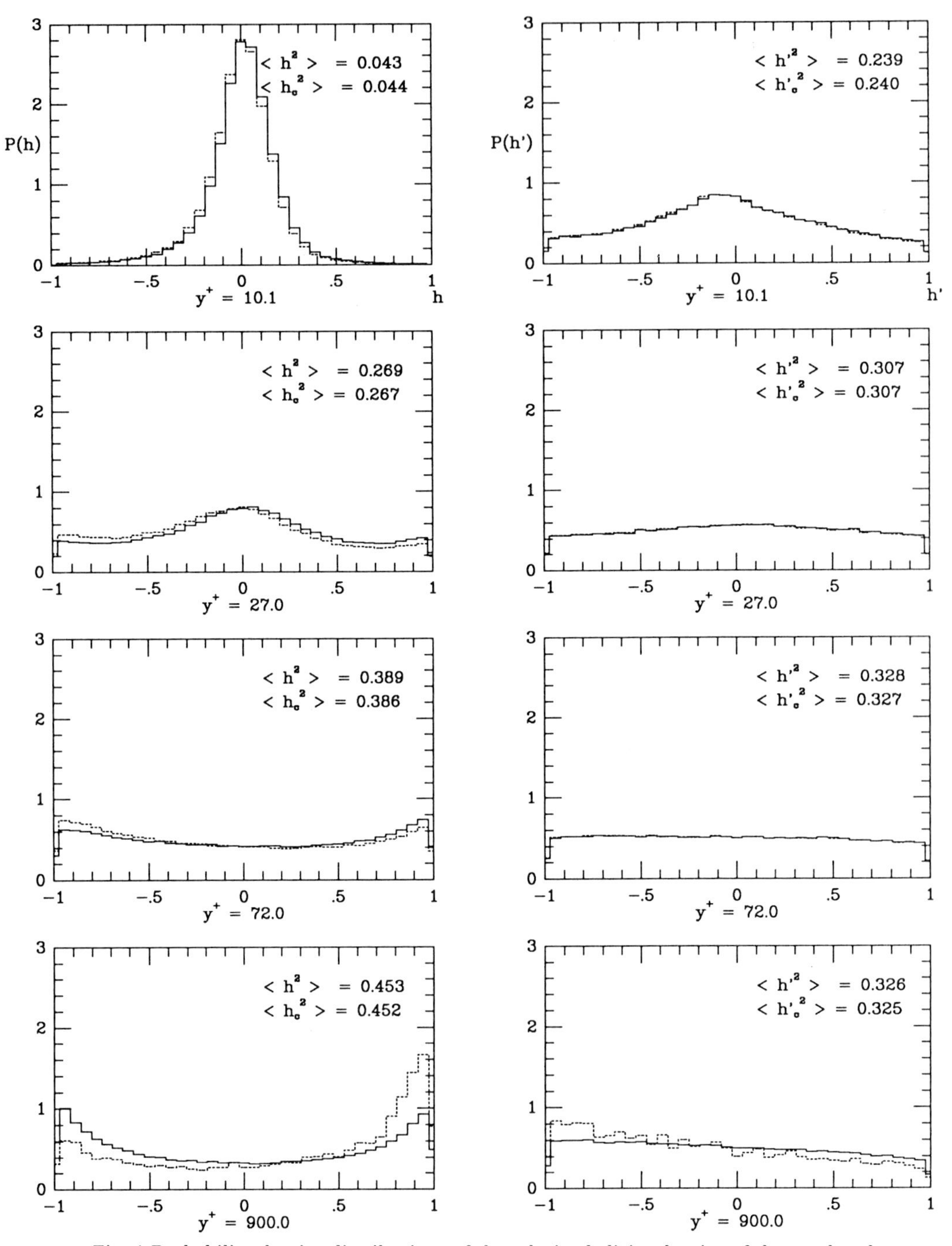

Fig. 1 Probability density distributions of the relative helicity density of the total and fluctuating velocity and vorticity fields h and h' in the boundary layer at $Re_\theta = 2688$. Corrected, ——; uncorrected, ····· (for all figures).

Rogers and Moin (1987) but in contrast to Pelz et al. (1985).

As Moffatt (1985) conjectured, it might be thought that, where the relative helicity density is large the instantaneous dissipation must be small based on the facts that, in the region of high dissipation near the wall, the velocity and vorticity vectors are nearly orthogonal, but the alignment becomes much greater where the mean dissipation rapidly decreases with distance from the wall. However, when the PDDs of h and h' are conditioned on the criterion that $\epsilon < 0.1\bar{\epsilon}$, as seen in figure 2, the shapes are only slightly changed compared to the unconditioned PDDs. Pelz et al. (1985) also found that, when the data at $y^+ > 15$ were conditioned on a low dissipation criterion, the alignment was hardly changed; Rogers and Moin (1987) found only a small increase in the alignment based on the low dissipation criterion at $y^+ = 41$. Little difference in the shapes of the PDDs compared to the unconditioned data was also found for the condition that $\epsilon > 3.5\bar{\epsilon}$, as can be seen in figure 2, with the exception that, at $y^+ = 72$, alignment is significantly lower conditioned on high dissipation compared to low dissipation.

3.2 Mixing Layer

In the two-stream mixing layer the rms value of the fluctuating streamwise vorticity component is about 40% larger than the rms values of the spanwise and normal components over the center part of the layer (Balint et al. (1988)). At the midplane of the layer studied here, the mean shear corresponds to that at about $y^+ = 80$ in the boundary layer described above, and other statistical properties including the enstrophy balance are similar. It could therefore plausibly be expected that the helicity characteristics in the mixing layer might be similar to those in the logarithmic region of the boundary layer. The PDDs of h and h' for positions at the midplane and on the low and high speed sides of the layer ($y/\theta = +1.36$ and -3.56 respectively) are shown in figure 3(a), where indeed it is seen that the shapes of the distributions are similar to those in the logarithmic region of the boundary layer. There is a slightly greater tendency for alignment at the midplane and at $y/\theta = +1.36$, as evidenced from $\overline{h_c^2}$, than at $y/\theta = -3.56$. At the latter position the rms streamwise vorticity is more in balance with the spanwise and normal components (Balint et al. (1988)). Just as for the boundary layer, conditioning on either low or high dissipation does not effect the shapes of these distributions very much, as seen for the mid-plane in figure 3(b).

In this flow the dominant coherent structures are large scale, weak, spanwise vortices which form by means of a Helmholtz instability downstream of the splitter plate and are basically orthogonal to the mean flow direction. Thus they should have very low relative helicity density. It is now well known, however, that in the "braid" between these spanwise vortices, counterrotating pairs of streamwise vortices are formed which are intensified by the strain rate field. These streamwise vortices should have signifigant relative helicity density and are likely to be the source of the observed tendency of the velocity and vorticity vectors to align, as seen in our data.

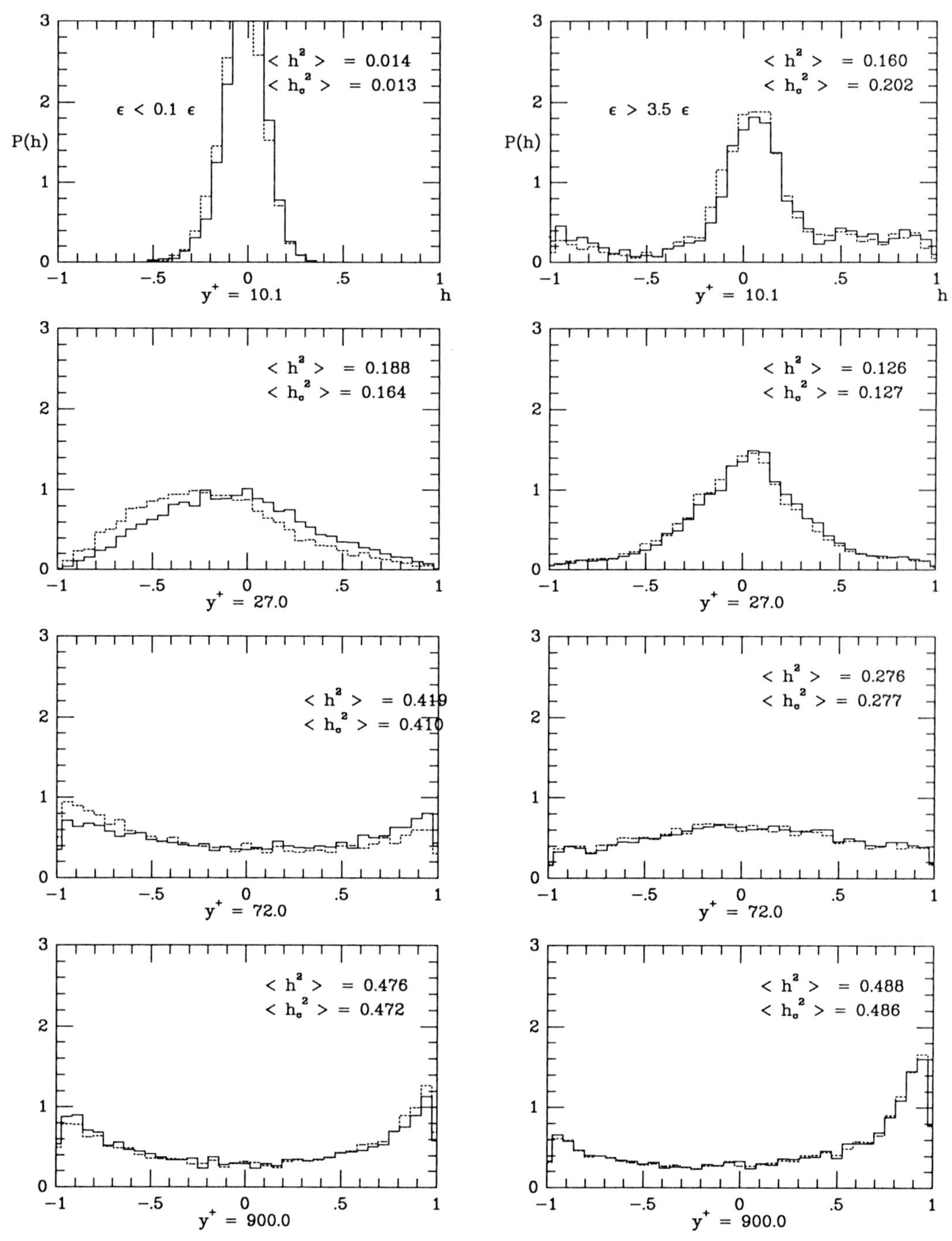

Fig. 2 Probability density distributions of h conditioned on the low and high instantaneous dissipation in the boundary layer at $Re_\theta = 2688$.

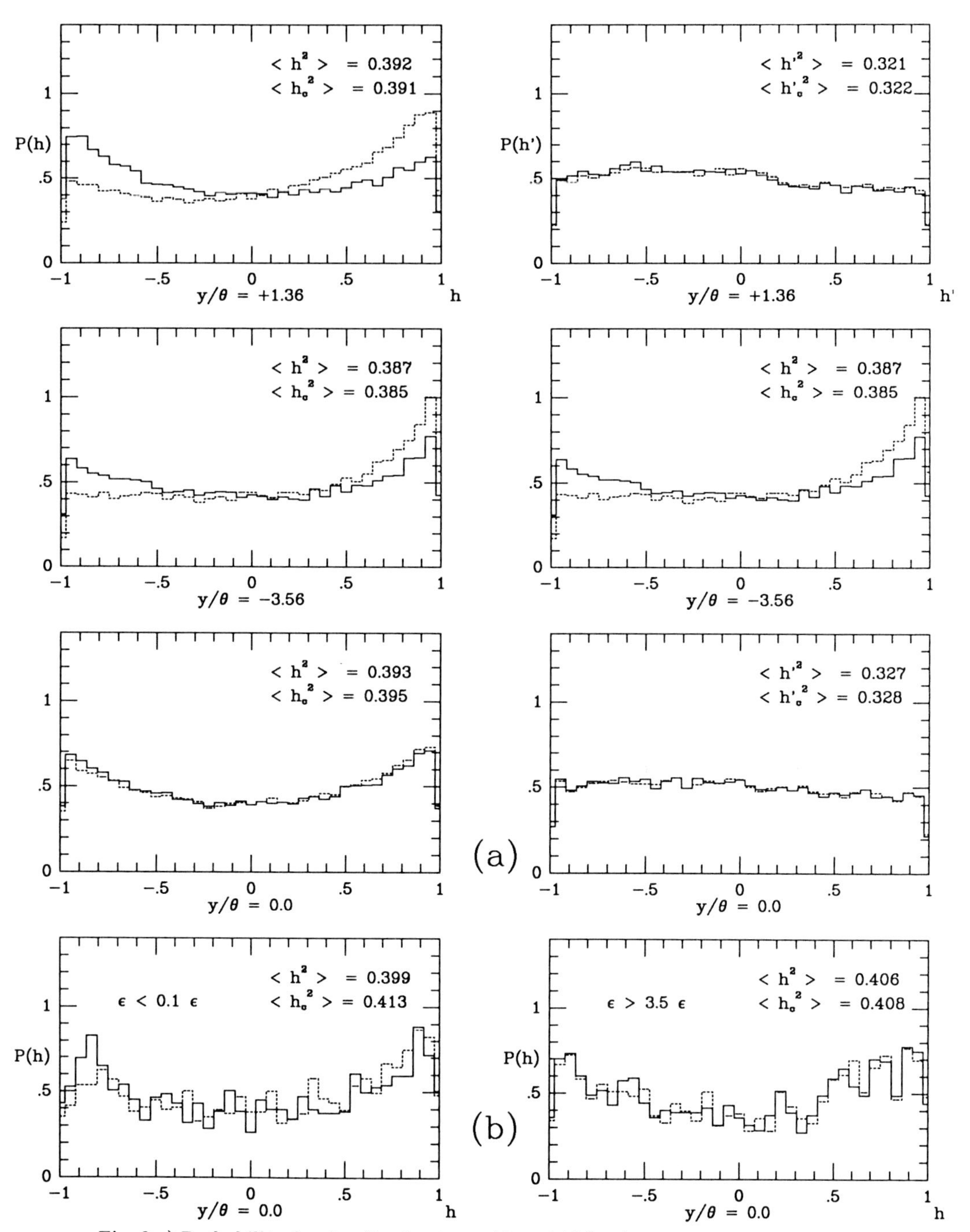

Fig. 3 a) Probability density distributions of h and h' for three positions in the mixing layer at $Re_\theta = 6406$. b) P(h) at the midplane $y/\theta = 0$ conditioned on low and high instantaneous dissipation.

3.3 Grid Flow

This flow, although not perfectly isotropic as indicated by the values of terms in the table below, had very similar characteristics to the grid flow studied by Kit et al. (1987). In fact the anisotropies are oddly similar in the two flows, although the wind tunnels and vorticity probes used were both different in the two cases. Figure 4(a) shows the PDDs of h and h' for the unconditioned data at two spanwise positions in the flow. To mimic true isotropic turbulence with no mean flow, a frame of reference convecting with the mean speed should be taken; thus the relative helicity density for the fluctuating field h' is the most relevant property. It is also possible in a laboratory grid flow, however, to determine h. For both h and h' it is evident that there is virtually no tendency for the vectors to align. The mean square values of h and h' are very close to those for two uncorrelated vector fields. The mean helicity in this flow was a result of very small angle misalignments of the arrays with the mean flow ($< 0.5^o$), but they had a large influence on the symmetry of these distributions, as seen in the figure. When the mean helicity was corrected, the asymmetry disappeared. This mean helicity was entirely due to the spurious residual mean velocities and vorticities. The very slight tendency for the two vector fields to align obtained here, as indicated by $\overline{h_c'^2} = 0.355$, is almost identical to the value of 0.359 found by Rogers and Moin (1987) and is a little less than the value of 0.380 found by Kit et al. (1987). The peak to minimum ratio is less than those found by Pelz et al. (1985). Data were taken at several spanwise positions, over the span of one mesh, in order to see if any residual wake from the grids remained. It is evident from the two positions shown in figure 4(a) that the measurements were far enough downstream for the grid wake to have become random.

There is a slightly smaller probability for the vectors to align when conditioned on either low or high dissipation compared to the unconditioned data, as seen in figure 4(b), this is in good agreement with the observations of Rogers and Moin (1987). Because there is no evidence that grid flow has a preferred orientation of organized vortices aligned with the mean velocity, this result is not suprising.

4 CONCLUSIONS

1. In the regions of turbulent shear flows where organized vortices are known to have axes at least partially aligned with the mean flow, the PPDs of h have shapes showing a preferred tendency for the velocity and vorticity vectors to be aligned. In the region of the flow where the total vorticity vector is nearly orthogonal to the mean flow, as in the sublayer of a boundary layer or channel flow, the PDDs show the opposite shape, as would be expected.

2. The conjecture that within coherent structures the relative helicity density should always be large is clearly not justified by these results. Coherent structures with vorticity axes nearly aligned with the mean velocity will have relatively high helicity densities; those that are orthogonal to the mean velocity,

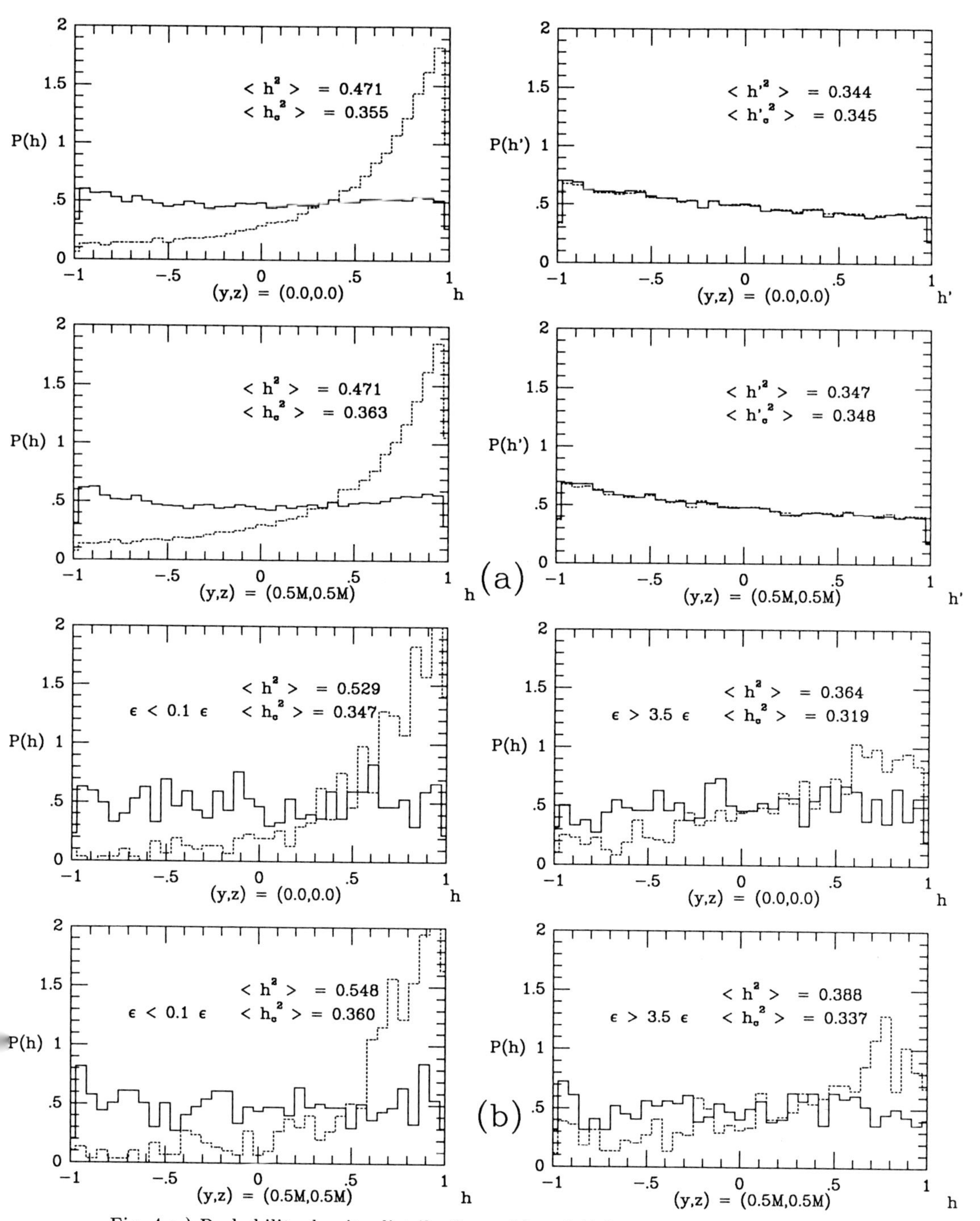

Fig. 4 a) Probability density distributions of h and h' for two transverse positions at $40M$ downstream of a $5.08cm$ mesh grid of round rods with 0.44 solidity. b) P(h) at the two transverse positions conditioned on low and high instantaneous dissipation.

as for example the large scale spanwise vortices in the mixing layer, will have very low relative helicity densities. Thus the prospects of modelling all coherent structures as Euler flows does not appear to be very bright.

3. The conjecture that large scale structures are a result of some universal intrinsic property of turbulent flow, even of homogeneous turbulence, rather than the result of shear layer instabilities, also does not appear justified from these results.
4. There appears to be little relationship between either low or high instantaneous dissipation and relative helicity density.
5. Asymmetries in experimentally determined PDDs of h and h' can be easily explained by small residual and spurious values of mean velocity and vorticity components which should be zero or negligible.

5 ACKNOWLEDGEMENTS

This work has been supported under the Department of Energy grant DEFG0588-ER13838. It was presented at its 7th Energy Sciences Symposium. We would like to thank Dr. Oscar Manley, the Basic Energy Sciences Program Director, for his continuing interest. Finally, we are grateful to Professors R. Adrian of the University of Illinois and N. Jones of Johns Hopkins University for making available to us their mixing layer and grid flow facilities, respectively.

6 REFERENCES

André, J.C. & Lesieur, M. 1977 *J. Fluid Mech.* **81**, 187.
Balint, J.-L. 1986 Thèse de Docteur d' Etat ès Sciences, Univ. de Lyon (France).
Balint, J.-L., Vukoslavčević, P. & Wallace, J.M. 1987 *Advances in Turbulence*, Springer-Verlag (ed. by G. Comte-Bellot & J. Mathieu), **459**.
Balint, J.-L., Wallace, J.M. & Vukoslavčević, P. to be published in *Proc. of 2nd European Turbulence Conf.* (1988), Springer-Verlag.
Comte-Bellot, G. & Corrsin, S. 1966 *J. Fluid Mech.* **25**(4), 657.
Hussain, A.K.M.F. 1986 *J. Fluid Mech.* **173**, 303.
Kerr, R.M. & Gibson, C.H. 1985 *Bull. Am. Phys. Soc.* **30**(10), 1733.
Kit, E., Tsinober, A., Balint, J.-L., Wallace, J.M. & Levich, E. 1987 *Phys. Fluids* **30**(11), 3323.
Kit, E., Tsinober, A., Teitel, M., Balint, J.-L., Wallace, J.M. & Levich, E. 1988 *Fluid Dyn. Res.* **3**, 289.
Kraichnan, R.H. 1973 *J. Fluid Mech.* **59**, 745.
Levich, E. & Tsinober, A. 1983 *Phys. Lett.* **93**A(6), 293.
Moffatt, H.K. 1969 *J. Fluid Mech.* **35**, 117.
Moffatt, H.K. 1985 *J. Fluid Mech.* **159**, 359.
Pelz, R.B., Yakhhot, V., Orszag, S.A, Shtilman, L. & Levich, E. 1985 *Phys. Rev.*

Lett. **54**(23), 2505.
Pelz, R.B., Shtilman, L. & Tsinober, A. 1986 *Phys. Fluids* **29**(11), 3506.
Rogers, M.M. & Moin, P. 1987 *Phys. Fluids* **30**(9), 2662.
Shtilman, L., Levich, E., Orszag, S.A., Pelz, R.B. & Tsinober, A. 1985 *Phys. Lett.* **113**A(1), 32.
Speziale, C.G. 1987 *ICASE Rep.* No. 87-69.
Tsinober, A. & Levich, E. 1983 *Phys. Lett.* **99**A(6,7), 321.
Vukoslavčević, P., Balint, J.-L. & Wallace, J.M. 1989 *Trans. ASME* I: *J. Fluids Engr.* **111**(2), 220.

Reduction of Nonlinearity and Energy Cascade in Helical and Nonhelical Turbulent Flows

LEONID SHTILMAN[1] & WOLFGANG POLIFKE[2]

[1]Tel-Aviv University, Department of Fluid Mechanics and Levich Insitute for PCH
[2]Levich Institute for PCH and Physics Department, The City College of CUNY, USA

ABSTRACT

A new initialization procedure is presented which allows to control both the energy and the helicity spectral density of the initial flow field. It is shown that large initial helicity leads to a reduced cascade of energy and inhibits the buildup of enstrophy. The mechanism of the reduction of nonlinearity observed in simulations of decaying isotropic turbulence is investigated. It is shown that Lamb vector $\vec{v}\times\vec{\omega}$ in Fourier space has a tendency to align with the wavevector $\vec{k}$. The ordering of the energy transfer is suggested to be a dynamically generic effect driving the reduction.

I. Introduction

Helicity, the scalar product of velocity $\vec{u}$ and vorticity $\vec{\omega}$ possesses a number of intriguing properties and has been the subject of a series of theoretical, numerical and experimental investigations. In inviscid flows with localized vorticity distributions, the total helicity $H = \int h d\vec{x} = \int \vec{v}\cdot\vec{\omega} d\vec{x}$ is invariant [1]. This follows from the equation of motion for H

$$\frac{d}{dt}H = \int_{\partial V} [-\vec{v}h + \vec{\omega}(p-\frac{v^2}{2})]\cdot\hat{n}\, dA + \nu\int_V (\vec{v}\cdot\nabla^2\vec{\omega}+\vec{\omega}\cdot\nabla^2\vec{v})d\vec{x} \tag{1}$$

if $\nu = 0$. It was shown in high Reynolds number numerical simulations of EDQN theory [2] that the presence of mean helicity slows down the energy transfer towards smaller scales. It was conjectured in a series of papers (see Ref 3 and references therein) that spontaneous fluctuations of helicity are intrinsic to turbulent flows. In this work we investigate various aspects of mean helicity in decaying isotropic turbulence by means of direct numerical simulation. In Section I we present a method of creating divergence-free Gaussian velocity fields with arbitrary spectral densities of energy and helicity, which then serve as initial fields for direct numerical simulations, and we investigate the effects of non-zero mean helicity on the energy transfer and related quantities in low Reynolds number turbulent flows. In Section II the role of helicity in the reduction of nonlinearity [4,5] is discussed. The work presented her constitutes a part of Ph.D. Thesis of one of the authors (see [8] for details)

I. Helicity and Energy Transfer

When setting up the initial flow field for a simulation of decaying turbulence, one frequently employs a random number generator to obtain the components $v_i(\vec{k})$ of the velocity field in Fourier space such that they yield an isotropic velocity field with Gaussian statistics and a certain energy spectrum $E(k)$. In particular, one obtains a divergence-free field ($\vec{k}\cdot\vec{v}(\vec{k}) = 0$) by projecting $\vec{v}(\vec{k})$ on the plane perpendicular to the wave vector $\vec{k}$. The reality constraint $\vec{v}(\vec{k}) = \vec{v}^{star}(-\vec{k})$ is satisfied implicitely by keeping only modes with $k_x \geqslant 0$ in the computation. The energy spectrum, defined as $E(k) = \frac{1}{2}\sum_{S(k)} |\vec{v}(\vec{k}')|^2$ and the average energy $E = \sum_k E(k)$ are determined by the appropriate choice of a 'shape function' $f(k)$ for the velocity amplitudes, $<|\vec{v}(\vec{k})|> = f(|\vec{k}|)$. The helicty spectral density $H(k)$ may be written as

$$H(k) = \sum_{S(k)} \vec{v}(\vec{k}) \cdot \vec{\omega}(-\vec{k}) = \sum_{S(k)} 2\vec{k}(\vec{R}(\vec{k}) \times \vec{I}(\vec{k})) \tag{2}$$

$\vec{R}(\vec{k})$ and $\vec{I}(\vec{k})$ denote the real and imaginary parts of the complex velocity vector $\vec{v}(\vec{k})$. As a consequence of the incompressibility constraint, the cross-product of $\vec{R}(\vec{k})$ and $\vec{I}(\vec{k})$ is always aligned with $\vec{k}$. This means that for given $R(\vec{k})$ and $I(\vec{k})$, i.e. given energy $E(\vec{k})$, the angle ϕ between $\vec{R}(\vec{k})$ and $\vec{I}(\vec{k})$ determines the helicity of mode $\vec{k}$. Therefore by controlling the angle ϕ between the real and the imaginary part of the velocity one can choose the initial helicity spectrum independently of the energy spectrum. The numerical implementation of this idea proceeds in a few straightforward steps [6]. Velocity fields created in this manner are - within statistical fluctuations - isotropic and Gaussian [6]. The energy and helicity spectra are now equal to

$$E(k) = 4\pi k^2 f(k)^2 <\chi^2> \quad H(k) = 8\pi k^3 f(k)^2 <\chi>^2 <\sin\Phi> \tag{3}$$

In order to study the effects of helicity on the dynamics of homogeneous turbulence, we have conducted a series of numerical simulations, employing a pseudospectral code with resolution 64^3. The initial fields for these runs were identical to each other in every respect except helicity. The energy spectrum was chosen to be $E(k) = Ak^4 e^{-2(k/k_{max})^2}$ with k_{max} equal to 4 and average initial energy E (determined by the the normalization constant A) equal to 2.8 . Other initial parameters are : kinematic viscosity ν equal to 0.015, turnover time τ equal to .45, integral scale equal to .64, Taylor Microscale equal to .44, Taylor Microscale Reynolds number R_λ equal to 45. During the early stage of the simulation, the energy, which initially is concentrated at small wavenumbers, spreads towards high wavenumbers without suffering significant dissipation. Consequently, the average enstrophy $\Omega = k^2 E(k)$ usually increases. One assumes that the flow is 'most developed' once the enstrophy starts to decay and the statistical characteristics of the velocity derivatives become typical of turbulence at moderate Reynolds numbers, i.e. a flatness of approximately 4 and a skew ness of approximately -0.5 .

Fig.1 shows the energy spectra for runs O_{zero}, accumulated at equal timeintervals $\delta t = 0.2$ for times $t = 0$ until $t = 2$, which is about four times the initial turnover time. The initial field for this run had zero helicity in all modes. In Fig. 2 we present the equivalent data for run O_{max}, a simulation with a maximally helical initial field, i.e. $H(k,t=0) = 2kE(k)$ for all wavenumbers. This flow is strongly Beltramized, we find that initially at 82 % of all gridpoints in physical space the cosine of the angle between velocity $\vec{v}$ and vorticity $\vec{\omega}$ is greater than 0.9 . From Fig. 2 we may conclude that the presence of high mean helicity slows down the flow of energy

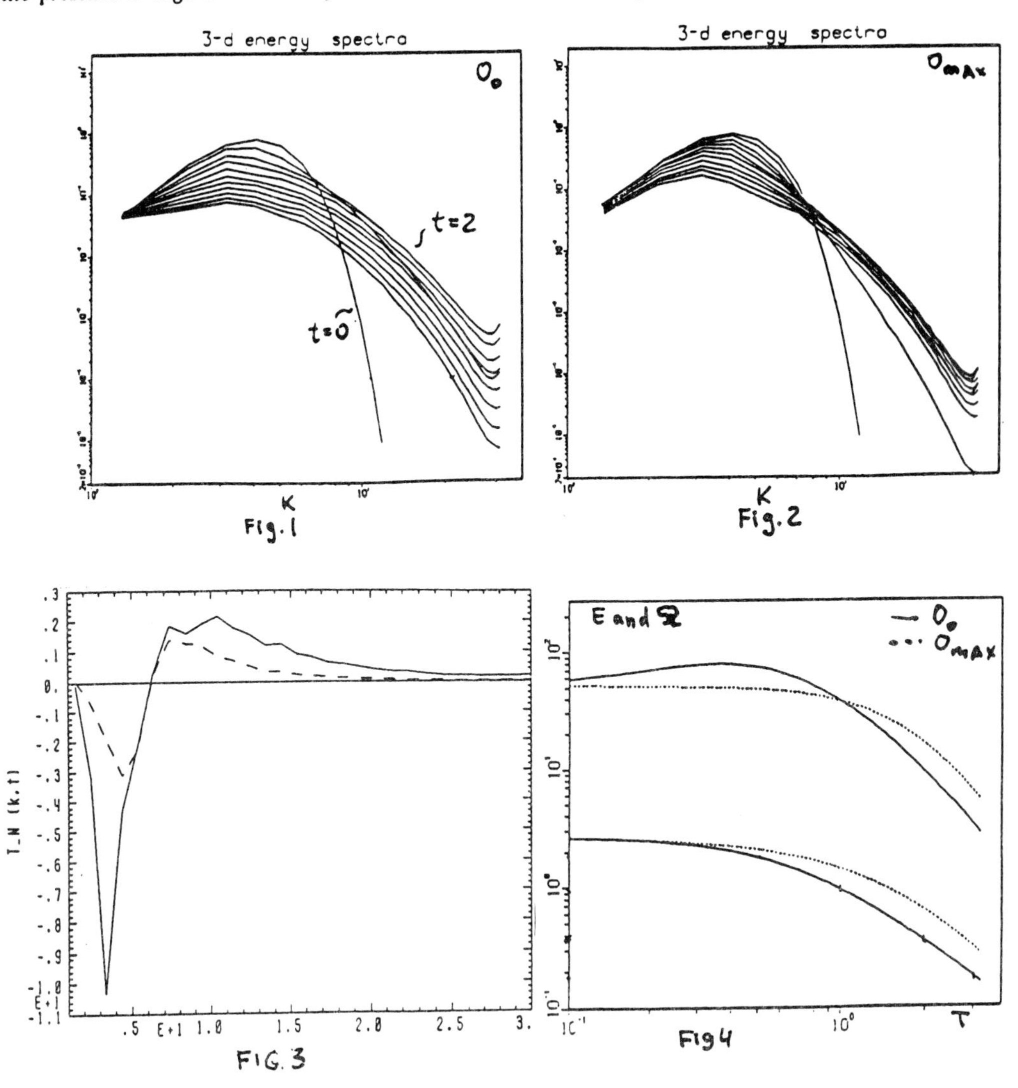

Fig.1 Fig.2

FIG.3 Fig4

towards higher wavenumbers. The time development of the energy spectrum is determined by

$$\left(\frac{d}{dt} + 2\nu k^2\right) E(k,t) = T(k,t) \tag{4}$$

where the transfer term T(k,t) can be written as

$$T(k,t) = 2\sum_{Sh} \mathrm{Real}[\vec{u}\times\vec{\omega}(\vec{k})\cdot\vec{u}\ \mathrm{star}(\vec{k})] \tag{5}$$

In Fig. 3 we plot the normalized transfer term $T_N(k,t) := T(k,t)\, L / (\nu E^{3/2})$ for runs O_{zero} and O_{max}, where L is the integral length scale at time t = 0.4 We see that the flow of energy out of the energy containing range is significantly reduced in the helical case. The difference in transfer strength can be observed for about two turnover times. However, at later times the decay rate of O_{max} is approximately equal to that of O_{zero}. This is illustrated in Fig. 4, which shows the time-behaviour of average energy and average enstrophy for both runs for times t = 0. up to t = 3.2. The time behaviour of helicity and of relative helicity $H_R = H / \sum_k 2kE(k)$ for run O_{max} is shown in Fig. 5. It is of inerest to assess what value of H_R is required to significantly influence the nonlinear dynamics of the decaying flow.

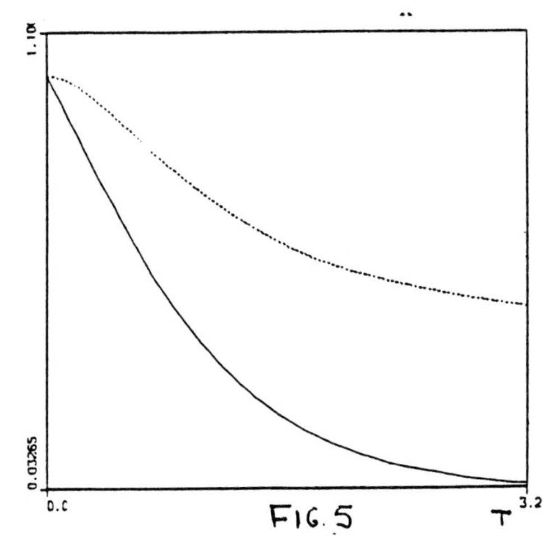

FIG. 5

In table I. we present data from runs O_{zero}, O_{max} and 4 other runs O_{20}, O_{40}, O_{60} and O_{80}, which had the same initial energy spectrum as O_{zero} and O_{max} and a helicity spectrum H(k) equal to 20, 40, 60 and 80 percent of the maximum value 2kE(k) for all wavenumbers. This was achieved by assigning maximum positive or negative helicity ($\phi = \pm\pi/2$) in varying proportions to the modes in Fourier space. In the first rows we list the time t, average energy E(t), average enstrophy $\Omega(t)$ and relative helicity $H_R(t)$. Recall that $E(t=0) = 2.8$ and $\Omega(t=0)=56$. Note that only for the maximally helical flow O_{max} there is no initial growth of enstrophy. In the last two rows we present quantities that seem adequate to measure the strength or effectiveness of the energy transfer in the turbulent flow: the normalized mean-square nonlinear term Q(t)

$$Q(t) := \overline{|\vec{u}\times\vec{\omega} - \nabla p|^2} \,/\, (\overline{u^2}\,\overline{\omega^2}) \tag{5}$$

and the sum S_T of the normalized transfer $T_N(t)$ over the energy containing modes (all k such that $T(k,t) < 0$):

$$S_T(t) := -\sum_{k=1}^{5} T_N(k,t)$$

Obviously, $S_T(t)$ indicates how rapidly the energy of the energy containing modes is being transferred to the small scales. Similarly, $Q(t)$ relates the average magnitude of the nonlinear term of the NSE, which is responsible for the energy transfer, to the average magnitude of energy and enstrophy. The data in table I indicate that the presence of non-zero mean helicity has significant consequences for the dynamics and nonlinearity of turbulent flows only if H_R is quite large. Finally we remark that the maximum Kolmogorov wavenumber k_D for run O_{max} was 26.6, compared with 28.9 for O_{zero}. As a consequence, run O_{max} is slightly better resolved, consider the upward turns of the energy spectrum in Fig. 1 and 2.

TABLE I

run	O_{zero}	O_{20}	O_{40}	O_{60}	O_{80}	O_{max}	O_{zero}	O_{max}	O_{zero}	O_{max}
t	0.4	0.4	0.4	0.4	0.4	0.4	1.2	1.2	2.0	2.0
E(t)	1.97	1.98	2.00	2.05	2.11	2.18	.745	1.19	.348	.622
$\Omega(t)$	77.7	76.9	74.3	68.0	59.13	48.7	27.6	32.4	9.30	16.2
$H_R(t)$	.001	.182	.334	.515	.719	.923	.015	.684	.038	.543
$R_\lambda(t)$	27.6	27.5	28.4	30.5	33.9	38.2	17.2	25.3	14.0	19.0
Q(t)	.135	.145	.134	.122	.093	.056	.142	.087	.141	.1
$S_T(t)$	20.7	20.1	19.8	17.0	13.6	7.9	19.23	14.0	19.3	17.2

II Reduction of Nonlinearity.

Recently velocity fields obtained from numerical simulations of the incompressible Navier-Stokes equations (NSE) were compared with Gaussian solenoidal vector fields that have the same instantaneous energy spectrum as the turbulent field ("corresponding Gaussian fields")[4,5]. It was observed that the normalized mean-square nonlinear term $Q = \langle |\vec{\lambda} - \nabla p|^2 \rangle / [\langle |\vec{v}|^2 \rangle \langle |\vec{\omega}|^2 \rangle]$ is about 60% of the value for the corresponding Gaussian field. Here $\vec{v}$ is the velocity, $\vec{\omega} = \nabla \times \vec{v}$ the vorticity, p is the pressure, and $\vec{\lambda} = \vec{u} \times \vec{\omega}$ is commonly referred to as the Lamb vector.

In Fig. 6 we present Q, the normalized mean-square of the nonlinear term for two simulations and the corresponding Gaussian fields at various times. Circles correspond to run s_r, a 64^3 simulation of decaying turbulence in a box with sidelength 2π with initial conditions similar to those of Kraichnan & Panda [4] : average energy $E \approx 1.1$, average enstrophy ≈ 9.0, average helicity $H \approx 0.07$, Reynolds number $R_\lambda \approx 45$, eddy turnover time ≈ 1.3 (computational time). Maximum enstrophy was reached at time $t \approx 0.9$, then $R_\lambda \approx 29$. Triangles denote Q for the corresponding Gaussian fields, i.e.

fields that have instantaneously the same energy spectrum as the simulation fields. We should remark here, that all quantities like kinematic pressure, energy, and helicity for random Gaussian fields are defined by analogy with the corresponding expression for the incompressible NSE; however, they have no physical meaning. At time t=0.9 Q of the simulation is 58 % of the corresponding Gaussian value. In order to elucidate the role of the alignment of velocity and vorticity in the reduction we employed the method of creating Gaussian fields with controlled helicity spectrum described in section I.

The crosses in Fig. 6 represent Q for maximally helical Gaussian (MHG) fields, i.e. $H(\vec{k}) = 2kE(\vec{k})$ for all $\vec{k}$. Again, the energy spectra of these MHG-fields are identical to those of run s_r (circles) at each moment. Initially Q for the MHG-fields is considerably less than Q for the simulation. Eventually however, the suppression of nonlinearity in the simulation becomes stronger than in the MHG-fields. The conclusion that must be drawn from this result is that the alignment of $\vec{v}$ and $\vec{\omega}$ cannot be the sole source of the drop of Q. However, it can enhance the suppression significantly. Using a MHG-field as an initial field for a simulation (run s_m, see squares in Fig. 6), we observe that Q in a simulation with high helicity ($H(t=0) \approx 4.6$) is at all times considerably less than for a simulation with small initial helicity. Note that for runs s_m the normalized mean-square of the nonlinear term is only 37% of the corresponding value for a random Gaussian field.

Let us consider the following equation

$$<|\nabla p|^2> = <\vec{\lambda}\cdot\nabla p> \tag{6}$$

The angular brackets denote here either spatial averaging or integration over Fourier space. Eqn. (1) can be verified by considering the expression for the kinematic pressure of a homogeneous incompressible flow field in Fourier space, i.e. $p(\vec{k}) = -i\vec{k}\cdot\vec{\lambda}(\vec{k})/\ k^2$. As an immediate consequence the second term on the right hand side can be rewritten as

$$<|\nabla p|^2> = <|\frac{\vec{k}}{k^2}\vec{k}\cdot\vec{\lambda}(\vec{k})|^2> = <R_{\vec{\lambda}}(\vec{k})^2\cos^2\xi_R> + <I_{\vec{\lambda}}(\vec{k})^2\cos^2\xi_I>$$

Here $\vec{R}_{\vec{\lambda}}(\vec{k})$ $(\vec{I}_{\vec{\lambda}}(\vec{k}))$ denotes the real (imaginary) part of the Fourier transform of the Lamb vector and ξ_R (ξ_I) the angle between $\vec{k}$ and $\vec{R}_{\vec{\lambda}}(\vec{k})$ $(\vec{I}_{\vec{\lambda}}(\vec{k}))$. Let us assume for the moment that the absolute value of the Lamb vector is statistically independent of its orientation in Fourier space. Let us further assume that the probability distribution functions (pdf) of ξ_R and ξ_I are identical and therefore

$$<\cos^2\xi_R> = <\cos^2\xi_I> = <\cos^2\xi>$$

Then (6) can be rewritten as

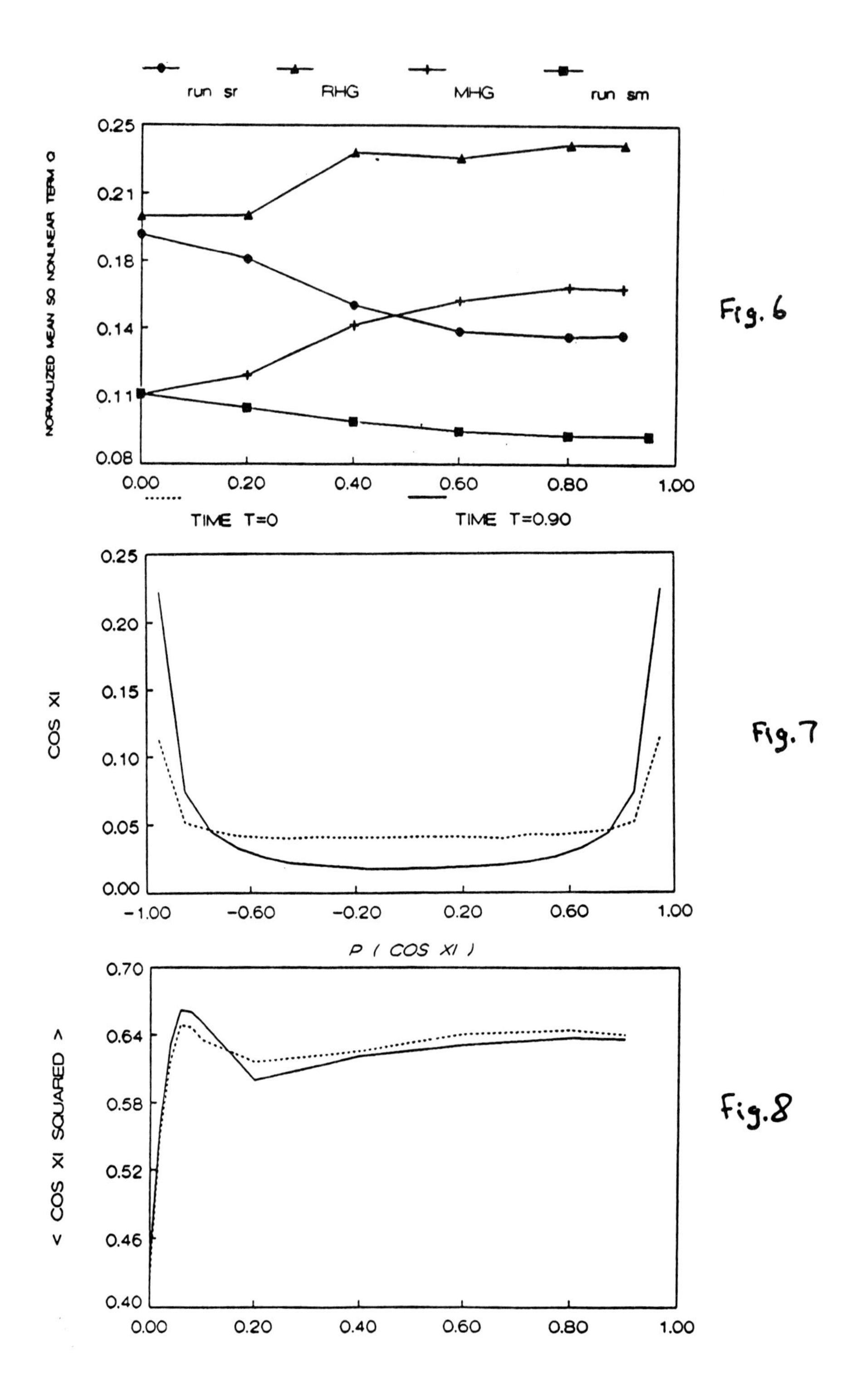
run sr
RHG
MHG
run sm
NORMALIZED MEAN SQ NONLINEAR TERM Q
0.25
0.21
0.18
0.14
0.11
0.08
0.00
0.20
0.40
0.60
0.80
1.00
Fig. 6
TIME T=0
TIME T=0.90
COS XI
0.25
0.20
0.15
0.10
0.05
0.00
-1.00
-0.60
-0.20
0.20
0.60
1.00
P (COS XI)
Fig. 7
< COS XI SQUARED >
0.70
0.64
0.58
0.52
0.46
0.40
0.00
0.20
0.40
0.60
0.80
1.00
Fig. 8

$$<|\vec{\lambda}-\nabla p|^2> \approx <|\vec{\lambda}|^2> - <|\vec{\lambda}|^2><\cos^2\xi> = <\lambda^2><\sin^2\xi> \quad (7)$$

We will comment below upon the validity of the assumptions made. The conclusion that we can draw from Eqn. (7) is that the suppression of nonlinearity may be related to a non-uniform distribution of the angles ξ_R and ξ_I, i.e. an anisotropic orientation of the Fourier transform of the Lamb vector. Analyzing data from several runs we found that the pdf's of ξ_R and ξ_I are always identical, we therefore present only plots related to ξ_R Fig. 8 shows the pdf of $\cos\xi_R$ for the run s_r at times $t = 0$ (dotted line) and $t = 0.9$ (solid line). Note that even initially $\cos\xi_R$ is distributed nonuniformly. It turns out that this is a consequence of incompressibility and the fact that $\vec{\omega}=\nabla\times\vec{v}$. This was checked by evaluating the pdf of cos ξ' for $\vec{\lambda}' = \vec{A}\times\vec{B}$ where $\vec{A}$ and $\vec{B}$ are Gaussian fields and i) $\vec{B}=\nabla\times\vec{A}$, $\vec{A}$ not divergence-free; ii) $\vec{A}$ and $\vec{B}$ independent and divergence-free; iii) $\vec{A}$ and $\vec{B}$ independent and not divergence-free. Only in case iii) the distribution of cos ξ was flat. Fig. 7 also shows clearly that the "horns" of the distribution grow considerably . In Fig. 8 we plot $<\cos^2\xi_R>$ at times $t = 0$ up to $t = 0.9$ for two runs : random initial helicity s_r (solid line) and maximum initial helicity s_m (dotted line). The first observation is that these two curves are almost identical, which support conjecture that there is no direct relation between helicity spectrum and the distribution of ksi. The second observation is that $<\cos^2\xi>$ reaches its maximum value at $t = 0.06$, which is long before the developed stage ($t \approx 0.9$) and long before Q experiences a significant drop. We therefore must conclude that the correlator $<R_{\vec{\lambda}}(\vec{k})^2\cos^2\xi_R>$ cannot be approximated by $<R_{\vec{\lambda}}(\vec{k})^2> <\cos^2\xi_R>$. The same is true for the imaginary part.

The results of numerical experiments presented above show that an enhanced alignment of the Lamb vector $\vec{\lambda}(\vec{k})$ and wavevector $\vec{k}$ is at least partly responsible for the reduction of nonlinear fluctuations. Let us consider a possible relation between this alignment and the cascade of energy. For isotropic turbulence the time development of the energy $E(k)$ of scales of length $2\pi/k$ is described by eqs.(4),(5) where the transfer-term $T(k,t)$ is based on triple velocity correlators. Examining transfer spectra from various runs we observed that at time $t = 0$, $T(k,t)$ is wildly fluctuating; there is obviously no net energy transfer between various regions of Fourier space. This is due to the statistics of the initial field: third order moments of a Gaussian random variable vanish in the mean. However, very soon the energy transfer becomes ordered, i.e. the energy containing modes lose energy ($T(k,t) < 0$), whereas the small scales receive energy. It is remarkable, that this ordering of $T(k,t)$ happens very fast and on the same timescale as the growth of $<\cos^2\xi>$. To illustrate this, we present in table 2 values of $T(k,t)$ for $k<7$ obtained from run s_r. Note that in this run most of the energy was initially contained in modes $k<4$.

TABLE 2

k	1.5	2.5	3.5	4.5	5.5	6.5
TIME						
0.00	0.33-2	-0.14-02	0.94-02	-0.11-01	-0.82-04	-0.53-04
0.02	0.20-02	-0.10-01	0.62-02	-0.29-02	0.13-01	0.33-02
0.04	0.57-03	-0.19-01	-0.21-01	0.54-02	0.26-01	0.66-02
0.06	-0.77-03	-0.28-01	-0.35-01	0.13-01	0.38-01	0.96-02

Obviously the organization of the energy transfer occurs within $t = 0.06$ (compare Fig. 8 and Table 2 !). The developed stage of the simulation (enstrophy maximal , skewness and flatness of velocity derivatives equal to -0.4 to -0.5 and 4 to 5, respectively [5]) is reached much later at time $t \approx 0.9$. We therefore argue that the suppression of nonlinear fluctuations is closely related to the organization of the energy transfer. In our opinion, the ordering of the energy transfer could be a dynamically generic effect that leads to the suppression of the nonlinear fluctuations in NSE-turbulence and other systems that exhibit organized flow of energy. The mechanism through which such a suppression is achieved, however, is by no means trivial and is dependent on the system under consideration. In decaying turbulence the reduction of the nonlinear interaction is clearly manifested in two geometrical properties - the alignment of the velocity and vorticity fields in physical space[4,10] and the enhanced alignment of $\vec{\lambda}(\vec{k})$ with $\vec{k}$ in Fourier space.

Acknowledgements

The authors thank Prof. Evgeny Levich for fruitful discussions. This work was performed under DOE grant DE-FG0288ER13837. All computations were performed at NMFECC at Lawrence Livermore National Laboratory and at NASA-Ames Center for Turbulence Research.

References

1. H.K.Moffatt, J.Fluid Mech. 35, p 117 (1969)
2. J.C. Andre & M. Lesieur, J. Fluid Mech. 81, pp 187-207(1977).
2. R. Kraichnan, J.Fluid.Mech. 59,745-752 (1973)
3. E. Levich , Phys. Rep. 151, 131-238 (1987)
4. R.H. Kraichnan and R. Panda, Phys. Fluids 31 , 2395 (1988)
5. L. Shtilman and W. Polifke, Phys.Fluid A 1(5) 778-780 (1989)
6. W.Polifke and L.Shtilman, Phys. Fluids A 1 (12)(1989)
7. R.B. Pelz, L. Shtilman and A. Tsinober, Phys. Fluids 29, 3506-3508 (1986)
8. W.Polifke, Ph.D. Thesis, The City College of CUNY (1990)

Vortex Filaments and Turbulence Theory[1]

ALEXANDRE J. CHORIN[2]

Department of Mathematics, University of California,
Berkeley, California 94720, USA

Introduction. Turbulence is described by the Navier-Stokes equations. Since the solution of the Navier-Stokes equations in a subsonic regime can be approximated by the motion of a collection of vortex blobs or vortex filaments [1,2,11], turbulence can be thought of as a many-vortex phenomenon, similar to other many-body problems in physics. The constraints imposed by the conservation properties of the Navier-Stokes equations, in particular the conservation of helicity, evoke similarities to another well-known statistical mechanical system with constraints a collection of polymers in a solution.

In this talk we shall present the main properties of vortex systems, explain some similarities to other physical systems, and present an analysis of the inertial scales of turbulence that explains some experimental facts and leads to a useful modeling technique. Some of the conclusions are quite unexpected.

The Random Vortex Equations. We begin by writing down the equation of motion for a vortex system. We assume for simplicity that the fluid is incompressible. The Navier-Stokes equations take the form

$$\partial_t \underline{\xi} + (\underline{u}\cdot\underline{\nabla})\underline{\xi} - (\underline{\xi}\cdot\nabla)\underline{u} = R^{-1}\Delta\underline{\xi} \tag{1a}$$

$$\text{div } \underline{u} = 0 , \quad \underline{\xi} = \text{curl } \underline{u} \tag{1b,c}$$

---

[1] Work partially supported by the Office of Energy Research, U.S. Department of Energy, under contract DE-AC03-76SF0098.

[2] Paper presented by T.F. Buttke

where $\underline{u}$ is the velocity, $\underline{\xi}$ is the vorticity, t is the time, $\underline{\nabla}$ is the differentiation vector, $\Delta = \underline{\nabla}\cdot\underline{\nabla}$, and R is the Reynolds number. R will be large or infinite. Equations (1b,c) can be inverted and yield

$$\underline{u} = K*\underline{\xi} \tag{2}$$

where $*$ denotes a convolution and K is the matrix

$$K = -\frac{1}{4\pi r^2}\begin{pmatrix} 0 & -x_3 & -x_2 \\ x_3 & 0 & -x_1 \\ x_2 & x_1 & 0 \end{pmatrix}$$

where $\underline{x} = (x_1,x_2,x_3)$ is the position vector and $r = |\underline{x}|$. Equation (2) is known as the Biot-Savart law. If $R^{-1} = 0$ equations (1) are known as Euler's equations.

Consider for simplicity the case of unbounded flow. Suppose the initial vorticity $\underline{\xi}(\underline{x}, t=0)$ is broken down into a sum of vector functions of small support (= "vortex blobs"),

$$\underline{\xi} = \sum_{i=1}^{N} \underline{\xi}_i \tag{3}$$

Let σ be a length characteristic of the support of $\underline{\xi}_i$. σ is the "cut-off length". Choose further a decomposition (3) for which $\underline{\xi}_i = \bar{\underline{\xi}}_i\Phi(\underline{x} - \underline{x}_i)$, $\underline{x}_i = (x_{1i},x_{2i},x_{3i})$, where $\Phi(\underline{x})$ is a function of small support, $\int\Phi dx = 1$, and $\bar{\underline{\xi}}_i$ is a constant. Equation (1) with $R^{-1} =$ $= 0$ can then be approximated by the following set of ordinary differential equations

$$d\underline{x}_i = (K*\underline{\xi})_{\underline{x}=\underline{x}_i}\, dt \quad , \tag{4a}$$

$$d\bar{\underline{\xi}}_i = ((\underline{\xi}\cdot\underline{\nabla})\underline{u})_{\underline{x}=\underline{x}_i}\, dt \quad , \tag{4b}$$

where one allows the "blobs" the move. The right-hand sides of equations (4) are known functions of the locations $\underline{x}_i$ and "strengths" $\bar{\underline{\xi}}_i$. The accuracy of the approximation depends on the choice of the

$\bar{\xi}_i$. The accuracy of the approximation depends on the choice of the decomposition (3) and thus on the form of ϕ. The convolution in (4a) can be rewritten as

$$(K*\xi)_{\underline{x}=\underline{x}_i} = \sum K_\sigma(\underline{x}-\underline{x}_i)\underline{\xi}_i \quad ,$$

where K_σ is a smoothed kernel, $K_\sigma = K*\phi$. The obvious constraint $\text{div}\ \underline{\xi} = \text{div curl}\ \underline{u} = 0$ can be imposed on (4b) explicitly and leads to representations by vortex tubes, or it can be left to take care of itself as a consequence of convergence; in this latter case, one can treat the blobs as independent vortex "arrows" or segments. These elements can be chosen so that their cross-sections remain circular. For details and convergence theorems, see [1,2,11].

Suppose now that $R^{-1} \neq 0$. Equations (1) then look like Fokker-Planck equations for the set of stochastic ordinary differential equations

$$d\underline{x}_i = \Sigma\, K_\sigma(\underline{x}-\underline{x}_i)\bar{\underline{\xi}}_i\ dt + R^{-1/2} d\underline{w}(t) \quad , \tag{5a}$$

$$d\underline{\xi}_i = ((\underline{\xi}\cdot\underline{\nabla})\underline{u})_{x=x_i}\ dt \tag{5b}$$

where $\underline{w}(t)$ is Brownian motion in three dimensional space. Equations (5) can be solved and sample the solution of equations (1). In the special case of two-dimensional flow $\bar{\underline{\xi}}_i$ = constant, $\underline{\xi}_i = (0,0,\xi_i)$, and the system (5) reduces to

$$d\underline{x}_i = \sum_{j\neq i} K_\sigma(\underline{x}-\underline{x}_i)\bar{\xi}_i + R^{-1/2}\ d\underline{w}(t)$$

where $\underline{w}(t)$ is two-dimensional Brownian motion, $K_\sigma = K*\phi$, ϕ has small support, $\underline{x}_i = (x_{1i},x_{2i})$, and

$$K = \frac{1}{2\pi}\begin{pmatrix} -\partial_y \\ \partial_x \end{pmatrix} \log r_{ij} \quad , \qquad r_{ij} = |\underline{x}_i-\underline{x}_j| \ .$$

The development below will be based on equation (5). In most practical applications, equations (5) have been treated by a fractional step formulation which is convenient but not necessary [11]. It can be shown that the addition of viscous dissipation and thus of a random forcing in equations (5) is similar to a quantization process in mechanics. In fact, equations (5) look like Langevin equations for a *low* temperature system (with temperature T proportional to viscosity, as in the Einstein relations of irreversible statistical mechanics). Note further the formal similarilty between equations (5) and the equations that describe the motion of polymers in a solution [7].

Numerical vortices and physical vortices. One aspect of equations (5) leads to endless confusion. The elements whose motion is described by these equations are not physical vortices (i.e., they are not real tornadoes, or pieces of hairpins that can be photographed in a laboratory). They are approximating computational elements. In particular, no one is suggesting that real physical vortices undergo Brownian motion. What is claimed, and indeed proved in the references mentioned above, is that an arbitrary flow can be approximated by these vortex filaments or vortex segments; this statement makes no claim as to the physical shape of the vortical parts of a given flow. For example, in two space dimensions, the motion of vortex sheets can perfectly well be approximated by round vortex blobs.

On the other hand, it is not claimed either that real vortex structures, which can be viewed as collective modes of motion for computational vortex clouds, are *never* tube-like. In each argument and in particular in each statistical mechanical argument, one has to distinguish carefully between what is claimed for computational vortices and what is claimed for physical vortices.

Some properties of tubular vortex structures. We begin by listing some properties of tubular vortex structures (computational elements or physical vortices when the latter happen to be tubular).

(i) Vortex tubes stretch. This stretching is very striking in numerical calculations, and there is no mathematical analysis that fully explains its mechanism. An interesting analysis in [9] shows that in a random isotropic flow lines stretch on the average, possibly quite slowly, but no clear inference can be drawn about vortex lines, since the latter are a small subset of all lines. The argument is interesting however because it links vortex stretching to information loss and thus to entropy, as one would expect from the polymer analogy. The increase in vortex length is measured by the increase in enstrophy

$$Z = \int |\underline{\xi}|^2 d\underline{x} \quad .$$

(ii) Vortex tubes absorb energy and transfer it from large to small scales, as a result of the stretching process. A quantitative analysis can be found in [6].

(iii) If a set of vortex tubes is stretching while energy is conserved, these tubes fold in a zig-zag manner and form "hairpins". A qualitative explanation runs as follows:

Consider a long vortex tube V of unit circulation and small, non-zero, approximately circular cross section. The kinetic energy T associated with the tube can be written as

$$T = \frac{1}{2}\int |\underline{u}|^2 dx = \frac{1}{8\pi}\int_V d\underline{x}\int_V d\underline{x}' \, \frac{\underline{\xi}(\underline{x})\cdot\underline{\xi}(\underline{x}')}{|\underline{x}-\underline{x}'|}$$

where $\underline{\xi} = \text{curl}\ \underline{u}$ is the vorticity. Suppose the support of V can be approximately covered by N circular cylinders I_i, $i = 1,\ldots,N$, of equal lengths ℓ and some radii r_i, $i = 1,\ldots,N$. T can then be approximated by $\tilde{T}$,

$$8\pi\tilde{T} = \sum_{i=1}^{N}\sum_{\substack{j=1 \\ j\neq i}}^{N} \tilde{T}_{ij} + \sum_{1}^{N} \tilde{T}_{ii} \qquad (6)$$

where

$$\tilde{T}_{ij} = \int_{I_i} d\underline{x}\int_{I_j} d\underline{x}' \, \frac{\underline{\xi}(\underline{x})\cdot\underline{\xi}(\underline{x}')}{|\underline{x}-\underline{x}'|}$$

Let $\underline{t}_i$ be a vector lying along the axis of the cylinder I_i, originating at the center of I_i, of length $|t_i| = \ell$, and pointing in the direction of $\underline{\xi}$ in I_i. If I_i and I_j are far from each other,

$$\tilde{T}_{ij} \sim \frac{\underline{t}_i \cdot \underline{t}_j}{|i-j|} \quad , \tag{7}$$

where $|i-j|$ is the distance between I_i and I_j. Pretend (7) holds approximately whenever $i \neq j$ — clearly a simplification. $\tilde{T}_{ii}$ (the case i=j) is a function of the radius r_i of I_i, and becomes infinite as $r_i \to 0$. Thus,

$$8\pi\tilde{T} \sim \sum_i \sum_{j\neq i} \frac{\underline{t}_i \cdot \underline{t}_j}{|i-j|} + \sum_i \tilde{T}_{ii}(r_i) \quad ,$$

where $d\tilde{T}_{ii}/dr_i < 0$.

Suppose now that the tube V is stretched by the velocity field that includes the velocity field that it itself induces, in such a way that its volume is preserved. Suppose that the support of the new stretched tube V can still be approximated by a collection of cylinders of lengths ℓ; their number N' will be larger than N, and most of their cross sections will be smaller than before. Thus the sum $\Sigma\tilde{T}_{ii}$ will increase because it will have more and larger entries. As the radii tend to zero, this sum will diverge. If $\tilde{T}$, the total energy, remains bounded, then many of the terms in the double sum over i,j must become negative, i.e., the tube must fold. Thus the vortex segments I_i arrange themselves in such a way that they shield the incipient singularity due to the singular Biot-Savart kernel. Note that this phenomenon has no analogue in two space dimensions, where there is no stretching and where "self-energies" can be safely subtracted from the total energy in defining a Hamiltonian. This shielding effect has some vague similarity to the shielding of a "bare" particle by a cloud of bosons in a renormalization of field theory. Nearby vortices have a reduced interaction because of the folding.

A simple one-dimensional cartoon of equation (6) is:

$$\tilde{T} = \sum_{i=1}^{N} \sum_{\substack{j=1 \\ j \neq i}}^{N} t_i t_j \frac{1}{|i-j|} + \sum \tilde{T}_{ii}$$

where N is finite and fixed, the t_i are Ising-like spins, i.e., vectors that can point either up ($t_i = 1$) or down ($t_i = -1$), $|i-j|$ is the distance between the position of t_i and t_j, and $\tilde{T}_{ii}$ is a function of a parameter r^{-1} that increases monotonically. The "spins" are located at the nodes of a regular lattice, most of whose nodes are empty. The "spins" can move to empty locations or flip (i.e., changes signs); r^{-1} increasing is interpreted as a stretching of the "spins". In [5] a sequence of spin configurations is constructed, such that the "energy" $\tilde{T}$ remains fixed. The $\tilde{T}_{ii}$ increase and thus the double sum must decrease; this requires the "spins" to bunch up as closely as possible on the lattice, with neighboring "spins" having opposite signs.

(iv) Vortex lines in a turbulent fluid are not spread uniformly in the fluid and tend to congregate. This phenomenon is known as "intermittency". In two space dimensions intermittency can be displayed numerically. Its occurrence can be made plausible by noting again the qualitative analogy between a plane vortex system and an Ising model in the plane; an Ising spin system at low temperature exhibits intermittency (i.e., a clustering of spins of like sign).

The most interesting manifestations of intermittency are, however, three-dimensional. In three space dimensions intermittency is a strong, dramatic effect. Its main source is the differential stretching of vortex lines. In addition, a shielding effect brings stretched portions close to each other. The result is a very uneven distribution of vorticity. As a consequence the enstrophy Z is dominated by the vorticity in a volume much smaller than the total volume available to the fluid.

Singularity formation, spectrum, dimension, and renormalization for tubular vortices. Several additional results can be obtained for tubular vortex structures and then applied to numerical algorithms.

Let ε be a small positive number; we define the ε-support of the vorticity to be the smallest set Λ_ε such that

$$\int_{\Lambda_\varepsilon} \xi^2 dx = (1-\varepsilon) \int_{\text{space}} \xi^2 dx \quad ;$$

i.e. Λ_ε is the set that carries most of the vorticity. Since ξ is stretching non-uniformly, and the vorticity that has been stretched most contributes most to $\int \xi^2 dx$, Λ_ε shrinks in size. If ξ is smooth initially, then initially the Hausdorff (= "fractal") dimension of Λ_ε, $\dim \Lambda_\varepsilon$, equals 3 for all $0 \le \varepsilon < 1$. As stretching occurs, and if there is no viscosity, it is conceivable that the dimension D of the limit

$$\lim_{\varepsilon \to 0} \lim_{t \to \infty} \Lambda_\varepsilon$$

is well defined for t large enough, with D less than three.

If this is indeed true, D is a measure of intermittency. The limiting set, if it exists, is the "essential support" of the vorticity. In the case of the model equation $u_t + (u^2)_x = 0$, $\xi = u_x$, one can readily see that $\lim \Lambda_\varepsilon$ exists, is the set of points where the shocks are located, and its dimension is zero.

This discussion raises the question whether the solutions of Euler's equations are smooth. The numerical evidence is that they are not, as is the experimental evidence. A numerical scaling analysis proceeds as follows: run an initial value problem for ξ, allowing $Z = \int |\xi|^2 d\underline{x}$ to grow until it becomes larger than some threshold Z_c at a time t_1. Then take out a portion V_1 (say an eighth) of the domain available to the fluid, making sure that $\xi \neq 0$ in V_1, rescale the length and time scales so that the energy is consistent with the velocity field in V_1. Run the smaller problem for a time $\tilde{t}_2$ until $Z > Z_c$ and repeat the process. Let t_i be the real time interval that corresponds to $\tilde{t}_i$. If $\sum_1^\infty \tilde{t}_i < +\infty$, then in the finite time $\sum_1^\infty \tilde{t}_i$ the original volume has an infinite Z. This calculation was done in [3], and, on a simplified lattice model, in [4], with the conclusion that Z does indeed blow up in finite time. The conclusion is of course non-rigorous, mostly because the boundary conditions at the edges of the sequence of shrinking volumes are artificial. The same calculation can be used to estimate the Hausdorff dimension D and yields $D \sim 2.35$.

The determination of the scaling law for the spectrum in this model is more difficult. If one removes the constraint of energy conservation the vortex tubes should, in thermal equilibrium, approximate a self-avoiding walk. If this is the case, one can readily see [7] that the spectrum in the inertial range is bounded by a constant times $k^{-5/3}$. One can then visualize the process of energy dissipation in the collection of filaments as a process in which folds form in reponse to stretching and are cancelled by pairing, leaving the large scale structure of the filament to approximate a self-avoiding walk.

This picture suggests an effective way of simplifying vortex calculations based on vortex segments [8]. As hairpins form they are removed. More generally, tightly folded vortex regions form in all turbulent flows and one could remove them as they begin to tighten. A similar idea has been used in the two dimensional case [10].

The direct applicability of these tubular models to turbulence, other than through the medium of computation, is rather unclear. There are flows in which real tubular vortices appear, form counterrotating pairs, and cancel through an elaborate hydrodynamical mechanism [1]. It is not clear at present how general this situation is and to what extent one can draw quantitative conclusions about it from the tubular models.

<u>Some qualitative properties of the inertial range</u>. The discussion above does have some qualitative consequences for theories of the inertial range.

The example of a flow dominated by filaments for which $D < 3$ and the inertial exponent γ is near 5/3 shows that there is no obvious basis for the belief that $D = 3$ corresponds to $\gamma = 5/3$. Indeed, the value $\gamma = 5/3$ cannot be understood without intermittency. If one considers a one-parameter family of turbulence models in which γ can be defined as a function of D, then it is likely that $\gamma(3) \geq 3$, since the member of the family that corresponds to D=3 is the one likely to be smooth. Moreover, unless the carriers of vorticity change topology quickly one should have $d\gamma/dD > 0$, i.e., an increase in intermittency decreases γ, contrary to what is often asserted. The reason is that as the support of a function decreases the support of its Fourier transform increases (see e.g. [7]). This conclusion is also supported by experiment [12] and by a spectral model [13].

It is also clear from the models above that there is no reason why energy should move across the range of scales in a local and orderly fashion. (In fact, it was shown in [6] that if it did, the spectrum would behave like k^{-2} in the inertial range.) More generally, the observed properties of turbulence are due to the complex interactions of localized structures in physical space; the statistical mechanics of those structures resemble the statistical mechanics of polymers; at high Reynolds numbers one reaches a state that resembles a critical state and the Kolmogorov exponent is a critical exponent that depends heavily on the presence of fluctuations.

References

[1] C. Anderson and C. Greengard, *Comm. Pure appl. Math.* (1989), in press.
[2] A. J. Chorin, *SIAM J. Sc. Stat. Comp.* 1 (1980), 1-21.
[3] A. J. Chorin, *Comm. Pure appl. Math.* 34 1981, 853-866.
[4] A. J. Chorin, *Comm. Pure. appl. Math* 39 (special issue)(1986), S47-S65.
[5] A. J. Chorin, Lattice vortex models and turbulence theory, in *Wave Motion*, Lax's 60th birthday volume, A.Chorin and A.Majda, editors (MSRI-Springer, 1987).
[6] A. J. Chorin, *Comm. Math. Phys.* 114 (1988), 167-176.
[7] A. J. Chorin, *Phys. Rev. Lett.* 60 (1988), 1947-1949.
[8] A. J. Chorin, Hairpin removal in vortex interactions, to appear.
[9] W. J. Cocke, *Phys. Fluids* 12 (1969), 2488-2492.
[10] C. Dritschel, Contour surgery, *J. Comp. Phys.* 77 (1988), 240-266.
[11] D. W. Long, Convergence of random vortex methods in three dimensions, preprint (Princeton, 1988).
[12] C. Meneveau, Ph.d. thesis (Yale, 1989).
[13] V. Yakhot, Z. S. She and S. Orszag, *Phys.Fluids* A1 (1989), 289-293.

Distortion of Material Surfaces in Steady and Unsteady Flows

N.A. MALIK

Department of Applied Mathematics and Theoretical Physics,
University of Cambridge, Silver Street, Cambridge CB3 9EW, UK

1. ABSTRACT

The stretching and distortion of surfaces in turbulent flows is an important mathematical problem in its own right, but it also has important applications in many areas, for example in mixing processes, chemical reactors and the propagation of flames.

We release material surfaces into simulated 3-d turbulent flow fields and we find that the surface area increases exponentially by a factor which depends upon the size of the inertial subrange. We also present results for higher moments of the surface area. Visualisations of material surfaces in a single flow field has proved particularly successful in demonstrating how structures within the flow distort different parts of the surface in very different ways to produce very fine scale details.

2. 3-D FLOW FIELD: KINEMATIC SIMULATION

We express the velocity as a finite sum of discrete Fourier modes ranging from k_1 to k_η $(k_1 < k_\eta)$ and $\eta = 2\pi/k_\eta$ is the Kolmogorov scale:

$$\mathbf{u}(\mathbf{x},t) = \sum_{m=1}^{M}\sum_{n=1}^{N}(\mathbf{a}_{mn} \times \hat{\boldsymbol{\kappa}}_n)\cos(\boldsymbol{\kappa}_n.\mathbf{x} + \omega_{mn}t) + (\mathbf{b}_{mn} \times \hat{\boldsymbol{\kappa}}_n)\sin(\boldsymbol{\kappa}_n.\mathbf{x} + \omega_{mn}t)$$

(This has been developed from earlier methods by Kraichnan (1970) and Drummond et.al. (1984)). We choose N discrete modes: $k_n = |\boldsymbol{\kappa}_n| = k_1+(n-1)k_\eta/(N-1)$, $n = 1,..,N$, and for each of these modes we pick M frequencies, ω_{mn}, from a distribution. The amplitudes and modes are determined by the energy spectrum $\mathcal{E}(k,\omega)$, but their directions are distributed uniformly on a unit sphere. We are interested in high Reynolds number turbulence so we have chosen the inertial subrange spectrum for the wavenumber part of the energy spectrum: $E(k) = \alpha_k\varepsilon^{2/3}k^{-5/3}$, where $\alpha_k = 1.5$ the Kolmogorov constant and ε is the rate of energy dissipation. The frequencies are picked from a Gaussian distribution. Hence $\mathcal{E}(k,\omega)$ is (see figure 1):

$$\mathcal{E}(k,\omega) = \alpha_k\varepsilon^{2/3}k^{-5/3}\exp\left[-\frac{(\omega - \overline{\omega}(k))^2}{2\sigma_\omega^2(k)}\right]/2\pi\sigma_\omega$$

with $\overline{\omega} = \sigma_\omega = \varepsilon^{1/3}k^{2/3}$. (Details of this method and an investigation of statistical properties of the flow is in Fung et. al. (1989)).

The most important feature of this technique is that the flow field has many scales of fluctuating motions which is particularly suitable for our work. It has been used successfully to investigate important aspects of the structure of turbulence, and 1-particle and two-particle relative dispersion (Fung et.al. (1989)).

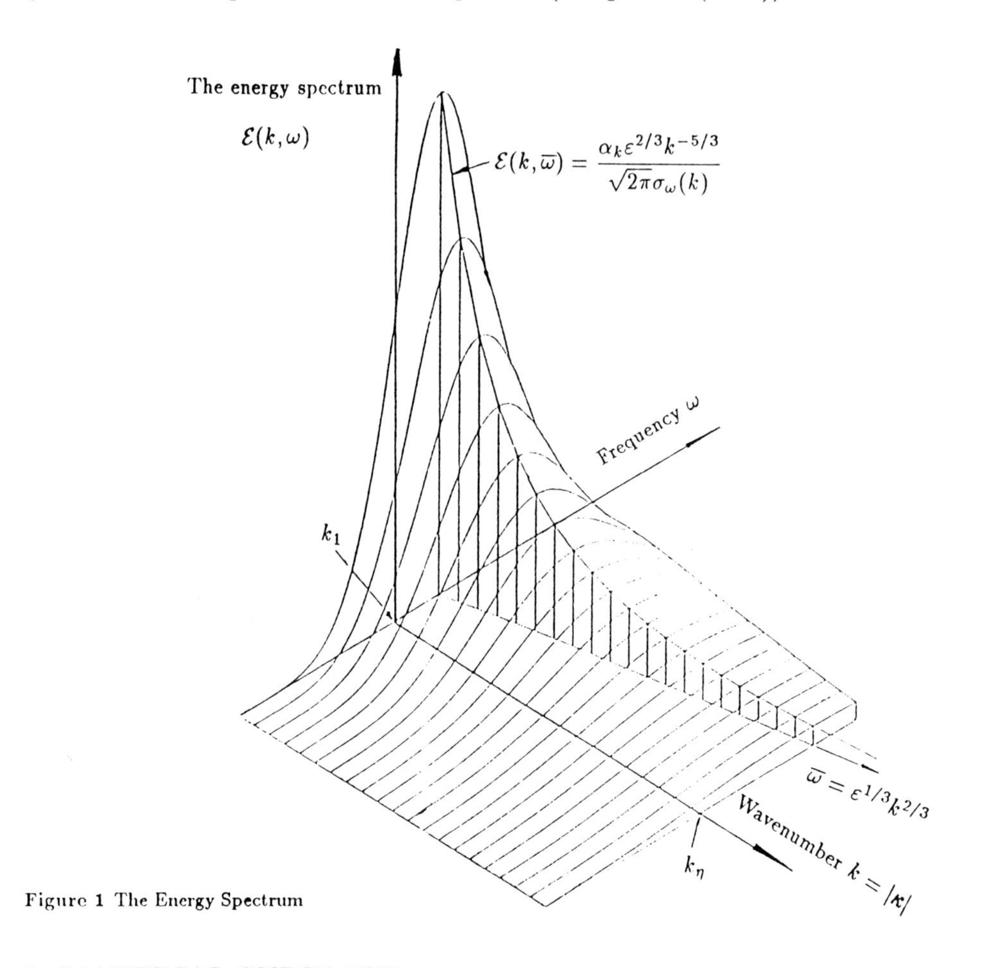

Figure 1 The Energy Spectrum

3. MATERIAL SURFACES

We use the Kinematic Simulation technique to investigate the growth of material surfaces in turbulent flows. We are interested in the effect on the growth of the surface by (1) varying the size of the inertial subrange and (2) by altering the time dependence of the flow field (steady and unsteady).

In the first set of simulations (3.1 below), we look at a surface in a single flow field realisation to gain insight into the way it folds and stretches and what scales of distortion appear. We release a regular square grid of thousands of particles, representing the material surface, in the $Z = 0$ plane and plot its configuration at regular timesteps. The spacing of the grid is initially 1/40th of respective Kolmogorov scale η i.e for $k_\eta = 10$, $\eta = 2\pi/10$ the initial spacing is $\Delta_0 = 2\pi/400$; and for $k_\eta = 20$, $\Delta_0 = 2\pi/800$. The integral length scale L is 1.2 and 1.1 respectively. In all our simulations the total energy is kept constant and the root mean square velocity fluctuation $u_i' = 1, \quad i = 1, 2, 3$.

In the second set of simulations(3.2), we extract statistical information about the growth of the surface and its higher moments. We average quantities over several flow field realisations.

Our computations are very intensive – we used the CRAY XMP/48 at the Rutherford Appleton Laboratory. When we came to represent a surface (see below), even limiting ourselves to an ensemble of just 15 flow fields, and a mesh of 81×81 for $k_\eta = 10$, and 161×161 for $k_\eta = 20$ (that is more than an integral length scale on each side) and for 151 timesteps it took almost 20 mins and 2.5 hours cpu respectively.

3.1 Visualisation of a Single Surface

We have obtained plots of the evolution in time of a material surface (Figures 2-5). In these simulations we were limited by computing expense to releasing material surfaces with dimensions of only $\eta \times \eta$, and to $M = 1$, i.e. a single frequency per mode. However, a much information can still be extracted from these results. The surfaces were represented by a grid of 41×41 points. We tracked the motion of each point and then we plotted the surface every 20th timestep ($0.2T$ where $T = L/u$ is an Eulerian timescale, and u is rms velocity fluctuation. $u = 1$ in all the simulations)

Figure 2 is for $1 < k < 10$ and $N = 35$ modes and the flow is unsteady;

Figure 3 is the corresponding result for the steady (or frozen) field, (i.e. $\omega = 0$, and $\mathbf{u}(\mathbf{x}, t) = \mathbf{u}(\mathbf{x}, 0)$: no time dependence);

Figure 4 is for $1 < k < 20$ and $N = 80$ modes and the flow is unsteady;

Figure 5 is the corresponding result for the steady field.

(Note that for longer times the graphics break down because of the severe contortions of some parts of the surface — these are not numerical errors. They are included because other parts of the surface and the overall structure of the surface is still good).

Some general features emerge from these graphics. Firstly, the surface is stretched, folded and twisted in very different ways in different parts of the surface. Such features have been associated with different structures in the flow field (Hunt et.al. 1988; Vassilicos 1989). Although the surface area increases exponentially (see section 3.2), the physical extent of the surface does not increase so dramatically. Most of the increase in surface area is taken up in folds and twists etc. In our simulations we start with a flat square surface of length η on each side and the final extent of the surface is of the order of 3η to 5η. This means that some of the structures on the surface that we see are actually *smaller* than the Kolmogorov scale. We do not model molecular diffusivity ($Pr = \infty$) so the Batchelor scale is 0. That is, if we were to let the surface continue to evolve indefinately then structures on the surface would continue to appear down to arbitrary small scales (Batchelor (1953)).

FIGURE 2. EVOLVING MATERIAL SURFACE: UNSTEADY FLOW

$1 \leq k \leq 10$

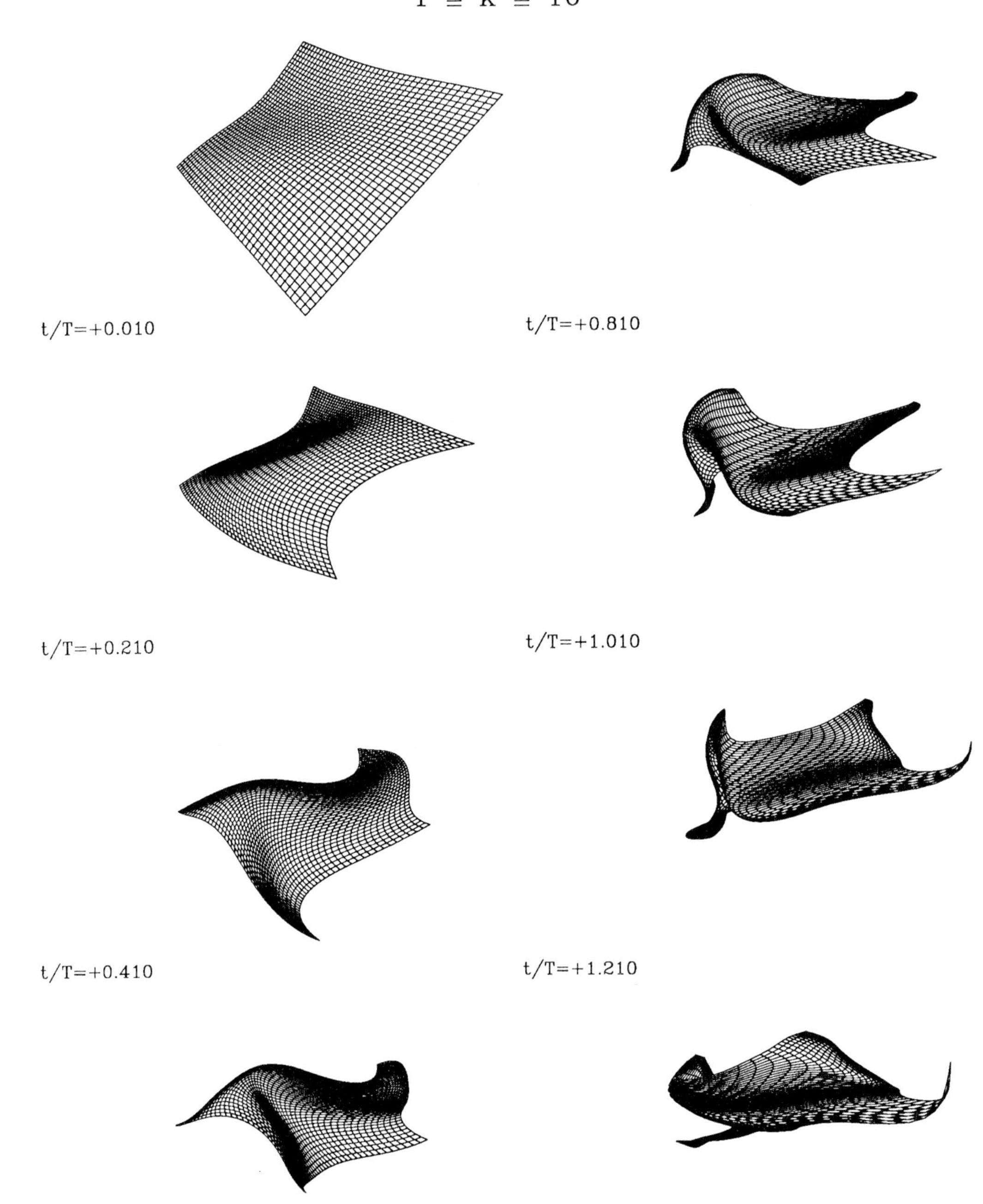

FIGURE 3. EVOLVING MATERIAL SURFACE: STEADY FLOW

$1 \leq k \leq 10$

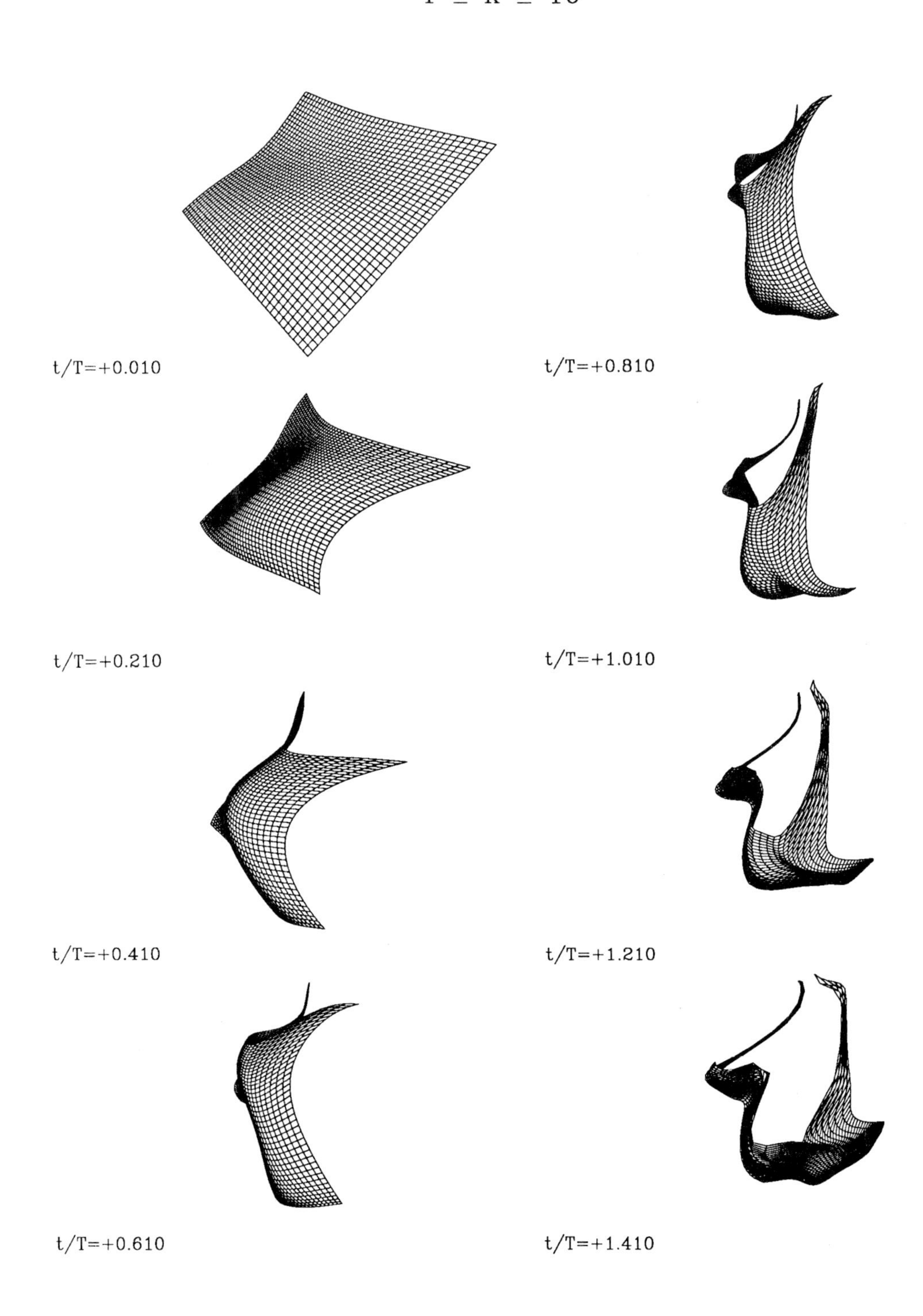

FIGURE 4. EVOLVING MATERIAL SURFACE: UNSTEADY FLOW

$1 \leq k \leq 20$

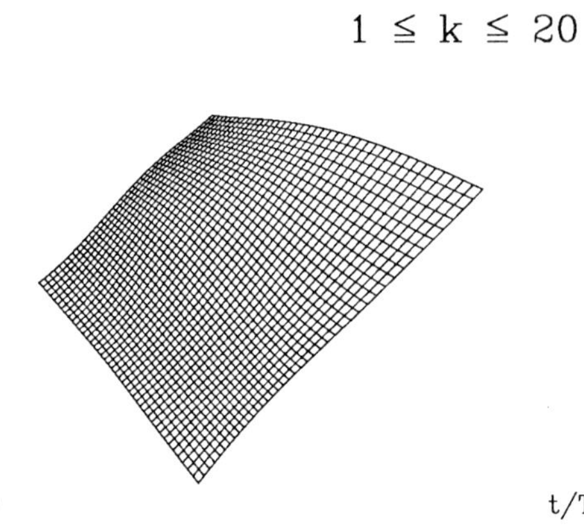

t/T=+0.010

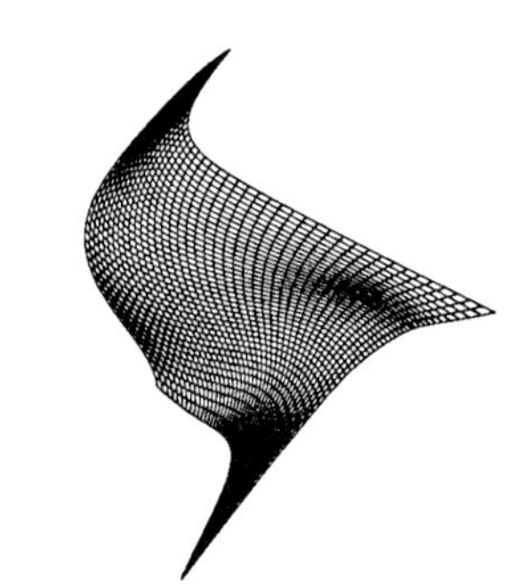

t/T=+0.210

t/T=+0.410

t/T=+0.610

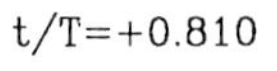

t/T=+0.810

t/T=+1.010

t/T=+1.210

t/T=+1.410

FIGURE 5. EVOLVING MATERIAL SURFACE: STEADY FLOW

$1 \leqq k \leqq 20$

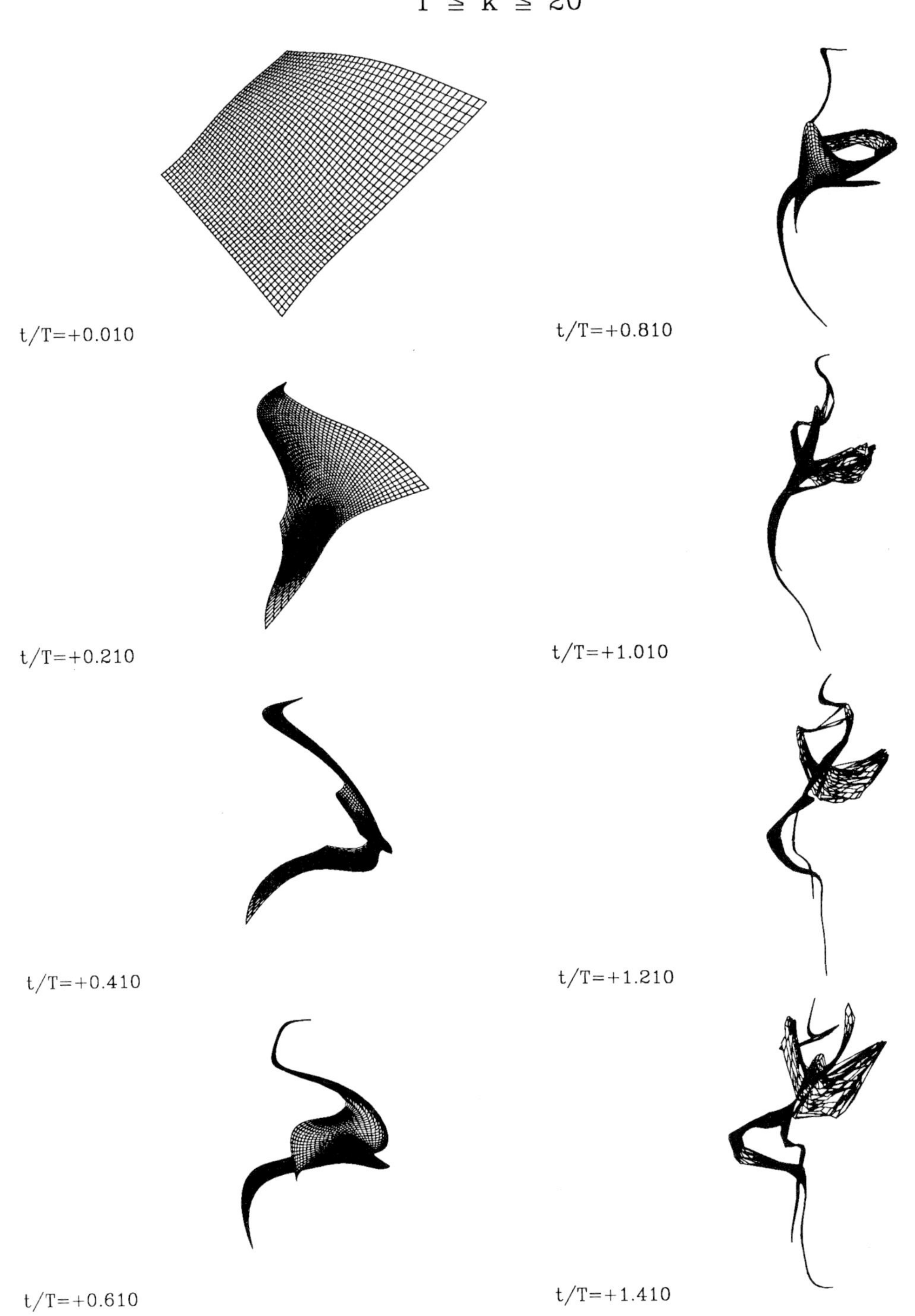

Secondly, increasing the inertial subrange also increases the rate at which the surface contorts. This is not surprising, but it emphasises again the important effects of the smallest scales of eddying motion, even though the energy contained in these scales is very small.

Thirdly, comparing the steady to the unsteady cases, it appears that in steady turbulence a surface becomes *more* convoluted. This suggests that the rapid oscillations in the flow smooths out small scale details on the material surface. In steady flows, mechanisms of straining, stretching and folding are persistent. For example a spiralling structure will continue to entangle a surface within it for much longer than in an unsteady flow where, after a short time, the flow structure in that part of the surface will have changed. Vassilicos (1989) has also found that in 2-D random flows the fractal dimension of a material line in a steady flow asymptotes to a higher value than in an unsteady flow.

These observations are fundamental in understanding the process of mixing where the small scale structure of the surface plays a decisive role. The basic process by which fluids mix is firstly that an initially marked material region is stretched and folded by the flow into thin sheets. The sheets come close together, especially in straining regions of the flow, and ultimately molecular diffusivity merges the sheets and the fluid is then well mixed. Although we do not model the effect of molecular diffusivity, the stretching and folding of the material surface is apparent in the plots.

3.2 Surface Statistics

We have measured the mean rate of growth of the surface area and its higher moments. We released a mesh of points representing a surface whose initial dimensions covered more than $L \times L$ so that the surface sampled the whole range of eddying motions. This was repeated for many randomly generated flow fields realisations. Because the simulations require large amounts of cpu time, we have been limited to relatively small ensembles — typically 15. However, the statistical variations are reduced because each single surface in a flow realisation covers at least a length scale square and consequently samples a large number of different flow structures within it – it is a 'mini-ensemble' of flow structures in itself.

Batchelor (1952) and subsequent workers (Cooke (1969); Orszag (1970)) showed that if you assume that there is no length scale governing the asymptotic behaviour of the probability distribution of a material surface, then at large times the area $S(t)$ of the surface grows exponentially:

$$S(t) = S_0 \exp(\gamma_1^s t)$$

where γ_1^s is the fractional rate of growth of an infinitesimal surface element.

Drummond and Muench (1989) introduced a set of numbers $\{\gamma_p^s; p = 1, 2, ..\}$ which are related to the time derivatives of the higher moments of $S(t)$:

$$\gamma_p^s = \frac{1}{p}\frac{d}{dt}\log(\langle S^p/S_0^p\rangle) \quad \text{for} \quad p > 0$$

and

$$\gamma_0^s = \frac{d}{dt}\langle \log(S/S_0) \rangle$$

[They also defined a corresponding set for material lines]. If the flow is homogeneous, isotropic stationary and time reversal invariant then under Batchelor's hypothesis, $\gamma_1^s = \gamma_2^s = \gamma_3^s = ...$etc. However, Drummond and Muench showed that this not the case in their flows.

We have run our simulations to obtain values for $\gamma_0^s, \gamma_1^s, \gamma_2^s, \gamma_3^s$.

Figure 6 shows log-linear plots for the first four moments; The flow fields have $1 < k < 10$, and are (a) unsteady and (b) steady.

Figure 7 is the same as for figure 6 except that $1 < k < 20$.

The slopes of the curves in figures 6 and 7 gives the values of γ_p^s. Table I below shows values of γ_0^s and γ_1^s and ratios γ_p^s/γ_0^s for $p = 1, 2, 3$ from our simulations for $k_\eta = 10$ and 20 (c.f. Drummond and Muench (1989)):

	$k_\eta = 10$ Unsteady	Steady	$k_\eta = 20$ Unsteady	Steady
γ_0^s	0.52	0.61	0.87	0.83
γ_1^s	0.54	0.63	0.87	0.83
γ_1^s/γ_0^s	1.04	1.03	1.00	1.00
γ_2^s/γ_0^s	1.08	1.07	1.00	1.00
γ_3^s/γ_0^s	1.10	1.11	1.02	1.01

Table I

We see the emergence of the characteristic exponential regimes in each case. The values of $\{\gamma_p^s\}$ increases with the size of the inertial subrange, but there is no clear difference between unsteady and steady flows.

However, we find in contrast to Drummond and Muench that $\gamma_1^s/\gamma_0^s \approx \gamma_2^s/\gamma_0^s \approx \gamma_3^s/\gamma_0^s \approx 1$, in line with Batchelor's hypothesis. In fact the increase in the size of the inertial subrange brings the ratios closer to unity (compare figures 6 and 7). Drummond and Muench used a similar technique to ours in representing the flow field, except that their flow fields had a particular length scale, $\ell_0 = 2\pi/k_0$ for some k_0, i.e. their energy spectrum $E(k)$ was proportional to $E(k) \propto \delta(k - k_0)$. The values for the ratios γ_p^s/γ_0^s which they obtained were significantly different from 1. These results are consistent only with the fact that if there is a preferred length scale in the flow then Batchelor's hypothesis is not expected to be valid and we expect to find the values of the ratios to be different from 1. However, as a wide range of length scales are introduced into the flow (as in real turbulence), we do expect Batchelor's hypothesis to be valid and the ratios to approach unity. Such is the case in our flows and we have shown that a range of wavenumbers of just $1 < k < 20$ is sufficient to bring these values to within 2% of unity (Table I). Thus all scales of motion above the Kolmogorov scale

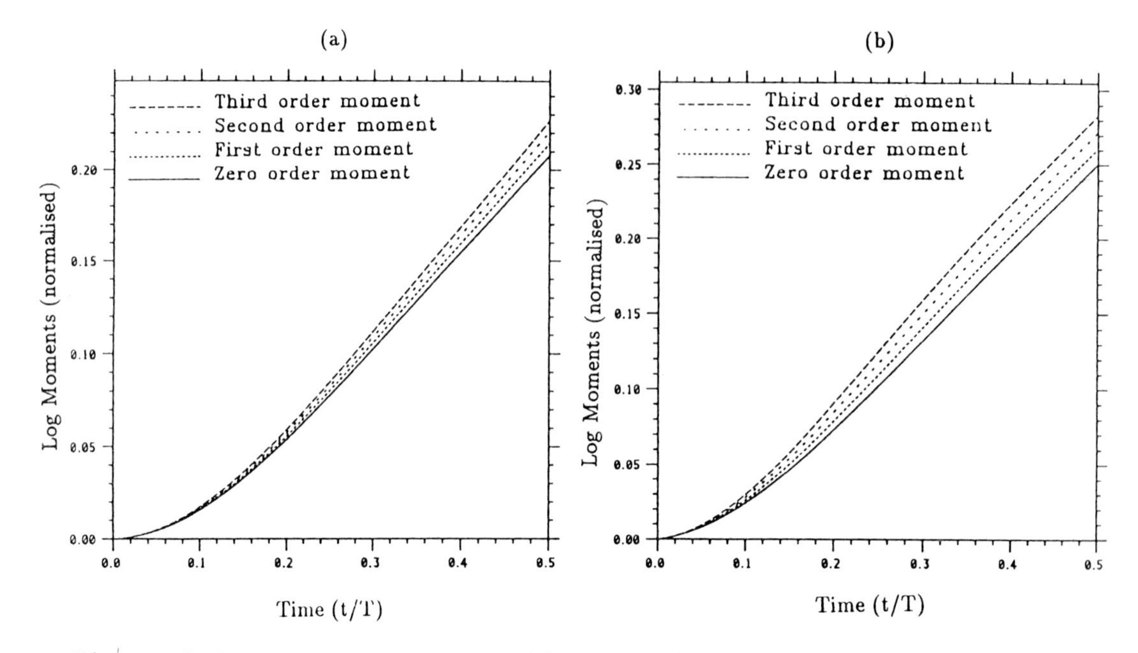

Figure 6 Surface growth in 3-D. $k_\eta = 10$; (a) Unsteady (b) Steady. Number of Fourier modes is 35. The ensemble size is 15 flow fields. Moments $\langle (S(t)/S(t=0))^p \rangle^{1/p}$.

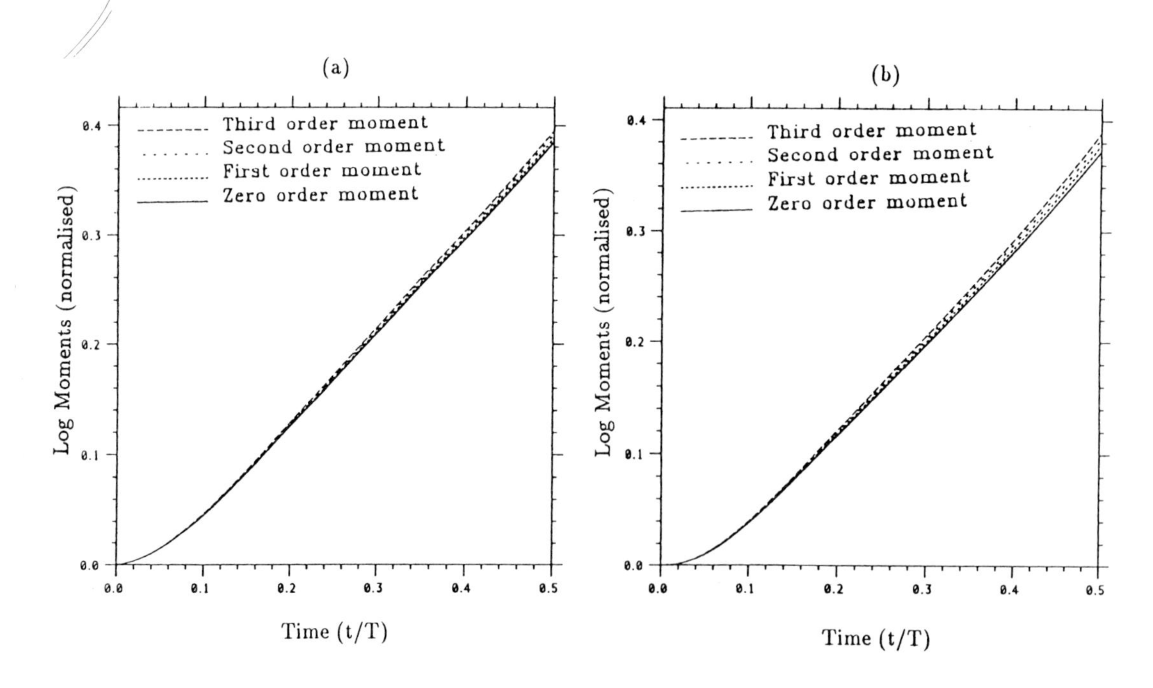

Figure 7 Surface growth in 3-D. $k_\eta = 20$; (a) Unsteady (b) Steady. Number of Fourier modes is 80. The ensemble size is 15 flow fields. Moments $\langle (S(t)/S(t=0))^p \rangle^{1/p}$.

are contributing significantly to the growth of the surface area – it is not just the smallest scales which are dominating the growth.

4. CONCLUSION

We have shown visualisations of surfaces deforming in 3-D random flows with explicit details of how such surfaces deform locally and at all scales. The effect of increasing the size of the inertial subrange is to increase the degree of distortion and to increase the rate of growth of the surface area. Furthermore, steady flow produces a higher degree of distortion than unsteady flow.

We have also found that the ratios $\{\gamma_p^s/\gamma_0^s\}$ are approximately equal to 1 to within 2% for $1 < k < 20$ which supports Batchelor's hypothesis that there is no length scale that governs the growth of the area i.e. that all length scales in the flow make a significant contribution to the growth of the surface area.

Acknowledgements I would like to thank my supervisor Dr. J.C.R Hunt for his constant guidance and help and also J.C. Vassilicos for his helpful comments on this paper. I would also like to thank S.E.R.C. for use of their CRAY XMP/48 at the Rutherford Appleton Laboratory.

REFERENCES

BATCHELOR G.K. (1953) "Small Scale Variation of Convected Quantities Like Temperature in Turbulent Fluid", J. Fluid Mech **5**, 113

COOKE W.J. (1969) "Turbulent Hydrodynamic Line Stretching: consequences of isotropy", Phys. Fluids **12**, 2448,

DRUMMOND I.T., DUANE S., and HORGAN R.R. (1984) "Scalar diffusion in simulated helical turbulence with molecular diffusivity", J. Fluid Mech. **138**,75,

DRUMMOND I.T. and MUENCH W. (1989) "Turbulent Stretching of Line and Surface Elements", IUTAM symposium 1989 (present volume)

FUNG J.C., HUNT J.C.R., MALIK N.A. and PERKINS R.J. (1989) (to be submitted to J. Fluid Mech.)

KRAICHNAN R.H. (1970) "Diffusion by a random velocity field", Phys. Fluid **13**,22,

HUNT J.C.R, MOIN P. and WRAY A.A. (1988) "Eddies, Stream and Convergence zones in turbulent flows", *NASA Report CTR-589.*

ORSZAG S.A. (1970) comments on "Turbulent Hydrodynamic Line Stretching: consequences of isotropy" Phys. Fluids **13**, 2203,

VASSILICOS J.C (1989) "On the geometry of lines in 2-D turbulence", in Proc. 2nd European Turb. Conf., ed. Springer Verlag.

Turbulent Stretching of Lines and Surfaces

IAN T. DRUMMOND & WOLFRAM H.P. MÜNCH

Department of Applied Mathematics and Theoretical Physics,
University of Cambridge, Silver Street, Cambridge CB3 9EW, UK

1. Introduction

Material line and surface elements transported in a turbulent velocity field increase in length or area at an exponential rate. We investigate how the stretching rates are related to the statistical properties of the velocity field both analytically and numerically in simple models of turbulence. In a Gaussian model the statistics exhibit time-reversal invariance. We demonstrate that, as pointed out by Kraichnan (1974), this leads to equality of line and area stretching rates. We also construct a model which violates the time reversal property and splits the values of the rates for lines and surfaces. The sign of the splitting depends on the sign of the time-reversal breakdown.

In section 2 we set out the theoretical background for line elements in turbulent flow and present the results of numerical simulations using a simple model turbulence[5,6]. In section 3 we discuss the stretching of surface elements and the relationship with the behavior of line elements. We introduce a numerical model turbulence lacking time-reversal invariance and illustrate how the stretching rates for lines and surfaces can be made to differ with respect to each other.

2. Line Element Stretching

Line element stretching was discussed by Batchelor[1] (1952) and by subsequent authors[2,3,4]. Although their analysis is substantially correct it is a little too restrictive. It is possible to define a sequence of stretching exponents by considering the behavior of various powers of l. For example we set

$$\gamma_p = \frac{1}{p}\frac{\frac{d}{dt} < l^p >}{< l^p >}, \tag{2.1}$$

for $p > 0$ and $\gamma_0 = \frac{d}{dt} < \log(l) >$, (note $\gamma_0 = \lim_{p\to 0}\gamma_p$). We can compute γ_p as a power series in the velocity field and the results exhibit a dependence on the value of p.

The evolution of a line element $\mathbf{l}(\mathbf{t})$, in a velocity field $\mathbf{u}(\mathbf{x},t)$ is

$$\frac{d}{dt}l_i(t) = W_{ij}(t)l_j(t) \tag{2.2}$$

where $W_{ij}(t) = u_{i,j}(\mathbf{X}(t), t)$ and $\mathbf{X}(t)$ is a solution of $\frac{d}{dt}\mathbf{x} = \mathbf{u}(\mathbf{x}, t)$. The solution of equation (2.2) has the form

$$l_i(t) = U_{ij}(t) l_j(0). \tag{2.3}$$

with $U(t) = \vec{T}\exp(\int_0^t dt' W(t'))$, where $\vec{T}$ denotes the time ordering. Of course,

$$\frac{d}{dt} U = W\ U. \tag{2.4}$$

It is convenient to split W into a symmetric part B and an antisymmetric part A. Introduce the orthogonal rotation matrix $R(t)$ which obeys

$$\frac{d}{dt} R = A\ R, \tag{2.5}$$

where $R = \vec{T}\exp(\int_0^t dt' A(t'))$. Now put

$$U = R\ M \tag{2.6}$$

and we have

$$\frac{d}{dt} M = V\ M, \tag{2.7}$$

where $V(t) = R^T(t)\ B(t)\ R(t)$, note that V(t) is a symmetric matrix and $M(t) = \vec{T}\exp(\int_0^t dt' V(t'))$. Now the orthogonality of $R(t)$ allows us to conclude that

$$l^p(t) = (\mathbf{l}^T(0) M^T(t) M(t) \mathbf{l}(0))^{\frac{p}{2}}. \tag{2.8}$$

If we evaluate the solution of equation (2.16) then we can write $M^T\ M = \mathbf{1} + C$, where C is $\mathrm{O}(\mathbf{u})$ and $\mathbf{1}$ is the unit matrix. Assuming $l_i(0)$ is a unit vector we expand (2.8); as the turbulenec is isotropic we lose nothing by averaging over the direction of $\mathbf{l}(0)$. In D - dimensional space, we find

$$\begin{aligned} l^p(t) = 1 &+ \frac{p}{2D} tr\ C \\ &+ (\frac{p}{2} - 1)\frac{p}{4D(D+2)}(2tr\ C^2 + (tr\ C)^2) \\ &+ (\frac{p}{2} - 1)(\frac{p}{2} - 2)\frac{p}{12D(D+2)(D+4)}(8tr\ C^3 + 6tr\ C^2 tr\ C + (tr\ C)^3) \\ &+ \cdots . \end{aligned} \tag{2.9}$$

Subsequently we average over the velocity field ensemble to obtain $< l^p(t) >$. The result is to $\mathrm{O}(\mathbf{u}^3)$

$$\begin{aligned} \gamma_p = \frac{1}{p}\frac{\frac{d}{dt} < l^p >}{< l^p >} &= \alpha_p \int_0^t dt' < V(t)V(t') > \\ &+ \beta_p \int_0^t dt' \int_0^{t'} dt" < V(t)V(t')V(t") > \\ &+ \cdots , \end{aligned} \tag{2.10}$$

where $\alpha_p = \frac{2}{D} + \frac{2}{D(D+2)}(\frac{p}{2}-1)$, $\beta_p = \frac{4}{D} + \frac{4}{D(D+2)}(\frac{p}{2}-1) + \frac{4}{D(D+2)(D+4)}(\frac{p}{2}-1)(\frac{p}{4}-2)$. To $O(\mathbf{u}^2)$ we see that

$$\gamma_p = \alpha_p \int_0^t dt' < tr\ B(t)B(t') > . \tag{2.11}$$

If we define an integral time scale τ_s by the formula

$$\tau_s = \frac{1}{< B(0)^2 >} \int_{-\infty}^{\infty} dt < tr\ B(t)B(0) >$$

and note that $< B(0)^2 >= \frac{1}{2}\Omega^2$, where Ω^2 is the mean square vorticity of the turbulence, then we have an estimate for γ_p, namely,

$$\gamma_p = \frac{1}{4}\alpha_p \Omega^2 \tau_s, \tag{2.12}$$

and we can evaluate the various γ_p in that approximation.
The turbulence was represented by an incompressible random velocity field which was chosen from a Gaussian distribution according to ideas of Kraichnan[5], 1970, and Drummond et al.[6], 1984. In order to make the calculations a little easier the autocorrelation function was chosen to be of a simple kind and was characterized by only one length- and timescale. However, the simulation does not depend for its success on this choice of spectrum for the turbulence or on the precise number of relevant time scales. All the computations were carried out at the Rutherford Laboratory, Abingdon, and at the Department of Applied Mathematics and Theoretical Physics, Cambridge. At the Rutherford Laboratory we used an AMT-DAP 510, a 32 by 32 array - processor. In Cambridge we used an AMT-DAP 610, a 64 by 64 array - processor.
The velocity field $\mathbf{u}(\mathbf{x}, t)$ is generated as a sum of Fourier components, each of which is determined by certain parameters distributed according to various probability distributions. A typical member of the velocity field ensemble in three dimensional space is then realized by

$$\begin{aligned}\mathbf{u}(\mathbf{x}, t) = a \sum_{n=1}^{N} &(\mathbf{f}^n \cos(\psi^n) - \mathbf{g}^n \wedge \hat{\mathbf{k}}^n \sin(\psi^n)) \wedge \mathbf{k}^n \cos(\mathbf{k}^n \cdot \mathbf{x} + \omega^n t + \phi^n) \\ &+ (\mathbf{g}^n \cos(\psi^n) + \mathbf{f}^n \wedge \hat{\mathbf{k}}^n \sin(\psi^n)) \wedge \mathbf{k}^n \sin(\mathbf{k}^n \cdot \mathbf{x} + \omega^n t + \phi^n)\end{aligned} \tag{2.13}$$

where $\mathbf{k}^n$ is distributed uniformly on a sphere with radius k_0, ω^n is chosen from a Gaussian distribution $P(\omega) = (2\pi\omega_0)^{-\frac{1}{2}} \exp -\frac{\omega^2}{2\omega_0}$, ψ^n is an adjustable helicity parameter which we set to $\psi^n = \psi$ for all n and $\psi \epsilon [0, \frac{\pi}{4}]$, $\mathbf{f}^n, \mathbf{g}^n$ are distributed uniformly and independently over the unit sphere, ϕ^n is distributed uniformly and independently between 0 and 2π, a is a normalisation factor $a = (\frac{3}{2N})^{\frac{1}{2}} \frac{u_0}{k_0}$. The model therefore has a timescale ω_0^{-1} and a lengthscale k_0^{-1}. The eddy circulation time is $(k_0 u_0)^{-1}$, where u_0 is the mean square velocity. The parameter Ω is $k_0 u_0$. In most of our simulation N was chosen to be 32 but other values of N have also been studied for the purposes of comparison.

The method of simulation comprises choosing a set of flows and following a number of particles distributed in different configurations in the flow in such a way that two nearest neighbors are separated by at least two correlation lengths of the flow field. Array-processors are particularly suited for these calculations as 1024 or 4096 particles could be followed at once. At each timestep the particle position and the first spatial derivatives were calculated and different l_p's were evaluated. As an integration procedure we used an algorithm based on ideas of Burlisch and Stoer[7].

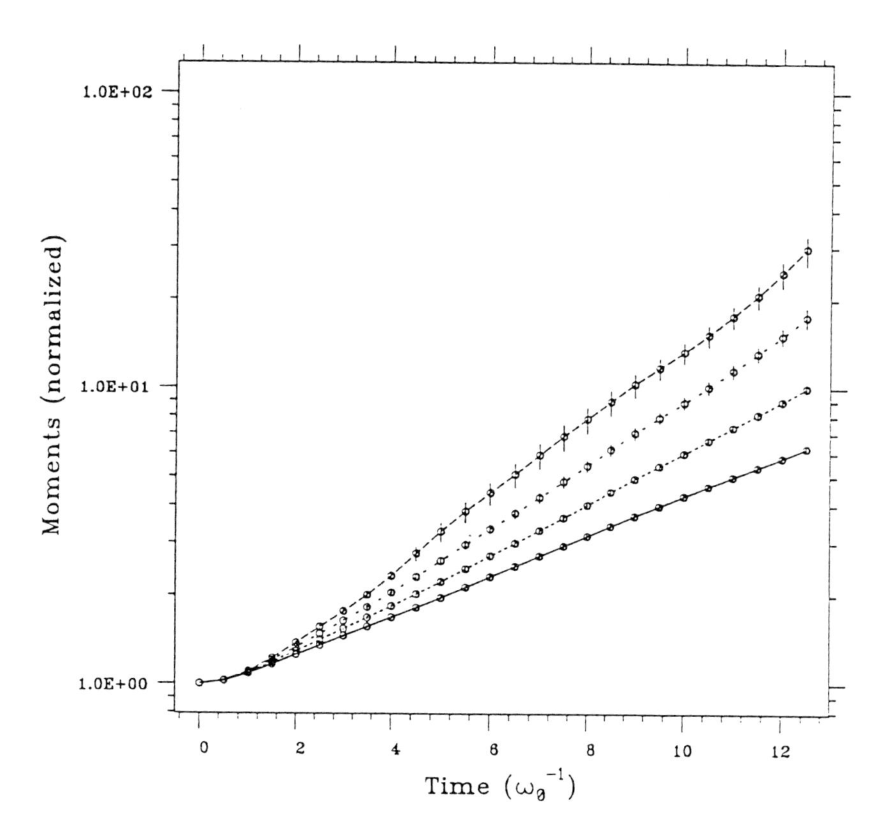

Figure 1: Line stretching in 3-D: the moments $< \frac{l(t)^p}{l(t=0)^p} >^{\frac{1}{p}}$, $u_0 k_0 = \omega_0 = 1$. The number of Fourier modes in the velocity field is 32, the ensemble size was 2000.

In figure 1 we show the different $< l_p >^{\frac{1}{p}} /p, p = \{1, 2, 3\}$ and $\exp(< log l >), p \to 0$, in 3 - D for $u_0 k_0 = \omega_0 = 1$. Clearly the results confirm the existence of the time independent γ_p. Table 1 compares the estimates of equation (2.12) with the numerical results. As the γ_p are just approximated to $O(\mathbf{u}^2)$, we do not expect these predictions to be very accurate.

Table 1: Comparison of the ratios $\gamma_p/\gamma_0, p = \{1,2,3\}$ from the numerical simulation with the estimates of equation (2.12) in 3 - D.

D=3	analytically	numerically: $(u_0k_0)/\omega_0 = 1$	$(u_0k_0)/\omega_0 = 2$	$(u_0k_0)/\omega_0 = 3$
γ_1/γ_0	1.166	1.26 ± 0.02	1.28 ± 0.02	1.32 ± 0.03
γ_2/γ_0	1.666	1.57 ± 0.02	1.63 ± 0.03	1.75 ± 0.06
γ_3/γ_0	1.833	1.89 ± 0.03	1.96 ± 0.05	2.16 ± 0.09

In the presence of helicity ($\psi = \frac{\pi}{4}$) the stretching moments did not change. As long as the velocity field ensemble does not contain odd - order correlation functions helicity does only affect the rotation matrix R of equation (2.11), i.e. the anti-symmetric part A of W of equation (2.6), instead of affecting the symmetric part which is important for the stretching of the line element. The general picture first proposed by Batchelor[1] then does hold although it is more complicated in detail since the existence of the distribution he proposed implies the equality of the γ_p and they are clearly distinct both on theoretical grounds and on the basis of the simulation.

3. Surface Element Stretching

For incompressible flow, surface element stretching only occurs in three and higher dimensions. We consider only D = 3. The theory of surface element stretching has been considered by different authors[1,2,3,4]. Our analysis is very close to that of Kraichnan[4] but we apply it to the whole sequence of exponents γ_p.
It is convenient to consider a triad of line elements $\mathbf{l}_1, \mathbf{l}_2, \mathbf{l}_3$ being transported along with the particles in the fluid, each evolving according to the rules exhibited in section 2, that is

$$l_{ai}(t) = U_{ij} l_{aj}(0). \tag{3.1}$$

We introduce the surface elements $\mathbf{L_a}$ (a=1,2,3), where

$$\mathbf{L_a} = \mathbf{l_b} \wedge \mathbf{l_c} \tag{3.2}$$

($\{a, b, c\}$ are a cyclic permutation of $\{1,2,3\}$). We have the result

$$\mathbf{L_a} \cdot \mathbf{l_b} = V \delta_{ab}, \tag{3.3}$$

where V is the volume of the parallelepiped spanned by $\mathbf{l}_1, \mathbf{l}_2, \mathbf{l}_3$. Because the flow is incompressible V is independent of time. This tells us that the time evolution of the surface elements proceeds according to

$$L_{ai}(t) = Q_{ij}(t) L_{ai}(0), \tag{3.4}$$

where

$$Q(t) = [U^{-1}(t)]^T. \tag{4.5}$$

The area of the plaquette a is $A_a = |\mathbf{L_a}|$ so

$$A_a^2 = \mathbf{L_a}^T(0)Q^T(t)Q(t)\mathbf{L_a}(0). \quad (4.6)$$

On the basis of equation (4.6), we see that $Q(t)$ is playing the role that $U(t)$ for line elements. We now show that, after directional averaging of L_a, which is appropriate in isotropic turbulence, $Q(t)$ and $Q^T(t)$ can be interchanged. This is most easily seen for A_a^2 ($p=2$ in the notation of section 2) itself. Directional averaging yields $A_a^2 = \frac{1}{3}tr\ Q^T(t)Q(t)$ but using the cyclic property of the trace, this becomes $A_a^2 = \frac{1}{3}trQ(t)Q^T(t)$, that is $A_a^2 = \frac{1}{3}tr\ [(U^{-1}(t))^T U^{-1}(t)]$. This means that we can now view $U^{-1}(t)$ as playing the role of $U(t)$ in the line element case. This allows us to relate the time reversal properties ofthe turbulent velocity field ensemble to the statistics of area element stretching. The point is that $U^{-1}(t_0)$ is the transformation matrix for a time reversed flow $\mathbf{u}(\mathbf{x},t) \rightarrow \mathbf{u}^T(\mathbf{x},t) = \pm\mathbf{u}(\mathbf{x},-t)$. Therefore, if the turbulent velocity field ensemble is time reversal invariant so that the reverse flow appears as frequently as the original flow it follows that $U^{-1}(t)$ has the same statistical properties as $U(t)$. This means thatγ_2 will have the same value both for both line elements (controlled by $U(t)$) and area elements (controlled by $U^{-1}(t)$). In fact it has been shown[8] that the argument can be extended to any power of A_a.

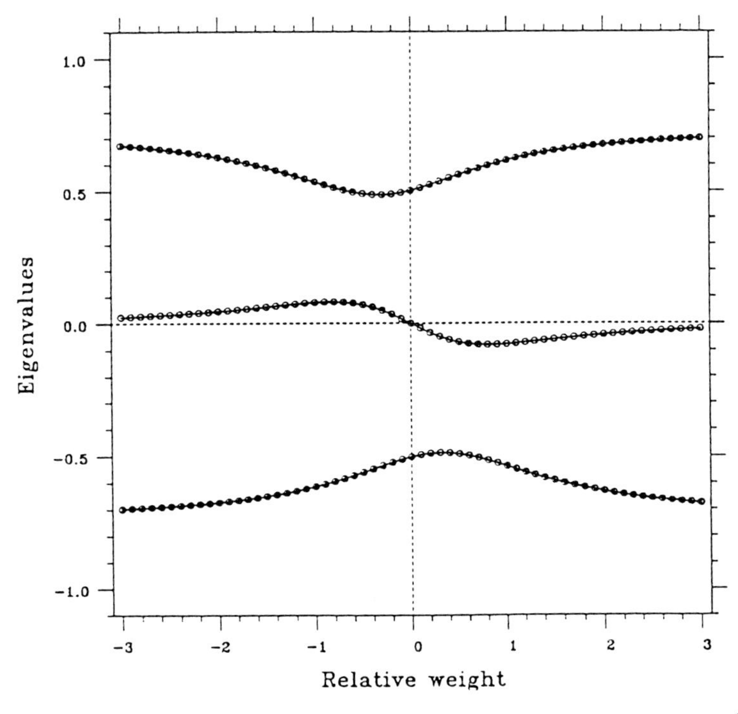

Figure 2: The eigenvalues of the symmetric rate of strain tensor $\frac{1}{2}(u_{i,j} + u_{j,i})$ in terms of the mean values $u_0 k_0$ as a function of λ, ($\mu = 1$).

The numerical simulations of the growth of the different moments of the surface area were carried out in an analog way to those of the line stretching moments. Particle positions and the normal to the surface were calculated at every timestep and the γ_p

were evaluated. As the velocity field of section 2 is time-reversal invariant we expect the stretching moments of area elements are distinct in that higher moments grow faster but we expect them to equal those in the previous section. The numerical results for γ_p, $\{p = 0, 1, 2, 3\}$ clearly support that the stretching exponents γ_p are different for area elements. Table 2 compares the stretching exponents calcultated from the surfaces with those calculated from the line. They are almost identical within the statistical errors.

Table 2: Comparison of the stretching exponents γ_p calcultated from the surface with those calculated from the line. ($\gamma_p^a = \gamma_p^{\text{surface}}$, $\gamma_p^l = \gamma_p^{\text{line}}$)

γ_0^a/γ_0^l	γ_1^a/γ_1^l	γ_2^a/γ_2^l	γ_3^a/γ_3^l
1.04 ± 0.01	1.00 ± 0.01	0.98 ± 0.02	0.97 ± 0.02

It is obvious that realistic turbulence (even when stationary) cannot be time-reversal invariant since it is generally associated with an energy cascade which proceeds in one direction from longer to shorter length scales. Clearly it is physically important to investigate the breakdown of time-reversal invariance. It follows from the analysis of the preceeding section that such a model is necessarily non-Gaussian.
Following a suggestion by H. K. Moffatt we construct a new velocity field $\mathbf{v}(\mathbf{x}, t)$ from the Gaussian field $\mathbf{u}(\mathbf{x}, t)$ by forming the combination

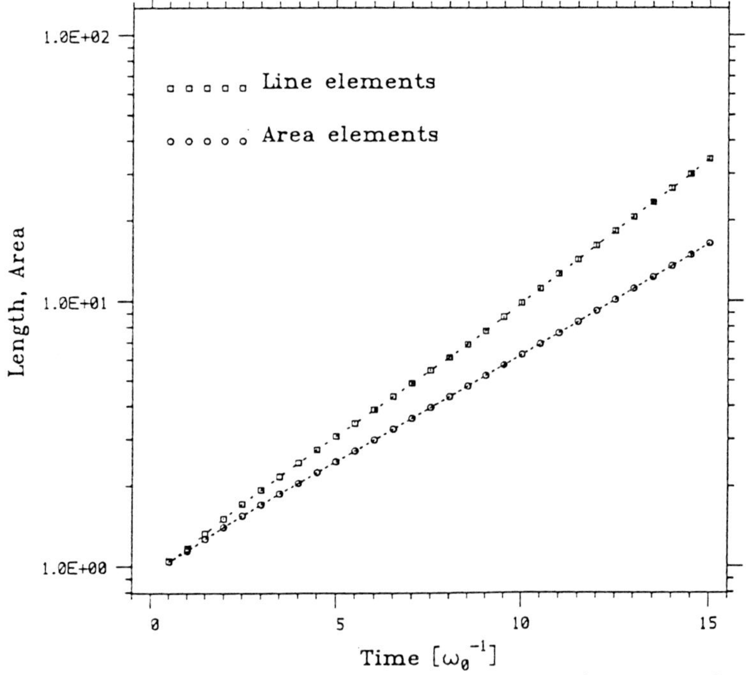

Figure 3: Line and surface element stretching as a function of time. For $\lambda = 0.7, \mu = 1$ lines grow faster than surfaces.

$$\mathbf{v}(\mathbf{x},t) = \mu\mathbf{u}(\mathbf{x},t) + \lambda[\mathbf{u}(\mathbf{x},t)\cdot\nabla\mathbf{u}(\mathbf{x},t)]_p$$
$$= \mu\mathbf{u}(\mathbf{x},t) + \lambda\mathbf{w}(\mathbf{x},t), \tag{4.7}$$

where $[\mathbf{a}(\mathbf{x})]_p$ means the divergenceless part of $\mathbf{a}(\mathbf{x})$, that is in a formal notation $[\mathbf{a}(\mathbf{x})]_p = \mathbf{a}(\mathbf{x}) - \nabla\frac{1}{\nabla^2}\nabla\cdot\mathbf{a}(\mathbf{x})$. This extra term is inspired by the quadratic term of the Navier-Stokes equations. The form of $\mathbf{v}(\mathbf{x},t)$ in terms the basic modes of $\mathbf{u}(\mathbf{x},t)$ is exhibited in reference 8.
Because of its quadratic structure it is clear that the time-reversal properties of the new term are different from those of the Gaussian field. Both velocity fields $\mathbf{u}(\mathbf{x},t)$ and $\mathbf{w}(\mathbf{x},t)$ are time-reversal invariant. They transform, however, differently under time-reversal. Whereas the velocity field $\mathbf{u}$ transforms according to $\mathbf{u}(\mathbf{x},t) \to \mathbf{u}^T(\mathbf{x},t) = -\mathbf{u}(\mathbf{x},t_0-t)$, the velocity field $\mathbf{w}$ transforms according to $\mathbf{w}(\mathbf{x},t) \to \mathbf{w}^T(\mathbf{x},t) = \mathbf{w}(\mathbf{x},t_0-t)$. Only the combined velocity field $\mathbf{v}$ lacks time-reversal invariance if the coefficients $\{\mu,\ \lambda\}$ are roughly of the same order of magnitude. For $\mu >> \lambda$ or $\mu << \lambda$ the velocity field $\mathbf{u}$ is approximately time-reversal invariant. One straight forward way to test this is to compute the ensemble average of the diagonalised form of the symmetric rate of strain tensor $S_{ij} = \frac{1}{2}(\mathbf{u}_{i,j}+\mathbf{u}_{j,i})$. For the Gaussian model we expect this to have the form

$$\begin{pmatrix} \gamma & 0 & 0 \\ 0 & 0 & 0 \\ 0 & 0 & -\gamma \end{pmatrix} \tag{4.8}$$

where a numerical evaluation yields $\gamma = (0.500 \pm 0.001)(u_0 k_0)$ for the velocity field $\mathbf{u}$. This is just the structure we would expect for a time-reversal situation since after an appropriate re-ordering of the axes the above form is invariant under a reversal of the sign of the elements of S_{ij}. When $\lambda, \mu \neq 0$ we can detect the breakdown of time-reversal invariance from the presence of a non-zero result for the intermediate eigenvalue. That is the expectation value of the diagonalised rate of strain tensor is now

$$\begin{pmatrix} \alpha & 0 & 0 \\ 0 & \beta & 0 \\ 0 & 0 & \gamma \end{pmatrix} \tag{4.9}$$

where $\alpha \geq \beta \geq \gamma$ and $\alpha+\beta+\gamma = 0$.
In Figure 2 we show α, β, and γ as a function of λ ($\mu = 1$) where the units on the y-axis are in terms of $u_0 k_0$. We see that β vanishes for $\lambda = 0$ and that $\lim_{\lambda\to\infty}\beta = 0$. The maximum breakdown occurs for $|\lambda| \approx 0.7$. The sign of the breakdown depends on the sign of λ. If $\lambda > 0$, $\beta \leq 0$ and line elements grow faster than surface elements. For $\lambda < 0$ we find $\beta > 0$ and surface elements expand at a greater rate than line elements. These results are illustrated in Figure 3 and Figure 4 where we plotted $< A >$ of a surface area element and $< l >$ of a line element. In Figure 3 we set $\lambda = 0.7$ and find that line elements grow faster than surface elements. In Figure 4 we set $\lambda = -0.7$ and see that surface elements grow faster than line elements.

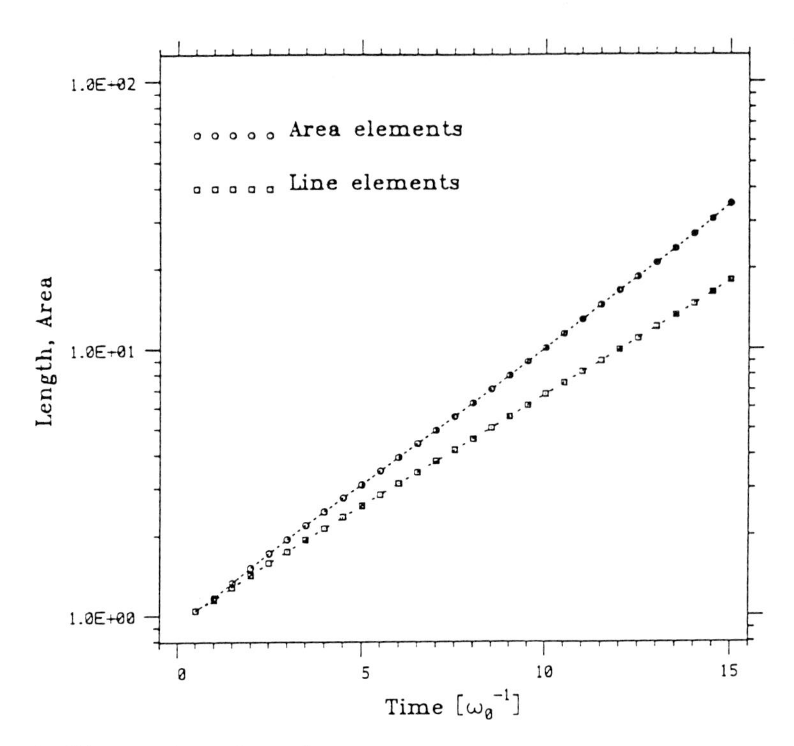

Figure 4: Line and surface element stretching as a function of time. For $\lambda = -0.7, \mu = 1$ surfaces grow faster than lines.

Acknowledgements

This work was supported by the EEC Twinning Contract ST2J-0029-1-F.

References

[1] G. K. Batchelor (1952),"The effect of homogeneous turbulence on material lines and surfaces", Proc. Roy. Soc. **A** 213, 349,
[2] W. J. Cocke (1969),"Turbulent hydrodynamic line stretching: consequences of isotropy", Phys. Fluids **12**,2448,
[3] S. A. Orzag (1970), Comments on "Turbulent hydrodynamic line stretching: consequences of isotropy", Phys. Fluids **13**, 2203,
[4] R. H. Kraichnan (1974),"Convection of a passive scalar by a quasi-uniform random straining field", J. Fluid Mech. **64**, 737,
[5] R. H. Kraichnan (1970),"Diffusion by a random velocity field", Phys. Fluids **13**, 22,
[6] I. T. Drummond, S. Duane, R. R. Horgan (1984),"Scalar diffusion in simulated helical turbulence with molecular diffusivity", J. Fluid Mech. **138**, 75,
[7] W. H. Press, B. P. Flannery, S. A. Teukolsky, W. T. Vetterling, "Numerical Recipes", Cambridge University Press (1986).
[8] I. T. Drummond, W. Münch, "Turbulent stretching of line and surface elements", submitted to J. Fluid Mech (1989).

Entangledness of Vortex Lines in Turbulent Flows

WOLFGANG POLIFKE & EVGENY LEVICH

Levich Institute for PCH, and Physics Department
City College of CUNY, New York, NY, 10031, USA

ABSTRACT

It is well known that the total helicity $H = \int_V \underline{v} \cdot \underline{\omega}\, dV$ of a localized vorticity distribution can be associated with the knottedness of the vortex lines. However, no absolute measure of topological complexity exists for open field structures. Berger and Field (J. Fluid Mech. 147, 1984) have demonstrated that the difference between the total helicity of the vorticity field $\underline{\omega}$ and that of a suitable reference field $\underline{\omega}'$ can provide a topologically meaningful and gauge-invariant measure of the entangledness of the vortex lines in an open subdomain D. In particular, choosing a potential reference field $\underline{\omega}' = \nabla\phi$, where ϕ is the harmonic function determined by Neumann boundary conditions $\underline{n} \cdot \nabla\phi = \underline{n} \cdot \underline{\omega}$ at ∂D, will yield a well defined measure that reduces to the usual volume integral of helicity for closed field structures.

In this study we discuss the properties of a related measure of local entangledness ; the relative helicity $H_R = \int_D \underline{v} \cdot (\underline{\omega} - \underline{\omega}')\, dV$. In an analysis of vorticity fields obtained from direct numerical simulations of homogeneous turbulent flows it is found that the vorticity field shows indeed a more complex structure in regions of high relative helicity. Also, statistical correlations between relative helicity and turbulent activity have been observed. Implications for some scenarios of turbulence are discussed briefly.

Introduction

The structure and topology of the vorticity field of turbulent flows has attracted considerable interest in recent years ; see for example Hussain (1986), Kida (1989), Levich (1987), Moffatt (1985) and Sagdeev *et. al.* (1986). In this study we will discuss the properties and possible usefulness of an analytical measure of the local 'entangledness' - rather than the global 'knottedness' in a strict topological sense - of vortex lines. Moreau (1961), Moffatt (1969) and Berger and Field (1984) have shown that the total helicity $H = \int_D \underline{v} \cdot \underline{\omega}\, dV$ of a localized vorticity distribution ($\underline{\omega} \cdot \underline{n} = 0$, where $\underline{n}$ is the unit normal to the surface of the domain D) is associated with the topological structure of the vorticity field $\underline{\omega} = \nabla \times \underline{v}$, i.e., the knots, twists and kinks of vortex lines and vortex tubes. The invariance of total helicity H for Euler flows can be understood as a manifestation of the fact that in inviscid flows the vortex lines move with the fluid and consequently their linkage properties are conserved.

However, the topology of open field structures, where vortex lines cross the surface of the volume under consideration, cannot be probed by the integral of helicity. This is because the linkage properties of vortex lines may depend on the field configuration outside D, see Fig. 1. A related problem is that of gauge-invariance : adding the gradient of a scalar field χ to the velocity, $\underline{v} \rightarrow \underline{v} + \nabla\chi$, changes the total helicity by an amount δH

$$\delta H = \int_D \nabla\chi \cdot \underline{\omega} \, dV = \int_{\partial D} \chi\underline{\omega} \cdot \underline{n} \, dA$$

In general, δH will not vanish if vortex lines cross the surface ∂D of the domain D. The vorticity field $\underline{\omega}$ and the linkage properties of the vortex lines, however, are obviously invariant under such gauge-transformations. Note that Galilean transformations and advection-effects by large scale eddies may be regarded as two important examples of gauge-transformations. Clearly, helicity cannot be a meaningful measure of the local entangledness of vortex lines.

Analytical Measures of Entangledness

Drawing on ideas originally developed in biology for the study of DNA structure (Fuller 1978), Berger and Field (1984) have argued that although there exists no absolute measure of knottedness for open field structures, it should be possible to define a topologically meaningful and gauge-invariant relative measure of topological complexity. We shall briefly recall the work of Berger and Field.

Let space V (periodic or unbounded) be divided into two simply connected regions D and $\overline{D}$ (the complement of D). For the case of unbounded volume we assume that all surface integrals 'at infinity' vanish. Consider two vorticity fields $\underline{\omega}^i$, i=1,2 , and the corresponding velocity fields $\underline{v}^i$, $\underline{\omega}^i = \nabla\times\underline{v}^i$. Outside of D the two vorticity fields are identical. We assume that the velocity fields $\underline{v}^i$ are solenoidal and continuous (C^0), the normal components of both velocity and vorticity across the boundary ∂D shall be continuous . Berger and Field have shown that the difference $\Delta H = H_1 - H_2$ of total helicities $H_i = \int_V \underline{\omega}^i \cdot \underline{v}^i \, dV$ depends only on the vorticity fields inside D. Rewrite

$$\Delta H = \int_V (\underline{v}^1 - \underline{v}^2) \cdot (\underline{\omega}^1 + \underline{\omega}^2) \, dV + \int_V (\underline{v}^2 \cdot \underline{\omega}^1 - \underline{v}^1 \cdot \underline{\omega}^2) \, dV$$

The second integral vanishes after an integration by parts. Note that $\nabla\times\underline{v}^1 = \nabla\times\underline{v}^2$ in $\overline{D}$, therefore

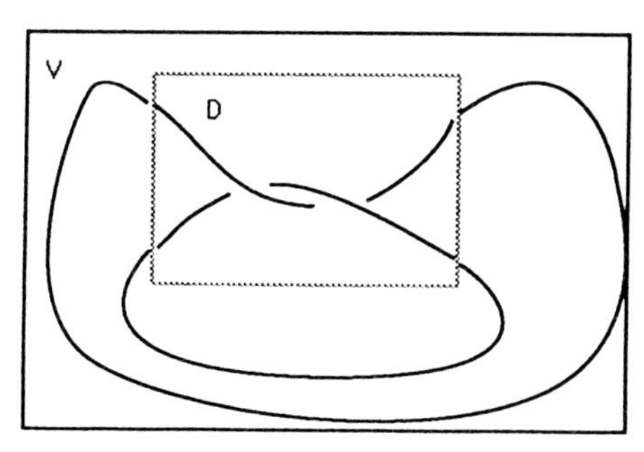

FIG. 1
KNOTTED VORTEX LINES

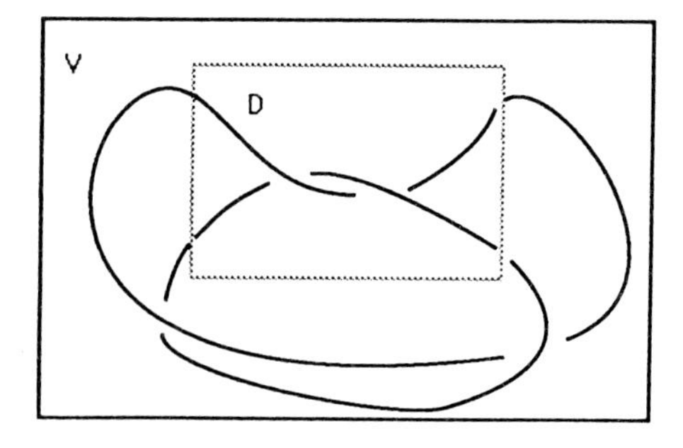

UNLINKED VORTEX LINES

outside D, $\underline{v}^1 - \underline{v}^2 = \nabla\zeta$ for some scalar ζ. Let us separate the first integral into contributions from D and $\bar{D}$:

$$\Delta H = \int_D (\underline{v}^1 - \underline{v}^2)\cdot(\underline{\omega}^1 + \underline{\omega}^2)\,dV + \int_{\bar{D}} \nabla\zeta\cdot(\underline{\omega}^1 + \underline{\omega}^2)\,dV$$

$$= \int_D (\underline{v}^1 - \underline{v}^2)\cdot(\underline{\omega}^1 + \underline{\omega}^2)\,dV - \int_{\partial D} \zeta(\underline{\omega}^1 + \underline{\omega}^2)\cdot\underline{n}\,dA$$

Expressing the velocities as a functional of the vorticities with the help of the Green's function,

$$\underline{v}^1 - \underline{v}^2 = \nabla\times\frac{1}{4\pi}\int_D \frac{\omega^1 - \omega^2}{r}\,d\underline{r}$$

we see that $\underline{v}^1 - \underline{v}^2$ and ζ and therefore also the difference ΔH in total helicities depend only on the vorticity fields inside D. Berger and Field also show that ΔH is gauge-invariant and thus argue that ΔH can be a meaningful measure of the topological complexity of the vorticity field $\underline{\omega}$ relative to an appropriate reference field $\underline{\omega}'$. The most suitable reference field is a potential field $\underline{\omega}' = \nabla\phi$ inside D, where ϕ is the harmonic function $\nabla^2\phi = 0$, uniquely determined by the Neumann boundary conditions $\nabla\phi\cdot\underline{n}\,|_{\partial D} = \underline{\omega}\cdot\underline{n}\,|_{\partial D}$. The potential field is the minimum enstrophy field for given boundary conditions and it is natural to assume that it is in some sense topologically simplest. Constructing $\underline{\omega}'$ with the help of a standard Poisson solver is a straightforward numerical task. Note, however, that the evaluation of the vectorpotential $\underline{v}'$ poses a formidable computational problem.

An alternative method of investigating the topological structure of open vortex tangles with the help of a potential reference field $\underline{\omega}' = \nabla\phi$ was developed by Kuz'min and Patashinsky (1985). Here a test field $\underline{\tilde{\omega}} = \underline{\omega} - \underline{\omega}'$ is set up in a spherical subdomain D, $\underline{\omega}'$ being again the gradient of a scalar ϕ, which is the solution of the appropriate Neumann problem. It is pointed out that $\underline{\omega}'$ does not contain any information about the structure of the vorticity field inside D, as it is completely determined by the vorticity field at the boundary ∂D. Kuz'min and Patashinsky then proceed to discuss the set of moments of $\underline{\tilde{\omega}}$, defined as

$$M^{(n)}_{i_1\cdots i_n}(x) \equiv \int_D r_{i_1} r_{i_2}\cdots r_{i_{(n-1)}}\tilde{\omega}_{i_n}(\underline{x} + \underline{r})\,d\underline{r}^3$$

and suggest that a comparison of field configurations by their moments will provide a classification scheme for coherent structures. Note that the test field $\underline{\tilde{\omega}}$ forms a localized 'blob' of vorticity, as the field lines of $\underline{\tilde{\omega}}$ do not cross the boundary. The entangledness of the vorticity field $\underline{\omega}$ will be reflected in the linkage of the field lines of the test field $\underline{\tilde{\omega}}$, see Fig. 2 for a schematic illustration. Therefore, the helicity $\tilde{H} \equiv \int_D \underline{\tilde{v}}\cdot\underline{\tilde{\omega}}\,dV$ of the test field should provide an alternative measure of entangledness.

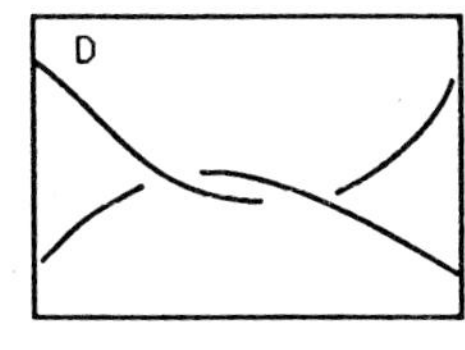

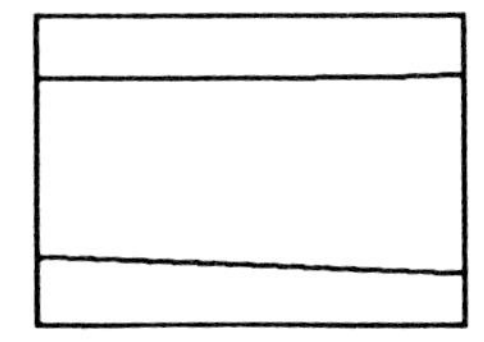

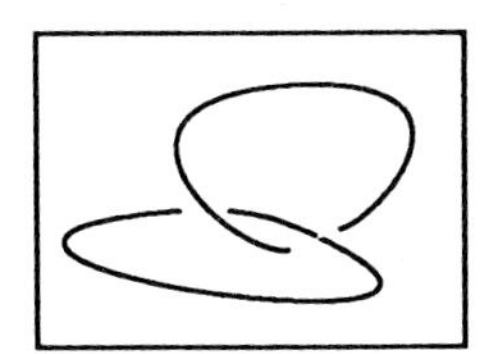

FIG. 2
VORTICITY FIELD $\underline{\omega}$ REFERENCE FIELD $\underline{\omega}'$ TEST FIELD $\underline{\tilde{\omega}}$

The boundary conditions for $\underline{\omega}'$ ensure that $\tilde{H}$ is invariant under a gauge-transformation $\underline{v} \rightarrow \underline{v} + \nabla\chi$:

$$\delta\tilde{H} = \int_D \nabla\chi \cdot \tilde{\underline{\omega}}\, dV = \int_{\partial D} \chi(\underline{\omega} - \underline{\omega}') \cdot \underline{n}\, dA = 0$$

$\tilde{H}$ - like the difference of total helicities ΔH - reduces to the usual integral of helicity for closed vortex structures, the potential reference field of which is identically zero. However, $\tilde{H}$ is in general not equal to ΔH (see below).

It was then proposed by Levich (1987), that a relative helicity H_R

$$H_R \equiv \int_D \underline{v} \cdot \tilde{\underline{\omega}}\, dV$$

would as well serve as measure of relative entangledness. The test field $\tilde{\underline{\omega}}$ is defined as above, which ensures gauge invariance. For computational purposes it is most advantageous that the evaluation of the relative helicity H_R does not require the knowledge of the vectorpotential $\underline{v}'$ of the reference field $\underline{\omega}'$. However, it is not clear what the topological significance of H_R really is. Obviously, $\tilde{H} = H_R - \delta$, where $\delta \equiv \int_D \underline{v}' \cdot \tilde{\underline{\omega}}\, dV$. Similarly, by dividing the domain of integration and regrouping the integrands we rewrite the difference ΔH of total helicities as

$$\begin{aligned}\Delta H &= \int_V (\underline{v}^1 \cdot \underline{\omega}^1 - \underline{v}^2 \cdot \underline{\omega}^2)\, dV \\ &= \int_D \underline{v}^1 \cdot (\underline{\omega}^1 - \underline{\omega}^2)\, dV + \int_D (\underline{v}^1 - \underline{v}^2) \cdot \underline{\omega}^2\, dV + \int_{\bar{D}} (\underline{v}^1 - \underline{v}^2) \cdot \underline{\omega}\, dV\end{aligned}$$

Recall that $\underline{\omega}^1 = \underline{\omega}$ in V , $\underline{\omega}^2 = \begin{cases} \underline{\omega} \text{ in } \bar{D} \\ \underline{\omega}' \text{ in } D \end{cases}$. After an integration by parts we obtain

$$\Delta H = H_R + \int_D \underline{v}' \cdot \tilde{\underline{\omega}}\, dV$$

We see that $\tilde{H} + \delta = H_R = \Delta H - \delta$. Unfortunately, the difference term $\delta = \int_D \underline{v}' \cdot \tilde{\underline{\omega}}\, dV$ will in general not vanish[1]. We argue, however, that because both $\tilde{H}$ and ΔH are reasonable measures of relative entangledness, and $\nabla^2 \underline{v}' = 0$ inside D, δ should be insignificant for most field configurations. This in turn implies that H_R may also serve as a measure of the local entangledness of vortex lines.

It seems that 'relative measures of entangledness' can be defined in a rather arbitrary fashion, given that they are gauge invariant and reduce to the integral of helicity for closed field structures. But of course, 'entangledness' as such is an intuitive notion and not mathematically well defined. Unlike the linkage properties of closed vortex lines, local 'entangledness' is not topologically invariant, and it will not be conserved for Euler flows. It is therefore perfectly acceptable that several measures of (relative) entangledness can be constructed. We hope that such measures will provide at least qualitative information about the local topological complexity of the vorticity field. One must realize, however, that vortex line configurations may exist that are considerably entangled but have little or no (relative) helicity (Incidentally, we do not expect the inverse to be true). Moffatt (1981) , for example, has pointed out that a Boromean ring configuration of vortex lines, although certainly

---

1 A. Frenkel, private communication

'entangled', has zero helicity. Yet simpler, cancellations due to left- and righthanded tangles cohabitant in one domain can also result in small values of helicity in regions where the vorticity field is considerably entangled. This does not render (relative) helicity useless as a measure of topological structure, remember that topological invariants provide in general only partial information about topological characteristics.

Numerical Results

We have investigated whether relative helicity H_R does correspond to the local entangledness of vortex lines in an analysis of turbulent velocity fields obtained from direct numerical simulations. We will present mostly results from two velocity fields : field HIE24 from a simulation of decaying homogeneous isotropic turbulence by Lee (1985) , and field C128U12 from a simulation of shear driven turbulence by Rogers (1986). Both simulations were performed on a 128^3 computational grid, the Taylor microscale Reynolds number R_λ was approximately equal to 80. The vorticity field $\underline{\omega}$ is computed from the velocity fields in Fourier space. When transforming back to physical space, the fields are (Fourier-) interpolated to a 256^3 computational grid. This is done to allow the computation of the potential reference field $\underline{\omega}' = \nabla\phi$ with satisfactory numerical accuracy. Note that in the shear flow case the mean vorticity due to the homogeneous shear is subtracted from the vorticity field. A standard Helmholtz solver is used to solve the Neumann problem for ϕ in cubical subdomains of varying sizes N_D. The potential vorticity $\underline{\omega}'$ is computed from ϕ with a second order finite difference scheme. This approach yields satisfactory precision in the calculation of the relative helicity H_R for domain sizes N_D of about 10^3 and greater. Furthermore, the fields of turbulent dissipation $\epsilon = 2e_{ij}e_{ij}$, vortex stretching $\sigma = e_{ij}e_{jk}e_{ki}$ and enstrophy production $\rho = \omega_i e_{ij}\omega_j$ are evaluated. Einstein summation is implied here, and $e_{ij} = \frac{1}{2}(\partial_j v_i + \partial_i v_j)$ is the stress tensor.

As the turbulent fields under study are homogeneous, it is initially not known where regions of high relative helicity are to be found. In order to locate such regions, we position a large number N_S (typically several thousand) of test-domains of a given size N_D inside the flow domain. For each domain, the total and relative helicities H and H_R and the volume integrals over the domain E_D , Ω_D , ϵ_D , σ_D , ρ_D of turbulent energy, enstrophy, dissipation, vortex stretching and enstrophy production are evaluated.

We then attempt to visualize the structure of the vorticity field by 3-D plots of vortex lines in domains where we expect to find very high (or low) topological complexity, as indicated by the 'topological charge' $T_C \equiv H_R / \sqrt{E_D \Omega_D}$ (Levich, 1987), i.e. the appropriately normalized value of H_R. A second order Runge-Kutta scheme is employed to integrate the vortex lines. The starting points for the integration of vortex lines were chosen to be the maxima of enstrophy in planes parallel to the y- and z-axis at various x-positions (and cyclic permutations of the axes). Then the integration advances in forward and reverse direction from the starting point until the vortex line reaches the domain boundary. Vortex lines are only plotted if the maximum of enstrophy does not lie on the boundary and if the average line strength is equal to or greater than the average vorticity inside D. It is found that although the vorticity field of these homogeneous flows is everywhere quite complicated, the vorticity field indeed tends to be more entangled in regions of high topological charge. For example, Figs. 3 and 4 show 3-D perspective plots of two domains with $N_D = 24^3$ which had the maximum topological charge T_C in a sample of 4096 domains taken from field HIE24. The two domains with lowest T_C from that sample are shown in Figs. 5 and 6 . However, beginning the vortex line integration at a fixed set of starting points, rather than at the extrema of vorticity, yields plots that appear to be equally entangled no matter what the relative helicity is. Discarding vortex lines of lesser strength brings little improvement. Clearly, a more detailed and sophisticated graphical representation of the vorticity field is called for. It might also be instructive to look at the structure of the reference field $\underline{\omega}'$ and the test field $\tilde{\underline{\omega}}$.

Another important question is whether high topological complexity T_C is mostly to be found in the active or inactive regions of turbulent flows. The relatively large number of domains sampled from each flow makes it possible to conduct a coarse statistical analysis. We have evaluated the linear correlation coefficient $r(X,Y)$, defined as

$$r(X,Y) = \frac{<X - <X>><Y - <Y>>}{\sqrt{<(X - <X>)^2><(Y - <Y>)^2>}}$$

for various pairs of domain parameters. Angular brackets $<\ldots>$ denote averaging over the sample of domains. The correlation coefficient is normalized to lie in the interval [-1 , +1], where the extrema correspond to complete positive or negative correlation. The results presented in table I show that relative helicity H_R correlates slightly - but statistically significantly - with ϵ, σ and ρ , i.e., with the active regions of the flow. It is appropriate to use the square of T_C, for otherwise symmetry between left- and righthanded tangles would obscure all correlations with quantities that are expected to be insensitive to handedness. For domain sizes larger than $N_D = 24$, the correlations weaken and become insignificant for N_D approximately equal to 60. This presumably indicates that taking the volume integrals over larger domains that contain both active and inactive regions amounts to a pre-averaging that destroys all statistical correlations. We have also gathered statistics from the initial field of the shear flow simulation and from a field from a simulation at lower Reynoldsnumber (C12 in Lee (1985), $R_\lambda \approx 20$). In both cases, no correlations were found. This was expected for the initial field, which is merely an amplitude field of uncorrelated modes without any structure. The absence of correlations between T_C and the active regions in the low Reynolds number case suggests that here viscous diffusion of vorticity is too strong to permit the existence of sufficiently well-defined and localized vortex tubes. It will be worthwhile to analyze turbulent flows at yet higher Reynolds number, once these come into reach of direct numerical simulation. No statistically significant correlations between the usual helicity and dissipation and vortex stretching or enstrophy production have been found. This is in agreement with results obtained by Rogers and Moin (1987). To further illustrate this point, we present in Fig. 7 contour plots of the normalized joint probability distribution

TABLE I

Field	N_S	N_D	$r<T_C^2, \epsilon>$	$r<T_C^2, \rho>$	$r<T_C^2, \sigma>$
HIE24	8000	12^3	0.20	0.17	-0.13
	4096	24^3	0.14	0.09	-0.09
C128U12	6859	12^3	0.16	0.13	-0.13
	1728	20^3	0.19	0.16	-0.17

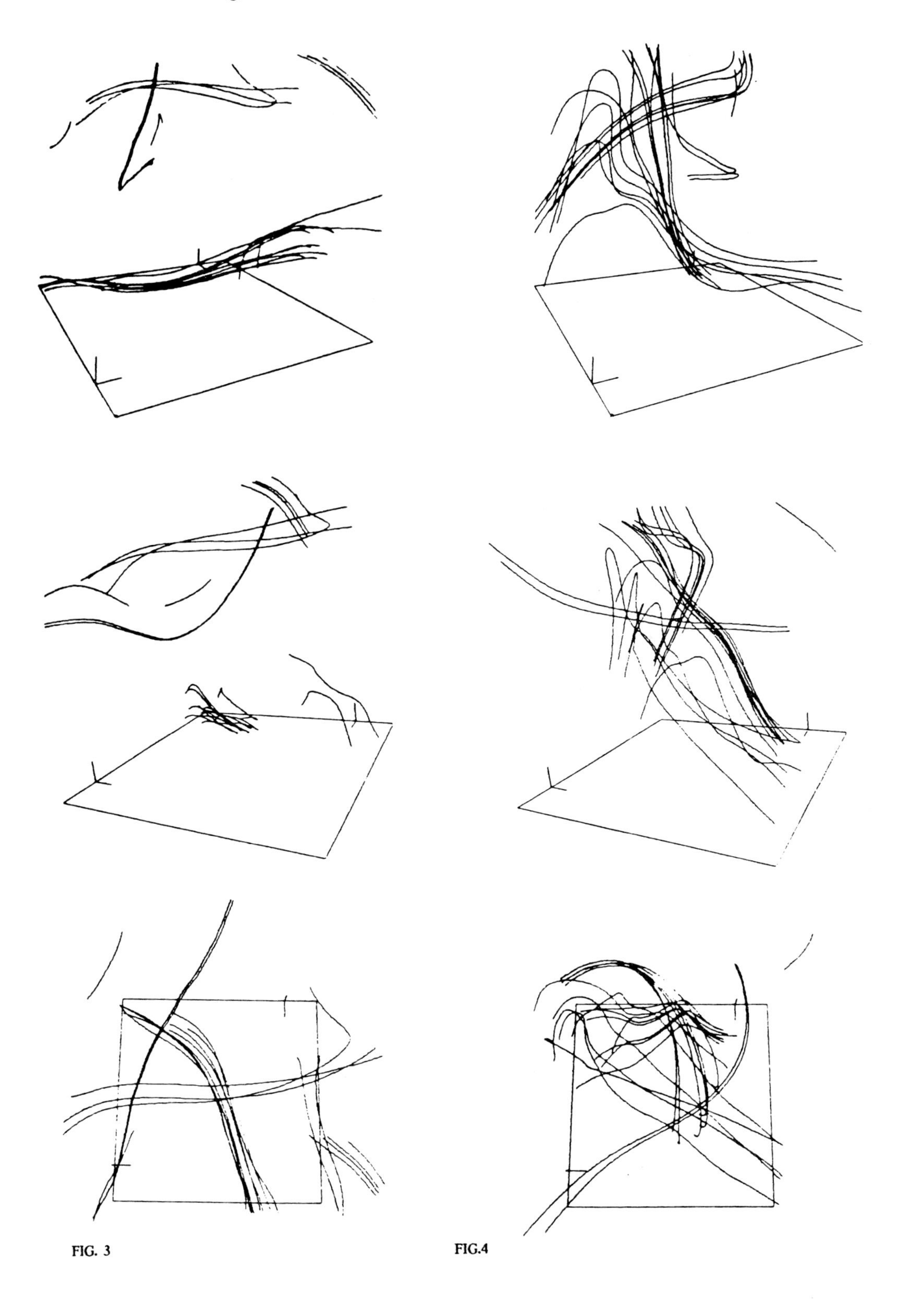

FIG. 3

FIG.4

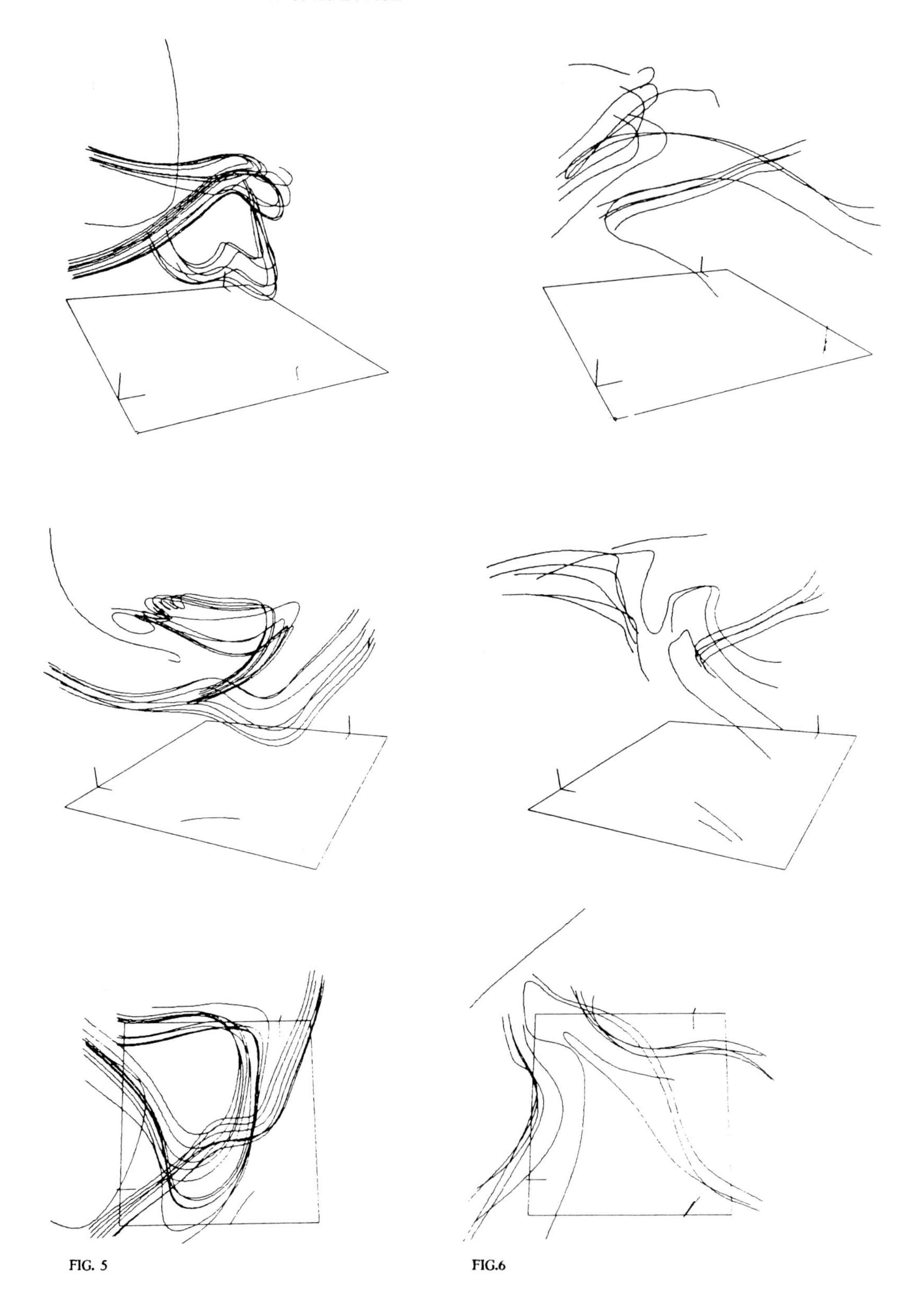

FIG. 5

FIG.6

function $P_{NJ} = pdf(T_C, \epsilon) / (pdf(T_C) pdf(\epsilon))$ obtained from a sample of 8000 domains with $N_D = 12^3$ from field HIE24. The normalization was chosen such that for uncorrelated quantities P_{NJ} is unity. Statistical fluctuations are very strong, because the number of sample domains is too small to adequately represent the long tails in the probability distribution function of dissipation. Nevertheless, one can discern a region of $P_{NJ} > 1.2$ for low dissipation (note that the peak of the pdf of dissipation lies at $\epsilon \approx 5$) and low topological charge. Correspondingly, there is a slight preference for T_C to be large where dissipation is high. The contour plot with the usual helicity , Fig. 7 b., apparently shows only statistical fluctuations.

As mentioned above, the linkage properties of vortex lines are preserved in inviscid flow. Assuming that invariants of Euler flows are 'almost invariant' in flows at high Reynolds numbers , scenarios of turbulence have been developed where it was suggested that flow structures associated with topological invariants will be long-lived, see, e.g. Moffatt (1985), Levich (1987) and Sagdeev *et. al.* (1986). There is no room here for an adequate discussion of these scenarios, but we want to point out that the observed positive correlation between high entangledness and dissipation indicates that vortex tangles are not long-lived.

Conclusions

We have argued that relative helicity $H_R = \int_D \underline{v} \cdot (\underline{\omega} - \underline{\omega}')\, dV$ is a gauge-invariant measure of the topological complexity of the vorticity $\underline{\omega}$ in a domain D relative to a potential reference field $\underline{\omega}'$. Gauge-invariance and the fact that the reference field is completely determined by the vorticity at the boundary of the domain implies that the value of H_R depends only on the vorticity inside D. Analyzing vorticity fields obtained from direct numerical simulations of homogeneous turbulent flows, we have found that the vortex lines indeed appear to be more entangled in regions of high relative helicity. Furthermore, slight but statistically significant correlations between H_R and turbulent dissipation, vortex stretching and enstrophy production have been observed. The correlations with entangledness as such might be stronger, because we may assume that there are regions with small relative helicity but considerable entangledness. It is remarkable that high entangledness seems to be found preferentially in regions where turbulent dissipation is high. This observation might invalidate the assumption that invariants of Euler flows are 'almost invariant' in flow at high Reynolds numbers, which some scenarios of turbulence are based on. However, at present the numerical evidence is weak and not unambiguous. We hope that simulations at higher Reynolds number, where the vortex lines and -tubes are more localised, will clarify the situation.

Acknowledgements

We would like to thank Dr. Michael M. Rogers, Dr. Robert D. Moser and Dr. Alan A. Wray for many stimulating discussions and computational advice. Without their support and continued interest this work could not have been brought to completion. Comments by Dr. Alexander Frenkel are greatly appreciated. The flow fields analyzed in this study were provided by Dr. Moon J. Lee and Dr. Michael M. Rogers. Computations were performed at NASA Ames Center for Turbulence Research and NMFECC at Lawrence Livermore National Laboratory. This work was performed under support of DOE grant DE-FG0288ER13837. Financial support from the Center for Turbulence Research is gratefully acknowledged.

BERGER, M.A. and FIELD, G.B. 1984 *J. Fluid Mech.*
FULLER, F.B. 1978 *Proc. Natl. Acad. Sci. USA* 75, 3557
HUSSAIN, F. 1986 *J. Fluid Mech.* 173, 303-356
KIDA, S., TAKAOKA, M. and HUSSAIN, F. 1989 *Phys. Fluids A* 1, (4), 630-632
KUZ'MIN, G.A. and PATASHINSKY, A.Z. 1985 *Phys. Lett.* 113A, 266-268
LEE, M.J. 1985 Ph. D. dissertation, Stanford University
LEVICH, E. 1987 *Phys. Rep.* 151, 129-238
MOFFATT, H.K. 1969 *J. Fluid Mech.* 35, 117-129
MOFFATT, H.K. 1981 *J. Fluid Mech.* 106, 27-47
MOFFATT, H.K. 1985 *J. Fluid Mech.* 159, 359-378
MOREAU, J.J. 1961 *C.R. Akad. Sci. Paris* 252, 2810
ROGERS, M.M. 1986 Ph. D. dissertation, Stanford University
ROGERS, M.M. and MOIN, P. 1987 *Phys. Fluids* 30 (9), 2662-2671
SAGDEEV, R.Z. , MOISEEV, S.S. , TUR, A.V. and YANOVSKII. V.V. 1986
Nonlinear Phenomena in Plasma Physics and Hydrodynamics , Mir Publishers, Moscow, 137-182

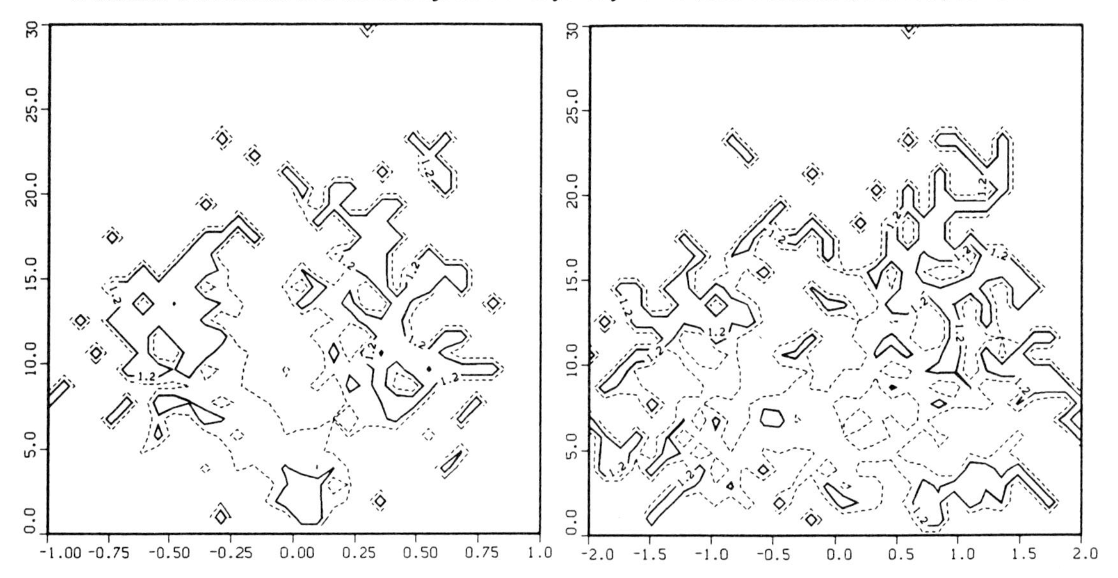

Fig. 7a
Normalized joint pdf of T_C vs. ϵ

Fig. 7b
Normalized joint pdf of $H_D/\sqrt{E_D \Omega_D}$ vs. ϵ

VIII: Inhomogeneous Turbulent and Convective Flows

Coherent Structures of Mixing Layers in Large-Eddy Simulation

P. COMTE, M. LESIEUR and Y. FOUILLET

Institut de Mécanique de Grenoble* B.P. 53 X, 38041 Grenoble Cedex, FRANCE

1 ABSTRACT

Three-dimensional temporally growing mixing layers are simulated by means of a pseudo-spectral code. Periodicity is assumed in streamwise and spanwise directions. High Reynolds numbers are reached with the aid of an appropriate subgrid-scale parameterization.

The initial condition results from the superposition of a three-dimensional small amplitude random noise upon a one-directional hyperbolic tangent basic flow (Fig. 1). The noise's spectral band-width is chosen broad enough to ensure that all of the instabilities of the basic flow are triggered. The calculation then continues without any forcing.

The evolution of the flow topology is surveyed using colour interactive graphic techniques: the first coherent structures to appear are Kelvin-Helmholtz vortices having the shape of spanwise billows corrugated due to three-dimensional unstable modes (Figs. 2 and 3). Couples of counter-rotating streamwise vortices form later on (Figs. 4 and 5). A possible scenario explaining their origin is proposed on the basis of our observations. With the aid of interactive visualizations, a spanwise spacing s between secondary vortices of same sign is measured and found to be very close to 2/3 of the streamwise spacing between the Kelvin-Helmholtz vortices, which is in excellent agreement with experiments.

2 INTRODUCTION

Quasi two-dimensional spatially-organized structures can be found in many high Reynolds number free shear flows, such as mixing layers [1]. They play an important part in momentum and temperature transport processes. In many cases, purely two-dimensional mechanisms allow one to understand most of the dynamics of these structures. For instance, reasonably intricate stability calculations account for formation [2] and pairing [3] of Kelvin-Helmholtz vortices in the case of an incompressible mixing layer; two-dimensional numerical simulations reported in [4] have shown such formation, using a small-amplitude white noise superimposed onto the basic hyperbolic tangent velocity profile, followed by three successive pairings and evidence for the turbulent character of these vortices was presented (unpredictability, broad-band kinetic energy and passive temperature spatial spectra). On the other hand, the afore-mentioned predictability study was interpreted in terms of progressive spanwise decorrelation of initially two-dimensional Kelvin-Helmholtz billows.

---

* Unité Mixte de Recherche numero 101 du C.N.R.S

In this paper, we simulate numerically a three-dimensional unforced mixing layer at high Reynolds number, by means of a pseudo-spectral method using a subgrid-scale parameterization, so expanding upon direct simulations reported in [5]. Both two- and three-dimensional instabilities are triggered by broad-band three-dimensional noise of small variance, superposed initially upon the basic flow. Evolution of the flow is followed with the aid of three-dimensional interactive visualization techniques. This enables us to see how coherent structures appear and evolve. On the basis of observations, we propose a scenario that might shed light on the origin of these coherent structures.

3 NUMERICAL METHODS

The large-eddy incompressible Navier-Stokes equations written in Fourier space are solved using pseudo-spectral techniques (collocation methods) in the form

$$\frac{\partial \hat{\mathbf{u}}}{\partial t} = F\left[F^{-1}(\hat{\mathbf{u}}) \times F^{-1}(\hat{\omega})\right] - i\mathbf{k}\, F\left[\frac{p}{\rho} + \frac{1}{2}|\mathbf{u}|^2\right] - \left[\nu + \nu_t(k|k_c)\right] \mathbf{k}^2\, \hat{\mathbf{u}} \qquad (1)$$

$$\mathbf{k}.\hat{\mathbf{u}} = 0 \quad , \qquad (2)$$

where $\hat{\omega} = i\, \mathbf{k} \times \hat{\mathbf{u}}$ is the vorticity in Fourier space ($i^2 = -1$), F is the direct fast Fourier transform operator, and $k = \sqrt{\mathbf{k}^2}$.

We solve simultaneously the following equation, satisfied by a passive temperature θ:

$$\frac{\partial \hat{\theta}}{\partial t} = -i\mathbf{k}.F\left[F^{-1}(\hat{\theta}).F^{-1}(\hat{\mathbf{u}})\right] - \left[\kappa + \kappa_t(k|k_c)\right] \mathbf{k}^2\, \hat{\theta} \quad . \qquad (3)$$

The spectral eddy-viscosity $\nu_t(k|k_c)$ and eddy-conductivity $\kappa_t(k|k_c)$ correspond to the Chollet-Lesieur subgrid-scale parameterization [6], k_c being the cut-off wave-number. The reader is referred to [7] for details concerning this parameterization.

The initial basic flow

$$\bar{u}(y) = U \tanh\,(2y/\delta_i) \qquad (4)$$

corresponds approximately to the mean velocity profile of a turbulent mixing layer between two quasi-parallel co-flowing streams of velocities U_1 and U_2, viewed in a frame translating downstream at the velocity $(U_1 + U_2)/2$. Thus we have $U = (U_1 - U_2)/2$ (cf. Fig. 1). δ_i is the initial vorticity thickness.

Onto this basic flow is superimposed a non-divergent isotropic pertubation of small amplitude, defined in Fourier space in such a way that its kinetic energy spectrum decreases exponentially for any wavenumber larger than an adjustable value k_i, and decreases following a given power law for $k < k_i$. Here, k_i is chosen as the most amplified streamwise wavenumber k_a predicted by linear instability theory. This perturbation is then modulated by a gaussian function defined in physical space by $f(y) = \exp\,[-0.5\,(y/\delta_i)^2]$, which limits its effects to the rotational zone of the basic flow. Finally, the filtered perturbation is rendered non-divergent by projection, in Fourier space, on the plane perpendicular to $\mathbf{k}$. It models, in a somewhat primitive manner, residual turbulence brought about by the splitter plate and responsible for the mixing layer's transition to turbulence.

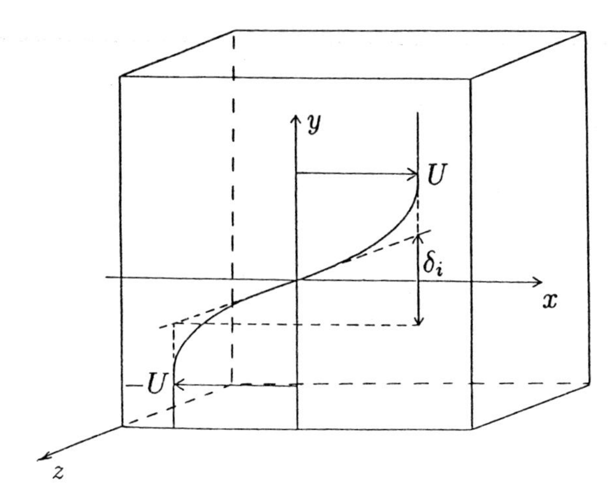

Figure 1: computational domain with basic velocity and passive temperature profile of a temporally growing mixing layer. U is half the velocity difference between the two streams. δ_i stands for the initial vorticity thickness of the layer.

Since the temperature θ is passive, the choice of its initial value is purely arbitrary. In order to visualize the mixing of the two flows, we have set $\theta(k_x, k_y, k_z, t = 0)$ equal to the streamwise velocity component $u(k_x, k_y, k_z, t = 0)$, of profile defined by (4), with a different realization of the perturbation having the same statistical properties.

With a spatial resolution of $64 \times 32 \times 32$ Fourier modes, the computational domain is, in physical space, a rectangular parallelepiped of sides L, $L/2$ and $L/2$ in the streamwise x, transverse y and spanwise z directions respectively. We also perform simulations with $64 \times 32 \times 64$ Fourier modes in a domain of size $L \times L/2 \times L$.

Periodic boundary conditions are applied in the streamwise and spanwise directions, while the flow is assumed quiescent for $y = \pm L/4$. The value of L is set to $m_i \lambda_a = 2\pi/k_a$, so that m_i Kelvin-Helmholtz billows of streamwise period λ_a are expected, $k_a \approx 0.4446\ (\delta_i/2)^{-1}$ being the most amplified mode predicted by linear stability theory [2].

4 COHERENT STRUCTURE OBSERVATION

We first run a large-eddy simulation in a domain of length $L = 2\ \lambda_a$, with an initial perturbation variance $< \hat{\mathbf{u}}'^2 >= 10^{-4}\ U^2$.

Fig. 2 shows the iso-surface $\theta = 0$ of the passive temperature at $t = 17\ \delta_i/U$, visualized with the aid of the interactive graphic software we have developed. This iso-surface corresponds to the virtual interface between the two streams of different velocities. As yet, no coherent structures are displayed. However, we can see that this interface rolls up at two different streamwise locations, where the development of a spanwise undulation is also visible. Consequently, two Kelvin-Helmholtz billows form, at roughly $t \approx 20\ \delta_i/U$, a value in good agreement with our previous two-dimensional simulations reported in [4].

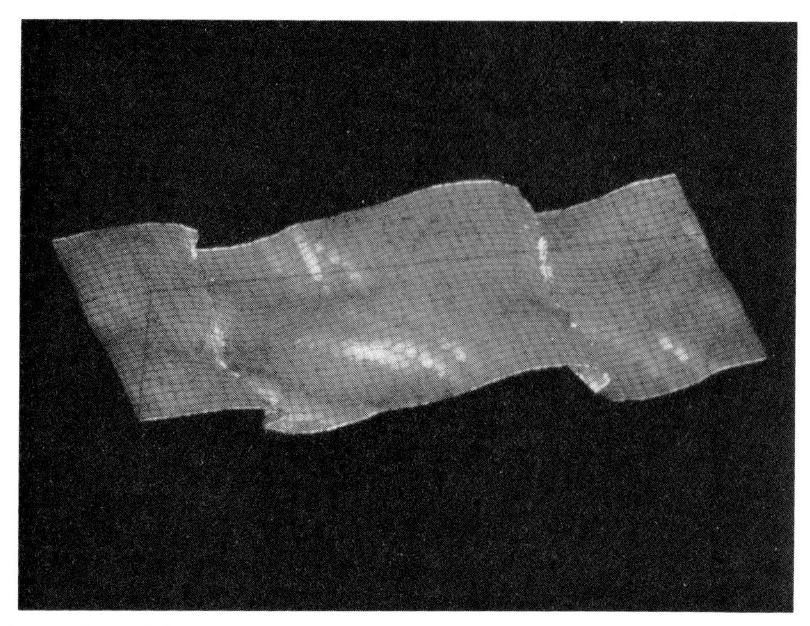

Figure 2: iso-surface of the passive temperature $\theta = 0$ which represents the interface between the two streams at $t = 17\ \delta_i/U$. The three-dimensional feature of the flow is already noticeable. The resolution is of $64 \times 32 \times 32$ Fourier modes. The size of the computational domain is ($L_x = 2\ \lambda_a$, $L_y = \lambda_a$, $L_z = \lambda_a$), λ_a being the wavelength of the most amplified streamwise mode.

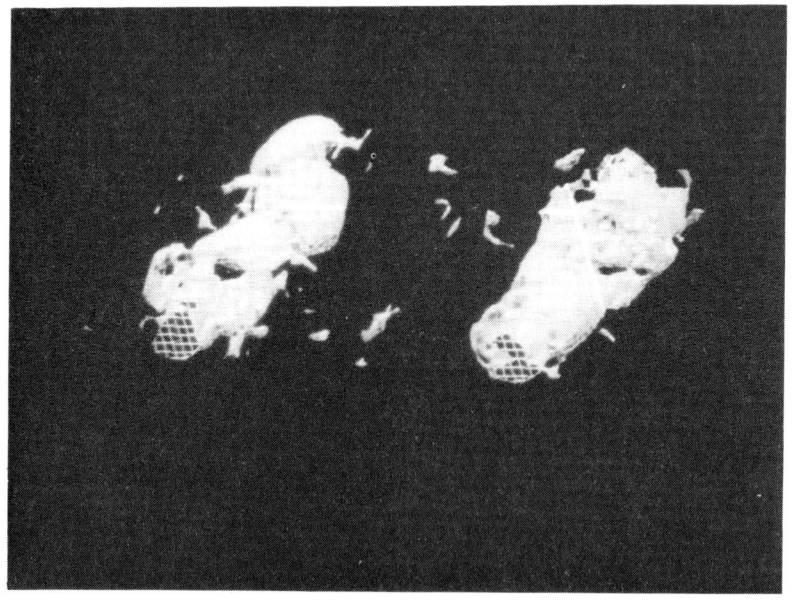

Figure 3: iso-contour of the spanwise vorticity component $\omega_z \approx 2U/\delta_i$ at $t = 26\ \delta_i/U$, for the same realization as before. The undulation of the Kelvin-Helmholtz billows appears clearly.

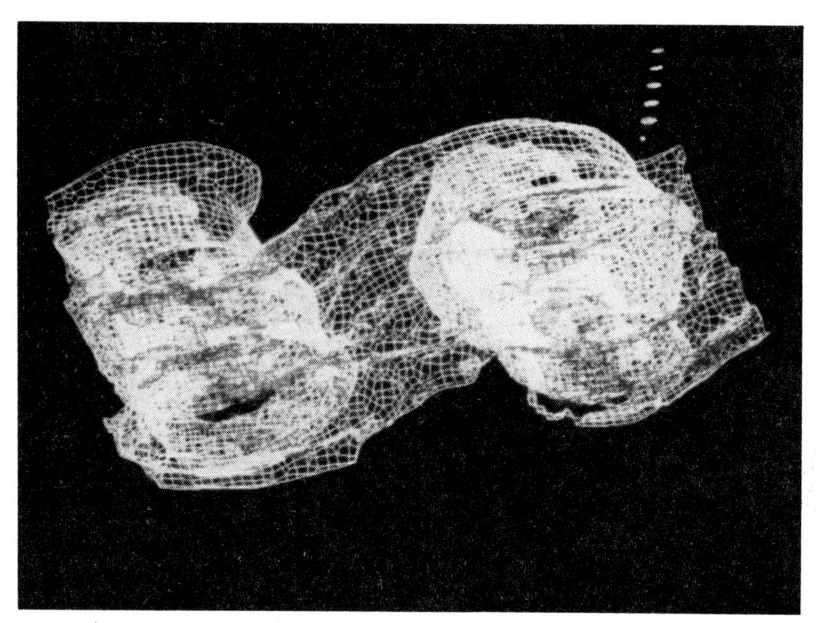

Figure 4: same calculation, at the same timestep. In dark: iso-contour of the streamwise vorticity component $\omega_x \approx \omega_i = 2\,U/\delta_i$ at $t = 26\,\delta_i/U$ (only positive vortices are displayed). The interface between the two streams, defined as in Fig. 2, appears in white.

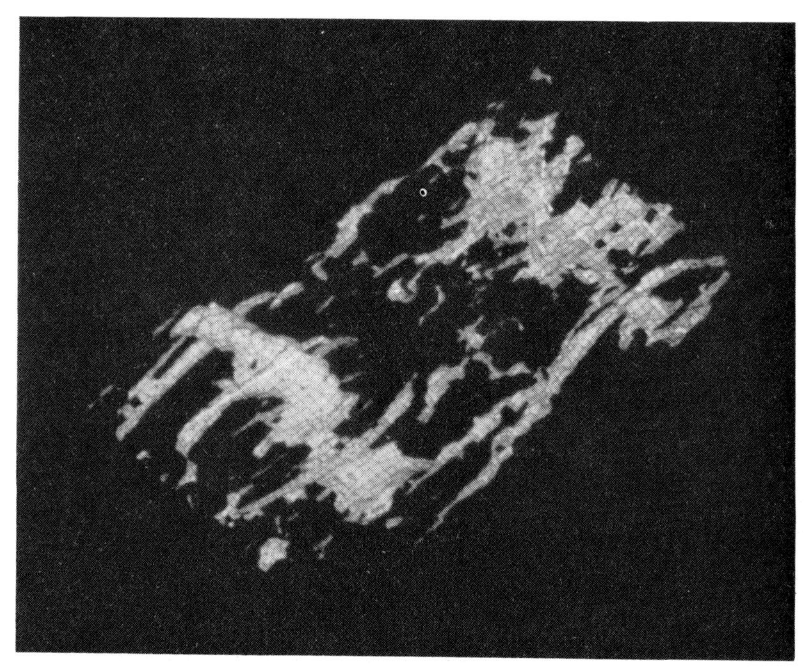

Figure 5: same calculation, at the same timestep. Iso-surface $\sqrt{\omega_x^2 + \omega_y^2} = \omega_i$. (See also Colour Plates)

Fig. 3 shows the iso-surface $\omega_z \approx 2U/\delta_i$ at $t = 26\ \delta_i/U$. Frow now on, we shall use the notation ω_i for the modulus $2U/\delta_i$ of the vorticity introduced initially by the basic flow, in the streamwise direction. The coherent structures thus displayed appear distorted by a sine spanwise mode of wavelength close to $L = \lambda_a$, the breadth of the box. This suggests the possible existence of a most amplified spanwise mode, whose modulus k_b would be close to the modulus k_a of the streamwise fundamental mode. This is in acceptable agreement with Pierrehumbert and Widnall's stability calculations [8], where a value $k_b \approx 1.5k_a$ was found. However, one must bear in mind that we have imposed spanwise periodicity of period L. We can also notice in Fig. 3, between the two vortical billows, small elongated filaments carrying relatively strong vorticity (of the order of ω_i).

Fig. 4 shows, still at $t = 26\ \delta_i/U$, the interface $\theta \approx 0$ between the two streams (in white), upon which is superimposed the iso-surface $\omega_x = \omega_i$. The latter shows intense secondary vortices, orientated streamwise. These can also be seen in Fig. 5, where the iso-surface $\sqrt{\omega_x^2 + \omega_y^2} = \omega_i$ is plotted.

We then run a large-eddy simulation in a domain of size $(L, L/2, L)$, with $L = 4\ \lambda_a$, at a resolution of $64 \times 32 \times 64$. Fig. 6 shows, at $t = 26\ \delta_i/U$, the iso-surface $\omega_x = \omega_i$.

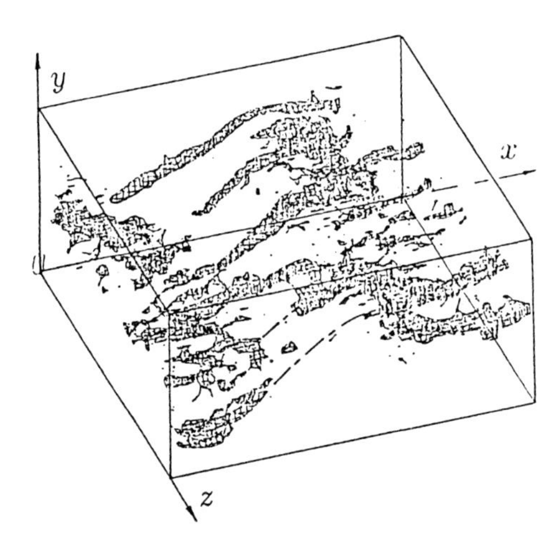

Figure 6: iso-surface $\omega_x = \omega_i$ at $t = 26\ \delta_i/U$, showing positive streamwise eddies. The dashed lines suggest possible vortex lines.

Streamwise secondary vortices of positive sign are displayed (analogous rib-like structures also appear when plotting the iso-surface $\omega_x = -\omega_i$). The computational domain is now broad enough to enable us to measure, with acceptable precision, a spanwise spacing s between these secondary vortices: in Fig. 6, it is possible to make out six interrupted streamwise vortices, suggesting a spanwise spacing s of about $L/6 = 2/3\ \lambda_a$, in excellent agreement with Bernal and Roshko's experiments [9]. This also suggests that these secondary vortices may result from the non-linear development of the most amplified spanwise unstable mode $k_b = 2\pi/s \approx 1.5k_a$ found by Pierrehumbert and Widnall [8].

Evolution with time of the peak values of streamwise vorticity has been plotted in Fig. 7: positive and negative contributions evolve in the same way, catching up on $\omega_i = 2U/\delta_i$ at $8\ \delta_i/U$ and levelling off at about 2.5 ω_i after 17 δ_i/U, the timestep which corresponds to Fig. 2. Before 17 δ_i/U, these peak values are located where the Kelvin-Helmholtz vortices roll up. Then, they concentrate into the secondary vortices as soon as these form.

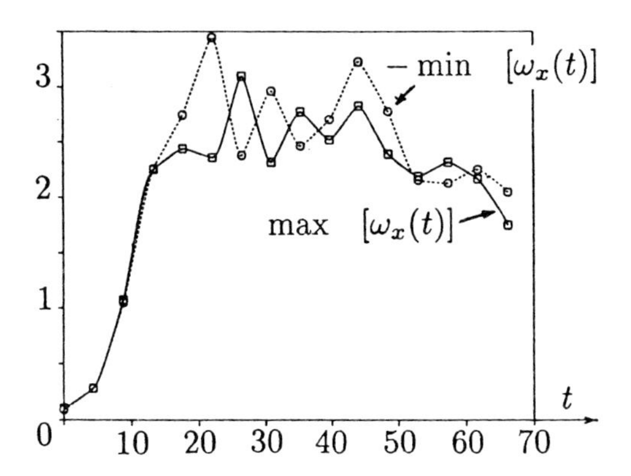

Figure 7: temporal evolution of the peak values of streamwise vorticity normalized by $\omega_i = 2\ U/\delta_i$. The unit of time is δ_i/U.

5 POSSIBLE INTERPRETATION

Let us now assume some hypotheses which seem to be consistent with the observations above:

(a) - Three-dimensional modes grow from the beginning of the evolution, together with two-dimensional instabilities.

(b) - During this growth, up to the formation of coherent structures, the most amplified modes (k_a streamwise and k_b spanwise) emerge.

(c) - Effects of three-dimensionality are mostly localized in the vicinity of high-vorticity regions, where Kelvin-Helmholtz billows will form: indeed, in Fig. 2, the interface between the two streams is considerably distorted three-dimensionally where it rolls up, whereas it looks almost two-dimensional in between (see for example the interface at $x = 0$ and $x = L$).

(d) - After roll-up, a two-dimensional process leading to formation of these vortices, streamlines in each longitudinal section xy can be schematized as in Fig. 7, with eigen-direction ξ and a saddle-point O.

(e) - According to (c), we can assume that three-dimensionality, which distorts Kelvin-Helmholtz billows (cf. Fig. 3), is negligible in the middle of the braids, so that saddle-points O remain on a straight stagnation line parallel to z. This is certainly valid as long as three-dimensionality remains weak in the vicinity of this line.

We are now in position to propose a plausible scenario explaining how secondary vortices form. Fig. 2 shows that Kelvin-Helmholtz vortices are about to form, whereas nothing suggests secondary vortices. For this reason, our scenario starts from a state where primary vortices are already formed and distorted by a spanwise mode k_b, by contrast with the scenario proposed by Lasheras and Choi [10].

Looking at Fig. 8, it is obvious that secondary vortices cannot result from ejection of vorticity from fundamental Kelvin-Helmholtz vortices, where an almost isotropic turbulence is likely to develop: the only effect of the streamlines drawn in Fig. 8 can be to attract fluid particles in the direction of the nearest fundamental vortex, without crossing the plane $(O, \xi^{\perp})$. Therefore, the only means of producing a secondary vortex starting from an almost two-dimensional state is to re-organize the small amount of spanwise vorticity contained in the braids. This can be done by k_b: let us consider (Fig. 9-a) a couple of vortex lines, one on either side of the stagnation line (O, z).

According to (c), the amplification rate of this mode is almost zero on (O, z) and maximal near the fundamental vortices. Evidence of such an amplification-rate variation is given by Meiburg's vortex-filament numerical simulations [11] (cf. Fig.2)

In any case, the vortex lines drawn on Fig. 8-a will undulate in phase with the Kelvin-Helmholtz vortices, with a smaller amplitude. However, this vortex-line undulation cannot, due to the strain field induced by fundamental vortices (Fig. 8), keep on growing in a symmetric way: growth towards the stagnation line will be considerably inhibited, while it will be favoured along ξ, in the other direction (Corcos' mechanism [12]) (Fig. 9-b).

Now the flow is strongly three-dimensional, even in the vicinity of the stagnation line. Hypothesis (e) is no longer valid: the stagnation line will be distorted three-dimensionally, in phase with k_b. Thus, the two vortex lines drawn in Fig. 9-b will merge to form a little vortex tube (Fig. 9-c), which keeps on being stretched, and hence strengthened, by Corcos' mechanism. Eventually, this hairpin vortex tube will wind up around the fundamental vortices (Fig. 9-d).

It is clear that the same process applies to all vortex lines initially localized in the braids: they will collapse onto the same vortex tube and feed it in vorticity.

6 CONCLUSION

We have proposed a possible scenario accounting for the formation of streamwise secondary vortices observed in mixing layers simulated numerically. By contrast with that proposed in [10], it is based on the fact that three-dimensionality appears first in the vicinity of the fundamental Kelvin-Helmholtz vortices, which has yet to be confirmed by higher resolution simulations. However, both scenarii agree about the basic mechanisms involved. In any case, reconstruction of chronology of the events leading to the formation of secondary vortices is of crucial interest: this is necessary to establish a relationship of causality between these events, the first stage towards understanding. High-resolution numerical simulations with frequent interactive graphic post-processing are in process at the moment. A video document is to be produced.

7 REFERENCES

[1] Brown, G.L. and Roshko, A., 1974 On density effects and large structure in two-dimensional mixing layer. *J. Fluid Mech.*, **64**, pp. 775-816.

[2] Michalke, A., 1964 On the inviscid instability of the hyperbolic-tangent velocity profile. *J. Fluid Mech.*, **19**, pp. 543-556.

[3] Kelly, R. E., 1967 On the stability of an inviscid shear layer which is periodic in space and time. *J. Fluid Mech.*, **27**, pp. 657-689.

[4] Lesieur, M. , Staquet, C. , Le Roy, P. and Comte, P. , 1988 The mixing layer and its coherence examined from the point of view of two-dimensional turbulence. *J. Fluid Mech.*, **192**, pp. 511-534.

[5] Metcalfe, R.W. and Hussain, F., 1989 Topology of coherent structures and flame sheets in reacting mixing layers. In these proceedings.

[6] Chollet, J.P. and Lesieur, M., 1981 Parameterization of small scales of three-dimensional isotropic turbulence utilizing spectral closures. *J. Atmos. Sci.*, **38**, pp. 2747-2757.

[7] Comte, P., Lesieur, M. and Fouillet, Y., 1988 Large-eddy simulation of a three-dimensional mixing layer. NATO-ARW on "New trends in non-linear dynamics and pattern forming phenomena: the geometry of non-equilibrium", Cargèse, France, NATO-ASI Series B (Physics), Plenum Press.

[8] Pierrehumbert, R.T. and Widnall, S.E, 1982 The two and three-dimensional instabilities of a spatially periodic shear layer. *J. Fluid Mech.*, **114**, pp. 59-82.

[9] Bernal, L.P. et Roshko, A., 1986 Streamwise vortex structure in plane mixing layers. *J. Fluid Mech.*, **170**, pp. 499-525.

[10] Lasheras, J.C. and H. Choi, 1988 Three-dimensional instability of a plane shear layer: an experimental study of the formation and evolution of streamwise vortices. *J. Fluid Mech.*, **189**, pp. 53-86.

[11] Meiburg, E. and Lasheras, J.C., 1989 Free shear flows; symmetries and topological transitions of the vorticity field. In these proceedings.

[12] Corcos, G.M., 1979 The mixing layer: deterministic models of a turbulent flow. *U.C. Berkeley, Mech. Engng. Rep.*, FM-79-2.

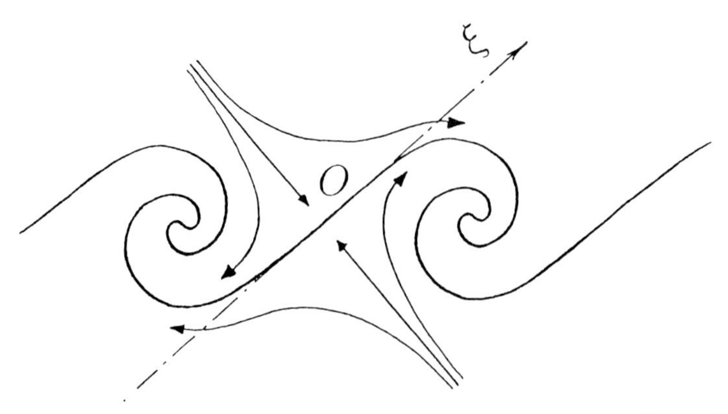

Figure 8 : two-dimensional schematic representation of streamlines in a mixing layer : Corcos' mechanism [11].

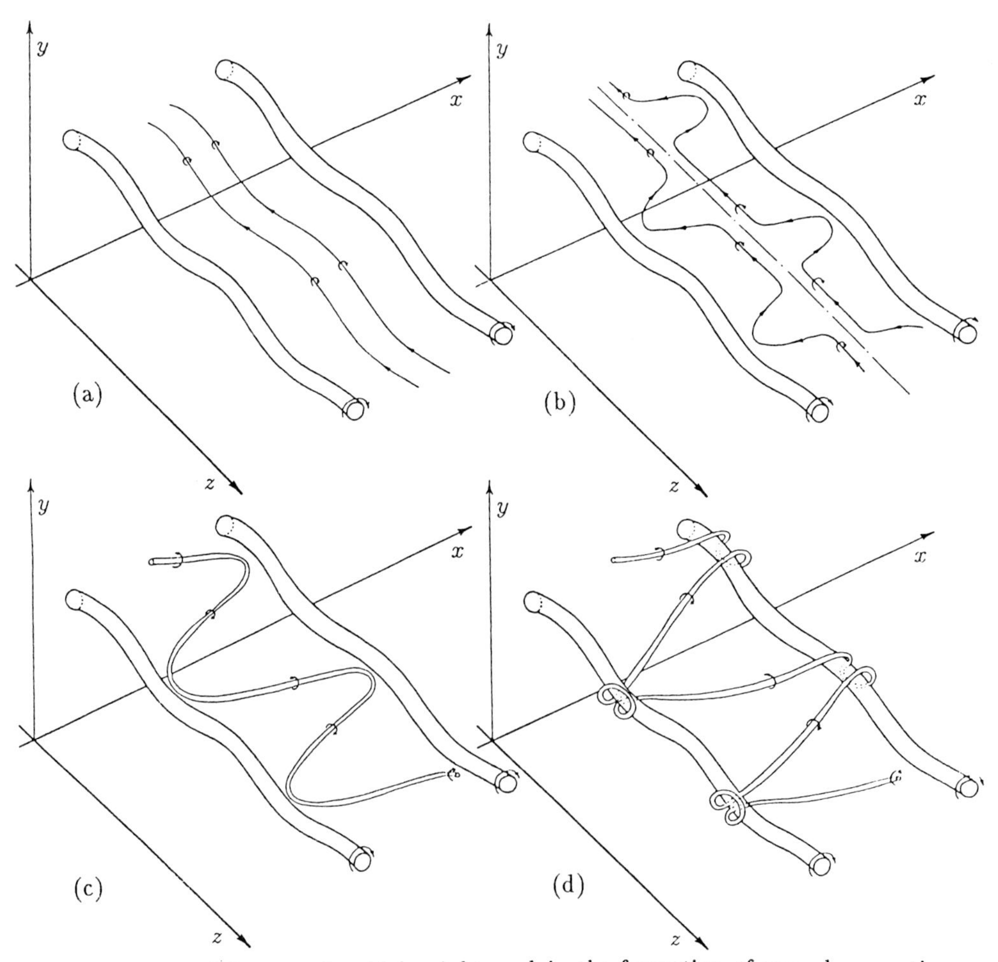

Figure 9 : possible scenario which might explain the formation of secondary vortices.

Topology of Coherent Structures and Flame Sheets in Reacting Mixing Layers

RALPH W. METCALFE AND FAZLE HUSSAIN

Department of Mechanical Engineering, University of Houston
Houston, TX 77204-4792, USA

1 ABSTRACT

We have performed three-dimensional direct numerical simulations of a temporally evolving mixing layer undergoing transition to turbulence with a binary, isothermal chemical reaction. With very low levels of spanwise coherent forcing, the spanwise coherent structures ("rolls"), otherwise kinked and convoluted, become nearly parallel, quasi-two-dimensional structures. Streamwise vortices ("ribs") develop simultaneously with the formation of rolls: there is an initial localization and reorientation of the largely spanwise braid vorticity into horseshoe-shaped ribbons which then become rod-like ribs. The ribs play a crucial role both in cross-stream transport and in wrinkling the flame sheet, producing enhanced mixing and chemical reaction, and removal of product. Under certain conditions, the ribs remain dynamically active coherent structures over times even longer than for the rolls.

2 INTRODUCTION

It has been clearly demonstrated experimentally (Brown & Roshko 1974; Hussain 1981) that large scale structures play a significant role in the dynamics of the evolution of free mixing layers in transitioning and turbulent flows even at high Reynolds numbers. Numerical simulations at more modest Reynolds numbers have validated these observations (Metcalfe et al. 1987a) and comparisons between the numerical simulations and experiments (Metcalfe et al. 1987b) have confirmed that there are important characteristic features of these structures. Recent laboratory experiments (Lasheras et al. 1986) have shed light on the complex interaction between the ribs and the rolls. We have been investigating certain aspects of the development of these structures by the use of direct numerical simulations.

3 METHODOLOGY

These are simulations of a temporally growing free mixing layer with a binary, single-step, irreversible, isothermal chemical reaction in an incompressible flow. The basic governing equations are the 3D Navier-Stokes equations with reaction-diffusion equations describing the interaction of the two initially segregated species. The reaction-diffusion equations are

$$\frac{\partial C_1}{\partial t} + \mathbf{u} \cdot \nabla C_1 = -\alpha C_1 C_2 + \kappa \nabla^2 C_1 ,$$
$$\frac{\partial C_2}{\partial t} + \mathbf{u} \cdot \nabla C_2 = -\alpha C_1 C_2 + \kappa \nabla^2 C_2 .$$

The reaction rate coefficient α and the species diffusivity κ are constants. Periodic boundary conditions are used in the streamwise and spanwise directions and free-slip conditions in the transverse direction. The simulations have been initialized using both the most unstable linear mode (fundamental) and its subharmonic at perturbation velocity amplitudes of about 1% of ΔU with a hyperbolic tangent mean velocity profile of the form

$$U(z) = \frac{\Delta U}{2} \tanh\left(\frac{z}{\delta_i}\right)$$

where ΔU is the velocity difference across the layer and δ_i is the initial mean vorticity thickness. The velocity field is nondimensionalized by ΔU, time by $\lambda_f/\Delta U$, where λ_f is the wavelength of the fundamental. The Reynolds number is defined by $Re = \Delta U \delta_i/\nu$, the Schmidt number $Sc = \nu/\kappa$, and the first and second Damköhler numbers $D_I = z_0 \alpha C_\infty/\Delta U$ and $D_{II} = z_0{}^2 \alpha C_\infty/\kappa$, which represent the ratio of the convection to the reaction time scales and diffusion to reaction time scales respectively. z_0 is the initial thickness of the reaction zone.

In addition to the 2D unstable modes, a low amplitude 3D spectrally broad-banded velocity perturbation was added to permit the growth of the most unstable three-dimensional disturbances as well. The initial reactant concentrations are defined by smooth functions to represent initially segregated species. Accurate pseudo-spectral numerical methods have been used to solve the equations. Details regarding the initialization procedure, equations of motion, and numerical methods are given elsewhere (Metcalfe et al. 1987a; Riley, Metcalfe & Orszag, 1986).

To aid in the visualization of the flow field, the computational domain has been extended by 50% in the streamwise and transverse directions (in which the flow field is periodic) in a number of the plots. This is particularly useful when the end of the computational domain cuts through a structure.

4 MODES OF INSTABILITY

The two dominant modes of instability in incompressible free mixing layers are the "rolls" and "ribs." The rolls dominate the early transition stages of the mixing layer evolution and generate a strain field which excites the secondary rib instability. The development and interaction of these modes have a significant effect on the dynamics of the transitioning flow field, cross-stream transport, flame sheet contortion and hence on the reaction rates between the initially segregated species and product generation. We have investigated certain important aspects of the topology of these structures which sheds some light on these issues.

The basic structure of the flow is shown in Figures 1a-c [subsequent figure references will be given in parentheses: (1a-c)]. These are constant value surfaces of ω_i^2 after the rollup of the fundamental and just before pairing. Three different vorticity levels are shown for the same structures at one instant so that the shape and strength of the vortical structures are apparent. Note that the structure shape and geometry can change drastically with the level of vorticity plotted. The essential dynamics of the rolls is 2D, but there are important 3D instability mechanisms that do significantly modify their structure. This is apparent especially (1c) where the spanwise cores (labelled **A1** and **A2**) can clearly be seen to be kinked. There are three important mechanisms which contribute to the kinking of the rolls: 1) spanwise and transverse perturbations enhanced by the background shear, 2) self- and mutual inductions, and 3) interaction with the ribs. These can subsequently introduce a spanwise variation in the relative phase of the vortex cores during the pairing process and thus reduce the spanwise coherence of the flow. The inherent three-dimensionality of the flow is also clear in the behavior of the braids where the vorticity quickly realigns from an initially dominant spanwise component ω_y to form the strong ribs (e.g. **B** and **C**, (1c)) dominated by the streamwise and transverse vorticity components ω_x and ω_z. The ribs vary significantly in amplitude, the strongest ribs (**B** and **C**) having a peak vorticity near that of the rolls (**A1** and **A2** (1c)). While they generally alternate in sign, this is not always the case, as ribs **D** and **E** are corotating (1a).

5 DYNAMICS OF RIB FORMATION

The formation of ribs in the braid region is initiated by basic instability mechanisms present in the shear flow as the rolls reach finite amplitude and nonlinear dynamics becomes important (Pierrehumbert & Widnall 1982). As the rolls begin to form, a significant strain field develops in the braid region which enhances the growth of streamwise vorticity (Lin & Corcos 1984). Experiments indicate that the 3D instabilities appear first in the braid regions (Lasheras et al. 1986), and this is consistent with our simulations.

The early stages of rib formation in the braid region are shown in Figures 2a-c. These are contour plots of ω_x on a y-z plane through the center of the braid region ($x = 16$, cf. (1a)) for $t = 0.8$, 1.6, and 2.4. Initially small perturbations (2a) grow into more intense, flattened, ribbon-like structures (2b) (as also noted in the simulations of Hussain et al. 1988) which then attain their more mature cylindrical shape (2c) under continued stretching. The enstrophy production ($\omega_i S_{ij} \omega_j$) representing stretching (maximum in the ribs) is plotted in Figure 3a, which is on spanwise plane $y = 24$ (cf. (1a)) passing through the center of the largest of the long ribs (**B**). During rollup and pairing, peak enstrophy production is at the middle of the braid, whereas as the rolls begin to saturate ($t \approx 4.8$), it decreases sharply in the braids (3b).

The evolution of rib structures can be seen in the time sequence of contour plots in Figures 4a-d at the transverse x-y plane corresponding to the initial position of symmetry of the mixing layer, z = 33. Initially, ribs form independently on both sides of the rolls, both in the braid region (labelled **L** in (1a, 3a)) and in the region between the merging rolls (labelled **S**) (4a). In this respect, these simulations may differ from laboratory experiments with spatially evolving flows in which upstream disturbances (natural or forced) may induce the ribs. As the rolls merge, region **S** contracts and then disappears (4b,c,d) into the subharmonic core while the braid region **L** is stretched. As the merging takes place (4d), the ribs in the braid region reach maximum strength while those in the merging cores undergo a complex interaction among themselves and the strong spanwise vorticity in the rolls. The dynamics of the rib generation process in regions **L** and **S** can be compared in Figures 4e-g, showing enstrophy production, dissipation, and helicity density ($u_i\omega_i$) in the same x-y plane as (4a-d).

There are two distinct regions in which ribs form as the rolls merge that involve significantly different dynamical behavior. Region **L** (1a, 3a) is that between the growing subharmonic rolls that has been studied extensively both experimentally (Bernal & Roshko 1986; Hussain & Zaman 1980) and numerically (Metcalfe et al. 1987a). However, region **S** is also subject to some very intense shear as the rolls merge which can also excite ribs by vortex stretching.

In the braid region **L**, some of the ribs which form, like **B**, maintain their coherence from their formation early in the rollup stage until well past the vortex pairing and saturation of the subharmonic, and are the most enduring structures in the flow. Other ribs, such as **D** and **E**, which are adjacent and have like-signed vorticity, appear to blend or fuse laterally into a single rib by a mechanism which must clearly be distinctly different from that governing the merging of counterrotating rib pairs. Strong ribs like **B** can generate high local dissipation in the braid region by pumping fluid from above and below the braid into a fluid stream moving in opposite directions (4f). This cross-stream transport creates a characteristic dual-peaked structure in the dissipation field.

In region **S**, the ribs rotate as the rolls merge, initially having the same orientation as the braid ribs (4b), rotating through the plane of symmetry (z = 0) (4c), and then tilting in the opposite direction (4d). Structures such as **C** have a peak vorticity that is actually higher than the braid ribs like **B** (4c), and the enstrophy production can also be much higher (4e), in this case larger by a factor 3. This is consistent with the experimental observations of Lasheras et al. (1986) who see evidence that ribs in this region remain dynamically active even after vortex pairing has begun. As the merging approaches saturation, the strain rate in this region decreases, and the intensity of the ribs like **C** is reduced while those in the braid region like **B** remain strong. Ribs in the **S** region also create high local dissipation peaks like the braid ribs (4f), and the characteristic change in the sign of helicity at the center of the rib is also exhibited (4g). Since the plane of Figure 4 passes approximately through the saddle point of the braid ribs (**B**, **D**, and **E**), the helicity is very nearly zero in the braid region (see also simulation analysis in Hussain, 1986). Due to the rib generation mechanism occurring here, helicity density by itself is not very useful in characterizing these important coherent structures. In fact, at the middle of the rib, where the helicity density vanishes, the enstrophy production is at its peak (cf. 4e, 4g). The existence of very intense ribs in the cores of pairing structures represents an important mechanism for enhancing mixing in the cores, and we are currently performing more detailed simulations to isolate these effects.

6 STRUCTURE OF THE FLAME SHEET

The rollup and pairing of the rolls enhance the chemical reaction by stretching the flame sheet. This can be seen in Figure 5a which shows where the intensity of the chemical reaction exceeds a threshold value. At this point, the vortical motion in the cores of the rolls is very important in continuing to lengthen the flame sheet and thereby enhance the reaction. However, the cores of the rolls soon become filled with reaction product, and flame shortening occurs. The reaction zone retreats from the core of the roll back toward the braid region (5b) where the ribs continue to bring fresh reactants together and to remove accumulated product toward the roll cores. Further enhancement of the reaction by the primary instability will require an additional vortex pairing. It should be noted that for temperature dependent reactions (for example if there is an Arrhenius factor in the reaction rate term), regions of high strain like the braids can also be regions of high scalar dissipation, and it has been argued theoretically by Linan (1979) and Peters & Williams (1983) that high scalar dissipation can be associated with flame extinction. This has been confirmed by numerical simulations by Givi et al. (1986). Thus, maximizing the strain rate or vortex stretching in the braids may not always maximize the reaction.

The ribs can affect the reaction rate in several ways. First, they tend to wrinkle the flame sheet which increases its area and hence the rate of product generation. This occurs both in the braids and in the merging rolls. In Figure 5a, the plane $x = 1$ cuts through one of the cores, showing how the very early stages of rib formation are beginning to convolute the flame sheet there. In the core, significant flame shortening effects have occurred by the time the subharmonic saturates (5b), so that additional mixing due to instabilities in the core will not enhance the reaction at this point. However, as D_I/D_{II} decreases, i.e. as scalar diffusivity decreases, flame shortening is delayed, so late stage mixing in the cores of the rolls can still be important for sufficiently small values of this parameter ratio. Flame sheet wrinkling can also be very significant in the braid region, especially for large values of D_{II}.

The ribs also augment the reaction rate in the braids by enhancing cross-stream transport. Figure 5c is a contour plot of the reaction zone at streamwise location $x = 10$ which is in the braid region, but not at the braid center. The strong convolution of the flame sheet is evident here, and it is clearly generated by the ribs (5d), whose locations are also indicated on the plot. The reaction rate reaches its maximum at the sides of the ribs where the cross-stream pumping of reacting species is the highest. Inside the rib itself, the reaction rate is much smaller as product tends to accumulate in the core of the rib vortex, inhibiting the reaction. As with the flame sheet wrinkling, the effectiveness of enhanced cross-stream transport in increasing reaction rate is strongly dependent on the Damköhler numbers of the flow. For example, decreasing D_{II} by a factor 10 results in much faster interdiffusion of species (5e). Consequently, the effect of the ribs, and thus the spanwise variation in reaction rate is substantially reduced.

7 ACKNOWLEDGEMENTS

The authors would like to thank Mr. C. F. Lee for help in performing the simulations and producing the plots. This work has been supported by the Office of Naval Research under Contract No. N00014-87-K-0670. Computer time has been supplied by NAS of the NASA Ames Research Center and by NASA Lewis Research Center.

8 REFERENCES

Bernal, L. P. & Roshko, A. 1986 *J. Fluid Mech.* **170**, 499-525.

Brown, G. L. & Roshko, A. 1974 *J. Fluid Mech.* **64**, 775-816.

Givi, P., W.H. Jou & R. W. Metcalfe 1986 *21st Symposium (International) on Combustion*?The Combustion Institute, 1251-1261.

Hussain, F. 1981 *Lect. Notes Phys.* **136**, Springer-Verlag 252-291.

Hussain, A.K.M.F. 1986 *J. Fluid Mech.*, **173**, 303-356.

Hussain, A.K.M.F. & Zaman, K.B.M.Q. 1980 *J. Fluid Mech.* **101**, 493-544.

Hussain, F., R. Moser, T. Colonius, P. Moin, and M. M. Rogers 1988. *Proceedings of the 1988 Summer Program, Center for Turbulence Research*, 49-56.

Lasheras, J. C., Cho, J. S. & Maxworthy, T. 1986 *J. Fluid Mech.* **172**, 231-258.

Linan, A. 1979 *Acta Astronautica* **1**, 1007-1039.

Lin, S.J. & Corcos, G. M. 1984 *J. Fluid Mech.* **141**, 130-178.

Metcalfe, R.W., S. A. Orszag, M.E. Brachet, S. Menon, & J. J. Riley 1987a *J. Fluid Mech.*, **184**, 207-143.

Metcalfe, R. W., F. Hussain, S. Menon, & M. Hayakawa, 1987b *Turbulent Shear Flows* **5**, 110-123.

Peters, N. & Williams, F.A. 1983 *AIAA Journal* **21**, 423-429.

Pierrehumbert, R. T. & Widnall, S. E. 1982 *J. Fluid Mech.* **114**, 50-82.

Riley, J. J., R. W. Metcalfe & S. A. Orszag 1986 *Physics of Fluids* **29**, 406-422.

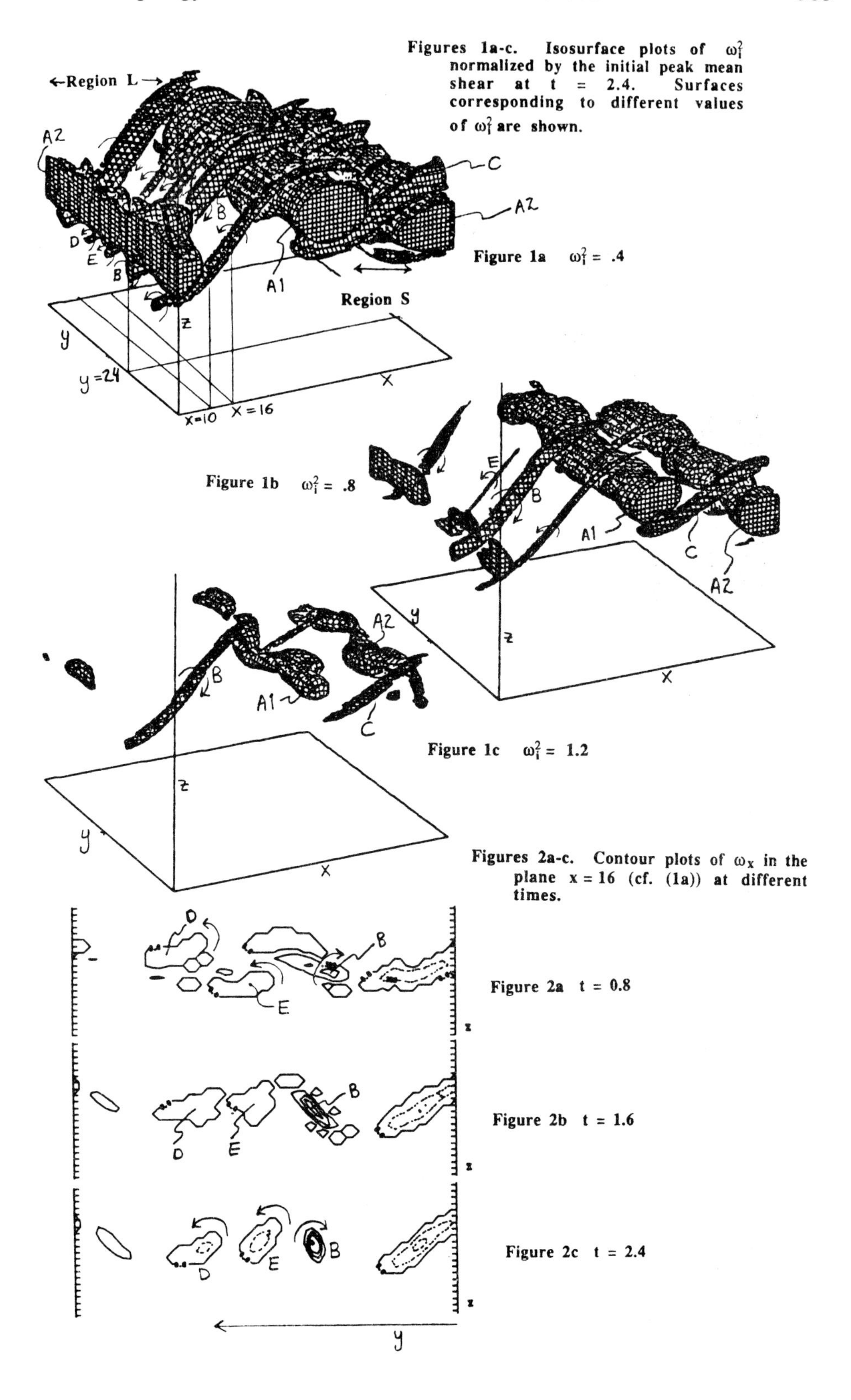

Figures 1a-c. Isosurface plots of ω_i^2 normalized by the initial peak mean shear at t = 2.4. Surfaces corresponding to different values of ω_i^2 are shown.

Figure 1a $\omega_i^2 = .4$

Figure 1b $\omega_i^2 = .8$

Figure 1c $\omega_i^2 = 1.2$

Figures 2a-c. Contour plots of ω_x in the plane x = 16 (cf. (1a)) at different times.

Figure 2a t = 0.8

Figure 2b t = 1.6

Figure 2c t = 2.4

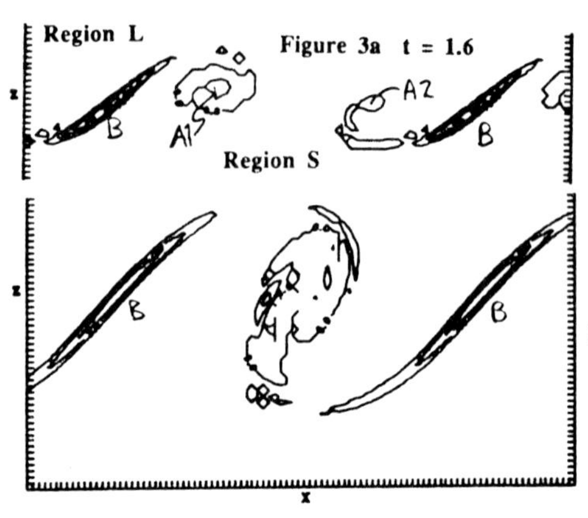

Figures 3a,b. Enstrophy production in the plane y = 24 at different times.

Figure 3b t = 4.8

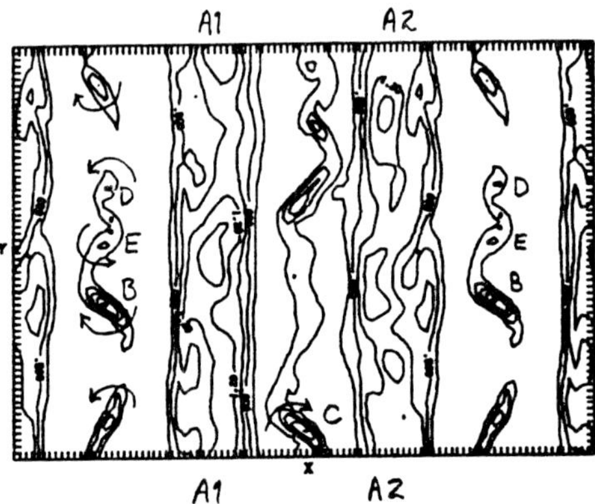

Figure 4a-d. ω_1^2 in the plane z = 33, the position of symmetry of the mean velocity profile, at different times.

Figure 4a t = 1.6

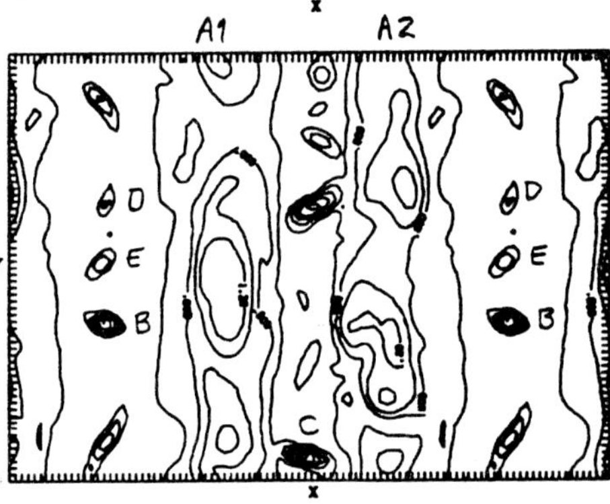

Figure 4b t = 2.4

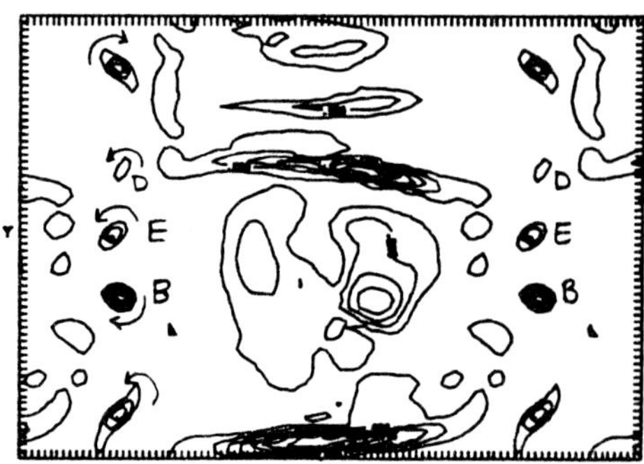

Figure 4c t = 3.2

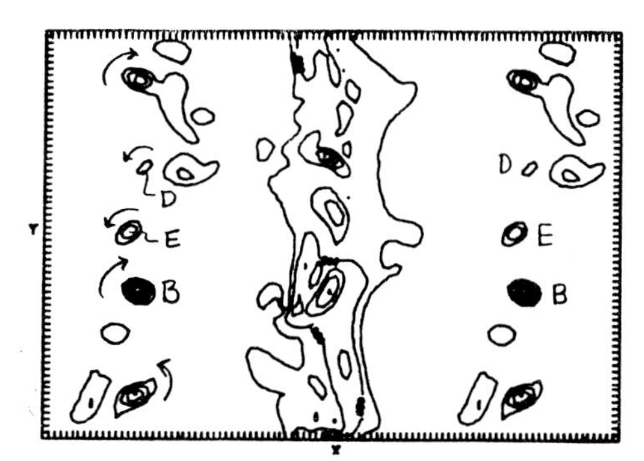

Figure 4d t = 4.0

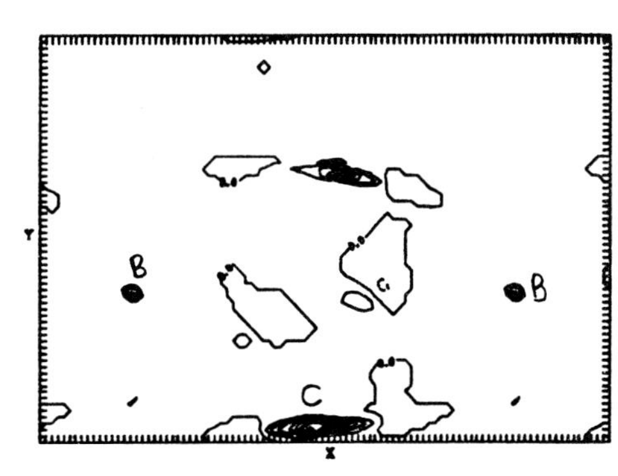

Figure 4e Enstrophy production in the plane z = 33 at t = 3.2.

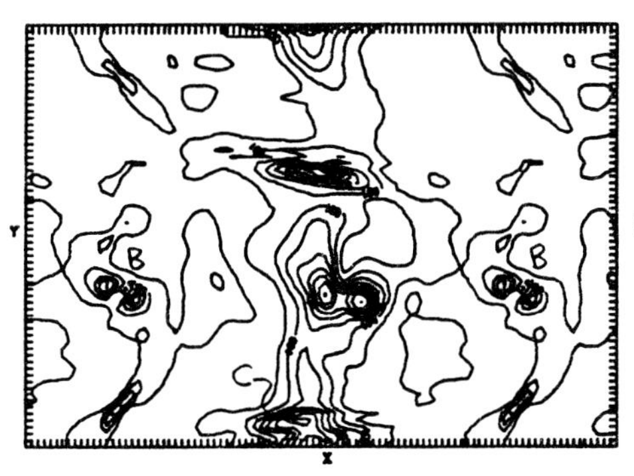

Figure 4f Turbulent dissipation in the plane z = 33 at t = 3.2.

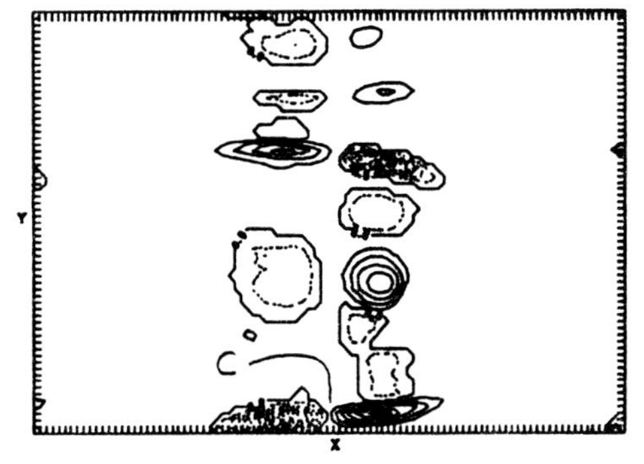

Figure 4g Helicity density in the plane z = 33 at t = 3.2.

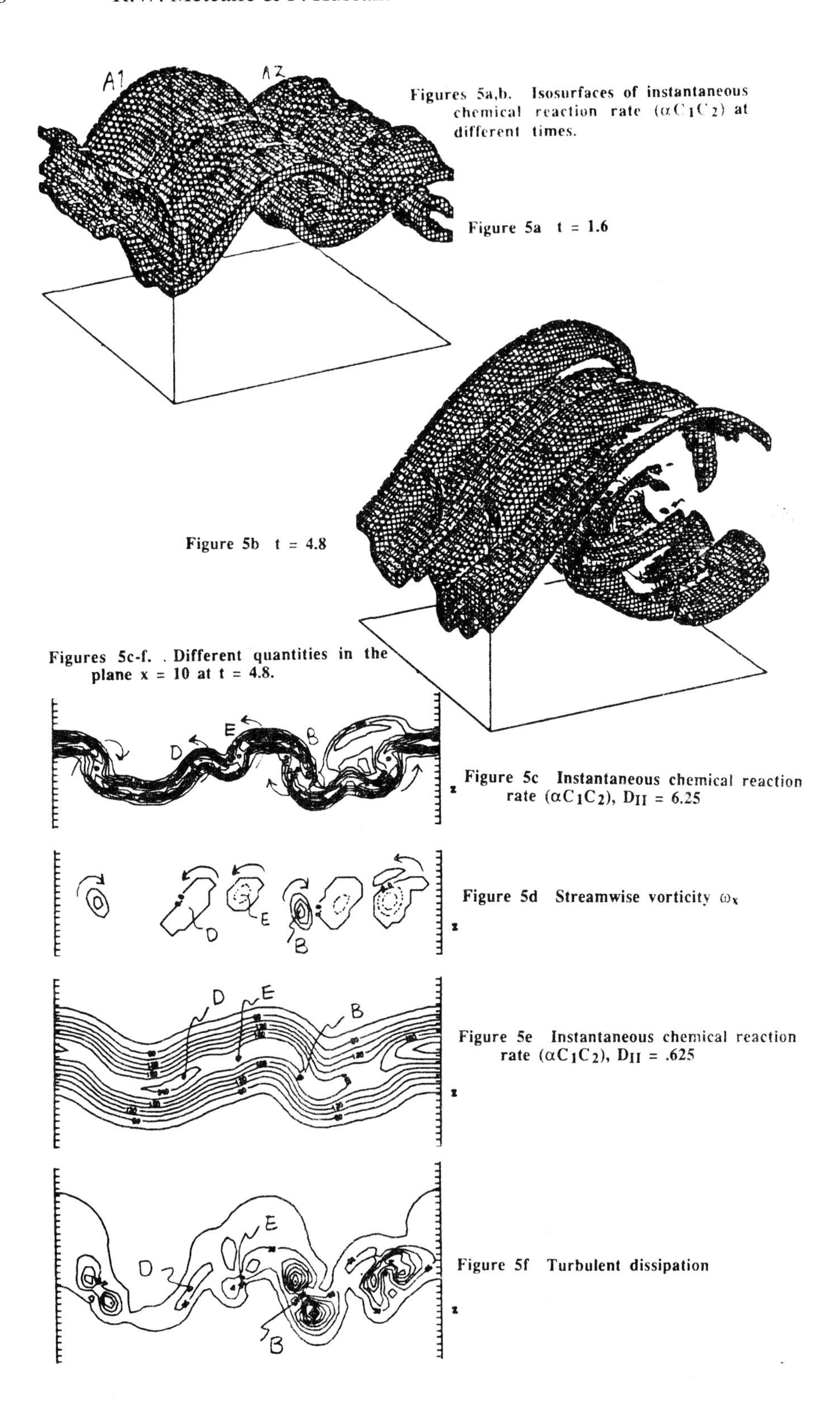

Figures 5a,b. Isosurfaces of instantaneous chemical reaction rate ($\alpha C_1 C_2$) at different times.

Figure 5a $t = 1.6$

Figure 5b $t = 4.8$

Figures 5c-f. Different quantities in the plane $x = 10$ at $t = 4.8$.

Figure 5c Instantaneous chemical reaction rate ($\alpha C_1 C_2$), $D_{II} = 6.25$

Figure 5d Streamwise vorticity ω_x

Figure 5e Instantaneous chemical reaction rate ($\alpha C_1 C_2$), $D_{II} = .625$

Figure 5f Turbulent dissipation

Topological Magnetoconvection

WAYNE ARTER

Culham Laboratory, Abingdon, Oxon OX14 3DB, UK

1 INTRODUCTION

Magnetoconvection (Proctor and Weiss, 1982) describes the convection of electrically conducting fluids or plasmas in the presence of magnetic fields. It finds numerous astrophysical applications, notably in the solar convection zone. Progress has been made mostly through analysis and direct numerical simulation: this review emphasises the latter, with a strong bias towards the author's own work.

Convection cells in the absence of magnetic field, easily studied in the laboratory, have topological properties that are far from trivial, see Section 2. Section 3 looks at how changes in flow topology affect the redistribution of a magnetic flux, initially imposed at right angles to the gravity vector. The Lorentz (magnetic) force is allowed to react back on the flow in some simulations.

2 RAYLEIGH-BÉNARD CONVECTION

2.1 Fixed Planform

The governing equations of Rayleigh-Bénard convection are given e.g. by Chandrasekhar (1961, Chap. II). For motions close to onset of convection, the flow field $\boldsymbol{u}(\boldsymbol{x})$ can be expressed by

$$\boldsymbol{u} = \text{curl curl}[F(x,y,z)\hat{\boldsymbol{z}}], \tag{1}$$

where to a first approximation

$$F(x,y,z) = f(x,y)g(z), \quad \nabla_2^2 f = -\lambda^2 f. \tag{2}$$

f is the planform function, that can be chosen so that the basic element of a periodic layer $(x-y)$ is a roll, rectangle, hexagon etc. $\boldsymbol{u}$ with separable F have closed streamlines because

$$I(x,y) = \int (\frac{\partial f}{\partial y}dx - \frac{\partial f}{\partial x}dy) \tag{3}$$

is an invariant for particle motion in the flow-field, and by a result due to Boltzmann and Larmor (see Whittaker, 1937, §119), the existence of one invariant implies a second, distinct constant of the motion (at least generically), see Arter (1983b) for examples.

At next order, F is not separable. It is more revealing to consider the general steady-state flow satisfying

$$div(\rho \boldsymbol{u}) = 0, \tag{4}$$

where $\rho(x, y, z)$ is the mass density. There is a close connection between flows satisfying (4) and Hamiltonian systems H with two degrees of freedom. Such Hamiltonians have trajectories that can be represented as streamlines of a flow (4) (Whittaker, 1937, §122). The Kerst-Haas-Whiteman construction (Whiteman, 1977) shows how to calculate H for a given $\boldsymbol{u}$. In a general curvilinear co-ordinate system (x^1, x^2, x^3) with infinitesimal volume J, find the contravariant components of the vector potential $\boldsymbol{\Phi}$ in the gauge where e.g. $\Phi_2 = 0$, i.e. find Φ_1 and Φ_3 such that

$$\rho u^1 = \frac{1}{J}\frac{\partial \Phi_1}{\partial x^2}, \rho u^3 = \frac{1}{J}\frac{\partial \Phi_3}{\partial x^2}, \tag{5}$$

then

$$H(x^1, p_1, x^3, p_3) = \Phi_3(x^1, x^2(p_1, x^3), x^3) - p_3, \tag{6}$$

where $p_1 = -\Phi_1$ and p_3 are the momenta conjugate to x^1 and x^3 respectively.

The simple construction (6) works when x^3 can be used to play the role of time, i.e., for cellular flows, if an angular co-ordinate can be found. In the case of the square cell, with

$$F = cos\pi x \, cos\pi y \, sin\pi z, \tag{7}$$

the axis can be taken to be $y = z = 1/2$, for flow interior to the triangular prism $(0 \le x \le 1/2, x \le y \le 1 - x, 0 \le z \le 1)$. The equivalent H is degenerate, since all streamlines close after one circuit about the axis. Birkhoff (1927, Chap. VIII) indicates that perturbations, such as are represented by the higher order corrections to F, should first open the streamlines so that they cover a surface. Arter (1983a) verified this, and showed that ergodic regions develop as the perturbations increase further in size.

The remarkable property of these topology changes is that they seem to have no dynamical significance. Direct numerical simulation of the governing equations using a fixed, square planform (Arter, 1985a) reveals no sudden changes in e.g. the heat transport Nu by the convection, see Figure 1. This is apparently because Nu is also controlled by thermal boundary layers near $z = 0$ and $z = 1$, the thicknesses of which are determined by the flow speed (for fixed viscosity and thermal diffusivity).

2.2 Shallow Tanks

The previous section assumed that convection took place in an infinite, periodic layer.

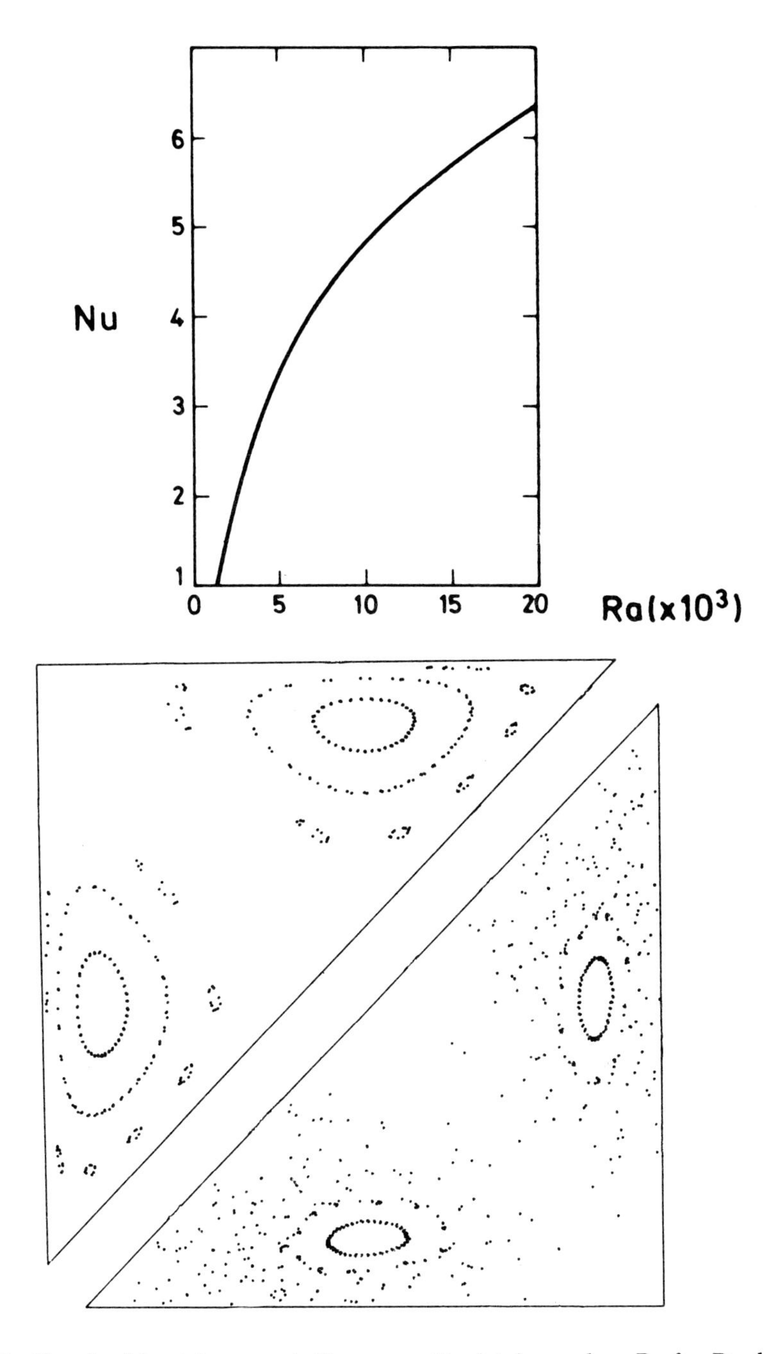

FIGURE 1. Graph of heat transport Nu versus Rayleigh number Ra for Rayleigh-Bénard convection with square planform at $\sigma = 1$. The Poincaré plots show intersections of streamlines of the convection with the midplane $z = 1/2$ for the region $\{0 \leq x \leq 1/2, x \leq y \leq 1-x\}$. The plot at left is for $Ra = 2000$; at right for $Ra = 10000$.

In the laboratory, the natural equivalent is a shallow tank, i.e. one with its horizontal dimensions much greater than its depth. Unfortunately, perfectly regular patterns can be hard or impossible to produce in such tanks, e.g. due to the appearance of defects.

With a view to elucidating these laboratory experiments, direct numerical simulations, employing no-slip boundary conditions, were undertaken, beginning with time-dependent 2-D calculations using a box of size 16×1 in $x - z$ co-ordinates (Arter et al., 1987). Linear theory, assuming periodicity in x, predicts (for a given Rayleigh number Ra and Prandtl number σ) that stable rolls occur for a finite range in wavenumber (Bolton et al., 1986). Simulations (at $Ra = 2000$ and $\sigma = 2.5$) when run for twelve turnover times bore out theory. However, over timescales approaching that for lateral diffusion, the solutions with more extreme wavenumbers developed an instability that eliminated, or added, a roll at a side-wall. The roll closest to either wall is a weak point in the pattern because the vertical velocity, while large on the "fluid" side has to vanish on the no-slip, wall side. When the instability acts to eliminate it, the roll's topology is preserved until its flow is greatly reduced, even though the roll gets very squashed.

Turning to 3-D calculations, in a $16 \times 11.5 \times 1$ box (Arter and Newell, 1988), intrinsically three-dimensional instabilities act to limit the range of wavenumbers still further. One such is the skew-varicose instability: simulations at $\sigma = 2.5$ show how its nonlinear development leads to the local destruction of two of the original rolls, usually consecutively, producing a defect, see Figure 2. The flow topology is seemingly preserved until the first roll disappears. In detail, a defect resembles the result of inserting a single, truncated, roll-like flow perpendicularly into a regular roll pattern. In sufficiently regular patterns, defects propagate to the sides where the component of motion with axis parallel to the wall is annihilated.

At $\sigma = 2.5$ with $Ra < 8500$, for initial conditions consisting of rolls, theory is largely borne out by simulation. The band of stable wave-numbers is slightly reduced by side-wall effects that stimulate the cross-roll instability at the lower wavenumber end, and is increased at the high wavenumber end because the skew-varicose instability has a wavelength parallel to the roll axes that is too long to fit our box. Less regular starting points, or regular, unstable initial conditions at $Ra = 8500$ lead to topologically more convoluted patterns that do not seem always to settle, in agreement with experiment.

At $Ra = 8500$, for example, irregular time-dependence persists for at least $2.5t_{Dh}$, where t_{Dh} is the lateral diffusion timescale. After initial transients, little happens on the turnover timescale: the computations confirm that turbulence is truly topological, i.e. it is determined by changes not within rolls or cells, but in the overall pattern.

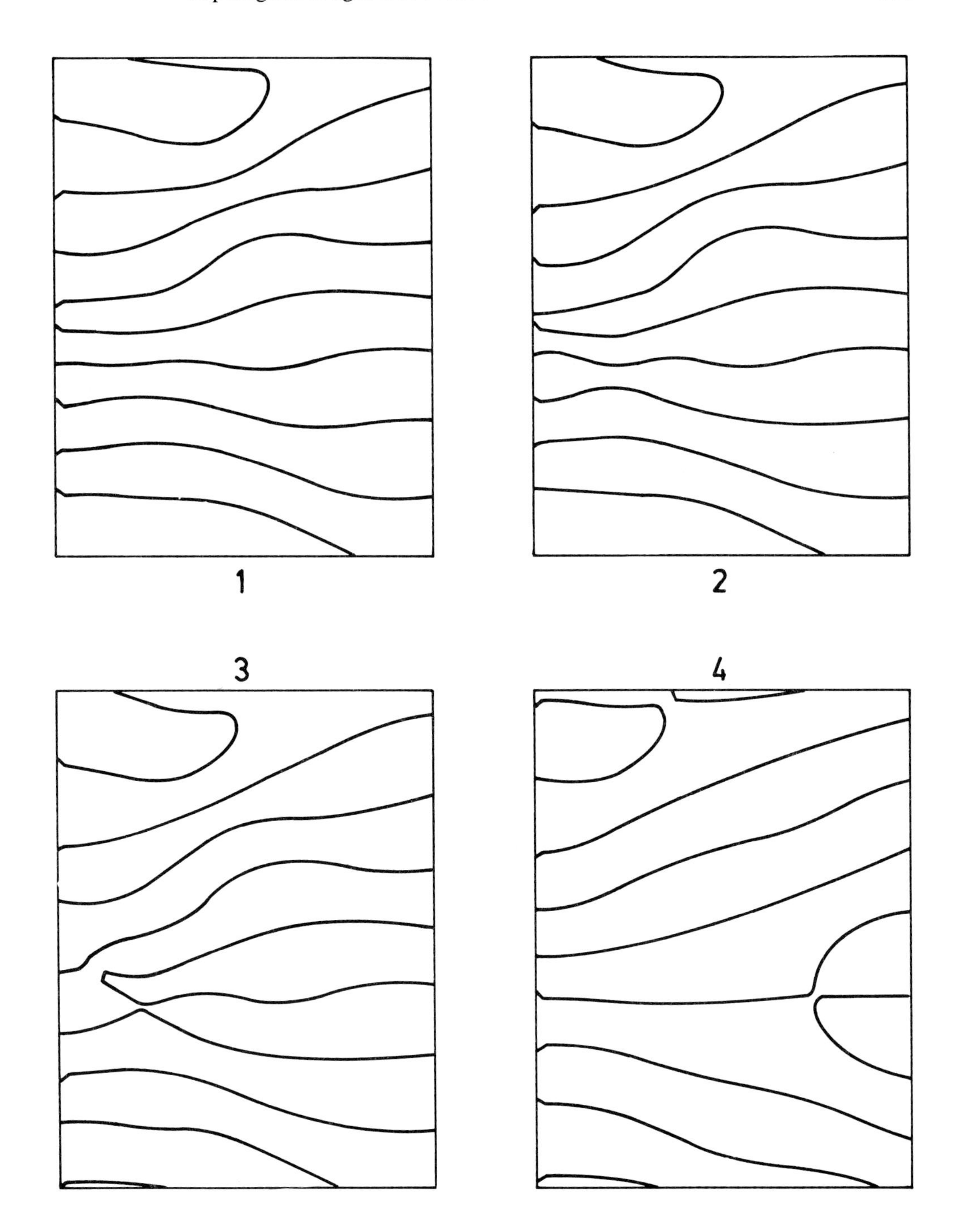

FIGURE 2. A sequence as time varies for $Ra = 8500, \sigma = 2.5$ showing the formation and initial motion of a defect in a $16 \times 11.5 \times 1$ tank. The contours join zeroes of the u_y- field for the sub-region $\{0 \leq x \leq 8, 3 \leq y \leq 13\}$ of the plane $z = 2/3$.

For example, over a time interval $O(t_{Dh})$ the flow field is dominated by foci, located in the corners of the box (viewed from above) that "push out" semi-circular patterns of rolls. The interactions between rolls emerging from different corners is the dynamic behind many pattern changes. While there is still some evidence for a preferred wavenumber, it must clearly be strongly influenced by the walls. At times, however, jostling cells appear to predominate. It remains unclear as to how well amplitude equations (e.g. Cross and Newell, 1984) describe the behaviour at such high Ra.

3 MAGNETOCONVECTION

3.1 Kinematics

The formal statement of the problem is to solve, for the magnetic field $\boldsymbol{B}$, the magnetic induction equation

$$\frac{\partial \boldsymbol{B}}{\partial t} = Rm\, curl(\boldsymbol{u} \times \boldsymbol{B}) + \nabla^2 \boldsymbol{B}, \tag{8}$$

where the magnetic Reynolds' number Rm is a dimensionless measure of the strength of the given flow pattern $\boldsymbol{u}(\boldsymbol{x})$. Initially $\boldsymbol{B} = (1,0,0)$, and the reflecting boundary conditions conserve the horizontal (x) magnetic flux (Arter, 1983b; Galloway and Proctor, 1983).

By taking $\boldsymbol{u}$ of the form (1) with

$$F(x, y, z) = (cos\, \pi x + cos\, \pi y + 1/2 cos\, \pi x\, cos\, \pi y) sin\, \pi z, \tag{9}$$

the flow is made to have a topological asymmetry (Drobyshevski and Yuferev, 1974): regions of rising flow ($u_z > 0$), e.g. centred at $x = y = 0$, are surrounded by falling fluid. Since F is separable, $\boldsymbol{u}$ has closed streamlines (Section 2.1).

As time increases in the numerical simulation of (8), the initially straight field-lines are dragged along with the flow, if $Rm >> 1$. An effect called flux enhancement occurs, where the flux is wound up into spirals, so that in the final steady state the positive x-component of field may be increased by a factor $O(\sqrt{Rm})$, with a compensating negative horizontal flux. The topological asymmetry is inessential for this effect (Arter, 1985b): it does however enable the near-complete vertical separation of regions of oppositely signed flux. Reconnection occurs in the region of falling flow, producing loops (due to symmetry) of flux, plus almost straight lines of positive field that are easily carried down. Negative flux occupies most of the top of the layer, while the positive flux is concentrated into a tube at the bottom, where the maximum value of the field is found, scaling as $Rm^{3/2}$.

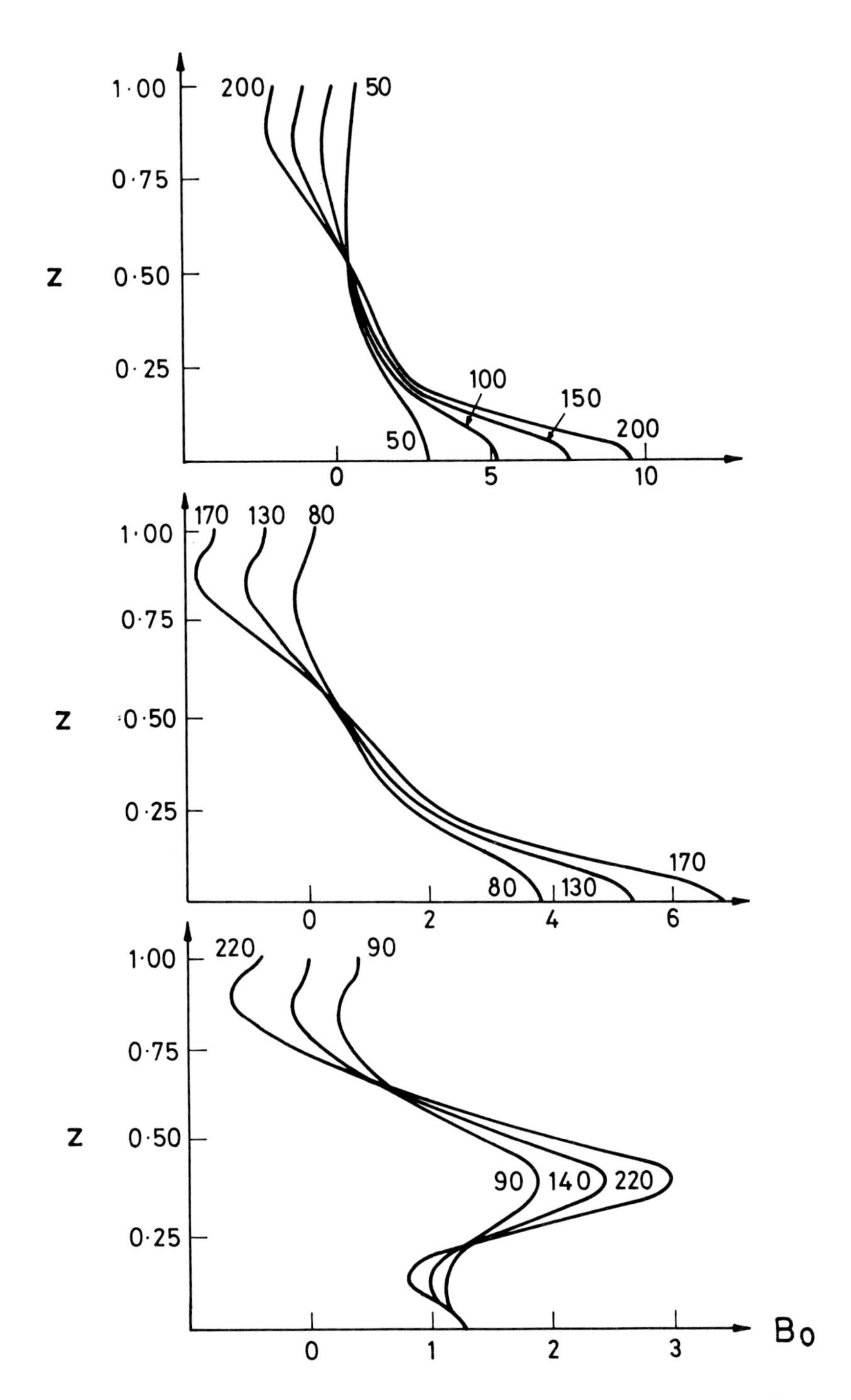

FIGURE 3. Graphs of B_o, horizontally averaged B_x, against z at steady state, for the flows described (top) by (9), (middle) by (10) with $c = 1/6$, and (bottom) by (10) with $c = 1$.

The topological complexity of $\boldsymbol{u}$ was increased by adding to F in (9), taking

$$F(x,y,z) = (\cos\pi x + \cos\pi y + 1/2\cos\pi x\cos\pi y)\sin\pi z - c[1/8(\cos 2\pi x + \cos 2\pi y) + \cos\pi x\cos\pi y]\sin 2\pi z. \quad (10)$$

Taking $c = 1/6$ opens out the streamlines, that to a first approximation now lie on surfaces. Nonetheless, the steady-state flux distribution is little changed, and the horizontally averaged flux distribution $B_o(z)$ is hardly altered, see Figure 3. Choosing $c = 1$ destroys nearly all the surfaces, but the qualitative properties of the equilibrium remain unchanged. However, flux now concentrates away from the bottom of the layer, since the stagnation point associated with flux-tube formation has moved up. Generally, the location and nature of the stagnation points of $\boldsymbol{u}$ is the key to where flux is concentrated.

3.2 Dynamics

For the last series of calculations, (8) is solved coupled to the convection equations, in a square planform. The dynamically determined $\boldsymbol{u}$ lack topological asymmetry. Figure 4 shows profiles of B_o obtained using different assumptions. At the top is the steady-state B_o for a fully coupled system, where $\boldsymbol{u}$ is driven by buoyancy forces and $\boldsymbol{B}$ acts back on $\boldsymbol{u}$ via the Lorentz force. Switching off the magnetic force leads to the middle set of profiles. Finally, the bottom graph indicates that most of the higher order spatial components of $\boldsymbol{u}$ are unimportant, since it was obtained kinematically, using a flow of form (1) with

$$F = \cos\pi x\cos\pi y\sin\pi z - 1/8(\cos 2\pi x + \cos 2\pi y)\sin 2\pi z. \quad (11)$$

The indications are that magnetic fluxes found kinematically, using simple $\boldsymbol{u}(\boldsymbol{x})$, may be very good approximations to dynamically consistent fields. It is necessary to note however, that if $\boldsymbol{B}$ is too strong, the convection may revert to rolls with axes aligned along the field.

4 CONCLUSION

The review has highlighted areas of magnetoconvection where the author's work has shown topology to play an important role – and some areas, where surprisingly, it seems to be unimportant. Flow topology ***is*** important for the generation of a large scale near-separation of flux of opposite signs, at least for parameters relevant to the solar convection. The turbulence seen in the laboratory, when convection occurs in shallow tanks at moderate Rayleigh number, ***is*** topological in nature.

Whether streamlines are closed or open has little effect on heat transport or magnetic flux concentration, however. This is because these are in some way boundary layer effects. There is one caveat, that the numerical simulations in these instances were performed using idealised, reflecting boundary conditions.

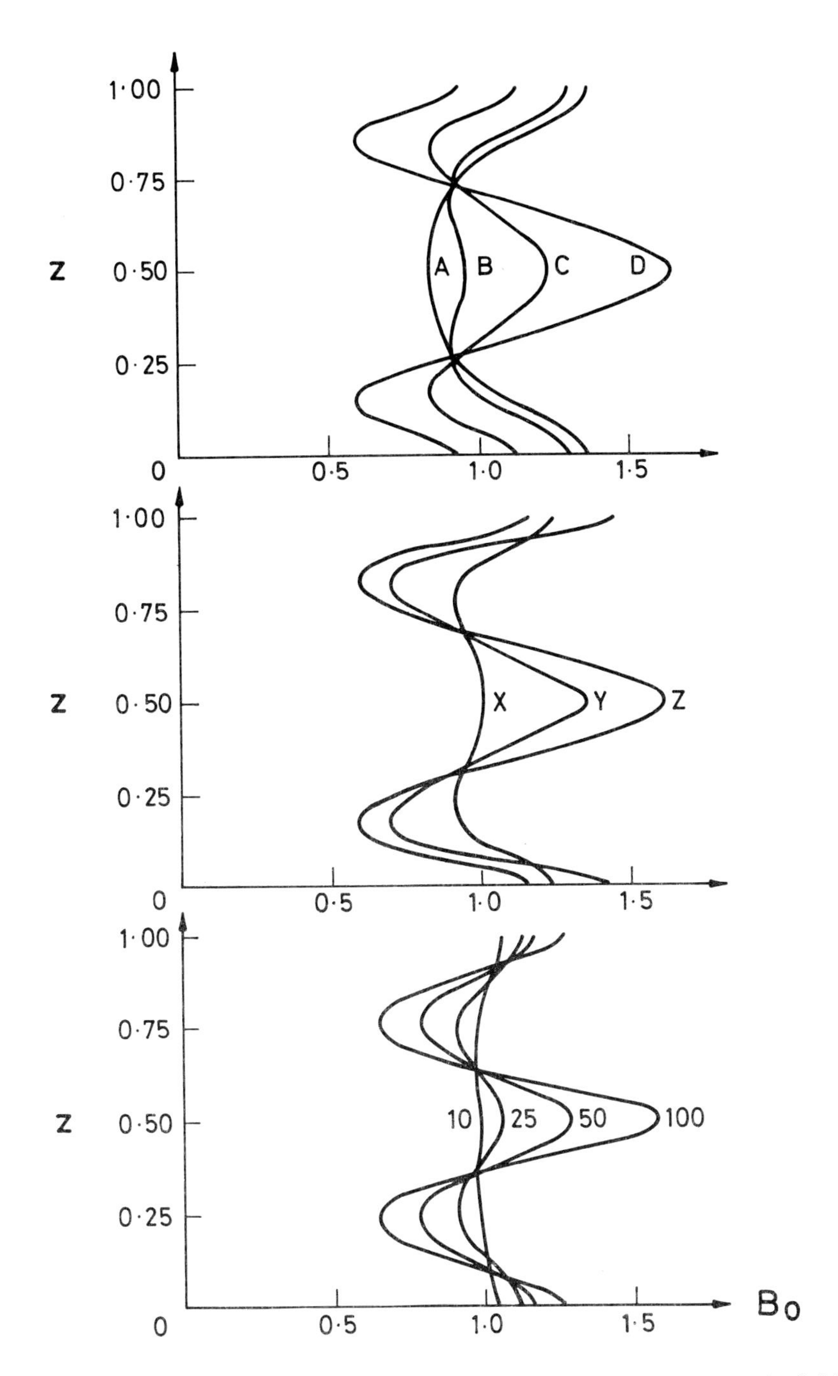

FIGURE 4. Graphs of $B_o(z)$ for (top) dynamically active magnetic field, (middle) passive $\boldsymbol{B}$, and (bottom) kinematic $\boldsymbol{B}$ for a model flow described in the text. The curves labelled A,B,C,D correspond to an effective Rm of 19, 27, 31 and 203 respectively; those labelled X,Y and Z to an Rm of 70, 116 and 134 respectively, and those at the bottom are labelled with Rm.

5 REFERENCES

W Arter 1983a "Ergodic streamlines in steady convection", *Phys Lett* **97A**, *171 -174*

W Arter 1983b "Magnetic-flux transport by a convecting layer – topological, geometrical and compressible phenomena", *J Fluid Mech* **132**, *25-48*

W Arter 1985a "Nonlinear Rayleigh-Bénard convection with square planform", *J Fluid Mech* **152**, *391-418*

W Arter 1985b "Magnetic-flux transport by a convecting layer including dynamical effects", *Geophys Astrophys Fluid Dyn* **31**, *311–344*

W Arter, A Bernoff and A C Newell 1987 "Wavenumber selection of convection rolls in a box", *Phys Fluids* **30**, *3840–3842*

W Arter and A C Newell 1988 "Numerical simulation of Rayleigh-Bénard convection in shallow tanks", *Phys Fluids* **31**, *2474-2485*

G D Birkhoff 1927 "Dynamical Systems", *American Math Society*

E W Bolton, F H Busse and R M Clever 1986 "Oscillatory instabilities of convection rolls at intermediate Prandtl numbers", *J Fluid Mech* **164**, *469–485*

S Chandrasekhar 1961 "Hydrodynamic and Hydromagnetic Stability", *Clarendon, Oxford*

M C Cross and A C Newell 1984 "Convection patterns in large aspect ratio systems", *Physica* **10D**, *299–328*

E M Drobyshevski and V S Yuferev 1974 "Topological pumping of magnetic flux by three-dimensional convection", *J Fluid Mech.* **65**, *33–44*

D J Galloway and M R E Proctor 1983 "The kinematics of hexagonal magnetoconvection", *Geophys Astrophys Fluid Dyn* **23**, *109–136*

M R E Proctor and N O Weiss 1982 "Magnetoconvection", *Rep Prog Phys* **45**, *1317–1379*

K J Whiteman 1977 "Invariants and stability in classical mechanics", *Rep Prog Phys* **40**, *1033–1069*

E T Whittaker 1937 "A Treatise on the Analytical Dynamics of Particles and Rigid Bodies", *Cambridge University*

The Nonlinear Break-up of a Twisted Magnetic Layer

FAUSTO CATTANEO[1], TZIHONG CHIUEH[2] & DAVID W. HUGHES[3]

[1]Joint Institute for Laboratory Astrophysics, University of Colorado, Boulder, CO 80309, USA
[2]Department of Astrophysical, Planetary and Atmospheric Sciences, University of Colorado, Boulder, CO 80309, USA
[3]Department of Applied Mathematics and Theoretical Physics, University of Cambridge, Silver Street, Cambridge CB3 9EW, UK

Motivated by problems concerning the storage and subsequent escape of the solar magnetic field we have studied how a magnetic layer embedded in a convectively stable atmosphere evolves due to axisymmetric instablities driven by magnetic buoyancy. The initial equilibrium consists of a toroidal field sheared by a weaker poloidal component. The key result of the paper is that the nature of the instability is greatly affected by the distribution of the poloidal field — of particular significance is the location of the *resonant surface* on which the poloidal field vanishes. A resonant surface close to the interface between the magnetised and field-free fluid leads to the localisation of the instability so that only a fraction of the magnetic region is disrupted by the motions. By contrast, a deeply seated resonant surface leads to the complete disruption of the layer and to the formation of large, helical magnetic fragments whose identity is preserved for long periods of time.

1. INTRODUCTION

One of the key problems of solar magnetohydrodynamics (MHD) is to explain the mechanism by which the large-scale, predominantly toroidal magnetic field escapes from the sun's interior, eventually to appear at the surface as sunspots or smaller magnetic elements. The rapid rise of isolated magnetic flux tubes through the convection zone, together with problems in matching the results of convection-zone dynamo models to the observed solar magnetic field, suggests that the bulk of the toroidal field is not in the convection zone itself but that it may be more readily accommodated in the convectively stable overshoot zone situated beneath the convection zone proper. A full discussion of this matter, along with the appropriate references, may be found in Cattaneo & Hughes (1988) (hereinafter referred to as I).

In I we modelled the escape of a magnetic field from the overshoot zone by considering the instabilities of a magnetic layer embedded in a convectively stable atmosphere. To simplify matters we considered the Cartesian analogue of the spherical problem with the y axis identified with the azimuthal direction, the equilibrium state being piecewise polytropic with a uniform "toroidal" field $(0, B, 0)$ sandwiched between two non-magnetic regions. The magnetic field, by virtue of its pressure, supports denser fluid above and consequently Rayleigh-Taylor type instabilities can ensue. We restricted our attention to axisymmetric $(\partial_y \equiv 0)$ modes which, for interfacial instabilities, are a good approximation to the most rapidly growing 3-dimensional modes (see I for a full discussion of this rather subtle issue). As a consequence the magnetic tension was identically zero and the field behaved essentially as a scalar. The most important result of I was that the shear flow arising from the primary Rayleigh-Taylor instabilities leads to secondary Kelvin-Helmholtz instabilities which wrap the gas into regions of strong vorticity and subsequently rapidly destroy the coherence of the magnetic layer (see, for example, Figure 8 of I).

This paper describes a natural extension of the work of I to take account of the fact that the field in the overshoot zone, although predominantly toroidal, will also have a weak poloidal component. The key question that we wish to address by this study is whether the addition of a poloidal ingredient leads to a notable change in the character of the instability. As in I we again work in Cartesian geometry, where a meridional slice of an axisymmetric magnetic field at the equator is transformed into a horizontal field whose direction changes with depth, $(B_x(z), B_y(z), 0)$ say. We again consider only axisymmetric motions although it should be noted that since the field is no longer unidirectional these no longer simply interchange magnetic field lines.

The addition of a poloidal field brings about a fundamental change in the nature of the problem with the magnetic tension now coming into play. The shearing of the field imparts a certain rigidity to the magnetised region, thereby hampering the development of the instability, the precise influence of the shear depending crucially on the manner in which it is distributed. Of particular significance in this field configuration is the possible existence of a so-called *resonant surface* on which the poloidal field vanishes — in axisymmetric geometry the field lines on the surface are purely azimuthal, in our Cartesian model they are purely in the y direction. As we shall see, the location of the resonant surface plays a crucial role in the evolution of the instability, determining not only the scale of the escaping field but also the structure of what remains of the magnetic layer after the instability has occurred.

The importance of resonant surfaces in MHD has long been recognised. For instance, resistive MHD instabilities are almost all related to the presence of resonant surfaces near which resistivity plays a crucial role and field lines reconnection occurs (see the review by White 1983). In ideal (diffusionless) MHD resonant surfaces again can become of importance if the instability is weak (Rosenbluth, Dagazian & Rutherford 1973).

In this paper we shall consider the results of just two numerical simulations in order to illustrate the very different nonlinear behaviour arising from equilibria with resonant surfaces at different depths. A much fuller account of this work, including detailed linear theory and further nonlinear simulations (with pictures in glorious technicolour), is to appear shortly (Cattaneo, Chiueh & Hughes 1990).

2. MATHEMATICAL FORMULATION

We consider the development of axisymmetric perturbations of an equilibrium state in which a magnetic layer supports an overlying layer of field-free gas. Both regions in themselves are convectively stable with the temperature and total pressure (gas + magnetic) everywhere continuous. The sole source of potential energy available for instability is thus the density jump, arising from the jump in gas pressure, at the magnetic interface. The equilibrium magnetic field is made up of a toroidal component sheared by a weaker poloidal ingredient so that the resulting magnetic field is predominantly azimuthal. As explained in the introduction, the analysis is simplified by adopting a Cartesian model where the coordinate axis y is identified with the azimuthal direction and where the restriction to axisymmetry is translated into the requirement that the solutions be independent of y.

The computational domain extends from $z = 0$ (top) to $z = d$ (bottom) and the magnetic field is confined initially below some depth z_t. In the region $z \geq z_t$ the equilibrium magnetic field $\mathbf{B}_e$ is horizontal and has constant magnitude B_0 but its orientation varies with depth so that it can be written as

$$\mathbf{B}_e = \begin{cases} 0 & 0 \leq z < z_t \\ B_0(B_x, \sqrt{1 - B_x^2}, 0) & z_t \leq z \leq 1. \end{cases} \tag{1}$$

In (1) the x and y components of $\mathbf{B}$ are identified with the poloidal and toroidal components respectively. In this paper we shall consider only the following quadratic distribution for the poloidal field:

$$B_x = \delta B \frac{(\xi - \xi_r)|\xi - \xi_r|}{(1-\xi_r)^2 + \xi_r^2}, \tag{2}$$

where $\xi = (z - z_t)/(1 - z_t)$ and δB measures the total excursion of B_x across the magnetic layer. The resonant surface on which B_x vanishes is given by $z = z_r$ ($\xi = \xi_r$). It is important to realise that in ideal MHD the topology of the resonant surface is preserved by axisymmetric (y-independent) motions and that, therefore, the regions above and below it cannot be brought into contact without reconnection of the field lines. As we shall see, this property has a profound effect on the development of the instability. Although the existence of a resonant surface in the Cartesian model might seem rather arbitrary, since a rotation of the coordinates about the z-axis swaps the "poloidal" and "toroidal" components, the uniqueness of the resonant surface is ensured by noting that the Cartesian model is constructed as a local representation of the system in spherical geometry where the poloidal and toroidal components are uniquely specified.

The fluid is assumed to be a perfect gas with constant shear viscosity μ, thermal conductivity K, magnetic diffusivity η and principal specific heats C_p and C_v. Initially the atmosphere is piecewise polytropic with a temperature distribution of the form $T = T_0 + \Delta z$. Choosing d as the unit of length and $d/\sqrt{RT_0}$ as the unit of time, where $R = C_p - C_v$, the evolution equations can be written as:

$$p = \rho T, \tag{3}$$

$$\partial_t \rho + \nabla \cdot \rho \mathbf{u} = 0, \tag{4}$$

$$\partial_t B + \nabla \cdot (B\mathbf{u} - v\mathbf{B}) = \tau C_k \nabla^2 B, \tag{5}$$

$$\partial_t A + \mathbf{u} \cdot \nabla A = \tau C_k \nabla^2 A, \tag{6}$$

$$\partial_t \rho \mathbf{u} + \nabla \cdot \rho \mathbf{u}\mathbf{u} = -\nabla p + (m+1)\theta \rho \hat{\mathbf{z}} + C_k \sigma (\nabla^2 \mathbf{u} + \frac{1}{3}\nabla(\nabla \cdot \mathbf{u})) - \frac{1}{\beta}\left(\nabla^2 A \nabla A + \frac{1}{2}\nabla B^2 + \nabla A \times \nabla B\right), \tag{7}$$

$$\rho(\partial_t T + \mathbf{u} \cdot \nabla T) + (\gamma - 1) p \nabla \cdot \mathbf{u} = \gamma C_k \nabla^2 T + (\gamma - 1) C_k (\sigma \phi_{ij} \partial_i u_j + \tau (\nabla \times \mathbf{B})^2 / \beta), \tag{8}$$

where $\phi_{ij} = \partial_i u_j + \partial_j u_i - \frac{2}{3}\delta_{ij} \nabla \cdot \mathbf{u}$, $\mathbf{B} = (-\partial_z A, B, \partial_x A)$ and $\mathbf{u} = (u, v, w)$.

The six dimensionless parameters are defined by:

$$\sigma = \frac{\mu C_p}{K}, \qquad \tau = \frac{\eta \rho_0 C_p}{K}, \qquad \beta = \frac{\mu_0 p_0}{B_0^2},$$

$$m = (\frac{g}{R\Delta} - 1), \qquad \theta = \frac{\Delta d}{T_0}, \qquad C_k = \frac{K}{\rho_0 C_p d \sqrt{RT_0}}.$$

We assume that the fields and motions are periodic in the x direction and adopt the following conditions on the horizontal boundaries:

$$T = 1 \,,\ A = 0 \quad \text{at} \quad z = 0 \quad \text{and} \quad T = 1 + \theta \,,\ A = A_0 \quad \text{at} \quad z = 1;$$

$$w = \partial_z u = \partial_z v = \partial_z B = 0 \quad \text{at} \quad z = 0, 1;$$

where the constant A_0 has a simple dependence on the excursion δB. These boundary conditions correspond to impenetrable, stress-free, perfectly conducting walls (thermally). It is easy to show that the conditions on A and B imply that both the toroidal and poloidal fluxes are conserved.

The numerical techniques used to solve (3)-(8) are described in detail in I.

3. RESULTS

In this section we study the nonlinear evolution of the instability and the effects of the resulting large-amplitude motions on the magnetic layer. We consider two particular initial profiles of the form (2). In case (a) $\xi_r = 0.05$, corresponding to a resonant surface very close to the top of the magnetic region; in case (b) $\xi_r = 0.95$ and the resonant layer lies just above the lower boundary.

As argued in I, and discussed briefly in the introduction, the nonlinear evolution of a purely toroidal field is controlled to a large extent by two factors; the tendency for light, magnetised fluid to rise, thereby contributing positive buoyancy work, and by the formation of vortices by a secondary Kelvin-Helmholtz instabilty whose interactions dominate the flow in the later stages of the instability. In order to understand the evolution in the present situation we must consider how each factor might be affected by a varying poloidal field.

Clearly the impact on the buoyancy force is minimal since in a compressible fluid buoyancy is a pressure effect and therefore depends only on the modulus of $\mathbf{B}$ — on the other hand the effects on the secondary instability are profound. When the field is purely toroidal neighbouring fluid elements transfer momentum amongst each other by viscous stresses (neglecting the stratification) and consequently no long range communication exists between distant regions of fluid on the dynamical timescale ($\ll$ viscous timescale). Two simple consequences of this fact are that the instability develops on the smallest horizontal scale compatible with dissipation and that the (velocity) shear at the interface between magnetised and unmagnetised fluid becomes unstable to the secondary Kelvin-Helmholtz instability. The only factor hindering the secondary instability is the stable stratification (Chandrasekhar 1961) whose effectiveness, however, can be greatly reduced if the Prandtl number is small (Zhan 1983 and references therein). A poloidal field on the other hand provides a very efficient mechanism for momentum transfer since Alfvén waves can now propagate across the layer. Distant fluid elements can thus be coupled with an effectiveness which depends on the (poloidal) field strength. A distribution of poloidal field can thus be regarded as a distribution in effective coupling between fluid particles. In this context the resonant surface is of special importance since in its neighbourhood the Alfvén crossing time (in the x–z plane) is large and hence the fluid is virtually uncoupled. Furthermore, so long as the topology of the resonant surface is preserved, fluid elements on opposite sides of it cannot be coupled in the above sense. It should be noted that the coupling, being mediated by the field lines, is not between physical locations but between actual Lagrangian fluid elements and therefore evolves with the instability.

Bearing these considerations in mind we inspect the numerical solutions for cases (*a*) and (*b*). The time evolution for these two cases is captured by Figures 1 and 2 which show density plots of the toroidal field intensity at four different times. It is apparent that the two cases differ strikingly in three important respects: (i) in the spatial scales of the instability, (ii) in the effectiveness of the instability at disrupting the magnetic layer, and (iii) in the structure of the magnetic "fragments" produced by the instability.

When the resonant layer is near the interface (case (*a*)), the instability develops on small horizontal scales and over a region of limited vertical extent. The penetration by the motions into the lower part of the layer is hindered by the increasing strength of the poloidal field and thus part of the layer escapes disruption. In Figure 1d remnants of the initial layer are visible, stabilised against further disturbances by the combined action of the poloidal field and of the lower boundary. When the resonant layer is near the lower boundary, as it is in case (*b*), the instability is initially more sluggish than in (*a*) because of the strong poloidal field at the top of the magnetic region. However, it develops on larger scales (in case (*a*) between 8 and 9 bumps can be distinguished at the magnetic interface whereas only 2 are present in case (*b*)) that reach down to the resonant surface and lead to the complete disruption of the layer. Very little field remains in the lower part of the domain in Figure 2d.

A useful quantity to make some of these notions more quantitative is afforded by $F[z;t]$, defined to be the fraction of the initial toroidal flux below some depth z, namely

$$F[z;t] = \frac{\int_z^1 B_y(\mathbf{x},t)d\mathbf{x}}{\int_z^1 B_y(\mathbf{x},0)d\mathbf{x}}. \tag{9}$$

Time = 2.27

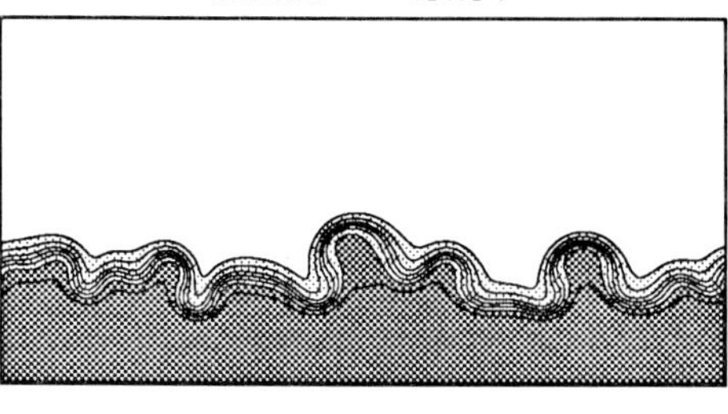

Time = 3.40

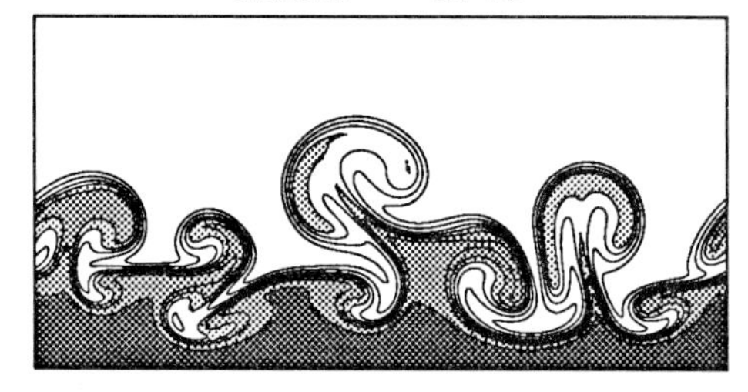

Time = 4.90

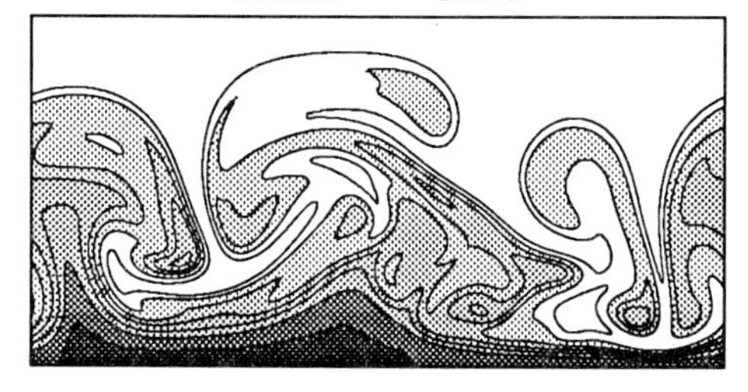

Time = 7.91

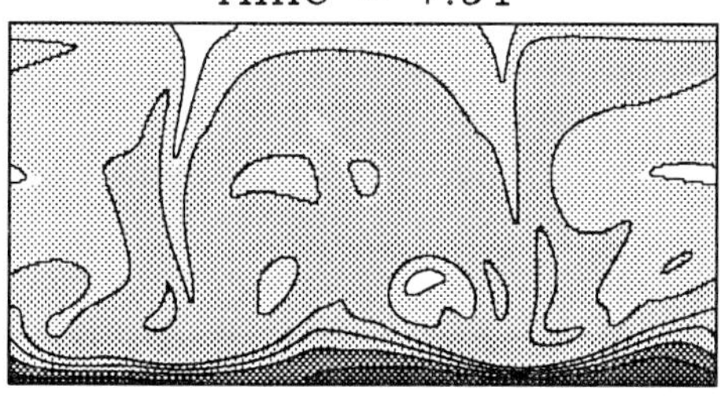

FIGURE 1. Grey-scale plot of the evolution of the toroidal field. The shading is uniform between adjacent contour lines with darker shading corresponding to more intense fields. The initial position of the resonant layer is $\xi_r = 0.05$.

Time = 2.27

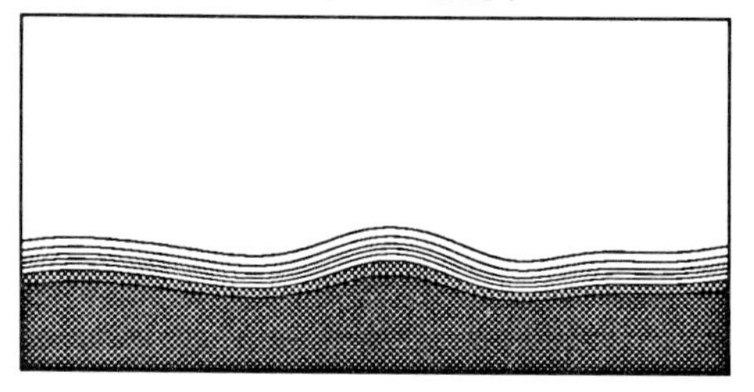

Time = 3.40

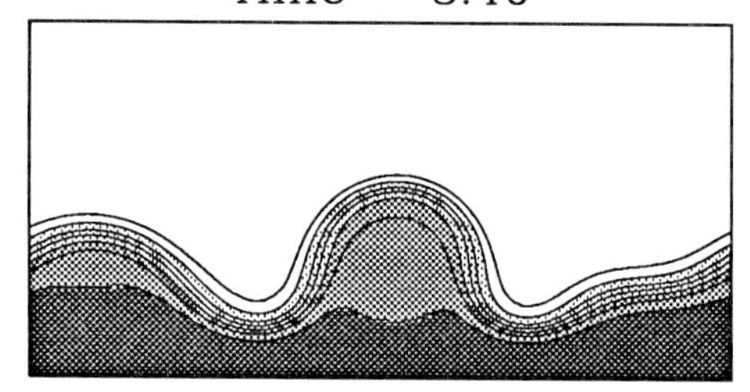

Time = 4.89

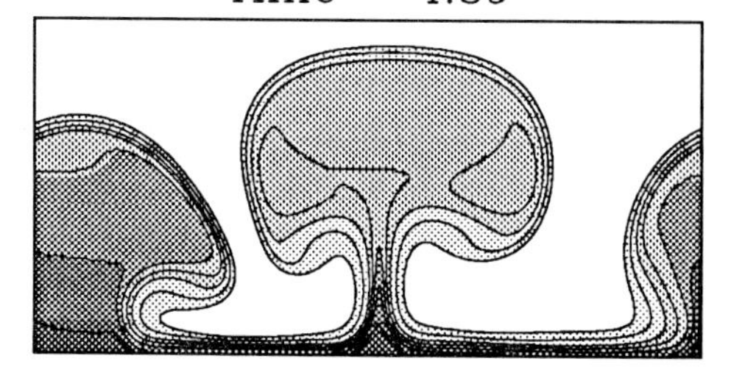

Time = 7.79

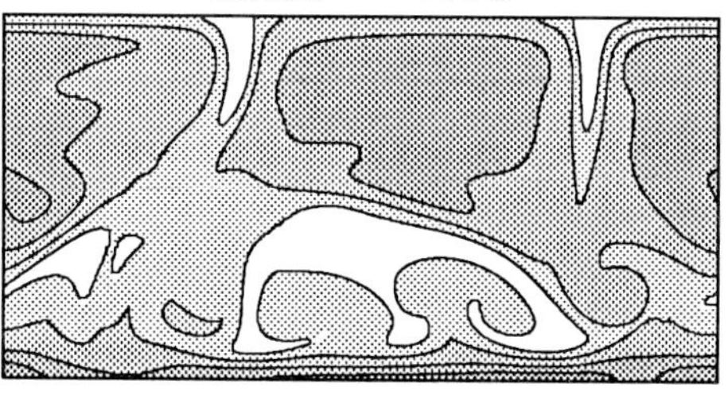

FIGURE 2. As above but with $\xi_r = 0.95$.
(See also Colour Plates)

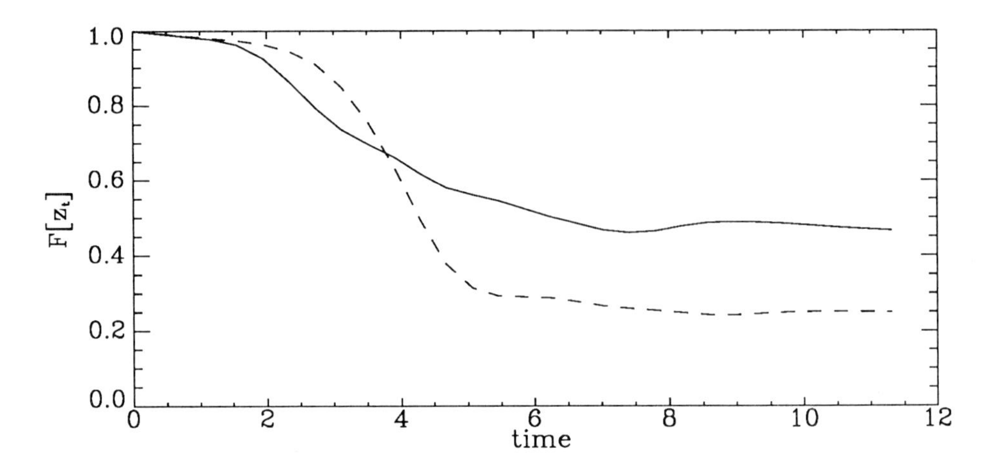

FIGURE 3. Fraction of the initial toroidal flux below z_t as a function of time. The solid line corresponds to case (a), the dashed line to case (b).

The graphs of F for $z = z_t$ shown in Figure 3 give a measure of the long term resilience of the layer against the instability. By the time most of the dynamically interesting phases have occurred, case (b) (resonance near the lower boundary) has about 20% of the initial toroidal flux in the region below z_t, whereas in case (a) over 50% of the initial flux still remains. The picture that emerges from these observations is one where the resonant layer plays a dual role; on the one hand layers with a resonant surface near the magnetic interface are more easily destabilised (in the sense that a weaker field can lead to instability), on the other hand for such cases the motions are confined to a shallow region near the interface and the disruption of the magnetic layer is limited. By comparison, a layer with a deep resonant surface is harder to destabilise but once instability is possible it leads to the complete disruption of the magnetic region.

The magnetic fragments resulting from the instability also differ greatly between the two cases. Besides the obvious difference in size we notice a more fundamental difference in evolution. In case (a) the small corrugations formed at the interface by the instability grow into the familiar mushrooms which become unstable to the secondary Kelvin-Helmholtz instability, leading to a very effective mixing of the ejected flux. In Figure 1d it is almost impossible to relate different magnetised regions in the upper part of the domain with the mushrooms at an earlier time. The homogenisation of the field occurs, mainly, because the ejected field is only weakly twisted and, hence, field lines can be interchanged efficiently. In case (*b*) the fragments produced by the instability retain their identity for the entire simulation, their enhanced stability arising from the distribution of poloidal field within the fragments themselves. Inspection of Figures 1 and 2 shows that when the initial gradient of B_x is positive (increasing downwards) the fragments produced tend to have the strongest poloidal field near their centres; contrariwise, when the gradient is negative the strongest poloidal field is found near the fragments' surfaces. Since the secondary instability is caused by the velocity shear between the rising magnetised fragment and the descending unmagnetised fluid a poloidal field concentrated at the interface is most effective at preventing the secondary instability. A measure of this effectiveness is provided by linear theory (Chandrasekhar 1961; however, see Chiueh & Zweibel 1987) which indicates that a velocity difference comparable to the parallel Alfvén speed is necessary for the secondary instablity to succeed. In a way it is as if the magnetic layer and the fragments have swapped stability; a deep resonant surface causes a substantial fraction of the magnetic layer to be destroyed but the fragments produced in the process are very stable to subsequent disruption; on the other hand, for a shallow resonant surface the instability leaves some of the initial layer intact but the fragments produced are strongly affected by further instabilities.

4. CONCLUSIONS

To recap, the key results of the paper are clearly illustrated by Figures 1 and 2 which show how the detailed nature of the evolution is affected by the position of the resonant surface. When the resonant surface is situated close to the magnetic interface the instability is localised to the top of the magnetic region leaving intact a significant portion of the initial layer. The field that does escape is rapidly distorted under the influence of secondary Kelvin-Helmholtz instabilities. On the other hand, when the resonant surface is deep-seated the instability acts over the entire magnetic region. The magnetic fragments that emerge are large and retain their coherence due to their helical structure.

The work in this paper has benefitted from numerous discussions with T. Bogdan, E. de Luca, R. Rosner, T. Tajima, J. Toomre and N. Weiss. It was partially supported by the National Science Foundation through grant ATM8506632, by the National Aeronautics and Space Administration through grants NSG-7511 and NAGW-91 and by Trinity College, Cambridge.

5. REFERENCES

CATTANEO, F. & HUGHES, D.W. 1988 The nonlinear breakup of a magnetic layer: instability to interchange modes. *J. Fluid Mech.* **196**, 323-344.

CATTANEO, F., CHIUEH, T. & HUGHES, D.W. 1990 Buoyancy driven instabilities and the nonlinear breakup of a magnetic layer. *J. Fluid Mech.* (submitted).

CHANDRASEKHAR, S. 1961 *Hydrodynamic and Hydromagnetic Stability.* Clarendon.

CHIUEH, T. & ZWEIBEL, E.G. 1987 The structure and dissipation of forced current sheets in the solar atmosphere. *Astrophys. J.* **317**, 900-917.

ROSENBLUTH, M.N., DAGAZIAN, R.Y. & RUTHERFORD, P.H. 1973 Nonlinear properties of the internal $m = 1$ kink instability in the cylindrical tokamak. *Phys. Fluids* **16**, 1894-1902.

WHITE, R. 1983 Resistive instabilities and field lines reconnection. In *Handbook of plasma physics. Vol I* (eds. A.A. Galeev and R.N. Sudan), pp. 611-676. North-Holland.

ZHAN, J-P. 1983 Instability and mixing processes in upper main sequence stars. In *Astrophysical processes in upper main sequence stars* (eds. A.N. Cox, S. Vauclair and J-P. Zhan), pp. 253-329. Geneva Observatory.

Free Shear Flows: Symmetries and Topological Transitions of the Vorticity Field

ECKART MEIBURG[1] and JUAN C. LASHERAS[2]

[1]Center for Fluid Mechanics. Division of Applied Mathematics
Brown University, Providence, RI 02912. USA

[2]Department of Mechanical Engineering.
University of Southern California, Los Angeles, CA 90089-1453. USA

ABSTRACT
On the basis of three-dimensional vortex dynamics simulations and flow visualization experiments, we analyze plane shear layers and wakes as well as axisymmetric jets with respect to their symmetry features and potential topological transitions of the vorticity field. The plane wake, as opposed to the shear layer, tends to form closed vortex loops. It can furthermore acquire different three-dimensional symmetries depending on the form of the initial perturbations. The jet shares common features with both of the above plane flows, which renders it an interesting object for flow control by means of variable initial conditions.

1 INTRODUCTION

Numerous investigations performed over more than two decades have tried to shed light on the characteristic features of turbulent free shear flows (for reviews see Roshko 1976, Townsend 1979, Cantwell 1981, Ho and Huerre 1984, Liu 1989). Flow visualization, quantitative measurements, and, more recently, numerical simulations have elucidated the two-dimensional aspects of the dominant large-scale structures. It now appears well-established that turbulent plane mixing layers and wakes as well as axisymmetric jets exhibit features which bear a certain similarity to those of corresponding laminar flows, namely spanwise rollers, Karman vortices, and vortex rings, respectively. At the same time, both experiments and simulations show that these coherent structures are far from being perfectly two-dimensional or axisymmetric. This is to be expected, since it is already known that the transition process from laminar to turbulent flow gives rise to three-dimensional effects. Searching for potential ways to actively or passively control the flow, it is natural to ask how strong a link exists between the complex three-dimensional structures observed in the transitional stages and those that dominate the fully turbulent state. For plane boundary layers, Gilbert and Kleiser (1988) have recently demonstrated by means of numerical simulation that structures typical of wall turbulence, such as hairpin vortices and low- and high speed streaks, exist already during the late stages of transition. If one assumes that free shear flows behave similarly, i.e. that the flow structures formed during the transitional process will affect the fully turbulent flow, it is tempting to analyze and classify the emergence of such structures as a function of specific initial or boundary conditions. It is well-known that the dominant features of turbulent flows are best characterized in terms of the vorticity variable, and hence we will try to relate the appearance of new structures to changes in the topology of the vorticity field. We will base our analysis of potential topological changes on our earlier numerical and experimental results for free shear flows (Ashurst and Meiburg 1988, Lasheras and Choi 1988, Meiburg and Lasheras 1988, Meiburg, Lasheras & Martin 1989).

The findings of many researchers in the field demonstrate that mixing layers, wakes, and axisymmetric jets share a number of important features: Their dominant instability modes are known to lead to two-dimensional or axisymmetric concentrations of vorticity, respectively, thus setting up a strong strain field between these emerging vortical structures. The extensional component of this strain in turn amplifies small three-dimensional perturbations, thus leading to the reorientation and stretching of predominantly two-dimensional or axisymmetric vorticity into the streamwise direction. Considering these common ingredients in the flow fields of interest, it seems natural to look for generic features in their transitional and turbulent stages. In the present study, we will investigate ways in which the vorticity fields of the above flow fields can undergo topological changes. We are especially interested in the formation of closed vortex loops through such topological transitions because of their potential significance for mixing enhancement. On the basis of earlier experimental and numerical results, we expect these vortex loops to be generated in planes containing significant amounts of streamwise vorticity as a result of the extensional strain mentioned above. Since the direction of the extensional strain is close to the flow direction, the closed vortex loops will have a self-induced velocity almost perpendicular to the flow direction, so that they can affect the spreading and mixing rates of the flow. In a different context, one spectacular example of the mixing that can be achieved by vortex rings misaligned with the flow direction is the blooming jet examined by Parekh and Reynolds (1987). As a result, the question of topological formation of closed vortex loops and possible generic aspects common to several classes of free shear flows appears not only to be of fundamental interest for the understanding of the evolution of transitional and turbulent flows. It furthermore has the potential of offering some insight into potential active or passive control mechanisms for mixing enhancement.

In the following, we will examine the plane shear layer, the plane wake, and the axisymmetric jet with respect to their potential for topological changes, and especially for the formation of closed vortex loops. We base our investigation on flow visualization experiments as well as on three-dimensional vortex dynamics simulations. After describing the numerical and experimental techniques in more detail (sections 2 and 3), we will first discuss the temporally growing plane shear layer (section 4). Section 5 will demonstrate the effect of introducing a second layer of vorticity of the opposite sign, i.e. the plane wake, and section 6 will discuss the case of the axisymmetric jet in more detail. Some conclusions will be drawn in section 7.

2 EXPERIMENTAL TECHNIQUES

While the flow visualization experiments for shear layer and wake were performed in water, the axisymmetric jet was investigated in a wind tunnel. Detailed descriptions of the experimental facilities can be found in Lasheras, Cho & Maxworthy (1986), Lasheras and Choi (1988), Meiburg and Lasheras (1988), as well as in Meiburg, Lasheras & Martin (1989), here we will limit the discussion to a brief summary of the techniques used.

2.1 Interface Visualization for Shear Layer and Wake

The test rig consists of a water channel in which two horizontal laminar water streams of equal or unequal velocities are produced, initially separated by a flat plate. We introduce a three-dimensional perturbation to this two-dimensional base flow by modulating the trailing edge of the splitter plate in a sinusoidal fashion. In the mixing layer, we utilize a splitter plate with a slightly corrugated trailing edge, whereas in the wake flow we also investigate the flow past a plate with an indented trailing edge (Fig. 1). To track the evolution of the interface separating both streams, the interface is visualized by inducing the fluorescence of fluorescine particles added to the lower stream with a set of spotlights. In this way, a shadow effect can be created by the corrugated interface, thereby showing its three-dimensionality. A large concentration of fluorescine particles results in a solid, opaque effect at the interface. Additional information about the temporal and spatial evolution of the interface is obtained by using laser induced fluorescence (LIF) cross-cuts of the wake.

2.1 Visualization of Co-flowing Jets

Our base flow consists of two laminar co-flowing jets of air. Acoustic forcing of the central jet leads to a sinusoidal variation in the strength of its boundary layer, which in turn causes a strongly periodic axisymmetric roll-up once the two streams have merged. The three-dimensional evolution is triggered by a sinusoidal corrugation of the nozzle lip in the circumferential direction. The nozzle has an average diameter of 5 cm, the corrugation amplitude is 1 mm. While the central jet has typical velocities of 40-50 cm/s, the outer jet, which merely serves to stabilize the flow pattern, typically has a velocity one order of magnitude smaller. The air that is to form the central jet is first guided through a tank of $TiCl_4$ and becomes saturated with $TiCl_4$ vapor. Subsequently, this saturated air enters a pre-chamber where it is mixed with small amounts of water vapor. The reaction $TiCl_4 + H_2O \dashrightarrow Ti_2O_4 + HCl$ results in the generation of submicron Ti_2O_4 particles, which can subsequently be visualized by laser light. Hence in sheets of laser light the central jet will appear as bright fluid within the dark ambient fluid.

3 NUMERICAL TECHNIQUE

The non-divergent nature of the velocity field in incompressible flows, along with the definition of vorticity, allow for a complete description of the kinematics of the flow in the form of the Biot-Savart law. Following the general concepts reviewed by Leonard (1985), vortex filaments are used for the representation of the vorticity field. A detailed description can be found in Ashurst and Meiburg (1988), hence we will only review the main points of the technique here. Each filament is represented by a number of node points along its centerline, through which a cubic spline is fitted to give it a smooth shape. The Biot-Savart law is evaluated assuming a smooth invariant vorticity distribution around the filament centerline. Incorporation of this vorticity distribution into the Biot-Savart law then allows us to obtain the velocity at any position by integrating over the arclength of all filaments in the flowfield. For the numerical simulation, we limit ourselves to the temporally growing problem, i.e., our flow is periodic in the streamwise direction. The Biot-Savart integration is carried out with second order accuracy both in space and in time by employing the predictor-corrector time-stepping scheme and the trapezoidal rule. As the flow develops a three-dimensional structure, the vortex filaments undergo considerable stretching. To maintain an adequate resolution, additional nodes are introduced, based on a criterion involving distance and curvature. Furthermore, the time-step is repeatedly reduced as local acceleration effects increase. The filament core radius decreases as its arclength increases to conserve its total volume.

4 THE PLANE SHEAR LAYER

Initially, we discretize into one layer of vortex filaments only the circulation related to the global velocity jump across the shear layer, thus assuming that the vorticity of the weaker one of the two boundary layers coming off the splitter plate is quickly canceled by part of the vorticity of opposite sign. To duplicate the experiment, we disturb the vorticity layer by one wave in the streamwise and one in the spanwise direction. The nature of these perturbations is such that they displace the filaments out of their initial plane. Fig. 2 shows the concentration of the spanwise vorticity into rollers and the reorientation and stretching of the vorticity residing in the braid region in between them. In agreement with the analysis by Lin and Corcos (1984), the final state of our simulation shows concentrated streamwise vortex tubes of alternating sign. If we analyze a schematic sketch of the vorticity field at this stage with respect to potential topological changes (Fig. 3), we recognize that a merging of the streamwise vortex tubes with the spanwise ones is likely. However, the formation of closed vortex loops is not possible, as none of the vortex lines have acquired a direction opposite to their initial one. There appears to be no likely mechanism that could achieve this in the present flowfield. Even two- or three-dimensional subharmonic perturbations leading to effects such as pairing or fractional pairing (Browand and Troutt 1985) do not appear to be able to promote the formation of closed vortex loops.

When analyzing the symmetry properties of the emerging flow, we have to differentiate between those of the velocity and the vorticity field, respectively. The two are related, of course, as the vorticity is obtained by differentiating the velocity field. There exists a three-dimensional symmetry of the velocity and vorticity fields with respect to those points at which initially both the streamwise and the spanwise perturbations had zero values. In addition, velocity and vorticity fields are symmetric with respect to the planes y=const. through the center between two streamwise vortices. However, when taking the direction of the vorticity into account, these symmetries might 5more properly be called antisymmetries.

If we include a second layer of vorticity of opposite sign and half the strength to also model the effect of the weaker boundary layer (Fig. 4), we obtain better agreement with the flow visualization experiment (Fig. 5). We furthermore immediately notice that the three-dimensional point symmetry of the velocity and vorticity fields is lost. The symmetry with respect to a plane y=const., however, remains. It is of interest to note that the change of the symmetry properties of the flow due to the inclusion of the second sign of vorticity appears to be very similar to the one we would obtain if we looked at spatially growing flows instead of the temporal problem.

5 THE PLANE WAKE

We now start the simulation with two equally strong vorticity layers of opposite signs. The evolution resulting from a streamwise and a spanwise perturbation wave displacing the filaments out of their own plane is shown in Fig. 6a and described in detail in Meiburg and Lasheras (1988). This numerical simulation corresponds to the flow past the splitter plate with the corrugated trailing edge (Fig. 7a). Again the Kelvin-Helmholtz instability leads to the formation of spanwise rollers with a strong strain field in between them. However, these Karman vortices are staggered now, and we observe an interaction of the evolving streamwise braid vorticity with the rollers of the opposite layer. This interaction represents the key difference between the plane shear layer and the wake, as it provides vorticity of the correct sign to complete the formation of a closed vortex loop (Fig. 8a). As we had shown in our earlier work (Meiburg and Lasheras 1988, 1989), a different configuration of closed vortex loops can be generated by imposing a horizontal spanwise perturbation instead of a vertical one, i.e., by indenting the trailing edge of the splitter plate instead of corrugating it (Figs. 6b, 7b, 8b). We thus notice that a strong strain field alone is not enough to cause the topological transition leading to the formation of closed vortex loops; presence of vorticity of the opposite sign represents another necessary ingredient.

With respect to the symmetry properties of the flow we notice for both of the above modes that planes at the center between two neighboring streamwise vortices remain symmetry planes, in line with our findings for the shear layer with both signs of vorticity. In addition, however, we now observe a further symmetry relationship between the upper and the lower vorticity layer. For the case of a vertical perturbation, this can best be described in the following way: if we flip the upper vorticity layer over, shift it half a wavelength in the streamwise direction and half a wavelength in the spanwise direction, then it becomes identical to the lower layer. For the case of the horizontal vorticity perturbation, the situation is slightly different: flipping the upper layer over and shifting it half a wavelength in the streamwise direction is enough to render it identical to the lower layer. These symmetry relationships are a consequence of the equal strengths of the two vorticity layers, which is why we did not observe a similar symmetry for the case of two unequal vorticity layers. They furthermore demonstrate how different initial perturbations to the two-dimensional base flow can establish three-dimensional flow fields of different symmetries.

6 THE AXISYMMETRIC JET

The roll-up of a single axisymmetric layer of vorticity subjected to a wavy streamwise perturbation as well as a corrugation in the circumferential direction is shown in Figs. 9 and 10. Details of the numerical simulation and the experiment are given in Meiburg, Lasheras & Martin (1989). There are similarities to the plane shear layer case in that strong streamwise vorticity is generated in the strain field between the concentrated vortices forming as a result of the primary jet instability. However, from the point of view of potential topological changes, the situation of the jet is now different from that of the shear layer or the wake. On the one hand, we have only one axisymmetric shear layer, on the other hand, however, a cut along a longitudinal plane containing the jet axis shows both signs of vorticity. We notice that, due to the axisymmetric nature of the problem, the radius of the highly strained braid region varies in the streamwise direction. At its smallest circumference, the legs of the streamwise vortex pair come quite close to each other, so that the 'pinching off' of a closed vortex loop does not appear to be as remote a possibility as for the plane shear layer having only one sign of vorticity. A cut-and connect event (e.g. Melander, 1989) involving the two nearly anti-parallel legs of the streamwise vortex pair would be able to create such a closed vortex loop involving both braid and ring vorticity. In addition, it seems possible that for strong initial perturbations, e.g a highly elliptical jet, opposite sides of the jet might start to interact and play a role similar to that of the two signs of vorticity in the plane wake, thus leading to closed vortex loops. This possibility still has to be investigated. The only symmetry property the jet exhibits is that of rotational symmetry around the jet axis.

A situation similar to that of the plane wake should emerge for the case of the co-flowing jet, if the inner and outer stream have equal velocities and the boundary layers emanating from the jet nozzle are thin compared to the jet diameter. We are currently analyzing this situation, and the results will be published elsewhere.

7 CONCLUSIONS

On the basis of flow visualization experiments and three-dimensional vortex dynamics simulations we have studied various free shear flows with respect to potential changes in the topology of their vorticity field, especially the formation of closed vortex loops. By comparing plane shear layers and wakes, we observe that in addition to a strain field, which reorients the perturbed two-dimensional vorticity into streamwise one, a necessary ingredient for the formation of closed vortex loops is the presence of vorticity of the opposite sign. The plane wake furthermore represents an example of a nominally two-dimensional flow that can acquire three-dimensional states of different symmetry as a function of different initial perturbations. The single axisymmetric shear layer behaves similarly to the plane shear layer in some ways. However, the axisymmetric nature of the extensional strain field results in a configuration that appears to be more favorable towards the formation of closed vortex loops than the plane shear layer. It hence appears possible that by carefully selecting the flow conditions such as the nozzle shape, one can modify the flow evolution considerably. Two co-flowing jets of equal strengths are expected to exhibit many of the same features observed in the plane wake.

This work is supported by the National Science Foundation under grant #MSM-8809438 (to EM), by DARPA under URI contract #N00014-86-K0754 to Brown University and by United Technologies Corporation (to JCL). The San Diego Supercomputer Center is providing computing time on its CRAY-X/MP.

8 REFERENCES

ASHURST, W.T. and MEIBURG, E. 1988 Three-dimensional shear layers via vortex dynamics. J. Fluid Mech. 189, p. 87.

BROWAND, F.K. and TROUTT, T.R. 1985 The turbulent mixing layer: geometry of large vortices. J. Fluid Mech. 158, p. 489.

CANTWELL, B.J. 1981 Organized motion in turbulent flow. Ann. Rev. Fluid Mech. 13, p.457.

GILBERT, N. and KLEISER, L. 1988 Near-wall phenomena in transition to turbulence. To appear in Proceedings of the International Seminar on Near-Wall Turbulence, Dubrovnik, Yugoslavia, Hemisphere Press.

HO, C.M. and HUERRE, P. 1984 Perturbed free shear layers. Ann. Rev. Fluid Mech. 16, p. 365.

LASHERAS, J.C., CHO, J.S. and MAXWORTHY, T. 1986 On the origin and evolution of streamwise vortical structures in a plane free shear-layer. J. Fluid Mech. 172, p. 231.

LASHERAS, J.C. and CHOI, H. 1988 Three-dimensional instability of a plane, free shear layer: an experimental study of the formation and evolution of streamwise vortices. J. Fluid Mech. 189, p. 53.

LEONARD, A. 1985 Computing three-dimensional incompressible flows with vortex elements. Ann. Rev. Fluid Mech. 17, p. 523.

LIN, S.J. and CORCOS, G.M. 1984 The mixing layer: deterministic models of a turbulent flow. Part 3: The effect of plane strain on the dynamics of streamwise vortices. J. Fluid Mech. 141, p. 139.

LIU, J.T.C. 1989 Coherent structures in transitional and turbulent free shear flows. Ann. Rev. Fluid Mech. 21, p.285.

MEIBURG, E., LASHERAS, J.C. and ASHURST, W.T. 1987 Topology of the vorticity field in three-dimensional shear layers and wakes. Fluid Dynamics Research 3, p. 140.

MEIBURG, E. and LASHERAS, J.C. 1988 Experimental and numerical investigation of the three-dimensional transition in plane wakes. J. Fluid Mech. 190, p.1.

MEIBURG, E., LASHERAS, J.C. and MARTIN, J.E. 1989 Experimental and numerical analysis of the three-dimensional evolution of an axisymmetric jet. To appear in Proceedings of the Seventh Symposium on Turbulent Shear Flows, Stanford, CA.

MELANDER, M. V. and HUSSAIN, F. 1989 Cut-and-connect of two antiparallel vortex tubes: a new cascade mechanism. Procedings of the 7th Symposium on Turbulent Shear Flows, Stanford, California.

PAREKH, D.E. and REYNOLDS, W.C. 1987 Bifurcating air jets at high subsonic speeds. Presentation at the Sixth Symposium on Turbulent Shear Flows, Toulouse, France.

ROSHKO, A. 1976 Structure of turbulent shear flows: a new look. AIAA J. 14, 1349.

TOWNSEND, A.A. 1979 Flow patterns of large eddies in a wake and in a boundary layer. J. Fluid Mech. 95, 515.

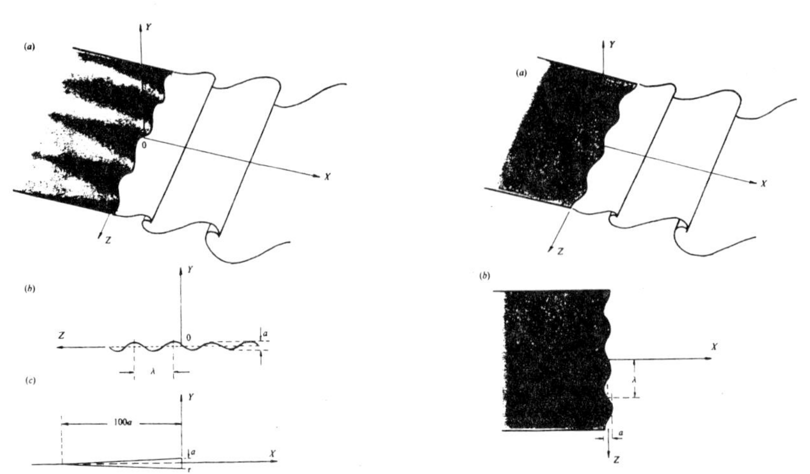

Figure. 1: The two types of splitter plates used in the experiment. a) A corrugated trailing edge that induces verticall oriented perturbation vorticity component. b) An indented trailing edge inducing horizontally oriented perturbation vorticity.

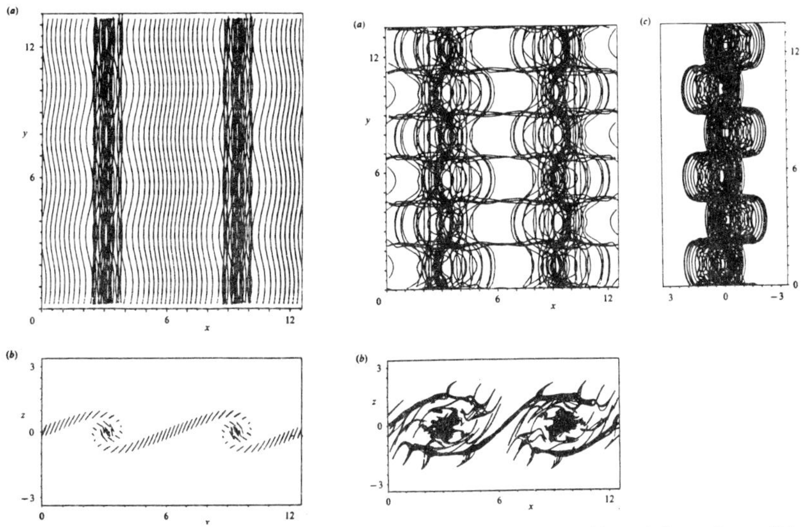

Fig. 2: The single plane shear layer at times 145 and 245. Shown are top, side, and front views of the vortex filament configurations.

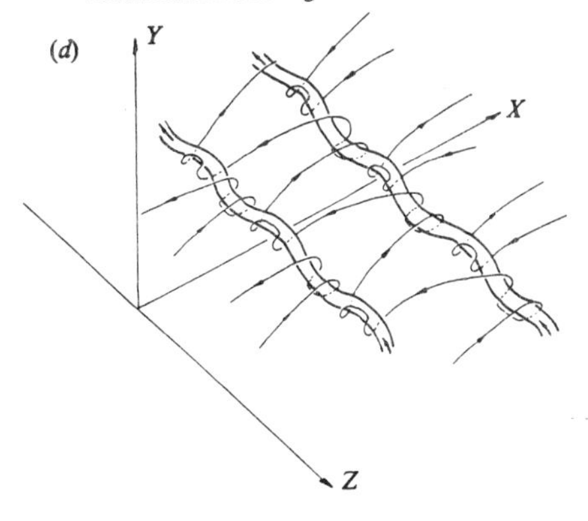

Figure. 3: Schematic representation of the vorticity field evolving in a three-dimensional plane shear layer. Note the existance of counter-rotating pairs of streamwise vortex tubes wrapping around the spanwise ones.

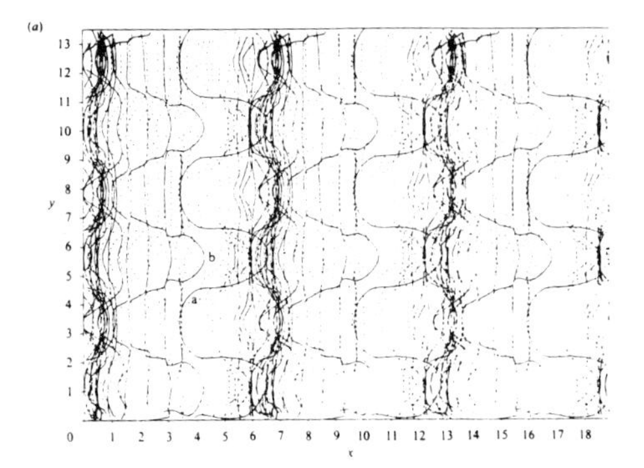

Figure. 4: Inclusion of a second weaker layer of opposite vorticity in the shear layer simulation changes the symmetry properties and leads to better agreement with the experiment. Shown is a top view of the vortex filaments at time 191.875.

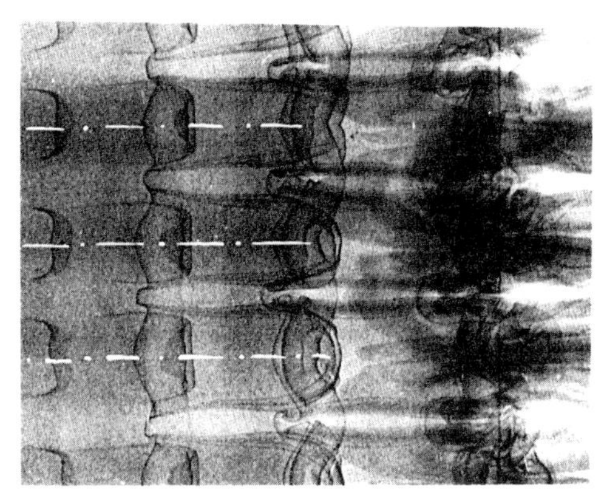

Figure. 5: Plan view flow visualization of the mixing layerunder the effect of a sinusoidal spanwise perturbation.

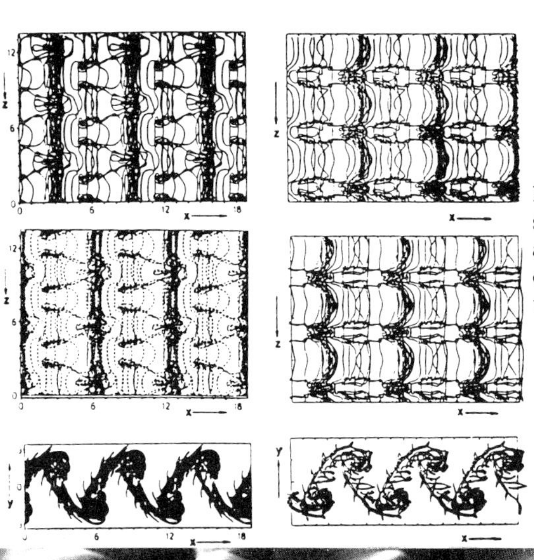

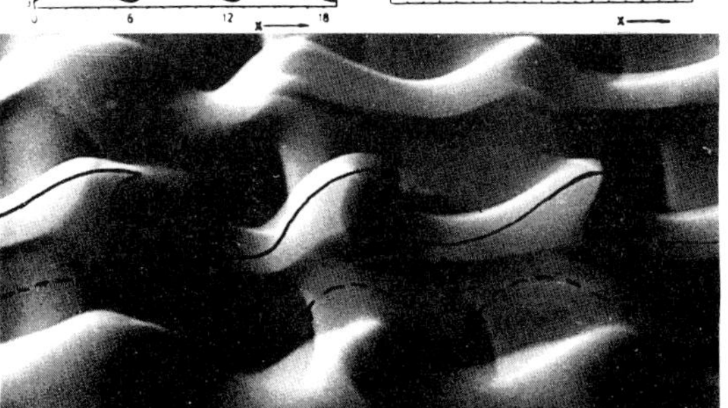

Figure. 6: Numerical simulation of a plane wake subjected to sinusoidal spanwise perturbations. Shown are top and side views of the vortex filament configurations at time 169 . a) Vertical perturbation vorticity. b) Horizontal perturbation vorticity.

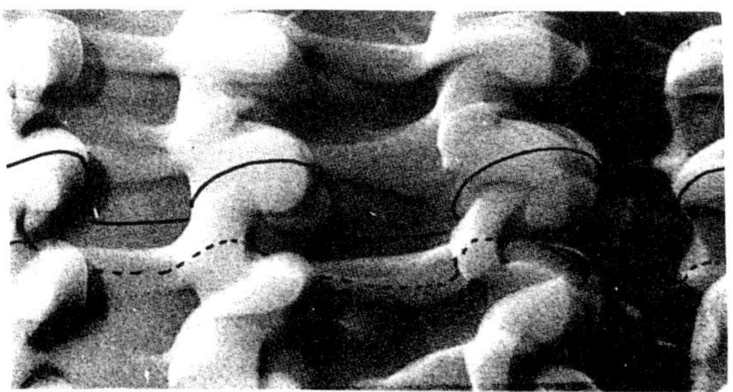

Figure. 7: Near-side perspective view of the flow visualization experiment of the plane wake. a) Corrugated trailing edge. b) Indented trailing edge. Note the different symmetry characteristics developed by the wake in each case.

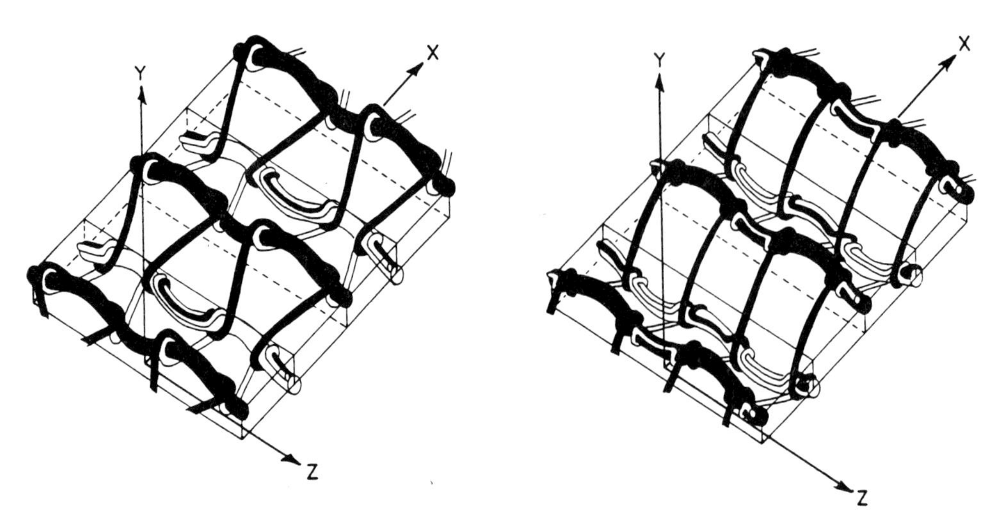

Figure. 8: Schematic representation of the vorticity field. Observe the formation of closed vortex loops. **a)** Vertical perturbation vorticity. **b)** Horizontal perturbation vorticity. The vortex filaments of the upper layer appear darker in the figure.

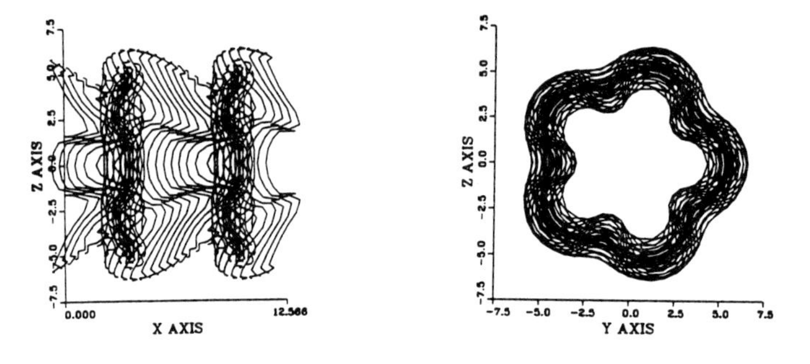

Figure. 9: Numerical simulation of a three-dimensional axisymmetric jet. Shown is a side view and a front view of the vortex filament configuration at time 14.22.

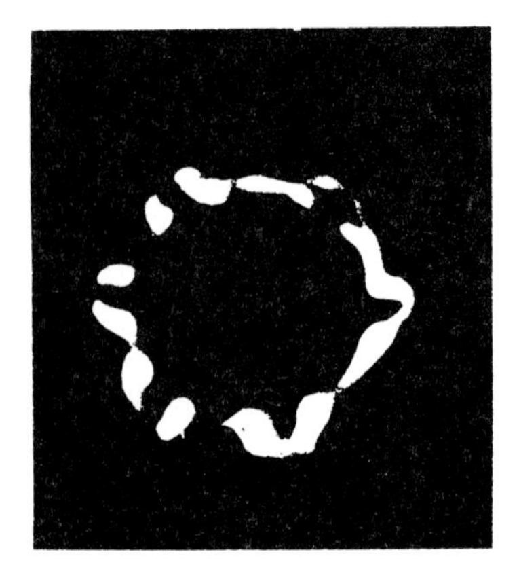

Figure. 10: Cross-cut visualization of the jet interface. For this experiment the jet was forced in the azimuthal direction with an almost sinusoidal wawe containing 6 wavelengths around.

Recirculation Zones in a Cylinder with Rotating Lid

K.G. ROESNER

Technische Hochschule Darmstadt, West Germany

SUMMARY

Experimental data are discussed for the flow of a fluid in the confined region of an upright circular cylinder with independently rotating top and bottom plate. The existence of recirculating zones near the axis of the cylinder is investigated by a variation of the geometrical and kinematic parameters of the problem. Two Reynolds numbers are chosen as similarity parameters of the problem based on the angular velocities of the rotating plates and the radius of the cylinder together with the aspect ratio of the cylinder describing the influence of the rotating lid and the bottom on the onset of secondary flows. The appearance of recirculating zones near the axis of the cylinder is studied when the lid is rotating with the bottom at rest and is compared with the case when the bottom is co–rotating with the lid at relatively low angular velocities. Especially the interaction between two recirculating zones on the axis is analysed for large aspect ratios. The question whether the recirculating zone has a geometrically closed or open character is investigated. From the experimental data diagrams are derived which show the region of existence of recirculating zones in cylindrical geometries for a wide range of the nondimensional parameters.

1 INTRODUCTION

The behavior of a fluid under rotation when the motion is induced by a rotating top or bottom plate was discussed by many authors starting with the famous paper by Theodore von Kármán (1921), followed by Cochran (1934), Batchelor (1951),

Stewartson (1953), Gregory, Stuart, and Walker (1955), Grohne (1955), Bar–Yoseph, Blech, and Solan (1979), Zandbergen and Dijkstra (1987), Szeri, Schneider, Labbe, and Kaufmann (1983), Szeri, Giron, Schneider, and Kaufmann (1983) and by many others. All these investigations are treating those geometrical situations where no secondary flow in the axial region is present. The first experimental proof for the appearance of a vortex breakdown phenomenon in the axial region of the cylinder was given by Vogel (1968). The experimental data show a recirculating zone of bubble–like character near the axis in the confined region of a circular cylinder. Vogel could measure the region of existence of such secondary motions for Reynolds numbers larger than 1000 and aspect ratios above 1.2. Because of the great practical relevance of this type of flow with respect to the mixing behavior of swirling gas flows in burning chambers and the calculation of the friction losses in pumping devices based on the centrifugal force field produced by rotating disks, an intensive research program was started in the past to understand totally the phenomenon of vortex breakdown in a cylindrical geometry. This challenging task has caused a hudge amount of research work with the aim to understand this problem from different points of view: As from the physical (experimental), and the mathematical (numerical and analytical) point of view. Other papers which are devoted to experimental results are due to Ronnenberg (1977), Escudier and Zehnder (1982), Escudier (1984), and Escudier (1988). Recently the access to powerful computers has opened the possibility to simulate the velocity field under the simplifying assumption of an axially symmetrical character of the flow. Among several other important contributions to this problem especially the results by Neitzel (1988), Lopez (1988a,b,c), Lugt and Gorski (1988), and Daube and Sørensen (1989) should be mentioned. The recent review article by Escudier (1988) which is devoted to the vortex breakdown phenomenon in different geometries summarizes all the theoretical and phenomenological approaches to this problem. It seems to be evident that after more than thirty years of research in this field still many questions remain open. For the bubble concept of the vortex breakdown phenomenon e.g. Lugt and Gorski (1988) have reported that the appearance of a stagnation point on the axis of the cylindrical flow field is not a necessary condition for the existence of vortex breakdown in a flowfield with swirl. Their numerical simulations have shown that the formation of a recirculating zone may be possible even without a stagnation point on the axis. From the experimental point of view such a type of vortical motion has not yet been found. On the other hand numerical calculations by Lopez (1988a,b,c) which are based on the assumption of a perfect rotational symmetry of the flowfield in the cylinder show good agreement with experiments for Reynolds numbers up to 1800. But the branching of the Navier–Stokes equations into time periodic solutions at higher Reynolds numbers which was observed in the present experiments could not be simulated numerically up to now, because this motion has a fully three–dimensional time dependent character. By the numerical simulation oscillations only in axial direction were found which could not yet be proved to exist by experiments. The question of the appearance of a stationary or time dependent flow in a circular cylinder can be decided numerically by using the three–dimensional time dependent Navier–Stokes equations. The apparent discrepancies between experimental results and numerical predictions are due to the assumptions made for an easy approach of the problem from the numerical point of view.

2 CYLINDER WITH ROTATING LID

A cylinder filled with a mixture of silicone oil and paraffin oil with radius R = 70 mm and a movable lid was used for the experiments. Three different sequences of experiments were performed: First, at a fixed Reynolds number $Re = \omega R^2/\nu = 1800$ (ω = angular velocity of the lid, R = radius of the cylinder, and ν = kinematic viscosity of the oil mixture) for different aspect ratios H/R (H = height of the lid above the bottom) the flow field was visualized by a disturbance–free method which is based on the laser–induced coloring technique described by Larsen and Roesner (1982). The ultraviolet laser light is transmitted through a transparent window in the center of the bottom and reacts with a photochromic organic compound dissolved in the oil. The upwards convected colored fluid follows in the steady state the stream surfaces and visualizes recirculating zones near the axis. Fig.1a shows only one recirculating zone at H/R = 1.68. A transition to a second recirculation region can be seen in Fig.1b where in the wake of the first recirculating zone a diffuse part of the colored flow field marks the region where the second recirculation zone should appear. If the aspect ratio is slightly raised to H/R = 1.99 (Fig.1c) a fully developed second recirculating zone can be seen. Raising the value of H/R to 2.16 (Fig.1d) gives a larger second egg–shaped recirculating zone. The shape of the first zone has a different character for larger values of the aspect ratio compared with the flat shape in Fig.1a. Still both the recirculating zones are separated. This situation changes at different Reynolds numbers and aspect ratios. Therefore a second set of experiments was performed and the results are shown in the sequence of Fig.2. The aspect ratio was chosen to be 2.5, while the Reynolds number is now changed gradually from 1928 (Fig.2a) to 2790 (Fig.2g). At Re = 1928 apart from the first recirculating zone another one is visualized in statu nascendi (Fig.2a). Increasing the Reynolds number to 1993 (Fig.2b) leads to a fully developed second egg–shaped region which contains a vortical motion. With further increase of the Reynolds number to 2021 (Fig.2c) this region becomes larger and approaches from behind the first recirculation zone which does not change very much its position relative to the bottom of the cylinder. If one enlarges more and more the angular velocity of the lid the second recirculating zone developes a thin funnel and enters the first recirculation zone from behind (Fig.2e). This funnel is elongated till it reaches the first stagnation zone. A further increase of the Reynolds number (Fig.2f) opens the second recirculation zone from behind and shows a strong merging of both the vortical motions. This process is indicated in Fig.2f and Fig.2g by a colored axial part of the flow field. In the experiment the entering of the colored liquid can be followed visually after having switched on the laser light. The merging of the two vortical recirculating zones in the steady state is similar to the unsteady behavior of two coaxially moving vortex rings one behind the other. The merging in the steady state will disappear for higher Reynolds numbers when one leaves the zone where such type of flow can exist. The flow then shows a srew–like character along the axial region. To reach a time periodic bahavior of the fluid one has to raise the aspect ratio to about 3.3. This ensures that two recirculating zones may exist which become unsteady. The sequence of pictures (Fig.3a–f) are taken for H/R = 3.3 in the range of Reynolds numbers between 2555 and 2718. Fig.3a–c show the development of two recirculating zones which represent stationary motion. The second recirculation zone gets very elongated and shows a tendency to be squeezed into two parts. This squeezing effect never led to a real separation into two smaller recirculation zones. Instead of this at the Reynolds number 2647 (Fig.3d) the upper part of the elongated recirculating zone and the first one, too, start to get an circumferential amplitude modulation which resembles the wavy mode of Taylor vortices around the axis of rotation. The second recirculation zone gets more and more distorted and looses completely rotational symmetry. The frequency varies with varying the Reynolds number and dies out when the Reynolds number is high enough. This time dependent character of the flow is observed only in an interval of Reynolds numbers which was predicted by Escudier (1984). This type of motion was up to now not yet simulated by numerical experiments even on the biggest computer available.

3 INFLUENCE OF A CO–ROTATING BOTTOM

If the top Reynolds number is close to the critical value where we expect to get a first recirculating zone a slight co–rotation of the bottom immediately creates the vortical motion in the axial region. On the other hand if we are slightly above the critical value of the top Reynolds number for the onset of a recirculating zone and we switch on a slow counterrotation of the bottom the recirculating zone immediately disappears. So one can conclude that a co–rotation of the bottom destabilizes the flow and leads to an immediate change of the flow field from a subcritical to a supercritical situation. Due to that fact the existence of recirculating zones could be proved for co–rotating top and bottom plate even for values of H/R = 0.8 where for the case of a bottom at rest no recirculation can appear. In the case of a counterrotation of lid and bottom the flow field is influenced just in the opposite way and does not create a recirculation zone.

4 CONCLUSION

The vortex breakdown phenomenon can be observed in cylindrical geometry for different kinematic conditions of the lid and the bottom motion. Fig.4 shows the regions of existence for one or two recirculating zones on the axis of a cylinder with a rotating lid. Below a Reynolds number of 1000 and an aspect ratio of 1.2 no recirculation zone exists. The dashed line indicates the region when both the recirculating zones merge. For other geometries as e.g. for an eccentrical or concentrical spherical gap the appearance of a vortex breakdown could also be observed by experiments due to Roesner, Viehl, Bar–Yoseph, and Solan (1989) and predicted by numerical simulations due to Bar–Yoseph, Seelig, Solan, and Roesner (1987). The question concerning the topological character of the flow field in the vicinity of the axis whether the recirculating region has a closed or open geometrical form can still not be answered in any situation.

5 REFERENCES

Bar–Yoseph, P., Blech, J. J. & Solan, A. 1979 Finite Element Solution of the Navier–Stokes Equations in Rotating Flow. TME–367, Faculty of Mechanical Engineering, Technion – Israel Institue of Technology, Haifa.

Bar–Yoseph, P., Seelig, S., Solan, A. & Roesner, K.G. 1987 Vortex breakdown in spherical gap. Phys. Fluids **30**, 1581.

Batchelor, K. G. 1951 Note on a class of solutions of the Navier–Stokes equations representing steady rotationally symmetric flow. Quart. Journ. Mech. and Appl. Math. **4**, 29.

Cochran, W.G. 1934 The flow due to a rotating disc. Proc. Cambridge Phil. Soc. **30**, 365.

Daube, O. & Sørensen, J. N. 1989 Simulation numérique de l'écoulement périodique axisymetrique dans une cavité cylindrique. C.R. Acad. Sci. Paris, t. **308**, Série II, 463.

Escudier, M. P. & Zehnder, N. 1982 Vortex–flow regimes. J. Fluid Mech. vol. **115**, 105.

Escudier, M. P. 1984 Observations of the flow produced in a cylindrical container by a rotating endwall. Exp. in Fluids, **2**, 189.

Escudier, M. P. 1988 Vortex Breakdown: Observations and Explanations. in: Progr. in Aerospace Sci., vol **25**, 189.

Gregory, N., Stuart, J. T. & Walker, W. S. 1955 On the stability of three–dimensional boundary layers with application to the flow due to a rotating disk. Phil. Trans. Roy. Soc. London, Ser. A, Vol. **943**, 155.

Grohne, D. 1955 Über die laminare Strömung in einer kreiszylindrischen Dose mit rotierendem Deckel. Nach. Akad. Wiss. Göttingen, Math.–Phys. Klasse, Nr. **12**, 263.

v. Kármán, Th. 1921 Über laminare und turbulente Reibung. ZAMM 1, 233.

Larsen, J. & Roesner, K. G. 1982 Optical Flow–Velocity Measurement in Irregularly Shaped Cavities. in: Recent Contrib. to Fluid Mechanics, Springer–Verlag, Berlin.

Lopez, J. M. 1988a Axisymmetric vortex breakdown in an enclosed cylinder flow. Proc. 11th Intern. Conf. on Num. Meth.in Fluid Dyn., Williamsburg, VA, Springer–Verlag, Berlin.

Lopez, J. M. 1988b Axisymmetric vortex breakdown, Part I: Confined swirling flow. A.R.L. Aero. Rep. 173, AR–004–572.

Lopez, J. M. 1988c Axisymmetric vortex breakdown, Part II: Physical Mechanisms. A.R.L. Aero. Rep. 174, AR–004–573.

Lugt, H. J. & Gorski, J. J. 1988 The "Bubble" Concept of Axisymmetric Vortex Breakdown With and Without Obstacles in the Vortex Core. DTRC–88/042.

Neitzel, G. P. 1988 Streak–line motion during steady and unsteady axisymmetric vortex breakdown. Phys. Fluids **31**, 958.

Roesner K. G., Viehl, M., Bar–Yoseph, P. & Solan, A. 1989 Asymptotic solutions of the Navier–Stokes equations for compressible fluids at low Mach numbers. in: Finite Approximations in Fluid Mechanics II, Friedr. Vieweg & Sohn, Braunschweig.

Ronnenberg, B. 1977 Ein selbstjustierendes 3–Komponenten Laserdoppleranemometer nach dem Vergleichsstrahlverfahren. Max–Planck–Institut für Strömungsforschung, Ber. 20/1977.

Stewartson, K. 1953 On the flow between two rotating coaxial disks. Proc. Cambridge Phil. Soc., 333.

Szeri, A. Z., Schneider, S. J., Labbe, F. & Kaufmann, H. N. 1983a Flow between rotating disks. Part1. Basic flow. J. Fluid Mech., vol. **134,** 103.

Szeri, A. Z., Giron, A., Schneider, S. J. & Kaufmann, H. N. 1983b Flow between rotating disks. Part 2. Stability. J. Fluid Mech., vol **134**, 133.

Vogel, H. U. 1968 Experimentelle Ergebnisse über die laminare Strömung in einem zylindrischen Gehäuse mit darin rotierender Scheibe. MPI für Strömungsforschung, Ber. 6.

Zandbergen P. J. & Dijkstra, D. 1987 von Karman swirling flows. Ann. Rev. Fluid Mech., **19**, 465.

Fig.1a H/R=1.68 Re=1800

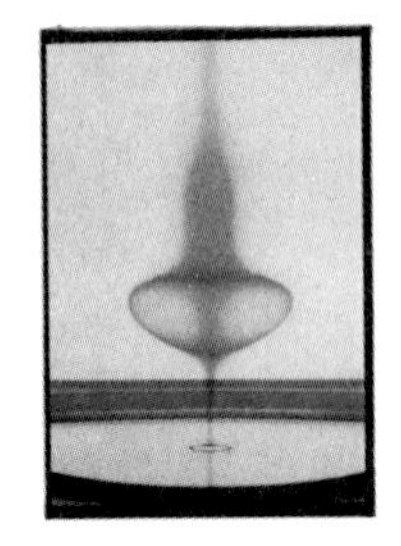

Fig.1b H/R=1.91 Re=1800

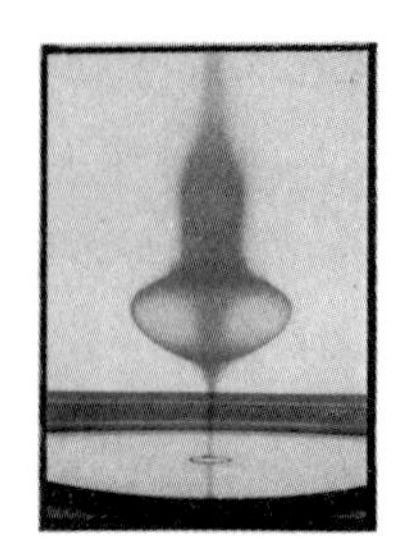

Fig.1c H/R=1.99 Re=1800

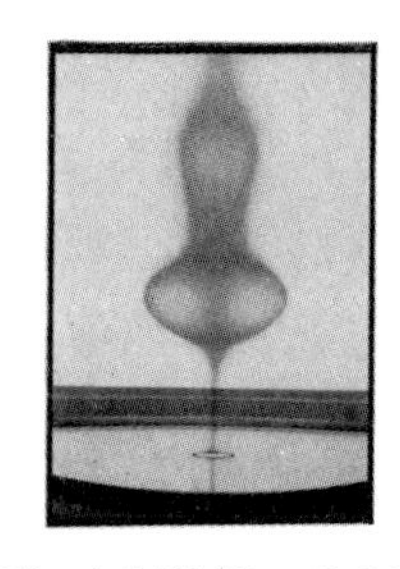

Fig.1d H/R=2.16 Re=1800

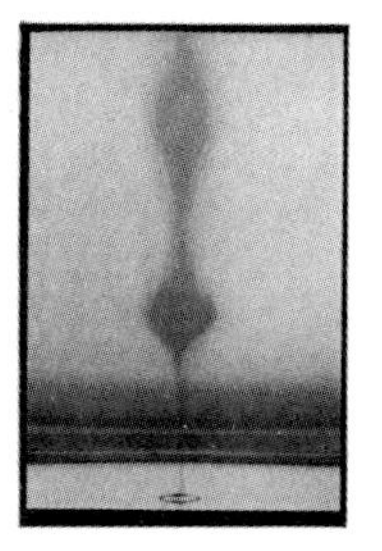

Fig.2a H/R=2.5 Re=1928

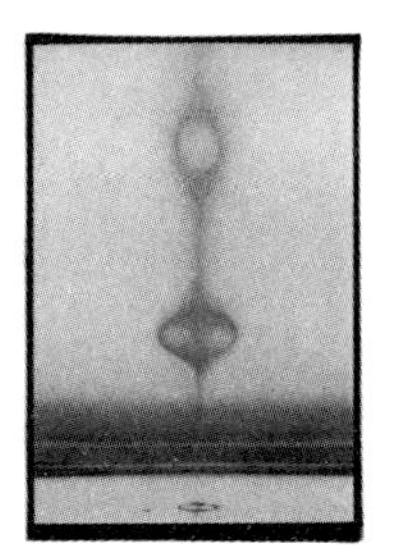

Fig.2b H/R=2.5 Re=1993

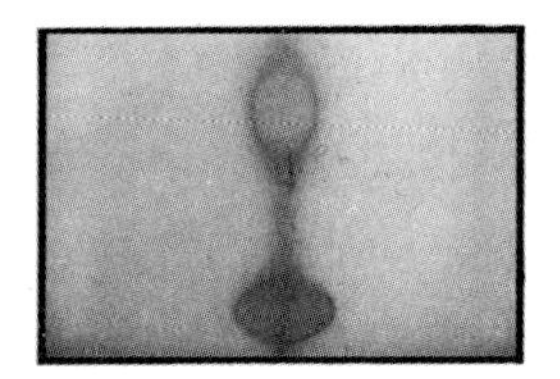

Fig.2c H/R=2.5 Re=2021

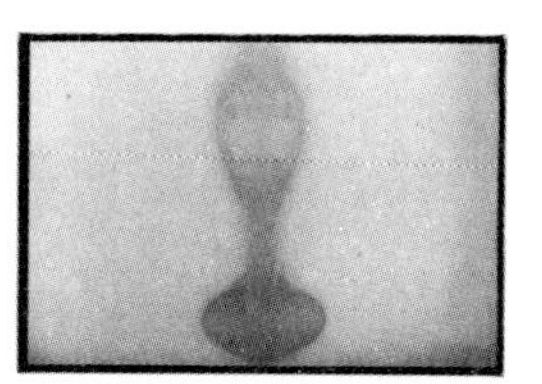

Fig.2d H/R=2.5 Re=2120

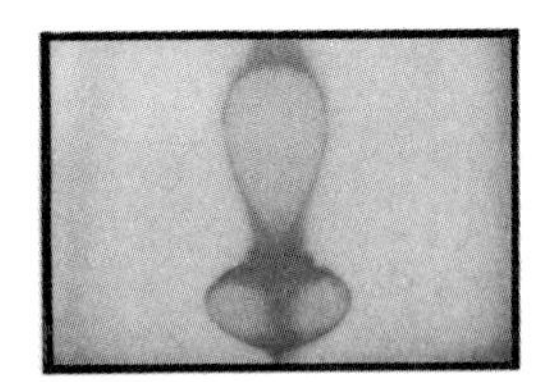

Fig.2e H/R=2.5 Re=2274

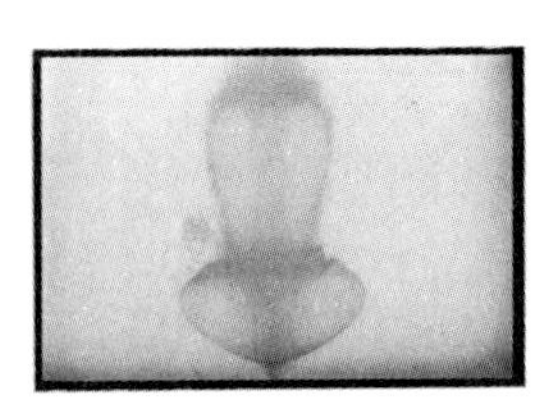

Fig.2f H/R=2.5 Re=2642

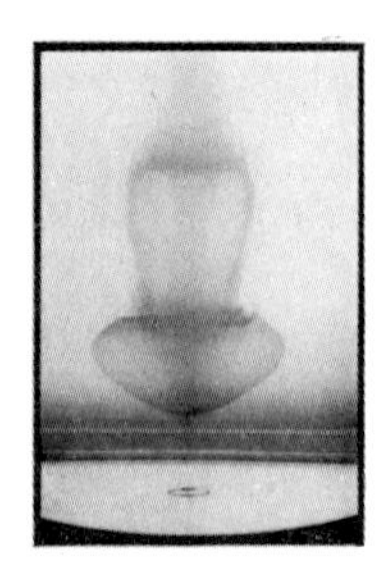

Fig.2g H/R=2.5 Re=2790

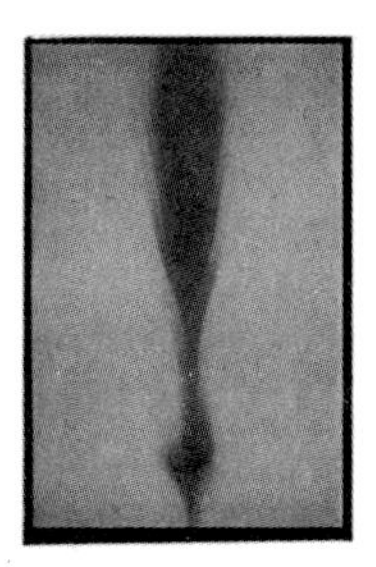

Fig.3a H/R=3.3
Re=2555

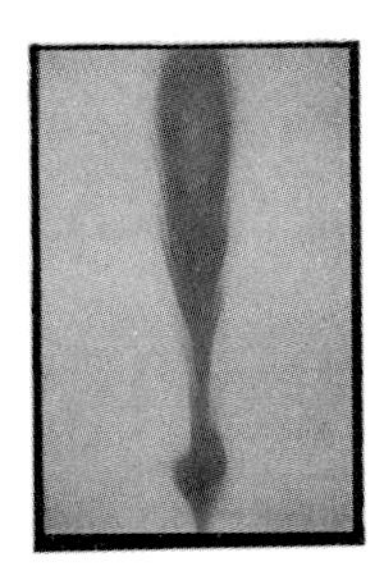

Fig.3b H/R=3.3
Re=2598

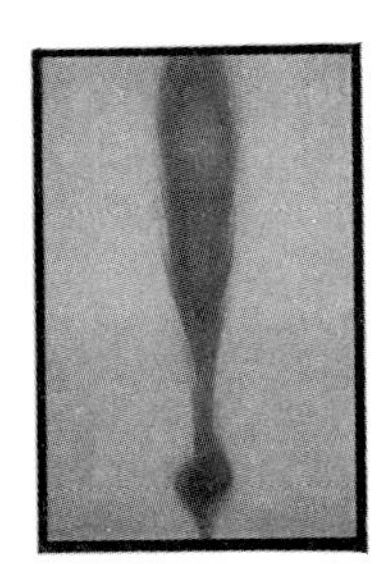

Fig.3c H/R=3.3
Re=2616

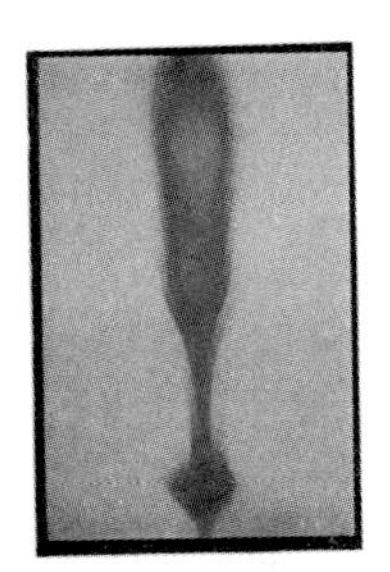

Fig.3d H/R=3.3
Re=2647

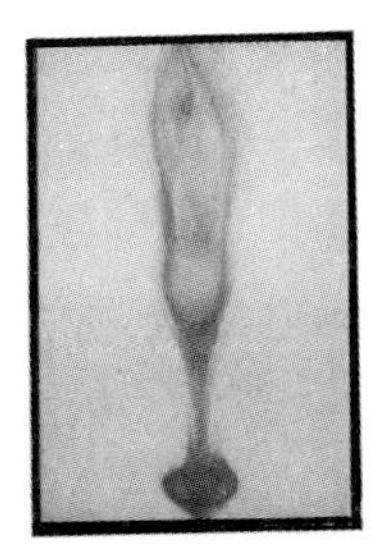

Fig.3e H/R=3.3
Re=2678

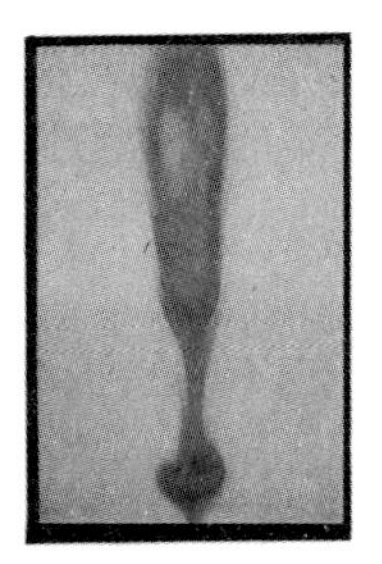

Fig.3f H/R=3.3
Re=2718

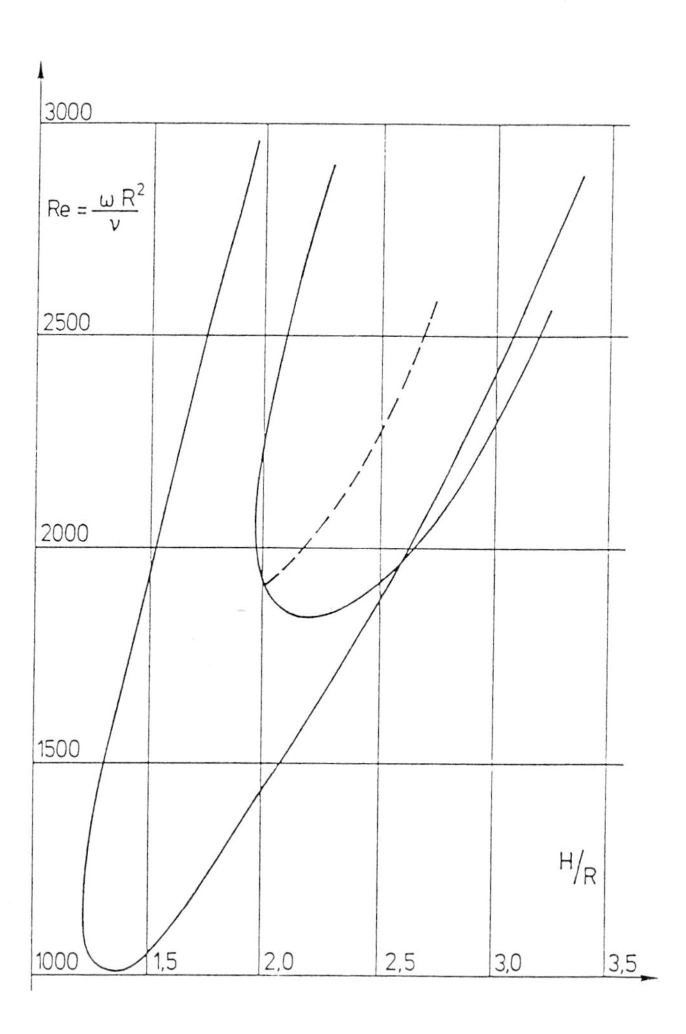

Fig.4 Regions of existence of one or two recirculating zones in a circular cylinder with rotating lid (– – – indicates merging of two recirculating zones)

Instabilities of Tertiary Flow in the Taylor-Couette System

E. WEISSHAAR, F.H. BUSSE & M. NAGATA

Physikalisches Institut, Univ. Bayreuth, Postfach 101251, 8580 Bayreuth, FRG

1 INTRODUCTION

Besides Rayleigh-Benard Convection, Taylor-Couette flow between differentially rotating cylinders is a well known system for the study of subsequent transitions from simple to more complex forms of fluid motion. In the limit of a small gap approximation between the two coaxial cylinders an additional symmetry is gained in comparison to the general Taylor system. Systems with high degrees of symmetries are especially useful for the theoretical investigation of transition to complex flows. While the basic solutions reflect the symmetry of the basic conditions, one or more symmetries are usually broken as bifurcations occur. Axisymmetric Taylor vortices are introduced at the first bifurcation as the Reynolds number is increased. In the secondary instability the axisymmetry is broken and a variety of three dimensional flows bifurcate from the axisymmetric flows depending on the position in the parameter space. Wavy vortices and twist vortices appear to be the most widely observed flows in small gap experiments (Andereck et al. 1983, 1986). In this paper twist solutions will be studied numerically by the same methods as used in the earlier work by Nagata (1986, 1988) and Nagata and Busse (1983).

2 PARAMETERS AND METHODS

We consider the flow in the narrow gap between two coaxial

cylinders rotating with the speeds $\Omega_1 R_1$ and $\Omega_2 R_2$, respectively. The gap $D = R_2 - R_1$ will be used as a length scale and D^2/ν as a time scale, where ν is the kinematic viscosity of the fluid. The small gap approximation allows us to introduce a cartesian coordinate system in a rotating frame with the unit vector $\vec{i}$ perpendicular to the plates, the unit vector $\vec{k}$ parallel to the axis of rotation and $\vec{j} = \vec{k} \times \vec{i}$. The dimensionless equations governing the flow are given by:

$$(1a) \qquad (\partial_t + \vec{u} \cdot \nabla)\vec{u} + \Omega \vec{k} \times \vec{u} = - \nabla \pi + \nabla^2 \vec{u}$$

$$(1b) \qquad \nabla \cdot \vec{u} = 0,$$

The following definitions are used,
the rotation parameter,

$$(2) \qquad \Omega = (\Omega_1 + \Omega_2) \; D^2/\nu$$

and the Reynolds number,

$$(3) \qquad R = (\Omega_1 - \Omega_2) \; D^2/2 \, \nu \; (R_1 + R_2)$$

We use a special small gap approximation defined by

$$(4) \qquad \frac{|\Omega_2 - \Omega_1|}{|\Omega_2 + \Omega_1|} \ll 1 \, , \qquad \frac{R_2 - R_1}{R_2 + R_1} \ll 1$$

The boundary conditions at the cylinder walls are given by

$$(5) \qquad \vec{u} = \mp \tfrac{1}{2} R \, \vec{j}$$

In the other directions periodic boundary conditions are used.

For the numerical investigation of the problem we proceed in two steps. In a first step, we calculate the finite amplitude solutions and in a second step we determine the stability of these solutions. The finite amplitude solutions are obtained numerically by a Galerkin method. As series expansions we use discrete Fourier series in axial and azimuthal direction with basic wavelengths $2\pi/\gamma$ and $2\pi/\alpha$ respectively. In radial direction Chandrasekhar functions and sine functions that satisfy the boundary conditions are used. The stability analysis starts from eqs. (1) linearized about the a steady finite amplitude solution. The infinitesimal disturbances are expanded in a series of the same form as for the finite amplitude solutions multiplied by a factor $\exp(i(dy+bz) + \sigma t)$ according to Floquet theory. d and b are Floquet parameters, y and z are the coordinates in azimuthal and axial directions, respectively. The resulting generalized eigenvalue problem yields the complex eigenvalue σ. If the real part of σ is positive, the finite amplitude solution is considered unstable owing to a disturbance with the wavenumbers b and d.

3 RESULTS

Twisted vortex flow is characterized by the external parameters R and Ω, by the axial wavenumber γ of the axisymmetric Taylor vortices from which the twisted vortices bifurcate, and by the azimuthal wavenumber α which corresponds to the strongest growing disturbance at the stability boundary of the axissymmetric vortices. Beyond the stability boundary the twisted vortices are assumed to prserve this azimuthal wave number as the Reynolds number is increased.

Two different symmetry classes of growing modes and finite amplitude solutions that become decoupled in the small gap limit can be distinguished: A first class, which will be called ordinary twists, is characterized by the property that the azimuthal structures of counter rotating Taylor vortices are in phase. The boundaries between the rolls are straight. The second class are the wavy twists. Here the boundaries are wavy and the azimuthal variations are out of phase on neighboring rolls.

Fig. 1 shows the neutral curves of ordinary and wavy twists

for the case $\gamma=3.1$. The neutral curves of the two types of instability intersect at $\Omega/R \approx 0.7$. The wavy twists are preferred at higher rotation rates, while the ordinary twists appear first at lower rotation rates. Everywhere in the region above the neutral curves steady finite amplitude solutions exist. Neither subcritical bifurcations nor Hopf bifurcations were found.

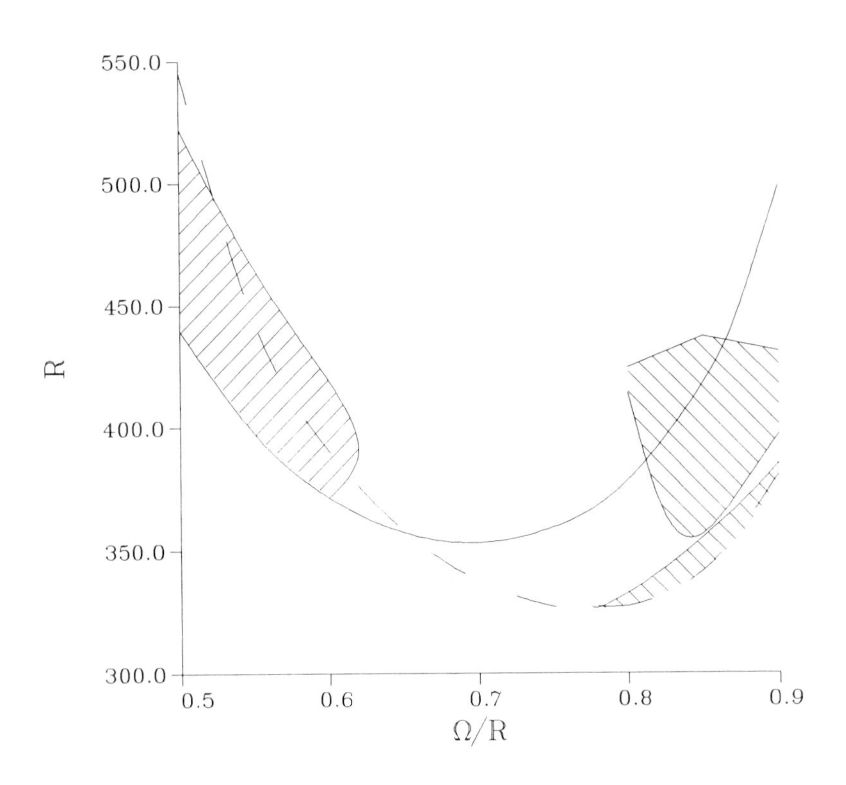

Fig. 1: Twist neutral curves for γ = 3.1
solid line: ordinary twists
broken line: wavy twists
hatched area: stable regime

The 3-dimensional solutions can be visualized by plotting the streamlines near the outer cylinder. Fig.2 and Fig. 3 show examples of ordinary twists and wavy twists, respec-

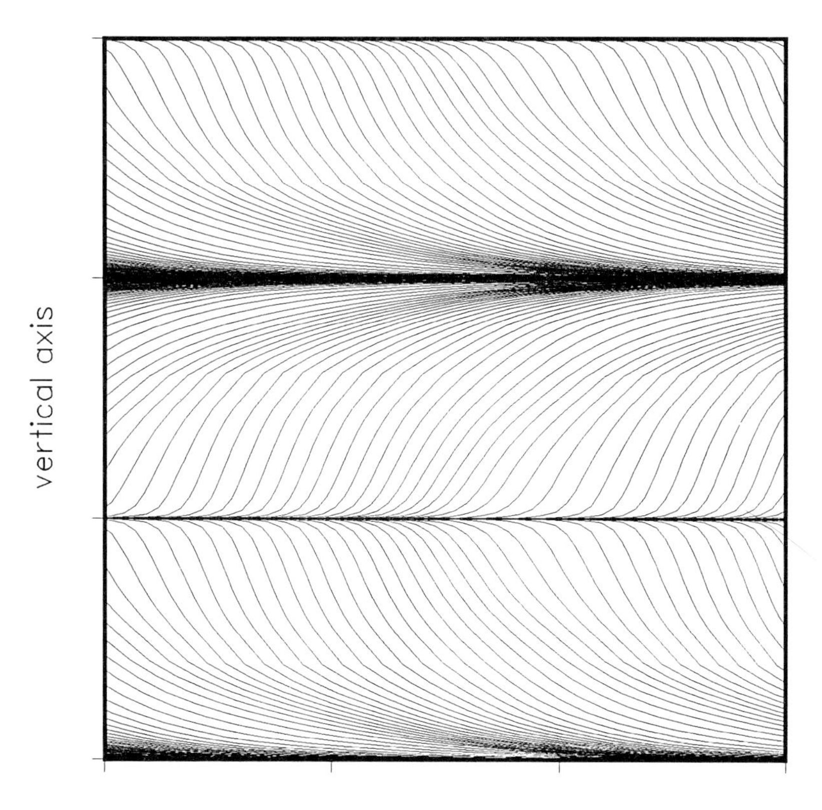

Fig. 2: ordinary twists.
R = 271.6, γ = 2.3, α = 2.63, Ω/R = 0.55

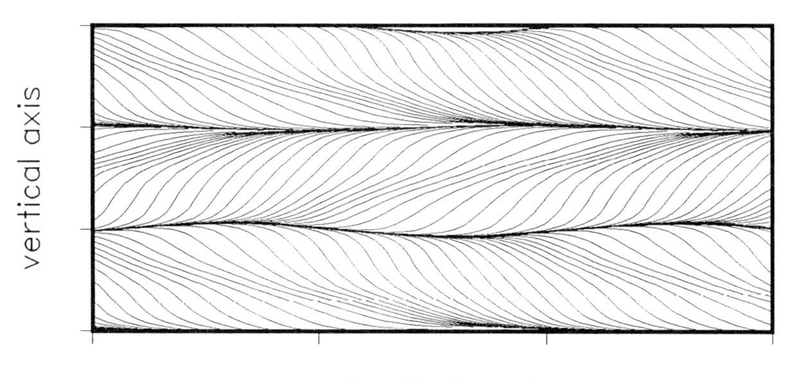

Fig. 3: wavy twists.
R = 640., γ = 2.8, α = 1.23, Ω/R = 0.90

tively. The plots cover $1\frac{1}{2}$ wavelengths in both, azimuthal and axial direction. The aspect ratio of the plots reflects the ratio between the wavelengths in the two directions. The difference between α and γ is more apparent for wavy twists than for ordinary twists. It becomes clear from the plots that the impression of twisted vortices arises from the fact that streamlines converge in regions which are twisted around the vortex.

Next, the stability of the 3-dimensional twist solutions is examined. Surprisingly, there is only a small regime of stable solutions that appears in the case of ordinary twists at low rotation rates $\Omega/R < 0.6$. The stable regime, which is also shown in Fig. 1, is bounded towards higher Reynolds number by an instability of the skewed varicose type. This means, that the most unstable mode results, when the Floquet parameters are both finite but arbitrary small at the onset of instability. The typical angle $\delta = \arctan(b/d)$ is about 60^0. All other solutions are unstable. The instabilities set in already at the twist neutral curve for arbitrary small parameters b. The properties of this wavelength changing instability are similar to the skewed varicose instability.

The wavy twists are stable in a regime at high rotation rates $\Omega/R > 0.8$ and for axial wave numbers $\gamma \geq 2.7$ with an interesting structure. It consists of two parts, which are separated by a strip of unstable solutions (see Fig 1). When γ is decreased the strip becomes thinner. At some critical γ the two stable regions meet in one point and fuse after γ is decreased further. Outside the stable regions wavelength changing instabilities in axial direction are dominant. At high Reynolds numbers the upper stable region is bounded by wavelength changing instabilities in azimuthal direction.

3 DISCUSSION

In their experiments Andereck et al. (1983) observe that twist vortices are in phase, which corresponds to the symmetry of our ordinary twists. The ordinary twists represent a class, which exhibits a stable range at low rotation rates. However, in experiments twists are also observed for higher rotation rates, where only wavy twists are stable according to our theory. On the other hand nonuniform "domain" states have also been observed in these experiments (Andereck and Baxter, 1988). The authors mention a side band instability for this regime and comment that "the system seems unable to chose a new wave length". It appears that the system is even more unstable in the small gap limit than in the experimentally realized finite gap with lower symmetry. A more detailed discussion of this and other aspects of the analysis will be presented in a future publication.

REFERENCES

C.D. Andereck, R. Dickman, H.L. Swinney (1983), Phys. Fluids, **26**, 1395-1401

C.D. Andereck, S.S. Liu, H.L. Swinney (1986), J. Fluid Mech, **164**, 155-183

C.D. Andereck, C.D. Baxter (1988) in: "Propagation in Systems Far From Equilibrium", Springer Verlag, Berlin 1988, 315-324

G.W. Baxter, C.D. Andereck (1986), Phys. Rev. Lett., **57**, 3046-3049

M. Nagata, F.H. Busse (1983) J. Fluid Mech., **135**, 1-26

M. Nagata (1986) J. Fluid Mech., **169**, 229-250

M. Nagata (1988) J. Fluid Mech., **188**, 585-598

3 DISCUSSION

In their experiments Andereck et al. (1983) observe that twist vortices are in phase, which corresponds to the symmetry of our ordinary twists. The ordinary twists represent a class, which exhibits a stable range at low rotation rates. However, in experiments twists are also observed for higher rotation rates, where only wavy twists are stable according to our theory. On the other hand nonuniform "domain" states have also been observed in these experiments (Andereck and Baxter, 1988). The authors mention a side band instability for this regime and comment that "the system seems unable to chose a new wave length". It appears that the system is even more unstable in the small gap limit than in the experimentally realized finite gap with lower symmetry. A more detailed discussion of this and other aspects of the analysis will be presented in a future publication.

REFERENCES:

C.D. Andereck, R. Dickman, H.L. Swinney (1983), Phys. Fluids, **26**, 1395-1401

C.D. Andereck, S.S. Liu, H.L. Swinney (1986), J. Fluid Mech, **164**, 155-183

C.D. Andereck, C.D. Baxter (1988) in: "Propagation in Systems Far From Equilibrium", Springer Verlag, Berlin 1988, 315-324

G.W. Baxter, C.D. Andereck (1986), Phys. Rev. Lett., **57**, 3046-3049

M. Nagata, F.H. Busse (1983) J. Fluid Mech., **135**, 1-26

M. Nagata (1986) J. Fluid Mech., **169**, 229-250

M. Nagata (1988) J. Fluid Mech., **188**, 585-598

Experimental and Numerical Investigation of Natural Convection in a Cube with Two Heated Side Walls

W.J. HILLER[1], ST. KOCH[1], T.A. KOWALEWSKI[1], G. DE VAHL DAVIS[2] & M. BEHNIA[2]

[1]Max-Planck-Institut für Strömungsforschung, Göttingen, FRG
[2]University of New South Wales, Kensington, N.S.W., Australia

1 INTRODUCTION

Experimental knowledge of natural convection within enclosures is still far from complete. Even stationary convective flow in a square box with differentially heated side walls, a problem which at first view seems to be closely related to the corresponding two-dimensional case, exhibits an intricate three-dimensional structure. The experimental simulation of two-dimensional plane flows is always a compromise as additional walls are needed to prevent the medium from lateral escape. These walls in their turn induce three-dimensional motion, as has been discussed in [1]. Even if by proper design of the size of the box the influence of these walls can be kept small in the centre region of the enclosure, the topological structure of the flow field is fundamentally changed. Problems like separation, stability of flow configurations, and onset of oscillatory motion can be discussed only by taking into account the three-dimensional character of the flow field [2].

Therefore, it appears quite useful to perform supplementary experiments to collect more details on the flow structure. For this purpose, a cube shaped box with two opposite, vertical walls kept at different temperatures and the other walls kept adiabatic appears very suitable, as a pronounced three-dimensional behaviour of the flow field is to be expected.

It has been already revealed in a former study [3] that the flow structure away from the vertical mid-plane of the cavity is strongly three-dimensional and, for the liquid used in that study, asymmetrical, and differs strikingly from existing two- and three-dimensional numerical results. At relatively low Rayleigh numbers ($Ra < 6 \cdot 10^4$) the core of the spiral vortex transporting fluid from the front and back walls (i.e. the adiabatic vertical walls) to the centre of the cavity is concave with respect to the heated wall. At increasing Rayleigh number a second spiral vortex twists off from the first vortex at approximately half the distance between the front and back walls and the centre of the cavity.

These effects may be due to failure to achieve a true adiabatic condition on the insulating vertical walls and are certainly affected by the temperature-dependent viscosity of the liquid. Unfortunately, there exist neither transparent fluids of constant material properties which are suited for liquid crystal tracers nor transparent, ideally adiabatic walls which are needed for the purpose of observation. Therefore, the experiments have to be accompanied by a numerical procedure able to take into account the experimental conditions, as most of the theoretical treatments of convective flow known to us have been performed by assuming a constant property Boussinesq fluid and ideal isothermal or adiabatic walls.

The present paper is a preliminary combined experimental and numerical study of three-dimensional natural convection aimed at revealing details of the flow structures, especially for higher Rayleigh numbers. The discrepancies between the observed flow structures and prior numerical results are under numerical investigation with respect to wall heat losses and the temperature dependence of the fluid properties.

2 DESCRIPTION OF THE PROBLEM

We consider the convective flow in a cubic box filled with a viscous heat-conducting liquid. The fluid viscosity and density are assumed to be only temperature dependent. Two opposite, vertical walls of the cube are isothermal at temperatures T_h (hot) and T_c (cold), respectively. The four other walls are isolators of finite thermal diffusivity α_w. Due to temperature gradients existing between the fluid inside the cavity and the surrounding atmosphere and also along the front and back walls, lid and floor of the box, a heat flux both through and along the walls comes into existence.

The origin of a rectangular Cartesian coordinate system is placed at an upper corner of the box as illustrated in Fig.1. Two non-dimensional parameters are chosen to compare the numerical and experimental results:

Rayleigh number,

$$(1)\quad Ra = \frac{g \cdot \beta \cdot d^3 \cdot (T_h - T_c)}{\alpha \nu}$$

and the Prandtl number,

$$(2)\quad Pr = \nu/\alpha \ .$$

In the above definitions, $g, d, T_h, T_c, \alpha, \beta, \nu$ denote gravitational acceleration, cavity dimension, vertical wall temperatures, thermal diffusivity, coefficient of thermal expansion and kinematic viscosity, respectively.

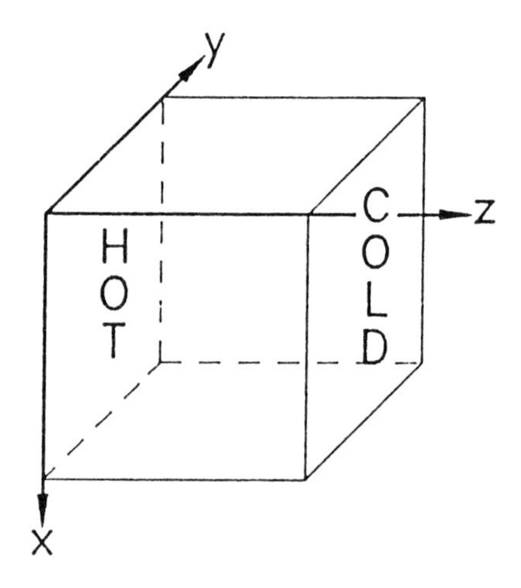

Fig. 1. The cube shaped enclosure and coordinate orientation. Hot wall at z = 0, cold wall at z = 1.

3 EXPERIMENTAL

3.1 Apparatus and Procedure

The apparatus used in the experiment is essentially the same as that used previously [3]. The convection cavity consists of a tube of square cross-section of internal dimension and length 38mm made from 8 mm thick Plexiglas. Both ends of the tube are closed with a black anodized copper plate of 60mm by 60mm, and 50mm thick. Each of these plates is maintained at constant temperature by a water flow which passes through internal passages in the plates. The temperature of the water is controlled by a thermostat. The temperature difference between the two walls was varied in the range from 2.5^{o}C to 18°C, whereas the mean temperature was kept at approx. 27°C. The temperatures were measured continuously by means of thermocouples and recorded by a Philips multichannel recorder. The temperature fluctuations with time observed on the end plates were below 0.1^{o}C. The room temperature was measured to be 22 ± 1°C.

The Rayleigh numbers ranged from 10^4 to 10^6. Aqueous solutions of glycerol of various concentrations were employed as flow media. The Prandtl number varied from 6900 (pure glycerol) to approximately 200 for a 70% glycerol-water mixture. Temperature and velocity fields were measured with help of liquid crystals suspended as small tracer particles in the liquid [3]. The visualization of temperature using liquid crystals is based on their temperature-dependent refractivity for the wavelength (colour) of visible light. If they are illuminated with white light, then the colour of the light reflected from the liquid crystals changes from red to blue if the temperature is raised (in a well defined temperature range which depends on the type of LC used). The flow was observed at different vertical (z-x) and horizontal (y-z) cross sections of the cavity using a light sheet technique.

3.2 Flow Structure

At first view, a characteristic unicellular flow pattern similar to the two-dimensional case is observed at moderate Rayleigh numbers $< 6 \cdot 10^4$ in the centre plane of the cavity (Fig. 2a), and the slope of the isotherms in the centre region of the channel is very small but still positive. The influence of the non-zero thermal conductivity of the top and bottom walls becomes visible by a deflection of the ends of the isotherms in such a way that close to these walls the z-component of the temperature gradient becomes more uniform. Estimates of the relative rates of heat transfer by convection to the surroundings and conduction along the Plexiglas walls suggest that this effect is due rather more to the latter than the former. The presence of the front and back walls induces axial flow components in the entire flow field with the result that the tracer particles in z-x planes no longer follow closed lines. They spiral outwards on the centre mid-plane of the cube and inwards in the neighbourhood of the front and back walls. Both regions are connected by a mean effective axial flow component which is directed to the centre of the cavity (for the inner region of the spirals) and to the front or back walls (for the

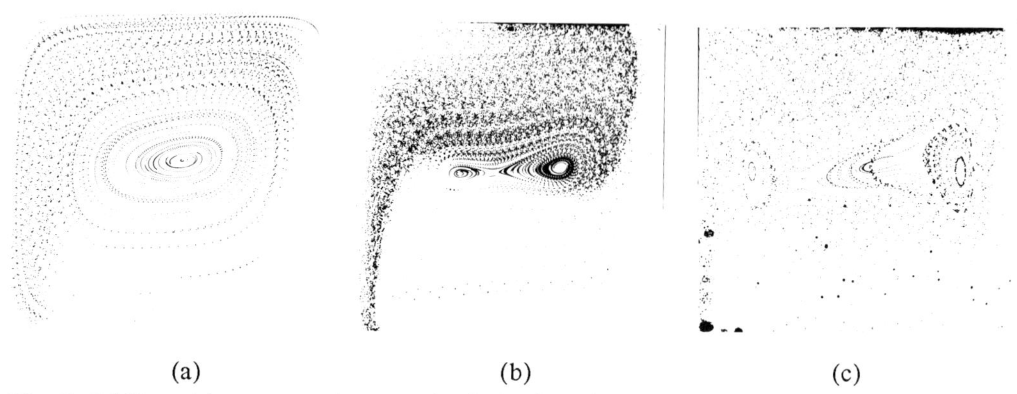

(a) (b) (c)

Fig. 2. B&W multiexposure photograph of the flow in the vertical centre plane (y = 0.5). Left side (z=0) is the hot wall; (a) Ra = $2 \cdot 10^4$, Pr = 6300; (b) Ra = $8 \cdot 10^4$, Pr = 6900; (c) Ra = $7.5 \cdot 10^5$, Pr = 225. See also colour plate, displaying additionally streamlines in a vertical plane y = 0.05. Red colour corresponds to ~27°C, blue to ~29°C.

outer region of the spirals). The ends of the spiral axis are shifted towards the heated side wall (Fig.3a). The flow field is symmetric with respect to the centre (y = 0.5) plane. If the Rayleigh number is raised above approximately $6.5 \cdot 10^4$ a negative slope of the isotherms appears in the middle region of the cubic enclosure which is caused by the increased convective heat transport (Fig.2b). As a consequence of the negative temperature gradient, the vortex splits up there into two vortices as in the two-dimensional case [4]. It seems as if part of the outer region of the original vortex twists off and forms a new spiral. Both spirals are clockwise when viewed in the direction of increasing y. For a Rayleigh number of 80,000 the points where the vortex splits are about midway between the centre vertical z-x plane and the front and back walls, respectively. On the front and back walls only one vortex is observed, and the slope of the isotherms is positive as is to be expected. Like in the one-vortex system the velocity component along the centre region of the two spirals is directed towards the vertical centre plane. There, the flow is deflected by an outward spiraling motion and finally returns back to the front wall and spirals inward. The temperature in the core of the spiral close to the cold wall is higher than the temperature in the core of the spiral close to the hot one. While the axis of the vortex near the hot wall is quite straight, the axis of the other one is considerably curved (Fig. 3b). Between these two configurations no hysteresis could be detected experimentally.

Further increase of the Rayleigh number results in a shift of the two vortices closer to the side walls z=0 and z=1, respectively, and the flow becomes strongly three-dimensional. This can be immediately deduced from the crossing of the particle traces visible in a vertical centre cross section when illuminated by a thin (2 mm) light sheet. The inner loops of the two spirals are strongly inclined with respect to the z-x plane in the neighbourhood of the front and back walls. For Ra = $7.5 \cdot 10^5$ on each side between the centre plane and the walls a third vortex is observed. Its pitch is so large that for the visualization in the experiment thick light sheets have to be applied. Fig. 2c shows a vertical cross section at y = 0.4 illuminated by a light sheet of 5mm. All spirals are turning clockwise. In the vicinity of the cores of the two spirals close to the hot and cold walls, respectively, the flow direction along their axes shows inward and outward components. There exist ringshaped regions around the vortex cores in which the tracer particles stay for several revolutions as if they were trapped. The flow observed at horizontal plane x = 0.5 has close to the front and back walls dominating velocity component into these walls (Fig. 3c).

The largest Rayleigh number realized up to now is about $3.2 \cdot 10^7$. In this case, water was used as a flow medium in order to avoid too high temperature differences T_h-T_c. The flow is still stationary and on the vertical mid-plane the two aforementioned vortices are also observed, with these cores now being very close to the corresponding side walls.

4 NUMERICAL MODELING

In conjunction with the experimental programme, a numerical study of this problem has commenced. This involves the solution of finite difference approximations to the equations of motion and energy. Two computer codes are in use. In one, the thermal conductivity, viscosity and specific heat of the fluid are assumed to be constant, while in the other this restriction is removed: polynomial functions of temperature are used to represent these properties. However, in the temperature range where the experiments are performed only viscosity varies significantly with temperature. The functional description of this dependence was done by fitting the approximating function to the measured values of the viscosity of the working fluid. The thermal conductivity and specific heat of the fluid were kept constant in the present calculations.

In each code the Boussinesq approximation is made: the density is assumed constant except in the buoyancy term of the equation of motion.

The constant property equations are well-known (e.g. [1,5]) and need not be stated here. The variable property equations, containing derivatives of the fluid properties, are extremely lengthy and will not be given here. They can be found in [6].

Also in each code, the thermal boundary conditions are sufficiently flexible to allow the imposition of arbitrary temperature, specified heat flux or specified heat transfer coefficient boundary conditions on each of the six surfaces of the box. In the present study, the two heated side walls are assumed to be isothermal. The non-adiabatic boundary conditions of the remaining four non-isothermal walls were approximated by assuming a constant heat flux through each wall. This heat flux was estimated by using the theory for the heat transfer through a thick, infinite wide plane plate of uniform conductivity into a gaseous environment. The ambient temperature was the room temperature (22°C). For the calculation of the heat flux coefficients the inner temperature of the wall was assumed to be constant. For the lid surface, this temperature was assumed to be T_h, for the floor surface T_c, and for the front and back walls $(T_h - T_c)/2$.

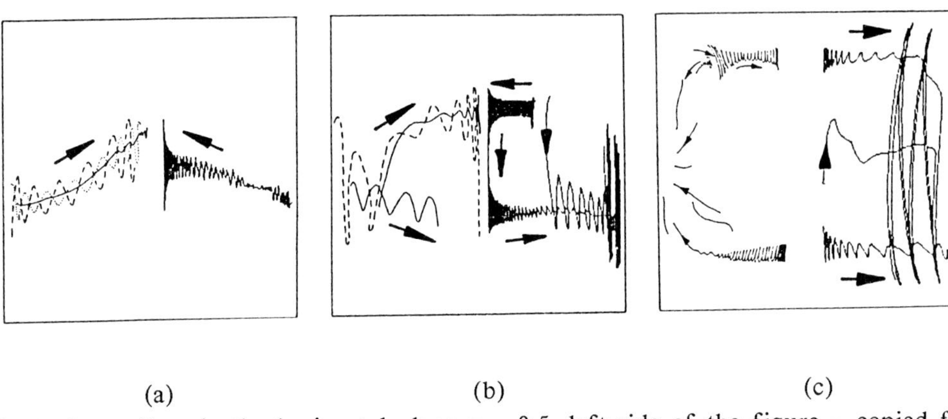

(a) (b) (c)

Fig. 3. Streamlines in the horizontal plane x = 0.5, left side of the figure - copied from the photographs, right - numerical simulation. (a) Ra = $2 \cdot 10^4$; (b) Ra = $8 \cdot 10^4$; (c) Ra = $7.5 \cdot 10^5$.

The two codes (variable and constant fluid properties) were developed for different purposes, but it is hoped that their parallel use in this work will enable a ready evaluation to be made of the effects of the temperature dependence of the fluid properties on the essential features of the flow. The motion of the fluid has been visualized from the computer solutions by the construction of particle tracks through the integration of the equations defining the velocity components (i.e., $\partial x/\partial t$ = u, etc.) The validity of the tracks has been verified in a number of ways including reduction of the time step and a reversal of all velocities to enable the track to be traversed in reverse. Initially all solutions were computed on a 21 by 21 by 21 mesh, chosen as a compromise between accuracy and cost. This mesh has been shown to be sufficient to yield errors of less that 2% for natural convection in air at values of Rayleigh number up to 10^5 [7]. The present computations are done for a much higher Prandtl number (and therefore much thicker boundary layers). It is confidently believed, therefore, that this mesh is more that adequate if a global description of the flow is of interest. However, the localization of singularities appeared to be more sensible of the mesh size, especially at higher values of Rayleigh number. Increasing the mesh size (up to 51x51x51) several test runs of the constant properties code were done to check how far the mesh size influences the form of the calculated tracks. It was found that for Ra > $4 \cdot 10^4$ a finer mesh of 31 by 31 by 31 must be employed if the code is used to detect local flow structures.

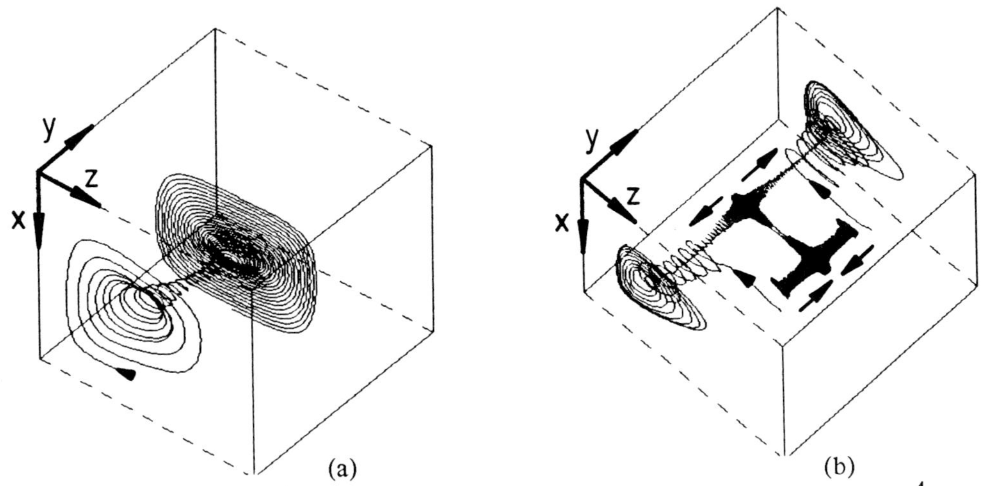

Fig. 4 Streamlines in perspective view - numerical simulation. (a) Ra = $2 \cdot 10^4$, single particle released near the front wall at the point (0.82,0.09,0.42); (b) Ra = $8 \cdot 10^4$, two particles released at the points (0.5,0.32,0.7) and (0.5,0.27,0.72) (other two released symmetrical to the centre plane y = 0.5). The first particle moves into the centre plane, then towards the hot wall and finally spirals towards the front wall. The second particle spirals directly to the front wall, merging with the spiral of the first one.

The variable viscosity of the fluid changes the symmetry of the flow with respect to the plane z = 0.5 (the flow at the hot side of the cavity is faster than at the cold one). Calculations performed with the variable property code give a very good quantitative description of the non-symmetrical velocity profiles in the centre region of the cavity (the calculated velocity components deviate within a few percent from the measurements reported in [3]). However, when comparing the tracks calculated with help of both codes it became obvious that the major influence on the flow structures is due to the thermal boundary conditions on the four non-isothermal walls, whereas variable fluid properties modify only slightly the calculated tracks. On the other hand, the huge storage needed for the variable property code limits (at the present time) its application to a 21x21x21 mesh. Therefore further numerical calculation were done with the constant property code, non-adiabatic thermal conditions on the side walls, and a uniform 31x31x31 mesh.

The preliminary numerical results which we have obtained reveal many of the same principal features of the flow: at low Rayleigh number, the movement of the fluid along a single spiral (Fig.4a) from the end walls (y = 0.1) towards the mid-plane (y = 0.5). Increasing Rayleigh number we observe a splitting of the single spiral into two in the region on either side of the mid-plane accompanying by a reversal of their direction. Also an additional singular point is detected for the spiral on the cold side of the cavity (Fig.4b). Such a point, observed in the experiments for higher Rayleigh number ($7.5 \cdot 10^5$), disappears in the numerical simulation made for this Rayleigh number.

There is a major discrepancy between the experimental and numerical results which we have not yet resolved: in the latter, the reversal of the direction of the spiral takes place at a Rayleigh number between 40,000 and 60,000 while in the former it has not been observed at 80,000 but at Ra = 750,000. Further experiments are being made in the intermediate range of Ra to identify the experimental critical value of Ra for this reversal. So far, we are unable to explain the discrepancy. Due to the unsolved experimental-numerical discrepancies we based our topological description of the flow exclusively on the experimental data.

5 DISCUSSION

The global topological structure of the one-vortex system (Ra = 20,000) is shown in Fig. 5a. The singularities in the corners and on the edges of the cube are either nodes or saddle points. They result from the v-component of the velocity generated by the front and back walls. Additionally on each of these walls there exists a focal point into which the streamlines from the wall spiral in, and on the vertical centre plane there is a third focal point from where the streamlines spiral out. The streamline connecting the front focal point with that of the centre plain shall only indicate that we have observed tracer particles which closely follow this path. For very low Rayleigh numbers there exist separation bubbles in the lower right front and back corners that disappear when the Rayleigh number increases.

The topological structure of the two-vortex system (Ra = 80,000) is shown in Fig. 5b. The singularities on the corners and the edges of the cube and those on the front and back walls remain unchanged, with exception of the singularity on the upper left edge where possibly a flow reversal occurs. In the center vertical plane two vortices are observed to which the two focal points separated by a saddle point correspond.

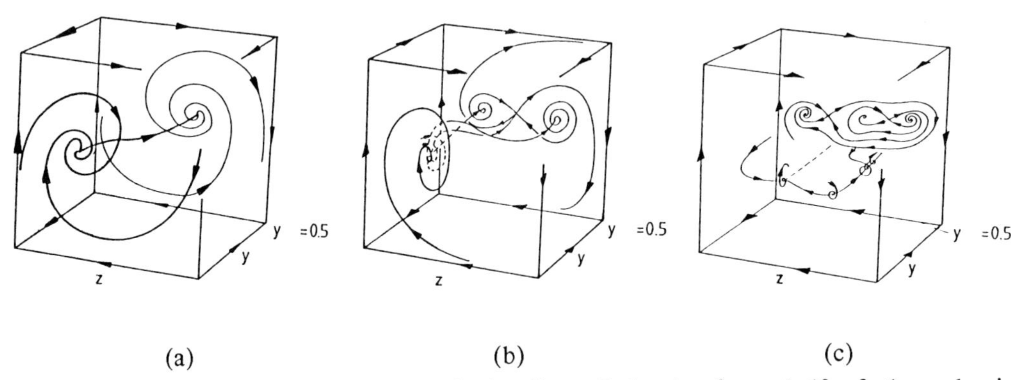

(a) (b) (c)

Fig. 5 Principal topological structure of the flow. Only the front half of the cube is displayed. (a) Ra = $2 \cdot 10^4$; (b) Ra = $8 \cdot 10^4$; (c) Ra = $7.5 \cdot 10^5$. (Arrows indicate the direction of streamlines).

The case where a third vortex becomes clearly visible (Ra = $7.5 \cdot 10^5$) is at the moment unsolved. In contrast to the one- and two vortex system, it is not even sure whether the final steady state, expected from the numerical results, has already been established. Some (but surely not all) of the additionally singular points have been found and could be defined. It seems that directly on the front and back walls still only one focal point still exists. That means that the topology on these walls remains unchanged. The interior of the cube, however, and in particular the regions close to the axes of the vortices where flow reversal with respect to the y-direction occurs, has not yet been fully analyzed. Therefore, the topological flow pattern proposed in Fig. 5c has only tentative character. Here, more detailed information is needed. The same is true for the embedding of the third vortex into the flow field. On the vertical center plane the aforementioned two vortices are still present, but the v-component of the vortex close to the hot wall has changed its sign.

Discrepancies between the calculations and observations (especially with respect to the flow direction along the axes of the spirals) seem not to be so serious as one would suspect at first. The net motion along a streamline into the y-direction is the result of alternating forward-backward movements. This also holds for the loops of the inner spirals which cross the positive and negative regions of the v components. The velocity distribution in the v-direction on a z-x plane (y = 0.5), loosely speaking, has a maximum in the left upper and right lower corners, and a minimum in the two other corners. In the centre of the plane there is a saddle point, and the velocity is zero on the walls. So, a small shift of the spiral axis due to unconvincing values of the fluid properties or deviations from the boundary conditions assumed, may deform the flow field. These speculations are supported by two facts: 1. The observed shape of the vortex axes deviates considerably from the calculated ones, and 2. the experimentally observed isotherms on the x-y planes are considerably shifted in the x-direction into the neighbourhood of the front and back walls (y = 0.1) in contrast to the calculated positions. This induced us to reformulate the boundary conditions for the heat flux, a work which is under way. As the numerical code enables us to prescribe the boundary temperature on the inner surfaces, it is planned to repeat the calculations by defining *a priori* the measured temperature distribution for all six walls.

The observed transition from one-vortex to two-vortex and also to three-vortex convection changes the number of critical points, i.e. local topological bifurcations occur [8]. Due to reversal of the direction of streamlines between singularities, global topological structures are also changed. The problem, whether one can define a critical number dividing these different topological structures, could not be solved. From the experimental point of view there are no sharply defined values for the Rayleigh number separating these structures. To answer these open questions there is a need to extend the present study by further experiments and improved numerical simulation.

6 LITERATURE

[1] Mallinson G.D., de Vahl Davis G., Three-dimensional natural convection in a box: a numerical study, J.Fluid Mech., **83**, pp.1-31, 1977.

[2] Kirchartz K.R., Dreidimensionale Konvektion in quaderförmigen Behältern, Habilitationsschrift, Universität Karlsruhe 1988.

[3] Hiller W.J., Koch St., Kowalewski T.A., Three-dimensional structures in laminar natural convection in a cube enclosure, Experimental Thermal and Fluid Scs. **2**, 34-44, 1989

[4] de Vahl Davis G., Laminar natural convection in an enclosed rectangular cavity, Int. J. Heat Mass Transfer, **11**, pp. 1675-1693, 1968.

[5] de Vahl Davis G., Finite difference methods for natural and mixed convection in enclosures, Heat Transfer 1982, vol.1, pp. 101-109, Hemisphere, Washington 1982.

[6] Yeah B.P., de Vahl Davis G., Behnia M., Program SOLCON 3D-solidification and convection of a viscous liquid in a box: Users'Manual. UNSW Report (in preparation)

[7] de Vahl Davis G., Convection of air in a square cavity: a bench mark numerical solution. Int. J. Num. Meth. Fluids **3**, 249-264, 1983.

[8] Dallmann U., Three-dimensional vortex structures and vorticity topology, Fluid Dyn. Res., **3**, pp.183-189,1988.

We are indebted to Peter Baumann from Max-Planck-Institut for his great help in displaying computer generated tracks.

HELICITY ASSOCIATED WITH FLOW AROUND FLUID LUMPS AND WITH INHOMOGENEOUS TURBULENCE

J. C. R. HUNT, DAMTP, University of Cambridge, England

and

F. HUSSAIN, Dept. of Mech. Eng., University of Houston, USA

SUMMARY

A brief review is given, with a new geometrical derivation, for the changes in velocity, vorticity and helicity of fluid elements and fluid volumes in inviscid flow. When a closed fluid volume V_b moves with a velocity $\mathbf{v}_b$ in a flow which at infinity has a velocity $\mathbf{U}_\infty$ and uniform vorticity $\boldsymbol{\Omega}$, it is shown that in general there is a net change ΔH_E in the integral of helicity H_E in the external region V_E *outside* the volume, i.e. $H_E = \int_{V_E} \mathbf{u}.\boldsymbol{\omega} dV \neq 0$ changes by ΔH_E where $\mathbf{u}$ and $\boldsymbol{\omega}$ are the velocity and vorticity fields. When the vorticity at infinity is weak (i.e. $|\boldsymbol{\Omega}| V_b^{1/3} \ll |\mathbf{U}_\infty - \mathbf{v}_b|$), the change in the external helicity integral, ΔH_e, is proportional to the dipole strength of V_b. For the case of volumes with two planes of symmetry parallel to their direction of motion (e.g. axisymmetric volumes), $\Delta H_E = -((\mathbf{v}_b - \mathbf{U}_\infty).\boldsymbol{\Omega}) V_b C_H$, where $C_H = (1 + C_M)/3$, and C_M is the added mass coefficient. So for a sphere moving along the axis of a rotating flow, $\Delta H_e = -\frac{1}{2} V_b\ (\mathbf{v}_b.\boldsymbol{\Omega})$, which is negative. Large values of the local helicity $(\mathbf{u}.\boldsymbol{\omega})$ are also generated by the flow around the volume when $(\mathbf{v}_b - \mathbf{U}_\infty).\boldsymbol{\Omega} = 0$, there is no net contribution to ΔH_e for symmetric volumes. These results are used to develop some new physical concepts about helicity in turbulent flow; in particular, it is suggested that vigorous entrainment at interfaces between turbulent and non-turbulent flow is associated with turbulence having relatively low fluctuations in local helicity.

1. REVIEW OF VORTICITY AND VELOCITY OF FLUID ELEMENTS

For inviscid adiabatic rotational flows, the vorticity $\omega(\mathbf{x}, t)$ and density $\rho(\mathbf{x}, t)$ at a point $\mathbf{x}$ and time t are related to the vorticity and density of the same fluid element located at $\mathbf{a}$ at time t_0, by Cauchy's result (expressed in suffix notation as) (Batchelor 1967).

$$\frac{\omega_i(\mathbf{x},t)}{\rho(\mathbf{x},t)} = \frac{\omega_j(\mathbf{a},t_0)}{\rho(\mathbf{a},t_0)} \left(\frac{\partial x_i}{\partial a_j}\right) \quad . \tag{1.1}$$

There is also the well-established theorem (see Serrin 1959, p.169; Weber 1868) for the change in the velocity $\mathbf{u}$ of the fluid element, namely

$$u_i(\mathbf{x},t) = u_j(\mathbf{a},t_0)\frac{\partial a_j}{\partial x_i} + \frac{\partial \Phi(\mathbf{x},t)}{\partial x_i} \quad , \tag{1.2}$$

where $\Phi(\mathbf{x}, t)$ is a scalar potential function of $\mathbf{x}$. This has been discovered and proved several times (e.g. Elsasser 1956), and applied to rapid distortion theory by Goldstein (1978), and Goldstein & Durbin (1980).

There is a simple geometrical and physical demonstration of the validity of (1.2), which we now give based on Kelvin's circulation theorem (see Fig. 1). Take the line integral along the path C_0, $\int \mathbf{u}.d\mathbf{l}$, between the point A at $\mathbf{a}$ and B at $\mathbf{b}$ at time t_0. After a time $(t-t_0)$ this loop has moved and been distorted. The points A and B have moved to A', B' and and C_0 changes to C_0'. Then consider an infinite number of possible partial circuits from B to A, e.g. C_n, and which are changed to C_n' from B' to A' at time t. The circulation theorem can be rewritten in terms of C_0 and any of these other partial circuits, i.e.

$$\int_{A'\ (C_0')}^{B'} \mathbf{u}.d\mathbf{l}(t) - \int_{A\ (C_0)}^{B} \mathbf{u}.d\mathbf{l}(t_0) = \int_{B\ (C_n)}^{A} \mathbf{u}.d\mathbf{l}(t_0) - \int_{B'\ (C_n')}^{A'} \mathbf{u}.d\mathbf{l}(t) \quad .$$

Since C_n or C_n' can be chosen arbitrarily and since a line integral that is independent of the path is only a function of its end points this difference can only be a scalar function Φ of the positions $\mathbf{x}$, $\mathbf{y}$ of A' and B'. Therefore

$$\int_{A'}^{B'} \mathbf{u}.d\mathbf{l}(t) = \int_{A}^{B} \mathbf{u}.d\mathbf{l}(t_0) + \Phi(\mathbf{y}) - \Phi(\mathbf{x}) \quad , \tag{1.3}$$

where $\Phi(\mathbf{x})$ is an unknown function of $\mathbf{x}$, t and t_0. It has the property that when $t = t_0$, $\Phi = 0$ for all $\mathbf{x}$.

Taking the limit of B' tending to A', and expressing the position of B' and B as $\mathbf{y} = \mathbf{x} + d\mathbf{x}(t)$ and $\mathbf{b} = \mathbf{a} + d\mathbf{a}(t_0) = \mathbf{a} + \frac{\partial \mathbf{a}}{\partial x_i} dx_i$, (1.3) reduces to

$$u_i(\mathbf{x},t)dx_i(t) = u_j(\mathbf{a},t_0)da_j(t_0) + \frac{\partial \Phi}{\partial x_k} dx_k(t) \quad . \tag{1.4}$$

Then, since each component of $d\mathbf{x}(t)$ can be taken as independent, (1.2) is recovered.

2. HELICITY OF FLUID ELEMENTS AND FLUID VOLUMES

The concept of the helicity density

$$h = \mathbf{u}.\omega(\mathbf{x},t), \tag{2.1}$$

and the helicity integral

$$H = \int (\mathbf{u}.\omega) dV \tag{2.2}$$

were introduced to fluid mechanics by Moffatt (1969). He considered this integral over the whole flow or over a region occupying part of the flow where the vorticity is non-zero and the region is such that it is bounded by a surface on which the normal component of vorticity is zero. Then H is conserved in inviscid flow where ρ is a unique function of pressure p (i.e. the fluid is barotropic) and any body forces are conservative. Obviously, h is not a conserved property and can indeed vary considerably in a turbulent flow.

Since then there have been many papers reporting on computations and measurements of h and H in various flows (Hussain 1986; Levich & Tsinober 1983; Rogers & Moin 1986; Andre & Lesieur 1977).

Almost no use has been made of the conservation condition (2.2) possibly because it has usually been interpreted as only being valid when the integral is taken over the whole region of the flow. (Moffatt's (1969) discussion of a blob of vorticity should be read carefully!). An important point of this note is to show how helicity integrals can and should be used within different regions of a flow. In §4 we consider in detail helicity in coherent structures.

From (1.4) and (1.1) the helicity density of a material element at time t can be expressed in term of its value at time t_0. Since

$$h(\mathbf{x},t) = \mathbf{u}.\omega(\mathbf{x},t) = \left[\frac{\rho(\mathbf{x},t)}{\rho(\mathbf{a},t_0)}\right] \mathbf{u}.\omega(\mathbf{a},t_0) + \boldsymbol{\omega}.\nabla\Phi \quad . \tag{2.3}$$

Note that $h(\mathbf{x},t)$ does not equal $h(\mathbf{a},t_0)$.

Now consider a set of non-overlapping volumes each of volume V_b, in which $\omega \neq 0$. Let the surface around each volume be S_b, such that $\omega.\hat{\mathbf{n}} = 0$, and let the flow within the space between these volumes be irrotational, i.e $\boldsymbol{\omega} = 0$ (see Fig 2) ($\hat{\mathbf{n}}$ is a unit vector).

Using (2.3), H_b at time t can be related to H_b of the same material volume at time t_0, since

$$H_b(t) = \int_{V_b} \frac{\rho(\mathbf{x},t)}{\rho(\mathbf{a},t_0)} (\mathbf{u}.\omega(\mathbf{a},t_0))\, d\mathbf{x} + \int_{V_b} \omega_k \frac{\partial\Phi}{\partial x_k} d\mathbf{x}$$

From the conservation of mass of each material element

$$\rho(\mathbf{x},t)d\mathbf{x} = \rho(\mathbf{a},t_0)d\mathbf{a}$$

and since $\nabla.\omega = 0$,

$$H_b(t) = \int_{V_b} (\mathbf{u}.\omega(\mathbf{a},t_0))\, d\mathbf{a} + \int_{S_b} \Phi(\omega.d\mathbf{S}) \tag{2.4}$$

where $d\mathbf{S}$ is the outward normal area element on S_b. Since $(\omega.d\mathbf{S}) = 0$ on S_b it follows that the helicity integral for each material volume as it moves around is constant, i.e.

$$H_b(t) = H_b(t_0) \tag{2.5}$$

In turbulent flows the total helicity integral H taken over the whole volume may or may not be zero. But the helicity integral H_b of *each* volume is conserved (if viscous diffusion is negligible), even though the interaction between the volumes may be strong, causing large changes of $\mathbf{u}$ and ω within the volumes. Also in the presence of any large scale irrotational straining of the flow H_b is conserved. In these interactions, the impulse $P_b = \frac{\rho}{2}\int(\mathbf{x}\wedge\omega)dV$ of each volume is generally changed (Hunt 1987).

The result (2.5) can be applied to flows with variation in density in space and time, provided that the fluid is barotropic within each volume and that any body forces per unit mass are conservative. In turbulent flows with significant density fluctuations it has become customary to use Favre-averaging (i.e. considering $\overline{\rho u_i u_i}$ as opposed to Reynolds-averaging (i.e. $\overline{u_i u_j}$) in analysing the dynamics (Favre 1969). The results (2.3)-(2.5) clearly suggest that, 'Favre' averaging does not lead to invariant integrals for vortical volumes. However, by considering helicity, it is possible to construct integrals that are invariant to barotropic density fluctuations.

3. HELICITY INDUCED OUTSIDE FLUID VOLUMES IN ROTATIONAL FLOWS

We consider the distribution of helicity in *steady inviscid rotational* flows *outside* closed volumes which are moving through the flow.

From this idealised flow some interesting suggestions emerge about helicity in turbulent flows. Consider a large volume of fluid in solid body rotation with vorticity $\boldsymbol{\Omega}$, into which a volume V_b with surface S_b is introduced. The volume moves steadily at a velocity $\mathbf{v}_b$. The volume might be a bubble or vortex ring, etc. It may or may not contain helicity itself (Fig. 3a). Let the diameter of the volume be 2a, which is of order $(V_b)^{1/3}$.

3.1 THE VOLUME MOVES PARALLEL TO THE ROTATION AXIS

The formal problem to be solved is to obtain the solution of

$$\frac{\partial\omega}{\partial t} + (\mathbf{u}.\nabla)\omega = (\omega.\nabla)\mathbf{u} \quad , \tag{3.1}$$

where $\nabla.\mathbf{u} = 0$ and $\omega = \nabla\wedge\mathbf{u}$, subject to $\mathbf{u}.\hat{\mathbf{n}} = \mathbf{v}_b.\hat{\mathbf{n}}$ on S_b, where the normal vector $\mathbf{n} = 0$ and $\hat{\mathbf{n}}$ is the unit normal vector. Far from the volume, as $|\mathbf{n}|/a \to \infty$, $\mathbf{u} \to \boldsymbol{\Omega}\wedge\mathbf{x}$. The solution for ω in terms of $\mathbf{u}$ is obtained by rewriting (3.1) in co-ordinates $\mathbf{x}'$ moving with the body where $\mathbf{x}' = \mathbf{x} - \mathbf{v}_b t$, where $\mathbf{x} = \mathbf{v}_b t$ is the location of a reference point (e.g. the centre) of the volume, and in terms of the relative velocity $\mathbf{u}' = \mathbf{u} - \mathbf{v}_b$, since the flow in these co-ordinates is steady, (3.1) becomes

$$(\mathbf{u}'.\nabla)\omega = (\omega.\nabla)\mathbf{u}' \tag{3.2a}$$

$$\text{where} \quad \mathbf{u}'.\hat{\mathbf{n}} = 0 \quad \text{on} \quad \mathbf{n} = 0, \tag{3.2b}$$

$$\text{and} \quad \left.\begin{array}{l}\mathbf{u}' \to \boldsymbol{\Omega}\wedge\mathbf{x}' - \mathbf{v}_b \\ \omega = \boldsymbol{\Omega}\end{array}\right\} \quad \text{as} \quad |\mathbf{n}|/a \to \infty \quad . \tag{3.2c}$$

We consider the solution where the background rotation is weak, compared with the gradient in the velocity field around the moving volume, so

$$\epsilon \ll 1, \qquad \text{where} \quad \epsilon = \frac{a\Omega}{v_b} \tag{3.3}$$

$$\text{and} \quad v_b = |\mathbf{v}_b|, \quad \Omega = |\mathbf{\Omega}|$$

For future reference

$$\hat{\mathbf{v}}_b = \frac{\mathbf{v}_b}{v_b} \quad . \tag{3.4}$$

The solutions for $\mathbf{u}^{'}$, $\boldsymbol{\omega}$ can be expanded as expansions in ϵ, normalised on v_b and a.

$$\left.\begin{aligned} \mathbf{u}^{'} &= v_b \left(\mathbf{u}^{'(0)} + \epsilon \mathbf{u}^{'(1)} + \ldots\right) \\ \boldsymbol{\omega} &= \frac{v_b}{a}\left(\boldsymbol{\omega}^{'(0)} + \epsilon \boldsymbol{\omega}^{'(1)} + \ldots\right) \end{aligned}\right\} \quad . \tag{3.5}$$

where as $|\mathbf{n}|/a \to \infty$,

$$\mathbf{u}^{'(0)} \to -\hat{\mathbf{v}}_b, \quad \mathbf{u}^{'(1)} \to \frac{\mathbf{\Omega} \wedge \mathbf{x}'}{a\Omega} \tag{3.6a}$$

$$\text{and} \quad \boldsymbol{\omega}^{'(0)} = 0, \quad \boldsymbol{\omega}^{'(1)} = \frac{\mathbf{\Omega}}{\Omega} \quad . \tag{3.6b}$$

Therefore, $\boldsymbol{\omega}^{'(0)}$ is zero everywhere and $\mathbf{u}^{'(0)}$ is an irrotational velocity field, with boundary conditions (3.2b) and (3.6a). The first order vorticity satisfies

$$\left(\mathbf{u}^{'(0)}.\nabla\right)\boldsymbol{\omega}^{'(1)} = \left(\boldsymbol{\omega}^{'(1)}.\nabla\right)\mathbf{u}^{'(0)} \quad . \tag{3.7}$$

Since $\boldsymbol{\omega}^{'(0)}$ is parallel to $\mathbf{u}^{'(0)}$ as $|\mathbf{n}| \to \infty$, (from 3.6a,b) it follows that the solution to (3.7) is

$$\boldsymbol{\omega}^{'(1)}(\mathbf{x}') = \alpha \mathbf{u}^{'(0)}(\mathbf{x}') \quad \forall\, \mathbf{x}'(|\mathbf{n}| \geq 0), \tag{3.8}$$

$$\text{where} \quad \alpha = -\frac{\mathbf{\Omega}.\mathbf{v}_b}{(v_b\Omega)} \quad . \tag{3.9}$$

(This a well known solution; see review by Hunt 1987.) Thence the helicity is given by

$$\begin{aligned} h = (\mathbf{u}.\boldsymbol{\omega}) &= v_b \frac{v_b}{a}\left[\left(\mathbf{u}^{'(0)} + \hat{\mathbf{v}}_b\right).\epsilon\alpha\mathbf{u}^{'(0)}\right] \\ &= -(\mathbf{\Omega}.\mathbf{v}_b)(\mathbf{u}^{'(0)} + \hat{\mathbf{v}}_b).\mathbf{u}^{'(0)} \quad . \end{aligned} \tag{3.10}$$

Around any volume the relative velocity $\mathbf{u}^{'}$ and therefore $\mathbf{u}^{'(0)}$ is parallel to the surface S_b. The tangential component of the velocity in fixed co-ordinates, $(\mathbf{u}^{'(0)} + \hat{\mathbf{v}}_b)$ has a component parallel to $\mathbf{u}'^{(0)}$ around bodies which are symmetrical to the flow direction. Therefore around the surface of these bodies h is negative if $\mathbf{\Omega}.\mathbf{v}_b$ is positive. Far upstream and downstream, on the stagnation streamline, $\mathbf{u}'^{(0)}$ is approximately equal to $-\mathbf{v}_b$, but $(\mathbf{u}'^{(0)} + \hat{\mathbf{v}}_b)$ has the same sign as $\mathbf{v}_b$ and is very small compared to unity. Therefore in these regions $(\mathbf{u}'^{(0)} + \hat{\mathbf{v}}_b).\mathbf{u}'^{(0)}$ is negative and h is positive if $\mathbf{\Omega}.\mathbf{v}_b$ is positive (Fig. 3a).

The external helicity integral H_E over the external volume V_E can be calculated from (3.10), for general flows satisfying (3.3). Since $\mathbf{u}'^{(0)}$ is irrotational, we can express $\mathbf{u}'^{(0)}$, $\mathbf{u}'^{(0)} + \hat{\mathbf{v}}_b$ in terms of a normalised potential ϕ as:

$$\mathbf{u}'^{(0)} + \hat{\mathbf{v}}_b = \nabla\phi, \tag{3.11}$$

where $\nabla^2\phi = 0$, and from (3.10)

$$H_E = \int_{V_E} h dV = -(\mathbf{\Omega}.\mathbf{v}_b) \int_{V_E} [\nabla\phi.\nabla\phi - \nabla.(\phi\hat{\mathbf{v}}_b)]\, dV$$

$$= -(\mathbf{\Omega}.\mathbf{v}_b) \left\{ \int_{S_\infty} [\phi\nabla\phi - \phi\hat{\mathbf{v}}_b]\, .\hat{\mathbf{n}} dS - \int_{S_b} [\phi\nabla\phi - \phi\hat{\mathbf{v}}_b]\, .\hat{\mathbf{n}} dS \right\} \quad , \tag{3.12a}$$

where S_∞ is an arbitrary surface far from V_b.

Since $(\nabla\phi.\hat{\mathbf{n}}) = (\hat{\mathbf{v}}_b.\hat{\mathbf{n}})$ on S_b, and $\nabla\phi \to O(r^{-3})$ as $r \to \infty$,

$$H_E = (\mathbf{\Omega}.\hat{\mathbf{v}}_b) \int_{S_\infty} (\hat{\mathbf{v}}_b.\hat{\mathbf{n}})\phi dS \quad \text{as} \quad r \to \infty \tag{3.12b}$$

$$= -\frac{4}{3}\pi C_1 (\mathbf{\Omega}.\mathbf{v}_b) V_b \quad ,$$

where C_1 is a constant depending on the dipole strength of the volume (Batchelor 1967, Chap. 6). In other words, H_E only depends on the effect of the volume on the flow in the far field.

For the class of shapes of volume V_b that have two planes of symmetry which are both parallel to $\mathbf{v}_b$, the dipole is aligned with the flow. Then the integral over the surface at infinity (3.12b) can be expressed in terms of the kinetic energy of the velocity field *outside* the volume $(\frac{1}{2}C_M \rho v_b^2 V_b)$ and the kinetic energy of the fictional velocity within the volume $(\frac{1}{2}\rho v_b^2 V_b)$, so that

$$C_1 = \frac{1}{4\pi}(1 + C_M) \quad , \tag{3.12c}$$

where

$$C_M = \frac{2}{V_b} \int_{V_E} (\nabla\phi)^2 dV = -\frac{2a}{V_b} \int_{S_b} \phi\hat{\mathbf{n}}.\hat{\mathbf{v}}_b dS \quad ,$$

(Batchelor 1967, pp.398-403). Then (3.12b) becomes

$$H_E = -C_H(\mathbf{\Omega}.\mathbf{v}_b)V_b, \qquad \text{where} \qquad C_H = (1 + C_M)/3 \quad . \tag{3.12d}$$

One implication of (3.12d) is that any elongated cylindrical shape moving parallel to itself (e.g. a needle) produces a negligible change in kinetic energy (i.e. $C_M = 0$), but because it displaces the flow it induces a change in helicity, with the coefficient $C_H = 1/3$.

However for volumes that significantly disturb the flow, C_M and therefore C_H are increased. For a sphere, $C_M = 1/2$, $C_H = 1/2$; for a circular cylinder moving normal to its axis, $C_M = 1$, $C_H = 2/3$, and for a disc moving normal to its axis, the high kinetic energy at the edges leads to $C_M = 8/3$, and thence $C_H = 11/9$.

Thus, from (3.10), (3.12) any symmetrical *fluid* volume moving with velocity $\mathbf{v}_b$ parallel to a weak rotational motion produces a net helicity integral

$$H = -(\mathbf{\Omega}.\mathbf{v}_b)V_b C_H + H_b \quad , \tag{3.13}$$

where H_b is the helicity integral *within* the volumes.

If there is a mean uniform motion $\mathbf{U}_\infty$ parallel to $\mathbf{\Omega}$ the helicity density is non-zero far from the volume, so the integral (3.12a) does not converge. However by considering the *change* in the net helicity integral, ΔH_E, defined by

$$\Delta H_E = \int_{V_E} (\omega.\mathbf{u} - \mathbf{\Omega}.\mathbf{U}_\infty)dV \quad ,$$

$$|\mathbf{v}_b - \mathbf{U}_\infty| >> \mathbf{\Omega} a \quad , \tag{3.14a}$$

and if the vorticity is weak enough that the same analysis as in (3.12) can be used, whence

$$\Delta H_E = ((\mathbf{U}_\infty - \mathbf{v}_b).\mathbf{\Omega})V_b C_H \quad . \tag{3.14b}$$

Note that (3.12) and (3.14) do *not* contradict the result that the helicity integral H is constant in inviscid flow. If we consider a finite volume V of fluid in rotation with $\mathbf{U}_\infty = 0$, then as each volume V_b enters it, the integral H in V is increased by ΔH_E (from Moffatt 1969), where ΔH_E is determined by the integral involving the flow across the entering surface S_V (Fig. 3a). It can be shown that

$$\Delta H_E = -\int\int_{S_V} \left\{ (\mathbf{u}.\hat{\mathbf{n}})(\mathbf{u}.\omega) + (\omega.\hat{\mathbf{n}}) \left[\frac{p}{\rho} - \frac{u^2}{2} \right] \right\} dS_V dt + H_b \quad , \tag{3.15}$$

where $\hat{\mathbf{n}}$ is the unit normal vector out of V, and the time is that taken for the volume to enter V. Note that $\frac{p}{\rho} = -\frac{u^2}{2} + v_b u_1$. For the case of weak rotation, this integral reduces exactly to (3.12a), which confirms the result.

In general it is simpler to estimate H from the local motion rather than from the production of H by motions across the boundary or by viscous processes, so we concentrate on the former direct approach.

3.2 VOLUMES MOVE PERPENDICULAR TO THE ROTATION AXIS

If in our model experiment, the volume V_b moves with velocity $\mathbf{v}_b = (v_b, 0, 0)$ *perpendicular* to the imposed vorticity $\mathbf{\Omega} = (0, \Omega, 0)$, the vorticity and velocity outside V_b is distorted. By again taking weak vorticity, i.e. $|\Omega| \ll \frac{v_b}{a}$, it is possible to use the result (1.1) (with the methods reviewed by Hunt (1987)).

Once again we analyse the change in ω by considering a steady flow with a steady incident velocity $(-\mathbf{v}_b)$ in a frame of reference moving with the volume. The weak vorticity is transported and distorted by the potential flow $\mathbf{u}'^{(0)}$ around the volume. Since ω is initially perpendicular to the approach flow $(-\mathbf{v}_b)$ it lies in surfaces where the 'drift' or 'time of flight' $T(\mathbf{x})$ (of the potential flow) is constant. These drift surfaces are material surfaces which are deformed as they are transported over the volume. Consequently the vortex lines remain within the surfaces and are therefore perpendicular to ∇T (Fig. 3b). Using the theory originated by Lighthill (1957), and later applied by Durbin (1981) and Hunt (1987), and using the fact that $\nabla T.\mathbf{u}'^{(0)} = 1$ (in normalised form), ω can be expressed in terms of ∇T and $\mathbf{u}'^{(0)}$ (to first order). Thence

$$\boldsymbol{\omega}.\mathbf{u} = -(\nabla T \wedge (\hat{\mathbf{x}}_1 \wedge \hat{\mathbf{x}}) \cos\theta . \mathbf{u}'^{(0)}) v_b \Omega \quad , \tag{3.16}$$

where $\cos\theta = \hat{\mathbf{R}}.\mathbf{\Omega}/\Omega$, $\hat{\mathbf{R}} = -(\hat{\mathbf{x}} \wedge \hat{\mathbf{x}}_1) \wedge \hat{\mathbf{x}}_1 = \hat{\mathbf{x}} - (\hat{\mathbf{x}}_1.\hat{\mathbf{x}})\hat{\mathbf{x}}_1$ and $\hat{\mathbf{x}}, \hat{\mathbf{x}}_1$ are unit vectors in the $\mathbf{x}$ and $\mathbf{x}_1$ directions.

Far downstream of the volume the drift surfaces, and therefore ω, become nearly parallel to the x-axis (assumed to be the axis of symmetry) (Hawthorne & Martin 1955). But the large component parallel to the x-axis, ω_1, has opposite signs above and below the plane $y = 0$ (or $\boldsymbol{\omega}.\hat{\mathbf{x}}_1 \gtrless 0$ for $\mathbf{x}.\mathbf{\Omega} \lessgtr 0$, respectively). The maximum positive and negative values of helicity above and below the $y = 0$ plane are given by

$$h \approx -v_b(\nabla T.\mathbf{\Omega}) \quad . \tag{3.17}$$

Since any fluid element approaching V_b on the stagnation line takes an infinite time T to reach the x-axis downwind of V_b, $T(R)$ and $\nabla T(R)$ tend to infinity as the distance R from the x-axis tends to zero. Therefore from (3.17) strong helicity variations are generated near the axis of the volume (and on its surface), and they persist far downstream of the volume. In the previous case of $\mathbf{v}_b$ parallel to $\mathbf{\Omega}$, there was no effect on h far downstream of the volume.

The next step is to consider the contribution to the net helicity integral H_E. From (3.17) it follows that over the area (e.g. A_∞) of any plane downwind of V_b,the integral of helicity is zero. Therefore to evaluate the integral of helicity around the volume H_E, we need only consider the integral near the volume. The component of vorticity parallel to the surface becomes very large as the vortex lines are wrapped around the surface.

Near the surface of V_b, where the distance normal to the surface $n = |\mathbf{n}|$ is small compared with a, the helicity (from (3.16)) is

$$h = \boldsymbol{\omega}.\mathbf{u} \approx O\left((\nabla T.\hat{\mathbf{n}})\,\Omega v_b\right) \propto \frac{\Omega a v_b}{n} \quad . \tag{3.18}$$

Because $h \propto \left(\frac{1}{n}\right)$ as $n \to 0$, H_E is a divergent volume integral. Suppose the integral is taken to a small distance $n_0(<< a)$ from the surface of V_b, then from the symmetry it is clear that $H_E(n > n_0) = \int_{V_E} (\boldsymbol{\omega}.\mathbf{u})dV$ can only be non-zero, if the flow around V_b is asymmetric.

Since $H_E = 0$ for a sphere moving perpendicular to weak vorticity it follows that the change in external helicity ΔH_E can be computed for a sphere moving with a velocity $\mathbf{v}_b$ in a general flow field $\mathbf{U}_\infty$ with vorticity $\boldsymbol{\Omega}$ provided (3.14) is satisfied and $a \ll \left(\frac{|\mathbf{U}_\infty - \mathbf{v}_b|}{\|\nabla \mathbf{U}_\infty\|}\right)$. The result is the same as (3.14b), viz:

$$\Delta H_E = \frac{1}{2}((\mathbf{U}_\infty - \mathbf{v}_b).\boldsymbol{\Omega})V_b \quad . \tag{3.19}$$

4. FURTHER COMMENT ON HELICITY IN TURBULENT FLOW

4.1 HELICITY STATISTICS

In this section we consider helicity in turbulent flows in the absence of any large-scale rotation or acceleration. In some flows it is of interest to distinguish between the *contributions* to the helicity density h from the mean velocity and vorticity $\overline{\mathbf{u}}, \overline{\boldsymbol{\omega}}$ and from all kinds of random fluctuations $\mathbf{u}_r$, $\boldsymbol{\omega}_r$.

It can be more revealing to focus on the large-scale, slowly-changing but randomly moving coherent structures in turbulent flows. In these flow regions we distinguish between the large-scale coherent velocity and vorticity fields $\mathbf{u}_c, \boldsymbol{\omega}_c$ which are correlated across the structure, and the incoherent random fields $\mathbf{u}_{rc}, \boldsymbol{\omega}_{rc}$, which include small-scale motions and (at high Reynolds number) the motions with highest vorticity (Hussain 1986). The mean fields $\overline{\mathbf{u}}, \overline{\boldsymbol{\omega}}$ are largely determined by $\mathbf{u}_c, \boldsymbol{\omega}_c$, but the random fields have contributions from the coherent and incoherent fields. (For convenience, $\mathbf{u}_c, \boldsymbol{\omega}_c, \mathbf{u}_{rc}, \boldsymbol{\omega}_{rc}$ are defined in a frame moving with the structure.)

Since the helicity is a product, the mean helicity $\overline{h}$, and for coherent sructures, the coherent helicity h_c, have contributions from the random and incoherent fields, respectively. Thus

$$\overline{h} = \overline{\mathbf{u}}.\overline{\boldsymbol{\omega}} + \overline{h_r}, \quad \text{where} \quad \overline{h_r} = \overline{\mathbf{u}_r.\boldsymbol{\omega}_r} \quad , \tag{4.1}$$

and

$$h_c = \mathbf{u}_c.\boldsymbol{\omega}_c + \{h_{rc}\}, \quad \text{where} \quad h_{rc} = \mathbf{u}_{rc}.\boldsymbol{\omega}_{rc} \quad , \tag{4.2}$$

where { } denotes averaging over an ensemble of similar coherent structures.

The r.m.s. of the fluctuations of h_r and h_{rc} about their mean values are also of interest, as defined by

$$h_r' = \left[\overline{\left(h_r - \overline{h}_r\right)^2}\right]^{\frac{1}{2}} \quad , h_{rc}' = \left\{\left((h_r) - \{h_{rc}\}\right)^2\right\}^{\frac{1}{2}} . \tag{4.3}$$

When h_r' is large but $\overline{h}_r$ is small, it means that $\mathbf{u}$ and $\boldsymbol{\omega}$ are aligned; but they are not on average in the same direction. Note that if the angle between the vectors $\mathbf{u}_r$ and $\boldsymbol{\omega}_r$ is θ, in isotropic turbulence $\overline{\cos^2\theta} = \frac{1}{3}$ (Rogers & Moin,1987).

From the idealised calculation of §3, we can estimate the possible magnitude of these helicity statistics for a flow field of distinct 'eddies' (denoted by t for 'tourbillon') moving randomly through a velocity field $\mathbf{U}$ with weak vorticity $\boldsymbol{\Omega}$. Λ is the proportion of the total volume occupied by the 'tourbillons', which could be buoyant thermals or vortex rings (Fig. 4a) and they move relative to $\mathbf{U}$ and (in their frame of reference) have approximately closed streamline surfaces around them. If they are energetic enough, they largely determine the random components of the whole flow. Therefore, these 'eddies' are localised volumes with a net motion in a particular direction producing a positive high velocity within the volume and upstream and downstream of the volume, but a negative velocity outside them at their 'equator'. In a frame moving with the mean velocity, there must be a slow net back flow between the eddies. [For details of buoyant thermals, see Hunt, Kaimal & Gaynor 1988.]

The mean helicity from these random motions $\overline{h}_r$ can be estimated from (3.19). Assuming that the eddies are on the average symmetrical, then there is no contribution to $\overline{h}_r$ from the flow external to eddies moving perpendicular to $\boldsymbol{\Omega}$. One contribution comes from the mean helicity of the 'tourbillons' $\{H_t\}$, i.e. the integral of $\int_{V_t}(\mathbf{U}+\mathbf{u}).(\boldsymbol{\Omega}+\boldsymbol{\omega})dV$ over *all the 'eddies'*. This produces a contribution $\overline{h}_t$ when averaged *over all space.* The other contribution comes from the motion outside those eddies moving parallel or anti-parallel to $\boldsymbol{\Omega}$. Therefore $\overline{h}_r \approx -C_{H_t}\Lambda(\boldsymbol{\Omega}.\mathbf{v}_t) + \overline{h}_t$ where the mean contribution from the flow within the 'eddies' is $\overline{h}_t \approx \frac{\Lambda\{H_t\}}{V_t}$. Typically the coefficient C_{H_t} varies between 1/3 and 1/2 as the eddy geometry varies from elongated to spherical.

This model applies to weakly rotating flow, if non-helical eddies (such as simple vortices and thermals) are generated so as to produce a turbulent velocity field whose *large scales* have skewed non-Gaussian probability distributions. For example, the external flow around buoyant thermals with a typical velocity w_* rising in a rotating system can contribute a mean helicity proportional to $(\boldsymbol{\Omega}.\mathrm{g})w_*$. However, within the thermals the helicity is likely to be of opposite sign caused by concentration of vorticity by entrainment.

Also fluctuations in helicity are produced by eddies moving both parallel or perpendicular to the mean rotation $\mathbf{\Omega}$. The analysis of §3.2 shows that there may be regions of rather large helicity fluctuation just outside eddies or coherent structures moving through a rotational large-scale flow. In cases where $(\mathbf{u}.\mathbf{\Omega})$ is large, these regions of high local helicity fluctuation are associated with regions of high vortex stretching and large dissipation.

4.2 HELICITY AND ENTRAINMENT

The main reasons given for considering helicity have been: that it is interesting because it is a conserved dynamical and topological quantity; that it *may* be an indicator of non-linear interactions, and that it indicates the potential for the generation of magnetic fields in conducting fluids. However information about helicity may have a more immediate practical use, because it is probably related to *entrainment* at a boundary between turbulent and non-turbulent motions, such as occur in jets, wakes, mixing layers, boundary layers, etc.

A surface S of a coherent structure lies between rotational and approximately irrotational flow, and is defined by $S(\mathbf{x}_s) = 0$ with unit normal vector $\hat{\mathbf{n}}$. Since vorticity is solenoidal and vortex sheets do not exist at such surfaces, the normal component of vorticity must be zero ($\boldsymbol{\omega}.\hat{\mathbf{n}} = 0$) (Fig. 4c). Therefore, $\boldsymbol{\omega}$ is parallel to the surface which implies that $\boldsymbol{\omega}_{rc}$ is also approximately parallel to S, since $|\boldsymbol{\omega}_{rc}|$ is large compared with $|\boldsymbol{\omega}_c|$. However, if the surface *moves* relative to the local large-scale flow, $\mathbf{u}_c$, there must be a component of the velocity fluctuation perpendicular to the surface, i.e $(\mathbf{u}_{rc}.\hat{\mathbf{n}}) \neq 0$.

Therefore if $|\mathbf{u}_{rc}.\boldsymbol{\omega}_{rc}|$ is small compared with $u_0'^2/\ell$, where ℓ is the integral scale of incoherent eddies and $u_0' = \{u_{rc}^2\}^{1/2}$, the surface S can move into the irrotational flow around the coherent structure, i.e. the criterion is that $\hat{h}_{rc} = |\mathbf{u}_{rc}.\boldsymbol{\omega}_{rc}|/(u_0'^2/\ell) << 1$. The mean rate of movement of such an interface, in some local frame (e.g. at the edge of a vortex or a jet), is called the boundary entrainment velocity (E_b). If E_b is defined in a frame moving with coherent velocity $\mathbf{u}_c$, then it must depend on the modulus of the local *helicity* of the incoherent turbulence as well as the magnitude of local velocity fluctuations u_0' and on the structure of large scales of motion (being independent of Reynold's number). Thus where $\hat{h}_{rc} \sim 1$, $\frac{E_b}{u_0'}$ is small and where $\hat{h}_c << 1$, $E_b/u_0' \sim 1$. (For a recent discussion of E_b see Turner 1986 and Hunt et al. 1983. Note that E_b may differ in magnitude and *sign* from other definitions of entrainment velocity.)

In the large scale structures of jets, $\hat{h}_{rc}$ is small and this is consistent with $\frac{E_b}{u_0'}$ being of order unity. In the coherent structures measured in mixing layers h_{rc}' has been measured and reviewed by Hussain (1986) (see Fig. 4d). At the stagnation point at the centre of the ribs, there is irrotational straining by the large scale motion, transporting external turbulence into the ribs and amplifying the local turbulence. Thus $\{\omega_{rc}^2\}$ and $\{\mathbf{u}_{rc}^2\}$ increase, but the helicity $(\mathbf{u}_{rc}.\boldsymbol{\omega}_{rc})$ of a fluid element or a fluid volume, remains approximately constant during the straining, so that the normalised helicity h is small. In this region the turbulence is strong enough that the boundary entrainment velocity E_b balances the mean straining velocity ($\approx u_0'$) (which opposes the spreading of the turbulent region), i.e. $\frac{E_b}{u_0'} \approx 1$. However, in the *rotational* regions of the vortex structures measurements show that $\hat{h}_{rc}$ is larger, and $\frac{E_b}{u_0'}$ is small.

Maxworthy (1974) shows how the entrainment at the surface of vortex rings is very weak ($\frac{E_b}{u_0'} \ll 1$), *and* that the helicity is large. In this case it is associated with wavelike motions along the vortex, in which motions perpendicular to the vortex lines are suppressed.

This hypothesis is consistent with a general dynamical argument. Where the average modulus of helicity is relatively large in a turbulent flow, the production of velocity fluctuations and vorticity diffusion is small, and therefore the entrainment is weak (such as in the 'roller' regions of a mixing layer). On the other hand, helicity of small-scale turbulence is weak in the presence of large-scale straining, where there is stretching of small-scale vorticity and strong velocity fluctuations normal to the small-scale vorticity, such as occurs in the region of low helicity density at the saddle points of mixing layers. Entrainment is large here because these large normal velocity fluctuations diffuse the mean vorticity, and because high gradients of amplified vorticity also amplify its viscous diffusion.

This preliminary examination of fluctuating helicity in coherent structures will be extended.

Acknowledgement

We are grateful for conversations with H. K. Moffatt, and S. R. Ramsay, who said he could not understand (1.2), and for criticisms from referees. This work was supported by Dr. O. Manley under DOE Grant DE - FG05 - 88ER13839.

References

André J. C. & Lesieur M. 1977 J. Fluid Mech. **81**, 187-207.

Batchelor G. K. 1967 Introduction to Fluid Mechanics, C.U.P.

Durbin, P.A. 1981 Quart. J. Mech. Appl. Math. **34**, 489-500.

Elsasser, W. M. 1956 Rev. Modern Physics **28**, 135-163.

Favre A. 1969 S.I.A.M. ; 237-266.

Goldstein, M. E. 1978 J. Fluid Mech. **89**, 433-468.

Goldstein, M. E. & Durbin, P. A. 1980 J. Fluid Mech. **98**, 473-508.

Hawthorne, W.R. & Martin, M.E. 1955 Proc. Roy. Soc. Ser.A **232**, 184-195.

Hunt, J.C.R. 1983 J. Fluid Mech. **61**, 625-706.

Hunt, J.C.R. 1987 Trans. Can. Soc. Mech. Eng. **11**, 21-35.

Hunt, J.C.R., Rottman, J.W. & Britter, R.E., 1984 Proc. IUTAM Symp., Atm. Disp. Heavy Gases. (Ed G. Oom & H. Tennekes), Springer, pp.361-395.

Hunt, J.C.R., Kaimal, J. & Gaynor, J.E., 1988 Quart. J. Roy. Met. Soc. **114**, 827-858.

Hussain, A.K.M.F. 1986 J. Fluid Mech. **173**, 303-356.

Ishii, K. & Hussain, A.K.M.F. 1989 Advances in Turbulence (2), Springer (in press).

Levich, E. & Tsinober, A. 1983 Phys. Lett. **93A**, pp.293-297.

Lighthill, M.J. 1956 J. Fluid Mech. **1**, 31-53.

Maxworthy 1974 J. Fluid Mech. **64**, 227-239.

Moffatt, H.K, 1969 J. Fluid Mech. **35**, 117-129.

Moffatt, H.K., 1978 Magnetic Field Generation in Electrically Conducting Fluids. Cambridge Univ. Press.

Rogers, M.M. & Moin. P. 1987 Phys. Fluids **30**, 2662-2671.

Serrin, J., 1959 Handbuch der Physik vol VIII/I, 125-263, Berlin.

Turner, J.S., 1986 J. Fluids Mech. **173**, 431-471.

Weber (1868) Über eine Transformation der hydrodynamischen Gleichungen. J. Reine Angew. Math. **68**, 286.

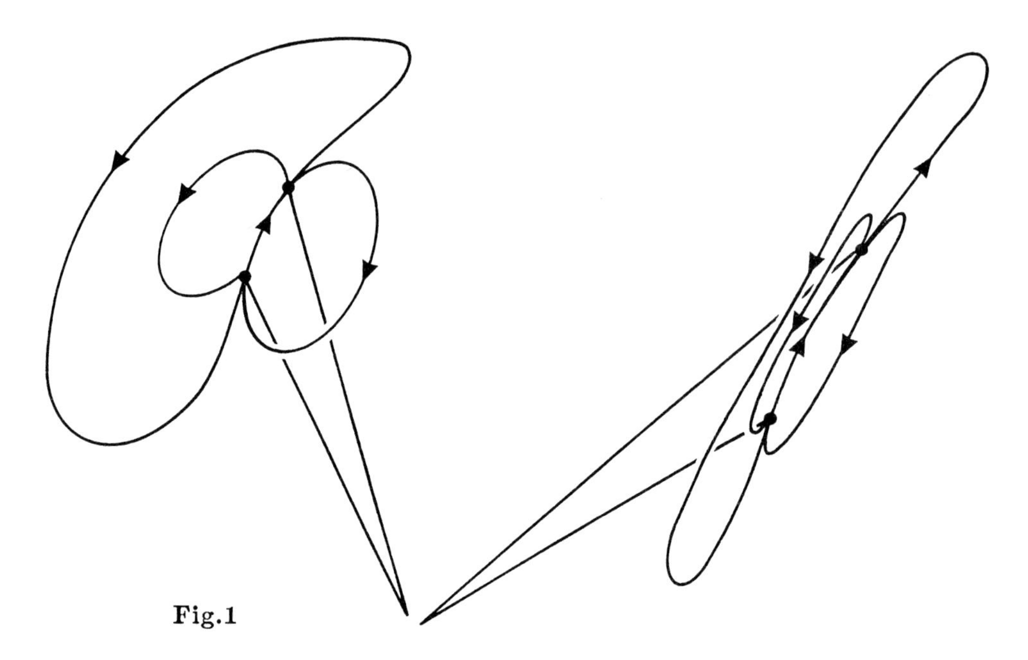

Fig.1

Showing how the material line C_0 between AB moves to C_0' between $A'B'$ and also how the 'partial' circuits $C_1, C_2, C_3 \ldots$ change to $C_1', C_2', C_3' \ldots$. Kelvin's theorem is taken round C_0 and each partial circuit.

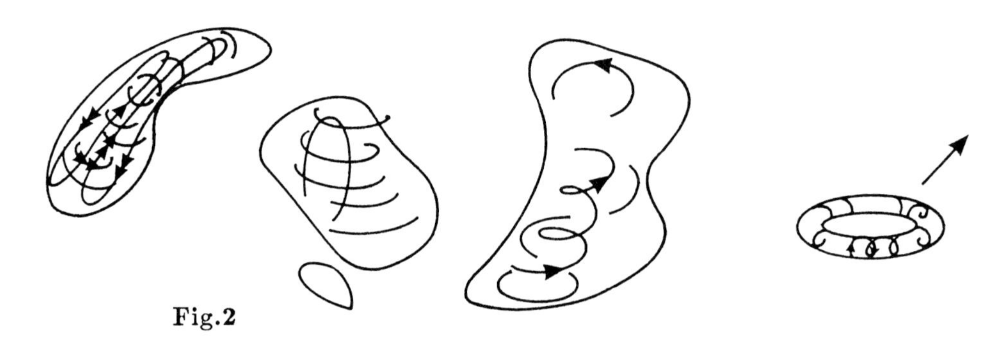

Fig.2

(a) Discrete non-overlapping volumes $V(n)$ of inviscid vortical flow separated by irrotational flow. The density of each $V(n)$ may differ but within $V(n)$ the flow is adiabatic and reversible. The helicity integral H_n is conserved for each volume.

Fig.3

(a) Volume V_b moving with velocity $\mathbf{v}_b$ parallel to a rotating flow. V_E is the external region round the volume. V is the total volume. S_V is the surface of the total volume.

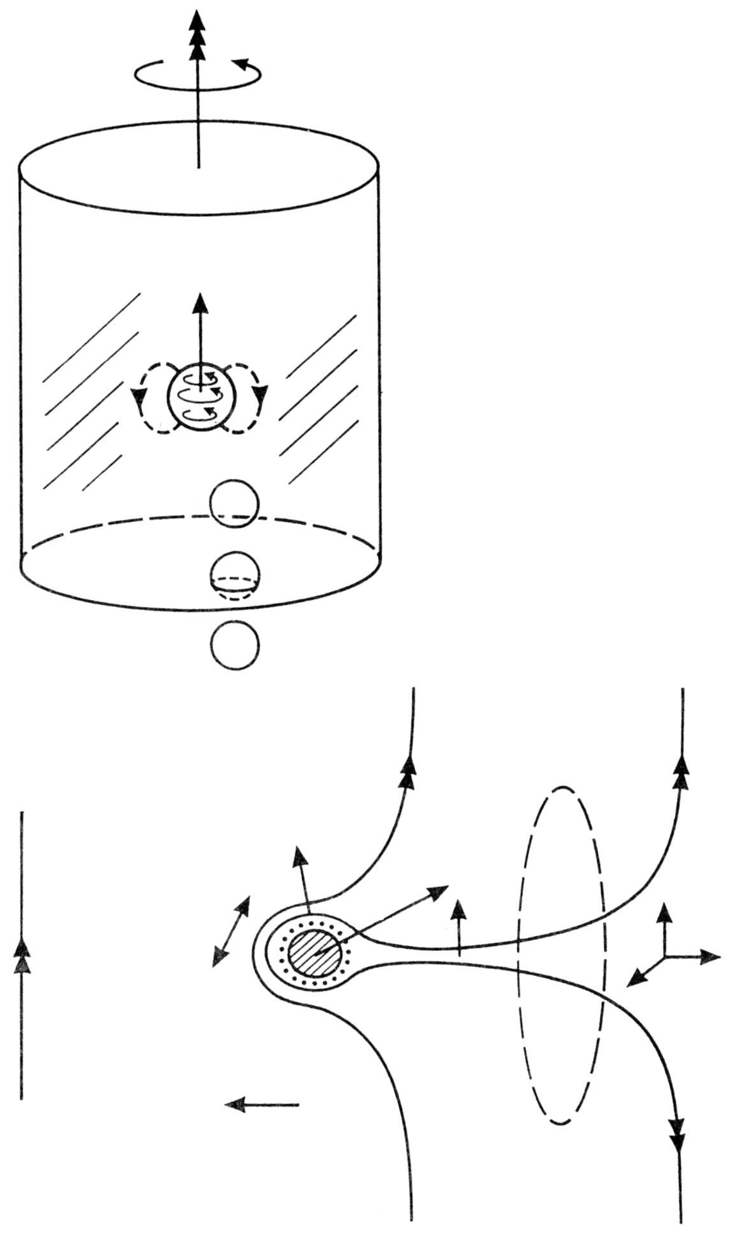

(b) Volume V_b moving with velocity $\mathbf{v}_b$ perpendicular to $\mathbf{\Omega}$. Note how as the vortex lines are distorted, they remain parallel to the surfaces where the 'drift' or time function $T(\mathbf{x})$ is constant.

Fig.4

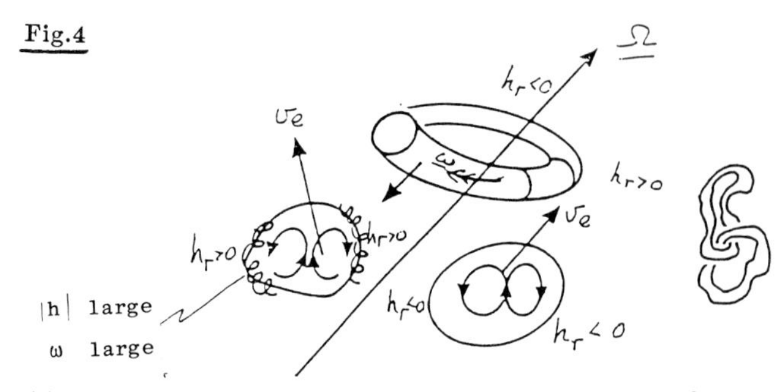

(a) Helicity fluctuations induced by various eddies moving in a large-scale vorticity field

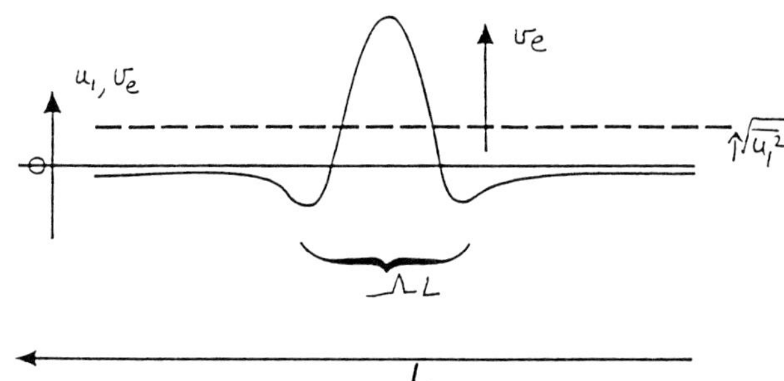

(b) Typical velocity profile associated with an eddy, to show how the number of 'eddies' per unit volume A is related to $\overline{u_1^3}, \overline{u_1^2}$.

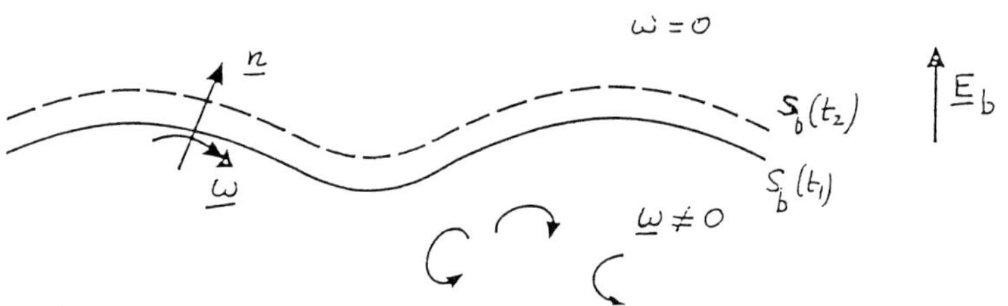

(c) The movement of an interface, at a velocity $\mathbf{E_b}$, between a turbulent region ($n < 0$) and a non-turbulent region ($n > 0$).

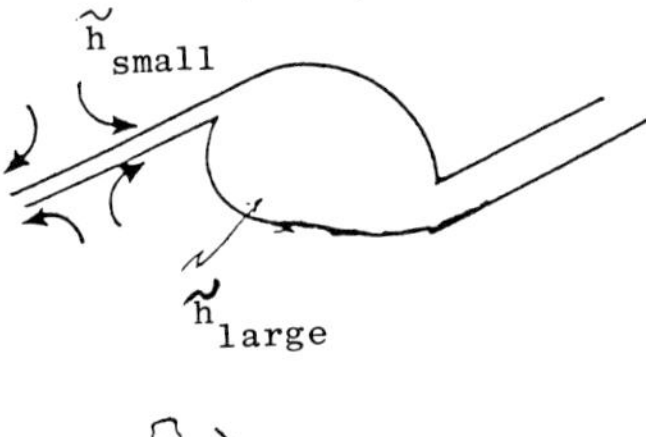

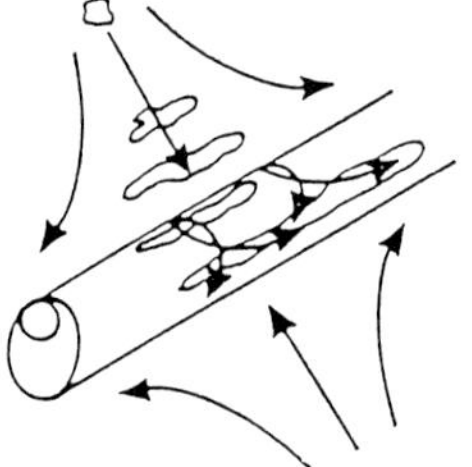

(d) The distortion of turbulence at the stagnation point on 'ribs' between vortices, and indications of regions of relatively low- and high-level helicity.

IX: Special Topics

Fluid Networks

HASSAN AREF*† AND THOMAS HERDTLE*

University of California, San Diego
La Jolla, CA 92093, USA

1. INTRODUCTION

The memoirs and treatises of Leonhard Euler from the 1750's belong to the foundations of fluid mechanics (see, for example, Truesdell 1954). The importance of these contributions is apparent, among other things, in the naming of the equations for inviscid flow after him. Equally well known, at least to mathematicians, is the seminal nature of Euler's work in the field of topology. His paper "Solutio problematis ad geometriam situs pertinentis" ("Solution of a problem pertaining to the geometry of position") on the puzzle of the Königsberg bridges appeared in 1736. His two papers on the polyhedral formula[1], that today bears his name, were written in 1752-3 but not published until 1758. We know from his correspondence with Christian Goldbach that he struggled with the proof of this formula (cf. Biggs, Lloyd & Wilson 1976).

The polyhedral formula relates the number of faces (F), vertices (V) and edges (E) in a network of polygons making up a polyhedron. In a more general form, applicable to any polygonal network, it may be written

$$F - E + V = \chi , \tag{1}$$

where the number χ is a characteristic of the surface on which the network is drawn. For example, $\chi=1$ for a finite network on the infinite plane, $\chi=0$ for a network on the surface of a torus or for a planar network subject to periodic boundary conditions, and $\chi=2$ for a network on the surface of sphere (in which case the network is "topologically equivalent" to a convex polyhedron, the original application of the result). The fundamental nature of (1) for topology comes about because the invariant it represents specifies a topological type

*) Affiliated with Department of Applied Mechanics and Engineering Science
†) Affiliated with Institute of Geophysics and Planetary Physics, and San Diego Supercomputer Center.

[1]) Euler, L. "Elementa doctrinae solidorum." *Novi Comm. Acad. Sci. Imp. Petropol.* **4**, 109-140; "Demonstratio nonnullarum insignium proprietatum quibus solida hedris planis inclusa sunt praedita." *ibid.* 140-160; both papers reproduced in *Opera Omnia* (1), vol. **26**, 72-108.

of the surface on which the network is drawn.

The IUTAM Symposium on *Topological Fluid Mechanics* is an appropriate forum in which to pursue an application of fluid mechanics where Eq.(1) has a central role. For this to occur the fluid must somehow be confined to a polygonal or polyhedral subset of space. There are several such examples, in particular if the fluid volume is locally one-dimensional as in systems of pipes or tubes. Euler's formula enters the discussion here in much the same way as one encounters it in circuit theory for electrical networks, and one might call such a situation a "codimension 2" fluid network. Our specific interest here will be the dynamics of a "co-dimension 1" network as realized by the faces of a foam. In so-called "polyhedral foam" the fluid is confined to a set of thin films that form a polyhedral, or in two dimensions polygonal, arrangement. The dynamics of such foams has been the subject of considerable work (see the recent review by Kraynik 1988) both stimulated by an intrinsic interest in the dynamics of fluid foams, and from a more general "morphological" vantage point, where the correspondence with cellular aggregates arising in other fields, such as botany and mineralogy, is pursued and elaborated (see the stimulating exposition by Weaire & Rivier 1984, and the essay by Smith 1952, for this point of view).

This paper presents computer simulation results that we have obtained for a two-dimensional polygonal foam. In order to appreciate these results, it is necessary to provide some background material, including a statement of the basic equations being solved and the algorithm used, and to specify some of the main issues in the field. Thus, §2 is devoted to a quick review of general results on fluid interfaces, bubbles, foams, cellular aggregates, etc., of particular relevance to later sections of the paper. This section concludes with the formulation of a mechanical model for the evolution of two-dimensional polygonal foams. In §2 we have relied heavily on the papers of Weaire & Rivier (1984) and their collaborators (Weaire & Kermode 1983a,b, 1984; Rivier 1983; Weaire, Fu & Kermode 1984; Wejchert, Weaire & Kermode 1986). Since we wish to focus on the "mechanical" aspects of foam evolution and, as far as possible, avoid detailed considerations of the physical chemistry of soap films, a large segment of the field having to do with the structure of individual soap films is hardly considered. For the reader interested in this material we recommend as a starting point the classic monograph by Mysels, Shinoda & Frankel (1959). The reader may also find the article by Rayleigh (1890) to be of interest[2].

In §3 we outline and discuss the numerical algorithm that we have constructed and used in conducting simulations of two-dimensional polygonal foams. Although the physical basis of our algorithm is identical to that used by Weaire and Kermode in the studies just cited, our implementation is different and more "global" than theirs in the sense that we update the entire state of the foam rather than performing a sequence of local adjustments to bubbles. We show and discuss examples of our computed results in §3. Topological

2) We may mention that there is much of fluid mechanical interest that goes on within a soap film, and that soap films have become interesting laboratory systems for studying two-dimensional hydrodynamics. The reader interested in this topic should consult the papers by Y. Couder and M. Gharib and their collaborators in the forthcoming proceedings of the conference *Advances in Fluid Turbulence* to appear as a special issue of *Physica D*. This recent work has a remarkable precedent in the experiments reported by Dewar (1923).

considerations re-enter here since rearrangements of polygonal cells will occur as the foam evolves, and the computational issues of such "topology changes" are briefly discussed. The main new results that we report have to do with the equilibration of the foam. In particular, when the foam is bounded completely or in part by rigid boundaries, there may be more than one asymptotic state depending on initial conditions. In §4 we discuss unresolved problems in the present work and various possible extensions.

2. REVIEW OF FOAM DYNAMICS

A suitable starting point for a review of the dynamics of foams is the work of Joseph-Antoine-Ferdinand Plateau whose treatise *Statique experiméntale et théorique des liquides soumis aux seules forces moléculaires,* summarizing research done over the course of a quarter century (1843-1869), and contained in eleven long scientific papers, appeared in 1873 (Nitsche 1974). Plateau relied, of course, on earlier work[3], in particular the formulation of what we now call the Young-Laplace law[4] for the pressure difference across a two-fluid interface:

$$\Delta p = \sigma \left(\frac{1}{R_1} + \frac{1}{R_2} \right) . \tag{2}$$

Here Δp is the jump in pressure at the interface, σ is the surface tension coefficient, and R_1, R_2 are the two principal radii of curvature of the interface. The result (2) says that the pressure jump at a two-fluid interface is proportional to the mean curvature of that interface. It is a central result in the well known connection between soap film dynamics and the mathematical theory of minimal surfaces. For a film Eq.(2) must be applied twice, once at each surface of the film.

On the basis of extensive observations Plateau abstracted the following two rules concerning the structure of interacting soap films (in three dimensions):

I. Soap films in equilibrium meeting at an edge form angles of 120°, i.e. three films always meet at an edge.

II. Vertices are formed by four films meeting at a point. The angle between the edges is the same as found by connecting the vertices of a regular tetrahedron to its centroid, i.e. the "Maraldi angle" γ for which $\cos\gamma = -\frac{1}{3}$.

In two dimensions law II does not pertain, and law I says that each vertex of the polygonal net is at the juncture of three edges meeting at 120°. The deduction of Plateau's rules from

[3]) He also relied on his family and several able assistants since he became irrevocably blind at about the time his seminal work on soap films started.

[4]) This law was stated by Young in his 1805 essay *Cohesion of Fluids,* and independently by Laplace in his treatise *Mécanique Céleste,* published in 1806.

(2), long surmised to be possible, has only yielded to full mathematical elucidation in recent times (see Almgren & Taylor 1976 for a popular account). There is a beautiful, elementary argument, due to Steiner, that three straight films attached to parallel pegs and meeting at a point must form angles of 120° (see Ch.VII, §5 of Courant & Robbins 1941; see also the discussion by Isenberg 1978).

For our purposes it is important to note that according to (2) the soap film between two bubbles is a circular arc in two dimensions. This feature allows a two-dimensional foam configuration to be specified by giving the locations of the vertices of the foam polygons and the pressure differences between bubbles in the foam, since by (2) these give the radii of the films making up the foam. In three dimensions such simplification is no longer possible, and each film is a minimal surface with boundaries that must be determined as part of the solution to the problem. Hence, fully three-dimensional foams are fundamentally very much more complicated than their two-dimensional counterparts.

Although we have little new to say about the three-dimensional problem, we mention the extensive investigations of Matzke (1945, 1946) on the statistics of edges and faces in polyhedral foams. In these experiments foams were produced by filling a dish with individually blown bubbles, so that a very uniform foam was produced as an initial condition. Some 25,000 bubbles were produced to make up the various foams used. Individual photographs were taken of many polyhedral cells within the foam, and comprehensive statistics were performed. Matzke & Nestler (1946) continued the work using foams with initially two species of bubbles, "large" and "small." For bubbles in the interior of the foam, well away from the walls of the container, Matzke (1946) quotes the results in Table I for a sample of 600 bubbles. The average number of faces per bubble on the basis of this data is 13.70, just under 14. This average number of faces per bubble, which arises also for other systems of polyhedral cells, notably plant cells as studied extensively by F.T. Lewis (1923; 1925; 1928a,b; 1943), leads to a famous "paradox" of the subject that relates to two papers of Kelvin (1887, 1894). We mention it here because it again underscores the relative simplicity of two dimensions versus three.

Table I. Statistics for 600 bubbles in foam experiments of Matzke (1946)

Faces per bubble	No. of bubbles	Vertices per face	No. of faces
11	2	4 (Quadrilaterals)	866
12	73	5 (Pentagons)	5503
13	179	6 (Hexagons)	1817
14	218	7 (Heptagons)	35
15	106		
16	20		
17	2		

Early workers had believed that in three dimensions the average bubble would be a dodecahedron. The argument started from the familiar stacking of spheres in which each sphere, of course, touches its twelve nearest neighbors. As the spheres are pressed into a polyhedral foam, the reasoning went, bubbles of twelve sides will be produced. In two dimensions the analogous argument works nicely: If a triangular lattice of identical circular bubbles is compressed, a hexagonal lattice of plane edges is produced. However, in three dimensions regular dodecahedra cannot be packed to fill space and some major readjustments must occur.

In his 1887 paper Kelvin started from the truncated octahedron, a tetrakaidecahedron that is one of the well known Archimedean semi-regular polyhedra, and that can be packed to fill space (cf. Coxeter 1973 or Williams 1979). He argued that by curving certain edges of this figure one would obtain a solution to the problem of polyhedral arrangement that minimizes surface area. Hence, this structure should arise by the natural minimization problem inherent in equilibration of a three-dimensional foam. This result fits in very nicely with the observed average of 14 for the number of faces per cell from experiments on foams. The main problem with it, apart from the substantial fluctuation levels apparent in data such as that of Table I, is that the Kelvin tetrakaidecahedra have faces that are either hexagons or quadrilaterals, and, as the data in Table I shows, the most common face in a real foam is a pentagon! The average number of edges per face from this data equals 5.124. Indeed, Matzke (1946) gives a detailed table of the number of sides per face observed in the 218 tetrakaidecahedra of his 600 bubble sample. He found just eight different combinations of sides per face, all of them containing at least four pentagons[5].

2.1 Bubble Statistics: Theoretical Developments

Kelvin's 1894 paper is aimed at providing a general argument leading to the average number of faces per cell without regard for detailed structure. It is basically a statistical analysis of a system subject to Euler's polyhedral formula. This is a fruitful route to follow both in two and three dimensions, and we pause to give some results obtainable by this method (cf. Weaire & Rivier, 1984): Consider the two-dimensional foam first. According to Plateau, for a stable foam three edges meet at each vertex[6]. Hence, since each edge connects two vertices, $3V=2E$. Let F_n denote the number of polygons with n edges. Then $\sum nF_n = 2E$, where the sum is over all numbers of edges $n \geq 3$. This says that counting the edges by faces double-counts, since each edge is shared by two faces. In Euler's formula (1) substitute for V in terms of E, and then for E in terms of $\sum nF_n$. Divide through by F and there results

---

[5]) Dormer (1980) summarizes the situation in these words: "Nothing is more revealing than the fact that Kelvin, a brilliant and tireless experimenter, having proposed the ideal shape for close-packed equal-sized bubbles, and having exhausted his ingenuity in the laboratory, was obliged to resort to the soldering iron. His 14-hedron was ultimately produced by the aid of a complete wire skeleton in which the position of every edge was *forced* upon the liquid films."

[6]) We shall say that the foam has the *coordination number* $z=3$. The graph theorist would use the terminology that the polygonal network of the foam is a regular graph of degree 3.

$$(6 - \frac{1}{F}\sum nF_n) = \frac{\chi}{F} \ . \tag{3}$$

The quantity

$$<n> \equiv \frac{1}{F}\sum nF_n \tag{4}$$

is the average number of edges per polygon since the distribution F_n is, of course, normalized by

$$\sum F_n = F \ . \tag{5}$$

Thus, as we let $F \to \infty$ we get

$$<n> = 6 \tag{6}$$

This is an exact result valid for a sufficiently large polygonal network of coordination number 3 on any surface. We conclude that in a two-dimensional foam the average number of edges per polygon is six.

In three dimensions Euler's formula states that[7]

$$C - F + E - V = -1 \tag{7}$$

where C is the number of cells. Since by Plateau's second rule each vertex is connected to 4 others in a three-dimensional foam, i.e. z=4, we have 4V=2E. Also, with the same notation for faces with n edges as above, we now have $\sum nF_n = 3E$ by Plateau's first rule. Finally, if we let C_f denote the number of cells with f faces, we have by analogy with (4) that

$$<f> C = \sum fC_f \tag{8}$$

where the sum is over $f \geq 4$. Since each face is shared by two cells $\sum fC_f = 2F$. Putting these results together we obtain

$$<f> = \frac{12}{6 - <n>} \ . \tag{9}$$

As derived (9) pertains to the foam as a whole[8]. It also holds for any individual bubble in the foam if $<f>$ is interpreted as the number of faces, F, of that bubble and $<n>$ is interpreted as the average of sides per face over those F faces. Weaire & Rivier (1984) go on to discuss and criticize various approximate analyses that yield $<f> \approx 14$ from Eq.(9).

[7]) We assume that all the cells are simple polyhedra, i.e. do not have holes. In any case the number on the right hand side of (7) is of order 1.

[8]) The data of Matzke (Table I) is in accord with (9), as it should be with $<f>=13.70$, $<n>=5.124$.

One such argument is reproduced in Isenberg's (1978) book. Here we turn to results on higher order statistics concentrating on the two-dimensional case.

Having established that $\langle n\rangle=6$ we consider the second moment of the distribution

$$p(n) = \lim_{F\to\infty} \frac{F_n}{F} , \tag{10}$$

i.e.,

$$\mu_2 = \sum (n-6)^2\, p(n) = \langle n^2\rangle - 36 , \tag{11}$$

where the last transformation follows from (6). Let $E_{m|n}$ designate the number of edges that bound a bubble with m edges from one with n edges. Then, clearly,

$$E_{m|n} = E_{n|m} \tag{12a}$$

and

$$\sum_m E_{m|n} = n\times F_n \tag{12b}$$

since the left hand side is the total number of edges of bubbles of any kind shared with an n-sided bubble, and for each n-sided bubble there are n such edges[9].

Aboav (1970) considered the quantity

$$\nu_n = \frac{\sum_m m\, E_{m|n}}{\sum_m E_{m|n}} , \tag{13}$$

which is the average number of edges among all bubbles that adjoin n-sided bubbles. Weaire (1974) noted the identity

$$\langle n\times(n-\nu_n)\rangle = 0 , \tag{14a}$$

or, using (11),

$$\langle n\nu_n\rangle = \mu_2 + 36 . \tag{14b}$$

Equations (14) follow from the transformations:

$$\sum_n n^2 F_n = \sum_n n \sum_m E_{n|m} = \sum_m \sum_n n\, E_{n|m} = \sum_m \nu_m \sum_n E_{n|m} = \sum_m m\nu_m F_m$$

[9]) A popular design of a soccer ball consists of $F_5=12$ pentagons and $F_6=20$ hexagons with each pentagon sharing all five edges with hexagons, and each hexagon sharing three edges with pentagons and three with hexagons. Thus, $E_{5|6} = E_{6|5} = 60$, $E_{6|6} = 60$, and all other $E_{m|n} = 0$. It is instructive to verify Eq.(12b) for the two cases n=5,6.

using (12b) and (13). Division by F and use of (10) now gives the result (14a).

Empirical evidence obtained from counting sides in some 3,000 grains of polycrystalline magnesium oxide suggested to Aboav (1970) a relation of the form

$$v_n = a + bn^{-1} \tag{16a}$$

often referred to as *Aboav's law*. If this is substituted into (14b), we find

$$b = \mu_2 + 36 - 6a \tag{16b}$$

The fit

$$v_n = a + (\mu_2 + 36 - 6a)n^{-1} \tag{16c}$$

with a=5 has frequently been found to match experimental data.

2.2 Bubble Dynamics: Von Neumann's Law

The results above pertain to the geometrical configuration of any instantaneous state of the foam. We next consider the formulation of a dynamical law describing the way in which the bubbles making up the foam change. Our main result, Eq.(19), was first obtained by von Neumann (1952) in a comment on the paper by Smith (1952) mentioned previously.

We return to Eq.(2). Let a given circular edge of a bubble subtend an angle α, i.e. let its arclength be $R\alpha$, where R is the radius of that edge. Then, since the pressure difference across the liquid film making up the edge is proportional to σ/R by Eq.(2), and the flow rate is proportional to the pressure difference (for discussion see Princen & Mason 1965), this edge contributes a flow due to diffusion out of the bubble that is proportional to α. If the bubble in question has n edges, the total flow out of that bubble is proportional to the sum of angles α_i, i=1,...,n, over all n edges. We reckon R positive if the edge is convex, negative if concave.

The rest is geometry. Draw the chords instead of the circular arcs delimiting the bubble to obtain a convex n-gon crudely representing the bubble. Let the angles of this n-gon be ϕ_i, i=1,...,n, with the indexing chosen such that angle ϕ_i appears at the vertex separating the circular edges of angle α_i and α_{i+1}.The sum of angles ϕ_i is just the sum of internal angles in a planar n-gon, i.e. $(n-2)\pi$. On the other hand, by Plateau's first rule

$$\frac{1}{2}\alpha_i + \frac{1}{2}\alpha_{i+1} + \phi_i = \frac{2\pi}{3} \tag{17}$$

because of well known geometrical relations between circular arcs and the angles spanned by chords and tangents associated with them. Sum (17) over all edges of the bubble to obtain

$$\sum_{i=1}^{n} \alpha_i = \frac{\pi}{3}(6-n) \tag{18}$$

This is proportional to the net flow *out* of the bubble. Thus, the rate of growth of area for a bubble of n edges is proportional to n – 6. This is *von Neumann's law:*

$$\frac{dA_n}{dt} = k(n-6) \tag{19}$$

Bubbles with more than six edges grow. Bubbles with less than six edges contract. Hexagonal bubbles do not change their area. The constant k in (19) is phenomenological. It should depend linearly on the surface tension, but must also depend on the permeability of the film making up the bubble wall and on the viscosity of the gas within the bubbles.

From what we have said in the derivation of (19) it would be possible, albeit inconvenient, for the growth rate k to depend on the bubble in question and on the time. Previous theoretical and numerical studies (Weaire & Kermode 1983a,b, 1984) have assumed that k in (19) is a constant. A recent experimental study by Glazier, Gross & Stavans (1987) finds that k is independent of time to better than 4%, but has a weak dependence on n. In our computer simulations we have taken k to be a constant.

It is of interest to generalize (19) to bubbles in the foam that are in contact with a rigid surface. We assume the rigid surface is coated by film. Plateau's rule I in this case states that the bubble edge makes a right angle with such a surface. When we compute the diffusive flux through the edges of a bubble delimited on one side by the rigid surface, the calculations proceed as before except for that surface, through which there is, of course, no diffusion. In this case the straight-sided polygon with the same vertices as the bubble is completed by extending the tangents to the rigid surface at the two points of intersection. At these points (17) is replaced by an equation of the form:

$$\frac{1}{2}\alpha_i + \phi_i = \frac{\pi}{2} \tag{20}$$

Summing (17) for i=2,...,n–1, and adding (20) for i=1 and i=n, we may express the sum of angles α_i corresponding to internal edges of the bubble in terms of n and the angle θ through which the tangent must be turned in passing from vertex n back to vertex 1. The end result for bubbles on the boundary is:

$$\frac{dA_n}{dt} = k\left\{\frac{\pi}{3}(n-5) + \theta\right\} \tag{21}$$

in place of (19). Thus, for a straight boundary, θ=0, and pentagonal bubbles are the ones that do not change their area when in contact with such a boundary (whereas in the interior of the foam hexagonal bubbles still have this property). Equation (21) has interesting consequences for the equilibria obtainable with a two-dimensional foam relaxing in containers of different shape.

3. COMPUTER SIMULATION OF EVOLVING FOAM

3.1 Computational Algorithm

Equations (19) may be made the basis for a numerical simulation of an evolving two-dimensional foam in the following way: We use as our independent variables the coordinates of the bubble vertices (x_i, y_i) and the pressures within bubbles p_i. From the vertex coordinates, a labelling of which vertices belong to which bubbles, and the information on edge radii contained in the pressures we can calculate the bubble areas. We can also obtain the angles between the edges that meet at a given vertex. Only two of these are independent. Consider all these formulae to have been written down.

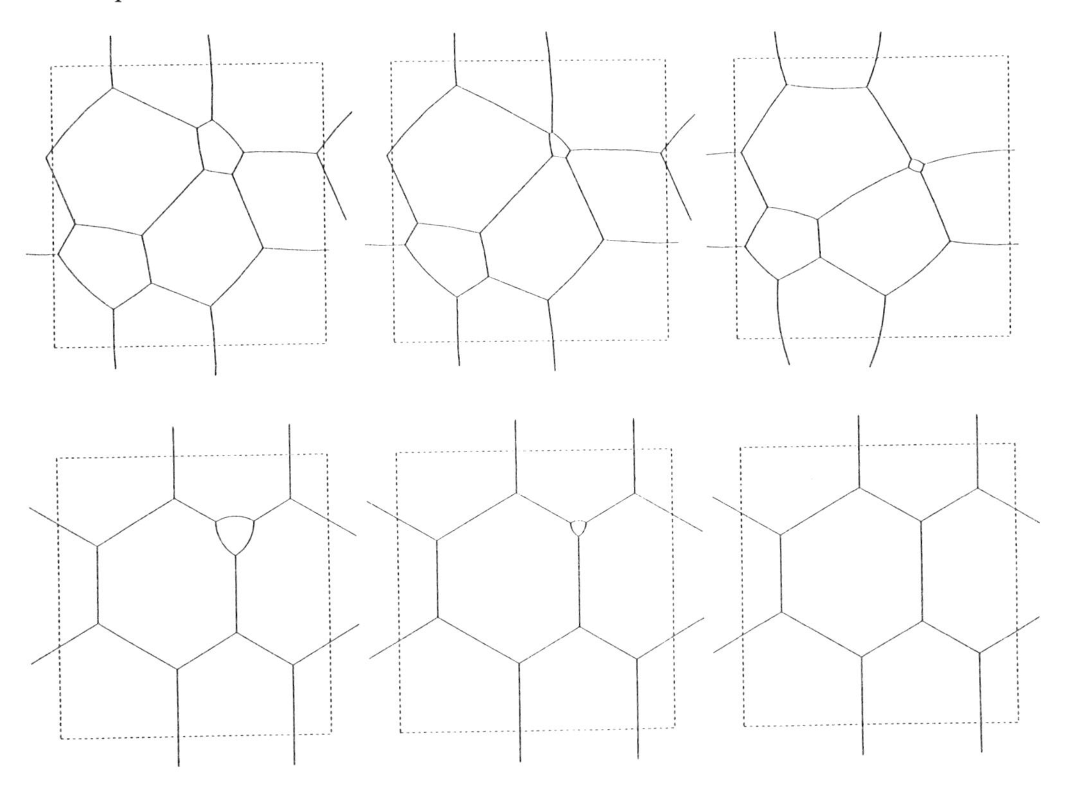

Figure 1: Topology changes in polygonal foam. Top row: T1 process; bottom row: T2 process.

We now differentiate the expressions for bubble areas and for angles between edges with respect to the vertex coordinates and pressures. We write the differentials of bubble areas, and for each vertex the differentials of two independent angles in terms of the differentials of the vertex coordinates and the bubble pressures. A system of linear equations results. Since the independent variables are two coordinates per vertex and one pressure per bubble, and the dependent variables are two angles per vertex (the third following from these) and one area per bubble, we see that the coefficient matrix of these equations is square. We turn this system into evolutionary equations for the foam by inverting it and using (19) to update the bubble areas while the vertex angles are kept fixed[10] at $2\pi/3$.

The matrix that must be inverted in this procedure is clearly singular. Any rigid displacement of all the vertices must correspond to a null vector, for example. In the numerical inversion of the matrix we therefore make use of a singular value decomposition (cf. Press et al. 1986).

In practice we consider two types of steps: "elastic relaxation" and "diffusion." In the former the areas are assumed to remain constant and the angles relax to $2\pi/3$. In the latter areas evolve according to von Neumann's law while angles remain fixed. If we were only taking infinitesimal steps, or exactly solving the differential equations, these two types would suffice. However, since we are taking finite computational steps using a matrix of derivatives evaluated for certain intermediate states, the elastic relaxation step incurs area changes and the diffusion step incurs angle changes. Hence, the steps need to be alternated. If after a diffusion step the error in angles is found to have grown to large, a relaxation step is invoked. Note that the relaxation step is not a single, predictive step but, due to the nonlinearity of the system, an iteration procedure.

Modifications to the above procedures for a bubble in contact with a rigid wall of known geometry are not difficult to include. This has been done for the simplest boundary shapes: (i) periodic boundaries in both x- and y-directions (referred to as doubly-periodic boundaries); (ii) a rigid, straight wall along y=0 and periodic boundary conditions in the x-direction; (iii) two straight, parallel rigid walls at y=0 and y=h and periodic boundary conditions in the x-direction ("channel"); (iv) a circular domain.

In what we have said so far there appears to be no intrinsic limitation on the size of the time step that can be taken. However, we soon realize that as the foam evolves some edges will continually diminish in size and must ultimately vanish altogether. When this situation arises in a simulation, a decision must be made on how to continue the evolution. Observations on real foams suggest that the connectivity of the bubbles making up the foam changes. This *topology change* can happen in one of two ways in two dimensions. Figure 1 illustrates both mechanisms. In each row temporal evolution progresses from left to right. In the top row a small pentagon (left panel) shrinks to the point where one side is very short (middle panel). A reconnection now takes place such that the small side is eliminated and replaced by a new side at a finite angle to it. This new side quickly grows to

[10]) We have also allowed in our code an elastic relaxation step in which the areas are kept fixed but the angles are forced to relax to $2\pi/3$. This is useful for allowing artificial, geometrically defined, initial conditions to relax to conceivable foam configurations, and for other purposes as discussed later. Physically the elastic relaxation is much more rapid than the diffusion. Hence, much of the time it is physically sensible to keep the angles fixed and allow only the areas to evolve.

considerable length (right panel). Notice how the sides in the small pentagon and in the larger hexagon above it, that are separated by the short side about to vanish, become connected and pull apart producing a small quadrilateral bubble and a larger pentagon that do not share a side. This type of process is referred to as *neighbor-switching* or as a *T1 process*. In a computer implementation a criterion for what constitutes a short side is established and instructions to reconnect and relax elastically are then called.

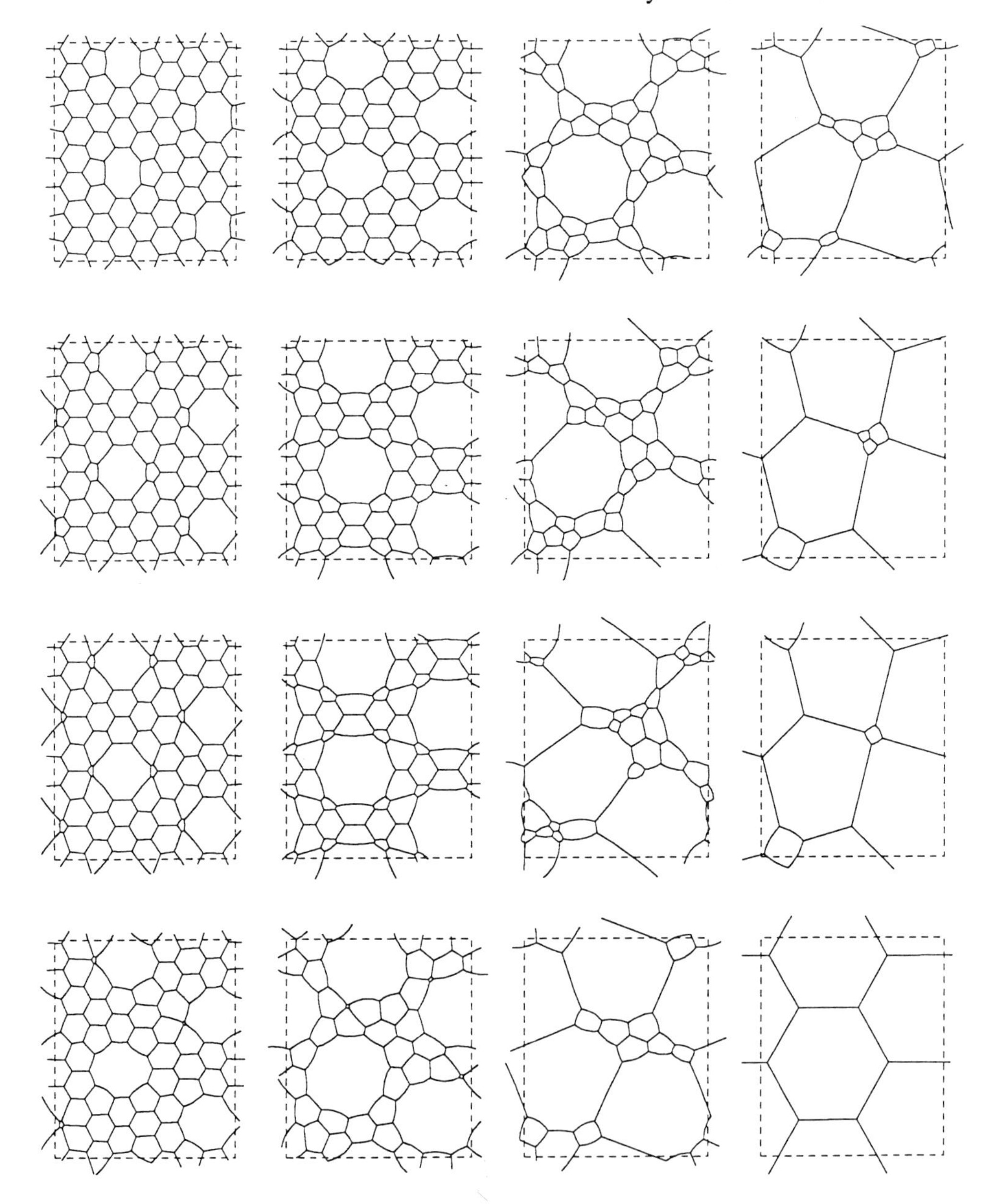

Figure 2: Equilibration of foam in a doubly periodic domain. Evolution proceeds top to bottom, left to right.

The bottom row of Fig.1 illustrates the second topology change mechanism, *cell disappearance,* also known as a *T2 process* in the literature. In this case a small triangular bubble (left panel) contracts uniformly (middle panel) and vanishes at a vertex (right panel). Clearly there is no possibility for a T1 process to intervene. Similar contraction of n-gons for n≥4 is not possible because of Plateau's rules. Schwarz (1964) has considered the analogous processes in three-dimensional foam.

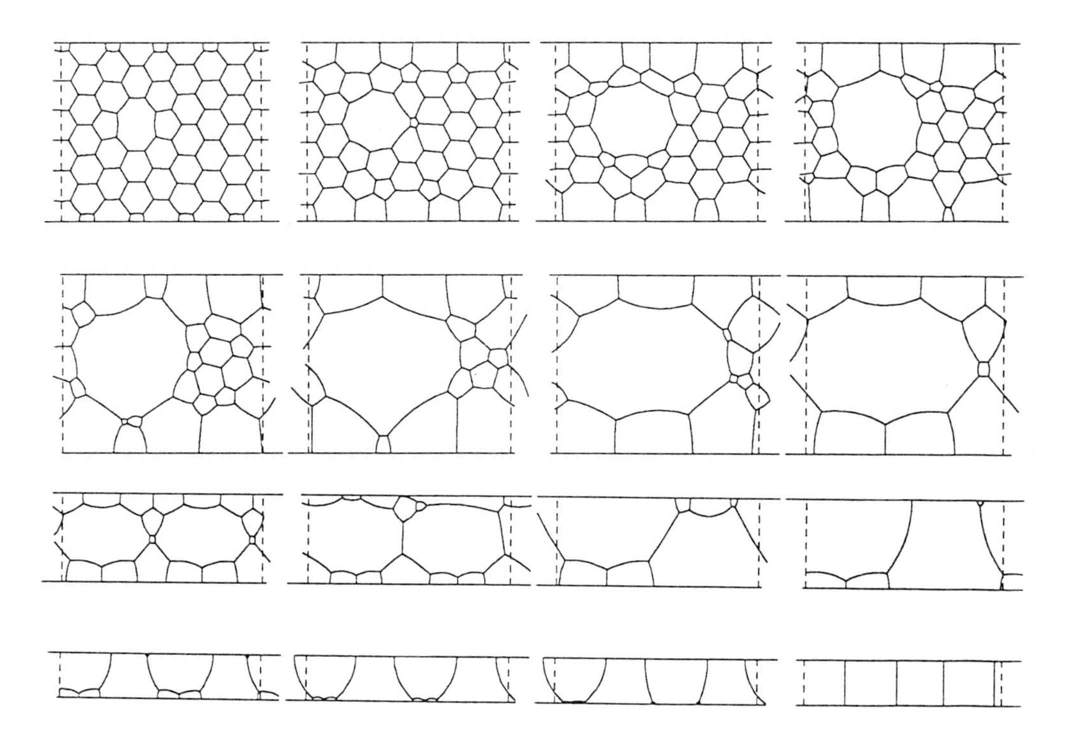

Figure 3: Equilibration of foam in a channel. The domain has periodic boundaries left and right, rigid boundaries top and bottom. Evolution proceeds left to right, top to bottom. The domain is doubled in the horizontal in both the third and fourth row.

The sources of numerical error that we mentioned above in connection with relaxation and diffusion steps become particularly important when one is close to a topology change. Clearly, just after a T1 or T2 process a relaxation step is required to bring all angles, and in particular the newly created angles, back to $2\pi/3$. Here problems sometimes occur in particular after a T1 process since some angles in the reconnected system may be very far from their equilibrium value. Furthermore, as other investigators have observed, T1 processes tend to be closely linked with T2 processes in the sense that a bubble having a very short side typically implies that the bubble itself is small, and will soon be reduced to a (small) triangular bubble that will disappear in a T2 process. This is an important point which has as its statistical counterpart *Lewis' law,* an empirical correlation between average bubble (or cell) size for n-gons, $\overline{A}_n$, and n itself. Lewis (1928b) found

$$\overline{A}_n = a\,(n-2) \tag{22}$$

Multiplying this by F_n and summing on n we obtain on the left hand side the total area of the foam, A. On the right hand side we get 4aF in view of Eq.(6), i.e., a=A/4F.

Two competing theories of (22) have been proposed. Beenakker (1987, 1988) suggests that it is an approximation, increasingly accurate for large n, that may be understood from a mechanistic coupling of the T1 and T2 processes. Rivier & Lissowski (1982; see also Rivier 1983), on the other hand, produce a derivation that suggests (22) is a general consequence of statistical mechanics for any network that fills the plane and has coordination number 3.

From a numerical simulation standpoint very small bubbles undergoing T1 processes are often difficult to handle since they require the elastic relaxation step to iterate many times, a manifestation of "stiffness" in the solution of this problem. A procedure that appears to help is the following: When a very small bubble, which is about to disappear anyway, needs to relax elastically, we let it relax with respect to its neighboring vertices keeping all other vertices fixed, and then let it complete the diffusion step. Once it disappears, we let the entire network relax elastically (which is now much easier without the small bubble).

3.2 Results of Numerical Experiments

We now describe simulation results that we have obtained using a code based on the considerations in §3.1. Several calculations have been performed for a foam equilibrating in a doubly periodic domain. An example is shown in Fig.2. The panels shown here and in Figs.3-5, were selected to be representative of the evolution and are not uniformly spaced in time. The initial condition was set such that a regular lattice of hexagonal bubbles had a periodic array of octagonal "imperfections." Associated with each octagon are two pentagonal bubbles. One advantage of computer simulations for this type of system is immediately apparent: The control over initial conditions that is achieved in Fig.2 is virtually unrealizable in the laboratory.

The simulation in Fig.2 shows the satisfying result that the foam ultimately equilibrates to a regular pattern of hexagons, which according to Eq.(19) is the only possibility. Obtaining the correct equilibrium state is an important consistency check on the calculations. For all cases tried we have not been able to obtain any other equilibrium than the regular hexagonal pattern for doubly periodic boundary conditions.

The situation changes dramatically when rigid boundaries are present. Figure 3 shows the evolution of an imperfection very similar to those in Fig.2 for a channel, i.e., rigid boundaries top and bottom, periodic boundaries left and right. As the flow evolves the large bubble in the center of the channel eventually spans the entire horizontal stretch from one periodic boundary to the other. At this stage our code requires that we double the horizontal dimension taking two replicas side by side. This is done for the first time in the third row of Fig.3, and again in the fourth row. The equilibrium consists of films spanning the channel from rigid wall to rigid wall. There are four such films because we started with one imperfection and doubled the horizontal dimension twice. Similar results are shown in experiments of Smith (1952, 1954).

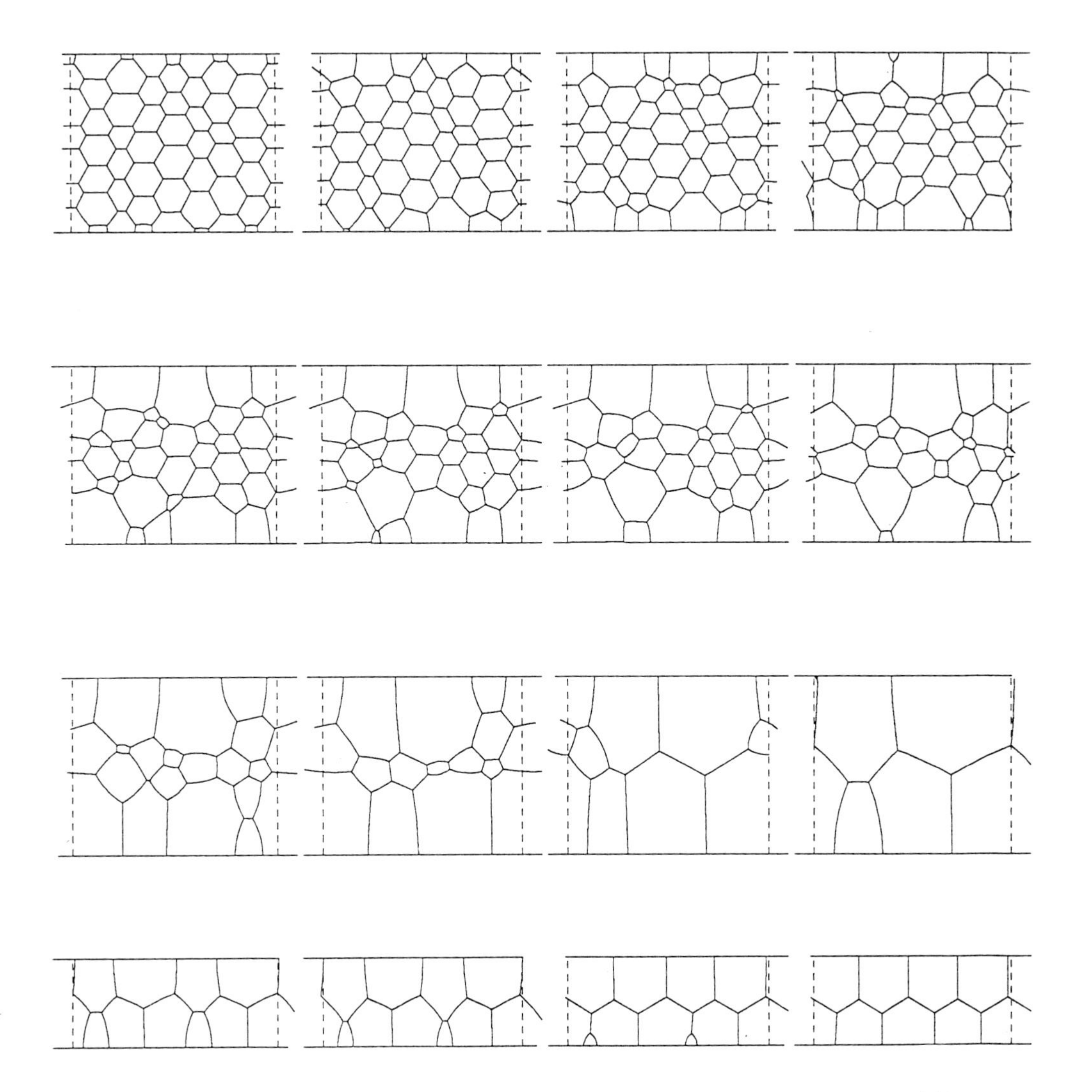

Figure 4: Equilibration of foam in a channel. The domain has periodic boundaries left and right, rigid boundaries top and bottom. Evolution proceeds left to right, top to bottom. The domain is doubled in the horizontal in the fourth row.

Figure 4 illustrates that for this boundary geometry there is more than one equilibrium configuration. In the calculation in Fig.4 the initial state did not have any designated imperfection. The internal bubbles were hexagonal of varying sizes. Again it eventually became necessary for computational reasons to double the domain in the horizontal direction (bottom row of Fig.4). The equilibrium for this case is different, consisting of two rows of pentagonal bubbles, one attached to either wall.

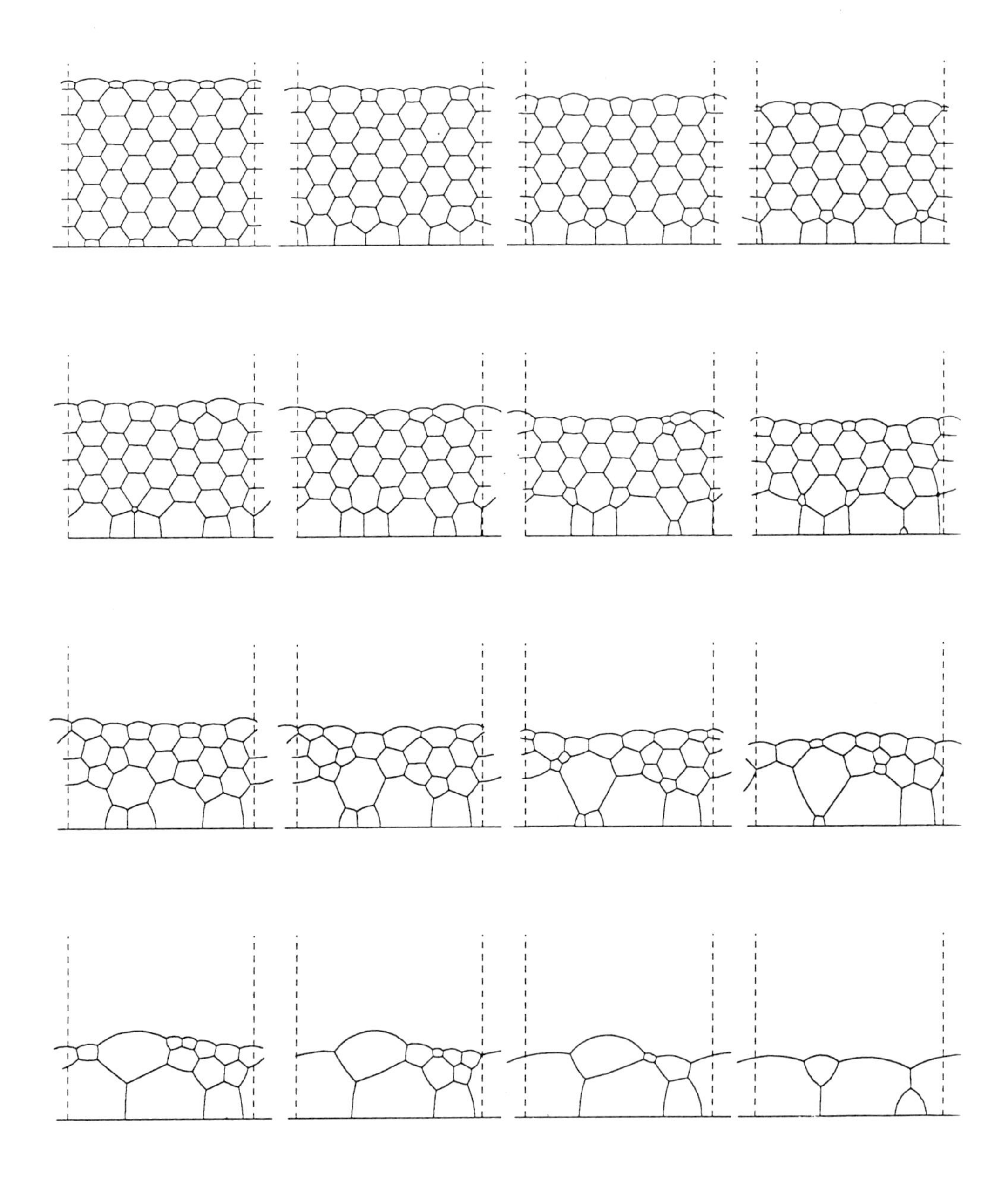

Figure 5: Equilibration of foam on a flat plate with a free surface.
The domain has periodic boundaries left and right.
There are no effects of gravity such as film draining.

It appears to be possible to have equilibria derived from that in the last panel of Fig.4 by inserting one or more rows of hexagons. However, we have not been able to find such equilibria in initial value calculations.

Figure 5 is a simulation of a foam with a free surface, represented computationally as an infinitely large bubble. Since the surface edges of the foam must always be convex, it follows that diffusion takes place from the foam to the space above and, thus, that the foam itself contracts. This is, indeed, what we see in Fig.5. By the end of the computation only a monolayer of bubbles survives. The triangular bubbles will eventually disappear via T2 processes, and the quadrilaterals must also contract. Hence, this model of two-dimensional foam dynamics predicts that a layer of foam with a free surface will collapse in a finite time (without consideration of effects of gravity such as film drainage).

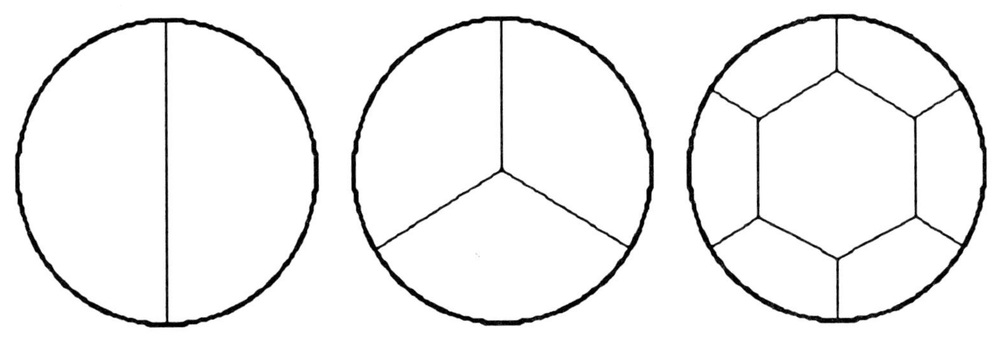

Figure 6: Steady states of few-bubble systems in a circle. All three are unstable.

It is interesting to speculate on the possible equilibria in other domains, for example, a foam confined in a shallow cylindrical container. Simulations for this case are in progress. From Eq.(21) we see that an n-gon situated on the wall must span an arc $\frac{\pi}{3}(5-n)$ to be stationary. The possibilities for n=2, 3 and 4 are shown in Fig.6. However, even though these states are possible in principle, it is highly unlikely they will ever be seen. For a slight displacement of any of the edges reaching the boundary will make the right hand side in (21) non-zero, and one of the boundary bubbles must contract. Thus, it appears that no stable equilibria exist for a two-dimensional foam in a circular container.

4. UNRESOLVED PROBLEMS, EXTENSIONS AND OUTLOOK

Considerable attention has been paid in the literature to the evolution of statistical measures of two-dimensional foams, such as the moment μ_2 introduced in §2. We have monitored this and other statistical quantities in our simulations. However, it does not appear to us at the present time that our simulated bubble aggregates are large enough to contain statistically significant information.

Weaire & Kermode (1983a,b, 1984) monitored Lewis' law (22) for their numerical experiments and found approximate agreement with a linear relation. In experiments by Glazier, Gross & Stavans (1987), using much larger numbers of bubbles, this relation is also seen approximately, at least for bubbles with more than 5 sides, although with considerable error bars. Aboav's law (16a) is convincingly verified in the experiments of Stavans & Glazier (1989), and more approximately in the numerics of Weaire & Kermode (1984). Our verifications of both Aboav's and Lewis' law are comparable to those in the earlier numerical work.

There appears to be some disagreement on the asymptotics of μ_2. Glazier, Gross & Stavans (1989) suggest that there is an intermediate regime in which μ_2 settles down to an asymptotic value of about 1.4. Weaire & Kermode (1984), on the other hand, do not seem to see a plateau in μ_2. In computer simulations it is easy to monitor μ_2 frequently and so considerable fluctuations occur in the time series since individual T1 and T2 events are visible in the record for small sample size. Our results for μ_2 are qualitatively similar to those of Weaire & Kermode (1984), but our peak values are higher, and since we follow the foam to equilibrium, our asymptotic value for doubly-periodic boundaries is zero.

The experiments of Glazier, Gross & Stavans (1989) suggest that the distribution p(n), Eq.(10), settles down to an asymptotic form peaked around n=6. There is some indication of similar behavior in the numerics of Weaire & Kermode (1984) but their data begins with a number of bubbles roughly equal to that at which the laboratory experiments end. Our results for p(n) are comparable to those of Weaire & Kermode (1984), but possibly even more affected by sample size and by our choice of initial conditions with few imperfections. Our numerical procedure appears to lead more directly to equilibration of the foam than the earlier implementation based on local adjustments used by Weaire & Kermode (1983, 1984). These authors do not show equilibration calculations.

We suggest that a fruitful quantity to monitor in future calculations is the distribution $E_{m|n}$ introduced in §2. This distribution seems suitable for kinetic considerations along the lines pursued by Beenakker (1986, 1987) albeit for a different distribution function.

Physical effects not embodied in von Neumann's model are accessible to simulations of the kind reported here. For example, one may consider the random rupturing of films in the foam. The analogies to evolution of biological cellular systems, that have led to so much direct work on foams, seem worth reviving, and we have begun exploring the role of such mechanisms as "cell division," the converse in a sense of rupturing. We are disturbed by the reports in the experimental work of Stavans & Glazier (1989) of deviations from Plateau's first rule as well as von Neumann's law. Nevertheless, the methods in §3 could incorporate these experimental observations.

A major impediment to progress of the simulations at the present is the scaling of the numerical work with the number of bubbles. The computer time for a step in our numerical model currently scales roughly as $N^{2.7}$, where N is the number of bubbles. We believe that substantial improvements are possible here, and we are currently working in this direction.

ACKNOWLEDGMENTS

We are indebted to A. Kraynik, K. Mysels and N. Rivier for comments on the original version of the manuscript.

This work was supported by NSF/PYI award MSM84-51107 with matching funds from Cray Research, Sun Microsystems and Apple Computer. T.H. acknowledges support of an ONR fellowship. We are indebted to the San Diego Supercomputer Center (SDSC) for the allocation of computing resources. SDSC is supported by NSF.

REFERENCES

Aboav, D.A. 1970 The arrangement of grains in a polycrystal. *Metallography* **3**, 383-390.

Almgren, F.J. & Taylor, J.E. 1976 The geometry of soap films and soap bubbles. *Scient. Amer.* **235**(1), 82-93.

Beenakker, C.W.J. 1986 Evolution of two-dimensional soap-film networks. *Phys. Rev. Lett.* **57**, 2454-2457.

Beenakker, C.W.J. 1987 Two-dimensional soap froths and polycrystalline networks: why are large cells many-sided? *Physica A* **147**, 256-267.

Beenakker, C.W.J. 1988 Numerical simulation of a coarsening two-dimensional network. *Phys. Rev. A* **37**, 1697-1702.

Biggs, N.L., Lloyd, E.K. & Wilson, R.J. 1976 *Graph Theory 1736-1936.* Clarendon Press, Oxford.

Courant, R. & H. Robbins 1941 *What is Mathematics? An Elementary Approach to Ideas and Methods.* Oxford University Press.

Coxeter, H.S.M. 1973 *Regular Polytopes.* Third edition. Dover Publications, New York.

Dewar, J. 1923 Soap films as detectors: stream lines and sound. *Proc. Roy. Inst.* **24**, 197-259. (Also in *Collected Papers of Sir James Dewar,* Lady Dewar ed., Cambridge University Press 1927, vol. II, pp. 1334-1379.)

Dormer, K.J. 1980 *Fundamental Tissue Geometry for Biologists.* Cambridge University Press.

Glazier, J.A., Gross, S.P. & Stavans, J. 1987 Dynamics of two-dimensional soap froths. *Phys. Rev. A* **36**, 306-312.

Isenberg, C. 1978 *The Science of Soap Films and Soap Bubbles.* Woodspring Press Ltd., Somerset, UK.

Kelvin, Lord (as Sir William Thomson) 1887 On the division of space with minimal partitional area. *Phil. Mag. (Fifth series)* **24**, 503-514.

Kelvin, Lord 1894 On the homogeneous division of space. *Proc. Roy. Soc. (London)* **55**, 1-16.

Kraynik, A.M. 1988 Foam flows. *Ann. Rev. Fluid Mech.* **20**, 325-357.

Lewis, F.T. 1923 The typical shape of polyhedral cells in vegetable parenchyma and the restoration of that shape following cell division. *Proc. Amer. Acad. Arts & Sci.* **58**, 537-552.

Lewis, F.T. 1925 A further study of the polyhedral shapes of cells. *Proc. Amer. Acad. Arts & Sci.* **61**, 1-34.

Lewis, F.T. 1928a The shape of cork cells: A simple demonstration that they are tetrakaidecahedral. *Science* **68**, 625-626.

Lewis, F.T. 1928b The correlation between cell division and the shapes and sizes of prismatic cells in the epidermis of Cucumis. *Anat. Rec.* **38**, 341-376

Lewis, F.T. 1943 A geometric accounting for diverse shapes of 14-hedral cells: the transition from dodecahedra to tetrakaidecahedra. *Amer. J. Bot.* **30**, 766-776.

Matzke, E.B. 1945 The three-dimensional shapes of bubbles in foams. *Proc. Nat. Acad. Sci. (USA)* **31**, 281-289.

Matzke, E.B. 1946 The three-dimensional shape of bubbles in foam - An analysis of the role of surface forces in three-dimensional cell shape determination. *Amer. J. Bot.* **33**, 58-80.

Matzke, E.B. & Nestler, J. 1946 Volume-shape relationships in variant foams. A further study of the role of surface forces in three-dimensional cell shape determination. *Amer. J. Bot.* **33**, 130-144.

Mysels, K.J., Shinoda, K. & Frankel, S. 1959 *Soap Films - Studies of their Thinning*. Pergamon Press, New York.

von Neumann, J. 1952 Discussion remark concerning paper of C. S. Smith, 'Grain shapes and other metallurgical applications of topology'. In *Metal Interfaces,* Amer. Soc. for Metals, Cleveland, Ohio, pp. 108-110.

Nitsche, J.C.C. 1974 Plateau's problems and their modern ramifications. *Amer. Math. Monthly* **81**, 945-968.

Press, W.H., Flannery, B.P., Teukolsky, S.A. & Vetterling, W.T. 1986 *Numerical Recipes, The Art of Scientific Programming*. Cambridge University Press.

Princen, H.M. & Mason, S.G. 1965 The permeability of soap films to gases. *J. Colloid Sci.* **20**, 353-375.

Rayleigh, Lord 1890 Foam. *Proc. Roy. Inst.* **13**, 85-97. (Also in *Scientific Papers,* Vol. III, Dover Publ., New York 1964, paper #169, pp. 351-362.)

Rivier, N. 1983 On the structure of random tissues or froths, and their evolution. *Phil. Mag. B* **47**(5), L45-L49.

Rivier, N. & Lissowski, A. 1982 On the correlation between sizes and shapes of cells in epithelial mosaics. *J. Phys. A: Math. Gen.* **15**, L143-L148.

Schwarz, H.W. 1964 Rearrangements in polyhedric foam. *Recueil* **84**, 771-781.

Smith, C.S. 1952 Grain shapes and other metallurgical applications of topology. In *Metal Interfaces,* Amer. Soc. Metals, Cleveland, pp. 65-108; edited version in C.S. Smith *A Search for Structure,* MIT Press, 1981, pp. 3-32.

Smith, C.S. 1954 The shape of things. *Scient. Amer.* **190** (1), 58-64.

Stavans, J. & Glazier, J.A. 1988 Soap froth revisited: dynamic scaling in the two-dimensional froth. *Phys. Rev. Lett.* **62**, 1318-1321.

Truesdell, C. 1954 *The Kinematics of Vorticity*. Indiana University Press, Bloomington.

Weaire, D. 1974 Some remarks on the arrangement of grains in a polycrystal. *Metallography* **7**, 157-160.

Weaire, D., Fu, T-L. & Kermode, J.P. 1984 On the shear elastic constant of a two-dimensional froth. *Phil. Mag. B* **54**, L39-L43.

Weaire, D. & Kermode, J.P. 1983a The evolution of the structure of a two-dimensional soap froth. *Phil. Mag. B* **47**, L29-L31.

Weaire, D. & Kermode, J.P. 1983b Computer simulation of two-dimensional soap froth I. Method and motivation. *Phil. Mag. B* **48**, 245-259.

Weaire, D. & Kermode, J.P. 1984 Computer simulation of two-dimensional soap froth II. Analysis of results. *Phil. Mag. B* **50**, 379-395.

Weaire, D. & Rivier, N. 1984 Soap, cells and statistics - Random patterns in two dimensions. *Contemp. Phys.* **25**, 59-99.

Wejchert, J., Weaire, D. & Kermode, J.P. 1986 Monte Carlo simulation of the evolution of a two-dimensional soap froth. *Phil. Mag. B* **53**, 15-24.

Williams, R. 1979 *The Geometrical Foundation of Natural Structure*. Dover Publications, New York.

* * *

Wavelet Analysis of Coherent Structures in Two-dimensional Turbulent Flows

MARIE FARGE

LMD-CNRS, Ecole Normale Supérieure,
24, rue Lhomond. 75231 Paris Cedex 5, France

MATTHIAS HOLSCHNEIDER

Centre de Physique Théorique, CNRS
Luminy, case 907. 13288 Marseille Cedex, France

JEAN-FRANCOIS COLONNA

Centre de Mathématiques Appliquées
Ecole Polytechnique. 91128 Palaiseau, France

1 INTRODUCTION

We will introduce a new technique, the so-called 'wavelet transform', which allows a decomposition of arbitrary functions in terms of space, scale and direction. This leads to a sort of local spectrum that we will apply to study the scale production in decaying two-dimensional turbulent flows obtained from numerical simulations.

2 TWO-DIMENSIONAL TURBULENCE AND COHERENT STRUCTURES

In the inviscid (inertial) range the dynamics of two-dimensional flows is governed by the conservation of both energy ($E=\langle v^2\rangle$) and enstrophy ($Z=\langle(\mathrm{rot}\ v)^2\rangle=k^2 E$, k being the mean wavenumber), which leads to a direct enstrophy cascade from the injection scale towards smaller scales, conjointly with an inverse energy cascade from the injection scale towards larger scales (Kraichnan 1967, Batchelor 1969) . Therefore the time evolution of such a flow will generate larger and larger scales in the velocity field v while exciting smaller and smaller scales in the vorticity field (rot v).

Numerical simulations of two-dimensional flows have revealed the presence of coherent structures *(Figure 1)*. Their study is essential for a better understanding of turbulence, because the flow dynamics results from their mutual interactions. They spontaneously emerge out of random-phase turbulent flows by a condensation of the vorticity field into vortex structures which confine most of the enstrophy . They exist at very different scales, all along the inertial range, and survive on time scales much longer than the eddy turn-over time characteristic of nonlinear transfers. They are observed both in the laboratory, e.g. Couder 1984, and in numerical experiments, e.g. McWilliams 1984. We are here considering the time evolution of two-dimensional decaying turbulent flows obtained by numerically integrating the Saint-Venant equations (two-dimensional Navier-Stokes equations with a free surface) from an initial random field in nonlinear balance equilibrium (Farge 1988, Farge and Sadourny 1989).

The Fourier spectral analysis gives the decomposition of energy in terms of scale, but does not take into account the coherence present in physical space and is therefore not suited to study coherent structures. The flow visualization reveals the presence and shape of the coherent structures *(Figure 1)* but does not give quantitative information on the turbulent cascades, i.e. the transfer of energy and enstrophy between different scales. We need a decomposition in terms of both scale and space to unify these two approaches. This is why we use the wavelet transform, which realizes a local scale decomposition and, due to its invertibility property, would allow us to separate the coherent structures from the background flow passively advected by them.

3 THE WAVELET TRANSFORM

The need for a Fourier transform localized in the signal has a long history, for instance in quantum mechanics to study coherent states. In 1946 Gabor (1946) devised the windowed Fourier transform, for which the analyzing functions are trigonometric functions, modulated by a Gaussian envelope of constant shape which is translated all along the signal. But this transform present several drawbacks, among others:

- a low-frequency cut-off and limitation on the localization *(Figure 2c)* due to the finite width of the Gaussian envelope,
- a high sensitivity to phase errors for the reconstruction of the signal due to the alternating character of the trigonometric series to be added,
- a periodicity condition on the signal as soon as we consider a discretized version of the transform.

The one-dimensional continuous wavelet transform also generates a two-dimensional phase space as the windowed Fourier transform, one parameter corresponds to translation, but the other corresponds here to a dilatation of the time scale instead of a frequency modulation. This technique was first developed by Grossmann and Morlet (1985) to analyze seismic data in geophysics. Lemarié and Meyer (1986), Daubechies (1988) and Jaffard (1989) found several orthonormal wavelet bases, which can be related to the pioneering works of Haar (1910), Franklin (1928) and Stromberg (1981). Today they play an important role in functional analysis and many developments are now encountered in numerical analysis (Beylkin et al. 1989, Perrier 1989). A fast wavelet transform algorithm, based on a factorization technique similar to those used for quadratic mirror filters, has been devised (Holschneider et al. 1988). Murenzi (1989) has recently extended the theory to the case of the two-dimensional wavelet transform adding the group of rotation to those of dilatation and translation already used for the one-dimensional wavelet transform. A general review on continuous and orthonormal wavelets is given in Combes, Grossmann and Tchamitchian (1989). Nowadays the main applications of wavelets are in acoustics (Kronland-Martinet et al. 1987, Saracco and Tchamitchian 1988), statistical mechanics, with for instance the analysis of multi-fractal objects related to phase transitions (Arneodo et al. 1988, Argoul et al. 1989), and image processing (Mallat 1989). Other applications can be found in quantum field theory, in relation to the renormalization problem (Battle 1987, Federbush 1987), in quantum mechanics (Daubechies et al. 1987, Paul 1989) and in vision theory (Duval-Destin and Menu 1989). Its use for turbulence was first developed for the case of one-dimensional signals (Farge and Rabreau 1988, Argoul et al. 1989) and then to the case of two-dimensional flows by Farge and Holschneider (1989).

The wavelet transform of a two-dimensional scalar field f uses a zero mean value function w, called wavelet, presenting only a few oscillations in a Gaussian envelope, that is translated (parameter b), dilated (parameter a) and rotated (operator R) in order to generate a family of wavelets $w_{a,R,b}$, such that:

analysis
$$\tilde{f}(a,R,b) = < w_{a,R,b} \mid f >, \tag{1}$$

with $w_{a,R,b}(x)=a^{-1}w(a^{-1}R^{-1}(x-b))$,

synthesis
$$f(x) = C \int a^{-3}da \int dm(R) \int w_{a,R,b}(x)\,\tilde{f}(a,R,b)\, d^2b, \tag{2}$$

dm(R) being the invariant measure of the rotation group and C a constant.

The wavelet transform also conserves energy (Parseval):

$$\int |f(x)|^2\, dx = \int a^{-3}da \int dm(R) \int |\tilde{f}(a,R,b)|^2\, d^2b. \tag{3}$$

The two-dimensional wavelet transform may be interpreted as a mathematical polarizing microscope: its magnification being a^{-1}, its polarization R and its position b, while the choice of the wavelet w controls the quality of the optics. We will use here the two-dimensional complex wavelet, called Morlet wavelet, of the form:

$$w(x)=e^{ik_0.x}\, e^{-(x.x)/2}, \tag{4}$$

with k_0 fixed frequency.

To analyze turbulent signals we prefer complex wavelets, because they allow a clear separation between the modulus, which corresponds to energy, and the phase, which shows more precisely the local regularity of the function (for the lines of constant phase point on the singularities). The family generated from the Morlet wavelet is non-orthogonal, therefore there is a redundancy of coefficients in wavelet space, which is actually very useful for analyzing and filtering the data. In the discrete case to reconstruct the signal we then have two possibilities, either to perform a triple integration on the dyadic grid *(Figure 3b)*, which corresponds to the only coefficients which are orthonormal and therefore independent, using a discretized version of (2), or a double integration on the complete grid *(Figure 3a)* according to:

$$f(x) = K \sum dm(R) \sum w_{a,R,x}(x)\,\tilde{f}(a,R,x)\, a^{-3}da, \tag{5}$$

K being a constant.

4 ITS APPLICATION TO DECAYING TWO-DIMENSIONAL TURBULENCE

Computing the two-dimensional complex wavelet transform of the vorticity field at a given time using the Morlet wavelet, we show *(Figure 4)* that the vortex cores contain very small-scale features compared to the background flow, especially when the vortices are strongly interacting, e.g. merging. This contradicts the classical interpretation of two-dimensional turbulence: the smallest scales are not the filamentary structures, which get developed at the vortex periphery under the action of the enstrophy cascade, but are mostly concentrated inside the vortex cores. This means that the nonlinear interactions may excite degrees of freedom internal to the vortices, and that dissipation, which is sensitive to the smallest scales, would act mostly on the internal structure of the vortices *(Figure 5)*.

Consequently, if the smallest scales of the flow do not correspond to the background flow but to the coherent structures, i.e. the vortices, dissipation should then be maximal there. This is confirmed when we plot the Laplacian of vorticity (Δ dissipation operator): the extrema are correlated with the vortex cores and both the vorticity and Laplacian of vorticity fields show the same spatial coherency *(Figure 5a)*. On the contrary, this correlation is lost if we consider now an iterated Laplacian of vorticity (Δ^n hyperdissipation operator, with here n=8): the spatial structure of this field is very different compared to the vorticity field and we do not see anymore the presence of coherent structures in the hyperdissipation field *(Figure 5b)*. It has been shown that the choice of a hyperdissipation instead of a dissipation operator presents the advantage of reducing the dissipation bandwidth in the small scales, the effect of dissipation becoming then more local in terms of scales, while not affecting the spectrum at the large scales (Basdevant 1981, Bennett and Haidvogel 1983). But would the use of an hyperdissipation change the spatial structure of the flow: would the lack of spatial correlation between the vorticity and the hyperdissipation fields we observe modify the dynamics of the flow and its topological properties? Where, in physical space, does dissipation act on the flow and how would it affect the coherent structures? These open questions seem to us essential and we will try to answer them by tracking the spectrum locally in space with the help of the wavelet transform.

We will also use the wavelet transform to separate coherent structures, whose dynamics drives the whole flow, from the background fluid which is passively advected by the vortices. For this, we will develop filtering techniques, such as 'skeleton extraction' based on a stationary phase hypothesis, and use the invertibility property of the wavelet transform (2) in order to reconstruct the vorticity field corresponding only to the dynamically active component of the flow. In doing this, our goal would be to reduce the number of degrees of freedom necessary to compute the evolution of two-dimensional turbulent flows, by projecting on wavelet modes instead of Fourier modes.

5 CONCLUSION

Computing the two-dimensional complex wavelet transform of the vorticity field, we have found that the vortex cores contain very small-scale features compared to the background flow. This contradicts the classical interpretation of two-dimensional turbulence: the smallest scales are not the filamentary structures, which get developed at the vortex periphery, but are concentrated inside the vortex cores. This means that the nonlinear interactions may excite degrees of freedom internal to the vortices.

Acknowledgments

We thank Ted Shepherd for useful comments.

REFERENCES

ARGOUL F., ARNEODO A., ELEZGARAY J. and GRASSEAU G. 1989
Wavelet transform of fractal aggregates
Phys. Lett. A, ***135,*** *n° 6/7, 327-336*

ARGOUL F., ARNEODO A., GRASSEAU G., GAGNE Y., HOPFINGER E. J., FRISCH U. 1989
Wavelet analysis of turbulence reveals the multifractal nature of the Richardson cascade
Nature, ***338,*** *n°6210, 51-53*

ARNEODO A., ARGOUL F., ELEZGARAY J. and GRASSEAU G. 1988
Wavelet transform analysis of fractals: applications to nonequilibrium phase transitions
International Conference on 'Nonlinear Dynamics', Bologna, ed. Turchetti, World Scientific

ARNEODO A., GRASSEAU G. and HOLSCHNEIDER M. 1988
Wavelet transform of multifractals
Phys. Rev., ***61,*** *n° 20, 2281-2284*

BATTLE G. 1987
Ondelettes and phase cluster expansion, a vindication
Comm. Math. Phys., ***109,*** *417-419*

BATCHELOR G. K. 1969
Computation of the energy spectrum in homogeneous two-dimensional turbulence
Phys. Fluids, Suppl. II, ***12,*** *233-239*

BASDEVANT C. 1981
Contribution à l'étude numérique et théorique de la turbulence bidimensionnelle
Thèse d'Etat, Université Paris VI

BENNETT A.F. and HAIDVOGEL D.B. 1983
Low-resolution numerical simulation of decaying two-dimensional turbulence
J. Atmos. Sci., ***40,*** *738-748*

BEYLKIN G., COIFMAN R. and ROKHLIN V. 1989
Fast wavelet transforms
Preprint, Math. Dep., Yale University

COMBES J.M., GROSSMANN A. and TCHAMITCHIAN P. 1989
Wavelets, time-frequency methods and phase space
Proceedings of the 1st International Conference on Wavelets, Marseille, 14-18 December 1987, Inverse Problems and Theoretical Imaging, Springer

COUDER Y. 1984
Two-dimensional grid turbulence in a thin liquid film
J. Physique Lett., ***45***, *8, 353-360*

DAUBECHIES I. 1988
Orthonormal bases of compactly supported wavelets
Comm. in Pure and Applied Math., ***49***, *909-996*

DAUBECHIES I., KLAUDER J.R. and PAUL T. 1987
Wiener measure for path integrals with affine kinematic variables
J. Math. Phys., ***28***

DUVAL-DESTIN M. and MENU J.P.1989
Wavelet transform: a new basic spatial operator for visual psychophysics
Submitted to Vision Research

ESCUDIE B., KRONLAND-MARTINET R., GROSSMANN A. and TORRESANI B. 1989
Analyse par ondelettes de signaux asymptotiques: emploi de la phase stationnaire
12ième Colloque du GRETSI, Juin 1989

FARGE M. 1988
Vortex motion in a rotating barotropic fluid layer
Fluid Dynamics Research, ***3***, *282-288*

FARGE M. and RABREAU G. 1988
Transformée en ondelettes pour détecter et analyser les structures cohérentes dans les écoulements turbulents bidimensionnels
C. R. Acad. Sci. Paris, ***307***, *série II, 1479-1486*

FARGE M. and SADOURNY R. 1989
Wave-vortex dynamics in rotating shallow water
J. Fluid Mech., ***206***, *433-462*

FARGE M. and HOLSCHNEIDER M. 1989
Wavelet analysis of coherent structures in two-dimensional turbulent flows
To appear in the proceedings of the 2nd International Conference on 'Wavelets', Marseille 29 May-2 June 1989, ed. Meyer, Springer

FEDERBUSH P. 1987
Quantum field theory in ninety minutes
Bull. Amer. Math. Soc., July

FRANKLIN P. 1928
A set of continuous orthogonal functions
Math. Annalen, ***100***, *522-529*

GABOR D. 1946
Theory of Communication
J. Inst. Electr. Engin., ***93***, *III, 429-457*

GROSSMANN A. and MORLET J. 1985
Decomposition of functions into wavelets of constant shape, and related transforms
'Mathematics and Physics', Lectures on recent results, World Scientific Publishing

HAAR A. 1910
Zür Theorie der Orthogonales Funktionensysteme
Math. Ann, ***69****, 336*

HOLSCHNEIDER M., KRONLAND-MARTINET R., MORLET J. and TCHAMITCHIAN P. 1988
The "algorithme à trous"
Submitted to IEEE

HOLSCHNEIDER M. 1988
On the wavelet transform of fractal objects
Stat. Phys., ***50****, n° 5/6*

JAFFARD S. 1989
Construction of wavelets on open sets
Proceedings of the 1st International Conference on Wavelets, Marseille, 14-18 December 1987, ed. Combes et al., Springer, 247-252

LEMARIE P.G. and MEYER Y. 1986
Ondelettes et bases Hilbertiennes
Rev. Mat. Ibero-americana, ***2****, 1*

KRAICHNAN R.H. 1967
Inertial ranges of two-dimensional turbulence
Phys. Fluids, ***10****, 1417-1423*

KRONLAND-MARTINET R., MORLET J. and GROSSMANN A. 1987
Analysis of sound patterns through wavelet transform
Int. J. Pattern Analysis and Artificial Intelligence, ***1****, n°2, 273-302*

MALLAT S.G. 1989
A theory for multiresolution signal decomposition: the wavelet representation
IEEE Trans. on Pattern Analysis and Machine Intelligence, ***II****, n° 7*

MURENZI R. 1989
Wavelet transforms associated to the N-dimensional Euclidean group with dilatations: signal in more than one dimension
Proceedings of the 1st International Conference on Wavelets, Marseille, 14-18 December 1987, ed. Combes et al., Springer, 239-246

PAUL T. 1989
Wavelets and path integrals
Proceedings of the 1st International Conference on Wavelets, Marseille, 14-18 December 1987, ed. Combes et al., Springer, 204-208

PERRIER V. 1989
Towards a method for solving partial differential equations using wavelet bases
Proceedings of the 1st International Conference on Wavelets, Marseille, 14-18 December 1987, ed. Combes et al., Springer, 269-283

PAUL T. 1989
Affine coherent states and the radial Schrödinger equation
Submitted to Annales de l'Institut Henri Poincaré

STROMBERG J. O. 1981
A modified Haar system and higher order spline systems
International Conference on 'Harmonic Analysis in honor of Antoni Zygmund', II, 475-494, ed. Beckner et al., Wadsworth

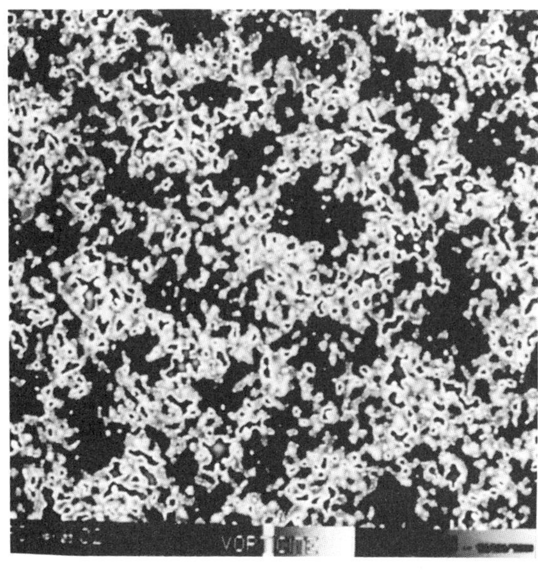

a. $t = 0\ s$

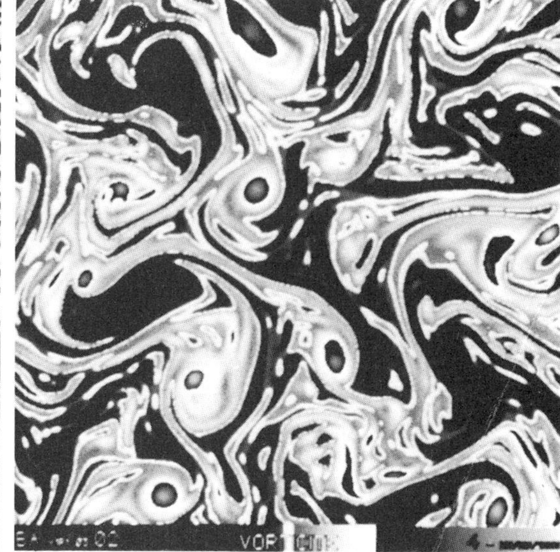

b. $t = 0.3\ 10^6\ s$

c. $t = 2.7\ 10^6\ s$

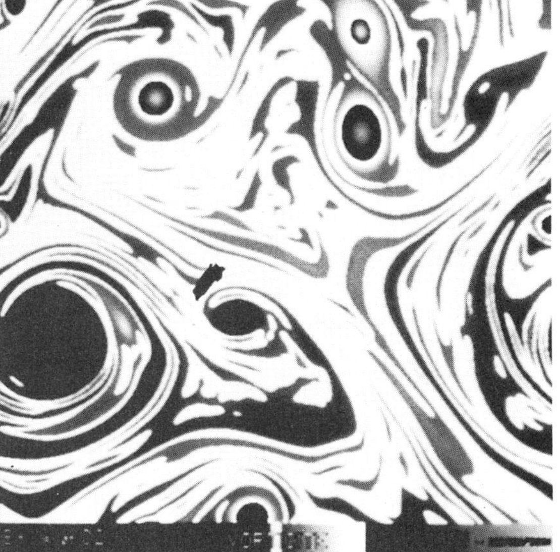

d. $t = 3.7\ 10^6\ s$

Figure 1

Time evolution of a
decaying two-dimensional turbulent flow
from random initial conditions in nonlinear balance equilibrium

(See also Colour Plates)

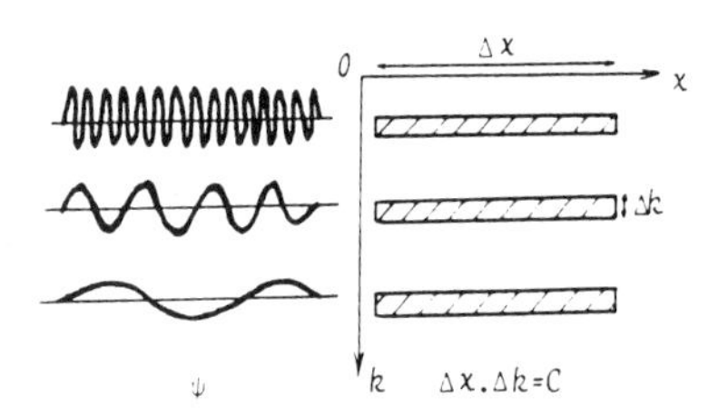

a. *ψ: trigonometric function*
No space localisation

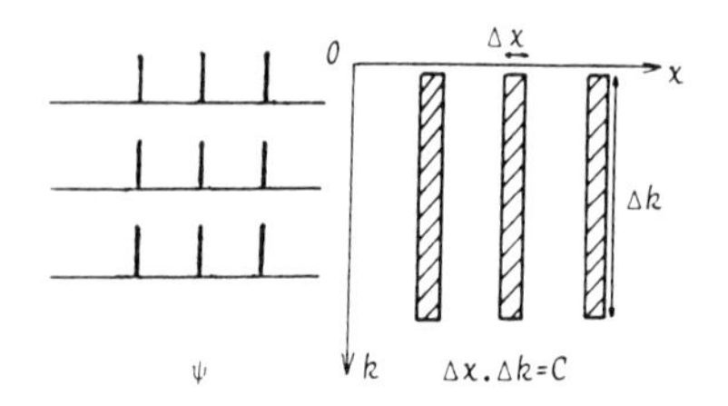

b. *ψ: Dirac pulse*
No scale localisation

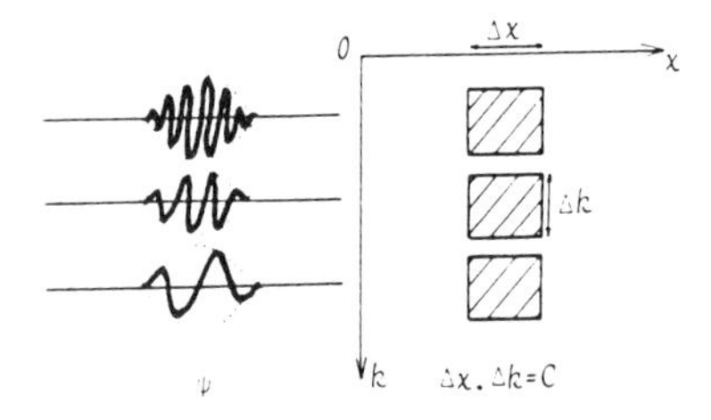

c. *ψ: trigonometric function with a constant Gaussian envelope whatever the scale*
Fixed space-scale localisation

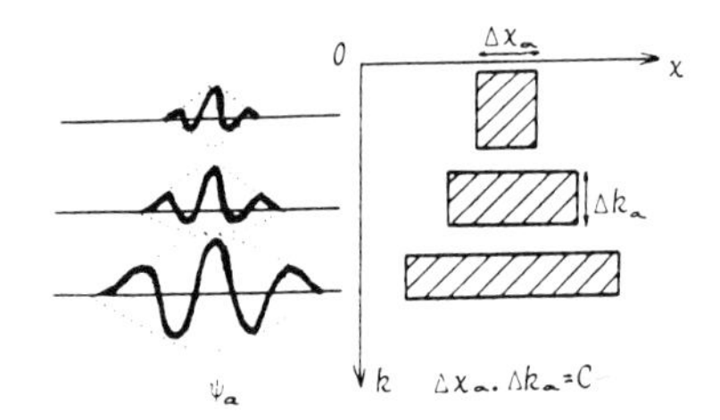

d. *ψ: wavelet dilated with a parameter a corresponding to the different scales*
Adaptative space-scale localisation

Figure 2

Space-scale diagram showing the limitation due to the uncertainty principle $\Delta x . \Delta k$ = constant for different analyzing function ψ

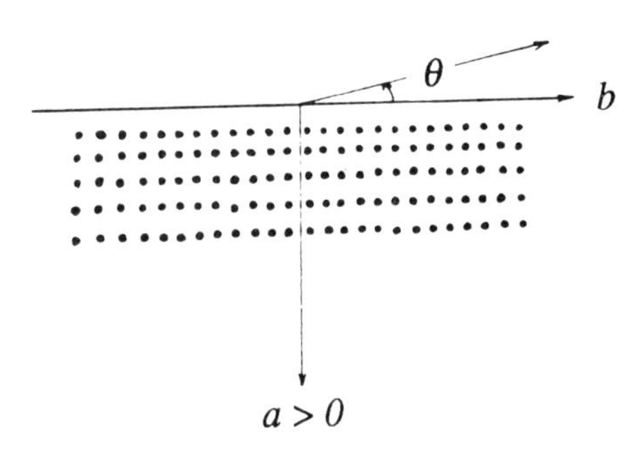

a. *Complete grid: $N_a . N_b$ non-independent coefficients*

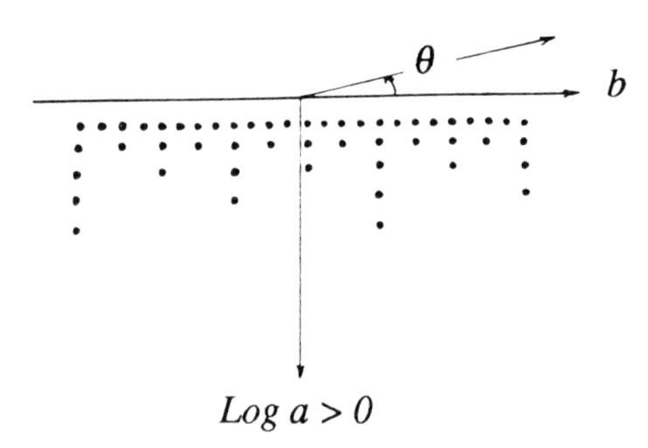

b. *Dyadic grid: $\log_2 N_a . N_b$ independent coefficients*

Figure 3

Wavelet coefficient space $\widetilde{f}(a,R,b)$

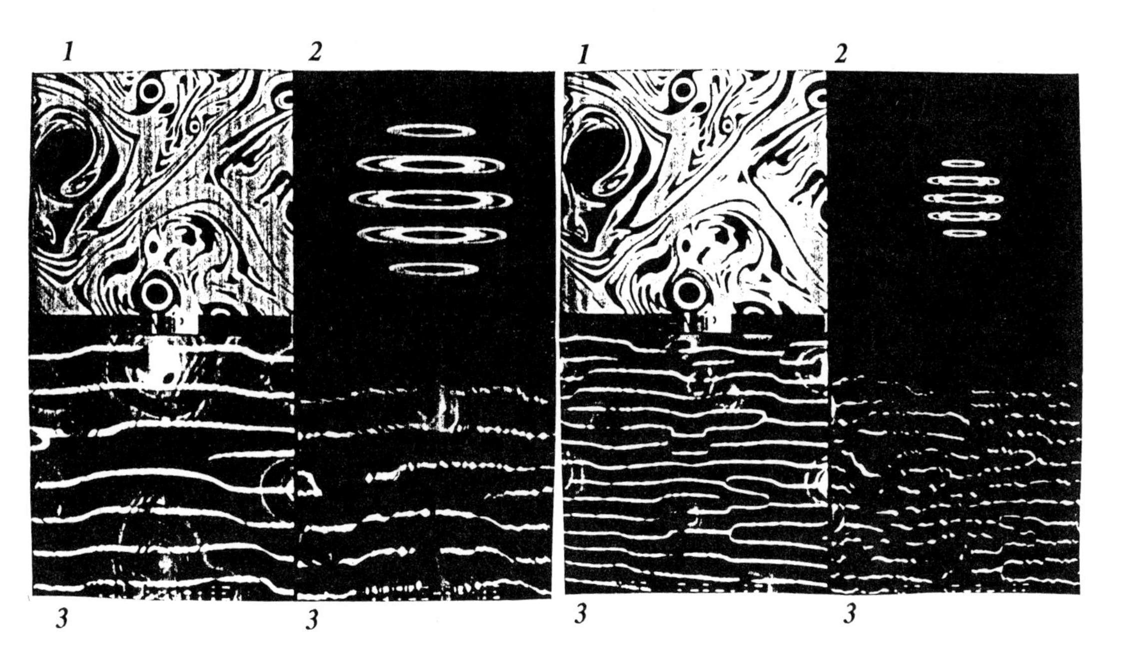

a. *Scale k=8 and angle α=0* *b.* *Scale k=16 and angle α=0*

1 ***Vorticity** field to be analyzed (cartographic view)*

2 *Analyzing **wavelet** at scale k and angle α=0*

3 *Superposition of :*
*-**vorticity** field*
(perspective view coded in luminance)
*-**modulus** of the wavelet transformat scale k and angle α=0*
(cartographic view color coded in the order: blue, red, magenta, green, cyan, yellow, white)
*-**phase** of the wavelet transform at scale k and angle α=0*
(isoline zero beingwhite)

Figure 4

Two-dimensional wavelet transform
of the vorticity field
(corresponding to the last time step shown on figure 1)

(See also Colour Plates)

a. *Vorticity ω field*

b. *Dissipation $\Delta\omega$ field: we recognize the underlying coherency of the vorticity field*

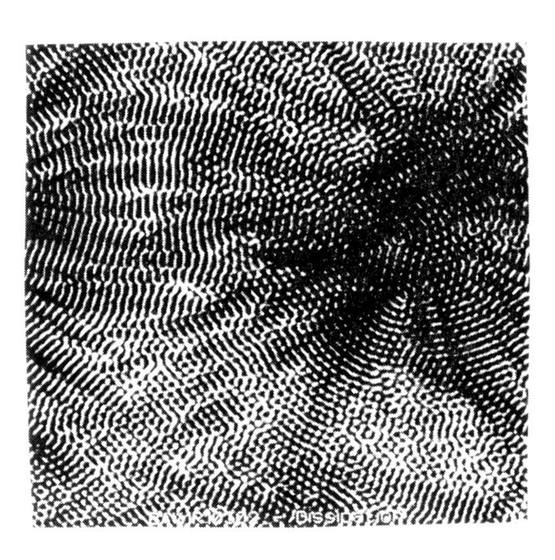

c. *Hyperdissipation $\Delta^8\omega$ field: we have lost the coherency present in the vorticity field*

Figure 5

Comparaison between the vorticity ω, dissipation $\Delta\omega$ and hyperdissipation $\Delta^8\omega$ fields

Structure Formation in Self-Gravitating Flows

J. LEORAT[1], T. PASSOT[2*] and A. POUQUET[3*]

1: Observatoire de Meudon, 92190- Meudon, France.
2: Department of Mathematics, University of Arizona, Tucson, 85721-AZ, USA.
3: NCAR, PO Box 3000, Boulder, CO 80303, USA.
*: on leave from Observatoire de Nice, BP139, 06003-Nice, France.

1 INTRODUCTION

Matter in the Universe is distributed in a wide range of structures, either in diffuse or condensed states. Uniformity of thermodynamical equilibrium is delayed through various dynamical phenomena were self gravitation plays a main role. Consider for example the stellar cycle which reprocesses the interstellar matter. In spite of noticeable observationnal advances, the fundamental processes leading to star formation, proceeding continuously in galaxies, are still poorly understood. Supersonic velocities are observed in the molecular clouds were stars form, and the Jeans linear stability analysis obviously fails to describe such flows. The life time of these clouds is large compared to their free-fall time, and it is interesting to examine the influence of turbulence on gravitational collapse in order to estimate the relevant dynamical time scale.
Theoretical knowledge in the field of supersonic turbulence is unfortunately not in a better state than for incompressible turbulence. Numerical simulations allow a quasi-experimental approach of self gravitating flows. In the first stage of the collapse, the interstellar medium may be considered as optically thin and one has to follow the evolution of an isothermal gas, using periodic boundary conditions appropriate to homogeneous flows.
To explore the parameter space and classify the numerical results, phenomenological considerations have proved to be usefull. We outline such a phenomenology in the following Section ; we then give a summary of the difficulties overcome in the numerical study and present some results; a conclusion is proposed in the last Section.

2 PHENOMENOLOGICAL CONSIDERATIONS

The linear instability analysis, which is valid only for vanishing velocities, leads to a statement in terms of scales, such as the Jeans length. Turbulence may induce stron modifications in this picture: on one hand, the large scales of the flow tend to tear the collapsing clumps and fragment them into stable pieces and, on the other hand, the smaller scales may form density fluctuations with such an amplitude that collapse could be initiated at scales smaller than the Jeans length.

In order to evaluate the impact of turbulence on gravitational collapse, it is convenient to compare characteristic time scales, such as the free fall time T^{FF}, the acoustic time T^{AC} (corresponding to the largest available scale) or the nonlinear turn-over time T^{NL} at a given scale. For example, Jeans instability occurs when the ratio

$$r^J = T^{AC}/T^{FF}$$

is greater than a number close to one. If this number r_J is smaller, the gas pressure is able to overcome the gravitational acceleration. We now tentatively generalize this criterion to turbulence, defining a "turbulent gravitational number",

$$r^* = T^{*NL}/T^{FF},$$

where the asterix recalls that this quantity has been evaluated at a characteristic scale of turbulence, such as the integral scale. In the case of supersonic turbulence ($T^{*NL}<T^{AC}$), the number r^* is small and eventually collapse is stopped when this quantity stays below a "critical gravitational number" of order unity.

A few remarks are here in order. First, the ratio T^{NL}/T^{FF} extends in fact along a complete spectrum and one would have to compare it with its critical value at all excited scales. Second, the relevant dynamical time is the transfer time of compressible energy and not the global turn-over time T^{NL}, since the gravitational acceleration acts only upon the divergent part of the flow. As the preceeding phenomenological arguments are checked with numerical simulations unable to exhibit a complete inertial range, these refinements need not to be examined here.

This phenomenology could be compared with the predictions of Bonazzola et al. (1987), who have suggested that turbulence may lead to an inversion of the Jeans criterium for gravitational instability. Note also that a modification of this criterium based on a "turbulent pressure" argument has been proposed by Chandrasekhar (1951).

3. NUMERICAL RESULTS

As recalled in the Introduction, the numerical study of self gravitating flows involves simulation of supersonic flows, including shocks and high density gradients. We use a Fourier spectral method (Gottlieb and Orszag 1977, Canuto et al. 1988), particularly efficient for homogeneous turbulence with periodic boundary conditions. It has been verified that this numerical method is able to cope with supersonic flows, with a typical CPU time of 5 minutes on a Cray 1 for a turn-over time when the maximal Mach number is three and the resolution is 256*256 (see for example, Léorat et al, 1984, Passot and Pouquet, 1988). The Poisson equation for the gravitational potential is directly solved, but some numerical difficulties related to self gravitation must be pointed out:

- the conservative formulation of the momentum equation must be avoided, since aliasing in the quadratic gravitational force gives rise to large scale pertubations and numerical instability.

-Fluctuations of density around zero tend to occur in the final stage of collapse. In the terms where density appears in the denominator (viscous term for example) or in a power law (if a polytropic equation of state is used), a small and uniform density background is introduced to prevent numerical troubles

The numerical procedure involves two successive steps: a stationary state of homogeneous turbulence is first achieved without gravitation, and thus external forcing is introduced. For simplicity reasons, it is a white noise in time and its spatial spectrum is a combination of two power laws (k^4 and k^{-4} at resp. large and small scales). In the second step, gravitation is set on, with a given number of Jeans masses in the computational box. This method allows to control the value of the gravitational number r* when gravitation begins to act, and to distinguish the consequences of turbulence from other questions related to its origin. Unforced runs produce results dependent on the initial conditions and since parameters are continuously evolving, the conclusions remain ambiguous.

To reduce the computing time, we have performed two-dimensional computations at a resolution of 128*128 There are generally 5 Jeans masses in the box, so that all wavelengths smaller than $\sqrt{5}$ are linearly unstable; a greater number of Jeans masses is paid by a smaller time step. It has been found that to get values of the gravitational parameter r* small enough to stop the collapse (below unity), the forcing spectrum must have a maximum around the wavenumber 20; white noise forcing at large scales only leads to a rather artificial supersonic turbulence state, with high density maxima able to initiate the collapse.

As a more complete report with comparative results has been submitted for publication (Léorat et al. 1988), we give here only a short account of a typical run, with r*=0.2, such that turbulence stabilizes the collapse.The turbulent Mach number (based on the r.m.s. velocity) is about 3. The figure below shows the evolution of the difference between the maximal and mean density (romax-1) and the one of the lowest wavenumbers, after gravitation has been set on in a statistically stationary turbulence (solenoidal forcing).

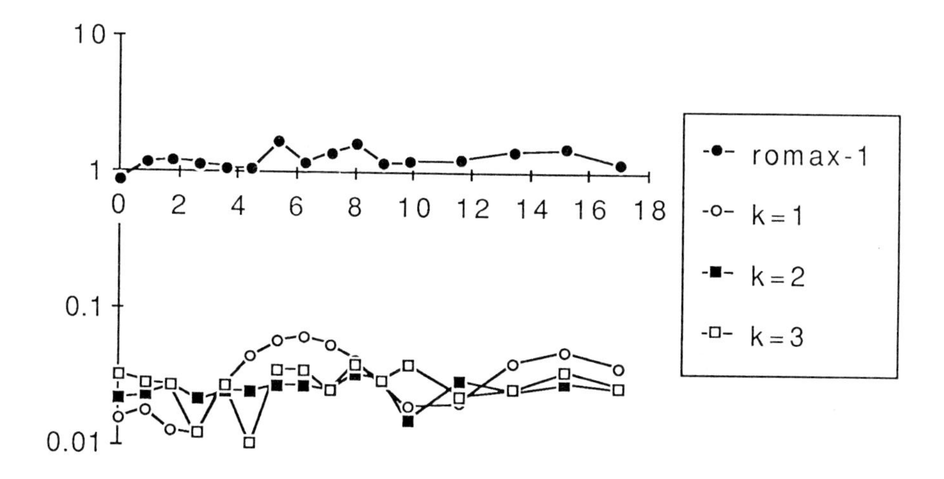

evolution of maximal density and some density modes over 17 free fall times

The final time of the computation is about 17 free fall times, and the final density contrast (ratio of the maxima density to its minimal value) is comparable to the one at the initial time. Time evolution of the density contrast provides a simple way to decide if collapse is suppressed by turbulence. Keeping the same forcing, if the number of Jeans masses is, say, 10, r* rises to 0.3, and it turns out that turbulent stabilization is lost: in about 6 free-fall times, the density contrast rises to 42 and computation must stop.

4. CONCLUSION

The numerical experiments have shown that supersonic turbulence has a strong impact on the evolution of self gravitating flows, being able to stop the gravitational collapse and they have confirmed the usefulness of the definition of the "gravitational number" r*. These runs, although consistent with the phenomenology of Section 2, are not a sufficient proof of the existence of a critical value of the gravitational number (around 0.25). In fact, they suffer from obvious shortcomings, which are often encountered in numerical fluid dynamics: the resolution is too low and artificial forcing is necessary to get statistical stationarity. The number of Jeans lengths would need to be larger in order to avoid the disturbing effect of the truncation at large scales; when there is only about one Jeans mass in the computation, gravitational collapse disappears even for values of the gravitational number greater than unity. To maintain a supersonic turbulent regime, forcing is relatively strong so that its spectral range cannot be entirely at large scales and overlaps the dissipation domain. Moreover, the results presented above are based on two-dimensional flows and one knows that their dynamics has some peculiarities. Three dimensional simulations are under way by P. Chantry and R. Grappin (Observatoire de Meudon) and the resolution problem is still more severe in this case. If supersonic turbulence cannot be sustained with a realistic time dependent forcing, one can study with some details the influence on gravitational collapse of basic flows, such as shearing, and the numerical approach is well suited to tackle this type of problems.
The main result of these investigations concerns the display of the structurating effect of turbulence on gravitational collapse. In absence of initial velocity, the collapse first enhances the primordial density pertubations (Jeans analysis applies), and the nonlinear step then gathers the various density peaks together in a single condensation, which minimizes energy in about one free-fall time. When turbulence is present at the beginning of the collapse, it prescribes its characteristic length and time scales so that the collapse of the whole mass may be inhibited. Thus the life time of clouds would be related to turbulence decay rather than to free fall and the turbulence spectral properties would have a major influence on the initial mass spectrum of stars. If it turns possible to show in a further step that collapse is itself partly able to feed the turbulence, a feed-back loop would thus be completed. The numerical approach will certainly continue to play a leading role in such a program.

REFERENCES

Bonazzola, S., Falgarone, E., Heyvaerts, J., Perault, M., Puget, J.-L., 1987. Astron. Astrophys., **172**, 293.
Canuto, C., Hussaini,Y., Quarteroni A., Zang T.A., 1988. "Spectral methods in fluid dynamics", Springer.
Chandrasekhar, S.: 1951, Proc. Roy. Soc., A**210**,26.
Gottlieb, D., Orszag, S., 1977, "Numerical Analysis of Spectral Methods", SIAM, Philadelphia.
Léorat, J., Pouquet, A. and Poyet, J.-P.: 1984, in 'Proceedings of Colloquium on Problems of Collapse and Numerical Relativity', eds. M. Bancel and M. Signore, p. 287, Reidel.
Léorat, J., Passot, T. and Pouquet, A., 1988, "Influence of supersonic turbulence on self-gravitating flows", submitted.
Passot, T. and Pouquet, A., 1987. J. Fluid Mech., **181**, 441.

Attractors in Problems of Magnetohydrodynamics

O.A. LADYZHENSKAYA

Leningrad Department of Steklov Mathematical Institute,
Academy of Sciences, USSR.

In my paper (Ladyzhensksaya, 1972), I introduced the set $\mathcal{M}$ of all limit states for some boundary value problems for the 2D Navier-Stokes equations in a bounded domain. The set $\mathcal{M}$ is the minimal global B-attractor of the problem. The approach is useful for the study of some problems of MHD for viscous fluids and for many other dissipative problems of partial differential equations (see my survey, Ladyzhenskaya 1987).

Here we formulate some results for the problem

$$v_t - \nu\triangle v + \sum_{k=1}^{2} v_k v_{x_k} + [H, \text{ rot } H] = -\nabla p + f, \tag{1}$$

$$H_t + \gamma \text{ rot rot } H + \text{ rot } [H, v] = \text{ rot } j, \tag{2}$$

$$\text{div } v = 0 \quad , \quad \text{div } H = 0, \tag{3}$$

$$v|_{\partial\Omega} = 0 \ , \quad (H, n)|_{\partial\Omega} = 0 \ , \quad (\text{rot } H, \tau)|_{\partial\Omega} = 0, \tag{4}$$

$$v|_{t=0} = \overset{\circ}{v} \quad , \quad H|_{t=0} = \overset{\circ}{H} \tag{5}$$

assuming that the cartesian components $v_k(\cdot)$ and $H_k(\cdot)(k = 1, 2, 3)$ of vector fields $v(\cdot)$ and $H(\cdot)$ depend only on $(x_1, x_2) \in \Omega$ and $t \in R^+ \equiv [0, \infty)$ and do not depend on x_3 (so $f(\cdot)$ and $j(\cdot)$ do not depend on x_3 either). We suppose that $\nu = \text{cst} > 0, \gamma = \text{cst} > 0$, Ω is a bounded simply-connected domain in R^2, $\partial\Omega$ is its boundary, n is the external normal to $\partial\Omega$, and τ is the tangent to $\partial\Omega$. Let also $(j, \tau)|_{\partial\Omega} = 0$ and $\partial\Omega$ be smooth enough i.e. $\partial\Omega \subset C^2$). We shall use the following notations:
$L_2(\Omega)$ is the set of all vector-fields $u(x), x \in \Omega$, with the finite norm
$\|u\| =: \left(\int_\Omega |u(x)|^2 dx_1 dx_2\right)^{\frac{1}{2}}$;

$L_2(\Omega)$ is the Hilbert space with the inner product $(u,v) =: \int_\Omega u(x)\cdot v(x)dx_1dx_2$

$$\begin{aligned}
W_2^l(\Omega) &=: \{u \in L_2(\Omega) \| \exists \partial_x^k u \in L_2(\Omega), |k| \le l\}, l \ge 1 \\
\mathcal{T}^l(\Omega) &=: \{u \in W_2^l(\Omega) \| \text{ div } u = 0\}, \\
\overset{\circ}{\mathcal{T}}{}_n^l(\Omega) &=: \{u \in \mathcal{T}^l(\Omega) \| (u,n)|_{\partial\Omega} = 0\}, \\
\overset{\circ}{\mathcal{T}}{}_\tau^l(\Omega) &=: \{u \in \mathcal{T}^l(\Omega) \| (u,\tau)|_{\partial\Omega} = 0\}, \\
\dot{\mathcal{T}}^\infty(\Omega) &=: \{u \in C^\infty(\Omega) \| \text{ div } u = 0 \text{ and } u \equiv 0 \text{ near } \partial\Omega,\} \\
G(\Omega) &=: \{u \in L_2(\Omega) \| u = \nabla q\ ,\ q \in W_2^1(\Omega)\}, \\
\overset{\circ}{\mathcal{T}}(\Omega) &=: L_2(\Omega) \ominus G(\Omega),
\end{aligned}$$

$\mathcal{T}(\Omega)$ is the closure in $L_2(\Omega)$ of the set $\mathcal{T}^1(\Omega)$. We shall use the following well-known facts (Ladyzhenskaya 1961, Bykhovskii & Smirnov 1960, Ladyzhenskaya & Solonnikov 1960):

THEOREM 1

1) $\dot{\mathcal{T}}^\infty(\Omega)$ is dense in $\overset{\circ}{\mathcal{T}}(\Omega)$.

2) $\overset{\circ}{\mathcal{T}}{}_\tau^l(\Omega)$ is dense in $\mathcal{T}(\Omega)$.

3) If $\partial\Omega \subset C^{l+1}$, $l \ge 1$, then

$$\overset{\circ}{\mathcal{T}}{}_q^l(\Omega) \underset{(\mathrm{rot})^{-1}}{\overset{\mathrm{rot}}{\Longleftrightarrow}} \overset{\circ}{\mathcal{T}}{}_n^{l-1}(\Omega)\ ,\ \overset{\circ}{\mathcal{T}}{}_n^l(\Omega) \underset{(\mathrm{rot})^{-1}}{\overset{\mathrm{rot}}{\Longleftrightarrow}} \mathcal{T}^{l-1}(\Omega)$$

and

$$c_1\|\text{rot } u\|^{l-1} \le \|u\|^{(l)} \le c_2\|\text{rot } u\|^{(l-1)}\ ,\ c_k > 0.$$

Here $\|\cdot\|^{(l)}$ is a norm in $W_2^l(\Omega)$ $\left(\text{for example} \|u\|^{(l)} = (\sum_{|k|\le l} \|\partial_x^k u\|^2)^{\frac{1}{2}}\right)$, $\overset{\circ}{\mathcal{T}}{}_n^{\circ} =:$ $\mathcal{T}$ and $\mathcal{T}^o =: \mathcal{T}$.

THEOREM 2

The problem

$$-\Delta u(x) = -\nabla q(x) + f(x),\ \ \text{div } u(x) = 0,\ x \in \Omega,\ u|_{\partial\Omega} = 0, \tag{6}$$

has a unique solution $u \in \overset{\circ}{\mathcal{T}}{}^2_n \cap \overset{\circ}{\mathcal{T}}{}^2_\tau$, $q \in W_2^1(\Omega)$ $(\int_\Omega q dx = 0)$ for any $f \in \overset{\circ}{\mathcal{T}}$ and

$$\|u\|^{(2)} + \|q\|^{(1)} \leq C\|f\| \tag{7}$$

The operator $A_1 = -P\triangle$, where P is the ortho-projector in $L_2(\Omega)$ onto $\overset{\circ}{\mathcal{T}}$, is an unbounded operator in Hilbert space $\overset{\circ}{\mathcal{T}}$. It is self-adjoint and positive definite on the domain $\mathcal{D}(A_1) = \overset{\circ}{\mathcal{T}}{}^2_n \cap \overset{\circ}{\mathcal{T}}{}^2_\tau$; its inverse A_1^{-1} gives the solution $u = A_1^{-1} f$ of the problem (6); A_1^{-1} is a completely continuous operator in $\overset{\circ}{\mathcal{T}}$.

THEOREM 3

The problem

$$\begin{aligned} \text{rot rot } u(x) &= f(x), \ \text{div } u(x) = 0, \ x \in \Omega, \\ (u, n)|_{\partial\Omega} &= 0, \ (\text{rot } u, \tau)|_{\partial\Omega} = 0, \end{aligned} \tag{8}$$

has a unique solution $u \in \mathcal{D}(A_2) =: \{v \| v \in \overset{\circ}{\mathcal{T}}{}^2_n, \ 0\text{rot } v \in \overset{\circ}{\mathcal{T}}{}^1_\tau\}$ for any $f \in \overset{\circ}{\mathcal{T}}$. The operator $A_2 =$ rot rot $: \mathcal{D}(A_2) \to \overset{\circ}{\mathcal{T}}$ is self-adjoint and positive definite; its inverse A_2^{-1} is completely a continuous operator in $\overset{\circ}{\mathcal{T}}$. The estimate

$$\|u\|^{(2)} \leq C\| \text{ rot rot } u\| \tag{9}$$

is true for any $u \in \mathcal{D}(A_2)$. The domain of definition $D(A_2^{\frac{1}{2}})$ for the operator $A_2^{\frac{1}{2}}$ coincides with $\overset{\circ}{\mathcal{T}}{}^1_n$. (Note that $\mathcal{D}(A_2) \overset{\text{rot}}{\Longleftrightarrow} \overset{\circ}{\mathcal{T}}{}^1_\tau(\Omega) \overset{\text{rot}}{\Longleftrightarrow} \overset{\circ}{\mathcal{T}}$).

Using Theorems 2 and 3 we transform problem (1)-(5) to the problem

$$\left. \begin{aligned} v_t + \nu A_1 v + P(v_k v_{x_k}) + P([H, \text{ rot } H]) &= f \\ H_t + \gamma A_2 H + P(\text{rot } [H, \vec{v}]) &= \text{ rot } j \\ v|_{t=0} = \overset{\circ}{v} \ , \ H|_{t=0} &= \overset{\circ}{H} \ , \end{aligned} \right\} \tag{10}$$

and choose $X = \overset{\circ}{\mathcal{T}} \times \overset{\circ}{\mathcal{T}}$ as the phase-space for pairs $\{v(\cdot, t); H(\cdot, t)\}$. The following theorem is true.

THEOREM 4

Suppose that f and j do not depend on $t, f \in \overset{\circ}{\gamma}{}^1_\tau(\Omega), j \in \overset{\circ}{\mathcal{T}} \ \tau^1(\Omega)$ and $\partial\Omega \subset C^2$. Then the problem (10) has a unique solution $\{v(\cdot, t; \overset{\circ}{v}, \overset{\circ}{H}); H(\cdot, t; \overset{\circ}{v}, \overset{\circ}{H})\} =:$

$V_t(\{\overset{\circ}{v}, \overset{\circ}{H}\})$ for any $\{\overset{\circ}{v}; \overset{\circ}{H}\} \in X = \overset{\circ}{\mathcal{T}} \times \overset{\circ}{\mathcal{T}}$ and all $t \geq 0$. The solving operators $V_t : X \to X, t \in R^+$, are bounded and continuous; they form a continuous semi-group $\{V_t, t \in R^+, X\}$. This semi-group has a bounded absorbing set – a ball $B_o \subset X$, and belongs to the class 1 (i.e. $V_t : X \to X$ is completely continuous for any $t > 0$). It has the minimal global B-attractor $\mathcal{M}$ – a compact, invariant, connected subset of X. The attractor $\mathcal{M}$ has a finite fractal dimension and a finite number of determining modes. The semi-group $\{V_t, t \in R^+, \mathcal{M}\}$ can be extended in a unique way to the full continuous group $\{V_t, t \in R, \mathcal{M}\}$.

These results are analogous to results proved earlier for the Navier-Stokes equations. The other results of Ladyzhenskaya (1972) concerning the full trajectories on $\mathcal{M}$ are also true for the problem (1)-(5). Instead of the boundary conditions (4), we can adapt other suitable boundary conditions (for example, the conditions of periodicity in x_k), or consider nonsimply-connected Ω and nonhomogeneous boundary conditions.

For fully 3-D problems of MHD we have only some conditional results (as even their global unique solvability for all $t \geq 0$ is unknown). Let us formulate one of these results:

THEOREM 5

Suppose that there is a bounded subset B of the space $Y = (\overset{\circ}{\mathcal{T}}{}^1_n \cap \overset{\circ}{\mathcal{T}}{}^1_\tau) \times \overset{\circ}{\mathcal{T}}{}^1_n$ for which the norms $\|\cdot\|_Y$ of all possible solutions of the 3-D problem (10) with $\{\overset{\circ}{v}; \overset{\circ}{H}\}$ belonging to B are bounded by a finite number c. Then the problem (10) has really the unique solution $\{v(\cdot, t; \overset{\circ}{v}, \overset{\circ}{H}); H(\cdot, t; \overset{\circ}{v}, \overset{\circ}{H})\} \equiv V_t(\{\overset{\circ}{v}; \overset{\circ}{H}\})$ for each $\{\overset{\circ}{v}; \overset{\circ}{H}\} \in B$ and all $t \geq 0$. So there is the evolution $V_t(B)$ in Y of the set B and it is possible to use the closure of the set $\cup_{t \geq 0} V_t(B)$ (in the norm $\|\cdot\|_Y$) as a complete metric phase-space X for the problem (10). The collection $\{V_t, t \in R^+, X\}$ forms a semi-group and all properties enumerated in theorem 4 are true for it.

The existence of a dynamo (for a fixed Ω, some forces $f(x) \not\equiv 0$ and $j \equiv 0$) depends on $\mathcal{M}$. If $\mathcal{M}$ does not lie completely in the subspace $X_1 = \overset{\circ}{\mathcal{T}} \times \{0\}$ of $X = \overset{\circ}{\mathcal{T}} \times \overset{\circ}{\mathcal{T}}$ then there is a dynamo. But we do not know "good" examples for which $\mathcal{M} \not\subset X_1$.

REFERENCES

Bykhovskii E.B. & Smirnov N.V. (1960) On orthogonal decomposition of space of vector-functions square-integrated on a domain and on some operators of vector-analysis. Trudy of Steklov Math. Institute, LIX.

Ladyzhenskaya O.A. & Solonnikov V.A. (1960) The solving of some nonstationary problems of MHD for viscous incompressible fluid. Trudy of Steklov Math. Institute, LIX, 115-173.

Ladyzhenskaya O.A. (1961) The mathematical theory of viscous incompressible flow, 2nd ed, Gordon and Breach, New York-London-Paris 1969. Mr 40# 7610.

Ladyzhenskaya O.A. (1972) A dynamical system generated by the Navier-Stokes equations. Translation in *J. of Soviet Math.*, (1975), **3**, 458-479.

Ladyzhenskaya O.A. (1987) On the determination of minimal global attractors for the Navier-Stokes and other partial differential equations. Translation in *Russian Math. Surveys*, **42:6***, 27-73.

Postscript: Knots for Three-dimensional Vector Fields

R.S. MACKAY

Mathematics Institute,
University of Warwick,
Coventry CV4 7AL, UK

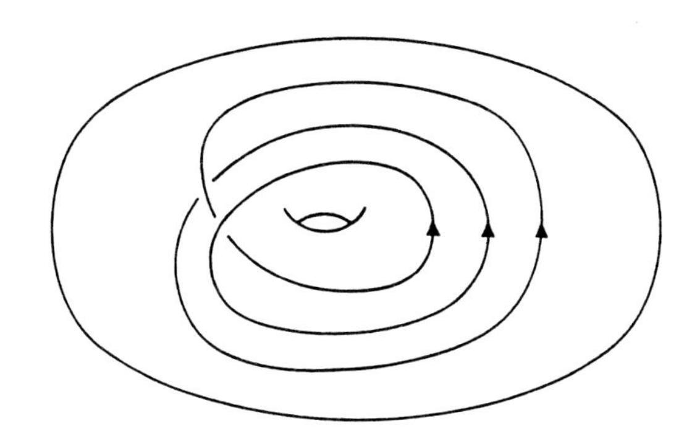

There is a growing set of results of the form: if a flow in three dimensions has closed orbits which are linked in a sufficiently non-trivial way then there must coexist certain other types of closed orbits [B,GST,LM]. For example, a flow in a solid torus with no equilibria and possessing a closed orbit as shown, possesses closed orbits of all periods [GST] (here the period is the number of revolutions around the hole). Lower bounds for the topological entropy can also be obtained. Associated with each link is a simplest flow possessing that link. Could such results be useful to fluid mechanics or MHD? The flow could represent a velocity field, or a vorticity field or a magnetic field. For background on knots in dynamical systems see [H]. The above results are based on the Nielsen-Thurston theory of isotopy classes of surface homeomorphisms [FLP].

[B] P L Boyland, An analog of Sharkovski's theorem for twist maps, in Hamiltonian Dynamical Systems, Contemp. Math. 81 (1988) 119-133, eds K R Meyer & D G Saari.

[FLP] A Fathi, F Laudenbach and V Poenaru, Travaux de Thurston sur les surfaces, Asterisque 66-67 (1979).

[GST] J M Gambaudo, S van Strien, C Tresser, Vers un ordre de Sarkovskii pour les plongements du disque preservant l'orientation, preprint.

[H] P Holmes, Knots and orbit genealogies in nonlinear oscillators, in New Directions in Dynamical Systems, T Bedford and J Swift (eds) (Cambridge, 1988).

[LM] J Llibre, R S MacKay, Rotation vectors and entropy for homeomorphisms of the torus, Erg. Th. Dyn. Sys., to appear.

List of Participants

Dr. J.J. Aly
Service d'Astrophysique
CEN-Saclay
F-91191 Gif-sur-Yvette CEDEX
FRANCE

Prof. H. Aref
IGPP A-025
UCSD
La Jolla CA 92093-0225
U.S.A.

Dr. W. Arter
Culham Laboratory
Abingdon
Oxon OX14 3DB
U.K.

Dr. K. Bajer
Uniwersytet Warszawski
Instytut Geofizyki
Ul. Pasteura 7
02-930 Warszawa
POLAND

Prof. Dr. P.G. Bakker
Dept. of Aerospace Engineering
Kluyverweg 1
2629 HS Delft
THE NETHERLANDS

Prof. G.K. Batchelor
D.A.M.T.P.
University of Cambridge
Silver Street
Cambridge CB3 9EW
U.K.

Dr. M.A. Berger
Dept. of Mathematics
University College London
Gower Street
London WC1E 6BT
U.K.

Prof. M.S. Berger
University of Massachusetts, Amherst
345 Elm Street
Northampton Mass 01060
U.S.A.

Dr. M.K. Bevir
UKAEA
Abingdon
Oxon OX14 3DB
U.K.

Prof. A. Bhattacharjee
Dept. Applied Physics
S.W. Mudd Building Rm 210
Columbia University
500 West 120th Street
New York NY 10027
U.S.A.

Dr. T.F. Buttke
Prog. Appld & Comput. Math
Princeton University
Princeton NJ 08544
U.S.A.

Dr. G.F. Carnevale
Mail Stop A-030
Scripps Inst. of Oceanography
UCSD
La Jolla CA-92093
U.S.A.

Prof. S. Childress
Courant Inst. Mathematical Science
New York University
251 Mercer Street
New York NY 10014
U.S.A.

Dr. M.S. Chong
Dept. Mechanical Engineering
University of Melbourne
Parkville
Victoria 3052
AUSTRALIA

Dr. P. Comte
Institut de Mécanique de Grenoble
BP 53
38041 Grenoble-Cedex
FRANCE

Dr. O.E. Coté
Geophysics & Space Dept. &
Air Force Office of Scientific Research
European Office Aerospace R. & D.
223/231 Old Marylebone Road
London NW1 5TH
UK

Prof. U. Dallmann
DLR-Institute for Theoretical
Fluid Mechanics
Bunsenstrasse 10
D 3400 Göttingen
F.R. GERMANY

Dr. R. Davies-Jones
National Severe Storms Laboratory
NOAA
1313 Halley Circle
Norman, Oklahoma 73069
U.S.A.

Dr. G. Doolen
Los Alamos National Laboratory
Los Alamos &
NM 87544
U.S.A.

Prof. Dr. T. Dracos
Fed. Inst. of Technology Zürich
Inst. Hydromechanics &
Water Resource Management
CH-8093 Zürich-Hönggerberg
SWITZERLAND

Dr. D. Dritschel
D.A.M.T.P.
University of Cambridge
Silver Street
Cambridge CB3 9EW
U.K.

Dr. B. Dubrulle
Observatoire Midi Pyrénées
14 Avenue E. Belin
31400 Toulouse
FRANCE

Dr. B. Eckhardt
FB Physik
Philipps Universität
Renthof 6
3550 Marburg
F.R. GERMANY

Mr. P. Enright
ALCAN International Ltd
Southam Road
Banbury, Oxon
U.K.

Dr. M. Farge
Ecole Normale Superieure
24 Rue Lhomond
75231 Paris Cedex 5
FRANCE

Prof. G.B. Field
Center for Astrophysics
60 Garden Street
Cambridge, MA 02138
U.S.A.

Mr. M.J. Filipiak
Department of Physics
The University of Edinburgh
JCMB, The King's Buildings
Mayfield Road
Edinburgh EH9 3JZ
U.K.

Dr. J.J. Finnigan
CSIRO Centre Environmental Mech.
GPO Box 821
Canberra Act 2601
AUSTRALIA

Dr. L. Fradkin
D.A.M.T.P.
University of Cambridge
Silver Street
Cambridge CB3 9EW
U.K.

Dr. U. Frisch
Observatoire de Nice
BP 139
06003 Nice Cedex
FRANCE

Dr. A. Gailitis
Physics Institute
Latvian SSR Academy of Sciences
Riga, Salaspils 1
U.S.S.R.

Prof. M. Gaster
Dept. of Engineering
University of Cambridge
Trumpington Street
Cambridge CB2 1PZ
U.K.

Prof. P. Germain
3 Avenue de Champaubert
75015 Paris
FRANCE

Prof. C.H. Gibson
Dept. of Mech. Engineering
R-011 UCSD
La Jolla CA 92093
U.S.A.

Dr. A.D. Gilbert
D.A.M.T.P.
University of Cambridge
Silver Street
Cambridge CB3 9EW
U.K.

Dr. C.G. Gimblett
UKAEA Culham Laboratory
Abingdon
Oxon OX14 3DB
U.K.

Dr. J.M. Greene
GA Technologies Inc.
PO Box 85608
San Diego CA 92138-5608
U.S.A.

Dr. A. Gyr
Fed. Inst. Technology Zürich
Inst. Hydromechanics &
Water Resource Management
CH-8039 Zürich-Hönggerberg
SWITZERLAND

Prof. H. Hasimoto
Faculty of Engineering
Hosei University
Kajinocho, Koganei
Tokyo 184
JAPAN

Dr. P. Haynes
D.A.M.T.P.
University of Cambridge
Silver Street
Cambridge CB3 9EW
U.K.

Mr. J.L. Helman
Dept. of Applied Physics
Stanford University
Durand Bldg., Room 370
Stanford CA 94305-4035
U.S.A.

Mr. T. Herdtle
Institute GPP
UCSD
A-025 La Jolla CA 92093
U.S.A.

Dr. W. Hiller
Max Planck Institute
Bunsenstr. 10
D-3400 Göttingen
F.R. GERMANY

Dr. D.W. Hughes
D.A.M.T.P.
University of Cambridge
Silver Street
Cambridge CB3 9EW
U.K.

Dr. A. Hunt
Head of Department, Fluid Mechanics
Schlumberger Cambridge Research Ltd.
High Cross, Madingley Road
Cambridge
U.K.

Dr. J.C.R. Hunt
D.A.M.T.P.
University of Cambridge
Silver Street
Cambridge CB3 9EW
U.K.

Prof. H.E. Huppert
D.A.M.T.P.
University of Cambridge
Silver Street
Cambridge CB3 9EW
U.K.

Prof. A.K.M.F. Hussain
Dept. of Mechanical Engineering
University of Houston
Houston TX 772-4-4792
U.S.A.

Dr. K. Ishii
ICFD
1-22-3 Haramachi
Meguro-ku, Tokyo 152
JAPAN

Dr. R. Iwatsu
ICFD
1-22-3 Haramachi
Meguro-ku, Tokyo 152
JAPAN

Dr. A.V. Jones
CEC Joint Research Centre
Ed. 65
21020 ISPRA (VA)
ITALY

Prof. T. Kambe
Dept. of Physics
University of Tokyo
Hongo, Bunkyo-ku
Tokyo 113
JAPAN

Dr. R.M. Kerr
National Center Atmospheric Res.
PO Box 3000 MMM
Boulder CO 80307
U.S.A.

Prof. S. Kida
Prog. Appld. & Computational Maths
218 Fine Hall, Princeton University
Princeton, New Jersey 08544
U.S.A.

Prof. R.M. Kiehn
Physics Department
University of Houston
Houston TX 77004
U.S.A.

Dr. Y. Kimura
Center for Nonlinear Studies
Los Alamos National Laboratory
Los Alamos NM 87544
U.S.A.

Dr. E.Kit
Faculty of Engineering
Tel-Aviv University
69978 Ramat-Aviv, Tel Aviv
ISRAEL

Dr. T.A. Kowalewski
Max-Planck-Institut
Bunsenstr. 10
3400 Göttingen
F.R. GERMANY

Prof. I. Kunin
Mechanical Engineering
4800 Calhoun Road
University of Houston
University Park
Houston TX 77004
U.S.A.

Prof. O.A. Ladyzhenskaya
191011 Leningrad D-11
Mathematical Institute
Fontanka 27
U.S.S.R.

Dr. J. Léorat
DEAC
Observatoire de Meudon
92190-Meudon (Hauts de Seine)
FRANCE

Dr. E. Levich
Benjamin Levich Institute for PCH
City College T202
Convent Ave. at 138th Street
New York NY 10031
U.S.A.

Dr. B.C. Low
High Altitude Observatory – NCAR
PO Box 3000
Boulder CO 80307
U.S.S.R.

Dr. R.S. MacKay
Mathematics Institute
University of Warwick
Coventry CV4 7AL
U.K.

Dr. M.E. McIntyre
D.A.M.T.P.
University of Cambridge
Silver Street
Cambridge CB3 9EW
U.K.

Mr. N. Malik
D.A.M.T.P.
University of Cambridge
Silver Street
Cambridge CB3 9EW
U.K.

Mr. N. Mattor
Culham Laboratory
Abingdon
Oxfordshire OX14 3DB
U.K.

Dr. S. Meacham
Rm. 54-1422
M.I.T.
77 Mass. Ave
Cambridge MA 02139
U.S.A.

Prof. E. Meiburg
Divn. of Appld. Math.
Box 1966 Brown University
Providence RI 02912
U.S.A.

Dr. A.J. Mestel
Dept. of Mathematics
Imperial College
Queen's Gate
London SW7 2BZ
U.K.

Prof. R.W. Metcalfe
Dept. Mech. Eng.
University of Houston
Houston TX 77204-4792
U.S.A.

Prof. H.K. Moffatt
D.A.M.T.P.
University of Cambridge
Silver Street
Cambridge CB3 9EW
U.K.

Prof. D.C. Montgomery
Dept. of Physics
Dartmouth College
Hanover NH 03755
U.S.A.

Mr. W.H.P. Münch
D.A.M.T.P.
University of Cambridge
Silver Street
Cambridge CB3 9EW
U.K.

Dr. B. Nicolaenko
Dept. of Mathematics
Arizona State University
Tempe AZ 85287
U.S.A.

Dr. J.W. Norris
Dept. of Mathematics
The University of Birmingham
PO Box 363
Birmingham B15 2TT
U.K.

Prof. A.M. Oboukhov
Institute of Atmospheric Physics
USSR Academy of Sciences
Pyzhevsky 3
109017 Moscow Zh-17
U.S.S.R.

Prof. J.M. Ottino
159 Goessmann Lab.
Dept. of Chemical Engineering
University Massachusetts, Amherst
Amherst MA 01003
U.S.A.

Dr. A. Otto
Ruhr-Universität Bochum
Theoretische Physik
Lehrstuhl IV
Universitätsstrasse 150
D-4630 Bochum 1
F.R. GERMANY

Prof. E.N. Parker
Laboratory for Astrophysics
933 East 56th Street
Chicago Ill 60637
U.S.A.

Prof. A.E. Perry
Dept. Mechanical Engineering
University of Melbourne
Parkville, Victoria 3052
AUSTRALIA

Dr. W. Polifke
Benjamin Levich Institute for PCH
City College T202
Convent Ave. at 138th Street
New York NY 10031
U.K.

Dr. H. Politano
Observatoire de Nice
BP 139
06003 Nice Cedex
FRANCE

Dr. L.M. Polvani
Dept. of Mathematics
M.I.T. Room 2-339
Cambridge MA 02139
U.S.A.

Prof. S.C. Prager
Department of Physics
University of Wisconsin – Madison
Madison
Wisconsin 53706
U.S.A.

Dr. M.R.E. Proctor
D.A.M.T.P.
University of Cambridge
Silver Street
Cambridge CB3 9EW
U.K.

Dr. A. Pumir
LPS, ENS
24 rue Lhomond
F-75231 Paris Cedex
FRANCE

Prof. K.-H. Rädler
Zentralinstitut für Astrophysik
Rosa-Luxembourg-Strasse 17A
1591 Potsdam-Babelsberg
D.R. GERMANY

Dott. R. Ricca
D.A.M.T.P.
University of Cambridge
Silver Street
Cambridge CB3 9EW
U.K.

Dr. J.P. Rivet
Observatoire de Nice
BP 139
06003 Nice Cedex
FRANCE

Mr. W. Roberts
Department of Physics
University of Edinburgh
JCMB, The King's Buildings
Mayfield Road
Edinburgh EH9 3JZ
U.K.

Prof. Dr. K.G. Roesner
Technische Hochscule Darmstadt
Institut für Mechanik
Hochschulstr. 1
D-6100 Darmstadt
F.R. GERMANY

Dr. S. Rogers
ALCAN International Ltd.
Southam Road
Banbury
Oxon
U.K.

Dr. A.A. Ruzmaikin
Instit. Terrestrial Magnetism
Ionosphere & Radio Wave Propagation
SU-142092 Troitsk
Moscow Region
U.S.S.R.

Dr. O.E. Sero-Guillaume
CNRS LEMTA UA 875
24-30 Rue Lionnois
54000 Nancy
FRANCE

Dr. T.G. Shepherd
Dept. of Physics
University of Toronto
Toronto M55 1A7
CANADA

Dr. S. Shirayama
ICFD
1-22-3 Haramachi
Meguro-ku, Tokyo 152
JAPAN

Prof. L. Shtilman
The Benjamin Levich Inst. of PCH
City College T202
Convent Ave/138th Street
New York 10031
U.S.A.

Prof. E. Siggia
Dept. de Physique
Ecole Normale Superieure
24 Rue Lhomond
75231 Paris, Cedex 05
FRANCE

Prof. A.M. Soward
School of Mathematics
The University
Newcastle upon Tyne NE1 7RU
U.K.

Prof. E.A. Spiegel
Dept. of Astronomy
Columbia University
New York NY 10027
U.S.A.

Dr. Wenhan Su
Fluid Mechanics Inst.
Beijing Univ. Aero- & Astronautics
Beijing
CHINA

Prof. N. Sugimoto
Dept. of Mechanical Engineering
Faculty of Engineering Science
Osaka University
Toyonaka, Osaka 560
JAPAN

Prof. P.L. Sulem
Observatoire de Nice
BP 139
06003 Nice Cedex
FRANCE

Prof. M. Tabor
Dept. Applied Physics
Columbia University
New York NY 10027
U.S.A.

Mr. M. Takaoka
Dept. of Physics
Faculty of Science
Kyoto University
Kyoto 606
JAPAN

Dr. J.B. Taylor
Institute for Fusion Studies
University of Texas at Austin
Austin TX 78712
U.S.A.

Prof. A. Tsinober
Dept. Fluid Mech. and Heat Transfer
Faculty of Engineering
Tel-Aviv University
Ramat-Aviv, Tel-Aviv 69978
ISRAEL

Prof. A.V. Tur
Institute for Space Research
USSR Academy of Sciences
Profosoyuznaya UL. 84/32
Moscow 117810
U.S.S.R.

Dr. G.K. Vallis
Dept. of Physics
UCSC
Santa Cruz CA 95064
U.S.A.

Mr. J.C. Vassilicos
D.A.M.T.P.
University of Cambridge
Silver Street
Cambridge CB3 9EW
U.K.

Prof. J.M. Wallace
Dept. of Mechanical Engineering
University of Maryland
College Park MD 20742
U.S.A.

Dr. I. Walton
Schlumberger Cambridge Research Ltd.
P.O. Box 153
Cambridge CB3 0HG
U.K.

Dr. W.E. Ward
D.A.M.T.P.
University of Cambridge
Silver Street
Cambridge CB3 9EW
U.K.

Mr. A.G. Watt
Department of Physics
University of Edinburgh
JCMB, The King's Buildings
Mayfield Road
Edinburgh EH9 3JZ
U.K.

Dr. E. Weisshaar
Universität Bayreuth
Physikalisches Institut
Postfach 101251
8580 Bayreuth
F.R. GERMANY

Drs. M.E.M. de Winkel
Dept. of Aerospace Engineering
Delft University of Technology
Kluyverweg 1
2629 HS Delft
THE NETHERLANDS

Dr. A.A. Wray
MS202A-1 Moffett Field
California CA 94035
U.S.A.

Prof. G.M. Zaslavsky
Space Research Institute
Profsoyuznaya 84
Moscow 117810
U.S.S.R.

Prof. E.G.Zweibel
APAS Dept. Box 391
University of Colorado
Boulder CO 80309
U.S.A.

Author Index

Author Index

PLATE SECTION

Plate 1 Ottino.

EXPERIMENT

COMPUTER SIMULATION

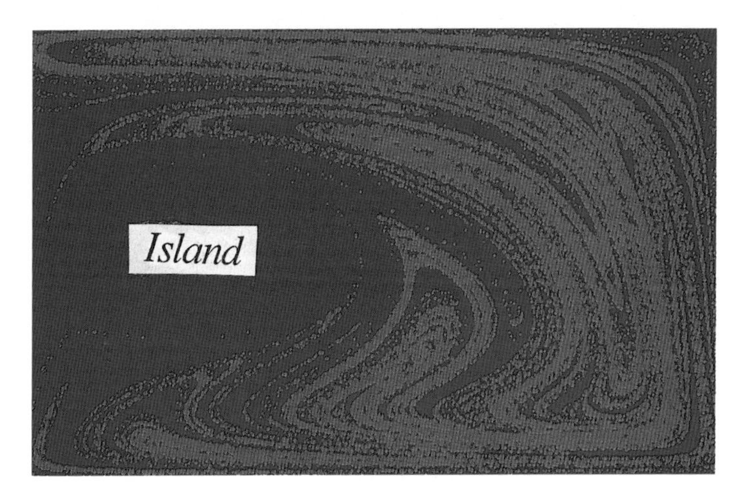

POINCARÉ SECTION

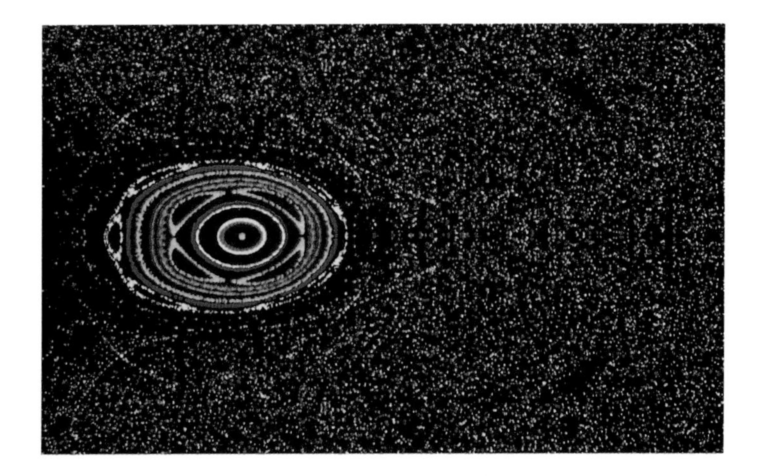

Figure 1. Comparison between a typical experiment, a computer simulation, and its corresponding Poincaré section, for a cavity flow (Leong 1989). The top two patterns are generated in just 8 periods. The Poincaré section, $O(10^3)$ periods, reveals a complex structure within the island, involving smaller islands, as well as a chaotic web. The rate of rotation in the smaller islands is very slow, and the green marker is largely unaffected by this structure.

See page 16.

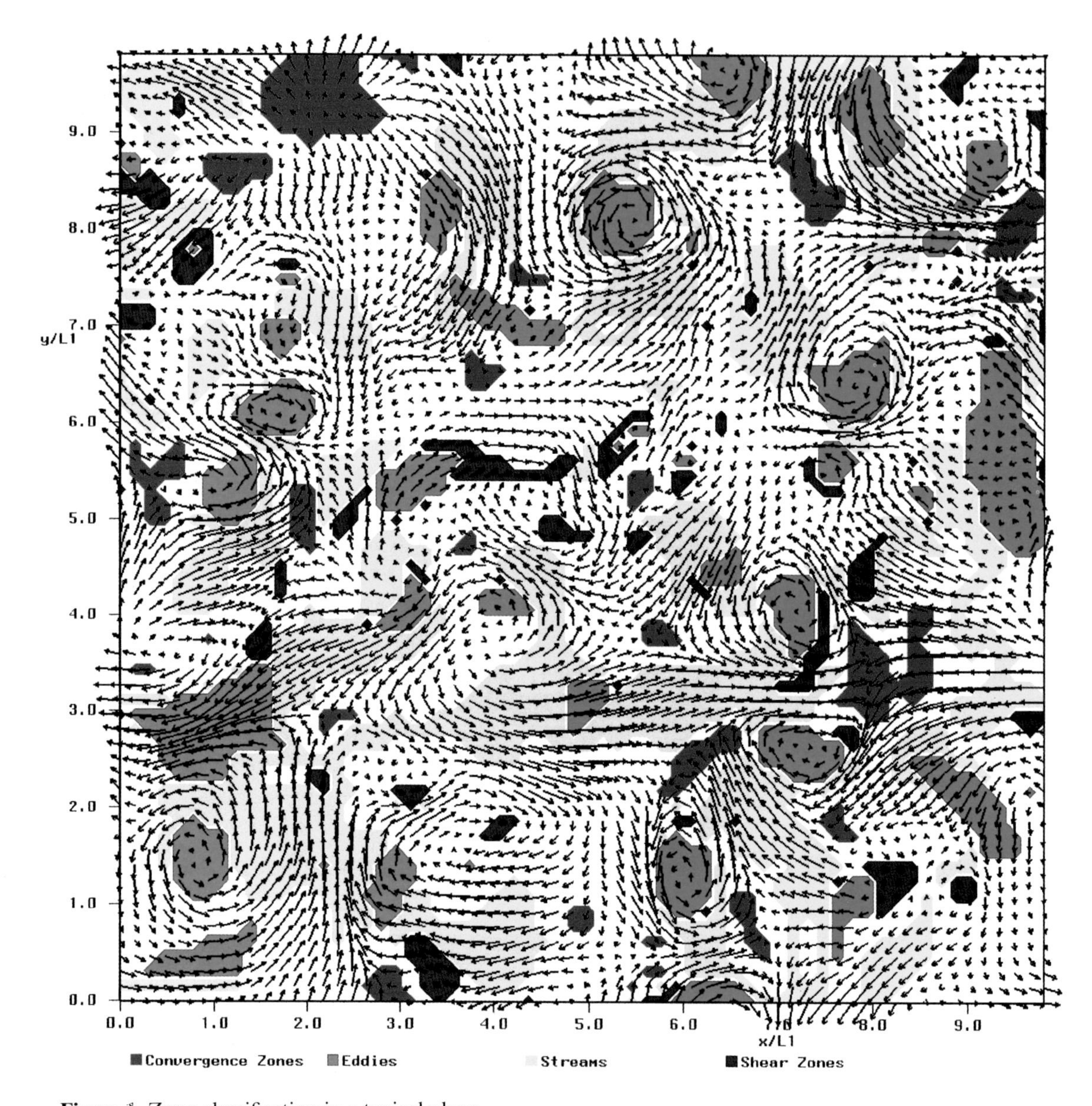

Figure 1. Zone classification in a typical plane.

See page 99

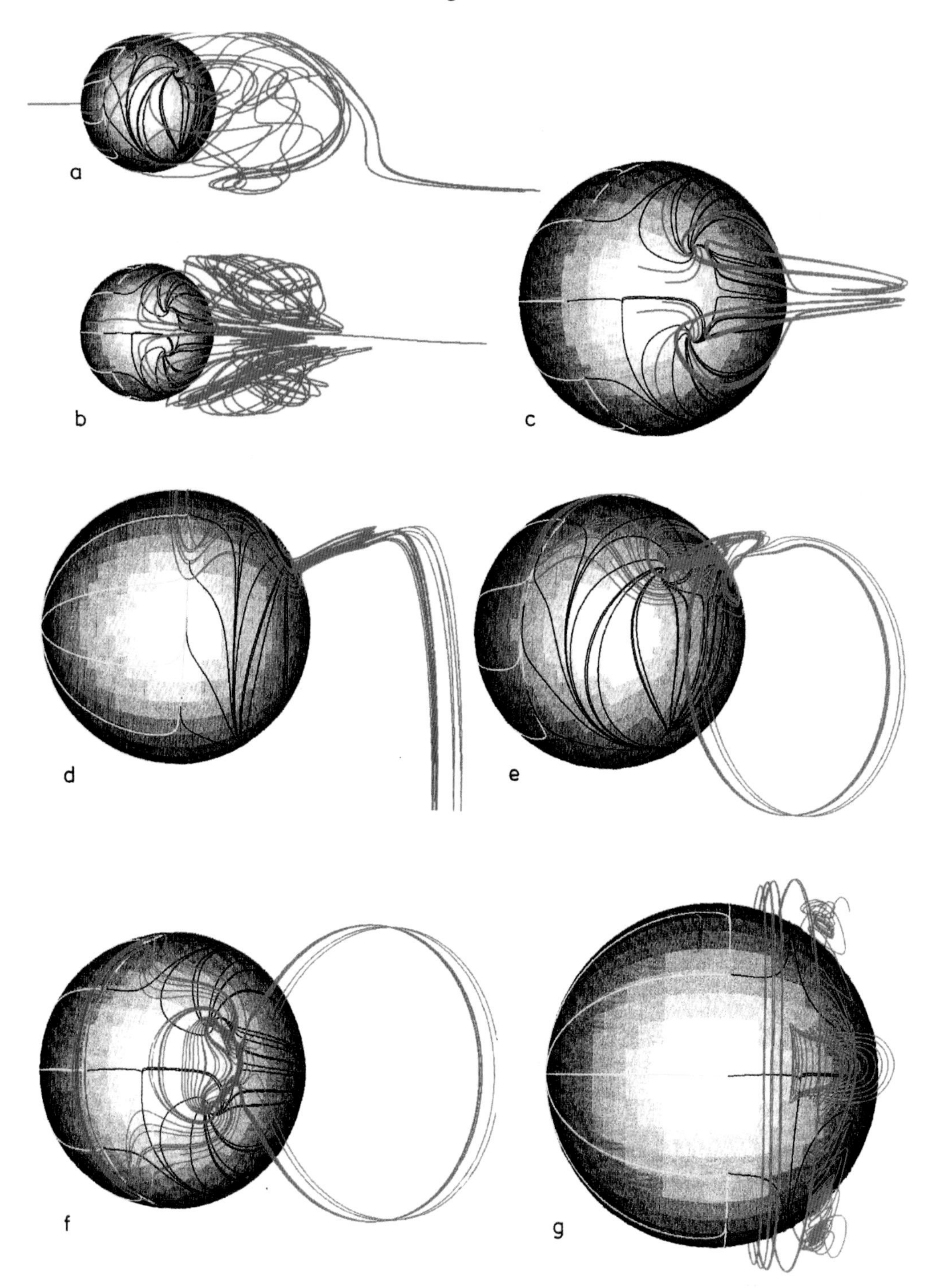

Skin friction lines and vorticity lines of 3-D flow simulation at *Re*=2000, *Ma*=0.40, streamlines starting at the front stagnation line (a), selected streamlines (b), streamlines starting at the foci (c), vorticity lines starting at the foci (d, e, f) and closed vorticity loops (g) which are getting folded in the separated flow region.

See page 383.

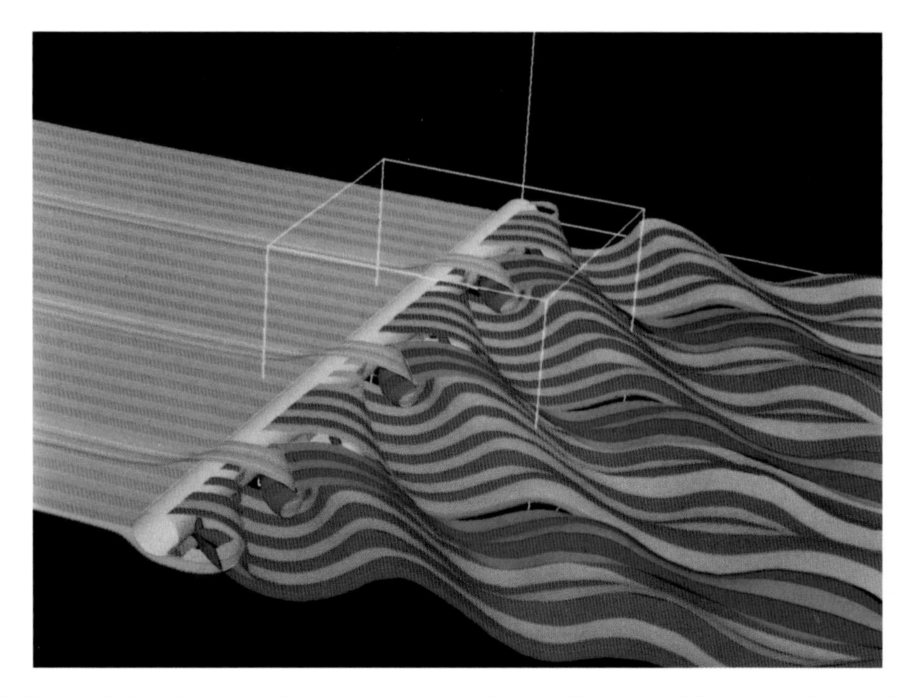

Figure 5: Topological surfaces showing time evolution of a two-dimensional flow around a circular cylinder.

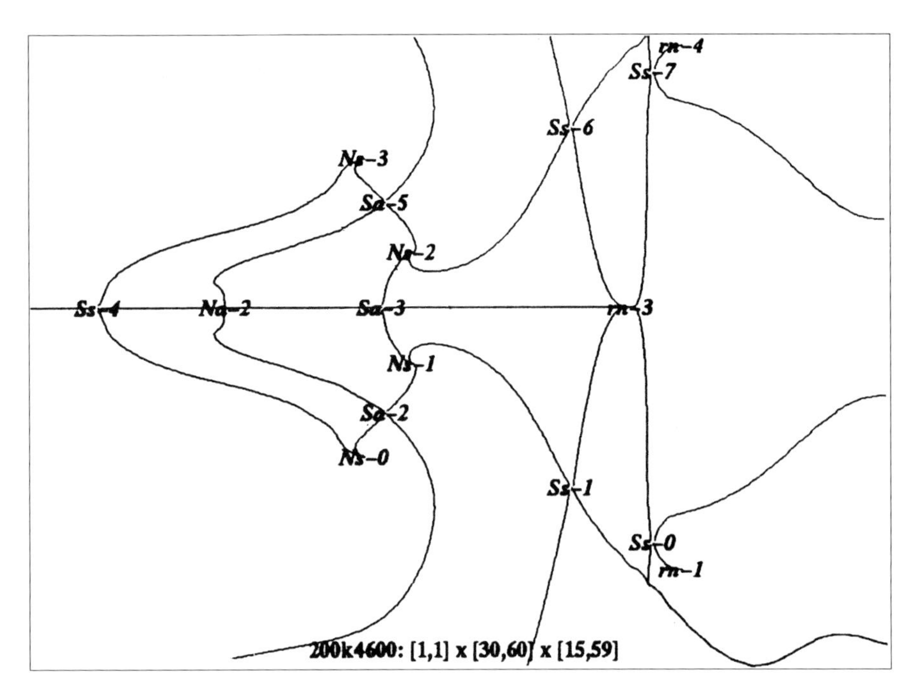

Figure 6: Skin friction curves and critical points from the analysis of the surface topology of a hemisphere cylinder.

See page 370.

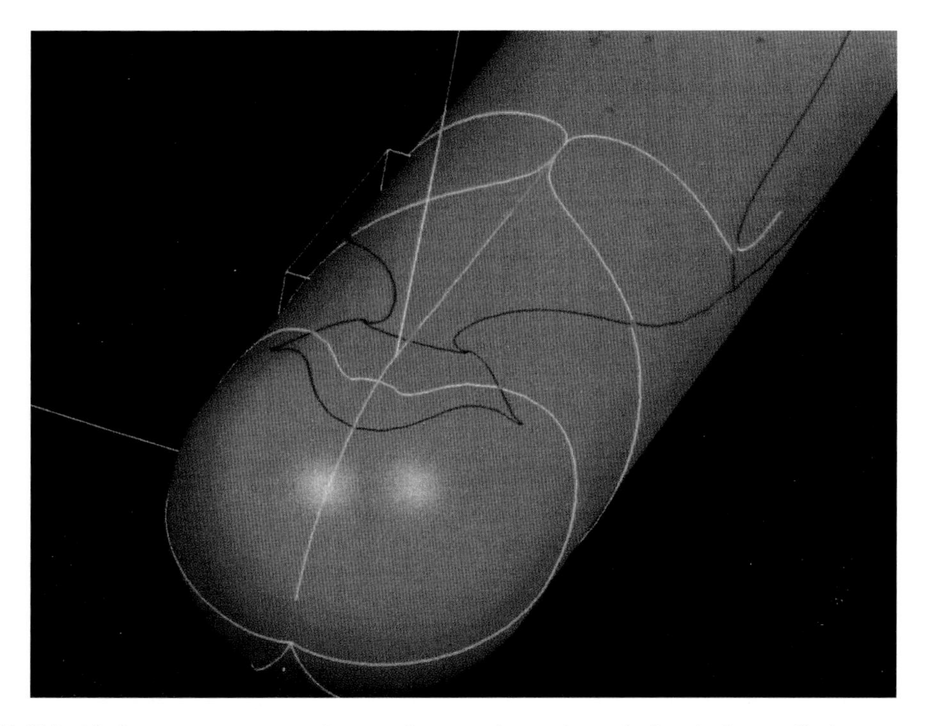

Figure 7: Skin friction curves representing topology on the surface of a hemisphere cylinder.

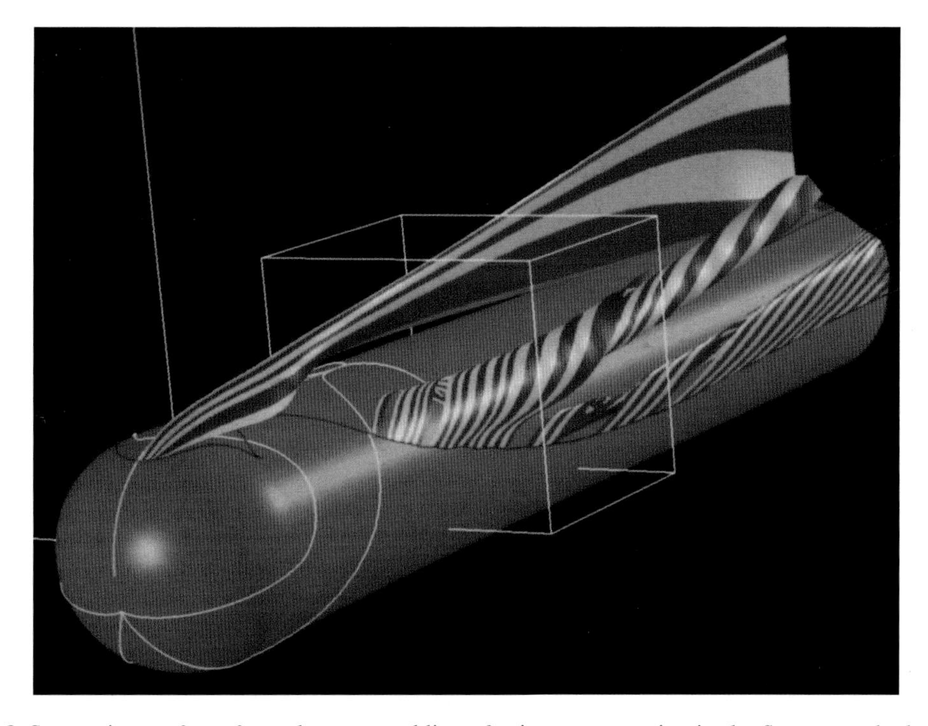

Figure 8: Separation surfaces from the nose and line of primary separation in the flow around a hemisphere cylinder.

See page 371.

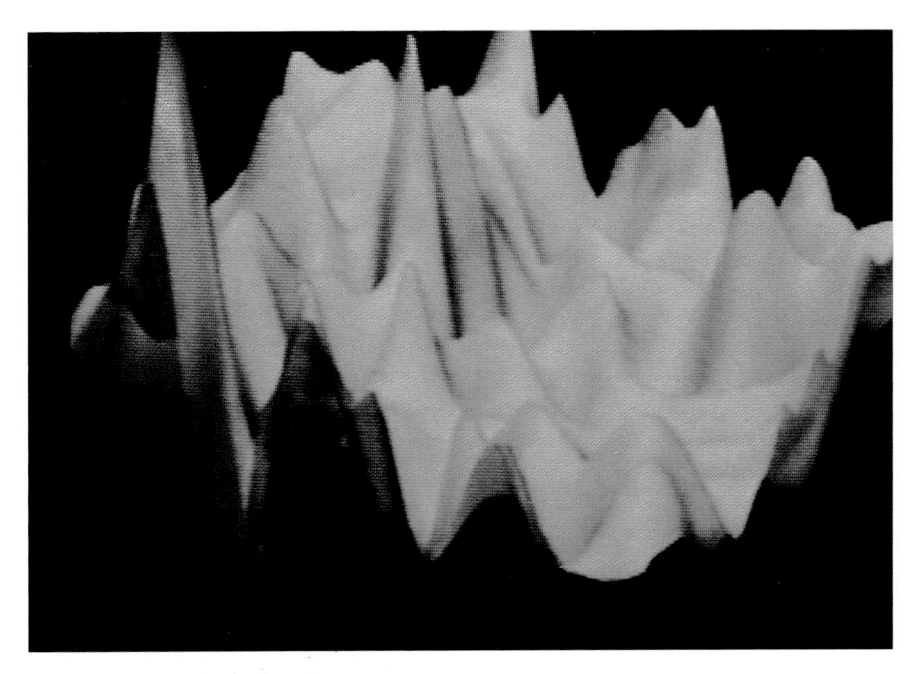

Figure 1a The 2-D vorticity distribution for a field during the burst, with the z-axis indicating the vorticity amplitude. Here R=36 and k_f =8. The scale of the periodic force controls the typical scale of vorticity structures.

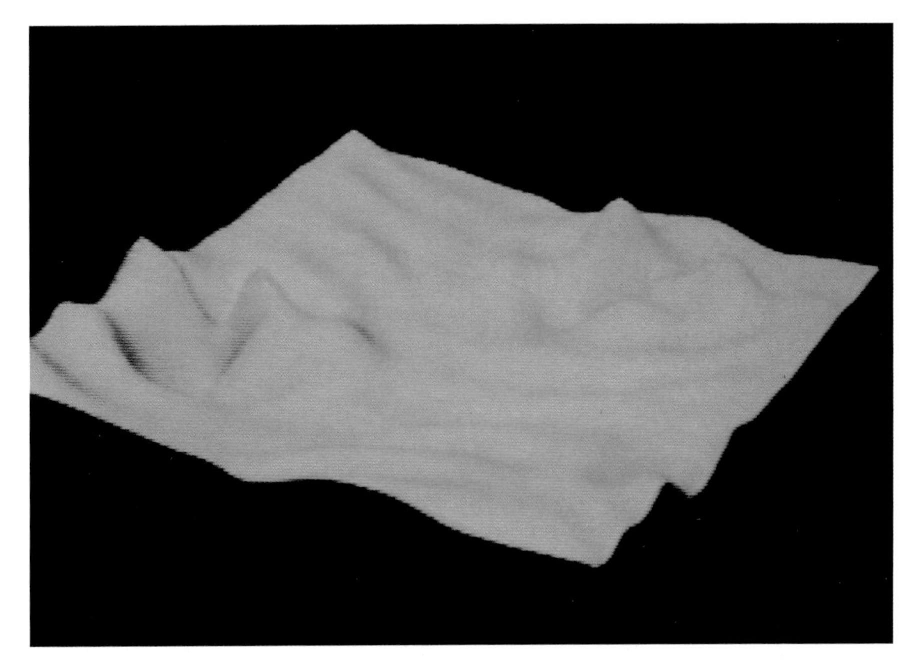

Figure 1b The 2-D vorticity distribution for a field between bursts for the same parameters. Notice the counter-rotating eddies, with tripoles within their core. The scale of the eddies is much larger than the forcing scale. Computations performed on ASU's Titan Graphics Computers.

See page 273.

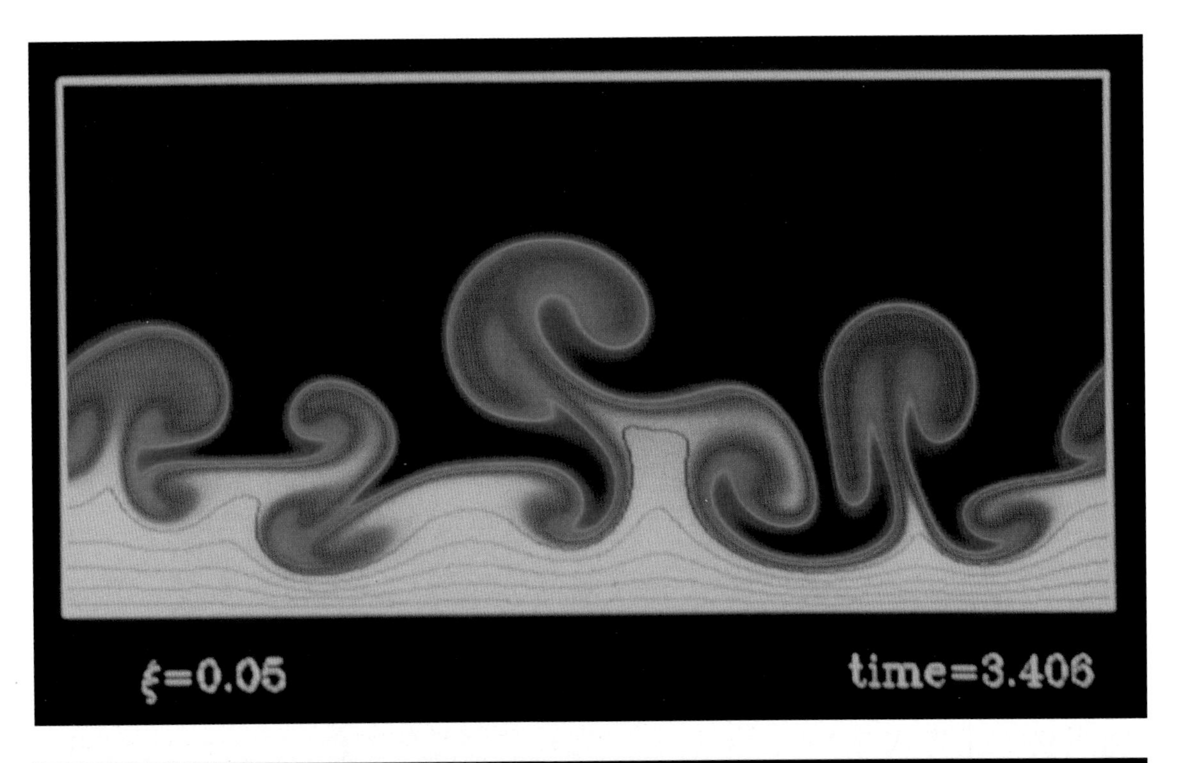

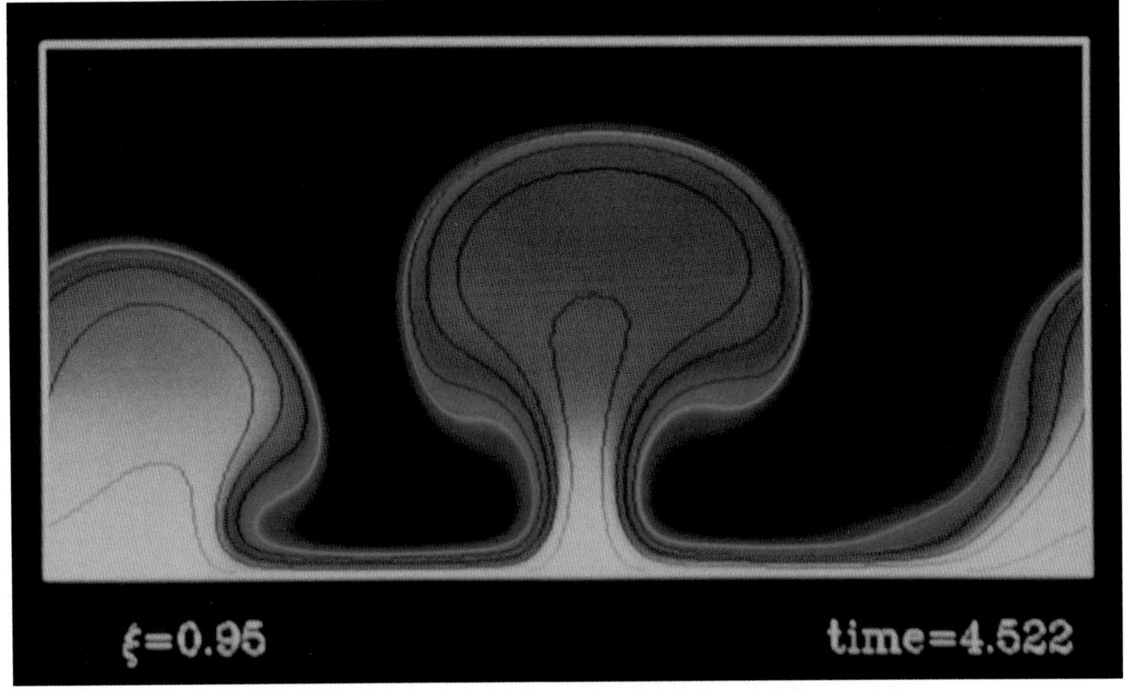

Snapshots of the evolution of a twisted magnetic field, as described by figures 1 and 2. The toroidal field is shown in colour, brighter regions corresponding to stronger field. Overlaid are the poloidal field lines.

See page 685.

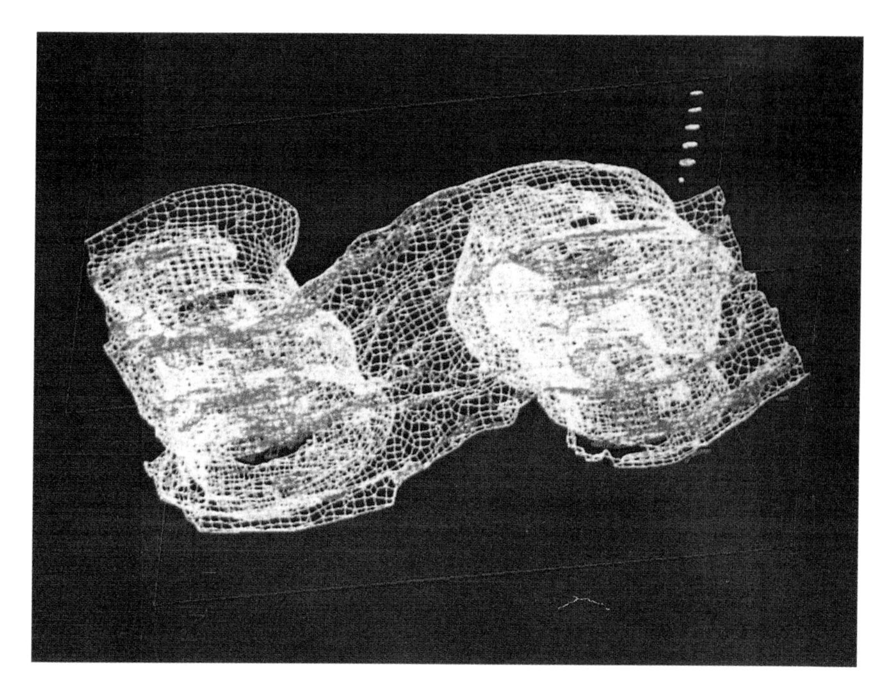

Figure 4: same calculation, at the same timestep. In red: iso-contour of the streamwise vorticity component $\omega_I \approx \omega_i = 2\ U/\delta_i$ at $t=35\ \delta_i/U$ (only positive vortices are displayed). The interface between the two streams, defined as in Fig. 2, appears in white.

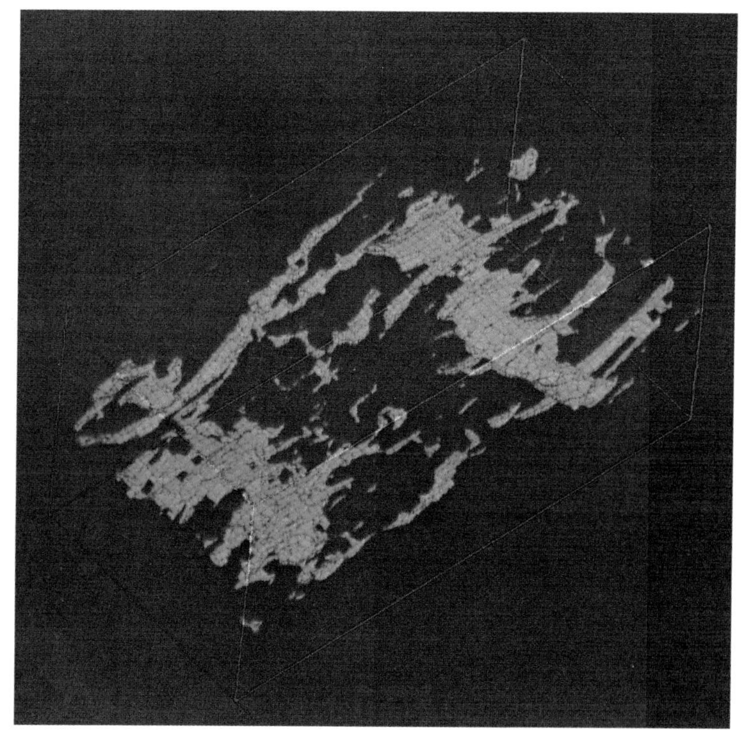

Figure 5: same calculation, at the same timestep. Iso-surface $\sqrt{\omega_x^2 + \omega_y^2} = \omega$

See page 653.

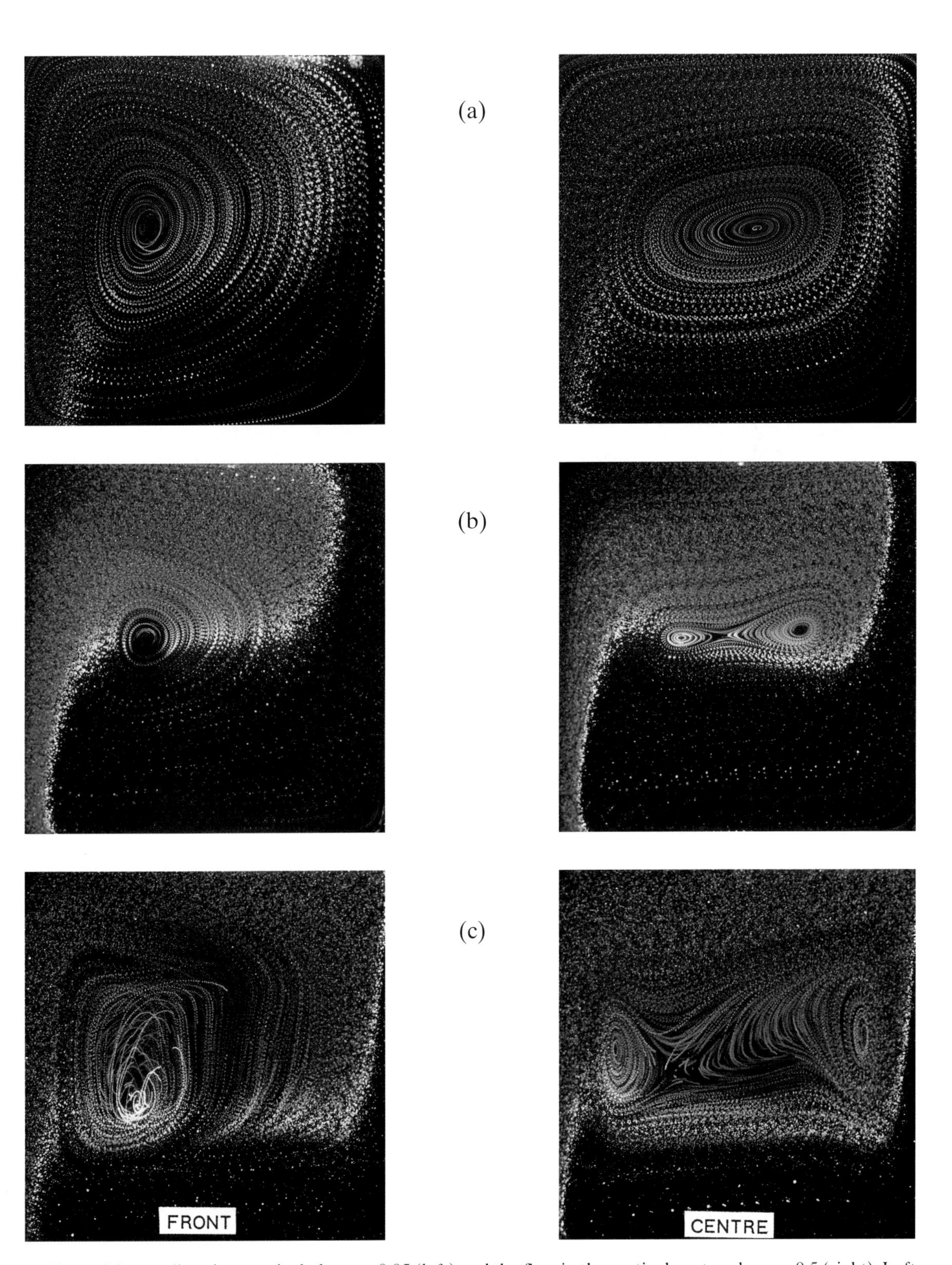

Figure 2 Streamlines in a vertical plane y=0.05 (left) and the flow in the vertical centre plane y=0.5 (right). Left side (z=0) is the hot wall; (a) $Ra=2.10^4$, Pr=6300; (b) $Ra=8.10^4$, Pr=6900; (c) $Ra=7.5\cdot10^5$, Pr=225. Red colour corresponds to ~27°C, blue to ~29°C.

See page 720.

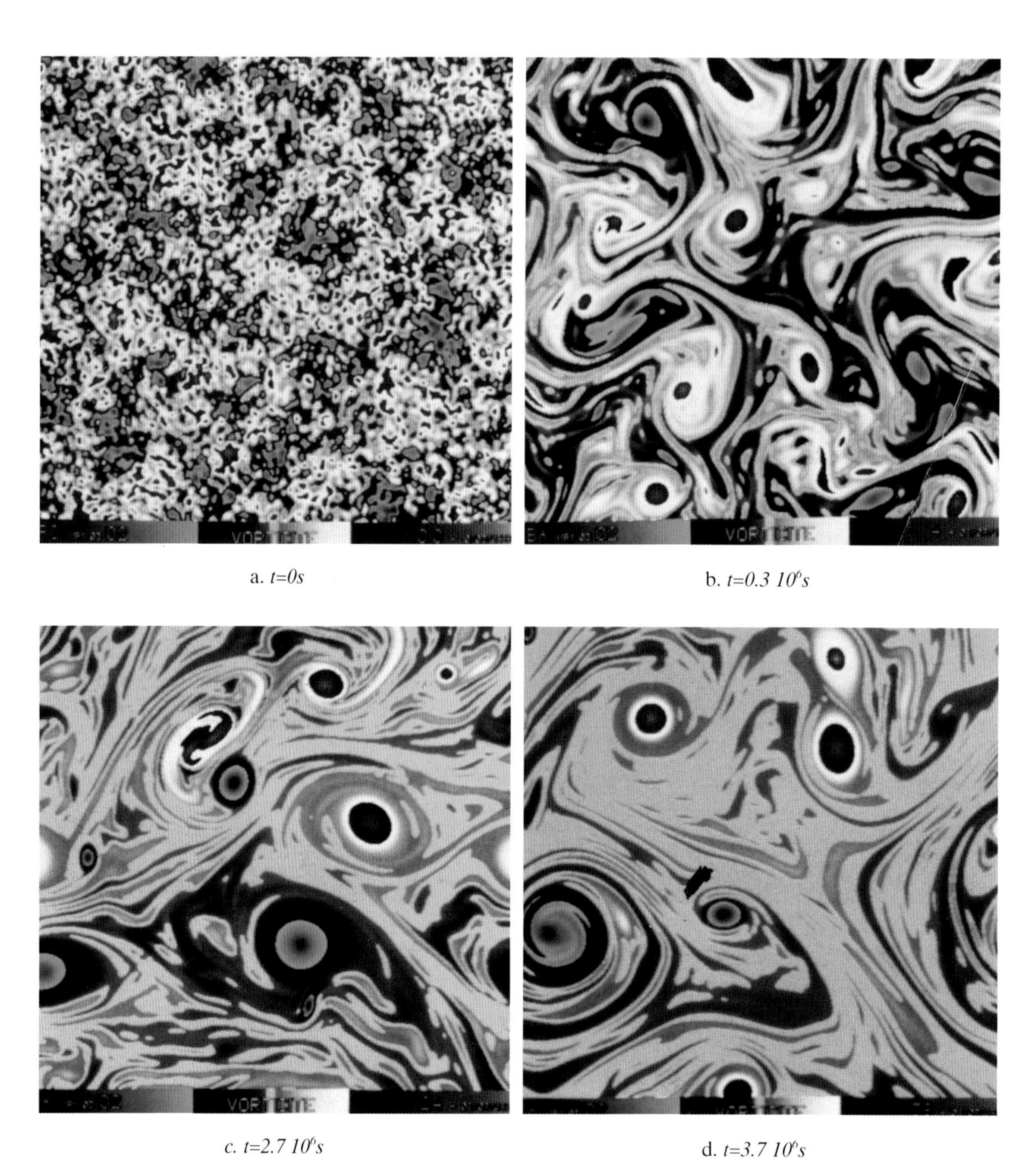

a. $t=0s$

b. $t=0.3\ 10^6 s$

c. $t=2.7\ 10^6 s$

d. $t=3.7\ 10^6 s$

Figure 1 Time evolution of a decaying two-dimensional turbulent flow from random initial conditions in nonlinear balance equilibrium

See page 773.

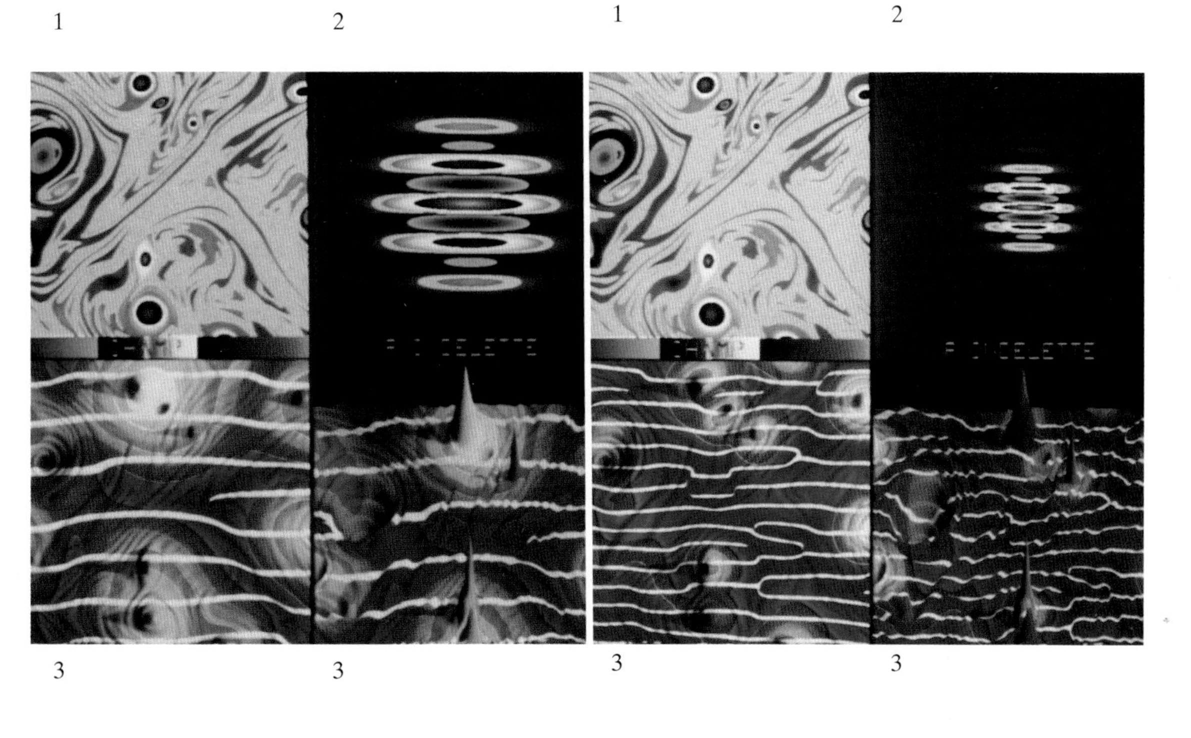

a. Scale k=8 and angle α=0

b. Scale k=16 and angle α=0

1. Vorticity field to be analyzed
   (cartographic view)

2. Analyzing wavelet at scale k and angle α=0

3. Superposition of:
   -vorticity field
   (perspective view coded in luminance)

   -modulus of the wavelet transformat scale k and angle α=0
   (cartographic view color coded in the order: blue, red, magenta, green, cyan, yellow, white)

   -phase of the wavelet transform at scale k and angle α=0
   (isoline zero being white)

Figure 4: Two-dimensional wavelet transform of the vorticity field
(corresponding to the last time step shown on figure 1)

See page 775.

Plate 12 Symposium Photograph

Symposium Photograph